Vegetable Crops

— Volume 4 —

THE EDITORS

Professor T.K. Bose obtained his M.Sc. (Ag.) and Ph.D. degree from Calcutta University, West Bengal, India. Trained at the Royal Botanic Garden, Kew, UK. Former Secretary of the Agri-Horticultural Society of India; Project Coordinator, All India Coordinated Floriculture Improvement Project (ICAR); Professor and Head, Department of Horticulture; Dean, Faculty of Agriculture; Director of Research, Bidhan Chandra Krishi Viswavidyalaya (BCKV), West Bengal; Member of Scientific Advisory Committees of ICAR, CSIR (Govt. of India) and Government of West Bengal. Awarded (D.Sc.) Honoris Causa from OUAT, Bhubaneshwar and UBKV, Coochbehar, India. Guided around 45 Ph.D. students, published more than 250 research papers and authored 30 books on horticultural crops.

Prof. M.G. Som served as former Vice-chancellor, Dean, Faculty of Agriculture and Professor of Vegetable Science, BCKV, West Bengal. Obtained his M.Sc. and Ph.D. degrees from ICAR-IARI, New Delhi. Received fellowship in 1997 and Plaque in 1998 from the Indian Society of Vegetable Science (ISVS). Served as FAO expert in Bangladesh for increasing the vegetable production. Acted as a member of governing body of the ICAR, New Delhi; member of Advisory Research Council, ICAR-IIVR, Varanasi, and ICAR-CIARI, Port Blair, Andaman. Published more than 100 research papers, written 8 books, 2 monographs and few bulletins and several popular articles. Guided 14 Ph.D. students and more than 50 M.Sc (Ag.) students. In 2019, ISVS felicitated Prof. Som for his outstanding contribution in the field of vegetable research and education.

Prof. Arup Chattopadhyay obtained his M.Sc. and Ph.D. degrees from BCKV, West Bengal. He has 23 years of experience in Research, Teaching and Extension. Acting as the Officer-in-Charge, AICRP on Vegetable Crops, and served as Professor for more than 8 years. He has guided 15 M.Sc. and 9 Ph.D. students in Vegetable Science. Published more than 150 research papers, authored 05 books, 07 book chapters, and many popular articles, folders, leaflets. Developed nine trait specific varieties/hybrids of different vegetable crops. First recipient of 'Dwarikanath Young Scientist Gold Medal Award' for the best Ph.D. thesis in Vegetable Science; 'Fellows' of ISVS, Varanasi, and CWSS, West Bengal. BCKV has been felicitated 'Lt. Amit Singh Memorial Award' by the ICAR for the Best coordinating centre on Vegetable Crops in India for outstanding contribution in vegetable research under his able leadership.

Prof. Tapan Kumar Maity has acclaimed professional career of more than 35 years in the Department of Vegetable Science, BCKV, West Bengal. Acted as Nodal Officer of Quality Control Laboratory for Horticultural Produce, and PI of AINRP on Onion and Garlic (ICAR). Served as the Head of the Department of Vegetable Science and the Dean, Faculty of Horticulture. Published more than 100 original research papers, contributed 13 book chapters, edited 3 books. Supervised 12 Ph.D. and 32 M.Sc. students. Acted as Member of the Working Group of the Agriculture Commission, Government of West Bengal, Member of the Monitoring Team of AINRP on Onion and Garlic for different centers; acting as consultant of National Horticulture Board, Government of India.

Prof. Jahangir Kabir obtained his M.Sc.(Ag) and Ph.D.(Horticulture) degrees from BCKV, West Bengal. He has 33 years of varied experience in Teaching and Research in Vegetable Crops and Post Harvest Physiology and Technology of Horticultural Crops. Acted as former Head of the Department of Post Harvest Technology, and served as Professor for more than 15 years. He has guided more than 30 M.Sc. and 10 Ph.D. students, authored more than 120 publications including books (edited), research papers, book chapters, review/technical and popular articles. Acted as reviewer of research papers, paper setter, and member of selection committee of Public Service Commission and several State and Central Universities.

Vegetable Crops
— Volume 4 —

— Senior Editor —

M.G. Som

Department of Vegetable Science
Faculty of Horticulture
Bidhan Chandra
Krishi Viswavidyalaya
West Bengal 741 252, India

— Editor Emeritus —

T.K. Bose

Faculty of Horticulture
Bidhan Chandra
Krishi Viswavidyalaya
West Bengal 741 252, India

— Edited by —

A. Chattopadhyay • T.K. Maity • J. Kabir

Faculty of Horticulture
Bidhan Chandra Krishi Viswavidyalaya
West Bengal 741 252, India

2022

Daya Publishing House®
A Division of

Astral International Pvt. Ltd.
New Delhi – 110002

Cataloging in Publication Data--DK
Courtesy: D.K. Agencies (P) Ltd. <docinfo@dkagencies.com>

Vegetable crops. Volume 4 / senior editor, M.G. Som ;
editor emeritus, T.K. Bose ; edited by A. Chattopadhyay,
T.K. Maity, J. Kabir. -- Fourth revised and illutrated
edition.
    pages cm
    ISBN 9789354616853 (HB)

    1. Vegetables.   I. Som, M. G., editor.

LCC SB320.9.V44 2021 | DDC 635   23

Published by   :   **Daya Publishing House®**
                                A Division of
                                Astral International Pvt. Ltd.
                                – ISO 9001:2015 Certified Company –
                                4736/23, Ansari Road, Darya Ganj
                                New Delhi-110 002
                                Ph. 011-43549197, 23278134
                                E-mail: info@astralint.com
                                Website: www.astralint.com

From the Desk of Editor Emeritus

The undersigned had initiated the publications entitled 'Fruits: Tropical and Sub-tropical, Vegetable Crops and 'Commercial Flowers' in one volume each in the 1980's in collaboration with co-editors, humbly expresses from the core of his heart that the updated revised editions of the above mentioned publications with colour illustrations in thirteen volumes which was planned in June, 2020, are being printed from October this year, a rare event in Horticultural Science has been possible by the blessing of Almighty God, with the cooperation of dedicated horticulturist as editors and contributors.

Kolkata

T.K. Bose

Editor Emeritus

From the Desk of
Senior Editor

The book entitled "Vegetable Crops in India" was first published as a single volume in 1986, afterwards revised and published in 1993 and 2003 in three volumes by the name "Vegetable Crops". The previous three editions proved to be very useful to the students, teachers and researchers across the world. Concerted research in production and productivity of vegetable crops during last decades necessitates further revision of the book with coloured illustrations as the fourth edition in four volumes.

The fourth volume of this edition includes chapters on cucurbits, bulbous vegetables and perennial vegetables. I am extremely happy to acknowledge the contributors for their dedicated efforts to make this volume very effective and meaningful. I would like to thank all the co-editors of this volume who tirelessly and sincerely participated in editing and proof reading of the manuscript. It is expected that this revised edition will act as a platform to enrich and strengthen the students and the researchers over their knowledge on vegetable crops.

I would like to specially appreciate Prof. T. K. Bose, the Editor Emeritus, who took active initiative and guided us to complete the fourth revised and illustrated edition of this book. It would have been impossible for us to put this book into table without his encouragement and guidance.

Despite of our best efforts, the editors are conscious about the lapses that might have crept in the manuscript, however, we would like to emphasis on the fact that these lapses or errors are unintentional but circumstantial.

Kolkata   **M.G. Som**
                                                                                                  Senior Editor

Preface to the Fourth Edition from Editors

The first edition of the book entitled 'Vegetable Crops in India' published in 1986 was greatly appreciated and proved valuable to the students, teachers, researchers and extension specialists in horticulture in general and vegetable science in particular, in different parts of the world. Then the book was revised twice in 1993 and 2003 with the title 'Vegetable Crops' in three volumes due to its increasing demand. All the previous editions, which contained updated scientific information of all commercially grown vegetable crops, were considered as valuable contribution on the subject. After about 17 years of the last revised publication in 2003, voluminous information pertaining to vegetable crop improvement, production, protection and postharvest management has been generated necessitating further revision in four volumes.

This particular edition has been enlarged by incorporation of up-to-date information on various aspects of vegetable crop production, protection and improvement particularly through biotechnological interventions with meaningful illustrations, and published in four volumes. Efforts have been made to cover as much information as possible. Fourth volume in this edition includes chapters on cucurbits, bulbous vegetables and perennial vegetables. Most of the contributors have taken care to review up-to-date research works and prepared the manuscript. The revision has been made by the editors and in few cases new contributors have been associated with some chapters.

We express our sincere thanks to all contributors, the Editor Emeritus Prof. T.K. Bose and the Senior Editor Prof. M.G. Som who sincerely participated in revision, editing and proof-reading of the manuscript and we appreciate their commendable efforts to bring out this publication. We express our sincere thank to Prof. P. Hazra for correction of first draft. We express our sincere gratitude to Dr. Praveen Kumar Maurya, Dr. Tridip Bhattacharjee, Mr. Swadesh Banerjee, Mr. Soham Hazra, Mr. Subhrajyoti Chatterjee, SK. Masudul Islam and Miss Debmala Mukherjee Ph.D. scholars, Bidhan Chandra Krishi Viswavidyalaya, India for their help and cooperation for arranging illustrations, necessary typing and correction of citations.

Apart from own collection of photographs, editors have also taken some photographs which are in public domain from different websites as far as possible. Therefore we acknowledge with enormous thanks to all State Agricultural Universities and Research Institutes under Indian Council of Agricultural Research, New Delhi, India; Public and Private International Agricultural Institutes across the world for taking photographs from their websites to make this book a valuable knowledge for the readers across the world. We think images can be powerful teaching tools, as illustrations to in-class lectures, or for studying concepts outside of the classroom. We also think identification of vegetable crops, their package of practices, cultivars, utilization, diseases and pests with the help of suitable illustrations in a consolidated manner will make a dent for the students and teachers of vegetable science.

Our thanks are due to Shri Prateek Mittal, Director of Astral International (P) Ltd., New Delhi for his keen interest in bringing out this publication.

Kolkata   *A. Chattopadhyay • T.K. Maity • J. Kabir*

Editors

Contents

List of Contributors

Banerjee, S.

Department of Vegetable Science, Faculty of Horticulture, Bidhan Chandra Krishi Viswavidyalaya, Mohanpur, 741252, Nadia, West Bengal, India

Chatterjee, R.

Department of Vegetable and Spice Crops, Faculty of Horticulture, Uttar Banga Krishi Viswavidyalaya, Pundibari 736165, Cooch Behar, West Bengal, India

Chatterjee, S.

Department of Horticulture, Centurion University of Technology and Management, Paralakhemundi 761211, Odisha, India

Dey, Shyam Sundar

Division of Vegetable Science, ICAR-Indian Agricultural Research Institute, Pusa 110012, New Delhi, India

Hazra, P.

Department of Vegetable Science, Faculty of Horticulture, Bidhan Chandra Krishi Viswavidyalaya, Mohanpur, 741252, Nadia, West Bengal, India

Islam, Sk Masudul

Department of Vegetable Science, Faculty of Horticulture, Bidhan Chandra Krishi Viswavidyalaya, Mohanpur, 741252, Nadia, West Bengal, India

Jana, J.C.

Department of Vegetable and Spice Crops, Faculty of Horticulture, Uttar Banga Krishi Viswavidyalaya, Pundibari 736165, Cooch Behar, West Bengal, India

Kabir, J.

Department of Postharvest Technology, Faculty of Horticulture, Bidhan Chandra Krishi Viswavidyalaya, Mohanpur 741252, Nadia, West Bengal, India

Maity, T.K.

Department of Vegetable Science, Faculty of Horticulture, Bidhan Chandra Krishi Viswavidyalaya, Mohanpur, 741252, Nadia, West Bengal, India

Malik, Ajaz Ahmed

Department of Vegetable Science, Sher-e-Kashmir University of Agricultural Sciences and Technology, Shalimar 190025, Kashmir, India

Maurya, P.K.

Department of Vegetable Science, Faculty of Horticulture, Bidhan Chandra Krishi Viswavidyalaya, Mohanpur, 741252, Nadia, West Bengal, India

Mohapatra, Priyadarshani P.

Department of Horticulture, College of Agriculture, Central Agricultural University, Kyrdemkulai 793103, Meghalaya, India

Mandal, J.

Department of Horticulture and Postharvest Technology, Palli Siksha Bhavan (Institute of Agriculture), Sriniketan 731236, Visva Bharati, West Bengal, India

Mukherjee, D.

Department of Vegetable Science, Faculty of Horticulture, Bidhan Chandra Krishi Viswavidyalaya, Mohanpur, 741252, Nadia, West Bengal, India

Munshi, A.D.

Division of Vegetable Science, ICAR-Indian Institute of Agricultural Research, Pusa 110012, New Delhi, India

Parthasarathy, V.A.

ICAR – Indian Institute of Spices Research, 32/482 C Narmada Nilayam, Bharathan Bazar, Chelavoor, Calicut 673571, Kerala, India

Sanyal, D.

Department of Horticulture and Postharvest Technology, Palli Siksha Bhavan (Institute of Agriculture), Sriniketan 731236, Visva Bharati, West Bengal, India

Selvakumar, R.

Division of Vegetable Science, ICAR-Indian Agricultural Research Institute, Pusa 110012, New Delhi, India

Seshadri, V.S.

Division of Vegetable Science, ICAR-Indian Agricultural Research Institute, Pusa 110012, New Delhi, India

Shinde, N.N.

Department of Horticulture, Vasantrao Naik Marathwada Krishi Vidyapeeth, Parbhani 431 402, Maharashtra, India

Sontakke, M.B.

Department of Horticulture, Vasantrao Naik Marathwada Krishi Vidyapeeth, Parbhani 431 402, Maharashtra, India

Sureja, Amish K.

Division of Vegetable Science, ICAR-Indian Agricultural Research Institute, Pusa 110012, New Delhi, India

①

CUCURBITS

V.S. Seshadri, A.D. Munshi, R. Selvakumar,
Amish K. Sureja, Shyam Sundar Dey,
V.A. Parthasarathy and P. Hazra

1.0 Introduction

Cucurbits form an important and a big group of vegetable crops cultivated extensively in the sub-tropical and tropical countries. This group consists of wide range of vegetable, either used as salad (cucumber) or for cooking (all the gourds) or for pickling (cucumber) or as dessert fruits (muskmelon and watermelon) or candied or preserved (ash gourd). They include mostly seed propagated ones, besides few perennials like chow-chow and ivy gourd (*kundsru*) (Table 1). They are of tremendous economic importance as food plants and it has been difficult to estimate or quantify it because of absence of reliable statistics of their area and production in India. They are extensively grown in mixed cropping of more than one kind, in long and meandering river-beds.

Table 1: List of Cultivated Cucurbits

Sl. No.	Common English Name	Hindi Name	Botanical Species	Chromosome Number (2n)
1	Bitter gourd	Karela	*Momordica charantia* L.	22
2	Spine gourd	Kakoda	*Momordica dioica* Roxb.	28
3	Sweet gourd (of Assam)	Kheksa	*Momordica cochinchinensis* Spreng.	28
4	Teasle gourd	Kakrol	*Momordica subangulata* subsp. renigera	56
5	Bottle gourd	Lauki or Ghiya	*Lagenaria siceraria* (Mol.) Standl.	22
6	Cucumber	Khira	*Cucumis sativus* L.	14
7	Indian squash (Round melon)	Tinda	*Praecitrullus fistulosus* Pang.	24
8	Muskmelon	Kharbuza	*Cucumis melo* L.	24
9	Snapmelon	Phoot	*Cucumis melo* var. *momordica*	24
10	Long or serpent melon	Kakri (of North India)	*Cucumis melo* var. *utilissimus* (syn. *C. melo* var. *flexuosus* (Naud.)	24
11	Pumpkin	Sitaphal or Kasiphal	*Cucurbita moschata* Duch ex Poir.	40
12	Squash	Chappan kaddu	*Cucurbita pepo* L.	40
13	Squash or pumpkin or winter squash	–	*Cucurbita maxima* Duch.	40
14	Ribbed or ridge gourd	Kali tori	*Luffa acutangula* (Roxb.) L.	26
15	Sponge gourd	Ghiatori	*Luffa cylindrica* Roem (syn. *L. aegyptiaca*)	26
16	Snake gourd	Chichinda	*Trichosanthes cucumerina* (syn. *T. anguina* L.)	24
17	Pointed gourd	Parwal	*Trichosanthes dioica* Roxb.	24

Sl. No.	Common English Name	Hindi Name	Botanical Species	Chromosome Number (2n)
18	Watermelon	Tarbuz	*Citrullus lanatus* (Thunb.) Matsumura and Nakai (syn. *C. vulgaris* Schrad)	22
19	Wax or ash gourd	Petha kaddu	*Benincasa hispida* (Thunb.) Cogn.	24
20	Ivy gourd or scarlet gourd	*Kundsru* or *Tondli*	*Coccinia cordifolia* (Voigt) L. (syn. *C. indica* W. and A)	24
21	Chow-chow or chayote	*Squash* or *ishkus*	*Sechium edule* (Jack) Sw.	28
22	Stuffing cucumber or slipper gourd	*Meetha karela*	*Cyclanthera pedata* Schrad.	32

2.0 Composition and Uses

2.1.0 Composition

The use of cucurbits as food plants is not primarily for calorie, mineral or vitamin values, since they are poor or only modest sources of these nutrients (Table 4). Fruit water content ranges from 86 (winter squash) to 95 per cent (cucumber), and caloric content per 100 g fresh material ranges from 15 kcal for cucumber to 46 kcal, on average, for winter squash (Ensminger *et al.*, 1983). The minor cucurbits addressed in this volume are comparable to the major cucurbits for water content and caloric value, *e.g.*, bottle gourd (*Lagenaria siceraria* (Molina.) Standley) 92 per cent water and 26 kcal/100 g fresh fruit weight and wax gourd [*Benincasa hispida* (Thunb.) Cogn.] 96 per cent water and 13 kcal/100 g fresh fruit weight (Ensminger *et al.*, 1983). There are few exceptions like bitter gourd richer in vitamin C, pumpkin containing high carotenoid pigments, *kakrol* (*Momordica dioica*) high in protein and chow-chow fairly high in calcium. Some cultivars of squashes and pumpkins are relatively high in energy and carbohydrates. For people lacking modern refrigeration ash gourd and pumpkin are ideal, because they can be stored for a long time to tide over unfavourable season.

2.1.1 Cucurbitacins

Biochemically the cucurbits are characterized by bitter principles, called 'cucurbitacins'. A systematic search for these substances in the family indicates that great majority of species contain bitter principles in some portion of the plant at some stage of development. Chemically cucurbitacins are tetracyclic triterpenes having extensive oxidation level. They occur in nature, free as glycosides or in complicated mixtures which have been shown to be of taxonomic significance.

Highest concentrations of cucurbitacins are found in fruits and roots, leaves being normally or slightly bitter. Bitter principles found in roots differ from those in fruit. Seedlings, especially radicles contain cucurbitacins, which are considered primary and hence significant. The pollen also carries bitter principles and hence when bitter pollen fertilizes non-bitter ovules, the resulting fruit will be bitter. This phenomenon which is very common in cucumber, bottle gourd, *etc.*, is called *metaxenia* and so consumers often try to remove bitterness from ends of the cucumber fruits. In a plant three situations can occur: (i) fruit and vegetative parts bitter, (ii) fruit not bitter but vegetative parts bitter and (iii) bitterness absent in both vegetative parts and fruits. Kano *et al.* (1997) postulated that in cucumber more cucurbitacin C, the bitter constituent, is synthesized in young, vigorous plant than in older, less vigorous plants. Cucumber beetle developed an extraordinary detoxification mechanism that enabled these insects to grow, develop and reproduce on highly toxic level of cucurbitacins. These beetles are attracted to feed on bitter plant organs. Interestingly, eggs produced by these beetles have substantial quantities of cucurbitacins that protect them against ant predators (Metcalf and Rhodes, 1990). One single dominant gene is responsible for the formation of bitter cucurbitacin compounds in *Lagenaria*, *Cucumis*, *Cucurbita* and *Citrullus*. Since cucurbitacin is attractive to cucumber beetle, resistance is achieved by selecting for reduced cucurbitacin content (Robinson *et al.*, 1976). Some kind of coevolutionary relationship between Cucurbitaceae and beetles (New World *Diabroticites* and Old World *Aulacophorini*) has been envisaged through the secondary plant substances such as oxygenated tetracyclic triterpenes (Metcalf *et al.*, 1983).

Many cucurbits have both bitter and non-bitter forms within the same species. In the bitter forms of *C. anguria*, the bitterness increases considerably as the fruit ripens. Cucurbitacins are among the most bitter substances known and are extremely toxic to mammals. In *C. anguria*, the main bitter principle is cucurbitacin B ($C_{32}H_{48}O_8$) with a much smaller amount of cucurbitacin D ($C_{30}H_{46}O_7$) and traces of cucurbitacins G and H. Toxicity studies showed the juice of the fruits to be highly toxic to rats (LD50 1.6 mg/kg). The toxicity is reported to be reduced more than 100-fold if the juice is first boiled. Studies on then larvicidal activity of aqueous, ethanolic and citric acid extracts from *C. anguria* on *Aedes aegypti*, the yellow fever and dengue fever mosquito, showed that concentrations of 0.5 mg/mL after 24 h exposure caused larval mortalities upto 40 per cent (Petrus, 2014).

The *Bi* gene confers bitterness to the whole plant (Huang *et al.*, 2009) and a genetic and biochemical study showed that *Bi* encodes a cucurbitadienol synthase that catalyses the first committed step in cucurbitacin C biosynthesis in cucumber (Shang *et al.*, 2014). Two bHLH (basic helix–loop–helix) transcription factors regulate cucurbitacin C biosynthesis by upregulating *Bi* in the leaves (*Bl*) and fruits (*Bt*) directly via binding in the E-box elements of the *Bi* promoter (Shang *et al.*, 2014). In cucumber, no selective sweep was associated with the *Bi* gene (Qi *et al.*, 2013), suggesting that loss of cucurbitacin biosynthesis in the whole plant would have been deleterious, even in cultivated cucurbits in terms of defence against generalist herbivores. Examination of different cucumber lines with varying bitterness has revealed that domestication of nonbitter cucumber occurred via the down regulation of *Bt*, either by mutation in its *cis* acting elements or by affecting the binding site (Qi *et al.*, 2013; Shang *et al.*, 2014; Zhou *et al.*, 2016). A genomic analysis of homologues comparing cucumber, melon and watermelon (producing cucurbitacin C, B and E, respectively) revealed a convergent domestication sweep at the *Bt* loci and the loss of bitterness is, in all cases, due to convergent mutations at this locus (Shang *et al.*, 2014; Zhou *et al.*, 2016). Selected mutations in an upstream regulatory gene controlling the expression of a given pathway in specific tissue may be common during domestication, as this can help avoid pleiotropic effects associated with modification of the pathway itself (Lenser and Theiben, 2013). Different type of cucurbitacin present in cucurbits have been shown in Table 2.

Table 2: Different Types of Cucurbitacin Present in Cucurbits

Cucurbits	Type of Cucurbitacin	Variety/Hybrid	Part	Reference
Watermelon	Cucurbitacin E	Warpaint	Fruit	Herrington *et al.* (1986)
		Hawkesbury	Fruit	Chambliss *et al.* (1968)
		Acc. 242	Fruit	Chambliss *et al.* (1968)
Cucumber	Cucurbitacin C	Bitter cucumber	Fruit	Enslin (1954)
	Cucurbitacin C	Hanzil	Fruit	Guha and Sen (1975)
	Cucurbitacin E	Sweet form	Fruit	Guha and Sen (1975)
Squash	Cucurbitacin E	Yellow straight-neck	Fruit	Rymal *et al.* (1984)
	Cucurbitacin E	Castleverde	Fruit	Rymal *et al.* (1984)
	Cucurbitacin E	Blackjack	Fruit	Hutt and Herrington (1985)

2.1.2 Seed Oils and Proteins

The nutritive value of cucurbit seeds is due to their high oil and protein content. The seed oil of Cucurbitaceae falls into 2 groups: one with the palmitic-oleic-linoleic acid composition, the other with conjugated triene acids, pumicic acid, α-oleostearic acid. The taxonomic significance of this difference is yet unclear. Seed proteins of Cucurbitaceae are comparable in nutritive value to those of *Leguminosae*, though like other plant proteins they tend to be low in lysine and sulphur containing amino acids. They are generally richer in methionine than those of the legumes and selective enrichment may be a possibility in some cases.

At least 150 fatty acids are known to occur as components of seed oils. They are of two categories, one with simple saturated and unsaturated fatty acids, the other with highly unsaturated fatty acids with conjugated double bonds. Genera like *Cucumis* and *Cucurbita* have usually ordinary unsaturated acids (oleic and linoleic), while other like *Momordica* and *Trichosanthes* are noted for acids with conjugated double bonds (pumicic and α-cleostearic).

Chang *et al.* (1996) reported that oil contents of the seeds of bitter gourd ranged from 41 to 45 per cent and the oils contained 63-68 per cent eleo-stearic acid and 22-27 per cent stearic acid. The ratio of stearic to eleostearicin seed oil is ten times greater than that in tung oil, an industrially important 'fast-drying oil' used in paints and varnishes. The defatted meals contained 52-61 per cent protein and would be a good source of methionine.

Cucurbita moschata seeds contained 40.27 per cent protein, 34.59 per cent crude oil, 13.79 per cent crude fibre and 4.45 per cent ash. Triglycerides were the major lipid fraction in seeds and oleic, linoleic and palmitic acids were the predominant fatty acids in seed oil (Aboul-Nasr *et al.,* 1997).

2.2.0 Uses

Cucumber is used as salad, pickle and also as cooked vegetables. The seeds are used in Ayurvedic preparations. Muskmelon is eaten raw. It prevents constipation and is nutritious. The fruits of other cucurbits (bottle gourd, bitter gourd, pointed gourd, spine gourd, coccinia, ash gourd, ridge gourd, sponge gourd and snake gourd) are mostly cooked in various ways. Bottle gourd and bitter gourd fruits are also used for making pickle.

Bottle gourd and ash gourd are used for preparation of sweets. The sweet meat of ash gourd is known as *petha*. It is also largely used for making confectionery. The leaves, and stem or shoots of pumpkin, bottle gourd and pointed gourd are cooked and used as vegetables. The hard shell of bottle gourd fruit is used for making musical instrument.

Extracts of bitter gourd fruits are used to cure rheumatism and disorders of liver and spleen. These are also recommended for curing diabetes and asthma. The fruits also have wormicidal properties. The roots, stems and leaves of spine gourd are used to cure skin disease, bronchitis and diabetes. Ash gourd is good for patient having a weak nervous system.

Luffa sponge is a lignocellulosic material composed mainly of cellulose (60 per cent), hemicelluloses (30 per cent) and lignin (10 per cent) (Rowell *et al.,* 2002; Mazali and Alves, 2005). The dry fibre of the mature fruits of sponge gourd is used for bathing purposes, cleaning utensils and also utilized by many countries commercially for making natural scrub. An industrial oil is extracted from its seed. Parkash *et al.* (2002) reported presence of *luffacyclin*, a ribosome inactivating peptide with antifungal activity against *Mycospharella arachidicola* and *Fusarium oxysporum* from *L. cylindrica* seeds. *Luffaculin 1*, a ribosome inactivating protein purified from *L. acutangula* is reported to possess anti-HIV1 activities (Jing *et al.,* 2008).

Two trypsin inhibitors and a ribosome-inactivating peptide (*luffangulin*) have been isolated from ridged gourd seeds. The glycoprotein '*luffaculin*', also isolated from the seeds, exhibits abortifacient, antitumor, ribosome-inactivating and immunomodulatory activities (Fernando and Grun, 2001). Rahman *et al.* (2013) also reported that *M. charantia* reduces serum sialic acid in type 2 diabetic patients, which results in the delay of the process of atherosclerosis. The antidiabetic effects of *M. charantia* (bitter melon) and its medicinal potency were also reported by Joseph and Jini (2013). The seed aril of *M. cochinchinensis* fruits is the richest source of lycopene and reported to contain even 70 times more lycopene than tomato (Ishida *et al.,* 2004; Rao and Rao, 2007). It is reported to have anticancer activities, particularly those against prostate cancer. This trait attracts the industry for the commercial utilization of the gac in its goal of developing various lycopene-rich health supplements. Health promoting molecules present in cucurbitaceae are given in Table 3.

Uses of Cucurbits

Different Recipes of Cucurbit Vegetables in Indian Cuisine

Table 3: Health Promoting Molecules Present in Cucurbitaceae

Cucurbits	Biomolecules	Uses
Cucumber	Alkaloid, 9-beta-methyl-19-norlanosta-5-ene glycoside, Steroid, Saponin and Tannin	Antiulcer, diuretic and anthelmintic
Pumpkin	Alkaloids, Flavonoids, and Palmitic, Oleic and Linoleic acids	Antidiabetic, antioxidant, anticarcinogenic, antiinflammatory
Watermelon	Lycopene, Cucurbitacin	Diuretic, cardiovascular activity, anti-inflammatory activity
Summer squash	Cucurbitacins, Avenasterol, Spinasterol, Triterpenoids, Sesquiterpenoids, Squalene, Tocopherols, Carotenoids	Antiandrogenic activity, immunological activity, antiviral activity, antifungal activity, cardiovascular activity, anti-inflammatory activity and hepatoprotective activity
Winter squash	Flavonoids, Polyphenolics, Saponins, Cucurbitaxanthin and α-tocopherol	Antidiabetic, antitumor, antihypertensive, antiinflammatory and immunomodulatory
Bottle gourd	β-carotene, Cucurbitacins, Saponins, Flavone-C-glycoside and Polyphenol	Antioxidant, antihepatotoxic, antitumor, immunoprotective and antipoliferative activity
Bitter gourd	Steroidal glycosides like Charantin, Insulinomimeticlectins and Alkaloids	Antidiabetic, antitumor
Ridge gourd	Flavonoids, Saponins, Luffangulin, Sapogenin, Oleanolic acid and Cucurbitacin	Diuretic, antioxidant, expectorant, laxative, purgative, hypoglycemic agent
Sponge gourd	Alkaloids, Saponins, Carotenoids, Terpenoids	Antioxidant, antimicrobial, anticancer, cardioprotective, gastroprotective, antidiabetic, hypolipidemic, hepato-protective
Snake gourd	Triterpenoids, Saponins, Cucurbitacins	Anti-inflammatory activity, ant-diabetic activity

3.0 Origin, Taxonomy and Cytogenetics

Being warm season crops, cucurbits are of tropical origin, mostly in Africa, tropical America and Asia, chiefly South East. Phylogeny of this large group of crops has not been studied adequately, possibly with an exception of *Cucurbita*. Recently, Chomick *et al.* (2019) reviewed in detail the origin, domestication and phylogenetic distribution of cucurbitaceae crops:

3.1.0 *Cucumis* (Cucumber and Muskmelon)

Cucumis belongs to the family Cucurbitaceae, subfamily Cucurbitoideae, and is currently placed in the tribe Benincaseae (Jeffrey, 2005). Taxonomy of *Cucumis* was first described by Linnaeus in 1753 (Ghebretinsae *et al.*, 2007). According to this classification, the genus *Cucumis* contains seven species, all of which were cultivated or economically useful. There have been a number of taxonomic placements of *Cucumis* since the work of Linnaeus (Pangalo, 1950; Jeffrey, 1962, 1967, 1980, 1990; Kirkbride, 1993; Schaefer, 2007).

Table 4: Composition of Nutrients in Cucurbitaceous Vegetables*

Composition	1	2	3	4	5	6	7	8	9	10	11	12	13	14	15
Moisture	92.17	90.87	91.60	91.24	95.17	94.50	94.63	93.78	92.96	93.52	92.80	92.78	92.40	91.73	91.85
Protein (g)	0.79	1.44	1.34	1.61	0.53	0.42	0.49	0.66	0.71	0.83	0.98	1.39	1.22	0.87	0.84
Ash (g)	0.70	0.86	0.81	0.88	0.36	0.34	0.13	0.38	0.54	0.52	0.52	0.58	0.47	0.47	0.58
Total Fat (g)	0.14	0.24	0.24	0.26	0.13	0.12	0.40	0.15	0.16	0.18	0.24	0.24	0.24	0.18	0.16
Dietary Fiber (g)	3.37	3.78	2.96	3.72	2.12	1.72	2.11	1.55	2.14	2.13	2.46	3.00	3.25	2.53	2.56
Carbohydrate (g)	2.52	2.82	2.53	2.29	1.68	2.53	2.25	3.47	3.48	2.82	3.01	2.01	2.41	4.22	4.00
Energy (KJ)	73	87	79	81	46	57	54	79	82	73	82	73	80	103	97
Thiamine (B$_1$)	0.03	0.05	0.06	0.06	0.03	0.03	0.03	0.01	0.02	0.02	0.02	0.04	0.04	0.03	0.03
Riboflavin (B$_2$)	0.01	0.04	0.04	0.04	0.01	0.01	0.01	0.03	0.01	0.01	0.01	0.02	0.02	0.02	0.03
Niacin (B$_3$)	0.12	0.27	0.29	0.30	0.14	0.14	0.14	0.23	0.35	0.35	0.36	0.19	0.51	0.44	0.41
Pantothenic Acid (B$_5$)	0.37	0.33	0.36	0.28	0.56	0.50	0.59	0.21	0.45	0.32	0.34	0.27	0.28	0.16	0.18
Total B$_6$	0.18	0.05	0.04	0.05	0.02	0.02	0.01	0.07	0.06	0.07	0.04	0.08	0.05	0.05	0.08
Biotin (B$_7$)	2.01	5.76	51.45	6.85	2.55	2.33	2.54	1.06	2.82	2.97	3.13	2.96	2.87	1.41	1.63
Total Folates (B$_9$)	14.11	60.28	50.87	60.03	41.99	49.59	46.31	63.03	16.84	14.67	18.77	48.68	50.13	31.60	24.14
Total Ascorbic Acid	11.41	46.53	0.06	54.30	4.33	4.54	4.54	20.21	6.11	6.21	6.24	17.62	21.08	7.29	8.04
Ergocalciferol (D$_2$)	1.35	1.92	1.90	1.83	0.74	0.70	0.60	5.46	1.26	1.36	1.59	6.25	6.06	1.07	1.40
Phylloquinones (K$_1$)	27.15	4.55	4.85	4.70	2.10	2.06	1.80	1.78	8.20	8.00	7.50	19.15	19.10	80.80	83.70
Lutein	18.29	282	244	313	28.88	29.64	28.98	8.56	2.22	4.86	3.19	170	157	394	161
Zeaxanthin	5.76	5.83	6.96	7.20	-	8.60	2.50	1.80	-	-	-	5.74	4.50	26.60	34.91
β-Carotene	-	122	126	130	44.05	47.13	44.82	1.57	5.33	4.80	5.55	134	498	363	149
Total Carotenoids	66.67	717	505	481	95.12	92.56	97.02	24.53	171	172	134	147	586	1319	1449
Magnesium (mg)	19.95	32.14	31.58	33.34	10.93	10.89	12.90	13.05	20.38	18.48	20.34	19.60	18.87	13.27	10.43
Manganese (mg)	0.09	0.25	0.23	0.50	0.13	0.11	0.15	0.17	0.08	0.09	0.07	0.16	0.12	0.08	0.07
Phosphorus (mg)	29.07	44.90	40.21	44.75	16.01	16.99	26.86	21.61	28.34	29.74	23.17	36.90	26.29	24.51	22.18

Composition	1	2	3	4	5	6	7	8	9	10	11	12	13	14	15
Potassium (mg)	372	326	282	356	124	116	171	120	183	198	185	198	167	186	253
Selenium (mg)	1.15	4.97	3.72	5.22	1.77	1.80	2.05	0.16	0.17	0.19	0.14	-	-	0.34	0.37
Sodium (mg)	0.77	13.09	12.59	11.16	1.46	1.52	1.35	1.28	6.33	6.11	8.16	1.53	0.18	5.21	8.81
Zinc (mg)	0.13	0.31	0.36	0.43	0.15	0.15	0.18	0.10	0.17	0.19	0.16	2.20	0.13	0.14	0.11
Calcium (mg)	19.39	21.36	16.27	17.62	15.42	15.05	16.64	18.64	16.39	19.25	21.98	34.39	37.12	24.10	23.06
Iron (mg)	0.47	1.15	1.08	1.28	0.26	0.28	0.34	0.48	0.46	0.59	0.45	0.38	0.29	0.29	0.36
Total polyphenols (mg)	10.47	49.76	50.46	52.98	34.76	31.83	38.20	5.04	14.10	13.92	14.56	1.76	1.60	15.21	15.25
Citric Acid (mg)	36.91	59.33	66.42	65.68	31.16	31.12	33.41	60.10	81.51	60.82	37.31	22.68	25.89	159	33.32
Mallic Acid (mg)	-	-	-	-	-	-	-	-	68.75	67.07	64.48	46.41	51.63	13.51	3.34
Succinic Acid (mg)	7.49	7.44	9.15	64.96	65.18	64.80	32.49	32.49	40.70	47.40	50.82	254	239.14	-	-
Fumaric Acid (mg)	0.57	1.41	1.51	1.42	0.16	0.14	0.12	-	4.13	3.50	4.40	0.56	0.41	1.13	1.35
Quinic Acid (mg)	-	-	-	-	-	-	-	37.30	4.97	3.38	5.70	90.46	92.82	37.98	-
Total Oxalate (mg)	4.89	45.40	48.83	43.37	3.75	2.53	1.70	1.55	12.13	10.49	11.55	7.72	8.15	57.33	41.22
Total Starch (g)	1.01	0.92	0.93	0.87	0.67	0.67	0.58	1.17	1.17	0.58	0.53	0.22	0.12	0.20	0.33

1. Ash gourd; 2. Bitter gourd, jagged, teeth ridges, elongate; 3. Bitter gourd, jagged, teeth ridges, short; 4. Bitter gourd, jagged, smooth ridges, elongate; 5. Bottle gourd, elongate, pale green; 6. Bottle gourd, round, pale green; 7. Bottle gourd, elongate, dark green; 8. Chow-chow; 9. Cucumber, green, elongate; 10. Cucumber, green, short; 11. Cucumber, orange, round; 12. Ivy gourd big; 13. Ivy gourd small; 14. Pumpkin, green, cylindrical; 15. Pumpkin, orange, round

Composition	16	17	18	19	20	21	22	23	24	25	26	27	28
Moisture (per cent)	85.82	94.99	94.27	94.81	94.92	94.35	94.41	92.83	93.15	92.97	91.84	94.54	95.33
Protein (g)	4.21	0.91	0.98	0.98	0.89	0.54	1.02	1.10	1.31	0.42	0.53	0.60	0.59
Ash (g)	2.24	0.44	0.53	0.42	0.45	0.41	0.51	0.93	1.08	0.52	0.48	0.13	0.12
Total Fat (g)	0.74	0.14	0.13	0.25	0.25	0.26	0.17	0.51	0.44	0.35	0.26	0.16	0.16
Dietary Fiber (g)	2.25	1.81	1.85	2.27	2.27	2.29	2.00	2.30	1.84	1.51	1.49	0.70	0.78
Carbohydrate (g)	4.75	1.72	2.24	1.27	1.23	2.15	1.90	2.33	2.20	4.24	5.40	3.86	3.02
Energy (KJ)	185	55	64	52	50	61	58	84	79	97	116	85	70
Thiamine (B_1) (mg)	0.07	0.02	0.02	0.03	0.03	0.03	0.02	0.05	0.03	0.01	0.01	0.02	0.02
Riboflavin (B_2) (mg)	0.13	0.01	0.01	0.03	0.03	0.02	0.03	0.09	0.02	0.01	0.02	0.02	0.02
Niacin (B_3) (mg)	1.49	0.20	0.21	0.34	0.33	0.33	0.56	1.03	0.42	0.41	0.43	0.28	0.30
Pantothenic Acid (B_5) (mg)	0.36	0.28	0.25	0.27	0.27	0.31	0.39	0.99	0.72	0.13	0.11	0.19	0.19
Total B_6	0.17	0.07	0.09	0.10	0.07	0.06	0.06	0.25	0.20	0.05	0.06	0.10	0.07
Biotin (B_7) (µg)	3.40	2.27	2.22	2.50	2.43	2.50	3.26	1.02	1.13	0.75	0.80	0.59	0.57
Total Folates (B_9) (µg)	33.82	29.26	27.36	18.34	16.52	17.74	43.23	18.85	21.50	22.31	20.23	5.88	5.55
Total Ascorbic Acid (µg)	12.33	5.42	8.10	2.72	2.85	2.30	14.20	15.78	16.71	22.76	21.32	13.26	11.45
Ergocalciferol (D_2) (µg)	3.19	0.37	0.34	2.67	3.08	3.12	2.76	0.40	0.38	4.41	2.33	0.56	0.56
Phylloquinones (K1) (µg)	243	11.23	11.57	8.35	8.40	8.30	2.73	41.05	53.28	5.70	1.74	2.10	2.80
Lutein (µg)	6139	129	127	38.9	39.60	32.75	8.19	87.86	294	20.17	22.48	914	981
Zeaxanthin (µg)	45.28	19.50	18.83	3.67	5.41	4.50	2.22	16.23	25.60	2.52	2.69	25.95	23.40
β-Carotene (µg)	1455	348	349	61.29	61.64	62.84	7.96	85.79	69.90	771	6.87	605	576
Total Carotenoids (µg)	8247	838	851	190	188	188	49.81	963	1177	925	92.15	4176	4300
Lycopene (µg)	-	-	-	-	-	-	-	-	-	-	-	1477	1257
Magnesium (mg)	84.21	16.15	17.66	18.70	21.70	15.07	18.96	15.41	10.82	11.62	9.81	9.91	7.42

Composition	16	17	18	19	20	21	22	23	24	25	26	27	28
Manganese (mg)	1.14	0.11	0.15	0.14	0.19	0.11	0.10	0.13	0.21	0.04	0.04	0.04	0.03
Phosphorus (mg)	64.54	33.06	39.25	23.27	31.03	21.33	30.37	21.38	32.03	17.28	13.09	11.33	8.09
Potassium (mg)	423	118	125	100	104	84.00	56.18	178	131	206	196	124	126
Selenium (mg)	1.38	0.59	-	-	-	-	0.22	0.21	0.30	0.88	1.35	-	-
Sodium (mg)	12.20	4.71	6.27	7.07	5.04	2.50	20.61	0.40	0.39	14.94	15.78	1.89	1.62
Zinc (mg)	0.90	0.22	0.26	0.14	0.20	0.11	0.20	0.29	0.27	0.09	0.09	0.10	0.07
Calcium (mg)	271	13.70	14.96	24.60	27.11	17.90	19.68	17.26	20.98	9.80	9.02	5.29	4.35
Iron (mg)	5.58	0.42	0.50	0.32	0.47	0.20	0.41	0.52	0.34	0.18	0.21	0.22	0.16
Total polyphenols (mg)	20.33	15.21	15.25	8.31	11.30	6.78	6.10	5.90	2.61	42.94	295	5.89	5.78
Citric Acid (mg)	7.74	152	112	110	123	126	119	47.81	47.81	5.71	9.08	47.35	52.65
Mallic Acid (mg)	14.35	12.36	45.56	54.35	68.18	62.38	63.86	2.32	2.32	1.32	2.94	7.27	7.55
Succinic Acid (mg)	54.07	-	-	-	-	-	-	55.30	55.30	82.54	63.45	-	-
Fumaric Acid (mg)	-	11.35	10.13	1.67	1.91	1.89	15.51	49.65	45.17	0.42	0.55	14.58	15.18
Quinic Acid (mg)	20.22	32.05	-	-	36.82	33.48	33.36	-	-	-	-	-	-
Total Oxalate (mg)	13.61	29.55	35.85	24.22	13.44	14.25	3.36	17.83	16.06	2.62	1.88	0.72	0.55
Total Starch (g)	2.10	1.40	1.19	1.20	1.19	1.18	1.07	1.61	1.37	2.99	2.36	3.66	3.00

16. Pumpkin leaves, tender 17. Ridge gourd; 18. Sponge gourd; 19. Snake gourd, long, pale green; 20. Snake gourd, long, dark green; 21. Snake gourd, short; 22. Tinda, tender; 23. Zucchini, green; 24. Zucchini, yellow; 25. Muskmelon, orange flesh; 26. Muskmelon, yellow flesh; 27. Watermelon, dark green; 28. Watermelon, pale green

* Longvah *et al.* (2017).

The genus comprises about 32 species distributed over two distinct geographic areas (i) South-East of Himalayas is an important region of Asiatic group with basic number of x=7 to which cucumber belongs, (ii) Africa group, comprising large part of Africa, Middle East, Central Asia extending to Pakistan and South Arabia. Mostly species with basic number x=12 are found in this region with few tetraploids and hexaploids. Muskmelon occurs in this region, besides West Indian Gherkin. Other wild species originating mostly from arid and/or semi-arid regions of Africa are cultivated as ornamental plants (*e.g.*, *C. dipsaceus* – "hedge hog gourd" and *C. myriocarpus* – "goose berry gourd") (Garcia-Mas *et al.*, 2004; Rubatzky and Yamaguchi, 1997).

The evolution of genus *Cucumis*, in particular the relationship between two basic number x=7 and x=12 is not clear. Whitaker (1933), Bhaduri and Bose (1947), Ayyangar (1967) assumed that species with 2n=24 chromosomes have arisen from species with 2n =14 chromosomes by fragmentation. Trivedi and Roy (1970) however, were of the opinion that by fusion of chromosomes with (sub) terminal centromere species with 14 chromosomes have arisen from species with 24 chromosomes. Roy and Singh (1974) related the degree of relationship between the species to the relative length of total chromatin of their chromosome complement and on the basis of chromosome structure, they considered *Cucumis melo* as a derived species.

Jeffrey (1980) classified the entire genus *Cucumis* into two subgenera:

1. Subgenus *Cucumis* x = 7. Three or four Sino-Himalayan species including *C. sativus*, *C. hystrix* (syn. *C. muriculatus*).

2. Subgenus *Melo* x = 12. Twenty five species mostly in tropical and South Africa.

    (i)   *anguria group-dioecious, monoecious* or andromonoecious perennials or annuals with yellowish or brown stripped browny fruit (20 spp.), closely related and partially fertile interspecific hybrids, *C. anguria*, *C. dipsaceus*, *C. prophetarum*, *C. myriocarpus* and *C. sacleuxii*.

    (ii) *metuliferus* group – monoecious annuals with red spiny fruits, *C. metuliferus*.

    (iii) melo group – monoecious, andromonoecious perennials or annuals with smooth fruits, *C. melo, C. sagittatus* (syn. *C. angolensis*), *C. dinteri*, *C. humifructus*.

    (iv) *hirsutus* – dioecious perennials with smooth orange fruit, *C. hirsutus*.

Taxonomic classification of cucumber (*C. sativus* L.) and melon (*C. melo* L.) in the family Cucurbitaceae (Chung *et al.*, 2006).

Cucurbitaceae (Family)

   Zanonioideae (Subfamily) Cucurbitoideae (Subfamily)

      Melothrieae (Tribe)

         Cucumis (Genus)

            Cucumis (Subgenus)

C. sativus L. (Species)

    var. *sativus*

    var. *hardwickii*

C. hystrix Chakr. (Species)

    Melo (Subgenus)

C. melo L. (Species)

    subsp. *agrestis*

    subsp. *melo*

Cucumber is thought to be indigenous to India, the evidence being more of a circumstantial nature, as the cucumber has never been found in wild state. De Candolle (1882) thought that cucumber has been cultivated for over 3000 years in India. *Cucumis hardwickii*, a small bitter cucumber with sparse and stiff spines has been found wild in the foothills of the Himalayas. The free hybridization with cultivated *sativus* with no reduction of fertility in F_2 generation (Deakin *et al.*, 1971) suggested that *C. hardwickii* be a feral form or more likely a progenitor of the cultivated cucumber. There is no evidence that unlike the other species like *Cucumis melo* and other feral species which have extensive distribution in Africa, cucumber has been found even in semi-wild state.

Muskmelon is said to be a native of tropical Africa more specifically in the eastern region, south of Sahara desert. Leippik (1966) considered that the phytogeographic distribution would point towards East Africa adjoining Arabian Peninsula and eastern part of Mediterranean and he relied on host-parasite relationship in respect of several diseases. These areas have the greatest concentration of several wild species (2n=24, 48, 72) that showed phylogenetically older characters. There are several primitive species with xerophytic and perennial habit and there are some tetraploids. Many of these species are good sources of resistance to pests and diseases (Leippik, 1967) and as also those from South Africa (Kroon *et al.*, 1979 and Dane *et al.*, 1980). Most of the species are tolerant to aphids and powdery mildew and *C. anguria*, *C. ficifolius* and *C. metuliferus* are resistant to nematodes and there are several species which carry resistance to bean spider mite and white fly. This promoted interest in the interspecific hybridization among the 24 (also 48) chromosome species. Studies by Deakin *et al.* (1971) showed that all the wild Africa species did not cross with *C. melo* either at diploid or tetraploid level. They grouped *Cucumis* spp. under 5 groups, where *C. sativus* and other 14 chromosome species stand a part in one group, and cultivated *melo* in another. The *anguria* group consisted of *C. longipes*, *C. africanus*, *C. dipsaceus*, *C. leptodermis*, *C. myriocarpus*, *C. prophetarum*, *C. zeyheri* and tetraploids (2n=48), *C. ficifolius* and *C. heptadactylus*. The fourth group consisted of *C. dinteri* and its conspecific *C. sagittatus*, *C. humifructus* and the fifth group wasty pified by *C. metuliferus*. The later investigations which included cytological studies by Dane *et al.* (1980) confirmed the conspecificity between *C. dinteri* and *C. sagittatus* and between *C. myriocarpus* and *C. leptodermis*. The cross-compatible group listed by them included annual species *C. africanus*, *C. anguria*, *C. dipsaceus*, *C. leptodermis* and *C. myriocarpus,* perennial ones *C. ficifolius* and *C. zeyheri,* perennial *C. aculeatus, C. heptadactylus* and *C. zeyheri* and perennial hexaploid *C. figarei.* Dubravee (1994) reported a new adventive species, *Cucumis metuliferus* in the Croatian flora. The vegetable is of interest as a vegetable with edible, decorative fruits. This species is used as a food plant in Africa and other continents.

Whitaker and Davis (1962) thought that Central Asia comprising some parts of southern Russia, Iran, Afghanistan, Pakistan and North West India may be regarded as a secondary centre of muskmelon. The huge diversity of this polymorphic species *C. melo* has been recognized as a complex of cultigenic species with free intercrossing possible. In fact, Russian botanists like Pangalo (1951) and Zhukovsky (1962) have raised *C. melo* to generic status of *Melo* sp. retaining cucumber under *Cucumis.* According to them, no African wild species could be suspected to be an ancestor of the cultivated or wild melons. But Alcazar's (1978) studies on alloenzyme variation on wild and cultivated species of *Cucumis* totaling 21, suggested India rather than Africa, as the area of origin of cultivated melons. Dane *et al.* (1980) on the basis of extensive cytogenetical studies of 43 wild species from South, East and West Africa, consisting of annuals and perennials, diploid, tetraploids and hexaploid, however, considered South Africa to be the primary gene centre of the genus.

All these bring out the wide and divergent observations, which make it difficult to arrive at a precise decision of the origin of the polymorphic taxon *C.melo.*

The third but less important cultivated species *Cucumis anguria* L. (2n=24), familiarly called as *West Indian Gherkin* (but sometimes erroneously referred to pickling cucumber) has been known in semi-wild state in West Indies and is now considered to be conspecific with *C. longipes* of Africa and probably became stabilized after its introduction by slave trade from Africa. The wild form was retained as var. *longipes* and the cultivated as var. *anguria* (Meeuse, 1958; Deakin *et al.,* 1971).

The most comprehensive placement of *Cucumis* was proposed by Kirkbride (1993). On the basis of his investigations, the genus *Cucumis* was divided into two subgenera with different geographical origin and basic chromosome numbers. Subgenus *Melo* (30 spp., n = 12) was originated in Africa and was partitioned into two sections (*Melo* and *Aculeatosi*), whereas subg. *Cucumis* (two spp., n = 7) was originated in Asia. However, this taxonomic treatment was challenged by the rediscovery of *C. hystrix,* a wild *Cucumis* species of Asian origin possessing 24 chromosomes. *C. hystrix* (2n = 24) was successfully crossed with cucumber (*C. sativus,* 2n= 14) (Chen *et al.,* 1997). A new species *C.* × *hytivus* was proposed in 2000 followed by chromosome doubling of the F_1 hybrid (Chen and Kirkbride, 2000). Recently, on the *Dicoelospermum, Mukia, Myrmecosicyos,* and *Oreosyce* have been transferred to *Cucumis,* resulting in 14 new combinations, two changes in status, and three new names (*Cucumis indicus, C. kirkbrideana,* and *C. oreosyce*). A complete morphological key to all these species now included in *Cucumis* was provided in Schaefer (2007).

Pitrat, Hanelt, and Hammer (2000) proposed 16 botanical groups or varieties in the two subspecies. The classification was further refined by Pitrat (2008) with 10 botanical groups for subspecies *melo* and five for subspecies *agrestis.* Burger *et al.* (2010) further modified Pitrat *et al.* (2000) scheme to the International Code of Nomenclature for Cultivated Plants (Brickell *et al.*, 2004) with 11 Groups for subspecies *melo* (*cantalupensis, reticulatus, adana, chandalak, ameri, inodorus, chate, flexuosus, dudaim, tibish* and *chito*), and five Groups in subspecies *agrestis* (*momordica, conomon, chinenesis, makuwa* and *acidulus*).

3.2.0 *Citrullus* (Watermelon)

The name of *Citrullus* was first coined by Forskal in the year 1775 but H. Schrader was the first who classified the genus systematically, which was adopted by the Eighth International Botanical Congress, 1954 to be included in the Nomina Conservanda (Fursa, 1972). Some of the morphological traits of taxonomic importance in various species of *Citrullus* genus are pollen structure, anatomy of fruits, seed structure, presence or absence of nectary flowers, characteristics of embryos, and variations in chromosome karyotypes (Robinson and Decker-Walters, 1999; Jarret and Newman, 2000).

De Candolle (1882) considered that there was much evidence of the possibility of watermelon being indigenous to tropical Africa, more specifically South. It is truly wild or native only in sandy dry areas of South Africa chiefly Kalahari Desert. Livingstone reported in 1857 in a book on his expedition in South Africa that in the desert of Kalahari, there were natural outgrowths of watermelon of various forms both edible and inedible. The African distribution of the genus has been recognized and Cogniaux and Harmes (1924) listed four species, *Citrullus vulgaris* which includes cultivated watermelon [now renamed *C. lanatus* Matsumura and Nakai], *C. colocynthis, C. ecirrhosus* and *C. naudinianus*. The species classification of *Citrullus* and their relationships have been studied in relation to cucurbitacin content or bitter principle by South African investigators Rehm *et al.* (1957) and Rehm (1960). They were of the opinion that there is one group of closely related species, *viz., C. vulgaris, C. colocynthis* and *C. ecirrhosus* with cucurbitacin E as the main bitter substance and the other group of isolated species, *viz., C. naudinianus* with cucurbitacin B together with E and their derivatives. From the results of morphological and cytogenetic studies by Shimotsuma (1963), it was observed that all these four species were cross-compatible with each other and several factors like geographical isolation, differences in flowering habit, genic differences and structural changes in chromosomes would have contributed to the maintenance of their identity.

Cultivated watermelon includes three subspecies: *C. lanatus* subsp. *lanatus* (Shrad. Ex Eckl. et Zeyh.), *C. lanatus* subsp. *vulgaris* (Shrad. Ex Eckl. et Zeyh.) Fursa, and *C. lanatus* subsp. *mucosospermus* Fursa (Levi *et al.*, 2001a, b). On the other hand, now all these three species are under the var. group *lanatus* (Jeffrey, 2001). Currently, the species *C. lanatus* (Thunb.) Matsum and Nakai) includes two botanical varieties, namely var. *lanatus* (Bailey) and var. *citroides* (Mansf). Cultivated watermelons belong to var. *lanatus* and have endocarps in wide ranging colours. The var. *citroides* is cultivated in southern Africa, and also called "Tsamma" or "citron" melon, whose rind is used as preservative in pickles (Whitaker and Davis, 1962; Fursa, 1972; Whitaker and Bemis, 1976; Burkill, 1985; Jarret *et al.*, 1997; Jeffrey, 2001). The citron fruits have green- or white-colored flesh and their taste varies from bland to bitter.

Seed production fields should be isolated from weedy citron types since these two botanical varieties cross readily (Wehner, 2008).

Pangalo (1930) and Goldhausen (1938) reported that bitter and non-bitter wild races of *vulgaris* should be distinguished and treated as C. *edulis* and C. *colocynthoides*. Bailey (1930) stated that *vulgaris* or cultivated watermelons can be divided into vars. *lanatus* and *citroides*, the latter comprising the citron or preserving melon producing fruits with hard in edible flesh and greenish or tan seeds. The species *Citrullus colocynthis* (Schrad.) is a perennial herb known as bitter apple and is a desert species with a rich history as a medicinal plant (Dane *et al.*, 2007). Whitaker (1933) considered C. *colocynthoides* as probable ancestor of watermelon. Morphologically the colocynth, which is found in India also and used in Ayurvedic medicine, has similar characters with C. *vulgaris*, but fruits are bitter and seed small. But African investigators Rehm *et al.* (1957) and Meeuse (1962) considered bitter forms C. *vulgaris* as the ancestor of cultivated watermelons. Shimotsuma (1963) also supported this view, on the basis of the same chromosome number, free crossability and their cohabitation throughout Africa and Asia. The 'citron' or preserving melon (var. *citroides* of Bailey) can be considered as intermediate stage between primitive bitter and cultivated sweet forms.

T.W. Whitaker considered C. *colocynthis* to be a likely ancestor of watermelon as it is morphologically similar to C. *lanatus*, and is freely intercrossable and produces fertile hybrids (Wehner, 2007). Dane *et al.* (2007) reported divergent lineages of C. *colocynthis* that are from tropical Asia and Africa, now widely distributed in the Saharo-Arabian phylogeographic region of Africa and in the Mediterranean region.

It is pertinent to mention here about the relationship of *tinda* or Indian squash or round melon, with watermelon. This is primarily an Indian vegetable crop with very limited cultivation in North and North-western regions. This was originally thought to be a subspecies of watermelon as C. *vulgaris* var. *fistulosus* and later raised to an independent species as C. *fistulosus*. The relationship with watermelon has been conclusively disproved by cytogenetical studies of Shimotsuma (1963) and Khoshoo (1955) nor is it related to muskmelon inspite of the same chromosome number (Khoshoo and Vij, 1963).

Since it is not at all found in Africa, the Russian botanist, (1944) had raised this taxon to a new independent genus *Praecitrullus* much earlier and Jeffrey (1980) concurred. The present accepted botanical name is *Praecitrullus fistulosus*. Genetic diversity of this crop is rather limited suggesting its partial or recent domestication.

3.3.0 *Cucurbita* (Pumpkin, Squash, Marrow)

This genus is important and better understood among the other cucurbits, because of extensive investigations on the phylogeny by Whitaker, Bemis and their associates. It is a comparatively closed group or system of about 27 species concentrated in the tropical regions of Central and South Africa and comprising both wild and cultivated species having the same chromosome number of 2n=40. From archaeological remains, it has been possible to trace man's association with these plants backward in time, for a maximum of 10000 years (Whitaker and Cutler, 1971). The three cultivated species C. *moschata*, C. *maxima*, C. *pepo* and the other two lesser known (in India) *C. ficifolia* and *C. mixta* were all selected by American Indians long before the European colonists and it was reported that the major food crops of the Americans (Aztec, Inca and Maya) were maize, bean and squash (Whitaker and Bemis, 1975).

Among the 27 species listed, five as mentioned above, were cultivated and the rest were mesophytic and xerophytic perennial. Numerical taxonomic (Bemis *et al.*, 1970) and cytogenetic studies (Whitaker and Bemis, 1975) show that *Cucurbita* species can bear ranged into 9 groups of 2 xerophytic and 7 mesophytic. *C. digitata*, *C. palmata*, *C. californica*, *C. cylindrata* and *C. cordata* are in group I and the second one consisted of *C. lundelliana*, *C. okeechobeensis* and *C. martinezii*, *C. mixta*, *C. palmeri*, *C. sororia* and *C. gracilor*, while *C. pepo* and *C. texana* were treated as conspecific. Similarly *C. maxima* and *C. andreana* are together in group V; *C. pedatifolia* and *C. foetidissima*, *C. moschata*, *C. ficifolia* and *C. ecuadorensis* are single species in groups VI, VII, VIII, IX and X, respectively, on the basis of phenograms. This grouping is closely corroborated if not wholly, by behavior of squash and gourd bees (Hurd *et al.*, 1971). *C. pedatifolia* is grouped with *C. ficifolia* instead of a group by itself and *C. ecuadorensis* is placed with *C. maxima* group instead of by itself. Thirdly, *C. mixta* has been placed in a group by itself instead of with *C. sororia*. The one interesting and remarkable close agreement was that in both the studies the five cultivated species came out as separate entities. The wild species separated geographically but genetically cross-compatible. They are viable from mesophytic to xerophytic and from annuals to perennials. While there are barriers to hybridization between the species of *Cucurbita*, no species or group is completely isolated from all other species.

Whitaker and Bemis (1975) assigned a key role to *C. moschata* as the nearest putative ancestor of the genus. They suggested *C. moschata* as a connecting link between the wild and remaining cultivated ones and that all the five cultivated species have been independently domesticated, *moschata* in Mexico and also in South America, *maxima* in Argentina, Bolivia, Chile and Peru, *ficifolia* in Central America and South Mexico, *pepo* probably in Mexico, north of Mexico city and *mixta* in South of Mexico city. In India *C. moschata* or the field pumpkin is more widely distributed than other species. The role of wild species in the evolution of cultivated species still remained obscure. In Mexico and Guatemala, there is a group of wild species closely allied inbreeding and other criteria to the cultivated forms. These include *C. lundelliana* and *C. martinezii* which are to some extent cross-compatible with the cultivars. They are vigorous, lush plants bearing small fruits with tough rinds, bitter, coarse and stringy flesh and abundant seeds.The evidence that the complex of 5 cultivated species derived from these small bitter gourds is suggestive but by no means decisive (Whitaker and Bemis, 1978). Lehmann (1996) examined with reference to the varietal assortment of *C. pepo*, *C. maxima*, *C. moschata*, *C. mixta* and *C. ficifolia*.

Paris (1996) outlined the history, diversity and distribution of summer squash (*C. pepo*). Summer squash is grown in many temperate and subtropical regions, ranking high in economic importance among vegetable crops world wide. A native of North America, it has been grown in Europe since the Renaissance. There are 6 extant groups of cultivars-cocozelle, crook neck, scallop, straight neck, vegetable marrow and *zucchini,* most of them existed for hundreds of years. The characteristics of these groups are discussed by Paris (1996). Their different fruit shapes determine their suitability for various methods of culinary preparation. The group differs in geographical distribution and economic importance. The *Zucchini* group, a relatively recent development, has undergone intensive breeding in the United States and Europe and is probably by far the most widely grown and economically important of the summer squashes.

Cytogenetically the species of *Cucurbita* show amazing uniformity of chromosome number (2n=40) and they are considered as secondary polyploids probably with a basic number of x=10 (Weilung, 1959). Groff (1966) studied the microsporogenesis of species and interspecific hybrids which supported Weilung's hypothesis of natural tetraploids. Interspecific hybridization in *Cucurbita* has been directed towards transfer of desirable traits. *C. lundelliana* has resistance to powdery mildew and *C. ecuadorensis* and *C. foetidissima* have resistance to several strains of cucurbit viruses. The transfer of single chromosome from the wild species to the genomes of cultivated species and inter-specific trisomics have been attempted (Bemis, 1973). Another interesting development is the attempt being made to produce an instant domesticated plant *C. foetidissima* a xerophytic perennial native to Northern Mexico. It is called 'buffalo gourd', suited to low rainfall regions producing an abundant crop of fruits containing seeds rich in oil and protein. It is propagated vegetatively by rooting at nodes and produces a large storage root rich in starch (Bemis *et al.*, 1978).

3.4.0 *Lagenaria* (Bottle Gourd)

It has one cultivated annual monoecious species and five wild perennial dioecious species, the latter confined to Africa and Madagascar. Tropical Africa remains as the primary gene pool for this species (Singh, 1990). The cultivated *L. siceraria* has been archaeologically documented, associated with man and argument of a single origin revolves round the fact that bottle gourd has long been considered as a monotypic genus, but Willis (1966) has listed six species.

Bottle gourd has been credited with a wide bihemispheric distribution. Schweinfurth has identified remains of *L. siceraria* from Egyptian tombs dated about 3000-3500 BC. Similarly bottle gourd remains have been identified in Spirit caves of Thailand (10000-6000 BC), in Mexico (7000-5000 BC), in Peru (4000-3000 BC) and in China (500 AD). Archaeological evidence estimates a time depth of at least 12000 years for *Lagenaria* remains, both in new and old worlds. A theory of trans-oceanic diffusion or drift of bottle gourd has been suggested to explain pre-Columbian distribution in tropical America. With the mass of conflicting evidence from historical and archaeological sources, it is not easy to arrive at a reasonably precise conclusion regarding its origin. Prevalence of Sanskrit equivalent *'alabu'* suggested long history of its cultivation in India.

Harris (1967) suggested that *Lagenaria* had a pan-tropical distribution and was independently domesticated in both old and new worlds, while Whitaker (1971) reaffirmed his hypothesis that bottle gourd was indigenous to tropical Africa (south of Equator) and has diffused by trans-oceanic drift or human transport. Richardson

(1972) concluded that (i) *Lagenaria* was not a monotypic genus with an ancient pantropical distribution, (ii) human utilization of *Lagenaria* was atleast 15000 years old in new world and 12000 years in old world, (iii) the early dates of use of *Lagenaria* were far too early convincingly to suggest trans-oceanic diffusion by man and (iv) *Lagenaria* was independently domesticated in the old and new worlds. Heiser (1980) supported the opinion of Whitaker, that of Africa as the origin of bottle gourd, though he conceded that there was no evidence really decisive enough to distinguish between Africa and America as the original home of the species.

Though hybridization with at least one perennial wild species *L. sphaerica* is known, the hybrid is sterile and possibility of transfer of desirable traits to cultivated species requires investigation.

3.5.0 *Momordica* (Bitter Gourd)

It is a large genus comprising nearly 23 species (Jeffrey, 1967) in Africa alone. The cultivated species are *M. charantia*, the familiar bitter gourd, or *karela* of India, *M. cochinchinensis*, the sweet gourd, or *golkakora* of Assam and NE States, *M. dioica*, the *kakrol* of Bengal, Bihar and Odisha tribal regions and *kartoli* of Thane district of Maharashtra, besides *M. balsamina* (balsam apple) and *M. cymbalaria* (syn. *M. tuberosa*). This genus is essentially a native of tropical regions of Asia, Polynesia besides tropical Africa and South America. There has been no systematics study on the origin of this gourd and its allied species which made Zeven and Zhukovsky (1975) to regard this genus as one of unidentified origin.

Schafer (2005) considers the genus *Momordica* to comprise 47 species including eight Asian species, which are all dioecious, and 39 African species of which 20 are dioecious and 19 monoecious. According to de Wilde and Duyfjes (2002), 10 species are reported in Southeast Asia, of which six occur in Malaysia and India, where *M. balsamina* L., *M. charantia* L., M. sub. *angulata* Blume (ssp. *renigera* (G. Don) W. J. de Wilde), and *M. cochinchinensis* (Lour.) Spreng. are common.

The monoecious group has $2n = 2x = 22$, whereas the dioecious group has $2n = 2x = 28$. In *M. balsamina*, $n = 11$ was recorded by Whitaker (1933) and McKay (1931), and in *M. charantia*, $2n = 22$ was recorded by McKay (1931) and Bhaduri and Bose (1947). Yasuhiro Cho *et al.* (2006) reported that most of the "cultivated kakrol" (most probably *M. subangulata* ssp. *renigera* and not *M. dioica* Roxb. as assigned by the authors) plants in Bangladesh were confirmed to be tetraploid ($2n = 4X = 56$).

The higher chromosome number in *M. dioica* ($2n=28$) compared to *M. charantia* or *M. balsamina* ($2n=22$) is suggested to be due to duplication rather than fragmentation. Natural polyploids ($2n=56$) have been recorded in *M. dioica*. Interspecific hybridization among *Momordica* species has been attempted by Roy *et al.* (1966) and Trivedi and Roy (1972). The crosses between *dioica* and *charantia* were unsuccessful as also between *dioica* and *balsamina*. The possible origin of *M. dioica* from *M. charantia* could not be fully explained (Trivedi and Roy, 1972). Somatic chromosome number and detailed karyotype analysis were carried out in six Indian *Momordica* species *viz.*, *M. balsamina*, *M. charantia*, *M. cochinchinensis*, *M. dioica*, *M. sahyadrica* and *M. cymbalaria* (syn. *Luffa cymbalaria*; a taxon of controversial taxonomic identity (Bharthi *et al.*, 2010, 2011) The somatic chromosome number $2n = 22$ was reconfirmed in monoecious species (*M. balsamina* and *M. charantia*). Out of four dioecious species, the chromosome number was reconfirmed in *M. cochinchinensis* ($2n = 28$), *M. dioica* ($2n = 28$) and *M. subangulata* subsp. *renigera* ($2n = 56$), while in *M. sahyadrica* ($2n = 28$) somatic chromosome number was reported for the first time. A new chromosome number of $2n = 18$ was reported in *M. cymbalaria* against its previous reports of $2n = 16, 22$. The karyotype analysis of all the species revealed significant numerical and structural variations of chromosomes and segmental allopolyploid origin of teasle gourd (*Momordica subangulata* subsp. *renigera*) through cyto-morphological evidence.

3.6.0 *Luffa* (Sponge Gourd and Ridge Gourd)

It is an essentially old world genus, consisting of two cultivated and two wild species besides only one wild new world species. It is rather difficult to assign with accuracy the indigenous area of *Luffa* species. They have along history of cultivation in the tropical countries of Asia and Africa. The name 'Luffa 'or' Loofah' is of Arabic origin because sponge characteristic has been described in Egyptian writings and Chinese name 'Szkua'- 'dish cloth gourd' or 'towel gourd' signifies its mention in early Chinese literature. Sanskrit name 'Koshataki' indicates its early cultivation in India.

The genus *Luffa* is comprised of seven species, four well-differentiated species from the Old World (*L. echinata* Roxb., *L. acutangula*, *L. aegyptiaca*, and *L. graveolens* Roxb.) and three species from the New World (*L. quinquefida* (Hook. and Arn.) Seem., *L. operculata* (L.) Cogn., and *L. astorii* Svens.) (Heiser and Schilling, 1990).

Cytogenetic investigations have been conducted by Dutt and Roy (1969, 1971) and Roy *et al*. (1970) among the two cultivated species *L. acutangula* (ridge or ribbed gourd) and *L. cylindrica* (sponge gourd) and two wild species *L. graveolens* and *L. echinata*. Chromosome counts in all the species were found to be the same (2n=26) and comparative morphology of the wild and cultivated species and chromosome pairing in interspecific hybrids, made Roy and its associates to suggest that *L. graveolens* as the prime species which has given rise to the two cultivated monoecious species *L. acutangula* and *L. cylindrica*. The sex expression, *i.e.*, monoecism and dioecism have, however, evolved independently because the advanced sex form of dioecism to be found only in wild primitive species *L. echinata*. Singh and Bhandari (1963) raised a hermaphrodite variety of *L. acutangula* to that of aspecific status *L. hermaphrodita*, which has not been accepted because of its easy crossability and fully fertile hybrids with *L. acutangula*.

In general, both way crosses between *L. acutangula* and *L. cylindrica* are successful in the sense that the F_1 hybrid plants were obtained (Thakur and Choudhury, 1967).

High pollen sterility existed in the interspecific hybrids and seed setting was poor. Fertility was restored when F_1 hybrid was crossed to either of the parents. Cytological studies by Pathak and Singh (1949) of the interspecific hybrids and later Trivedi and Roy (1976) found that each chromosome of haploid complement of the two species are sufficiently homologous but non-homologous segments are present in normally pairing chromosomes of F_1 hybrids between two species.

Without artificial pollination, however, the two species do not hybridize, in view of the sterility barriers, which are effective in preventing exchange of genes between the two species.

3.7.0 *Trichosanthes* (Snake Gourd and Pointed Gourd)

It is a large genus principally of Indo-Malay and distribution with about 44 species of which 22 occur in India (Chakravarty, 1982). It is quite diverse in seed-coat anatomy (Singh and Dathan, 1976) and poorly known taxonomically.

The centre of origin of *Trichosanthes* is not precisely known but most of the authors agree India or Indo-Malayan region as the original home. The name 'petola' or 'patola' signifying snake gourd in the Malay Peninsula, Moluccas and the Philippines Island, is possibly of Sanskrit origin indicating its Indian nativity. De Candolle wrote that the species of *Trichosanthes* are all of old world and considered Indian origin as the most probable one, especially in the case of pointed gourd (or *parwal*) *T. dioica*.

The genus has undergone great changes in its species classification in the hands of several workers. The classification at species level was done mainly by Kundu (1942) and Chakravarty (1982). Kundu divided the genus into 2 sections: *Eutrichosanthes* and *Pseudotrichosanthes*, former containing 23 species and latter three only. But Chakravarty described only 22 species and 3 species have been described in Flora Malayasiana (Backer and Brink, 1963) and 4 under Flora of Japan (Ohuri, 1965). But similarity or remoteness of these species with Indian one, has not been studied. *T. bracteata* is by far the commonest species in India, while *T. cucumerina* has extensive range. The third one which has principal Indian distribution is *T. dioica*. The cultivated snake gourd originally designated as *T. anguina* but crossability studies with *T. cucumerina* have brought out the near synonymy between these two species. Hence the older epithet *T. cucumerina* has been adopted. Roy *et al.* (1972) studied *T. lobata* also in the crossability studies and opined that all the three did not differ from each other.

Cytological studies have shown that *T. anguina* and *T. dioica* have 2n=22 chromosomes, while polyploidy series of 22, 44, 66 chromosome have been recorded in *T. palmate* (Ayyangar, 1949; Varghese, 1971; Varghese, 1972 and Singh and Roy, 1973). Sarkar *et al.* (1987) reported that cultivated species *T. anguina* and *T. dioica* and wild species. *T. cucumerina* and *T. palmate* had 2n=22 chromosomes with a graded karyotype consisting of metacentric to sub metacentric medium sized chromosomes.

3.8.0 *Benincasa* (Wax Gourd)

It is a monotypic genus and the only cultivated species is *B. hispida* (syn. *T. cerifera*), commonly known as wax or ash gourd (*petha*). It is unknown in the wild. It is indigenous to Asian subtropics. De Candolle found it wild on these a shore of Java and it has spread northwards to Japan and also to Central America and West Indies. The chromosome number is 2n = 24 (Verghese, 1973). Two botanical forms have been recognized in Japan (Kimura and Sugiyama, 1961). One is called typical which is characterized by velvety testa and marginal band around the seeds, while this characteristic absent in the other form is called *T. emarginata*. The waxy coat of the fruit ensures outstanding storage quality. It has Sanskrit equivalent 'kooshmanda' as well as Persian 'pazadaba', which signify its antiquity.

3.9.0 *Coccinia* (Ivy Gourd)

Coccinia is placed in the tribe *Benincaseae* and comprises about 30 species, all of them confined to tropical Africa with the exception of *Coccinia grandis* (Imbumi, 2004). This typical dioecious species of India is extensively cultivated and has Sanskrit equivalent '*bimba*' taking it to pre-Christian era. In Ghana, a monoecious species is cultivated, while in Ethiopia *C. abysinnica* has edible tuberous roots. Anchote (*Coccinia abyssinica* (Lam.) Cogn.) is cultivated mainly for its tuberous fleshy rootstock in Ethiopia, but its young shoots are eaten as a cooked vegetable. In Malawi children may eat the fruits of *Coccinia rehmannii* Cogn., although they are said to cause sore eyes; the starchy tuber is eaten after roasting. The leaves of *Coccinia trilobata* (Cogn.) C. Jeffrey are used as a famine food, occasionally used as a relish by some tribes in Kenya, *e.g.* in the Mbeere district. The fruits are said to be toxic (Boonkerd, 1993; Imbumi, 2004).

3.10.0 *Sechium* (Chow-chow or Chayote)

It is the only cultivated Cucurbit in the sub-family Sicyoideae. This genus is characterized by one-seeded fruit and polycolporate spiny pollen grains. The cultivated species *Sechium edule* is a native of humid tropical regions of Central America and the Caribbean. Though originally considered as monotypic genus without wild relatives but recently eight wild species have been noted in Central America, opening the way of study of their relationships (Newstrom, 1990). This species and all their wild relatives were native to the New World (Whitaker and Davis, 1962). Taxonomic field studies revealed the presence of at least 2 wild species, *S. compositum* and *S. hintonii* in Mexico and Guatemala (Newstrom, 1987). Verghese (1973) has recorded 2n=28 as the chromosome number, while the earlier reports recorded 2n = 24. After America colonization, chayote spread rapidly to all tropical areas of the New World and became a popular staple item in the diet for people of the Old World (Whitaker and Davis, 1962). Today, chayote is cultivated throughout tropical and subtropical regions of the world (Newstrom, 1990).

4.0 Botany

The family Cucurbitaceae is a moderately large one, comprising about 117 genera and 825 species distributed in warmer regions of both the hemispheres (Jeffrey, 1983) and in India Chakravarty (1982) estimated that 36 genera and around 100 species have been described. The taxonomic classification of Cucurbitaceae has been dealt with differently by various botanists. Jeffrey (1980) of Kew Gardens revised his earlier classification, collecting information from palynology, seed-coat anatomy and phytochemistry. He divided the family into two sub-families (i) Zanonioideae and (ii) Cucurbitoideae and the latter was subdivided into eight tribes: (i) Melothrieae - *Cucumis*; (ii) Schizopeponeae; (iii) Joliffieae - *Momordica*, (iv) Trichosantheae - *Trichosanthes*; (v) Benincaseae - *Coccinia, Benincasa, Lagenaria, Citrullus, Luffa*; (vi)

Wild Species of Cucurbits

Citrullus colocynthis

Citrullus ecirrhosus

Cucurbita okeechobeensis subsp. *martinezii*

Cucurbita texana

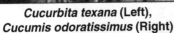

Cucurbita texana (Left),
Cucumis odoratissimus (Right)

Cucurbita maxima subsp. *andreana*

Cucurbita foetidissima

Wild Species of Cucurbits

Momordica balsamina   *Momordica anigosantha*   *Momordica bovinii*

Acanthosicyos naudinianus   *Cucumis metuliferous*   *Cucumis anguria*

Cucumis dipsaceus   *Cucumis myriocarpus*

Cucumis sativus var. *hardwickii*   *Luffa echinata*

Cucurbiteae - *Cucurbita*; (vii) Cyclanthereae - *Cyclanthera*; (viii) Sicyoeae - *Sechium*. Chakravarty (1982), however, classified the family into three tribes: (i) Cucumerineae comprising the genera *Trichosanthes, Lagenaria, Luffa, Benincasa, Momordica, Cucumis, Citrullus* and *Coccinia, etc.* (ii) Orthospermeae consisting *Cyclanthera* and *Dicaelospermum*, and (iii) Zanonieae including *Sechium* and others.

The gourd family has a distinct set of morphological features that easily distinguish it from other plant families. Generally, the plants have a fairly long tap root with lateral roots, confined to to player of 60 cm, in some cases as in *Cucurbita* taproot going down to 170 to 180 cm. Hence, these are adapted to grow in river-beds to utilize subterranean moisture and also some of them (mostly semi-wild) have xerophytic habit. The stems are branched (3 to 8), prostrate or climbing the branches cover large areas, some like *Cucurbita* or *Lagenaria* spread 9 to 10 metres under abundant moisture conditions, nodes usually root when they touch the soil. There are some short internode cultivars with erect habit of growth like 'bush' squashed of *Cucurbita pepo* as also genetic stocks of bush types of muskmelon and watermelon. Leaves are simple, mostly 3 to 5 lobed, variously shaped, palmate, cordate or reniform. Some have deeply lobed or nearly pinnatified as in *Citrullus*. Tendrils are borne on the axils of leaves, simple in *Cucumis*, simple or bifid in *Citrullus, Cucurbita, Luffa, Sechium* and *Lagenaria* and they are absent in 'bush' or erect types.

Inflorescence is axillary, solitary or clustered or racemose. Flowers mostly unisexual, large and showy, mostly monoecious (staminate and pistillate flowers separately in the same plant) while dioecious, andromonoecious and hermaphrodite forms are also met with. The cultivated forms of cucumber are mostly monoecious, while gynoecious (pistillate flowers only) stocks have been developed for producing F_1 hybrids but in muskmelon cultivated forms are mostly andromonoecious (bisexual flowers and staminate flowers in separate nodes). In *Lagenaria, Cucurbita, Citrullus, Luffa, Momordica, Trichosanthes*, mostly monoecious forms are met with. However, the dioecious forms (staminate and pistillate flowers in separate plants) that are in cultivation, are pointed gourd or *Trichosanthes dioica*, ivy gourd or *Coccinia grandis*, kakrol or *Momordica dioica* and feral forms like *Luffa echinata, Citrullus naudinianus, etc.* Staminate flowers mostly in long pedicels are borne singly, but cluster in *Cucumis* and in racemes in *Luffa* and have campanulate, showy corolla, calyx forming a perianth tube, calyx lobes alternating with corolla lobes. Filaments free, two stamens having two locules (thecae) and one unilocular. Pistillate flowers are borne singly in short peduncles, 1 to 5 carpels usually three, thick short style terminated by 3 bilobed or divided papillate stigma (pistillate and hermaphrodite flowers similar). Usually staminate and pistillate (or bisexual) flowers are borne in different axils, while in some like *Sechium and Luffa* may be on the same node. Hermaphrodite form which bears only bisexual flowers, like 'Satputia' cultivar of *Luffa acutangula*, is rare.

Fruit is essentially a (inferior) berry, even though called a *pepo*, because of hard and tough rind (when completely mature) as in bottle gourd or summer squash (*Cucurbita pepo*). The epithet 'gourd' actually refers to this character, even though it is applied to other fleshy fruits like bitter gourd or snake gourd, whose skin does not become hard or tough when ripe. The fruit peduncle is characteristic in *Cucurbita* species, 5 to 8 angular, ridged in *pepo*, terete and soft in *maxima* and 5 to 8 ridged and flared near the base of the fruit in *moschata*. The seeds are borne with parietal placentation. The edible portion is placentae in cucumber and watermelon, while in muskmelon and pumpkin mostly pericarp with very little mesocarp is edible. In ridge and sponge gourds endocarp is edible which later becomes fibrous and spongy and in ash gourd also endocarp is fleshy and edible.

Seeds are usually many, with characteristic shapes and colours as in *Citrullus* and varying in size. In chow-chow it is monocarpellary, single seeded and viviparous. Seeds are usually exendospermous and seed-coat anatomy is characteristic in several genera. The functional seed-coat is exotestal and the mature exotesta consists of three layers, *viz.*, seed epidermis, seed hypodermis and a sclerenchymatous layer (Singh, 1971; Singh and Dathan, 1973, 1974 and 1976).

5.0 Flowering, Sex Mechanism, Pollination and Fruit Set

5.1.0 Floral Biology and Pollination Behaviour

The flowers of different genera and species of cucurbits vary in size, shape and colour but are similar in general morphology. With regard to sex expression, the cucurbits generally fall into two groups (1) Monoecious (staminate and pistillate flowers separately on the same plant) and (2) Dioecious (with separate staminate and pistillate flowering plants). Flowers are borne in the axils of the leaves. The ratio of staminate and pistillate flowers varies between species and cultivars. But the staminate flowers are always more in number than pistillate in all monoecious cucurbits. Staminate flowers are single in *Lagenaria*, *Cucurbita*, *Citrullus*, and *Momordica* and in clusters in *Cucumis* and *Luffa*. In genus *Sechium* both staminate and pistillate flowers are borne on the same node of the stem while in other cucurbits both are borne on separate nodes. The staminate and pistillate flowering nodes are very near in genus *Luffa*. Staminate flowers are borne mostly on long pedicles. The flowers of *Citrullus* are smaller and less showy than those of *Cucumis*. In the genus *Cucurbita*, flowers are bright yellow, large and showy.

Different sex forms in cucurbits are given below.

1. **Hermaphrodite**: Primitive form where only bisexual or perfect flowers are produced. As for example 'Satputia' cultivar of ridge gourd, cucumber and muskmelon.

2. **Monoecious**: Where staminate and pistillate flowers are separately produced in the same plant. As found in cucumber, bottle gourd, bitter gourd, watermelon, pumpkin, squash, ash gourd, ridge gourd, sponge gourd, snake gourd *etc.*

3. **Andromonoecious**: Staminate and perfect flowers produced separately in the same plant. For example, watermelon (in some cultivars), muskmelon (dessert type).

4. **Gynomonoecious**: Pure pistillate and perfect (hermaphrodite) flowers are produced separately in the same plant (genetic stock of cucumber).

5. **Gynoecious**: Where only pistillate flowers are produced in a plant. Example, genetic stock of cucumber, it has been evolved from gynomonoecious sex.

6. **Trimonoecious or gynoandromonoecious**: Where staminate, pistillate and perfect flowers are produced in the same plant in varying proportion (genetic stock of bitter gourd).

7. **Androecious**: Where only staminate flowers are produced.

8. **Dioecious**: Where staminate and pistillate flowers are produced separately in separate plants, *e.g.* parwal (pointed gourd), ivy gourd, *kakrol* (*Momordica dioica*), etc.

Floral biology of important cucurbits has been shown in Table 5.

5.1.1 Flowering

Flowering in cucurbits normally starts in about 40-45 days after sowing depending upon the weather condition. The node number of the first flower is important and more so the node number of the first female or perfect flower. The latter is an index of earliness of the cultivar and there are variations among the cultivars in a given species in this regard.

The sequence of flowering follows a set pattern, namely the first 4 to 6 flowering nodes would bear staminate flowers and later pistillate flowers appear in few nodes in the main axis and secondary branches in cycles. In a typical monoecious sex form, the numbers of male flowers produced are in far greater proportion than the pistillate flowers. This is called 'sex ratio', which may range from 25 to 30:1 to 15:1, the latter condition is advantageous and economical, because it results in greater number of pistillate flowers per plant, consequently higher fruit set and yield. Sex ratio is highly sensitive to environment and high N, long days and high temperature generally promote greater number of male flowers. The alternating sequence continues until fruits set and developing fruits in a plant in a large measure determine the production of pistillate/perfect flowers (in muskmelon), further down in the vine. In case of those crops where immature fruits are harvested at vegetable (or edible) stage, this kind of inhibitory mechanism will not be perceptible, but in melons, pumpkins, *etc.* (where ripe fruits are harvested), this will be very obvious. Even if more pistillate or perfect flowers are produced in the vine, fruits may also set, but these may not develop fully, or shed in immature condition. The vine strikes physiological balance at the threshold limit of maximum fruits that it can carry to maturity. That is why the number of fruits per vine in a seed crop will be less (4 or 5) than in a vegetable crop (12 to 15), in a plant like bottle gourd.

Hossain *et al.* (1990) observed in a local cultivar of bottle gourd that staminate flowers appeared about 7 days earlier than pistillate flowers and the staminate: pistillate ratio was 9:1.

5.2.0 Sex Form

As indicated earlier, the sex form in cucurbits showed a wide range of variation. Primarily, hermaphrodite is considered as a primitive form, where only bisexual

Table 5: Floral Biology of Important Cucurbits

Cucurbits	Flower Colour	Floral Habit	Anthesis	Pollen Dehiscence Time	Duration	
					Pollen Fertility	Stigma Receptivity
Muskmelon	Yellow	Monoecious, andromonoecious	5.00-8.00 hr	5.00-8.30 hr	5.00-14.00 hr	2 hr before to 2-3 hr after anthesis
Watermelon	Greenish yellow	Monoecious, andromonoecious	6.00-8.30 hr	6.00-8.30 hr	7.00-11.00 hr	2 hr before to 3 hr after anthesis
Bottle gourd	White gourd	Monoecious, andromonoecious	11.00-16.00 hr	13.00-14.30 hr	On the day of anthesis till next morning	36 hr before to 60 hr after anthesis
Bitter gourd	Yellow	Monoecious	6.30-10.30 hr	7.30-10.30 hr	6.00-12.00 hr	One day before and one day after anthesis
Snake gourd	White	Monoecious	18.00-21.00hr	Shortly before anthesis	10 hr before 40 hr after dehiscence	7hr before to 50hr after anthsis
Ridge gourd	Yellow	Monoecious	15.00-18.00 hr	15.00-18.00 hr	On the day of anthesis to till 2-3 days after anthesis in spring season and 1 day in rainy season	6 hr before to 80 hr after anthesis
Sponge gourd	Yellow	Monoecious, dioecious	5.00-8.00 hr	5.00-8.00 hr	On the day of anthesis	10 hr before to 100 hr after anthesis
Cucumber	Yellow	Monoecious, andromonoecious, gynoecious, hermaphrodite	5.30-7.30 hr	4.30-7.00 hr	Up to 14 hr	12 hr before to 6-7 hr after anthesis
Summer Squash	Yellow	Monoecious	4.00-10.00 hr	4.00-11.00 hr	16 hr after anthesis	2 hr before and 10 hr after anthesis
Pumpkin	Creamy white to deep orange yellow	Monoecious	4.00-8.00 hr	4.00-8.00 hr	16 hr after anthesis	2 hr before and 8 hr after anthesis

Source: Kalloo, G. (1994) Vegetable Breeding. Volume (I, II, III) Combined Edition., Panima Educational Agency, New Delhi, pp. 26-27.

Sex Forms of Cucurbits

Gynoecism in Cucurbits

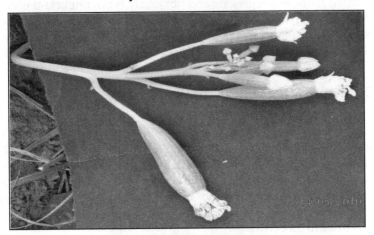

Hermaphroditism in Cucurbits

Dioecism in Cucurbits

or perfect flowers are produced. This is met with in '*Satputia*' cultivar of ribbed gourd, while hermaphrodite genetic stocks in cucumber and muskmelon are also recorded. The advanced form is monoecious where staminate and pistillate flowers are separately produced in the same plant, as in cucumber, bottle gourd, bitter gourd, watermelon, pumpkin, squash, ash gourd, ridge and sponge gourds, snake gourd, *etc*. In muskmelon, there is a variation in that pistillate flowers are replaced by perfect flowers and this sex form is called andromonoecious. Even though major dessert muskmelons are andromonoecious, non-dessert forms like long or serpent melon, 'dosakaya', 'phoot' (snap melon), 'vellarikkai', 'kachri', *etc*., are monoecious. Similarly, andromonoecious sex form is met within some watermelon cultivars in varying proportion. There are also other sex forms like gynomonoecious where pistillate and perfect flowers are produced separately in the same plant. This is met within cucumber, more as a genetic stock or a segregate from the crosses between the two sexes. This sex is comparatively unstable in muskmelon. Gynoecious sex form, where only pistillate flowers are produced in a plant is important in cucumber and has been evolved from the gynomonoecious sex. This has been perpetuated and maintained as a true breeding stock, through artificial induction of staminate flowers by sprays of plant regulators. There is another sex form, which is highly unstable and not breeding true is the trimonoecious or gynoandromonoecious condition, where staminate, pistillate and perfect flowers are produced in the same plant in varying proportion. Androecious sex form where only staminate flowers are produced is rarely recorded more as a segregate and obviously is not important economically. Then there is dioecious sexform where staminate flowers and pistillate flowers are produced in separate plants and pointed gourd (*parwal*), ivy gourd (*kundsru*), *kartoli* or *kakrol* (*Momordica dioica*) are some of the examples in cultivated cucurbits, while feral forms like *Luffa echinata*, *Citrullus naudinianus* and *Cucumis asper* are dioecious.

5.3.0 Cytology of Sex Differentiation

In *Coccinia grandis* (syn. *C. indica*), the first report on sex mechanism and sex chromosomes was made by Kumar and Deodikar (1940), Chakravarty (1948) and Kumar and Visweswariah (1952). Later, detailed studies by Roy and Roy (1971), Roy (1974) and Roy and Bithi Dutt (1977) have clearly brought out the X/Y mechanism of sex determination. Karyotype studies and comparison of chromosome morphology have established the heteromorphy of male and the presence of XY mechanism. The male is heterogametic with Y chromosome. It was also found that polyploidy was not a barrier to dioecism.

$22A + XX$ female	$33A + XXX$ female
$22A + XY$ male	$44A + XXX$ female
$33A + XYY$ male	$44A + XXXY$ female
$33A + XXY$ male	(A-autosomes)

Development of male plant in various polyploidy series, in the presence of single Y chromosome by the suppression of the effects of $2X$ and $3X$ chromosomes, was an evidence of a strong male determining role of Y.

In other dioecious cucurbit species such a distinct differentiation has not been noted. In *T.dioica* there is no heteromorphy in the male and female complements, however, microsporogenesis shows in regard to a pair of chromosomes distinct patterns of abnormality which could be correlated with dioecy (Patel, 1952; Roy *et al.*, 1982). In *L. echinata* and *M. dioica* differentiation of sex is entirely genic or genetical without any cytological evidence of heterogamety.

5.4.0 Genetical Mechanism

Most of the Angiosperm species typically produce bisexual flower which favours self pollination. However, throughout evolution numerous biochemical, physiological and morphological mechanisms have evolved to promote out-crossing *viz.*, self-incompatibility, dichogamy, heterostyly, unisexuality, *etc.* Unisexual flower development shows the most extreme morphological alterations which can take two general forms, monoecy and dioecy (Renner and Ricklefs, 1995). The sex types of the members of Cucurbitaceae plants can be identified early in their vegetative growth stage. The relative numbers of male and female flowers determines the sex of the plant (Dellaporta and Calderon, 1994; Saito *et al.*, 2007). The Cucurbitaceae is one of the handful of families where unisexual flower production is widespread, including an estimated 32 per cent dioecious genera (Renner and Ricklefs, 1995). Most of the Cucurbitaceae species are monoecious, many are dioecious and only few are hermaphrodite (Roy and Saran, 1990). In addition to the standard classification of monoecy and dioecy, cucurbit species can exhibit diverse range of various combinations of sex types such as gynomonoecious (separate pistillate and bisexual flower in the same plant), gynodioecious (separate plant producing pistillate and bisexual flower), andromonoecious (separate staminate and bisexual flower in the same plant) or androdioecious (separate plant producing staminate and bisexual flower). Of these different types of sex types and their combinations, androdioecy is considered as the most rare sex system and this sex system has so far been documented only in three species including one, *Schizopepon bryoniaefolius* Maxim belonging to Cucurbitaceae (Akimoto *et al.*, 1999).

Sex types of different members of the Cucurbitaceae family are highly plastic and typically controlled by small number of key genes. Extensive genetical studies on sex mechanism have been conducted in different cucurbits and have considerable importance from a theoretical stand point as well as in practical breeding. The two major genes that regulate phenotypic sex expression in Cucurbitaceae are the *F* and *M* genes (Saito *et al.*, 2007; Shiber *et al.*, 2008). Genotypically, gynoecious lines (*M_F_*) produce only females; monoecious lines (*M_ff*) produce males and females; andromonoecious lines (*mmff*) produce males and bisexuals; and hermaphrodite lines (*mmF_*) produce only bisexual flowers. The dominant *F* gene regulates the formation of female flowers, while the interaction between the dominant *M* and *F* genes confirm the production of female flowers. The recessive *m* alleles (*mm*) produce bisexual flowers. Androecious phenotypes that produce only male flowers have been reported in cucumber and muskmelon plants. The phenotypic expression of androecious lines can be ascribed to the presence of the recessive *A* and *F* genes in the form of *aaff*.

Dioecious species are typically characterized by an *XY* types where one sex is homogametic or homomorphic (*XX*) and the other heterogametic or heteromorphic (*XY*). The *XY* system may represent specific genes or distinct sex chromosome. Among the cucurbits, sex chromosomes have been verified in *Coccinia indica* (Roy and Saran, 1990; Guha *et al.*, 2004).

5.4.1 Cucumber

Gynoecism is conditioned by a single dominant gene, subject to considerable influence from modifying genes and environmental factors. These have been variously designated as *Acr* (Shifriss, 1961), *acr^F* (Kubicki, 1965), *st* (Galun, 1961), *etc.* Kubicki (1969a, b, c and d) suggested multiple alleles at this locus and established that different gynoecious lines (inbreds) differed in the degree of female expression and their response to GA_3 application and postulated another major gene that interacts with the previous gene, to increase the degree of sex expression. Two major genes, *F* and *M*, control sex expression phenotypes in cucumber (Galun, 1961; Shifriss, 1961; Kubicki, 1969e). Yamasaki *et al.* (2002) opined that sex determination in cucumber plants is genetically controlled by the *F* and *M* loci. The *F* and *M* loci interact to produce three different sex phenotypes *viz.*, gynoecious (*M-F-*), monoecious (*M-ff*), andromonoecious (*mmff*) and hermaphrodite *mmF-*) phenotypes. The F gene is relatively dominant and promotes the production of female flowers. The *M* gene acts on the production of female flowers, and plants with recessive m alleles (*mm*) produce bisexual flowers. In addition to these genotypes, androecious cucumbers that produce only male flowers have been reported (Galun, 1961). The *A* gene acts downstream of the *F* gene, and plants become androecious when both *A* and *F* are recessive (*aaff*).

Galun and his associates (Frankel and Galun, 1977) propounded that gene *M* controls a trigger mechanism that permits only stamen or pistil development. Flowers of *M*/plants are, therefore, unisexual and flowers of *m/m* plants may be hermaphrodite. The second type of gene controls the flowering pattern which in monoecious cucumber is (in the main shoot) an initial strong male tendency which gradually changes into a female tendency. One gene of this type *'st'* brings femaleness closer to the base and another *'a'* has the opposite effect. The different genotypes in this scheme are androecious (*M/-st⁻/st⁻a/a*), monoecious (*M/-st⁻/st⁻, A-*), gynoecious (*M/-st/st, A/-*), andromonoecious (*m/mst⁻, A/-*) and hermaphrodite (*m/mst/ st, A/-*). M is completely dominant but there is no complete dominance in *st* gene, *st⁻/st* heterozygote shows an intermediate phenotype and it sexpression is strongly affected by environment. The gene *'a'* is fully expressed only in *M/-st⁻/st⁻* genotypes.

Trebitsh *et al.* (1997) reported that sex determination in cucumber (*C. sativus* L.) is controlled largely by three genes: *F*, *m*, and *a*. The *F* and *m* loci interact to produce monoecious (*M_f_*) or gynoecious (*M_F_*) sex phenotypes. Ethylene and factors that induce ethylene biosynthesis, such as 1-aminocyclopropane-1-carboxylate (*ACC*) and auxin, also enhance female sex expression. They also suggested that *CS-ACS1G* is closely linked to the *F* locus and may play a pivotal role in the determination of sex in cucumber flowers.

Yamasaki *et al*. (2002) determined the relationship between the genotype and ethylene production in gynoecious (cv. Higan-fushinari), monoecious (cv. Otone No. 1) and andromonoecious (cv. Lemon) cucumbers. The levels of ethylene production and the accumulation of CS-ACS2 mRNA in andromonoecious cucumber plants did not differ from those of monoecious plants and were less than those of the gynoecious plants. Ethylene inhibited stamen development in monoecious cucumber but not in andromonoecious one. To investigate the action mechanism of ethylene in the induction of femaleness of cucumber flowers, they isolated three ethylene-receptor-related genes *viz*., *CS-ETR1, CS-ETR2* and *CS-ERS*. Ethylene substantially increased the accumulation of *CS-ETR2, CS-ERS* and *CS-ACS2* mRNA in monoecious cucumber plants, but not in andromonoecious plants. These results suggest that ethylene responses in andromonoecious cucumber plants are reduced compared to those in monoecious plants. This is the first evidence that ethylene signals may mediate the product of the *M* locus to inhibit stamen development in cucumber. The andromonoecious line provides novel material to study the function of the *M* locus during sex determination of flowers in cucumber plants. Knopf and Trebitsh (2006) isolated the *Cs-ACS1* (ACS, 1-aminocyclopropane-1-carboxylate synthase) gene that encodes the rate-limiting enzyme in the ethylene biosynthetic pathway. They proposed that *Cs-ACS1* is present in a single copy in monoecious (*ffMM*) plants whereas gynoecious plants (*FFMM*) contain an additional copy *Cs-ACS1G* that was mapped to the *F* locus. To study the origin of *Cs-ACS1G*, Knopf and Trebitsh (2006) cloned and analysed both the gynoecious-specific Cs-ACS1G gene and the non-sex-specific *Cs-ACS1* gene. Their results indicated that *Cs-ACS1G* was the result of a relatively recent gene duplication and recombination, between *Cs-ACS1* and a branched-chain amino acid transaminase (*BCAT*) gene.

5.4.2 Muskmelon

The original hypothesis of Poole and Grimball (1939) is based on digenic inheritance of sex form. Hermaphrodite is designated as double recessive with monoecious having the two dominant genes. The andromonoecious (*+a*) and gynomonoecious (*g+*) were heterozygotes, conforming to 9: 3: 3: 1 segregating pattern. They considered that environmental factors can modify gynomonoecious into gynoecious or trimonoecious. Kubicki (1969e) thought that four pairs of independently segregating genes were responsible for various forms of sex expression. Kubicki (1966) described a method for producing 100 per cent genetically pure gynoecious lines. Such lines are derived from a cross of hermaphrodite (*aagg*) with a monoecious (*AAGG*) line and isolating gynoecious (*AAgg*) and hermaphrodite (*aagg*) lines. Crossing of gynoecious (*AAgg*) and hermaphrodite (*aagg*) lines produce all gynoecious (*Aagg*) plants. Gynoecious lines are generally maintained by applying 200-250 ppm silver thiosulphate.

Rowe (1969) concluded that gynoecious sex expression was controlled by multiple modifying genes in addition to the major genotype *A-gg*. Studies made at the Indian Agricultural Research Institute, New Delhi have confirmed the high instability of gynoecism in muskmelon, and also of the gynomonoecious condition. The possibility of modifier at '*g*' locus has been proposed to explain the varying degrees of segregation among gynoecious, gynomonoecious and hermaphrodite crosses (Shinde and Seshadri, 1983; Magdum and Seshadri, 1983).

Later, the gynoecious line WI 998 of Wisconsin (the United States) was also studied by More and Seshadri (1988) in F_2 generation for the cross of the gynoecious and monoecious lines, which were found to be segregated in a 3:1 ratio.

In melon, Boualem *et al.* (2009) showed that the transition from monoecy to andromonoecy resulted from a mutation in 1-aminocyclopropane-1-carboxylic acid synthase (*ACS*) gene, *CmACS-7*.

5.4.3 Watermelon

Inheritance of sex forms in watermelon is not well described. The sex form of watermelon is basically monoecious but four sex expression responses *viz.*, monoecism, gynoecism, andromonoecism, and hermaphrodite are found in different watermelon forms (Jie *et al.*, 2017). The monoecious or andromonoecious sex form is governed by a single pair of genes (Rosa, 1928; Poole and Grimball, 1945; Rudich and Zamski, 1985). However, Sugiyama *et al.* (1998) have developed hermaphroditic flower-bearing accession of andromonoecious watermelon that is expected to be useful in developing a high female bearing genetic stock. They also noticed the role of silver thiosulfate (STS) for sex modification.

Ji *et al.* (2015) made five pairs of crosses of watermelon plants with different sex forms and grew progeny in two different seasons to investigate the inheritance of sex forms and the seasonal effect on sex expression. They showed that environmental factors had no effect on sex forms, but they affect sex expression on individual flowers as more pistillate flowers were observed in spring than in autumn. The study suggested that short photoperiod and low temperatures promote formation of pistillate flowers in watermelon. In the F_2 population of the cross of andromonoecious (*SL3H* or *AKKZW*) × monoecious (*XHB*), the segregation ratio is 9 monoecious: 3 trimonoecious: 4 andromonoecious, and the segregation ratio in BC_1P_1 (F_1 × andromonoecious parent) is 1 monoecious: 1 trimonoecious: 2 andromonoecious. The segregation ratio in the F_2 population of the gynoecious (*XHBGM*) × monoecious (*XHB*) is 3 monoecious: 1 gynoecious whereas the segregation ratio in the BC_1P_1 (F_1 × gynoecious parent) is 1 monoecious: 1 gynoecious. The segregation ratio in the F_2 population of gynoecious × andromonecious cross is 27 monoecious: 12 andromonoecious: 9 gynoecious: 9 trimonoecious: 4 hermaphroditic: 3 gynomonoecious. The segregation ratio in the BC_1P_1 population (F_1 × gynoecious) is 1 monoecious: 1 gynoecious whereas the segregation ratio in the BC_1P_2 (F_1 × andromonoecious) is 1 monoecious: 1 trimonoecious: 2 andromonoecious. Taken together, the results suggested that three recessive alleles, andromonoecious (a), gynoecious (gy) and trimonoecious (tm) control the sex forms in watermelon, and a allele is epistatic to the tm allele. The following phenotype-genotype relationships were proposed for each of the sex forms in watermelon: monoecious, *A_Gy_Tm_*; trimonoecious, *A_Gy_tmtm*; andromonoecious, *aaGy_Tm_* or *aaGy_tmtm*; gynoecious, *A_gygyTm_*; gynomonoecious, *A_gygytmtm*; and hermaphroditic, *aagygyTm_* or *aagygytmtm*.

5.4.4 *Luffa*

Extensive investigations have been conducted in India. Singh *et al.* (1948) came to the conclusion that two gene loci are involved, each associated with multiple allelomorphic series (*A-aa?* and *G-gg?*). The genotypes were monoecious (*AG*), andromonoecious (*a'G*), androecious (*aG*), gynomonoecious (*Ag'*), gynoecious (*Ag*) and perfector hermaphrodite (*ag, a'g, a'g', ag'*). But Richharia (1948) assumed two independent genes one of which determines the inheritance of both sexes, whereas other controls the female sex only. The first has been found to be epistatic in action and in the absence of both, the individual has perfect flowers. He obtained F_2 segregation, 12 monoecious: 3, gynoecious: 1 hermaphrodite. Choudhury and Thakur (1965) studied intervarietal and interspecific crosses in *Luffa*. They postulated that sex expression was controlled by two independent suppress or genes '*A*' and '*G*', the former suppressing male organs in solitary flower and the latter suppressing the femaleness in the racemes. Roy and his associates (Roy *et al.*, 1975) thought that *A*, *a'a* and *G*, *g'g* controlled the appearance of male and female sexes allowing for weak suppression of androecium and gynoecium in different phenotypes. The monoecious sex form has been assigned more than one genotypic expression as *AAGG, AaGG, Aa'Gg, a'a'GG*, and *aa'Gg'* gynoecious as *AAgg, Aa'gg, Aa'g'g* and androecious as *aaGG, aa'GG* and *a'a'Gg'*. The studies of Roy *et al.* (1970) have brought out an interesting variation regarding hermaphrodite sex form 'Satputia' of *L. acutangula*. The hermaphrodite sex though considered primitive, has been considered here, as a derived one from monoecious species *L. graveolens*. Inheritance of hermaphroditism in ridge gourd (*Luffa acutangula* Roxb.) was studied by Karmakar *et al.* (2012). In F_2 generation, the observed distribution of plant phenotypes fitted to the expected ratio of 9:3:3:1 (monoecious: andromonoecious: gynoecious: hermaphrodite), 3:1 (solitary *vs* cluster) and 15:1 (ridge *vs* non-ridge) for hermaphroditism, bearing habit and fruit surface morphology, respectively. The segregation pattern suggested digenic recessive control of hermaphroditism while monogenic recessive and duplicate recessive control was observed for cluster bearing habit and non-ridge fruit surface, respectively.

5.4.5 *Momordica*

The appearance of various intermediate sex forms like andromonoecious, gynoecious and trimonoecious, in colchicines treated plants of *Momordica charantia*, but remaining as diploids, is an interesting phenomenon reported for the first time (Trivedi and Roy, 1973). Probably the hidden genetic mechanism has been exposed, but more information could not be gathered, since all these plants of intermediate sex forms were found to be sterile. Though monoecious is the predominant sex form in bitter gourd, however, gynoecious sex form has been reported from India (Behera *et al.*, 2006). Gynoecious sex form has been reported (Ram *et al.*, 2002; Behera *et al.*, 2006) in Indian germplasm that is governed by a single recessive gene "*gy-1*" (Ram *et al.*, 2006; Behera *et al.*, 2009), whereas Iwamoto and Ishida (2006) reported that gynoecious sex expression in bitter gourd is partially dominant. A predominantly gynoecious line one predominately gynoecious line in bitter gourd (INGR 12014) with high female: male ratio (5:1 to 7:1) was reported by Behera *et al.* (2012).

5.4.6 *Cucurbita*

Cucurbita species are monoecious and fairly stable with occasional perfect flowers as reported by Hayase (1956) in *Cucurbita moschata*. However, gynoecious sex form has been recorded in buffalo gourd (*Cucurbita foetidissima*), a semi-arid cucurbit native to Mexico.

5.4.7 *Benincasa*

Monoecy is the most prevalent sex form for this cucurbitaceous vegetable. "Andromon-6," the andromonoecious line, has been reported by Singh *et al.* (1996) in segregating lines of bottle gourd that is recessive in nature.

5.4.8 Cucurbits with Dioecious Sex Form

Most of the cultivated cucurbits are monoecious, although some important cucurbit taxa are dioecious, that is, *Trichosanthes dioica, Coccinia indica, Momordica dioica, Momordica cochinchinensis* and *Momordica subangulata* subsp. *renigera*. Besides these cultivated cucurbits, there are also a few feral species that contribute dioecious sex forms: *Luffa echinata, Melothria heterophylla, Edgeria darjeelingensis, etc.* (Seshadri, 1986).

5.5.0 Sex Modification

A majority of cucurbits are monoecious, and the sex ratio (male:female) ranges from 25–30:1 to 15:1. The sex ratio is influenced by environmental factors. High nitrogen content in the soil, long days, and high temperature favours maleness. Besides environmental factors, endogenous levels of auxins, gibberellins, ethylene, and abscisic acid also determine the sex ratio and sequence of flowering. A primordium can form either a female or a male flower, and it can be manipulated by addition or deletion of auxins. Exogenous application of plant growth regulators can alter sex form, if applied at 2–4 leaf stage.

The principle in sex modification in cucurbits lies in altering the sequence of flowering and sex ratio. Short day treatment (9 h light) stimulated early development of female flowers on lower nodes in cucumber, long melon, snap melon, ridge gourd, bitter gourd and round gourd and increased the number of female flowers by 61 to 290 per cent when compared with natural day and long day (16 h light). Long day, on the other hand stimulated maleness and caused significant reduction of pistillate flowers (Bose and Ghosh, 1975). Besides environmental factors, exogenous application of plant regulators can alter the sex ratio and sequence if applied at 2-4 true leaf stage, the critical stage at which the suppression or promotion of either sex. Hence, modification of sex to desired direction has to be manipulated by exogenous application of growth regulators once, twice or even thrice at regular intervals. High ethylene level is favourable to female sex expression and it is suggested that it promotes the formation of the ovary in cucumber, muskmelon and summer squash but it affects the early evolution of ethylene. Gibberellins play a key role in promoting male sex expression and is antagonistic to that of ethylene and abscisic acid (Rudich, 1983). To promote greater number of pistillate flowers and alter the sex ratio in desired direction, several chemicals have been tried and found useful. In cucumber, Maleic hydrazide (MH) at 50 to 100 ppm, GA_3 5 to 10 ppm, 2-chloroethylphosphonic acid (commercially called ethrel, or ethephon or CEPA) 150 to 250 ppm; in watermelon, tri-iodobenzoic acid (TIBA) 25 to 50 ppm, boron 3 ppm; in bottle gourd, boron 3 ppm and calcium 5 ppm and in sponge gourd, ethrel 250 ppm have been found useful. These bring about increased yield in terms of number of fruits per plant, but the individual fruit set is slightly reduced, especially in watermelon. (Choudhury, 1966, 1979) This kind of sex modification is more useful and practicable, in crops like cucumber and bottle gourd, where continuous and simultaneous flowering, fruit set and fruit picking takes place. Gynoecious lines of cucumber are maintained by induction of male flowers through sprays of GA_3/GA_7 at 1500 to 2000 ppm. Now-a-days silver nitrate at 300-400 ppm and silver thiosulphate at 250-300 ppm has been found to bring about this same modification. However due to harmful effects of Silver Nitrate and GA_3 the Silver Thiosulphate is most oreferred because of its safe and persistent effect to produce abundant male flowers. Kasrawi (1988) noticed that largest numbers of staminate flowers were produced on parthenocarpic cucumber plants (cv. Diyala) sprayed twice with 300 ppm $AgNo_3$ at weekly intervals, with the initial spray given at the first true-leaf stage. Al Xin *et al.* (2000) reported induction of perfect flowers in cucumber at lower nodes by spraying silver nitrate at 200-250 mg/1 at 3-leaf and 4-leaf stages.

Table 6: Role of Plant Growth Regulators in Modification of different Sex Forms in Cucurbits

Growth Regulators	Functions	References
Gibberellic acid	Maleness in bitter gourd	Wien (1997)
	Maleness in *Cucumis* and *Cucurbita* spp.	Sedghi *et al.* (2008), Arabsalmani *et al.* (2012)
	Femaleness in bitter melon	Ghani *et al.* (2013)
Auxins	Femaleness in *Cucumis* and *Cucurbita* spp.	Sulochanamma (2001)
Ethephon	Femaleness	Arabsalmani *et al.* (2012)
α-/γ-Naphthaleneacetic acid	Femaleness in bitter gourd and pumpkin	Sedghi *et al.* (2008), Ghani *et al.* (2013)
Brassinosteroids	Femaleness in watermelon	Susila *et al.* (2010)
Ethrel	Maleness in watermelon	Devaraju *et al.* (2002), Susila *et al.* (2010)
	Femaleness in gherkin	Ghani *et al.* (2013)
Paclobutrazol	Femaleness in watermelon	Susila *et al.* (2010)
β-Indoleacetic acid	Femaleness in pumpkin	Ntui *et al.* (2007)
Cycocel	Maleness in bitter gourd	Wang and Zeng (1996)
	Femaleness in gherkin	Devaraju *et al.* (2002)
Mepiquat chloride	Femaleness in gherkin	Devaraju *et al.* (2002)

Application of ethrel increased the number of female flowers, gave uniform fruit growth and increased the yield in cucumber (Vysochin *et al.*, 1985). Dibutyl phthalate 0.01 per cent and dipotassium phthalate 0.2 to 0.5 per cent were effective in increasing the number of pistillate flowers on the main and lateral shoots in cucumber (Egorov *et al.*, 1987). El-Ghamriny *et al.* (1988) confirmed that in cucumber sex differentiation takes place at 2-leaf stage and this is the best internal condition for the application of growth regulators. Singh and Choudhury (1989) obtained highest number of fruits/vine in cucumber cv. Khira Poosa (6.2-7.2) and bottle gourd cv. Pusa Summer Prolific Long (15.4-18.7) with ethephon at 50 and 100 ppm and in watermelon *cv.* Sugar Baby (3.0-3.4) with MH at 50 ppm or B at 2 ppm. Das *et al.* (1995) reported in green house trial with cucumber cultivars Chinese Green, Pusa Sanyog and Poinsette that application of NAA (30 or 100 ppm) or ethrel (250 or 500 ppm) twice at 4 to 5 leaf stage and just before bud initiation resulted in production of female flowers on lower nodes and early flowering. Foliar application of cephalexin altered the phenology of cucumber plants, shorting the vegetative phase and extending the flowering phase. It suppressed the production of male flowers but increased the number of female flowers, thereby decreasing the ratio of male: female flowers (Das and Basu, 1998).

Triazole chemical, bitertanol at 1000 ppm sprayed on to cucumber seedlings was found to be promising for producing stronger seedlings with more pistillate flowers per plant without influencing fruit yield and quality (Lee *et al.*, 1999).

Xia *et al.* (1999) found that one or two applications of 1000 µl ethylene/l increased early yields of cucumber *cv.* Zhongnong No. 8. Higher concentrations or more frequent application leads to lower numbers of female flowers, later flowering and lower yield.

Basu *et al.* (1998) reported that flowering was promoted in bitter gourd following foliar application with lower concentrations of penicillin (50 and 100 mg/l). It also promoted fruit set and development and also fruit quality.

Helmy *et al.* (1996) conducted a study to determine the effect of ethrel, salinity, mefluidide and chilling on flowering and modifying sex expression of cucumber plants. Ethrel at 25, 50 and 75 mg/l increased the total number of female flowers with 13.98, 17.49 and 22.38 per cent, respectively compared to control (Figure 1). Ethrel inhibited male flowers by 50.57, 47.13 and 59.77 per cent, respectively. Similar trend had been obtained with mefluidide treatment (15 mg/l).The effect of different treatments on cucumber yields and its relation in female flowers were illustrated in Figure 2. It is obvious that ethrel at 25 and 50 mg/l increased total yield perplant compared to control, similar trend has been obtained by mefluidide treatments at 5mg/l and chilling exposure (7 days) to10°C.

In summer squash (*Cucurbita pepo*), Edelstein *et al.* (1985a) reported that one spraying with ethephon at 500 ppm delayed male flowering by about 16 days in cv. Table Queen while in cv. Vegetable Spaghetti two sprays with 500 ppm (at 2-leaf and 4-leaf stages) were required to obtain similar effects. Sheveleva and Sutulova (1987) noticed that ethephon and CCC stimulated pistil at flower formation and increased productivity in summer squash.Spraying on seedlings of *Cucurbita pepo* cv. Amcobella 94 with 225 and 300 ppm ethrel (ethephon) markedly reduced the number of male flowers and increased the number of female flowers within the first

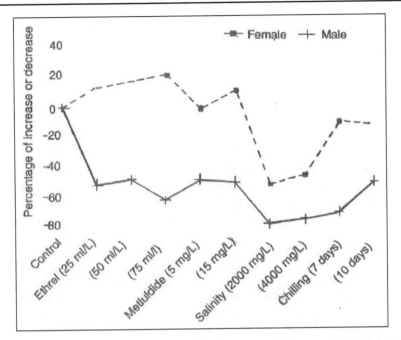

Figure 1: Percentage of Increase and/or Decrease of Accumulated Male and Female Flowers Relative to Control (Helmy *et al.*, 1996).

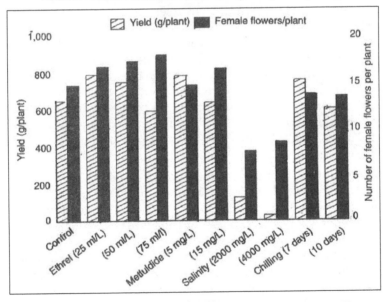

Figure 2: Effect of different Treatments on Accumulative Female Flowers and Total Yield per Plant of Cucumber Plants (Helmy *et al.*, 1996).

15-20 days from anthesis and increased early and total yield, number of fruits/plant and fruit weight, compared with unsprayed controls (Gad *et al.*, 1993).

Arora *et al.* (1985) reported that MH at 50 mg/1 to bottle gourd and at 50 mg/1 to round melon, ethephon at 100 mg/1 to sponge gourd and at 250 mg/1 to summer squash, and GA$_3$ at 25 mg/1 to watermelon were most effective in lowering the male: female ratio, setting additional fruit and producing highest fruit yield as compared to the untreated controls. In watermelon, male: female ratio was lowest with ethephon at 250 mg/1 while fruits per plant and yield were highest with GA$_3$ 25 mg/1. According to Baruah and Das (1997), spraying on bottle gourd plants (cv. Kiyari Lao) at 2-true leaf stage with NAA at 25 ppm or MH at 50ppm produced the best yields (5.48 and 4.86 kg/plant, respectively).

In bitter gourd it has been reported by Yonemori and Fujieda (1985) that 25 ppm BA was the best treatment for differentiation of female flowers followed by 200 ppm ethephon. Wang and Zeng (1997a) noticed that in *M. charantia* GA$_3$ delayed appearance of the first male flower but promoted appearance of the first female flower. At low concentration, GA$_3$ increased the number of female flowers and the ratio of female: male flowers, CCC at 50 200 mg/1 showed a masculinizing effect, while at 500 mg/1 it promoted femaleness. They (1997b) further investigated the effect of phytohormones and polyamines on sexual differentiation of *M. charantia* and suggested that the variation in spermidine content might be related to the initiation of development of female flowers, while the rise in putrescine content could be related to the initiation of male flowers. The response of different cucurbit species and different sex forms within a taxon, to various kinds of exogenous sprays of plant regulators is explained by 'optimum curve hypothesis' by Friedlander *et al.* (1977). This hypothesis is based on a scheme where sex expression (node number) may be considered as a function of abscisic acid and ethylene concentration and described by an optimum curve. The location of each sex type of a crop plant is determined according to the growth substance content of its shoot tips and its response to exogenous treatment. Low gibberellin content in the gynoecious cucumber under short day condition seems to be correlated with a marked response to external addition of gibberellin. On the other hand, in plants having a high gibberellin concentration, addition of this growth substance caused only a small change in sex expression. It is propounded that the node number as a function of endogenous gibberellin concentrations may be suggested as a saturation curve.

Verma *et al.* (1986) found that application of ethrel at 100 ppm at 2-leaf stage increased fruiting and yield in pumpkin (*C. moschata*). Ethrel at 250 and 100 ppm significantly increased the number of fruits/vine and yield/ha of pumpkin, because of a reduction in sex ratio resulting in more fruits/plant GA$_3$ at 10 and 25 ppm improved vine growth and seed content (Arora *et al.*, 1989). GA$_3$ tended to increase the sex ratio (staminate-pistillale) of flowers of pumpkin whereas NAA and ethrel reduced it (Das and Das, 1995). GA$_3$ at 100 ppm however, produced the most compact plants and highest fruit yields of 588.33 q/ha compared with 298.33 q/ha in the control (Das and Das, 1996). Spraying MH @ 150 ppm or CCC @ 500 ppm or GA$_3$ @ 30 ppm to the seedlings of bottle gourd at two-leaf stage and second after four days of 1st spray (Moniruzzaman *et al.*, 2019) and seed soaking in 0.5 per cent borax (0.05 per cent boron) solution for 12 hours and spraying of Ethrel 100 ppm or GA 10 ppm at two and four true leaf stages enhanced the yield in bottle gourd (Ansari *et al.*, 2018). Mia *et al.* (2014) reported that application of CEPA at 150 ppm and NAA at 50 ppm was found to be the best treatments for reducing sex ratio by increasing the female flowers by suppressing the male ones, and resulted higher yield. Ethephon at 300 ppm was found superior to its other levels including control for increasing number of female flowers and inhibiting male flowers and reducing male: female sex ratio in cucumber. Dhakal *et al.* (2019) and Acharya *et al.* (2020) observed that the spray of GA$_3$ 50-100 ppm increases growth, number of male flowers and weight of fruits. Ethrel 400-500 ppm increased number of female flowers, enhanced maturity cycle and improve sex ratio by suppressing the male flowers. Auxin 50-100 ppm has great influence on growth as well as ethrel improves the yield attributing parameters in cucurbits.

Parthenocarpy (seedless fruit) is an economically important yield- and quality-related trait in cucumber. Parthenocarpy is regulated by endogenous plant growth regulators (*e.g.*, diffusible auxin, IAA), and their balance is dramatically influenced by 252 cucumber environment (Kim *et al.*, 1994). Phenotypic selection has, however, resulted in the development of parthenocarpic hybrids (More and Budgujar, 2002) and genetic stocks (Sztangret *et al.*, 2004). It is clear that parthenocarpy is genetically controlled, but there is little agreement regarding the number and type of gene action involved. Hawthorn and Wellington (1930) and Meshcherov and Juldasheva (1974) suggested that parthenocarpy is recessive and controlled by a single gene. Kvasnikov *et al.* (1970), however, proposed that many incompletely recessive genes control parthenocarpy. Pike and Peterson (1969) simultaneously proposed that a single dominate gene expressing incomplete dominance controls parthenocarpy in cucumber. Results of de Ponti and Garretsen (1976) and El-Shawaf and Baker (1981a and b) indicate that parthenocarpy may be quantitatively inherited in this species. In fact, studies by Sun *et al.* (2006a and b) indicate that the genetics of parthenocarpy are complex. Generation means analyses in cross-progeny derived from elite processing cucumber lines indicated gene action generally could not be adequately explained by a simple additive-dominance model. Moreover, the analysis of F$_3$ families indicated that more than five genes control parthenocarpy, and that growing environment and epistatic interactions dramatically influence trait expression.

5.6.0 Sex Regulation by Environmental Factors

It has been proved earlier that environmental factors affect fruit development and sex expression in a number of crop species. However, with a few exceptions (*L. cylindrica, Lagenaria siceraria*), cucurbits are day-neutral plants, that is, flowering is a photo-insensitive process in cucurbits, but sex expression of flowers is affected by day length and temperatures (Nitsch *et al.*, 1952). In general, the number of pistillate flowers increased under short-day conditions and staminate flowers under long-day conditions for monoecious cultivars of cucumber (Matsuo, 1968), but *Cucumis hardwickii* (progenitor of *Cucumis sativus*) is an obligatory short-day plant. The effect of photoperiod can be evaluated by close examination of flowering shoots on which the first female flower appears, although different genotypes do not separate on the basis of node location. Usually at first, a male flush appears at the proximal region of flowering shoots, followed by a mix flush of male and female flowers at higher nodes, and only a female flush tends to appear at the distal portion of flowering shoots. Transition of phase in cucurbits depends on the photoperiod and the kind of species along with other external factors (Nitsch *et al.*, 1952; Saito and Ito, 1964).

Temperature and photoperiod are often associated together, and they are supposed to influence sex determination in association. Separate studies were conducted by Staub *et al.* (1987) in cucumber and more number of staminate flowers was obtained on the plants that were grown at 30°C compared to those on plants kept at 16°C. In brief, low temperature can promote femaleness and high temperature facilitates maleness in cucumber (Fukushima *et al.*, 1968; Cantliffe, 1981; Miao *et al.*, 2011). Low light intensity accompanied low temperature and short-day condition and high light intensity accompanied high temperature, so this matter of study was analyzed and found to produce more number of staminate flowers in a series of gynoecious processing cucumber at light intensity of 17,200 lx than 8,600, 12,900 and 25,800 lx, while a gynoecious inbred line showed no significant effect of light intensity (Cantliffe, 1981). The combined effects of light and air temperature (photothermal ratio). Wang *et al.* (2014) clearly indicated that the total number of female nodes significantly increase by high photothermal ratio, and this is opposite for the number of male nodes in three cultivars of cucumber when compared with the plants grown under low photothermal ratio.

Miao *et al.* (2011) investigated how low temperature alters the sex expression of monoeceous cucumbers (*C. sativus* L.). Plants were grown under different day/ night temperature regimens, 28°C/18°C (12/12 h), 18°C/12°C, 28°C/12°C, and 28°C/ (6 h 18°C + 6 h 12°C). It was found that plant femaleness is highest in the 28°C/(6 h 18°C + 6 h 12°C) treatment.

5.7.0 Physiology of Sex Expression

In the organ culture experiments of Galun *et al.* (1962), it has been convincingly demonstrated in cucumber that a given primordium can form either a pistille or staminate flower and has been found to be manipulated by addition or deletion of auxin (IAA). Determination of sex occurs early in the growth of primordium and is susceptible to control through environmental and chemical treatments and environmental control functions through the agency of growth metabolism of the plant which affects determination by establishing various auxin levels in the neighbourhood of differentiating primordia.

Heslop-Harrison (1963) propounded a kind of 'gene-operon' system to explain the numerous processes concerning the sequential initiation of the organs of the flower, like the one in monoecious cucumber. It is a kind of system which reflects the function of a relay system of gene activation operating through the agency of specific 'inducers' with short intercellular ranges. From several studies it has been possible to speculate on the functions and characteristics of 'organ forming' substances and the function may be supposed to control specifically the release of blocks of genetic information or perhaps more accurately to set in train, the processes leading to the step by step release of such information in defined cell lineages. Super imposed on the above mentioned short range systems governing differentiation in the flowering apex, there may be a generalized form of control, mediated through auxin metabolism of the whole plant. This auxin mediated control may operate in two ways; in conjunction with the determination of individual floral members in the early life of primordium and later in determining the balance of growth between gynoecium and androecium. Galun (1983) concluded that genetic, environmental and chemical factors are involved in the control of stamens and ovary differentiation in the cucumber floral bud. Three to four major genes coupled with modifiers, environmental factors like day-length and temperature and almost all major groups of growth regulators, all in conjunction determine the sex differentiation in cucumber plant. The transition from vegetative to reproductive phase was accompanied by a significant increase in the level of GA_3 in both *T. dioica* and *T. cucumerina* (Sarkar and Datta, 1990).

5.8.0 Pollination

Anthesis, pollen dehiscence and fruit set in cucurbits are influenced by environmental factors. Usually fruit set takes place in the early morning between 6 a.m. and 8 a.m. in the months of March and April, in crops like cucumber, pumpkin, muskmelon and watermelon. Optimum temperature during this period would range between 55°F (12.8°C) and 65°F (18.3°C). There are other cucurbits which flower later in the day and fruits set at higher temperature of mid-day as in bottle gourd and ridge gourd. In snake gourd and pointed gourd, night temperature favours anthesis during nights and fruits set in the early hours of the morning when the insects visit them. Hossain *et al.* (1990) reported that in bottle gourd both staminate and pistillate flowers opened between 16.30 and 18.00 h and closed after 14.00 h. The staminate and pistillate flowers required 10 and 6.5 days, respectively for anthesis from date of initiation. Natural pollination supplemented with hand pollination was most successful (59.26 per cent) for fruit set. Pollen production in different genera is variable. In watermelon and pumpkin, pollen production is good while in crops like muskmelon the pollen production is scanty and also the pollen is sticky due to the oily film surrounding the pollen grains. Hence mechanical scratching of pollen grain becomes necessary in hand pollination which bees accomplish much more efficiently in an involuntary way.

Pollen grains of the cultivated genera are generally described as tricolporate oblate to sub-oblate and occasionally spinose grains (Ayyangar, 1976). Nair and Kapoor (1974) described in detail the variations found in different cultivars of cultivated species. An interesting phenomenon reported by Ayyanga (1976) was that in the two dioecious species *Coccinia grandis* (syn. *C. indica*) and *Trichosanthes dioica*, distinct differences in the size ranges of pollen grains, inferring that two distinctive categories of pollen grains may be male determining and female determining.

The very fact that staminate and pistillate flowers are separate within the same plant, imposes a situation conductive to cross-pollination. This group of crops are classified as cross-pollinated but it may be worth emphasizing that this condition does not preclude natural self-pollination taking place within the same plant. The extent of cross-pollination ranges from 60 to 80 per cent depending upon the environment and visitation by the insect pollinators. In fact in the case of muskmelon, the andromonoecious condition (due to presence of perfect flowers beside the staminate ones) favours a higher degree of natural self-pollination than in the monoecious condition. It is obvious that insects do not distinguish flowers from the same or different plants at the time of involuntary pollination. In dioecious crops, provision of at least 10 to 15 per cent male plants will ensure good fruitset and yield.

The pollination is entomophilous and bees mostly *Apis florea, Apis dorsata, Nomioide* ssp., *Halictine* sp., are the main pollinating agents. There are also beetles like *Conpophilus* sp., moths like *Planidia* sp., *Pygargonia* sp., etc. To ensure the visit of pollinating insects, the flowers in most cucurbits are showy in colour, big in size and staminate flowers are produced in greater abundance than the pistillate ones. For maximum fruit set, fruit and seed yield, one bee colony per half hectare would be required. Man has not been as efficient as bee in pollinating cucurbit flower and in most cases, controlled hand pollination (whether selfing or crossing) result in lower percentage of successful fertilization. Entomophilous pollination imposes the necessity of isolation when seeds of more than one cultivar of a given species are multiplied. While it is reported that bees can fly as far as 8 to 16 km in search of nectar, for practical purposes 0.8 to 1.0 km is considered adequate isolation between two cultivars of the same species.

Where isolation facilities are not available and where seeds of two or more cultivars of the same species are to be multiplied, a kind of barrier system with 3 or 4 cucurbits, can be adopted providing minimum isolation.

Cucumber-1	Watermelon-1	Bottle gourd-1	Squash-1
Bottle gourd-2	Squash-2	Cucumber-2	Watermelon-2
Cucumber-3	Watermelon-3	Bottle gourd-3	Squash-3
Bottle gourd-4	Squash-4	Cucumber-4	Watermelon-4

In this system, in an acre or 1.5 acres 16 plots are divided with minimum area of 9×30 metres to be assigned to each cultivar. Sowing of 4 cultivars of each of four cucurbits has to be done efficiently to ensure fullstand at the time of flowering without any gaps in any plot. Further, adjustment of sowing has to be carefully done, so that all the four cultivars in all the species flower simultaneously. The main principle here is that when bees or insects carrying pollen of one cultivar of one species will have a barrier of another species and the pollen of the former will be wasted. The barrier between the two cultivars of the same species is effective if full stand and simultaneous flowering are ensured. Under this system, natural self (or sib) pollinations take place within the cultivar without the need or controlled hand (or self) pollination. This is a practical method which ensures large number of fruits available for selection, at the same time avoiding cross-pollination between the two cultivars of the same species. The plot areas of each cultivar should be big enough and the population within the plot should be large enough, so that the bees donot skip or overfly from one cultivar to another cultivar of the same species. Even if occasional cross-pollination were to take place, it would not exceed more than one per cent and to eliminate this, a row of plants on the outer periphery of each plot may be taken as border plants to discard their fruits.

6.0 Cultivars

6.1.0 Cucumber

Under *Cucumis sativus*, no sub divisions are usually recognized but Chakravarty (1982) distinguished two botanical varieties var. *sativus* and var. *sikkimensis*, on the basis of higher number of leaves and higher number of placentae in the latter.

Cucumber cultivars are usually classified on the basis of how they are used, fresh market (slicer) and pickling. In general, fruits of slicers are larger than pickling cultivars and they develop a darker and heavier skin with uniformly cylindrical shape. Present day pickling cultivars tend to develop shorter vines than fresh market ones and they are more prolific. 'Balam Khira' of Saharanpur (U.P.) region is akin to pickling types of the West. The fruit of pickling cultivars are smaller and the seed cavity develops more slowly, an important characteristic affecting suitable for storage in brine. In pickling cultivars, a thin skin with a lighter colour is preferred.

Cucumber cultivars are also classified according to spine colour, a trait associated with a gene controlling skin colour of fully mature fruits (at the time of seed extraction). The cultivars with white spine retain their green colour longer in immature condition and become dull yellow when fully ripe. All the slicing cultivars of the Western Europe and the U.S.A., are white spined. Glass house cucumbers of Europe have different characteristics as English types, are large ones and the Russian cultivars have short, thick and rough netted brown skin. In France, thick large cultivar of irregular shape are grown for cosmetic purposes, while in the East and Near East (Asia) slender smooth skin types are popular and in Japan consumers prefer small sized slicer cultivars.

Black spined cultivars develop a dull orange colour at full maturity and tend to turn colour prematurely under high temperature conditions both in the field and during transport. Black spined cucumbers have been preferred over white spined pickling cultivars because they develop a more attractive colour after prolonged storage in brine.

Significant advance has been made in recent years to produce hybrid cultivars on commercial scale. The hybrid incorporate such characteristics as high yield potential and prolonged harvests as in glass house cultivars of Western Europe, relatively concentrated fruit set and maturity as in pickling cultivars of the U.S.A. for machine harvest and resistance to a broad range of cucumber disease. Multiple disease resistant hybrids both in pickling and slicing kinds and for glass house and open field culture are now available. All the present day hybrids are gynoecious, parthenocarpic (in the case of glass house hybrids of Western Europe) and multiple disease resistant.

6.1.1 Cucumber Cultivars Released in India

Japanese Long Green

It is released by ICAR-Indian Agricultural Research Institute (IARI) Regional Station, Katrain (Kullu Valley). A temperate cultivar, suited to hills and lower hills, extra early with 45 days maturity, fruits 30 to 40 cm long, flesh is light green and crisp.

Straight Eight

Early cultivar suited to hills, white spined, fruit medium long, thick, straight with round end, colour medium green, released by IARI Regional Station, Katrain (Kullu Valley).

Pusa Sanyog (F₁ Hybrid)

An early and high yielding hybrid, released by Indian Agricultural Research Institute (IARI) Regional Station, Katrain (Kullu Valley). It is a hybrid between Japanese gynoecious line and Green Long Naples. Its fruits are 28 to 30 cm long, cylindrical and dark green with yellow stripes, crisp flesh, maturity in 50 days.

Calypso (F₁ Hybrid)

It is an early maturing, gynoecious pickling type F_1 hybrid introduced from USA and takes 50-55 days to reach harvestable maturity. The fruits are cylindrical, 2.0-2.5 cm thick and 8-10 cm long. The rind is medium green and bears white spines.

Pusa Uday

It is released by ICAR-IARI, New Delhi. Fruits are medium in size (13-15 cm long), light green in colour with whitish green stripes, straight, non-prickled and soft skinned. It is suitable for cultivation both in spring-summer and rainy seasons. Maturity in 48-52 days. Average yield is 15 t/ha.

Pusa Parthenocarpic Cucumber-6

It is released by ICAR-IARI, New Delhi. It is the first extra early (40-45 days for first fruit harvest) improved variety of parthenocarpic gynoecious cucumber suitable for cultivation in protected condition. Fruits are attractive, uniform, dark green, glossy, cylindrical, straight, slightly ribbed, non-hairy, non-warty, slightly striped at blossom end and has tender skin and crispy flesh. Average fruit yield is 126 t/ha (1260 kg/100 m²) during winter season.

Pusa Gynoecious Cucumber Hybrid-18

It is the Gynoecy based hybrid released by ICAR-Indian Agricultural Research Institute , New Delhi for cultivation in Zone I [Jammu & Kashmir (J&K), Himachal Pradesh and Uttarakhand]. Fruits are attractive green in colour with mild whitish green stripes originating from the blossom end and brownish green blotchy patches present near the stem end. Fruits are 18-20 cm long having soft skin, crispy and tender flesh with average fruit weight 200-210 g. Average yield is 24.0 t/ha.

Priya

It is a hybrid cucumber of Indo-American Hybrid-Seeds, commercially available in India. The fruit is about 25 cm long, dark green, bitter-free and slicing type. It is tolerant to mildew, mosaic, anthracnose and scab disease.

Poinsette

An American introduction multiplied by National Seeds Corporation. The fruits are dark green, 20 to 25 cm long. Originally developed at Charleston (South Carolina), U.S.A., it carries resistance to downy mildew, powdery mildew, anthracnose and angular leaf spot.

Kashi Nutan (F₁ Hybrid)

It was released by ICAR-IIVR, Varanasi. It is an early maturing hybrid having cylindrical long light green colour with mottling at peduncle side of fruit. It is resistant to downy mildew and is suitable for rainy and summer season. Average yield is 17 t/ha.

CO 1

It was released from Tamil Nadu Agricultural University, Coimbatore. It is a selection from a local type of Kanyakumari District. The fruits are long (60-65 cm), attractive, slightly curved, tapering towards stalk end. As a tender vegetable for salad it yields 15 t/ha in a crop duration of 100 days.

Harith (F₁ Hybrid)

It is released by KAU, Vellanikara. It is suitable for open and rain shelter conditions. Fruits are light green, 18.67 cm long and 260 g wt. Potential yield is 10.28 t/ha.

Shubra (F₁ Hybrid)

It is released by KAU, Vellanikara. It is suitable for open and rain shelter conditions. Fruits are greenish white, 14.4 cm long and weigh 275 g. Potential yield is 10.23 t/ha.

KPCH 1 (F₁ Hybrid)

It is released by KAU, Vellanikara. This F₁ hybrid is generated from a cross between parthenocarpic lines, suitable for polyhouse cultivation, early maturing with long dark green fruits and moderately resistant to downy mildew. It is suitable for humid tropics of Kerala. Average yield is 1.0 t/100 m².

Punjab Naveen

It is released from PAU, Ludhiana. The fruits are uniform cylindrical fruit with attractive light green colour and smooth surface. The fruits are bitter free, having soft seeds at edible maturity and are very crispy. It takes 68 days from the transplanting to harvesting. The cultivar is excellent in taste, appearance, colour, size and texture and its average yield is 7 t/acre.

Punjab Kheera-1

It is released from PAU, Ludhiana. The plants are vigorous, bearing 1-2 fruits per node. It is suitable for poly-net house only. Flowers are parthenocarpic and fruits are dark green, seedless, bitter free, medium sized (125 g), 13-15 cm long and do not require peeling. First fruit picking is possible after 45 and 60 days of sowing in September and January sown crop, respectively. Its average yield is 30 t/acre and 37 t/acre in September and January sown crop, respectively.

Pant Sankar Khira-1 (F₁ Hybrid)

It was released from G.B.U.A. and T., Pant Nagar. The fruits are about 20 cm long, cylindrical and green with light stripes. Vine length is about 120 cm. It takes 50 days to first picking, the yield potential is 20 t/ha.

Pant Khira-1

It was released from G.B.U.A. and T., Pant Nagar. The fruits are long (20 cm), cylindrical with light white stripes. The fruits attain first picking stage in 50-60 days. Recommended seed rate is 4 kg/ha. The yield potential is 15 t/ha.

Pant Parthenocarpic Khira-2

This parthenocarpic cucumber cultivar was released from G.B.U.A. and T., Pant Nagar. It is suitable for polyhouse cultivation. No. of female flower per vine is around 550. Average fruit wt. is 630 g and yield potential is 21 t/ha.

Pant Parthenocarpic Khira-3

It is a parthenocarpic cucumber cultivar suitable for polyhouse cultivation and is released from G.B.U.A. and T., Pant Nagar. Number of female flower per vine is 465. Average fruit weight is 415 g and yield potential is 19.92 t/ha.

Himangi

It is released by MPKV, Rahuri. The fruits are white in colour and resistant to bronzing, suitable for *kharif* season. The average yield is 18 t/ha.

Phule Shubangi

It is released by MPKV, Rahuri. Its fruits are straight, light green in colour, fruit surface smooth with trichomes. It is tolerant to powdery mildew and is suitable for *kharif* season. Average yield is 18 t/ha.

Phule Prachi

It is released by MPKV, Rahuri. The fruits are yellowish white in colour and suitable for green house cultivation. Average fruit weight is 600 g and yield potential is 36 t/ha.

Swarna Ageti

Released from ICAR-Research Complex for Eastern Region, Ranchi, Patna. Fruits are cylindrical long (150-200 g), green and without prominent placental hollowness. Sowing time is July-August and February-March and the first harvest takes place 45-50 days after sowing. It is tolerant to powdery mildew. Yield is about 30.0-35.5 t/ha.

Swarna Sheetal

Released from ICAR-Research Complex for Eastern Region, Ranchi, Patna. Fruit is cylindrical long 200-250 g, greenish white and without prominent hollowness. Sowing time is February-March and the first harvest takes place 60-65 days after sowing. It is tolerant to powdery mildew disease. Yield is about 20-30 t/ha.

Swarna Poorna

Released from ICAR-Research Complex for Eastern Region, Ranchi, Patna. Fruits are cylindrical with approximately 300 g weight, generally light green in colour and without prominent hollowness. Time of sowing is July-August and February-March. First harvest takes place 55-60 days after sowing. It is tolerant to powdery mildew disease. Yield is about 30-35 t/ha.

There are several local cultivars grown in different regions of India. In Poona region, there is a cultivar called 'Poona Khira' producing small-sized, pale green fruit. In Saharanpur region (U.P.), there is a cultivar called 'Balam Khira', similar to picking type of the U.S.A. In North Bengal, a cultivar called Darjeeling or Sikkim is known. In some parts of Maharashtra and in South India, the real cucumber is called *kakri* which should not be confused with *kakri* of North India.

6.1.2 Promising Cultivars from other Countries

Moviri (F₁ Hybrid)

Bred at the Moscow branch of the Vavilov Institute of Plant Industry from a cross between line 598 and the monoecious line 1-29N, is suitable for growing under glass or plastic. The plants are tall with many side shoots. The fruits are dark green with white spines, weight 90-120 g and attain a length of 11-13 cm. Moviri is resistant to *Cladosporium cucumerium*, bacterial disease and cucumber mosaic virus (Pyzhen-kov *et al.*, 1988).

Askon

Turchenkov *et al.* (1992) reported from Russia a cucumber cultivar Askon. This is predominantly gynoecious type, with a high degree of parthenocarpy, gives consistently high yields without insect pollination. It was bred by individual plant selection from K2842.

Kid

Kid is a mid-season type (48-50 days from emergence to first harvest) of Russia, suitable for pickling and is resistant to downy mildew (Migina and Kuzmitskaya, 1998).

Redlands Long White

The cultivar is developed from a cross between Green Gem and Crystal Apple is adapted to growing conditions in Queensland, Australia. The fruits are white and smooth with crisp and white flesh (Saranah and Harrington, 1985).

Guloi 82

From Korea Democratic People's Republic, a female F₁ hybrid of cucumber, Guloi 82, was obtained by crossing a female line with a complex variety as male. It is early maturing and produces high yields under different systems of cultivation, including the field, plastic tunnel and green house (Chan Giand Su Yon, 1995).

Martina

Martina is a new small-fruited cucumber hybrid from Bulgaria, obtained by crossing the gynoecious line 81066 with the monoecious line V8l. It produces dark-green warty fruits with good texture, fine flavour, and suitable for canning. The growth period from emergence to fruiting is 44-46 days. It is resistant to powdery mildew (Khristova, 1997).

Renesansa

In Yugoslavia, a F₁ hybrid of salad cucumber Renesansa has been developed from a cross between the gynoecious line 135 and the monoecious line MM76 (the

Cultivars/Hybrids of Cucumber

Pusa Parthenocarpic Cucumber-6

Poinsette

Pusa Gynoecious Cucumber Hybrid-18

Pusa Barkha

Swarna Sheetal

Swarna Ageti

Swarna Poorna

Cultivars/Hybrids of Cucumber

Malini

Padmini

Amira

Armenian

Burpless

Bush Champion

County Fair

Dasher II

English

Fanfare

Cultivars/Hybrids of Cucumber

Garden

Gherkins

Gold Standard

Green Sleeves

Japanese

Kirby

Lemon

Marketmore

Lemon Crystal

Miniature White

Cultivars/Hybrids of Cucumber

National Pickling

Orient Express

Persian

Pickle Bush

Regal

Salad Bush

Saladin

Sambar

Cultivars/Hybrids of Cucumber

Slicemore

Spacemaster

Straight Eight

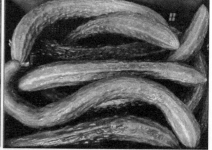

Suo Long

Sweet Slice

Sweet Success

Tender Green

Burpless Diva

hybrid being of the gynoecious type). The hybrid is suitable for growing under unprotected and protected cultivation (Stankovic, 1992).

Chernenko (1989) reported from Russia, F_1 hybrids Brigadnyi, Cherol, G707, G711, G697 and G736 which belong to various maturity groups, disease resistance and suitable for pickling.

According to description made by Suchkova (1995), the Russian F_1 hybrid cucumber green house cultivars *viz.*, Villina, Yuventa, Reddo, Korelek and Izumrud are all predominantly gynoecious. The first 2 produce smooth fruits with a mean length of 22-27 cm. The fruits of the other hybrids have warty skin and vary in length from 14-16cm in Izumrud (suitable for pickling) and 16-18 cm in Korelek to 18-20 cm in Reddo.

Some more Russian F_1 hybrid parthenocarpic cucumber cultivars of the gynoecious type are Sapfir, Vesnyanka, Effekt, Korolek, Alisa, Moskovskii,Yubileinyi, El'f, Sharm and Vavilon (Suchkova,1997).

From the Netherlands, cucumber cv. Ventura has been reported to be the best performed cultivars in an evaluation trial (Uffelen, 1989). Parinova (1989) reported an early cv. Nadeyhnyi from Russia which requires 4 days for emergence to first harvest. In Siberia, frost resistant high yielding cvs. Krepysh and Uchitelskii, can be grown without protection (Saraev, 1991).

A survey of cucumber cultivars grown in North Carolina, USA revealed that most popular pickling cucumber was Calypso, followed by Primepak and Regal, and the most popular fresh-market cultivar was Dasher II, followed by Poinsette 76, Centurion, Marketmore, Supersett and Revenue (Schultheis, 1992). EI-Aidy (1991) tested 17 cultivars of cucumber in Egypt and highest total yields were obtained from Katla, Cordito, Tetenyi F_1, Budai and Csemege F_1 and Peteta F_1 in the winter season and Rewa, Nagah and Picabillo in the early summer season.

In lower Bavaria, Germany, 12 parthenocarpic processing cucumber cultivars were compared by Arold (1998). Marketable yield was highest in cv. RS94108 and it was lowest in cv. Bj1799.

A new cucumber cv. Svetlyachok was bred in western Siberia by Variety × line hybridization between disease-resistant forms. It is a mid-early field cultivar, which begins to fruit 44 days after emergence, gives a total yield of 30 t/ha and marketable yield of 25 t/ha (Vysochin, 1998).

In Korea Republic, Pusan Daemok 1 was selected as a promising rootstock for grafting of cucumber (Kim *et al.*, 1998). Total and marketable yields of Pusan Daemok 1 in protected cultivation were 14 and 11 per cent higher, respectively, than the Deruderu root-stock on the market.

6.2.0 Muskmelon

The species *Cucumis melo* is a large polymorphic taxon, encompassing a large number of botanical and horticultural varieties or groups. It includes both dessert as well as cooking and salad types used like cucumber. Naudin (1859) an African botanist divided this species into several botanical varieties adopting a trinomial classification. Even though this classification is not now botanically recognized, because of absence of distinct botanical variation and of free intercrossability, the botanical epithets used by him are still in current usage.

C. melo var. *agrestis*: Plant with slender vines, small flowers and inedible fruits.

var. *cultura*: Plant with stouter vines, flowers large with edible fruits.

var. *reticulatus*: Netted melons, fruits medium size, surface strongly netted often length wise, rind fleshy with tough skin. Most American cultivars like cantaloupe and Persian melons belong to this group.

var. *cantaloupensis*: Cantaloupe melons, fruits warty, scaly and rough with hard skin, warted surface distinct, netting absent, still cultivated in parts of Europe.

var. *inodorus*: Called winter melons of the USA, comprising Honeydew (green fleshed) Casaba and Crenshaw. Fruits with little musky odour, ripening late and keeping into winter, fruit surface usually smooth, leaves light green or medium green.

var. *flexuosus*: Snake, or serpent or longmelon, fruits are long and slender 2.5-5.0 cm in diameter, 45-90 cm long. It is the same as *C. melo* var. *utilissimus* of India, known as *kakri* (of North India), tar, *etc.*, and eaten like salad cucumber.

var. *conomon*: Pickling melon, fruit smooth, glabrous, without musky odour, variously shaped.

var. *chito*: Mango lemon or lemon cucumber, fruit small with the size of an ordinary orange, variously described as vegetable orange or melon apple, *etc.*

var. *dudaim*: Fruits small like chito, surface marbled with rich brown, very fragrant, grown for ornamental purpose.

Apart from these botanical varieties one more is recognized, namely *Cucumis melo* var. *momordica* which is the *phoont* of North and Eastern India or called as snap melon, whose fruit skin burst or cracks on maturity. The rind is yellow, flesh fluffy and sourish.

This classification of muskmelon is no longer officially recognized but nevertheless it brings for the large variation met within this species. Keeping muskmelon with cucumber in one genus *Cucumis* has been questioned in relation to the large diversity of *melo* forms, complete absence of intercrossability, both natural or artificial between *sativus* and *melo* and differing chromosome numbers. The Russian botanists like Pangolo (1951) and Zhukovsky (1962) placed the muskmelon in an independent genus *Melo*. Three sections Eumelo, Melonosides and Bubalion have been recognized by them.

Eumelo consists of: *M. microcarpus, M. adzhur*

M. cassaba, M. adana

M. cantalupa, M. flexuosus

M. chandalak, M. ameri and *M. zard*

Melonosides consists of: (Chinese melons) *M. chinensis, M. conomon*

M. monoclinus

Bubalion consists of: (Weed species) *M. agrestis* and *M. figari*

Most of the Central Asian cultivars like those of Afghanistan and Uzbekistan are long duration types grown during dry period from April-May to October. These are generally oblong or round, netted or smooth skinned, green, orange or white fleshed. 'Sarda' melon available in North Indian cities during September-October is one such variety imported from Afghanistan.

Horticulturally, the muskmelon and cantaloupe (of the USA) differ somewhat in physical characteristics and regional adaptation. Today the 'cantaloupe' simply refers to cultivars that are highly uniform in overall netting (corky tissue on the rind) with relatively indistinct ribs or vein tracts. Internally, flesh is thick, salmon-orange in colour with characteristic flavor and the seed cavity is small and dry. Muskmelon cultivars on the other hand have a stronger aroma, juicier flesh and larger seed cavity.

In the USA most cantaloupe cultivars have been developed to meet requirements for packing in crates and transport to long distance. Winter melons of the USA conversely require a long growing season at relatively higher temperatures under semi-arid conditions and are suited to limited storage and shipment after harvest.

Among the leading cantaloupe cultivars, Hale's Best, Edisto, Campo, Jacumba, Top Mark, Perlita, Wescan, Planters Jumbo, *etc.*, can be mentioned. In the Persian group, Persian Large and Small are recognized. Among the long duration winter melons, Honey Dew, Casaba, Golden Beauty, Crenshaw and Santa Claus are important. There are also melons grown in glass houses in Europe as Charentias in France and smaller smooth skinned ones in Japan.

Some of the improved cultivars of muskmelon evolved by different Institutes and Agricultural Universities in India are described below:

Arka Rajhans

It is released by ICAR-Indian Institute of Horticultural Research (IIHR), Hessarghatta for Southern region. It is a mid-season cultivar bearing large oval fruits weighing above one kg with fine net. The flesh is white and sweet and fruit has transportable quality, said to be tolerant to powdery mildew.

Arka Jeet

It is released by ICAR-IIHR, Hessarghatta, Bengaluru. An early cultivar very similar to 'Lucknow Safeda' of U.P. and a selection from 'Bati' strain of U.P. The fruits are flat, small, weighing 300 to 500 g, orange to orange-brown skin, white flesh, big seed cavity, very sweet (12 to 14 per cent TSS).

Arka Siri

It is released by ICAR-IIHR, Hessarghatta, Bengaluru. It is cantaloupe type fruit weighing about 1kg each, with an appealing pattern of netting, plus green sutures on creamish orange rind background, sweet (TSS 12 per cent), dark-orange flesh and a strong musky aroma. Yield is 20- 25 t/ha.

Pusa Sharbati

It is released by ICAR-IARI, New Delhi. It is a derivative from the cross between Kutana and Resistant No.6 of the USA. An early cultivar maturing in 85 days, with round fruits having netted skin. Salmon-orange flesh is firm and thick with small seed cavity, moderately sweet (11 to12 per cent TSS).

Pusa Madhuras

It is released by ICAR-IARI, New Delhi. It is a mid-season selection from a Rajasthan collection with roundish flat fruits weighing one kg or slightly more. The skin is pale green, sparsely netted with dark green stripes and salmon-orange flesh, juicy and sweet (12 to 14 per cent TSS). Keeping quality is poor.

Pusa Madhurima

It is released by ICAR-IARI, New Delhi. This is a unique shaped muskmelon cultivar with high yield (22.45 t/ha) and increased shelf life. Its fruit is ovate to obovate shape with average weight of 775 gram. Fruits get ready for harvest in about 80 days. The rind colour of fruit is creamish yellow with green sutures. Fruit flesh is thick, green, juicy and crispy with medium musky flavour and high sweetness (TSS 12 °Brix). Fruit surface is grooved with moderate netting and is slipable at maturity. Fruits attain nipple shape at peduncle end. Its leaves are weakly lobed and shows andromonoecious sex expression.

Pusa Sarda (Sarda melon)

It is released by ICAR-IARI, New Delhi. This is a first cultivar of Sarda melon from public sector in India which can be grown in net-house under north Indian plains. Its fruit are golden yellow, roundish to elongated globe shaped with average weight of 1.1 kg. Fruits get ready for harvest in about 85-90 days. Fruit flesh is thick, greenish white, and very crispy with high sweetness (TSS 13.60 °B). It can be stored for 15 to 20 days. Its average yield is 5.44 t/1000 m^2 under net-house.

Kashi Madhu

It is released by ICAR-IIVR, Varanasi. Plants have medium vine and leaves sparsely lobed and dark green, fruits are round, with open prominent green sutures, weight 650-725 g, half slip in nature, thin rind, smooth and pale yellow at maturity, flesh salmon orange (mango colour), thick, very juicy, T.S.S. 13-14 per cent and seeds are loosely packed in the seed cavity. Postharvest life is long with good transportability, tolerant to powdery and downy mildew, medium maturity and yield 20-27 t/ha.

Hara Madhu

It is released by Punjab Agricultural University, India. It is a late cultivar from a local collection of Haryana. The fruit is globose with dark green stripes, weighing one kg; flesh is light green, juicy, very sweet (with 12 to 15 per cent TSS). Keeping quality is poor.

Punjab Sunehri

It is released by Punjab Agricultural University, India. It is a derivative from the cross between Hara Madhu and Edisto, early maturing, pale green, thick skin, flesh salmon-orange, thick with moderate sweetness (11 to 12 per cent TSS).

Punjab Hybrid

It is released by Punjab Agricultural University, India. It is a F_1 hybrid between a male sterile line (ms_1) and Hara Madhu. It is an early maturing hybrid with orange flesh and netted skin. The fruits are nearly round, rind is light yellow, sutured and moderately netted. The fruits are invariably sweet. Flesh is thick, orange and juicy. Average TSS is 12 per cent. Fruits develop full slip stage. Average fruit weight is one kg and yield is 16 t/ha.

MH-27

It is released by Punjab Agricultural University, India. Fruit is round, light yellow, sutured and netted. Flesh is thick, salmon orange, medium juicy with 12.5 per cent TSS. The fruits develop 'full slip' stage. Its first picking is after 63 days of transplanting. Average fruit weight is 860 g. It is tolerant to wilt and root knot nematodes. The yield is 8.75 t/acre. It has long shelf life and suitable for distant transportation.

MH-51

It is released by Punjab Agricultural University, India. Fruits are round and netted with green sutures. Fruit flesh is thick, salmon orange, medium juicy and flavorsome with 12.2 per cent TSS. Its average fruit weight is 890 g. It is an early maturing hybrid and can be harvested after 62 days of transplanting. Its average fruit yield is 8.9 t/acre.

Durgapura Madhu

It is recommended by Department of Agriculture, Jaipur, India. A very early cultivar confined to Jaipur region of Rajasthan. The fruits are oblong, weighing 500 to 600g, pale green rind, light green flesh with dry texture, very sweet (13 to 14 per cent TSS), seed cavity big.

Sona (Cantaloupe)

A hybrid from Indo-American Hybrid Seeds in India. Fruits are closely netted, slightly ribbed, and orange cream coloured. It is tolerant to powdery mildew and downy mildew and possesses good keeping quality.

Swarna (Cantaloupe)

This is also a hybrid from Indo-American Hybrid Seeds. Fruit is yellowish orange in colour with very sweet, dark orange flesh inside. It can withstand long distance transport.

Gujarat Muskmelon-3

It is released by AAU, Gujarat. The fruits are oblong shape and it has soft, attracting green flesh textured with sweet taste and good aroma having 12 per cent TSS. Average yield is 12.7 t/ha.

There are several local cultivars in muskmelon grown in different regions of India, among the noted dessert cultivars are Lucknow Safeda, Baghpat melon of Meerut, Jaunpuri Netted, Mau melon of Azamgarh, *etc.*, of U.P., Tonk melon of Rajasthan, Kharri melon of Hoshangabad (M.P.), Goose melon and Jamor Neel of Northern Maharashtra, Sharbat-e-Anar, Bathesa of Andhra Pradesh are some of the cultivars extensively grown in river-beds and uplands. Besides, nondessert types like, *photo* (snap melon),'kachri' or 'kachauri' (which is cooked) of North India, 'dosakaya', 'budamkaya' cooking types of Andhra Pradesh, but called as cucumber, vellarikkai - a salad cucumber type of Tamil Nadu, all these belong to *Cucumis melo*, even though not considered or used as muskmelon. Similar is the case of *kakri* or *tar* of North India, otherwise called long or snake or serpent melon.

6.2.1 Promising Cultivars from other Countries

Allrora is a new cultivar adapted to growing conditions in the south eastern USA. It was selected from the cross between Southland × PI140471, followed by backcrossing and disease screening. The fruits are 15.2-17.8 cm in diameter with an average weight of 1.9 kg. The flesh is thick, orange and takes 70-75 days to mature and compares favourably with other cultivars in yield. It is resistant to *Pseudoperonospora cubensis, Sphaerotheca fuliginea* and *Didymella bryoniae* (Cosper and Norton, 1985; Norton *et al.*, 1985). In varietal trials, Hill (1996) recorded the highest yield in cv. Tenerife (Canary type).

The cultivars Qi Tian-I was selected as a natural mutant from the population of the land-race Tie Ba Qing in China. It has 65 days' growth period from emergence to maturity and yields 25.5 t/ha (Zhu *et al.*, 1993). Lanmi is a mid season F_1 hybrid from China (Zhang and Zhang, 1998). The fruits are yellow, oblong of around 3 kg weight with a fine netting when ripe. Other high yielding Chinese cultivar developed are Mitianguan (Wei *et al.*, 1995) and Xinmi No.13 (Li and Li, 1998).

Aisophy is a newly developed netted cultivar of Japan, resistant to *Fusarium* wilt and powdery mildew. The cultivar has high brix value and good shelf-life (Ogawa *et al.*, 1995).

Galia F_1 hybrid was produced in Israel for high yield and excellent fruit quality (Karchi *et al.*, 2000).

It has been reported from Italy that long storage life and high yielding cultivars are Nun 0326 (28.8 t/ha) and Pegaso (27.1 t/ha) (Quiattrucci and Conti, 1997); early cultivars are Perseo and John and late cultivars are Orlando and Century (Benomi *et al.*, 1999). Other cultivars with good performance are Baggio, Calipso, Supermarket and Avalon (Benomi *et al.*, 1999).

In Spain cultivars RQ2000 and Sancho were most appreciated for their external appearance and fruit quality (Pardo *et al.*, 2000).

In Panama, cultivars Cristobal and Durango have been reported to produce export quality fruits with high yield and good tolerance to pests and diseases (Carranza *et al.*, 1998).

Cultivars/Hybrids of Muskmelon

Kashi Madhu

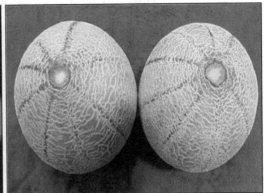

Arka Siri

Hara Madhu

Pusa Madhurima

Arka Jeet

Arka Rajhans

Different Market Segments of Muskmelon

Canary Yellow   Galia   Honey Dews   Charentai

Cantaloupe   Madhuras   Kajari

In South Korea, Kim *et al.* (1984) reported hybrid cultivar Poongmi (Choonge 3 × Kurume 2) for green house cultivation. It bears well-netted fruit with a high TSS.

Oriental Pickling Melon

KAU Vishal

It has medium to large cylindrical shaped fruits, crop duration 70-75 days and is less susceptible to pests and diseases. It is suitable for warm humid tropics of Kerala. Yield is 33 t/ha.

Some other varieties and local types are Soubhaghya, Valakkavu local (PS), Mudicode and Arunima.

6.3.0 Watermelon

Asahi Yamato

It is a mid-season Japanese introduction, released by ICAR-IARI. It produces medium sized fruits averaging 6 to 8 kg. The rind colour is light green with deep pink flesh (TSS 11 to 13 per cent). The fruits ripen in 95 days.

Sugar Baby

It is an early season American introduction, released by ICAR-IARI, New Delhi. The fruit is slightly smaller in size, weighing 3 to 5 kg, round in shape, having bluish black rind and deep pink flesh (TSS 11 to 13 per cent) and small seeds. The fruits ripen in 85 days.

Pusa Bedana

It is a triploid seedless watermelon developed from the cross Tetra 2 (tetraploid) × Pusa Rasal (diploid). Its fruits are medium green in colour each weighing 3-4 kg. Released by ICAR-IARI, New Delhi.

Arka Jyoti

It is a mid-season F_1 hybrid (IHR-20 × Crimson Sweet), released by ICAR-IIHR, Bengaluru. The fruits are round, weighing 6 to 8 kg. The rind colour is light green with dark green stripes and flesh colour crimson (11 to 13 per cent TSS).

Arka Manik

It is released by ICAR-IIHR, Bengaluru. It is resistant to powdery mildew and tolerant to anthracnose. The fruits are round to oval with green rind and dull green stripes. The flesh is deep red, very sweet (12-13 per cent TSS). The average fruit weight is 6 kg. It stands well in transport and storage.

Arka Akash

It is F_1 hybrid between IIHR-60-1 × Arka Manik, released by ICAR-IIHR, Bengaluru. Fruits are oval round with high TSS (11-12°Brix). Average yield is 90-100 t/ha.

Arka Aiswarya

It is F$_1$ hybrid between IIHR-86-3 × Arka Manik, released by ICAR-IIHR, Bengaluru. Fruits are oblong with high TSS (10-11°Brix). Average yield is 80-90 t/ha.

Arka Muthu

It is a selection from IIHR-81-1-1 and released by ICAR-IIHR, Bengaluru. It has unique character of dwarf vine (vine length 1.2 m), shorter internodal length and early maturing type (75-80 days). It has round to oval fruits with dark green stripes and deep red flesh. Average fruit weight is 2.5-3 kg with TSS ranging from 12 to 14 °Brix. Fruit yield is 85 to 90 t/ha.

Arka Madhura

It is a triploid seedless watermelon hybrid between Tetra-1 × Arka Manik and released by ICAR-IIHR, Bengaluru. Fruits are round with dark green rind colour with light green broad stripes. Crimson red flesh is with pleasant aroma and without seeds, TSS 14° Brix. Average fruit weight is 6 kg. Yield is 50-60 t/ha in 100-110 days duration.

Thar Manak

This cultivar is released from CIAH, Bikaner and is developed through selection from the local land races found in arid region. It is very early in fruiting. First marketable harvesting is 75-80 days after sowing. The yield potential is 50-80 t/ha under arid conditions.

AHW-19

It is released from CIAH, Bikaner. It is medium-early maturing cultivar, produces 3.0-3.5 fruits per vine, flesh dark pink, solid (firm) with good eating quality and taste having 8.0 to 8.4 per cent TSS. Average yield is 46-50 t/ha and tolerates high temperature.

AHW-65

It is released from CIAH, Bikaner. It is very early maturing cultivar, produces 3-4 mature fruits per vine and gives yield of 37-40 t/ha. The flesh is delicious, pink, solid (firm) having 8.0-8.5 per cent TSS.

Improved Shipper

It is released by Punjab Agricultural University, India. It is an introduction from the USA having a big-sized fruit weighing 8 to 9 kg. The fruit is dark green with moderate sweetness (8 to 9 per cent TSS).

Special No.1

It is released by Punjab Agricultural University, India. Fruits are round and small with red flesh and red seeds. It is early in maturity. The average TSS is slightly lower than Shipper.

Durgapura Meetha

It is released by Agricultural Research Station, Durgapura, Rajasthan. A late cultivar maturing in 125 days, fruits round with light green. Rind is thick with good keeping quality, flesh sweet, TSS around 11 per cent with dark red colour, average fruit weight 6 to 8 kg, seed with black tip and margin.

Durgapura Kesar

It is released by Agricultural Research Station, Durgapura, Rajasthan. It is also a late cultivar, fruit weight 4-5 kg. Skin green with stripes, flesh yellow in colour, moderately sweet, seeds are large.

PKM-1

It is a selection from a local type and is released by HC and RI, Tamil Nadu Agricultural University, Periyakulam. It is suitable for arid, semi-arid and irrigated conditions. The skin is dark green with attractive pink flesh. Fruits are bigger in size with dark green skin and pinkish red flesh. Each fruit weighs 3-4 kg. The crop duration is 120 to 135 days. The season for cultivation is June-October and December-April. It yields 36-38 t/ha.

Shonima

This red fleshed seedless triploid hybrid is released from KAU, Vellanikkara. The rind colour is dark green with light green stripes. Average fruit weight is 3.92 kg.

Swarna

This bright yellow fleshed seedless triploid hybrid is released from KAU, Vellanikkara. The rind colour is green with yellow stripes. Average weight is 3.18 kg.

There are several cultivars, locally grown which are named after the region in which they are grown, such as Farrukhabadi, Moradabadi, and Faizabadi of U.P. Most of them have fruits with darkgreen colour or pale green with black stripes, oblong to round shape weighing 8 to10 kg with thick rind. The flesh colour would be variable from pale pink to pink with big flat black to brown seeds. In Rajasthan, Mateera cultivar is grown in rainy season around Bikaner. There is a local cultivar of Yamuna river-bed called Katagolan whose flesh is not sweet but it keeps well for over 2 to 3 months at ambient temperature during months of July-September. The flesh becomes dry and fibrous instead of fermentation and rotting.

Arka Sumet was considered to be the most promising cultivar for the southern dry region of Karnataka, India (Anjanappa *et al.*, 1999).

Madhu, Milan and Mohini are some of the hybrids of Indo-American Hybrid Seeds.

Madhu

Fruits are light green with dark green stripes, of excellent flesh quality and mature in 85 days.

Milan

It is an early maturing hybrid with tough rind and smooth red sweet flesh. It is tolerant to Fusarium wilt and has good shipping quality.

Mohini

Fruits are elongated with bright red, crisp flesh and good for shipping. It is tolerant to Fusarium wilt.

All-Jubilant and All-Producer

Norton *et al.* (1985) from USA reported these two cultivars of watermelon. All-Jubilant has large, symmetrically elongate fruit, while that of All-Producer is round to oblong-round. Both are high yielding and resistant to race 2 of *Colletotrichum lagenarium, Fusarium oxysporum* f.sp. *niveum* and *Didymella bryoniae*.

Red-N-Sweet

This watermelon has been reported from Louisiana USA by Lancaster *et al.* (1987). It is an induced mutant, having vermilion flesh and comparable to or better than standard Louisiana cultivars for plant vigour, disease resistance and yield.

Wuchazao

Wuchazao is an unbranched dwarf, early cultivar, developed from Anurechax Xiaozi 3 by Li (1986) in China. Its vine length is only 1.39 m, the fruits are round and pale green with dark green stripes, weighing 3.6 kg.

Moodeungsansoobak

A watermelon cultivar from Republic of Korea which is late maturing, high yielder and with large fruit of good quality (Han *et al.*, 1986).

Hungaria (H-8) and Napsugar (sun beam)

Both Hungaria 8 (H-8) and Napsugar are cultivars from Hungary. Hungaria (H-8) is an early cultivar of excellent quality producing higher yields than Sugar Baby. Napsugar has bright lemon-coloured sweet flesh and is suitable for small scale commercial land garden cultivation (Matonffy, 1987).

Vodlei

The most useful traits of this cultivar from Russia are earliness, uniform ripening, good flavor and multiple resistance to diseases (Yurina, 1990).

Watermelon cultivars of the USA have spread to different countries. Charleston Gray, one of the older cultivars of USA, grown in Gulf countries and characterized by its oblong to cylindrical shape, is a leading cultivar suited to long distance transportation and carries resistance to fusarium wilt and anthracnose. Likewise Congo, is another popular transportable cultivar resistant to anthracnose. Klondike types with black stripes from California and Crimson Sweet another resistant cultivar to two diseases are being grown extensively. Japanese cultivars especially F_1 hybrids and seedless triploids are grown in Taiwan and Philippines and exported to Hongkong and Singapore. There are cream to yellow fleshed cultivars like Yamato

Cultivars/Hybrids of Watermelon

Arka Akash

 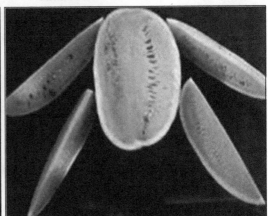

Arka Aiswarya

Arka Madhura   **Arka Manik**

Cultivars/Hybrids of Watermelon

Pusa Bedana

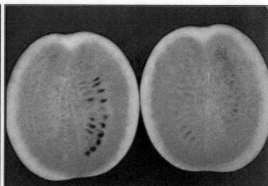

Arka Muthu

Sugar Baby   **Yellow Watermelon**

Cultivars/Hybrids of Watermelon

Ice Box

Improved Shipper

Asahi Yamato

Arka Jyoti

Durgapura Meetha

Cream, Shin Yamato, which satisfy the discriminating connoisseurs. There are small ice-box varieties weighing 1 to 2 kg like Takii Gem and New Hampshire Midget.

Watermelon cultivars, Bykovskii22, Yubileinyi72, Bodrost, and Tselnolistnyi 215 (Sincha, 1985) and Kamyzyakskii and Latos (Dyutin and Bicherev, 1985) with uniform ripening suitable for mechanical harvesting have been developed. Tekanovich and Fursa (1986) produced promising hybrids of bush form.

The yield of a new watermelon hybrid cv. Hainong 6 developed in China ranged from 1.49 to 4.76 t/667m². The fruit was elliptical with dark green stripes and red flesh with an average weight of 10 kg. The soluble solids content was 9-13 per cent. It is highly resistant to *Fusarium* wilt and anthracnose (Wan *et al.,* 1998).

6.4.0 Pumpkin and Squashes

Squashes botanically comprise all the three species *Cucurbita pepo, C. maxima* and *C. moschata*. There has been a lot confusion in the common epithets like squash, pumpkin, marrow, *etc.*, and more than one common name is ascribed to any one botanical species. Horticulturally, however, squash cultivars are commonly divided into two classes, 'bush' and 'vining'. Bush cultivars produce stems with greatly shortened internodes and set fruits in close succession. Generally, the fruits are ready for picking in about six weeks and are picked tender within a few days after pollination. These bush squashes do not store well and are commonly called as 'summer squash' under the species *C. pepo*.

Vining cultivars produce large plants with one or more long stems, covering the ground to a distance of 6 metres. Usually, the fruits are allowed to full maturity and stored for extended periods. Winter squashes are included in this group, besides the common pumpkin, orange, pale orange on yellow fleshed *kashiphal* of India. Winter squashes belong to *C. maxima*. There are also vining cultivars in *C. pepo* and all the pumpkin cultivars (*C. moschata*) are vining types.

Most of the improvement work in the USA is in *C. pepo*. There are several cultivars like Scallop with fruits of semi-globe, somewhat tart shaped with scallop margin, Black Zucchini with cylindrical smooth dark green fruits with greenish white flesh, Early Prolific, Straight-neck, Yellow Summer Crook-neck, *etc.* In winter squash, Boston Marrow, Butter Cup, Green Hubbard, Winter Banana are some of the well known cultivars. The most popular cultivar in C. *moschata* is the Butternut squash with fruits relatively small, cylindrical with pronounced bulb surrounding the seed cavity and bright orange flesh.

In India, pumpkin (*C. moschata*) or variously called as *kashiphal* or *seethaphal* or *lalkaddu* is extensively grown in different parts of the country mainly because of its storage capacity. The familiar ones have pale reddish brown skin, ribbed or smooth when completely ripe and greenish or greenish yellow skinned ones are also harvested and marketed. The shapes met with are mostly oblong, round, sometimes oval and the average weight will not be less than 10 kg and may even go to 20 kg. The flesh colour ranges from pale yellow to crimson and flesh thickness often varies widely. The pumpkin is a common vegetable grown in all parts of India (barring Himalayan Hills) and wide variation in fruit characteristics is encountered.

Some of the released cultivars are listed below:

Arka Suryamukhi (C. maxima)

It is released by ICAR-IIHR, Bengaluru. It is a selection from a foreign introduction. The fruits are small (1 kg), round with flat ends and with deep orange skin. Flesh is firm, orange and flavoured. Maturity is at 100 days under South Indian conditions.

Arka Chandan (C. moschata)

It is released by ICAR-IIHR, Bengaluru. A selection from Rajasthan collection of medium sized flat fruits weighing 2-3 kg with depressed polar ends, rind colour light brown with creamy patches at maturity, fruits mature in 125 days.

Pusa Vishwas (C. moschata)

It is released from ICAR-IARI, New Delhi. Fruits are spherical oval, medium in size (4-5 kg). Fruit flesh is thick and golden yellow in colour. Fruit maturity is in 120 days. Average yield is 35-40 t/ha (Sirohi *et al.*, 1991).

Pusa Vikas (C. moschata)

It is released from ICAR-IARI, New Delhi. Vines are semi-dwarf (about 2.5m long), leaves soft with light green or yellow spots. Fruits are flattish round, small in size (2 to 2.5 kg) with yellow flesh. Fruit maturity is in 110 days. Average yield is 25 t/ha.

Pusa Hybrid 1 (C. moschata)

It is released from ICAR-IARI, New Delhi. Fruits are round, medium in size, weighing 4.75 kg, flesh golden yellow. About 25 per cent higher yield than Pusa Vishwas. Yield 50 t/ha.

Kashi Harit (C. moschata)

This cultivar is derived from the cross between NDPK-24 × PKM through pedigree selection and released from ICAR-IIVR, Varanasi. Vines are short, leaves dark green with white spots. Fruits are green, spherical, weight 2.5-3.0 kg at green stage; yield of 30-35 t/ha in 65 days of crop duration.

Kashi Shishir F₁ Hybrid (C. moschata)

It is released from IIVR, Varanasi. It is an early maturing hybrid with small round and mottled green fruits (2-2.25 kg) having 3-4 fruits/plant. It can be grown in both the season summer as well as *Kharif*. Fruit yield is 38-42 t/ha.

Thar Kavi

It is developed through hybridization followed by selection from the segregating population of CM16 × CM19 at Central Horticultural Experiment Station, CIAH, Vejalpur, Gujarat. Fruits are very small in size with yellowish spots over the fruit surface and greenish lines at floral end of the fruit. It is moderately resistant to fruit fly, powdery mildew and pumpkin mosaic virus under field conditions. Average yield is 7-8 kg per plant.

CO 1 (C. moschata)

It is a selection from local type and is released from TNAU, Coimbatore. The fruits are flattened at the base measuring 34 cm length and 26.0 cm girth. The flesh thickness is 4 to 5 cm. The seed content is less (1.2 per cent) and each plant bears 6 to 7 fruits/plant. Each fruit weighs about 7.0 kg. The first harvest can be done in 115 days. The crop duration is 180 days and yields 25-30 t/ha, with good cooking quality.

CO 2 (C. moschata)

It is another selection from a local type and is released from TNAU, Coimbatore. The vines are moderately vigorous, less spreading than Co.1 and adopts well for high density planting at 2.2 × 1.9 m spacing. Vines flower in 55 days after sowing and first harvest can be had in 95-100 days. Fruits are small sized, slightly ridged with bright orange colored flesh. Each fruits weigh on an average of 1.5 to 1.8 kg. The crop duration is 135 days with a mean yield of 22.65 t/ha.

Ambili

It is a selection from a local material and is released from Kerala Agricultural University, Vellanikkara. The fruits are flattish-round, medium sized, each weighing 5.0-6.0 kg and with shallow furrows on the surface. Immature fruit colour is green and turns tan at maturity. It takes 130 days to reach maturity. Average yield is 30 t/ha.

PAU Magaz Kadoo-1

Released from PAU, Ludhiana. It is an edible seeded cultivar. Its seeds are hull-less (without testa) and can be used as 'Magaz' and snacks. Its fruits are of medium size, round and turn golden yellow at maturity. Its seed yield is 2.9 q/acre.

Punjab Samrat

It is released from PAU, Ludhiana. It is a selection from an exotic material introduced from Russia. The fruits are spherical, small sized and average weight is 2.5 kg. Fruit skin is mottled green when immature and orange-brown at maturity. The flesh colour is golden-yellow and is rich in β-carotene. Average yield is 41 t/ha.

Punjab Nawab

It is released from PAU, Ludhiana. This variety is suitable for rainy season as it is tolerant to white fly transmitted yellow vein mosaic disease. The average yield is 16 t/acre.

PPH-1 hybrid

It is released from PAU, Ludhiana. Vines of the hybrid are dwarf, internodal length is short and leaves are dark green. The fruits are small, round, mottled-green at immature stage and mottled-brown at mature stage. Fruit cavity is small and flesh is golden yellow. It is extra-early in maturity and yields 50 t/ha.

PPH-2 Hybrid (PAU, Ludhiana)

It is released from PAU, Ludhiana. Vines are dwarf, internodal length is short and leaves are green. The fruits are small, round, light green at immature stage and smooth-brown at mature stage. Fruit cavity is small and flesh is golden yellow. It is extra-early in maturity and yields 55 t/ha.

Azad Pumpkin-1 (C. moschata)

It is released by C.S.A.U.A. and T., Kanpur. Fruit green medium broken white pattern, spherical. Average yield 45-50 t/ha.

Narendra Agrim (C. moschata)

It is released from N.D.U.A. and T., Faizabad. Fruits are small, round and stripe less dark green. It is suitable for summer crop. Early in maturity (55 days), average yield is 30-40 t/ha.

Narendra Amrit (C. moschata)

It is released from N.D.U.A. and T., Faizabad. Fruits are round in shape and light green mottled in colour. It is suitable for summer crop, mature fruit weight is 5-6 kg.

Narendra Abhooshan (C. moschata)

It is released from N.D.U.A. and T., Faizabad. Fruits are in round shape with dark green striped and highly attractive. Average fruit yield is 70 t/ha.

Anand Pumpkin-1 (C. moschata)

Released from AAU, Anand, Gujarat. The fruits are medium sized, globular in shape with deep yellow flesh colour. The fruits of this cultivar possess higher total soluble solids, carotene, total soluble sugars and protein. Average yield is 24.4 t/ha.

Hamdan and Qasim (C. moschata)

The *Cucurbita moschata* cultivars Hamdan and Qasim were selected in Saudi Arabia from the only marketed cultivars, which is known as local or Egyptian pumpkin. Hamdan has small fruits of uniform shape which mature 122 days after planting. The skin is cream coloured and the flesh is light orange. Fruits are easy to peel and have a smooth texture. Hamdan produces an average of 3 fruits/plant, compared with 1 for local cultivar and yields about 11 per cent more than the local cultivars (Ibrahim *et al.*, 1996).

Khersonskaya (C. maxima)

Khersonskaya, a drought resistant cultivar released from Russia, produces dark grey, round and flattened fruit, weighing on an average 7.5 kg with light grey stripes, flesh sweet and juicy. It is resistant to *Colletotrichum lagenarium* and is moderately resistant to powdery mildew (Sokolov, 1984).

Chung *et al.* (1998) studied the morphological and ecological characteristics of 24 native Korean cultivars of *C. moschata* and observed that Yangku was an early maturing cultivar; Haenam, Saugju and Namhal produced large fruits; and Jeju 2 and Ulreung had fruits with dark orange flesh.

Types of Pumpkin

Variability in Shape, Size and Colour of Pumpkin

Types of Pumpkin

Variability in Shape, Size and Colour of Pumpkin in India

Exotic Cultivars of Pumpkin

Green Goblin

Blue Jarrahdale

Lumina

Cinderella

Fairytale

Baby Boo

Orange Hubbard

Blue Hubbard

Peanut

Striped Cushaw

White Cushaw

Turk's Turban

Indian Cultivars of Pumpkin

Anand Pumpkin-1

Swarna Amrit

Arka Chandan

Arka Suryamukhi

Kashi Harit

Kashi Shubhangi

Indian Cultivars of Pumpkin

Narendra Agrim   Narendra Amrit

Narendra Abhooshan   Pusa Vishwas

Narendra Upcar

Rahman *et al.* (1995) from Bangladesh studied for *Cucurbita moschata* cultivars and noted highest leaf area, number of female flowers and yield per plant in cultivar CC_4 but it had the lowest male: female ratio.

In pumpkin, Rana *et al.* (1986) identified Butter Ball 3 as a promising parent for yield and HP37 for earliness. In an evaluation trial, Devi *et al.* (1989) found that accession CM15 was the earliest, with the first flower emerging after 56 days while CM 20 had a wide sex ratio of 50.8:1 and gave highest yield.

South African pumpkin cultivar Star 7001 has a bush habit, which makes weed control easier and promotes improved utilization of nutrients. It is suitable for mechanical harvesting and gives consistently good yields of high quality fruit with thick flesh and deep orange colour (Anon, 1994).

Kue Hyon *et al.* (1998) developed semi-bush type Putho bag squash (*C. moschata*) lines,"Wonye403" and "Wonye404" from interspecific hybrids between courgette (*C. pepo*) cv. Ford, Zhucchini and squash (*C. moschata*) cv. Puthobag by backcrossing with latter parent.

Rios *et al.* (1998) studied 10 genotypes of pumpkin in Cuba and considered that the cv. P-l523B, P-1300 and P-130E to have potential for use in both dry and wet seasons.

Alice is a squash cultivar (*C. maxima*) with bush type growth habit (*Bugene*) recommended for cultivation in Brazil. It was selected from an F_6 population of Coroax Zapallode Tronco, flowers 45-50days after germination and produces 25-30 tonnes of fruits/ha (Vecchia *et al.*, 1991).

In Cheju province of Korea Republic, the summer season production of sweet pumpkin (*C. maxima*) cultivars (Evis, Hamaguri, Miyako, Yako and Bamhobak) was investigated and it was found that marketable yield was highest in Evis (28.9 t/ha). Evis fruits contained the highest contents of β-carotene and carbohydrate (Cho *et al.*, 1997).

Nakagawara (2000) reported from Japan that a pumpkin cv. Kachiwari, developed from local types is suitable for cultivation under low fertilizer application, without pesticides. The vines were robust, large and resistant to diseases, especially powdery mildew. Even if stored over the entire winter the fruit retains excellent flavour, with high starch and sugar concentrations.

The Cultivars of *C. pepo* are Described Below

Early Yellow Prolific (C. pepo)

It is released by IARI Regional Station, Katrain. An early bush type, fruit medium sized, warted, tapering towards stem end, skin light yellow turning orange-yellow on maturity. Flesh tender at which stage it is consumed as vegetable.

Australian Green (C. pepo)

It is released by IARI Regional Station, Katrain. An introduction, very early bush type, fruits dark green with longitudinal white stripes, 25 to 30 cm long, very tender at edible stage.

Pusa Pasand (C. pepo)

It is released from ICAR-IARI, New Delhi. It is early improved flattish round variety of summer squash for spring summer season cultivation under open and off-season winter cultivation under protected condition. Its fruits are attractive light green, shiny, uniform, flattish round, 70-80 g with tender flesh. It has continuous and concentrated fruit setting. First harvesting is 45-50 days after sowing in spring summer season. Average fruit yield is 16.3 t/ha during spring summer season.

Pusa Alankar (C. pepo)

It is released by IARI Regional Station, Katrain. It is an early maturing F_1 hybrid between EC207050 and 51-1-8 (a derivative from cross between Chappan Kaddu and Early Yellow Prolific), having uniform dark green fruits with light coloured stripes. The flesh is tender and fruits mature in 45 to 50 days.

Kashi Subhangi (C. pepo)

It is released from ICAR-IIVR, Varanasi. It is a medium maturing bushy cultivar bearing 8-10 medium size elongated dark green shining fruit with average fruit weight (800-900 g), resistant to downy mildew. It is most suitable for autumn and also can be grown in spring season. Higher yield two to three times (60-65 t/ha) over check cultivar.

Punjab Chappan Kaddoo No. 1 (C. pepo)

It is a selection from the local material and is released from PAU, Ludhiana. The plants are bushy with thick foliage. Female flowers appear at a very early stage of plant growth and sometimes even before the male flowers. The fruits are attractive, light green, disc shaped, mildly ribbed with flat stem-end. First picking is possible 60 days after sowing. Average fruit weight is 80 g and yield is 24 t/ha. It is tolerant to downy and powdery mildews and red pumpkin beetle.

Patty Pan (C. pepo)

An introduction from the USA, a bush type, producing disc shaped, chalky white and attractive fruits. A short duration cultivar of 85 days.

Orangetti and Go-Getti (C. pepo)

These two semi-bush type cultivars have been reported by Paris *et al.* (1985) from Israel. The skin of cv. Go-Getti is green and yellow-orange when immature and becomes uniformly yellow-orange at maturity. Cultivars orangetti has orange fruit skin and orange flesh of excellent flavour.

Bareqet (Hybrid) (C. pepo)

A hybrid from Israel, high yielding, with dark green, non-tapering uniform fruits (Paris *et al.*, 1986).

In Korea Republic, *Cucurbita pepo* lines and cultivars of P84051, P84057, YSC10-IS, P.M. Zucchini, Caserta and Zucchini showed good parthenocarpic ability (Om and Hong, 1989).

Cultivars of Summer Squash

Early Yellow Prolific

Austrlian Green

Pusa Pasand

Pusa Alankar

Patty Pan

Punjab Chappan Kadu No-1

6.5.0 Bottle Gourd

The fruits of bottle gourd show immense variation in shape and size ranging from flattened or discoidal forms to bottle like forms, to long club shapes, sometimes nearly a metre long. In India, conical, round, club and 'tumri' shapes are recognized.

Pusa Summer Prolific Long

It is released from ICAR-IARI, New Delhi. It has prolific fruiting, fruits 40 to 50 cm long, pale green in colour, 10 to 15 fruits per plant. Suitable for spring summer sowing.

Pusa Summer Prolific Round

It is released from ICAR-IARI, New Delhi. It has prolific fruiting, fruits round 15 to 18 cm in girth when young and green in colour, heavy yielder.

Pusa Meghdoot

It is released from ICAR-IARI, New Delhi. A F_1 hybrid between Pusa Summer Prolific Long and Selection-2. Fruits are long and light green.

Pusa Manjari

It is released from ICAR-IARI, New Delhi. A F_1 hybrid between Pusa Summer Prolific Round and Selection-11. Fruits are round and light green.

Pusa Naveen

It is released from ICAR-IARI, New Delhi. It is free from crook necked fruits. Its fruits are cylindrical, straight and each weighing 550 g. It is suitable for cultivation in spring-summer and rainy seasons. Average yield is 35-40 t/ha.

Pusa Santushti

It is released from ICAR-IARI, New Delhi. It set fruits under low temperature (10-12°C) as well as high temperature (35-40°C). Fruits are attractive green, smooth, pear shaped, fruit length 18.50 cm, fruit diameter 12.40 cm, fruit weight 0.8-1.0 kg. Maturity is in 55-60 days. Average yield is 26 t/ha.

Pusa Samridhi

It is released from ICAR-IARI, New Delhi. Fruits are long without neck. Maturity is in 50-55 days. Average yield is 27 t/ha.

Pusa Sandesh

It is released from ICAR-IARI, New Delhi. Vines are medium to long, fruits are attractive green, round, deep oblate, medium size, weighing 600 g, first picking in 55-60 days in *kharif* and 60-65 days in summer. It is suitable for both, summer and *kharif* seasons. Yield is 30 t/ha.

Pusa Hybrid 3

It is released from ICAR-IARI, New Delhi. Fruits are green, slightly club shaped without neck, suitable for easy packing and long distance transportation, first picking in 50-55 days. Average fruit weight is 1.0 kg and yield is 42 t/ha in spring-summer season and 47 t/ha in *kharif* season.

Arka Nutan

It is released from ICAR-IIHR, Bangalore. Early flowering and first female flower appears at 9th node. It takes 45 days for the first female flower appearance and 56 days for first picking of fruits. Fruits are green, cylindrical and tender. It is resistant to Gummy stem blight. Yield is 46 t/ha.

Arka Shreyas

It is released from ICAR-IIHR, Bangalore. First female flower appears at 15th node. It takes 48 days for the first female flower appearance and 60 days for first picking of fruits. Fruits are green, club shaped and tender. It is resistant to Gummy stem blight. It has good shelf life with less weight loss, firmness and colour retention up to 10 days of storage under RT. Yield is 48 t/ha.

Arka Ganga

It is released from ICAR-IIHR, Bangalore. Early flowering and first female flower appears at 12th node. It takes 47 days for the first female flower appearance and 56 days for first picking of fruits. Fruits are green, oblong/oval and tender. It is resistant to Gummy stem blight. It has good shelf life with less weight loss, firmness and colour retention up to 10 days of storage under RT. Yield is 60 t/ha.

Arka Bahar

It is released from ICAR-IIHR, Bangalore. It is a pure line selection from IIHR-20A. Fruits are light green, shining, medium long, straight without crookneck with average weight of 1 kg. It is tolerant to blossom end rot. Yield is 40-45 t/ha.

Kashi Kiran (VRBG-4)

It is released from ICAR-IIVR, Varanasi. Light green attractive round fruits, tolerant to downy mildew. Fruit weight 600-700 g. Number of fruits 13-14/plant, average yield is 45-48 t/ha.

Kashi Kundal (VRBOG-16)

It is released from ICAR-IIVR, Varanasi. It has attractive pear shaped green fruits, medium in size, suitable for July to September sowing. Number of fruits per plant is 12-14 with average fruit weight of 1.3 to 1.5 kg. It is resistant to downy mildew. Yield is 47 t/ha.

Kashi Kirti (VRBOG-63-02)

It is released from ICAR-IIVR, Varanasi. Fruits are green, small, cylindrical shape (Gutka type). It is resistant to downy mildew. It is early maturing and suitable for distant marketing and transportation due to better postharvest life.

Kashi Ganga

It is released from ICAR-IIVR, Varanasi. This is an early cultivar derived from the cross IC-92465 × DVBG-151. Fruits are light green, length 30 cm, diameter 7 cm, fruit weight 800-900 g and yield 48-55 t/ha. It is tolerant to anthracnose and suitable for rainy and summer season cultivation.

Kashi Bahar (F₁)

It is released from ICAR-IIVR, Varanasi. This is a long fruited hybrid with green vine and vigorous growth, fruit straight, light green, length 30-32 cm and average weight 780-850 g and yield 50-55 t/ha. It is suitable for rainy and summer season cultivation. It is tolerant to anthracnose, downy mildew and Cercospora leaf spot under field conditions.

Thar Samridhi (AHLS Round 1)

It is released from CIAH, Bikaner. The fruits are attractive, shiny light green, oblate-round, internal flesh is white and solid, weighing 450-750 g each with average yield of 4.83-5.78 kg per plant. The fruit yield potential is 24-30 t/ha.

CO 1

It is a selection from a germplasm collection and is released from TNAU, Coimbatore. Fruits are round at the base with a prominent bottleneck at the top, medium sized, attractive light green in color with a mean weight of 2.03 kg. The crop duration is 135 days with a yield of 36.0 t/ha.

CO BoGH1

It is released by HC and RI, TNAU, Coimbatore. It is F_1 hybrid developed by crossing NDBG 121 × Arka Bahar. Fruits are cylindrical, without crook neck, medium sized (0.95-1.00 kg) and suitable for nuclear family. It is suitable for bower system of cultivation. The crop duration is 100 – 110 days with a yield of 79 t/ha.

Punjab Komal

It is released from PAU, Ludhiana and is developed from the cross between local collections LC 11 and LC 5. The fruits are pear shaped, medium sized each weighing about 625 g, light green and pubescent. First picking is possible 70 days after sowing. Average yield is 50 t/ha. It is tolerant to CMV.

Punjab Long

It is released from PAU, Ludhiana. The plants are vigorous, profusely branched bearing cylindrical light green and shinning fruits. It is suitable for packaging and long distance marketing. Average yield is 18 t/acre.

Punjab Barkat

It is released from PAU, Ludhiana. It is developed by selection from a local material collected from Khanna in Ludhiana district. Its plants are vigorous and profusely branched. Fruits are about 27 cm long, shining light green, cylindrical and tender. First fruit picking is possible 40 days after transplanting. Fruits are suitable for packaging and distant transportation. It is moderately resistant to mosaic disease. Average fruit yield is 55 t/ha.

Punjab Bahar

It is released from PAU, Ludhiana. Its vines are medium long and pubescent. Fruits are nearly round, medium sized, light green, shining and pubescent. Its vines bear average of 9 to 10 fruits. Its average yield is 22 t/acre.

Pant Sankar Lauki-1

It is released from G.B.P.U.A. and T., Pantnagar. The fruits are intermediate sized long and somewhat cylindrical (about 35 cm long). The fruits are green. Vine length is about 5.5 m. The first picking is possible in about 60 days. It is suitable for planting in the plains as well as in the hills. The yield potential is 40 t/ha.

Pant Sankar Lauki-2

It is released from G.B.P.U.A. and T., Pantnagar. The fruits are about 40 cm long, club shaped with smooth green colour. The first green fruit is in harvest is suitable for plains and hills. It can be sown from March to July in plains and April to May in the hills. The yield potential is 40 t/ha.

Pant Lauki-4

It is released from G.B.P.U.A. and T., Pantnagar. This is the medium duration and high yielding variety. It has long fruits (40 cm) of light green colour with light strips having hairs. Yield potential is about 30 t/ha.

Narendra Bottle Gourd Hybrid-4

It is released from N.D.U.A. and T., Faizabad. Fruits are cylindrical, uniform and attractive. Early maturity (50-55 days) and prolific bearer with yield potential of 70 t/ha in summer, 100 t/ha in rainy and early winter seasons.

Narendra Rashmi (NDBG-1)

It is released from N.D.U.A. and T., Faizabad. Fruits are long bottle shape, suitable for cultivation in summer and rainy seasons. Early maturity (60 days), average fruit yield is 30-40 t/ha.

Narendra Dharidar (NDBG-208-1)

It is released from N.D.U.A. and T., Faizabad. Fruits are long bottle shape with striped green suitable for summer seasons. Average fruit yield is 35 t/ha.

Narendra Jyoti (NDBG-104)

It is released from N.D.U.A. and T., Faizabad. Fruits are long slender and attractive. Early in maturity (60 days). Average fruit yield during summer 40 t/ha and 60 t/ha during rainy seasons.

Narendra Shishir (NDBG-202)

It is released from N.D.U.A. and T., Faizabad. It is winter type, round fruited variety and has peculiar multifid/podate leaf shape. It has resistance against downy mildew, powdery mildew and viral disease complex. Average yield is 85 t/ha.

Types of Bottle Gourd

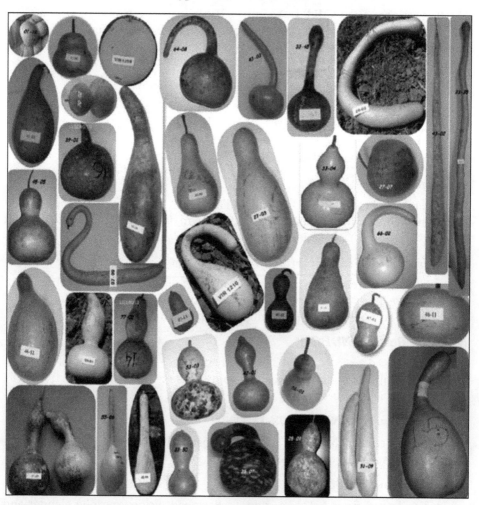

Variability in Shape, Size and Colour of Bottle Gourd
(Dhillon *et al.*, 2017b)

Cultivars of Bottle Gourd

Kashi Ganga

Anand Bottle Gourd-1

Jorabotta

Kashi Bahar

Arka Shreyas

Pusa Santushti

Narendra Madhuri (NDBG-505)

It is released from N.D.U.A. and T., Faizabad. Fruits are attractive with round shape, it produces highly palatable cooked vegetable. Average yield is 100 t/ha.

Narendra Shivani (NDBG- 403)

It is released from N.D.U.A. and T., Faizabad. It is winter type prolific bearer very long slender fruited fit for kitchen garden purpose. Average yield is 130 t/ha.

Narendra Bottle Gourd (NDBG-132)

It is released from N.D.U.A. and T., Faizabad. Fruits are long slender shape, it is an early cultivar requires 60 days for first picking in spring/summer crop. Average yield is 30 t/ha.

Samrat

It is released from MPKV, Rahuri. It is developed by selection from a local material. Fruits are 30-40 cm long, green in colour, cylindrical in shape. Good for box packing, suitable for *Kharif* and summer seasons.

Azad Nutan

It is released from CSAUAT, Kanpur. The fruits are attractive, shining light green and medium long. It is recommended for cultivation in both spring-summer and rainy seasons. First picking is possible 55-60 days after sowing. Average fruit weight is 1.0-1.5 kg.

Kalyanpur Long Green

It is released from CSAUAT, Kanpur. It is developed by selection from a local material. The fruits are long with tapering blossom end. Average yield is 30 t/ha.

Some cultivars of Madhya Pradesh have been described by Amarchandra and Parikh (1969) like White Surat, Hot Season Long White, 'Sarmiwali Safaid Lambee', Doodhi Long White and All Seasons Long, *etc.*

Provvidenti (1995) suggested that bottle gourd cv. Cow Leg should be grown in the USA to supply fruit to oriental markets. It is multiviral resistant and commonly grown in Taiwan.

6.6.0 Bitter Gourd

Pusa Do Mausmi

It is released from ICAR-IARI, New Delhi. It is a selection from local collection suitable for spring summer and rainy seasons. The fruits reach edible stage in about 55 days from sowing, fruits dark green, long medium thick, club shaped with 7-8 continuous ridges,18 cm long at edible stage, 8-10 fruits weigh one kg.

Pusa Vishesh

It is released from ICAR-IARI, New Delhi. It is a dwarf vine cultivar suitable for planting in high plant density. Fruits are glossy green, medium long and thick, suitable as vegetable for pickling and dehydration.

Pusa Hybrid-1

It is released from ICAR-IARI, New Delhi. Fruits are medium long, medium thick, glossy-green, suitable for pickling and dehydration. First picking is in 55-60 days. It is suitable for growing in spring summer season. Yield is 20 t/ha.

Pusa Hybrid-2

It is released from ICAR-IARI, New Delhi. Its fruits are medium long and thick, glossy-green, suitable for pickling and dehydration. First picking starts at 55-60 days after seed sowing. Average yield is 19-20 t/ha.

Coimbatore Long

A selection by National Seeds Corporation, fruits long, tender, white in colour, suitable for rainy season.

Coimbatore Long White

It is a selection from the local material. The fruits are extra-long (up to 60 cm) and white in colour. Average yield is 15 t/ha.

CO 1

It is a selection from a local type collected from Thudiyalur (Long Green) and is released from HC and RI, TNAU, Coimbatore. The fruit is dark green, medium long (25-30 cm) and thick (6-8 cm) with characteristic warts. Each vine produces 20-22 fruits each weighing 100120 g. Vines flower in 45-50 days and first harvest can be had in 55-60 days after sowing. Each fruit contains 24 to 30 seeds weighing 6 to 8 g. The total crop duration is 115 days with 6-8 harvests. It yields 14.4 t/ha.

MDU-1

It is released from AC and RI, TNAU, Madurai. It is an induced mutant developed by gamma irradiation to local cultivar (MC 103). It is early in flowering (60 days) with a sex ratio of 1:20 of female and male flowers. The fruits are long with a mean length of 40.34 cm and a girth of 17.54 cm and each fruit weighs 410 g on an average. Fruits contain less seeds. Each vine yields on an average of 16.66 fruits. It yields 32.19 t/ha.

COBgoH 1

It is a F_1 hybrid developed by a crossing MC-84 × MDU-1 and is released from HC and RI, TNAU, Coimbatore. It recorded an average yield of 44.4 tonnes/ha. The average individual fruit weight is 300 g. The potential yield goes up to 51 t/ha. June-Sept (*Kharif*) and Dec-March (*Rabi*) are ideal season for cultivation of this cultivar. Fruits at vegetable maturity are useful for making fried curies, porial and stuffed food. It has high momordicine content (2.99 mg/g). The crop duration is 115120 days.

Arka Harit

A selection from Rajasthan collection and is released by ICAR-IIHR, Bengaluru. The fruits are attractive, spindle shaped with green colour, small in size with smooth regular ribs.

VK-1 Priya

A selection from Kerala Agricultural University, Vellanikkara. The fruits are extra long, around 40 cm, first picking in 61 days and heavy bearing, 50 fruits per plant.

Kashi Urvasi

It is released from ICAR-IIVR, Varanasi. This cultivar has been derived from the cross IC-85650B × IC-44435A, having dark green and long fruits, mild projection, length 16-18 cm, fruit weight 90-110 g and yield 20-22 t/ha. This is suitable for cultivation under both rainy and summer seasons.

Pant Karela-1

It is released from GBPUAT, Pantanagar. The vine length is about 2 m. Fruits are thick, about 15 cm long with tapering ends. It takes about 55 days to first harvest. It is suitable for planting in the hills. The yield potential is 15 t/ha.

Hirkani

It is released from MPKV, Rahuri. Fruits are dark green, 15-20 cm long, prominent prickles on fruits, fruits are tapering both ends. It is suitable for *Kharif* and summer seasons. Average yield is 23 t/ha.

Phlue Green Gold

It is released from MPKV, Rahuri. Fruits are dark green, 25-30 cm long with prickles, tolerant to downy mildew, suitable for Kharif and summer seasons. Average yield is 23 t/ha.

Phule Ujwala

It is released from MPKV, Rahuri. Fruits are dark green, medium in length and prickled, excellent shelf-life and suitable for long distance market, tolerant to downy mildew. Average fruit yield is 14.7 t/ha.

Phule Priyanka (F₁ Hybrid)

Fruits are dark green, length 20-22 cm, prominent prickles on fruits and it is tolerant to downy mildew. Average yield is about 20 t/ha.

Punjab Kareli-1

It is a selection from the local material collected from Khanna of Ludhiana district and is released from Punjab Agricultural University, Ludhiana. Its vines are long; and leaves are green, smooth and serrated. Fruits are long (30 cm), thin and ridged. First picking is possible 65 days after sowing. Average fruit yield is 17.5 t/ha.

Punjab-14

It is a selection from the local material and is released from Punjab Agricultural University, Ludhiana. The vines are small, average fruit weight is 35 g and have light green skin colour. It is suitable for sowing during spring and rainy seasons. Average yield is about 12.5 t/ha.

Types of Bitter Gourd

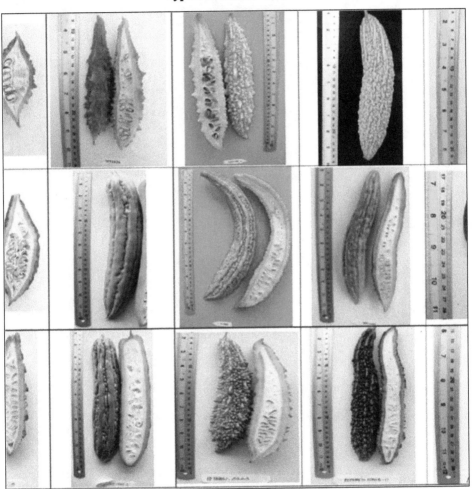

Variability in Shape, Size and Colour of Bitter Gourd

Types of Bitter Gourd

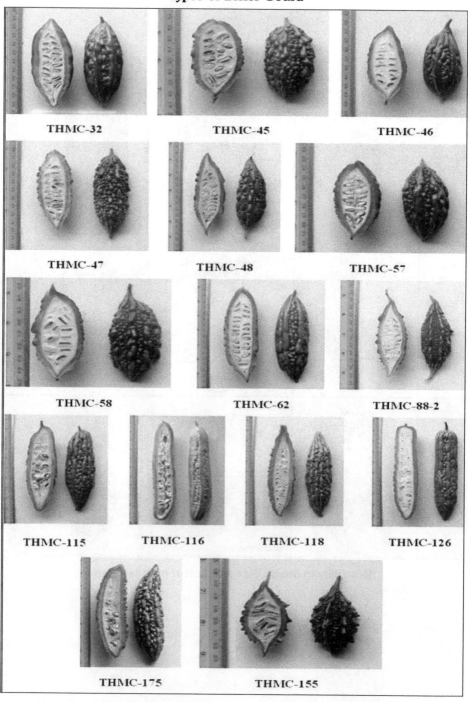

Fruit Type of 15 Earliest Staminate-Flowering Inbred Lines of Bitter Gourd
(Dhillon *et al.*, 2017a)

Types of Bitter Gourd

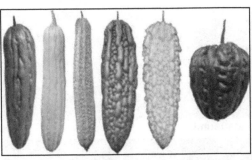

Thailand (Left), Chinese (Middle), and Taiwan (Right) Types Bitter Gourd

Vietnam (Left), and Philippine (Right) Types Bitter Gourd

Indonesian (Lleft), and South Asian (Right) Types Bitter Gourd
(Dhillon *et al.*, 2017b)

Cultivars/Hybrids of Indian Bitter Gourd

Pusa Do Mausumi

Arka Harit

Pusa Aushadhi-1

Pusa Hybrid-4

Pusa Rasdar

Pusa Purvi

Swarna Yamini

Pusa Hybrid-1

Pusa Hybrid-2

Punjab Karela-15

It is released from Punjab Agricultural University, Ludhiana. Fruits are dark green and with mat appearance. It is moderately resistant to yellow mosaic disease of bitter gourd. The average yield is 5.1 t/acre.

Punjab Jhaar Karela-1

It is released from Punjab Agricultural University, Ludhiana. Fruits are attractive green, tender, spindle shaped and suitable for cooking by chopping. It is resistant to root knot nematode and virus diseases. Its average yield is 3.5 t/acre.

In Gujarat, there are cultivars called Pride of Surat, Pride of Gujarat whose fruits are small, roundish, whitish green in colour with a few white dabs on thin skin. Each fruit weighs 8-10g only.

Reddy *et al.* (1995) studied the performance of thirteen cultivars of *M. charantia* in Karnataka, India and recorded highest fruit yields (4763.37 and 5430.04 kg/ha) in local cultivars Siddavanahalli and Bellary, respectively.

Evaluation of bitter gourd genotype by Thakur *et al.* (1996) revealed that Pusa Vishesh was the earliest cultivar for days to harvest and days to 50 per cent harvest. Phule BG6 had the heaviest fruits and HK 96 gave the highest value for fruits/plant.

Bitter gourd F_1 hybrid Xiang Kugua-1 is an early maturing, high yielding and of good quality, developed in Hunan Vegetable Research Institute, China (Xue and Huang, 1996).

Cuilii I is an early maturing and high yielding bitter gourd F_1 hybrid suitable for cultivation in southern China in spring and autumn. This hybrid has been derived by crossing early gynoecious line 19 with Jiang Xuan 105 (Zhou *et al.*, 1997).

6.7.0 Ridge Gourd

There is wide variation in shape and length of fruits. In South India, longer fruits are preferred and some are called yard long ribbed gourd, while in North India shorter fruits are preferred.

Pusa Nasdar

It is released from ICAR-IARI, New Delhi. It is a selection from Madhya Pradesh collection. It is an early maturing cultivar producing light green coloured club shaped fruits, 15 to 20 per vine. Flowering starts in 60 days after sowing.

Pusa Nutan

It is released from ICAR-IARI, New Delhi. Fruits are long (25-30 cm), straight, attractive green, average fruit weight 105 g, flesh is tender. It is suitable for spring summer and *kharif* seasons. Maturity is in 45-50 days. Average yield is 16 t/ha.

Arka Prasan

It is an open pollinated cultivar developed by inbred selection from the segregating germplasm, IIHR-53 and is released from ICAR-IIHR, Bangalore. It is an early cultivar (42-45 days for first picking) having green, long, tender fruits with excellent cooking quality. Yield is 26 t/ha in 120-135 days.

Arka Vikram

It is released from ICAR-IIHR, Bangalore. Hybrid developed by crossing the inbreds, IIHR-6-1-1 x IIHR-53-1-3. Early flowering hybrid (46 days for first picking) with green, long, tender fruits having excellent cooking quality. Yield is 34 t/ha in 120-135 days.

Arka Sumeet

It is released from ICAR-IIHR, Bangalore and developed by pedigree method of breeding between IIHR-54 × IIHR-18 followed by selection. It has lush green and tender long fruits (50-65 cm) with prominent ridges and delicate aroma. Good transport and cooking qualities. Yield is 50 t/ha.

Arka Sujat

It is released from ICAR-IIHR, Bangalore and is developed by pedigree method of breeding between IIHR-54 × IIHR-18 followed by selection. Fruits are lush green and tender, medium long (35-45 cm) with prominent ridges and delicate aroma. Good transport and cooking qualities. Yield 53 t/ha.

Kashi Jyoti

It is released from ICAR-IIVR, Varanasi. It has light green fruits of 100-150 g with yield of 15-18 t/ha, resistant to sponge gourd mosaic virus, tolerant to downy mildew and root knot nematode.

Kashi Shivaani

It is released from ICAR-IIVR, Varanasi. Fruits are green, long straight (20-30 cm and may increase on bower up to 40 cm) with a diameter of 3-4 cm. Average fruit weight ranges from 100 g to 150 g. Fruit is ready for harvest after 50-60 days from the date of sowing. The yield potential is 18-20 t/ha. It is resistant to anthracnose and tolerant to downy and powdery mildew disease under field condition.

Thar Karni

It is released from CIAH, Bikaner. Fruits are cylindrical with 10 shallow ridges and light green colour. It is early in harvesting and takes 51-55 days to first picking from sowing. It is resistant to mosaic disease and melon fruit fly under field condition. It sets fruit at high temperature during April-May under hot arid condition of Rajasthan. Average fruit yield is 10-15 kg/plant and 15 t/ha.

CO-1

It is released by Tamil Nadu Agricultural University, Coimbatore. Moderately vigorous, 10 to 12 fruits per vine weighing 3 to 4 kg per vine. The fruits are long 60 to 75 cm and 30 cm in girth, first harvest is taken in 55 days and it is completed by 125 days.

CO 2

It is a selection from a germplasm collection and released from HC and RI, TNAU, Coimbatore. Each vine produces 8 to 10 fruits weighing 9 to 10 kg. The

Cultivars of Ridge Gourd

Arka Sumeet   Arka Prasan

Pusa Nutan   Swarna Uphar

Fruit Shape Variation of *Luffa* sp.
a-c: *Luffa cylindrica*, d-f : *Luffa acutangula*,
g: *Luffa hermaphrodita* (Dhillon *et al.*, 2017b).

fruits are green, very long (1 m) and fleshy. The crop duration is 120 days with a yield of 25 t/ha.

PKM 1

It is an induced mutant from the type H 160 and is released from HC and RI, TNAU, Periyakulam. The fruits are dark-green 60-70 cm long with shallow grooves. They are characterized by very broad terminal end than the basal end. The plants are tolerant to pumpkin beetle, fruit fly and leaf spot. It yields 25-28 t/ha in a duration of 160 days.

KRH 1

It is a hybrid from MS-LA 101 × LA 102 and released from KAU, Vellanikara. It yields 7.41 kg/plant.

GJRGH-1

It is released from J.A.U., Junagadh, Gujarat. The fruits are long in size with green colour. Suitable for growing in *kharif* season. Average yield is 11 t/ha.

Hisar Kalitori

It is released from HAU, Hisar, Haryana. It is an early long thin and straight fruited cultivar, tolerant to powdery mildew, suitable for rain fed areas. Average yield 8-10 t/ha.

Pant Torai-1

It is released from GBPUA and T, Pantnagar. The main shoot is 5 m long. Fruits are 15-20 cm long and club shaped. It takes about 65 days to first harvest. It is specifically suitable for rainy season. The yield potential is about 10 t/ha.

Gujarat Anand Ridge Gourd-1

Released from GAU, Anand. This variety is medium sized and elliptical shape with green fruit skin colour. The variety has less mosaic and downy mildew disease reaction. Average yield is 15 t/ha.

Haritha

Long, Green fruited, and high yielding cultivar.

Deepthi

It has been developed by a single plant selection from the local material from KAU, Vellanikara. Fruits are green with intermediate lusture and tapered stem end, medium sized (23 cm long and 15 cm in girth) with finely wrinkled surface. First picking is possible 53 days after sowing. Average fruit weight is 165 g. It possesses field resistance to mosaic virus and downy mildew diseases.

Konkan Harita

It is an early maturing selection from the local material and is released from KKVP, Dapoli. First picking is possible 45 days after sowing. The fruits are dark green, straight, 30-45 cm long and tapered at both the ends. Average yield is 17 t/ha.

Phule Sucheta

It is released from MPKV, Rahuri. Fruit colour is green with prominent ridges on fruit, tolerant to downy mildew and powdery mildew under field conditions. Average yield is 12 t/ha.

Swarna Manjhari

Released from ICAR-Research Complex for Eastern Region, Ranchi, Jharkhand. Fruits are elongated, long and highly ridged, pulp soft and contain less fibre. Resistant to powdery mildew disease. Yield is about 18-20 t/ha.

Swarna Uphar

Released from ICAR-Research Complex for Eastern Region, Ranchi, Jharkhand. Fruits are elongated in shape, long ridged at edible stage, pulp soft and contain less fibre Resistant to powdery mildew disease. Yield – 20-30 t/ha.

Luffa acutangula var. *hermaphrodita*, called Satputia' is having hermaphrodite in sex form and produces smaller fruits in clusters.

Kashi Khushi (Satputia)

It is released from ICAR-IIVR, Varanasi. It has hermaphrodite flower with cluster bearing fruiting habit. It is an early maturing cultivar having light green fruit with ten dark superficial and continuous longitudinal ridges. Suitable for *kharif* season and can be also grown in spring season. It yields 130-150 fruits/plant, and average fruit yield is 12.7 t/ha.

In China, a new *Luffa acutangula* F_1 hybrid, Feng Keng was developed from the cross KR91-1-1 × 2-1-1. It is early maturing, vigorous, produces dark green fruits, 62 cm long with an average yield of 23.7 t/ha (Chen *et al.*,1996).

6.8.0 Sponge Gourd

Pusa Chikni

It is a selection from Bihar collection and is released from ICAR-IARI, New Delhi. It is an early fruiting cultivar, flowering in about 45 days. The fruits are smooth and dark green colour, more or less cylindrical, 15 to 20 fruits per vine, suitable or both spring-summer and rainy seasons.

Pusa Supriya

It is released from ICAR-IARI, New Delhi. Fruits are pale green, smooth, 15-20 cm long, straight and slightly curved at the stem end, pointed distal end, non-hairy, flesh tender, suitable for spring summer and *kharif* season. Fruits become ready for picking at 50-55 days after sowing in spring summer and 44-48 days after sowing in *kharif* season. Average yield is 13-14 t/ha.

Pusa Sneha

It is released from ICAR-IARI, New Delhi. It was developed by selection from Pusa Supriya. The fruits are dark green, 20-25 cm long and straight. It is suitable for cultivation in both spring summer and *kharif* seasons. Maturity is in 50-55 days.

First picking is possible 50-55 days after sowing in spring summer season and 40-45 days in *kharif* season. Average fruit weight is 120 g and yield is 11-12 t/ha.

Pusa Shrestha (F₁)

It is released from ICAR-IARI, New Delhi. Fruits have desirable marketable attributes and are attractive, uniform, green, elongated, cylindrical, with superficial ribs, smooth texture, thick skin and white flesh. Early (45-50 days for first fruit harvest) maturing. Average fruit length is 27 cm and girth is 13 cm. Average fruit weight is 120 g. Average yield is 20.0 t/ha during spring summer season.

Kashi Rakshita

It is released from ICAR-IIVR, Varanasi. Average number of fruits/plant ranges from 12-14. Fruits are dark green and become ready to harvest at 48-52 days after sowing. It is resistant to disease downy mildew and sponge gourd mosaic virus under field condition.

Kashi Shreya

It is released from ICAR-IIVR, Varanasi. Plants starts flowering at 45 days after planting. Fruits are cylindrical and dark green. Fruit diameter is 3-3.75 cm and fruit weight 100 to 150 g. Fruits start maturing after 105-115 days. It is resistant to downy mildew, powdery mildew and sponge gourd mosaic virus under field condition. It is tolerant to leaf minor, fruit fly and red pumpkin beetle. Suitable for river bed cultivation.

Kashi Divya

It is released from ICAR-IIVR, Varanasi. It has medium sized, attractive and cylindrical fruit with high yield potential and is suitable for distant marketing. It can be grown successfully in rainy and summer seasons and performs better at 25-30°C at medium humidity. Yield is 25-27 t/ha.

Kashi Saumya (VRSGH-3) F₁ Hybrid

It is a medium maturing hybrid released from ICAR-IIVR, Varanasi. It has attractive dark green medium fruits and is suitable for river bed cultivation. It is resistant to sponge gourd mosaic virus and tolerant to downy and powdery mildew under field condition. Yield is 18-19 t/ha.

Thar Tapish

It is released from CIAH, Bikaner. Its fruits are dark green, straight with slightly curved neck and shining lusture. Average yield is 14 t/ha (1.3 kg/plant). It is suitable both for rainy and summer season cultivation. It is tolerant to high temperature and abiotic stresses of hot arid agro-climate as spring summer season crop.

Azad Taroi -1

It is released from CSAUAT, Kanpur. Early, fruits are green, smooth. Average yield is 12-15 t/ha.

Cultivars of Sponge Gourd

Pusa Shrestha

Swarna Prabha

Kashi Khushi

Kashi Divya

Azad Taroi -2

It is released from CSAUAT, Kanpur. It has profuse fruiting, suitable for summer and rainy seasons. Average yield is 14-16 t/ha.

PSG-9

It is a selection from the local material, released by Punjab Agricultural University, Ludhiana. It is suitable for planting in February-March in north Indian plains. Its fruits are smooth, cylindrical, long (15-18 cm) and dark green. It takes 60 days from sowing to reach maturity. Average yield is 16 t/ha.

Swarna Prabha

Fruits are light greenish, pulp soft and contain less fibre. Yield is about 25-30 t/ha. It is tolerant to powdery mildew disease. Released from ICAR-Research Complex for Eastern Region, Ranchi, Jharkhand.

Phule Prajakta

It is released from Mahatma Phule Krishi Vidyapeeth, Rahuri. Fruits are medium green, straight, slender, tapering at ends and flesh colour pure white. Average yield is 14-15 t/ha.

Phule Komal (F$_1$ Hybrid)

It is released from Mahatma Phule Krishi Vidyapeeth (MPKV), Rahuri. Fruits are attractive, shiny green, cylindrical fruits, average weight of fruit 111.93 g. It is moderately resistant to powdery mildew, downy mildew diseases and pests such as fruit fly, leaf miner, white fly and thrips under field condition. Average fruit yield is 21-22 t/ha.

An extremely long cultivar Yizhangquang, introduced from Zhejiang province of China produces many female flowers. Fruits grow up to 1 m long at 20 days after flowering and a single plant produces 50 kg fruits in a season (Zhang, 1989).

6.9.0 Indian Squash (Tinda)/Round Melon

Pusa Raunak

It is released by ICAR-IARI, New Delhi. It is an early maturing cultivar of round melon for spring summer season cultivation for North Indian plains. Fruits become ready for first harvesting in 55-60 days after sowing. It produces 8-10 fruits per vine. Young fruits at marketable stage are attractive green, shiny, uniform, flattish round in shape, 5 cm in diameter. Flesh is white, tender, less-seeded and has good cooking quality. Each fruit is medium in size and weighs 60 g at marketable stage. The seeds are black in colour with ridged border. Average fruit yield 7.59 t/ha during spring summer season.

Arka Tinda

It is released by ICAR-IIHR, Bangalore. It is developed by hybridization between Rajasthan Local and T8 (Punjab) followed by pedigree method of selection.

Cultivars of Round Melon and Long Melon

Pusa Raunak

Arka Tinda

Pusa Utkarsh

Fruits are round with light green shining skin covered with soft hair when young and tender (60 g). Yield is 10 t/ha.

Hisar Tinda (HT-10)

It is released by HAU, Hisar. It is an early and high yielding, fruits round medium in size tender and tolerant to downy mildew and root rot wilt. Average yield 7.5-10 t/ha.

S-48

It is released by PAU, Ludhiana. It is an early maturing variety developed by selection from a local material. Plants are vigorous 75-100 cm long. Leaves are deeply lobed and light green. Fruits are medium sized (50 g), flat round, pubescent, light green and shinning. Flesh is white. First picking is possible 60 days after sowing. Average fruit weight is 50 g and yield is 6.5 t/ha.

Punjab Tinda-1

It is released by PAU, Ludhiana. It is an early maturing cultivar and is suitable for sowing in spring season. Fruits are round, shining, green, pubescent, white fleshed with average fruit weight of 60 g (immature stage). First picking is possible 54 days after sowing. Its average yield is 7.2 t/acre.

6.10.0 Long Melon

Pusa Utkarsh

It is released from ICAR-IARI, New Delhi. It is an early maturing cultivar for spring summer season cultivation under North Indian plains. Fruits are ready for first harvest in 45-50 days after sowing in spring summer season. Fruits are slightly curved, medium long (length 50 cm), thin (diameter 2.4 cm), light green, smooth non-prominent ridges, shiny with tender skin, crispy flesh and free from bitterness. Each fruit weighs 130-145g at marketable stage. Average fruit yield 29.2 t/ha during spring summer season.

Arka Sheetal

It is released from ICAR-IIHR, Bangalore. The fruits are medium long (length 22cm and girth 7.8 cm) with light green skin colour covered with soft hair. Each fruit weighs 90-100 g at marketable stage. It is free from bitter principles with crisp texture. Seeds light tan color. Average yield is 35 t/ha.

Thar Sheetal

Fruits are light green coloured and tender at edible stage. It is moderately resistant against melon fruit fly. Suitable for culinary and salad purpose. Average yield is 13-14 t/ha.

AHC-2

It is released by CIAH, Bikaner. Its fruits are light green skin without furrows. It is very early maturing cultivar bearing uniform, medium long fruits. Fruits become ready for harvest in 53-55 days after sowing and harvesting continues up to 95-110 days. About 12-15 tender fruits can be harvested giving a yield of 4 kg per vine and 17-20 t/ha under arid situations.

AHC-13

It is released by CIAH, Bikaner. It is very early and highly productive heat tolerant cultivar with profuse hermaphrodite flowers. For slicing, the fruits can be harvested at very early stage (3-6 days after fertilization). First harvest can be obtained 50 days after sowing and harvesting continue up to 95-100 days. About 20-25 fruits are borne per vine. On an average 2.15 kg tender fruits can be harvested per vine giving a yield of 8.5-12.5 t/ha.

Pant Kakri-1

Released by G.B.U.A. and T., Pant Nagar. The vines are vigorous with long light green straight fruits. It is free from common diseases and insects. The first green immature fruit picking is possible in 50 days after sowing. Seed to seed stage is about 90 days. Yield potential is 30 t/ha.

Punjab Long Melon-1

It is released by PAU, Ludhiana. It is an early maturing selection from a local material collected from Hoshiarpur, Punjab. Vines are long with angled, hairy and light green stem. Leaf petiole is large, cylindrical and light green. The fruits are long, thin and light green. Average yield is 21 t/ha.

6.11.0 Snap Melon

Pusa Shandar

It is released from ICAR-IARI, New Delhi. Its fruits are oblong, medium in size with creamy white skin, thick and light pink flesh. Average yield is 38 t/ha. Maturity in 45-50 days.

Types of Snap Melon

Variability in Shape, Size and Colour of Snap Melon

AHS-10

It is released from ICAR-CIAH, Bikaner. Fruits are oblong and medium in size (900 g), flesh whitish pink and sweet in taste having 4.5-5.0 per cent TSS. Fruits can be harvested 68 days after sowing. It bears 4.0-4.5 fruits vine each giving yield of 22-23 t/ha under arid conditions.

AHS- 82

It is released from ICAR-CIAH, Bikaner. The flesh is light pink, sweet having 4.3-4.9 per cent TSS. Fruit harvest starts 67-70 days after sowing, each vine bears 4.5-5.0 fruits giving an yield of 24-25 t/ha.

6.12.0 Ash Gourd

Pusa Ujjwal

It is released from ICAR-IARI, New Delhi. Fruits are oblong, ellipsoid, rind greenish white while flesh is white with average fruit weight of 7.0 kg. Its fruits are ideal for long distance transportation. Average yield is 48-50 t/ha in *kharif* season and 41-42 t/ha in summer season.

Pusa Sabzi Petha

It is released from ICAR-IARI, New Delhi. This cultivar is identified for *Kharif* cultivation in Zone VIII (Karnataka, Tamil Nadu and Kerala). Its vines are medium long (average length 7.0m) and fruits are cylindrical. It requires 100-110 days for first fruit maturity. The average flesh thickness of fruits is 6.40 cm. Average yield is 36.5 t/ha and fruit weight is 3.5 kg. Suitable for culinary purpose.

Pusa Urmi

It is released from ICAR-IARI, New Delhi. It is suitable for cultivation in *kharif* season. Its fruits are oblong-ellipsoid with greenish white rind and white flesh. Days to first harvest is after 115-120 days of sowing. Average fruit weight is 10.0 kg. Average yield is 40 t/ha. It is suitable for processing.

CO-1

Released by Tamil Nadu Agricultural University, Coimbatore. It is a selection from a local type from Tamil Nadu with crop duration of 150 days. Fruits are globular weighing 5 to 6 kg with less seeds. Six to eight fruits are borne in a vine. Moderately resistant to pests and diseases. It yields 20-25 t/ha with good cooking quality.

CO-2

Released by Tamil Nadu Agricultural University, Coimbatore. Early maturing (120days), fruits small weighing 3kg, long spherical shape, less seeds (around 200-300 seeds), light green coloured flesh, higher yielder than CO-1. It is suitable for both kitchen garden and commercial cultivation.

TNAU Ash Gourd Hybrid CO 1

Released from TNAU, Coimbatore. It is an F_1 hybrid between PAG 3 × CO 2.

Cultivars and Types of Ash Gourd

Pusa Sabzi Petha

Pusa Urmi-1   **Pusa Urmi-2**

Variability in Shape, Size and Colour of Ash Gourd (Dhillon *et al.*, 2017b)

Fruits are oblong and medium sized and suitable for small family. Duration is 130 – 135 days. Average yield is 90 t/ha.

Kashi Surbhi

It is released from ICAR-IIVR, Varanasi. Fruits are oblong, ellipsoid, rind greenish white, flesh white. Average fruit weight 10-12 kg. Fruits are suitable for long distance transportation. It has yield potential of 60-70 t/ha (*Kharif* season) and 51-55 t/ha (summer season).

Kashi Ujwal

It is released from ICAR-IIVR, Varanasi. Fruits are globular, each weighing 10-12 kg and less seeded. Yield potential is 60 t/ha in 130-140 days of crop duration. Suitable for candy/petha making and is recommended for Tamil Nadu.

Kashi Dhawal

It is released from ICAR-IIVR, Varanasi. This cultivar is derived from a local collection. The fruits are oblong, flesh white, thickness 8.5-8.7 cm, seed arrangements linear, average weight 11-12 kg. This is suitable for preparation of Petha sweets due to high flesh recovery. Crop duration 120 days and yield is 55-60 t/ha.

Indu

It is developed by selection from the local material collected from Koyilandi in Kozhikode. High yielding cultivar with good flesh thickness released from KAU, Pattambi. Yield potential is 24.5 t/ha with mean fruit weight of 4.82 kg.

Thara

Released by Kerala Agricultural University. It is a small fruited (1.22 kg) cultivar suited for homestead and commercial cultivation. Average yield is 22 t/ha.

Shakthi

It is a selection from the local material and is released from APAU, Hyderabad. The fruits are cylindrical. It takes 140-150 days from sowing to marketable maturity. Average yield is 30-35 t/ha.

PAG-3

It is a selection from the local material and is released from PAU, Ludhiana. It is suitable for cultivation in north Indian plains in February-March and May-June. The fruits are attractive, globular and medium sized. The immature fruits are hairy but smooth and waxy at maturity. It takes 145 days from sowing to reach maturity. Average fruit weight is 10-12 kg and yield is 55 t/ha.

6.13.0 Snake Gourd

CO-1

It is a pure line selection from local and is released from TNAU, Coimbatore. It is an early maturing cultivar, first fruit coming to harvest in 70 days. Fruits are long between 160 and 180 cm in length, dark green with white stripes, flesh light green 10 to 12 fruits are borne in a vine weighing 4 to 5 kg.

CO 2

It is a pureline selection from a local type of Coimbatore district and is released from TNAU, Coimbatore. The fruits are short and stoutThe fruit is light greenish white and each weighs 400-600g. The crop duration is 105 days. The cultivar does not require pandal. The vines are less spreading and hence seeds can be sown at close spacing of 1.5 × 1.5 m. It yields on an average of 36 t/ha.

PKM 1

Released from HC and RI, TNAU, Periyakulam. It is an induced mutant from H 375. The vines are vigorous growing. Fruit color is dark green with white stripes on outer side and light green inside, with a mean fruit weight of 700 g. The fruits are extra long (180-200 cm) and suitable for growing all through the year. It yields on an average of 25.5 t/ha in crop duration of 145 days.

MDU 1

It is a F_1 hybrid between Panripudal and Selection-1 from Thaniyamangalam and is released from AC and RI, TNAU, Madurai. It is an easily flowering type (84 days) with a sex ratio of 1:38. It produces 13 fruits per vine weighing 7.15 kg with an average yield of 31.75 t/ha in a crop duration of 145 days. The fruits are medium long (66.94 cm) with white stripes under green background. Each fruit on an average weighs 55 g. The fruits are fairly rich in vitamin-A (44.4 mg/100 g) and very low in fibre content (0.6 per cent).

PLR (SG) 1

Released from Vegetable Research Station, TNAU, Palur. It is a pure line selection from white long type. This cultivar is suitable for cultivation under irrigated conditions only. Excellent cooking quality due to less fibre and high flesh content and does not twist due to maturity. This is having a yield potential of 35 – 40 t/ha with 30.50 per cent increase over CO 1. This cultivar can be cultivated during June – September, November – March and April – May.

PLR (SG) 2

Released from Vegetable Research Station, TNAU, Palur. Fruits are plumpy, fleshy with attractive white colour with less fibre content. It has excellent cooking quality. Average single fruit weighs 600g. Short fruit enables easy handling and long distance transport.

Kaumudi

It is high yielding cultivar released from Kerala Agricultural University. Fruits are long white, with average fruit length of one metre. Suitable for growing in acid alluvial soils of Kerala. Average yield is 50 t/ha.

Cultivar and Types of Snake Gourd

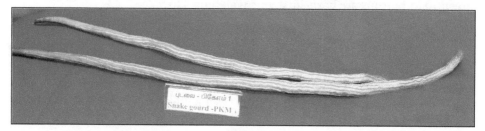

PKM-1

**Variability in Shape, Size, Colour and Striation of Snake Gourd
(Dhillon *et al.*, 2017b)**

Baby

It is a high yielding cultivar released from the Kerala Agricultural University with small, uniformly white coloured fruits and average fruit weight of 474 g. The crop starts yielding in about 55 days from sowing and it has a potential yield of 57 t/ha.

Harithasree

Released from Kerala Agricultural University. It is a high yielding cultivar having green fruits with white stripes for areas where green fruits are preferred.

Manusree

It is an early maturing cultivar developed by selection from a local material and released from Kerala Agricultural University, ARS, Mannuthi. Fruits are attractive, medium long (65-70 cm), white with green stripes at the pedicel end. First picking is possible 55 days after sowing. Average fruit weight is 750-800 g and yield potential is 60 t/ha.

TA 19 (KAU, Vellanikkara)

It is developed by selection from a local material and released from Kerala Agricultural University, Vellanikkara. It takes 65-70 days to reach marketable maturity. The fruits are about 60 cm long, light green with white stripes at styler end when immature. Average fruit weight is about 600 g.

Konkan Sweta

It is a selection from a local material and is released from KKV, Dapoli. The fruits are medium long (90-100 cm) and whitish-green. Average yield is 15-20 t/ha.

Phule Vaibhav

It is released from MPKV, Rahuri. Fruit is white colour with green strips and smooth surface, tolerant to downy mildew under field condition. Average yield is 28 t/ha.

There are cultivars, whose fruits reach a length of about 160 to 180 cm with variations in stripe sand colour (pale or dark green) of the fruit.

6.14.0 Perennial Cucurbits

In vegetatively propagated cucurbits like pointed gourd (*parwal*) and ivy gourd (*kundsru* or *tondli*), clonal variations exist and different shapes, sizes and markings of fruits are known. Most of these clones go by names of the localities where they are grown. In pointed gourd the major types are (i)10-13 cm long, dark green with white stripes, (ii) 10-16 cm long thick, dark green with very faint stripes and pale green in colour, (iii) small 5-8 cm long roundish dark green and stripped, and (iv) small tapering at the ends, green and stripped. These are grown mostly in Bihar, Bengal and Eastern U.P. In M.P. Pale White Oval cultivar is popular.

In chow- chow, variations in fruit shape and colour occur and cultivars like Round White, Long White, Pointed Green, Board Green, Oval Green are met with. In

Bangalore region, two types, green and the other creamy green are grown. In North-East India, these types are met with, besides minor variations in the fruit shape.

In ivy gourd (*kundsru*), there are several clones grown in different parts of India. In South India, mainly in Tamil Nadu, Andhra Pradesh and Karnataka, fruits are small, thinner in size with green ground colour and longitudinal white stripes. In Western India, clones have medium sized fruits, around 5 cm long and thin, dark green or light green with scattered small longitudinal white stripes. From Bihar one clone has been identified with bigger fruits, longer than 5 cm and thicker than one's middle finger. There are cultivars whose fruits can be used as salads, while there are some which can be pickled. Bitter forms are also sometimes encountered.

6.14.1 Pointed Gourd

Arka Neelachal Kirti

It is released from ICAR-IIHR, Bangalore. Fruits are spindle shaped, dark green in colour with 3-4 fragmented stripes of cream colour and medium sized (10.45 × 3.5 cm) weighing around 40-45g with solid core. It has a yield potential of 15 t/ha.

Kashi Amulya

It is released from ICAR-IIVR, Varanasi. It is less seeded (5-8 seed/fruit as compared to 25-30 seeds in normal cultivar, suitable for confectionary purpose, more fleshy, attractive light green fruit with sparsely distributed white stripes. Yield is 20-22 t/ha.

Kashi Suphal

It is released from ICAR-IIVR, Varanasi. Fruits are attractive light green with mild stripes, fleshy with soft seed, long duration of fruit retention in plant, better keeping quality suitable for culinary purpose and sweet making. Yield is 18-20 t/ha.

Kashi Alankar

It is released from ICAR-IIVR, Varanasi and is developed through clonal selection. Fruits are light green in colour and devoid of any white stripe with 6-7cm in length and 2-3 in diameter having an average fruit weight of 25-27g. Plants of this cultivar are capable of producing 120-130 fruits/vine. The average yield this cultivar is ranged from 18-19 t/ha.

Swarna Rekha

It is developed by clonal selection from the local material and is released from Horticulture and Agro-forestry Research Programme, Ranchi. The fruits are elongated (8-10 cm long) and tapered at both ends, striped and green. Average yield is 19 t/ha.

Swarna Alaukik

It is developed by clonal selection from the local material and is released from Horticulture and Agro-forestry Research Programme, Ranchi. The fruits are elongated (5-8 cm long) with blunt ends, light green and suitable for making sweets. The yield potential is 23 t/ha.

**Variability in Shape, Size and Striation of Pointed Gourd Germplasm
Maintained at BCKV, West Bengal, India**

Locally Adopted Types of Pointed Gourd

Cultivars of Pointed Gourd

Swarna Rekha   Swarna Suruchi

Kashi Alankar   Swarna Alaukik

BCPG-3   BCPG-4   BCPG-5

Faizabad Parwal-1

It is developed by clonal selection from the local material and is released from NDUAT, Faizabad. The fruits are round and green. Average yield is 16 t/ha.

Faizabad Parwal-3

It is developed by clonal selection from the local material and is released from NDUAT, Faizabad. The fruits are spindle shaped, green and striped. Average yield is 13.5 t/ha.

Faizabad Parwal-4

It is developed by clonal selection from the local material and is released from NDUAT, Faizabad. The fruits are spindle shaped and light green. Average yield is 11 t/ha.

BCPG-3

This cultivar has been developed through clonal selection at BCKV, West Bengal. Early plant vigour good and medium viny in nature. Stem shape angular, tendril branched and coiled. Leaves serrated round and medium in size (6.5-7cm), pubescent, intermediate and sparse. Fruit spindle shaped, medium curved, fruit skin primary color light green with white alternate stripe. Average fruit length 7.30 cm, diameter 3.5 cm, fruit weight 34.0 g. Average yield is 520 q/ha upon hand pollination.

BCPG-4

This cultivar has been developed through clonal selection at BCKV, West Bengal. Early plant vigour good and medium viny in nature. Stem shape angular and pubescent, tendril branched and coiled. Leaves serrated oblong and medium in size (7.5-8cm), pubescent, intermediate and sparse. Fruit spindle shaped, fruit skin primary color dark green with white alternate stripe having rough skin. Average fruit length 8.78 cm, diameter 3.52 cm, fruit weight 44.0 g. Average yield is 580 q/ha upon hand pollination.

BCPG-5

This cultivar has been developed through clonal selection at BCKV, West Bengal. Early plant vigour good and medium viny in nature. Stem shape angular and pubescent, tendril branched and coiled. Leaves serrated oblong and medium in size (8-9 cm), pubescent, intermediate and sparse. Fruit long shaped, fruit skin primary color dark green with prominent white alternate stripe having smooth skin. Average fruit length 9.50 cm, diameter 3.75 cm, fruit weight 46.0 g. Average yield is 550 q/ha upon hand pollination.

6.14.2 Ivy Gourd

CO 1

Clonal selection from Anaikatti type and is released from TNAU, Coimbatore. Fruits are long, green with white stripes, less seeded and sweet (4.5° Brix). It is suitable for culinary purpose with yield of 83 t/year.

Arka Neelachal Kunkhi

It is released from ICAR-IIHR, Bangalore. It is a dual-purpose (salad as well as cooked) early cultivar. Fruits are extra-long (8.39 cm), weighing around 15-20 g, uniform, cylindrical with attractive stripes. It produces around 800 fruits in a season with yield potential of 15-20 t/ha.

Arka Neelachal Sabuja

It is released from ICAR-IIHR, Bangalore. The plants are very vigorous (>10 m long) and produce high biomass. Fruits are dark green in appearance with fractured stripe and conical in shape. It gives 70-80 harvest per season (10-11 months) and yield is up to 20-25 t/ha.

Thar Sundari (AHIG-01)

It is developed through single plant selection and is released from CIAH, Bikaner. It can be cultivated *kharif* and spring seasons. It is tolerant to high temperature (40-42 °C).

Sulabha

It is a clonal selection from the local material and is released from Kerala Agricultural University, Vellanikkara. The fruits are set parthenocarpically and the first picking is possible 45-50 days after planting. The fruits are long (9.5 cm), cylindrical and pale green with continuous striations. It is suitable for salad making, pickles and cooking. Average fruit weight is 18 g and average yield is 19-20 kg per plant.

Indira Kundru-05

Released from IGKV, Raipur, Chhattisgarh. Fruits are oblong size, green colour with white strips Suitable for table purpose. Average yield is 101.9 t//ha.

Indira Kundru-35

Released from IGKY, Raipur, Chhattisgarh. It has long sized and light green fruits. It is suitable for table purpose. Average yield is 93 t/ha.

6.14.3 Spine Gourd

Arka Neelachal Shree

It is released from ICAR-IIHR, Bangalore. It is developed through selection.

Arka Neelanchal Shanti (Hybrid)

It is released from ICAR-IIHR, Bangalore. It is developed through hybridization between spine gourd and teasle gourd. The cultivar exhibits plant and flower morphology more similar to teasle gourd while its fruit morphology is more close to spine gourd. It is high yielding (15-16 kg/vine) with medium sized fruit (20 g), moderately tolerant to fruit borer, anthracnose and downy mildew.

Cultivars of Ivy Gourd and Types of Teasle Gourd

Arka Neelachal Kunkhi

Arka Neelachal Sabuja

Indira Kundru-5

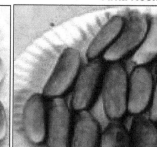

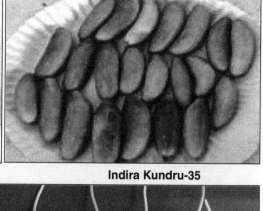

Indira Kundru-35

Variability in Teasle Gourd Germplasm Maintained at BCKV, West Bengal, India

6.14.4 Teasle Gourd

Arka Neelachal Gaurav

It is released from ICAR-IIHR, Bangalore. Fruits are attractive, uniform lush green round-oval fruit with soft seed and high-quality edible portion for culinary purposes and soft seeded. It is tolerant to downy mildew and anthracnose. It yields 18-20 t/ha.

7.0 Soil and Climate

7.1.0 Soil

A well drained soil of loamy type is preferred for cucurbits. Lighter soils which warm quickly in spring are usually utilized, for early yields and in heavier soil, vine growth will be more and fruits are late maturing. In sandy river-beds, alluvial substrata and subterranean moisture of river streams support the cucurbits. The soils should not crack in summer and should not be waterlogged in rainy season. It is necessary that soils should be fertile, provided with organic matter.

All the cucurbits are sensitive to acid soils. Below pH 5.5, no cucurbit can be successfully grown and most of the cucurbits prefer a soil pH between 6.0 and 7.0. Muskmelon is slightly tolerant to soil acidity, while the other cucurbits like cucumber, pumpkin and squashes fall under the intermediate or moderate category. Likewise, alkaline soils with high salt concentration are unsuitable. There is no cucurbit which can tolerate high salt concentration like those of beet and spinach and watermelon is comparatively more tolerant to medium salt concentration with Ec_e values ($Ec_e \times 1000$) ranging from 10 to 4.

Soil temperature is a determining factor for quick germination and early maturity and production. The minimum should not go below 10°C and maximum beyond 25°C, the optimum range is around 18-22°C. In river-beds, the sand remains comparatively warmer and does not cool quickly as in garden soil. Further, sandy river-beds have moisture beneath and warm up quickly in spring. That is there as on why cucurbits in river-beds survive low temperature periods of winter month sand produce early crops in spring.

Soil moisture is important for rapid growth and it should be atleast 10 to 15 per cent above the permanent wilting point. Even though primitive and wild cucurbits are xerophytic in character, nevertheless, cucurbits require good moisture but not excessive moisture during its vegetative growth. Rainy season cucurbits are mostly unirrigated.

Depth of the soil is an important consideration, because in perennial cucurbits, the soil has to support the vines for atleast 3 years as in chow-chow or ivy gourd, *etc.* Pit system of growing in pandals or trellis or arbours are adopted in these cucurbits so that deep soil can support the vine for longer period.

7.2.0 Climate

Cucurbits are essentially warm season crops grown mainly in tropical and subtropical regions. In temperate countries, forcing of cucumber is done mainly in glass houses where temperature, humidity, light and CO_2 are controlled. Similarly, river-bed cultivation of cucurbits (Diara cultivation) in North, Western and Central India during winter months, is a kind of indigenous system of vegetable forcing practiced with minimum capital outlay, but with maximum risks of growing in low temperature.

The growth requirements of cucurbits are generally, long period of warm preferably dry weather with abundant sunshine. They are not adapted to resist even light frosts and will have to be provided with at least partial protection if grown during winter months. There are few specific cucurbits which donot stand extreme dry weather but prefer moderate humid conditions, like chow-chow, pointed and snake gourds. But excessive humid weather will promote diseases like downy mildew, anthracnose and virus diseases and pests like fruit fly. For good fruit quality and sweetness in muskmelon and watermelons, dry weather during fruit development is necessary and cool nights and warm days are ideal for accumulation of sugars in the fruits.

The average temperature for growth would be around 30-35°C with maximum ranging around 40°C and minimum between 20 and 25°C. Most of the cucurbits germinate well when the day temperature is above 25°C. Seed germinates relatively poorly at temperatures below 15°C and such temperatures also retard root growth. For normal growth, they require optimum average monthly temperature from 25 to 30°C. Cucumber prefers slightly lower temperature than other cucurbits and hence it is extensively grown in subtropical regions. Temperature above 45°C causes vines to wilt temporarily resulting in sunburning of fruits and reduced shelf life.

Maximum temperatures of 21°C in the day and 16°C at night increase retention of female flowers and fruit set in cucumber, squash, and pumpkin (Maynard and Hochmuth, 2007). Development of female flowers in cucurbitaceous plants is reduced when the plants are subjected to high diurnal and night temperatures above 30°C and 18°C, respectively. Abortion of female flowers and young fruits is also high when the plants are subjected to temperature extremities.

The opening of flowers and anthesis requires optimum temperature and relative humidity (Agbagwa *et al.,* 2007; Welbaum, 2014). Most of the cucurbitaceous flowers will open at temperatures >10°C. For instance, flowers of pumpkins and squashes will open at >10°C, cucumber and watermelon at approximately 16°C, and muskmelon and cantaloupe at approximately 18°C. The optimum temperature for anthesis and pollen dehiscence is between 13°C and 18°C in pumpkin, cucumber, muskmelon, and watermelon.

Melon requires warm and dry weather with plenty of sunshine for growth and production. The optimum temperature range is 18°C–28°C, growth being severely retarded below 12°C. Melon easily withstands several hours per day of very high temperatures, up to 40°C. In snake melon, stem elongation was found to be greater under short 8 h days than under 16 h days. High humidity will reduce growth, adversely affect fruit quality, and encourage leaf diseases.

All cucurbits are sensitive to frost. Almost no growth takes place at temperatures below 15°C, but is rapid between 18°C and 30°C. Plants grow more luxuriantly at higher temperatures. However, relatively low temperatures and short daylight periods promote the formation of more female, in relation to male flowers; while at high temperatures (35°C or more) only male flowers may be formed. There is poor pollination of female flowers when temperatures drop below 10°C for pumpkins and squashes, 15°C for watermelons and cucumbers and 20°C for muskmelons (also called spanspek or cantaloupe); higher temperatures tend to promote pollination. High atmospheric humidity, especially in the later growing stages, and particularly for muskmelons and watermelons, favour the development of fungal diseases. Prolonged cool, cloudy or moist weather over the flowering season will often reduce bee activity, and result in poor pollination of flowers and a reduced fruit-set.

Table 7: Sensitivity to Days and Night Temperature

Crop	Germination Temperature (°C)	Soil Temperature (°C)
Cucumber	28-32.2	20-30
Muskmelon	28-32.2	25-30
Watermelon	28-32.2	25-30
Bitter gourd	28-30	30-35
Bottle gourd	28-30	30-35
Sponge gourd	26-28	30-35
Ridge gourd	16-28	30-35
Squash	28-32.2	28-30

In trials with cucumber cv. Alma-Alinshii, Obshatko and Shabalina (1984) opined that the time of fruiting was related to early temperature conditions, those raised at relatively low temperatures had lower requirements for both temperature and light than plants raised at relatively higher temperature. Kano *et al.* (2000) noticed that the occurrence of bitter fruit in cucumber (cv. Kagafutokyuri) was higher in the plots with a lower air temperature. Leaf weight was lower and total-N, amino acid-N and protein contents were higher in the plots with a lower air temperature. The melons, on the other hand, require tropical climate with fairly high temperature of 35 to 40°C during fruit development. Cool nights and warm days are ideal for accumulation of sugars in the fruits. If the nights are warm in summer as in some parts of North India, maturity is hastened. Infact, some of the sweetest melons are produced in Central Asia during May-September because of prevalent of warm dry weather and cool nights. Sarda muskmelons of Afghanistan and Uzbekistan are highly sweet and adapted to this kind of climatic conditions. In the arid Imperial Valley of California (USA), cantaloupes are grown during spring-summer and also autumn seasons and these are short duration types maturing with 90 days. Medany *et al.* (1999) observed that fruit growth rate (FGR) was highest in both autumn and spring season in the non-shaded treatment with a night set point of 18°C. In the Negev Desert, Israel, Huyskens *et al.* (1992) observed that germination of bitter gourd was optimum at temperatures between 25 and 35°C. Huyskens *et al.* (1993) reported from Germany, maximum germination (50 per cent) in *L. acutangula* at 35°C and at 8, 12 and 45°C germination was completely inhibited. Partial removal of the seed coat increased the percentage of germination while vernalization and exposure to salinity ≥mM NaCl reduced it. Flowering pattern was also affected by planting season with significantly more female flowers being produced in spring summer under long days and high temperatures than in autumn-winter under short day sand low temperatures in bitter gourd (Huyskens *et al.*, 1992) and ridge gourd (Huyskens *et al.*, 1993). A day length of at least 16 h (long-day) promotes male flower development in cucumbers, whereas a short-day length of at most 9 h at moderate light intensity promotes female flower development.

Other cucurbits, namely gourds are grown mostly in summer as well as in rainy season (depending on the rainfall). In the latter season, vine growth and spread will be very extensive. In South and Central India, most of them are grown round the year. In the hills, cucumber and summer squash are grown in summer (April-August) and in lower hills, two crops are taken depending on the altitude and location. Snake gourd and pointed gourd require hot and humid weather.

The perennial cucurbits like pointed gourds (parwal) are sensitive to frost and cold temperature, especially below 5°C. They over winter and the roots remain dormant underneath the soil. They can be sprouted in spring with irrigation. Chow-chow requires moderate climate with good humidity as prevalent in Bengaluru, lower hills of Darjeeling, North-East India, Koraput region of Odisha, *etc.* Chow-chow does not stand extreme dry winds (*loo*) during summer and frost in winter. In the case of ivy gourd (*kundsru*) it is confined to warmer regions of Maharashtra, Andhra Pradesh, Gujarat, Karnataka, *etc.*

8.0 Cultivation

8.1.0 Seed Treatment

Seed treatment for improving germination and yield is very much effective in cucurbitaceous crops. In cucumber, a pre-sowing seed soaking in l-cyanoethyl-5, 6-nitrobenzimidazole increased yields by 11.5-15.0 per cent (Sovetkina *et al.*, 1984). Solanki and Joshi (1985) obtained better germination percentage (92 per cent) in four years' old seeds of cucumber with the treatment of 12 hours pre-soaking in KH_2PO_4 (1-3 per cent) solution. Squash cvs. Hoyo, Camote and Arizona did not germinate at 8°C. Pre-soaking of seeds for 48 hours lowered the threshold temperature to 8°C for three cultivars (Bravo and Venegas, 1984). Seeds of summer squash soaked in 500 ppm Cu, 2000 ppm Mn or 2000 ppm Zn gave the highest germination rates, greatest plant growth and highest yield (Abed and Sharabash, 1985). Germination of bitter gourd seeds was found to be enhanced by soaking in low concentrations (100-200 ppm) of IAA, GA_3 and CCC (Sharma and Govil, 1985). Presowing treatment of bitter gourd seeds with 1 per cent KNO_3 increased germination percentage (Devi and Selvaraj, 1996). Priming seeds of *Lagenaria siceraria* seeds for 3 days in the presence of KNO_3 or thiourea (0.5, 1 or 3 per cent) increased germination rate and decreased the mean time required for germination (Yoo *et al.*, 1996a).

Ilamanova (1987) reported that soaking of muskmelon seeds in Co, Mn, Cu or Zn solutions caused earlier flowering, and induced a greater number of male and female flowers in cv. Ameri (early) and Gulyabi Oranzhevaya. Co, Zn and especially Mn were found to increase the percentage of female relative to male flowers. Lebedeva and Pal'm (1986) reported that soaking of cucumber seeds in Cu, B and Zn solutions extended the growing period and produced a greater leaf assimilation area. Seed enrichment with B or Cu was recommended for early warm springs and with Zn for late cool springs in Russia.

Devi and Selvaraj (1995) recorded highest mean germination percentage of seeds of *B. hispida* cv. CO-1 (88.7 per cent) and *L. acutangula* cv. CO-1 (92.7 per cent) with arappu powder (*Albizia amara*) pelleting at 500g/kg seed.

Chen *et al.* (1999) noticed that five strains of bacteria, *CN11, CN31, CN45, CV116* and *CV129*, largely promoted growth of cucumber seedlings, including increasing incidence of emergence, bringing forward the time of emergence, increasing seedling height, fresh weight and dry weight.

Treatment of spine gourd tubers with thiourea 1.0 per cent for 12 h was recorded best for breaking the dormancy and percentage of sprouted tuber was 87.5 per cent (Panda *et al.*, 1994).

8.2.0 Land Preparation

Bring soil to a fine tilth after 2-3 ploughings and harrowings.

1. Broadcast 25 t/ha of well decomposed FYM after first ploughing. The FYM may be enriched with *Trichoderma*.
2. Prepare 60 cm wide and 15 cm high beds across slope (can be done using tractor drawn bed former) with 1.5 to 2 m spacing between beds (bed center) for muskmelon and 2.5 to 3 m spacing between beds (bed center) for watermelon.

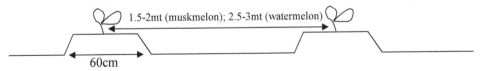

1.5-2mt (muskmelon); 2.5-3mt (watermelon)

60cm

3. As basal dose apply 50 per cent recommended N, 100 per cent of recommended P_2O_5 and K_2O in shallow furrows over the bed, 5 cm below the soil surface.
4. Lay inline drip laterals with 30 cm emitter spacing and 2 L/hr discharge over the beds. Flush drip laterals and check for leakages and blocks.
5. Cover the bed with reflective plastic mulch (black – silver), 30 micron thickness and 4 ft width over each bed with the silver side on the top. This helps in water saving, reducing evaporation losses, maintains soil temperature, reduces weeds and repels pests.
6. Make 10 cm diameter holes for seedling transplanting in the plastic mulch on the center of the bed with a spacing of 50 cm between holes for muskmelon and 60 cm for watermelon.

8.3.0 Sowing and Planting

Being a tropical group of vegetables, these are grown essentially in summer and rainy season. Most of them are sown during winter to take an early spring summer crop. In South and Central India, where winter is neither severe nor long, these are grown almost round the year. Melons are grown only in those seasons, when weather is dry and warm during fruit development. Generally, in North India early sowing is done in river-beds in November and extending to February in garden lands. For the rainy season crop of other cucurbits, the sowing is done in June-July. In South and Central India, sowing is done in October-November and the crops mature in March-April. The season of growing is mainly determined by the rainfall and temperature conditions. It is preferable to avoid the fruits of cucurbits maturing during spells of rainy weather of heavy downpour. Hence the rainy season crop matures in September and October in North India, when south-west monsoon has receded. Likewise, in South India the fruits mature in February-March when north-east monsoon has receded. In Western India, cucumber sowing extends from September to February. In plains of North-East India most of the cucurbits are grown from November to March when the weather is comparatively dry. In the hills of North India, the sowing season starts from April-May and the crop matures in August and hence the crops, especially cucumber and squashes are attacked by mildew. In North-West arid Rajasthan like Bikaner, a rainy season crop of watermelon (Mateera cultivar) is taken because of absence of heavy rainfall during August-September.

Pumpkins are sown in Tamil Nadu in India during August and also in December-January. Snake gourd is generally sown in July and ash gourd is raised in two seasons, *viz.*, December-March and July September. It is, therefore, a common sight to notice one or the other cucurbit being available in the urban city markets round the year.

Yield of *M. dioica* was recorded highest from planting on 1st February (Islam *et al.*, 1994).

Most of the cucurbits are seed propagated and *in situ* sowing is practised. In some cases where early crop is desired, seeds can be sown in alkathene bags and germinated under protected cover from low temperature. The seedlings are transplanted from the bags at 2-true-leaf stage. This practice is prevalent in Punjab, especially in the case of muskmelon and it can also be done in the hills to get early crop in July. Normally, the cucurbits donot stand transplantation beyond this stage due to injury to tap root. There is considerable saving in seed quantity, nearly 50 to 60 per cent as compared to *in situ* sowing. Waterer (2000) reported that in pumpkins, transplanting did not improve and establishment over direct sowing but enhanced yields without influencing crop maturity.

C. pepo cv. Dixie, Brown *et al.* (1996b) reported from USA that fruit yield was significantly higher with transplanting than with direct sowing. Various systems of sowing have adopted depending on the season, crop and system of cultivation. In garden soil, furrows are made at 1 to 1.5 m spacing in the case of cucumber, tinda and bitter gourd, 2 to 2.5 m in the case of muskmelon, 3 to 4 m in the case of watermelon, bottle gourd, pumpkin, ash gourd, *etc.*

Staub *et al.* (1992) reported that in cucumber, number and weight of fruits/ ha increased with increasing plant density, fruit weight/plant decreased with increasing density. A spacing of 60cm between plants has been recommended in muskmelon (Prabhakar *et al.*, 1985). Usually, the sowing is done on the top of the sides of furrows and the vines are allowed to trail on the ground, especially in summer season. The hills where sowing is done are spaced at a distance of 0.5 to 0.75 m. In some regions, bed system is in vogue where the seeds are sown on the periphery of beds. The essential feature is that irrigation water should not spoil the developing fruits while the vines spread between the rows. In some cases, raised beds or mounds are made to facilitate drainage and seeds are sown into centre of the raised beds, especially in heavy rainfall regions. Pit system is practised during rainy season and vines of snake grourd, ivy gourd, *parwal*, sponge or ridge gourds are trained over in trellises, arbours or pandals at a height of 1.5-2 metres. The pits spaced at 1.5-2 m, are dug at about a metre depth and well manured with FYM. Five to six seeds are sown in a pit and finally two to three vines are trained on supports. In small homestead gardens, bottle gourd, ridge gourds, *etc.*, are trailed over thatched hut sand walls.

Arora and Malik (1989) found that in *L. acutangula* cv. Pusa Nasdar, a spacing of 9 plants per bed (2 × 4 m) and pruning to 6 primary branches gave the longest plants with the highest number of secondary branches. It resulted in early appearance of pistillate flowers, highest number of weight of fruits in both the early and total yields. Fruits yield was generally higher in the rainy than in the summer season. Huyskens *et al.* (1993) obtained a high yield of 44.5-47.3 tonnes/ha in *L. acutangula* for plants trained on trellises at planting densities of 10000 or 20000 plants/ha in Germany.

Yadav *et al.* (1989) obtained the highest yield of pointed gourd cv. FP-4 at 1.5 × 1.5m spacing and by training the vines in bower system. In pointed gourd total and early fruit yields were reported highest (101.71 and 169.82 q/ha, respectively) when plants were spaced 0.60m apart in rows (Pandit *et al.*, 1997).

Yield of *M. dioica* cv. Kathali were found to be highest (26.2 t/ha) with spacing at 3.0 m in Bangladesh (Islam *et al.*, 1994). In spine gourd, Puzari (1997) obtained greatest yield in main (10.7 t/ha) and ratoon crop (7.3 t/ha) from largest tubers and from the narrowest spacing of 1m apart.

Bottle gourd cv. Arka Bahar gave the highest average yield of 384.55 q/ha at a spacing of 300 × 45 cm with 1 plant/hill (Vishnu and Prabhakar, 1987).

In Deccan plateau, especially in Telengana region, India of black clay loamy soil, some cucurbits like ribbed gourd are grown in pits, under entirely rainfed condition. Pits are spaced 1.5 m either way in June-July and pits are manured 5-10 kg of cattle manure. If the soil is sandy loam, the crop is raised during October-November. The 'Nakka Dosakaya', the cucumber-like-muskmelon used for cooking, is also grown in such regions.

The effects of plant density and training method of Japanese white spined cucumber were investigated by Young Hah *et al.* (1995) and obtained highest yield (380t/ha) with 3 lateral shoots at a density of 45000 plants/ha.

Investigation with pumpkin (*C. pepo*) by Reiners and Riggs (1997) resulted in a significantly greater yield in cv. Howden and Wizard in irrigated plots at closer spacing (from 1.2 m to 0.3 m), whereas there was no significant increase in yield without irrigation. Field studies were conducted by Reiners and Riggs (1999) on 2 pumpkin (*C. pepo*) cultivars, Howden (vining-type growth habit) and Wizard (semi-bush growth habit), at 2 locations (Geneva, New York, and near Schoharie, New York, USA) to determine the effect of plant population and row width on marketable yield. Increasing plant populations from 2990 to 8960 plants/ha resulted in significantly greater fruit numbers and yields at both locations and for both cultivars, with yields of 49.0 t/ha and 61.4 t/ha at Geneva and Schoharie, respectively, at the highest population. Average fruit size declined at the highest populations. Increasing row width from 1.8 to 3.6 m resulted in a slight but significant decrease in number of fruits/ha with no effect on other yield parameters. At one location (Schoharie), the effect of row width on yield and number of fruits/ha depended on the population. At low populations, row width did not influence yield or fruit number; at high populations, wide rows produced lower yield and fewer fruits than narrow rows.The results demonstrate that growers may increase pumpkin yield by increasing plant populations but should use narrower row widths and wider in-row spacing.

From Cuba, Cortes and Hernandez (1996) reported highest yield at a spacing of 3 × 2 m in pumpkin (*C. moschata*) cultivar Mariuchaon a red ferrallitic soil. The yield was influenced by number of plants/hole, being significantly lower with 1 than with 2, 3 or 4 plants/hole.

From Australia, Botwright *et al.* (1998) reported that in *C. maxima* marketable yield increased to a maximum of 18 t/ha at 1.1 plants/m^2 and declined at higher densities of increased numbers of undersized fruits.

General seed rate and planting distance of some cucurbits are given in Table 8.

Paris *et al.* (1986) reported that a density of 44,444 plants/ha gave the highest yield of courgette. It produced 51 per cent more fruits, 66 per cent more marketable yield and 70 per cent more grade A yield than at 11,111 plants/ha.

Table 8: Seed Rate and Planting Distance of some Cucurbits

Crops	Planting Distance		
	Seed Rate (kg/ha)	Row to Row (m)	Hill to Hill (cm)
Watermelon	3.5 to 4.5	2.5 to 3.5	75 to 90
Round melon (Tinda)	3.5 to 4.5	1.5 to 2.0	45 to 60
Muskmelon	2.5 to 3.5	1.5 to 2.0	45 to 60
Long melon (*Kakri*)	2.0 to 2.5	1.5 to 2.5	45 to 60
Cucumber	1.5 to 2.0	1.5 to 2.0	45 to 60
Bottle gourd	4.0 to 4.5	2.5 to 3.5	60 to 75

Crops	Planting Distance		
	Seed Rate (kg/ha)	Row to Row (m)	Hill to Hill (cm)
Bitter gourd	4.0 to 4.5	2.0 to 2.5	45 to 60
Ridge gourd	3.5 to 4.0	2.5 to 3.0	60 to 75
Sponge gourd	2.5 to 3.0	2.5 to 3.0	60 to 75
Snake gourd	5.0 to 6.0	2.0 to 2.5	60 to 75
Ash gourd	5.0 to 6.0	2.5 to 3.0	60 to 75
Pumpkin	6.0 to 7.0	3.0 to 3.5	60 to 75
Summer squash	6.0 to 7.0	0.6 to 0.75	60 to 75

Source: Premnath (1976).

Edelstein *et al.* (1989) reported that in *C. pepo*, yield was not affected significantly by plant population but there was negative relationship between plant density and number of fruits/plant. In plants of the vine type, increased plant density decreased the ratio of large to small fruits, whereas in bush type plants, this ratio was almost unaffected.

In Israel, a high yield of 40-45 t/ha was achieved in bitter gourd at densities of 10,000 and 20,000 plants/ha and at both growing regimes of ground and trellises (Huyskens *et al.*, 1992).

Davis (1994) developed a luffa sponge gourd (*Luffa aegyptiaca*) production practices for a cool, temperate climate. According to him highest marketable yields were obtained when plants were spaced 30.5 cm apart in the row and the first 4 lateral shoots were removed (76900 gourds/ha). Plants spaced 91cm apart produced gourds with the largest diameter (mean 8.6 cm), whereas plants spaced 30.5 cm apart produced the highest yields of gourds with bath sponge diameters (5.1-7.6 cm). Plant spaced 91 cm apart and topped at node 6, produced fruits with high fibre density, strong fibres and excellent visual appeal, but yields were low.

8.3.1 Sowing System

In cucurbitaceous vegetables following systems are adopted for seeding.

8.3.1.1 Shallow Pit Method

In this method, shallow pits of 60 × 60 × 45 cm size are dug at a distance of 120-150 cm. The pits are left open for a week for partial solarization. Then each pit is filled with mixture of sufficient soil and 4-5 kg well decomposed FYM or compost and part of fertilizers like urea @ 50 g, SSP @ 100 g, and MOP @ 80 g. In the area where termite is problematic, carbofuran is added @ 1-1.5 g per pit. All the manures and fertilizers are mixed properly and filled in the pits. After filling the pits, circular basins are made. Generally, 3-4 seeds per hill at 2-3 cm depth are placed and they are covered with fine soil and FYM in the ratio of 50:50.

8.3.1.2 Raised Bed Method

According to nature of crop, ridges and furrows are prepared manually or mechanically in well prepared field at the distance of 1.5-3.5 m. Generally, depth of trench is kept 20-30 cm with the width of 40-50 cm. During rainy season, excess water drained easily by these trenches. On these beds (near to trench) pits of 30 × 30 × 30 cm size are dug and filled with well decomposed FYM @ 4-5 kg per pit. Generally, 3-4 g carbofuran mixed in soil and filled upto the height of 10 cm above ground.

8.3.1.3 Deep Pit Method

Deep pit method is quite common to grow cucurbits in riverbeds. In this method circular pits of 60-75 cm size and 1-1.5 m of depth are dug at the distance of 120-150 cm. The pits are filled with 10-12 kg FYM, 80 g urea, 100 g, SSP and 70-80 g MOP. Soil may be added with other nutrients as per the requirement.

8.3.1.4 Mound Method

In this method, 15-20 cm raised mounds are generally prepared. In each mound, 4-6 seeds are shown at a depth of 2-3 cm.

8.3.1.5 Hill and Channel System

In North Indian plains, during spring summer season, hill and channel system of sowing/planting has been recommended. In this method, channels are prepared from East to West direction and hills are prepared in the bund of the channels on the Northern slope. Generally 2 or 3 seeds are sown on each hill and only one plant is retained after the seedlings attain 2-4 true leaf stage. The channel to channel distance and hill to hill distance depends upon the growth of the vine of the cucurbits as sown in the table above. Generally vines are allowed to trail in between the two channels. This system has advantage as irrigation and manure and fertilizer application is done only in the channels which saves water and fertilizers. Since irrigation is applied only in the channels and not in between the channels, there is less chances of rotting of fruit.

However, machan/arbour system is followed mostly in Eastern and Southern India where rainfall is high and hill and channel system is not effective. In north Indian plains, many farmers adopt this system of planting during rainy season to get a good quality of produce.

8.4.0 Propagation by Grafting

Seven diverse rootstocks (*C. moschata, C. maxima* and *Sicyos angulatuss*) for cucumber grafting were evaluated in Germany using an apical and a tongue grafting method (Kell *et al.,* 2000). Yields of tongue grafted cultivars were much higher than those of apical grafted cultivars. A promising rootstock of cucumber, Pusan Daemok 1 was selected for grafting in Korea Rupublic (Kim *et al.,* 1998).

8.4.1 Cucurbit Grafting

The primary motive for grafting cucurbits is to avoid damage caused by soil borne pests and pathogens when genetic or chemical approaches for disease management are not available (Oda, 2002a; b). After methyl bromide fumigation was phased out, researchers and producers began to use grafting as an alternative method to avoid soil borne pests and diseases. Grafting of fruit bearing vegetables is used to control soil borne wilt diseases caused by *Fusarium* and *Verticillium* (Lee, 1994; Yetioir *et al.*, 2003), to increase low temperature tolerance (Tachibana, 1989), salinity tolerance (Behboudian *et al.*, 1986), and to increase plant growth and fruit yield by enhancing water and nutrient uptake (Lee, 1994; Oda, 1995; Ruiz *et al.*, 1996, 1997), improve fruit yield, and quality (Shimada and Moritani, 1977; Romero *et al.*, 1997; Oda, 2002a;b; Trionfetti-Nisini *et al.*, 2002; Lee and Oda, 2003; Rivero *et al.*, 2003, Hang *et al.*, 2005).

Research on cucurbit grafting began in the 1920s with the use of *Cucurbita moschata* as a rootstock for watermelon. Initial reports by Tateishi (1927) and Sato and Takamatsu (1930) described several studies and grafting applications at Kyusyu University. Tateishi (1931) reported on grafting of watermelon on to *Cucurbita moschata* rootstocks, a technique that was well known at the time. However, bottle gourd and wax gourd [*Benincasa hispida* (Thunb.) Cogn.] have became preferred rootstocks for watermelon grafting on account of their resistance to fusarium wilt, their high affinity for watermelon and the highly stable growth of the plants after grafting (Sato and Takamatsu, 1930; Matsumoto, 1931; Tateishi, 1931; Kijima, 1933; Murata and Ohara, 1936; Kijima, 1938). Pumpkin and squash, figleaf gourd, cucumber, and bottle gourd were reported as compatible with cantaloupe (*Cucumis melo* L. var. *cantalupensis*) in 1931 by Matsumoto. Matsumoto (1931) reported that *Cucurbita* spp. and melon are compatible with cucumber, and Imazu (1949) noted that *Cucurbita moschata*, bottle gourd, wax gourd, and luffa were also compatible with cucumber. Cucumber grafting started in Japan around 1960 to strengthen low-temperature tolerance and fusarium wilt resistance (Fujieda, 1994).

The main purpose for grafting watermelon has been to induce resistance to soil borne pathogens, such as root-knot nematodes, *Fusarium oxysporum* f. sp. *niveum*, and *Verticillium spp.* (Davis *et al.*, 2008; Lee *et al.*, 2010; Louws *et al.*, 2010). Watermelon has been grafted mainly onto Bottle Gourd (*L. siceraria*), Pumpkin (*Cucurbita pepo* L.), Squash (*C. moschata* Duch.), and interspecific hybrid Squash (*C. maxima* Duch. × *C. moschata* Duch.) (King *et al.*, 2010; Lee *et al.*, 2010). In an initial study on watermelon grafting, 'Crimson Tide' was grafted onto 10 different rootstocks and all grafted plants exhibited resistance to *Fusarium oxysporum* f. sp. *niveum*. Grafted plants had significantly greater vegetative growth, fruit yield and quality, except in cases of graft incompatibility (Yetioir *et al.*, 2003). A recent report showed that grafting melon onto squash interspecific hybrids can provide resistance not only to *F. oxysporum* f. sp. *melonis* race 1,2, but also to *Didymella bryoniae* (Fuckel) Rehm, the causal agent of gummy stem blight (Crino *et al.*, 2007). Among the rootstocks, squash and interspecific hybrids have been largely used for their vigorous root system and extreme temperature and *Fusarium* resistance (Lee *et al.*, 2010; Ling *et al.*, 2013; Miguel *et al.*, 2004). The rootstock 'Shintoza' and 'Super Shintoza', a *C. maxima* × *C. moschata* hybrid, provided resistance to watermelon when cultivated in Fusarium infested soils; the rootstocks increased fruit size and yield compared to those of nongrafted plants (Álvarez-Hernández *et al.*, 2015; Miguel *et al.*, 2004). Among these, the 'Super Shintoza' rootstocks also provide tolerance to *Verticillium* wilt, reducing the microsclerotia incidence, survival structure of *Verticillium* spp., and maintenance of fruit yield (Dabirian *et al.*, 2017).

Table 9: Rootstocks Used in Cucurbit Grafting

Rootstock	Targets	References
Cucumis melo: var. *flexuosus*, var. *conomon*, var. *agrestis*, Asian land races	Improved tolerance to *Monosporascus* vine decline	Fita *et al.* (2007); Jang *et al.* (2016)
	Tolerance to nematodes	Ito *et al.* (2014)
	Fruit quality	Condurso *et al.* (2012); Verzera *et al.* (2014)
	Salt tolerance	Dasgan *et al.* (2015)
	Tolerance to *Fusarium oxysporum*	Lee and Oda (2003)
Cucumis metuliferus	Management of *Meloidogyne incognita*	Siguenza *et al.* (2005); Kokalis-Burelle and Rosskopf (2011); Guan *et al.* (2014); Punithaveni *et al.* (2015)
Cucumis africanus, *Cucumis myriocarpus*, *Cucumis ficifolius*, *Cumis anguria*	Management of *M. incognita*	Pofu *et al.* (2011); Pofu *et al.* (2013)
Cucumis maxima × *Cucumis moschata*	Tolerance to *Fusarium* wilt	Trionfetti Nisini *et al.* (2002)
	Improved tolerance to vine decline diseases	Jifon and Crosby (2008)
	Improved tolerance to salinity	Orsini *et al.* (2013); Rouphael *et al.* (2012)
	Improved tolerance to boron	Edelstein *et al.* (2011)
	Improved tolerance to Fusarium wilt and fruit quality	Zhou *et al.* (2014)
	Improved response in nematode-infested soil	Goreta Ban *et al.* (2014)
	Management of *Verticillium*	Buller *et al.* (2013)
	Higher yield, larger fruit	Zhang *et al.* (2014)
	Water use efficiency	Kivi *et al.* (2014)
	Nitrogen use efficiency	Salar *et al.* (2015)
	Didymella bryoniae tolerance	Silva *et al.* (2012)
C. ficifolia	*Macrophomina* wilt tolerance	Cohen *et al.* (2012)
	Improved cold tolerance	Bulder *et al.* (1991); Zhou *et al.* (2009); Li *et al.* (2014a)
	Improved salt tolerance	Huang *et al.* (2010)
	Increase yield	Cheshmehmanesh *et al.* (2004); Hoyos Echebarria *et al.* (2001); Hernandez-Gonzalez *et al.* (2014)
C. moschata	*M. incognita* tolerance	Punithaveni *et al.* (2015)
	Improved salt tolerance	Liu *et al.* (2012)
	Tolerance to *F. oxysporum*	Traka-Mavrona *et al.* (2000)
	Tolerance to high pH	Roosta and Karimi (2012)
Cucurbita pepo	Improved vegetative growth (but reduced yield)	Bekhradi *et al.* (2011)
Cucurbita argyrosperma	Increased yield	Hernandez-Gonzalez *et al.* (2014)

Rootstock	Targets	References
Cucurbita martinezii	Grafting incompatibility	Huh *et al.* (2003)
Lagenaria siceraria	Tolerance to Fusarium wilt	Yetisir *et al.* (2007); Karaca *et al.* (2012)
	Tolerance to flooding	Yetisir *et al.* (2006)
	Improved tolerance to *Phytophthora capsici*	Kousik *et al.* (2012)
	Tolerance to salinity	Yang *et al.* (2013)
	Management of Verticillum wilt	Buller *et al.* (2013)
	Tolerance to nematodes	Ozarslandan *et al.* (2011)
	Tolerance to powdery mildew	Kousik *et al.* (2008)
	Fruit volatiles, quality and yield	Petropoulos *et al.* (2012); Guler *et al.* (2014)
	Plant development and fruit quality	Petropoulos *et al.* (2014); Bekhradi *et al.* (2011)
Luffa cylindrica	Improved heat tolerance	Li *et al.* (2014a, b); Yetisir and Sari (2004); Galatti *et al.* (2013)
Sicyos angulatus	Improved cold tolerance	Zhang *et al.* (2008); Bulder *et al.* (1991)
Citrullus spp.	Reduced negative impact on fruit quality	Huh *et al.* (2003); Selvi *et al.* (2013)
	M. incognita tolerance	Punithaveni *et al.* (2015)
Citrullus lanatus var. *citroides*	Tolerance to *Fusarium* wilt	Keinath and Hassell (2014)
	Tolerance to nematodes	Thies *et al.* (2010); Thies *et al.* (2015a)
	Melon necrotic spot virus tolerance	Huitron *et al.* (2007)
Benincasa hispida	Tolerance to *Fusarium* wilt	Trionfetti Nisini *et al.* (2002)
	Tolerance to nematodes	Ito *et al.* (2014); Aminand Mona (2014); Galatti *et al.* (2013)
	Tolerance to *Verticillum* wilt	Wimer *et al.* (2014)
Trichosanthes cucumerina	Tolerance to nematodes	Ito *et al.* (2014)
Sicana odorifera	Tolerance to nematodes	Ito *et al.* (2014)
Momordicacharantia	Tolerance to nematodes	El-Eslamboly and Deabes (2014)

Takahashi and Kawagoe (1971) found that certain rootstocks reduced *phytophthora* blight on cucumbers, and Wang *et al.* (2004) showed improved cucumber tolerance to this disease through grafting. It is probable that improved tolerance to *phytophthora* blight is all grafting can achieve, since no cucurbit compatible rootstock with genetic resistance for *P. capsici* has been reported (Babadoost, 2005). Increased tolerance to viral complexes (*CMV, WMV-II, PRSV* and *ZYMV*) of grafted seedless watermelons compared to nongrafted controls was reported by Wang *et al.* (2002). Another report demonstrated that grafted melon plants had improved tolerance to Melon Necrotic Spot Virus (*MNSV*) (Wang *et al.*, 2002). Similarly, in grafted tomato, plants were more tolerant to *Tomato Yellow Leaf Curl Virus* (Rivero *et al.*, 2003). It is presumed that the increased tolerance is due to increases in vigor, photosynthesis, chlorophyll content, and/or peroxidase activity associated with the grafted plants. It will be interesting to determine if new virus-resistant rootstocks, such as Ling and Levi's (2007) *ZYMV*-Florida strain-resistant bottle gourd rootstocks will confer virus resistance to the scion.

Grafted watermelon seedlings under low temperature stress have higher antioxidants and antioxidative enzyme activities in leaves than self-rooted watermelon seedlings (Liu *et al.*, 2003a; Liu *et al.*, 2004a). Nongrafted control watermelon and watermelon grafted onto wax gourd showed greater resistance to drought stress than watermelon grafted onto bottle gourd (Sakata *et al.*, 2007). Yang *et al.* (2006) demonstrated that grafted cucumber plants have higher net photosynthesis, stomatal conductance, and intercellular CO_2 concentrations under NaCl stress than self-rooted plants. Under salt stress, watermelon grafted onto salt-tolerant rootstock demonstrated improved vigor and yield, and maintained good fruit quality compared to watermelon-watermelon-grafted controls (Liu *et al.*, 2003b; 2004b, c). Increased photosynthesis can often be realized under less than optimal growing conditions, such as weak sunlight and low CO_2 content in solar green houses during winter months, allowing grafted plants to produce higher yields and sometimes improved fruit quality (Xu *et al.*, 2005; Xu *et al.*, 2006; Zhu *et al.*, 2006).

Cucumber plants grafted onto pumpkin rootstocks had 27 per cent more marketable fruit per plant than self-rooted cucumber (Seong *et al.*, 2003). A squash interspecific hybrid rootstock increased watermelon yield and increased fruit size (Miguel *et al.*, 2004). Yetisir and Sari (2003) and Yetisir *et al.* (2003) tested multiple rootstocks (*Cucurbita moschata*, *Cucurbita maxima*, squash interspecific hybrids, and bottle gourd) for effects on yield. Wax gourd-grafted and watermelon-grafted watermelon are affected by soil conditions, and often show decreased yield in areas of low fertility or moisture (Sakata *et al.*, 2007).

8.5.0 Pruning

Pruning is an important operation in cucurbitaceous vegetable production not only for high quality fresh harvest but also for maximum quality seed production. A plant of this group bears so many branches but all are not able to bear fruits. So, it is necessary to remove the branches from main stem for better flowering and fruiting. In case of cucumber, single stem should be kept up to 30 cm of plant height. In case of gourds, pruning must be done upto main stem of trellis.

8.6.0 Training

There are mainly four training systems that are adopted for training in cucurbitaceous vegetables.

8.6.1 Bower System

Bower is a criss-cross netting structure which provides a suitable platform/ condition expressing the full potential of growth and fruiting. Training of plant on bower system is beneficial practice for getting high yield of uniform quality

Grafting Methods of Cucurbits

Different Steps of Splice Grafting

Different Steps of Side Grafting

Different Steps of Approach Grafting

Different Steps of Hole Insertion Grafting

fruits of cucurbits. Since cucurbits are viny in nature, they bear more number of fruits for longer period on supporting structure. It has been observed that if vines are allowed on the ground, nearly 25-30 per cent less yield has been recorded and 8-10 per cent fruits become unmarketable due to misshaped and discolored. The produces obtained by bower system get steady market price and export quality throughout the growing period.

8.6.2 Kniffin System

Kniffin system is another important system by which quality fruit can be harvested for longer period. This system is prepared with the help of bamboo poles or wooden poles and connected by 16 gauge G.I. wire. The wooden poles are fixed in the soil at 45 cm depth and 5 m, apart along the length of rows. For strengthening each kniffin, 16 gauge G.I.wire stretched and fastened on the poles at 45 cm, 90 cm, and 135 cm height above the ground. On this system flowering and fruiting start very early and due to hanging of fruits freely length increase many folds.

8.6.3 Bush System

Under bush system, individual plants are trained or allowed to grow on waste bushes of dry cotton plants, pigeon pea plants, sarkanda, *etc*. These supporters are fixed near the plants and vines are allowed to develop.

8.6.4 Ground System

Earlier spreading of vines on the ground was the common practice to take the cucurbits crops. During summer season still this practice is quite useful because vines, tender leaves, flowers, and fruits flourish well by getting sufficient moisture and there is also saving.

8.7.0 Irrigation

In spring-summer crop, frequency of irrigation is very important, while in rainy season crop, irrigation may not be necessary at all, if rainfall is well distributed between July and September. Soil moisture is an important factor governing good germination in cucurbits. More often in those kinds like bottle gourd and bitter gourd, whose seed coat is thick, germination has to be ensured by mulching to prevent moisture evaporation. Generally, sprouted seeds are planted in spring-summer season and adequate moisture has to be maintained at the time of emergence. Usually pits, ridges or beds are irrigated a day or two prior to planting of seeds and the next irrigation preferably light, is given 4 or 5 days after sowing of seeds. This irrigation should not flood the hills where seeds have been sown and crust formation of the top soil should be prevented. Infact, mulching of the hills after sowing will be advisable to reduce moisture loss at the time of emergence and to prevent crust formation. It is always necessary to keep the moisture well maintained at the root zone, to promote rapid tap root development. Even if top soil is comparatively dry, the vine growth will not be affected, and the subterranean moisture at the tap root zone will be able to keep the vine vigorous.In fact, frequent irrigation does real harm to the vines. While no ready recommendation is possible, irrigation once in 5 or 6 days will be necessary depending upon the soil, location,

Growing of Cucumber in Open Field and Protected Structure

Field Growing of Watermelon

Field Growing of Muskmelon

Field Growing of Roundmelon

Field Growing of Bottle Gourd

Field Growing of Bitter Gourd

Field Growing of Ridge Gourd

Field Growing of Sponge Gourd

Field Growing of Ash Gourd

AVT-II on Ash gourd

Local Check

RCAG-15

Field Growing of Snake Gourd

Field Growing and Marketing of Pumpkin

Field Growing and Marketing of Pointed Gourd

Field Growing and Marketing of Teasle Gourd

Field Growing and Marketing of Ivy Gourd

temperature, *etc.* In river-bed system pot watering is practised in the early stages of germination and it is stopped after the roots touch the subterranean moisture region. Frequent irrigation also promotes excessive vegetative growth, especially in heavy soils, which needs to be avoided. Further, one precaution needs to be taken that application of water should be restricted to the base of the plant or root zone. In other words, irrigation water should not wet the vines or vegetative parts, especially when flowering, fruit set and fruit development is in progress. Frequent wetting of stems, leaves and developing fruits will promote diseases and rotting of fruits especially in melons. As far as possible the beds or inter-row spaces should be kept dry so that developing fruits are not damaged.

In rainy season crop, this is not possible and hence incidence of anthracnose and fruit fly are severe if microclimate near the vines is humid. Some of the rainy season crops like bitter gourd, sponge or bottle gourd are therefore, trailed over supports to prevent rotting of fruits.

Usually, irrigation frequency is reduced, when the fruits reach near maturity and completely stopped in the last stages of harvest. Especially in pumpkins, ash gourd and melons, irrigation frequency is reduced during harvesting stage. It is a common belief that melon fruits would be affected, if or these were irrigated too frequently.

In trials with melon cv. Ogen grown in green houses and plastic tunnels, drip irrigation was compared with sprinkler irrigation by Paunel *et al.* (1984) in Romania. They found that energy saving with the former system was upto 62 per cent greater than the latter one, and weed and disease incidences were lower. Also the crop could be fertilized via drip irrigation system, less water was used and the average yield was 21.4 t/ha compared with 17.6 t/ha with sprinkler irrigation. Bhella (1985) reported that early and total marketable yields of trickle irrigation were significantly higher than yields of the unirrigated control. However, fruits produced under trickle irrigation had significantly lower soluble solids than under unirrigated control. The effect of planting method on yield was non-significant.

In studies with cucumber cv. Monastyrski in Poland, cavities were found in very young fruits which were suggested to low soil moisture and high N-nutrition conditions (Elkner, 1985).

Profirev and Laptev (1986) found that more adventitious roots were formed, their ramification and length increased and consequently productivity was raised in watermelon when the crop was irrigated with magnetised water. In another trial, response of watermelon cv. Charleston Gray to factorial combinations of trickle irrigation or no irrigation and black polythene mulch or no mulch was studied by Bhella (1988) and it revealed from the results that greatest stem growth, early and total yield of the crop were obtained when plants were grown with polythene mulch in combination with trickle irrigation.

Drip irrigation is more suitable for cucumber cultivation than sprinkler irrigation. Drip irrigation in combination with a film mulch accelerated seedling development, thus leading to earlier yields and prolonged harvest periods (Kunzelmann and Paschold, 1999).

The water consumption of summer squash was determined by water table lysimeters in the Botucatu region (latitude 22°51'S, longitude 48°27'W and altitude 786 m), Sao Paulo, Brazil (Klosowski *et al.*, 1999). The crop had higher water consumption during the blossom and fruit development stages. Total water consumption, was 231.52 mm, in a crop season development of 70 days. The crop coefficient obtained, varied between 0.68 and 1.96.

8.8.0 Manuring and Fertilization

All the gourds, melons, *etc.*, respond well to manuring and fertilizer application. The doses of fertilizers depend upon the soil type, climate and system of cultivation. It is difficult to be specific about fertilizer recommendations because of variation in soil types, soil fertility and system of cultivation. Well rotten farm yard manure at the rate of 150 to 200 quintals per hectare is applied to the field at the time of preparation of land. This is supplemented by full dose of super phosphate and potash before sowing and one-half dose of nitrogen at the time of vining and the other 10-15days later. In cucurbits, excessive nitrogen and consequently enormous vine growth require to be avoided. In general, high N under high temperature conditions promote maleness in flowering and number of female/perfect flowers per vine gets reduced resulting in low fruit set and low yield. In fact, excessive vine growth can be pruned manually to promote higher female to male ratio. This is not practicable in commercial growing and the best way to control the vine growth within reasonable limits is by adjusting fertilizer doses and frequency of irrigation. It is better to complete all the fertilizer applications just before the fruit set. Barring melons and pumpkins, where fruits are harvested at full maturity, in other cucurbits there is simultaneous flowering, fruit set and fruit picking, the vines will have to remain vigorous, so that larger numbers of fruits are harvested per vine.

Another consideration is the leaching of nutrients in the sandy loam soils and especially in river-beds. This can be partially remedied by foliar sprays of urea and it can be mixed with insecticides. Further, there are preliminary indications that some Cucurbit leaves do fix atmospheric nitrogen and some micro-organisms in the phyllosphere of most of the cucurbits, have been associated with active N_2 fixing properties (Sen *et al.*, 1983). This explains the higher nitrogen content of leaves than in the soils, on which they grow.

8.8.1 Cucumber

The deficiency of N results in marked reduction in growth of vines with leaves turning pale yellow-green in colour, the stem becomes thin and woody. P deficiency causes stunted growth and reduction in leaf size while chlorosis, bronzing and marginal scorch appears on the leaves of K deficient plants.

In heated green house, trials with cv. F_1 TSKhA-211$_2$, V'Yukov (1984) applied simple or compound fertilizer in solution to the soil surface or by sub-soil drip irrigation. Sub-soil application resulted in better plant development and a 20:16:10 (N: P: K) fertilizer gave higher yield than simple fertilizer applied by the same method. Results of solution culture studies in cucumber by Sen Yan *et al.* (1996) showed that vegetative growth was best with 100 per cent NO_3^--N,when NH_4^+-N was 25 per cent or 50 per cent of total N, reproductive growth was promoted and the percentage of female flowers and the yield/plant were increased compared with plants receiving 100 per cent NO_3^--N. Zhang *et al.* (1991) found that yield of cucumber cv. Jin Za No. 1 was improved by increasing the proportion of nitrogen as NH_4 form upto 50 per cent.

Hanna and Adams (1993) reported from USA that for staked cucumber 600-800 lb NPK (13-13-13)/acre pre-planting was needed to increase yields significantly on less fertile soils. Weekly foliar fertilization with a seaweed extract containing 9 per cent N, 9 per cent P and 7 per cent K at 0.2 gallons/acre for 8 weeks increased total yields.

Florescu *et al.* (1991) observed that total yields of cucumber (cv. Plishka) was heaviest in unheated green house with 30t/ha urban-waste compost. Fruits grown with compost had higher contents of vitamin C, carbohydrates, K and Mg, less dry matter and less acidity than fruits grown with FYM.

Schon and Compton (1997) concluded that under Florida conditions, when cucumber was grown in rock wool using a double-stem training technique, N and P should be provided at 275 and 75mg/l, respectively, to minimize depletion of these nutrients from the growing medium. Average yield of cucumber cv. Himangi were highest with 150 kg N+50 kg P+50 kg K/ha in Maharashtra, India (Patil *et al.*, 1998).

Cole and Lesaint (1978) and EI-Beheidi *et al.* (1978) reported that spraying cucumber plants with 0.01 per cent, borax, zinc sulphate or manganese sulphate stimulated growth, advanced flowering, enhanced production of male and female flowers and improved seed yield and quality. Seed yields and carbohydrate, nitrogen and protein contents of seed were highest with Zn treatment. EI-Behairy *et al.* (1997) stressed the importance of reducing the Zn concentration in the nutrient solution to give better uptake of P and Mn in cucumbers for better growth in NFT.

Nutritional diagnosis using petiole nitrate levels was successful in cucumber when petioles of leaves at the 14-16th node were used. With rapid growth, the optimum levels of nitrates in petioles were 800-1200 ppm at the early stage of harvest, 200-400 ppm at mid-harvest and 100-300 ppm during the late stage of harvest. With slower growth, the optimum levels was 1000-1200 ppm for all stages (Roppongi, 1991).

In cucumber, Tuncay *et al.* (1999) reported significant correlations between leaf nutrient contents and fruit diameter, fruit length, TSS, titratable acidity, pH, dry matter, fruit firmness and colour as quality characteristics. K had significant positive direct effects on all of the quality traits with the exception of dry matter, which was affected positively by P. Leaf Ca content had negative direct effects on all of the quality traits determined in the study.

High total N and amino acid-N levels in the leaves induce bitterness in leaves and fruits of cucumber by promoting N metabolism, which in turn favours the enzymatic synthesis of cucurbitacin C, the bitter factor (Kano *et al.*, 1999).

Nitrogen fertilization is important for vegetative growth of plants and needs to be managed properly to achieve a balance between generative and vegetative growth. High nitrogen fertilization leads to excessive vegetative growth and reduction in reproductive growth. It also delays and/or reduces female flower development in cucurbitaceous plants (Abd El-Fattah and Sorial, 2000; Agbaje *et al.*, 2012). Potassium fertilization and the application of biofertilizers can increase female flower production in cucumber and squash plants (Abd El-Fattah and Sorial, 2000).

Umamaheswarappa *et al.* (2005) reported that nitrogen application showed a significant effect on number of days required for initiation of earliness in flowering, number of days required for first fruit set whereas phosphorus application also exhibited positive effect on flower initiation, development whereas potassium application had no substantial effect on flowering, fruit set of cucumber in cv. Poinsette.

Jilani *et al.* (2009) stated that application of NPK fertilizer (100-50-50) in cucumber induced earliness in flowering and fruiting, least days for flowering, fruit setting, maturity, Arshad *et al.* (2014) concluded that the application of NPK as fertigation resulted in early flowering, fruiting with maximum number of fruits per plant, more weight, length of fruit and higher yield in cucumber.

Anjanappa *et al.* (2012) revealed that the plants applied with 75 per cent RDF + 75 per cent FYM + Azotobacter + Phospobacteria + Trichoderma resulted in earliness and higher productivity of cucumber crop.

Mohan *et al.* (2016) described that 60 per cent each of RDF and vermicompost along with Azotobacter, Trichoderma and PSB were found to be superior among all the combinations of organic, inorganic and biofertilizer sources of nutrients for characters, namely minimum number of days to 50 per cent flowering, average fruit length, fruit weight, edible fruit count and maximum edible fruit yield in cucumber.

8.8.2 Watermelon

Sundstrom *et al.* (1983) recorded the maximum yield by applying 112 kg N/ha at a soil pH of 6.0. Khristov (1984) suggested the optimum economic rate at 140 kg N and 120 kg P_2O_5/ha. Som *et al.* (1986) recorded maximum growth and yield by application of 40 kg each of N and P_2O_5/ha. Alikhan *et al.* (1986) reported that fruit quality of watermelon was improved by spraying of 3 ppm B, 3 ppm Mo and 20 ppm Ca, twice at 2 and 4-leaf stages. Oga and Umekwe (2013) suggested that NPK fertilizer significantly affected the vine length, flowering, fruiting and marketable yield in watermelon. Audi *et al.* (2013) indicated that application of 6.0 ton ha⁻¹ cattle manure + 405 kg ha⁻¹ CAN, in all cases, either in combination or alone, resulted in a significant increase in qualitative and quantitative parameters which led to higher yield in watermelon.

8.8.3 Muskmelon

Prabhakar *et al.* (1985) obtained highest yield (63.82 q/ha) in cv. Hara Madhu by applying 100 kg N and 60 kg P/ha. Application of 60 kg K/ha was found to increase the yield by another 16 per cent. Bhella and Wilcox (1986) reported that highest total yields were produced with 67 kg N/ha pre-plant plus 50 to 100 ppm post-plant N fertigation. In cv. Hara Madhu, Sidhu *et al.* (1984) obtained highest yield (123.51

q/ha) by spraying Agromin at the rate of 2 kg/ha (containing Zn, Cu, Mn, Mg, B and Mo) at 2-leaf stage and again at 6-leaf stage, the control yielded 104.64 q/ha.

8.8.4 Bottle Gourd

Singh *et al.* (2012) revealed results of three years experiment that bottle gourd responded to the application of vermicompost @ 2.5 t + 50 per cent RDF, which registered higher fruits yield over other nutrients combination. In Southern Italy, Damato *et al.* (1998) obtained highest marketable yield of *L. siceraria* cv. Locale with N fertilizer applications of 100 or 150 kg/ha (average of 20.4 t/ha in 1994 and 15.2 t/ha in 1995). The 150 kg N/ha treatment also reduced the time taken to achieve 25 per cent of marketable yield by 5 to 6 days. Das *et al.* (2015) concluded that equal amount of N+ Organic sources + *Azotobacter* and PSB gave maximum primary branch count, fruit count, average fruit weight and fruit size in bottle gourd.

8.8.5 Bitter Gourd

According to Rajput and Gautam (1995), seed yields were highest in bitter gourd with 60 kg N and with 80 kg P/ha. Application of 40 kg N, 30 kg P and 60 kg K produced highest seed yield in cv. MDU-1 (Vijayakumar *et al.*, 1995). In Peshawar, Pakistan, highest yield (24.90 t/ha) was recorded in bittergourd with 80 kg N/ha (Ali *et al.*, 1995).

Samdyan *et al.* (1994) reported that 75 kg N/ha + 50 mg MH/I gave the thickest and heaviest rind and thickest flesh, where as 25 kg N/ha + 50 kg MH/l gave the highest rind: flesh ratio. The highest flesh weight and drymatter, ascorbic acid and TSS contents were obtained with 50 kg N/ha + 250 mg Cycocel/1. Prasad *et al.* (2009) concluded that the application of inorganic nitrogen but in combination with the azotobacter + PSB biofertilizers resulted in healthy and maximum yield in bitter gourd.

8.8.6 Indian Squash (Tinda)

Natchathra *et al.* (2017) revealed that 75 per cent of NPK along with vermicompost @ 2.5 t ha^{-1} combined with azospirillum and phosphobacteria @ 2 kg ha^{-1} in tinda had good potential to promote and improve growth parameters due to an effective and alternative source of macro and micronutrients in crop of tinda.

8.8.7 Squash

Reiners and Riggs (1997) reported from USA that marketable yield of pumpkin (*C. pepo*) was not affected by 3 rates of applied N (67, 112 or 157 kg/ha). Stephano *et al.* (1995) from Sao Paulo, Brazil reported that yields of summer squash (cv. Caserta) were highest with NPK (4-14-8,100 g/plant) + humus (1kg/hill) + B (1g borax/hill or 0.03 per cent borax applied twice as a foliar spray).

8.8.8 Pointed Gourd (Parwal)

Das *et al.* (1987) observed that plant growth and yield increased with rising N: P rates, with the maximum average early yield (45.9 q/ha) and total yield (138.8 q/ha) being obtained at 90:60 kg/ha.

In parwal, Misra *et al.* (1994) obtained highest yield (114.70 and 68.73 q/ha in the first and second season respectively) and largest fruits with 150 kg N + 80 kg P$_2$0$_5$+80 kg K$_2$0/ha. Nayak *et al.* (2016) stated that recommended dose of chemical fertilizer which was applied to pointed gourd in conjunction with biofertilizer and vermicompost in presence of lime, resulted in improvement the quality of the produce (fruits) without hampering the yield potential of the crop.

8.8.9 Spine Gourd

Goswami and Sharma (1997) reported that fruit yield of spine gourd (*M. dioica*) increased as both P (from 0 to 60 kg P$_2$0$_5$/ha) and K rates (from 0 to 75 kg K$_2$O/ha) increased but there was no significant interaction between P and K. K rate had no effect on fruit ascorbic acid content but this was highest when P was applied at 40 kg P$_2$0$_5$/ha.

8.8.10 Sponge Gourd

In cv. Pusa Chikni, application of N at 50 kg + P at 20 kg/ha in the summer season and N at 25 kg + P at 40 kg/ha in the winter season produced the highest female flowers of 15.1/plant and 21.8/plant, respectively (Arora and Siyag, 1989).

8.8.11 Ridge Gourd

Arora *et al.* (1995) recorded highest vitamin C (ascorbic acid) and total soluble solids content with 60 kg N/ha and highest fruit drymatter (DM) content with 90 kg N/ha.

Sreenivas *et al.* (2000) recommended vermicompost at 10 t/ha+N: P: K at 50:25:25 kg/ha for optimizing yield and quality of ridge gourd.

8.8.12 Pumpkin

Swiader and Al-Redhaiman (1998) suggested that near maximum fruit yields of dryland and fertigated pumpkins can be expected when petiole-sap NO$_3$-N concentrations during early fruiting are ≈900-1500 µg/ml and those during fruit enlargement and ripening are ≈500-700 µg/ml.

The role of micronutrients in modification of sex ratio in some cucurbits has been studied by several authors. Foliar application of Ca at 20 ppm significantly increased fruit yield which was attributed to be due to increase in hermaphrodite flowers. Besides, Ca 20 ppm and boron 4 ppm improved fruit quality, especially TSS content, ascorbic acid content, flesh appearance and juiciness in muskmelon (Randhawa and Singh, 1970). In watermelon, boron and molybdenum at 3 ppm and calcium at 20 ppm proved to be effective in inducing greater number of female flowers and increasing fruiset and fruit yield (Choudhury, 1966). In bottle gourd, Choudhury (1979) reported the effect of boron at 2-3 ppm as most significant in increasing the number of female flowers, total number of flowers and fruit yield.

8.9.0 Rotations

In garden land areas, in contrast to river-beds, there are definite systems of growing cucurbits in rotation and mixed cropping. In Eastern Uttar Pradesh, melons are grown from February to May, as a pure crop in rotation after potato or garden pea and followed by maize in Varanasi and Jaunpur districts. In Lucknow region, Lucknow Safed a muskmelon is grown after paddy during January to May. In Eastern India also, bottle gourd and pumpkin are grown in rice fallows after 'aman' paddy. In some districts of Andhra Pradesh, watermelons are grown after rice from December to April. In rice fallows of Cuddapah district of Andhra Pradesh, there is a small well defined pocket of growing muskmelon between February-May, the fruits of which reach the market after the river-bed crops are over. Such crops require adequate irrigation facilities during summer months.

Mixed cropping of melons ininitial stages of sugarcane is done in Eastern U.P. and in Bhatinda region of Punjab mixed cropping with cotton is prevalent. Usually cucurbits are sown in relay system just before digging of potatoes in late January or early February in North India and crops like bottle gourd and pumpkin derive residuary benefit of heavy fertilizers applied to potato. *Phoot* or snapmelon is invariably grown as mixed crop along with sorghum or maize during rainy season.

Studies on the influence of different companion crops on growth and yield of pointed gourd showed that yields were highest (105.62-113.11q/ha) with spinach beet and lowest with tomatoes (38.29-43.56 q/ha) while net returns were highest from pointed gourd with onions followed by spinach beet (Maity *et al.*,1995).

The valley of Ujarras, Costa Rica, is one of the most important regions of chayote production in the world where it is grown in mixecropping systems with other vegetables (tomatoes, *Capsicum, Phaseolus,* cabbages and cucumbers) (Gambao *et al.*, 1996).

8.10.0 Intercutlure

The cucurbits donot require much attention by way of interculture. In the early stages before they start vining, the beds, ridges, *etc.*, require to be kept free from weeds. At the time of top dressing of nitrogenous fertilizers, weeding and earthing up are done. When the vines start spreading, weeding in between the rows or ridges, becomes unnecessary since vine growth can smoother the weed. Further, in snake gourd and perennial cucurbits like chow-chow, or ivy gourd, vines have to be trained over trellises or supports, weeding of the pits are alone necessary. Under the canopy of vine growth in pandals, weed growth down below, is nearly arrested. The only attention is that big weed has to be manually pulled out, without disturbing the vines, at later stages. Vertical training increased cucumber yields and quality. The increased yields were attributed to increased fruit set and development to marketable size and reduction in fruit rot (Hanna and Adams, 1993).

Staking trial on cucumber cv. Bhaktapur Local for off season production in Nepal revealed that on average staking with jute string gave 5.6 and 29 per cent more marketable fruits than the farmers 'practice of staking (bamboo sticks or tree branches)' and no staking, respectively (Jaiswal *et al.*, 1997).

8.11.0 Weed Control

Application of 2 kg alachlor/ha, 4 kg chloramben or 4 kg asulam/ha was effective to control the weeds in *C. maxima* (Rahman *et al.*, 1985); Scheffer and Wood (1985) also suggested alachlor at 1.5 kg/ha in *C. maxima*.

Application of 0.8 kg a.i. treflan/ha 18 days before sowing cucumbers gave effective control of weeds and increased yields (Dyachenko and Zhovner, 1986).

In a trial on different herbicides in muskmelon, Sandhu *et al.* (1986) reported that either 1.2 kg fluchloralin as pre-sowing treatment, 0.48 kg fluchloralin + 0.5 kg nitrofen as pre-emergence or 1.25 kg nitrofen/ha were effective for weed control.

In pointed gourd crop yield was greatest (11.8 t/ha) with gramoxone (paraquat) at 1.0 l/ha, followed by hand weeding and fernoxone (2, 4-D) at 1.0 l/ ha (Chattopadhyay *et al.*, 1997).

Field trials were conducted at Lafayette, USA by Birge *et al.* (1996) to assess the efficacy of herbicides, plastic mulch and cover crops for weed control in transplanted pumpkins cv. Howden. Results indicated that weed control persisted for ≤ 4weeks where ethalfluralin (at 1.68 kg/ha) had been applied before transplanting pumpkins, but crop vigour was reduced for ≥ 4 weeks. Excellent control with no crop injury was achieved in the black plastic mulch treatment.

8.12.0 Mulching

Vigour, yield, earliness and quality in melon cv. Honeydew were increased with mulching and the soil temperature at 15 cm under mulch was higher and uniform, resulting in strong growth (Izquierdo and Menendez, 1980). However, Basky (1984) advocated the use of blue and transparent plastic mulch along with application of a gride x EG318 and atplus 411 oil sprays for seed cucumber with a view to decrease the incidence of virus. The treatment reduced the incidence by 83 per cent compared with the untreated plots at the end of the season. A black mulch was less effective in this respect. Yang (1984) observed that plastic mulch was effective in inhibiting the upward movement of soil salts, resulting in better seedling survival and improved yields in watermelon.

The effect of using red or black plastic mulch on the yield of cucumber cultivar Vista Alegre was studied in Brazil by Campos De Araujo *et al.* (1992). The red plastic mulch treatment produced the best yield of 60.27 t/ha. The yield with black plastic mulch was 47.03 t/ha and with no mulch 42.33 t/ha. According to Farias-Larios *et al.* (1994) however, clear plastic mulches increased yield and reduced number of days to flowering and first harvest.

In USA, Brown and Channel-Butcher (1999) reported that flat surface soil with black plastic mulch + spun bonded polyester (FBPM + SPE) row cover and Fl at surface soil with black plastic mulch (FBPM) treatments increased total watermelon (cv. AU Producer) yields by 105 per cent and 111 per cent over those produced on flat surface bare soil (FBS). Treatments of hilled surface soil with black plastic mulch + spun bonded polyester (HBPM + SPE) and flat surface soil with black plastic mulch + spun bonded polyester (FBPM + SPE) produced a higher number of early-matured melons than those on bare soil. It (HBPM + SPE and FBPM + SPE treatments) caused 55 per cent and 32 per cent of the total number of fruits produced to mature 2-3 weeks earlier than those grown FBS.

In Poland, Siwek and Kunicki (1998) reported that transparent and black mulches gave 76.3 per cent and 66.1 per cent increase yield over unmulched control when cucumber cv. Marinda F_1 is grown with protected low tunnels.

In pumpkins, Waterer (2000) reported from Canada that plastic' mulches improved stand establishment and fruit yields relative to non-mulched control. Clear mulch was superior to black plastic in some cases.

In field trials by Ibarra and Flores (1997) in Mexico, *C. pepo* cv. Gray Zucchini was sown directly and watermelon cv. Charleston Gray was transplanted into beds in early June with or without black plastic mulch and/or introduction of row covers 5 days later. The greatest benefit was obtained with black plastic mulch alone which increased yields of cv. Charleston Gray by 208 per cent compared with untreated control (13.6 t/ha) and of cv. Gray Zucchini by 177 per cent compared with control (13.1 t/ha) and also reduced the period to harvest by 9 and 2 days, respectively.

Comparison between silver spray mulches and polyethylene film mulches in delaying the onset of aphid-transmitted virus diseases in Zucchini squash (*C. pepo*) revealed that water soluble bio-degradable silver spray may be advantageous over polyethylene films because they may be incorporated into the soil at the end of the season, rather than requiring removal and disposal in and fill (Summers *et al.*, 1995). In an experiment in Germany, cucumber cultivar Stimora was planted on two biodegradable black mulching foils (Walocomp and Ecoflex) and a polyethylene (PE) and mulch-free control by Weber (2000) and observed that yields were highest on PE foil, followed by Ecoflex. He concluded that while yields on biodegradable foils are slightly lower than on PE foils, they are still significantly higher than under cultivation without foil. Brown *et al.* (1996) observed that reflective mulches increase yield, reduce aphids and delay infection of mosaic virus in summer squash (*C. pepo*). Mansour *et al.* (2000) reported from Jordan that aluminium foil whether surface mulched or as boards reduced aphid-borne mosaic virus disease incidence and symptom severity at early stages of squash (*C. pepo*) growth and eventually increased the total yields of squash with a higher marketable fruit yield as compared to the control treatment.

8.13.0 River-bed System

In India, growing of cucurbitaceous vegetables in river-beds or river basins constitute a distinct type of farming. These areas are familiarly called *'diara'* land in UP and Bihar. The river-beds of Yamuna, Ganges, Gomti, Saryu and other distributaries in Haryana, Uttar Pradesh and Bihar; Banas river-bed in Tonk district of Rajasthan; Narmada, Tawa and Tapti river-beds of Madhya Pradesh and Maharashtra; Sabarmati, Panam, Vartak and Orsung of Gujarat; Tungabhadra, Krishna, Hundri, Pennar river beds of Andhra Pradesh are some of the important areas where cucurbits like bottle gourd, pumpkin, melons, *etc.*, are extensively grown. Even in Kerala, in Pamba and Manimala river-bed bitter gourd, snake gourd, *etc.*, are grown. These river-beds are formed and subject to alluvion and diluvion action of perennial Himalayan rivers and due to inundation caused by swollen rivers during South-West monsoon. Fresh silt and clay deposits received every year, during the monsoon months, especially in Himalayan Rivers, makes these lands suitable for growing vegetable crops, literally on sand. Even though upper layers of land seem unsuitable for growing crops, the subterranean moisture seeped from adjacent river streams, makes it possible to grow early crops. This system is unconnected with any other crop rotation and cucurbits are specially adapted to this system of growing due to their long tap root system. It can be treated as a kind of vegetable forcing where in the cucurbits are grown under sub-normal conditions, literally on sand, during winter months from November-February, especially in North and North-western India.

The system consists of digging trenches at 2-3 metres spacing or pits at 4 metres spacing (if the soil moisture is below 2 metres) after the cessation of South-West monsoon in late October. Most of the cucurbits are sown in November-December. Before sowing, the trenches are manured with FYM or any other organic decomposed waste or oilcakes. Especially in North-western India, where winter temperatures in December-January go down to 1 to 2°C, the protection is provided by planting of grass stubbles (probably of *Saccharum* sp.). This protection has three-fold uses: (i) it checks the sand drifting on the dug-up trenches and covering the hills sown with seeds, (ii) it provides partial though insufficient protection against chilly winds, and (iii) this grass is available for spreading over the sand when the vines grow and cover the sand. It helps to prevent the sand being blown off with the vines, especially when '*loo*' or hot summer winds sweep these areas in May. Due to prevalent low temperature, sprouted seeds are sown in trenches or pits in November-December and mixed cropping of several cucurbits like muskmelon, watermelon, pumpkin, bottle gourd, ridge and sponge gourds, is practised. In Bihar, pointed gourd is also grown in river-beds, where rooted cuttings are transplanted. Plants are pot watered in the initial stages, until roots touch the water regime down below. In some parts of Andhra Pradesh, germinated young seedlings are transplanted. After 25 to 30 days of sowing, depending on the growth and weather conditions, top-dressing in two split doses is done, especially of nitrogenous fertilizers like urea or fertilizer mixture. This top dressing is applied away from the plants in shallow side trenches. Cakes like groundnut cake in Maharashtra and Andhra Pradesh or cast or cake in Gujarat are also applied.

When the vines grow, they are spread over the sand and before that trenches are levelled up and the stubbles of grass are spread over the sand on the inter spaces between the rows of plants. The melon fruits come to the market in February-March, in Andhra Pradesh, slightly later in April in Maharashtra and Rajasthan, extending to June in West Uttar Pradesh and Haryana. Other cucurbits come earlier because they are picked in edible (vegetable) maturity stage like bottle gourd in March. The pumpkin is the last to arrive, since it is of longer duration. In years of unseasonal rainfall or floods in the rivers, the crop gets damaged, and resowing has to be done. The mixed cropping has some advantages in that it gives river-bed farmer continuous income from March to June and cushions losses or failure of any one crop. But diseases like mildews and viruses can spread from one cucurbit to another and their control becomes difficult.

River bed culture of melons has been developed through the native ingenuity of the farmers. They keep the seeds of the fruits which are found sweet in the current season, unmindful and unwary of the fact that most of these fruits are produced by cross pollination before selection of the fruit for seed extraction. That is the reason why the fruits coming from river-beds are of undependable quality, especially in sweetness and flesh colour, of which urban consumers often complain. Since each farmer keeps his own seeds, there is no guarantee that a cultivar will remain pure in river-bed culture for more than one season. There is a blessing in disguise amidst this chaotic situation of mixture of cultivars of muskmelon and watermelons in river-beds. It has enabled the perpetuation of natural variability and there has been a continuous process of recombinations and selection in voluntarily promoted by the farmers. Another interesting feature is that despite the effects of cross pollination on the maintenance of varietal purity, there are still recognizable typical phenotypes maintained by the farmers in each melon growing region or pocket. This has been made possible because of some degree of natural self-pollination that is taking place without conscious effect of the farmers. This has enabled perpetuation of some characteristics cultivars, which have stable external characteristics but of variable internal quality.

The yield from river-beds of different crops is naturally variable depending upon the location and the number of cucurbits constituting mixed cropping. The river-bed system is comparatively free from major disease epidemics. Powdery mildew and downy mildew have now become serious in several river-beds. Stray incidence of *fusarium* wilt, besides anthracnose and angular leaf spot is also noted. Virus diseases have been found to be common and three or four strains with a wide host range, like *CMV*, melon mild mosaic have been noted. The pests like aphids and red pumpkin beetle are usually noted in early stages of the crop. The fruit fly incidence is more in pointed gourd and mite infestation increases in arid situations, when the day temperature rises to 40°C. Non-pathogenic diseases, especially mineral deficiencies, are also prevalent in some situations. Absence of rich subsoil, silt or alluvium beneath the sandy layer and leaching of nutrients due to sandy substrata sometimes cause deficiencies of macro- and micro-nutrients.

9.0 Growing Cucurbits under Protection

In Western Europe, Japan and other countries, summer months (April-September) are usually cloudy with frequent spells of rains, making it difficult to grow cucumbers, especially salad or slicing cucumbers. Hence protected cultivation has come into vogue and green house culture of cucumbers is a highly specialized system of vegetable forcing. In this culture, temperature, humidity, light and carbon dioxide inside the glass house are controlled to provide ideal conditions under computerized microprocessor systems. High initial capital outlay due to sophisticated control of green house environment, long duration of the crop from December to July or early August and intensive care and attention needed in the management, all contribute to production of high quality cucumber of uniform dark green cylindrical fruits with minimum spines and parthenocarpic development.

In Japan, the green house industry has achieved a phenomenal increase during the last four decades. Production is mostly confined to southern Islands, with about 25 per cent located near Tokyo and 40 per cent in southern Islands of Kyushu and

Shikoku. The vast majority of the houses are covered with plastic (polyvinyl chloride or polyethylene) sheets, usually called 'polytunnels'. Most of them are heated with hot air from oil or kerosene fuelled heaters. Holdings are usually small, about 0.5 to 1 acre and cucumbers and melons are usually grown besides tomatoes, strawberries, eggplant and capsicums (peppers). There are both permanent and temporary plastic houses. In the latter case, after the double cropping of rice fields, growers install in expensive pipe frame supports over the harvested paddy fields.

These plastic houses cater to winter production of vegetables, while in permanent plastic houses, cantaloupes, watermelons, *etc.*, are grown and they are cropped for nearly 9 months in a year. In plastic house, 5.5 plants/m² and 2-row arrangement with triangular configuration increased the yield of cucumber substantially (Kasrawi, 1989).

9.1.0 Off-season Cultivation of Cucurbits

Cucurbit crops muskmelon, round melon, long melon, cucumber, bitter gourd and bottle ground *etc.*, if produced one or one and half month early than the normal season during summer, often command a greater price on the market.Accordingly, low-cost polyhouse technology to grow off-season nursery of cucurbitaceous vegetables for raising extra early crop in spring-summer season in north Indian plains was developed by Sirohi (1994). The crop can be taken one to one and a half months in advance than normal method of direct seed sowing and farmers can get a remunerative price by selling the product during off season. The utilization of plastic low tunnels (row covers) and plastic mulching for off-season muskmelon, watermelon and summer squash production is a common practice in USA, Israel and in some European countries for fetching high price of the produce. Row covers or low tunnels are flexible transparent covering that are installed over the rows or individual beds of transplanted vegetables to enhance plant growth by warming the air around the plants in the open field during winter. They can also warm the soil and protect the plants from hails, cold wind, injury and advance the crop by 30 to 45 days than the normal season. Standardisation agrotechnique for off-season nursery raising and cultivation of cucurbits under high to medium cost polyhouse, low plastic tunnels with drip irrigation, fertigation schedule was reported by Singh *et al.* (2007, 2010).

River bed cultivation is an old practice for production of certain cucurbits in off-season in our country, but area for river bed cultivation is very limited.

9.1.1 Growing Structures

Campiotti *et al.* (1991) reported highest cumulative yield for both cucumber and Zucchini (*C. pepo*) in Italy under non-heated green house with PVC covering in the spring and double layer of polyethylene (PE) covering in autumn. For cucumber production in climatic conditions similar to South West Ontario, Canada, the double inflated polyethylene film green house is strongly recommended, because there is no loss of productivity in comparison with a green house clad with single-glass (glass-clad), while great savings on initial investment and energy use are achieved (Papadopoulos and Hao, 1997).

A study was carried out in Turkey by Uzun *et al.* (1999) to determine the effects of perforated and standard plastic covering materials and different mulching materials on the growth and yield of cucumber cultivars Cengelkoy and Hylarcs F_1. In Cengelkoy, the highest early yield was obtained in low tunnels covered with plastic having 40 perforations/m^2 and combined with plastic mulch, while in Hylares, low plastic tunnels with 40 perforations/m^2 and straw mulch gave the highest value. Three parthenocarpic cucumber cultivars (Marinda F_1, a pickling type with warted, coarse spined fruits, Othello F_1, a pickling type with soft-spined fruits and Gracius F_1, a slicing type with smooth fruits) were grown in an experiment in the open field near Karkow Poland by Siwek and Capecka (1999) under nonwoven polypropylene (PP) cover and in a polyethylene (PE) tunnel. Vegetative growth was greatest in plants in the tunnel where the thermal conditions were best. Early and total marketable yields were highest under the PE tunnel for all the cultivars. Yields under the PP cover were lower but exceeded those in the open field several fold. Yields were highest from Othello which was slightly earlier than Marinda. In a green house trial in France, when cucumber cv. Regina was grown with misting system to maintain relative humidity at 70 per cent, the total class 1 and class 2 fruits were increased from 33.3 to 42.3kg/m^2 and plant losses were decreased (Pelletier, 1989). In plastic house under Egyptian conditions, total and early yields per plant and per m^2 of cucumber F_1 hybrids were recorded for each month of the year. Overall, the highest prices and early and total yields were obtained in April and May from Cordito, followed by Maram (El-Doweny *et al.*, 1993).

9.2.0 Growing Environment

Temperatures maintained are 18.9°C during night, 21.1°C during day and 26.6°C during ventilation. Most of the glass houses are fitted with heating systems both ducted warm air systems and hot water systems. Temperature maintenance and opening of ventilators according to atmospheric conditions are all computerized.

A green house trial was conducted by Krug and Liebig (1991) to investigate the effect of soil temperature on vegetative growth and fruit production of cucumber cv. Pepinex. From the results, the optimum soil temperature recommended for the 1st (1 week after planting), 2nd (until anthesis) and 3rd (until harvesting) growth phases 24, 16-18 and 15-18°C, respectively, and the minimum recommended temperatures for the 3 phases were 16, 14-16 and 12-18°C, respectively.

Lu *et al.* (1996) suggested from the photosynthesis rate of cucumber that during the coldest period in winter, a high day time temperature should be maintained in the solar plastic green house as this is beneficial for growth and yield formation. Further, reducing the plant density will increase the photosynthetic productivity of cucumber.

Uffelen (1986) compared the effects of night/day temperature regimes of 21/21°C or 17/25°C and observed that cucumber per node was almost double with lower day temperature. Short branches with many fruitlets may develop and pruning of these shoots has been suggested as a means of correcting the balance between growths and cropping. A hot house trial with cucumber, cv. Venture, set on substrate mat at different day/night temperature (of 26/20°C, 21/21°C and 18/22.5°C) revealed that fruit numbers/plant were highest with regimes of 23/20°C or 21/21°C. The highest day temperature advanced the time of harvest and improved shelf life, but reduced average fruit weight (Uffelen, 1989). Carbon dioxide enrichment has been found to be beneficial in increasing the yield of cucumber. The decomposition of straw will give out CO_2 but enrichment is also done at 1000 ppm from half an hour before sunrise to one and a half hours before sunset.

Heij and Uffelen (1984) obtained the highest yield of cucumber in Netherlands when the plants were grown in green house at a CO_2 concentration of 790 ppm, both at two different dates of planting during the first 6 weeks and first 11 weeks of harvesting, respectively.

Glas (1985) reported that early yield was highest with the early sown plants with normal CO_2 concentration. Nederhoff (1987) was of the view that regulated CO_2 supply according to computer controlled economically optimal programme was effective in 30 per cent higher yield over a continuous supply of CO_2. In Norway, the yield of cucumber cvs. Cordoba and Mustang increased almost linearly with the level of illumination and a combination of lighting at 16 klx, spacing at 0.4 m and CO_2 enrichment at 700 ppm gave the highest cumulative yield (Grimstad, 1991).

The use of supplementary CO_2 advanced harvest date and increased yield (Slobbe, 1965). Maintaining the CO_2 content in the green house at 0.1 per cent was found to increase the yield of cucumber by about 10 kg/m^2 (Anon., 1963). Uffelen (1971) reported that higher CO_2 concentration (0.21 per cent) promoted early yield by 20 per cent and with 0.07 per cent CO_2 by 11 per cent compared with plants which had not received any supplementary O_2.

In China, Yang Yang (1998) observed that cucumber (cv. Chang chun Mici) plants grown in a solar green house with CO_2 supplementation (1200 mg/kg) over 30 days increased plant height, stem diameter and leaf area by 18.5 per cent, 11.4 per cent and 31.4 per cent, respectively compared with the controls. It also increased the photosynthesis rate, decreased the light compensation point and had no detrimental effect of CO_2 supplementation on fruit quality.

9.3.0 Media, Fertigation and Irrigation

In green house experiment with soilless media in Egypt by Abou Hadid *et al.* (1994) cucumber cv. Corona F$_1$ seedlings (seeds sown on 1 July, 1990) were grown in plastic seedling trays (3 ×3 × 3 cm cells) or rock wool containers (7 × 7 × 7 cm) containing peat moss mixed with fine washed sand and/or vermiculite or with perlite and vermiculite. All media contained equal parts (by volume) of each component. Two irrigation methods (sub-irrigation and spray irrigation) were also compared. Of the growing medium, peat+vermiculite and peat+sand+vermiculite generally resulted in the best seedling quality. Rock wool containers gave superior results to plastic trays for all parameters. Sub-irrigation resulted in greater average stem length and diameter than spray irrigation, but other parameters were not significantly affected. In another trial in Egypt, cucumber (cv. Cordito) grown on agricultural wastes (rice straw, legume residues or a mixture of both) covered with a layer of clay or sand increased plant height, leaf area, plant dry weight, early and total yield and also improved fruit quality in green house, compared with plants grown in clay or sand (Salama and Mohammedien, 1996).

In China, using organic soilless cultivation technique, cucumbers were grown in 2 rows on 48cm wide beds of vermiculite and gravel (in equal proportions) with a built-in spray system. Autumn/winter crops yielded 4.5 t/667 m² and winter/spring crops yielded 8.5 t/667 m² (Wei Qiang *et al.*, 1998). In green house trials in Poland, the yield of cucumber cv. Aramis 1 was highest on mushroom compost: peat mixture at ratio of 1:2 (Martyniak-Przybyszewska and Weirbicka, 1996). Sawan *et al.* (1999) indicated from green house experiment that sawdust can be used as a substitute for high percentages of peat in media for cucumber seedling production. Nutrient contents (N, P, K, Fe, Mn, Zn and Cu) and heavy metals contents (Pb, Cd, Ni, Cr and Co) were determined in the cucumber fruits. In a green house trial of cucumber grown in perlite, Altunlu *et al.* (1999) from Turkey concluded that the N concentration of the nutrient solution should not be more than 200 ppm and the K concentration should be between 200-300 ppm for profitable cucumber production. From Italy, Garibaldi *et al.* (1995) reported in green house trials that potassium hydrogen phosphate, applied at weekly intervals at a rate of 10g/1 gave satisfactory and consistent control of *Sphaerotheca fuliginea* in Zucchini (Marrows).

Cucumber cv. Picapello, grown in a sandy green house soil in Saudi Arabia, recorded highest numbers (>30) and weight (>2 kg) of total and marketable fruits per plant with 150 kg P_2O_5/donum (1000 m²) when applied preplant or 2 weeks after planting (Mohamoud, 1995a). Sulphur application had little effect on yield and quality of cucumber (Mohamoud, 1995b).

Ruiz and Romero (2000) reported from Spain that for improved cucumber cultivation 8 g P per sq.m. gave maximum yield while 16 g P per sq.m. increased the metabolism and efficient utilization of N.

'Blue print' system of culture has now widely been used and production under protection is of greatest importance in temperate latitudes between 40 and 55°N of oceanic Europe. The seedlings are raised in (6 cm) peat pots in December and the plants are set out in small beds on top of straw bales. These beds are made from a mixture of sterilized soil and horse manure or a mixture of equal parts of steam sterilized loam and stable manure. Only sufficient bed material is required to accommodate plants which have been raised in peat pots. The peat was saturated with a liquid feed of stock solution (containing 681 g of potassium nitrate, 567 g ammonium nitrate in 4.5 litre of water) and diluted in 1:200. Liquid feeding consisted of 283 g ammonium nitrate in 4.5 litres applied 1:200 three times per week, during first three weeks after planting. Later, during 3-6 weeks after planting 595 g of ammonium nitrate in 4.5 litres applied 1:200 at every watering. Thereafter, stock solution consisted of 480 g ammonium nitrate, 1.13 kg potassium nitrate and 113 g magnesium sulphate in 4.5 litres water applied at 1:200, three times per week. Cucumber cv. Camir explants grown in soilless culture with a circulating nutrient solution containing 50 mg/1 produced the highest yield. As the concentration of N increased from 70 to 100 mg/1, growth and yield declined and Ca deficiency symptoms appeared, although the nutrient solution contained 200 ppm Ca. This apparently occurred because high N concentrations damaged the roots and affected their ability to takeup Ca (Schacht *et al.*, 1992).

Chartzoulakis and Michelakis (1990) observed that fruit size was not affected by irrigation system but number of fruit/plant was significantly higher with porous plastic tube irrigation system of green house cucumber. Water use efficiency for harvested yield was highest with drip and lowest with furrow system with 27.7 and 16.8 kg/m^2 water, respectively.

On the basis of the results in green house experiment of drip irrigation of short-fruit cucumbers in Bulgaria, Kireva and Kalcheva (1997) recommended that irrigation rate of about 19 mm and 4000 m^3/ha under conditions of water deficiency yielded 8813 kg/decare (0.1 ha). In solar green house in China, Xin Yuan *et al.* (1999) recorded that yield increased with increasing rates of irrigation in cucumber but quality slightly decreased. Water use efficiency (yield/irrigation quantity) decreased with increasing rates of irrigation.

Benott and Ceustermans (1997) carried out a green house study to compare the traditional NFT (nutrient film technique) system with poly urethane (PUR) block, the NFT-drip hybrid system and the DFT (deep flow technique) drip hybrid system with cucumber (cv. Cardita). The lowest productivity was found with the NFT drip hybrid system due to in sufficient nutrient solution supply (1.8 1/ha).

9.4.0 Training and other Cultural Practices

There are several systems of training and pruning of cucumbers in the green houses: (i) archway system where spacing of 46 cm apart to give 3800 plants per acre, (ii) inclined cordons planted 54 cm apart in rows alternately at 2.14 m and 76 cm bedcentres,and (iii) vertical cordon where spacing of 54 cm apart in rows at 1.52 mm centres to give 4800 plants per acre. There are two systems of pruning, (i) English or moderate systems where first five laterals are removed while the other laterals and sublaterals are pinched off at second leaf stage and then sublaterals are trimmed at 14 days intervals and (ii) Dutch (minimum) systems where first five laterals are removed and the next five are topped at leaf, the next five at 2 leaves and the next at 3 leaves except for the top two which are stopped when they hang halfway down to the beds. The tips are pinched out of sublaterals at 14 days intervals.

Ganikhozhdacva (1984) studied the different training systems in cucumber for early yield in unheated green houses and reported that of three systems tried in the plastic-clad green houses during the spring season, training the plants with one stem produced the highest early and total yields. While Hanna and Adams (1987) described that vertically trained (staked) plants produced double marketable yield in slicing cucumber types over vining plants.

In green house trials with autumn cucumber, Uffclen (1984) from Netherlands obtained best results with a double 'V' training system, with stems 50 cm apart, 1 mainstem and a planting distance of 56 cm. In an experiment the effects of plant densities from 0.7 to 3.1 plants/m^2 on production and mean fruit weight were investigated by Bakker and Vooren (1984). Plants were trained according to the umbrella system (V-system) or as cordons. There were significant differences in production and mean fruit weight between the two training systems. The effect of density on production and mean fruit weight decreased with time. Bakker and Vooren (1985) suggested 'V' system of training and high density (3.1/m²) for maximum yield in cucumber but the mean fruit weight decreased under green house cultivation.

Lin and Zhu (1998) described a method of umbrella style training for modem green house cucumbers in China. The growing tip of the main stem is removed when 12-15 leaves have formed and the side shoots produced are trained into an umbrella shape. Yields of 17.86 t/667m² were obtained with this method.

Most of the melons are grafted to specially bred pumpkin or gourd rootstock primarily to control fusarium wilt but also to increase tolerance to lower soil temperature. Ten thousand transplants per acre are set out in twin rows on raised and fumigated beds, covered with plastic mulch and equipped with trickle irrigation tubes. Plants are trained on a single system and trained onto an upright trellis system. Melons require a lot of sunlight and a minimum night temperature of 15.5°C. When the vines begin to flower bees are allowed inside for pollination. As the fruits enlarge, plastic cords or slings are tied to support their increasing weight. Japanese consumers are willing to pay extremely high prices for an admittedly luxury cantaloupe, rather than import even at much lesser price. This allows a monetary return to the green house grower sufficient to make the production of crops such as melons or strawberries economically feasible and provides the incentive behind the rapid development of this industry in Japan today. Ethephon (250 to 450 ppm) applied at seedling stage induced femaleness (pistillate flowers) on the main stem that led to greater fruit population in cucumber cv. Beit Alpha grown in green house (Al-Masoum and Al-Masri, 1999). Transparent polyethylene mulch was found to be effective for early yield of cucumber in glass house (Tiizel and Giil, 1991).

9.5.0 Plant Protection in Greenhouse

The major problem in green house is the plant protection. Since plant protection is a major problem in green house cultivation, biological control measures of important pests and diseases have been discussed hereunder. Diseases and pests of cucurbits, in general, have also been summarized separately.

Biological control of thrips (*Thrips tabaci* and *Frankliniella occidentalis*) on seedless cucumber in green houses by predatory phytoseud mite *Amblyseius cucumeris* (*Neoseiulus cucumeris*) has been reported by Gillespie (1989). Bennison and Jacobson (1991) reported a controlled release system (CRS) of introducing the predatory mite *Amblyseius cucumeris* as a biological control agent against the thrips (*Frankliniella occidentalis*) on cucumber in green houses. Biological control of *Thrips tabaci* by the phytoseuds, *A. cucumeris* and A. *barkeri* was studied by Brodsgaard and Hansen (1992) in Denmark on green house cucumbers. Successful control was achieved with *N. cucumeris* alone but not with *A. barkeri* alone or with both species. In Czech Republic, assessment of predatory mite *A. barkeri* as a control agent for thrips (*Thrips tabaci* and *Frankliniella occidentalis*) indicated satisfactory result with an early application of predators (total of 3500 predators/plant) where both thrips species were present (Jarosik and Pliva, 1995).

In France, the arrival of *Frankliniella occidentalis* in 1986 ended the use of biological control agents on green house cucumber. Original formulations of *Amblyseius cucumeris* were ineffective against high thrips populations. A new release formulation was presented, which provides great numbers of *N. cucumeris* over a minimum period of 6 weeks (Cheyrias, 1998). Predatory anthocorid, *Orius majusculus* was reported as potential biological control agents against thrips *Frankliniella occidentalis* on cucumber in green houses in France (Grasselly *et al.*, 1995). Castane *et al.* (1997) reported of biological control of green house cucumber pests (*Trtaleurodes vaporariorum, Frankliniella occidentalis* and *Aphis gossypii*) by use of predatory mirid bug (*Dicyphus tamaninii*) in the Mediterranean region. Chat-Locussol *et al.* (1998) reported that although *Amblyseius cucumeris* resulted in better control but the release strategy used was inefficient. *A. degenerans* was also tested and a new control strategy resulted in improved control by both species.

The biological control agents of *Aphis gossypii* on cucumbers that were used in green houses in Germany were the parasitoid *Aphidius matricariae* and the predatory cecidomyiid *Aphidoletes aphidimyza* (Albert, 1995). In plastic tunnels in Bologna, Italy, Burgio *et al.* (1997) found that wild predators (*Cocinellids* and *Chrysoperla carnea*) were essential for full aphid (*Aphis gossypii*) control in cucumber. Siffnis and Ei-khawass (1996) reported success of biological control of aphid (*Aphis gossypii*) with parasitoid *Aphidius colemani* on green house grown cucumber in Netherlands.

The use of banker plants (*Eleusine coracana*) and the parasitoid *Aphidius colemani* had positive effects during 1997 on the control of *A. gossypii* in melons, courgettes (Martin *et al.*, 1998).

A bioactive substance extracted from *Rheum undulatum* controls cucumber powdery mildew diseases (*Sphaerotheca fuliginea*) grown in a polyhouse (Su Bong *et al.*, 1996).

Biological control agents *Gliocladium virens* G8728 and *Pseudomonas putida* Pf_3, were compatible with each other and successfully colonized in cucumber rhizospheres which contributed to along term inhibition of cucumber fusarium wilt (*F.oxysporum* f. sp. *cucumerinum*) of about 80 days with the efficacy greater than that obtained by any individual strains treatment under green house conditions (Bae *et al.*, 1995).

Green house studies were carried out in Austria by Walzer and Schausberger (2000) to investigate biological control of spider mites (*Tetranychus urticae* and *T. cinnabarinus*) by *Phytoseiulus persimilis* and *Neoseiulus californicus*. Results showed that the use of both predatory mites in crops with a short growing season (cucumber) did not increase spider mites control compared to the single use of *P. persimilis*. In pot plants, *N. californicus* should be present before the occurrence of spider mites, while *P. persimilis* should be used when required.

In plastic houses experiment at central Jordan Valley, the fungicides ridomil (metalaxyl) and trimiltox forte (copper oxychloride + mancozeb) were found to be most effective in decreasing angular leaf spot (*Pseudomonas syringae* pv. *lachrymans*) disease (Khlaif, 1995). Reduction in yield due to infection ranged from 31 to 64 per cent depending on cultivar. R-C-17 was the most susceptible of the tested cultivars and Shorouk was the least susceptible (Khlaif, 1995). Application of dichlorvos 3 times at 10 days interval by spraying, fogging or smoking in cucumber grown under plastic film house controlled 98-100 per cent melon aphids (*Aphis gossypii*) and fruits can be picked for human consumption 5 days after spraying (ChunHce *et al.*, 1998).

Fusarium solani f. sp. *cucurbitae* causes serious damage in green house cucumbers (*Cucumis sativus*) in Greece. Combined treatments of calcium cyanamide, solarization and shredded wheat straw gave the greatest reduction of soil populations of the pathogen (Bourbos *et al.*, 1997).

Production of early green house cucumbers is often accompanied by physiological problems in root. Roots become brown, the cortex decays and plants may wilt on sunny days; however, new roots eventually form as laterals (Vlugt, 1989a). In an attempt to prevent physiological root death of cucumber grown in nutrient film technique in glass house, lower plant density reduced the extent to which the plants shaded each other and it allowed more assimilates to become available to roots. An alternative approach was the growing of an extra side shoot, which would also increase the assimilate supply to the roots (Vlugt, 1989b).

9.6.0 Quality and Productivity

Cutting of fruit starts in April and extends to late July and early August. The quality of the fruits is uniformly of highest grade. One important characteristic is the parthenocarpic development which ensures exact cylindrical shape of the fruit without any goose-neck, crook-neck shape, which make it unmarketable. Hence no pollinators are allowed inside the green house. Investigation on the effect of bagging on the quality of bitter gourd fruits grown in plastic houses showed that the bagging did improve dry weight percentage of pericarp, percentage of prime grade fruits, the bitter value awarded by panel test and colour (Kuo *et al.*, 1999). The productivity under protected cultivation is by far the highest of any system in agriculture. Gynoecious hybrids where each flowering node produces a fruit and which ensures sustaining yields over along period of 4 months, have brought in a high degree sophistication in cucumber production. The yield range is 200 to 250 t/ha, while the maximum can reach as high as 350 tonnes.

In Germany, Leber and Heck (1991) compared 10 cultivars of cucumber and obtained highest yield with cv. Birgit (26.1 kg/m^2). But in terms of quality cv. Girola was best producing 87.7 per cent of class I crop while cv. Birgit was resistant to mildew. Uffelcn *et al.* (1991) considered cv. Jessica, Optica and Dugan to be best glass house cucumber cultivar suitable for autumn culture in Netherlands. More *et al.* (1990) recommended cucumber cv. Poinsette for protected cultivation in North India during winter months. The result of green house trials from first crop of 12 cucumber cultivars (topped and non-topped) in Belguim revealed that highest yields were obtained from Cardita (>9 kg/m^2) (Broeck, 1997).

9.7.0 Fruit Development

Cucurbit fruit grows exponentially for a period after fruit set and then the growth rate slows. The increase in fruit size after pollination is largely a result of cell expansion rather than an increase in the number of cells. Cucurbits can be divided into two major groups based on whether the fruit are harvested when immature-summer squash and cucumbers or mature all types of melons, winter squash, pumpkins, gourds. Cucumbers and summer squash are harvested during the period of rapid growth. They may be ready for harvest as early as 3 days after pollination, depending on the market requirements. In the other crops, fruits typically reach their full size about 2–3 weeks after pollination and take another 3 or more weeks to mature to a harvestable stage. During this time, seeds develop to maturity and sweetness, flavor, and color develop in the fruit. The rind toughens, becomes less permeable to water, and in the case of muskmelon, develops corky netting (Loy, 2004). A color change may occur, either subtle, as in the change from pale green to yellowish in muskmelon; or just on the portion of the fruit near the ground, as in a yellow ground spot of a watermelon; or across the entire fruit surface, as in pumpkin. The size of the mature fruit is influenced by genetics, the environment, and plant conditions during development of the pistillate flower and fruit. Conditions that reduce the amount of assimilate available tend to decrease the size of individual fruit. Increased plant density, greater numbers of fruit per plant, and reduced water supply tend to decrease fruit size. In muskmelon and watermelon, the soluble solids content of the fruit is an important measure of quality. Like fruit size, soluble solids tend to be lower under conditions that reduce assimilate level. High night temperatures, reduced leaf area, increased numbers of fruit per plant, and increased plant density can all reduce soluble solids (Wien, 1997). In contrast to its affect on fruit size, reduced water supply can increase fruit-soluble solids (Bhalla, 1971)

10.0 Harvesting and Yield

10.1.0 Harvesting

Generally, the cucurbit fruits mature faster, immediately after fruit set. In such kinds like summer squash, Indian squash melon (tinda), bitter gourd, small (pickling) cucumber and long (or serpent) melon, fruits reach optimum edible maturity stage within a week or slightly later. Promotion of fruit development in cucumber by spraying the receptacles of female flowers once with thidiazuron (2 ppm) + GA_3 (100 ppm) mixture has been reported by Yang *et al.* (1992). It markedly increased the fruit yield of the first harvest from 2.81 kg/20 plants in the untreated control to 4.99 kg/20 plants. In other kinds like sponge and ridge gourds, bottle gourd, snake gourd, pointed gourd, *etc.*, picking is done in about 12 to 15 or even 20 days after fruit set. In muskmelon and watermelon, where fruits have to be picked at full maturity stage, ready to be consumed as dessert fruit, usually 30-40 days interval would be necessary, watermelon taking slightly longer time. In pumpkin and ash gourd, where the fruits have to reach their full size and fruits have to store well, the picking of fruits gets delayed until 100-120 days after sowing.

Fruits of cucumber attain edible maturity within a week from anthesis of female flowers. Picking of fruits as the right edible maturity stage is dependent upon individual kinds and cultivars. In salad or slicing cucumber, dark green skin colour should not turn into brownish yellow or russeting and white spine colour will also been useful indication for edible maturity. Further, over mature fruit would show carpel separation in transverse section of the fruits. Optimum length of the fruit will be around 20-25cm at edible maturity stage, depending upon the cultivars. In small fruited types, which have pale green skin colour, the length of the fruit would be around 8-12cm. Here also fruits turning yellow will be overmature and show carpel separation in cross-section. In western countries, there are definite standards of size for pickling cultivars and they are pickled as whole fruit or as slices in brine. Hence most of the pickling cultivars are adapted to machine or once-over harvest. An American harvester suitable for mounting on a 75 KW tractor, was tested under Italian conditions for mechanical harvesting of pickling cucumber. Preliminary results indicate a work rate of 1.5-3 h/ha, with yields of about 13 t/ha (Siviero and Chillemi, 1992).

In muskmelon, there are two groups of cultivars which behave distinctly. In one group, the fruits when mature slips out easily from the vine with a little pressure or jerk or if not, it will remain separated the next day. This is called *full slip* stage. Most of the cultivars like Lucknow Safeda, Durgapura Madhu, cantaloupes of the USA behave in this way. In netted cantaloupe, green colour of vein tract between the netting changes into yellow, netting becomes pale or dirty white. The cantaloupe can be picked at ¾ slip stage for long distance transportation. There is another group of melons like Honey Dew and Casaba, the fruits of which do not separate at maturity and they have to be picked on the basis of external colour. In some Indian cultivars, green stripes on the skin begin to turn yellow which is an indication of full maturity. The long or serpent melon (*kakri*) is picked at tender stage before the fruit skin turns pale yellow and carpel separation is noticed in cross section. In *phoot* or snap melon, the green rind turns yellow or orange at full maturity. The quality characteristics in muskmelon are several and varied, namely thick skin and netting suitable for long distance hauling, juicer flesh with at least 10 per cent TSS (for Indian palate) while thick and dry flesh with salmon-orange colour of cantaloupe suited to western taste, small seed cavity in contrast to large seed cavity met within Indian cultivars, higher aroma in Indian cultivars in contrast to characteristic cantaloupe flavour, *etc*. Muskmelon is a climacteric fruit, which ripens during transit and storage and hence it is harvested before it is fully ripe so that it will reach the consumer at full ripe condition. Juicier the flesh is, the less will be the keeping quality of the whole fruit. Firmer flesh stands transport well. The sweetness is entirely due to reducing and non-reducing sugars chiefly glucose, fructose, sucrose, mannose, *etc*. Salmon-orange colour flesh contains carotenoid pigments, green contain chlorophyll and white flesh have none.

Ethylene plays a major role in the ripening of climacteric fruit such as melons (*Cucumis melo*). However, the importance of other modes of regulation is emerging and demonstrates the inter play between ethylene-dependent and ethylene-independent events during ripening, including at the molecular level. Inhibition of ethylene production in transgenic cantaloupe fruit has helped to isolate ethylene-regulated genes such as those involved in aroma synthesis and to understand the precise role of ethylene during development, ripening and postharvest life (Lelievre *et al.*, 2000). An engineered antisense *ACC* oxidase line with 99 per cent inhibition of ethylene production exhibited a strong but reversible inhibition of fruit ripening in melon (Guis *et al.*, 1998).

Watermelon harvest maturity is identified by yellowing of the ground spot and wilting of tendrils near the place of fruit attachment. Sugar levels do not increase in muskmelon or watermelon after harvest (Wien, 1997). In watermelon, maturity is judged by taking into consideration of several factors: (i) dull sound when the fruit is thumped, in contrast to metallic sound, (ii) withering of tendril at the fruit axil, (iii) ground spot (where the fruit touches ground) turning yellow, and (iv) the rind of the ripe melon yields to pressure. Cumulatively all these criteria will help to judge the fruit for picking. A sharp knife should be used to cut melons from the vines; melons pulled from the vine may crack open. Harvested fruit is windrowed to nearby roadways, often located 10 beds apart. The major point in watermelon is the ability of the fruit to withstand transport, especially with cracking or bursting. In most of the native Indian cultivars, thicker skin help in this respect, but tough skin, even though thinner will contribute to better transportability. Cylindrical or oblong shape like that of Charleston Gray is suitable for transportation in trucks. Most of the cultivars have deep pink or pink or pale pink flesh colour with slightly reddish tinge, containing largely lycopene and anthocyanin pigments. The colour of the flesh has no relationship to the degree of sweetness, even pale pink flesh can be sweeter than deep pink flesh in some cultivars. The average sweetness will record around 9 to 10 per cent TSS. Uniform development of pink colour from centre to rind is necessary. More often in Indian cultivars, 'white heart' seed at the central portion, shows poor quality. Similarly, hollow heart and fibrous flesh signify over maturity.

In bottle gourd, fruits are harvested at tender stage when they grow to one-third to half. Fruits attain edible maturity 10–12 days after anthesis and are judged by pressing the fruit skin and noting pubescence persisting on the skin. At edible maturity, seeds are soft. Seeds become hard and the flesh turns coarse and dry during aging. Tender fruits with a cylindrical shape are preferred in market. Harvesting starts 55–60 days after sowing and is done at 3–4 days intervals. Seed should be soft, if examined in transverse section. Hence smaller fruits fetch better price in the market. Bottle gourd cv. Pusa Summer Prolific Long fruits were found edible, tender and green from 12 to 24 days after anthesis. Ascorbic acid content was low, but increased at an intermediate stage of fruit development and decreased thereafter. Chlorophyll content decreased during fruit development, but carotenoid content increased (Bhatnagar and Sharma, 1994).

In the case of sponge and ribbed gourds, the flesh should not turn fibrous and picking should be done earlier. There are long fruited types of ridge gourd in South India and tenderness and not the size would decide edible maturity.

In bitter gourd, the fruits should not turn yellow or yellowish orange during transport and hence fruits are picked at green stage. Similarly, in snake gourd also, fruits should not turn yellowish, and tender fruits are heavier than mature fruits. The ash gourd fruit becomes smooth and covered with a bluish white wax bloom when ripe and the fruits should not be bruished at the time of harvest.

In the case of pumpkin, even though immature fruits do find a ready sale in nearby market, it is always better that complete mature fruits are harvested when the vines start drying. The fruits should have good storage capacity and it is always better to reduce the frequency of irrigation before fruit picking. The greenish yellow colour of the skin should turn to pale brown and inner flesh orange colour according to the cultivar.

Ash gourds are mature when the stems connecting the fruit to the vine begin to shrivel. Cut fruits from the vines carefully using pruning shears or a sharp knife leaving 3-4 inch of stem attached. Snapping the stems from the vines results in many broken or missing "handles." The fruits can be harvested at different stages depending on the purpose for which it will be used. Normally, green fruits are ready for harvest within 45–60 days~ matured ones coated with powdery substance are harvested between 80 and 90 days after sowing.

In sponge gourd, the crop is ready for harvest in about 60 days after sowing. Both crops are picked at immature tender stage. Fruits attain marketable maturity 5–7 days after anthesis. Overmature fruits will be fibrous and are unfit for consumption. To avoid overmaturity, picking is done at 3–4 days interval. Harvested fruits are packed in baskets to avoid injury and can be kept for 3–4 days in a cool atmosphere.

In pointed gourd, the tender fruits are harvested before seeds become hard. In ivy gourd (*kundsru*), tender fruits containing immature seeds are picked. Parthenocarpic cultivars are sometimes met with in this gourd. In chow-chow, the fruit easily separates at the time of harvesting, skin colour should remain pale green or green according to the cultivar.

10.2.0 Yield

The yield of cucurbits varies according to the system of cultivation, cultivar, season and several other factors. Approximate estimation would be:

In open pollinated cultivars the average yield is as follows:

- ☆ Cucumber: 12 to 14 tonnes/ha
- ☆ Muskmelon: 12 to 15 tonnes/ha
- ☆ Watermelon: 25-30 tonnes/ha
- ☆ Bottle gourd: 30 to 40 tonnes/ha
- ☆ Sponge and ridge gourd: 12 to 15 tonnes/ha
- ☆ Bitter gourd: 12 to 15 tonnes/ha
- ☆ Snake gourd: 10 to 12 tonnes/ha
- ☆ Ash gourd: 15 to 25 tonnes/ha
- ☆ Pumpkin: 25 to 30 tonnes/ha
- ☆ Indian squash/Squash melon (tinda): 10 to 12 tonnes/ha

In F_1 hybrids the average yield is as follows:

- ☆ Cucumber: 20 to 25 tonnes/ha
- ☆ Muskmelon: 20 to 25 tonnes/ha
- ☆ Watermelon: 35-40 tonnes/ha
- ☆ Bottle gourd: 40 to 45 tonnes/ha
- ☆ Sponge and ridge gourd: 18 to 22 tonnes/ha

☆ Bitter gourd: 18 to 22 tonnes/ha

☆ Ash gourd: 25 to 35 tonnes/ha

☆ Pumpkin: 35 to 45 tonnes/ha

11.0 Post-harvest Management and Storage

In rainy season crops or crops damaged by summer rains, fly attacked fruits would be considerable which will have to be culled out before sending to market. In cucurbits like cucumber, bitter gourd, goose-neck, crook-neck shaped fruits produced specially at the fag end of the harvest, would have less market value. These fruits are produced because of imperfect pollination and fertilization and also due to fly attack. Anthracnose affected fruit, especially in watermelon and bottle gourd, are of poor quality.

Fruit growth of *Luffa cylindrica* followed an 'S' pattern and the size of mature fruit was positively correlated with that of the ovary (Yang *et al.,* 2000). Deformed fruits were formed as a result of low temperatures, reduced water supply during fertilization or mechanical injury to young fruits (Yang *et al.,* 2000).

The other consideration in harvesting of cucurbits is the distance of the markets. Most of the cucurbits donot stand long transportation. In fact long or serpent melon, Indian squash or round melon (tinda), cucumber, bottle gourd, *etc.,* have to be sold out in nearby urban markets.

Muskmelon fruits for storage should be cooled to 10°C–15°C immediately after harvesting to retard ripening. Storage for 10–15 days at 3°C–4°C (90 per cent relative humidity) is possible, but lower temperatures can cause chilling injury. "Honeydew" and other winter melon fruits can be stored at 10°C–15°C for longer periods, some cultivars up to 90 days. Heavily netted melon fruits (*e.g.,* the Mediterranean "Galia" and the American "Western Shipper") are relatively resistant to handling and transport. Muskmelons, especially, can be hauled to long distance metropolitan markets of Mumbai (from Andhra Pradesh), Calcutta (from U.P.), Delhi (from Rajasthan, Lucknow, *etc.*).Watermelons from U.P. and A.P. are transported to Kolkata markets. Because of thick skinned nature, *parwal* or pointed gourd can be transported to distant markets. Pumpkin, ash gourd stand storage well under ambient temperature conditions.

Watermelons are not adapted to long-term storage. Normally, the upper limit of suitable storage is about 3 weeks. However, this will vary from variety to variety. Storage for more than 2 weeks triggers a loss in flesh crispness. Storing melons for several weeks at room temperature will result in poor flavour. However, when a fruit is held just a few days at warmer temperatures, the flesh colour tends to intensify. Sugar content does not change after harvest. Watermelons' flesh tends to lose its red color if held too long at temperatures below 10°C. Watermelons may lose crispness and colour in prolonged storage. They should be held at 10°C–15°C and 90 per cent relative humidity. Sugar content does not change after harvest, but flavour may be improved because of a drop in acidity of slightly immature melons. Chilling damage will occur after several days below 5°C. The resulting pits in the rind will be invaded by decay-causing organisms.

The harvested ash gourd fruits can be stored for several weeks in ambient conditions. It can be kept for 23 months in temperatures from 10°C to 12°C and 50–75 per cent relative humidity. Avoid cutting and bruising the ash gourds when handling them.

Most of the smaller fruit kinds like pointed gourd, ivy gourd, bitter gourd, cucumber, and Indian squash or round melon are packed in baskets and transported. Muskmelon and watermelon are transported in trucks without any individual packing and are liable to bruises and damage during transit. The river bed farmers sell their produce to transport contractor sand *'ahrtiyas'* or mandi agents who advance funds to them for cultivation. Being highly perishable, marketing of cucurbits involves risks and losses and hence cultivation of these vegetables are more concentrated around metropolitan cities. Some far flung river-bed growing melons and some specialized regions like cucumber of Poona region, or of Darjeeling hills, bitter gourd from Konkan districts of Maharashtra, *etc.*, supply cucurbitaceous vegetable to distant markets.

Since manual grading of cucumber is time consuming, tedious and costly, image processing systems for grading of cucumber has been developed in Korea Republic (Sung and Lee, 1996). Lin *et al.* (1993) opined whole fruit imaging was suitable for estimating fruit chlorophyll content and quantifying colour intensity in cucumber. There is potentiality of using this technique in cucumber sorting and prediction of shelf-life of cucumber.

Most of the cucurbits in India are not stored, except pumpkin and ash gourd and no cold storage facility is available in India for these vegetables. Those vegetables generally have short storage life for few weeks under 10°C and 60 to 70 per cent R.H. Fruit decay largely due to *Penicillium* varied with storage temperature and weight of fruits packed. It could be reduced significantly by pre-treating cucumbers with Imazalil (Laa *et al.*, 1995). Shelf-life of cucumber was negatively correlated with fruit N, P, B, Mn and particularly Ca content, but was positively correlated with N: Ca, Fe: Ca and Mg: Ca ratios (Lin, 1991).

According to Lee and Kang (1998), squash fruits (*C. moschata*) stored at 12°C showed no chilling injury (CI) symptoms after 20 days of storage but did not show any increase in marketability due to accelerated quality deterioration after 10 days in storage. For fruits stored at 2 and 5°C, CA treatments of 1 per cent CO_2 + 1 per cent O_2 and 3 per cent CO_2 + 1 per cent O_2 were most effective at reducing CI symptoms for 10 days at 2°C and 20 days at 5°C. At 10°C, these CA treatments reduced the development of CI symptoms to 5 per cent after 20 days of storage.

Laamin *et al.* (1995) reported that modified atmosphere storage of cucumbers significantly reduced weight loss, maintained fruit colour, firmness and acidity and doubled the shelf-life. Park and Kang (1998) observed that ceramic film of 0.02mm thickness was the best for maintaining fruit quality during modified atmosphere storage of cucumber. Shelf-life of about 2 weeks at 4°C has been reported by Huyskens *et al.* (1993) in cucumber. At 12°C, water loss, yellowing and degradation of soluble proteins were inhibited in cucumber by wrapping in polyethylene film (Zi De *et al.*, 1996). Bhatnagar and Sharma (1997) noticed that Summer Long Green and Rainy Green could be stored for upto 12 days at 25°C. Ascorbic acid content decreased significantly with increased storage duration, at 25°C this degradation was greater than at 5°C.

Cucumber fruits can be stored at 12.5°C without risks to fruit quality and it developed symptoms of chilling injury when stored at 2.5°C (Rab and Ishtiaq, 1996). Maezawa and Akimoto (1996) suggested that electrical conductivity and electrolytic leakage should be considered as an indicator of chilling injury of vegetables like cucumbers.

Keeping quality of cucumber fruits can be predicted by biological age (C_{ba}) which is evaluated by photochemical yield (ϕ_{psu}), photochemical quenching (q/p) and a photosystem 1 (K_e) (Schouten *et al.*, 1997).

Buta *et al.* (1997) concluded from their experiment with marrows (*C. pepo*) that fourier transform infra red (FTIR) spectroscopy is a rapid way to detect changes in chilled tissues before the eventual appearance of visible symptoms.

Kumar *et al.* (1988) found that by dipping in benzyladenine at 50mg/l, fruits of ridge gourd cv. HRQ-14 could be stored for 9 days at 0.5°C with least physiological weight loss, lowest decay and the best fruit quality. The cucurbits are also not usually processed. For meeting needs of defence forces, canning of bitter gourd, tinda, *etc.*, is done. There is a well developed cucumber pickle industry in the USA and specific pickling cultivars have been developed for processing in brine. In India, ash gourd is preserved in sugar syrup, and *'petha'* is the common native confectionary prepared from ash gourd in North India. Bitter gourd slices are sometimes dried. It has been observed by Waskar *et al.* (1999) that bottle gourd fruits (cv. Samrat) packed in polyethylene bags + CFB boxes and stored in a cool chamber (20.16-21.18°C and 90.25-94.50 per cent RH) had the longest storage life (28 days).

12.0 Perennial and Minor Cucurbits

12.1.0 Pointed Gourd (*Trichosanthes dioica*)

In North Bihar and in eastern Uttar Pradesh, it is extensively grown in *'diara'* land sand in Assam, Bengal, Odisha, *etc.*, it is also grown in loamy soils and in places with hot and humid climate. The main precaution is to prevent water-logging. Being a vegetatively propagated crop, there are several clones with distinct characters met within different states. Seed propagation is avoided because (a) germination is poor, (b) about 50 per cent plants are male and non-bearing, and (c) flowering takes nearly two years.

Cuttings are planted in two systems. In one system, cuttings of 60 cm long or more from one year old plants are taken in October when the plants complete fruiting and vines are mature. These cuttings are coiled in the shape of a ring and planted directly in the hills of prepared land or nursery. Here a number of nodes are covered by the soil and rooted cuttings are planted in February-March in permanent places. In the other system, furrows are dug 30 cm deep and after filling with organic manure, 60 cm long cuttings are planted 15 cm deep keeping both ends exposed. In root sucker propagation, the smaller root suckers at the nodes of creepers are uprooted in October and planted in hills. At the time of planting it is to be ensured that 10 to 15 per cent of the cuttings should be from male plants for adequate fruit set. Dash *et al.* (2000) advocated 15 female plants around one male plant. Further increasing the female plant population beyond 15 around one male resulted in significant reduction in the yield per hectare.

Basal application of farm yard manure @ 20-25 tonnes per hectare and the recommended dose of fertilizers @ 120 to 150 kg N, 60-75 kg P and 60-75 kg K per hectare can be applied, depending upon the soil and location. In some areas like in eastern U.P. and West Bengal, growers do staking. During summer, frequent irrigation may be necessary. Drying of flowers (1-2 days after anthesis) and yellowing and drying of fruits (5-7 days after anthesis) of pointed gourd were most common during summer (April-May). Mehta *et al.* (1999) attributed this flower drying and fruit drying and yellowing to the lack of pollination and fertilization during summer months. In garden land conditions, the crop is grown as a perennial for three to four years with ratooning practised in late October. However, in river-beds, the crop is planted every year. The crop starts fruiting from February in Bengal and later in other states like Bihar. The fruiting continues upto south-west monsoon when another flush of flowering begins. It may continue until October. The fruits are harvested when they are green and tender. In the first year, the yield would be around 75-80 quintals per hectare, while in the second year it may increase to 140-150 quintals. Artificial pollination can increase yield up to 500-600 quintals per hectare.

12.2.0 Chow-chow (*Sechium edule*)

In chow-chow, propagation is by planting the whole fruit in specially prepared pits. Though pandals or trellises can be erected for the vines to trail over at a height of 2 metres in homestead gardens, the vines are allowed to climb on small trees. The single seed is located near the broad end of the pear shaped fruit. The vines start flowering in about 3-4 months and production is continuous. In Bangalore regions, planting is done all round the year, but preferably in July. Planting is done in pits or basins of 45-60 cm diameter which are spaced at 1.8 × 2.4 m. The pits are well manured before planting with 10 to 15 kg of cattle manure per pit and sometimes seedlings raised in pots are transplanted. They are staked to reach the trellis and in Bangalore conditions fruiting is all round the year. A well grown plant of about one year yields 500-600 fruits per year. The fruit weighs 200-450 g. In North East India, fruits are slightly bigger. An acre yield of 80 to 100 quintals has been recorded in state of Meghalaya. The crop is semi-perennial, lasting 3-4 years. In Black Sea littoral of Georgia, a yield upto 126 t/ha was recorded (Rossinskii *et al.*, 1986). Storage is recommended at from 5 to 10°C and 85 per cent to 95 per cent RH. The respiration rate is low (4 mg/kg/hr), and room cooling is normally adequate. Chayote can be stored from 4 to 6 weeks (Stephens, 2003).

12.3.0 Ivy Gourd (*Coccinia grandis*)

It is a semi-perennial crop of 4-5 years, yielding fruits in summer and rainy season. Stem cuttings, 25 to 30 cm long, 1.5 cm to 2 cm thick are used for planting. Thicker cuttings sprout earlier. They are planted in basins of 60 cm diameter, spaced at 2 m and cuttings are planted at 3 cm deep in July or February. About 10 per cent male plants have to be planted to ensure good fruit set. Some clones produce parthenocarpic fruits while some bitter fruits too. Bowers are erected at about 1.5 metre height. Fruiting starts in 10 to 12 weeks after planting. At least once a year, the pits should be manured. In south and Central India, fruiting is round the year, while in North India, fruiting terminates when the temperature comes down in November. A vine can yield 200-350 fruits, weighing 3 to 4 kg going up to 8 kg which works out to about 120 quintals per hectare.

12.4.0 *Momordica* spp.

There are four dioecious species of *Momordica, viz., cochinchinensis* (Lour.) Spreng, *dioica* Roxb., *subangulata* Blume subsp. *renigera* (G. Don) (De Wilde and Duyfjes) and *sahyadrica* (Joseph and Antony) which are grown in a limited way in some parts of India, but are often considered synonymous because of confusing vernacular names like *kakoda, keksa, kakrol, bhat karela* etc., signifying either of the four species. *M. cochinchinensis* is called *'golka koda'* or 'sweet gourd', grown in North Bengal, North East India while *M. dioica* is called, *'ghee karela'* , grown in tribal areas of Bihar, West Bengal, Odisha and Thane district of Maharashtra. *M. subangulata* subsp. *renigera* is called *'kakrol'* grown in West Bengal, Odisha, Assam and Tripura. These are all dioecious perennials, tuberous rooted, but differ in fruit and seed character. Fruit of *M. cochinchinensis* are ovate 10-15 cm, pointed, orange or red, densely covered with conical spines, very fleshy orange-red pulp containing numerous seeds, ovoid and brownish black in colour. It grows in warm humid weather and tuberous roots are planted in pits spaced 120 cm apart. The vines are trained in bowers and 5-10 per cent of male plants are provided for good fruit set. Flowering starts in April and fruiting ends in October-November. The plants remain dormant in winter. The yield is 30-50 fruits per plant weighing around 4 kg. The tubers are left *in situ* and they overwinter (Sadhu and Chakraborty, 1980).

The cultivation of *M. dioica* is somewhat similar with *M. subangulata* subsp. *renigera*. Fruits of *M. dioica* are smaller (10-15 g weight), 2.5-5 cm long, ovoid ellipsoidal, densely echinate with soft spines, while fruits of *M. subangulata* are 5.0 -7.5 cm long, 60-100 g in weight, round or ovoid ellipsoid or oblong in shape with soft spines. Crossing experiments indicated that sex is controlled by a single factor, with heterozygous males and homozygous recessive females in *M. dioica* (Hossain *et al.*, 1996) and *M. cochinchinensis* (Sanwal *et al.*, 2011).

Ali *et al.* (1991) from Japan reported that seeds of *M. dioica* sown in vermiculite germinated at 30°C when the seed coat was removed but not at 20 or 25°C. Seed originated plants had a 1:1 ratio of male and female plants. The plants were successfully propagated by vine cuttings with or without rooting hormone and use of ethephon on the male plant was not effective in converting the sex. Application of silver nitrate at a concentration of 300-600 mg/1 to female plant was effective in inducing bisexual flowers. The pollen of the bisexual flowers was effective in producing fruits and seeds in female flowers of the same as well as different plants but produced no fruit on self-pollination. Foliar sprays with $AgNO_3$ (400 ppm) at preflowering stage could induce 70 per cent –90 per cent hermaphrodite flowers in *M. dioica* (Rajput *et al.*, 1994). Application of 500 mg/L $AgNO_3$ on female plants produced the maximum proportion of induced hermaphrodite flowers in *M. cochinchinensis*, and the pollen viability was similar to that of a normal male plant (Sanwal *et al.*, 2011).

Shekhawat *et al.* (2011) developed an *in vitro* propagation method for female plants of *M. dioica* (Roxb.) using nodal segments on Murashige and Skoog's (MS) agar-gelled medium + 2.0 mg L⁻¹ 6-benzylaminopurine (BAP) + 0.1 mg L⁻¹ indole-3 acetic acid (IAA). The cultures were amplified on MS medium with 1.0 mg L⁻¹ BAP + 0.1 mg L⁻¹ IAA and shoots multiplied by subculturing of shoot clump on MS medium + 0.5 mg L⁻¹ BAP + 0.1 mg L⁻¹ IAA. Root formation was done on half-strength MS medium + 2.0 mg L⁻¹ indole-3 butyric acid (IBA) (89 per cent success rate) and green house hardened plantlets transferred to the field.

Aileni *et al.* (2009) attempted a protocol for *in vitro* propagation of *Momordica tuberosa* (Cogn) Roxb. using nodal segments and shoot apex on MS medium and reported the highest regeneration efficiency when MS medium is supplemented with 4.40 µM BA combined with 4.60 µM Kn. The BA at 13.30 µM induced regeneration from shoot apex cultures. Microshoots were rooted onto MS medium supplemented with 4.90 µM IBA and regenerated plants established with 90 per cent survival rate. Park *et al.* (2012) used IBA for treating vine cuttings and observed better responses.

12.5.0 Gherkin/Gooseberry Gourd (*Cucumis anguria*)

C. anguria is synonymous to *C. longipes* Hook.f. (1871) of African origin, and it occurs wild in East and southern Africa. It has bitter fruits, but occasionally nonbitter types occur. Seeds were taken to the Americas with the slave trade, where the cultivated West Indian gherkin was developed. This edible, nonbitter type spread through the Caribbean, parts of Latin America, and the southern United States. It can now be found in a semiwild state as an escape from cultivation, and in some cases, it appears to be an element of the indigenous flora. Wild and cultivated types of *C. anguria* differ in bitterness of the fruits but also in the length of fruit spines (longer in wild forms). Wild types have been distinguished as var. *longipes* (Hook.f.) A. Meeuse or var. *longaculeatus* J.H. Kirkbride, cultivated ones as var. *anguria*. However, plants with short-spined fruits are often naturalized in tropical America and rarely in Africa (Kirkbride, 1993).

Temperatures during the growing season range from 15°C to 35°C. *C. anguria* is intolerant of frosts and cold temperatures. The culture and agronomic requirements are similar to those of the common garden cucumber. In cultivation, the plants should be trailed. The application of organic manure and NPK fertilizer is beneficial. Irrigation can be given in periods of drought.

West Indian gherkin is propagated by seed, which requires light for germination. Seeds are sown in pockets of three to four at a spacing of 30 cm in the row and 100–150 cm between rows. The seed requirement is 2.5–4.5 kg/ha. In the growing season, the period from seeding to first harvest is 2–2.5 months. Plants continue to flower and set fruit for several months. For leaf production, the same cultural practices can be followed (Fernandes, 2011).

As the fruits are preferred for pickling, they are harvested in the immature stage, while still green. If grown for leaves, these can be picked many times during several months. A single plant can produce 50 or more fruits. No statistics on fruit or leaf yield are reported. The yield potential is probably higher than for pickling cucumbers. The fruits can be kept for a few days at room temperature; the leaves should be consumed or marketed within a day.

West Indian gherkin is quite resistant to pests and diseases. It displays varying degrees of natural resistance to pathogens and insects, such as the cucumber green mottle mosaic virus, root-knot nematodes, powdery mildew, and green house white fly. The fruits are seldom parasitized by fruit fly larvae, which attack most other cucurbit species in southern Africa.

12.6.0 *Cyclanthera*

This is an annual cucurbit introduced from South America, called *meetha karela* (*Cyclanthera pedata*) grown in western Himalayan regions. It is grown on arches, pandals or bowers in the hills of U.P., upto about 300m above mean sea level. It is primarily grown for local consumption. It is a monoecious cucurbit with smooth fruits, oblong with narrow base, green when young, white when mature, containing 8 to 10 seeds in each fruit. Singh *et al.* (1972) introduced a cultivar which bore flowers in 43 to 52 days and fruits in 80 days during September to October. Four to six green pickings are possible.

13.0 Disorders, Diseases and Pests

13.1.0 Disorders

13.1.1 Blossom-end Rot

The blossom end of the fruit develops a dark leathery appearance. In severe cases the entire end of the fruit turns black and rots. It is associated with insufficient calcium uptake and alternating periods of wet and dry soil.

Maintain constant soil moisture, apply calcium fertilizers and avoid excessive use of nitrogen. Practise drip irrigation for water management.

13.1.2 Pillow

It is a fruit disorder of processing cucumber due to low calcium level in the tissue. In this disorder, an abnormal white styofoam like porous textured tissue is formed in the mesocarp of the fleshy harvested fruits. Vascular tissue with some pillow areas may collapse and become necrotic.

13.1.3 Light Belly Color

In this disorder, the undersurface of cucumber fruit remains light in color instead of turning dark green. It commonly occurs on fruit lying on cool and moist soil.

It can be partially controlled by avoiding use of excessive nitrogen and avoiding luxuriant vine growth.

13.1.4 Measles

Small brown spots appear scattered over the smooth-skinned fruit surface of melons and cucumbers. The spots are superficial and do not penetrate beyond the outer epidermal layers of the fruit. These spots also may occur on leaves and stems.

This disorder is associated with environmental conditions favouring guttation. The guttation droplets develop high concentrations of salts which burn the fruit epidermis. Measles spots occur where a guttation droplet had formed.

Reduce the frequency and duration of irrigation as fruit approach maturity in fall-harvested crops. Irrigation reduction at the later stages of fruit development has not shown any adverse effects on fruit size and soluble solid content.

13.1.5 Rind Necrosis

It mostly occurs in cantaloupe or watermelon. Symptoms appear as dead, hard, dry reddish-brown to brown spots or patches of tissue in the fruit rind. The affected areas vary in size from 3mm spots to extensive dead areas throughout the entire rind. In cantaloupe, circular, water-soaked depressions develop on the fruit surface and the dead tissue may extend into the fruit flesh. In watermelon, symptoms are not visible from the outside and are rarely found in the flesh.

The environmental conditions causing stress on the plants may trigger the onset of this disorder. Susceptibility to rind necrosis varies among varieties.

Avoid drought stress in melons. Genetic tolerance to this disorder has been identified in watermelon.

13.1.6 Hollow Heart

In watermelon, internal cracks appear in the fruit flesh due to accelerated growth in response to ideal growing conditions. Both genetic factors and growing conditions are responsible for this disorder. It is associated with conditions that result in poor pollination (enough pollination to set the fruit but not enough to fertilize a high percentage of the ovules) followed by conditions favouring rapid fruit growth (too much fertility, water and high temperatures).

Avoid watermelon varieties having a tendency to exhibit hollow heart. Implement good practices for irrigation and fertilization.

13.1.7 Sunscald

Papery white areas appear on fruit surface. Symptoms develop during hot summer weather when fruit are suddenly exposed to direct sunlight.

It can be minimised by maintaining strong vine growth to ensure the fruit is well covered.

13.1.8 Leaf Silvering

It is a disorder of summer squash which occurs due to moisture scarcity. The leaves become silver coloured and contain less chlorophyll. Photosynthesis is hampered in the silvered leaves.

13.1.9 Unfruitfulness in Pointed Gourd

Pointed gourd is a dioecious cucurbit in which the female plants produce the fruit and male plants act as a pollen donor. Adequate number of male plants should be planted depending upon the population of female plants to ensure sufficient pollination, fertilization and good fruit set. A common problem is met with where

Physiological Disorders of Cucurbits

Splitting (Left), and Blossom End Rot (Right) of Watermelon

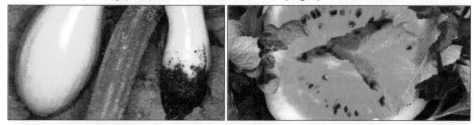

Blossom End Rot (Left) of Squash, and Hollow Heart (Right) of Watermelon

Light Belly Colour on Cucumber (Left), and Measles on Honeydew Melon (Right)

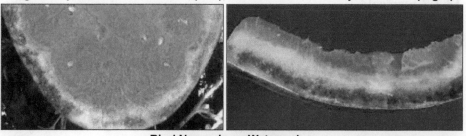

Rind Necrosis on Watermelon

Sunsclading on Lumpkin (Left), and Unfruitfulness in Pointed Gourd (Right)

pistillate flowers in female plants are shed due to lack of pollination and fertilization. In some cases, ovary of the unfertilized flower may grow a bit due to parthenocarpic stimulation which also abscises after a few days.

Male plants must be grown in the field along with the female plants at the rates of 10-12 male plants per 100 female plants to ensure adequate pollination and fruit set. Hand pollination may be done to improve the fruit set. Hand pollination to the female flowers should be done in the early morning hours because stigma receptivity decreases with an advancement of the day.

13.1.10 Delay in Fruit Ripening

In muskmelon and watermelon, delay in ripening is sometime associated with less sweetness and cracking of fruits which occur due to high moisture level and temperature fluctuation at ripening stage.

Irrigation should be stopped at the ripening stage to hasten ripening. Sowing time should be adjusted in such a way that fruits ripe in hot and rainless condition which hastens ripening and also improve sweetness of the fruits.

13.2.0 Diseases

A large number of diseases affect the cucurbits at different stages of growth. The fungicidal control recommendations given below are purely tentative, since they are liable to be changed by newer formulations from time to time.

13.2.1 Fungal Diseases

13.2.1.1 Powdery Mildew

Powdery mildew affects almost all cucurbits under field and green house conditions. The disease is widely distributed and destructive among cucurbits in most areas of the world and can be a major production problem causing yield losses of 30 per cent –50 per cent (El-Naggar *et al.*, 2012). It is caused by *Golovinomyces cichoracearum* (synonym = *Erysiphe cichoracearum*) and *Podosphaera xanthii* (synonym = *Sphaerotheca fuliginea*). It very often becomes severe in warm rain free growing region. Symptoms first appear as white or fluffy, somewhat circular patches or pale yellow spots on the stems, petioles and undersurface of the leaves. Gradually the spots enlarge, conidia are produced from affected tissue and the spots have a powdery appearance. In severe cases, these spread, coalesce and cover both the surfaces of the leaves and spread also to the petioles, stem *etc*. Powdery mass on the leaves decreases the photosynthetic rate (Queiroga *et al.*, 2008), causing reduction in plant growth, premature foliage loss, and consequently reduction in yield. Severely attacked leaves become brown, papery and shriveled and defoliation may occur. Fruits of the affected plants do not develop fully and remain small. The yield loss is proportional to the severity of the disease and the length of time that plants have been infected (Mossler and Nesheim, 2005). If this disease is not controlled in time, symptoms can be severe enough to cause extensive premature defoliation of older leaves and wipe out the crop (Nunez-Palenius *et al.*, 2012). This disease occurs mostly on cucumber, muskmelon, pumpkin, bottle gourd, *etc*. In some parts of South India, powdery mildew is found to occur especially on watermelon, which is not affected in North India. There appears to be several biotypes and the extent of specialization has not been investigated in India.

Powdery mildew causing fungi are obligate parasites which require a living host to survive. They overwinter on weeds and are carried for long distances by air currents. The most favorable conditions for disease development are high relative humidity of more than 70 per cent (Ali *et al.*, 2013) and 20-27°C temperature. Dry conditions favour colonization, sporulation and dispersal of conidia. Development of disease is favoured by vigorous plant growth, moderate temperatures, low light and dew.

In earlier years, this disease used to be controlled by sulphur dusting, but most of the cucurbits are susceptible to sulphur injury, especially when it is done during hot days. Modern fungicides like cosan, calixin, *etc.*, have been found useful in controlling this disease. Weed control and better field sanitation practices also help in controlling powdery mildew. Folicure or Depacozile 1ml/litre or meptyl dinocap @ 1.5 ml/litre of water controls the disease, if sprays are given when the first initial symptoms appear. The sprays will have to be repeated at least thrice, at 5-6 days interval. In bottle gourd, powdery mildew disease caused by *Sphaerotheca fuliginea* and *Erysiphe cichoracearum* can be controlled well by application of bavistin. It was also less phytotoxic (Ratnam *et al.*, 1985). White mustard oil at 1 per cent was also found effective (Stoyka *et al.*, 2014). It is necessary that drenching of leaves should be thoroughly done, especially when the vines are in full growth. Application of a fugan (pyrazophos) and baycor 300 EC (bitertanol) proved very effective in controlling powdery mildew of green house cucumber (Qvarnstrom, 1989). The disease spreads quickly as spores are carried by winds. Although synthetic fungicides are effective against powdery mildew on cucurbitaceae, caused by *Erysiphe cichoracearum* and/or *Sphaerotheca fuliginea*, their waiting periods are not compatible with the harvesting of these crops and further more sulphur is phytotoxic to cucumber. Studies were therefore carried out in Bologna, Italy by Collina (1996) to evaluate the efficacy of some natural material and plant products which are nonphytotoxic on powdery mildew on courgette (cv. Storr's Green) and cucumber (cv. Jezzer) in the green house and courgette (cv. Storr's Green) in the field. The efficacy of sodium bicarbonate, tween-20, pinolene (a pine resin derivative), mineral oil, monopotassium phosphate (potassium dihydrogen phosphate) and rape oil (cv.Canola), alone and combination, were compared with that of the standard, sulphur. The addition of sodium bicarbonate improved the efficacy of tween-20, pinolene and mineral oil in the green house. In the field, the most effective treatments, after sulphur, were sodium bicarbonate + mineral oil or pinolene. Gamil (1995) reported that soaking seeds in acetyl salicylic acid at 2.5 or 5 m M for 24 h induced resistance to natural infection by *Sphaerotheca fuliginea*. Spraying with similar concentrations of aspirin at the 1st true leaf stage reduced disease severity. According to Sharma and Kumar (1999) for best control of powdery mildew of cucumber, the most economic fungicide was Karathane (dinocap). Chlorothalonil was toxic to cucumber and sulphur was unprofitable.The results of experiment conducted by Mosa (1997) provide evidence that phosphate salt (K_2HPO_4) could induce resistance and provide a curative effect against cucumber powdery mildew. The biocompatible products AQ10 (hyper parasitic fungus *Ampelomyces quisqualis*), JMS stylet-oil (medium viscosity mineral oil), M-pede (potassium salt of fatty acids) and Kaligreen (82 per cent potassium bicarbonate) suppressed powdery mildew (*Sphaerotheca fusa*) on winter squash, muskmelon and pumpkin and increased yield compared with non-treated plants under field conditions in New York, USA (Mc Grath and Shishkoff, 1999). Fujiwara *et al.* (2000) indicated that the electrolyzed anode-side water (AW) obtained by electrolysis of dilute aqueous KCl solution can be an alternative to chemical fungicides for powdery mildew in cucumber. Biological control involves the use of fungal spores of *Ampelomyces quisqualis* Ces., which parasitize and destroy the powdery mildew. Similarly, bacteria Bacillus subtilis and fungus *Sporothrix flocculosa* (syn. *Pseudozyma flocculosa*) gave promising results (Nunez-Palenius *et al.*, 2012). Moreover, foliar sprays of chlorite mica clay that contains silicon have also shown suppression of powdery mildew in cucumber (Ehret *et al.*, 2001).

Considerable effort has been made in developing cultivars resistant to sulphur injury and resistant to the attack of the causal organism. Sulphur resistant cultivars are known in California in cantaloupes, like SR-91 and V-1. Breeding for resistance to powdery mildew in cantaloupe in Imperial Valley of the USA and cucumber in the east coast of the USA was successful, from the collections introduced from India (Pl79376), especially from Kathiawar (Gujarat) region and in cucumber from Puerto Rico collections. Now there are several cultivars developed in cucumber and muskmelon (cantaloupe) resistant to different races of powdery mildew. Lebeda and Kristova (1996) from Czech Republic reported high level of field resistance to powdery mildew (*Erysiphe cichoracearum*) disease in *C. pepo* cultivars, Acceste F_1, Albina Ambassador F_1, CU-235, Elite F_1, Goldfinger, Parmanta F_1 and Seneca Hybrid F_1. In pumpkin and squashes, *Cucurbita lundelliana* has been identified as a good source of resistance to powdery mildew. Under Indian conditions, Campo, a cantaloupe from California has been found resistant to an unidentified race at Delhi. In cantaloupe, sources of powdery mildew resistance are PI79376, PI124111 and PI134198, moderately resistant sources Georgia-47, race-1 resistant PMR-45, race-2 resistant Seminole and highly resistant to race-1 and race-2 are Planter's Jumbo. In cucumber, PI197087, PI120815 are resistance sources. Resistance is governed by one gene in cultivars, Palmetto and Ashley, 2 genes in Poinsette and Cherokee and 3 genes in PI197087 (Barnes, 1961, 1966). Arka Manik, watermelon cultivar from ICAR-IIHR carries high resistance to powdery mildew under Bangalore conditions.

Biological control of cucumber powdery mildew disease by mycoparasite, *Verticillium lecanii* together with use of resistant cultivar Flamingo has been reported by Verhaar and Hijwegen (1994) from Netherlands.

13.2.1.2 Downy Mildew

This is caused by the fungus *Pseudopernospora cubensis*. It is a disease affecting most of the cucurbits like cucumber, muskmelon, ridge gourd, bottle gourd, *etc.* It is prevalent in areas of fog, heavy dew, high humidity, especially when summer rains occur regularly (Keinath, 2014). Sporangia are transmitted between fields by wind. Within fields sporangia are spread by air currents, splashing water, workers and/or equipment. High temperatures of >35°C are not favourable for development of downy mildew disease. But disease development progresses if cool night temperature of 15–20°C is prevalent. The disease is characterised by formation of yellow, more or less angular spots on the upper surface of leaves, while purplish spores appear on the upper surface of leaves. The disease spreads rapidly killing the plant quickly through rapid defoliation. The affected foliage in the early plant development stage causes a reduction in photosynthetic activity that result in stunted plant growth and yield reduction, especially in cucumber (Colucci and Holmes, 2010). Hence, the control by fungicidal sprays has to be taken up at the initial stage itself. In contrast to powdery mildew, spores of downy mildew are dark purplish gray and appear only on the underside of the leaves (McGrath, 2006). Symptoms on watermelon and cantaloupe are not as distinctive as on cucumber and squash and are mostly mistaken for other diseases such as anthracnose, target spot, Alternaria leaf spot, or gummy stem blight (Colucci and Holmes, 2010). The symptoms of downy mildew infection are variable in different cucurbit crops, for example, the lesions are angular and are limited by the leaf veins on cucumber and squash, whereas these are typically irrregularly shaped on the foliage of watermelon and cantaloupe and turn brown rapidly. As the disease progresses, the lesions expand and multiply, causing the field to have a brown and brittle look (Celetti and Roddy, 2010). Subsequently, the lesions coalesce, become necrotic and continue to expand until the leaf dies. Severe infection leads to defoliation, stunting of plants and poor fruit development.

In Belarus, this destructive disease of cucumber has been observed since 1985. Presowing disinfection by physical method using a device for bactericidal irradiation of seed (λ=254 nm) as an alternative to chemical treatment was developed (Zherdetskaya and Levashenko, 1996).

The sprays of Dithane 45 @ 0.75 - 1.5 kg per hectare if given early and repeated 2 or 3 times can control the disease effectively though not completely. Early planting can also avoid the disease (Keinath, 2014). Biocontrol agent *B. subtilis* is available in a wettable powder formulation that can be used for downy mildew control (Fravel, 1999). However, Keinath (2014) reported that no organic-approved fungicides or biofungicides prevent or control cucurbit downy mildew. Yang (1985) recommended the application of sugar solution (1 per cent) along with urea (1 per cent) every 10 days to control the disease. Golyshin *et al.* (1992) recommended dimethomorph + mancozeb or a mixture of dimethomorph + daconil (chlorothalonil) (3-5 applications) for *Peronosporosis* disease control in cucumber. The disease is serious in muskmelon in Punjab and on ridge gourd in West Bengal. Ethaboxam 25 per cent wettable powder (a derivative of aminothiazole carboxamide) is a fungicide developed in Korea Republic, effectively controlled cucumber downy mildew caused by *Pseudoperonospora cubensis* (Kim, 1999).

Adequate spacing between plants should be provided to reduce the canopy density. Grow varieties with genetic resistance. The Indian cucumber cultivar Bangalore was the starting point of development of resistant cultivars in cucumber in the USA. Later, Chinese Long and PI197097 have been used. Resistant cultivars are several like Ashley, Poinsette, *etc.* In cantaloupe, sources of resistance are Seminole and PI124112, moderately resistant Georgia-47 and resistant Edisto-47.

In other cucurbits, resistance breeding has not made much headway. There are several cultivars in cucumber and in cantaloupe which carry resistance to both the mildews like Chipper, Sumter in cucumber, Perlita and Planter's Jumbo in cantaloupes (Sitterly, 1972). Thakur *et al.* (1996) recorded lowest disease index for downy mildew (*P. cubensis*) incidence in bitter gourd genotype BL240. Cucumber cultivars Chitradurga and Kuknoor showed resistance to downy mildew (Reddy *et al.*, 1997).

A study was conducted to determine the most effective treatment for controlling downy and powdery mildew (caused by *Pseudoperonospora cubensis* and *Erysiphe cichoracearum*, respectively) on bitter gourd (*Momordica charantia*), and among all treatments, eight sprays of 0.3 per cent copper oxychloride + 0.3 per cent wettable sulfur at 10-day intervals from 30 days after crop sowing were the most effective for the control of downy and powdery mildew diseases of bitter gourd during the rainy season, recording the highest yields and economic returns (Memane and Khetmalas, 2003).

The fungicidal mixture metalaxyl + mancozeb was highly effective in controlling downy mildew in both dry and rainy seasons, while Chlorothalonil + methyl hyphenate was effective only during the dry season (Santos *et al.*, 2004). ICI A5504, a beta-methoxyacrylate compound, is particularly effective on Cucurbitaceae, providing unique control to both *P. cubensis* and *Sphaerotheca fuliginea* causing powdery mildew.

Different intensities of the disease in melon were achieved by spraying the following fungicide mixtures: methyl thiophenate (thiophenate-methyl) + Chlorothalonil or metalaxyl + mancozeb (Santos *et al.*, 2003).

Potassium enrichment reduces incidence of downy mildew in cucumber. Follow a 3-year crop rotation to reduce soil borne inoculum and field sanitation by burning the crop debris after harvest (Saha, 2002).

Abou-Hadid *et al.* (2003) reported that *Trichoderma harzianum* and *Trichoderma hamatum* were the most effective antagonists against the pathogens of powdery or downy mildew disease.

Pusa Hybrid-1 and Arka Chandan of pumpkin; Punjab Chappan Kaddu-1 of summer squash; Pusa Hybrid-3 and Pusa Summer Prolific Round and Pusa Summer Prolific Long of bottle gourd; BL-240, Hybrid BTH-7, BTH-165, Phule Green, and RHR BGH 1 of bitter gourd; IIHR-8 of sponge gourd; Arka Sujat of ridge gourd; Poinsette and Priya of cucumber; Arka Manik of watermelon; and Punjab Rasila and Pusa Madhuras of muskmelon cultivar are tolerant to downy mildew disease (Thamburaj and Singh, 2005).

13.2.1.3 Anthracnose

The disease is very serious in watermelon, bottle gourd, cucumber, snake gourd, *etc.* and is caused by the fungus *Colletotrichum* sp. which is both air- and seed-borne disease that occur on most cucurbits during warm and moist seasons. In case of cucumber and muskmelon, reddish brown dry leaf spots are formed which often coalesce and cause shriveling and death of leaf. Lesions on petioles and stems are water soaked and yellowish. The leaf spots on watermelon are black and foliage presents a scorched appearance. Under moist conditions, the lesions are dotted with pink conidia. It spreads to fruits of bottle gourd and watermelon in severe cases of incidence. This is another disease promoted by high humidity and moist weather and is severe in rainy seasons and in regions of heavy summer rains as in West Bengal. Seed borne anthracnose infection results in drooping and wilting of cotyledons, and lesions may appear on the stem near the soil line.

This fungus spread by infected crop debris, splashing rain, overhead irrigation, insects, field workers and equipment. The disease development is favoured by warm and humid weather. Optimum temperature for disease development is 24°C.

Adopt crop rotation with non-cucurbit crops, avoid overhead irrigation, maintain sanitation of field, workers and equipment and use resistant varieties. The disease can be controlled by repeated spraying at 5 to 7 days interval with Indofil M-45 @ 0.2 per cent. Soaking the seed in hot water at 57.2°C for 20 minutes is also effective to check seed borne inoculum. Two distinct races (and perhaps as many as seven), which vary in their ability to infect a range of cucurbit genera, species, and cultivars, have been reported (Goldberg, 2004). Carry out deep ploughing immediately after the final harvest to destroy all infected cucurbit plants and debris in the field (Palenchar *et al.*, 2012). Copper products and biological control agent, *B. subtilis* strain, QST713, have also been recommended (Li, 2014). Shimizu *et al.* (2009) reported that endophytic *Streptomyces* sp. strain, MBCu-56, has strong potential for controlling cucumber anthracnose.

In USA, high degree of resistance to all the three races occurring in cucumber has been found in Pl197087 with some kind of multigenic inheritance. Poinsette is a highly resistant cultivar, with moderately resistant sources in PI175111. In watermelon, resistance to *races* 1 and 3 has been noted in African Citron-8 and to *race* 2 in African Citron W-695. Charleston Gray, the leading watermelon cultivar and Congo are resistant to *race* 1 and 3 (Sitterly, 1972).

13.2.1.4 Fusarium Wilt

It is a common disease affecting several cucurbits like watermelon, muskmelon, bottle gourd, *etc.* Even though cucumber and squashes are affected by fusarium root rot, they are not affected by *Fusarium* wilt. The causal organism has been identified as *F. oxysporum* with sub-species *niveum* affecting watermelon, *melonis* on muskmelon and *cucumerinum* (Owen) affects cucumber. Many *Fusarium* wilt pathogens including *F. oxysporum* f. sp. *niveum* are capable of being seed borne, although the extent of the contamination varies widely (Egel and Martyn, 2013). There are also reports that *F. solani* is causing wilt in muskmelon and bottle gourd. On the other hand, Fusarium crown and foot rot [*Fusarium solani* f. sp. *cucurbitae* (Snyder and Hansen)] causes root rot in squashes. The early symptoms appear as wilting of leaves, while the entire plant may wilt and die with the passage of time. The rot develops initially as water-soaked, light-coloured areas that progressively turn darker. The fungus is generally confined to the crown area of the plant and the infection starts in the cortex of the root. In young seedlings, cotyledons droop and wither. In older plants, leaves wilt suddenly and vascular bundles in the collar region become yellow or brown. There is some kind of relationship of soil temperature with disease resistance as in cantaloupe at 27°C and in watermelon at 32-33°C. Information on physiological specialization under Indian conditions is not available. High nitrogen, especially ammoniacal form, less than 25 per cent soil moisture, and light, sandy, and slightly acidic soils (pH 5–5.5) favour development of Fusarium wilt disease (Zitter, 1998). The fungus survives as chlamydospores in the soil and in plant debris and are disseminated in soil and plant debris during cultivation practices in field, in irrigation water, by wind-blown soil and workers.

Grow resistant varieties. Practice soil solarisation, maintain a soil pH of 6.5, follow equipment and worker sanitation and use of a NO_3 nitrogen source. Chemical control of such soil-borne and root diseases is very difficult. The disease can be checked to some extent by drenching the soil with captan or hexocap or thiride 0.2 per cent to 0.3 per cent solution. This should be repeated so that spread of the disease can be checked.

Sun and Huang (1984) reported that amendment of loamy soil with S-H mixture (4.4 per cent bagasse, 8.4 per cent rice husk, 4.25 per cent oyster shell powder, 8.25 per cent urea, 1.01 per cent KNO_3, 13.16 per cent calcium superphosphate and 60.5 per cent mineral ash) was highly effective in reducing the percentage survival of *F.oxysporum* f.sp. *niveum* in watermelon cultivation.

Singh *et al.* (1999) observed that two chitinolytic bacteria strains, *Paenibacillus* sp. 300 and *Streptomyces* sp. 385, suppressed *Fusarium* wilt of cucumber caused by *Fusarium oxysporum* f.sp. *cucumerinum* in nonsterile, soilless potting medium.

Watermelon when grafted on bottle gourd cv. Renshi, a *Fusarium* wilt resistant cultivar, the incidence of the disease in watermelon was uncommon. Besides, the graft compatibility with watermelon is good and the growth, quality and cropping characteristics of watermelon grafted onto Renshi are similar to those on other rootstocks (Matsuo *et al.*, 1985).

Sources of wilt resistance in watermelon have been from African citrons. Charleston Gray cultivar carries resistance to *Fusarium* wilt also. Iowa Belle and Garrisonain are other resistant cultivars. There are several watermelon cultivars like Crimson Sweet, Sweet Princess, Jubilee, *etc.*, which carry resistance to both *Fusarium* wilt and anthracnose (Sitterly, 1972). Arka Manik of ICAR-IIHR has been found to be moderately tolerant to anthracnose under Bangalore conditions.

In melon, Kuchkarov *et al.* (1982) identified Kurume 1 to be the most resistant cultivar to *Fusarium oxysporum*. Akbarov (1982) reported that a family from the cross IckzlUzbekskii 331 x Kurume 1 showed promise for breeding for resistance to *F. oxysporum* in melons. U.C. PMR 45 and U.C. Top Mark are two *Fusarium* wilt resistant muskmelon breeding lines (Zink and Gubler, l987a, b).

13.2.1.5 Gummy Stem Blight

It is caused by the fungus *Didymella bryoniae* (anamorph: *Phoma cucurbitacearum*) which is a foliar pathogen infecting all cucurbit species and having worldwide distribution. Young seedlings show damp-off symptoms. On older plants, circular, dark tan to black spots appear on leaves and are surrounded by a yellow halo. As the disease progresses, these lesions dry, crack and fall out, which is called "shot-holing." Wilt starts at leaf margins and progresses toward the center, resulting in leaf blight. Infected stems develop cankers and produce a red or brown and gummy exudate. In severe cases, the infected stems may be girdled, resulting in vine death. Small, water-soaked, oval or circular spots to large necrotic panels appear on fruits. The blossom end of fruit may soften and turn brown or green. Numerous small black fruiting bodies (pycnidia or pseudothecia) develop within the infected leaf or stem tissue or within the lesions on fruits.

The fungus can be seed-borne and overwinters on infected crop debris of cucurbit crops and cucurbit volunteer plants. The optimum temperature range for infection is 20–25°C. Moderate temperatures, high humidity and wet weather favours disease development.

Avoid overhead irrigation, follow minimum 3 years crop rotation with non-susceptible non-cucurbit crops, keep the fields weed-free, practise soil sterilization, maintain a strict sanitation, grow resistant varieties and use treated seed. Use recommended fungicides. M17 is a monoecious, pickling *Cucumis sativus* inbred line with moderate resistance to gummy stem blight (*Didymella bryoniae*) (Wehner *et al.*, 1996). Ten days after transplanting copper oxy-chloride or copper hydroxide paste should be applied to the collar region to avoid wilt and gummy stem blight especially during *kharif* season.

13.2.1.6 Monosporascus Wilt

It is caused by *Monosporascus cannonballus* which is distributed worldwide. Initial symptoms include stunting and poor plant growth. The older leaves turn chlorotic, wilt, collapse and most of the canopy may be killed. Tan to reddish-brown lesions appear on the roots and the root system may become necrotic, resulting in plant death. Large, black perithecia are visible on the dead roots. Fruit of diseased plants are smaller, cracked, abscise from the pedicle before ripening and have reduced sugar content.

The disease is spread by the movement of infested soil or infected plant material. Rise in the soil temperature is favourable for the perithecia formation in the roots. Ascospores of the fungus help in long-term survival.

Management of *Monosporascus cannonballus* is difficult due to its heat tolerance and thick-walled resting structures. Adopt crop rotation with non-cucurbit crops, avoid planting melons and watermelons in infested fields, avoid excessive irrigation, destroy the infested roots and use resistant rootstocks. Methyl bromide fumigation of soil prior to sowing, the main control method for *Monosporascus cannonballus* in melons in the Arava Rift Valley of Israel is prohibited and thus Cohen *et al.* (2000a) high lighted the necessities of alternative strategies and discussed other approaches for the *Monosporascus* wilt control, including breeding for resistance, grafting melon plants onto resistant *Cucurbita* and melon rootstocks,changes in irrigation schemes, improved soil solarization, chemical control with fungicides and the use of other soil fumigants. Cohen *et al.* (2000b) observed that *Monosporascus* sudden wilt incidence on grafted plants to be significantly lower than on non-grafted plants. Melons grown in the daily irrigated plots, initial wilt symptoms appeared 47 days after planting, whilst in the less frequently irrigated plots, initial wilt symptoms appeared 60 days after planting and the plants did not collapse totally. The fungicide fluazinam was effective in both, inhibiting pathogen growth in culture and suppressing disease in the field.

13.2.1.7 Scab

The fungus *Cladosporium cucumerinum* is most common in cucumber, but also can affect pumpkin, squash and cantaloupe. It is prevalent in subtropical and temperate countries and is virulent in cool growing seasons accompanied by intense fog and dew. The disease appears on leaves, petioles, stems, and fruits (Ogorek *et al.*, 2012). Numerous water-soaked spots appear on leaves and runners, which eventually turn gray to white and become angular, often with yellow margins. The center of the spots could then drop out to give irregular-shaped holes in the leaves. The spots might ooze a gummy substance, which could then be invaded by secondary rotting bacteria, and these spots cause foul smell (Yuan 1989; Watson and Napier, 2009).

Spread of pathogen and development of the disease are favoured by low night temperature (21–24°C). Higher temperatures inhibit disease development. The conidia are dispersed by wind, insects, farming equipment, and workers. Grow resistant varieties. Some of the resistant cultivars of cucumber developed are Highmoor and Wisconsin SMR-15. Cucumber scab became a serious disease of cucumber in Jilin, China. The highest incidence reached 97.4 per cent and yield loss was maximum when the infection occurred during the seedling stage. Among the fungicide tested MBC (carbendazim) was the most effective in controlling the disease (Yuan, 1989). Photodynamic dyes (Bengal rose, toluidine blue, and methylene blue) systemically protect cucumber plants from cucurbit scab. It was found that all dyes at 0.5–200 µM significantly suppressed symptoms of the disease (Aver Yanov *et al.*, 2011).

13.2.1.8 Other Fungal Diseases

There are other fungal diseases of minor importance like angular leaf spot caused by *Pseudomonas lachrymans* in cucumber, alternaria leaf spot caused by *Alternaria cucumerina*, especially in cantaloupe and damping off caused by *Rhizoctonia solani* and *Phythium* sp. in cucumber.

Suhag *et al.* (1985) recorded effective control of *Cercospora* leaf spot disease in bottle gourd by using dithane M-45.

Control of cucumber damping-off caused by *R. solani* and *Pythium* spp. requires an integrated approach and can be achieved using the binucleate *Rhizoctonia* sp. isolate CG combined with the fungicide metalaxyl (Cubeta and Echandi, 1991). Chitosan has potential for the control of phythium root rot of cucumber when used as an amendment in nutrient solution of hydroponic systems (El-Ghaouth *et al.*, 1994). Roberts *et al.* (1997) indicated the capability of biological control of damping-off of cucumber caused by *Phythium ultimum* with a root-colonization deficient strain of *Escherichia coli*, S17R1.

Chen *et al.* (1999) reported that growth promoting bacteria (CN 11, CN 31, CN 45, CN 116 and CNJ 29) reduces the incidence of disease *Rhizoctonia* rot and damping-off caused by *Rhizoctonia solani* AG4, *Pythium aphanidermatum* and *P.ultimum* on cucumber.

13.2.1.9 Virus Diseases

There are a large number of viruses which cause much damage to different cucurbits all over India and world. A complex of viruses infects cucurbits, and over thirty viruses have been reported (Zitter *et al.*, 1996). Identified strains may be several, occurring single or in mixtures especially in river-beds. The leaves show mottling mosaic, crinkling and twisting and shortened internodes and flowering is adversely affected. There are some which are partially seed transmitted, some are transmitted through insect vectors like aphids and white flies and some are even mechanically transmitted.

The majority of these viruses cause huge losses in cucurbit production. The important viruses are cucumber mosaic virus (*CMV*), squash mosaic virus (*SqMV*), watermelon mosaic virus (*WMV*), zucchini yellow mosaic virus (*ZYMV*), and papaya ring spot virus (*PRSV*) (Tobias and Tulipan, 2002). These viruses are transmitted by aphids except *SqMV*, which is spread by beetles and is seed borne. The symptoms caused by different cucurbit viruses are very similar, and it is very difficult or even impossible to identify the causal virus. Some strains are reported to over-winter in alternate hosts. In recent years, Tomato Leaf Curl New Delhi Virus (*ToLCNDV*) has become a serious problem in many cucurbits in India and abroad.

Studies at IARI have identified (i) cucumber mosaic virus (cucumis virus 1), (ii) watermelon mosaic virus (poty-virus group),(iii) tobacco virus group which is sap transmissible (without vectors), comprising cucumis virus 2B, cucumis virus 2C and cucumis virus 3,(iv) non-sap transmissible but white fly transmissible, characterized by yellow vein mosaic, and (v) non-classified consisting Kakri mosaic virus and tori (*Luffa*) mosaic virus (Vasudeva and Lal, 1943; Capoor and Varma, 1948 :Vasudeva *et al.*,1949; Mitra and Nariani, 1965; Shankar *et al.*, 1969; Shankar *et al.*, 1972 and Mitra and Nariani, 1975). From Coimbatore, two isolates of melon mosaic virus, bitter gourd mosaic virus transmitted by aphids have been reported (Jagannathan and Ramakrishnan, 1971; Nagarajan and Ramakrishnan, 1971) and from West Bengal, nine different isolates on pumpkin have been identified (Ghosh and Mukhopadhyay, 1979) comprising bottle gourd mosaic virus, cucurbit latent virus, watermelon mosaic virus-I, cucurbit mosaic virus and two newly designated viruses, pumpkin venation mosaic virus and pumpkin mild mosaic virus. Likewise, there are detailed reports from Gorakhpur (U.P.) on cucumber and watermelon mosaic virus (Bhargava, 1951; Bhargava and Joshi, 1960). In a survey of river-bed growing different cucurbits, cucumber green mosaic virus, seed-borne muskmelon mosaic virus, melon mild mosaic and melon ring spot virus have been reported.

The complete control of virus diseases has not been possible. One way of checking the spread is to dissuade the farmers not to collect seeds from virus infected plants, especially in the case of mosaic occurring in squash. Some viruses are thermosensitive and get inactivated due to high temperature occurring in summer and sometimes virus affected plants do recover partially or temporarily.

Pink and Walkey (1985) compared the *CMV* disease resistance to the different cultivars of *C. pepo* namely, Cinderella, Cobham Bush Green and Gold Rush at different temperature and light intensities. Symptoms severity in all three cultivars decreased with increased temperature but only in Cinderella and Cobham Bush Green when the light intensity was raised. In general, Cinderella was most resistant and Gold Rush most susceptible, except at 25°C when most plants were symptomless.

Snyder *et al.* (1993) found that *PYG* squash cultivars are only tolerant of watermelon mosaic vims (*WMV*) and if planted with a susceptible cultivar such as Lemon Drop then low percentages of marketable fruit could result. Prevention of mosaic virus diseases of squash to the acceptable level was possible when JMS stylet oil was combined with aluminium foil mulch (Mansour, 1997).

In the cucurbits grown in rainy season, insect vectors like white fly spread the viruses rapidly. The only solution to control viruses is the development of virus resistant cultivars. In cucumber, Tokyo Long Green and Chinese Long are good sources of resistance to cucumber mosaic virus and some of the important resistant cultivars developed are SMR-18 Wisconsin and Table Green. In cantaloupe, some resistant sources to *CMV* and *WMV* have been identified. In *Cucurbita*, 12 species were found resistant to *CMV* or *TRSV*, 7 were immune to both WMV_1 and WMV_2, 7 resistant to *TmRSV*, 3 tolerant to *SqMV* and immunity to *ByMV* and *TmRSV* were found in cultivated forms (Provvidenti, 1983). Cow Leg is a multiviral resistant cultivar of bottle gourd commonly grown in Taiwan (Provvidenti, 1995). In Spain an accession of the wild Asiatic *Cucumis melo* sp. *agrestic* showed resistance to melon yellows virus (*MYV*) infection, a white fly transmitted clostero virus (Nuez *et al.*, 1999).

Field vacuuming was found to be as effective as insecticide applications for the control of *B.tabaci* in melons (Weintraub and Horowitz, 1999).

Thakur *et al.* (1996) found no yellow mosaic symptoms in genotypes BG14-4, BL240, BG14, HK12 and Palwal Sel-1 of bitter gourd.

13.2.1.10 Cucumber Mosaic Virus (CMV)

Several aphid species transmit this virus in a non-persistent manner. It has an extensive host range. Most of the cucurbits are susceptible to CMV. Symptoms first appear on younger leaves, which curl downward and become mottled, distorted and reduced in size. Plants may become stunted, have short internodes and youngest leaves have a rosette-like appearance. If infection occurs after flowering, vine growth may not be reduced, but there may be mottling and distortion of fruits.

Control the vectors, weeds, use reflective mulches, maintain sanitation in field, destroy the affected crop residues and grow resistant varieties.

13.2.1.11 Cucurbit Aphid-Borne Yellow Virus (CABYV)

Several aphid species transmit this virus in a persistent manner. It has world wide distribution. Leaves become chlorotic, leathery and brittle, while the mid-vein and primary veins remain green (interveinal chlorosis). Plants are stinted and there is flower abortion. Cucurbits are the primary hosts of *CABYV*. Alternate host crops are lettuce and fodder beet.

Control the vectors, weeds, maintain sanitation in field, destroy the affected crop residues and grow resistant varieties. Use of silver reflective plastic mulches can help to repel the aphids.

13.2.1.12 Squash Mosaic Virus (SqMV)

Striped cucumber beetle (*Acalymma* spp.) and spotted cucumber beetle (*Diabrotica* spp.) are the vectors for this virus. It is also transmitted by seed and mechanically by workers and equipment. It mostly affects muskmelon, pumpkin and squash.

Melon and Squash: Green vein banding, appearance of yellow spots, vein clearing and/or blistering are the initial symptoms on leaves. In severe cases, leaves are distorted, with marginal projections from the veins giving a fringe-like appearance to the leaf margin. Infected plants become stunted with fewer branches and fruit. Fruit show mild mottling to severe deformation. Infected fruits of netted melon types may not form netting.

Cucumber: Symptoms on leaves include chlorotic spots with an upward leaf curl, vein-clearing or a yellow vein-banding.

Control the vectors, weeds, maintain sanitation in field, destroy the affected crop residues and grow resistant varieties. Use virus free seeds or seedlings.

13.2.1.13 Cucumber Green Mottle Mosaic Virus (CGMMV)

It is transmitted mechanically *via* equipment and workers and also by seed. It affects bottle gourd, cucumber, pumpkin, squash, muskmelon and watermelon. Vein-clearing and crumpling of young leaves are the initial symptoms. Other symptoms include mild to severe leaf distortion, light and dark green mottling, yellow or silver leaf flecks and stunting. Mature leaves may become bleached. Chlorotic or silver spots or streaks appear on fruits and later fruits become distorted.

Grow resistant varieties, use healthy seed and resistant rootstocks. Adopt good sanitation practices.

13.2.1.14 Gemini Viruses

Several Gemini viruses affect cucurbits world wide, *viz. Cucurbit Leaf Crumple Virus (CuLCrV), Loofa Yellow Mosaic Virus (LYMV), Melon Chlorotic Leaf Curl Virus (MCLCuV), Pumpkin Yellow Vein Mosaic Virus (PYVMV), Squash Leaf Curl Virus (SLCV), Squash Mild Leaf Curl Virus (SMLCV), Squash Leaf Curl China Virus (SLCYNV), Tomato Leaf Curl New Delhi Virus (ToLCNDV), Watermelon Chlorotic Stunt Virus (WmCSV), Watermelon Curly Mottle Virus (WCMV), etc.* Of these, *ToLCNDV, PYVMV* and *WCMV* are serious problem of cucurbit crops in India. All these gemini viruses are transmitted by whitefly (*Bemisia tabaci*). Cucumber is least affected by gemini viruses whereas pumpkin, squash, Luffa, muskmelon, watermelon are most affected. Symptoms include upward curling of leaf margins, foliar stunting, chlorosis, interveinal mottling, vein clearing and thick, distorted veins. Flowers

Diseases of Cucurbits

Downy Mildew Infection of Ridge Gourd (Left), and Cucumber (Middle and right)

Downy Mildew Infection in Bitter Gourd (Left), Pointed Gourd (Middle), and Watermelon (Right)

Powdery Mildew Infection in Cucumber (Left), and Pumpkin (Right)

Powdery Mildew Infection in Ridge Gourd (Left), and Squash (Right)

Diseases of Cucurbits

Collar and Vine Rot of Pointed Gourd

Anthracnose Infection in Cucumber

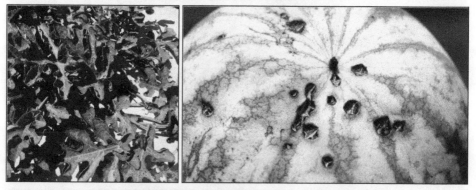

Anthracnose Infection in Watermelon

Angular Leaf Spot in Cucumber (Left), and *Alternaria* Leaf Blight in Watermelon (Right)

Diseases of Cucurbits

**Phytophthora Leaf Blight in Bottle Gourd (Left),
Cucumber (Middle), and Sclerotinia Infection in Cucumber (Right)**

Gummy Stem Blight Infection in Bottle Gourd

Sclerotinia Infection in Bottle Gourd

Diseases of Cucurbits

Bacterial Wilt of Summer Squash (Left), and Verticillium Wilt in Watermelon (Right)

Cucumber Mosaic Cucumovirus (CMV) Infection in Pumpkin

CMV Infection in Squash (Left), and Pumpkin (Right)

CMV in Muskmelon (Left), and Cucumber (Right)

Diseases of Cucurbits

Watermelon Mosaic Potyvirus Infection in Summer Squash

Watermelon Mosaic Potyvirus Infection in Pumpkin

Papaya Ringspot Potyvirus in Summer Squash

Diseases of Cucurbits

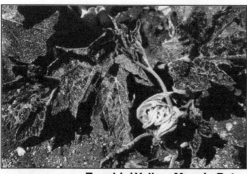

Zucchini Yellow Mosaic Potyvirus in Summer Squash

Zucchini Yellow Mosaic Potyvirus in Cucumber (Left), and
Tomato Leaf Curl New Delhi Virus Infection in Sponge Gourd (Right)

Squash Mosaic Comovirus in Muskmelon

Tobacco Ringspot Nepovirus in Cucumber (Left), and
Aster Yellows Phytoplasma in Summer Squash (Right)

Diseases of Cucurbits

Symptoms Caused by some Cucurbit Viruses

(A) mosaic on a leaf and fruit of a melon plant infected by *Watermelon mosaic virus* (WMV); (B) Severe mosaic and deformation on leaves and fruits of a zucchini squash plant infected by *Zucchini yellow mosaic virus* (ZYMV); (C) Mosaic and deformation on leaves and young fruit of a zucchini squash plant infected by *Cucumber mosaic virus* (CMV); (D) Yellowing of older leaves of a melon plant infected by *Cucurbit aphid-borne yellows virus* (CABYV); (E) Vein clearing on a leaf of a cucumber plant infected by *Cucumber vein yellowing virus* (CVYV); (F) Severe yellow mosaic on a watermelon plant infected by WmCSV; (G) Necrotic spots on a leaf of a cucumber plant infected by *Melon necrotic spot virus* (MNSV); and (H) Vein banding on a leaf of a melon plant infected by *Squash mosaic virus* (SqMV) (Lecoq and Desbiez, 2012).

Diseases of Cucurbits

**Blossom Blight of Teasle Gourd with
Emerging Sporangiola (A) Initial Stage (B) Mature Stage.**

Anthracnose Infection in Pointed Gourd

***Phytophthora* Fruit Rot Infection in Pointed Gourd**

***Rhizoctonia* Leaf and Vine Rot Infection in Pointed Gourd**

of infected plants are small and fail to develop normally. Affected fruits exhibit chlorotic blotches and are deformed.

Recently *Tomato Leaf Curl New Delhi Virus* (*ToLCNDV*: genus *Begomovirus*, family Geminiviridae), the causal virus of tomato leaf curl has been reported to be associated with sponge gourd, cucumber and muskmelon and causing a yield loss up to 100 per cent in north Indian plains during *kharif* season under epidemic condition (Sohrab *et al.*, 2003). The diseased plant is characterized by yellow spots appearing on newly emerging leaves, followed by a mosaic appearance and upward curling of the upper leaves. In cases of severe attack, the leaves of the plant are small and distorted and misshapen fruits are produced.

Sponge gourd line DSG-6 and DSG-7 were found to be highly resistant to ToLCNDV (Munshi *et al.*, 2012, Islam *et al.*, 2010). Saha *et al.* (2013) cloned and characterized the NBS-LRR encoding resistance gene candidates from DSG-6.

13.2.1.15 Poty Viruses

The major poty viruses affecting cucurbits are *Papaya ringspot virus* (PRSV; formerly *Watermelon Mosaic Virus-1*), *Watermelon mosaic virus* (WMV; formerly *Watermelon Mosaic Virus-2*) and *Zucchini Yellow Mosaic Virus* (ZYMV) and are transmitted in a non-persistent manner by several species of aphids. A typical symptom of cucurbit poty viruses is "shoe-string", a narrowing of tendril-like appearance of leaves.

PRSV: Symptoms on leaves include vein-clearing, development of a light to dark green mosaic, followed by distortion and deep leaf serration. In cucumber, leaves are distorted along the margins. Pumpkin and squash leaves assume a "shoe-string" appearance. In muskmelon, there is blistering of young leaves. In watermelon, the growing terminals stands erect and new leaves are reduced in size. Affeced fruits become blotchy and deformed. Concentric ring-spot patterns appear on the rind of watermelon fruit.

WMV: Symptoms on leaves include chlorosis of leaf veins, development of a green mosaic and leaves become deformed and blistered. In severe cases, leaf tissue surrounding the major veins has a "shoe string" appearance.

ZYMV: The infected leaves are yellow with severe mosaic symptoms and exhibit blistering and "shoe-string." Plants become stunted having uneven fruit colour and fruit malformation.

Control the aphid vectors, weeds, maintain sanitation in field, destroy the affected crop residues and grow resistant varieties. Use of silver reflective plastic mulches can help to repel the aphids.

13.2.1.16 Tospo Viruses

The major tospo viruses affecting cucurbits in Asia are *Groundnut Bud Necrosis Virus* (GBNV), *Melon Yellow Spot Virus* (MYSV), *Watermelon Bud Necrosis Virus* (WBNV) and *Watermelon Silver Mottle Virus* (WSMV) and are transmitted by several thrips species. Symptoms include bronzing and chlorotic spotting of leaves, leaf deformation, mosaic, die-back and plant stunting. Fruit symptoms include chlorotic ring spots on young fruit and necrotic lesions and sometimes cracking on older fruit.

Control the thrips vectors, weeds, maintain sanitation in field, destroy the affected crop residues and grow resistant varieties.

13.3.0 Pest

Several insect pests attack cucurbit vegetables. Among them, fruit fly is the major pest and causes significant economic loss. Aphids and white flies are the vectors of several viruses affecting cucurbits. Red pumpkin beetle is a problem mostly in seedling and early stages of plant growth in many cucurbit vegetables. The chemical insecticides control recommendations given below are purely tentative, since they are liable to be changed by newer formulations from time to time.

13.3.1 Red Pumpkin Beetle

These are brightly coloured elongated small beetles of orange-red colour. The beetles (*Aulacophora foveicollis*) attack most of the cucurbits at seedling stages, especially at cotyledonary leaf stage. They make holes in cotyledonary leaves. Severe damage is caused at this stage, although they attack the vines in the grown up stage also. Muskmelon, bottlegourd, pumpkin, cucumber and watermelon are attacked mostly with the exception of bitter gourd.

Mechanically collect and destroy pumpkin beetles if the pest incidence is low. They remain sluggish before sunrise. Hence, catch them before sunrise and destroy. As per the direction of Central Insecticide Board, red pumpkin beetle can be controlled by spraying of Dichlrovos 76 per cent EC @ 500 a.i or Tricholfon 50 per cent EC at 500 a.i. Spray the crop with Carbosulfan @ 2 ml/litre of water either very early in the morning or very late in the evening as during day the useful insects like honey bees are active in the crop for pollination and fruit setting. Soil application of Furadon granules at 15 kg per hectare before sowing the seeds controls the insects. Soil application of neem oil cake @ 20 kg/ha is effective in killing the pest larvae.

13.3.2 Aphids

These small green insects (*Aphis* sp.) damage the plants by sucking the leaf sap. In young stage, cotyledonary leaves crinkle and in severe cases the plants wither. In grown up vines, the leaves turn yellow and plant loses its vigour and yield. Both *Aphis gossypii*, melon aphid and *Myzus persicae*, the green peach aphid attack the cucurbits. They suck the sap from stems, leaves, and other tender plant parts by piercing with their mouth part and are also vector of many viruses.

Use reflective mulches to repel aphids. The aphids can be easily controlled by spraying cypermethrin (0.01 per cent) or acetamiprid (0.01 per cent) or bifenthrin (0.01 per cent) or malathion (0.05 per cent). Spraying of 1.5 per cent fish oil soap can also be used to control aphids. Usually sprays against aphids and beetles can be combined. Spray against aphids has to be done sufficiently early before the attack becomes severe, since aphid transmitted viruses rapidly spread. The aphids should be killed by using a detergent and vegetable oil solution before destroying the old crops to avoid winged adult aphids spreading virus to nearby crops. Some breeding work has been taken up in California to breed aphid resistant cultivars in cantaloupe which can provide indirect control of the virus.

13.3.3 Fruit Fly

This is by far the serious pest of cucurbits, defying chemical control. Maggots of this fly (*Bactrocera cucurbitae*) causes severe damage to young developing fruits. The adult fly lays eggs below the skin of the young ovaries. The eggs hatch into maggots which feed inside the fruits and cause rotting. The fly attack is severe on muskmelon, watermelon, bitter gourd, Indian squash melon (*tinda*), pointed gourd, long melon *etc*. The fly attack is severe, especially after summer rains when the humidity is high. It prefers to infest young and soft-skinned fruits.

There is no direct control of maggots because they are inside the developing fruits. Bagging of fruits, field sanitation, protein baits, cue-lure traps, host plant resistance, biological control, and new generation insecticides can be used for management of fruit fly. The adult flies can be controlled by using light traps in the night and poison baits. The fruit fly can be controlled by Cyantraniliprole 10.26 per cent OD @ 90 a.i. as per CIB, Government of India. Application of fenitrothion (0.025 per cent) in combination with protein hydrolysate (0.25 per cent) also reduces the fruit fly damage. Furthermore, Bait Application technique (BAT) by spraying liquid of 0.1 per cent insecticide and 10 per cent jaggery or 10 per cent ripe banana at 200 spots/ha or Erect Cue-lure (para pheromone trap) 3 per acre to attract and trap male fruit flies. Infestation can be controlled by baits containing sex attractants like protein hydrolysate 0.5 kg + 1.25 kg of 50 per cent Malathion + 200 g of molasses. Spinosad is also quite effectively used as a toxicant with protein bait. Now a days, other sex attractants are also being tried to control this pest. Cue-lure traps and some chemicals like methyl eugenol are also effective. Use of a repellent NSKE 4 per cent is used to enhance trapping and luring in bait spots. Spray of thiodan @ 6 ml per 4.5 litre of water also partially, checks the fly incidence (Pawar *et al.*, 1984). Reddy (1997) reported that triazophos was the most effective insecticide against cucurbit fruit fly. In controlling this pest, individual efforts by a single farmer will not be effective and cooperative efforts by all the farmers in river-beds or other areas growing different cucurbits, would be necessary. Another precaution of spraying insecticides at the time of flowering, when the fly attack is severe, is to ensure that there is no repellant action to pollinating insects like bees from insecticidal sprays. The affected fruits should be regularly pinched off and buried in a pit. Pumpkin is comparatively free from fly attack because the skin of the young ovaries of *Cucurbita* sp. is quite tough towards egg laying adults. *Opius fletcheri* Silv. is a dominant parasitoid of *Bactrocera cucurbitae*. The fungus *Gliocladium virens* is effective against the adult fruit fly.

Some work on breeding for resistance was carried out at IIHR, in muskmelon and bitter gourd. *Cucumis sagittatus* has been found to be immune to fly attack. *Cucumis melo* var. *callosus* has been reported to be resistant (Chelliah and Sambandam, 1972).

Mites are serious in watermelon and muskmelon, especially during severe summer weather. These tiny arachnids are seen on the under surface of the leaves. Spray of diazinon 0.03 per cent or lebaycid 0.05 per cent is effective.

Pests of Cucurbits

Red Pumpkin Beetle (Left), and Fruit Fly Infested Fruit (Middle and right)

Adult of Male (Left), and Female (Right) Fruit Fly

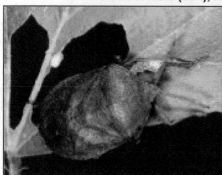

Stink Bug (Left), and White Fly (Right)

**Green Peach Aphids (Left), Banded Blister Beetle (Middle), and
Two Spotted Spider Mite (Right) Infestation**

Pests of Cucurbits

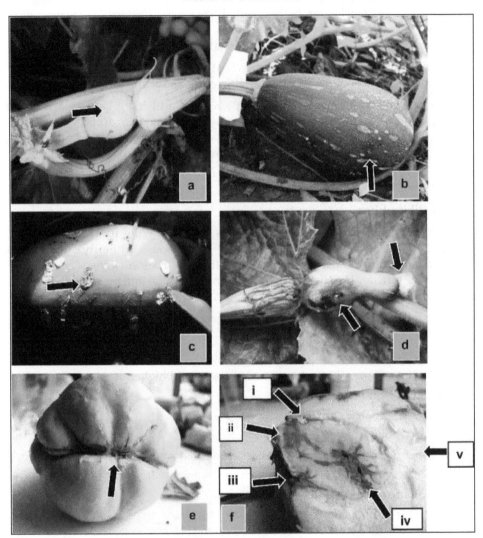

Damage Due to Fruit Flies on Various Fruits of Cucurbits.

Oviposition site on young fruit (a) and on maturing fruit (b) of *Cucurbita moschata*; oviposition site on maturing fruit (c) of *Cucumeropsis mannii*; fallen and decaying young fruit (d) of *Cucurbita moschata* showing larval activity; oviposition site on maturing opened fruit (e) and on dissected fruit (f) of *Sechium edule* showing: (i) eggs, (ii) larvae, (iii) feeding galleries, (iv) seed damage and (v) fruit pulp (Mokam *et al.*, 2018).

Besides, there are other pests like cutworms, thrips, jassids, *Epilachna* beetle which are not of major importance but are of consequence in localized regions. Biological control of thrips (*Frankliniella occidentalis*) on cucumbers using the predatory mites *Amblyseius barkeri* and *A. cucumeris* (*Neoseiulus cucumeris*) in Denmark and *Amblyseius cucumeris* and *A. limonicus* in Netherlands has been reported by Borrergaad (1988) and Houten (1996).

Fruit fly pheromone traps (10 traps/ha) may be placed along with baits of malathion for control of fruit fly during flowering and fruiting phase.

For fruit fly control, crush pumpkin 1 Kg and add 100 g Jaggery and 10 ml malathion or spinosad and keep in the plot (4-6 places/acre). Adults of fruit fly get attracted to the fermenting pumpkin and lay eggs and get killed. Repeat 2-3 times in cropping season. Alternately, erect cue-lure traps at 10 traps/acre to annihilate male flies or spray deltamethrin 1 ml/L+ 1 per cent Jaggery at fruit formation/ ripening stage.

13.3.4 White Flies

Among different species of white flies, tobacco white fly (*Bemisia tabaci*), silverleaf white fly (*Bemisia argentifolii*) and green house white fly (*Trialeurodes vaporariorum*) are the most serious in cucurbit crops. They suck the plant sap and transmit the viruses. The silver leaf white fly injects a toxin into the plant that causes slivering/whitening of the leaves.

The following measures can be taken to manage white fly and white fly-borne viruses:

1. Avoid excessive use of plant growth promoting hormones
2. Use of N-fertilizer at optimum doses, avoid higher doses
3. Application of K-fertilizer
4. Avoid dense cropping
5. Remove and destroy virus-infected plants
6. Clean cultivation, remove alternate weed hosts from the surrounding area
7. Use reflective polythene mulch to repel white fly
8. Raise nursery under insect-proof net
9. Plant tall barrier crops like maize, sorghum or pearl millet on border. Seeds for border rows have to be sown at least 2-3 weeks before sowing/ transplanting main crop.
10. Intercropping with non-susceptible crops
11. White fly in the early stages of population development can be managed by hosing down with water sprays
12. Use yellow stick traps to detect and monitor the activities of white flies in the field
13. White fly is difficult to control by insecticidal application. Rotation of following insecticides mixed with soap solution/sticker is recommended. Spay should cover the bottom side of foliages.

☆ Cyantraniliprole 10.26 per cent OD at 90 g ai/ha

☆ Imidacloprid 70.00 per cent WG, Imidacloprid 17.80 per cent SL at 35 g ai/ha

☆ Thiamethoxam 25.00 per cent WG at 50 g ai/ha

☆ Spiromesifen 22.90 per cent SC at 150 g ai/ha

☆ Diafenthiuron 50.00 per cent WP at 300 g ai/ha

☆ Afidopyropen 50 g/L DC at 50 g ai/ha

☆ Tolfenpyrad 15.00 per cent EC at 150 g ai/ha

☆ Buprofezin 25.00 per cent SC, Buprofezin 70.00 per cent DF at 250 g ai/ha

☆ Flonicamid 50.00 per cent WG at 75 g ai/ha

13.3.5 Mites

Two-spotted spider mite (*Tetranychus urticae*) is the common mite spp. attacking cucurbits. They feed on the underside of the leaves by piercing and sucking the leaf surface. The initial symptoms appear as tiny, light spots on the leaves which later turn brown. Severe infestation causes mottling of leaves with silvery-yellow appearance and premature leaf fall. Acaricides like spiromesifen are quite effective in controlling the mites.

13.3.6 Nematodes

The cucurbits are highly susceptible to nematode infestation, *viz.*, root-knot nematode, *Meloidogyne incognitoa crita*. Poor growth and stunted plants are the usual symptoms. Muskmelon, cucumber, pumpkins, *etc.*, are severely affected. Soil fumigation with fumigant nematicides or application of nemagon and long rotations can control the nematodes. Non-fumigant nematicides like thionazin, oxamyl, ethoprophos and carbofuran can also be used. Seed treatment with 6 per cent W/W of benfurocarb was effective in controlling *Meloidogyne* spp. in bottle gourd (Darekar *et al.*, 1989). Green manuring of soil with *Tagetes* or *Xanthium* leaf powder followed by application of *Verbesina* and *Artemesia* reduces population of *M. incognita* in cultivation of *C. pepo* (Sharma *et al.*, 1985).

Soil solarization and farm yard manure (40 t/ha) reduced the incidence of *Meloidogyne javanica* on cucumber roots in Isfahan, Iran (Nasr-Esfahana and Ahmadi, 1997). Soil fumigation and weed-free fallow periods help to control root-knot nematode.

Resistance to *Meloidogyne incognito* has been observed in wild species *Cucumis dipsaceus* and *C. anguria* and in *C. sativus* cultivars, Fem Cap, *Rozental* Tsepellin, Superator and Kuc-Vo-Kha-Bakh (Udalova and Prikhod'Ko, 1985).

Some work has been initiated to locate resistant sources for transfer into cultivated varieties, *Cucumis metuliferus* is one, which has high resistance to nematode and some attempts are successful in interspecific hybridization with *Cucumis melo*.

14.0 Seed Production

In cucurbitaceous vegetables (barring asexually propagated ones), seed crop is not anyway different from vegetable crops.The only difference is that instead of picking fruits at vegetable maturity, they are allowed to mature in the plant itself, so that seeds are extracted in full maturity. In crops like bottle gourd, cucumber, bitter gourd, summer squash, sponge and ridge gourds, *etc.*, the fruits are allowed to ripen fully and even dry as in the case of bottle gourd, summer squash and sponge and ridge gourds. In crops like muskmelon, watermelon and pumpkin, seed maturity coincides with edible maturity and so there is no difference in picking stage. The other point that is to be considered is the choosing of the right season. Seed crop should always be raised in seasons which remain dry at the time of seed maturity and seed extraction. In other words, summer season is preferred over rainy season for raising seed crops, since in crops like bottle gourd or bitter gourd, whose seeds have thick-seed-coat, seeds do not dry properly in humid weather.

The fact that staminate and pistillate flowers are separate in the same plant and consequent entomophilous condition favour cross-pollination, even though natural self-pollination is not precluded. This situation, therefore, emphasizes the need for isolation between two cultivars of the same species, when seeds are to be raised. Where isolation facilities are available, 0.5 to 1.0 km can be considered as adequate isolation, provided the intervening space has some other crops, if not other genera of cucurbits. Generally, crossing between the cultivars of the same species is possible like between muskmelon and long melon (*kakri*), *phoont* (snap melon) and other non-dessert types. But interspecific crossing even between closely related species like *L. acutangula* and *L. cylindrica* or between *Cucurbita* sp., is very negligible, if not rare. Intergenetic crossing is not also successful and that is there as on why barrier system of minimum isolation is sometimes adopted as explained earlier.

The cross pollinating nature of this group of crops naturally promotes variation even within a pure cultivar. Complete homozygosity will not be prevalent even in pure cultivars. There is always a need to constantly watch the performance of the cultivar for seed multiplication. Even in breeder's seed and foundation seed multiplication plots, selection of desirable and true-to-type plants is necessary. In cross pollinated crops, genetic drift or deterioration occurs during seed production if selection (specifically mass selection) is not enforced. Selection of desirable plants and removal of off-type variants are absolutely necessary year after year to maintain purity of the cultivar.

Further, this selection of desirable plants is possible long after the pollination is over, especially in melons, where fruit quality of individual plant is judged and flesh colour and other characteristics are verified. In other words, pollination precedes selection unlike in onion and cole crops. In cucurbits, off-type variants, if present, do take part in pollination earlier which will however, be identified at the time of fruit maturity for seed extraction. The spine colour in cucumber, netting and strips in rind of muskmelon stripes and other patterns in rind of watermelon, can help early detection of the variants, but not before the pollen contaminating other true-to-type plants. Further, this contamination due to cross-pollination will show up and segregate continuously in succeeding 2 or 3 generations of seed production. This handicap can be partly mitigated by resorting to continuous roguing, 2 or 3 times during vegetative or flowering phases.

Being entomophilous crops, provision of adequate pollination agency, namely bee colony is necessary. Two hives per hectare would be necessary. The effort should be to get maximum fruit set in the early stages of pistillate/perfect flower phase and earlier set of fruits, called crown set, develop to full size and produce plump seeds. If the earlier set is missed, fruits maturing later are poorly sized and developed and may not be sweet (in the case of melons). In a crop like bottle gourd or summer squash, 4 to 5 fruits per vine carried to seed maturity would be adequate enough to give a good seed yield. Korzcniewska *et al.* (2000) reported that seed yield of 400 ppm ethephon treated plants was approximately half that of the untreated control.

In case of fleshy fruits like cucumber, muskmelon, watermelon, *etc.*, seeds have to be scooped out when the fruits are fully mature and ripe. In melon, fruit quality characters in respect of flesh colour, *etc.*, are verified and a minimum index of 10 per cent TSS is fixed as standard of consumer acceptability. All fruits scoring less than 9 per cent should be outright rejected and fruits scoring between 9 and 10 per cent TSS or more may be retained, if better fruits have been obtained in earlier harvests. In Brazil, Alvarenga *et al.* (1984) studied the influence of age and postharvest storage of watermelon cv. Charleston Gray on seed quality and opined that best quality in relation to germination, vigour, dry weight and moisture content was generally in seeds from fruits harvested 35 and 45 days after anthesis and stored for 4 days. In melons, seeds of all fruits cannot be extracted on one day but will be spread over to 2 or not more than 3 pickings. Any fruit showing different flesh colour or external rind character or spine colour (in the case of cucumber) should be completely rejected and all the fruits of the off-type plants should be discarded. Nandesh *et al.* (1996) assessed seed quality of cucumber in terms of germination, field emergence and vigour and concluded that quality was highest in fruits harvested 40 days after anthesis followed by 15 days of postharvest ripening. In watermelon, seed characters of a particular cultivar are also diagnostic and can be identified at the time of seed extraction. Yoo (1996) reported that in *L. siceraria*, seed germination percentage increased with fruit maturity; seeds from 70-80 days after anthesis fruits had the highest percentage of germination (99 per cent). After-ripening of fruits also increased percentage of germination; seeds from fruits after-ripened for 30 and 40 days had the highest percentage of germination (99 per cent).

In the case of dry fruits like bottle gourd, sponge and ridge gourds, summer squash, seeds are extracted when the fruits dry and seeds rattle inside the shell. The shells have to be broken to extract the seeds and clean them.

In other cases, where seeds are mixed with pulp or placentae, the seeds are scooped out with little pulp attached as in muskmelon, cucumber, watermelon, *etc.*, and collected in a barrel. Seeds can be washed after they are rubbed with sand or ash to remove pulp. Otherwise, the pulp surrounding the seeds can be allowed to ferment for 48 hours, when the pulp can be easily separated. A rapid method of separating the seed from pulp is by acid treatment. Twenty five to thirty ml of hydrochloric acid for 5 kg of pulp containing seeds or about 8 to 10ml of commercial sulphuric acid can be used. The seeds can be washed free from the pulp in about 20 to 30 minutes. Then the seeds have to be washed thoroughly to remove excess acid and ill-developed seeds floating are discarded. The seeds are then dried in sun and drying has to be uniformly done. For safe storage, the moisture content of seeds should be below 8 per cent. Dry seeds should be stored in containers and placed in a cool room with provision for dehumidification and with protection against rat and vermin.

Munshi and Tomar (2013), Munshi and Sureja (2013) and Munshi *et al.* (2013) discussed in detail the techniques and basic requirements of seed production of open pollinated varieties of cucurbitaceous vegetable crops.

14.1.0 Basic Requirement of Seed Production of Cucurbits

The basic requirement of seed production is availability of improved varieties/ hybrids and their demand among the farmers. Preferably the variety should be realised and notified for certified seed production, however, any kind/variety can be multiplied for Truthfully Labeled Seed (TFL).

14.2.0 Isolation Requirements

The cucurbits are cross pollinated in nature and honey bees are major pollinator, thus for pure seed production an isolation distance all around seed field is necessary to separate it from fields of other varieties, fields of the same variety not confirming to varietal purity requirement. The isolation distance of 400 m for Certified seeds and 800 m for Foundation seeds and at least 1000 m isolation is required for breeder seed production.

It is important to mention that muskmelon, long melon and snap melon (phoot) can cross with each other. Similarly, species of genus *Cucurbita* have the risk of crossability among them. The cucumber can cross easily with its wild relative *Cucumis hardwickii* found in wild form in sub-mountainous regions of Himalayas.

14.3.0 Choice of Season and Areas of Seed Production

Seed crop should be raised in such a seasons which remain dry at the time of seed maturity and seed extraction.

Locations are also important in seed production with reference to seed yield and quality of seed. To harness the advantage of climate, private sector seed companies are organizing their seed production in these areas. The Punjab and Haryana, U.P, Jalna (Aurangabad) in Maharashtra, Ranibenur and around Bengaluru in Karnataka, Nandyal Valley in A.P., are the main areas of seed production for muskmelon and cucumber.

14.4.0 Roguing

Seed crop is to be monitored at various stages of crop growth for removal of off-type and obviously should be carried out before flowering to avoid natural cross-pollination. However, fruit set and complete fruit development stages are also important. Minimum four rogueing should be done at the following stages.

14.5.0 Stages of Roguing

1. **Before flowering**: Off types are detected on the basis of vegetative characters like vine growth, foliage morphology, colour *etc*.

2. **Flowering stage**: Early and late varieties can be easily identified on the basis of sex expression and sex ratio. Development of fruits at lower node which is an indicator for earliness should also be taken into consideration.

3. **Fruit developing stage**: Trueness to type of developing fruit is checked and on the basis off-type plants are rogued out. Fruit shape, colour, presence of spines, spines colour, colour of ripen fruit (green, yellow, white or orange) should be taken into consideration.

4. **Maturity stage**: At this stage it is essential to remove vine showing late maturity of fruits in the early variety and vice versa. Mosaic affected plants, which is seed borne should be carefully removed and destroyed.

The crop-wise main features described here under for effective rogueing.

☆ **Muskmelon**: Fruit shape, colour, rind colour, skin (netted/plain), flesh colour (orange/red), TSS and cavity size.

☆ **Watermelon**: Fruit shape, colour, rind colour, flesh colour (red/yellow/white).

☆ **Long melon**: Fruit shape, colour, bitterness.

☆ **Cucumber**: Fruit shape, colour, presence of spines, spines colour, colour of ripen fruit (green, yellow, white or orange)

☆ **Pumpkin (Squash, summer)**: Fruit shape, colour, flesh colour.

☆ **Gourds (Bottle gourd, Bitter gourd and Luffa etc.)**: Fruit shape, colour, stripe, neck *etc*.

14.6.0 Maturity of Fruit

Cucurbits take fairly long time to attain harvestable maturity. The maximum period is required in crops like pumpkin, ash gourd and watermelon, however, muskmelon, round melon, cucumber and bitter gourd take relatively less time. The maturity also influenced by the environmental factors and crop management (trailing *etc*.). The maturity period is shorter in summer season than rainy season. Harvestable maturity of few varieties of cucurbits in India have been shown in Table 10.

Besides days to maturity, some other parameters like change in colour also used as criteria of maturity index.

☆ **Cucumber**: Fruit turn pale yellow to golden yellow or brownish and attached with plant.

☆ **Pumpkin**: Fruit reddens and seeds inside the shell breaks readily from pulp.

☆ **Muskmelon**: Full slip stage.

Table 10: Harvestable Maturity of Cucurbits in India

Sl.No.	Crop	Variety	Period in Days (Seed to seed)
1.	Bitter gourd	Pusa Vishesh, Pusa Do Mausami	65 days
2.	Bottle gourd	Pusa Santusthi, Pusa Samriddhi, Pusa Naveen	75 to 80 days
3.	Pumpkin	Pusa Vishwas, Pusa Vikas	90 days (Rainy season)100 days
4.	Ash gourd	Pusa Ujjwal	90 days
5.	Cucumber	Pusa Uday, Poinsette	70 days
6.	Watermelon	Sugar Baby	85 days
7.	Muskmelon	Pusa Madhuras, Hara Madhu	80 days

☆ **Watermelon:** Fruits are ready for harvest when they reach edible maturity, fruit colour change from green/white to pale yellow of underside of the fruit.

☆ **Bitter gourd and snake gourd:** Fruit turn to bright yellow.

☆ **Bottle gourd:** At maturity fruit colour fade to straw green or pale yellow.

☆ **Luffa:** Complete drying/fruit turn to brown colour.

14.7.0 Choice of Season and Areas of Seed Production

Seed crop should be raised in such a seasons which remain dry at the time of seed maturity and seed extraction. Rainy season is preferred over summer season for raising seed crop.

Locations are also important in seed production with reference to seed yield and quality of seed. To harness the advantage of climate, private sector seed companies are organizing their seed production in these areas. The Jalna (Aurangabad) in Maharashtra, Ranibenur and around Bangalore in Karnataka, Nandyal Valley in A.P., are the main areas of seed production. Areas of major cucurbits seed production regions in India have been shown in Table 11.

Table 11: Major Cucurbits Seed Production Regions in India

Sl.No.	Crop	Areas	Region
1	Muskmelon, Long melon	Punjab and Haryana	Northern Region
2	Bitter gourd	Eastern U.P., Faizabad and Jaunpur	Northern Region
3	Cucumber, Muskmelon and Bottle gourd	Azamgarh, Ballia and Gonda in U.P., Jalna (Maharashtra)	Northern Region / Central Region
4	Pumpkin and Ridge gourd	West Bengal (South West)	Eastern Region
5	Sponge gourd, Watermelon, Cucumber	Telengana, Kurnool and Vijayawada (North A.P.) Ranibenur (Karnataka)	Southern Region / Southern Region
6	Watermelon, Bottle gourd and Bitter gourd	Costal districts of A.P.	Southern Region

14.8.0 Seed Extraction

There are two method of seed extraction employed in cucurbits.

(a) **Dry method:** The dried fruits are cut from one side and the seeds come out from the fruit *e.g.* sponge gourd, ridge gourd, snake gourd.

(b) **Wet method:** This method is employed for seeds extraction of cucumber, muskmelon, watermelon, ash gourd, bitter gourd, round melon and long melon. The fruit of cucumber and bitter gourd, summer squash and long melon are cut longitudinally and seed is scooped out while fruit of muskmelon and pumpkin are cut into two pieces and seed is scooped out from cavity.

Dry Method of Seed Extraction in Sponge Gourd (Left) and Ridge Gourd (Right).

Wet Seed Extraction Method of Cucumber.

However, in case of watermelon and ash gourd whole central portion are manually scooped out and macerated to separate the seed from pulp. In wet method, the seed extraction done by three ways:

(i) **Mechanical Extraction:** In this method the fruits are cut into pieces and macerated by machine. The seeds are separated out from pulp by floating with water. This method is quick, less expensive and seeds retain good lusture, but require good amount of water. This method is applicable in bottle gourd, watermelon, round melon and ash gourd.

(ii) **Natural Fermentation:** The scooped material kept in wooden/plastic or steel vessel for 48 hours at room temperature and stirred 2-3 times and then seed is washed thoroughly with water 2-3 times. The main problem with this method is discolouration and poor lusture of seed.

(iii) **Chemical Extraction:** 25-30 ml. of HCL or 8-10 ml. of commercial H_2SO_4 added per 5 kg of pulp and some quantity of water is mixed, stirring of pulp is done to enhance to separation and left for 30 minutes. The impurities will float and seed will sink. The seed should be washed thoroughly with clean water. This is quick method but accuracy of acid and time is important

14.9.0 Seed Yield

Seed yield depends upon the crop, variety, location, season and management of the seed crops. Average seed yield of different cucurbits is given below.

Cucumber 110-130 kg/ha, Bottle gourd, 200-250 kg/ha, Bitter gourd 150-200 kg/ha, Sponge gourd 200-250 kg/ha, Pumpkin 250-300 kg/ha, Muskmelon 150-160 kg/ha, Watermelon 250-300 kg/ha.

15.0 Hybrid Seed Production

Cucurbitaceous crops being cross-pollinated in nature exhibit wide range of variability with respect to yield and other morphological characters. They are distinct group where sex mechanism is unique and can be easily manipulated for easy and economic production of hybrid seeds. Further, low inbreeding depression, high heterosis percentage with respect to yield and economically desirable traits, favourarable genetic system like male sterility, gynoecy, comparatively easy technique of emasculation and pollination, availability of various morphological markers and chemical/growth regulators for sex expression and modification, advantage of getting large number of seeds from a single cross and low seed rate requirement per unit area can give distinct advantage in commercial exploitation of heterosis in cucurbits.

15.1.0 Techniques of Hybrid Seed Production

The recent advances in hybrid seed production in cucurbits has been reviewed by Robinson (2000), Sirohi (2004), Munshi (2013) and recently by (Munshi *et al.*, 2017), Tomar *et al.* (2017) and Thakur *et al.* (2016).

15.1.1 Bagging of Female Flowers and Hand Pollination

A day prior to anthesis, the female flowers before they open on the plants of female parent are covered by butter paper bags or buds of female flowers are tied with rubber band or clip specially where flower is large *e.g.* pumpkin. In the afternoon of the same day, unopened male flowers on the plants of male parent are protected from pollen contamination by bagging or tying the petals. In the next day morning (7-11 a.m.), pollen is applied on the stigma of the protected female flowers. After hand pollination, the female flowers are again bagged or petals are tied. This practice is applicable in most of the cucurbits except muskmelon where andromonoecious sex form is predominant. F_1 seed is collected from the mature fruits harvested from the plants of female parent. In muskmelon, utilization of monoecious lines has been advocated.

Satish (2005) reported that there is no significant differences among planting ratio for seed yield and seed attributes in bitter gourd. Kushwaha and Pandey (1998) concluded that 4:1 planting ratio for hybrid seed production in bottle gourd is better for good yield and quality. Sharma *et al.* (2004) observed that 3:1 ratio is suitable for hybrid seed production of cucumber.

**Bagging of Female and Male Flowers Followed by
Hand Pollination and Tagging of Female Flower.**

15.1.2 Emasculation and Hand Pollination

This method is practiced in muskmelon where andromonoecious sex form is predominant. In the plants of female parent, the hermaphrodite flowers are emasculated a day prior to anthesis and protected by bagging. The male flowers on the plants of intended male parent are also protected. Pollination is done in the early morning of next day by taking pollen from protected male flowers of male parent. Hybrid seed is collected from the mature fruits harvested from the plants of female parent.

Since muskmelon, which is andromonoecious in sex form, hand pollination for production of F_1 hybrid seeds involves emasculation of perfect flowers. Naturally, the cost of hybrid seeds is high but the value of hybrids is quite superior such that the USA imports hybrid seeds of andromonoecious cantaloupe parents from

Taiwan, Israel, Chile and Mexico, where labour cost is comparatively less. However, according to Munger (1942), muskmelon fruit contain large number of seeds and hybrid seeds can be manually produced by emasculation and pollination. About 3000 viable seeds are required to produce seedlings for one acre. These seeds can be obtained from at the most 10 fruits, as even small size fruit yields more than 300 seeds. Only ten successful pollinations are required to produce hybrid seeds sufficient for planting in an acre. Hybrid seed can also be produced by using andromonoecious line having small number of male flowers. In such case, male flowers were removed before anthesis at two or three days interval. However, total elimination of male flowers was difficult to achieve and this results in contamination of hybrid seed.

15.1.3 Use of Monoecy

Considering the problems associated with the use of male sterile gene(s) for production of muskmelon hybrid (maintenance of single recessive gene in heterozygous condition and problem of identifying and rogueing of 50 per cent male fertile plants from the female row at the time of flowering and also to overcome the andro-monoecious forms), the scientists at ICAR-IARI, New Delhi took keen interest in developing true breeding monoecious lines *viz.*, *M1, M2, M3* and *M4*. Of various combinations developed, M3 × Durgapura Madhu *i.e.* Pusa Rasraj was found to be outstanding in performance and has been released for commercial cultivation.

The andromonoecious nature of muskmelon involves tedious process of emasculation of perfect flower in female line for hybrid seed production. Though mostly it is andromonoecious, some genotypes show monoecious sex expression. Monoecious lines reduced the extent of self pollination in female parents, when seed production is done under open pollination with adjacent rows of male and female parents. Since use of monoecy exclude emasculation it can reduce the time required for a given number of pollination by 50 per cent and enhance fruit set by 40-70 per cent as compared to 5-10 per cent in andromonoecious parent. The problem in using monoecy in hybrid seed production is undesirable linkage between genes controlling monoecious sex expression on the fruit shape, consequently F_1 combinations with round fruits cannot be easily obtained. The problem was evident in Pusa Rasraj (*M3* hybrid) which was not acceptable commercially because of its undesirable fruit shape and poor external appearance.

15.1.4 Pinching of Male Flower Bud

This method is the most simple and economical and can easily be adopted by the growers who can identify male or female flower bud. This technique is practicable in those cucurbits like watermelon, bottle gourd, bitter gourd, *etc.*, where male flowers are produced singly in long pedicels, to facilitate pinching at bud stage.

F_1 hybrid production can economically be done on large scale by pinching all the male flower bud before opening from the female parent and allowing the male parent to grow side by side of female parent for natural cross-pollination. One row of male parent can be sown after every three rows of female parent as suggested by Choudhary and Singh (1971) for producing F_1 hybrid seeds on large scale in bottle gourd and pumpkin. All the fruits set in female parent would be necessarily through cross-pollination by insects. For this, as a precaution, there should not be a single male bud in female parent as it will promote self or sib-pollination within the female parent. Anthesis of the flowers of bottle gourd is in afternoon, the pinching operation should therefore be done in the forenoon. Isolation distance between different varieties should be kept about 400 metres. As the male flowers in bottle gourd, pumpkin and squash are quite big, showy, having long pedicels and less in number, the pinching operation can easily be performed. In muskmelon, cucumber and *Luffa*, where male flowers are small and produced in clusters or in racemes, pinching of male buds will not be effective and hence this technique will not workout satisfactorily. The precaution to be taken is not to leave a single male bud in female parent because it will promote self-or sib-pollination within the female parent for maximum fruit set and seed yield, availability of pollinator is pre-requisite. One medium sized bee colony per hectare would be enough in seed production block.

15.1.5 Utilization of Male Sterility

First recessive male sterile gene *ms-1* in muskmelon was reported by Bohn and Whitaker (1949). Since then at least four additional male sterile recessive alleles *viz.*, *ms-2* (Bohn and Principe, 1962), *ms-3* (McCreight and Elmstrom, 1984), *ms-4* (Pitrat, 1990) and *ms-5* (Lecouviour *et al.*, 1990) had been identified. The phenotypes of *ms-4* plant is different from *ms-1*, *ms-2*, *ms-3* and *ms-5* (McCreight, 1993). Sterility in all these male sterile mutants is monogenic recessive. Male sterile line *ms-1* identified by Bohn and Whitaker (1949) had proved boon to the hybrid seed industry. In summer squash (Eisa and Munger, 1968), winter squash and watermelon (Watts, 1962) were the first to report male sterile line. The line showed good nicking ability when used in a series of hybrid combination. Punjab Hybrid, a F_1 between MS-1 × Hara Madhu was released in Punjab (India) and subsequently at national level in 1984. This hybrid still occupies a larger percentage of muskmelon growing area. Genic male sterility in muskmelon is maintained under heterozygous (*Msms*) condition in isolation block by crossing with the recessive (*msms*) parent every year. For commercial seed production, heterozygous seed stock (*Msms*) is grown which segregates into 50 per cent heterozygous male fertile (*Msms*) and 50 per cent homozygous male sterile plants (*msms*). The male fertile plants are removed from the population before flowering where as the sterile plants are kept for hybrid seed production. For maintenance of sterility, a male sterile line is sown in an isolated field. At flowering, each plant is examined and marked as male sterile or male fertile. An adequate supply of bees is provided. After the fruit set, all male fertile plants are rogued before the fruit becomes mature and seed is harvested only from male sterile plants. These seeds, which will segregates as male sterile and male fertile plant in 1:1 ratio serves as a stock seed to repeat the cycle. For hybrid seed production, the same seed is planted at the ratio of 3:1 (female: male). Male fertile plants are identified and rogued at the appearance of first male flower. Hybrid seed is harvested from the fruits which develop on male sterile plants. To get maximum hybrid seed, six seedlings per hill are planted so that sufficient population is maintained after rogueing the fertile plant. Commercial exploitation of genic male sterility is handicapped by identification of male fertile plants before flowering.

Hence male sterility has not been exploited in any of the cucurbits for commercial hybrid seed production except Punjab Hybrid 1 by Punjab Agricultural University, Ludhiana. However, male sterile line *ms-3* in muskmelon had been found superior to *ms-1* and *ms-2* due to the fact that male sterile plants are very easy to identify and possess superior horticultural traits. In male sterile line *ms-5*, male flower buds abort pre-maturely, hence male sterile plants can easily be identified. Use of markers gene to simplify the procedure of identification and hybrid seed production was reported by Foster (1968) by utilizing glabrous seedling markers which was controlled by single recessive gene. This could eliminate the tedious method of identification of male sterile plant and keep down the cost of hybrid seed production. Efforts were made by Mishra (1981) to incorporate male sterility and genetic markers in monoecious lines to facilitate the identification and rogueing those seedlings on germination. It was expected that male sterility coupled with marker in monoecious sex form would promote greater cross pollination for the production of hybrid seed. But 100 per cent open pollinated pure seed could not be produced on field scale even with complex female parent monoecious + genic male sterility + marker gene, because male sterility was being simply inherited recessive character, could not be maintained under homozygous condition. The poor linkage of genetic marker with ms line was suggested by Sandha and Lal (1999). Functional male sterility in muskmelon was induced by applying two sprays of 0.3 per cent FW 450 (Sodium alpha, beta dichloroiso-butyrate). However, this method had not been exploited commercially for the production of F_1 hybrid. In China, a genetic male sterile (*ms*) watermelon line G17AB, which shows no major morphological differences between male sterile and male fertile plants has been used as the maternal parent of F_1 seed production. It has been crossed with lines carrying the glabrous male sterility gene (*gms*) to obtain a new male sterile type (Zhang and Wang, 1990). Preliminary studies of a dwarf, male-sterile watermelon (DMSW) by HeXun *et al.* (1998) showed that male-sterility was controlled by a pair of recessive nuclear genes. The dwarf gene, named *dw-3* was different from the 3 known watermelon dwarf genes (*dw-1*, *dw-1ˢ* and *dw-2*). Pollen aborted completely in male-sterile plants. The DMSW plants were 1.5m tall, with fewer leaf lobes than normal plants, a trait which could be identified when 2-3 true leaves have developed. Dyutin and Sokolov (1990) found a spontaneous mutant in cv. Kamyzyakskii to have male sterility controlled by a sin gle recessive gene designated *ms2*. Unlike other mutants of this type, it had normal seed production and could be used for obtaining hybrid seed by open pollination.

Table 12: Male Sterile Lines Reported in Cucurbits for HSP

Muskmelon	*ms-1*	Bohn *et al.* (1949)
	ms-2	Bohn *et al.* (1962)
	ms-3	McCreight *et al.* (1984)
	ms-4	Pitrat *et al.* (1990)
	ms-5	Lecouviour *et al.* (1990)
Watermelon	*msg* (male sterile mutant)	Watts *et al.* (1992)

A male sterile line with rudimentary male flowers was reported first time in ridge gourd (Pradeepkumar *et al.*, 2007) and was maintained through micropropagation. Further (Pradeepkumar *et al.*, 2012) first time reported of cytoplasmically controlled male sterility (CMS) in ridge gourd where two dominant male fertility restorer nuclear genes with complementary gene action governing the restoration of male fertility. Recessive mutants are reported in watermelon. Linkage of *ms* gene with delayed-green (*dg*) seedling marker gene and glabrous seedling marker was reported in watermelon. Monogenic recessive gene for male sterility is reported in cucumber and summer squash. However, the scope for utilization of male sterility for improvement of cucumber and summer squash is limited because of availability of gynoecious lines and availability of sex regulating mechanism using growth regulators, respectively (Rai *et al.*, 2004).

Table 13: Male Sterile Line with Marker Gene

Male Sterile Line with Marker		
MSDG-1 and MSDG-2	Male sterile with delayed green seedling marker	Zhang *et al.* (1996)
93JMSB-1, 93JMSB-1-1, and 93JMSF3-2	Juvenile albino seedling male sterile line	Zhang *et al.* (1996)

15.1.6 Utilization of Gynoecious Lines

Gynoecy condition where all the flowering nodes in the primary, secondary and tertiary branches bear pistillate flowers in the leaf axils, is so far the most important sex form which has made phenomenal exploitation of hybrid vigour possible in cucumber, bitter gourd and also in muskmelon.

Cucumber is the crop most extensively studied in the Cucurbitaceae for the production of hybrid seed. Among the many types of cucumber cultivars, it is possible to find gynoecious genotypes, *i.e.* plants that only have female flowers. Parthenocarpy, or the development of fruits without fertilization and seed formation, is another important trait available for cucumber breeding. Gynoecious cultivars with parthenocarpic fruits are usually preferred for green house production because of their higher yields and ease in crop management.

The commercial production of gynoecious cucumber seed was made possible only when it was discovered that gynoecious inbreds could self reproduce if a growth regulator is applied to induce male flower formation (Robinson, 1999). Peterson and Anhder (1960) for the first time discovered the effect of gibberellic acid (1500-2000 ppm) on promotion of male flower formation in gynoecious cucumber. A problem observed with gibberellic acid application is that different gynoecious lines vary in response to GA application and, in some cases, the number of induced male flowers was not sufficient for hybrid seed production. Additionally, GA applications typically cause excessive stem elongation or malformed male flowers (Robinson, 2000). Because of the erratic male flower induction by use of gibberellic acid, application of silver compound such as silver nitrate (250-400 ppm) is done to induce male flowers. Silver ions inhibit ethylene action and thus promote male flower formation in gynoecious cucumber plants (Beyer, 1976). However, due to phytotoxic effects of silver nitrate such as burning of plants, silver thiosulphate (400 ppm) is now widely used by seed producers for the maintenance of gynoecious cucumber lines. It induces male flowering of cucumber plants over a longer period and is less phytotoxic compared to silver nitrate. Hybrids of cucumber are produced mainly by crossing gynoecious lines with monoecious lines. Though, other systems of producing gynoecious hybrid seed such as gynoecious x gynoecious have been proposed but gynoecious × monoecious hybrids are still the most widely grown. The hybrids produced by the cross of a gynoecious and monoecious line resulted in hybrid vigor and a high degree of female sex expression, with uniform and concentrated fruit formation, which was especially advantageous for mechanical harvest (Robinson, 1999, 2000). Most of the commercial hybrids based on gynoecious cucumber lines are a blend of gynoecious hybrid and monoecious seed. About 10 per cent blending of monoecious genotype seed with gynoecious hybrid seed has been advocated (Peterson and De Zeew, 1967). This practice improves pollination, which is required for fruit set in genotypes that are not parthenocarpic. However, it has the disadvantage of affecting uniformity, which is one of the principal advantages of hybrid cultivar production (Robinson, 2000). In parthenocarpic gynoecious F_1 hybrid, blending is not required as fruits develop without pollination and this is advantageous for production of cucumber F_1 hybrid. In addition, homozygous gynoecious hybrid seed has been produced by crossing two gynoecious lines after one has been treated with a growth regulator to induce male flowers (Robinson, 1999). When two gynoecious inbred lines homozygous for the gene F are crossed, the resulting F_1 hybrid is homozygous for F. These hybrids are more stable for gynoecious sex expression compared to hybrids produced by crossing gynoecious and monoecious lines. In the case of hybrids heterozygous for the gene F, some environments such as high temperature and long days may promote the development of male flowers, which is less likely in gynoecious × gynoecious hybrids. The stability of gynoecious sex expression is especially important for gynoecious parthenocarpic F_1 hybrids used for green house production. These hybrids produce long, seedless fruits in the absence of pollination. However, when female flowers of parthenocarpic gynoecious hybrids are pollinated by hand the formation of seeds enlarges the fruits at the blossom end, affecting their shape and quality. Using homozygous gynoecious hybrids reduces the chance of pollination and misshapen fruit because these plants produce no pollen (Robinson, 2000).

The gynoecious trait in cucumber is determined by a single dominant gene "F". Since these plants have only female flowers, hybrid seeds may be produced using gynoecious maternal lines without the requirement for male flower emasculation. It is observed that gynoecious sex forms are stable only under moderate regimes of temperature and photoperiodic conditions. From USA, multiple disease resistant gynoecious inbred lines of cucumber $Gy4$ and $Gy5$ with high combining ability have been reported (Lower et al., 1991; Wehner et al., 1991). These lines are capable of producing high yielding hybrids when crossed to monoeceous inbred lines. However when the temperature exceeds beyond 30°C the stability gynoecious sex expression of these is affected. Unfortunately, the temperate gynoecious lines are unstable for gynoecy under high temperature and long photoperiodic conditions because of their thermo-specific response for gynoecious stability. That is why the gynoecy in cucumber did not receive much attention in the tropical countries and there is a paramount need to develop suitable hybrids, which may be utilized on commercial scale especially in the north Indian plains. Keeping in view the facts, at IARI New Delhi, inheritance traits gynoecious sex expression in cucumber was studied by Pati *et al.* (2015) and Jat *et al.* (2018) and they also confirmed monogenic dominant nature of the traits. Further, efforts have been made during recent past towards developing gynoecious sex forms under our tropical and indigenously adapted background and two tropical gynoecious lines DGC-102 and DGC 103 having stable true gynoecious sex even at temperatures around 35°C to 40°C have been developed at IARI (Munshi *et al.*, 2018). These are being exploited in heterosis breeding programme. Identification of SSR markers (*SSR13251, SSR11798, SSR15516*) closely linked with gynoecious trait covering all 7 chromosomes in cucumber has been reported for further effective introgression of the trait in monoecious cultivar (Jat *et al.*, 2018).

A gynoecious based hybrid Pusa Gynoecious Hybrid 18, the first notified gynoecy based hybrid in cucumber from Public Sector of India (Anonymous, 2020) was developed by utilizing tropical gynoecious line DGC-102 at ICAR-IARI, New Delhi.

However, for green house/polyhouse cultivation hybrid seeds of gynoecious parthenocarpic F_1 hybrids are produced through hand pollination by crossing two gynoecious parthenocarpic parents (one parent phenotypically modified as male by using silver thiosulpuate as growth regulator). The gynoecious parthenocarpic F_1 hybrids will produce seedless fruit without pollination which is most important requirement for polyhouse since there is no provision of honey bees inside polyhouse.

The utilisation of gynoecy is economical and easier for exploiting hybrid vigour in many cucurbits including cucumber (Pati *et al.*, 2015; Jat *et al.*, 2018) and bitter gourd (Behera, 2004). Gynoecious line in bitter gourd has been reported in recent past in India by Ram *et al.* (2002) from Varanasi and Behera *et al.* (2006). Behera *et al.* (2009) and Dey *et al.* (2010) from New Delhi reported two gynoecious lines *viz.* PVGy-201 in the back ground of Pusa Vishesh and PDMGy-201 in the background of Pusa Do Mausami. They were maintained and utilized in the hybrid breeding programme. Induction and morphological characterization of hermaphrodite flowers in a gynoecious line of bitter gourd by silver nitrate, gibberellic acid and silver thiosulfate was reported by Behera *et al.* (2010-2011) and Mishra *et al.* (2014). The inheritance pattern of gynoecy in bitter gourd was further confirmed by Misra *et al.* (2015). Use of gynoecious line as one parent in hybrid breeding showed positive impact in terms of yield and earliness (Dey *et al.*, 2010; Gangadhara Rao *et al.*, 2018) in bitter gourd. Gynoecious lines in bitter gourd from its feral form, *Momordica charantia* var. *muricata* (small fruited type) was isolated. The segregation pattern in F_2 suggested that gynoecious sex form in bitter gourd was governed by single recessive gene coined as *gy-1* gene (Behera *et al.*, 2009). Since gynoecious parent has only female flowers, the open pollinated seeds it produces will be F_1 hybrid. It will reduce the cost of male flower pinching and hand pollination. Two gynoecious lines,

DBGY-201 and *DBGY-202* were crossed with two monoecious cultivars, "Pusa Do Mausami" and "Pusa Vishesh" and their F_2 populations were observed to determine the inheritance pattern of gynoecious sex form. Another experiment was conducted to determine magnitude of heterosis by utilizing nine inbred lines including one gynoecious line, *DBGY-201* in diallel (without reciprocal) mating system. All the gynoecious hybrids showed significant heterosis in a desirable direction for traits like sex ratio (male:female), days to first picking, number of fruits per plant, yield per plant, and vine length. The gynoecious hybrids *DBGY-201* × Pusa Vishesh and *DBGY-201* × Pusa Do Mausami were important for early harvest (<"50 days after sowing) whereas *DBGY-201* × Priya and *DBGY-201* × Arka Harit were promising for higher yield and yield contributing characters (Behera *et al.*, 2009).

A gynoecious based hybrid of bitter gourd Pusa Hybrid-4, the first notified Gynoecious based hybrid from India by Public Sector (Anonymous, 2018) was developed by IARI, New Delhi. Two gynoecious lines (IIHRBTGy-491 and IIHRBTGy-492) were also identified in bitter gourd (Varalakshmi *et al.*, 2014).

In muskmelon by using gynoecy in heterosis breeding, the tedious emasculation and identification and rogueing of male fertile plants from the mixed population is avoided. Probably, for this reason, Frankel and Galun (1977) and Loy *et al.* (1979) advocated the use of gynoecious lines in hybrid seed production. The exact genetic make up of gynoecism in muskmelon is not yet understood due to influence of temperature and photoperiod for its expression. This breeds true and produces all pistillate flowers under short day and low temperature conditions, but when grown at high temperature and long day it becomes gynomonoecious (Kubicki, 1969e). Wisconsin 998 was the first gynoecious line in muskmelon developed by Peterson *et al.* (1983). Wisconsin 998, when used in hybrids seed production, has exhibited good combining ability for yield and earliness (Lal and Dhaliwal, 1993). A cross between W1 998 × Punjab Sunheri (MHL-10) was found promising and was released for commercial cultivation in Punjab during 1995. It was not only early and high yielding but also had good shipping quality (Lal and Dhaliwal, 1996; Dhaliwal and Lal, 1996). Procedure for commercial hybrid seed production of MHL-10 was discussed in detail by Lal (1995). Unlike cucumber, gynoecious lines of muskmelon could not be maintained by GA, however, perfect flower induction was reported with the application of 5-Methyl-7-Chloro-4-ethoxy carboxy- methoxy 2, 1, 3, benzothio-diazole (MCEB) at the fourth true leaf stage and following hand pollination, selfed seeds were produced (Byers *et al.*, 1972). Use of $AgNO_3$ @ 100-200 ppm (Owen *et al.*, 1980) induced production of 12 perfect flowers on the first 20 nodes, which can be used to self or sib for maintenance of gynoecious lines. Silver thiosulphate (More and Sheshadri, 1987) had also been suggested for inducing maleness. More *et al.* (1987) had developed three true breeding gynoecious line *viz.* 86-104, 105 and 118 and by subsequent selfing and selection for good horticultural characters and stability for gynoecy, finally seven lines namely GH 3-2, 5E-1, 7-7, 4D, 5D, 6C-4 and 6E-7 were selected (More *et al.*, 1991). Several cross combinations were made by the above but none of them were found to be commercially acceptable from quality point of view having low T.S.S. The utilization of gynoecious lines for the production of hybrid seed in muskmelon suffers with various types of limitation and therefore further needs to search the possibility to utilize the existing gynoecious lines for the production of hybrid.

Inheritance of hermaphrodism of the cross of ridge gourd Pusa Nutan and Satputia Long (DSat-4) revealed hemaphrodism is governed by 2 recessive genes. On the basis of the information for the first time gynoecious lines were developed by crossing *Luffa acutangula* and *Luffa hermaphorodita* (Munshi *et al.*, 2010-11). On the basis of the information gathered from inheritance of sex form in *Luffa acutangula* for the first ime gynoecious lines were developed by crossing *Luffa acutangula* and *Luffa hermaphorodita* and 10 gynoecious lines were evaluated on the basis of colour, shape and size. DRGGL-8 line showing true gynoecious character found to be most promising and utilized in hybrid breeding programmee (Anonymous, 2019).

Table 14: Gynoecious Lines Reported in Cucurbits for HSP

Cucurbits	Lines	Sources
Bittergourd	DBGy-201and DBGy-202	Behera *et al.* (2002)
	IIHRBTGy-491, IIHRBTGy-492	Varalakshmi *et al.* (2014)
Muskmelon	Wisconsin 998 (WI 998)	Peterson *et al.* (1983)
	86-104, 105 and 118	More *et al.* (1987)

15.1.7 Use of Morphological Markers

Use of morphological markers for hybrid seed production was first reported by Foster (1963) by utilizing glabrous seedling markers in muskmelon which was controlled by single recessive gene. This could eliminate the tedius method of identification of male sterile plant and keep down the cost of hybrid seed production. In watermelon hybrid seed production was facilitated by with single gene recessive delayed green seedling marker (*dg*) linked *ms* male sterile line, Juvenile albino Seedling Marker (*ja*) and yellow Spotted dominant marker (*Sp*) were reported and utilized in hybrid seed production (Watts, 1962; Rhodes, 1986; Zhang *et al.*, 1996). Ideally, seedling markers that are useful for hybrid seed production should be controlled by a single recessive gene and incorporated into the seed parent because the abnormal phenotype(s) conferred by the recessive gene will not be expressed in the hybrid. Furthermore, off-types in the seed parent, resulting from outcrosses during reproduction and mixes during seed handling, can be recognized before pollination.

Various morphological markers can be exploited for hybrid seed production of cucurbits:

   i) Glabrous seedling in watermelon and muskmelon

   ii) Yellow leaf (monogenic recessive) in muskmelon

   iii) Non-lobed leaf in watermelon (monogenic recessive)

   iv) Segmented leaf in bottle gourd (PBOG 54) (monogenic dominant)

   v) Silvery patches on leaf of pumpkin (monogenic dominant)

   vi) Protuberant and non-protuberant behaviour in bitter gourd

vii) Pubescence on bottle gourd fruit

viii) White flowered bitter gourd

  ix) Striping and non-striping in pointed gourd

   x) White seeded sponge gourd

Foster (1968) in a series of experiments highlighted the importance of monoecious parents in producing hybrids by cheaper method of hybrid seed production in muskmelon. He showed by using a marker gene, glabrous that field crossing could produce 30-35 per cent F_1 hybrid seed using andromonoecious lines and 60 per cent using monoecious lines and 75 per cent when male sterility and monoecism were combined with marker gene in seed parents. Marker character would facilitate rogueing of self/sib-pollinated seedlings occurring as mixture in hybrid population. Male sterility has not been of much of an advantage in melon hybrids as indicated earlier in hybrid seed production.

15.1.8 Use of Growth Regulators and Chemicals for Temporary Suppression of Male Flowers and Open Pollination

Use of growth regulator and chemical is yet another method of hybrid seed production by the use of chemicals. With the advent of growth regulating substances, very significant results have been obtained with regard to sex modification in cucurbits. It has now been possible to prove that the two true-leaf stage is the most responsive stage for application of chemicals for sex modification. Specific chemicals are known to induce femaleness or maleness as desired. In cucurbits like bottle gourd, pumpkin and squash female flowers can be increased by the application of ethrel (2-chloroethyl-phosphoric acid) at the rate of 200-300 ppm at two true-leaf and four-leaf stage. Ethrel helps in suppressing the staminate flowers and initiating pistillate flowers successively in the first few flowering nodes on the female parent. The row of male parent is grown by the side of female parent and natural cross pollination is allowed. In the absence of insect pollinators, hand pollination is possible when two sexes are separate. Precaution should be taken that during natural cross pollination, there should not be any other variety except the parents of hybrid. Four to five fruits set at initial nodes containing hybrid seed would give sufficient seed yield. Complete suppression of male flowers in summer squash can be achieved by the application of higher concentration of ethrel at 400-500 ppm, applied twice, which has made hybrid seed production comparatively easier (Taylor, 1983). However, Edelstein *et al.* (1985b) suggested four applications of ethephon in summer squash (as ethrel 1500-600 ppm) at 2, 4 and 6 leaf stages and again a month later at fruitset. Ethephon treatments delayed the appearance of staminate flowers by 28 days compared with the beginning of pistillate flowering. The period was long enough for honeybees to produce 400-500 kg/ha of hybrid seeds.

There are different compounds that induce of femaleness in cucurbits. Exogenous application growth regulators *viz.* MH (100-200 ppm) or Ethrel or Ethephon (150-200 ppm) for temporarily suppression of male flower in the female line of monoecious plants has been commercially adopted for producing F_1 hybrids. Ethrel releases ethylene gas, which has been used in hybrid seed production. The main practical advantage of ethrel is its persistent effect on some species. The response to ethrel varies among species and cultivars. For example, in monoecious cucumber, it can induce the formation of only female flowers for enough time to permit hybrid seed production by open pollination. Ethrel is less effective in promoting femaleness in melon and watermelon. When used in hybrid seed production of monoecious cucumbers, ethrel is applied at 2-4 true leaf stage. These applications suppress formation of male flowers during the pollination period. Because of the importance of avoiding self fertilization in hybrid seed production, the plants should be checked periodically and any male flowers that appear should be removed before anthesis. Once development of male flowers in the female line has been suppressed, bees can be used for pollination. Five to six beehives per hectare are recommended to ensure successful pollination. In the field, the proportion of female and male plants is usually 3 to 4 female plants per each male plant. An important practical implication of using ethephon in hybrid seed production of cucumber is that the plants must be very uniform when the product is applied. This requires good agronomic management of the crop, which includes adequate soil preparation, sowing, irrigation, fertilization, and weed and pest control. If the plants are uneven in growth at the time of ethephon application, some plants will receive the product at differing stages of development causing those plants to produce male flowers that results in undesired pollen contamination.

In muskmelon also, application of ethrel at 2-3 true leaf stage induced temporary suppression of male flower and produced pistillate flower at early node, thus the ethrel sprayed plants behave as female line at an early stage. Foliar application of ethephon induced temporary gynoecious stage (averaging 7-19 days, during which development of male bud was inhibited and bees could be used to pollinate flowers). The method of hybrid seed production was the selection of any two good combination on the basis of heterosis percentage out of which female line should be monoecious and planting them in the ratio at 4:2 (monoecious : pollen parent), where formers was sprayed with growth regulator. The monoecious line would produce female flowers at early nodes and at the same time, the corresponding pollen parents produce male flowers which hybridize the female flower of the female line. Pollinator rows would be destroyed after fruit set. The first few fruits produced in the female lines would be F_1 hybrid. However, according to More and Sheshadri (1975, 1998) even phenotypically, a temporary change of sequence of flowers attempted in monoecious seed parents through exogenous application of 2 chloro-ethyl-phosphonic acid had shown the distinct possibility of inducing perfect and female flowers at earlier nodes (five to six) in andromonoecious and monoecious lines. Its practical use on larger scale hybrid seed production under open field yet to be is worked out.

In monoecious cucurbits commercial seed production under open field condition in handicapped because of huge mixture of self or sibbed seeds in the hybrid seeds.

15.1.9 Hybrid Seed Production of Cucurbits under Protected Structures

The lack of sufficient isolation, insect's vector, diseases and a virus free environment in the production of disease free, healthy and genetically pure seed for commercial cultivation are the major challenges in quality hybrid seed production of cucurbit vegetable crops. Compared to open filed condition, protected cultivation can fetches higher seed yield with better quality (Tomar and Jat, 2015). Insect's vectors and viral diseases are the most devastating problems for quality seed production in most of the vegetable crops grown under open fields, and if the insect vectors are checked by protected structures the use of pesticides will automatically reduce. The seed production in summer season is affected by sudden increase in temperature and severe infestation of mottle mosaic virus and other insect pests in rainy season; against which still there is no effective and reliable management measure. The change in climatic conditions like unseasonal rains during April- June, increased temperature drastically reduced the seed yield and quality even in the summer season crop. Raising seed crop in insect-proof net house can overcome these problems by protecting the crop from various insect vectors and unfavourable climatic conditions for quality hybrid seed production of cucurbits. The major interest is to grow virus free seed crops and protection against major insect/pests as compared to open field condition (Jat *et al.*, 2015; Jat *et al.*, 2016). Semi-climate controlled green house is suitable for hybrid seed production of, bitter gourd, summer squash and parthenocarpic cucumber hybrids. Seed yield of such crops can be 3-4 times more compared to their open filed cultivation (Kaddi. 2014; Kalyanrao *et al.*, 2012; Jat *et al.*, 2017). Similarly, naturally ventilated green house is also suitable for hybrid seed production, where the seed yield is usually 2-3 times more over open field, but the cost of seed production is only 1/3 of the seed produced under semi-climate controlled green house condition (Kalyanrao *et al.*, 2014; Singh and Tomar, 2015). A comparative evaluation of hybrid seed production of bitter gourd was carried out in rainy and spring-summer season under net house. The results showed high pollen viability and stigma receptivity in both the seasons, however, higher fruit yield, fruit weight were achieved in rainy season whereas the fruit setting percentage and seed quality were superior in the spring-summer season. The last week of April to first week of May in spring summer, third and fourth week of September in rainy season were found to be better for hybrid seed production than later weeks of pollination in both the seasons (Sandra *et al.*, 2018).

Effect of plant growth regulators on sex expression, fruit setting, seed yield and quality in the parental lines for hybrid seed production in bitter gourd Pusa Hybrid 1 and Pusa Hybrid 2 was carried out (Nagamani *et al.*, 2015). In manually pollinated flowers, plants sprayed with GA_3 @ 50 ppm had higher fruit and seed setting, fruit weight and hybrid seed yield. All the growth regulators had positive influence on vegetative, flowering and fruit traits in both the seasons but effect of growth regulators were more evident in rainy than spring summer season. GA_3 @ 50 ppm, NAA @ 200 ppm, etherel @ 50 ppm were found most effective for enhancement in vegetative growth, fruit and seed yield for hybrid seed production of bitter gourd (Nagamani *et al.*, 2015).

The major advantages of hybrid vegetables seed production under protected conditions are:

1. Higher seed yield (generally 2-4 times more) and seed quality as compared to open field.

2. Requirement of isolation distance in cross pollinated vegetables can be minimized.

3. Problem of synchronization of flowering can be minimized.

4. Maximum plant population can be maintained.

5. Seed production under adverse climatic conditions is possible.

6. Training, pruning and hand pollination practices are very easily manageable under protected conditions compared with to field seed crop.

7. Emasculation of female parents is not required as there are no insect pollinators.

8. Seed crops will not be damaged by un-seasonal rains at the time of their maturity and seed viability, seed vigour could be extended through better nutrient management under protected conditions.

16.0 Crop Improvement

Cucurbitaceous vegetables form a distinct group among the cross-pollinated vegetable crops, in respect of methods of improvement. Besides the wide ranging sex forms and sex expression, which favour outbreeding, they are not completely cross-pollinated and as stated earlier; natural self-pollination does take place even though in a less degree. Another characteristic of this group of crops is that they do not suffer much from inbreeding depression or do so only negligible, in contrast to other cross-pollinated crops like cabbage, onion, carrot, *etc.* This phenomenon allows to breed cultivars through pureline or single plant selection. In fact, some earlier cultivars bred by farmers and breeders in the beginning of this century were through pureline selection. It does not, however, mean that mass selection is unsuitable, but it has got its value.

16.1.0 Breeding Objectives and Selection Criteria

The objectives of cucurbits breeding are highly crop specific but in general following aspects are taken into consideration during breeding programme.

1. **Sex ratio:** A high female to male ratio desirable.

2. **Growth habit:** Preference should be given to medium vine characteristics.

3. **Maturity:** The number of node at which first pistillate (or hermaphrodite) flower appears, gives a fair indication of earliness, especially in those kinds where immature fruits are edible.

4. **Fruit shape:** Spherical or oblong in watermelon; spherical, flat or oblong in muskmelon; round, long, cylindrical, club shaped in bottle gourd; round, oblong, spherical shaped in pumpkin. A shape having high flesh recovery per volume, especially in muskmelon and pumpkin would be desirable. In salad or slicing cucumber, uniform long cylindrical shape without neck is desirable.

5. **Fruit surface:** Attractive dark green with minimum of prickles in slicing cucumber. Netting and thick skin are useful in transportable quality in

muskmelon and toughness of skin helps in watermelon for long distance haulage. In bitter gourd, continuous ridges and blunt tubercles are preferred. A non-ridged fruit skin is attractive in pumpkin. Sparse hairs persisting on the skin of bottle gourd and round melon indicate tenderness of fruit and edible maturity.

6. **Flesh colour and thickness:** The flesh colour in muskmelon may be white or greenish or salmon-orange while in watermelon it can be pink or pale pink or deep pink and in pumpkin yellow, yellowish orange or deep yellow colour is met with. The flesh should be thick with small seed cavity in muskmelon and pumpkin. In cucumber, carpel separation showing hollow spots at edible maturity is undesirable. In most of the gourds where immature fruits are harvested like bottle gourd, ridge gourd, sponge gourd, round melon, snake gourd, *etc.*, the flesh should not become fibrous quickly in optimum sized fruits at edible maturity. Carotenoids produce pigmentation which results in the range of whites, creams, yellows, and oranges in the flesh color of squash and pumpkin. The expression of this flesh color is conditioned by the particular carotenoid type and concentrations which are influenced by genetic and environmental factors. The goal of the cucurbit breeding should be to investigate the various levels of carotenoids present and identify molecular markers associated with these carotenoids within pumpkin, summer and winter squash.

7. **Sweetness**: The sweetness in melon is judged as total soluble solids content read by squeezing the juice in a hand refractometer, subject to varying conditions of environment, 10 per cent TSS may be taken as a minimum standard for selection of fruits to meet consumer acceptance.

8. **Bitterness**: Bitterness in fruits of cucumber, bottle gourd, ridge gourd, *etc.* should be eliminated.

9. **Flavor**: High aroma in muskmelon is preferred by consumers.

10. **Higher early yield and total yield**: The yield in terms of weight and number of fruits of marketable quality will be the deciding criterion in selection. Misshapen and poorly developed fruits should be rejected. Early yield, especially of higher percentage of total yield is a desirable character.

11. **Seed**: Fewer and smaller seeds are desirable in watermelon. In cucumber, bottle gourd, bitter gourd, parwal, tinda *etc.* Fruits should not produce mature seeds at edible maturity.

    Wider adaptability, cold tolerance and photoinsensitivity.

12. Resistance/tolerance against biotic and abiotic stresses.

13. Parthenocarpic gynoecious and indeterminate varieties for protected cultivation especially in cucumber.

14. Bunching fruit habit producing multiple pistillate flowers on individual nodes for harvesting finger size fruits to suit whole fruit canning for export especially in case of pickling cucumber.

16.2.0 Breeding Methods

16.2.1 Commonly Adopted Methods for Cucurbits Breeding

Indian subcontinent is the centre of origin for a number of wild and cultivated cucurbits. It is primary centre of origin for a number of wild and cultivated cucurbits like cucumber, snake gourd, ridge gourd, smooth gourd, pointed gourd, bitter gourd and bottle gourd. By utilizing different selection methods like mass selection, Individual plant selection, inbred selection, hybridization followed by selection, clonal selection and back cross methods, several high yielding varieties and F_1 hybrids have been developed in cucurbit vegetables.

16.2.2 Mass Selection

The best individual plants are selected on the basis of phenotypic characters and their open pollinated seeds are mixed for raising the succeeding generations. The selection is repeated in uniform growing condition before treating the population as a new variety. Thus this selection scheme is generally known as 'phenotypic recurrent selection'. The size of the base population and selection intensity in each generation are dependent on the crop, character under selection, *etc*. There is no control over the pollination. Since mass selection is made exclusively on the phenotypic performance of the mother plant without any progeny test, the success of selection depends largely on the heritability of the character under selection. This method is effective in improving simply inherited and highly heritable qualitative characters like sugar content in watermelon and muskmelon, disease resistance *etc*.

16.2.3 Individual Plant Selection

The species under the family Cucurbitaceae, though cross-pollinated in nature, generally do not show significant loss in vigour due to inbreeding. In these crops, inbreeding and individual plant selection through pedigree breeding can be practised as effectively as in self-pollinated crops. The three important uses of inbreeding are to attain uniformity in plant characters, to improve yield by individual plant selection and to recombine suitable inbred lines. The base population should be heterozygous and non-uniform materials. The plants are selected only on individual basis and unlike mass selection; the seeds of the selected plants are not bulked together even if they are equally good and look similar. This method of selection is based on the principle that the actual worth of the selected plant may be masked by environmental variation and therefore, the only way to recognize it is by growing the progeny of each selected individual plant for re-evaluation. In cucurbits, the progenies of individual plant for evaluation can be developed by selfing because they do not show loss of vigour due to inbreeding. Homozygosity for the concerned characters can also be attained in the individuals of the progeny by such selfings. After necessary evaluation, the best selection may be treated as a new variety. It is the chief breeding method followed in India for developing several varieties in different cucurbits. Example, Cucumber (Pusa Uday, Pusa Barkha, Pusa Long Green, Sheetal, Phule Shubhangi), Muskmelon (Pusa Madhurima, RM-43, Durgapura Madhu, Arka Rajhans, Arka Jeet, Pusa Madhuras, Lucknow Safeda, Annamalai, MH-10), Watermelon (Durgapura Meetha, Durgapura Kesar), Pumpkin (C-1, C-2,

CM-14, Solan Badami), Summer squash (Pusa Pasand, Punjab Chappan Kaddu-1, Early Yellow Prolific), Winter squash (Arka Suryamukhi), Bitter gourd (Coimbatore Long, Pusa Do Mousami, Pusa Aushadhi, Pusa Purvi, Pusa Rasdar, Coimbatore Long White, Coimbatore Green, Preethi, Priyanka, Hirkani, Phule Green, Konkan Tara, Arka Harit, VK-1-Priya, CO-1, MC-23, Pusa Vishes, Punjab BG-6, Kalyanpur Sona), Ridge gourd (Pusa Nutan, Pusa Nasdar, CO-1, CO-2, Konkan Harita, Punjab Sadabahar), Sponge gourd (Pusa Sneha, Pusa Supriya, Kalyanpur Long Green, Kashi Rakshita, Kashi Divya, Kashi Shreya), Bottle gourd (Pusa Summer Prolific Long, Pusa Summer Prolific Round, Arka Bahar, Pusa Naveen, Punjab Round, CO-1, Azad Nutan, Kalyanpur Long Green, Samrat, Pusa Sandesh, Pusa Samriddhi, Pusa Santushti), Wax gourd (Pusa Ujwal, Pusa Sabzi Petha, CO-1, KAU Local, CO-2, Mudliar, APAU Shakthi), Snake gourd (CO-1, CO-2, TA-19, PKM-1, Konkan Sweta, APAU Sweta), Long melon (Pusa Utkarsh, Punjab Long Melon-1, Arka Sheetal), Round melon (Pusa Raunak, Hisar Tinda, Punjab Tinda, Arka Tinda).

16.2.4 Recurrent Selection

This population improvement scheme is designed to concentrate favourable genes scattered among number of individuals in the base population. This is based on selection of superior plants followed by controlled mating to produce the new base population. This method is effective in improving yield and other quantitative characters by enhancing additive genetic variance in the population. Even though inbreeding depression would be negligible or nearly absent in the parental lines, interference of natural self- and sib-pollination is major disadvantage of handicap in these crops for this method.

16.2.5 Inbred Selection

Inbreeding is the mating of closely related individuals, either by selfing or sib mating. Inbreeding increases homozygosity and reduces the proportion of heterozygosity in the population. Recessive, lethal and deleterious genes usually remain concealed in the heterozygous system of the cross pollinated crops which are increased in proportion upon rendering to homozygosity by inbreeding, thus resulting reduction in vigour-called inbreeding depression. Inbreeding is the way to get the necessary alleles in homozygous condition. Although, the total allele frequency remains the same in the population, the proportion of phenotypes in the population differs, *i.e.*, variation occurs in the population which favours selection. At the same time, uniformity in the population is attained. Cucurbitaceous vegetable crops do not show loss of vigour due to inbreeding. For this reason, individual selection of inbreds can be practised and maximum uniformity can be attained in several cucurbits namely muskmelon, pumpkin *etc.* Example, Muskmelon (Hara Madhu), Pumpkin (Arka Chandan, Pusa Viswas, Pusa Vikas), Summer squash (Punjab Chappan Kaddu).

16.2.6 Hybridization followed by Selection

Controlled hybridization between selected parental lines followed either by selfings and pedigree breeding or open pollination and selection is also effective in developing improved varieties because as a result of gene recombination in the progeny following a cross, it is possible to select desirable segregates. Hybridization refers to crossing between individuals having unequal genetic constitution. In the F_1 hybrid, the process of meiosis involves formation of gametes with all kinds of chromosome and consequently gene combinations. Besides, as a result of crossing over with the associated exchange of identical segments between chromatids of homologous chromosomes in meiosis, genes located in the same chromosome undergo recombination. As a result, the progeny display new combination of characters not found in the parents. So, hybridization aims at combining in a genotype the useful characters inherent in each parent separately. That is why; it can be called combination breeding. In other words, genetic variability is created in the population by hybridization for efficient selection of the desirable type. If a character is governed by small number of genes, their recombination is relatively easy, but recombining the quantitative characters is very difficult as they are governed by multiple genes (polygene). So, successful outcome of hybridization is to a great extent dependent on how well the inheritance pattern of the concerned character is understood which actually imply the genetic process occurring in the segregating progenies of the hybrid population. Good number of varieties has been developed in India following this method of breeding. Example. Muskmelon (Pusa Sharbati: Kutana × PMR-6; Punjab Sunheri: Hara Madhu × Edisto; Hisar Madhur: Pusa Sharbati × 75-34; Punjab Rasila: WMR-29 × Hara Madhu), Watermelon (Arka Manik: IIHR-21 x Crimson Sweet), Cucumber (Himangi: Poinsette × Kalyanpur Ageti), Bottle gourd (Punjab Komal: LC-11 × LC-5), Ridge gourd (Arka Sujat: IIHR 54 × IIHR 18; Arka Sumeet: IIHR 54 × IIHR 24).

16.2.7 Inter-specific Hybridization

When intra-specific hybridization fails to produce desirable recombinants, distant hybridization is practised particularly to introgress very specific characters like disease resistance from the wild relatives of the concerned cultivated species. *Cucumis sativus* × *C. trigonus* (source of fruit fly resistance), *C. sativus* × *C. hardwicki* in cucumber; *Cucumis melo* × *C. figarei*, *C. melo* × *C. zeyheri*, *C. melo* × *C. meeusii*, *C. melo* × *C. myriocarpus* (resistance to *Fusarium oxysporum* ssp. *melonis* and cucumber green mottle mosaic virus) and *C. melo* × *C. africanus* (*CGMMV* resistance) in melons; *Luffa acutangula* × *L. cylindrica*, *Luffa acutangula* × *L. graveolens* in ridge gourd, *C. pepo* × *C. moschata* for development of semi-bushy character, *C. maxima* × *C. moschata* to develop squash bug and squash vine borer resistance are the examples of some successful inter-specific hybridization in cucurbits. F_1 hybrids can be obtained from most interspecific crossings, usually with difficulty and crossability depends on the particular accessions of the two species used as parents however, such hybrids are normally highly sterile because of impaired ability of the staminate flowers to produce functional pollens. Inter-specific crosses among different species is generally done by repeated pollination during the bud and flowering stage although percentage fruit set and embryo size were greatly influenced by the pollen parent and/or variety. However, the use of *in vitro* culture of the immature embryo by embryo rescue technique is generally employed for obtaining the inter-specific F_1 hybrid.

Table 15: Cucurbit Varieties Developed through different Breeding Methods

Crop	Variety	Breeding Method
Cucumber	Poinsette	Introduction
	Sheetal	Individual plant selection
	Phule Shuvangi	Individual plant selection
	Pusa Uday	Individual plant selection
Bottle gourd	Pusa Naveen	Individual plant selection
	Punjab Long	Individual plant selection
	Pusa Sandesh	Individual plant selection
	Pusa Samridhi	Individual plant selection
	Pusa Santushti	Individual plant selection
Bitter gourd	Pusa Do Mausami	Individual plant selection
	Pusa Vishesh	Individual plant selection
	Coimbatore Long	Individual plant selection
	Arka Harit	Individual plant selection
	Priya	Individual plant selection
	NDBG-1	Individual plant selection
	MDU-1	Mutation Breeding (Gamma ray mutant of MC-103)
	Kalianpur Sona	Individual plant selection
Ridge gourd	Pusa Nasdar	Individual plant selection
	Pusa Nutan	Individual plant selection

Practical Difficulty of this Breeding Method

i) Huge requirement of space in the field for the evaluation of a reasonably large segregating population to identify the desirable one becomes the main practical difficulty of this breeding method.

ii) In distant hybridization, segregates with unacceptable horticultural traits usually predominate in the F_2 generation of the inter-specific crosses.

iii) The undesirable traits of the wild parental species are often dominant and consequently desirable segregates in the F_2 population may often be small.

In the context of modern concept of breeding of superior populations rather than individual plants, methods of breeding undergo suitable modifications. From pureline and mass selection methods, several techniques have been developed, combining the virtues of both the methods. In a mass pedigree system of selection, Andrus and Bohn (1967) obtained several advantages in cantaloupe breeding. They advocated that inbreeding be withheld until after some period of mass selection with or perhaps without conscious selection. The principle here is to allow natural simulated selection with a resultant opportunity to allow for more genetic recombinations. Emphasis in melon breeding, being dessert fruits, should be on getting a cultivar that has dependable quality of a standard grade rather than outstandingly sweet or extraordinary flavour. By inbreeding and pureline selection we can select a cultivar with uniformity in one character, but only by sacrificing some other indispensable quality characteristic. Invariably, more homozygous inbreds lack earliness, productivity or one or the other quality trait. These point out the lack of genetic balance of quality traits which can overcome environmental sensitivity. Thus far, in no case, a wholly satisfactory combination of horticultural characters in an apparently homozygous line has been obtained. The behavioural characteristics which result in good overall performance are based on polygenic systems and it is not uncommon to harvest two fruits of muskmelon or watermelon from the same vine on different days with varying quality of sweetness (TSS 13 per cent and 10 per cent). It is rather difficult to insist on pre-established dimensions in arithmetical terms in an improved melon cultivar and hence a margin of ±2 per cent in TSS content of fruits within the same plant even in a pure bred cultivar is allowed.

Xu Bin (1999) described the main techniques for cultivating watermelons for breeding purpose. Selection of pure (99 per cent) parental lines of watermelon is the essential first stage. Selection of a site in a district where cucurbits are not grown as crops is ideal, but if that is not possible, it is necessary to ensure that the site is isolated from other cucurbit crops by 500-1000m. Planting at 18000 plants/ha is recommended and male plants should be planted 10 days earlier than female plants so that pollen is readily available when the female plants commence flowering. The second and third developed female flowers are selected and hand pollinated in the early morning, after which they can be covered with a pollination bag. If the flower has been pollinated, a row of coloured hairs is visible on the main vein of the female lower. Fruits should be harvested later than the commercial crop to permit maximum development of the seeds. Seeds should be dried to a maximum water content of 8 per cent. Seed can then be trailed in rows 1.5m apart and 20cm spacing within the rows.

Koutsika-Sotiriou *et al.* (2004) also suggested three cycles of pedigree breeding and selfing to develop improved genetic stock of different *Cucurbita* species including *C. moschata*. Having acceded to this proposition, four generations of consecutive selfings and selections produced marked improvement for fruit yield/plant, marketable fruits/plant and carotene content of the pulp showing 89.4, 85.9 and 54.2 per cent increase, respectively over the base population and mode- rate improvement ranging between 7.8 to 22.7 per cent for polar and equatorial diameter of fruits, fruit weight, 100-seed weight and total sugar content of the pulp (Mandal, 2006). A dwarf variety of pumpkin with short vines 'Non-vine I' has been identified which was considered to be a GA- related mutant, and the dwarf trait results from the failure of normal internode cell elongation (Cao *et al.*, 2005). Munshi and Alvarez (2004), Sirohi *et al.* (2005), Munshi and Chaudhary (2012, 2014), and Munshi *et al.* (2010, 2019, 2020) discussed in details about the recent advances in breeding of different cucurbitaceous crops.

16.2.8 Resistance Breeding

The genetic balance is the principle of the 'recombination cross technique' adopted by Barnes (1966) in breeding multiple disease resistant cultivars and hybrids of cucumber. In this method, plants having desirable complementary characters were crossed in F$_3$ to F$_5$ generations. The entire complex consisting of widely divergent multigenic characteristics was made available for selecting better recombination than available in back-cross method of breeding. Another excellent method of combining sources of disease resistance in *Cucurbita* was described by Rhodes (1959). This method consisted of an interbreeding population or gene pool with *C. pepo, C. mixta, C. maxima* and *C. moschata*. Plants from this pool are used in a series of bridging crosses using *C. lundelliana* as a bridge species to transfer species genes between incompatible species. Rhodes (1959) utilized this method of transfer the bush habit of growth from *C. pepo* to *C.moschata* and powdery mildew resistance from *C. lundelliana* to the previously mentioned gene pool. Sitterly (1972) obtained from this pool of multiple resistance to powdery mildew, downy mildew and squash mosaic virus in *C. pepo*. Snyder *et al.* (1993) reported from USA that precocious yellow gene (*pyg*) squash (*C. pepo*) cultivars offer several advantages: fruits turn yellow earlier making them more suitable for baby vegetable markets, the number of female flowers are increased leading to increased yield and fruits retain a yellow colour after infection with watermelon mosaic 2 potyvirus (WMV). One draw back is that *PYG* confers a yellow instead of green peduncle which sometimes causes problem in marketing. *PYG* cultivars are not resistant to WMY but mask the symptoms of green streaks and blotches on the fruit under a yellow colour. *PYG* cultivars are Multipik, Superset, Superpik, XPH1636, *etc.*

Breeding cucurbits for disease resistance has resulted in several accomplishments, some of which are not recorded even in cereal crops or commercial crops. As individual diseases are conquered by breeding, it is practical to combine the genes for resistance to several specific diseases into multiple disease resistant cultivars. One of the earliest success was achieved by combining scab and cucumber mosaic resistance in Wisconsin SMR pickle lines (Walker and Pierson, 1955). Barnes (1961, 1966) has developed several inbred and hybrid cultivars carrying resistance to six diseases. In the slicers, F$_1$ Cherokee and F$_1$ Gemini and in pickling F$_1$ Explorer and F$_1$ Southern Cross are some of his multiple resistant hybrids. Feral species of *Cucurbita* are found resistant to many diseases *viz., C. lundelliana* and *C. martinezii* are resistant to powdery mildew (Rhodes, 1964; Contin and Munger, 1977) and different viruses such as watermelon mosaic, cucumber masaic virus and Zucchini yellow mosaic virus (Provvidenti *et al.*, 1978; Maluf *et al.*, 1986; Robinson *et al.*, 1988), *C. equadorensis* is resistant to papaya ringspot virus (Herrington *et al.*, 1989) and *C. andreana, C. lundelliana, C. equadorensis* and *C. foetidissima* are resistant to bacterial wilt caused by *Erwinia trachiiphila* (Watterson *et al.*, 1971).

Further, significantly progress has recently been achieved by development of cucumber Wisconsin 2757 by Peterson *et al.* (1982), in which resistance to the diseases has been incorporated, *viz.*, powdery mildew, downy mildew, anthracnose, angular leaf spot, mosaic, bacterial wilt, scab, Fusarium wilt and Target leaf spot, an outstanding achievement in the area of resistance breeding. M17 originated from the cross of the southern USA cv. Chipper with the northern USA cv.Wisconsin SMR18. F$_5$ lines cycled through five generation of selection for gummy stem blight resistance followed by selfing to produce M17. It has moderate to high resistance to other common cucumber diseases (Wehner *et al.*, 1996). Multiple resistant lines in cantaloupe have been developed carrying resistance to powdery and downy mildew, *Alternaria* and watermelon mosaic virus (Thomas and Web, 1981). To breed short fruited cucumbers resistant to downy mildew, highly resistant accession PI179676 and susceptible G-3 were crossed by Angelov and Krasteva (2000). Highly resistant plants from the BC_1 generation were selected and self-pollinated. Subsequently, selection for high resistance and self-pollination was conducted for several generations. The percentage of highly resistant plants increased in each generation from BC_1F_2 (28.2 per cent) through the BC_1F_7 (98.7 per cent).

Ivanova (1986) investigated sources of resistance to powdery mildew among melons in Uzbekistan and noticed highest resistance in Super Market, Perlita, Kurume 1 and Joknean. Kenigsbuch and Cohen (1987) found that inheritance of resistance to powdery mildew in muskmelon in PI124111 to race is conferred by a single dominant gene and to race 2 by an incompletely dominant gene. The downy mildew resistance in muskmelon line of MR-1 is conferred by 2 incompletely dominant genes designated *Pc1* and *Pc2* (Thomas *et al.*, 1988).

Powdery mildew resistance of *C. martinezii*, controlled by a single dominant gene with some modifications by minor gene was successfully transferred to the *C. moschata* cultivar "Wonye 402" through interspecific hybridization (Cho *et al.*, 2004c). Two powdery mildew oriental squash (*C. moschata*) cultivars "Mansu" and "Chensu" were developed at Korea through interspecific hybridization between *C. moschata* local cultivar "Jecheonjaerae" × *C. martinezii* followed by selection of powdery mildew resistant and non-bitter genotype from BC_2F_1 population which was selfed twice then backcrossed to the *C. moschata* local cultivar "Jecheonjaerae" and finally selfed thrice (Cho *et al.*, 2004a, 2004b).

Study on the genetics of powdery mildew resistance in cucumber by Pershin *et al.* (1988a) revealed that it was controlled by one major dominant and 2 semidominant independent genes. In order to analyze the inheritance of powdery mildew resistance Morishita *et al.* (2000) crossed PI 197088-5 (resistant) parent and Sharp One (susceptible) in Japan. They, however, found that resistance was controlled by two recessive genes, and plants with *aabb* show resistance at both 26 and 20°C, and plants with *aaBB* or *aaBb* showed resistance at 26°C and concluded that PI 197088-5 is a novel gene source for powdery mildew resistance in winter cucumbers in Japan. According to Provvidenti (1987) Zucchini yellow mosaic virus (ZYMG) in cucumber is controlled by a single recessive gene, designated *Zym*. Results from the study of Pershin *et al.* (1988b) indicated that resistance to downy mildew in cucumber was controlled by at least 3 major genes. Different accessions of *C. moschata viz.*, Minina, originating from Portugal (Gilbert *et al.*, 1993), Nigerian Local (Provvidenti *et al.*, 1997; Brown *et al.*, 2003), BGH 1934, BGH 1937 and BGH 1943 (Moura *et al.*, 2005) and wild species *C. ecuadorensis* (Paran *et al.*, 1989; Provvidenti *et al.*, 1997) are known to carry resistance against this virus. Transfer of ZYMV resistance from the resistant *C. moschata* cultivars and *C. ecuadorensis* to the varieties of *C. moschata*, *C. pepo* and *C. maxima* was successful in some cases resulting in the development of new cultivars showing high level of resistance (Provvidenti *et al.*, 1997; Desbiez *et al.*, 2003). No RAPD markers linked to ZYMV resistance were found (Brown *et al.*, 2003).

A full length cDNA clone of the RNA genome of the ZYMV was constructed down stream from a bacteriophage T 7 RNA polymerase promoter, a single extra guanosine residue not present in ZYMV RNA was added to the 5' and 3' ends. Capped (m7GppG) ZYMV RNA transcripts were infectious in 10 of 91 *C. pepo* test plants and uncapped RNA transcripts were not infectious (Gal-On *et al.*, 1991). The 3' terminal sequences of the ZYMV genome from Beijing, China was characterized by RT-PCR from the total RNA of infected leaves and cloned into the pMD18-T vector, and the sequences were 1269bp long and contained a coat protein (CP) gene, which consisted of 837 nucleotides encoding 279 amino acids (Zhang *et al.*, 2006).

Inheritance a resistance of downy mildew in ridge gourd could be explained by the additive dominance model, involving at least 4 additive genes. Narrow sense heritability was 69.63 per cent (Hua and Sen, 1999). The resistance to cucumber mosaic virus in pumpkin cv. Cinderella is controlled by 2 recessive genes (Pink *et al.*, 1985; Pink, 1987). Paris *et al.* (1988) observed from their investigation that ZYMV in *C. moschata* was controlled by a single dominant gene. Lin and Fang (2000) reported that resistance to cucumber blight was controlled by at least 3 pairs of genes. The narrow sense heritability was 78 per cent.

The genetics of resistance to *Cucumber mosaic virus* (CMV) in *Cucumis sativus* var. *hardwickii* R. Alef, the wild progenitor of cultivated cucumber was assessed by challenge inoculation and by natural infection of CMV. Among the 31 genotypes of *C. sativus* var. *hardwickii* collected from 21 locations in India the lowest mean per cent disease intensity (PDI) was recorded in IC-277048 (6.33 per cent) while the highest PDI was observed in IC331631 (75.33 per cent). A chi-square test of frequency distribution based on mean PDI in F_2 progenies revealed that CMV resistance in *C. sativus* var. *hardwickii* was controlled by a single recessive gene (Munshi *et al.*, 2008)

Based on screening against natural infection and challenge inoculation study and PCR detection of *Tomato Leaf Curl New Delhi Virus* (Islam *et al.*, 2010) two lines DSG-6 and DSG-7 were found highly resistant to the virus. The frequency distribution of F_2 based on chi-square test revealed monogenic dominant nature of resistance of the disease. The funding was further confirmed by Identification of SRAP markers linked to the single dominant resistance gene (Islam *et al.*, 2011). Further DSG-6 (IC0588956; INGR12013), was registered with NBPGR, New Delhi (Munshi *et al.*, 2012). More and Varma (1991) and More *et al.* (1992) found 13 collections out of 187 studied, as field tolerant to *Cucumber Green Mottled Mosaic Virus* (CGMMV) and by further screening under artificial inoculation condition with a pure isolate of CGMMV they found *Cucumis figarei* (GBNR 1804), *C. myriocarpus-1* (GBNR 1976), *C. myriocorpus-3* (GBNR 1007), *C. africans-1* (17048 J 14), *C. africans-2* (GBNR 1984), *C. meeusii* (GBNR 1800), *C. ficifolius* (GBNR 1801) and *C. zeyheri-2* (GBNR 1053) were resistant to CGMMV. Immature nature of CGMMV resistance of *C. figarei, C. ficifolinus, C. meensii, f. africanus-1 and C. zeyheri-2* was confirmed by back-inoculation technique and transmission electron microscopic study. Resistance to CGMMV is governed by recessive polygenes (Rajamony *et al.*, 1990; More and Varma 1991). Several CGMMV and downy mildew resistance lines DVRM-1, DVRM-2, DMDR-1, DMDR-2 and DMDR-3 were developed by Munshi and Verma (2000) by crossing Phoot (*C. melo* var. *momordica* (resistant) x M4 (susceptible) and FM1 (resistant) x Pusa Madhuras (susceptible) and further screened and selected for yield, disease reaction and horticulturally desirable characters (Anonymous, 1999, 2000, 2001). Two of the above resistant lines DMDR-1 and DMDR-2 were identified as resistance source against downy mildew and CGMMV at XXIIth group meeting of vegetable research held at Indian Institute of Vegetable Research Varanasi (U.P.) India (Anonymous, 2001).

Quemada (1998) described the use of coat protein technology to develop virus-resistant in cucurbits. A transgenic yellow crook neck squash (*C. pepo*), ZW20, expressing coat proteins give resistance to Zucchini yellow mosaic potyvirus and watermelon mosaic 2 potyvirus. Use of biotechnological tools for improvement is discussed later in the chapter.

Dhillon and Sharma (1989) suggested that cage evaluation may be used in screening for resistance to red pumpkin beetle. It has been found that *C. moschata* cultivars resistant to leaf silvering disorder had non-mottled leaves indicating a possible relationship between the genetically controlled silver leaf mottling and expression of the silver leaf disorder due to feeding of *Bemisia argentifolii* (Wessel-Beaver and Katzir, 2000).

Resistance to *M. javanica* in cucumber is conferred by the newly discovered *mj* gene and *M. javanica* resistance and the 17 other traits controlled by simple genes were evaluated in green house or field (Clinton, North Carolina, USA) trials in four inbred lines and crosses by Walters and Wehner (1998). None of the 17 genes were linked with *mj*. Cucumber breeders interested in nematode resistance should be able to incorporate the trait into elite breeding lines without breaking linkages with 17 genes studied.

16.2.9 Heterosis Breeding

The phenomenal success in heterosis breeding in cucumber and to a lesser extent in summer squash, is entirely due to manipulation of sex expression in the desired direction. The genetic control of sex mechanism in cucumber, especially of the gynoecious sex form, has made it possible to exploit heterosis in cucumber, muskmelon and bitter gourd. Thus, gynoecious hybrids in combination with gynoecious and monoecious parents had enormous potential for earliness, yield and yield contributing traits in cucumber (Pati *et al.*, 2015 and Jat *et al.*, 2016) and in bitter gourd (Alhariri *et al.*, 2018).

16.2.10 Manifestation of Heterosis in Cucurbits

☆ Earliness with respect to days to first fruit harvest.

☆ Uniform maturity.

☆ High female to male ratio especially in gynoecious hybrids.

☆ Uniform attractive fruit with good colour (internal and external) and appearance as per the consumer preference of different regions.

☆ Thick skinned fruit, suitable for long distance transportation.

☆ Good quality fruit like small cavity size, thick flesh in muskmelon and pumpkin.

☆ Fruits free from bitterness in cucumber, bottle gourd and ridge gourd.

☆ High aroma and crispy flesh in muskmelon.

☆ Tender and crispy flesh in cucumber.

☆ Fewer and smaller seeds in watermelon, ash gourd, bottle gourd, bitter gourd and cucumber.

☆ Disease and pest resistance. Resistance to many diseases is governded by single dominant gene.

☆ Drought tolerance in watermelon.

☆ Wider adaptability, particularly in muskmelon, watermelon and cucumber.

☆ Higher early and total yield with respect to number of fruits and fruit weight.

Hybrid cucumber and squash are not that important for mature fruit yield, as for earliness, number of fruits and external attributes of uniformity of size and shape, especially in cucumber slicers, attractive colour, flesh texture and other quality traits. Most of the hybrids carry multiple resistance to diseases due to the fact that dominant genes control resistance to some diseases.

Heterosis for high number of fruits, bearing at each flowering node, would mean maintenance of plant vigour to a very high degree which has been accomplished in glass house culture of protected environment. Not only high yield but sustaining one which is spread over nearly four months, would make the glass house culture competitive to produce superior quality fruits of uniformity and of attractive colour and shape. Earliness is another character which is manifested in the hybrids, besides lesser percentage of unmarketable culls. The latter character is important in picking hybrids because mis-shaped fruits are rejected at the time of grading for processing. In glass house culture of Western Europe, parthenocarpy and multiple disease resistance have been incorporated in the gynoecious hybrids. But in the open cultivation, parthenocarpy is not preferred because pollinators cannot be eliminated. Hence a mixture of seed pollinizing monoeceous cultivar is provided to an extent of 5-10 per cent. However, in pickling hybrids, machine harvesting has come into vogue requiring development of cultivars with a concentrated fruit set. Multiple pistillate lines producing more than one fruit at each flowering node are being attempted in pickling types to suit once-over harvest.

Leijon and Olsson (1999) reported from Sweden that recently parthenocarpic hybrids have gained ground in cucumber. Besides yield, improvement of optical quality characteristics like straight fruits with rounded ends, a smooth skin of homogeneous green colour and a light green-yellow flesh are of interest. An important goal in breeding for parthenocarpic varieties is a small seed room to avoid the cucumber losing structure during processing. Texture should be firm and crisp and the flavour should be without bitterness.

In muskmelon, heterosis breeding has different requirements. Being dessert fruits, quality characteristics of the hybrids should be superior in respect of uniformity and stability over pure bred cultivar. Earlier maturity is another major advantage in the acceptance of melon hybrids. It has been reported in cantaloupe hybrids that additive more than non-additive factors are responsible for the heterosis of most of the characters. The andromonoecious condition of muskmelon necessitates emasculation. Foster (1968) in a series of experiments highlighted the importance of monoeceous parents in producing hybrids by cheaper method of hybrid seed production. He showed by using a glabrous marker gene, field crossing could produce 30-35 per cent F_1 hybrid seed using andromonoccious lines and 60 per cent using monoeceous lines and 75 per cent when male sterility and monoecism were combined with marker gene in seed parents. Marker character would facilitate rogueing of self/sib-pollinated seedlings occurring as mixture in hybrid population. Male sterility has not been of much of an advantage in melon hybrids as indicated earlier in hybrid seed production.

Galaev (1986) found that F_1 hybrid seedlings in watermelon exhibited heterosis exceeding the parental forms and his data were supporting the theory that heterosis is expressed where the parental form has well developed roots and the maternal form has well developed stems and leaves.

Borisov and Krylov (1986) reported a new triple hybrid of cucumber, TSKhA 2693 from the then Soviet Union which is suitable for winter-spring cultivation, gives high yield, produces more female flowers on main stem and has better storage quality. Tarakanov *et al.* (1986) reported the use of complex material forms, designated as SMF-NF, developed at the Vegetable Experimental Station of the Timiryazev Agricultural Academy in Moscow for the production of triple hybrids of cucumber. The SMF-NF forms exceeded Natsufushinari in shade tolerance, salt resistance, growth rate, earliness and by 22 per cent seed yield. The quality and yield of cucumber hybrids using female (gynoecious) parents and bisexual male parents were indistinguishable from those obtained using near isotenic gynoecious male parents with pollen production induced by silver nitrate treatment (Staub *et al.*, 1986). The recent cross of cucumber and *Cucumis hystrix* made in 1997 in USA represented a break through and the F_1 hybrid is both male and female sterile. Fertility has been restored by using *C. hystrix* as the seed parent (Chen and Adelberg, 2000).

Dyutin and Puchkov (1996) outlined new trends in breeding *C. pepo* of the marrow and custard types in Russia and gave a short description of new varieties Sosnovskii (courgette type) and Tabolinskii (with fruit of the custard type). Particular attention is paid to the production of F_1 hybrid varieties and information is given on a source of male sterility found in the custard type Englischer Gelber, with sterility controlled by a single recessive gene. The variety crosses readily with other types of *C. pepo* and thus can be used in heterosis breeding. Plants are found in the marrow variety Oranzhevyi in which the male flowers were normal but the female ones wilted after opening and set no fruit. Hybrid analysis showed that this androecy was controlled by a single recessive gene.

Orangetti is a hybrid derived by backcrossing the donor parent precocious Fordhook Zucchini to Vegetable Spaghetti, with the donar parent contributing the dominant genes for bush habit (possibly Bu) and colouration (*B. D*, L_1 and L_2). The combination of the dominant alleles L_1 and L_2 results in the intense colouration of Orangetti while the Ballele produces its orange hue (Paris, 1993).

In summer squash (*C. pepo*), earliness and high number of fruits per plant were the advantages in hybrid. Application of ethephon or ethrel to young plants at 2-3-leaf stage would suppress male flower formation for 2-3 weeks and complete suppression can be achieved at higher concentrations of 600 ppm applied twice at 2- and 4-leaf stage has made hybrid squash seed production comparatively easier and nearly 56 per cent of total squash seed produced in the USA is of F_1 hybrids. Shifriss (1988) developed a method for synthesis of genetic females and their use in hybrid seed production in squash. It is applicable to other monoeceous species of plants.

Likewise, heterosis breeding has potentialities in Asiatic cucurbits as well, like bottle gourd, bitter gourd, pumpkin, Indian squash melon, ridge gourd, sponge gourd, *etc.* (Sirohi, 2004). Studies on combining ability and estimation of genetic components of variation and heterosis has been conducted by various workers *viz.* Munshi and Sirohi (1993, 1994a, 1994b), Dey *et al.* (2009, 2010, 2011 and 2012), Rao *et al.* (2017), Behera (2004) in bitter gourd; Munshi *et al.* (2005, 2006a, 2006b),Tiwari *et al.* (2010a, 2010b, 2013), Reddy *et al.* (2014), Pati *et al.* (2015a, 2015b), Jat *et al.* (2015, 2016, 2017), Bhutia *et al.* (2017, 2018) in cucumber; Karmakar *et al.* (2013, 2014), Singh *et al.* (2017, 2019), Sarker *et al.* (2015) and Varalaksmi *et al.* (2019) in ridge gourd; Kumar *et al.* (2012), Tyagi *et al.* (2010) in sponge gourd; Singh *et al.* (2011), Singh *et al.* (2012), Shinde *et al.* (2016), Janaranjani *et al.* (2016), Doloe *et al.* (2018) Mishra *et al.* (2019) in bottle gourd; Sureja *et al.* (2006, 2010) in ash gourd; Munshi and Verma (1998, 1999), Moon *et al.* (2002, 2003, 2004), Sashikumar and Pichaimuthu (2016), Selem (2019) in muskmelon, Kale and Seshadri (1981), Reddy (1987), Bansal *et al.* (2002), Gabriele and Wehner (2004), Omran *et al.* (2012), Sapovadiya *et al.* (2013) and recently Singh *et al.* (2020) in watermelon; and Mohanty and Mishra (1999), Mohanty (2001), Pandey *et al.* (2010), Tamilselvi and Jansirani (2016), Nagar and Sureja (2017), Nagar *et al.* (2017a,b,c) in pumpkin.

All of them advocated the importance of heterosis breeding for effective utilization of non-additive gene action for the improvement of important yield contributing taits like earliness, number of fruits per plant and yield per plant. However, majority of them observed the predominant role of partial dominance and additive gene action in the improvrment of fruit size especially fruit weight and fruit lengh through effective selection. Further, Munshi and Verma (1999) in muskmelon, Munshi *et al.* (2006), Tiwari *et al.* (2013) in cucumber and Singh *et al.* (2019) in ridge gourd observed that in most of the cases *per se* performance of parents bears direct reflection of their respective *gca* effects, ie parents showing highest *gca* effect for a character were also observed as good performer with respect to that particular character. A comparison of *sca* effects of the crosses and *gca* effects of the parents involved indicated that *gca* effects were reflected in the *sca* effects of the cross combination. It is apparent that in almost all the hybrids which showed best *sca* effects, the parental lines involved were at least one of the outstanding parental lines for which also had high *gca* effect for one or more characters contributing towards yield which indicated that there was strong tendency of transnlitting the higher gain from parents to offspring (Venna *et al.*, 2000, Bairagi *et al.*, 2001 and Gulamuddin *et al.*, 2002).

In some heterosis studies in different cucurbits, partial dominance and additive gene action has been noted to be predominant for quality traits especially minerals, pigments and T.S.S which point out the possibility of development of synthetic through the development of several inbred lines. However, due to negligible inbreeding depression, interference of natural self- and sib-pollination in early synthetic population, would be major disadvantage and handicap in these crops for producing synthetics.

The strategy for breeding F_1 hybrid pumpkin is to develop parental lines through self-pollination (Sarý *et al.*, 2006). Hazra (2006) advocated development of parental lines by inbred selection because no reduction in fruit, seed and fruit quality characters was evident even after four generations of successive selfings coupled with selection. Selection of divergent parents based on fruit weight, leaf area, seed weight, and fruit yield per plant may be useful for heterosis breeding in pumpkin (Kale *et al.*, 2002). Commercial hybrids were produced based on parental vine type: short × short vine and short × long vine. Marketable yield was highest in the short × long hybrids; flesh thickness however, was greatest in the open-pollinated control variety, intermediate in the short × short hybrids and lowest in the short × long hybrids (Maynard *et al.*, 2001).

New germplasm has also been developed by interspecific crosses among *C. moschata, C. pepo, C. maxima* and *C. argyrosperma*. The F_1 plants exhibited heterosis and disease resistance during vegetative growth, and showed a wide variation in botanical traits and metaxenia effects were also observed on the taste and colour of the pumpkins (Cheng *et al.*, 2002).

In vegetatively propagated cucurbits like pointed gourd, F_1 hybrids can be produced by crossing different clones and high yielding F_1s can then be clonally multiplied. Unlike in seed propagated cucurbits, there is obviously no necessity of producing F_1 hybrid seeds every year in these crops.

16.2.11 Seedless Watermelon

Seedless watermelon is one classic example of the practical use of colchicine and application of autopolyploidy in crop improvement. The interference of crunchy seeds at the time of eating slices of watermelon has long been recognized as a disadvantage not encountered in other fruits like muskmelon. Seedlessness has been induced by (3x) triploidy which is both female and male sterile. The seeds of seedless cultivar will have to be produced every year like a F_1 hybrid and the tetraploid (4x=44) is used as female parent with diploid (2x=22) as male in the cross. There are several steps in producing seedless watermelon fruits (Kihara, 1951; Kondo, 1955).

16.2.11.1 Production of Tetraploid Lines

Application of colchicine 0.2 per cent (to 0.4 per cent) to growing points of young seedlings (at 1 to 2 true-leaf stage), for 2 to 3 days successively, gives satisfactory results. The treatment is done under controlled conditions avoiding direct sunlight. The growth of tetraploids (44 chromosomes) will be slow and terminal bud should be carefully promoted to grow. The 4x plants are characterized by gigas characters

244 | *Vegetable Crops, Volume 4*

like thick and broad leaves and bigger stomata. When the plants start flowering, the pollen grains may be tested for fertility and also the size of pollen grains of 4x plants will be bigger. The fruiting will be less in first generation with very few seeds, around 30 to 40 due to high pollen sterility. The seeds will be broad and bigger. The 4x plants should be checked for chromosome number, both in somatic cells and in meiosis. There would not be 11 quadrivalents, but it will range from 4 to 10 with varying number of univalents, bivalents and trivalents.

Maintenance of tetraploid lines at stable level is important and continuous selection for improvement of quality and vigour in tetraploid lines has to be done for atleast 4 or 5 generations. Chance reversion to diploidy can be detected by numerous small seeds in the fruits. Tetraploid fruits have characteristic large blossom scar. Higher fertility during continuous selection is possibly related to lesser multivalent frequency, which probably leads to the true breeding behaviour. Attainment of superior fruit quality (firm flesh, deep colour, stable TSS content of 12 to 13 per cent) in tetraploids is an important requirement for production of high quality seedless fruits.

The need for caution in seed extraction of tetraploids is to be noted. Fermentation of the pulp should be avoided which will affect the seed germination. Similarly, at the time of sowing of seeds of tetraploids, overnight presoaking should not be done and they will have to be germinated under controlled conditions ensuring slow and continuous absorption of moisture.

16.2.11.2 Production of 3x Seeds

It can be done by (i) hand pollination, the female tetraploid by diploid male or (ii) in open fields with adjacent rows of tetraploid female and diploid male, pinching of male buds of tetraploid plants leaving the pollination to insects, or (iii) leaving the pollination and fruit set to natural condition of insect pollination. The last method involves interference of self- or sib-pollinated seeds in tetraploid lines among the cross-pollinated triploid seeds. In 1:1 planting, the percentage of 4x seeds (by sib/selfs) under open pollinated conditions comes to about 30 and the triploid and tetraploid seeds cannot be distinguished. The production of 3x seeds is, therefore, expensive partly because of low seed content of tetraploid lines and mixture of 4x seeds. Seed men get not more than 4.5 to 6.8 kg per acre of triploid seeds under open pollinated conditions.

Ne Smith and Duval (2001) recorded that greatest number of triploid fruit per unit land area was in the harvest row 3.0 m from the pollinizer row. When the distance from the pollinizer row was 6.0 m or greater, triploid fruit numbers diminished substantially (Fig. 3). Yield estimates made each year using the fruit density data suggested that a 1 pollinizer : 4 triploid ratio gave the maximum total triploid fruit yield per hectare for 1.5 m row spacings (Fig. 4). These results should prove useful in designing field planting strategies to optimize triploid watermelon production.

Liu (1998) observed no obvious differences in the fruit set, yield, flesh soluble solids content and the number of coloured aborted seeds in triploid watermelons when different diploid pollinators were used.

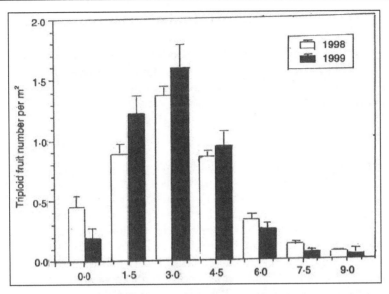

Distance from pollinizer row (m)

Figure 3: Fruit Number of 'Genesis' Triploid Watermelon as a Function of Distance from Pollinizer Row during 1998 and 1999 at Griffin, Georgia, USA.
The pollinizer was 'Ferrari'. Each vertical bar represents one 1.5m wide row.
The 0 distance represents the location of the pollinizer row.
Standard errors are depicted by vertical lines (Ne Smith and Duval, 2001).

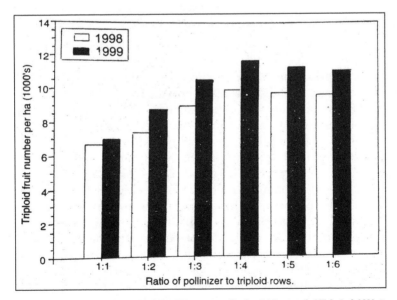

Ratio of pollinizer to triploid rows.

Figure 4: Estimates of Yield (Total fruit number/ha) of 'Genesis' Triploid Watermelon in Response to different Pollinizer: Triploid row ratios during 1998 and 1999.
The pollinizer was 'Ferrari'. Estimates are base on 1.5 m row width
(NeSmith and Duval, 2001).

16.2.11.3 Germination of 3x Seeds

This is important because of high cost of production of 3x seeds. Poor germination is caused by hard seed-coat. Seeds will have to be sown in flat or peat cups with moisture carefully maintained at a temperature of 24-29°C in the first 72 hours. Naturally, germination of seedlings under controlled conditions and their subsequent transplantation increase the cost of production of seedless fruits. Transplantation has to be done in 2-true-leaf stage.

16.2.11.4 Production of 3x Fruits

There will be no fruit set in triploid plants, if there are only 3x plants in the field. Even a few tetraploid plants occurring as a mixture would not be enough for good fruit set. Stimulation of fruit development has to be brought about by 2x pollen through bee pollination. Hence a pollinizer, a diploid cultivar has to be planted in the field. Another difficulty arises at the time of fruiting, when 2x, 3x and 4x fruits have to be distinguished. Hence at the time of choice of parents for crossing 4x plants with 2x male, contrasting fruit characters are chosen. If 4x plants produce fruits with light green rind and stripes and 2x male chosen, would produce fruits of dark green rind, the 3x fruits will be have intermediate green rind with stripes. The 3x fruits will have triangular configuration, in contrast to round fruits of 4x and 2x plants. The 3x plants are of long duration, fruits maturing in 120 to 130 days in contrast to 85-90 days in diploid cultivars. The firm texture of the rind is an attractive feature, but there may be few empty ovular cavities in the fruit. Sweetness will be uniformly distributed from the rind to the centre. Triploids are sterile both on male and female sides and unequal chromosomal distribution in meiosis causes pollen and ovular abortion. However, few ovules may develop into a hardened seed-coat and these are called 'ex-embryonate' or 'false' seeds.

The credit for production of seedless watermelons goes to Japanese scientists who were the first to discover this possibility. Kihara (1951) gave an exceptionally clear account of various steps in production of triploid fruits and later publications by Kondo (1955) and Shimotsuma (1961) provide an analysis of several problems encountered in commercial exploitation. Japanese triploids were produced mainly from Fumin tetra-ploid stocks, Asahi 4x and Ogon stocks. Most of them are of medium maturity and fruit flesh was crisp in texture in some and soft in other. The Vegetable Breeding Laboratory of USDA in Charleston (SC) in the USA has developed tetraploid lines, Tetra-1, Tetra-2, Tetra-3 and Tetra-4. These lines have been used by Eigsti for commercial production of seeds of seedless watermelons. Tetra-2 line has also been used as female parent at IARI for production of seedless watermelon cultivar 'Pusa Bedana' (Seshadri *et al.*, 1972). Andrus *et al.* (1970) discussed several problems of production of seedless watermelons under conditions of USA. Due to long duration and slow maturity, production of seedless watermelons has shifted to Taiwan from South Japan and seedless fruits are exported to Hong Kong and Singapore. It has been possible to transfer a recessive gene (*ms*) for male sterility from a diploid to a tetraploid *Citrullus lanatus* line through a series of controlled pollination. Progeny testing and morphological characteristics confirmed the development of tetraploid lines with *Ms* gene and it is suggested by Love *et al.*

(1986) that this gene transfer system may be useful in the improvement of tetraploid cultivars and in the production of seedless triploid watermelon. Resistance to fusarium wilt disease has been incorporated in tetraploid lines in the USA and Japan so that triploids carry resistance.

In view of several problems leading to high cost of production of triploid seeds and triploid fruits, Japanese scientists attempted at diploid level through gamma irradiation to get segmental interchange or reciprocal translocation stocks in watermelon. Such changes exhibit partial sterility of pollen grains and ovules. It is generally recognized that fertility decreases as the number of interchanges increase and in a watermelon plant, a ring of 22 chromosomes would be completely sterile. The ultimate aim of this work is to obtain two interchange homozygous stocks, whose hybrids would give a ring of 22 chromosomes (Oka *et al.*, 1967: Shimotsuma, 1968).

Gray and Elmstrom (1991) described a novel production of seedless watermelon seed. The process involves cloning desirable tetraploid watermelon parental lines. The method makes possible the use of these self sterile tetraploid parental lines in the production of triploid seed.

16.2.12 Breeding for High Pigments and Bioactive Compounds

Inheritance of flavonoid, carotenoids and chlorophyll pigments has been studied extensively in melon. The pigments accumulating in rinds of different melon genotypes during fruit development were analyzed. It was found that melon rind colour is based on different combinations of chlorophyll, carotenoids and flavonoids according to the cultivar tested and their ratios changed during fruit maturation. Moreover, in "canary yellow" type melons, naringenin chalcone, a yellow flavonoid pigment previously unknown to occur in melons, has been identified as the major fruit colorant in mature rinds. Naringenin chalcone is also prominent in other melon types, occurring together with carotenoids (mainly β-carotene) and chlorophyll. Both chlorophyll and carotenoid pigments segregate jointly in an F_2 population originating from a cross between a yellow canary line and a line with green rind. In contrast, the content of naringenin, chalcone segregates as a monogenic trait independently to carotenoids and chlorophyll (Tadmor *et al.*, 2010).

In melon, α-carotene and β-carotene are found in the fruit. There is an opportunity through traditional plant breeding to increase the amount of these and other carotenes to improve the nutrition of melon. Some carotenes can be observed in melon fruit as an orange colour (the higher the orange colour the higher the concentration of β-carotene). Genetic knowledge regarding the inheritance (how genes are transmitted from generation to generation) about these compounds is critical for effective and efficient plant improvement through breeding. Very little is known about the inheritance of α-carotene and β-carotene in melon. However, several genes were identified which control β-carotene in fruit and are related to the colour of the fruit. These genes can now be used to breed higher level of β-carotene in melon varieties. Further (Cuevas *et al.*, 2008) suggest that accumulation of β-carotene in melon is under complex genetic control, where epistasis plays an important role in trait expression. Although the inheritance of quantity of β-carotene and fruit maturity is complex, introgression of early fruit maturity genes from Chinese germplasm might solve the problem. Such introgression may lead to increased yield potential in US market types while retaining relatively high β-carotene fruit content (*i.e.*, orange mesocarp), if stringent, multiple location and early generation family selection (F_{3-4}) is practised for fruit maturity with concomitant selection for quantity of β-carotene (Cuevas *et al.*, 2010b).

Mapping of genetic loci that regulate quantity of β-carotene in fruit of US Western Shipping melon (*Cucumis melo* L.) was studied by (Cuevas *et al.*, 2008). The results suggest that accumulation of β-carotene in melon is under complex genetic control. This study provides the initial step for defining the genetic control of QbetaC in melon leading to the development of varieties with enhanced β-carotene content.

Identification of molecular markers and orange flesh colour associated with major QTL for high β-carotene content in muskmelon was reported by Napier (2006). Presence or absence β-carotene in muskmelon is controlled by two genes, green flesh *gf* and white flesh *wf*. In its dominant form the *wf* gene is responsible for orange flesh colour; however, the epistatic interactions of *gf* and *wf* can create three flesh colours: orange, white and green were used to examine the relationships of β-carotene content, flesh colour, and flesh colour intensity. Bulk segregent analysis was used with RAPD markers to identify molecular markers associated with high β-carotene content. Flesh color and flesh colour intensity both had significant relationships with β-carotene content. A single QTL was also found to be linked with the *WF* locus. The identified QTL can be used to screen potential breeding lines for high β-carotene. It was also confirmed that the visual ratings of flesh colour intensity can be reliably used to select high β-carotene content melons.

A distinct group of cultivars, known as *Xishuangbanna gourd*, is classified as *C. sativus* var. *xishuangbannesis* Qi and Yuan is grown by the Hani people of southwestern China at elevations of 1000 m or higher, these land races are largely unknown outside of eastern Asia. The rind of this variety is white, yellow or brownish-orange, sometimes with distinct netting, smooth, with yellow to orange flesh colour rich in provitamin A carotenes and used by the breeder to introgress of β-carotene in to the commercial cultivars. Three varieties namely Early Orange Mass 400 (EOM 400), Early Orange Mass 402 (EOM 402) and Late Orange Mass 404 (EOM 404), were developed derived from the cross between U.S pickling cucumber lines (*Cucumis sativus* L. var. *sativus*) and orange fruited Xishuangbanna cucumber (*Cucumis sativus* var. *xishuangbannanesis* Qi et Yuan) (XIS) PI 509549, from Peoples Republic of China (Simon and Navazio, 1997). Further the Inheritance of β-carotene-associated flesh colour in cucumber was studied (Cuevas *et al.*, 2010a) and found mesocarp and endocarp F_2 segregation adequately fit a 15:1 [low β-carotene (0.01–0.34 lg g^{-1}): high-β-carotene (1.90–2.72 lg g^{-1})] and 3:1 (low-β-carotene: high β-carotene) ratio, respectively. Likewise, segregation of carotene concentration in mesocarp and endocarp tissues in BC_1P_2 progeny adequately fit a 3:1 (lowb-carotene: high-β-carotene) and 1:1 (low-β-carotene: high-β-carotene) ratio, respectively. Progeny segregations indicate that two recessive genes control the b-carotene content in the mesocarp, while one recessive gene controls β-carotene content in the endocarp.

A large number of orange fleshed cucumber accessions were collected from Mizoram of North-east region of India which had significantly high carotenoid content than Chinese type and thus, appeared to be promising for utilization in breeding programme. A unique accession namely IC420405, showing orange flesh (mesocarp and endocarp) colour and lemon yellow skin colour was identified in 2012 at ICAR-NBPGR, New Delhi (Ranjan *et al.*, 2019) and subsequently registered with ICAR-NBPGR, New Delhi (INGR18029).

Morphological characters, anatomical evidences, trends of distribution and uses suggest the prevalence of diversity of orange fleshed cucumbers in north-east region of India. The germplasm having common character *viz.*, orange flesh colour in adjoining regions of Myanmar and southern China are indicative of close linkages of continuous variation in this region (Ranjan *et al.*, 2019).

The '*B*' gene, in *C. pepo* which conditions precocious yellow fruit pigmentation before anthesis produce deep yellow coloured fruit with high vitamin A content (Shifriss, 1981). It is chlorophyll deficiency character in different plant parts and controlled by two unlinked genes, *B1* and *B2* (Shifriss, 1991). This gene from *C. pepo* was successfully transferred to *C. moschata* (Paris *et al.*, 1985). Later, Shifriss *et al.* (1989) developed two different *B* lines of *C. moschata* after gene transfer from *C. pepo* and *C. maxima viz.*, NJ-B (*B1* gene from *C. pepo*) and IL-B (*B2* gene from *C. maxima*). Inheritance pattern of the *B* genes was further studied by Shifriss (1993) in the F_2 population of the *C. moschata* cross IL-B × NJ-B, where the female parent is thought to have the constitution *B1+ B1+ B2B2* and the male parent *B1B1 B2+ B2+*. The breeding programme led to the development of two phenotypically distinct lines.

16.2.13 Breeding for Quality

Attractive light green/green uniformly long cylindrical fruits with smooth surface without prominent spines, prickles, and crookneck and crispy with tender flesh are the primary objective of breeding for the quality character in cucumber. Fruits free from carpel separation, bitterness and less seed at edible maturity should be other objectives of the breeder. Cultivars differ in the time required for their fruits to develop from the optimal size for market to oversized fruits of little value, and 'Marketer' has become important because it produces a large proportion of fruits of marketable stage. Effort were also made in the development of cultivar with bunching fruit habit producing multiple pistillate flowers on individual nodes for harvesting finger size fruits to suit whole fruit canning for export. Newly developed multiple branching cultivars (*e.g.* 'little leaf' types) produce more fruits at the same stage of maturity, which is desirable for once-over harvesting systems. Besides, pickling cucumber cultivars have also been bred to withstand carpel separation in order to prevent bloating during the brining process. Recent breeding efforts towards increasing yield and multiple branching have focused on the wild var. *hardwickii*.

A set of 81 recombinant inbred lines (RIL) derived from Group *Cantalupensis* U.S. Western Shipper market type germplasm was examined in two locations [Wisconsin (Wisc.) and California (Calif.), USA] for two years to identify quantitative trait loci (QTL) associated with β-carotene accumulation in mature fruit. Three hundred fifty-eight melon simple sequence repeats (SSR), 191 cucumber expressed sequence tag (EST), and 42 cucumber EST-SSR markers were evaluated to enhance saturation of a resident 181-point map. Additionally, genomic information from diverse plant species was used to isolate partial nucleotide sequences of eight putative genes coding for carotenoid biosynthesis enzymes, identify single nucleotide polymorphisms (SNP), and perform candidate gene analysis. Mapping parent analyses detected 64 SSR polymorphisms, seven SNP using cucumber EST and four SNP in putative carotenoid candidate genes, and these markers were used to create a moderately saturated 256-point RIL-based map [104 SSR, 7 CAPS, 4 SNP, 140 dominant markers and one morphological trait (a) spanning 12 linkage groups (LG)] for β-carotene QTL analysis. Eight QTL were detected in this two-location RIL evaluation that were distributed across four LG that explained a significant portion of the associated phenotypic variation for β-carotene accumulation (R_2 = 8 to 31.0 per cent). Broad sense heritabilities for β-carotene accumulation obtained from RIL grown in Wisc. and Calif. were 0.56 and 0.68, respectively, and 0.62 combining both locations. Although genotypes x environment interactions were confirmed in two-year experiments, relative RIL performance rankings remained consistent. QTL map positions were not uniformly associated with putative carotenoid genes. One QTL (β-car 6.1) interval was located 10 cM from a β-carotene hydroxylase gene, and this region was colinear with previously reported QTL for color pigmentation. These results suggest that accumulation of β-carotene in melon is under complex genetic control, where epistasis plays an important role in trait expression. This study provides the initial step for defining the genetic control of β-carotene accumulation in melon leading to the development of varieties with enhanced β-carotene content (Cuevas *et al.*, 2008).

16.2.14 Mutation Breeding

In cucumber, bushy, compact and determinate mutants have been obtained, from an inbred line of cv. Borszczagowski by treatment with 0.6-0.1 per cent ethylencimine (Kubicki *et al.*, l986a; Kubicki *et al.*, l986b; Soltysiak *et al.*, 1986). Seeds of cucumber cv. M15 were treated with 0.05 per cent ethylencimine for 21 h, and selection for resistance to *Meloidogyne* spp. revealed that 7.6 per cent of plants were resistant (Udalova and Prikhod'Ko, 1986).

In *Momordica charantia*, gamma irradiation and EMS (ethylmethane sulfonate) treatment led to an increased genetic variability in quantitative characters (Mallaiah and Zafar, 1986). The M_1 progeny derived from radiation mutagenesis, which are controlled by single recessive genes. One variety, MDU 1, developed as a result of gamma raditation (seeds treatment) of the landrace MC-103 (Rajasasekharan and Shanmugavelu, 1984). Likewise, the white bitter gourd mutant "Pusa Do Mausami" (white-fruited type) was developed through spontaneous mutation from the natural population "Pusa Do Mausmi" (green-fruited type) at the Indian Agricultural Research Institute.

In *Trichosanthes anguina*, Datta (1987) isolated a mutant with yellow striped, yellow fruit in the M_2, following treatments with 18KR X-rays of seeds from a variety with white-striped, white fruit. The mutant was superior to its parent for fruit size and yield and content of punicic acid (a desirable industrial chemical) in its seed oil.

A space mutation breeding using dragon fruit cucumber was carried out and obtained an excellent variation of white cucumber strain "055-33-6-1-249" after five generations of mutated line selection (Li *et al.*, 2016). The cucumber varieties space "96-1" (Zhang *et al.*, 2005), "Hang Yan 1" (Li *et al.*, 2009) and "HangYu 1" (Wang *et al.*, 2016) were also obtained by space mutation breeding.

17.0 Biotechnology

17.1.0 Cucumber

Among the cucurbitaceous vegetables, marked progress has been made on biotechnology of cucumber, muskmelon and watermelon. The results of investigations and the achievement obtained so far have been reviewed by Parthasarathy *et al.* (2000).

17.1.1 Tissue Culture

Tissue culture research in cucumber has resulted in the development of protocols for micropropagation, embryo, anther and protoplast culture and somatic embryogenesis.

17.1.1.1 Micropropagation

Direct regeneration without intermediate callus phase is an ideal system of micropropagation as it does not produce any somaclonal variants. Navratilova (1987) standardized *in vitro* micropropagation of 6 breeding lines by culturing apical meristems of 9-day old seedlings on MS medium with 0.01 mg IBA, 0.1 mg BA and 10 mg ascorbic acid (all per litre). Proliferating meristems were transferred to fresh medium after 3 weeks, and subsequently individual developing plantlets were subcultured on the same medium every 3 weeks. Some differences were noted between lines in the capacity for callus formation, growth rate, production of secondary buds and formation of floral buds on the regenerated plants. Continuous *in vitro* culture was successful if light and humidity were controlled. Hisajima *et al.* (1989) induced multiple shoots using BA (2.55 µM) from the seeds of cv. Mizunasu cultured *in vitro*. Shoots were continuously multiplied using a combination of BA and IBA.

Misra and Bhatnagar (1995) developed an efficient protocol for direct shoot regeneration from cucumber (cv. Japanese Long Green), leaf explants. Regeneration of plantlets cotyledon (Lou *et al.*, 1996), leaf (Kuijpers *et al.*, 1996), immature embryos (Custers *et al.*, 1988) from cultivars of cucumber has been reported and the protocols have been standardized.

Lou *et al.* (1996) conducted a research on how the growth hormones and sucrose efficiently works in the *in vitro* regeneration of cucumber from different explants. Highest multiplication rates of normal embryos were achieved with 2,4-D at 4 and 8 µM. Subsequent results demonstrated that elevated sucrose concentrations could also change the morphogenic regeneration patterns inducible by IAA. In the presence of IAA at a given concentration (57.1 µM) and with sucrose at 131 µM, only adventitious shoot formation occurred, whereas 263 µM sucrose induced both shoot organogenesis and somatic embryogenesis. Selvaraj *et al.* (2006) studied the *in vitro* method of plant regeneration from cucumber using hypocotyl plantlets. For callus induction they used the media containing MS media, sucrose 87.64 µM, agar 8 gl⁻¹, 2,4-D 3.62 µM and BA 2.22 µM for callus induction. They get success of 25 new shoots from callus, the media composition for conversion of callus to embryogenesis like MS media added with 8.88 µM BA, 2.5 µM zeatin and also added the coconut water for subsequent subcultures. It was concluded that the media containing NAA, BA, Zeatin and L-glutamine at 1.34, 8.88, 0.91 and 136.85 µM respectively works efficiently in development of shoot buds (75.6 per cent). Abu-Romman *et al.* (2015) evaluated the role of zeatin in the *in vitro* shoot multiplication of cucumber. Nodal segments of 10 days old seedlings are shifted to the MS based supplemented with 0.5 mgl⁻¹ BAP, 1 mgl⁻¹ KIN, 2 mgl⁻¹ TDZ, 3 mgl⁻¹ Zeatin. They tried with all kinds of cytokinins but their experiments showing that kinetin is the most accepted cytokinin for *in vitro* shoot multiplication. The media containing MS and kinetin 1 mgl⁻¹ showing highest percentage of shoot multiplication (83 per cent). Per explant they got 7.93 shoots, but here they got the negative response from the media fortified with BAP and zeatin. So they came to conclusion that compared to the zeatin and BAP, kinetin role in the *in vitro* regeneration of cucumber through nodal segments will responding more. Jesmin *et al.* (2016) attempted micropropagation using the germinated seeds. Explants from germinated seedlings were cultured on MS medium supplemented with individual treatments of different auxins (2,4-dichloro-phenoxyacetic acid (2,4-D), α naphthalene acetic acid (NAA) or cytokinins (benzyl aminopurine (BAP). The optimum medium for callus induction from leave, stem and cotyledon explants was MS medium supplemented with 0.5 mgl⁻¹ BAP added with 1.0 mgl-1 NAA. Priyanka *et al.* (2019) reported the efficient protocols for *in vitro* plant regeneration of cucumber from hypocotyl explants with standard explant like hypocotyl (10 mm in length) excised from 7-day-old *in vitro*-raised seedlings with media composition of MS medium containing different concentrations of BAP (0-5.0 mgl⁻¹). BAP at a concentration of 2.0 mgl⁻¹ was found to be the most effective (79 per cent) for callus induction.

17.1.1.2 In vitro Flowering

In vitro flowering is extremely use in generation advancement under aseptic condition and has potential to get 4-6 generation within a single year. This technique has been used widely in different grain crops and its application in vegetable crops is very rare. Induction *in vitro* flowering from shoot tip explants of cucumber (*Cucumis sativus* L. cv. 'Green long') was reported by Sangeetha *et al.* (2014). *In vitro* flowering is an alternative breeding tool for generating hybrid *Cucumis* spp. as it is able to overcome limitations caused by interspecific incompatibility. This study describes an efficient method for *in vitro* flowering from shoot tip explants of cucumber (*Cucumis sativus* L.). They used different concentrations of 6-benzylaminopurine (BAP; 0.5–2.5 mg/L) alone or in combination with 0.5 mgl⁻¹ kinetin (KIN). For *in vitro* flowering, shoots were cultured on MS medium supplemented with 0.5 mgl⁻¹ BAP and different concentrations of sucrose.

17.1.1.3 *Embryo Culture*

The success of embryo culture depends on the stage of embryo development as well as supplementation with auxin and cytokinins. Visser and Franken (1979) described an embryo culture technique from tetraploid × diploid cucumber crosses which developed normally in tetraploid plants but the seeds were only partially filled and rarely germinated. The culture technique enabled the largest embryos (4mm or longer) to be grown successfully. The resulting plants were all triploids.

In vitro Propagation and Flowering of Cucumber

Micropropagated Plantlets (MP) Cultured on MS PGR-free Medium (a), and with Developed Male Flowers (b). Female Flowers (Kielkowska and Havey, 2012).

Mackiewicz *et al.* (1998) obtained triploids by embryo rescue following crosses between diploid and tetraploid parents. Ananalys of the effects of triploidy on some morphological traits in 3 lines of different origins confirmed that the effects depended on the genotype, although some traits were common. The new traits appeared in the triploids were parthenocarpy; increased fruit size, stronger femaleness, seedlessness, new classes of pollen size, a marked decreased in pollen fertility, appearance of pollen with 4-6 pollen tube holes, and large flowers. Malepszy *et al.* (1998) isolated and cultured late heart stage triploid embryos from reciprocal tetraploid × diploid crosses of 3 cucumber lines (411, 412 and 413) and found a high variation of development, depending on the lines, and cross direction. According to them the five main deficiencies that decreased the chance of obtaining triploid plants are inability of embryos to germinate, non-rooting plantlets, albinism, mixoploidy and non-survival on transfer to soil. The lowest rate of failure occurred when the tetraploid was the female parent. Depending on the line, 11.7-45.8 per cent of embryos obtained in the 4x × 2x direction developed into plants.Triploid plants did not show visible abnormalities.

Successful inter-specific hybridization between *C. sativus* × *C. hystrix* was attempted by Chen *et al.* (1997) through controlled pollination and subsequent embryo rescue. Controlled crossing resulted in fruit containing embryos which were excised and rescued on a Murashige and Skoog solid medium. Hybrid plants were morphologically uniform. The multiple branching habit, densely brown hairs (especially on corolla and pistil), orange-yellow collora, and ovate fruit of F_1 hybrid plants were similar to that of the *C. hystrix* paternal parent, while appearance of the first pistillate flower was more similar to that of *C. sativus* maternal parent than to *C. hystrix*, staminate flower appearance was mid-parent in occurrence. Based on this study they have reported a new synthetic species, *C. hytivus* J.F. Chen and J. H. Kirkbr. (2n = 38), obtained by doubling the number of chromosomes in the sterile interspecific F_1 hybrid (2n = 19) between *C. sativus* L. (2n = 14) and *C. hystrix* Chakr. (2n = 24), was described (Chen *et al.*, 2000). Beharav *et al.* (1995) studied the *in vitro* embryo culture of the hybrid between *C. melo* × *Cucumis metuliferus*. They pollinated the plants in controlled condition and flower from 14 and 17 days after pollination collected and cultured in MS media supplemented with casein hydrolysate, m-inositol, thiamine-HCl, pyridoxin-HCl, nicotinic acid, sucrose, IAA, kinetin and GA_3 at different concentration like 1 gl^{-1}, 0.1 gl^{-1}, 1 gl^{-1}, 1 gl^{-1}, 1 gl^{-1}, 35 gl^{-1}, 0.02 mg^{-1}, (0.125, 0.5, 1 mg^{-1}) and (0 to 0.5 mg^{-1}), respectively. They got the better response from embryo taken at 17 days after pollination. After embryo germination for the embryo development purpose they alter the media with 0.5 mg^{-1} kinetin found effective. Similarly embryo extracted from the 14 days after

Table 16: Successful Inter-specific Hybridization among different Species of *Cucumis*

Sl.No.	Cross	Results	Source
1	*C. sagittatus* × *C. melo*	Embryo only	Deakin *et al.* (1971)
2	*C. metuliferus* × *C. melo*	Embryo only	Fassuliotis (1977)
3	*C. sativus* × *C. melo*	Globular stage embryo only	Niemirowicz-Szczytt and Kubicki (1979)
4	*C. metuliferus* × *C.melo*	Fertile F_1	Norton and Granberry (1980)
5	*C. prophetarum* × *C. melo*	Fruits with unviable seeds	Singh and Yadava (1984)
6	*C. zeyheri* × *C. sativus*	Fruits with unviable seeds	Custers and Den Nijs (1986)
7	*C. sativus* × *C. metuliferus*	Embryo only	Franken *et al.* (1988)
8	*C. melo* × *C. metuliferus*	Embryo only	Soria *et al.* (1990)
9	*C. sativus* × *C. hystrix*	Sterile plants (2n and 4n)	Chen *et al.* (1997b)
10	*C. hystrix* × *C. sativus*	Fertile plants (4n)	Chen *et al.* (1998)
11	*C. anguria* × *C. zeyheri*	Embryonic callus regenerated plants	Skálová *et al.* (2007)
12	*C. sativus* × *C. melo*	Embryonic callus only	Skálová *et al.* (2007)
13	*C. sativus* × *C. metuliferus*	Embryonic callus only	Skálová *et al.* (2007)

pollination gives better response from the media containing 0.5 mg^{-1} and 0.5 mg^{-1} Kinetin and GA$_3$, respectively. The Table 16 enlists the susscessful interspecific hybridization in *Cucumis* by different workers.

17.1.1.4 Development of Haploids and Doubled Haploids

Development of doubled haploids have been used widely in creating hiomozygosity within a single generation, creation of mapping population, broadening the genetic base and create diversity in the existing gene pool. In cucumber this technique has been used widely and the different approaches successfully practised in cucumber are summarized as below:

Gynogenesis and Doubled Haploid Production of Cucumber

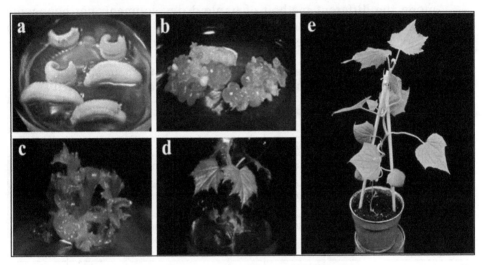

Gynogenesis and Regeneration of Plantlets.

a: Unpollinated ovary culture; b: Embryo like structure; c: Shoot like structure and shoot formation; d: Elongated shoot and roots (complete plantlet); e: Transplanted plantlet (Sorntip *et al.*, 2017).

Table 17: Status of Haploids and Doubled Haploids in Cucurbits*

Technique Employed	Donor Plant Parts	Stage of Culture	Specific Treatment	Success Rate	Researchers
Gynogenesis	Unfertilised ovule/ovary	6 hour before anthesis	Heat treatment of 35 °C for 4-5 days	18.4 per cent	Ge´mesne´ *et al.* (1997); Ge´mes-Juha´sz *et al.* (2002)
Gynogenesis/ Parthenocarpic embryogenesis	Ovaries pollinated with irradiated pollens	Embryo rescue 10-25 days after pollination	Gamma irradiation @ 200-500 rad	-	Troung-Andre (1988); Niemirowicz-Szczytt and Dumas de Vaulx (1989); Sauton (1989)

Technique Employed	Donor Plant Parts	Stage of Culture	Specific Treatment	Success Rate	Researchers
Androgenesis	Isolated microspores	Late uninucleate stage	Heat shock treatment at 32-35 °C	15-25 embryos/ plate	Niemirowicz-Szczytt (1994)
Androgenesis	Anther culture	Small buds 3-7 days before anthesis	Both cold and heat shock treatment	-	Sztangret Wiœniewska *et al.* (2006); Lofti *et al.* (1999); Lofti *et al.* (1997)

* Dey *et al.* (2020).

Among the different method of haploid development, parthenogenic development of haploids through pollination with irradiated pollen was found to be most successful in cucumber and reported widely by several workers.

Table 18: Comparative Analysis of Development of a Homozygous Material through DH Technology and Conventional Pedigree Inbreeding Method (PIM)*

Activity	Progeny	Genetic Composition	Time Line (Year)	
			DH	PIM
Crossing of two desired parents	F_1 hybrids	Heterozygous	-	-
Development of doubled haploids/Selfing	DH1/F_2	Homozygous/heterogeneous and heterozygous	1.0	1.0
Evaluation of DH plants and selection/selfing	DH2/F_3	Homozygous/heterogeneous and heterozygous	2.0	2.0
Multilocation trial after attending homozygosity and homogeneity	DH3/F_7	Homozygous/mostly homo-zygous and homogeneous	3.0	8.0
Development of commercial variety	DH$_4$/F_{10-12}	Homozygous/mostly homo-zygous and homogeneous	4.0	10.0-13.0

* Dey *et al.* (2020).

The homozygous materials developed within 2-3 years using DH technology can be readily used as one parent development of F_1 hybrids or they may be recommended as such ass a commercial variety if performed well under multi-location trial. Therefore, the time required is drastically reduced through development of DH in cucumber. Besides, the DH based populations are ideal in molecular mapping of complex traits. However, development of RILs (recombinant inbred lines) requires several years (5-8 yrs) for their use in molecular mapping. Therefore, DH technology offers very attractive option to accelerate the breeding methodologies in cucumber.

Development of doubled haploids from the F_1 hybrids offers a unique opportunity to create novel genotype and more diversity from an existing population. This is mainly because of one cycle of meiosis and recombination when the donor (microspore/megaspore) for the DH development is taken from the F_1

hybrids of two distantly related genotypes. Moreover, this technology is the only option to develop completely homozygous and stable traits like gynoecism and parthenocarpy without going for marker assisted backcross breeding. Moreover, there is possibility that all the developed DH are diverse from each other while possessing the desirable traits like gynoecy, parthenocapy and tolerance to different biotic stresses. Lofti *et al.* (2003) demonstrated that it is possible to develop completely homozygous lines with multiple virus resistance through induction of haploids from F_1 and their subsequent diplodization in *C. melo*. It is further explained in the following figure for its possible application in cucumber.

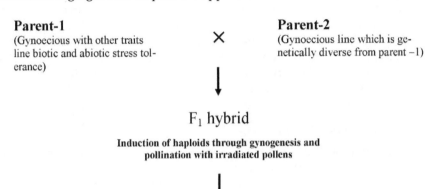

Parent-1
(Gynoecious with other traits line biotic and abiotic stress tolerance)

×

Parent-2
(Gynoecious line which is genetically diverse from parent −1)

F_1 hybrid

Induction of haploids through gynogenesis and pollination with irradiated pollens

Development of large number of DH lines through chromosome doubling of the haploids

i. They will be very diverse because of one cycle of meiosis in F_1.
ii. All the DH lines will be gynoecious in nature.
iii. Large number of novel gynoecious DH genotypes with complete homozygosity will be created.
iv. All the developed DH lines can be readily used in breeding programme and genetic studies.
v. Large number of new genotypes with biotic abd abiotic stress tolerance will be created which are not present in natural population.

Figure 5: Doubled Haploid (DH) based Approaches are Useful in Germplasm Innovation in Cucumber

Embryo rescue has been used to rescue haploid embryos. The production of haploid embryos depended on season. Caglar and Abak (1999) induced haploid cucumber embryos by pollination with irradiated pollen and cultured on E20A medium under aseptic conditions. They found that the regenerative ability of embryos was higher at advanced stages (60 per cent in the first year and 80 per cent in the second year) than at globular stage. These embryos also produced haploid plants rapidly (in 3.5 days).

17.1.1.5 Anther Culture

In 1958 Aalders isolated haploid embryos from normal seeds of cucumber, while Swaminathan and Singh (1958) observed a haploid shoot of watermelon on a diploid plant grown from an X-ray treated seed. The procedures for chromosome doubling of cucurbit haploids are mainly based on colchicine treatment (Çaðlar and Abak, 1997, Nikolova and Niemirowicz-Szczytt, 1996, Sari and Abak 1996, Yetisir and Sari, 2003), a method that is constantly being revised (Lim and Earle, 2008, 2009). Çaðlar and Abak (1997) analysed the influence of colchicine concentration and treatment duration on chromosome doubling of cucumber haploids *in vitro* and noted that the highest efficiency of chromosome doubling (60 per cent on average) was obtained with the immersion of plantlets in 0.5 per cent colchicine solution for four hours. Another chromosome doubling method designed for cucumber employs is the direct regeneration of plants from haploid leaf explants (Niemirowicz-Szczytt *et al.*, 1995), which showed to be more effective than colchicine treatment (Sztangret-Wiœniewska *et al.*, 2006). Faris *et al.* (2000) used explants from the first and second haploid leaf in order to directly regenerate DHs of cucumber. Plants were regenerated from both types of leaves (388 and 210, respectively), however, all plants derived from the first leaf were haploids, whereas the second leaf yielded 70.5 per cent haploids, 28.2 per cent diploids and 1.3 per cent mixoploids.

17.1.1.6 Protoplast Culture

The first demonstration of the operation *in vitro* of a mechanism conferring resistance to plant virus infection and underlining the potential value of the use of protoplasts in studying resistant as well as susceptible virus/host combinations was reported by Coutts and Wood (1977). Using protoplasts derived from a resistant (China-K) and susceptible (Ashley) cucumber cultivars they found that 48 h after inoculation with CMV strain Price's No.6, neither infectious RNA nor nucleoprotein could be recovered from the China-K protoplasts whereas infectivity was recovered from identically treated protoplasts from Ashley leaves.

An efficient method was developed for obtaining callus from cotyledon, hypocotyl and leaf protoplasts. Of the factors studied that affected protoplast isolation and culture, the key step for reproducibility of results and high planting efficiency was preculture of the material from which the protoplasts were to be isolated. The frequency of callus giving rise to plantlets was on an average 1.5 per cent with a maximum of 12.5 per cent (Garcia Sogo *et al.*, 1991). Garcia Sogo *et al.* (1992) also studied protoplast isolation, formation of microcalli and culture of the microcalli in media containing NAA/BA or 2,4-D/BA to form embryogenic callus. Nearly all the embryogenic calli produced normal cotyledonary embryoids at an average rate of 1-3 per callus, and the embryoids readily grew on basal medium without growth regulators.

Colijn Hooyrnans and Bouwer (1987) isolated protoplasts from cotyledons, hypocotyls and seedling leaves and cultured on MS medium with added sucrose, mannitol, NAA and 2iP. *Amorphous* callus from this medium formed embryoids when transferred to hormone-free MS medium. Embryoid growth was stimulated by reduction of sucrose from 3 per cent to 2 per cent and shoots formed in 2 per cent of the embryoids. A method is described for the isolation of protoplasts from cotyledons of cucumber (*Cucumis sativus* L.) by Dons and Bouwer (1986). After isolation, protoplasts were embedded in a mixture of agarose and Murashige and Skoog medium supplemented with 250 mg/L trypton, 2 per cent sucrose, 5 µM naphthalene acetic acid and 15 µM 2iP. The culture of the protoplasts was improved by the application of an agarose-disc culture procedure. The embedded protoplasts were plated in small (100 µL) droplets in petri dishes to which, after gelling of the agarose, liquid medium was added. For comparison, protoplasts were also cultured according to the agarose-bead procedure. The plating efficiency ranged from 50 to 80 per cent if the protoplasts were plated at high density (105 protoplasts per mL). In agarose-bead culture, divisions were induced at a lower rate. At lower densities the plating efficiency was dramatically decreased. Growth of microcalli was determined by homogenizing culture samples and measuring their density spectrophotometrically. The growth rate of the developing cell clusters was much higher in agarose-disc cultures as compared with bead-type cultures. It is concluded that for cucumber protoplasts the agarose-disc culture procedure provided optimal conditions for both the initiation of cell division and the growth of microcalli. A procedure describing the isolation and culture of protoplasts from cotyledons and leaves of *Cucumis sativus* was reported by Colijn-Hooymans *et al.* (1988). They grew the plants used for protoplast isolation under *in vitro* condition. Protoplasts were plated according to different procedures, which were compared on the basis of the observed plating efficiency. Protoplasts cultured in agarose solidified medium (either by the bead-type culture or the disc-type culture) showed plating efficiencies of up to 50 per cent. In contrast, division frequency was low (<2 per cent) when protoplasts were cultured in liquid medium. Frequent removal of the medium and addition of fresh medium with lower osmolarity and different concentrations of hormones resulted in the production of yellow capable of regeneration. Shoots were obtained from leaf protoplasts approx. 3 months after protoplast isolation. It was possible to induce regeneration in calli subcultured for 1 year. The transient gene expression system using plant protoplasts has become widely used for high-throughput analysis and functional characterization of genes. Huang *et al.* (2013) investigated protoplast isolation and green fluorescent protein (GFP) transient transfection and their main affecting factors, such as mannitol concentration in enzymolysis solution, enzymolysis time, and polyethylene glycol (PEG) concentration and transfection time, on 'xintaimici' cucumber. The results showed that when the enzyme solution had 1.5 per cent cellulase R-10 (w/v), 0.4 per cent macerozyme R-10 (w/v), 0.4 M mannitol, 20 mM 2-morpholinoethanesulfonic acid, 10 mM $CaCl_2$, 0.1 per cent bovine serum albumin, and was at pH 5.8 with an enzymolysis time of 8 h, the protoplast yield was 6–7 × 10^6/g fresh weight. Viability was about 90 per cent. When the concentration of PEG4000 was 20 per cent and transfection time was 20 or 30 min, transformation efficiency was greater than50 per cent and the green fluorescent signal could be detected in the cytoplasm, chloroplasts, and plasma membrane.

17.1.1.7 *Somatic Hybrids*

Protoplast fusion and somatic hybridization have been reported in different species of *Cucumis* and also in species of different genera. In *Cucumis* species, somatic hybridization by protoplast fusion was described between *C. sativus* and *C. melo* (electrofusion: Jarl *et al.*, 1995), *C. metuliferus* and *C. melo* (electrofusion:

Debeaujon, Branchard, 1990; chemical fusion: Roig *et al.*, 1986), *C. melo* and *C. anguria* (electrofusion: Dabauza *et al.*, 1998), and *C. melo* and *C. myriocarpus* (electrofusion: Bordas *et al.*, 1998). Mesophyll and callus protoplasts in two types of fusion were tested and compared: chemical fusion by polyethylene glycol (PEG 6000) and electrofusion (using apparatus ECM 2001). Protoplasts were viable and undamaged after isolation and fusion experiments. The hybrid products after fusion contained rich vacuolar systems and many chloroplasts, that is, characteristics of both fusion partners (Navrátilová *et al.*, 2006a, b). Polyethylene glycol (PEG) is a fusogen introduced in 1974 by Kao and Michayluk to increase the frequency of the fused protoplasts of lucerne (*Medicago sativa*). Fusing partners are mixed (mesophylland callus, mesophyll and mesophyll) and treated with PEG of different molecular mass (1,500–6,000) at a concentration range of 15–45 per cent for 15–30 min, to the final concentration of 106 protoplasts per ml (Navrátilová, 2004). Chemical fusion was performed using 33 per cent PEG for 15 min (Christey *et al.*, 1991; Navrátilová *et al.*, 2006a). Electrofusion was realized in electroporation chamber for 400–500 µl of volume with the 3, 2 mm electrode distance and following parameters: protoplast alignment 30 V AC and 1 pulse 90 V DC of length 80 µs (Navrátilová *et al.*, 2006a). The obtained calluses grew for several months until two years of *in vitro* culture. Many experiments showed that plant regeneration after isolation or fusion is very difficult (Gajdová *et al.*, 2004, 2007a, b), however, recent past Ondøej *et al.* (2009a) reported somatic pro-embryo formation derived from cucumber (*C. sativus*) protoplasts. Tang and Punja (1989) used leaves of *C. sativus* (Gy14) and *C. metuliferus* (PI292190) from 15 to 21 day-old seedlings raised from embryos cultured on MS medium, as protoplast source. Fusion was achieved by mixing suspensions in polyethylene glycol (PEG8000) solution and culturing protoplasts on MS medium with growth regulators (2,4-D, BAP and NAA) added in various concentrations. Fusion frequencies were estimated at 2-6 per cent and microcalli were obtained after 28 days.

Intergeneric fusions and somatic hybrids have also been attempted. Zhang and Liu (1998) observed that division of isolated protoplasts from the cotyledons of cucumber cultivars Erzaozi and Mici was inhibited when treated with 10mg rhodamine 6G (R6G) per litre for 15min. Cotyledonary protoplasts of China squash (*Cucurbita moschata*) and Malabar gourd (*Cucurbita ficifolia*) were inhibited after treatment with 1 and 1.3 mM iodoacetic acid per litre, respectively for 10 min. Fused protoplasts resumed division when treated with 25-35 per cent PEG (6000) for 5-10 min. Calli were obtained from the fusion combinations of *C. sativus* with *C. moschata* and *C. ficifolia* and some somatic hybrids were identified.

17.1.1.8 Somatic Embryogenesis

A system for efficient regeneration from diploid and haploid donor plants using by mesophyll material was developed by Dirks and Buggenum (1991). Only lines giving high regeneration frequencies on somatic embryogenesis were suitable for protoplast culture. Zhang *et al.* (1998) regenerated plant lets from calli which originated from mesophyll protoplasts and were cultured in solid MS medium supplemented with 1.0 mg BA and 0.05 mg NAA/l. Numerous calli were obtained from four kinds of intraspecific combinations of cucumber, some of which differentiated root sand/or were identified as somatic hybrid tissues according to isoenzyme analyses.

Somatic embryogenesis was induced in various explants on a medium consisting of MS salts supplemented with growth regulators. Embryoids matured on the same medium without 2, 4-D and developed into normal plants (Trulson and Shahin, 1986). Various explants like cotyledon, hypocotyl, leaves, radicles, shoots and zygotic embryos have been successfully used for production of callus for use in suspension cultures, embryogenesis and plantlet regeneration. The method has also been used to induce triploids (Zhang and Lu, 1995).

17.1.1.9 Somaclonal Variation

Plader *et al.* (1998) studied somaclonal variation in Borszczagowski line of *Cucumis sativus* for five regeneration systems, namely micropropagation (MP), direct leaf callus regeneration (DLR), leaf callus regeneration (LCR), recurrent leaf callus regeneration (RLCR) and direct protoplast regeneration (DPR). The highest frequency of change arose through DPR (90 per cent of lines) and RLCR (42.8 per cent), as opposed to 5.9 per cent with LCR. Tetraploids were produced only in the case of LCR (4.7 per cent) and RLCR (28 per cent).

17.1.2 Biochemical and Molecular Markers

17.1.2.1 Isozymes

Esquinas (1981) applied isozyme analysis to study the genetic variation among Spanish cultivars. He defined 15 enzyme coding loci among which the total number of allozyme was 22. Chen *et al.* (1997) analyzed isozymes of eight enzymes in *Cucumis hystrix* and two cultivated *Cucumis* species (*C. sativus* and *C. melo*) electrophoretically to investigate the biosystematics of these three species. Cluster analysis using data from six enzymes indicated that considerable genetic distance existed between *C. hystrix* and melon and between *C. hystrix* and cucumber. Isozymes have also been used by Puchalski *et al.* (1978) and Robinson *et al.* (1979) for establishing genetic relationship.

17.1.2.2 Molecular Markers

In cucumber, Kennard *et al.* (1994) constructed the genetic map utilizing RAPD, RFLP, Isozyme and phenotypic markers. Other genetic maps in cucumber developed by Serquen *et al.* (1997), Park *et al.* (2000), Staub and Serquen (2000), Bradeen *et al.* (2001), and Fazio *et al.* (2003b). The most dense map (contains 347 markers) of cucumber was developed by Park *et al.* (2000) spanning 816 cM, which is within the range of the length of the cucumber genome (750-1,000 cM) estimated by Staub and Meglic (1993).

Linkage maps have been constructed for *Cucumis sativus* (Kennard *et al.*, 1994, Serquen *et al.*, 1997). Dijkhuizen *et al.* (1996) used two sets of cucumber germplasm to determine the potential use of restriction fragment length polymorphisms (RFLPs) for estimating genetic relationships. Sixteen accessions [(15 from domesticated varieties of *sativus* and one from feral variety *hardwickii* (PII83967)] of diverse origins were used to assess RFLP variation in cucumber, and to determine if genetic relationships based on RFLPs were similar to those obtained by isozyme analysis.

Kennard *et al.* (1994) constructed a genetic map with RFLP, RAPD, isozyme, morphological and disease resistance markers spanning 766cM on 10 linkage groups for a cross within the cultivated cucumber. Earliness and fruit yield and quality components were investigated in progeny derived from the cross *C. sativus* var. *sativus* GY14 × *C. sativus* var. *hardwickii* PII83967 (Dijkuizen and Staub, 1999). A molecular marker map constructed from F_2 individuals was used to identify QTL for each trait examined and to assess the consistency of QTL over years (1991-92) and planting density (29000 and 58000 plants/ha), QTL affecting earliness, fruit yield and shape were identified. The traits examined were less affected.

Lv *et al.* (2012) fingerprinted 3345 accessions from worldwide cucumber collections with 23 highly polymorphic SSR markers.

The sex phenotype of cucumber is determined by three types of genes, namely, *F*, *M* and *A*, and their linkage markers have been widely reported. For example, the ACC synthase gene (*CsACS1* gene) markers are closely linked to F-locus (Ye *et al.*, 2000). SSR markers SSR19914 (3.2 cM), SSR23487 (0.28 cM), SCAR markers SCAR123 (0.94 cM), SRAP markers ME23SA4 (17.8 cM) and SCAP markers SCAP123 (0.94 cM) are linked to the M gene (Shi *et al.*, 2009).

Park *et al.* (2000) mapped the *ZYMV* resistance gene, *Zym* in the population TMG-1 × ST-8 and identified flanking marker E15/M47-F-197 (at 2.2 cM) and E14/M50-F-69, E14/M50-F-140, E15/M 50-F-221 (at 5.2 cM). In an other population scab resistance gene (*Ccu*) mapped with flanked marker CMTC51 and E14/M19-F-158-P2 at 0.5 and 1.9 cM (Bradeen *et al.*, 2001). The gene *dm* confers resistance for downy mildew disease mapped and BC5191100 identified as a flanking marker at 9.9 cM (Horejsi *et al.*, 2000). Bradeen *et al.* (2001) mapped the downy mildew resistance gene (*dm*) in the population G-421 × H-19 and WI-1983 × ST-8 utilizing RAPD, RFLP and AFLP markers.

Trebitsh *et al.* (1997) mapped gynoecious sex expression conditioned by a single major dominant locus *F*, which is involved with the regulation of bud primordial ethylene levels (higher ethylene levels support pistillate flower production). It was reported that *F* co-segregated with ACC synthase gene mapped in the population of Gy-14 (gynoecious) × PI-183967 (monoecious).

Bradeen *et al.* (2001) found a narrow linkage group in the population G-421 × H-19 and WI 1983 × ST8 for the little leaf (*ll*) gene and identified RAPD flanking marker (BC551) at 0.6 cM and another flanking marker (E14-M62-224 and OP-W7-2) for gene *ll* identified at 2.6 cM and 4.8 cM, respectively in the population G-421 × H-19 (Fazio *et al.*, 2003b). Bradeen *et al.* (2001) and Bradeen *et al.* (2003b) mapped the gene *de* (determinate plant growth habit) and identified RFLP flanking markers at 3.1 cM and 0.8 cM.

17.1.2.3 Genetic Transformation

Agrobacterium mediated genetic transformation is the most widely used method in *C. sativus*. Raharjo and Punja (1996) used *A. tumefaciens* strain EHA 105 carrying binary vector pMOG196 for transforming petiole explant of pickling cv. Endeavor. Similarly, binary vector was used for transforming hypocotyl (Nishibayashi *et al.*, 1996) and cotyledon (Chee and Slightom, 1991). Raharjo *et al.* (1996) transformed petiole explant of cucumber cv. Endeavor using three *A. tumefaciens* strains.

Neomycin phosphotransferse (*npt*) gene is most commonly used as selectable marker in transformation studies (Raharjo and Punja, 1996, Raharjo *et al.*, 1996). Microprojectile bombardment (Kodama *et al.*, 1993), protoplast transformation (Burza *et al.*, 1995) and microinjection (Dong *et al.*, 1993) methods have also proved successful.

Genetic transformation in *C. sativus* has been used for imparting resistance to viral and fungal diseases. Nishibayashi *et al.* (1996) introduced the *cmv-o* coat protein gene into cucumber plants, using a *Ti Agrobacterium*-mediated transformation system, with the aim of producing cucumber plants with cucumber mosaic cucumo-virus (*CMV*) resistance. Tabei *et al.* (1997) introduced a rice chitinasec cDNA (*RCC2*) under the control of the *CaMV 35S* promoter into cucumber by *A.tumefaciens*-mediated transformation. Regenerated plants were tested for resistance to *Botrytis cinerea* by infection with agar discs containing conidia. Some 75 per cent of regenerated shoots exhibited higher resistance than controls. Aartrijk (1990) tested genetically transformed cucumber plants expressing the cucumber mosaic cucumovirus (*CMV*) coat protein gene in the field for resistance to *CMV*. In a plot planted with transformed and untransformed plants, ca. 23 per cent of the transgenic plants became infected, compared with ca.70 per cent of non-transformed plants. Apart from *CMV* resistance in Poinsette and Marketmore cultivars in transgenic plants, fruit yield and vegetative growth were better in transgenic plants (Gonsalves *et al.*, 1992).

17.1.2.4 Genomics of Cucumber

The first cucumber used for sequencing is a northern China fresh market type inbred line '9930'. The 9930 draft genome sequence (Version 1.0, Huang *et al.*, 2009) was assembled using a combination of traditional Sanger and Solexa Next-gen sequencing technologies. Huang *et al.* (2009) reported 243.5 Mbp of sequence in the 9930 V1.0 assembly. The 9930 Version 1.0 draft genome assembly was based on 3.9-fold coverage of Sanger reads and 68.3-fold coverage of Illumina GA reads (average read length 42–53 bp). The genetic map used for anchoring scaffolds was developed using 77 recombinant inbred lines from a cross between the cultivated cucumber Gy14 and the wild cucumber (*C. sativus* var. *hardwickii*) line PI 183967 (Ren *et al.*, 2009) which only had 581cM map length in 7 linkage groups. During development of Version 2.0, Li *et al.* (2011a) employed 5.23G additional Illumina GAII sequencing data from 9930 and PI 183967; they also improved the prediction of protein-coding genes with evidence from RNA-Seq reads from 10 tissues of 9930. While Version V2.0 is a significant improvement over V1.0, the issue of mis-assemblies of many scaffolds in Version 1.0 was largely unaddressed in Version 2.0, which was found in several genetic mapping studies (*e.g.*, Li *et al.*, 2011b; Miao *et al.*, 2012; Yang *et al.*, 2013; Zhou *et al.*, 2015). In addition to 9930, the draft genome for the North American pickling cucumber inbred line *Gy14* was also developed (Yang *et al.*, 2012; http://www.phytozome.org).

Genetic Transformation in Cucumber

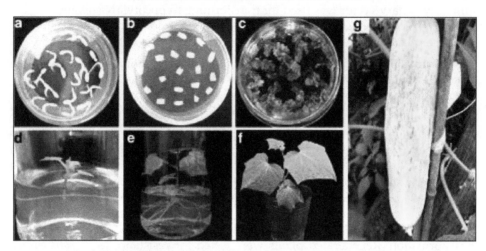

pCXSN-CsDIR16(±) genetic transformation of cucumber 'D9320'.

a: Seed germination; b: Co-culture; c: Screening culture; d: Plant regeneration; e: Resistant seedlings taking root; f: Regeneration of resistant seedlings; g: Seed of transgenic plants (Liu *et al.*, 2018).

17.1.2.5 QTL Mapping

Mapping of quantitative trait loci (QTLs) in cucurbits was first initiated in cucumber fruit traits by Kennard and Havey (1995). Among these traits, fruit weight was found to be in the vicinity of fruit diameter (Kennard and Havey, 1995; Serquen *et al.*, 1997), and multiple lateral branching (MLB) and fruit diameter were located in the same genomic region (Fazio *et al.*, 2003a). Four QTLs (LOD > 3) for sex expression were detected, as well as four for MLB, two for earliness, and two for fruit length (Serquen *et al.*, 1997).

The *Xishuangbanna cucumber* (XIS) is a semi-wild landrace that is endemic to the tropical south west China and surrounding regions with some unique traits such as tolerance to low light, large fruit size and heavy fruit weight, as well as orange flesh color in mature fruits that are very useful for cucumber breeding. Bo *et al.* (2015) conducted QTL mapping of domestication-related traits with recombinant inbred lines derived from the cross of the XIS cucumber with a cultivated cucumber inbred line. Comparative analysis of orders of common marker loci or marker-anchored draft genome scaffolds among the wild, semi-wild, and cultivated cucumber genetic maps revealed that the XIS cucumber shares the major chromosomal rearrangements in chromosomes 4, 5 and 7 between the wild and cultivated cucumbers suggesting the origin of the XIS cucumber through diversification selection after cucumber domestication

Dogimont *et al.* (2000) detected seven QTLs for resistance to cucumber mosaic cucumovirus (*CMV*). More recent past, Perchepied *et al.* (2005) mapped the *Fom* 1.2 gene that confer polygenic partial resistance to the *Fusarium* wilt race 1.2

using Isabelle (resistant) and Védrantais (susceptible) as parents. Nine QTLs were identified in five linkage groups which explained 41–64 per cent of the observed phenotypic variation.

Zhao (2011) used bulked segregant analysis SRAP and BSA technology to perform QTL analysis of cucumber neck length. Polymorphic bands were detected in nine markers, and two QTLs might affect the length of fruit stalk.

The application of SSR markers between subgynoecious S-2-98 and monoecious M95 bulks constructed from BC1 plants has identified three QTLs: *sg3.1, sg6.1* and *sg6.2*. The major QTL *sg3.1* accounts for 54.6 per cent of the phenotypic variation. PCR-based-markers from the SNP profile was developed to indicate that *sg3.1* was delimited to a 799-kb genomic region (Bu *et al.*, 2016).

QTL analysis revealed that 11 QTLs underlying fruit size variation and flowering time in the semi-wild Xishuangbanna cucumber, *FS5.2* played the major roles in determining the round fruit shape (characteristic of the WI7167 XIS cucumber) (Pan *et al.*, 2017a). Moreover, the round-fruited WI7239 had a 161-bp deletion in the first exon of CsSUN. A marker derived from this deletion was also mapped at the peak location of FS1.2 in QTL analysis (Pan *et al.*, 2017b).

17.2.0 Muskmelon

17.2.1 Tissue Culture

17.2.1.1 Micropropagation

Niedz *et al.* (1989) found cotyledonary explants of 4 day old muskmelon cv. Hale's Best Jumbo seedlings showed maximum initiation of shoot buds when cultured onto a modified MS medium supplemented with 5µM BA and cultured at 25-29°C under low light intensity. Culture of axillary buds (Spetsidis *et al.*, 1996), cotyledons (Guis *et al.*, 1997), hypocotyl stem and petiole (Tabei *et al.*, 1991) gave rise to plantlets.

Lee *et al.* (1995) cultured nodes of various melon cultivars from node explants *in vitro* on MS medium supplemented with BA (0.2-1.0ppm). Shoot growth was retarded by the addition of NAA, but callus formation was promoted. Shooting and shoot growth were also promoted in the presence of kinetin (0.1-1.0ppm).

Ren *et al.* (2012) investigated the different existing regeneration media and optimized the medium composition for an elite honeydew diploid breeding line, "150", using cotyledonary explants. Four combinations of three different plant growth regulators, 6-benzyladenine, abscisic acid, and indole-3-acetic acid (IAA), were tested in the shoot regeneration media. The presence of IAA in the medium caused the cotyledon explants to curl away from the medium. Omission of IAA from the culture media eliminated this problem and did not impact the shoot regeneration capacity of the cotyledonary explants. They had also estimated the ploidy of regenerated plants using flow cytometry, and 50–60 per cent were found to be polyploid (tetraploid or mixoploid). However, contrary to other studies, these polyploid plants did not show major morphological differences compared to the

In vitro Clonal Propagation of Muskmelon

A: Germination of seeds on MS medium containing 0.5 mg/l GA$_3$; B: Multiple shoot formation on MS + 1.0 mg/l BA; C: Rooting of in vitro shoots on MS + 0.1 mg/l NAA; D: A plantlet established in garden soil and vermicompost (2:1); E: Mature plant under field condition after 60 days of transplantation (Rahman *et al.*, 2012).

diploid plants. This explains the occurrence of somaclonal variants in the couse of *in vitro* culture of melons.

17.2.1.2 Embryo Culture

Embryo culture has also proved very useful to produce plants from seeds which failed to germinate under normal condition. Abak *et al.* (1996) investigated embryo induction in 18 melon (*Cucumis melo*) cultivars belonging to 3 botanical varieties (vars. *Dulain, odorus, reticulatus* and *cantalupensis*) pollinated with irradiated pollen. Embryos were successfully obtained from 14 genotypes by *in vitro* embryo rescue and regenerated into plantlets. These were treated with colchicine (0.5 per cent) for 2 h and then cloned to produce dihaploid plants which were acclimatized in the green house. Dore *et al.* (1995) pollinated with irradiated pollen, followed by embryo rescue to induce haploid plants.

Ezura *et al.* (1993) crossed diploid (2x=24; GSl and AdS1) and tetraploid (4x=48; GE46 and AsAd9) melon plants reciprocally. In the crosses of diploid × tetraploid plants, fruits contained three types of seed: (i) seeds containing abnormal embryos in which the cotyledon did not develop completely, (ii) seeds containing endosperm which propagates a typically, and (iii) empty seeds. In the crosses of tetraploid × diploid plants, fruits contained only empty seeds. When abnormal embryos excised from dry seeds were germinated *in vitro* in MS medium, about 20 per cent of them grew into normal plantlets and they were found to be triploid (3x=36). Haploid plants of *Cucumis melo* var. *reticulatus*, var. *inodorus* and of their hybrids were first obtained *in vitro* by culturing ovary slices removed from unpollinated hermaphrodite flowers (Ficcadenti *et al.*, 1999).

17.2.1.3 Haploid Production in Muskmelon

Doubled haploids (DH) have contributed to breeding programs for diverse crops. However, early attempts to produce haploids of melon or other cucurbit crops by anther or ovule culture were not successful in melon (Dryanovska and Ilieva, 1983). The first reported success in obtaining haploid or DH melons was achieved by rescue of parthenogenetic embryos induced by pollination with irradiated pollen (Sauton and Dumas de Vaulx, 1987). Ficcadenti *et al.* (1999) conducted study on haploids induction using gynogenesis in muskmelon. They selected the *Cucumis melo* L var. *reticulatus*, var. *inodorus* and their hybrids. Explants for their experiment were unpollinated ovaries of hermaphrodite flower. Under aseptic condition sliced ovaries are cultured in the media containing MS media supplemented with Thidiazuron 0.09 µM for four days. After few days of incubation explants shifted to MS media containing additional plant growth hormones like NAA 0.27 µM and BA 0.88 µM. From that cultures embryo like proliferations starts emerging after four to six weeks. They got germination from 82 cultured ovaries. They also got better response of the ovaries from the line 7-82, which belongs to the var. reticulatus shows 36.4 per cent of germination among all the cultured genotypes. For confirmation of obtained plants are haploids or germinated from somatic cells by using newly growing root tips for chromosome analysis. Through their findings they got 15.4 per cent of haploids, 23.1 per cent of diploids and 61.5 per cent mixoploiods. At

Production of Double Haploid in Muskmelon

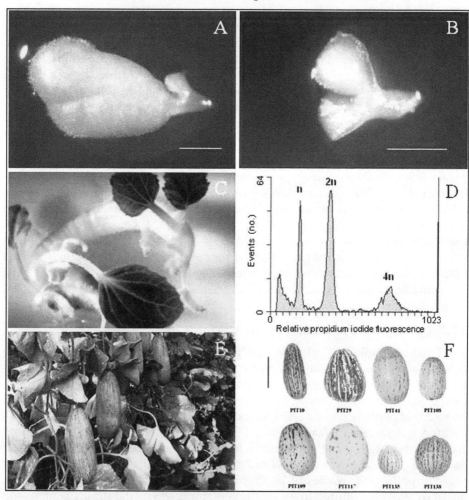

Melon doubled haploid lines produced through *in situ*-induced parthenogenesis. Parthenogenic embryos, 3 weeks after pollination with γ-irradiated pollen, with abnormal and asymmetric cotyledons [A–B (bar = 1 mm)]. Developed plantlet 3 weeks after in vitro culture (C). Histogram of flow cytometric analysis of DNA content of nuclei released from leaves of a haploid line treated with colchicine (D). Perpetuation of doubled haploid plants by self-pollination (E). Fruit variability from several doubled haploid lines derived from the PI 1613175 × 'Piel de Sapo' hybrid (F) (Gonzalo *et al.*, 2011).

last they came up with conclusion that genotype specificity place a major role in the *in-vitro* gynogenesis in muskmelon.

Lotfi *et al.* (2003) have developed improved procedures for recovery of haploid and doubled haploid (DH) melon plants, using hybrids derived from crosses of lines with multiple virus resistance. Seeds formed after pollination with irradiated pollen were cultured in liquid medium for 10 days before excision of the embryos for further culture. This made it easier to identify the seeds containing parthenogenetic embryos, thereby reducing the effort required and increasing the percentage of plants recovered. The plants obtained (approximately 175) were transferred to a green house for evaluation. Three fertile lines were identified, and selfed seeds were obtained for evaluating virus resistance. Flow cytometry of leaf tissues showed that two of these lines were spontaneous DH and the third was a mixoploid containing haploid and diploid cells. The other plants remained sterile through the flowering stage. Flow cytometry of 20 sterile plants showed that all were haploid. Attempts to induce chromosome doubling by applying colchicine to green house grown plants were unsuccessful. Shoot tips from the haploid plants were used to establish new *in vitro* cultures. *In vitro* treatment of 167 micropropagated haploid shoots with colchicine produced 10 diploid plants as well as 100 mixoploid plants. Pollen from male flowers that formed *in vitro* on the colchicine-treated plants was examined. High percentages of viable pollen that stained with acetocarmine were found not only in the diploids but also in >60 per cent of the plants scored as mixoploid or haploid by flow cytometry.

17.2.1.4 Protoplast Culture

Leaf protoplast of *C. melo* developed embryos and plantlets (Debcaujon and Branchard, 1992). Protoplasts have been isolated from cotyledons, hypocotyls and leaves. Debeaujon and Branchard (1988) isolated protoplasts from cv. Cantaloupe Charentais by enzymatic digestion; yields were 2×10^6/g from cotyledons and 4×10^6/g from leaves. Microcalli were produced on modified B5 medium supplemented with growth regulator sand those derived from cotyledon protoplasts gave rise to embryoids and small plantlets on supplemented modified MS medium.

Bordas *et al.* (1991a) studied the effect of several factors on electric field strength with a view to determining suitable conditions for an electrofusion system with protoplasts from cotyledons of *Citrullus colocynthis* and *Cucumis melo*.The effectiveness of the field applied depended on size and design of the fusion chamber, the fusion medium and the type of protoplast. There was an upper threshold for effective voltages. Yamaguchi and Shiga (1993) electrofused protoplasts of melon and pumpkin. Leaves from the 188 regenerants obtained from calli were analyzed for electrophoretic banding pattern of acid phosphatase; 33 putative somatic hybrids were selected.

Debeaujon and Branchard (1991) achieved regeneration of plants from muskmelon protoplasts through organogenesis or somatic embryogenesis, depending on the media sequence employed. They concluded that the morphogenetic competence appeared to be genotype-dependent.

17.2.1.5 Somatic Embryogenesis

Tabei (1997) studied various factors and conditions involved in plant tissue and cell culture techniques that influenced regeneration from plant tissue and found that cotyledonary explants of mature seeds and young seedlings possess high ability for shoot organogenesis and embryogenesis. Moreover, shoot organogenesis and embryogenesis of melon and cucumber were controlled by the concentrations of auxin in the regeneration media. Induction of somatic embryo genesis and plant regeneration is dependent on genotypes, light and growth regulators in the culture media. Cotyledonary and leaf explants produce high percentage of shoots (Bordas, 1991b). Branchard and Chateau (1988) described a method in which somatic embryos were produced from cotyledon callus cultivated on solid medium. Plants formed quickly and regularly if the medium contained both BA and 2,4-D.

17.2.1.6 Somaclonal Variation

Ezura and Oosawa (1994) examined the chromosome numbers in cells of zygotic embryo-derived callus, somatic embryos and root tips of regenerated plantlets in the cultivars Prince and Sunday Aki. In callus cells, ploidy levels ranging from haploidy to non-haploidy were observed. A small number of aneuploidy cells were observed at each ploidy level. In somatic embryo cells, diploidy, tetraploidy and octoploidy was observed at frequencies of 50, 38.5 and 11.5 per cent respectively. Ezura et al. (1992) studied morphological variation in leaf and fruit among plants derived from somatic embryos. When regenerated plants were cultured in a green house, 52 per cent showed leaves with asinus and 31 per cent of them set flat fruits. Somaclonal variation was also reported by Branchard et al. (1988).

17.2.2.0 Biochemical and Molecular Markers

Molecular marker technology and linkage maps were developed relatively late in melon compared with other major crop species. Restriction fragment length polymorphisms (RFLPs) were reported from the second part of the 1990s (Baudracco-Arnas and Pitrat, 1996; Oliver et al., 2001); simple sequence repeats (SSRs) some years later (Danin-Poleg et al., 2001; Ritschel et al., 2004; Gonzalo et al., 2005; Fernandez-Silva et al., 2008; Fukino et al., 2008); and finally single nucleotide polymorphisms (SNPs) more recent past (Deleu et al., 2009). First genetic map of melon using molecular marker (RAPD and RFLP) were constructed by Baudarcco-Arnas and Pitrat (1996) and second map with 188 predominantly AFLP markers was developed by Wang et al. (1997). The first molecular map did not cover the 12 melon chromosomes. The first maps covering the 12 chromosomes were reported by Oliver et al. (2001) and Périn et al. (2002a). Melon genome was mapped using 400 markers (Oliver et al., 2001). A high-density composite map from RIL population was constructed by Périn et al. (2002) consisting of 668 AFLP, IMA (inter micro satellite amplification) and phenotypic markers, which produced 12 linkage group. Gonzalo et al. (2005) generated another composite map of melon using SSR markers.

Dense linkage maps based on SSRs (Gonzalo et al., 2005; Fukino et al., 2008; Fernandez-Silva et al., 2008) and SNPs (Deleu et al., 2009) have also been developed. Linkage group (LG) nomenclature was not uniform, making difficult the comparison between different linkage maps. In 2005, Périn *et al.* (2002a) nomenclature was adopted by the international community and melon LG are named in Roman numbers from LG I to XII. An integrated map, including most of the published maps, has been constructed by the ICuGI and contains over 1,500 markers.

Puchalski *et al.* (1978) based on analysis of esterase, peroxidase and leucine aminopeptidase patterns from 24 species reported that: (i) the zymograins of *C. myriocarpus* and *C. leptodermis* were identical and similar to those of *C. africanus, C. anguria, C. meeusii* and *C. zeyheri*, (ii) on the basic differences in esterase patterns, *C. longipes* did not appear to be the direct progenitor of *C. anguria*, (iii) *C. angolensis, C. dinteri, C. sagittatus* and *C. asper* appeared to be closely related, (iv) *C. dipsaceus* and *C. heptadactylus* were distinct from all other species studied, (v) *C.trigonus* was similar to muskmelon and appeared to be a subspecies of *C. melo*, (vi) *C. hirsutus* and *C. humifructus* had some esterase and peroxidase bands in common with *C. melo*, (vii) *C. sativus* and *C. hardwickii*, both with 14 chromosomes, had isozymes distinct from all species with 24 or 48 chromosomes,and (viii) *C. hardwickii* appeared to be a subspecies of *C. sativus*, since its zymogram was similar to that of some cucumber varieties. Genetic linkage among cultivars *C. melo* (Staub *et al.*, 1998) and species belonging to different genera (Sujatha and Seshadri, 1989) has also been established through isozymes.

Apart from determining linkage, molecular markers have been used for selection of disease resistance. The RFLP analyses and hybridizations with different satellite DNAs and rDNA probes confirmed the division of the genus *Cucumis* into an African and an Asian subgenus and revealed that *C. melo* of the African subgenus is more closely related to *C. sativus* and *C. hardwickii* of the Asian subgenus than to *C. anguria* which also belongs to the African subgenus (Zentgraf *et al.*, 1992).

Wolff *et al.* (1996) identified 2 RAPD markers linked to susceptibility allele (*Fom 2*) for *Fusarium* wilt in melon. This marker was observed in susceptible cultivars but absent in resistant ones.

Wechter *et al.* (1995) tagged *Fom-2* gene for *Fusarium* wilt disease resistance gene using RAPD molecular markers caused by *Fusarium oxysporum* f. sp. *melonis* in a population MR-1 × AY and a flanking marker 596 identified at 2 cM. Later, this marker was converted into SCAR marker (Wechter *et al.*, 1998; Wang *et al.*, 2000). Two flanked marker AAC/CAT1 and ACC/CAT1 was found at 1.7 and 3.3 cM for *Fom-2* in the population MR-1 × AY (Wang *et al.*, 2000). Same gene was also mapped using the RAPD marker and two flanking markers (E07-1.3 and G17-1.0) identified from the cross of Vedrantais × PI 161375 (Baudarcco-Arnas and Pitrat, 1996).

The gene *nsv*, conferring resistance to melon necrotic spot virus, was first mapped by Morales *et al.* (2002). *nsv* was subsequently fine-mapped using 408 F_2 plants and 2727 backcross progeny to a single BAC clone with cosegregating marker 52K20sp6 (Morales *et al.*, 2005).

The gene, *Prv*, controls resistance to potyvirus which is a debilitating disease in melon. *Prv* was tightly linked to NBS47-3 (Brotman *et al.*, 2002) which is flanked at 1.4cM by *Fom-1* and NBS47-3 on each side (Brotman *et al.*, 2005). The *zucchini yellow mosaic virus (ZYMV)* resistance gene, *Zym-1*, was mapped in both melon (Danin-Poleg *et al.*, 2002) and cucumber (Park *et al.*, 2000).

Brotman *et al.* (2002) mapped the *Fusarium* wilt resitance gene *Fom-1* (races 0 and 2) and identified flanking marker, NBS 47-3 and K–1180 at 2.6 cM and 7.0 cM apart, respectively. Molecular markers are also identified for several virus resistance genes in melon. Morales *et al.* (2002) mapped gene *nsv*, responsible for melon necrotic spot virus and identified flanking marker OPD08-0.80, and CTA/ACG115, CTA/ACG120 at 4.4 and 1.5 cM. Two flanking marker X15L and M29 were identified at 0.25 and 2.0 cM for gene *nsv* from a population of PI 161375 × Pinoyonet Piel de Sapo for melon necrotic spot virus resitance (Morales *et al.*, 2003). Touyama *et al.* (2000) also mapped the same gene and identified the C2 marker at 2.0 cM. The gene *Zym-1*, conferring the resistance for *ZYMV* (zucchini yellow mosaic virus) was mapped in melon using the SSR marker CMAG36 (Danin-Poleg *et al.*, 2002). A gene *(Vat)* conferring the resistance against melon aphid *(Aphis gossypii)* have been mapped and RFLP flanked markers NSB 2 and AC 39 identified at 31 and 6.4 cM (Klingler *et al.*, 2001)

Danin-Poleg *et al.* (2002) mapped pH locus, which confer flesh acidity utilizing SSR marker and identified flanked marker CMAT141 at 1.7cM and another marker 269–0.9 at 8.7 cM on the other side. Perin *et al.* (1998) identified marker (EIU5 at 6 cM) for *p* gene that confer five carpels in flower. Silberstein *et al.* (2003) mapped the gene *a*, which confers a monoecious phenotype with stamenless flowers, as opposed to hermaphrodite flowers in homozygous recessive state and a flanking RFLP marker CS-DH 21 identified at 7 cM. Other important horticulture traits *i.e.*, non-yellowing, long shelf-life (Touyama *et al.*, 2000), and *ms-3* (GMS gene, Park and Crosby, 2004) was mapped in melon.

Joobeur *et al.* (2004) cloned *Fom-2* gene conferring the résistance against *Fusarium* wilt in melon. The colonization of melon/cotton aphid was inhibited by *Vat* and this process resists the transmission of unrelated cucumber mosaic virus *(CMV)* and poty viruses. Pauquet *et al.* (2004) reported that cloned *Vat* gene (6 kb in size with three introns and encodes a protein of 1473 amino acids) belongs to the coiled-coil (CC) NBS/LRR class of plant disease resistance genes. They also validated and confirm the function of *Vat* gene using susceptible melon varieties. The cloned gene conferring the resistance can be introduced to susceptible lines. The introduction of resistance genes should be through transformation or traditional breeding aided by MAS. Wang *et al.* (2000) used *Fom-2* marker for screening of 45 melon genotypes for Fusarium wilt resistance and markers correctly predicted disease phenotypes in the population. However, Zheng and Wolff (2000) stated that RAPD markers can not be relied for screening of the target phenotype.

17.2.2.1 Melon ESTs

Prior to the establishment of ICuGI in 2005, only several thousand ESTs were available in public domains. One of the major goals of ICuGI was to sequence ~100,000 ESTs from different melon genotypes and tissues. In the mean time, the Spanish Melon Genomics Project (MELOGEN) reported the generation of ~30,000

ESTs from eight normalized cDNA libraries prepared from tissues of fruits, roots, leaves, pathogen-infected roots and cotyledons derived from four different genotypes: *agrestis* pat81, "Piel de Sapo" Pinyonet torpedo, cantaloupe C-35, and "Piel de Sapo" T-111 (Gonzalez-Ibeas *et al.*, 2007). These ESTs were assembled into 16,637 unigenes, among which 6,023 were contigs and 10,614 were singletons. A total of 1,052 potential SSRs and 356 SNPs were identified from this EST collection (Gonzalez-Ibeas *et al.*, 2007). Omid *et al.* (2007) also reported ~1,800 ESTs from phloem-sap of melon cultivar Hales Best Jumbo.

Recently ICuGI released ~94,000 ESTs (Table 9-1; http://www.icugi.org), which represents a significant addition to the current melon EST and functional genomic resources. These ESTs were generated from fruits, flowers, and callus of four different genotypes: Dulce, Vedrantais, PI161375, and "Piel de Sapo" T111, as well as melon necrotic spot virus (MNSV)-infected leaf, root, and cotyledon of "Piel de Sapo" T111. To date, a total of 129,067 melon ESTs have been generated. All these ESTs, together with 173 mRNA sequences from GenBank, were assembled into 24,444 unigenes with an average length of 776.7 bp, comprising 11,653 contigs with an average length of 972 bp and 12,791 singletons of 598.7 bp. Furthermore, over 4,000 SSRs were also identified from the melon EST collection (ICuGI, http://www.icugi.org). These SSRs are potential valuable markers for melon breeding programs and part of them have been used to construct high density genetic maps (Deleu *et al.*, 2009; Harel-Beja *et al.*, 2010).

17.2.2.2 Genetic Transformation

The most efficient method of genetic transformation in *Cucumis melo* is *Agrobacterium*-mediated. Valles and Lasa (1994) co cultivated cotyledon explants with disarmed *Agrobacterium tumefaciens* strain LBA4404, containing the binary vector plasmid pBI121.l. Gonsalves *et al.* (1994) also used binary vector plasmid containing genes for *nptll, gus* and the CMV-WL coat protein (*cp*) for transforming cotyledon explants. The transformed plants contained all 3 genes. Microprojectile, electroporation, and pollination methods also proved successful in transforming muskmelons and *npt* gene is commonly used selectable marker.

17.2.2.3 Gene Transfer for Specific Traits

The genetic transformation in *Cucumis melo* has been used primarily to transfer resistance to viral and fungal diseases and understanding the role of ethylene biosynthesis.

The first successful genetic transformation was reported by Fang and Grumet (1990) who introduced the kanamycin resistance gene NPT-II into the cultivar 'Hale's Best Jumbo' mediated by *Agrobacterium tumefaciens*. Since then, more reports on melon transformation have been published and include reports using *A. tumefaciens*, biolistic particle bombardment (Nunez-Palenius *et al.*, 2008), and also transient expressions assays (Nieto *et al.*, 2006). Several difficulties have been found in melon transformation, such as the generation of tetraploid plants and a low frequency of transformation events (Guis *et al.*, 2000).

Yoshioka *et al.* (1992) transferred the gene for cucumber mosaic cucumovirus coat protein (*CMV-CP*) via *A. tumefaciens* to *C. melo*. Similarly, Fang and Grumet (1993) engineered three versions of the zucchini yellow mosaic potyvirus (*ZYMV*) coat protein gene for expression in plants, the full-length coat protein sequence, the conserved core portion of the gene and an antisense version.

Fang and Grumet (1993) introduced the full-length *zucchini yellow mosaic potyvirus* (*ZYMV*) coat protein into melon plants, and transgenic plants expressing the coat protein were found to exhibit apparent immunity to *ZYMV* infection (*i.e.*, there was neither observed symptom nor virus accumulation in infected transgenic plants). Similarly, when *CMV* coat proteins were expressed in melon, transgenic plants did not develop symptoms when inoculated with 1µg/ml virus and showed delayed symptoms when inoculum was 10 µg/ml (Yoshioka *et al.*, 1993).

Bull *et al.* (1993) used genetic transformation in cucumber to study the regulation of *Cucumis melo* ethylene-forming enzyme gene *mella*, which encode sac DNA (pMEL1) known to be expressed in fruit.

Ayub *et al.* (1996) developed transgenic Cantaloupe Charentais melon sexpressing an antisense *ace* oxidase gene, which catalyses the last step of ethylene biosynthesis. Ethylene production of transfenic fruit was <1 per cent of control untransformed fruit, and the ripening process was blocked both on and off the vine.

Bordas *et al.* (1997) transferred *hall* gene, a salt tolerance gene isolated from yeast (*Saccharomyces cerevisiae*), to melon. Although root and vegetative growth were reduced, transgenic *hall*-positive plants consistently showed a higher level of tolerance than control *hall*-negative plants.

Dahmani-Mardas *et al.* (2010) engineered the melon fruit for extended shelf life using TILLING approach. In plants, the most common techniques to produce altered or loss of function mutations are T-DNA or transposon insertional mutagenesis and RNAi. TILLING (Targeting Induced Local Lesions IN Genomes) combines advantages of random chemical mutagenesis and high throughput mutation discovery methods. To establish a reverse genetics platform in melon and to identify novel alleles of agronomic importance, Dahmani-Mardas *et al.* (2010) set up a melon TILLING platform and performed a screen for mutations in genes that control fruit ripening. Fruit ripening is a process characterized by a number of biochemical and physiological processes that alter fruit firmness, colour, flavour, aroma, and texture. Ripening patterns are conserved across fleshy fruit species. Characterizations of homologous genes involved in fruit ripening in different species suggest that genetic mechanisms are also conserved. Two missense mutations, L124F and G194D, of the ethylene biosynthetic enzyme, ACC oxidase 1, were identified and the mutant plants were characterized with respect to fruit maturation. The L124F mutation is a conservative mutation occurring away from the enzyme active site and thus was predicted to not affect ethylene production and thus fruit ripening. In contrast, G194D modification occurs in a highly conserved amino acid position predicted, by crystallographic analysis, to affect the enzymatic activity. Phenotypic analysis of the G194D mutant fruit showed complete delayed ripening and yellowing with improved shelf life and, as predicted, the L124F mutation did not have an effect.

17.2.2.4 In vitro Conservation

Shimonishi *et al.* (1991) cryopreserved somatic embryos of melon (*C. melo*) in LN after controlled desiccation (RH of 50-65 per cent) and preculture with 10mg/l ABA. Sixty five per cent of the embryos survived storage in LN when desiccated at RH of 60 per cent. Ogawa *et al.* (1997) successfully cryopreserved tissue-cultured shoot primordia of melon (*C. melo* cv. Prince Melon) in liquid nitrogen (LN) using a slow pre-freezing method.

The production of melon haploid plants and double haploids by chromosome doubling with colchicine has also been implemented by inducing *in situ* gynogenetic haploid parthenogenesis by pollination with either gamma or X-ray-irradiated pollen (Sauton and Dumas de Vaulx, 1987). The original protocol has been adopted and improved successfully in numerous melon cultivars (Cuny and Roudot, 1991; Cuny *et al.*, 1993; Yetisir and Sari, 2003; Nunez-Palenius *et al.*, 2006; Lim and Earle, 2009). Breeders usually develop double haploid genotypes to rapidly fix interesting alleles and to obtain inbred lines. Double haploid lines have also been used in basic melon research as genetic mapping and quantitative trait loci (QTL) analysis (Monforte *et al.*, 2004; Gonzalo *et al.*, 2005).

17.2.2.5 QTL Mapping

Melon fruit size is under a complex polygenic control. QTL for this trait have been described in a number of works (Monforte *et al.*, 2004; Zalapa *et al.*, 2007; Paris *et al.*, 2008). These QTL were defined in experimental populations, so they can not be transferred directly into breeding programs. Melon fruit shape is also under complex control. QTL have been detected in several works (Périn *et al.*, 2002b; Monforte *et al.*, 2004; Paris *et al.*, 2008). Four QTL have been detected in the same positions in two independent works (Périn *et al.*, 2002b; Monforte *et al.*, 2004), suggesting that QTL effects for this trait may be quite consistent, and therefore the genetic manipulation of fruit shape would be feasible based on QTL analysis. Eduardo *et al.* (2007) used an introgression line genomic library from the cross between a Spanish 'Piel de Sapo' cultivar and the Korean accession PI 161375 (Eduardo *et al.*, 2005) to introgress several fruit morphology QTL from PI 161375 into the 'Piel de Sapo' genetic background. The effects of some of these QTL were consistent across different locations and genetic backgrounds (Moreno *et al.*, 2008a; Fernandez-Silva *et al.*, 2010), demonstrating that the manipulation of melon fruit morphology is possible by combining QTL analysis and introgression line development.

Dogimont *et al.* (2000) analyzed the QTLs utilizing RIL population of Vedrantais × PI-161375 for cucumber mosaic cucumovirus (*CMV*). Seven QTLs were detected to carry the resistance to *CMV*.

In melon, Monforte *et al.* (2004) found that the heritability of fruit shape (FS) was comparatively higher than fruit weight (FW) and sugar content (SSC) in agreement with Perin *et al.* (2002b). Eight QTLs for FS were reported and among them were *fs1.1*, *fs9.1*, and *fs11.1*. Eduardo *et al.* (2004) generated three near-isogenic lines (NILs) from PI 161375 × Pinyonet Piel de Sapo to verify the effect of *fs1.1*, *fs9.1* and *fs11.1* on fruit shape. Perin *et al.* (2002a) mapped the QTL for fruit length (*fl*) and fruit shape (*fs*) and 8 QTLs were reported for fruit shape (*fs1.1*, *fs2.1*, *fs8.2*). It was concluded that fruit shape is generally determined during early ovary development stage as QTLs for fruit and ovary shape co-segregated. QTLs for fruit ethylene production was mapped and 4 QTLs (*eth1.1, eth2.1, eth3.1* and *eth111.1*) detected, which are tightly linked to ethylene receptor gene ERS1 (Perin *et al.*, 2002b). QTLs conferring disease resistance in melon have also been identified. More recent past, Perchepied *et al.* (2005) identified the QTLs for *Fusarium* wilt resistance gene. They used Isabelle as resistant and Védrantais as susceptible parents and mapped the *Fom 1.2* gene (*Fusarium* wilt race 1.2) that confer polygenic partial resistance and detected 9 QTLs in five linkage groups.

17.2.2.6 Genomics of Melon

The melon genome sequence was completed in 2012 in the frame work of the MELONOMICS project (http://www.melonomics.net) (Garcia-Mas *et al.*, 2012). The melon line selected for sequencing the genome was the double haploid line DHL92, which was generated from the cross between the agrestis melon type PI 161375 (SC) and the inodorus type "Piel de sapo T111" (PS). The melon genome was sequenced using a shotgun strategy with 454 single reads complemented with paired-end sequences obtained from 3, 8 and 20kb libraries. Additionally, 53,203 BAC-end sequences (BES) (Gonzalez *et al.*, 2010) were used to improve the genome scaffolding. The melon genome sequence v3.5 contained 1,594 scaffolds and 29,865 contigs, totalling 375Mb of sequence with a N50 scaffold size of 4.7 Mb. This represents 82.5 per cent of the 454Mb reported genome size (Arumuganathan and Earle, 1991). Mining the EST database has allowed the identification of new sesquiterpene synthases involved in volatile sesquiterpene biosynthesis in melon rinds, which are responsible of particular aromas in melon varieties (Portnoy *et al.*, 2008). The vast majority of genes known to be involved in ethylene biosynthesis, perception, response, and cell wall degradation can also be found in the melon EST database (Ezura and Owino, 2008; Moreno *et al.*, 2008b). To date, two melon microarrays have been described, the first one containing 3,000 unigenes mainly from fruit tissues (www.icugi.org), and a more recent past microarray containing 17,000 unigenes also enriched with ESTs from fruit tissues (Mascarell-Creus *et al.*, 2009).

In the recent years, two targeting induced local lesions in genomes (TILLING) platforms have been obtained in melon for functional genomics studies, one in the inodorus genetic background (González-To *et al.*, 2011), and another one in the 'Charentais' genetic background (Bendahmane *et al.*, 2017). The second TILLING platform has allowed the identification of mutants in the melon sex determination genes *a* and *g* (Boualem *et al.*, 2008; Martin *et al.*, 2009).

17.2.2.7 Melon Phylome

A phylogenomic analysis was performed after reconstructing more than 22,000 phylogenetic trees between melon and 22 other dicots, monocots, mosses and algae protein-coding genes. The phylogenetic trees can be accessed at http://phylomedb. org. The melon genome has been recently included in the plant comparative genomics PLAZA 3.0 resource (Proost *et al.*, 2015), a database that includes 37 plant genomes and allows browsing the annotated genomes, gene families and phylogenetic trees in a comprehensive way.

17.2.2.8 Melon Comparative Genomics and Retrotransposons

Melon exhibits substantial natural variation especially in fruit ripening physiology, including both climacteric (ethylene-producing) and non-climacteric types. At least 19 horticultural subgroups and six major groups of melon have been identified. The melon genome comprises 12 chromosomes, and its genome size was estimated to be ~454 Mb based on the nuclear DNA content. This is larger than the genomes of other cucurbit plants such as *Cucumis sativus* (7 chromosomes, 367 Mb) and *Citrullus lanatus* (11 chromosomes, 425 Mb). The first reported whole genome sequence of melon was that of the experimental line designated DHL92 (Garcia-Mas *et al.*, 2012). A genomic DNA sequence of 417 Mb was published in the latest version of the DHL92 genome reference (CM3.6.1), of which 337 or 79.6 Mb was the actual nucleotide sequence or ambiguous bases (*e.g.*, NNN), respectively. In addition, 29,980 protein-coding genes have been reported in the genome annotation CM4.0. he DHL92 genome reference has been utilized for supporting transcriptome analyses as well as quantitative trait loci (QTL) studies of important agricultural traits, including fruit ripening, fruit morphology, and disease resistance (Yano *et al.*, 2020). However, third generation sequencing technologies (*e.g.*, PacBio RSII/sequel and Oxford Nanopore Technology [ONT]), implemented with single molecule sequencing that can generate long reads (*e.g.* >10 kb) emerged as an alternative approach. Yano *et al.* (2020) assembled the whole genome sequence of the semi-climacteric Japanese cultivar "Earl's Favorite Harukei-3" (Harukei-3; var. *reticulatus*) by coupling ultra-long ONT sequencing (R9.4.1 + R10 flow cells) with Bionano optical mapping and Illumina mate pair sequencing. This melon shows moderate climacteric ripening behavior, although the rate of ripening is less than that of the well-studied cultivar Charentais (var. *cantalupensis*). ONT RNA-seq-based gene prediction coupled with other methods, such as ab initio prediction, also identified 33,829 protein-coding genes whose protein BUSCO benchmark value was 1372 (95.3 per cent).

Melon is usually described as producing sweet fruit; however, Harukei-3 produces considerably sweeter fruit than other melon accessions if it is grown in the appropriate seasons. Yano *et al.* (2020) reported Oxford Nanopore-based high-grade genome reference in the semi-climacteric cultivar Harukei-3 (378 Mb + 33,829 protein-coding genes), with an update of tissue-wide RNA-seq atlas in the Melonet-DB database. Comparison between Harukei-3 and DHL92, the first published melon genome, enabled identification of 24,758 one-to-one orthologue gene pairs, whereas others were candidates of copy number variation or presence/absence polymorphisms (PAPs). Further comparison based on 10 melon genome assemblies identified genome-wide PAPs of 415 retrotransposon Gag-like sequences. Of these, 160 showed fruit ripening-inducible expression, with 59.4 per cent of the neighboring genes showing similar expression patterns ($r > 0.8$). These results suggested that retrotransposons contributed to the modification of gene expression during diversification of melon genomes, and may affect fruit ripening-inducible gene expression.

17.2.2.9 Genome Editing for Melon Improvement

Genome editing tools have the potential to modify genomic sequences with accuracy. Some of these tools are: homologous recombination (HR), targeted induced local lesions in the genome (TILLING), zinc finger nucleases (ZFN), transcription activator-like effector nucleases (TALENs), or clustered regularly interspaced short palindromic repeats associated to nuclease Cas9 (CRISPR/Cas9). ZFN, TALENs and CRISPR/Cas9 are site-specific nucleases. The CRISPR/Cas9 genome editing tool was developed in 2013, and in comparison with other genome editing tools has better efficacy, efficiency, versatility and is simpler (Bortesi and Fischer 2015). CRISPR/Cas9 system cleaves a specific region of DNA by the Cas9 nuclease, which is guided by a 20-nt sequence named RNA-guide (gRNA). The association between Cas9 and gRNA, and the presence of a conserved protospacer-adjacent motif (PAM) downstream of the target DNA (typically NGG), allows a precise editing of DNA target sequences. The endonuclease domain induces DNA double-strand breaks (DSB), which can be repaired by either nonhomologous end-joining (NHEJ) or homology-directed repair (HDR) generating insertions and deletions events (INDELs) and substitutions. Within the *Cucurbitaceae* family, this genome editing technique has only been reported as successfully applied in cucumber and watermelon.

Hooghvorst *et al.* (2019) reported the use of genome editing approach for the knockout of phytoene desaturase gene. The phytoene desaturase gene of melon (*CmPDS*) was selected as target for the CRISPR/Cas9 system with two designed gRNAs, targeting exons 1 and 2. A construct (pHSE-CmPDS) carrying both gRNAs and the Cas9 protein were delivered by PEG-mediated transformation in protoplasts. Mutations were detected in protoplasts for both gRNAs. Subsequently, *Agrobacterium*-mediated transformation of cotyledonary explants was carried out, and fully albino and chimeric albino plants were successfully regenerated. A regeneration efficiency of 71 per cent of transformed plants was achieved from cotyledonary explants, a 39 per cent of genetic transformed plants were successful gene edited, and finally, a 42–45 per cent of mutation rate was detected by Sanger analysis. In melon protoplasts and plants most mutations were substitutions (91 per cent), followed by insertions (7 per cent) and deletions (2 per cent). Hooghvorst *et al.* (2019) established a CRISPR/Cas9-mediated genome editing protocol which is efficient and feasible in melon, generating multi-allelic mutations in both genomic target sites of the *CmPDS* gene showing an albino phenotype easily detectable after only few weeks after *Agrobacterium*-mediated transformation.

17.3.0 Watermelon

Biotechnological investigation in watermelon has been aimed at mass producing the triploid watermelon plants through micropropagation, identification and characterization of germplasm and gene tagging using molecular markers and gene transfer against viral and fungal diseases.

17.3.1 Tissue Culture

17.3.1.1 Micropropagation

Micropropagation has been widely attempted to produce triploid watermelon plants. Various explants such as cotyledon (Song *et al.*, 1988; Srivastava *et al.*, 1989;

Tissue Culture-based Regeneration System for Watermelon

Several Steps of the High Efficiency Regeneration System of Watermelon using Cotyledonary Nodes as Explants A, Five-day-old seedlings on propagation medium; B and C, cotyledonary node explants prepared for regeneration; D, regenerated shoots; E, elongated shoots after 1 week in elongation medium; F, rooted shoots after 1 week in rooting medium; G, rooted shoots after 3 weeks in rooting medium; H, regenerated plants ready for transfer to soil; I, regenerated plants after transfer to soil (Wang *et al.*, 2013).

Szala, 1995; Hao and Wang, 1998), hypocotyls (Srivastava *et al.*, 1989; shoot tips (Compton *et al.*, 1993; Desamero *et al.*, 1993) Yamamoto and Matsumoto, 1994) and axillary buds (Gao *et al.*, 1988; Tang *et al.*, 1994) have been used.

Tao *et al.* (1991) found that of 4 MS-based media tested, media containing ammonium sulphate and ammonium nitrate at half strength gave the best results, with are generation success rate of 52.9 per cent. Song *et al.* (1988) found that bud explants comprising part of the cotyledon, taken from sterile seedlings produced from triploid seeds and cultured on MS medium supplemented with BA at 2 mg/1, proliferated to form multiple bud clusters. The plants grew well, producing good quality fruits with high total sugar and ascorbicacid contents. Szalai (1995) propagated watermelon F_1 hybrid Szigetcsepi 5 *in vitro* from cotyledon explants using MS medium containing 0.2 mg 2, 4-D/l to induce callus in 4 weeks. Plantlets began to differentiate from callus on hormone free medium in the ninth week and completely regenerated by the 16th week. There generation capability of watermelons appeared to depend on genotype rather than on plant growth regulator concentrations. Hao and Wang (1998) developed an effective system for vegetative propagation using cotyledons from 5 watermelon cultivars.

Compton *et al.* (1993) used shoot-tip explants from 21-day-old aseptically germinated *Citrullus lanatus* seedlings which were incubated on solidified MS medium containing test concentrations BA and kinetin (each at 0, 1, 5 or 10 µM), and thidiazuron (TDZ 0, 0.1, 1 or 5µM) for 8 weeks. Approximately 1.5-2.8 times more axillary shoots formed at the optimum BA level (1µM) compared to the best TDZ (0.1 µM) or kinetin (10 µM) concentrations. Ren *et al.* (2000) found that herate of adventitious bud induction from the terminal buds was higher than that from the cotyledons; cotyledons from 4 to 5 day old seedlings, which had just changed from yellow to light green, produced the highest numbers of adventitious buds.

According to Thomas *et al.* (2000) hyper hydricity or glassiness was a frequent problem during the micropropagation of triploid watermelon cv. Arka Manik and this was influenced by level of BA, explant, medium and vesse laeration.

Abdollahi *et al.* (2015) reported the development of haploid embryos from anthers culture in watermelon. They studied with the effect of different pre-treatments like heat and cold shock treatments, induction media supplemented with wheat ovaries and phytohormones role in induction of embryo like structures from anther culture. Aseptically grown Charleston Gray and Crimson Sweet cultivars taken as experimental materials, they studied the heat shock pre-treatments of anthers like 30°C for one week, two days and a control. Meanwhile they also treated the anthers with cold treatment at 4 °C for two and five days, 4 °C for 4 and 8 days and a control. For initiation of callus from anthers using the protocol like MS media added with 10 wheat ovaries, 2, 4-D and BAP. They got highest percentage of callus from the media containing MS media, 3 per cent sucrose, 2.0 µM 2, 4-D, 1.5 µM BAP, 0.8 per cent agar and inclusion of wheat ovaries for boosting the growth. From their results showing that 4 °C for 4 and 2 days, 30 °C for 7 and 2 days cold and heat shock pre-treatments of anthers responding well from the media enriched with MS media added with BAP 2.22 µM, 3 per cent and 0.8 per cent sucrose and agar from the Crimson sweet cultivar. Out of 12 regenerated plantlets they got 10 haploids.

17.3.2 Molecular Markers

Molecular markers such as RAPD and SSRs have been used to study the genetic diversity, hybrid seed purity and character tag the genes. Several small linkage maps have been constructed in watermelon based on isozymes (Navot and Zamir, 1986; Navot *et al.*, 1990), RAPD and ISSR (Hashizume *et al.*, 1996, 2003; Hawkins *et al.*, 2001; Levi *et al.*, 2001c, 2002; Zhang *et al.*, 2004), and AFLP and SRAP (Levi *et al.*, 2006) markers. Genome mapping of watermelon is very recent and first map was construction by Hashizume *et al.* (1996) utilizing 62 markers (RFLP and RAPD) which produced 11 linkage groups. The other population utilized for map the watermelon genome (Fan *et al.*, 2000; Hawkins *et al.*, 2001; Levi *et al.*, 2001c, 2004; Zhang *et al.*, 2004) and high density map was released using 554 RAPD, RFLP, ISSR, isozyme and phenotypic markers (Hashizume *et al.*, 2003).

However, the most dense and useful map based on a large number of EST-SSR (Levi *et al.*, 2009) and SSR markers have been constructed by Xu *et al.* (2010). This map was constructed using a RIL population (F_2S_8) derived from a cross between the *Fusarium* wilt resistant PI 296341 (*C. lanatus* subsp. *citroides*) and the elite Chinese line 97103 (*C. lanatus* var. *lanatus*). Overall, in the genetic mapping experiments RAPD, ISSR, and SSR primers produced low polymorphisms while SRAP primer produced higher polymorphism among watermelon heirloom cultivars (Levi *et al.*, 2001c, 2004, 2006).

17.3.2.1 Marker for Disease Resistance

Fusarium wilt (caused by *Fusarium oxysporum* f. sp. *niveum*) race 1 resistance gene (*Fom-1*) was mapped based on the F_2 segregating population and flanking RAPD marker OPP01/700 placed at 3.0 cM (Xu *et al.*, 1999). Test cross population for *Fusraium* wilt race 1 resistance gene (*Fom-1*) mapped using 114 AFLP marker (Levi *et al.*, 2006). Hashizume *et al.* (1996) mapped the gene (*gs*) responsible for exocarp colour in the population H-7 × SA-1 and identified RAPD flanking marker (R1217A and R1280B) near 5 cM. Watermelon cold tolerance gene (*sly+*) mapped and flanking marker (OPG12/1950) identified at 6.98 cM (Xu *et al.*, 1998; Fan *et al.*, 2000).

17.3.2.1.1 Genetic Relationship

The first genetic map of watermelon was constructed by Navot and Zamir (1986) in a segregating population of *C. lanatus* × *C. colocynthis*, and later extended by Navot *et al.* (1990) into seven linkage groups covering a length of 354 cM.

RAPD has been widely used to study the genetic diversity (Shin *et al.*, 1995; Lee *et al.*, 1996). Shin *et al.* (1995) subjected a total of 39 genotypes of watermelon to RAPD analysis and, based on the polymorphisms present, these were grouped into sub-populations. Levi *et al.* (2000) estimated the genetic relatedness among 34 plants of introduction accessions (PIs) of the genus *Citrullus* (of these, 30PIs are known to have disease resistance) and five watermelon cultivars, using 28 RAPD primers. These primers produced 662 RAPD markers that could be scored with high confidence. Levi *et al.* (2001a, b) constructed a linkage map of 17 linkage groups using 155 RAPD markers and a sequenced characterized amplified region (SCAR) marker covering 1,295 cM in a backcross population [PI 296341 (*C. lanatus* var. *citroides*) ×

New Hampshire Midget (NHM, *C. lanatus* var. *lanatus*)] × NHM. Another linkage map was constructed by Levi *et al.* (2002) using a testcross population [Griffin 14113 (*C. lanatus* var. *citroides*) × NHM] × PI 386015 (*C. colocynthis*) with 141 RAPD, a SCAR, and 27 ISSR markers segregating in 25 linkage groups covering a total distance of 1,166 cM. Lately, they added another 114 AFLP markers to this map (Levi *et al.*, 2006).

Lee *et al.* (1996) generated RAPD markers by 15 arbitrary decamers used to determine the frequency of DNA polymorphism in 39 watermelon germplasm. In conclusion, RAPD assays can be used for providing alternative markers for identifying genotypes and quantitative characteristics (such as sweetness) in watermelon. Hashizume *et al.* (1996) constructed a linkage map for watermelon on the basis of RAPD, ribosomal DNA RFLP, isozyme, and morphological markers in a segregating population of 78 individuals derived from a backcrossing programme involving the cultivated inbredline H-7 and SA-I, a wild form from southern Africa, with the latter genotype being there current (male) parent.

Jarret *et al.* (1997) used simple sequence repeat (SSR) length polymorphisms to examine genetic relatedness among watermelon accessions.

Although a number of genetic studies were conducted in watermelon (Levi *et al.*, 2001a, 2002, 2006; Wechter *et al.*, 2008), only limited genetic data are available for genes that could confer disease and pest resistance in watermelon (Zhang *et al.*, 2004; Ling *et al.*, 2009; Harris *et al.*, 2009a, b).

Xu *et al.* (2010) have constructed an extensive map, based on 555SSR and EST-SSR markers that have proven quite useful as a watermelon genome representation. This map consists of 11 linkage groups with a total coverage of 739cM and the average distance between markers of 1.33 cM. Several linkage maps were constructed for watermelon using a BC_1 population (Levi *et al.*, 2001c), a testcross population (Levi *et al.*, 2002, 2006), an F_2 population (Hashizume *et al.*, 2003) and a recombinant inbred line (RIL) population (Zhang *et al.*, 2004; Xu *et al.*, 2010). The genetic maps and mapping populations (Levi *et al.*, 2006) have been useful for mapping the eukaryotic elongation factor "*eIF4E*" gene linked to *ZYMV* resistance in watermelon and for identification of markers linked to this resistance (Ling *et al.*, 2009; Harris *et al.*, 2010a).

17.3.2.1.2 Identification of Hybridity and Linkage

Ferreira *et al.* (2000) estimated the natural out crossing rate of a partially andromonoecious segregating population (with 53.3 per cent monoecious and 46.7 per cent andromonoecious plants) of watermelon. Twelve material families, each composed of 23 individuals plus the corresponding seed parents, were genotyped with RAPD markers. To determine the genetic purity of hybrid seeds, Hashizume *et al.* (1993) employed RAPD to discriminate parents from their hybrids using a single primer for each reaction.

17.3.3 Genetic Transformation

The use of biotechnological tools for transfer of genes particularly for viral diseases and fungal disease has been reported. Genetic transformation and plant regeneration of watermelon using *Agrobacterium tumefaciens* was successfully

Generation of Transgenic Watermelon Resistance to Cucumber Mosaic Virus

Growth Stages of Transgenic Watermelon.

A. Adventitious shoots regenerated from the cut edge of the cotyledon after 3 weeks. B. Enlarged adventitious shoots. C. Elongated plantlets. D. Regenerated shoot graftedonto rootstock. E. Watermelon fruit set in transgenic T0 plant. F. Watermelon fruit set from self-pollinated T1 plant (Liu *et al.*, 2016b).

reported. Srivastava *et al.* (1991) regenerated plants from cotyledon discs of *C. vulgaris* cv. Melitopol'skii cultured with A. *tumefaciens* carrying vector GV3850 Neo before transfer to selective medium. *A. rhizogenes* with vector AK320Neo + pRi15834 was infectious on hypocotyl segments. Roots produced by the derived callus were cultured on selective MS medium. Callus giving a positive response in aneomycin phospho-transferase II assay were used for plant regeneration. Phenotypic and enzymatic studies indicated that the chimaeric kanamycin resistance gene was expressed in material regenerated from cotyledondiscs. Choi *et al.* (1994) induced adventitious shoots on the proximal cut edges of different cotyledonary explants of watermelon cultivars Sweet Gem and Gold Medal cultured on MS medium with 1mg BA/l. Light (16-h photoperiod, about $7W/m^2$ cold-white fluorescent lamps) was essential for shoot formation. Cotyledonary explants of Sweet Gem were cocultured with the disarmed *A. tumefaciens* strain LBA4404, harbouring a binary vector pBI121 carrying the CaMV35S promoter-*GUS* gene fusion used as are porter gene and NOS promoter-*npt* gene as a positives election marker, for 48 h on MS medium with l mg BA/l and 200 μM β-hydroxyaceto syringone. Explants were then transferred to medium with 1mg BA, 250 mg carbenicillin and 100 mg kinetin/1 and cultured in the light. Adventitious shoots formed on the explants after 4 weeks of culture.

To ease public concern regarding genetically modified organisms (GMOs), a direct gene transfer without any use of antibiotics was developed by Chinese scientists who initially conducted their experiments in watermelon (Tao *et al.*, 1996; Chen *et al.*, 1998; Xiao *et al.*, 1999a–c). This technology involves simply injecting the source DNA into target plant tissues, and subsequently identifying transgenic progeny based on phenotype. This was first reported by Tao *et al.* (1996) who injected cashew DNA into watermelon plants. In another experiment, Xiao *et al.* (1999b) soaked watermelon embryos in bottle gourd total genomic DNA. Watermelon plants susceptible to *Fusarium* wilt were utilized, while DNA was extracted from resistant bottle gourd plants (Xiao *et al.*, 1999a).

Table 19: Genetic Transformation in Cucurbitaceae

Gene	Transformation Method	Type of Explant	Horticultural Traits	References
Cucumber				
CMV-C coat-protein	Cucumber mosaic virus	Leaf	CMV resistance	Gonsalves *et al.* (1992)
CMV-O coat-protein	Agrobacterium tumefaciens	Leaf/cotyledon	CMV/ZYMV resistance	Nishibayashi *et al.* (1996a, b)
Gus	A. tumefaciens	Hypocotyl	GUS positive	Nishibayashi *et al.* (1996a, b)
RCC2	A. tumefaciens	Leaf	Gray mold resistance	Kishimoto *et al.* (2002)
Thaumatin II	A. tumefaciens	Leaf/fruit	Change in fruit taste	Szwacka *et al.* (2002)
SOD	A. tumefaciens	Leaf/fruit	Increase in SOD activity	Lee *et al.* (2003)

Gene	Transformation Method	Type of Explant	Horticultural Traits	References
DHN10	A. tumefaciens	Leaf/cotyledon/hypocotyl	Chilling tolerance	Yin et al. (2004)
pDefH9	A. tumefaciens	Leaf/fruit	Parthenocarpic fruits	Yin et al. (2006)
CBF1	A. tumefaciens	Leaf	Chilling tolerance	Gupta et al. (2012)
PAC	A. tumefaciens	Leaf	Increase in β-carotene	Jang et al. (2016)
NOA1	A. tumefaciens	Leaf	Chilling tolerance	Liu et al. (2016c)
Watermelon				
gus and nptII	A. tumefaciens/biolistics	Cotyledon	Antibiotic resistance	Compton et al. (1993)
Gus	A. tumefaciens	Cotyledon	GUS positive	Choi et al. (1994)
Phosphomannose isomerase	A. tumefaciens	Cotyledon	Positive selection	Reed et al. (2001)
WMV-2 coat protein gene	A. tumefaciens	Leaf	WMV2 resistance	Wang et al. (2003)
HAL1 yeast gene	A. tumefaciens	Cotyledon	Salt resistance	Ellul et al. (2003)
hpt	A. tumefaciens	Leaf	Antibiotic resistance	Akashi et al. (2005)
ZYMV, WMV, CMV coat protein	A. tumefaciens	Cotyledon	ZYMV, WMV, CMV resistance	Sheng-Niao et al. (2005)
ZYMV, PRSV W coat protein	A. tumefaciens	Cotyledon	ZYMV, PRSV W resistance	Yu et al. (2011)
WSMoV coat protein	A. tumefaciens	Cotyledon	WSMoV, resistance	Huang et al. (2011)
Pti4	A. tumefaciens	Cotyledon	Transcription factor domain	Juan Li et al. (2012)
WSMoV, CGMMV, CMV coat protein	A. tumefaciens	Cotyledon	WSMoV, CGMMV, CMV resistance	Lin et al. (2012)
Coat protein	A. tumefaciens	Cotyledon	Leaf	Liu et al. (2016b)
Melon				
CMV	A. tumefaciens	Cotyledon	CMV resistance	Yoshioka et al. (1993)
ZYMV	A. tumefaciens	Cotyledon	ZYMV resistance	Fang and Grumet (1993)
CMV-WL coat protein	A. tumefaciens/biolistics	Cotyledon	CMV resistance	Gonsalves et al. (1994)
CMV/WMV/ZYMV	A. tumefaciens	Leaf	ZYMV, WMV, CMV resistance	Clough (1995)
Antisense ACC oxidase	A. tumefaciens	Cotyledon	Slow ripening	Ayub et al. (1996)
ZYMV, WMV, CMV coat protein	A. tumefaciens	Cotyledon	ZYMV, WMV, CMV resistance	Fuchs et al. (1997)

Gene	Transformation Method	Type of Explant	Horticultural Traits	References
HAL1 yeast gene	A. tumefaciens	Leaf/cotyledon	Salt resistance	Bordas et al. (1998) and Serrano et al. (1998)
S-adenosyl methionine hydrolase gene	A. tumefaciens	–	Slow ripening	Clendennen et al. (1999)
Bar gene	Potyvirus-vector inoculation	Direct inoculation on whole plant	Herbicide resistance	Shiboleth et al. (2001)
Polyribozyme genes	A. tumefaciens	Cotyledon	WMV2, ZYMV resistance	Huttner et al. (2001)
Photorespiratory eR genes	A. tumefaciens	Cotyledon	Downy mildew resistance	Taler et al. (2004)
Apple ACC oxidase (ethylene biosynthesis)	A. tumefaciens	Cotyledon	Slow ripening	Silva et al. (2004)
Petunia ACC synthase	A. tumefaciens	Cotyledon	Earlier floral development and fruit set	Papadopoulou et al. (2005)

CBF1, c-repeat binding factor-1; CGMMV, cucumber green mottle mosaic virus; CMV, cucumber mosaic virus; DHN10, gene encoding a Solanum sogarandium dehydrin with 10 kDa; NOA1, nitric oxide associated 1; PAC, phytoene synthase-2a carotene desaturase; pDefH9, Antirrhinum majus deficiens homologue 9 promoter; PRSV W, papaya ringspot virus type W; RCC2, a rice chitinase cDNA; SOD, superoxide dismutase; WMV, watermelon mosaic virus; WSMoV, watermelon silver mottle virus; ZYMV, zucchini yellow mosaic virus.

17.3.3.1 Virus Resistance

Zucchini yellow mosaic virus ($ZYMV$) is a member of the potyvirus group, causing devastating epidemics in commercial cucurbits worldwide. Development of virus-resistant cultivars by classical breeding or by the introduction of viral nucleic acid sequences into the plant genome is difficult and slow (Gal On *et al.*, 2000). The use of ribozyme genes to protect plants against viruses provides an alternative to the technologies currently used for protecting crops against viruses, based on the concept of pathogen derived resistance. Huttner *et al.* (1999) used ribozyme genes to protect melon plants against 2 potyviruses: $WMV2$ (water melon mosaic virus) and $ZYMV$ (zucchini yellow mosaic virus). Different poly ribozyme genes were designed, built and introduced into melon plants. Transgenic melon plants containing a resistance gene were obtained and their progeny was challenged by the appropriate virus. Most of the genes tested conferred some degree of resistance to the viruses in green house trials.

Another alternative method of cross protection is currently being studied with a genetically engineered clone of attenuated viral cDNA of $ZYMV$. A full-length infectious clone of an attenuated $ZYMV$-$AG1$ isolate was constructed and tested successfully by particle bombardment (Gal On *et al.*, 2000). Infection of cucurbits with the engineered $ZYMV$-AGI strain changed the symptoms dramatically from severe to water melon. The engineered virus was found to be stable and sofar no revertant virus has been found during several passages and long periods of incubation (Gal On *et al.*, 2000).

17.3.3.2 Wilt Resistance

Gene transfer *via* pollen-tube pathway for anti-*Fusarium* wilt in watermelon was reported by Chen *et al.* (1998). Total DNA was extracted from leaves of *Fusarium oxysporum* f. sp. *niveum*-resistant squash cv. Pai Chu Tso, digested with *EcoRi* and *HindIII* and ligated to a phosphorylated GUS plasmid containing the CaMV35S promoter. Two µl of squash DNA was injected into the ovaries of *Citrullus lanatus* F_1 hybrid cv. Pink Orchid at 24, 48 and 72 h following hand pollination GUS activity was confirmed using Southern blotting. Some 200 transformed seedlings were obtained and of these, 10 were wilt resistant. Xiao *et al.* (1999 a, b) extracted DNA of bottle gourd (*Lagenaria siceraria*) and used to treat embryos of watermelon by soaking them in a 500 m µg/cm^3 solution for 15 h. Plants regenerated from treated embryos showed differences in fruit characteristics from the original watermelon parent. The watermelon parent was highly susceptible to wilt caused by *Fusarium oxysporum* f. sp. *niveum* while the bottle gourd parent was resistant. Some of the plants from treated embryos were moderately susceptible to the disease but most were moderately or highly resistant.

Genetic analysis and chromosome mapping of resistance to *Fusarium oxysporum* f. sp. *niveum* (*FON*) race 1 and race 2 in watermelon detected one major QTL on chromosome 1 for *FON* race 1 resistance with a LOD of 13.2 that explained 48.1 per cent of the phenotypic variation. Two QTLs of *FON* race 2 resistance on chromosomes 9 and 10 were also discovered based on the high-density integrated genetic map constructed. The nearest molecular markers should be useful for marker-assisted selection of *FON* race 1 and race 2 resistance (Ren *et al.*, 2015).

17.3.3.3 QTL Mapping

In watermelon, fruit traits were mostly targeted for QTL mapping. The F_2 segregating population from the cross between 97103 (higher soluble solids content, thin rind, susceptible to *Fusarium* wilt) × PI-296341 (low soluble solids content, thick rind, resistance to *Fusarium* wilt) mapped using RAPD and SSR markers (Fan *et al.*, 2000). They detected 4 QTLs for total soluble solids, 2 for rind thickness, 5 for rind hardness, 3 for fruit weight and 6 for seed weight. Hashizume *et al.* (2003) evaluated the 60 BC population from H-7 × SA-1 cross and detected the QTLs for total soluble solids, red flesh colour, yellow flesh colour and rind hardness.

17.3.3.4 Genomics of Watermelon

Xu *et al.* (2010) initiated the sequencing of the watermelon genome (International Watermelon Genomics Initiative). To accomplish this sequencing project, Solexa's Sequencing-By-Synthesis technology was applied in Watermelon Whole Genome Sequencing (WWGS). Watermelon line 97103, which is the paternal parent of Jing Xin No.1, was sequenced. Until now, a total of 49.6 Gb high-quality base pairs of 97103 have been generated, which is about 115.42 fold coverage of the genome. K-mer depth distribution of the sequenced reads was used to estimate the genome size of watermelon. The estimated genome size is 421.75 Mb (Guo *et al.*, 2013). Additional 16 watermelon lines with different fruit qualities and resistance to diseases and pests were selected and used for shallow sequencing and SNP discovery. As in previous studies using DNA markers (Jarret *et al.*, 1997; Levi *et al.*, 2001b, 2001c), a large number of SNPs was discovered among the *C. lanatus* var. *lanatus* (watermelon cultivars) versus the wild type *C. lanatus* var. *citroides* (cow watermelon) while lower genetic diversity existed within the sub species.

Levi *et al.* (2006) isolated RNA from watermelon fruits at early, maturing, and ripe stages (Davis *et al.*, 2006), and constructed normalized cDNA libraries that were subtracted by hybridization with leaf cDNA. Eight thousand eight hundred cDNA clones of the watermelon flesh subtraction library were sequenced and ESTs associated with fruit setting, development, and ripening were identified. These 8,800 ESTs were assembled into 4,770 EST-unigenes. These EST-unigenes can be found at the International Cucurbit Genomics Initiative (ICuGI) website (http://www.icugi.org). About 29 per cent of the 4,770 watermelon EST-unigenes had no detectable homologous gene sequences to any other genomes or protein sequences in GenBank (Levi *et al.*, 2006; Wechter *et al.*, 2008).

Guo *et al.* (2019) reported resequencing of 414 watermelon accessions for identification of fruit quality traits. They assembled an improved genome sequence of the watermelon cultivar '97103' using PacBio long reads combined with BioNano optical and Hi-C chromatin interaction maps. We then resequenced the genomes of 414 watermelon accessions representing all seven extant *Citrullus* species and performed population genomic analyses and genome-wide association studies (GWAS) for several important fruit quality traits. Population genomic analyses reveal the evolutionary history of *Citrullus*, suggesting independent evolutions in *Citrullus amarus* and the lineage containing *Citrullus lanatus* and *Citrullus mucosospermus*. Their findings indicated that different loci affecting watermelon fruit size have been under selection during speciation, domestication and improvement. A non-bitter allele, arising in the progenitor of sweet watermelon, is largely fixed in *C. lanatus*. Selection for flesh sweetness started in the progenitor of *C. lanatus* and continues through modern breeding on loci controlling raffinose catabolism and sugar transport. Fruit flesh coloration and sugar accumulation might have co-evolved through shared genetic components including a sugar transporter gene.

Diversification after domestication has led cultivated watermelons to exhibit diverse fruit flesh colors, including red, yellow, and orange. Recently, there has been increased interest in red-fleshed watermelons because they contain the antioxidant *cis*-isomeric lycopene. Subburaj *et al.* (2019) performed whole genome resequencing (WGRS) of 24 watermelons with different flesh colors to identify single-nucleotide polymorphisms (SNPs) related to high lycopene content. The resequencing data revealed 203,894–279,412 SNPs from read mapping between inbred lines and the 97103 reference genome. In total, 295,065 filtered SNPs were identified, which had an average polymorphism information content of 0.297. Most of these SNPs were intergenic (90.1 per cent) and possessed a transversion (Tv) rate of 31.64 per cent. Overall, 2,369 SNPs were chosen at 0.5 Mb physical intervals to analyze genetic diversity across the 24 inbred lines. A neighbor-joining dendrogram and principal coordinate analysis (PCA) based on the 2,369 SNPs revealed that the 24 inbred lines could be grouped into high and low lycopene-type watermelons. In addition, we analyzed SNPs that could discriminate high lycopene content, red-fleshed watermelon from low lycopene, yellow or orange watermelon inbred lines.

Li *et al.* (2020) identified the genomic regions controlling scarlet red flesh color in watermelon. They constructed a high density genetic map based on an F_2 population from the coral red-fleshed line ZXG01478 and scarlet red-fleshed line 14CB11. A single dominant gene, Y^{scr}, produces the scarlet red flesh color rather than the coral red flesh color in watermelon. They postulated two high-density genetic maps and whole-genome variation detection aided by genome resequencing were first map the flesh color locus Y^{scr} to a small region on chromosome 6 based on two independent populations derived from two scarlet red-fleshed lines and two coral red-fleshed lines. Two major quantitative trait loci located in the same genomic regions were identified in the F_2 and BC_1P_2 populations and explained 90.36 per cent and 75.1 per cent of the phenotypic variation in flesh color, respectively. Based on the genetic variation in the two parental lines, newly developed PCR-based markers narrowed the Y^{scr} region to 40 Kb. Of the five putative genes in this region, four encoded glycine-rich cell wall structural proteins, which implied that a new regulatory mechanism might occur between scarlet red- and coral red-fleshed in watermelon.

17.4.0 *Cucurbita maxima*

Biotechnological work on *C. maxima* is limited to micropropagation. Islam *et al.* (1992) induced multiple shoots from shoot tip explants derived from 15-day-old aseptically grown seedlings of the squash hybrid *C. maxima* × *C. moschata* on MS medium. The number of shoots/explant was improved by the addition of BA (1 mg/1) and NAA (0.1 mg/l) to the MS medium. Further improvements were obtained with the addition of case in hydrolysate. Regenerated shoots rooted well on half-strength MS medium supplemented with IBA and NAA at 0.1 mg/l each. Seo *et al.* (1991) developed protocols for (i) micropropagation system for rootstock plants and (ii) embryo rescue technique allowing interspecific hybridization. Some of the characteristics of interspccific hybrids obtained were also investigated. Embryo germination in *C. ficifolia* was favourable on MS medium containing 6 per cent sucrose, and successful multipleshoots could be obtained by reculturing shoot tips from developing in MS media supplemented with 0.5-1.0 mg BNI. Transplantable plantlets were obtained when shoottips were cultured on the MS media containing 1.0 mg BA and 0.5 mg IBA or 0.5-3.0 mg IAA/1 in combination. Percentage of fruitset and embryo size was greatly influenced by pollen parent and/or variety. Rahman *et al.* (1993) could regenerate plant from internode segments of *C. maxima* × *C. moschata* hybrid on MS medium. Ex-planted internodal sections were induced to develop multiple shoots through direct regeneration without an intervening callus phase. The highest frequency of shoot bud formation was obtained when MS medium was supplemented with 4.4 µM benzyladenine and 0.54 µM NAA. Only one subculture of shoot bud-producing explants to the same medium was required for the development of shoots from shoot buds. Rooting of *in vitro* regenerated shoots obtained in half strength MS medium with 0.54 µM NAA.

17.4.1 Metabolomic Profiling of *C. maxima*

Cucurbita maxima is one of the major pumpkin species. Similar to melon, the fruit development of pumpkin undergoes five typical stages: young fruits, expanding, premature, mature, and post maturing stages. Open female flowers are pollinated during the day time, and the fruit is characterized as being at developmental day 0 after pollination. Ten days (10 d) after pollination, the fruit has a small volume and has reached the young fruit stage. A rapid increase of size and accumulation of metabolites subsequently occurs about twenty days (20 d) after flowering, and the pumpkin fruit enters the expanding stage. Forty days (40 d) after flowering, the fruit is recognized to be fully mature. During fruit maturation, fruit qualities such as color, nutrient content, sweetness, flavor, and texture are formed. Sweetness is determined by the sugar content, with higher sweetness levels being more acceptable by consumers. Carotenoids are one of the major nutrients in pumpkin fruits and provide the orange color for *C. maxima* fruit. Starch content is a major attribute of *C. maxima* fruit texture, and higher starch and dry matter content are preferred by consumers. Fruit qualities such as sweetness, fruit color and texture show great alteration during development. At the initial expanding stage, the fruit accumulates carotenoids, sweetness, starch and dry matter. Once the fruit reaches its full size, the ripening process is initiated and major biochemical changes occur in the maturing fruit which are associated with further dramatic changes in color, sweetness and fruit texture (Huang *et al.*, 2019).

RNA sequencing (RNA-seq) technology is widely applied for rapid gene discovery and molecular regulation studies. In the context of developmental studies, the technology could provide insights into functional genes and signal pathways involved in fruit ripening, with or without genomic reference information. Huang *et al.* (2019) reported the first transcriptome of this species is characterized, using fruit pulp at five different stages of the fruit's development. Combining the metabolic profiles during fruit development, the data provides an overview of the transcriptional characteristics and signaling pathways involved in fruit quality formation and fruit ripening regulation. Fruit transcriptional profiles offer helpful information for identifying valuable genes responsible for fruit quality formation, and can be used to assist in improvements of fruit breeding outcomes. Fruit transcriptome of *C. maxima* at five stages throughout development was assembled to elucidate the molecular regulation of fruit development. Almost 18 billion nucleotide bases were sequenced in total, and 48,471 unigenes were detected. A total of 32,397 (66.8 per cent) unigenes were identified to be differentially expressed. They found there was a correlation between ripening-associated transcripts and metabolites and the functions of regulating genes. KEGG analysis showed there are multiple transcripts enriched in starch, sugar, carotenoid, plant hormone signal transduction and pectin pathways and several pathways regulating quality formation were identified. Candidate genes involving in sugar, starch, pectin, fruit softening and carotenoid metabolism in fruit were firstly identified for the species of *C. maxima*.

17.4.2 Genetic Mapping

The high content of carotenoids, sugars, dry matter, vitamins and minerals makes the fruit of winter squash (*Cucurbita maxima* Duchesne) a valuable fresh-market vegetable and an interesting material for the food industry. Due to their nutritional value, long shelf-life and health protective properties, winter squash fruits have gained increased interest from researchers in recent years. Kazminska *et al.* (2018) reported the genetic mapping approach to map the ovary colour locus and to identify the quantitative trait loci (QTLs) for high carotenoid content and flesh colour. An F_6 recombinant inbred line (RIL) mapping population was developed and used for evaluations of ovary colour, carotenoid content and fruit flesh colour. SSR markers and DArTseq genotyping-by-sequencing were used to construct an advanced genetic map that consisted of 1824 molecular markers distributed across linkage groups corresponding to 20 chromosomes of C. maxima. Total map length was 2208 cM and the average distance between markers was 1.21 cM. The locus affecting ovary colour was mapped at the end of chromosome 14. The identified QTLs for carotenoid content in the fruit and fruit flesh colour shared locations on chromosomes 2, 4 and 14. QTLs on chromosomes 2 and 4 were the most meaningful. A correlation was clearly confirmed between fruit flesh colour as described by the chroma value and carotenoid content in the fruit. A high-density genetic map of C. *maxima* with mapped loci for important fruit quality traits is a valuable resource for winter squash improvement programmes.

17.5.0 *Cucurbita moschata*

In C. *moschata*, researches on micropropagation, somatic embryogenesis and molecular markers have been reported.

17.5.1 Tissue Culture

17.5.1.1 *Micropropagation*

Interspecific crosses among C. *moschata*, C. *pepo*, C. *maxima* and C. *aryrosperma* could be done by repeated pollination during the bud and flowering stage (Chen *et al.*, 2002) although percentage fruit set and embryo size were greatly influenced by the pollen parent and/or variety (Seo *et al.*, 1991). However, several researchers advocated the use of *in vitro* culture for obtaining the interspecific F_1 hybrid. The interspecific crosses C. *pepo* × C. *moschata* required embryo cultures to obtain interspecific hybrids (Munger, 1990).

Islam *et al.* (1992) cultured shoot tip explants derived from 15-day-old aseptically grown seedlings of the squash hybrid C. *maxima* × C. *moschata* and induced them to produce multiple shoots on MS medium. The number of shoots/explant was improved by the addition of BA (1 mg/l) and NAA (0.1 mg/l) to the MS medium. Further improvements were obtained with the addition of case in hydrolysate. Regenerated shoots rooted well on half-strength MS medium supplemented with IBA and NAA at 0.1 mg/l each. Seo *et al.* (1991) developed method for micropropagation system for rootstock plants and for embryo rescue allowing interspecific hybridization. Some of the characteristics of interspecific hybrids obtained were also investigated. No differentiation of shoots and roots occurred in hypocotyl segment culture. Transplantable plantlets were obtained when shoottips were cultured on the MS medium containing 1.0 mg BA and 0.5 mg IBA or 0.5-3.0 mg IAA/l in combination. Percentage of fruit set and embryo size were greatly incluenced by pollen parent and/or variety. The interspecific crosses, *C. maxima* Myako × *C. moschata* Butter Bush, *C. moschata* PM142 × *C. maxima* Myako, *C. moschata* × Shindosa and Shindosa × *C. moschata* PM142 require embryo culture to obtain interspecific hybrids. Male (*C. maxima* × *C. moschata*) and female (Shindosa × *C. mixta*) flowering *in vitro* was observed. The interspecific hybrid obtained from the cross *C. maxima* × *C. moschata* had a high incidence of female flowers. Zhao *et al.* (1998) cultured shoot segments of *C. moschata*, 1 cm in length, with axillary buds on MS medium supplemented with 0-10 mg PP333 [paclobutrazol]/l and 1.0 mg 6-BA + 0.5 mg IAA/l. PP333 at 0.05-0.5 mg/l inhibited shoot and petiole growth and increased leaf colour intensity. Survival rate after root formation and transplanting was high (94.8 per cent) with 0.1 mg PP333/l. An efficient *in vitro* micropropagation protocol was developed for direct shoot growth using shoot tips of 5 day old explants of interspecific *Cucurbita* hybrid (Sarowar *et al.*, 2003). The best condition for shoot growth was with 3 mg/l 6-benzyladenine (BA) in MS medium and the shoots of 1.0-1.5 cm length were rooted most effectively in 1.0 mg/l indole-3-butyric acid-supplemented MS medium.

17.5.1.2 Somatic Embryogenesis

Kwack and Fujieda (1988) excised ovules of *C. moschata* excised from ovaries at anthesis and pretreated at 5°C for 2 days and cultured on a 50 per cent MS medium containing 30 g sucrose/l. The embryosacs degenerated and nucellar cells gave rise to proembryos which developed further to give embryos. A few of these showed normal morphology. Upon transfer to the same medium containing 5 g sucrose and 8 g agar/l, most embryos produced callus, but a few produced normal shoots and roots. The regenerated plants were diploid (2n=40) or tetraploid (2n=80).

17.5.2 Molecular Markers

Leclerc (1987) detected an ovel family of tandemly repetitive elements by *Hind III* digestion of *C. pepo* genomic DNA. Homologous elements were detected in *C. moschata* and *C. foetidissima*, those of *C. moschata* being more homologous to those of *C. pepo* than were those of *C. foetidissima*.

Wilson *et al.* (1992) carried out cladistics analysis of 86 chloroplast DNA restriction site mutations among 30 samples representing 15 species of *Cucurbita*. Their study indicated that annual species of the genus are derived from perennials. The pattern of variation supports three species groups as monophyletic:

   (i) *C. fraterna, C. pepo* and *C. texana,*

   (ii) *C. lundeliana, C. maritinezii, C. mixta, C. moschata* and *C. sororia,* and (iii) *C. foetidissima* and *C. pedatifolia.*

Domesticated samples representing subspecies of *C. pepo* are divided into two concordant groups, one of which is allied to wild-types referable to *C. texana* and *C. fraterna*. The data failed to resolve relationships among cultivars of *C. moschata* and *C. mixta* and their association to the wild *C. sororia*. The South American domesticated, *C. maxima*, and its companion weed, *C. andreana*, show close affinity and alliance to *C. equadorensis*.

Somatic Embryogenesis of *Cucurbita moschata*

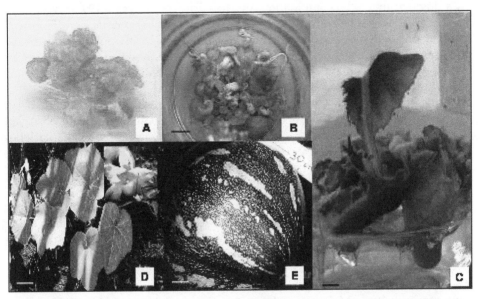

High Frequency Somatic Embryogenesis of *Cucurbita moschata* cv. Sello de Oro.

(A) Embryogenic callus formed after 16 weeks of culture on basal MS medium supplemented with 0.5 mg/l 2,4-D. (B) Regenerated squash plantlets from somatic embryos cultured on regeneration medium. (C) Squash plants cultured on basal MS medium. (D) Plant acclimated in the field with flowers. (E) Squash fruits with normal seeds (Valdez-Melara *et al.*, 2009).

RAPD markers have been used by some workers. Jeon *et al.* (1994) generated RAPD markers by 6 out of 50 arbitrary 10-mer primers were effective in discriminating among 9 *C. moschata* and 6 *C. pepo* cultivars. The 6 primers produced 64 useful RAPD markers of 350 to 6000 bp. Some 11 per cent of the markers were common to both *C. moschata* and *C. pepo*, where as 62.5 per cent were specific to one or other species. The rest were polymorphic in either or both species. The average dissimilarity coefficient matrix of markers was 5.84 between *C. moschata* and *C. pepo*, 3.41 between two *C.moschata* cultivars and 2.90 between two *C. pepo* cultivars. Fifteen *Cucurbita* cultivars screened were distinguished from each other by a single primer or by combinations of 2 to 3 primers. Lee *et al.* (1995) screened 70 arbitrary 10-mcr primers in an F_2 population of *C. pepo* × *C. moschata*, and 15 showed polymorphism in parental DNAs. These were used to identify polymorphismin 40 F_2 plants. Some 47 segregating RAPD markers were analysed and 5 linkag groups were constructed containing 28 RAPD marker loci. Youn and Chung (1998) used RAPD analysis to investigate genetic relationships among 22 native varieties of *C. moschata* from all regions of South Korea. Of the 35 primers used, only 11 revealed clearand reproducible banding patterns; 16 polymorphic bands were observed among 86 bands. Based on the presence of polymorphic bands, the 22 varieties were divided into 3 groups by cluster analysis. It was concluded that there was no relationship between fruit morphology and genetic relatedness, where as genetic relatedness was closely related to geographical distribution.

Gwanama *et al.* (2000) used RAPD markers to analyse the amount of genetic diversity in *C. moschata* landraces grown in south-central Africa and to classify the land races to assistin selection of parent genotypes for improvement of fruit characteristics. Cluster analysis, based on 39 polymorphic and 105 monomorphic DNA fragments amplified by 16 primers, was used to show relationships among 31 genotypes obtained from Zambia and Malawi. The analysis revealed four clusters, with genotypes from Malawi mainly clustering in three clusters while all genotypes from Zambia and three from Malawi clustered in one cluster. The pairwise mean genetic distance was 0.32 ± 0.04 for samples from Malawi and 0.26 ± 0.04 for samples from Zambia.

Gong *et al.* (2008a) have also used 205SSR markers to construct a linkage map of *C. moschata* with a marker density of 7.6 cM. Furthermore, they reported that 72 of the 76 common SSR markers between *C. pepo* and *C. moschata* were located in homologous linkage groups, with largely conserved orders and distances between markers. In *C. moschata*, the *Gr* locus for green versus tan fruit coloration was 12.5 cM from an SSR marker (Gong *et al.,* 2008a).

Brown and Mayers (2002) attempted the QTL mapping in squash for fruit shape and leaf indentation utilizing RAPD markers from interspecfic cross between *C. pepo* × *C. moschata* and detected the QTLs for both traits.

17.5.3 Transcriptomics

Guo *et al.* (2018) reported the transcriptome profiling of pumpkin for powdery mildew resistance. The genus Cucurbita is composed of several species, including the cultivated *C. moschata* (*Cucurbita moschata* Duch.), *C. pepo* (*Cucurbita pepo* L.), *C.*

maxima (*Cucurbita maxima* Duch.) and several wild species. *Cucurbita moschata* is an economically important species that is cultivated worldwide. Pumpkins are valued for their fruit and seeds and are rich in nutrients such as vitamins, amino acids, flavonoids, phenolics and carbohydrates. Cucurbit powdery mildew (PM), mainly caused by *Podosphaera xanthii* (formerly *Sphaerotheca fuliginea*), is a serious biotrophic pathogen disease in field and green house cucurbit crops worldwide. PM slows plant growth, and it causes premature desiccation of the leaves and a consequent reduction in the quality and marketability of the fruits. Breeding for PM resistance is the most desirable strategy to control this disease by means of resistant cultivars. Next-generation sequencing (NGS) based RNA sequencing for transcriptome methods (RNA-Seq) have been proven to be an effective method to analyze functional gene variation, and NGS has dramatically improved the speed and efficiency of gene discovery. Guo *et al.* (2018) studied the gene expression differences in PM-treated plants (harvested at 24 h and 48 h after inoculation) and untreated (control) plants of inbred line "112–2" using RNA sequencing (RNA-Seq). The inbred line "112–2" has been purified over 8 consecutive generations of self-pollination and shows high resistance to PM. More than 7600 transcripts were examined in pumpkin leaves, and 3129 and 3080 differentially expressed genes (DEGs) were identified in inbred line "112–2" at 24 and 48 hours post inoculation (hpi), respectively. Based on the KEGG (Kyoto Encyclopedia of Genes and Genomes) pathway database and GO (Gene Ontology) database, a complex regulatory network for PM resistance that may involve hormone signal transduction pathways, transcription factors and defense responses was revealed at the transcription level. In addition, the expression profiles of 16 selected genes were analyzed using quantitative RT-PCR. Among these genes, the transcript levels of 6 DEGs, including bHLH87 (Basic Helix-loop-helix transcription factor), ERF014 (Ethylene response factor), WRKY21 (WRKY domain), HSF (heat stress transcription factor A), MLO3 (Mildew Locus O), and SGT1 (Suppressor of G-Two Allele of Skp1), in PM-resistant "112–2" were found to be significantly up- or down-regulated both before 9 hpi and at 24 hpi or 48 hpi; this behavior differed from that observed in the PM-susceptible material (cultivar "Jiujiangjiaoding"). The transcriptome data provided novel insights into the response of *Cucurbita moschata* to PM stress and are expected to be highly useful for dissecting PM defense mechanisms in this major vegetable and for improving pumpkin breeding with enhanced resistance to PM.

17.6.0 *Cucurbita pepo*

Among the *Cucurbita* species, systematic investigations on various aspects of biotechnology have been carried out in *C. pepo*.

17.6.1 Tissue Culture

17.6.1.1 Micropropagation

Juretic *et al.* (1989) obtained plantlets from seedling shoottip explants and somatic embryos by culturing single-node explants on MS medium either without hormones or with IAA and/or benzyladeninc (BA). Average shoot length was greater on MS medium with 2.9 µM IAA and 0.2 µM BA when derived from seedling explants, whilst shoots derived from somatic embryos grew better on MS medium with 5.7 µM IAA.

17.6.1.2 Somatic Embryogenesis

Research on somatic embryogenesis has been reported since 1968 (Schroeder, 1968). Various explants such as hypocotyl (Jelaska, 1972, 1974), Calluslines (Krsnik-Rasol *et al.*, 1982), and long term embryogenic cellculture (Jelaska *et al.*, 1986) have been used for studying somatic embryogenesis in *C. pepo*. Tarasenko and Uspenskii (1996) induced indirect somatic embryogenesis using the first 2-3 leaves of variety Odesskii cultured on supplemented MS medium. The embryogenic callus, which was compact, shiny and structured, consisted of alternating sectors of large vacuolized and small meristematic-type cells. Predominantly the latter cells formed various types of protrusions on thecallus surface. The embryoids developed asynchronously, and various structural disturbances were seen in the course of somatic embryogenesis in comparison with the corresponding stages of zygotic embryo development.

Kintzios *et al.* (1998) studied the effect of light on the induction, development and maturation of somatic embryos, after prolonged incubation of cultures for *in vitro* conservation purposes. Cotyledon and leaf explants were cultured on solid MS medium supplemented with various plant growth regulators for the induction of callus and somatic embryos. Further embryoid development to the torpedo-shape stage and embryoid maturation was significantly affected by exposure of the cultures to light during the induction phase. Juretic and Jelaska (1991) induced embryogenic callus derived from hypocotyl segments and maintained for 15 years on MS medium supplemented with the auxins IBA (4.9 µM), 2,4-D (4.5 µM) or IAA (5.7 µM). As embryo maturation and development of adult plants failed on induction medium, small embryogenic clumps and individually isolated embryos were subcultured. The best result was achieved with a line that had been induced and maintained for 15 years on MS with IBA. In the IBA line, out of 100 embryos, 77 developed into plantlets on MS medium supplemented with 11.4 µM IAA.

Chee (1991) cultured somatic embryos organized from shoot apex-derived callus on MS medium supplemented with 1.2 mg 2,4,5-T, 0.8 mg BA and 0.1 mg kinetin/l. Immature somatic embryos developed into plantlets after transfer to MS medium with 0.05 mg NAA and 0.05 mg kinetin/l. Regenerated plants appeared morphologically normal and setfruits with seeds which germinated normally. Chee (1992) initiated embryogenic callus tissues from cotyledons of mature seeds. MS medium supplemented with either 22.7 µM 2, 4-Dora combination of 4.7 µ M2, 415-T, 4 µM BA and 0.5 µM kinetin. Clusters of somatic embryos were found in callus tissue. Maturation of these somatic embryos was achieved by transfer of embryo genic callus tissues to MS supplemented with 0.5 µM NAA and 0.25 µM kinetin. Regenerated mature plants were morphologically normal and set fruits containing seeds that germinated normally. Gonsalves *et al.* (1995) regenerated *via* somatic embryogenesis using cotyledons excised from germinated or non-germinated seeds.

17.6.1.3 Anther Culture

The first haploid plants of *Cucurbita pepo* L. (summer squash) were obtained by Dumas de Valux and Chambonnet in 1986 using a non-fertilised ovule culture. Hayase (1954) obtained a haploid plant of winter squash (*Cucurbita maxima* Duch. ex Lam.) by pollination with *Cucurbita moschata* Duch. ex Poir. Dryanovska (1985) cultured ovaries and anthers of various *C. pepo* varieties at various stages development on modified media containing various growth regulators and vitamins. Callus formed from all species on all media, irrespective of the stage of development of the ovaries and stamens, 7- 10 days after planting. The best callus development was seen on MS medium and from ovaries. After 30- 40 days the calluses were transferred to a second medium but no root formation occurred. Diploid, haploid and aneuploidy metaphases were observed in the callus tissue. Kurtar *et al.* (1999) determined the effects of different sucrose and 2,4-D combinations on plant propagationin 4 squash cultivars by means of anther culture on solid MS medium with 2 rates of sucrose (120 and 150 g/1) and 4 of 2, 4-D (5,7.5,10 and 12.5 mg/1). Anthers at the uninucleate microspore stage without filament excised from surface-sterilized male buds were placed on the media. Three plants were obtained from Sakiz with 120 g/1sucrose + 5 mg/1 2,4-D combination. Cytological and morphological studies indicated that all plants were diploid (2n=2x=40).

Metwally *et al.* (1998) studied the effects of sucrose and 2, 4-D on the induction of haploid plants of summer squash cv. Eskandarani by anther culture.Root tips from 20 plantlets were cytologically examined under a light microscope.Results revealed 10 diploid (2n=2x=40) and 10 haploid (2n=x=20) plants.

Most *in vitro* studies have focused on obtaining haploid plants from *C. pepo*, using both ovule (Metwally *et al.*, 1998a; Shalaby, 2007) and anther culture (Metwally *et al.*, 1998b). More recently, pollen irradiation has been used to induce haploid plants in *C. pepo* (Kurtar *et al.*, 2002), *C. maxima* (Kurtar and Balkaya, 2010) and *C. moschata* (Kurtar *et al.*, 2009).

Table 20: Status of Tissue Culture in Cucurbitaceae

Cucurbits	Medium	Explants	References
Cucumis sativus L.	MS + BAP (2.5-5 µm)	Cotyledon	Hisajima and Arai (1989)
	MS + BAP (1.0 µM) + Casein hydrolysate (200)	Node	Ahamad and Anis (2005)
	500 gamma radiation, Co 60 X- rays source	Parthenogenic embryo	Claveria *et al.* (2005)
	MS + Sucrose (87.64 µM) + agar (0.8 per cent) + 2,4-D (3.62 µM) + BAP (2.22 µM)	Hypocotyl	Selvaraj *et al.* (2006)
	MS + BAP (0.4 µm)	Shoot tip	Mohammadi and Siveritepe (2007)
	Primers (OP-C10, OP-G14, OP-H05, OP-Y03 and OP-AT01)	Somatic embryo	Elmeer *et al.* (2009)
	MS + BAP (1.5 mg/l)	Node	Firoz Alam *et al.* (2015)

Cucurbits	Medium	Explants	References
	MS + BAP (0.5 mg/l) + NAA (1.0)	Stem	Jesmine and Mian (2016)
	MS + Kinetine (6.0 µm)	Stem fragments	Kie³kowska and Havey (2011)
	MS + 2,4-D (5) + TDZ (0.1)	Leaf	Usman et al. (2011)
	MS + BAP (1.0 mg/l)	Shoot tip	Sangeetha et al. (2011)
	MS + Orange juice	Cuttings	Ikram-ulhaq et al. (2013)
	Kanamycin resistance and green fluorescent protein (GFP) fluorescence,	Cotyledon	Nanasato et al. (2013)
Cucumis melo L.	MS + IBA (5.0 µM)+ BAP (5.0 µM)+ 25-29°C + light intensity (5-30 µmolm⁻²s⁻²)	Cotyledon	Randall et al. (1989)
	MS + 2,4-D (5)+ TDZ (0.1)	Cotyledon	Grey et al. (1993)
	MS + BAP (8.0mg/l)	Node	Keng and Hoong (2005)
	MS + BAP (1.0 mg/l) + IBA (0.1) + GA₃ (0.3)	Shoot tip	Huda and Sikdar (2006)
	MS + BAP + 2.0-iP	Cotyledon	Chovelon et al. (2008)
	MS + BA (2.0 mg/l) + IAA (0.2)	Cotyledon	Li et al. (2011)
	MS + BAP (1.0)	Leaf node	Rahaman et al. (2012)
	MS + IAA (0.5) + BAP (2.0mg/l)	Shoot tip	Venkateshwaralu (2012)
	MS + BAP (2.0 mg/l)	Node	Parvin et al. (2013)
	MS + BAP (2.0 mg/l)	Shoot tip	Faria et al. (2013)
Cucumis figarei	MS + BAP (1.0 mg/l) + ABA (1.0 or 2.0)	Cotyledon	Yutaka et al. (1998)
Cucumis metuliferus	MS +BAP (1.0) + IAA. (0.2)	Cotyledon	Yutaka et al. (1998)
Cucumis hystrix	MS + Sucrose (30 g) + myoinositol (0.1 g) + Agargel plus (5 g) + IBA (1.7 µM) + Kinetin (0.5 µM) + GA₃ (0.3 µM)	Shoot tip	Compton et al. (2001)
Cucumis angurea	MS + BAP (1 mg/l) +NAA (0.2)+ L-glutamine (20)	Node	Margareate (2014)
Cucumis trigonus	MS + BA (1.0 mg/l) + IAA (0.25)	Leaf	Satapathy et al. (2014)
Cucumis melo var. utillisimus	MS + BAP (1.0 mg/l) + Adenine sulphate (15)	Node	Venkateshawaralu et al. (2010)
Momordica charantia	MS + BAP (2.0 mg/l) + NAA (0.2)	Node	Sultana et al. (2003)
	MS + BAP (2 mg/l) + NAA (0.2) + Sucrose 30 gl-1 + Agar 7.0 gl-1+ pH (5.5- 6.0)	Node	Sultana et al. (2005)
	MS + 2,4-D (1.0)	Leaf	Thiruvengadam et al. (2006)
	MS + BAP (1.5mg/l)	Leaf	Saima Malik et al. (2007)
	MS and Gamboge + NAA (3.0 µm) + TDZ (1.0 µm) + Putrecine (1.0 µm)	Petiole	Thiruvengadam et al. (2012b)
	½ MS + BAP (0.5 mg/l)	Node	Verma et al. (2014)

Cucurbits	Medium	Explants	References
Momordica cochinchinensis	MS agar gelled + 2, 4-D (2.0) + Coconut milk (15 per cent v/v)	Node	Debnath *et al.* (2013b)
Momordica balsamina	MS + BAP (2.0 mg/l) + NAA (0.2)	Node	Thakur *et al.* (2011)
Momordica cymbalarica	MA + BAP (2.0 mg/l)	Node	Devi *et al.* (2017)
Momordica sahyadrica	MS + BAP	Seedlings	Rajashekharan*et al.* (2012)
Momordica dioica	MSHP + AdSO$_4$ (80 ppm) + BAP (10 ppm) + IBA (5 ppm) + myoinositol (100) + Agar-Agar (0.8 per cent) + Sucrose (3 per cent)	Node	Kulkarni (1999)
	MS + BAP (2.0 mg/l) + NAA (0.5)	Cotyledon	Hoque *et al.* (2000)
	MS + BAP (1.0 mg/l) + NAA (0.1)	Cotyledon	Nabi *et al.* (2002a)
	MS + BAP (1.0 mg/l) + NAA (0.1)	Cotyledon	Nabi *et al.* (2002b)
	MS + IBA (1.0)	Healthy shoots	Ghive *et al.* (2006a)
	MS + AdSO$_4$ (70/80) + BAP (1.0) + NAA (1.0)	Node	Ghive *et al.* (2006b)
	MS + IBA (10.8) + NAA (1.08) + GA$_3$ (0.54)	Immature embryo	Hoque *et al.* (2007).
	MS + 2,4-D (2.2 μm) + L- glutamine (0.5 μm)	Petiole	Thiruvengadam *et al.* (2007)
	MS + 2, 4-D (1.0 mg/l) + BAP (2.0)	Leaf	Devendra *et al.* (2009)
	MS + BAP (0.1 mg/l) + NAA (0.1) + Sucrose (30 g/l w/v)	Internode	Karim and Ahamad (2010)
	MS + BAP (1.5 mg/l)	Cotyledon	Karim (2011)
	MS + BAP (1.0)	Cotyledon	Karim and Ullah (2011)
	MS + BAP (2.0 mg/l) + IAA (0.1)	Node	Shekhawat *et al.* (2011)
	MS + BAP (0.6 μm) + Casein hydrolysate (200)	Node	Rai *et al.* (2012)
	MS + 2,4-D (2.0) + BAP (1.0 mg/l)	Node	Mustapha *et al.* (2012)
	MS + BAP (4.44 and 8.88 μm)	Cotyledon	Arekar (2012)
	MS (0.7 per cent agar solidified) + BAP (0.5 μm)	Encapsulated Shoot tip	Thiruvengadam*et al.* (2012a)
	MS + BAP (1.0 mg/l)	Cotyledon	Karim (2013)
	MS + BAP (2.0 mg/l) + L- glutamic (2.0)	Node	Mustapha *et al.* (2013)
	MS + 2,4-D (3.3 μm) + Putrescine (0.5 μm)	Leaf	Thiruvengadam *et al.* (2013)
	MS + 2,4-D (2.0) + BAP (0.5 mg/l)/ Coconut milk (15 per cent v/v).	Node	Debnath *et al.* (2013a)
	MS + BAP (1.0 mg/l) + NAA (0.2)	Node	Jadhav (2015)
	MS + 2,4-D (2.0) + BAP (2.0mg/l)	Leaf	Raju *et al.* (2015)

Cucurbits	Medium	Explants	References
	MS + NB6 + BAP (0.5+0.5)	Node	Patel and Kalpesh (2015)
	MS + BAP (3.0 mg/l) + NAA (0.5)	Leaf	Swamy et al. (2015)
	MS + BAP (1.5 mg/l)	Node	Jamathia (2016)
	MS + BAP (0.5 mg/l) + IAA (0.1) + Ascorbic acid (50) +Adenine sulphate, Citric Acid, L-arginine (25)	Node	Choudhary et al. (2017)
	MS + BAP (1.0 mg/l) + NAA (1.0)	Node	Kapadia (2018)
Luffa acutangula	MS + BAP (1.5 mg/l) + NAA (1.0)	Cotyledon	Zohura et al. (2013)
	2, 4 – D + TDZ –(2.0)	Leaf	Moideen and Prabha (2013)
	MS + 2, 4-D + TDZ (1.5)	Petiole	Moideen and Prabha (2014)
	MS + BAP (1.0 mg/l) + Zeatin (0.2) + NAA (0.2) + 2,4-D (0.6) + Picloram (0.1) +AdS (20)	Cotyledon	Umamaheshwari et al. (2014)
	MS + BAP (2.0 mg/l) + NAA (0.2)	Petiole	Vellivella et al. (2016)
Luffa cylindrica	MS salts + B5 + BAP (10 µM)	Cotyledon	Singh et al. (2011)
	MS + BAP (1.5 mg/l) + NAA (1.0)	Leaf	Srivastava and Roy (2012)
Citrullus lanatus	MS + BAP (3.0 µM) +2iP (3.0 µM)	Cotyledon	Chaturvedi et al. (2001)
	MS + BAP (1.0 mg/l)	Shoot tip	Compton and Grey (1992)
	Agrobacterium tumefaciens LBA4404 + vector pBI121 + r gene β–glucuronidase (gus) + neomycin phosphotransferase (nptII)	Cotyledon	Dabauza et al. (1997)
	MS + BAP (5.0 mg/l)+ IAA (0.1)	Shoot tip	Khalekuzzama et al. (2012)
	MS + 2, 4-D (1.0)	Cotyledon	Khatun et al. (2010a)
	MS + BAP (1.0 mg/l) + NAA (0.2)	Node	Khatun et al. (2010b)
	MS + BAP (1 mg/l) + coconut water (10 per cent)	Cotyledon	Krug et al. (2005)
	MS + 2, 4-D (2.5)	Leaf	Sultana et al. (2004)
	MS + BAP (20.0 µM)	Cotyledon	Suratman et al. (2009)
	MS + BAP (0.5)	Shoot tip	Vedat et al. (2002)
Citrullus colocynthis	MS + 2,4-D (1.5) + BAP (1.0mg/l)	Leaf	Savitha et al. (2010)
	MS + Kn (2.0) + TDZ (1.0)	Leaf	Shashtree et al. (2014)
	MS + IAA (2.0) + IBA (1.5)	Cotyledon	Ram and Shashtri (2015)
Cucurbita moschata	Callus induction medium (CIM) + 2,4-D (0.5 or 3.5)	Cotyledon	Valdez-Melara et al. (2009)
Interspecific Cucurbita hybrid	MS + BAP (3.0 mg/l)	Shoot tip	Sarowar et al. (2003)

Cucurbits	Medium	Explants	References
Cucurbita pepo	2,4,5-T (4.7 µm)+ BAP (4.0 µm) + Kinetine(0.5 µm)	Cotyledon	Paula (1992)
	MS + 2,4,5-T (1.2) + BAP (0. 8 mg/l) + Kinetin (0.1)	Shoot tip	Paula *et al.* (1990)
	MS + Thidiazuron (0.5)	Hypocotyl	Pal *et al.* (2007)
Cucurbita maxima	MS + BAP (3.0 mg/l)	Shoot tip	Mahazabin (2008)
	MS + BAP (2.0 mg/l)	Node	Hoque *et al.* (2008)
Cucurbita ficifolia	MS + Zeatin (1.0) + IAA (0.1)	Cotyledon	Kim *et al.* (2010)
Trichosanthes cucumerina L.	MS + BAP (1.0 mg/l)+ NAA (0.1)	Shoot tip	Devendra *et al.* (2008)
	Kinetin (0.1) and BAP (2.0mg/l)	Cotyledonary node	Kawale and Choudhary (2009)
Trichosanthes dioica	MS + BAP (1.0 mg/l) + NAA (0.2)	Shoot tip	Abdul-Awal *et al.* (2005)
	MS + BAP (1.0 mg/l)	Cotyledon	Malex *et al.* (2010)
	MS + BAP (2.5 mg/l)	Node	Komal (2011b)
	Semi solid MS + Coconut milk (15 per cent)	Node	Komal (2011c)
	MA + BAP (2.0 mg/l) + NAA (0.3)	Node	Komal (2011a)
	MS + BAP (0.5 mg/l) + 2,4-D (0.5)	Leaf	Sourab *et al.* (2017)
Benincasa hispida	MS + BAP (1–6 µM) + NAA, 0.2 and 0.5 µM	Cotyledon	Thomas *et al.* (2004)
	MS + BAP (1.5 mg/l)	Shoot tip	Haque *et al.* (2008)
	MS + BAP (1.5 mg/l) + GA$_3$ (0.2)	Shoot tip	Kausar *et al.* (2013)
Coccinia abyssinica	5 per cent NaOC with 10 Minutes	Leaf	Guma *et al.* (2015)
Sechium edule	MS + BAP (0.1 mg/l)	Stem part	Abdelnour *et al.* (2002)

17.6.2 Genetic Transformation

Among cucurbits, research on the development of transgenics is well reported in case of *C. pepo*. Katavic *et al.* (1991) studied transformation by *A. rhizogenes* (wild-type strains 8196 (manno pine type) and 15834 (agro pine type) of intact seedlings grown *in vivo*, and 6-8-day-old excised cotyledons cultured in a xenic conditions. Transformed (hairy) roots were successfully induced only on the excised cotyledons with strain 8196, while intact seedlings failed to form hairyroots with either of the 2 bacterial strains. A xeni chairy-root cultures established on MS medium without hormones grew vigorously. Toppi *et al.* (1997) transformed two-week-old in *vitro*-grown intact plants and cotyledons (detached and undetached from the mother-plant) by *A. rhizogenes* strain NCPPB1855, grown for 48 h at 25°C on YMB medium. The transformed roots were successfully grown in liquid MS medium without plant growth regulators for an indefinite number of transfers. Genetic transformation of root DNA was confirmed by Southern analysis performed with a *rolABC* probe and a virprobe.

The main area of genetic transformation research in *C. pepo* is towards engineering coat protein genes for resistance to virus diseases. Transgenic squash cultivars incorporating single or multiple virus coat protein (*CP*) genes were among the first genetically engineered crop plants to be grown on a commercial scale. The first of these, the yellow crookneck hybrid Freedom II, commercialized in 1995 (Fuchs *et al.*, 1998), was resistant to *ZYMV* and WMV-2. Lines expressing *CP* genes of *CMV*, *ZYMV* and *WMV* have now been developed, and are resistant to these three viruses (Fuchs and Gonsalves, 2007). In 2005, transgenic squash accounted for 12 per cent of the summer squash acreage in the US (Fuchs and Gonsalves, 2007).

Pang *et al.* (2000) made the first report on the development of transgenic squash that are resistant to *SqMV* were produced and crossed with non-transgenic squash. Further green house, screenhouse and field tests were carried out on R_1 plants from three independent lines that showed susceptible, recovery or resistant phenotypes after inoculations with *SqMV*. Nearly all inoculated plants of the resistant line (SqMV-127) were resistant under green house and field conditions and less so under green house conditions.

17.6.2.1 Field Performance

Quemada (1996) genetically engineered the transgenic squashline ZW20 for resistance to zucchini yellow mosaic potyvirus and watermelon mosaic 2 potyvirus. Fuchs and Gonsalves (1995) investigated the resistance of three transgenic yellow crook neck squash lines expressing the zucchini yellow mosaic potyvirus (*ZYMV*) and/or the watermelon mosaic 2 potyvirus (*WMV2*) coatprotein (*cp*) genes under field conditions at Geneva. Resistance was evaluated under high disease pressure achieved by mechanical inoculations of *ZYMV* and *WMV2* and natural challenge inoculations by aphid vectors. Assessment of infection rates was based on visual monitoring of symptom development, enzyme-linked immunosorbent assays (ELISA), infectivity assays and analysis of virus occurrence in fruits. Horticultural performance was evaluated by counting the number of mature fruits per plant and estimating their fresh fruit weight. The transgenic line *ZW20B* expressing both the *ZYMV* and *WMV2 CP* genes showed excellent resistance in that none of the plants developed severe foliar symptoms, although localized chlorotic dots or blotches appeared on some leaves.

Tricoli *et al.* (1995) produced transgenic inbred squash lines containing various combinations of the cucumber mosaic cucumovirus (*CMV*), watermelon mosaic 2 potyvirus (*MMV2*) or zucchini yellow mosaic potyvirus (*ZYMV*) coat protein (*cp*) genes using *Agrobacterium*-mediated transformation. Progeny from lines transformed with single or multiple *CP* gene constructs were tested for virus resistance underfield conditions, and exhibited varying levels of resistance to infection by *CMV*, *WMV2* and *ZYMV*. Most transgenic lines remained non-symptomatic throughout the growing seasons and produced marketable fruits, while otherlines showed a delay in the on set of symptoms and/or a reduction in symptom severity. A few lines failed to display any level of resistance. Depending on the *cp* gene used, 40 to 95 per cent of the transgenic lines containing single *cp* constructs of either *CMV*, *WMV2* or *ZYMV* were resistant to the virus from which the *cp* gene was derived.

Arce Ochoa *et al.* (1995) tested Pavo, a commercially grown, virus-susceptible squash (*Cucurbita pepo*) hybrid, and two experimental virus-resistant transgenic hybrids, XPH-1719 and XPH-1739, for field performance. The two transgenic hybrids posses the desired fruit and plant characteristics of their parental line, Pavo, plus resistance to zucchini yellow mosaic potyvirus and watermelon mosaic 2 potyvirus (XPH-1719) and resistance to zucchini yellow mosaic potyvirus, watermelon mosaic 2 potyvirus and cucumber mosaic cucumovirus (XPH1739). Percent emergence and days to flowering were similar among the three hybrids. XPH1719 and XPH1739 were equally effective in producing a high percentage of quality marketable fruit and yield.

Fuchs *et al.* (1998) tested transgenic squash containing the coat protein (*cp*) gene of the aphid transmissible strain WL of cucumber mosaic cucumovirus (*CMV*) under field conditions to determine if they would assist the spread of the aphid non-transmissible strain C of *CMV*, possibly through heterologous encapsidation and recombination. Transgenic squash were resistant, although the latter occasionally developed chlorotic blotches on lower leaves. Transgenic squashline ZW-20, one of the parents of commercialized cultivar Freedom II, which expresses the *cp* genes of the aphid transmissible strains FL of zucchini yellow mosaic (*ZYMV*) and watermelon mosaic 2 (*WMV2*) potyviruses was also tested. Results indicated that transgenic plants expressing *CP* genes of aphid transmissible strains of *CMV*, *ZYMV*, and *WMV2* are unlikely to mediate the spread of aphid non-transmissible strains of *CMV*.

17.6.3 Molecular Markers

In squash, Lee *et al.* (1995) constructed a map using 28 RAPD markers. Another genetic map in squash was developed by Brown and Myers (2002).

Zraidi *et al.* (2007) constructed the first consensus map of *C. pepo* using a combination of genetic markers including RAPD, amplified fragment length polymorphism (AFLP), simple sequence repeat (SSR) and morphological traits in two different populations. A total of 332 and 323 markers, respectively, mapped in the two populations, were spread over 21 linkage groups and estimated to cover 2,200 cM of the genome. The *C. pepo* map was updated using SSR markers, expanding the map to 86 per cent genome coverage, representing 20 linkage groups and a map density of 2.9 cM (Gong *et al.*, 2008b). Meyers (2002) included five morphological traits, but only two identified by specific genes, and none were within 15 cM of a RAPD marker. Gong *et al.* (2008b) reported four SSR markers closely linked to the *h* locus (hull-less seed) and one SSR marker linked to the *Bu* locus (bush growth habit) in *C. pepo*.

Brown and Myers (2002) mapped the mature fruit colour intensity utilizing the inter-specific cross between *Cucurbita pepo* × *C. moschata* with RAPD markers and identified the G17700 at 9.7 cM.

17.6.4 Genomics of *Cucurbita* spp.

The inbreeding line MU-CU-16, belonging to the Zucchini morphotype of the *subsp.pepo* of *Cucurbita pepo*, was chosen to obtain the whole genome sequence. A whole-genome shotgun strategy based on Illumina sequencing, using a combination of 2 pair-end and 2 mate-pair (3 Kb and 7 Kb) libraries, was employed. The first sequencing assay produced 441 and 173 million paired-end and mate-pair reads, respectively. A preliminary assembly was then performed with a set of selected clean reads. Assembly was done with SOAP denovo2 (Li *et al.*, 2010). Assembly was also improved by splitting the scaffolds generated by SOAP denovo2 using REAPR (Hunt *et al.*, 2013), a tool that precisely identifies errors in genome assemblies without the need for a reference sequence, and scaffolding again with SSPACE, a tool for scaffolding pre-assembled contigs that was reported to show higher N50 value compared with common de novo assemblers, like SOAPdenovo2 (Boetzer *et al.*, 2011). The assembly was improved with Gapcloser (Li *et al.*, 2010), yielding an assembly of 247Mb with contig and scaffold N50 of 57.8 Kb and 0.24 Mb, respectively. The *C. pepo* genome assembly can be considered of good quality compared with other cucurbit genomes (Huang *et al.*, 2009; Garcia-Mas *et al.*, 2012; Guo *et al.*, 2013).

17.6.5 Comparative Genomics of Cucurbitaceae

Yang *et al.* (2012) investigated genetic differentiation between *C. sativus* var. *sativus* and the wild *C. sativus* var. *hardwickii* by comparative fluorescence *in situ* hybridization analysis of pachytene chromosomes with selected markers from the genetic map and draft genome assembly. After sequencing the cucumber genome, Huang *et al.* (2009) proposed that five cucumber chromosomes arose from a fusion of ten ancestral chromosomes after divergence from *C. melo*. The authors reported that 348/522 (66.7 per cent) melon genetic markers and 136/232 (58.6 per cent) watermelon genetic markers were aligned on the cucumber chromosomes. The comparison revealed cucumber chromosome 7 corresponds to melon chromosome 1 and watermelon group 7. Li *et al.* (2011a) constructed a consensus melon linkage map derived from two previous genetic maps with the largest number of cross-species cucumber molecular markers and identified that melon chromosome 1 was syntenic with cucumber chromosome 7. Garcia-Mas *et al.* (2012) compared an alignment of melon and cucumber genomes synteny to detect shorter regions of rearrangements that were not previously noted, to confirm most of the previously reported ancestral fusions of five melon chromosome pairs in cucumber and several inter- and intra-chromosome rearrangements. Garcia-Mas *et al.* (2012) identified 19,377 one-to-oneortholog pairs between melon and cucumber, yielding 497 orthologous syntenic blocks. Guo *et al.* (2013) analyzed the syntenic relationships between watermelon, cucumber, melon and grape to identify 3543 orthologous relationships covering 60 per cent of the watermelon genome. This study further resolved complicated syntenic patterns using detailed chromosome-to-chromosome relationships within the Cucurbitaceae family and identified orthologous chromosomes between watermelon, cucumber and melon.

17.6.6 Whole Genome Resequencing

Cucurbita pepo contains two cultivated subspecies, each of which encompasses four fruit-shape morphotypes (cultivar groups). The Pumpkin, Vegetable Marrow, Cocozelle, and Zucchini Groups are of subsp. *pepo* and the Acorn, Crookneck, Scallop, and Straightneck Groups are of subsp. *ovifera*. Recently, a de novo assembly of the *C. pepo* subsp. *pepo* Zucchini genome was published, providing insights into its evolution. Aliki *et al.* (2019) performed the whole genome resequening of *C. pepo* morphotypes and identified genomic regions of important horticulturally economic traits. They reported first time whole-genome resequencing of the four subsp. *pepo* (Pumpkin, Vegetable Marrow, Cocozelle, green Zucchini, and yellow Zucchini) morphotypes and three of the subsp. *ovifera* (Acorn, Crookneck, and Scallop) morphotypes. A high-depth resequencing approach was followed, using the BGISEQ-500 platform that enables the identification of rare variants, with an average of 33.5X. Approximately 94.5 per cent of the clean reads were mapped against the reference *Zucchini* genome. In total, 3,823,977 high confidence single-nucleotide polymorphisms (SNPs) were identified. Within each accession, SNPs varied from 636,918 in green Zucchini to 2,656,513 in Crookneck, and were distributed homogeneously along the chromosomes. Clear differences between subspecies *pepo* and *ovifera* in genetic variation and linkage disequilibrium are highlighted. In fact, comparison between subspecies *pepo* and *ovifera* indicated 5710 genes (22.5 per cent) with Fst > 0.80 and 1059 genes (4.1 per cent) with Fst = 1.00 as potential candidate genes that were fixed during the independent evolution and domestication of the two subspecies. Linkage disequilibrium was greater in subsp. *ovifera* than in subsp. *pepo*, perhaps reflective of the earlier differentiation of morphotypes within subsp. *ovifera*.

17.6.7 Bioinformatic Databases for Cucurbitaceae

To facilitate the usage and application of genomic resources in cucurbits, a family-wide cucurbit genomics database (CuGenDB, http://www.icugi.org) and several species-specific databases have been developed to integrate genomic sequences and annotation data. Genome sequences of three major cucurbit crops, cucumber, melon and watermelon, haven been published (Huang *et al.*, 2009; Garcia-Mas *et al.*, 2012; Guo *et al.*, 2013), while squash and pumpkin genomes have been generated but not officially published yet. Squash mainly has resources of transcriptomes and genetic maps (Esteras *et al.*, 2012a; Blanca *et al.*, 2011a), as well as an online draft version of zucchini genome at CucurbiGen (https://cucurbigene. upv.es). Pumpkin currently has small RNAs (sRNAs) available at CuGenDB, which is used in an evolutionary study of microRNAs in vascular plants (Chavez Montes *et al.*, 2014).

Table 21: Bioinformatic Databases for Cucumber, Melon, Watermelon and Squash/Pumpkin

Crop	Database	Website
Cucumber (*Cucumis sativus*)	Cucumber Genome Database	http://cucumber.genomics.org.cn
	CuGenDB	http://www.icugi.org
Melon (*Cucumis melo*)	MELONOMICS	https://melonomics.net/
	MeloGene	https://melogene.upv.es/
Watermleon (*Citrullus lanatus*)	Watermelon Genome Database	http://www.iwgi.org
	CuGenDB	http://www.icugi.org
Squash and pumpkin (*Cucurbita spp.*)	CucurbiGen	https://cucurbigene.upv.es
	CuGenDB	http://www.icugi.org

17.7.0 Bitter Gourd

17.7.1 Tissue Culture

17.7.1.1 Micropropagation

First tissue culture with multiple shoot induction of bitter gourd was reported by Agarwal and Kamal (2004) and Yang et al. (2004). Agarwal and Kamal (2004) used MS medium, whereas Yang et al. (2004) induced 90.0 per cent shoots on MS medium having 4.0 mg/l 6-BA +2.0 mg/l kinetin to induce green callus with 66.7 per cent frequency on MS medium with 5 mg/l zeatin and 0.5 mg/l kinetin with proliferation coefficient of 5–6. Similarly, Sikdar et al. (2005) induced multiple shoots on immature cotyledonary nodes of lines BGGB1 and BGGB14 on MS medium using GA_3, IAA, IBA, NAA and KIN, BAP. The shoots grew for 5–6 weeks on medium containing 2 mg/l BAP+0.1 mg/l IAA +2 mg/L GA_3 to induce adventitious shoot proliferation on immature cotyledonary nodes with 80–84 shoots, after 6 weeks of culture.

Al Munsur MAZ et al. (2007) noticed that root tips were better in the regeneration of callus compared to leaf explants on MS medium with 2.0 mg/l BAP and 0.3 mg/l NAA. They noted >65 per cent shoot regeneration on leaf segments and root tips on medium containing 2.0 or 2.5 mg/l BAP + 0.2 mg/l IAA each. Al Munsur MAZ et al. (2009) used root and nodes explants of bitter gourd cultured on MS medium to induce callus on 1.0 mg/l 2, 4-D and 1.0 mg/l BAP with 75.00 per cent shoot regeneration. However, Thiruvengadam et al. (2012 a, b) induced shoot regeneration on internodal explants induced calli of bitter gourd on MS medium with 4.0 lM TDZ, 1.5 lM 2,4-D + 0.07 mM L-glutamine induced 96.5 per cent (48 shoots per explant). Verma et al. (2014) obtained maximum shoot length on ABG-6 medium (½ MS medium with 0.5 mg/ml BAP after third subculture). Saglam (2017) obtained callus-induced somatic embryogenesis developed on stem explants of accession Silifke genus.

17.7.2 Molecular Markers

Studies on genetic divergence based on molecular markers in M. charantia were started in the last decade. In one of the earliest reports, Dey et al. (2006) conducted a study on molecular diversity in 38 bitter gourd lines using RAPD markers. A total of 208 markers were generated of which 76 (36.50 per cent) were polymorphic, and the number of bands per primer was 7.17 out of which 2.62 were polymorphic. In the same population, 15 inter-simple sequence repeat (ISSR) primers were used to reveal the extent of diversity which produced a total of 125 markers, 94 (74.8 per cent) of which were polymorphic (Singh et al., 2007). Behera et al. (2008a) reported a range of 3–15 RAPD amplicons per RAPD primer with an average of 2.6 amplicons per primer in bitter gourd. In another study, Behera et al. (2008b) analyzed genetic relationships among 38 bitter gourd accessions with the aid of 29 RAPD, 15 ISSR and six AFLP markers. Using RAPD and ISSR markers, Dalamu et al. (2012) assessed genotypic variation among 50 indigenous and exotic bitter gourd genotypes. Polymorphic microsatellite markers were developed by Wang et al. (2010) for M. charantia to investigate the genetic diversity and population structure within and

In vitro Propagation of Bitter Gourd

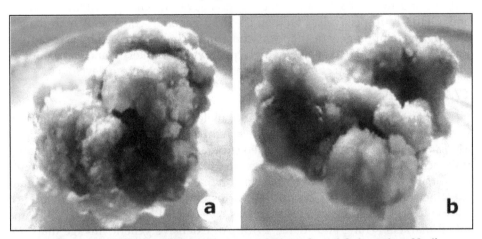

Callus Induced from different Explants of Bitter Gourd Cultured on Media
Containing different Concentrations of Growth Regulators: a. Nodal segments
at 1.0 mg/l 2,4 D and 1.0 mg/l BAP (16 DAC). b. Root segments at
1.0 mg/l 2,4 D and 0.50 mg/l BAP (38 DAC).

Shoot regeneration from nodal segments callus media containing at 1.0 mg/l BAP
(a) and 1.0 mg/l 2,4 D (b) after 45 days of culture (Al Mansur *et al.*, 2009).

between *M. charantia* and its four related species (*Cucurbita pepo* L., *Luffa cylindrica* L., *Lagenaria siceraria* L. and *Cucumis sativus* L.). Cui *et al.* (2018) identified a total of 188, 091 and 167, 160 SSR motifs in the genomes of the bitter gourd lines 'Dali-11' and 'OHB3-1,' respectively, based on the draft genome sequence. A high-density and high-resolution genetic map was constructed in bitter gourd by Gangadhara Rao *et al.* (2018), and a total of 2013 high-quality SNP markers binned to 20 linkage groups (LG) spanning a cumulative distance of 2329.2 cM were developed.

17.7.3 Genetic and QTL Mapping

The first genetic linkage map of bitter melon derived from an inter-botanical variety cross between Taiwan White, *Momordica charantia* var. *charantia*, and CBM12, *M. charantia* var. *muricata was* developed (Kole *et al.*, 2012). Besides, 12 quantitative trait loci (QTLs) controlling five polygenic fruit traits including length, diameter, weight, number, and yield were detected on five linkage groups that individually explained 11.1 to 39.7 per cent of the corresponding total phenotypic variance. A year later, the second genetic map was constructed based on SSR, AFLP, and SRAP markers (Wang and Xiang, 2013). An extensive genetic linkage map was constructed using F_2 progenies. The map included 194 loci on 11 chromosomes consisting of 26 expressed sequence tag (EST)-SSR loci, 28 SSR loci, 124 AFLP loci, and 16 SRAP loci. This map covered 1005.9 cM with 12 linkage groups. A total of 43 QTLs, with a single QTL associated with 5.1–33.1 per cent phenotypic variance, were identified on nine chromosomes for 13 horticultural traits. One QTL cluster region was detected on linkage group (LG)-5 which contained the most important QTLs with high contributions to phenotypic variance (5.8–25.4 per cent) (Wang and Xiang, 2013). The QTL/gene *gy/fffn/ffn*, controlling sex expression involved in gynoecy, first female flower node, and female flower number was detected. Particularly, two QTLs/genes, *Fwa/Wr* and *w*, were found to be responsible for fruit epidermal structure and white immature fruit color, respectively. Similarly, Wang and Xiang (2013) identified three QTLs (*fffn4.1*, *fff5.1* and fffn9.1) two QTLs in two different locations by Cui *et al.* (2018) in bitter gourd.

In bitter gourd, RAD-seq or GBS was used to genotype about a hundred of F_2 or $F_{2,3}$ individuals for developing linkage maps (Urasaki *et al.*, 2017; Cui *et al.*, 2018; Gangadhara Rao *et al.*, 2018). In these studies, RAD-seq tags or sequence reads were mapped to the genome sequence for calling SNPs or indels. This process avoids sequence differences among redundant regions, which were not derived from alleles. On the other hand, there was a strategy for genotyping by RAD-seq or GBS without reference genome and linkage maps were actually developed in bitter gourd (Matsumura *et al.*, 2014; Gangadhara Rao *et al.*, 2018). As quantitative traits, sex ratio (number of female flowers in a plant), first node of female flowers, days to first female flower, and several traits for fruits size were scored for mapping (Wang and Xiang, 2013; Cui *et al.*, 2018); Gangadhara Rao *et al.*, 2018). These traits were directly effective to yield of bitter gourd fruits.

17.7.4 Bitter Gourd Genomics

Recently, a draft genome sequence of the bitter gourd line *Momordica charantia* OHB3-1 was reported, with a scaffold-level genome assembly of 285.5 Mb and

45,859 protein-coding genes annotated by ab initio prediction. Cui *et al.* (2020) performed the whole genome resequencing of bitter gourd and revealed the basis of genetic diversity and evolution of bitter gourd. Bitter gourd (*Momordica charantia*) is a popular cultivated vegetable in Asian and African countries. To reveal the characteristics of the genomic structure, evolutionary trajectory, and genetic basis underlying the domestication of bitter gourd, Cui *et al.* (2020) performed whole-genome sequencing of the cultivar Dali-11 and the wild small-fruited line TR and resequencing of 187 bitter gourd germplasms from 16 countries. The major gene clusters (*Bi* clusters) for the biosynthesis of cucurbitane triterpenoids, which confer a bitter taste, are highly conserved in cucumber, melon, and watermelon. Comparative analysis among cucurbit genomes revealed that the *Bi* cluster involved in cucurbitane triterpenoid biosynthesis is absent in bitter gourd. Phylogenetic analysis revealed that the TR group, including 21 bitter gourd germplasms, may belong to a new species or subspecies independent from *M. charantia*. Furthermore, we found that the remaining 166 *M. charantia* germplasms are geographically differentiated, and we identified 710, 412, and 290 candidate domestication genes in the South Asia, Southeast Asia, and China populations, respectively.

18.0 References

Aartrijk, J. Van (1990) In: *Plant Molecular Biology Manual* (Eds. J.L.Slightom, P.P. Chee, D.Gonsalves, J.J.J. Nijkamp,and Plas, L.H.W. Vander), Kluwer Academic Publisher, Dordrecht, Netherlands, pp. 201-206.

Abak, K., Sari, N., Paksoy, M., Yimaz, H. and Tunali, C. (1996) *Turkish J. Agri. Forest.*, **20**: 425-430.

Abdelnour, A., Ramirez, C. and Engelmann, F. (2002) *Agronom ameso Americana*, **13**: 147-151.

Abdul-Awal, S. M., Alam, J. Md., Ali, R. Md., Hassan, N. Md., Basunia, S. R. and Rehaman, S. M. M. (2005) *Biotech.*, **4**: 221-224.

Abed, T.A. and Sharabash, M.T.M. (1985) *Agric. Sci., Cairo*, **22**: 473-486.

Abou-Hadid, A. F., El-Beltagy, A.S., Youssef, S.M. and Gaafer, S.A. (1994) *Egyptian J. Hort.*, **21**: 203-211.

Aboul-Nasr, M.H., Ramadan, B.R. and El-Dengawy, RA (1997) *Assiut. J. Agric. Sci.*, **28**: 163-172.

Abu-Romman, S.M., Al-Hadid, K.A. and Arabiyyat, A.R. (2015) *J. Agric. Sci.*, **7**: 159.

Acharya, S.K., Thakar, C., Brahmbhatt, J.H. and Joshi, N. (2020) *Int. J. Pharmacogn. Phytochem.*, **9**(4): 540-544.

Adams, P. (1978) *Grower*, **89**: 199- 201.

Agarwal, M. and Kamal, R. (2004) *J. Med. Arom. Plant Sci.*, **26**: 318–323.

Agbagwa, I.O., Ndukwu, B.C. and Mensah, S.I. (2007) *Turk. J. Bot.*, **31**: 451–458.

Ahmad, N. and Anis, M. (2005) *Turk. J. Bot.*, **29**: 237-240.

Aileni, M., Kota, S.R., Kokkirala, V.R., Umate, P. and Abbagani, S. (2009) *J. Herbs, Spices, Medicinal Plants*, **15**(2): 141–148.

Akashi, K., Morikawa, K. and Yokota, A. (2005) *Plant Biotechnol.*, **22**: 13–18.

Akbarov, S.A. (1982) *Tr. Uzb.NIIOvoshche-bakhch, Kul'tur Kartofelya*, No. 20, pp. 16-20

Akimoto, J., Fukuhara, T. and Kikuzawa, K. (1999) *Amer. J. Bot.*, **86**: 880-886.

Al Munsur, M.A.Z., Haque, M.S., Nasiruddin, K.M. and Hasan, M.J. (2007) *Prog. Agric.*, **18**(2): 1–9.

Al Munsur, M.A.Z., Haque, M.S., Nasiruddin, K.M. and Hossain, M.S. (2009) *Plant Tissue Cult. Biotechnol.*, **19**(1): 45–52.

Alam, M.Z. and Quadir, M.A. (1986) *Punjab-Vegetable Grower*, **21**: 32-34.Albert, R. (1995) *Gartenbau Magazin*, **4**: 32-34.

Alcazar, J.T. Esquinas (1978) *Diss. Abs. Inter.*, **38**: 10.

Ali, M., Hannan, A., Shafi, J., Ahmad, W., Ayyub, C.M., Asad, S., Abbas, H.T. and Sarwar, M.A. (2013) *Int. J. Adv. Res.*, **10**(1): 17–20.

Ali, M., Okubo, H., Fujii, T. and Fujieda, K. (1991) *Sci. Hort.*, **47**: 335-343.

Ali, N., Rehman, M. and Hussain, S.A. (1995) *Sarhad J. Agric.*, **11**: 585-589.

Alikhan, S., Reddy, N.T. and Reddy, E.N. (1986) *South Indian Hort.*, **34**: 46-49.

Alin Xin., He, C.Z., Peng, Z.C. Zhu, L., Shi, Z.Y. and He, G.C. (2000) *Hunan Agric. Univ.*, **26**: 93-96.

Al Munsur, M.A.Z., Haque, M.S., Nasiruddin, K.M. and Hossain, M.S. (2009) *Plant Tissue Cult. Biotech.*, **19**(1): 45-52.

Al-Masoum, A.A. and Al-Masri, H. (1999) *Egyptian J. Hort.*, **26**: 229-236.

Altunlu, H., Gul, A. and Tunc, A. (1999) *Acta Hortic.*, **49**: 377-382.

Alvarenga, E.M., Silva, R.F., Araujo, E.F. and Cardoso, A.A. (1984) *Horticulturae Brasileira*, **2**: 5-8.

Amarchandra and Parikh, H.S. (1969) *Farm. J.* (Dec.), pp. 19-21.

Amin, A.W. and Mona, A.W. (2014) *Pak. J. Nematol.*, **32**: 51–58.

An, K.B., Lim, J.W., Sac, M.W., Khee, H.C., Yu, C.J., Kwon, K.C. and Park, H.Y. (1995) *RDA J. Agric.Sci. Hort.*, **37**: 356-362.

Andrus, C.F. and Bohn, G.W. (1967) *Proc. Amer. Soc. Hort. Sci.*, **90**: 209-222.

Andrus, C.F., Seshadri, V.S. and Grimball, P.C. (1970) *Tech. Bull.1425*, U.S. Department of Agriculture, Washington, D.C.

Angelov, D. and Krasteva, L. (2000) *Acta Hortic.*, **510**: 135-137.

Anjanappa, M., Reddy, S.S., Muralli, K., Krishnappa, K.S. and Pitchaimuthu, M. (1999) *Karnataka J. Agric. Sci.*, **12**: 246-247.

Anonymous (1963) *Printed in Gartner Tindende*, **79**: 453.

Anonymous (1994) *Seedling News*, No. 1, pp. 1-2.

Anonymous (2018) IARI Annual Report – 2017-18. ICAR-Indian Agricultural Research Institute, New Delhi – 110012, P. 29.

Anonymous (2019) IARI Annual Report – 2018-19. ICAR-Indian Agricultural Research Institute, New Delhi – 110012, P. 37.

Anonymous (2020) IARI Annual Report – 2019. ICAR-Indian Agricultural Research Institute, New Delhi – 110012, P. 34.

Ansari, A.M. and Chowdhary, B.M. (2018) *Int. J. Pharmacogn. Phytochem.*, **1**: 202-206.

Arabsalmanik, K., Jalali, A.H. and Hasanpour, J. (2012) *Int. J. Agric. Sci.*, **2**(7): 605–612.

Arce Ochoa, J.P., Dainello, F., Pike, L.M. and Drews, D. (1995) *HortScience*, **30**: 492 - 493.

Arekar, A.R., Janhavi, A., Arekar, S.S., Barve and Paratkar, G.T. (2012) *J. Appl. Nat. Sci.*, **4**: 297-303.

Arold, G. (1998) *Gemüse (Munchen)*, **34**: 354-358.

Arora, S.K. and Malik, I.J. (1989) *Haryana J. Hort. Sci.*, **18**: 99-105.

Arora, S.K. and Siyag, S. (1989) *Haryana J. Hort. Sci.*, **18**: 106-112.

Arora, S.K., Pandita, M.L., Partap, P.S. and Sidhu, A.S. (1985) *J. Amer.Soc. Hort. Sci.*, **110**: 142-145.

Arora, S.K., Singh, Y. and Pandita, M.L. (1995) *Haryana J. Hort. Sci*, **24**: 144-147.

Arora, S.K., Vashistha, R.N. and Partap, P.S. (1989) *Res. Develop. Rep.*, **6**: 31-34.

Arumuganathan, K. and Earle, ED. (1991) *Plant Mol. Biol. Report.*, **9**: 208–218.

Aver Yanov, A., Lapikova, V., Pasechnik, T. and Baker, C.J. (2011) *Crop Protection*, **30**: 925-930.

Ayub, A., Guis, M., Ben-Amor, M., Gillet, L., Roustan, J.P., Laiche, A., Bonzaycn, M. and Pech, J.C. (1996) *Nat. Biotech.*, **14**: 862-866.

Ayyangar, K. and Rangaswami (1949) *Proc. 36ᵗʰ Indian Sci. Cong.*, Part III, P. 137.

Ayyangar, K. and Rangaswami (1976) *Adv. in Pollen Spore Res.*, **II**: 54-59.

Ayyangar, K. and Rangaswarni (1967) *Bull. Natl. Ins. Sci. India*, **34**: 380-396.

Backer, C.A. and Brink, R.C. (1963) *Flora of Java I*, Noordhoff, Netherlands.

Bae, Y.S., Shim, C.K., Park, C.S. and Kim, H.K. (1995) *Korean J. Plant Path.*, **11**: 287-291.

Bailey, L.H. (1930) *Gentes Herb., Ithaca*, **2**: 175-186.

Bakker, J.C. and Vooren, J. van De (1984) *Acta Hortic.*, **156**: 48.

Banerjee, S.N. and Chatterjee, B.K. (1955) *Proc. Indian Acad. Sci.* (B), **41**: 227-239.

Bansal, R., Sooch, B.S. and Dhall, R.K. (2002) *Environ. Ecol.*, **20**(4): 976-979.

Barnes, W.C. (1961) *Proc. Amer.Soc. Hort. Sci.*, **77**: 417-423.

Barnes, W.C. (1966) *Proc. Amer. Soc. Hort. Sci.*, **89**: 390-393.

Baruah, G.K.S. and Das, R.K. (1997) *Ann. Agric. Res.*, **18**: 371-374.

Basky, Z. (1984) *Protec. Ecol.*, **7**: 243-248.

Basu, P.S., Das, S. and Banerjee, S. (1998) *Plant Physiol. Biochem.*, **25**: 28-31.

Baudracco-Arnas, S. and Pitrat, M. (1996) *Theor. Appl. Genet.*, **93**: 57–64.

Behboudian, N.M., Walker, R.R. and Torokfalvy, E. (1986) *Sci. Hort.*, **29**: 251–261.

Behera, T.K. (2004) In: Hybrid Vegetable Development (Eds. P.K. Singh, S.K. Dasgupta, S.K. Tripathi), The Haworth Press. New York, USA, pp. 217– 221.

Behera, T.K., Dey, S.S. and Sirohi, P.S. (2002) *Indian J. Genet.*, **66:** 61-62.

Behera, T.K., Dey, S.S., Munshi, A.D., Gaikwad, A.B., Pal, A. and Singh, I. (2009) *Sci. Hort.*, **120:** 130-133.

Behera, T.K., Gaikwad, A.B., Singh, A.K. and Staub, J.E. (2008a) *J. Sci. Food. Agric.*, **88** (4): 733–737

Behera, T.K., Jat, G.S. and Dev, B. (2015). In: MTC on Entrepreneurship development to ensure quality vegetable seed production for making the country nutritionally secure from 10-17th December, 2015 in the Division of vegetable Science, IARI, New Delhi, pp. 46-50.

Behera, T.K., Pal, A. and Munshi, A.D. (2012) *Indian J. Plant Genet. Resour.*, **25**(3): 322-323.

Behera, T.K., Singh, A.K. and Staub, J.E. (2008b) *Sci. Hort.*, **115:** 209–217.

Bekhradi, F., Kashi, A. and Delshad, M. (2011) *Int. J. Plant Prod.*, **5:** 105–110.

Belfort, C.C., Haag, H.P. and Minami, K. (1986) *Anaisde Escola Superior de agriculture Liuzde Queeiroz*, **43:** 465-377.

Bemis, W.P., Berry, J.W, Webber, C.W. and Whitaker, T.W (1978) *HortScience*, **13:** 235-240.

Bemis, W.P., Rhodes, AM., Whitaker, T.W. and Canner, S.G. (1970) *Amer. J. Bot.*, **57:** 404-412.

Bemis, W.P. (1973) *Genet. Res. Camb.*, **21:** 221-228.

Bennison, J.A. and Jacobson, R. (1991) *Mededelingen van de Faculteit Landbouwwetenschappen, Rijksuniversiteit Gent.*, **56:** 251-258.

Benoit, F. and Ceustcrrnans, N. (1997) *Proeftuinnieuws*, **7:** 32-33.

Berenyi, B. (1996) *Acta Agronomica Ovariensis*, **38:** 91-97.

Berezhnova, V.V. and Agzamova, N. (1976) *Agrokhimiya*, **4:** 43-45.

Beyer, E.M. Jr. (1976) *HortScience*, **11**(3): 195–196. Bhaduri, P.N. and Bose, P.C. (1947) *J. Genet.*, **48:** 237-256.

Bhalla, P.R. (1971) *Physiol. Plant.*, **24:** 106–111.

Bharathi, L.K., Munshi, A.D., Vinod, Chandrashekaran, S., Behera, T.K., Das, A.B., Joseph John, K. and Vishalnath (2011) *J. Genet.*, **90**(1): 21-30.

Bharathi, L.K., Vinod, Munshi, A.D., Behera, T.K., Chandrashekaran, Kattukunnel, J.J., Das, A.B. and Vishalnath (2010) *Euphytica*, **176**(1): 79-85.

Bhargava, K.S. (1951) *Ann. Appl. Biol.*, **28:** 377-388.

Bhargava, K.S. and Joshi, R.D. (1960) *Curr. Sci.*, **29:** 443-444. Bhatnagar, D.K. and Sharma, N.K. (1994) *Res. Develop. Rep.*, **11:** 34-37.

Bhatnagar, D.K. and Sharma, N.K. (1997) *Haryana Agric. Univ. J. Res.*, **27:** 15-18.

Bhella, H.S. (1985) *J. Amer. Soc. Hort. Sci.*, **110:** 793-796.

Bhella, H.S. (1988) *HortScience*, **23:** 123-125.

Bhella, H.S. and Wilcon, G.E. (1986) *HortScience*, **21:** 86-88.

Birge, Z.K., Weller, S.C. and Daniels, D.D. (1996) In: *Proc. North Central Weed Sci. Soc.*, St. Louis, Missouri, USA, 10-12 December, 1996, Vol. 51, pp. 153-156.

Bohn, G.W and Whitaker, T.W (1949) *Proc. Amer. Soc. Hort. Sci.*, **53**: 309-314.

Bohn, G.W. and Principe, J.A. (1964) *J. Heredity*, **55**: 211-215.

Bonomi, L., Testoni, A., Lupi, D., Galasso, L. and Sarzi, A.A. (1999) *Informatore Agrario*, **55**: 39-44.

Boonkerd, T., Na Songkhla, B. and Thephuttee, W. (1993) In: *Plant Resources of South-East Asia No 8. Vegetables* (Eds. J.S. Siemonsma and Piluek Kasem), PudocScientific Publishers, Wageningen, Netherlands, pp. 150–151.

Bordas, M., Garcia-Sogo, B., Dabauza, M., Roig, L.A. and Moreno, V. (1991b) *Actas de horticultura VII Journadas de seleccion y mejora de plantas horticolas, Communicacioneses, Pontevedra*, 17-19 *de Septiembre de* 1991, pp. 149-157.

Bordas, M., González-Candelas, L., Dabauza, M., Ramón, D. and Moreno, V. (1998) *Plant Sci.*, **132**: 179–190.

Bordas, M., Montesinos, C., Dabauza, M., Salvador, A., Roig, L.A., Serrano, R. and Moreno, V. (1997) *Transgenic Res.*, **6**: 41-50.

Bordas, M., Moreno, V. and Roig, L.A. (1991a) *Cucurbit Genet. Coop. Rep.*, **14**: 71-73.

Borisov, A.V. and Krylov, O.N. (1986) In: Progress, *PriemyOvoshehevod, Selektsiii Semenovod, Ovoshch. Kul'tur*, Moscow, pp. 28-31.

Borrergaard, S. (1988) *Gartner Tidende*, **104**: 1214-1215. Bortesi, L. and Fischer, R. (2015) *Biotechnol. Adv.*, **33**: 41–52.

Bose, T.K. and Ghosh, M.S. (1975) *Indian J. Agric. Sci.*, **45**: 487-492. Bose, T.K. and Nitsch, J.P. (1970) *Physiol. Plant.*, **23**: 1206-1211.

Botwright, T., Mendham, N. and Chung, B. (1998) *Aust. J. Exp. Agric.*, **38**: 195-200.

Boualem, A., Fergany, M., Fernandez, R., Troadec, C., Martin, A., Morin, H., Sari, M.A., Collin. F., Flowers, J.M., Pitrat, M., Purugganan, M.D., Dogimont, C. and Bendahmane, A. (2008) *Science*, **321**: 836–838.

Bougie, G., Ferrari, R. and Nicoli, G. (1997) *Bollettino, dell Istitutodi Entomologia "Guido Grandidella Universita Degli Studidi Bologna*, No. 51, pp. 171-178.

Bourbos, V.A., Skoudridakis, M.T., Darakis, G.A. and Koulizakis, M. (1997) *Crop Protection*, **16**: 383-386.

Bradley, G.A., Fleming, J.W. and Mayes, R.L. (1961) *Bull. Arkans Agric. Exp. Stat.*, **643**: 23.

Branchard, M. and Chatcau, M. (1988) *C.R. Acad. Sci.*, 307:777-780.

Branchard, M., Chateau, M., Megnegneau, B. and Debeanjon, I. (1998) In : *Proc. Eucarpia meeting on cucurbit genetics and breeding*. Avignon-Montfavert, France, May 31-June 2, 1988, pp. 133-136.

Bravo, A. and Venegas, X. (1984) *Cienciae InvestigcionAgraria*, **11**: 135-140.

Brickell, C.D., Baum, B.R., Hetterscheid, W.L.A., Leslie, A. C., McNeill, J., Trehane, P. and Wiersema, J.H. (2004) *Acta Hortic.*, **647**: 28–30.

Brodsgaard, H.F. and Hansen, L.S. (1992) *Biocontrol Sci. and Technol.*, **2**: 215-223.

Broeck, L. and Van Den (1997) *Proeftuinnieuws*; **7**: 45.

Brotman, Y., Kovalski, I., Dogimont, C., Pitrat, M., Portnoy, V., Katzir, N. and Perl-Treves, R. (2005) *Theor. Appl. Genet.*, **110**: 337–345.

Brotman, Y., Silberstein, L., Kovalski, I., Perin, C., Dogimont, C., Pitrat, M., Klingler, J., Thompson, G.A. and Perl-Treves, R. (2002) *Theor. Appl. Genet.*, **104**: 1055–1063.

Brown, J.E. and Channell-Dutcher, C. (1999) *J. Veg. Crop Prod.*, **5**: 67-71.

Brown, J.E., Yates, R.P., Channelb-Butcher, C. and West, M.S. (1996a) *J. Veg. Crop Prod.*, **2**: 51-55.

Brown, J.E., Yates, R.P., Stevens, C., Khan, V.A. and Witt, J.B. (1996b) *J. Veg. Crop Prod.*, **2**: 55-60.

Brown, R.N., Bolanos, H.A., Myers, J.R. and Jahn, M.M. (2003) *Euphytica*, **129**: 253-258.

Bulder, H. A. M., DenNijs, A. P. M., Speek, E. J., Van Hasselt, P. R., and Kuiper, P. J. C. (1991a) *J. Plant Physiol.*, **138**: 661–666.

Bulder, H.A.M., Speek, E. J., Van Hasselt, P. R., and Kuiper, P. J. C. (1991b) *J. Plant Physiol.*, **138**: 655–660.

Bull, J.H., Lasserre, E., Brame, S. and Pech, J.C. (1993) In: *Proc. Internal. Symp. Cellular and Molecular Aspects of Biosynthesis and Action of the Plant Hormone Ethylene* (Eds. J.C. Pech, A. Laiche, and C. Balague), Agen, France, August 31-September 4, 1991, pp. 94-95.

Buller, S., Inglis, D. and Miles, C. (2013) *HortScience.*, **48**: 1003–1009.

Burger, Y., Paris, H.S., Cohen, R., Katzir, N., Tadmor, Y., Lewinsohn, E. and Schaffer, A.A. (2010) *Hortic. Rev.*, **36**: 165–198.

Burkill, H.M. (1985) Vol 1, 2nd edn. Royal Botanic Gardens, Kew, UK.

Burza, W., Wochniak, P., Wroblewski, T. and Malepszy, S. (1995) *J. Appl. Genet.*, **36**: 1-10.

Buta, J.G., Qi, Ling and Wang, C.Y. (1997) *Environ. Exptl. Bot.*, **38**: 1-6.

Byers, R.E., Baker, L.R., Sell, H.M., Herner, R.C. and Dilley, D.R. (1972) *Proc. Nat. Acad. Sci. USA*, **69**: 717.

Çaðlar, G. and Abak K. (1997) *Cucurbit Genetics Coop. Rep.*, **20**: 21-23.Caglar, G. and Abak, K. (1999) *Turkish J Agri. Forest*, **23**: 283-290.

Campiotti, C.A., Rocchi, P., Salice, M.F. and 'Taggi, R. (1991) *Acta Hortic.*, **287**: 443-450.

Campos De Araujo, J.A., Campos De Araujo, S.M., Castel Lane, P.D. and Siqueira,

C.F.M. (1992) In: *Proc. of a conference heldin Granada* (Eds. N.C. Prados and

J.H. Rodriguez), *Spain*, on 3-8 May, 1992 Madrid, Spain, Comite Espanolde Plasticosen Agricultura, pp. 108-113.

Cao, J.S., Yu, H.F., Ye, W.Z., Yu, X.L., Liu, L.C., Wang, Y.Q. and Xiang, X. (2005) *J. Hort. Sci. Bio.*, **80**: 29-31.

Capoor, S.P. and Verma, P.M. (1948) *Curr. Sci.*, **17**: 274-275.

Carranza, L., Osorio, N. and Rios, D. (1998) *Ciencia Agropecuaria Instituto de luvestigacion Agropecuaria-de-Panama*, No. 9, pp. 37-38.

Castane, C., Alomar, O. and Riudavets, J. (1997) In: *Proc. of the meeting at Tenerife* (Eds. R.Albajes, and A.Cameroeds), Canary Island, 3-6 November, 1997, **20**: 237-240.

Celetti, M. and Roddy, E. (2010) http://www.omafra.gov.on.ca/english/crops/facts/10-065.htm.

Chakravarty, A.K. (1948) *Proc. Ind. Acad. Sci., Sec. B.*, **27**: 74-86.

Chakravarty, H.L. (1982) *Fascicles of flora of India - II Cucurbitaceae*. Botanical Survey of India, p. 136.

Chang, M.K., Conkerton, E.J., Chapital, D.C., Wan, P.J., Vadhwa, O.P. and Spicers, J.M. (1996) *J. Amer. Oil Chemist Soc.*, **73**: 263-265.

ChanGi, R. and Su Yon, K. (1995) *Acta of Academy of Agric. Sci. of the Democratic Peoples Republic*, Korea, No. 1, pp. 72-77.

Chartzoulakis, K. and Michelakis, N. (1990) *Acta Hortic.*, **278**: 237-243.

Chat-Locussol, I., Rivenez, M.O. and Javoy, M. (1998) In: *First transnational workshop on biological. Integrated and rational control: status and perspectives with regard to regional and European experiences, Lille, France, 21-23 January*, 1998, Loos-en-Gohelle, France, Service Regionalde la Protectiondes Vegetaux, NolrdPas-Dc-Calis (1998), pp. 19-20.

Chattopadhyay, A., Maity, T.K. and Paria, N.C. (1997) *Crop Res.* (Hisar), **14**: 133-135.

Chaturvedi, R. and Bhatnagar, S. P. (2001) *In-Vitro Cell. Dev. Biol. Plant.*, **37**: 255-258.

Chee, P.P. (1991) *Plant Cell Rep.*, **9**: 620-622.

Chee, P.P. (1992) *HortScience*, **27**: 59-60.

Chee, P.P. and Slightom, J.L. (1991) *J. Amer. Soc. Hort. Sci.*, **116**: 1098-1102.

Chelliah, S. and Sambandarn, C.N. (1972) *Final Report, Pl480 Scheme*, Annamalai University, Chidambaram.

Chen, J. and Kirkbride, J.H. (2000) *Brittonia*, **52**: 315–319

Chen, J., Staub, J.E. and Tashiro, Y. (1997) *Euphytica*, **96**: 413–419.

Chen, J.F. and Adelberg (2000) *HortScience*, **35**: 11-15.

Chen, J.F. and Kirkbride, J. (2000) *Brittonia*, **52**: 315–319.

Chen, J.F. and Staub, J.E. (1997) *Cucurbit Genet. Coop. Rep.*, **20**: 24-26.

Chen, J.F., Adelberg, J.W., Staub, J.E., Skorupska, H.T. and Rhodes, B.B. (1998) In: Cucurbitaceae 98 — Evaluation and enhancement of Cucurbit germplasm (Ed. J. McCreight), ASHS Press, Alexandria, Va. pp. 336–339.

Chen, J.F., Isshiki, S., Tashiro, Y. and Miyazaki, S. (1997a) *Euphytica*, **97**: 139-141.

Chen, J.F., Luo, X.D., Qian, C.T., Jahn, M.M., Staub, J.E., Zhuang, F.Y., Lou, Q.F. and Ren, G. (2004b) *Theor. Appl. Genet.*, **108**: 1343-1348.

Chen, J.F., S. Isshiki, Y. Tashiro, and Miyazaki, S. (1995) *J. Jpn. Soc. Hort. Sci.*, **64**: 264-265.

Chen, J.F., Staub, J.E. and Jiang, J.M. (1998) *Genet. Resour. Crop Evol.*, **45**: 301–305.

Chen, J.F., Staub, J.E., Tashiro, Y., Isshiki, S. and Miyazaki, S. (1997b) *Euphytica*, **96**: 413–419.

Chen, J.F., Zhuang, F.Y., Liu, X.A. and Qian, C.T. (2004a) *Can. J. Bot.*, **82**: 16-21.

Chen, Q.H., Huang, T., Zhuo, Q.Y., He, X.Z. and Lin, T.E. (1996) *China Vegetables*, No. 2, pp. 7-8.

Chen, W.S., Chiu, C.C., Liu, H.Y., Lee, T.L., Cheng, J.T., Lin, C.C., Wu, Y.L. and Chang, H.Y (1998) *Biochem. Mol. Biol. Inter.*, **46**: 1201-1209.

Chen, X.B., Zhang, B.X., Lou, B.G. and Li, J. Y. (1999) *J. Zhejiang Univ. (Agric. and Life Sci.)*, **25**: 578-582.

Cheng, Y.A., Zhang, B.K., Zhang, E.H. and Zhao, Z.L. (2002) *Rep. Cucurbit Genet. Coop.*, **25**: 56-57.

Chernenko, O.L. (1989) In: *Selektsiyai semenovodstvoovoslichnykhibakhchevykh Kul'tur*, USSR, pp. 153-157.

Cheshmehmanesh, A.A., Kashi, A., Moshrefi, M.M. and Khososi, M. (2004) *Seed and Plant*, **19**: 447–456.

Cheyrias, J.M. (1998) In: *First transnational workshop on biological, integrated and rational control: status and perspectives with regard to regional and European experiences, Lille, France, 21-23 January, 1998*, Loos-en-Gohclle, France, Service Regionalde la Protection des Vegetaux, Nolrd Pas-De-Calis (1998), pp. 21-22.

Cho, M.C., Heo, Y.C., Kim, J.S., Om, Y.H., Mok, I.G., Hong, K.H. and Park, H.G. (2004a) *Korean J. Breed.*, **36**: 111-112.

Cho, M.C., Heo, Y.C., Kim, J.S., Om, Y.H., Mok, I.G., Hong, K.H. and Park, H.G. (2004b) *Korean J. Breed.*, **36**: 113-114.

Cho, M.C., Om, Y.H., Heo, Y.C., Kim, J.S. and Park, H.G. (2004c) *Korean J. Breed.*, **36**: 271-275.

Cho, Y.D., Kang, S.G.and Chung, J. (1997) *RDA J. Hort.Sci.*, **39**: 33-38.

Choi, P.S., Soh, W.Y., Kim, Y.S., Yoo, O.J. and Liu, J.R. (1994) *Plant Cell Rep.*, **13**: 344–348.

Chomicki, G., Schaefer, H. and Renner, S.S (2019) *New Phytologist.* https://doi.org/10.1111/nph.16015

Choudhary, S.K., Patel, A.K., Harish, Shekhawat, S., Narpat, S. and Shekhawat (2017) *Physiol. Mol. Biol. Plants.*, **23**: 713–722.

Choudhury, B. (1966) 18th *Proc.Int. Hort. Cong.*, Vol. I. Abst. No. 191. Choudhury, B. (1979) *Vegetables* (6th Rev. Edition), National Book Trust, Delhi.

Choudhury, B. and Singh, B. (1971) *Indian Hort.*, **16**: 15-16.

Choudhury, B. and Thakur, M.R. (1965) *Indian J. Genet.*, **25**: 188-197.

Chovelon, V., Restier, V., Dogimont, C. and Aarrouf, J. (2008) *Proceedings of the IXth EUCARPIA meeting on genetics and breeding of Cucurbitaceae* (Ed. M. Pitrat),INRA, Avignon (France), May 21-24th, 2008.

Christey, M.C., Makaroff, C.A. and Earle, E.D. (1991) *Theor. Appl. Genet.*, **83**: 210–208.

Chun Hee, L., Kyung Ae, S. and Seong Gun, C. (1998) *RDA J. Crop Protection*, **40**: 124-128.

Chung, HD., Youn, S.J. and Choi, Y.J. (1998) *J. Korean Soc. Hort. Sci.*, **39**: 377-384.

Chung, S.M., J.E. Staub, and Chen, J.F. (2006) *Genome*, **49**: 219-229.

Claveria, E., Garcia-Mas, J. and Dolcet-Sanjuon, R. (2005) *J. Amer. Soc. Hort. Sci.*, **130**: 555-560.

Clendennen, S.K., Kellogg, J.A., Wolff, K.A., Matsumura, W., Peters, S., Vanwinkle, J.E., Copes, B., Pieper, M., Kramer, M.G. (1999) In: *Biology and Biotechnology of the Plant Hormone Ethylene II*. Springer, (Eds. A.K. Kanellis, C., Chang, H. Klee, A.B. Bleecker, J.C. Pech, and D. Grierson), Springer, Netherlands, Dordrecht, pp. 371–379, http://dx.doi.org/10.1007/978-94-011-4453-7_68.

Clough, G.H. (1995) *Plant Dis.*, **79**: 1107.

Cogniaux, A. and Harms, H. (1924) *Cucurbitaceae – Cucurbitaceae et Cucumerinae*, Pflanzenreich, Berlin.

Cohen, R., Pivonia, S., Burger, Y, Edelstein, M., Gamlicl, A and Katan, J. (2000b) *Acta Hortic.*, **510**: 143-147.

Cohen, R., Pivonia, S., Burger, Y., Edelstein, M., Gamliel, A. and Katan, J. (2000a) *Plant Disease*, **84**: 496-505.

Cohen, R., Pivonia, S., Burger, Y., Edelstein, M., Gamliel, A. and Katan, J. (2012) *Plant Dis.*, **84**: 496–505.

Cole, Y. and Lesaint, C. (1978) *C.R. Acad. Agri. France*, **64**: 787-792.

Colijn-Hooymans, C.M. and Bouwer, R. (1987) *Acta Botanica Neerlandica*, **36**: 331-332.

Collina, M. (1996) *Colture Protette*, **25**: 39-42.

Colucci, S.J. and Holmes, G.J. (2010) *American Phytopathol. Soc.*, pp. 1–8. Compton, M. E. and Gray, D. G. (1992) *Proc. Fla. State Hort. Soc.*, **105**: 352-354.

Compton, M.E., Brenda, L., Pierson and Staub, J.E. (2001) *Plant Cell, Tissue and Organ Culture,* **64**: 63–67.

Compton, M.E., Gray, D.J. and Elmstrom, G.W. (1993) *Plant Cell Tissue Organ Cult.*, **33**: 211-217.

Compton, M.E., Gray, D.J., Hiebert, E. and Lin, C.M. (1993) *HortScience*, **28**: 498.

Condurso, C., Verzera, A., Dima, G., Tripodi, G., Crinò, P., Paratore, A. and Romano, D. (2012) *Sci. Hort.*, **148**: 9-16.

Contin, M. and Munger, H.M. (1977) *HortScience*, **12**: 397.

Cortes, S. and Hernandez, A. (1996) *Cultivos Tropicales*, **17**: 44-47.

Cosper, R.D. and Norton, J.D. (1985) *HortScience*, **20**: 656.

Coutts, R.H.A. and Wood, K.R. (1997) *FEMS Microbiology Letters*, **1**: 152-159.

Crin'o, P., Lo Bianco, C., Rouphael, Y., Colla, G., Saccardo, F. and Paratore, A. (2007) *HortScience*, **42**: 521–525.

Cubeta, M.A. and Echandi, E. (1991) *Biological Control*, **1**: 227-236.

Cuevas, H.E., Song, H., Staub, J. E. and Simon, P.W. (2010a) *Euphytica*, **171**: 301–311.

Cuevas, H.E., Staub, J.E. and Simon, P.W. (2010b) *Euphytica*, **173**: 129–140.

Cuevas, H.E., Staub, J.E., Simon, P.W., Zalapa, J.E. and McCreight, J.D. (2008) *Appl Genet.*, **117**(8): 1345-1359.

Cui, J., Luo, S., Niu, Y., Huang, R., Wen, Q. and Su, J. (2018) *Front. Plant Sci.*, **9**: 477.

Cui, J., Yang, Y. and Luo, S. (2020) *Hortic. Res.*, **7**: 85.

Custers, J.B.M. and Bergervoet, J.H.W. (1998) *Prophyta*, **42**: 150. Custers, J.B.M. and Den Nijs, A.P.M. (1986) *Euphytica*, **35**: 639–647.

Dabauza, M., Bordas, M., Salvador, A. and Roig, L. A. (1997) *Plant Cell Reports*, **16**: 888–892.

Dabauza, M., Gonzáles-Candelas, L., Bordas, M., Roig, L.A., Ramón, D., Moreno, V. (1998) *Plant Cell, Tissue and Organ Culture*, **52**: 123–131.

Dahmani-Mardas, F., Troadec, C., Boualem, A., Le´ve^que, S. and Alsadon, A.A. (2010) *PLoS ONE*, **5**(12): e15776.

Dalamu, Behera, T.K., Gaikwad, A.B., Saxena, S., Bharadwaj, C. and Munshi, A.D. (2012) *Austral. J. Crop Sci.*, **6**(2): 261–267.

Damato, G., Manolio, G. and Bianco, V.V. (1998) *Acta Hortic.*, **467**: 295-303.

Dane, F., Denna, D.W. and Tsuchiya, T. (1980) *Z. Pflanzenzuchtg.*, **85**: 89-109.

Dane, F., Liu, J. and Zhang, C. (2007) *Genet. Resour. Crop Evol.*, **54**: 327–336.

Danin-Poleg, Y., Tadmor, Y., Tzuri, G., Reis, N., Hirschberg, J. and Katzir N. (2002) *Euphytica*, **125**: 373–384.

Darekar, K.S., Mhase, N.L. and Shelke, S.S. (1989) *International Nematology Network Newsletter*, **6**: 14-16.

Das, B.C. and Das, T.K. (1995) *Orissa J. Hort.*, **23**: 87-91.

Das, B.C. and Das, T.K. (1996) *Orissa J. Hort.*, **24**: 74-78.

Das, M.K., Maity, T.K., Som, M.G. (1987) *Veg. Sci.*, **14**: 18-26.

Das, R., Dey, S.C. and Dutta, T.C. (1995) *J. Assam Sci. Soc.*, **37**: 182-184.

Das, S. and Basu, P.S. (1998) *Indian J. Plant Physiol.*, **3**: 201-204.

Dasgan, H.Y., Balacheva, E., Yetis¸ir, H., Yarsi, G., Altuntas, O., Akhoundnejad, Y. and Coban, A. (2015) *Procedia Environmental Sci.*, **29**: 268.

Dash, S.K., Saheb, S.K.and Tripathy, M.K. (2000) *Madras Agric. J.*, **86**: 282-286.

Datta, S.K. (1987) *Crop Improvement*, **14**: 182-184.

Davis, J.M. (1994) *HortScience*, **29**: 263-297.

De Candolle, A. (1882) *Origin of Cultivated Plants*, NewYork.

De Wilde, W.J.J.O. and Duyfjes, B.E.E. (2002) *Bot. Z.*, **87**: 132–148.

Deakin, J.R., Bohn, G.W. and Whitaker, T.W. (1971) *Econ. Bot.*, **25**: 195-211.

Debeaujon, I. and Branchard, M. (1988) *Cucurbitaceae 88, Proc. Eucarpia meeting on cucurbit genetics and breeding, Avignon-Montfavet.*, Franch, May 31-June 2, 1988.

Debeaujon, I. and Branchard, M. (1991) *Acta Hortic.*, **289**: 225-226.

Debeaujon, I. and Branchard, M. (1992) *Plant Cell Rep.*, **12**: 37-40.

Debnath, B., Sinha, S. and Sinha, R.K. (2013a) *Indian J. Plant Sci.*, **2**: 2319-3824.

Debnath, B., Sinha, S. and Sinha, R.K. (2013b) *Indian J. Plant Sci.*, **2**(3): 43-47.

Deleu, W., Esteras, C., Roig, C., Gonzalez-To, M., Fernandez-Silva, I., Gonzalez-Ibeas, D., Blanca, J., Aranda, M.A., Arus, P., Nuez, F., Monforte, A.J., Pico, M.B. and Garcia-Mas J. (2009) *BMC Plant Biol.*, **9**: 90.

Desamero, N.V., Adelberg, J.W., Hale, A., Young, R.E. and Rhodes, B.B. (1993) *Plant Cell Tissue Organ Cult.*, **33**: 265-271.

Desbiez, C., Gal-On, A., Girard, M., Wipf-Scheibel, C. and Lecoq, H. (2003) *Phytopathology*, **93**: 1478-1484.

Devaraju, M.A.N., Reddy, K.C., Sivakumar, M. and Deepu, D.D. (2002) *Curr. Res. Univ. Agric. Sci., Bangalore*, **31**(2): 16–17.

Devendra, N.K., Rajanna, L., Sheetal, C. and Seetharam, Y. N. (2008) *Plant Tiss. Cult. Biotechnol.*, **18**:103-111.

Devendra, N.K., Subhash, B. and Seetharam, Y. N. (2009) *American-Eurasian J. Sustainable Agric.*, **3**(4): 743-748.

Devi, D.S., Balakrishnan, R., Paulas, D., Subbiah, R. and Natarajan, S. (1989) *South Indian Hort.*, **37**: 274-276.

Devi, J.R. and Selvaraj, J.A. (1995) *Madras Agric. J.*, **82**: 75-77.

Devi, J.R. and Selvaraj, J.A. (1996) *Seed Res.*, **22**: 64-65.

Devi, T., Rajasree, V., Premalakshmi, V. and Hemapra, K. (2017) *Int. J. Curr. Microbiol. App. Sci.*, **6**: 2392-2402.

Dey, S.S, Behera, T.K., Munshi, A.D. and Anand Pal (2010) *Euphytica*, **173**: 37–47.

Dey, S.S., Behera, T.K., Bhatia, R., Munshi, A.D. (2020) In: *Accelerated Plant Breeding*, Volume 2 (Vegetable Crops), Springer Nature, Switzerland AG 2020, pp. 271-299.

Dey, S.S., Behera, T.K., Munshi, A.D., Rakshit, S. and Bhatia, R. (2012) *Indian J. Hortic.*, **69**(4): 523-529.

Dey, S.S., Behera, T.K., Pal, A. and Munshi, A.D. (2005) *Veg. Sci.*, **32**(2): 173-176.

Dey, S.S., Behera, T.K., Munshi, A.D. and Pal, A. (2009) *Euphytica*, **173**(1): 37-47.\

Dey, S.S., Singh, A.K., Chandel, D. and Behera, T.K. (2006) *Sci. Hort.*, **109**: 21–28.

Dhaliwal, M.S. and Lal, T. (1996) *Indian J.Genet.Plant Breed.*, **56**(2): 207-213.

Dhesi, N.S., Padda, D.S.and Malik, B.S. (1966) *Indian J. Hort.*, **23**: 169-171.

Dhesi, N.S., Padda, D.S. and Malik, B.S. (1964) *Punjab Hort. J.*, **4**: 45-47.

Dhillon, Narinder P.S., Phethin, Supunsa, Sanguansil, Supannika and McCreight, James D. (2017a) *Pak. J. Agric. Sci.*, **54**(1): 27-33.

Dhillon, Narinder P.S., Sanguansil, Supannika, Singh, Sheo Pujan, Masud, Mohammed Abu Taher, Kumar, Prashant, Bharathi, Latchumi Kanthan, Yetioir, Halit, Huang, Rukui, Canh, Doan Xuan, and McCreight, James D. (2017b) In: *Genetics and Genomics of Cucurbitaceae* (Ed. R. Grumet *et al.*), Plant Genetics and Genomics: Crops and Models, DOI 10.1007/7397_2016_24.

Dijkuizen, A. and Staub, J.E. (1999) *Cucurbit Genet. Coop. Res.*, **22**: 8-10.

Dijkuizen, A., Kennard, W.C., Havey, M.J. and Staub, J.E. (1996) *Euphytica*, **90**: 79-87.

Dirks, R. and Buggenum, M. van. (1991) *Physiol. Plant.*, **82**: 1-18.

Dogimont, C., Leconte, L., Perin, C., Thabuis, A., Lecoq, H. and Pitrat, M. (2000) *Acta Hort.,* **510**: 391–398.

Doloi, N., Patel, J.N. and Acharya, R.R. (2018) Vegetos, **31**: 1. doi: 10.4172/2229-4473.1000382

Dong, W., Chen, L. and Wu, Y. (1993) *Acta Horticulturae Sinica*, **20**: 155-160.

Dons, J.J.M. and Bouwer R (1986) In: International Atomic Energy Agency, Vienna (Austria); Food and Agriculture Organization of the United Nations, Rome (Italy); Proceedings series, IAEA, Vienna (Austria), pp. 497-504.

Dore, C., Boulidard, L., Sauton, A., Rode, J.C., Cuny, F., Niemirowicz-Szezytt, K., Sari, N. and Dumas-de-Valux (1995) *Acta Hortic.,* **392**: 123-128.

Dryanovska, O.A. (1985) *C.R. Acad. Bulenceare Des.Sci.*, **38**: 1243-1244.

Dryanovska, O.A. and Ilieva I.N. (1983) *C R Acad. Bulg. Sci.,* **36**: 1107–1110.

Dubey, G.S. and Nariani, T.K. (1975) *Indian Phytopath.*, **28**: 118-119.

Dubravec, K. (1994) *Poljoprivrcdna Znanstvena Smotra*, **59**: 425-428.

Dumas, de Vaulx R, Chambonnet, D. and Pochard, E. (1981) *Agronomie* **1:** 859–864.

Dutt, B. and Roy, R.P. (1969) *Genetica*, **40**: 7-18.

Dutt, B. and Roy, R.P. (1971) *Genetica*, **42**: 139-156.

D'yachcnko, I.I. and Zhovner, I.M. (1986) *Kartofel'i Ovoshchi*, No. 2, P. 38.

Dyutin, K.E. and Bicherev, V.A. (1985) In: *Prob. Oroshaem Ovosheheuodi bakhcheuod Astrakhan*, USSR, pp. 37-40.

Dyutin, K.E. and Puchkov, M. Yu (1996) *Kartofeli Ovoshchi*, No.5, P. 25.

Dyutin, K.E. and Sokolov, S.D. (1990) *Tsitologiyai Genetika;* **24**: 56-57.

Edelstein, M., Nerson, H., Nadler, K. and Burger, Y. (1989) *Hassadelt*, **70**: 398-400.

Edelstein, M., Nerson, H., Paris, H.S., Karchi, Z., Burger, Y. and Zohar, R. (1985a) *Hassadch*, **65**: 2416-2419.

Edelstein, M., Paris, H.S., Nerson, H., Karchi, Z. and Burger, Y. (1985b) *Rep. Cucurbit Genet. Coop.*, USA.

Edelstein, M., Ben-Hur, M., Leib, L. and Plaut, Z. (2011a) *Israel J. Plant Sci.*, **59**: 207–215.

Edelstein, M., Ben-Hur, M., and Plaut, Z. (2011b) *J. Amer. Soc. For Hort. Sci.*, **132**: 484–491.

Eduardo, I., Arús, P. and Monforte, A.J. (2007) *J. Amer. Soc. Hort. Sci.*, **132**: 80–89.

Egel, D.S. and Martyn, R.D. (2013) *The American Phytopathological Society*, pp. 1–11.

Egorov, I.V., Sutulova, V.I., L'Vova, I.N. and Grarnenitskaya, VN. (1987) *Fiziologiya Rastenii*, **34**: 400-405.

Ehret, D.L., Koch, C., Menzies, J., Sholberg, P. and Garland, T. (2001) *HortScience*, **36**: 934–936.

El-Aidy, F. (1991) *Acta Hor.*, **287**: 281-288.

El-Behairy, U.A.A., Abou-Hadid, A.F. and Burrage, S.W. (1997) In: *Proc.* 9th *International Congress on Soilless Culture. St. Helier, Jersey*, Channel Islands, 12-19April, 1997, Wageningen, Netherlands, pp. 157-166.

El-Beheidi, M., EI-Mansi. A.A., Abd-Alla, T.K. and Hewedy, A.M. (1978) *Res. Bull. Fac. Agric Ain Shams University*, No. 909, P. 18.

El-Doweny, H.H., Hashem, M.M. and Abou-Hadid, A.F. (1993) *Egyptian J. Hort.*, **19**: 13-22.

El-Eslamboly, A.A.S. and Deabes, A.A.A. (2014) *Minufiya J. Agric. Res.*, **39**: 1109–1129.

El-Ghamriny, E.A., Singh, N. and Vema, V.K. (1988) *Curr. Sci.*, **57**: 263-265.

El-Ghaouth, A., Arul, J., Grenier, J., Benhamou, N., Asselin, A. and Belanger, R. (1994) *Phytopath.*, **84**: 313-320.

Eisa, H.M. and Munger, H.M. (1968) *Proc. Amer. Soc. Hort. Sci.*, **92**: 473-479.

Elkner, K. (1985) *Acta Agrobotanica*, **35**: 61-68.

Ellul, P., Rios, G., Atares, A., Roig, L.A., Serrano, R. and Moreno, V. (2003) *Theor. Appl. Genet.*, **107**: 462–469.

Elmeer, K.M.S., Thomas, F., Gallagher and Hennerty, M.J. (2009) *African J. Biotech.*, **8**: 3219-3222.

El-Naggar, M., El-Deeh, H. and Ragab, S. (2012) *Pakistan Journal of Agriculture: Agricultural Engineering Veterinary Sciences*, **28**: 52–61.

El-Shawaf, I.I.S. and Baker, L.R. (1981a) *J. Amer. Soc. Hort. Sci.*, **106**: 359-364.

El-Shawaf, I.I.S. and Baker, L.R. (1981b) *J. Amer. Soc. Hort. Sci.*, **106**: 365-370.

Ensminger, A.H., Ensminger, M.E., Konlande, J.E. and Robson, J.R.K. (1983) In: *Foods and Nutrition Encyclopedia*, Vol. 1. Clovis: Pergus Press.

Esquinas, J.T. (1981) *Kulturplanza*, **24**: 337-362.

Ezura, H. and Oosawa, K. (1994) *Plant Cell Rep*; **14**: 107-111.

Ezura, H., Amagai, H. and Oosawa, K. (1993) *Japanese J. Breed.*, **43**: 193-199.

Ezura, V.H., Amagai, H., Yoshioka, K. and Oosawa, K. (1992) *Japanese J. Breed.*, **42**: 137-144.

Fang, G.W. and Grumet, R. (1993) *Mol. Plant Microbe Interact.*, **6**: 358-367.

Fang, X.J., Gu, X.F., Han, X. and Xhang, T.M. (1996) *China Vegetables*, **1**: 12-14.

Faria, A. L., Tanziman, A., Karim, R., Islam, R., and Hossain, M. (2013) *J. Genetic Env. Resour. Conservation*, **1**: 247-253.

Farias-Larios, J., Guzmaqn, S. and Michel, A.C. (1994) *Biological Agric. Hort.*, **10**: 303-306.

Faris, N.M., Rakoczy-Trojanowska, M., Malepszy S. and Niemirowicz-Szczytt, K. (2000) *Progr. Biotechnol., Food Biotechnology*, **17**: 49-54.

Fassuliotis, G. (1977) *J. Amer. Soc. Hort. Sci.*, **102**: 336–339.

Fazio, G., Staub, J.E. and Stevens, M.R. (2003a) *Theor. Appl. Genet.*, **107**: 864–874. Fellner, M. and Lebeda, A. (1998) *Biol. Plant.*, **41**: 11-24.

Fernandes, C.A. (2011) *M.Sc. Thesis*, University of Massachusetts Amherst. Fernandez-Silva,I.,Moreno, E.,Essafi,A.*et al*.(2010)*Theor. Appl. Genet.*,**121**:931–940.

Fernando, L. N. and Grün, I. U. (2001) *Flavour and Fragrance J.*, **16**(4): 289–293.

Ferreira, M.A.J. da F., Vencovsky, R., Vieira, M.L.C. and De-Queiroz, M.A. (2000) *Acta Hort.*, **510**: 47-54.

Ficcadenti, N., Veronese, P., Sestili, S., Crino, P., Lucretti, S., Schiavi, M. and Saccardo, F. (1995) *J. Genet. Breed.*, **49**: 359–364.

Ficcadenti, N., Sestili, S., Annibali, S., Marco, M. de and Schiavi, M. (1999) *J. Genet. Breed.*, **53**: 255-257.

Fita, A., Pico, B., Roig, C. and Nuez, F. (2007) *J. Hortic. Sci. Biotechnol.*, **82**: 184–190.

Firoz Alam, M., Ruhul, A., Ekhlas, U.M. and Sudhan (2015) *Int. Res. J. Biol. Sci.*, **4**: 48-52.

Florescu, E., Ciofu, R., Vajiala, M., Dumitru, M. and Berinde, J. (1991) *Lucrari Stuntifice, Institutul Agronomic Nicoiae Balcescu Bucuresti, Seria B, Horticultura*, **34**: 17-29.

Foster, R.E (1963) *J. Hered.*, **54**: 113-114.

Foster, R.E. (1968) *J. Hered.*, **59**: 205-207.

Frankel, R. and Galun, E. (1977) In: *Pollination Mechanisms, Reproduction and Plant Breeding*. Heidelberg: Springer-Verlag. pp. 104-157.

Franken, J., Custers, J.B.M. and Bino, R.J. (1988) *Plant Breed.*, **100**: 150–153.

Fravel, D.R., Rhodes, D.J. and Larkin, R.P. (1999) In: *Integrated Pest and Disease Management in Green house Crops*, (Eds. R. Albajes, M.L. Gullino, J.C. van Lenteren and Y. Elad, Dordrecht), Kluwer, pp. 365–76.

Friedlander, M., Atsmon, D. and Galun, E. (1977) *Plant Cell Physiol.*, **18**: 681-691.

Fuchs, M. and Gonsalves, D. (1995) *BioTechnol.*, **13**: 1466-1473.

Fuchs, M., Klas, F.E., Mc Ferson, J.R. and Gonsalves, D. (1998) *Transgenic Res.*, **7**: 449-462.

Fuchs, M., McFerson, J.R., Tricoli, D.M., McMaster, J.R., Deng, R.Z., Boeshore, M.L., Reynolds, J.F., Russell, P.F., Quemada, H.D. and Gonsalves, D. (1997) *Mol. Breed.*, **3**: 279–290.

Fujieda, K. (1994) In: *XXIVth International Horticultural Congress Publication Committee*, Horticulture in Japan, Asakura, Tokyo, pp. 69–73.

Fujiwara, K., Doi, R., Ii Moto, M. and Yano, A (2000) *Environment Control in Biology*, **38**: 33-38.

Fukino, N., Ohara, T., Monforte, A.J., Sugiyama, M., Sakata, Y., Kunihisa, M. and Matsumoto, S. (2008b) *Theor. Appl. Genet.*, **118**: 165–175.

Fukino, N., Sugiyama, M., Ohara, T., Sainoki, H., Kubo, N., Hirai, M., Matsumoto, S. and Sakata, Y. (2008a) In: *Proc IXth EUCARPIA Meeting on Genetics and Breeding of Cucurbitaceae*. Avignon, France, pp. 505–509.

Fursa, T.B. (1972) *Bot. Z.*, **57**(1): 31–41.

Gaableman, W.H. (1974) In : *Proceedings of XIXth International Horticulture Congress*. **I** (II): 419-426.

Gad, A.A., Alsadon, A.A. and Wahdan, H.M. (1993) *Ann. Agric. Sci.* (Cairo), **38**: 251-259.

Gajdová, J., Lebeda, A. and Navrátilová, B. (2004) In: *Proceedings of Cucurbitaceae 2004, the8th EUCARPIA Meeting on Cucurbit Genetics and Breeding*. Palack University in Olomouc, Olomouc, pp. 441–454.

Gajdová, J., Navrátilová. B., Smolná. J., Lebeda. A. (2007a) *Acta Hortic.*, **731**: 89–94.

Gajdová. J., Navrátilová. B., Smolná. J., Lebeda. A. (2007b) *J. Appl. Bot. Food Qual.*, **81**: 1–6.

Galaev, M.Kh. (1986) *Nauchno-Teknicheskii Byulleten Vsesoyuznogo Ordena Lenina i*

Ordena Druzby Narodov Nauchno-Issledovatel'skogo Instituta Rastenievodstva imeni *N.I. Vavilova*, No. 165, pp.77-78.

Gal-On, A, Katsir, P. and Wang, Y.Z. (2000) *Acta Hortic.*, **510**: 343-347.

Gal-On, A., Antignus, Y., Rosner, A. and Raccah, B. (1991) *J. Gen. Virol.*, **72**: 2639-2643.

Galatti, F.S., Franco, A.J., Ito, L.A., Charlo, H.C.O., Gaion, L.A. and Braz, L.T. (2013) *Ceres*, **60**: 432-436.

Galun, E. (1961) *Genetica*, **32**: 134-163.

Galun, E. (1983) *Proc. Conference on the Biology and Chemistry of Cucurbitaceae, Cornell University, Ithaca*, New York, August, 1980.

Galun, E., Jung, Y. and Lang, A. (1962) *Nature*, **194**: 596.

Gamboa, W., Philipp, D. and Pohlan, J. (1996) *Tropenlandwirt*, **97**: 85-94.

Gangadhara Rao, P., Behera, T.K., Gaikwad, A.B., Munshi, A.D., Jat, G.S. and Boopalakrishnan, G. (2018) *Front. Plant Sci.*, **9**: 1555.

Gangadhara Rao, P., Behera, T.K., Munshi, A.D. and Dev, B. (2018) *Indian J. Hortic.*, **74(2)**: 227-232.

Ganikhozhdaeva, R.A (1984) *Nauchnye Trudy Tashkentskii Sel skokhozyaistvennyi Institue*, No. 105, pp. 64-69.

Gao, X.Y., Wang, Z., Lin, X.Y., Lua, S.L., Wang, Y.Y., Wang, F.H. and Sun, B.L. (1988) *Scientia Agricultura Sinica*, **21**: 74-80.

Garcia Sogo, B., Dabauza, M., Bordas, M., Roig, L.A. and Moreno, V. (1991) VIII *Jornadas de selecciony mejora de plantas horticolas*, Comunicaciones, Pontevedra, pp. 133-147.

Garcia-Mas, J., Benjak, A., Sanseverino, W., Bourgeois, M., Mir, G., Gonzalez, V.M., *et al.* (2012) *Proc. Natl. Acad. Sci.*, **109**: 11872–11877.

Garcia-Mas, J., Monforte, A.J. and Arus, P. (2004) *Plant Syst. Evol.*, **248**: 191–203.

GarciaSogo, B., Debauza, M., Roig, L.A. and Moreno, V. (1992) *Cucurbit Genet. Coop. Rep.*, **15**: 40-44.

Garibaldi, A., Minuto, G. and Gullino, M.L. (1995) In: *47th Inter. Symp. on crop Protection, Gent, Belgium, 9 May, 1995*, Mededelingen-Facultieit Landbouwkundigeen Toegepaste Biologische Wetens-chappen, Universiteit Genet, **60**: 317-320.

Garnil, N.A.M. (1995) *Ann. Agric. Sci., Moshtohor*, **33**: 681-691.

Ge´mes-Juha´sz, A., Balogh, P., Ferenczy, A., Kristo´f, Z. (2002) *Plant Cell Rep.*, **21**: 105–111.

Ge´mesne´-Juha´sz, A., Venczel, G., Balogh, P. (1997) *Acta Hortic.*, **447**: 623–625.

Ghani, M.A., Amjad, M., Iqbal, Q., Nawaz, A., Ahmad, T., Hafeez, O.B.A. and Abbas, M. (2013) *Pak. J. Life Soc. Sci.*, **11**(3): 218–224.

Ghebretinsae, A.G., Thulin, M. and Barber, J.C. (2007) *Amer. J. Bot.*, **94**: 1256–1266.

Ghive, D.V., Raut, N.W. and Ghorade, R.B. (2006a) *Int. J. Plant Sci.*, **1**(2): 266-268.

Ghive, D.V., Ghorade, R.B., Khedekar, R.P., Jeughale, G.S. and Raut, N.W. (2006b) *Asian J. Biol. Sci.*, **1**: 146-148.

Ghosh, S.K. and Mukhopadhyay, S. (1979) *Phytopath. Z.*, **94**: 172-184.

Gillespie, D.R. (1989) *Amlelyseius cucumeris Entomophaga*, **34**: 185-192.

Glas, R. (1985) *Groenten Fruit*, **40**: 42-43.

Goldberg, N.P. (2004) Anthracnose of cucurbits. Guide H-247, New Mexico State University, USA. P. 2.

Goldhauscn, M. (1938) *C.R. Doklady Acad. Sci.*, USSR, **20**: 595-597.

Golyshin, N.M., Maslova, A.A. and Goneharova, T.F. (1992) *Zashchita Rastenii (Moskva)*, No. 12, p. 13.

Gong, L., Pachner, M., Kalai, K. and Lelley, T. (2008a) *Genome*, **51**: 878–887.

Gong, L., Stift, G., Kofler, R., Pachner, M. and Lelley, T. (2008b) *Theor. Appl. Genet.*, **117**: 37–48.

Gonsalves, C., Xue, B., Yepes, M., Fuchs, M., Ling, K., Namba, S., Chee, P., Slightom, J.L. and Gonsalves, D. (1994) *J. Amer. Soc. Hortic. Sci.*, **119**(2): 345-355.

Gonsalves, D., Chee, P., Provvidenti, R., Seem, R. and Slightom, J.L. (1992) *Biotechnology*, **10**: 1562-1570.

Gonsalves. C., Xue, Bao Di and Gonsalves, D. (1995) *HortScience*, **30**: 1295-1297.

González, V., Aventín, N., Centeno, E. and Puigdomènech, P. (2014) *BMC Genomics*, **15**: 1131.

Gonzalez, V.M., Rodriguez-Moreno, L., Centeno E, Benjak, A., Garcia-Mas, J., Puigdomenech P., *et al.* (2010) *BMC Genomics*, **11**: 618.

Gonzalez-Ibeas, D., Blanca, J., Roig, C., González-To, M. and Picó, B. (2007) *BMC Genomics*, **8**: 306.

Gonzalo, Maria Jose , Claveria, Elisabet , Monforte, Antonio J. and Dolcet-Sanjuan, Ramon (2011) *J. Amer. Soc. Hort. Sci.*, **136**(2):145–154.

Goreta Ban, S., Zanic, K. Dumicic, G., Raspudic, E., Vuletin Selak, G. and Ban, D. (2014) *Chilean J. Agric. Res.*, **74**: 29–34.

Goswami, R.K. and Sharma, S. (1997) *Hortic. J.*, **10**: 101-106.

Grassclly, D., Millot, P. and Aiauzct (1995) *Infos (Paris)*, **110**: 34-37.

Gray, D. J., McColley. D. W. and Michel, E. (1993) *J. Amer. Soc. Sci.*, **118**: 425-432.

Gray, D.J.and Elmstrom, G.W. (1991) *United States Patent*, YS 5007198, P. 4.

Grimstad, S.O. (1991) *Norsk. Land bruks for Sking.*, **5**: 333-341.

Groff, D.W. (1966) *Ph.D. Dissertation*, University of Arizona, Tucson.

Guan, W., Zhao, X., Dickson, D.W., Mendes, M.L. and Thies, J. (2014) *HortScience*, **49**: 1046- 1051.

Guis, M., Amor, M.D., Botondi, R., Ayub, R., Latchi, A., Bouzayen, M. and Pech, J.C. (1998) In: *Proc. of XV EUCARPIA Congress* (Eds. Scarascia, Mugnozza, G.T., Porceddu, E. and Pagnotta, M.A.), Viterbo, Italy.

Guis, M., Botondi, R., Ben Amor, M., Ayub, R., Bouzayen, M., Pech, J.C. and Guis, M., Latche, A., Pech, J.C. and Roustan, J.P. (1997) *Sci. Hort.*, **69**: 199-206.

Guis, M., Botondi, R., Ben-Amor, M., Ayub, R., Bouzayen, M., Pech, J.C. and Latche A. (1997) *J. Amer. Soc. Hort. Sci.*, **122**: 748–751.

GuJamuddin, Nazeer, A. and Ahmed, N. (2002) *Appl. Biol. Res.*, **4**(1-2): 31-38.

Guler, Z., Candir, E., Yetis¸ir, H., Karaca, F. and Solmaz, I. (2014) *J. Hort. Sci. Biotechnol.*, **89**: 448–452.

Guma, T. B., Kahia, J., Justus, O. and Peter, N. K. (2015) *Int. J. Res. Dev. Pharmacy and Life Sci.*, **4**: 1427-1433.

Guo, S., Zhang, J., Sun, H., Salse, J., Lucas, W.J., Zhang, H. *et al.* (2013) *Nat Genet.*, **45**: 51–58.

Guo, W.L., Chen, B.H., Chen, X.J., Guo, Y.Y., Yang, H.L. and Li, X.Z. (2018) *PLoS ONE*, **13**(1): e0190175.

Guo, S., Zhao, S. and Sun, H. (2019) *Nat Genet.*, **51**: 1616–1623.

Guo, Z.L. and Xu, Z. (1994) *China Vegetables*, **3**: 50.

Gwanama, C., Labuschagne, M.T. and Botha, A.M. (2000) *Euphytica*, **113**: 19-24.

Guha, A., Sinha, R.K. and Sinha, S. (2004) *Caryologia*, **57**: 117-120.

Gupta, N., Rathore, M., Goyary, D., Khare, N., Anandhan, S., Pande, V. and Ahmed, Z. (2012) *Biol. Plant.*, **56**: 57–63.

Gusmini, G. and Wehner, T.C. (2004) *Cucurbit Genet. Coop. Rep.*, **27**: 43-44.

Gwanama, C., Labuschagnc, M.T. and Botha, AM. (2000) *Euphytica*, **113**: 19-24.

Halsey, L.H. (1959-60) *Proc. Fla. St. Hort. Sci.*, **72:** 131-135.

Han, Taili, Feng Le Rongand and Shen Pei Rong (1999) *China Vegetables*, **2**: 27-28.

Han, K.P., Choi, S.K., Lee, D.K., Park, H.S. and Lee, C.H. (l986) *Res. Rep. Rural Develop. Administ.*, Horticulture, Korea Republic, **28**: 32-36.

Hang, S.D., Zhao, Y.P., Wang, G.Y. and Song, G.Y. (2005) *Vegetable Grafting*, China Agriculture Press, Beijng, China.

Hanna, H.Y. and Adams, A.J. (1993) *Bulletin-Louisiana Agric. Exp. Station*, **844**: 18.

Hanna, H.Y. and Adams, J. (1987) *HortScience*, **22**: 32-34.

Hao, L.H. and Wang, H.M. (1998) *Acta Agric. Boreali Sinica*, **13**: 112-115.

Haque, M.E., Sarkar, M.A.R., Mahmud, M.A., Rezwana, D. and Sikdar, B. (2008) *J. Bio-Sci.*, **16**: 67-71.

Harel-Beja, R., Tzuri, G., Portnoy, V., Lotan-Pompan, M., Lev, S., *et al.* (2010) *Theor. Appl. Genet.*, **121**: 511–533.

Harris, D.R. (1967) *Geographical Review*, **57**: 90-107.

Harris, K.R., Ling, K.S., Wechter, W.P. and Levi, A. (2009a) *J. Amer. Soc. Hort. Sci.*, **134**: 529–534.

Harris, K.R., Wechter, W.P. and Levi, A. (2009b) *J. Amer. Soc. Hort. Sci.*, **134**: 649–657.

Hashizume, T., Sato, T. and Hirai, M. (1993) *Japanese J. Breed.*, **43**: 367-375.

Hashizume, T., Shimamoto, I. and Hirai, M. (2003) *Theor. Appl. Genet.*, **106**: 779–785.

Hashizume, T., Shimamoto, I., Harushima, Y., Yui, M., Sato, T., Imai, T. and Hirai, M. (1996) *Euphytica*, **90**: 265-273.

Hawkins, L.K., Dane, F., Kubisiak, T.L., Rhodes, B.B. and Jarret, R.L. (2001) *J. Amer. Soc. Hort. Sci.*, **126**: 344–350.

Hazra, P. (2006) In: *Proceedings of the National Seminar on Cucurbits* (Eds. H.H. Ram,

H.P. Singh), College of Agriculture, Pantnagar, Uttaranchal, India, pp. 183-194.

Heij, G. and Uffclen, J.A.M. Van (1984) *Acta Hortic.*, **162**: 29-36.

Heiser, C.B. (1980) The Gourd Book, University of Oklahoma Press, Norman. Heiser, C.B., Schilling, E.E. (1990) In: *Biology and Utilization of Cucurbitaceae* (Eds. D.M.Bates, R.W. Robinson and Jeffrey), Cornell University Press, USA. pp. 120-133.

Helmy, Y.I., Singer, S.M., EI-Abd, S.O. and Abou-Hadid (1996) *Acta Hort.*, **434**: 361-366.

Hernández-González, Z., Sahagun-Castellanos, J., Espinosa-Robles, P., Colinas-Leon, M.T. and Rodriguez-Perez, J.E. (2014) *Revista Fitotecnia Mexicana*, 37: 41–47.

Herrington, M.E., Byth, D.E., Teakle, D.S. and Brown, P.J. (1989) *Aus. J. Exp. Agric.*, **29**: 253-259.

Heslop-Harrison, J. (1963) In: Meri stemsand Differentiation, Brook haven Symposia in Biology, No.16.

HeXun, H., Xiao Qi, Z., Zhen Chcng,W., Qing Huai, L. and Xi, L. (1998) *Scientia Hort.*, **74**: 175-181.

Hill, D.E. (1996) Agricultural Experiment Station, No. 933, P. 11.

Hisajima, S. and Yujl, A. (1989) *Japan. J. Trop. Agr.*, **33**: 1-5.

Hisajima, S., Aria, Y., Namwongprom, K. and Subhadrabandhu, S. (1989) *Japanese J. Trop. Agric.*, **33**: 1-5.

Hooghvorst, I., López-Cristoffanini, C. and Nogués, S. (2019) *Sci Rep.*, **9**: 17077.

Hooymansa, C.R., Bouwera. W., Orczykb J.J. and Dons, M. (1988) *Plant Science*, **57**: 63-71.

Hoque, A., Islam, R. and Arima, S. (2000) *Phytomorphology*, **50**: 267-272.

Hoque, M., Hossain, S., Alam, S., Arima and Islam, R. (2007) *Plant Tissue Cult. Biotech.*, **17**: 29-36.

Hossain, M.A., Islam, M. and Ali, M. (1996) *Euphytica*, **90**: 121–125.

Hossain, M.A., Sharfuddin, A.F.M., Monda, I.M.F. and Aditya, D.K. (1990) *Bangladesh J. Crop Sci.*, **1**: 53-56.

Houten, Y.M. and Van (1996) *Bulletin of BISROJ*, **19**: 59-62.

Hoyos Echebarría, P., Fernández, J.A., Martínez, P.F. and Castilla, N. (2001) *Acta Hortic.*, **559**: 139–143.

Hua, X.W. and Sen, X.D. (1999) *J. South China Agri. Univ.*, **20**: 20-23.

Huang, H., Wang, Z., Cheng, J., Zhao, W., Li, X., Wang, H., Zhang, Z. and Sui, X. (2013) *Scientia Hort.*, **150**: 206–212.

Huang, H., Yu, T., Li, J., *et al.* (2019) *J. Plant Biol.*, **62**: 203–216.

Huang, S., Li, R., Zhang, Z., Li, L., Gu, X., Fan, W., Lucas, W.J., Wang, X., Xie, B. and Ni, P. (2009) *Nat. Genet.*, **41**: 1275–1281.

Huang, Y., Bie, Z.L., He, S.P., Hua, B., Zhen, A. and Liu, Z.X. (2010) *Env. Exp. Bot.*, **69**: 32–38.

Huang, Y.C., Chiang, C.H., Li, C.M. and Yu, T.A. (2011) *Plant Cell Tissue Organ Cult.*, **106**: 21–29.

Huda, A.K.M.N. and Sikdar, B. (2006) *Plant Tissue Cult. Biotech.*, **16**: 31-36.

Huh, Y.C., Om, Y.H. and Lee, J.M. 2003. *Acta Hortic.*, **588**: 127–132.

Huitrón, M.V., Diaz, M., Diánez, F. and Camacho, F. (2007). *J. Food, Agric. Env.*, **5**: 344–348.

Hurd, P.D. Jr., Linsley, E.G. and Whitaker, T.W. (1971) *Evolution*, **25**: 218-234.

Huttner, E., Tucker, W., Vermeulen, A., Ignart, F., Sawyer, B. and Birch, R. (2001) *Curr. Issues Mol. Biol.*, **3**: 27–34.

Huyskens, S., Mendlinger, S., Benzioni, A. and Ventura, M. (1992) *J. Hort. Sci.*, **67**: 259-264.

Ibarra, L. and Flores, VJ. (1997) *Agrociencia*, **31**: 9-14.

Ibrahim, AM., Al-Sulirnan, A.I. and Al-Zcir, K.A. (1996) *HortScience*, **31**: 889-890.

Ikram-ul Haq, Arshad, Z. and Taseer (2013) *Int. J. Sci. Res.*, **6**: 81-84.

Ilarnanova, R.I. (1987) Turkmenistan SSR Ylymlar Akademijasynyn Habarlary, Biologik Ylymlaryn, No. 1, pp. 19-26.

Iluyskens, S., Mendlinger, S., Benzioni, A. and Ventura, M. (1993) *J. Hortic. Sci.*, **68**: 989-994.

Ito, L.A., Gaion L.A., Galatti, F.S., Leila, T., Jaime, B. and Santos, M. (2014) *Hortic. Bras.*, **32**: 297-302.

Imazu, T. (1949) *J. Jpn. Soc. Hort. Sci.* **18**: 6–42.

Imbumi, M.D. (2004) *Coccinia grandis* (L.) Voigt. In: *PROTA 2: Vegetables/Légumes* (Eds. G.J.H. Grubben, O.A. Denton) [CD-Rom], PROTA, Wageningen, The Netherlands.

Ishida, K.B., Turner, C., Chapman, H.M. and Mckeon, A.T. (2004) *J. Agric. Food Chem.*, **52**: 274–279.

Islam, A.K.M.R., Joarder, O.I., Rahman, S. M. and Hossain, M. (1992) *Indian J. Hort.*, **49**: 249-252.

Islam, M.O., Fakir, M.S.A. and Hossain, M.A. (1994) *Punjab Vegetable Grower*, **29**: 20-21.

Islam, S., Munshi, A.D, Verma, M., Arya, L., Mandal, B., Behera, T.K., Kumar, R. and Lal, S.K. (2011) *J. Hortic. Sci. Biotech.*, **86**(6): 661-667.

Islam, S., Munshi, A.D., Mandal, B., Kumar, R. and Behera, T.K. (2010) *Euphytica*, **174**(1): 83-89.

Ivanova, V.A.E. (1986) Nauchno – Teknicheskii Byulleten Vsesoyuznogo Ordena Leninai Ordena Druzby Narodov Nauc/1110 0 lssledovatel'skogo lnstituta Rastonievodstvaimeni, N.I. Vavilova, No. 165, pp. 77-78.

Izquierdo, J.A. and Menendez, R. (1980) Investigaciones Agronomicas, Centrode Investigae iones Agricolas' Alberto Boerger', **1**: 57-61.

Jadhav, S., Parmar, L.D. and Chauhan, R.M. (2015) 4th *International Conference on Agriculture and Horticulture,* **4**: 2.

Jagannathan, T. and Ramakrishnan, K. (1971) *Madras Agric. J.*, **58**: 331-337.

Jaiswal, J.P., Bhattaraj, S.P. and Subedi, P.P. (1997) Working paper – Lumle Agricultural Research Centre, No.97/48, P.6.

Jamatia, D. (2016) A thesis submitted to *Maharana Pratap University of Agricultural and Technology*, Udaipur, India.

Janaranjani, K.G., Kanthaswamy, V. and Kumar, S.R. (2016) *Int. J. Veg. Sci.*, **22**(5): 490-515.

Jang, H.A., Utomo, S.D., Kwon, S.Y., Ha, Sun-Hwa, Xing-guo, Y. and Choi, P.S. (2016) *J. Plant Biotechnol.*, **43**:341–346.

Jang, Y., Mun, B., Do, K., Um, Y. and Chun, C. (2014) *Hort. Environ. Biotechnol.*, **55**: 387-396.

Jarl, C.I., Bokelmann, G.S. and De Haas, J.M. (1995) *Plant Cell Tissue Org. Cult.*, **43**: 259-265.

Jarosik, V. and Pliva, J. (1995) *Acta Societatis Zoological Bohemicae*, **59**: 117-186.

Jarret, R.L. and Newman, M. (2000) *Genet. Resour. Crop Evol.*, **47**: 215–222.

Jarret, R.L., Merrick, L.C., Holms, T., Evans, J. and Aradhya, M.K. (1997) *Genome*, **40**: 433-441.

Jasmine, R. and Mian, M.A.K. (2016) *J. Biosci. Agric. Res.*, **9**: 796-803.

Jassal, N.S. and Nandpuri, K.S. (1972) *J. Res. Punjab Agric.Univ.* Ludhiana, **91**: 242-247.

Jat, G.S, Singh, B., Tomar, B.S., Singh, J., Ram, H. and Kumar, M. (2016) *J. Nat. Appl. Sci.*, **8**(4): 2111-2115.

Jat, G.S. (2011) *M.Sc. Thesis*, Indian Agricultural Research Institute, New Delhi-110012.

Jat, G.S., Munshi, A.D., Behera, T.K. and Tomar, B.S. (2016) *Indian J. Agric. Sci.*, **86**(3): 399–403.

Jat, G.S., Munshi, A.D., Behera, T.K., Choudhary, H. and Dev, B. (2015) *Indian J. Hort.*, **72**(4): 494-499.

Jat, G.S., Munshi, A.D., Behera, T.K., Singh, A.K. and Kumari, S. (2017a) *Chem. Sci. Rev. Lett.*, **6**(22): 1075-1079.

Jat, G.S., Singh, B, Tomar, B.S., Muthukumar, P. and Kumar, M. (2017b) *Indian Hortic.*, **62**(2): 34-37.

Jat, G.S., Singh, B., Tomar, B.S. and Ram, H. (2015) *Seed Res.*, **42**(2): 139-145.

Jat, G.S., Munshi, A.D., Behera, T.K and Bharadwaj, C. (2018) *Int. J. Hortic. Sci.*, **12**(2): 193-197.

Jeffrey, C. (1962) *Kew Bull.*, **15**: 337–371.

Jeffrey, C. (1967) *Kew Bull*, **20**: 417–426.

Jeffrey, C. (1980) *Bot. J. Linn. Soc.*, **81**: 233–247.

Jeffrey, C. (1983) In: *Proc. Conference on the Biology and Chemistry of Cucurbitacae*, Cornell University, Ithaca, New York, August 1980.

Jeffrey, C. (1990) In: Biology and Utilisation of Cucurbitaceae (Eds. D.M. Bates, R.W. Robinson, and C. Jeffrey), Cornell University Press, Ithaca, NY, USA, pp. 3–9.

Jeffrey, C. (2001) Cucurbitacae (Citrullus. In: *Mansfeld's Encyclopedia of Agricultural and Horticultural Crops (except ornamentals)* (Ed. P.Hanelt),Springer, New York, NY, USA, pp. 1533–1537.

Jeffrey, C. (2005) *Bot. Zhurn.*, **90**: 332–335.

Jelaska, S. (1972) *Planta*, **31**: 257-261.

Jelaska, S. (1974) *Physiol. Plantarum*, **31**: 257-261.

Jelaska, S., Juretick, B., Krsnik-Rasol, M., Papes, D. and Bosnjak, V. (1986) *Proc.

IAPTC Cong., Minneapoli*s, p. 106.

Jeon, H.J., Been, C.G., Hong, K.H., Om, Y.H. and Kim, B.D. (1994) *J. Korean Soc. Hort. Sci.*, **35**: 449-456.

Ji, G., Zang, J., Gong, G. and Xu, Y. (2015) *Sci Hort.*, **193**: 367-373.

Jie, Z., Jianting, S., Gaojie, J., Haiying, Z., Guoyi, G., Shaogui, G., Yi, R., Jianguang, F., Shouwei, T. and Yong, X. (2017) *Hort. Plant J.*, **3**: 91-100.

Jian Hua, Z. and Xin Hua, F. (1995) *Plant Physiol. Commun.*, **31**: 334-346.

Jifon, J.L. and Crosby, K.M. (2008) *Acta Hortic.*, **782**: 329–333.

Jing, L., Xiao Min, H., Gao Hong, Z., Ming Huang, C. and Yong Tang, Z. (2008) *Chin. J. Nat. Med.*, **6**(5): 372–376.

Joseph, B. and Jini, D. (2013) *Clinical Intervention in Aging*, **2**(2): 219–236.

Juan Li, J., Tang, Y., Qin, Y., Li, X. and Li, H. (2012) *Afr. J. Biotechnol.*, **11**: 6450–6456.

Juretic, B. and Jelaska, S. (1991) *Plant Cell Rep.*, **9**: 623-626.

Juretic, B., Katavic, V. and Jelaska, S. (1989) *Acta Botanica Croatica*, **48**: 27-34.

Kaddi G, Tomar B.S, Singh, B. and Kumar, S. (2014) *Indian J. Agr. Sci.*, **84**(5): 624–627.

Kale, P.B. and Seshadri, V.S. (1981) *Veg. Sci.*, **8**: 15-24.

Kale, V.S., Patil, B.R., Bindu, S. and Paithankar, D.H. (2002) *J. Soils Crops*, **12**: 213-216.

Kalloo, G. (1994) Vegetable Breeding. Combined Edn. (Vol I, II, III), Panima Educational Book Agency, New Delhi, pp. 26-27.

Kalyanrao, Tomar, B.S. and Singh, B. (2012) *Seed Res.*, **40**(2): 139–144.

Kalyanrao, Tomar, B.S. and Singh, B. (2014) *Indian J.l Hortic.*, **71**(3): 428–432.

Kano, Y., Goto, H., Fukuda, H. and Ishimoto, K. (2000) *Environ. Contr. Biol.*, **38**: 55-62.

Kano, Y., Yamabe, M. and Ishimoto, K. (1997) *J. Japanese Soc. Hort. Sci.*, **66**: 321-329.

Kano, Y., Yamabe, M., Ishimoto, K. and Fukuda, H. (1999) *J. Japanese Soc. Hort. Sci.*, **68**: 391-396.

Kao, K.N. and Michayluk, M.R. (1974) *Planta*, **115**: 355–367.

Kapadia, C., Patel, N., Patel, N. and Ahmad, T. (2018) *J. Experimental Biol. and Agric. Sci.*, **6**: 599–605.

Karaca, F., Yetisir, H., Solmaz, İ., Andir, E., Kurt, Ş., Sariand, N. and Guler, Z. (2012) *Turk. J. Agric. For.*, **36**: 167–177.

Karchi, Z., Katzir, N. and Paris, H.S. (2000) *Acta Hortic.*, **510**: 13-17.

Karim, M.A. (2013) *J. Life Sci. Technol.*, **1** : doi: 10.12720/jolst.1.1.79-83.

Karim, M.A. and Ahmed, S.U. (2010) *Int. J. Environ. Sci. Dev.*, **1**: 10-14.

Karim, M.A. and Ullah, M.A. (2011) *In-vitro* regeneration of teasle gourd. 2nd *Int. Conference on Biotech. Food Sci.*, **7**: 144-148.

Karim, M.A. and Ullah, M.A. (2011) IPCBEE, Vol. 7, IACSIT Press, Singapore. Karmakar, P., Munshi, A.D, Behera, T.K., Kumar, R. and Sureja, A.K. (2012) Extended Summary Vol 3, 3rd International Congress, Nov. 26-30, 2012, New Delhi. pp. 1038-1039.

Karmakar, P., Munshi, A.D., Behera, T.K. and Sureja, A.K. (2014) *Indian J. Hort.*, **71**(1): 61-66.

Kasrawi, M. (1989) *J. Horti. Sci.*, **64**: 573-579.

Kasrawi, M.A. (1988) *Dirasal*, **15**: 69-78.

Katavic, V., Jelaska, S., Bakran Peticioii, T. and David, C. (1991) *Plant Cell Tissue Org. Cult.*, **24**: 35-42.

Kausar, M., Parvin, S., Haque, M.E., Khalekuzzaman, M., Sikdar, B. and Islam, M.A. (2013) *J. Life Earth Sci.*, **8**: 17-20.

Kawale, M.V. and Choudhary, A.D. (2009) *Indian J. Plant Physiol.*, **14**: 116-123.

KaŸmiñska, K., Hallmann, E. and Rusaczonek, A. (2018) *Mol. Breed.*, **38**: 114.

Keinath, A.P. (2014) Clemson University Extension Information Leaflet, p. 90.

Keinath, A.P. and Hassell, R.L. (2014) *Plant Dis.*, 98: 255-266.

Kell, K., Jaksch, T. and Schmidt, U. (2000) *Gemuse (Munchen)*, **36**: 60-62.

Keng, C.L. and Hoong, L.K. (2005) *Biotechnology*, **4**: 354-357.

Kennard, W.C. and Havey, M.J. (1995) *Theor. Appl. Genet.*, **91**: 53–61.

Kennard, W.C., Poetter, K., Dijkhuizen, A., Meglie, V., Staub, J.E. and Havey, M.J. (1994) *Theor. Appl. Genet.*, **89**: 42-48.

Khalekuzzaman, M., Khatun, M.M. Rashid, M.H., Sheikh, M.I., Sharmin, S.A. and Alam, I. (2012) *Braz. Arch. Biol. Tech.*, **55**: 335-340.

Khalifa, H. (1995) *Pure Appl. Sci.*, **22**: 1201-1208.

Khatun, M.M., Hossain, M.S., Haque, M.A. and Khalekuzzaman, M. (2010b) *J. Bangladesh Agril. Univ.*, **8**: 203–206.

Khatun, M.M., Hossain, M.S., Khalekuzzaman, M., Rownaq, A. and Rahman, M. (2010a) *Int. J. Sustain. Crop Prod.*, **5**: 25-29.

Khoshoo, T.N. (1995) *Curr. Sci.*, **24**: 377-378.

Khoshoo, T.N. and Vij, S.P. (1963) *Caryologia*, **16**: 544-522.

Khristov, B. (1984) *Pochvoznanie I Agrokhimiya*, **19**: 94-99.

Khristov, K.H. (1997) *Rasteniev dni Nauki*, **34**: 52-54.

Kielkowska, A. and Havey, M.J. (2012) *Plant Cell Tiss. Organ. Cult.*, **109**:73-82.

Kihara, H. (1951) *Proc. Amer. Soc. Hort. Sci.*, **58**: 217-230.

Kijima, J. (1933) *J. Okitsu Hort. Soc.*, **29**: 111–115.

Kijima, J. (1938) *J. Okitsu Hort. Soc.*, **34**: 57–71.

Kim, D.S., Park, H.C., Chun, S.J., Yu, S.H., Choi, K.J., Oh, J.H., Shin, K.H., Koh, Y.J.

Kim, B.S., Hahm, Y.I. and Chung, B.K. (1999) *Plant Pathol. J.*, **15**: 48-52.

Kim, H.T., Kang, N.J. and Kang, K.Y. (1998) *RDA J. Hort. Sci.*, **40**: 158-161.

Kim, I.S., Yoo, K.C., Fujieda, K. and Okubo, H. (1994) *J. Kor. Soc. Hort. Sci.*, **35**: 195-200.

Kim, K., Chang Kil Kim and Jeung-Sul Han (2010) *Plant Biotechnol. Rep.*, **4**: 101–107.

Kim, M.S., Ban, C.D., Kang, K.Y. and Kim, H.T. (1984) Research Reports Office of the Rural Development, S. Korean, **26**: 73-75.

Kimura, K. and Sugiyama (1961) *Japanese J. Bot.*, **36**: 68-70.

Kintzios, S.E., Hiureas, G., Shortsianitis, E., Sereti, E., Blouhos, P., Manos, C., Makri, O., Taravira, N., Drossopulos, J.B., Holevas, C.D. and Drew, R.A. (1998) *Acta Hort.*, **461**: 427-432.

Kireva, R., and Kaleheva, S. (1997) *Rateniev dni Nauki*, **34**: 53-56.

Kirkbride, J.H. (1993) Parkway, Boone, NC, USA.

Kishimoto, K., Nishizawa, Y., Tabei, Y., Hibi, T., Nakajima, M. and Akutsu, K. (2002) *Plant Sci.*, **162**: 655–662.

Kivi, S.S., Salehi, R. and Liaghat, A. (2014) *Acta Hortic.*,**1038**: 573–578.

Klosowski, E.S., Lunardi, D.M.C. and Sandanielo, A. (1999) *Revista Brasieira de Engenhart Agricola e ambiental*, **3**: 409-412.

Knysh, A.N. and VVakulenko, R.I. (1976) *Agrokhimiya*, No. 2, pp. 100-101.

Kodama, H., Irifune, K., Kamada, H. and Morikawa, J. (1993) *Transfenic Res.*, **2**: 147-152.

Kokalis-Burelle, N. and Rosskopf, E.N. (2011) *J. Nematol.*, **43**: 166-171.

Kole, C., Olukolu, B.A., Kole, P., Rao, V.K., Bajpai, A., Backiyarani, S., Singh, J., Elanchezhian, R. and Abbott Albert, G. (2012) *J. Plant Sci. Mol. Breed.*, **1**: 1.

Komal, R. (2011a) *Biotechnol. Bioinf. Bioeng.*, **1**: 59-62.Komal, R. (2011b) *Indian J. Agric. Res.*, **45**: 140-145.

Komal, R. (2011c) *African J. Biotech.*, **10**: 9-12.

Kondo, N. (1955) Studies on the triploid watermelon. The Institute for Breeding Research, Tokyo Agricultural University.

Korzeniewaka, A., Galecka, T. and Niemirowiez-Syzyytt (2000) *Acta Hortic.*, **510**: 269-271.

Kousik, C.S., Donahoo, R.S. and Hassell, R. (2012) *Crop Prot.*, **39**: 18–25.

Kousik, C.S., Levi, A., Ling, K.S. and Wechter, W.P. (2008) *HortScience*, **43**: 1359–1364.

Koutsika-Sotiriou, M., Traka-Mavrona, E., Tsivelikas, A.L., Mpardas, G., Mpeis, A. and Klonari, E. (2004) In: Progress in Cucurbit Genetics and Breeding Research, Proc. of Cucurbitaceae 2004 (Eds. A.Lebeda, and H.S.Paris), the 8th EUCARPIA meeting on Cucurbit Genetics and Breeding, Olomouc, Czech Republic, July 12-17, 2004, pp. 163-167.

Kroon, G.H., Custers, J.B.N., Kho, Y.O., Den Nijs, A.P.M. and Varekamp, H.Q. (1979) *Euphytica*, **28**: 723-728.

Krsnik-Rasol, M., Ja1aska, S. and Sermzm, D. (1982) *Acta Bot Croatica*, **41**: 33-39.

Krug, H. and Liebig, H.P. (1991) *Acta Hortic.*, **287**: 427-430.

Krug, M.G.Z., Stipp, L.C.L., Rodriguez, A.P.M. and Mendes, B.M.J. (2005) *Pesq. Agropec. Bras., Brasília*, **40**: 861-865.

Kubicki, B. (1965) *Genet. Polonica*, **6**: 241-250.

Kubicki, B. (1966) *Genet. Polonica*, **7**: 27-29. Kubicki, B. (1969a) *Genet. Polonica*, **10**: 23-68.

Kubicki, B. (1969b) *Genet. Polonica*, **10**: 69-86.

Kubicki, B. (1969c) *Genet. Polonica*, **10**: 87-99.

Kubicki, B. (1969d) *Genet. Polonica*, **10**: 123-143.

Kubicki, B. (1969e) *Genet. Polonica*, **10**: 145-163.

Kubicki, B., Sollysiak, U. and Korezcniewska. A. (1986a) *Genet. Polonica*, **27**: 273-287.

Kubicki, B., Soltysiak. U. and Korezcniewska. A. (1986b) *Genet. Polonica*, **27**: 289-298.

Kuchkarov, S.K., Shchukina, A.S. and Pcstsova. S.T. (1982) Uzbekskngo Nauclno-issledovatel' skogo Instituta Ovoshche - bakhchevykh Kul 'tur Kartcfelva, No. 20. pp. 32-34.

KucHyon, H., YoungHyun, O., YoonChan, Hco and KwanDal, KO. (1998) *RDA J. Hort. Sci.*, **40**: 166-168.

Kuijpers, A.M., Bouman, H. and De'Klerk. G.J. (1996) *Plant Cell Tissue Organ Cult.*, **46**: 81-83.

Kulkarni, G.R. (1999) *Ph.D. Thesis,* submitted to Mahatma Phule Krishi Vidyapeeth, Rahuri, India.

Kumar, P., Shaunak, I., Thakur, A.K. and Srivastava, D.K. (2017) *J. Genetics Genomes,* **1**: 102.

Kumar, J., Arora, S.K. and Mehra, R. (1988) *Crop Res.*, **1**: 124-127.Kumar, L.S.S. and Deodikar, G.B. (1940) *Curr. Sci.*, **9**: 128-130.

Kumar, L.S.S. and Viseveshwaraiah, S. (1952) *Nature*, **170**: 330-331.

Kundu, B.C. (1942) *J. Bombay Nat. Hist. Soc.*, **43**: 362-368.

Kunzelmann, G. and Paschold, P.J. (1999) *Gemiise (Miinchen)*, **35**: 232-235.

Kuo, C.T., Lee, T.C. and Tsai, P.L. (1999) *J. Chinese Soc. Hort. Sci.*, **45**: 152-158.

Kurtar, E. S., Uzun, S. and Escndal, E. (1999) Ondokuzmayis Universitesi, Ziraat Fakultesi Dergisi, **14**: 33-45.

Kvasnikov, B.V., Rogova, N.T., Tarakonova, S.I. and Ignatova, I. (1970) *Trudy-po-Prikladnoi-Botanike-Genetiki-ISelektsii*, **42**: 45-57.

Kwack, S.N. and Fujieda, K. (1988) *J. Japanese Soc. Hort. Sci.*, **57**: 34-42.

KynEl, M. and Jasa, B. (1965) *First commmuicatio*ns, SI, vgs, sk, zemed. v. Brne, Rada H., N0. 3, pp. 417-424.

Laamin, M., Ait-Oubahou, A. and Benichou, M. (1995) In: *Proc. of an International Symposium* Agadir (Eds. A.Ait-Oubahou and M.El-Olmani), Morocco, January, 16-21, 1994.

Lal, T. and Dhaliwal, M.S. (1993) In: *Symp. on Heterosis Breeding in crop plants: Theory and Application,* Ludhiana, India, pp. 34-35.

Lal, T. (1995) *Punjab Vegetable Grower,* **30**: 47-49.

Lancaster, D.K., Johnson, C.E. and Young, W.A. (1987) *Louisiana Agriculture*, **30**: 7.

Laxuman, S.A., Patil, P.M., Salimath, P.R., Dharmatti, A.S., Byadgi and Nirmalayenagi (2012) *Karnataka J. Agric. Sci.*, **25**(1): 9-13.

Lcippik, E. (1966) *Euphytica*, **15**: 323-328.

Lebeda, A. and Kristova, E. (1996) *Genet. Resour. Crop Evol.*, **43**: 79-84.

Lebedeva, A.T. (1996) *Karlofel I Ovoshchii*, **4**: 16-18.

Lebedeva, L.A. and Pal'm, G.G. (1986) In: *Biologicheskii Krugovorot Veshchestv v Zemledem Kazan* USSR, 122-127.

Leber, B. and Heck, M. (1991) *Gemiise (Miinchen)*, **27**: 144-147.

Leclerc, R.F. (1987) Diss. Abstr. *Internatl. B-Sciences-and-Engineering*, **47**: 1, 4773B.

Lecoq, Herve'and Desbiez, Ce'cile (2012) In: *Advances in Virus Research*, Volume 84, DOI: 10.1016/B978-0-12-394314-9.00003-8, pp. 67-126.

Lecouviour, M., Pitrat, M. and Risser, G. (1990) *Cucurbit Genet. Coop. Rep.*, **13**: 10-13.

Lee, C.H., Son, K.A. and Chcon, S.G. (1998) RDA *J. Crop Protection*, **40**: 124-128.

Lee, H.S., Kwon, E.J., Kwon, S.Y., Jeong, Y.J., Lee, E.M., Jo, M.H., Kim, H.S., Woo, I.S., Shinmyo, A., Yoshida, K. and Kwak, S.S. (2003) *Mol. Breed.*, **11**: 213–220.

Lee, J. M. (1994) *Hort. Sci.*, **29**: 235–239.

Lee, J. M. and Oda, M. (2003) *Hortic. Rev.*, **28**: 61–124.

Lee, J.M., Bae, E.J., Lee, C. W., Kwong, S.S. (1999) *Acta Hortic.*, **484**: 125-132.

Lee, K. and Kang, Y.J. (1998) *J. Korean Soc. Hort. Sci.*, **39**: 402-407.

Lee, S.J., Shin, J.S., Park, K.W. and Hong, Y.P. (1996) *Theor. Appl. Genet*, **92**: 719-725.

Lee, W.S., Lee, H.Y., Know, H.J., Hwang, H.S., M., H.A. and Hong, S.Y. (1995) *J. Korean Soc. Hort. Sci.*, **36**: 199-210.

Lee, Y.H., Icon, H.J., Hong, K.H. and Kim, B.D. (1995) *J. Korean Soc. Hort. Sci.*, **36**: 323-330.

Lehmann, L. (1996) *Sveriges Utsadesforenings Tidskrift*, **106**: 97-107.

Leijon, S. and Olsson, K. (1999) *Sveriges Utsadesfarenings Tidskrift*, **109**: 35-40.

Leippik, E. (1967) *Adu Frontiers of Pl. Sci.*, **19**: 43-50.

Lelievre, J.M., Amor, M.B., Flores, B., Gomez, M.C., E1-Yahyaoui, F., Chatenct, C. Du, Perin, C. Hernandez, J.A., Romojaro, F., Latche, A., Bouzayen, M., Pitrat, M., Dogimont, C. and Pech, I.C. (2000) *Acta Hort.*, **510**: 499-506.

Lenser, T. and Theiben, G. (2013) *Trends Plant Sci.*, **18**: 704–714.

Levi, A., Davis, A.R., Hernandez, A., Wechter, W.P. and Thimmapuram, J. (2006) *Plant Cell Rept.*, **25**: 1233–1245.

Levi, A., Thomas, C.E., Keinath, A.P. and Wchner, T.C. (2000) *Acta Hortic.*, **510**: 385-390.

Levi, A., Thomas, C.E., Keinath, A.P. and Wehner, T.C. (2001a) *Genet. Resour. Crop. Evol.*, **48**: 559–566.

Levi, A., Thomas, C.E., Wehner, T.C. and Zhang, X. (2001b) *HortScience*, **36**: 1096–1101.

Levi, A., Thomas, C.E., Zhang, X., Joobeur, T., Dean, R.A, Wehner, T.C. and Carle, B.R. (2001c) *J. Amer. Soc. Hort. Sci.*, **126**: 730–737.

Levi, A., Thomas, C.E., Joobeur, T., Zhang, X. and Davis, A. (2002) *Theor. Appl. Genet.*, **105**: 555–563.

Levi, A., Thomas, C.E., Newman, M., Reddy, O.U.K., Zhang, X. and Xu, Y. (2004) *J. Am. Soc. Hort. Sci.*, **129**: 553–558.

Levi, A., Thomas, C.E., Thies, J.A., Simmons, A.M., Ling, K. and Harrison Jr, H.F. (2006) *HortScience*, **41**: 463–464.

Levi, A., Wechter, P. and Davis, A. (2009) *Plant Genet. Resour.*, **7**: 16–32.

Lewinsohn, E. and Schaffer, A.A. (2010) In: *Horticulture Reviews*(Ed. J. Janick), Wiley-Blackwell, **36**: 165–198.

Li, C.H., Li ,Y.S., Bai, L.Q., Zhang, T.Y., He, C.X., Yan ,Y. and Yu, X.C. (2014a) *Physiologia Plantarum*, **151**: 406– 422.

Li, H., Liu, S.S., Yi, C.Y., Wang, F., Zhou, J., Xia X.J. and Yu J.Q. (2014b) *Plant, Cell Env.*, **37**: 2768– 2780.

Li, H.N. (2014) Anthracnose of cucumber. http://www.ct.gov/caes

Li, H., Wang, F., Chen, X.J., Shi, K., Xia, X.J., Considine, M.J., Yu, J.Q. and Zhou, Y.H. (2014c) *Physiologia Plantarum*, **152**: 571–584.

Li, J., Li, X.M., Qin, Y.G., Tang, Y., Wang, Ma, C. and Li, H.X. (2011) *Afr. J. Biotech.*, **10**: 9760-9765.

Li, N., Shang, J., Wang, J. and Zhou, D. (2020) *Front. Plant Sci.*, **11**: 116.

Li, R., Fan, W., Tian, G., Zhu, H., He, L. and Cai, J. (2010) *Nature*, **463**: 311–317.

Li, X.E., Liu, F.Y., Huang, Y., Kong, Q.S., Wan, Z.J., Li, X. and Bie, Z. (2012) *Acta Hortic.*, **1086**: 59-64.

Li, Y.H.,Yang, L.M., Pathak, M., Li, D.W., He, X.M. and Weng, Y. (2011b) *Theor. Appl. Genet.*, **123**: 973–983.

Li, Z., Zhang, Z., Yan, P., Huang, S., Fei, Z. and Lin, K. (2011a) *BMC Genomics*, **12**: 540.

Li, J., Li, X.M., Qin, Y.G., Tang, Y., Wang, Ma, C. and Li, H.X. (2011b) *Afr. J. Biotech.*, **10**: 9760-9765.

Li, N., Shang, J., Wang, J. and Zhou, D. (2020) *Front. Plant Sci.*, **11**: 116.

Li, Y.H., Yang, L.M., Pathak, M., Li, D.W., He, X.M. and Weng, Y. (2011c) *Theor. Appl. Genet.*, **123**: 973–983.

Li, Z., Zhang, Z., Yan, P., Huang, S., Fei, Z. and Lin, K. (2011d) *BMC Genomics*, **12**: 540.

Lim, W. and Earle, E.D. (2008) *Plant Cell Tiss. Org. Cult.*, **95**: 115-124.

Lim, W. and Earle, E.D. (2009) *Plant Cell Tiss. Org. Cult.*, **98**: 351-356.

Lin, C.Y., Ku, H.M., Chiang, Y.H., Ho, H.Y., Yu, T.A. and Jan, F.J. (2012) *Transgenic Res.*, **21**: 983–993.

Ling, K., Harris, K.R., Meyer, J.D., Levi, A., Guner, N., Wehner, T.C., Bendahmane, A. and Havey, M.J. (2009) *Theor. Appl. Genet.*, **120**: 191–200.

Ling, K.S., Levi, A. (2007) *HortScience*, **42**: 1124–1126.

Liu, Chunhong, Qin,Zhiwei, Zhou, Xiuyan, Xin, Ming, Wang, Chunhua, Liu, Dong and Li, Shengnan (2018) *BMC Plant Biol.*, **18**:16 DOI 10.1186/s12870-018-1236-2.

Liu, G., Liu, L.W., Gong, Y.Q., Wang, Y., Yu, F., Shen, H. and Gui, W. (2007a) *Seed Sci. Technol.*, **35**: 476–485.

Liu, H.Y., Zhu, Z.J., Lu, G.H. and Qian, Q.Q. (2003a) *Scientia Agriculturae Sinica*, **36**: 1325–1329.

Liu, J.L., Song, X.H., Zhou, S.C. and Wang, J.S. (2016a) *Mol. Plant Breed.*, **4**: 23–26.

Liu, L., Gu, Q., Ijaz, R., Zhang, J. and Ye, Z. (2016b) *Sci. Hort.*, **205**: 32–38.

Liu, L.W., Wang, Y., Gong, Y.Q., Zhao, T.M., Liu, G., Li, X.Y. and Yu, F.M. (2007b) *Sci. Hortic.*, **115**: 7–12.

Liu, X., Liu, B., Xue, S., Cai, Y., Qi, W., Jian, C., Xu, S., Wang, T. and Ren, H. (2016c) *Front. Plant Sci.*, **7**. https://doi.org/10.3389/fpls.2016.01652.

Liu, Y.Q., Liu, S.Q. and Wang, H.B. (2004a) *Agri Sci.*, **4**: 30–31.

Liu, Y.Q., Liu, S.Q. and Wang, H.B. (2004b) *Acta Agriculturae* Shanghai, **20**: 62–64.

Liu, Y.Q., Liu, S.Q., Yang, F.J. and Li, D.F. (2003b) *Acta Agriculturae Boreali-Occidentalia Sinica*, **12**: 105–108.

Liu, W. (1998) *Fruit Sci.*, **15**: 336-339.

Lofti, M., Kashi, A. and Onsinejad, R. (1997) *Acta Hort.*, **492**: 323–328 .

Longvah, T., Ananthan, R., Bhaskarachary, K. and Venkaiah, K. (2017) Indian Food Composition Tables, National Institute of Nutrition, Hyderabad.

Lou, H., Obam Okeyu, P., Tamaki, M. and Kako, S. (1996) *J. Hort. Sci.*, **71**: 497-502.

Love, S.L., Rhodes, B.B. and Nugenl, P.E. (1986) *Euphytica*, **35**: 633-638.

Lower, R.L. Wehhner, T.C. and Jenkins, S.F.Jr. (1991) *HortScience*, **26**: 77-78.

Loy, J.B. (2004) *Crit. Rev. Plant Sci.*, **23**(4): 337–363.

Loy, J.B., Natti, T.A., Zack, C.D. and Fritts, S.K. (1979) *Amer. Soc. Hort. Sci.*, **104**: 100-101.

Lu, Y.H., Shen, Y.M., Liu. K.C. and Pang, J.A. (1996) *J. Shandong Agril. Univ.*, **27**: 39-43.

Luo, S.B., Zhou, W.B. and Nakashima, T. (1996) *China Vegetables*, **5**: 25-27.

Lv, J., Qi, J.J., Shi, Q.X., Shen, D., Zhang, S.P. and Shao, G.J. (2012) *PLoS One*, **7**: e46919.

Ma DcHua, Hou, F. and Lu, S.Z. (1998) *China Vegetables*, **2**: I7-19.

Ma DeHua, Lu, ShuZhcn and Zhen Rong (1997) *China Vegetables*, **6**: 21-23.

Mackicwicz, H.O., Malepsz, S., Sarreb, D.A. and Narkicwicz, M. (1998) *Gartenbauwissenschafi*, **63**: 125-129.

Maezawa, S. and Akimoto, K. (1996) *Research Bulletin of the Faculty of Agriculture*, Gifu University, No. 61, pp. 81-86.

Magdum, M.B. and Seshadri, V.S. (1983) *Proc. 15th International Congress of Genetics*, Part II. p. 562.

Mahazabin, F., Parvez, S. and Alam, M.F. (2008) *J. Biosci.*, **16**: 59-65.

Maity, T.K., Ghosh, B.K. and Som, M.G. (1995) In: *Proceedings National Symposium on Sustainable Agriculture in Sub-humid Zone*. (Eds. M.K. Dasgupta, D.C. Ghosh, D. Das Gupta, D.K. Majumdar, G.N. Chattopadhyay, P.K. Ganguli, P.S. Munsi and D.Bhatacharya), March 3-5, I995, pp. 180-184.

Malcpszy, S., Sarreb. D.A., Mackicwicz, H.O. and Narkicwicz, M. (1998) *Gartenbauwissensehaft*, **63**: 34-37.

Malex, M.A., Khanam, D., Khatun, M., Molla, M.H. and Mannan, M.A. (2010) *Bangladesh J. Agric. Res.*, **35**: 135-142.

Malik, B.S. (1965) *Indian J. Agron.*, **10**: 266-270.

Mallaiah, B. and Nizam, Z. (1986) *J. Indian Bot. Soc.*, **65**: 80.

Maluf, W.R., Moura, W.D.E.M., Silva, I.S.D.A. and Castelo-Branco, M. (1986) *Revista Brasileira de Genética*, **9**: 161-167.

Mandal, A. (2006) *Ph.D. Thesis*, Bidhan Chandra Krishi Viswavidyalaya, West Bengal, India, p. 110.

Manjunatha (2009) *Ph.D. Thesis. Indian Agricultural Research Institute*, New Delhi-110012.

Mansour, A.N. (1997) *Dirasat Agrie. Sci.*, **24**: 146-151.

Mansour, A.N., Akkawi, M. and Al-Musa, A. (2000) *Dirasat Agric. Sci.*, **27**: 1-9.

Maragal, S.Y., Singh, A.K., Behera, T.K. and Munshi, A.D. (2018a) *Veg.Sci.*, **45**(1): 38-41.

Maragal, S.Y., Singh, A.K., Behera, T.K., Munshi, A.D. and Sabir, N. (2018b) *IndianJ. Agric. Sci.*, **88**(11): 1801–1803.

Margaret, S., Maheswari, U., Ambethkar, Vasudevan, Sivanandhan and Selvaraj (2014) *Int. J. Innov. Res. Sci. Engg. Technol.*, **6**: 13876-13881.

Martin, A., Troadec, C., Boualem, A., Rajab, M., Fernandez, R., Morin, H., Pitrat, M., Dogimont, C. and Bendahmane, A. (2009) *Nature*, **461**: 1135–1138.

Martin, C., Schoen, L. and Arrufat, A. (1998) In: *First transnational workshop on biological, integrated and rational control status and perspectives with regard to regional and European experiences*, Lille, France, 21-23 January, 1998, Loos-en-Gohelle, France, Service Regional de la Protection des Vegetaux, Nolrd Pas-De-Calis (1998), pp. 31-32.

Martyniak-Przybvszewska, B. and Wierbicka, B. (1996) *Zeszyty Problemowe Postepow NaukRoImczyclz*, **429**: 237-240.

Mascarell-Creus, A., Cañizares, J., Vilarrasa-Blasi, J., Mora-García, S. and Blanca, J. (2009) *BMC Genomics*, **10**: 467.

Matonify, B. (1987) *Kerleszet es Syoleszet*, **36**: 6.

Matoria, G.R. and Khandelwal, R.C. (1999) *J. Appl. Hort.*, **1**: 139-141.

Matsumoto, S. (1931) Jissaiengei, Jissaiengeisya, Tokyo, Japan. **11**: 288–291.

Matsumura, H., Miyagi, N., Taniai, N., Fukushima, M., Tarora, K., Shudo, A. and Urasaki, N. (2014) *PLoS ONE*, **9**(1): e87138

Matsuo, S., Ishiuchi, D. and Kohyama, T. (1985) *Bull. Veg. Ornament. Crops Res. Sta.* (Kurume), **8**: 1-21.

Maynard, D.N. and Hochmuth, G.J. (2007) *Knott's Handbook for Vegetable Growers*. John Wiley and Sons Inc., Hoboken, NJ.

Maynard, D.N., Carle, R.B., Elmstrom, G.W., Ledesma, N. and Campbell, R.J. (2001) *Proc. Int.Am. Soc. Trop. Hort.*, **44**: 1-4.

McCreight, J. D. (1984) *Rep. Cucurbit Genet. Coop.*, **6**: 48.

McCreight, J.D. and Elstorm, G.W. (1984) *HortScience*, **19**: 268-270.

McCreight, J.D., Nerson, H. and Grumet, R. (1993) In: *Genetic Improvement of Vegetable Crops* (Eds. G. Kalloo, and B.O. Bergh), Pergamon press, Oxford, N.Y, pp. 267-294.

McGrath, G.T. and Shishkoff, N. (1999) *Crop Protec.*, **18**: 471-478.

McGrath, M.T. (2006) Update on managing downy mildew in cucurbits.McKay, J.W. (1931) *Univ. CA. Publ. Bot.*, **16**: 339.

Meeuse, A.D.J. (1962) *Bothalia*, **8**: 1-111.

Meeuse, A.D.J. (1958) *Blumea Suppl.*, **4**: 196-204.

Mehta, M.L., Singh, D.P. and Sachan, S.C.P. (1999) *Adv. Hort. Forestry*, **6**: 111-118.

Mehta, Y.R. (1959) In: *Vegetable Growing*, Uttar Pradesh Bureau of Agric. Information, Lucknow, India.

Memane, S.A. and Khetmalas, M.B. (2003) *J. Maharashtra Agric. Univ.*, **28**(3): 283–284.

Meshcherov, E.T. and Juldasheva, L.W. (1974) *Trudy-poPrikladnoi-Botanike-Genetiki-I-Selektsii*, **51**: 204-213.

Metcalf, R.I., Metcalf, R.A. and Rhodes, A.M. (1983) In: *Proc. Conference on the Biology and Chemistry of Cucurbitaceae Cornell University*, Ithaca, New York. August, 1980.

Metcalf, R.L., RhodeS, A.M (1990) In: *Biology and utilization of the Cucurbitaceae* (Eds. D.M. Bates, R.W. Robinson, and C.Jeffrey), Ithaca: Cornell University Press, pp. 167-182.

Metwally, E.I, Moustafa, S. A., El-Sawy, B. I. and Shalaby, T. A. (1998) *Plant Cell Tissue Org.*, **52**: 171-176.

Mgdany, M.A. Wadid, M.M., Abou Hadid, A.F. (1999) *Acta Hortic.*, **491**: 107-112.

Mia, B.M., Islam, M.S., Miah, M.Y., Das, M.R. and Khan, H.I. (2014) *Pak. J. Biol. Sci.*, **17**(3): 408-413.

Miao, H., Zhang, S.P., Wang, X.W., Zhang, Z.H. and Li, M. (2012) *Euphytica*, **172**: 167–176.

Migina, O.N. and Kuz'Mi1skuya, G.A. (1998) *Kartfel'i Ovoshchi*, **5**: 23.

Miguel, A. (2004) Proc. Fifth International Conference on Alternatives to Methyl Bromide. Lisbon, pp. 151–156.

Mishra, J.P. (1981) *Ph.D. Thesis*, Kanpur University, Kanpur, India.

Mishra, S., Pandey, S., Kumar, N., Pandey, V.P. and Singh, T. (2019) *J. Pharmacog. Phytochem.*, **8**(1): 29-38.

Misra, A.K. and Bhatnagar, SP. (1995) *Phytomorphology*, **45**: 47-55.

Misra, R.S., Kumar, R., Sirohi, R.K. and Misra, S.P. (1994) *Recent Hort.*, **1**: 58-60.

Mitra, P.K. and Nariani, T.K. (1965) *Indian Phytopath.*, **16**: 260-267.

Mohammadi, J. and Sivritepe, N. (2007) *J. Biol. Sci.*, **7**: 653-657.

Mohamoud, S.H. (1995a) *Assiut. J. Agric. Sci.*, **26**: 85-91.

Mohamoud, S.H. (1995b) *Assiut. J. Agric. Sci.*, **26**: 93-99.

Mohanty, B.K. (2001) *Indian Agric.*, **44**(3-4): 157-163.

Mohanty, B.K. and Mishra, R.S. (1999) *Indian J. Genet.*, **59**(4): 505-510.

Moideen, R.S. and Prabha, L.A. (2013) *Int. J. Sci. Res.*, pp. 2319-7064.

Moideen, R.S. and Prabha, L.A. (2014) *Int. J. Pharm Bio. Sci.*, **5**: 925–933.

Mokam, D.G., Djiéto-Lordon, C., Bilong Bilong, C.F. and Lumaret, J.P. (2018) *Afric. Entomol.*, **26**(2): 317–332.

Monforte, A.J., Oliver, M. and Gonzalo, M.J. (2004) *Theor. Appl. Genet.*, **108**: 750–758.

Mongkolporn, O., Dokmaihom, Y., Kanchana-udomkan, C. and Pakdeevaraporn, P. (2004) *J. Hortic. Sci. Biotechnol.*, **79**: 449–451.

Morales, M., Luis-Arteaga, M., Alvarez, J.M., Colcet-Sanjuan, R., Monfort, A., Arus, P. and Garcia-Mas, J. (2002) *J. Am. Soc. Hort. Sci.*, **127**: 540–544.

Morales, M., Orjeda, G., Nieto, C., van Leeuwen, H. and Monfort, A. (2005) *Theor. Appl. Genet.*, **111**: 914–922.

More, T.A. and Budgujar, C.D. (2002) *Acta Hortic.*, **588**: 255-260.

More, T.A. and Seshadri, V.S (1975) *Veg. Sci.*, **2**: 37-44.

More, T.A. and Seshadri, V.S. (1987) *Veg Sci.*, **14**: 138-142.

More, T.A. and Seshadri, V.S. (1998) In: *Cucurbits* (Eds. N.M. Nayar, and T.A. More), Oxford and IBH Publishing Co Pvt. Ltd. New Delhi, pp. 39-66.

More, T.A. and Seshadri, V.S. (1998) In: *Cucurbits* (Eds. N.M. Nayar, and T.A. More), Oxford and IBH Publishing Co. Pvt. Ltd. New Delhi, pp. 169-186.

More, T.A., Chandra, P. and Singh, J.K. (1990) *Indian J. Agric. Sci.*, **60**: 356-357.

More, T.A., Mishra, J.P., Sheshadri, V.S., Doshi, S.P. and Sharma, J.C. (1987) *Ann. Agril. Res.*, **8**: 237-242.

More, T.A., Sharma, S.C. and Mishra, J.P. (1991) In: *Golden Jubilee Symposium on Genetic Research and Education: Current Trends and Next Fifty Years of ISGPB*, New Delhi, Feb. 12-15, pp. 610-611.

Moreno, E., Fernandez-Silva, I. and Eduardo, I. (2008a) In: *Proceedings of the IXth Eucarpia Meeting on Genetics and Breeding of Cucurbitaceae* (Ed. M.Pitrat), Avignon, France, pp. 101–108.

Morishita, M., Sugiyama, K. and Saito, T. (2000) *Acta Hortic.*, **521**: 83-90.

Mosa, A.A. (1997) *Ann. Agric. Sci.* (Cairo), **42**: 241-255.

Mossler, M.A. and Nesheim, O.N. (2005) Florida Crop/Pest Management Profile: Squash. Electronic Data Information Source of UF/IFAS Extension (EDIS). CIR 1265. February 3, 2005. http://edis.ifas.ufl.edu/.

Moura M da C.C.L., Zerbini, F.M., da Silva, D.J.H. and de Queiróz, M.A. (2005) *Horticultura Brasileira*, **23**: 206-210.

Mukhamedkhanova, F.S., Faiziev, S.H.M., Engalycheva, S.S. and Abdukarimov, A.A. (1995) *Tsitologiya l Genetika,* **29**: 31-35.

Munger, H.M. (1990) *Rep. Cucurbit Genet. Coop.,* **13**: 4. Munger, H.M. (1942) *J. Am. Soc. Hortic. Sci.,* **40**: 405-410.

Munshi, A.D. and Sureja, A.K. (2013) In: Kalia, P. and Yadav, R.K. (Eds.) Entrepreneurship Development Programme (EDP) on Vegetable Seed Production Technology, Organized by Division of Vegetable Science and ZTM and BPD Unit, IARI, New Delhi, pp. 26-32.

Munshi, A.D and Verma, V.K. (1999) *Indian J. Agric. Sci.,* **69**(3): 214-216.

Munshi, A.D, Kumar, R. and Panda, B. (2006) *Indian J. Agric. Sci.,* **76**(12): 750-752.

Munshi, A.D. (2012) In: *Breeding for Productivity and Industry Suitable Food Colorants and Bioactive Health Compound in Vegetable Crops* (Eds. P. Kalia, and T.K. Behera), *Conventional and Hi-tech Cutting Approaches.* Winter School Sponsored by Indian Council of Agricultural Research (ICAR), Division of Vegetable Science, I.A.R.I, New Delhi, pp. 173-184.

Munshi, A.D. and Alvarez, J.M. (2004) *J. New Seeds,* **6**(4): 323-361.

Munshi, A.D. and Choudhary, H. (2012) In: *Breeding for Productivity and Industry Suitable Food Colorants and Bioactive Health Compound in Vegetable Crops* (Eds. P. Kalia, and T.K. Behera), *Conventional and Hi-tech Cutting Approaches.* Winter School Sponsored by Indian Council of Agricultural Research (ICAR), Division of Vegetable Science, I.A.R.I, New Delhi, pp. 78-90.

Munshi, A.D. and Choudhary, H. (2014) In: Peter, K.V. and Hazra, P. (Eds.). *Handbook of Vegetables Volume III,* Studium Press. pp. 271-310.

Munshi, A.D. and Kumar, R. (2007) *Indian Hort.,* **52**(2): 19-21.

Munshi, A.D. and Tomar, B.S. (2013) In: Kalia, P. and Behera, T.K. (Eds.). *Vegetable Seed Production Technques and Post –Harvest Handling of Seedsi.* Study tour cum training Programme sponsored by Food and Agriculture Organisation. Division of Vegetable Science, I.A.R.I, New Delhi, P. 21-26.

Munshi, A.D. and Verma, V.K. (1997) *Veg. Sci.,* **24**: 103-106.

Munshi, A.D. and Verma, V.K. (1998) *Veg. Sci.,* **25**: 93-94.

Munshi, A.D. and Verma, V.K. (1999) *Indian J. Agric. Sci.,* **69**(3): 214-216.

Munshi, A.D. and Verma, V.K. (2000) *Indian J. Plant Genet. Resour.,* **13**(1): 72-74.

Munshi, A.D., Behera, T.K., Sureja, A.K. and Kumar, R. (2011) *Rep. Cucurbit Genet. Coop.,* **33-34**: 57-59.

Munshi, A.D., Behera, T.K., Jat, G.S., Dey, S.S. and Singh, J. (2020) In: Training manual of CAAST on Genomics for improvement of horticultural crops (Eds. T.K. Behera, A.K. Goswami and Gograj Singh Jat), organized at Division of Vegetable Science, ICAR-IARI, New Delhi, pp. 165-171.

Munshi, A.D., Behera, T.K., Sureja, A.K., Jat, G.S. and Singh, J. (2015) In: MTC on Entrepreneurship development to ensure quality vegetable seed production for making the country nutritionally secure from 10- 17th December, 2015, Division of Vegetable Science, ICAR-IARI, New Delhi. pp. 20-23.

Munshi, A.D., Behera, T.K., Sureja, A.K., Singh, A.K., Singh, B., Tomar, B.S. and Singh, J. (2019) *New Age Protec. Cult.*, **5**(1): 49-51.

Munshi, A.D., Behera, T.K., Sureja, A.K., Singh, B., Singh, A.K. and Tomar, B.S. (2018) *ICAR News*, **24**(2): 22-23.

Munshi, A.D., Dey, S.S. and Singh, J. (2019) In: *Compendium ICAR Sponsored* Winter School on Breeding and genomic tools for stress resistance in vegetable crops (Eds.B.S. Tomar, M. Mangal, A. Srivatava, G.S. Jat, J. Singh and Y.A. Lyngdoh), Division of Vegetable Science, ICAR-IARI, New Delhi, pp. 43-50.

Munshi, A.D., Islam, S., Kumar, R., Mandal, B., Behera, T.K. and Sureja, A.K. (2012) *Indian J. Plant Genet. Resour.*, **25**(3): 321-322.

Munshi, A.D., Kumar, R. and Panda, B. (2006) *Indian J. Agric. Sci.*, **76**(12): 750-752.

Munshi, A.D., Panda, B., Mandal, B., Bist, I.S., Rao, E.S. and Kumar, R. (2008) *Euphytica,* **164:** 501-507.

Munshi, A.D., Sureja, A.K. and Saha, P. (2013) In: *National Training on Quality Seed Production of Vegetable Crops* (Eds. V. Pandey, D.K. Srivastava, M.K. Vishakarma,

M. Kumar and M.P. Yadav), Organised by National Seed Research and TrainingCentre (Ministry of Agriculture, Govt of India), Varanasi, pp. 66-72.

Munshi, A.D., Sureja, A.K., Kumar, R. and Karmakar, P. (2010) In: *Designing Nutraceutical and Food Colorant Rich Vegetable Crop Plants: Conventional and Molecular Approaches* (Eds. P.Kalia and T.K. Behera), Winter School Sponsored by Indian Council of Agricultural Research (ICAR) (15th October to 4th November, 2010). Division of Vegetable Science, I.A.R.I, New Delhi, pp. 98-108.

Munshi, A.D., Tomar, B.S., Jat, G.S. and Singh, J. (2017) Quality seed production of open pollinated varieties and F_1 hybrids in cucurbitaceous vegetables. In: ICAR sponsored short Course Advances in variety maintenance and quality seed production for entrepreneurship" pp. 107-125.

Mustafa, Md., Swamy, T.N., Raju, S. and Peer, S.N. (2013) *Int. J. Biosci.*, **3**: 8-12.

Mustafa, Md., Swamy, T.N., Raju, S., Mohammad, S.K. and Suresh, V. (2012) *Int. J. Pharm. Bio. Sci.*, **3**: 92–96.

Nabi, S.A., Rashid, M.M., Al-Amin, M. and Rasul, M.G. (2002a) *Plant Tissue Cult.*, **12**: 173-183.

Nabi, S.A., Rasul, M.G., Al-Amin, M., Rasheed, M. M., Ozaki, Y. and Okubo, H. (2002b) *J. Faculty Agric.*, **46**: 303-309.

Nagamani, S., Basu, S., Singh, S., Lal, S. K., Behera, T.K., Chakrabarty, S.K. and Talukdar, A. (2015) *Indian J. Agric. Sci.*, **85**(9): 1185-1191.

Nagar, A. and Sureja, A.K. (2017) *Ann. Agric. Res. New Series*, **38**(1): 109-112.

Nagar, A., Sureja, A.K., Kar, A. Bhardwaj, R., Gopala Krishnan, S. and Munshi, A.D. (2017a) *Chem. Sci. Rev. Lett.*, **6**(21): 574-580.

Nagar, A., Sureja, A.K., Kumar, S., Munshi, A.D., Gopala Krishnan, S. and Bhardwaj, R. (2017b) *Vegetos*, **30**(Special-1): 81-86.

Nagar, A., Sureja, A.K., Munshi, A.D., Bhardwaj, R., Kumar, S. and Tomar, B.S. (2017c) *Indian J. Agric. Sci.*, **87**(11): 1519–23.

Nagarajan, K. and Ramakrishnan, K. (1971) *Proc. Indian Acad. Sci.*, pp. 30-35.

Nair, P.K.K. and Kapoor, S.K. (1974) Pollen Morphology of Indian Vegetable Crops. *Glimpses in Plant Research*, **2**: 106-201.

Nakagawara, T. (2000) *Crop Prod.*, **3**: 113-118.

Nanasato, Y., Konagaya, K., Okuzaki, A., Tsuda, M. and Tabei, Y. (2013) *Plant Biotechnol. Rep.*, **7**: 267–276.

Nandesh, Javaregowda, S. and Ramegowd (1996) *Seed Res.*, **23**: 113-115.

Napier, A.B. (2006) *M.Sc. Thesis*, Texas A and M University, USA.

Nasr-Esfahana, M. and Ahmadi, A.R. (1997) *App. Entom. Phytopath.*, **65**: 18-20.

Naudin, C. (1859) *Ann. Sci. Natl. Sci. Bot.*, **11**: 5-87.

Navot, N. and Zamir, D. (1986) *Theor. Appl. Genet.*, **72**: 274–278.

Navot, N. and Zamir, D. (1987) *Plant Syst. Evol.*, **156**: 61–67.

Navot, N., Sarfatti, M. and Zamir, D. (1990) *J. Hered.*, **81**: 162–165.

Navrátilová, B., Greplová, M., Gajdová, J. and Skálová, D. (2006a) In: Sborník: Nové poznatky z genetiky a šäachtenia poånohospodársk ch rastlín, November 14–15, 2006. Pieš any, SK, pp. 125–126.

Navrátilová, B. (2004) *Hortic. Sci. (Prague)*, **31**: 140–157.

Navratilova, B. (1987) *Bulletin, Vvzkumny a Slechtitelsky Ustav Zelinarslty Olomouc*, **31**: 35-44.

Navrátilová, B., Greplová, M., Vyvadilová, M., Klíma, M., Gajdová, J., Skálová, D. (2006b) *Acta Hortic.*, **725**: 801–805.

Nederhoff, E. (1987) *Groenten en Fruit*, **42**: 40-43.

NeSmith, D.C. and Duval, L.R. (2001) *HortScience*, **36**: 60-61.

Newstrom, L.E. (1987) Diss. *Abst. Intern., B. (Sciences and Engineering)*, **48**: 1226B-1227B.

Newstrom, L.E. (1990) In: *Biology and Utilisation of Cucurbitaceae* (Eds. D.M.Bates, R.W. Robinson, and C. Jeffrey), Cornell University Press, pp. 141-149.

NHB Statistics (2015) www.nhb.org. National Horticulture Board, Gurgaon, India. Niedz, R.P., Schiller-Smith, S., Dunbar, K.B., Stephens, C.T. and Murakishi, H.H. (1989) *Plant Cell Tissue Organ Cult.*, **18**: 313-319.

Niemirowicz-Szczytt, K. and B. Kubicki. (1979) *Genetica Polonica*, **20**: 117–125.

Niemirowicz-Szczytt, K., Dumas de Vaulx, R. (1989) *Rept. Cucurbit Genet. Coop.*, **12**: 24–25.

Niemirowicz-Szczytt, K., Faris, N.M., Nikolova, V., Rakoczy-Trojanowska, M. and Malepszy S. (1995) In: G. Lester (Ed.). Cucurbitaceae' 94, pp. 169-171.

Nikolova, V. and Niemirowicz-Szczytt, K. (1996) *Acta Soc. Bot. Pol.*, **65**: 311-317.

Nishibayashi, S., Hayakawa, T., Nakajima, T., Suzuki, M. and Kaneko, H. (1996) *Theor. Appl. Genet.*, **93**: 672-678.

Nishibayashi, S., Kaneko, H. and Hayakawa, T. (1996b) *Plant Cell Rep.*, **15**: 809–814.

Norton, J.D. and Granberry, D.M. (1980) *J. Amer. Soc. Hort. Sci.*, **105**: 174–180.

Norton, J.D., Casper, R.D., Smith, D.A. and Rymal, K.S. (1985) *Cir Agri. Exp. Sta.*, Aubum University, Alabama, USA, No. 280, p. 11.

Norton, J.D., Cosper, R.D., Smith, D.A. and Rymal, K.S. (1985) *HortScience*, **20**: 955-956.

Ntui, V.O., Ugoh, E.A., Udensi, O. and Enok, L.N. (2007) *J. Food Agric. Environ.*, **5**(2): 211–214.

Nuez, F., Pico, B., Iglesias, A., Esteva, J. and Juarez, M. (1999) *European J. Plant Path.*, **105**: 453-464.

Nunez-Palenius H.G., Hopkins, D. and Cantliffe, D.J. (2012) Powdery mildew of cucurbits in Florida.

Obshatko, L.A. and Shabalina, L.P. (1984) In: *1 Termonezistent nort i Produktivonst Sel' skokhozyaistvennykh Rastenii Petrozavodsk*, USSR, pp. 113-119.

Oda, M. (1995) *JARQ*, **29**: 187–198

Oda, M. (2002a) *Sci. Rep. Agric. Biol. Sci.*, Osaka Pref. Univ., **53**: 1–5.

Oda, M. (2002b) *Sci. Rep. Agric. and Biol. Sci.*, Osaka Pref. Univ., **54**: 49–72.

Ogawa, R., Ishikawa, M., Niwata, E. and Oosawa, K. (1997) *Plant Cell Tissue Organ Cult.*, **49**: 171-177.

Ogawa, R., Sugahara, S., Ito, H., Kawai, H., Sakamori, M., Aoyagi, M. and Sakurai, Y. (1995) *Res. Bull. Aitchi Ken Agricultural Research Centre*, No. 27, pp. 175-180.

Ogorek, R., Lejman, A., Pusz, W., Mi uch, A. and Miodynska, P. (2012) **19**: 80–85.

Ohuri, J. (1965) *Flora of Japan*, Natural Science Museum, Tokyo.

Oka, H., Watanabe, T. and Nishiyuna, I. (1967) *Canad. J. Genet. Cytol.*, **9**: 482-489.

Oliver, M., Garcia-Mas, J., Cardús, M., Pueyo, N., López-Sesé, A.I., Arroyo, M., Gómez-Paniagua, H., Arús, P. and de Vicente C. (2001) *Genome*, **44**: 836–845.

Om, Y.H. and Hong, K.H. (1989) *Res. Rep. Rural Develop. Administ.*, Horticulture, Korea Republic, **31**: 30-33.

Omid, A., Keilin, T., Glass, A., Leshkowitz, D. and Wolf, S. (2007) *J. Exp. Bot.*, **58**: 3645–3656.

Omran, S.A.L., Ramadan, W.A.E. and Mostafa, Y.A.M. (2012) *J. Plant Production*, Mansoura Univ., 3(12): 3139–3148.

Ondøej, V., Navrátilová, B., Lebeda, A. (2009a) *Plant Cell, Tissue Org. Cult.*, **96**: 229–234.

Orsini, F., Sanoubar, R., Oztekin, G.B., Kappel, N., Tepecik, M., Quacquarelli, C., Tuzel, Y., Bona, S. and Gianquinto, G. (2013) *Functional Plant Bio.*, **40**: 628–636.

Ossom, E.M., Igbokwe, J.R. and Rhykerd, C.L. (1998) *Proc. Indian Acad. Sci.*, **105**: 169-176.

Owen, K.W., Peterson, C.E. and Tolla, G.E. (1980) *HortScience*, **18**: 116.

Ozarslandan, A., Sogut, M.A., Yetisir, H. and Elekcioglu, I.H. (2011) *Turkish J. Ent.*, **35**: 687–697.

Padda, D.S., Malik, B. and Kumar, J.C. (1969) *Indian J. Hort.*, **25**: 173-175

Pal, S.P., Alam, I., Anisuzzaman, M., Sarker, K.K., Sharmin, S.A. and Alam, M.F. (2007) *Turk. J. Agric.*, **31**: 63-70.

Palenchar, J., Treadwell, D.D., Datnoff, L.E., Gevens, A.J. and Vallad, G.E. (2012) Cucumber anthracnose in Florida. p. 266.

Papadopoulou, E., Little, H.A., Hammar, S.A. and Grumet, R. (2005) *Sexual Plant Reprod.*, **18**: 131–142.

Paran, I., Shifriss, C. and Raccah, B. (1989) *Euphytica,* **42**: 227-232.Paris, H.S. (1993) *HortTechnology*, **3**: 95-97.

Paris, H.S. (1996) *HortTechnology*, **6**: 16-l3.

Paris, H.S., Cohen, Burger, Y. and Yoseph, R. (1988) *Euphytica*, **37**: 27-29.

Paris, M.K., Zalapa, J.E. and McCreight, J.D. (2008) *Mol. Breed,* **22**: 405–419.

Paris, H.S., Nerson, H. and Karchi, Z. (1986) *J. Hort. Sci.*, **61**: 295-301.

Paris, H.S., Nerson. J. and Zass, N. (1986) *HortScience*, **21**: 1036-1037.

Paris, J.S., Edelstein, M., Nerson, H., Burger, Y., Karchi, Z. and Lozer, D. (1985) *Hassadeh*, **66**: 254-256.

Park, K.W. and Kang, H.M. (1998) *J. Korean Soc. Hort. Sci.*, **39**: 397-101.

Parkash, A., Ng, T.B. and Tso, W.W. (2002) *Peptides*, **23**(6): 1019–1024.

Parks, S.E., Murray, C.T. and Gale, D.L. (2012) *Exp. Agr.*, **49**: 234–243.

Parthasarathy, V.A., Upadhyay, A. and Karun, A. (2000) In: *Biotechnology of Horticultural Crops* (Eds. V.A. Parthasarathy, T.K. Bose, and P.C. Deka), Vol. 2, Naya Prokash, Calcutta, pp. 188-218.

Parvin, S., Kausar, M., Haque, E., Khalekuzzaman, Sikdar, B. and Islam, M.A. (2013) *Rajshahi Univ. J. Life Earth Agric. Sci.*, **41**: 71-77.

Patel, G.I. (1952) *Curr. Sci*, **21**: 343-344.

Patel, M.G. and Kalpesh, B.I. (2015) *J. Med. Plants Stud.,* **3**: 82-88.

Pathak, G.N. and Singh, S.N. (1949) *Indian J. Genet.*, **9**: 18-26.

Pati, K., Munshi, A.D., Behera, T.K. and Sureja, A.K. (2015) *Indian J. Agric. Sci.,* **85**(12): 1609-1613.

Pati, K., Munshi, A.D., Behera, T.K., Kumar, R. and Karmakar, P. (2015) *Genetika*, **47**(1): 349-356.

Patil, S.D., Keskar, B.G. and Lawande, K.E. (1998) *J. Soils Crops*, **8**: 11-15.

Paunel, I., Giorgota, M., Jileu, M. and Mitrache, D. (1984) Anale Inistitutal de Cereetari Pentru degumicullura si Floricultura, Vidra, V11: 333-341.

Pelletier, B. (1989) *Infos* (Paris), **57**: 6-8.

Perchepied, L., Dogimont, C. and Pitrat, M. (2005) *Theor. Appl. Genet.*, **111**: 65–74.

Périn, C., Hagen, L.S. and De Conto, V. (2002a) *Theor. Appl. Genet.,* **104**: 1017–1034.

Périn, C., Hagen, L.S. and Giovinazzo, N. (2002b) *Mol. Genet. Genom.*, **266**: 933–941.

Pershin, A.F., Medvedeva, N.I. and Medvedev, A.V. (1988) *Genetika*, **24**: 484-493.

Peterson, C.E. and Anhder, L.D. (1960) *Science,* **131**: 1673-1674.

Peterson, C.E. and Dezeew, D.J. (1967) *Spartan Dawn Quart. Bull. Mich. Agr. Exp. Sta.*, **46**: 267-273.

Peterson, C.E., Owensk, W. and Rowe, P.R. (1983) *HortScience*, **18**: 116.

Peterson, C.E., Williams, P.H., Palmer, M. and Louward, P. (1982) *HortScience*, **17**:268.

Petropoulos, S.A., Khah, E.M. and Passam, H.C. (2012) *Int. J. Plant Prod.*, **6**: 481–491.

Petropoulos, S.A., Olympios, C., Ropokis, A., Vlachou, G., Ntatsi, G., Paraskevopoulos, A. and Passam, H.C. (2014) *J. Agric. Sci. Technol.*, **16**: 873–885.

Petrus, A.J.A. (2014) *J. Chem.*, **30**(1): 149–154.

Pike, L.M. and Peterson, C.E. (1969) *Euphytica*, **18**: 101-105.

Pink, D.A.C. (1987) *Ann. Appl. Biol.*, **111**(2): 425-432.

Pink, D.A.C. and Walkey, D.G.A. (1985) *J. Agril. Sci.*, **104**: 325-329.

Pink, D.A.C., Carter, P.J. and Walkey, D.G.A. (1985) In: 36th Annual Rep. 1985. *Nat. Veg. Res. Sla.*, Wellsbourne, U.K., p. 46.

Pitrat, M. (1990) *Curcurbit Genet. Coop. Rep.*, **13**: 58.

Pitrat, M., and Besombes, D. (2008) In: Cucurbitaceae 2008, IXth EUCARPIA meeting on genetics and breeding of Cucurbitaceae (Ed. M. Pitrat), INRA, Avignon, France, pp. 135–142.

Pitrat, M., Chauvet, M. and Foury C. (1999) In: 1st *Int. Symp. on Cucurbits* (Eds. K. Abak and S. Buyukalaca), ISHS Adana, Turkey, pp. 21–28.

Pitrat, M., Hanelt, P. and Hammer, K. (2000) *Acta Hortic.*, **510**: 29–36.

Plader, W., Malepszy, S., Burza, W. and Rusinowski, Z. (1998) *Euphytica*, **103**: 9-15.

Pofu, K.M., Mashela, P.W. and Mafeo, T.P. (2013) *Acta Horti.*, **1007**: 807–812.

Pofu, K.M., Mashela, P.W. and Mphosi, M.S. (2011) *African J. Biotechnol.*, **10**: 8790–8793.

Poole, C.F. and Grimball, P.C. (1939) *J. Heredity*, **30**: 21-25.

Poole, C.F. and Grimball, P.C. (1945) *J. Agri. Res.*, **71**: 533-552.

Prabhakar, B.S., Srinivas, K. and Shukla, V. (1985) *Prog Hort.*, **71**: 533-552.

Pradeepkumar, T., Hegade, V.C., Kannan, D., Sujatha, R., George, T.E. and Nirmaladevi, S. (2013) *Sci. Hort.*, **144**(6): 60-64.

Pradeepkumar, T., Sujatha, R., Krishnaprasad, B.T. and Johnkutty, I. (2007) *Cucurbit Genet. Coop. Rep.*, **30**: 60-63.

Preethi, G., Anjanappa, M., Ramachandra, R.K. and Vishnuvardhana (2019) *Int. J. Curr. Microbiol. App. Sci.*, **8**(3): 925-932.

Premnath (1976) In: *Vegetables for the Tropical Region*, ICAR, New Delhi.

Profirev, N.P. and Laptev, V.N. (1986) In: *Problemy Oroshaemogo Owshvhewd I Bakhchevodastva Astrakhan*, USSR, 59-62.

Proost, S., Van Bel, M., Vaneechoutte, D., Van de Peer, Y., Inzé, D. and Mueller-Roeber, B. (2015) *Nucleic Acids Res.*, **43**: 81.

Provvidenti, R. (1983) Proc. conference an Biology and chemistry of the Cucurbitaceae, Cornell University, Ithaca, New York, August, 1980.

Provvidenti, R. (1987) *HortScience*, **22**: 102-103.

Provvidenti, R. (1995) *Cucurbit Genet. Coop. Rep.*, **18**: 65-67.

Provvidenti, R., Robinson, R.W. and Munger, H.M. (1978) *Plant Dis. Rept.*, **62**: 326-329.

Puchalski, J.T., Robinson, R.W. and Shail, J.W. (1978) *Cucurbit Genet. Coop. Rep.*, **1**: 39.

Punja, Z.K. and Raharjo. S.H.T. (1996) *Plant Dis.*, **80**: 999-1005.

Punithaveni,V. (2015) *Ph.D. (Hort.) Thesis,* Tamil Nadu Agricultural Univ., Coimbatore.

Punithaveni, V., Jansirani, P. and Sivakumar, M. (2015) *Electronic J. Plant Breeding.*, **6**: 486-492.

Puzari, N.N. (1997) *Ann. Agric. Res*, **18**: 508-509.

Pyzhcnkov, V.I., Kosarcva, G.l. and Davidich, N.1 (1988) *Selektsiya 1 Semekavodstvo, Moscow*, **4**: 40-41.

Qi, J., Liu, X., Shen, D., Miao, H., Xie, B., Li, X., Zeng, P., Wang, S., Shang, Y. and Gu, X. (2013) *Nat. Genet.*, **45**: 1510–1515.

Quanrucci, M. and Conii, D. (1997) *Informatore Agrario*, **53**: 53-56.

Queiroga, R., Puiatti, M., Fontes, P.C.R. and Cecon, P.R. (2008) *Horticultura Brasileira*, **26**: 209–215.

Quemada, H. (1998) In: Ives, C.L. and Bedfnrc, B.M. (Eds.), *Agricultural biotechnology in international development*, Wallingford, UK, CAB Intemational, pp. 147-160.

Quemada, L.L. (1994) Food safety evaluation. Proc. OECD-sponsored workshop, 12-15 September, 1994. Oxford, UK. pp. 71-79.

Qvamstrom, K. (1989) *Vaxl Skyddsnotiser*, **53**: 54-57.

Rab, A. and Ishtiaq, M. (1996) *Sarhad J. Agric.*, **12**: 123-126.

Raharjo, S.H.T. and Punja, Z.K. (1996) *Plant Disease*, **80**: 999-1015.,

Rahaujo, S.H.T. and Punja, Z.K. (1992) *Cucurbit Genet. Coop Rep.*, **15**: 35-39.

Rahaxjo, S.H.T., Hernandez, M.O., Zang, Y.Y. and Punza, Z.K. (1996) *Plant Cell Rep.*, **15**: 591-596.

Rahman, A.H.M.M. (2013) *J. Med. Plants Stud.*, **1**(3): 118–125.

Rahman, H., Shahinozzaman, M., Karim, M. R., Hoque, A., Hossain, M. M. and Rafiul, A. K. M. (2012) *Int. J. Agron. Agric. Res.*, **2**: 47-52.

Rahman, Ma, Alamgir, A.N.M. and Khan, M.A.A. (1995) *J. Asiat. Soc. Bangladesh Sci.*, **21**: 227-232.

Rahman, S.M., Hossain, M., Islam, R. and Joarder, O.I. (1993) *Curr. Sci.*, **65**: 562-564.

Rahman, A., James, J.K. and Moflimcr, J. (1985) In: Proceedings, New Zealand Weed and Pest Control Conference, Hastings, New Zealand, pp. 135-138.

Rai, G.K., Singh, M., Rai, N.P., Bhardwaj, D.R. and Kumar, S. (2012) *Physiol. Mol. Biol. Plants*, **18**: 273-280.

Rai, M., Singh, M., Kumar, S., Pandey, S., Singh, B. and Ram, D. (2004) In: Souvenir–National Symposium on Harnessing Heterosis in Crop Plants, March 13-15, 2004, Varanasi, India. pp. 110-118.

Rajasekharan, K.R. and Shanmugavelu, K.G. (1984) *South Indian Hort.*, **32**: 47.

Rajasekharan, P.E., Bhaskaran, S., John, J.K., Koshy, Eapen, P. and Antony, V.T. (2012) *J. Biotech.*, **7**: 50-56.

Rajkumar (1962) *Kheti*, **14**: 13-15.

Rajput, A.L. and Gautam, G.P. (1995) *J. Recent Adv. A.I.S* 10 pp CI, **10**: 87-88.

Rajput, J.C., Parulekar, Y.R., Sawant, S.S. and Jamadagni, B.M. (1994) *Curr. Sci.*, **66**: 779.

Raju, S., Chithkari, R., Bylla, P. and Mustafa, Md. (2015) *J. Exp. Biol. Agric. Sci.*, **3**: 407-414.

Ram, D., Kumar, S., Banerjee, M.K., Singh, B. and Singh, S. (2002) *Cucurbit Genet. Coop. Rep.*, **25**: 65-66.

Ram, D., Kumar, S., Singh, M., Rai, M. and Kalloo, G. (2006) *J. Hered.*, **97**: 294-295.

Rama Krishna, D. and Shashtri, T. (2015) *Asian J. Biotech.*, **7**(2): 88-95.

Rana, T.K., Vahistha, R.N. and Pandita, M.L. (1986) *Haryana J. Hort. Sci.*, **15**: 71-75.

Randall, P., Niedz, L., Smith, S.S., Kerry, B., Dunbar, Christine, T. and Haeey, H. (1989) *Plant Cell, Tissue Org. Cult.*, **18**: 313-319.

Randhawa, K.S. and Singh, K. (1970) *Plant Sci.*, **2**: 118-122.

Ranjan, P., Gangopadhyay, K.K., Bhardwaj, R., Pandey, C.D., Srivastava, R., Bag, M.K., Prasad, T.V., Meena, B.L., Dutta, M., Bansal, K.C., Pal, A.K., Koley, T.K., Pandey, S., Singh, B., Munshi, A.D. and Harish, G.D. (2019) *Indian J. Plant Genet. Resour.*, **32**(3): 437.

Ranjan, P., Pandey, A., Munshi, A.D., Bhardwaj, R., Gangopadhyay, K.K., Malav, P.K, Pandey, C.D., Pradheep, K., Tomar, B.S. and Kumar, A. (2019) *Genet. Resour. Crop. Evol.*, **66**: 1217–1230.

Rao, A.V. and Rao, L.G. (2007) *Pharmacol. Res.*, **55**(3): 207–216.

Ratnam, C.V., Pandit, S.V. and Rao, K.C. (1985) *Pesticides*, **19**: 42, 51.

Reddy, A.V. (1997) *Annals of Agricultural Research*, **18**: 252-254.

Reddy, B.S., Thammaiah, N., Patil, R.V. and Nandihalli, B.S. (1995) *Adv. Agric. Res., India*, **4**: 103408.

Reddy, B.S., Thammaiah, N., Patil, R.V. and Nandihalli, B.S. and Patil, D.R. (1997) *Adv. Agric. Res.*, **7**: 175-177.

Reddy, V.V.P., Rao, M.R. and Reddy, C.R. (1987) *Veg. Sci.*, **14**: 152-160.

Reed, J., Privalle, L., Powell, M.L., Meghji, M., Dawson, J., Dunder, E., Sutthe, J., Wenck, A., Launis, K., Kramer, C., Chang, Y.-F., Hansen, G. and Wright, M. (2001) *In-Vitro Cell. Dev. Biol. Plant*, **37**: 127–132.

Rehm, S. (1960) *Ergebnisse den Biologie,* **22**: 108-136.

Rehm, S., Enslin, P.R., Mceusc, A.D.J. and Wcssels, I.H. (1957) *J. Sci. Food Agric.,* **8**: 678-680.

Reine, S. and Riggs, D.I.M. (1999) *HortScience,* **34**: 1076-1078.

Reiners, S. and Riggs, D.I.M. (1997) *HortScience,* **32**: 1037-1039.

Rekhi, S.S., Nandpuri, K.S. and Singh, H. (1968) *J. Res. Ludhiana,* **5**: 199-202.

Ren, C.M, Dong, Y.Y., Hong, Y.H. and Zhao, Y. (2000) *J. Hunan Agril. Univ.,* **26**: 50-53.

Ren, Y., Bang, H. and Gould, J. (2013). *In Vitro Cell. Dev. Biol. - Plant,* **49**: 223–229. https://doi.org/10.1007/s11627-012-9482-8

Ren, Y., Mc Gregor, C., Zhang, Y., Gong, G., Zhang, H., Guo, S., Sun, H., Cai, W., Zhang, J., and Xu, Y. (2014) *BMC Plant Biol.,* **14**: 33.

Ren, Y., Zhang, Z., Liu, J., Staub, J.E., Han, Y., Cheng, Z., Li, X., *et al.* (2009) *PLoS ONE,* **4**: e5795.

Renner, S.S. (2017) *PhytoKeys,* **85**: 87–94.

Renner, S.S and Ricklefs, R.E. (1995) *Amer J. Bot.,* **82**: 596-606.

Rhodes, A.M. (1959) *Proc. Amer. Soc. Hort Sci.,* **74**: 546452.

Rhodes, A.M. (1964) *Plant Disease Reporter,* **48**: 54-55.

Richardson, J.B. (1972) *Econ. Bot.,* **16**: 266-273.

Richharia, R.H. (1948) *Curr. Sci.,* **17**: 358.

Rios, H., Fernandez, A., Casanova, E. (1998) *Cultivos Tropicales,* **19**: 33-35.

Rivero, R.M., Ruiz, J.M., and Romero, L. (2003) *Sci. Technol.,* **1**: 70–74.

Roberts, D.P., Dery, P.D., Hebbar, P.K., Mao, W. and Lumsden, R.D. (1997) *J. Phytopathology,* **145**: 383-388.

Robinson, R.W. (1999) *J. New Seeds,* **1**: 1-47.

Robinson, R.W. (2000) In: Hybrid Seed Production in Vegetables, Rationale and Methods in Selected Species (Ed. A.S.Basra), Food Products Press, pp. 1-47.

Robinson, R.W. and Decker-Walters, D. (1999) Curcurbits. CAB International, Wallingford, Oxon, OX108DE, U.K.

Robinson, R.W., Munger, H.M. and Whitaker, T.W. (1976) *HortScience,* **11**: 554-568.

Robinson, R.W., Weeden, N.F. and Provvidenti, R. (1988) *Rep. Cucurbit Genet. Coop.,* **11**: 74-75.

Robinson, R.W. (2000) Rationale and Methods to Produce Hybrid Cucurbit Seed, pp. 1- 47. In: Hybrid Seed Production in Vegetables, Rationale and Methods in Selected Species, A.S. Basra (ed.). Food Products Press. 135 p.

Roig, L.A., Roche, M.V., Orts, M.C., Zubeldia, L., Moreno, V. (1986) *Rep. Cucurbit Genet. Coop.,* **9**: 74–77.

Romero, L., Belakbir, A., Ragala, L. and Ruiz, M. (1997) *Soil Sci. Plant Nutr.,* **43**: 855–862.

Roosta, H.R. and Karimi, H.R. (2012) *J. Plant Nut.,* **35**: 1843–1852.

Roppongi, K. (1991) *Bull. Saitama Hortic. Exp. Stn.* (Japan), **5**: 1-15.

Rouphael, Y., Cardarelli, M., Rea, E. and Colla, G. (2012) *Photosynthetica,* **50**: 180–188.

Roy, R.P. and Saran, S. (1990) In: *Biology and Utilization of the Cucurbitaceae* (Eds. D.M. Bates, R.W. Robinson and C. Jeffrey), Cornell Univ. Press, Ithaca, NY, USA, pp. 251-268.

Rubatzky, V.E. and Yamaguchi, M. (1997) Chapman and Hall, New York, NY, USA.

Ruggieri, A., Labopin, M., Bacigalupo, A., Gülbas, Z., Koc, Y. and Blaise, D.V. (2018) *Sci. Rep.,* **8**: 80-88.

Ruiz, J.M., Belakbir, A. and Romero, L. (1996) *J. Plant Physiol.,* **149**: 400–404.

Ruiz, J.M., Belakbir, A. Lopez-Cantarero, A., and Romero, L. (1997) *Sci. Hortic.,* **71**: 113–123.

Saglam, S. (2017) *Sci. Bull. Sr. F. Biotechnol.,* **21**: 46–50.

Saha, D., Rana, R.S., Sureja, A.K., Verma, M., Arya, L. and Munshi, A.D. (2013) *Physiol. Mol. Plant Pathol.,* **81**: 107-117.

Saima Malik, Zia, M., Riaz-ur-Rehaman, and Choudhary, M.F. (2007) *Pak. J. Biol. Sci.,* **10**: 4118-4122.

Sakata, Y., Takayoshi, O. and Mitsuhiro, S. (2007) *Acta Hort.,* **731**: 159–170.

Salar, N., Salehi, R. and Delshad, M. (2015) *Acta Horti.,* **1086:** 225–230.

Sandha, M.S. and Lal, T. (1999) *Veg. Sci.,* **26**(1): 1-5.

Sandhu. K.S., Bal, P.S. and Randhawa, K.S. (1986) *J. Res. Punjab Agril. Univ.,* India, **23**: 49-53.

Sandra, N., Basu, S. and Behera, T.K. (2018) *Indian J. Hort.,* **75**(2): 245-251.

Sangeetha, P. and Venkatachalam, P. (2014) *In Vitro Cellular and Developmental Biology-Plant,* **50**: 242-248.

Sangeetha, P. and Venkatachalau, P. (2011) *Plant Cell Biotechnol. Mol. Biol.,* **12**:1-4.

Sanwal, S.K., Kozak, M., Kumar, S., Singh, B. and Deka, B.C. (2011) *Acta Physiologiae Plantarum,* **33**: 1991–1996.

Sapovadiya, M.H., Dhaduk, H.L., Mehta, D.R. and Patel, N.B. (2013) *Progress. Res.,* **8**(2): 217-220.

Saraev (1991) *Kartofel' I Ovoshchi,* **3**: 22-29.

Saranah, S. and Harrington, M.E. (1985) *HortScience,* **20**: 1145.

Sari, N. and Abak, K. (1996) *Turk. J Agric. For.,* **20**: 555-559.

Sar , N., Yetisir, H. and Bal, U. (2006) In: Global Science Books (Ed. J.A. Teixeia da Silva), London, pp. 376-384.

Sarkar, D. De and Datta, K.B. (1990) *Bangladesh J. Bot.,* **19**: 1-4.

Sarkar, D. de, Datta, K.B. and Sen, R. (1987) *Cytologia,* **52**: 405-417.

Sarkar, M., Singh, D.K., Lohani, M., Das, A.K. and Ojha, S. (2015) *Int. J. Agric. Environ. Biotechnol.,* **8**(1): 153-161.

Sarowar, S., Oh, H. Y., Hyung, N. I., Min, B. W., Harn, C. H., Yang, S. K., Ok, S. H. and Shin, J.S. (2003) *Plant Cell, Tissue Org. Cult.*, **75**: 179–182.

Satapathy, G. and Thirunavoukkarasu, M. (2014) *Int. J. Pure App. Biosci.*, **2**: 139-146.

Sato, N. and Takamatsu, T. (1930) *Nogyo Sekai*, **25**: 24–28. (in Japanese).

Saurabh, S., Prasada, D., Ambarish, S. and Vidyarthi (2017) *J. Crop Sci. Biotech.*, **20**: 81-87.

Sauton, A. (1989) *Rep. Cucurbit Genet. Coop.*, **12**: 22–23.

Savitha, R., Shasthree, T., Sudhakar and Mallaiah, B. (2010) *Int. J. Pharma Bio Sci.*, **1**: 1-8.

Sawan, O.M., Eissa, A.M., Abou-Hadid, A.F. (1999) *Acta Hort.*, **491**: 369-376.

Schacht H., Exner. M. and Schenk, M. (1991) *Gartenbauwissenschaft*, **57**: 238-242.

Schaefer, H. (2007) *Blumea*, **52**: 165–177.

Schafer, H. (2005) *Cucurbit Network* News, **12**: 5.

Scheffer, J.J.C. and Wood, R.J. (1939) In: Proceedings, New Zealand Weed and Pest Control Conference. Hastings, New Zealand, pp. 139-141.

Schon, M.K. and Compton, M.P. (1997) *HortScience*, **7**: 33-38.

Schouten, R.E., Otma. E.C., Kooten. O. Van and Tuskens, I.M.M. (1997) *Postharvest Biol. Tech.*, **12**: 175-181.

Schroeder, C.A. (1968) *Bot. Gaz.*, **129**: 374-276.

Schultheis, J.R. (1992) *Rep. Cucurbit Genet. Coop.*, **15**: 9-10.

Sedghi, M., Gholipouri, A. and Seyed Sharifi, R. (2008) *Notulae Botanicae Horti Agrobotanici Cluj-Napoca*, **36**: 80–84.

Selim, M.A.M (2019) *Egypt. J. Agric. Res.*, **97**(1): 317-342.

Selvi, N.A.T., Pugalendhi, L. and Sivakumar, M. (2013) *Asian J. Hort.*, **8**: 720–725.

Selvaraj, N., Vasudevan, A., Manickavasagam, M. and Ganapathi, A. (2006) *Biologia Plantarum*, **50**: 123-126.

Sen, S.P., Sen, M., Sengupta, B., Somchoudhury, M. and Ghosh, P.D. (1983) *Proc. Conference on the Biology and Chemistry of Cucurbitaceae*, Cornell University,Itheca, New York, August, 1989.

SenYan, S., WeiMin, D. and HuiNing, L. (1996) *Acta Horticulturae Sinica*, **23**: 49-53.

Seo, Y.K., Paek, K.Y., Hwang, J.K. and Cho, Y.H. (1991) *J. Korean Soc. Hort. Sci.*, **32**: 137-145.

Seong, K. C., Moon, J. M., Lee, S. G., Kang, Y. G., Kim, K. Y. and Seo, H. D. (2003) *J. Kor. Soc. Hort. Sci.*, **44**: 478–482.

Serquen, F.C., Bacher, J. and Staub, J.E. (I997) *Mol. Breed.*, **3**: 257-268.

Serrano, R., Culiañz-Maciá, F.A. and Moreno, V. (1998) *Sci. Hort.*, **78**: 261–269.

Seshadri, V.S., Sharma, J.C. and Choudhury, B. (1972) *Indian Hort.*, **16**: 21-22.

Shang, Y., Ma, Y., Zhou, Y., Zhang, H., Duan, L., Chen, H., Zeng, J., Zhou, Q., Wang, S. and Gu, W. (2014) *Science*, **346**: 1084–1088.

Shankar, G., Nariani, T.K. and Namprakash (1969) *Indian J. Microscopy in Life Sci.*, Calcutta.

Shankar, G., Nariani, T.K. and Namprakash (1972) *J. Microbiol.*, **12**: 154-165.

Sharma, C., Trivedi, P.C. and Tiagi, B. (1985) *Int. Nematol. Net. Newsl.*, University of Rajasthan, India. **3**: 7-9.

Sharma, C.B. and Shukla, V. (1972) *Proc. Third Int. Symp. Trop. Subtrop. Hort.*, Bangalore, p. 45.

Sharma, J.P. and Kumar, S. (1999) *Indian J. Agri. Sci.*, **69**: 678-679.

Shashikumar, K.T. and Pitchaimuthu, M. (2016) *Int. J Agric. Sci. Res.*, **6**(2): 341-348.

Shasthree, T., Ramakrishna, D., Imran, M.A. and Chandrashekar, Ch. (2014) *J. Herbs, Spices Med. Plants*, **20**: 235-244.

Shekhawat, M.S., Shekhawat, N.S., Harish, Kheta, R., Phulwaria, M. and Gupta, A.K. (2011) *J. Crop Sci. Biotech.*, **14**: 133-137.

Sheng-Niao, N., Xue-Sen, H., Sek-Man, W., Jia-Lin, Y., Fu-Xing, Z., Da-Wei, L., Sheng-You, W., Guang-Ming, Z. and Fan-Sheng, S. (2005) *Chin. J. Agric. Biotechnol.*, **2**: 179–185.

Shiboleth, Y.M., Arazi, T., Wang, Y. and Gal-On, A. (2001) *J. Biotechnol.*, **92**: 37–46.

Shifriss, O, Volin, R.B. and Williams, T.V. (1989) *Rep. Cucurbit Genet. Coop.*, **12**: 75-78.

Shifriss, O. (1961) *Proc. Amer. Soc. Hort. Sci.*, **67**: 479-486.

Shifriss, O. (1981) *J. Amer. Soc. Hortic. Sci.*, **106**: 220- 232.

Shifriss, O. (1988) *Biotechnol. Adv.*, **6**: 72.

Shifriss, O. (1991) *Rep. Cucurbit Genet. Coop.*, **14**: 116-122.

Shifriss, O. (1993) *Rep. Cucurbit Genet. Coop.*, **16**: 64-67.

Shimada, N. and Moritani, M. (1977) *J. Japan Soc. Soil Sci. Plant Nutr.*, **48**: 396–401.

Shimizu, M., Yazawa, S. and Ushijima, Y. (2009) *J. Gen. Plant Pathol.*, **75**(1): 27–36.

Shimoinishi, K., Ishikawa, M., Suzuki, S. and Oosawa, K. (1991) *Japanese J. Breed.*, **41**: 347-351.

Shimotsuma, M. (1961) *Seiken Ziho.*, **12**: 75-84.

Shimotsuma, M. (1963) *Seiken Ziho.*, **15**: 24-34.

Shimotsuma. M. (1968) *Seikhen Ziho.*, **20**: 47-53.

Shin, J.S., Lee, S.J. and Park, K.W. (1995) *Korean J. Breed.*, **27**: 194-207.

Shinde, S., Supe, V.S. and Gaikwad, S.S. (2016) *Asian J. Sci. Technol.*, **7**(5): 2846-2849.

Shinde, N.N. and Seshadri, V.S. (1983) Proc. 15th International congress of Genetics, Pt II, P. 613.

Shrivastava, D.K., Andrianov, V.M. and Piruzyau, L.Z.S. (1985) *Fiziologiva Rastenii.*, **35**: 1243-1247.

Siguenza, C. Schochow, M., Turini, T. and Ploeg, A. (2005) *J. Nematol.*, **37**: 276–280.

Siffnis, M.J. van and Ei-Khawass, K.A.M.H. (1996) In: Proc.Of the meeting Integrated Control in glass house (Ed. J.C. van Lenteren), held in Vienna, Austra, 20-25May, 1996, Bulletin 01LB/SROP **19**: 159-162.

Sikdar, B., Shafiullah, M., Chowdhury, A.R., Sharmin, N., Nahar, S. and Joarder, O.I. (2005) *Biotechnology*, **4**: 149–152.

Silva, E.de.S.da., Palangana, F.C., Goto, R., Furtado, E.L. and Fernandes, D.M. (2012) *Summa Phytopathologica*, **38**: 139–143.

Silva, J.A., da Costa, T.S., Lucchetta, L., Marini, L.J., Zanuzo, M.R., Nora, L., Nora,F.R., Twyman, R.M. and Rombaldi, C.V. (2004) *Postharvest Biol. Technol.*, **32**: 263–268.

Simon, P.W. and Navazio, J.P. (1997) *HortScience* **32**(1): 144-145.

Sincha, K.P. (1985) In: *I Puti intensi*, bakhchevod. V. Volgogr Zavolzhe. Mytishchi, USSR, pp. 64-67.

Singh, A.K. and Yadava, K.S. (1984) *Plant Syst. Evol.*, **147**: 237–252.Singh, A.K. and Roy, R.P. (1973) *Sci. Cult*, **39**: 505-506.

Singh, A.K. (1990) In: Bates, D.M., Robinson, R.W. and Jeffrey, C. (Eds.), Biology and Utilisation of Cucurbitaceae, Cornell University Press, pp. 10-28.

Singh, A.K., Behera, T.K., Chandel, D., Sharma, P. and Singh, N.K. (2007) *J. Hort. Sci. Biotechnol.*, **82**(2): 217–222.

Singh, B. and Tomar, B.S. (2015) *Indian J. Agric. Sci.*, **85**(10): 3-11.

Singh, B., Kumar, M., Singh, V. and Mehto, S.P. (2007) *Acta Hort.*, **742**(11): 85-87.

Singh, B., Tomar, B.S. and Hasan, M. (2010) *Acta Hort.*, **871**(37): 279-282.

Singh, B.P., Dadlani, S.A. and Mittal, SK. (1972) *Indian Hort.*, **17**: 24.

Singh, D. (1971) *J. Indian Bot. Soc.*, **50A**: 208-215.

Singh, D. and Dathan, A.S.R. (1973) *Phytomorphology*, **22**: 29-45.

Singh, D. and Dathan, A.S.R. (1974) *New Botanist*, **1**: 8-22.

Singh, D. and Dathan, A.S.R. (1976) *J. Indian Bot. Soc.*, **55**: 160-168.

Singh, H.B., Ramanujam, S. and Pal. B.P. (1948) *Nature*, **161**: 775-776.

Singh, J., Munshi, A.D., Sureja, A.K., Shrivastava, S., and Tomar, B.S. (2019) *Indian J. Agric. Sci.*, **89**(11): 1959–1963.

Singh, K. and Randhawa, K.S. (1968) *Progress. Hort.*, **2**: 45-50.

Singh, M.N., Misra, A.K. and Bhatnagar, SP. (1996) *Phytomorphology*, **46**: 395-402.

Singh, N., Raj, A., Sharma, A. and Kumar, P. (2011) *Reviewed Proceedings of National Seminar on Internet: Applications in Research*, pp. 36-40.

Singh, P.P., Singh, S.P. and Pandey, S. (2011) *Veg. Sci.*, **38**(1): 39-43.

Singh, R.K. and Chowdhury, B. (1989) *Indian J. Hort.*, **46**: 215-221.

Singh, S.K., Upadhyay, A.K., Pandey J. and Pandey, A.K. (2012) *Ann. Hortic.*, **5**(2): 246-251.

Singh, G., Singh, R. and Singh, D. (2020) *Veg. Sci.*, **47**(1): 7-15.

Singh, D. and Bhandari, M.M. (1963) *Baileya,* **2.**

Singh, P.P., Shin, Yong Chul, Park. Chang Seuk and Chung Younghyun (1999) *Phytapath.,* **89:** 92-99.

Sirohi, P.S (2004) In: Souvneir - National Symposium on Harnessing Heterosis in Crop Plants, Varanasi, India, pp. 121-129.

Sirohi, P.S. (1994) *Indian Hort.,* **38**(4): 21-23.

Sirohi, P.S. and Munshi, A.D. (2009) In: *Vegetable Variety Development and Evaluation.* (Eds. P.Kalia, S.Joshi,T.K. Behera, and S.Pandey), Training sponsored by Food and Agriculture Organization (15th September to 14th October, 2009), Division of Vegetable Science, I.A.R.I, New Delhi, pp. 41-52.

Sirohi, P.S., Choudhury, B. and Kalda, T.S. (1991) *Indian Hort.,* **36:** 24-26.

Sirohi, P.S., Munshi, A.D., Kumar, G. and Behera, T.K. (2005) In: *Plant Genetic Resources: Indian Perspective,* (Ed. B.S. Dhillon), Narosa Publishers, New Delhi, pp. 34-58.

Sitterly, W.R. (1972) *Ann. Rev. Phytopath.,* **10:** 471-490.

Siviero, P. and Chillemi, G. (1992) *Informatore Agrario,* **48:** 31-34.

Siwek, P. and Capecka. E. (1999) *Folia Horticulturae,* **11:** 33-42.

Siwek, P. and Kunicki, E. (1998) *Roczniki Akademii Rolniczej w Poznaniu, Pgrodnictwo,* No. 27, pp. 277-283.

Skálová, D., Dziechciarková, M., Lebeda, A., Navrátilová, B. and Køístková, E. (2007) *Acta Hortic.,* **731:** 77-82.

Slobbe (1965) *Tuinderij,* **4:** 954-955.

Snyder, R.G., Kellebrew, F. and Fox, J.A. (1993) *HortTechnology,* **3:** 421-423.

Sodhu, A.S., Hooda, R.S., Pandita, M.L. and Singh, K.P. (1984) *Veg. Sci.,* **11:** 94-99.

Sohrab, S.S., Mandal, B., Pant, R.P. and Varma, A. (2003) *Plant Disease,* **87:** 1148.

Sokolov, D.K. (1984) *Kartofel i Ovoshchi,* **2:** 35.

Solanki, S.S. and Joshi, R.P. (1985) *Progress. Hort.,* **17:** 122-124.

Soltysiak, U., Kubicki, B. and Korzeniewska, A. (1986) *Genetica Polonica,* **27:** 299-308.

Som, M.G., Biswas, D. and Maity, T.K. (1986) Abst. 22nd Int. Hort. Congr., California, No. 509.

Song, P.L., Pen, W.Z. and Yang, Y.I-I. (1988) *Acta Scientiarum Naturalium Universitatis Normalis Hunanensis,* **11:** 340-345.

Sood, N.K. and Misra, U.S. (1965) *J. Agric. Coll.,* Gwalior, **7:** 9-13.

Sood, N.K., Kaushik, U.K. and Rathore, V.S. (1972) *Indian J. Hort.,* **29:** 111-113.

Soria, C., Gomez-Guillamon, M.L., Esteva, J. and Nuez, F. (1990) *Cucurbit Genet. Coop. Rpt.,* **13:** 31–33.

Sorntip, A., Poolsawat, O., Kativat, C. and Tantasawat, P.A. (2017) *Can. J. Plant Sci.,* https://doi.org/10.1139/cjps-2017-0112.

Sovetkina, V.E., Shashenkova, D.kh. and Dement'eva, F.I. (1984) In: *Ispol 'zovanie Regulyatorov Rostai Polimernykh Materialov v Ovoshchevodstve*, 3-6.

Spetsidis, N., Sapountzakis, G. and Tsafiaris, A.S. (1996) *Cucurbit Genet. Coop. Ret.*, **19**: 63-65.

Sreenivas, C., Muralidhar, S. and Rao, M.S. (2000) *Ann. Agric. Res.*, **21**: 262-266.

Srivastava, A. and Roy, S. (2012) *Int. J. Pharm. Bio. Sci.*, **3**: 526-531.

Srivastava, D.K., Andrianov, V.M. and Piruzyan, E.s. (1991) In: Proc. of the International Seminar on New Frontiers in Horticulture, organized by Indo-American Hybrid seeds, Bangalore, India, 25-28 November, 1990, Current plant science and biotechnology in agriculture, **12**: 127-130.

Srivastava, D.R., Andrianov, V.M. and Piruzyan, E.S. (1989) *Plant Cell Rep.*, **8**: 300-302.

Stankovic, L., Marmkovic, N., Zdravkovic, N. and Damjanovic, M. (1992) *Savremena Pofioprivsreda*, **40**: 106-110.

Staub, J.E., Balgooyen, B. and Tolla, G.E. (1986) *HortScience*, **21**: 510-512.

Stephens, J.M. (2003) Chayote. HS579. Hort. Sci. Dept., Fla. Coop. Ext. Serv., IFAS, Univ. Fla., Gainesville. http://edis.ifas.ufl.edu/MV046.

Stoyka, M., Tsvetana, L., Nikolay, V. and Georgi, V. (2014) *Turk. J. of Agric. Nat. Sci.*, **2**: 1707–1712.

Stzangret, J., Wronka, J., Galecka, T., Korzeniewska, A. and Niemirowicz-Szczytt, K. (2004) In: Progress in cucurbit genetics and breeding research (Eds. A. Lebeda, and H.S. Paris), Proceedings of Cucurbitaceae 2004, pp. 411-414.

Subburaj, S., Lee, K., Jeon, Y., Tu, L., Son, G. and Choi, S. (2019) *PLoS ONE*, **14**(10): e0223441.

Sulochanamma, B.N. (2001) *J. Res., ANGRAU*, **29**: 91–93.

Sultana, R.S., and Miha, B.M.A. (2003) *J. Biol. Sci.*, **3**: 1134-1139.

Sultana, R.S., Miha, B.M.A., Rahaman, M.M. and Mollah, M.U. (2005) *J. Biol. Sci.*, **5**: 781-785.

Sun, Z., Lower, R.L. and Staub, J.E. (2006a) *Euphytica*, **138**: 333-341.

Sun, Z., Lower, R.L. and Staub, J.E. (2006b) *Plant Breed.*, **125**: 277-280.

Sun, Z., Lower, R.L., Chung, S.M. and Staub, J.E. (2006c) *Plant Breed.*, **125**: 281-287.

Suratman, F., Huyop, F. and Parveez, G.K.A. (2009) *Biotechnology*, **8**: 393-404.

Sureja, A.K., Sirohi, P.S., Behera, T.K. and Mohapatra, T. (2006) *J. Hort. Sci. Biotech.*, **81**(1): 33-38.

Sureja, A.K., Sirohi, P.S., Patel, V.B. and Mahure, H.R. (2010) *Indian J. Hort.*, **67**(Special Issue): 170-173.

Susila, T., Reddy, S.A., Rajkumar, M., Padmaja, A. and Rao, P.V. (2010) *J. Hortic. Sci. Ornamental Plants*, **2**(1): 19–23.

Swaminathan, M.S. and Singh, M.P. (1958) *Curr. Sci.*, **27**: 63-64.

Swamy, T.N., Bylla, P., Suresh, V. and Mustafa, Md. (2015) *Sci. Res. Report.*, **5**: 177-180.

Swiadcr, J.M. and Al-Redhaiman, K. (1998) *J. Veg. Crop Prod.*, **4**: 45-56.

Swiadcr, J.M., Sipp, SK. and Brown, R.E. (1994) *J. Amer. Soc. Hort. Sci.*, **119**: 414-419.

Szalai, J. (1995) *HortScience*, **27**: 111-113.

Sztangret-Wiœniewska, J., Ga ecka, T., Korzeniewska, A., Marzec, L., Ko akowska, G. and Piskurewicz, U. (2006) In: *Proc. Cucurbitaceae* 2006 (Ed. G.J. Holmes), Raleigh, North Carolina, USA. pp. 515-526.

Szwacka, M., Krzymowska, M., Osuch, A., Kowalczyk, M.E. and Malepszy, S. (2002) *Acta Physiol. Plant*, **24**: 173–185.

Tabei, Y. (1997) *Bull. Nat. Inst. Agrobiol. Resour.*, **11**: 100-107.

Tabei, Y., Kanno, T. and Nishio, T. (1991) *Plant Cell Rep.*, **10**: 225-229.

Tabei, Y., Kitude, Nishizawa, Y., Kikuchi, N., Kayano, T., Hibi, T. and Akutsu, K. (1997) *Plant Cell Rep.*, **17**: 159-164.

Tabei, Y., Nishio, T. and Kanno. T. (1992) *J. Japanese Soc. Hort. Sci.*, **61**: 225-229.

Tachibana, S. (1989) *J. Jpn. Soc. Hor. Sci.*, **58**: 333–337.

Tadmor, Y., Burger, J., Yaakov, I., Feder, A., Libhaber, S.E., Portnoy, V., Meir, A., Tzuri, G., Sa'ar, U., Rogachev, I., Aharoni, A., Abeliovich, H., Schaffer, A.A., Lewinsohn, E. and Katzir, N. (2010) *J. Agric. Food Chem.*, **58**: 10722–10728.

Takahashi, H. and Kawagoe, H. (1971) *Agr. Hort.*, **46**: 1581–1584. (In Japanese).

Talekar, N.S., Vaddoria, M.A. and Kalkaran, G. V. (2013) *Progress. Res.*, **8**: 650-653.

Taler, D., Galperin, M., Benjamin, I., Cohen, Y. and Kenigsbuch, D. (2004) *Plant Cell*, **16:** 172–184.

Tamilselvi, N.A. and Jansirani, P. (2016) *Int. J. Veg. Sci.*, **22**(2): 170-182.

Tang, S.H., Liao, Y.F. and Xu, R.C. (1994) *J. Southwest Agric. Univ.*, **16**: 540-542.

Tang, T.A. and Punja, Z.K. (1989) *Cucurbit Genet Coop. Rep.*, **12**: 29-34.

Tao, Q., Chang, Y.L., Wang, J., Chen, H., Islam-Faridi, M.N., Scheuring, C., Wang, B., Stelly, D.M. and Zhang, H.B. (2001) *Genetics*, **158**: 1711–1724.

Tao, Y.M., Zhou, M.Y. and Xia, 7.11. (1991) *J. Shanghai Agric. College*, **9**: 213-216.

Tarakanov, G.I, Borisov, A.V. and Krylov, O.N. (1986) In: *Pragressivnye Priemy Ovoshchevodstve, slektsii i semenovodstve ovoschenykh Kul 'tur*, Moscow, USSR, pp. 23-27.

Tarasenko, L.V. and Uspenskii, G. B. (1996) *Tsitologiya I Genetika*, **30**: 19-23.

Tateishi, K. (1931) *Jissaiengei*, **11**: 283-282. (In Japanese).

Taylor, A.D. (1983) In: *Proc. Conference on the Biology and Chemistry of Cucurbitaceae*, Cornell University, Ithaca, New York, August, 1980.

Tekheinovich, G.A. and Fursa, T.B. (1986) *Sbornik Nauclmykh Trudovpo Prikladnoi Botanike, Genetike i Selcktsii*, **101**: 38-44.

Thakur, G.S., Sharma, R., Sanodiya, Pandey, M., Baghel, R., Gupta, A. and Bisen, P.S. (2011) *African J. Biotech*, **10**: 15808-15812.

Thakur, J.C., Khattra, A.S. and Dhanju, K.C. (1996) *Punjab Vegetable Grower*, 31: 25-28.

Thakur, M.R. and Choudhury, B. (1967) *Indian J. Hort.*, **24**: 87-94.

Thakur, P., Dash, S.P. and Kumar, K. (2016) *J. Agroecol. Nat. Resour. Manag.*, **3**(3): 220-224.

Thamburaj, S. and Singh, N. (2005) Vegetable, tuber crops and spices. Indian Council of Agricultural Research, New Delhi, pp. 10–75.

Thies, J.A., Ariss, J.J., Hassell, R.L., Buckner, S. and Levi A. (2015) *HortScience*, **50**: 4–8.

Thies, J.A., Ariss, J.J., Hassell, R.L., Olson, S., Kousik, C.S. and Levi, A. (2010) *Plant Dis.*, **94**: 1195 1199.

Thiruvengadam, M., Chung M-III, and Sechun Chun (2012b) *J. Med. Plants Res.*, **6**: 3579-3585.

Thiruvengadam, M., Jeyakumar, J. J., Kamaraj, M., Lee, Y.J. and Chung, I.M. (2013) *Aust. J. Crop Sci.*, **7**: 969-977.

Thiruvengadam, M., Mohamed, S.V., Yanf, C.H. and Jayabalan, N. (2006) *Sci. Hortic.*, **109**: 123-129.

Thiruvengadam, M., Praveen, N. and Chung, I.M. (2012a) *Aust. J. Crop. Sci.*, **6**(6): 1094–1100.

Thiruvengadam, M., Praveen, N. and Chung, I.M. (2012b) *Afr. J. Biotechnol.*, **11**(32): 8218–8224.

Thiruvengadam, M., Rekha, K. T., Jayabalan, N., Praveen, N., Kim, E. H. and Chung,I. M. (2013) *Aust. J. Crop Sci.*, **7**: 449-453.

Thiruvengadam, M., Rekha, S.K. and Yang, C.H. (2007) *Funct. Plant Sci. Biotechnol.*, **1** : 200-206.

Thomas, C.E. and Webb, R.E. (1981) *HortScience*, **16**: 96.

Thomas, C.E., Cohe, Y. and McCreight, J.D., Jourdain, E.L. and Cohen, S. (1988) *Plant Disease*, **72**: 33-35.

Thomas, P., Mythili, J.B. and Shivashankara, K.S. (2000) *J. Hort. Sci. Biotech.*, **75**: 19-25.

Thomas, T.D. and Sreejesh, K.R. (2004) *Sci. Hortic.*, **100**: 359–367.

Tiizel, Y. and Giil, A. (1991) *Plasticulture*, **91**: 37-40.

Tiwari, J.K., Munshi, A.D., Kumar, R. and and Sharma, R.K. (2010a) *Veg. Sci.*, **37**(1): 81-83.

Tiwari, J. K., Munshi, A.D., Kumar, R. and Sureja, A.K. (2010b) *Indian J. Hortic.*, **67**(2): 197-201.

Tiwari, J.K., Munshi, A.D., Kumar, R, Sureja, A.K., and Sharma, R.K. (2013) *Indian J. Hortic.*, **70**(1): 135-138.

Tobias, I. and Tulipan M. (2002) *Novenyvedelem*, **38**: 23–27.

Tomar, B.S., Jat, G.S. and Singh, J. (2017) Advances in Quality Seed Production of Vegetable Crops. In: CAFT 2017-18, Advances in Hybrids Seed Production of Vegetable Crops.

Tomar, B.S., Sharma, B.B., Jat, G.S., Munshi, A.D., Behera, T.K., Yadav, R.K., Saha, P. and Prakash, C. (2018) Pusa Vegetable Varieties for Nutrition and Health. Division of Vegetable Science, IARI, New Delhi. TB-ICN: 187/2018, p. 32.

Toppi, L. S. di., Pecchioni, N. and Durante, M. (1997) *Plant Cell Tissue Org. Cult.,* **51**: 89-93.

Traka-Mavrona, E., Koutsika-Sotiriou, M. and Pritsa, T. (2000) *Sci. Hortic.,* 83: 353–362.

Trionfetti-Nisini, P., Colla, G., Granati, E., Temperini, O., Crino, P. and Saccardo, F. (2002) *Sci. Hortic.* **93**: 281–288.

Triooli, D.M., Camey, K.J., Russell, P.F., McMaster, J.R., Groff, D.W., Hadden, K.C., Himmel, P.T., Hubbard, J.P., Boeshore, M.L. and Quemada, H.D. (1995) *BioTechnol.,* **13**: 1458-1465.

Trivedi, R.N. and Roy, R.P. (1970) *Cytologia,* **35**: 561-569.

Trivedi, R.N. and Roy, R.P. (1972) *Genetica,* **43**: 282-291.

Trivedi, R.N. and Roy, R.P. (1973) *Cytologia,* **38**: 317-325.Trivedi, R.N. and Roy, R.P. (1976) *Genet. Iber.,* **28**: 83-106.

Troung-Andre, I. (1988) Proc. Eucarpia Meet Cucurbit. Avignon-Monfavet, France, pp. 143–144.

Trulson, A.J. and Shahin, E.A. (1986) *Plant Sci.,* **47**: 35-43.

Tuncay, O.I., DO. Over, Rget, M.E., Gul, A. and Budak. N. (1999) In: *Improved Crop Quality by nutrient management,* (Eds. D. Anac, and P.Marlin-Prevel), pp. 193-195.

Turchenkov, G.L., Avdeev, Xu, I., Alekseeva, V.S. and Sheherbinin, N. (1992) *Kartofel I Ovoshchi,* No. 1, p. 9.

Tyagi, S.V.S, Sharma, P., Siddiqui, S.A. and Khandelwal, R.C. (2010) *Int. J. Veg. Sci.,* **16**: 267–277.

Udalova, V.B. and Prikhod'Ko, V.F. (1986) *Byulleten Vsesoyuznogo Instituta Gel 'mintologii imeni K.I. Skryabina,* No. 45, pp. 63-67.

Udalova, V.B. and Prikhod'Ko, VF. (1985) *Byulleten Vsesoyuznogo Instituta Gel 'mintologii imeni K.I. Skryabina,* No. 41, pp. 67-70.

Uffelen, J. (1971) *Groenten en Fruit,* **27**: 455.

Uffelen, J.A.M. Van (1984) *Groenten en Fruit,* **40**: 43-45.

Uffelen, J.A.M. Van (1986) *Groenten en Fruit,* **41**: 48-49.

Uffelen, J.A.M. Van (1989) Groenten en Fruit, **45**: 38-39.

Umamaheswari, C., Ambethkar, A., Margaret, F.S. and Selvaraj, N. (2014) *Inter. J. Current Biotech.,* **2**: 7-13.

Urasaki, N., Takagi, H., Natsume, S., Uemura, A., Taniai, N., Miyagi, N., Fukushima, M., Suzuki, S., Tarora, K., Tamaki, M., Sakamoto, M., Terauchi, R. and Matsumura, H. (2017) *DNA Res.,* **24**(1): 51–58.

Usman, Hussain, Z. and Fatima, B. (2011) *Pak. J. Bot.,* **43**: 1283-1293.

Utffelen, J.A.M. Van, Hogendonk, L. and Steenbergen, P. (1991) *Groenten en Fruit,* **47**: 12-13.

Uzun, S., Reksen, A. and Alan, R. (1999) Ondokumayis Universitesi, Ziraat Fakultesi Dergisi, **14**: 166-179.

V'yukov, A. (1984) *Byulletin Vsesoyuznogo Nauchno Issledovate'skogo Instituta Udobreniil Agropochvovedeniva*, No. 64. pp. 34-36.

Valdez-Melara, M., García, A., Delgado, M., Andres, M., Gatica-Arias and Ramírez-Fonseca, P. (2009) *Rev. Biol. Trop.*, **57**: 119-127.

Valles, M.P. and Lasa, J.M. (1994) *Plant Cell Rep.*, **10**: 225-229.

Varalakshmi, B., Pitchaimuthu, M. and Rao, E.S. (2019) *J. Hortic. Sci.*, **14**(1): 48-57.

Varalakshmi, B., Pitchaimuthu, M., Rao, E.S., Krishnamurthy, D., Suchitha, Y. and Manjunath, K.S.S. (2014) In: *International Bitter gourd Conference* (BiG2014) organized by AVRDC at ICRISAT, Hyderabad in March, 2014, P. 36.

Varghese, B.M. (1972) *Genetica*, **43**: 292-301.

Varghese, B.M. (1971) *Cytologia*, **36**: 205-209.

Varghese, B.M. (1973) *Curr. Sci.* **42**: 30.

Vasudeva, R.S. and Lal, TB. (1943) *Indian J. Agric. Sci.*, **13**: 182-192.

Vasudeva, R.S., Raychoudhuri. S.P. and Singh, I. (1949) *Indian Phytopath.*, **2**: 181-185.

Vecchia, P.T., Della, Takazaki. P.E., Terenciano, A. and Costa, C.P. (1991) *Horticultura Brasiliera*, **9**: 27.

Vedat, P., Onay, A., Yildirim, H., Adiyaman, F., Ifiikalan, C. and Bafiaran, D. (2002) *Turk J. Bio.*, **27**: 101-105.

Velivela, Y., Narra, M., Ellendula, R., Kota, S. and Abbagani, S. (2016) *J. Appl. Biol. Biotech.*, **4**: 41-45.

Vemla, T.S., Singh, R.V. and Sharma, S.C. (2000) *Indian J. Hortic.*, **57**(2): 144-147.

Venkateshwaralu, M. (2010) *Int. J. Plant Protection*, **3**: 107-110.

Venkateshwaralu, M. (2012) *Int. J. Pharma Bio Sci.*, **3**: 645-652.

Verhaar, M.A. and Hijwegen, T. (1994) In: Proc. Of the International Conference at the Occasion of the 75th Anniversary of the Wageningen Agricultural University (Eds.P.C.Struik, , W.J., Vscdenberg, J.A. Renkema, and J.E Parlevliet), 8 June-1 July I993, Dordrecht, Netherlands, Kluwer Academic Publishers, pp.373-374.

Verma, A.K., Kumar, M., Tarafdar, S., Singh, R. and Takur, S. (2014) *Int. J. Plant, Animal and Environ. Sci.*, **4**: 275-280.

Verma, V.K., Singh, N. and Choudhury, B. (1986) *South Indian Hort.*, 34: 105-111.

Verzera, A., Dima, G., Tripodi, G., Condurso, C., Crinò, P., Romano, D., Mazzaglia, A., Lanza, C.M., Restuccia, C. and Paratore, A. (2014) *Sci. Hortic.*, 169: 118-124.

Vijayakumar, A., Arunachalam, M. and Pandian. I.R.S. (1995) *South Indian Hort.*, **43**: 103-105.

Vishnu, S. and Prabhakar, B.S. (1987) *South Indian Hort.*, **35**: 453-454.

Visser, D.L. and Franken, J. (1979) *Cucurbit Genet. Coop. Rep.*, **2**: 46-47.

Vlugt, J.L.F. Van Der (1989a) *Norwegian J. Agric. Sci.*, **3**: 265-274.

Vlugt, J.L.F. Van Der (1989b) *Norwegian J. Agric. Sci.*, **3**: 275-279.

Vysochin, V.G. (1998) *Kartofel i Ovoshchi*, No. 2, p. 42.

Vysochin, V.G., Sankin, L.S. and Tulupov, Yu Ku (1985) *Sibirskii Véstnik Sellskokhozyaistvennoi Nauk*i No. 3, pp. 103-105, 124.

Walker, J.C. and Pierson, Cf. (1955) *Phytopathology*, **45**: 451-453.

Walters, S.A. and Wehner, T.C. (1998) *HortScience*, **33**: 1050-1052.

Walzer, A. and Schausberger, P. (2000) *Forderungsdienst*, **48**: 50-51.

Wan, S.Q., Zheng, H. and Xu, M. (1998) *China vegetables*, **6**: 24-26.

Wang, H.R., Ru, S.J., Wang, L.P. and Feng, Z.M. (2004) *Acta Agriculturae Zhejiangensis*, **16**: 336–339.

Wang, H.Z., Zhao, P.J., Xu, J.C., Zhao, H. and Zhang, H.S. (2003) *Yi Chuan Xue Bao*, **30**: 70–75.

Wang, J., Zhang, D.W. and Fang, Q. (2002) *J. Anhui Agr. Univ.*, **29**: 336–339.

Wang, Q.M. and Zeng, G.W. (1996) *Journal of Zhejiang Agriculture University*, **22**: 541–546.

Wang, Q.M. and Zeng, G.W. (1997a) *J. Zhejiang Agric. Univ.*, **27**: 555-556.

Wang, Q.M. and Zeng, G.W. (1997b) *Acta Horticulturae Sinica*, **24**: 28-52.

Wang, S.Z., Pan, L., Hu, K., Chen, C. and Ding, Y. (2010) *Amer. J. Bot.*, **97**: e75–e78.

Wang, X., Shang, L. and Luan, F. (2013) *Pak. J. Bot.*, **45**(1): 145-150.

Wang, Z. and Xiang, C. (2013) *Euphytica*, **193**: 235–250.

Waskar, D.P., Yadav, B.B., Garande, V.K. (1999) *Indian J. Agric. Res.*, **33**: 287-292.

Waterer, D. (2000) *Canadian J. Plant Sci.*, **80**: 385-388.

Watson, A. and Napier, Y.T. (2009) *Primefact*, 832.

Watterson, J.C., Williams, P.H. and Durbin, R.D. (1971) *Plant Disease Reporter*, **55**: 816-819.

Watts, V.M. (1962) *Proc. Amen Soc. Hort. Sci.*, **81**: 498-505.

Watts, V.M. (1992) *Proc. Amer. Soc. Hort. Sci.*, **91**: 579-583.

Weber, C. (2000) *Gemuse (Munchen)*, **36**: 30-32.

Wechter, W.P., Levi, A., Harris, K.R., Davis, A.R., Fei, Z.J., Giovannoni, J.J., Trebitsh, T., Salman, A., Hernandez, A., Thimmapuram, J., Tadmor, Y., Portnoy, V. and Katzir, N. (2008) *BMC Genomics*, **9**: 275–282.

Wehner, T.C. (2007) North Carolina State University, Raleigh, NC, USA, pp. 381–418. Wehner, T.C. (2008) In: Prohens, J. and Nuez, F. (Eds.), Watermelon, Springer, New York, NY, USA, pp. 381–418.

Wehner, T.C., Jenkins, S.F.Jr. and Lower, R.L. (1991) *HortScience*, **26**: 78-79.

Whitaker, T.W. (1933) *Bot. Gaz.*, **94**: 780–790.

Whitaker, T.W. and Bemis, W.B. (1976) Longman, London, UK, pp. 64–69.

Whitaker, T.W. and Davis, G.N. (1962) Interscience, New York, NY, USA.

Wien, H.C. (1997) In: *The Physiology of Vegetable Crops* (Ed. H.C. Wien), CAB International, New York, pp. 345–386.

Wien, H.C. (2007) In: *The Physiology of Vegetable Crops* (Ed. H.C. Wien), CAB International, New York, pp. 345–386.

Wimer, J.A., Miles, C.A. and Inglis, D.A. (2014) *Acta Hortic.*, **1085**: 363–364.

Xanthopoulou, A., Montero-Pau, J. and Mellidou, I. (2019) *Hortic Res.*, **6**: 94.

Xu, C.Q., Li, T. L. and Qi, H. Y. (2006) *Acta Hort. Sinica*, **33**: 773–778.

Xu, S. L., Chen, Q. Y., Li, S. H., Zhang, L. L., Gao, J. S., and Wang, H. L. (2005) *J. Fruit Sci.*, **22**: 514–518.

Xu, Y., Guo, S., Zhang, H., Fei, Z., Zhang, H., Ren, Y., Zhao, H., Lv, G., Gong, G., Kou, Q., Zou, X., Wang, H. and Hou, W. (2010) *J. Amer. Soc. Hortic. Sci.*, pp. 125–128.

Xu, Y., Ouyang, X.X., Zhang, H.Y., Kang, G.B., Wang, Y.J. and Chen, H. (1999) *Acta Bot. Sin.*, **41**: 952–955.

Yamaguchi, J. and Shiga, T. (1993) *Japanese J. Breed.*, **43**: 173-182.

Yamasaki, S., Fujii, N., Takahashi, H., Mizusawa, H. and Matsuura, S. (2002) *Acta Hort.*, **588**: 309-312.

Yamamoto. Y. and Matsumoto, O. (1994) *J. Japanese Soc. Hort. Sci.*, **63**: 67-72.

Yang.C. (1985) *Acta Hort. Sinica*, **12**: 101-106.

Yang, L.F., Zhu, Y.L., Hu, C.M., Liu, Z.L. and Zhang, G.W. (2006) *Acta Botanica Boreali-occidentalia Sinica*, **26**: 1195–1200.

Yang, L.M., Koo, D.H., Li, Y.H., Zhang, X.J., Luan, F.S. and Havey, M.J., Jiang, J. and Weng, Y. (2012) *Plant*, **71**: 895–906.

Yang, L.M., Li, D.W., Li, Y.H., Gu, X.F., Huang, S.W. and Garcia-Mas, J. (2013) *BMC Plant Biol.*, **13**: 53.

Yang, M., Zhao, M., Zeng, Y., Lan, L. and Chen, F. (2004) *High Technol. Lett.*, **10**(1): 44–48.

Yang, Y. (1998) *Beijing Agricultural Science*, **16**: 23-26.

Yang, Y.G., Zhang, N.Z., Xie, J.C., Xu, W.J., Liu, K.Q., Chen, Y.J. and Fan, S.Y. (2000) *Acta Agriculturae Universitatis Jiangxiensis*, **22**: 66-69.

Yang, Y.J., Lu, X.M., Yan, B., Li, B., Sun, J., Guo, S.R. and Tezuka, T. (2013) *J. Plant Physiol.*,**170**: 653–661.

Yang, Y.Z., Lin, D.C. and Guo. Z.Y. (1992) *Sci. Hort.*, **50**: 47-51.

Yang, Y.Z. (1984) *Ningxia Agril. Sci. Tech.*, **2**: 27-29.

Yano, R., Ariizumi, T., Nonaka, S., Kawazu, Y., Zhong, S. and Mueller, L. (2020) *Commun. Biol.*, **3**: 432.

Yetisir, H., Kurt, S., Sari, N. and Tok, F. M. (2006) *Australian J. Expt. Agric.*, **43**: 1269–1274.

Yetisir, H., Kurt, S., Sari, N. and Tok, F.M. (2007) *Turk. J. Agr. For.*, **31**: 381-388.

Yetisir, H. and Sari, N. (2003) *Sci. Hort.*, **98**: 277-283.

Yetisir, H. and Sari, N. (2003) *Australian J. Expt. Agri.*, **43**: 1269-1274.

Yetisir, H. and Sari, N. (2004) *J. Agric. Turk.*, **28**: 231–237.

Yin, Z., Malinowski, R., Ziolkowska, A., Sommer, H., Plcader, W. and Malepszy, S. (2006) *Cell. Mol. Biol. Lett.*, **11**: 279–290.

Yin, Z., Pawlowicz, I., Bartoszewski, G., Malinowski, R., Malepszy, S. and Rorat, T. (2004) *Cell. Mol. Biol. Lett.*, **9**: 891–902.

Yonemori, S. and Fujieda, K. (1985) Science Bulletin, College of Agriculture, University of Ryukyus, Okinawa, Japan, No. 32, pp. 189-192.

Yoo, K.C., Kim, J.H., Yeoung, Y. and Lee, S.H. (1996a) *J. Korean Soc. Hort. Sci.*, **37**: 42-46.

Yoo, K.C., Kim, J.H., Yeoung, Y. and Lee, S.H. (1996b) *J. Korean Soc. Hort. Sci.*, **37**: 197-209.

Yoshioka, K., Hanada, K., Harada, T., Minobe, Y. and Oosawa K. (1993) *Jpn. J. Breed.*, **43**: 629–634.

Yu, T.A., Chiang, C.H., Wu, H.W., Li, C.M., Yang, C.F., Chen, J.H., Chen, Y.W. and Yeh, S.D. (2011) *Plant Cell Rep.*, **30**: 359–371.

Yue Jin, X. and RuLai, R. (1998) *China Vegetables*, **6**: 3-5.Yurina, O.V. (1990) *Selektsiyai Semenovodstvo*, **3**: 20-22.

Yutaka, T., Tomohiro, Y., Toshikazu, M., and Takeshi, O. (1998) *JARQ*, **32**: 281-286.

Zalapa, J.E., Staub, J.E. and Mc Creight, J.D. (2007) *Theor. Appl. Genet.*, **114**: 1185–1201.

Zentgraf, U., Kind, K. and Hemleben, V. (1992) *Acta Botanica Neerlandica*, **41**: 397-406.

Zeven, A.C. and Zhukovsky. P.M. (1975) Dictionary of Cultivated Plants and their centres of Diversity, Wageningen.

Zhang, F.M., Liu, B.Z. and Liu, Z.T. (1991) In: *Proc. of international symposium on applied technology of green house* (Ed. B.Z. Liu), 7-10 October, 1991 Beijing, China, Knowledge Publishing House, pp. 216-221.

Zhang, Q. and Zhang, Q. (1998) *China Vegetables*, **5**: 31-32.

Zhang, R., Xu, Y., Yi, K., Zhang, H., Lie, L., Gong, G. and Levi, A. (2004) *J. Amer. Soc. Hort. Sci.*, **129**: 237–243.

Zhang, S. (1989) *Zuowu Pinzlzong Zivuan*, No. 4, P. 7.

Zhang, S.P., Gu, X.F. and Wang, Y. (2014) *Acta Horticulturae Sinica*, **33**: 1231–1236.

Zhang, X.G. and Liu, P.Y. (1998) *J. Southwest Agric. Univ.*, **20**: 293-297.

Zhang, X.G., Chen, J.F. and Liu, P.Y. (1998) *Adv. Hort.*, **2**: 434-438.

Zhang, X.P., Rhodes, B.B., Baird, W.V., Skorupska, H.T. and Bridges W.C. (1996) *HortScience*, **31**(1): 123–126.

Zhang, X.P. and Wang, M. (1990) *Rep. Cucurbit Genet. Coop.*, **13**: 45-46.

Zhang, X.Y., Fu, Q., Zhu, H. and Wang, H. (2014) *China Vegetables*, **6**: 13–19.

Zhang, Y.J., Li, M.F., Xiang, N. and Li, G.F. (2006) *Acta Phytophylacica Sinica*, **33**: 32-36.

Zhang, C.M. and Lu, J. (1995) *Acta Agriculturae Shanghai*, **11**: 31-36.

Zhao, D.G., Sun, H.Y. and Ren, G.S. (1998) *China Vegetables*, **2**: 20-21.

Zhao, J. P., Bi, K. H., Jiang, X. M. and Bai, X. F. (1998) *Plant Physiol. Commun.*, **29**: 435-437.

Zherdetskaya, T.N. and Levashenko. G.I. (1996) *Zashchita Karantin Rastenii*, **4**: 43.

Zhou, Q., Miao, H., Li, S., Zhang, S., Wang, Y. and Weng, Y. (2015) *Mol. Plant*, **8**: 961–963.

Zhou, W.B., Luo, S.S.B. and Luo, T.N. (1997) *China Vegetables*, **3**: 19-20.

Zhou, X.Z., Wu, Y.F., Chen, S., Chen, Y., Zhang, W.G., Sun, X.T. and Zhao, Y.J. (2014) *HortScience*, **49**: 1365–1369.

Zhou, Y.H., Zhou, J., Huang, L.F., Ding, X.T., Shi, K. and Yu, J.Q. (2009) *J. Plant Res.*, **122**: 529–540.

Zhou, Y., Ma, Y., Zeng, J., Duan, L., Xue, X., Wang, H., Lin, T., Liu, Z., Zeng, K. and Zhong, Y. (2016) *Nature Plants*, **2**: 16183.

Zhu, C.E., Yan, Q.G., Che, E.Z. and Qian, G.Y. (1993) *Chinese Vegetables*, **2**: 3-4.

Zhu, J., Bie, Z.L., Huang, Y. and Han, X.Y. (2006) *China Vegetables*, **9**: 24–25.

Zhukovsky, P.M. (1962) *Cultivated Plants and their Relatives* (Translation by P.S. Hudson), Bucks.

ZiDe, Zhang, JunLian, Ma and RucJing, Z. (1996) *I. Hebei Agric. Univ.*, **19**: 4044.

Zink, F.W. and Gubler, W.D. (1987a) *HortScience*, **22**: 1342.

Zink, F.W. and Gubler, W.D. (1987b) *Hortscience*, **22**: 1372.

Zitter, T.A. (1998) Fusarium diseases of cucurbits. http://vegetablemdonline.ppath.cornell.edu/factsheets/Cucurbits_Fusarium.htm.

Zitter, T.A., Hopkins, D.L. and Thomas, C.E. (1996) APS Press, St. Paul, MN, USA. Zohura, F.T., Haque, M. E., Islam, M. A., Khalekuzzaman, M. and Sikdar, B. (2013) *Int. J. Sci. Technol. Res.*, **2**(9): 33-37.

Zraidi, A, Stift, G., Pachner, M., Shojaeiyan, A., Gong, L. and Lelley, T. (2007) *Mol. Breed.*, **20**: 375–388.

ONION

J. Mandal , T.K. Maity, N.N. Shinde,
M.B. Sontakke, J. Kabir and P. Munshi

Bulb Crops

This bulb group consists of a wide range of underground vegetables like onions, garlic, leek, welsh onion, shallot, chive, *etc*. They are the members of genus *Allium*, which is tropically well adapted to suit different ecosystems. Among the bulb crops, onion and garlic are the most important vegetables, extensively grown throughout the tropics. Both the crops are used as spices and condiments all over the world. These are among the old cultivated species grown in India. Utility of them as vegetable drugs was mentioned in our ancient literatures like 'Charaka Samhita' around 600 B.C. (Ray and Gupta, 1980) and 'Susruta Samhita' of 3rd and 4th century A.D. (Ray *et al.*, 1980).

Distribution

Alliums are widely distributed throughout the temperate northern hemisphere of the globe. In the new world, there are 80 species mostly in western states, extending to Mexico and Guatemala. The old world species are abundant in Europe, especially in Russia, North Africa and Asia. Only the species *A. schoenoprasum* is common both in North America and Eurasia and this hardy species is exceptional in its range of distribution– it extends from Arctic down to tropical latitudes in Ethiopia and Sri Lanka (Ceylon). It is assumed that India, Pakistan and Bangladesh, are the areas where it is most likely that onion started its tropical domesticated existence and is the probable route taken by onion to reach other parts of tropics (Currah and Proctor, 1990).

Zeven and Zhukovsky (1975) recorded *Allium* species in different centres of diversity.

Allium chinense, A. fistulosum, A. ledebournianum, A. macrostemon, A. nipponicum, A. sativum, A. ramosum, A. schoenoprasum (chive), *A. tuberosum* (Chinense chive) Chinese-Japanese centre.

Allium ampeloprasum – Levant garlic, 2n = 16, perennial sweet leek in Kashmir, tropical South Asian centre or Hindusthan centre.

Allium cepa and *A. sativum* originated in the region of Central Asia (Pakistan, Afghanistan, Iran, southern Russia, western China).

A. ampeloprasum, A. ascalonicum, A. kurrat, A. porrum and *A. sativum* – near Eastern Saudi Arabia, Iraq, Turkey, Georgia, Russia. They further identified the secondary centre (Mediterranean) for *A. cepa* and *A. sativum* and European and Siberian centre for *A. ampeloprasum* and *A. scorodoprasum*.

There are more than 70 species (belonging to 8 genera) of edible Alliaceae vegetables in China. In *Allium*, garlic, Welsh onion, Chinese chives and onion rank the first four places, respectively. Garlic has long been cultivated in China both for its edible parts of tender seedlings, flower stalks and bulbs, and for its medicinal effects. To date, 148 accessions of garlic have been collected. Onion has been cultivated for only 90 years and, to date, 54 accessions and 9 clones of 3 varieties have been collected. The main areas for garlic and onion production centre are the central and eastern parts of China (Feng *et al.*, 1997).

The genus *Allium* is one of the largest genera of monocots comprising currently more than 900 species widely distributed on the northern hemisphere from the dry subtropics to the boreal zone. Especially, the region from the Mediterranean area to southwest and central Asia is characterized by high species diversity (Fritsch and Friesen 2002; Fritsch *et al.*, 2010; Fritsch and Abbasi 2013). The bulb onion (*Allium cepa* L.) is grown on all continents except Antarctica (Havey, 2018)

Fritsch *et al.* (2001) reported that *A. vavilovii* and a new Iranian species *Allium asarense* are the closest among the known relatives of the common onion *A. cepa*.

Taxonomy

The genus as a whole is assigned to the family Liliaceae by some taxonomists (Polunin, 1969) to which it belongs on account of the superior ovary and to the Amaryllidaceae by other (Traub, 1968) because of the umbellate inflorescence.

Others have opted for a subfamily – Allioideae – of the Liliaceae (McCollum, 1976) and others again favour a separate family altogether, Alliaceae – within which can be placed all those other genera, *e.g.*, *Agapanthus, Brodiaea, Triteleia, Tulbaghia* which have superior ovaries and umbellate inflorescence. The Angiosperm Phylogeny Group (APG) reassessed the taxonomic position of this genus and finally *Allium* was placed in the Amaryllidaceae family (APG III 2009).

The number of species in the genus has varied from as low as the thirty one described by Linnaeus in 1753 (Traub, 1968) to as high as 1100 (Stearn, 1944). Some of the species are highly variable and difficult to classify unambiguously. *A. schoenoprasum* (chives) has been grown and sold in the USA under 12 different specific names and the ornamental species *A. senescens* distributed under an even larger list over 32 different species (Moore, 1954). The number of synonyms and wrong classifications has greatly been reduced and estimates are down to around 600 species. However, according to Govaerts *et al.* (2013) and Keusgen *et al.* (2011) the genus *Allium* L. is one of the largest monocotyledonous genera with c. 900 species distributed world-wide.

Vvedensky (1935) grouped 228 species from the then USSR into 9 sections and Stearn (1944) later proposed that the old world species be provisionally classified into 14 separate sections. The new world *Alliums* from North America were organized by Ownbey and Aase (1955) into 9 alliances which were later reduced down to 8 (Saghir *et al.*, 1966). Traub (1968) grouped American species under 4 sections to put them on the same basis of classifications as the old world species, which were later revised into subgenera, sections and subsections.

The three subgenera are based on natural groupings of significant relationships between basic chromosome number and orientation of vascular bundles in the leaf blade. These are *Amerallium* with x = 7 (with a few sp. x = 8 or x = 9), *Nectaroscordum* (all with x = 8) and *Allium* (mainly x = 8, few x = 9 or 10). Within the subgenera there are XVII sections and their subsections with grouping of species on a phenological basis which takes into account of heritable characteristics, such as anatomy, chemical composition, method of propagation, colour, taste, *etc.* in addition to morphological variation. Sections I-IV includes the North American species with x = 7 and subsections are based on the Origin Alliances of Saghir *et al.* (1966). The remaining sections (V-XVII) include all other species

Decorative Alliums

They embrace a wide range of colours, form and several of the more familiar ones described by Moore (1954) and Mathew (1973) as ornamental species are:

Allium azureum (2n = 2x = 16) bright blue

A. curmum (2n = 2x = 14) pink/red nodding onion

A. flavum (2x = 16) highly fragrant yellow pendulous inflorescences

A. moley (2x = 14) bright yellow

A. karataviense (2x = 18) dwarf, preferred in rock gardens

A. pulchellum (2x = 16) purple, scented, pendulous inflorescence

A. neapolitanum (2x = 14) white flowered, *A. senescens* (4x = 32) pink flowered

A. subhirsutum (2x = 14), *A. zebdanense* (2x =18) white flowered

A. roseum (2x = 16) white flowered

A. triquetrum (2x =18) vigorous white flowered (Jones, 1983)

Edible Alliums

Allium cepa (2n = 2x = 16): It has been in use as vegetable for even 5000 years. The ancestral species in unknown but there are several closely related wild diploids of Central Asia (*A. vavilovii, A. oschaninii, A. pskemense, A. galanthum*) which are near relative and all of which form interspecific hybrids with *A. cepa*.

A. cepa var. *ascalonicum* (2n= 2x = 16): Shallot. It is a perennial onion, which rarely produced seed and which is perpetuated each year by replanting some of the bulbs formed in clusters on the surface of the soil.

A. cepa var. *aggregatum* (2n = 2x = 16): Potato onion. Underground onion or multiplier onion is also known as Egyptian ground onion. It grows as closely packed clusters of bulbs underground rather than on the surface like shallot.

A. cepa var. *viviparum/proliferum* (2n = 2x = 16): Tree onion or Egyptian tree onion. It is a *cepa* × *fistulosum* interspecific hybrid (Bozzini, 1964). It is a viviparous plant that grows as a perennial bulb underground, having leaves similar to onion and produces clusters of bulblets at the top of the stem in place of inflorescence. A few flowers may form with bulblets which are sterile, not widely grown. The tree onion is noted for resistance to pests and diseases. Puizina and Papes (1999) reported that top, tree or Egyptian onion, *A. proliferum* 2n = 2x = 16, is a minor, vegetatively propagated garden crop in Europe, North America and north East Asia. Several clones of the top onion were found to be locally cultivated in the region of south Croatia under the name Ljutika-talijanka (Puizina and Papes, 1999).

Different Species of Allium

A. chinense   *A. fistulosum*

A. schoenoprasum

A. tuberosum

A. ampeloprasum

Different Species of Allium

A. cepa var. *aggregatum*

A. cepa var. *viviparum/proliferum*

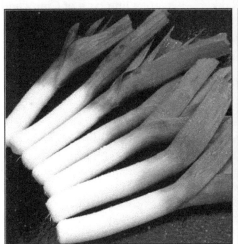

A. kurrat   **A. porrum**

A. sativum (2n = 2x = 16): Second most important *Allium* species. Perennial species are producing a much divided bulb consisting of several cloves covered by a thin white skin. It differs from *cepa* in that the leaf bases do not store food, but mature as dry scales enclosing cloves. The cloves themselves are well developed axillary buds within the thin foliage leaves. *A. sativum* is sexually sterile diploid unknown in the wild. Some authors consider *A. longicuspis* widespread in Central Asia, equally viviparous and sterile, to be its wild ancestor. Under domestication *A. sativum* has become a vegetatively propagated plant exclusively (should not be confused with great headed garlic *A. ampeloprasum*).

A. fistulosum (2n = 2x = 16): Welsh onion or Japanese bunching onion. No special connection with Wales of the U.K., perennial bunching type grown for its edible tops and long leaf bases, very common in China and Japanese gardens.

A. porrum (2n = 4x = 32): Leek. It is considered as a cultivated form of wild *A. ampeloprasum* which is common in the Mediterranean and south-west Asia, a biennial, edible leaf bases, green leaves, a much lesser tendency to form bulbs than *A. ampeloprasum*.

A. kurrat (2n = 4x = 32): Kurrat. Another cultivated form of *ampeloprasum* grown for green leaves, common in Egypt and Near East.

A. ampeloprasum (2n = 4x = 32, 6x = 48): Great headed garlic. Well-known in North America and Europe, used as substitute for garlic in cooking, larger bulb than garlic leaves like the leek, and also produces an inflorescence.

A. schoenoprasum (2x = 16, 24, 32): Chive. These are hardy bunching perennial herbs, widely grown and favoured for their hollow green leaves. It is propagated by root division; tolerant to extreme cold and drought.

A. tuberosum (2n = 4x = 32): Chinese chives of eastern Asia. These are grown for green leaves.

A. chinense (2n = 2x = 16, 4x = 32): Rakkyo. It is a pickling type of China and Japan and South Asia.

Besides these, there are also other wild species like *A. ursinum*, wild garlic, *A. victorialis*, broad-leaved species eaten in Russia.

ONION

1.0 Introduction

Onions have an extensive culinary, dietary, therapeutic, trading, income and employment generation value. Unlike other vegetables, it can be kept for a fairly long period. Onion is second only to tomato in their importance as a vegetable in the tropics. The demand for onions is worldwide. Onions are found on most markets of the world throughout the year and can be grown under wide range of agro-climatic conditions. Irrespective of price, the demand for onion remains almost constant in the market as it is essential ingradient of the cusine of many regions. Onion is primarily used to enhance flavour of different recipes.

2.0 Area, Production and Export

Onion is commercially cultivated about one hundred fifty countries of the world with production of 96.8 million tonnes in 2018. Asia contributed 67.5 per cent of world onion production. Out of which India contributed 22.81 per cent and China contributed 25.6 per cent, and thus two countries together contributed about half of the world onion production (FAOSTAT, 2020).

Table 1: Onion (Dry) Production in the World (tonnes)*

Continents	Year				
	2014	2015	2016	2017	2018
Asia	58494431	59607763	62647931	64397201	65319706
Africa	10950174	12148517	11757956	12230005	12453010
Americas	9401746	9597072	10368934	10494407	9768450
Europe	10054738	9789575	9799307	10099363	8953700
Oceania	256058	314950	264562	263251	278953
World	89157147	91457876	94838690	97484228	96773819

* FAOSTAT (2020).

Table 2: Top Ten Onion (Dry) Producing Countries of the World (2018)*

Country	Production (tonnes)
China	24775344
India	22071000
USA	3284420
Egypt	2958324
Iran	2406718
Pakistan	2119675
Turkey	1930695
Bangladesh	1737714
Russian Federation	1642106
Mexico	1572608

* FAOSTAT (2020).

Area wise top five onion growing countries in the world are India, China, Nigeria, Bangladesh and Indonesia. However, India occupies second position in total onion production after China. USA, Egypt and Iran are the next top onion growing countries but are much behind the two Asian giant. However, the productivity of onion in India (16.8 t/ha) is very low in comparison to Guyana (73.1 t/ha), USA (62.3 t/ha), Korea (57.6 t/ha), Taiwan (54.2 t/ha) and Australia (54.1 t/ha) (FAOSTAT, 2020).

Table 3: Area and Production of Onion in India*

Year	Area ('000 ha)	Production ('000 MT)	Productivity (kg/ha)
2015-16	1320	20931	15856.8
2016-17	1306	22427	17172.3
2017-18	1285	23262	18102.7

* Ministry of Agriculture and Farmers' Welfare (2018).

Table 4: State-wise Area, Production and Productivity of Onion in India (2017-18)*

States	Area ('000 ha)	Production ('000 MT)	Productivity (MT/ha)
Andhra Pradesh	42.00	915.73	21.80
Assam	8.34	80.37	9.64
Bihar	53.77	1240.59	23.07
Chhattisgarh	25.54	421.21	16.49
Gujarat	22.49	546.20	24.29
Haryana	29.93	701.50	23.44
Himachal Pradesh	2.69	52.19	19.43
Jammu and Kashmir	3.10	57.96	18.73
Jharkhand	17.16	289.04	16.84
Karnataka	195.28	2986.59	15.29
Kerala	0.03	0.31	10.00
Madhya Pradesh	150.87	3701.01	24.53
Maharashtra	507.96	8854.09	17.43
Manipur	0.56	6.84	12.25
Meghalaya	0.56	5.06	9.05
Mizoram	2.09	7.93	3.79
Nagaland	0.70	7.20	10.22
Odisha	33.47	379.34	11.33
Punjab	9.36	214.55	22.91
Rajasthan	64.76	996.73	15.39
Sikkim	0.62	35.00	56.45
Tamil Nadu	28.36	301.14	10.62
Telangana	17.97	326.59	18.18
Tripura	0.16	1.05	6.46
Uttar Pradesh	26.85	439.64	16.37
Uttarakhand	4.30	44.09	10.26
West Bengal	35.20	633.60	18.00
Others	0.88	16.77	19.15
India	1284.99	23262.33	18.10

* Ministry of Agriculture and Farmers' Welfare (2018).

India produced 23262.33 thousand metric tons of onion from 1284.99 thousand ha in 2017-18. During this period, the annual demand for onion was 20531 thousand metric tons (Table 3 and 4). Among the total onion production in India, about 71 per cent is used for domestic consumption, 20 per cent goes as waste during storage and handling, 5 per cent is used for export, 3 per cent for processing and 1 per cent bulbs are used for seed production (Singh *et al.*, 2017). As a major vegetable crop of India, it is mainly grown in the states like Maharashtra, Madhya Pradesh, Karnataka, Bihar, Rajasthan, Andhra Pradesh, Haryana, West Bengal, Gujarat and Uttar Pradesh. Maharashtra has the pre-eminent position in respect of onion production accounting for more than 39.5 per cent of area and 38.1 per cent of production. However, the top productivity of onion has been found in Sikkim (56.45), Madhya Pradesh (24.53), Gujarat (24.29), Haryana (23.44) and Bihar (23.07).

Table 5: Total Export of Fresh Onion from India*

Year	Quantity (MT)	Value (Rs. Lakhs)
2014-15	12,38,102.60	2,30,054.14
2015-16	13,82,959.54	3,09,720.85
2016-17	24,15,739.06	3,10,606.44
2017-18	15,88,985.72	3,08,882.22
2018-19	21,83,766.42	3,46,887.38

* APEDA, Ministry of Commerce and Industry, GOI.

Table 6: Top Ten Fresh Onion Importing Countries from India (2018-19)*

Importing Countries	Quantity (MT)	Value (Rs. Lakhs)
Bangladesh	578111.71	105814.01
Malaysia	332450.83	51769.83
United Arab Emirates	258492.15	37305.47
Sri Lanka	229711.81	35899.72
Nepal	139494.86	14634.60
Saudi Arab	77045.21	11387.68
Kuwait	74715.20	10980.46
Qatar	75293.05	10730.98
Oman	74739.22	10584.35
Indonesia	62272.61	9430.08
Other Countries	281439.77	48350.20
Total	2183766.42	346887.38
Per cent Share of Top 10 Countries	87.11	86.06

* DGCIS, Ministry of Commerce and Industry, GOI.

The requirement of onion is almost constant throughout the year and availability of fresh onion is limited to 7 or 8 months and there is lean periods when prices shoot up because of poor storage conditions available in the country. There are three main seasons of onion production namely (i) *kharif* crop (ii) late *kharif* (iii) *rabi* crop. About 55-60 per cent of onion comes from *Rabi* season and 40-45 per cent from *Kharif* and late *Kharif* season.

Table 7: List of Major Onion Exporting Countries in the World (2019)*

Exporter	Quantity (MT)	Value (US$ Milllion)
Netherland	10,12,150.00	451.00
Mexico	4,02,927.00	379.00
India	10,25,005.00	310.00
China	6,69,383.00	240.00
Spain	4,56,009.00	227.00
Egypt	4,20,454.00	179.00
USA	1,49,302.00	167.00
Peru	2,22,228.00	99.00
Poland	1,83,767.00	82.00
New Zealand	1,36,775.00	81.00

* APEDA, Ministry of Commerce and Industry, GOI.

Because of its high export potential, it comes under cash crop apart from vegetable. Besides meeting domestic need, a sizeable quantity of onion is being exported from India. India's recent export of onion during 2018-2019 is 21, 83,766.42 MT which valued of 3, 46,887.38 lakhs (Table 5).

About 170 countries of the world cultivate onions for domestic use while some also grow onions for trade. About 8 per cent of the harvested onion is internationally traded (Sen Nag, 2017). In 2019, India is the world's largest onion exporter by volume and third largest by value after the Netherlands and Mexico (Table 7). India sells onions to the neighbouring countries like Bangladesh, Malaysia, Sri Lanka, Indonesia and Nepal and also to the Gulf States. China exports a lot to Japan, Vietnam, Malaysia and Thailand, but Russia is also a major recipient of Chinese onions. Mexican onions almost exclusively go to the USA. US exports tend to go to Canada and Mexico, and to Taiwan and Japan.

In 2018, the Netherlands followed by India, China, Spain, Mexico, Egypt and the USA represented the largest exporters of onion and shallots, together achieving 67 per cent of total exports. In value terms, the Netherlands, Mexico and the USA were the countries with the highest levels of exports, with a combined 44 per cent share of global exports. These countries were followed by Spain, India, China, Egypt, Peru, Poland, France, Germany and New Zealand, which together accounted for a further 40 per cent. About 6.6M tonnes of onion and shallots were exported worldwide. The global onion and shallot market revenue amounted to $43.1B (Global Trade, 2019).

3.0 Composition and Uses

3.1.0 Composition

The species of the genus *Allium* are very important crops for human health. They contain many health beneficial substances, such as polyphenols (especially flavonoids), sulphur compounds, vitamins, mineral substances and substances with antioxidant activity (Lenkova *et al.*, 2016). Importance of onion is greatly increasing and now it has become second most medicinal and horticultural crop after tomatoes. It has multiple functional compounds including organosulphur, anthocyanins, flavonoids, quercetin, kaempferol and polyphenols (Arshad *et al.*, 2017). In onions secondary metabolites occur in two main classes; flavonoids and organosulfur compounds. Onion bulb is rich in minerals like phosphorus and calcium, and carbohydrates. It also contains protein and vitamin C. The composition of onion bulb has been presented in Table 8.

Table 8: Composition of Onion Bulb (per 100 g of edible portion)*

Moisture	86.8 g	Vitamin C	11 mg
Carbohydrates	11.0 g	Calcium	180 mg
Protein	1.2 g	Phosphorus	50 mg
Fibre	0.6 g	Iron	0.7 mg
Minerals	0.4 g	Nictotinic acid	0.4 mg
Thiamine	0.08 mg	Riboflavin	0.01 mg

* Aykroyd (1963).

Bajaj *et al.* (1980) studied the chemical composition of bulbs of 12 onion cultivars and observed that the dry matter varied from 10.66 to 14.80 per cent, TSS 41.5 to 74.0 per cent, reducing sugars 12 to 22.25 per cent and phenols 1.75 to 2.95 per cent on dry matter basis. According to Song *et al.* (1997) high protein and amino acid contents were present in bunching onion, high protein and poor amino acid contents in green onions with a long fleshy sheath, good amino acid and poor protein contents in types with short fleshy sheaths, and poor protein and amino acid contents in swollen fleshy sheath types. Schulz *et al.* (1998) reported that major volatile components detected in onion were 2-methyl-2-pentenal, (E)-propenyl disulfide, methyl propyl trisulfide and propanethiol, whereas dipropyl trisulfide, dipropyl disulfide and (E)-propenyl propyl disulfide predominated in leek oils. In accordance with the higher amount of leek chromosomes in the cell nucleus, the percentages of the measured sulphur volatiles in the hybrid material *Allium cepa* × *Allium porrum* corresponded more to the leek than to the onion flavour profile.

Lenkova *et al.* (2016) compared the total phenolic content and antioxidant activity of garlic (*Allium sativum* L.), chives (*Allium schoenoprasum* L.), ramson (*Allium ursinum* L.) and red, yellow and white onion (*Allium cepa* L.). Total polyphenols content was determined in the range 444.3-1591 mg/kg. Total polyphenols content in the observed crops declined in the following order: chives > red onion > garlic > yellow onion > ramson > white onion. The value of antioxidant activity ranged

12.29-76.57 per cent. Antioxidant activity observed in crops declined in the following order: chives > ramson > red onion > yellow onion > garlic > white onion.

Onion bulbs accumulate significant amounts of flavonoids, such as quercetin derivatives in yellow and anthocyanins in red bulbs. Quercetin levels tend to be highest in red and yellow onions and lowest in white onions. Amounts of quercetin in onions vary with bulb color, type, and variety. However, information on absolute amount of anthocyanins in red onions is inadequate. Ferreres *et al.* (1999) and Clifford (2000) reported that up to 250 mg/kg anthocyanins present in red onion. Onion dry skins show that values extending from a minimum (109/100 g) to a maximum (219/100 g). The total quercetin concentration in onion is from traces in white varieties to 2.5-3 mmol/kg of fresh weight in red varieties (Arshad *et al.*, 2013). There are four predominant forms of Q in onions exist in: Q3G (quercetin-3-O-glucoside), Q3, 42 G (quercetin-3, 4-di-O-glucoside), Q aglycone and Q 42 G (quercetin-4-Oglucoside). Among fruits and vegetables, onion rated maximum in quercetin (Q) content (Ko *et al.*, 2011). Bulb type and color vary the concentration of (Q) in onions and quercetin is spread typically in the outer rings and skins (Lombard *et al.*, 2005). It is reported that boiling and frying process show an overall loss of 25 per cent in quercetin glucosides in case of onions.

Flavonoids and flavonols are present in high levels in *Allium* vegetables. Flavonoid ranges from <0.03 to >1 g/kg in the edible portions of *Allium* vegetables. Higher levels of flavonoids 2 ± 10 g/kg are present in onion skins rather than the edible portion. The most significant flavonoids were quercetin aglycone, quercetin diglucoside and quercetin 40-glucoside and kaempferol monoglycoside or isorhamnetin monoglycoside in some cases (Leighton *et al.*, 1992). After onion processing, a huge amount of onion skins remains useless regardless. Onion skin contains higher levels of flavonoids. Prakash *et al.* (2007) reported that dry peel of onion (consider as a waste product) contain major flavonoids which includes quercetin glycoside which is an effective antioxidants against oxidative stress.

Various anthocyanins have identified in onion: cyanidin 3-O-(300-O-b-glucopyranosyl-600-O-malonyl-b-glucopyranoside)-40-O-b-glucopyranoside, cyanidin 7-O-(300-O-b-glucopyranosyl-600-O-malonyl-b-glucopyranoside)-40-O-b-glucopyranoside, cyanidin 3,40-di-O-b-glucopyranoside, cyanidin 40-O-b-glucoside, peonidin 3-O-(600-O-malonyl-b-glucopyranoside)-5-O-b glucopyranoside and peonidin 3-O-(600-O-malonyl-b-glucopyranoside) were present in minute amounts from pigmented parts of red onion (Perez Gregorio *et al.*, 2010). Additionally, four anthocyanins with the same novel 4-substituted aglycone, carboxypyranocyanidin, were isolated from methanolic extracts of red onion. The structures of two of them were identified as 5-carboxypyranocyanidin 3-O-(6″-O-malonylb-glucopyranoside and 5-carboxypyranocyanidin 3-O-b-glucopyranoside (Fossen *et al.*, 2003).

3.1.1 Pungency

The pungency in onion odour is formed by enzymatic reaction only when tissues are damaged. The pungency in onion is due to volatile oil allyl-propyl disulphide. Pungency varies with cultivar, growing conditions, stage of maturity and storage conditions. Pungency is maximum just before top fall in the field. Skin colour is due to the presence of quercetin. From an investigation in USA, Kopsell and Randle (1997) reported that pungency was reduced in some cultivars grown under high selenium (Se), indicating that although sulphur (S) uptake was enhanced, S metabolism in the flavour precursor biosynthetic pathway was affected. Hamilton *et al.* (1997) observed that there was significant variation in the pyruvic acid content among clones in high sulphur (S) treatment, but not in the low-S treatment. They also noted that S content in leaf and bulb tissues were significantly lower in the low-S treatment and indicated S deficiency. Bakr and Gawish (1998) found that the sulphur fertilization increased the contents of all volatile components and the horizontal solar tunnel drier with a maximum temperature of 60°C was beneficial to maintain volatile odour with good colour of onion slices.

Onion possesses organosulphur compounds and carbohydrates that provide unique flavour and health-enhancing characteristics. Significant phenotypic correlations have been reported among soluble solids content (SSC), total dry matter, pungency, and onion-induced *in vitro* antiplatelet activity. The results further indicate and that it will be difficult to develop onion population with lower pungency and high *in vitro* antiplatelet activity; however, the strong genetic and phenotypic correlation between *in vitro* antiplatelet acitivity and high soluble solids content (SSC) are beneficial for the health functionality of onion (Galmarini and Goldman, 2001).

Flavour intensity in onions is influenced by the genetic potential of a cultivar and the environment in which the cultivar grows. Expression of onion flavour is dominated by organ sulphur compounds and modified by simple and complex sugars. Flavour intensity can be increased, within cultivars, by low irrigation rates, high temperatures during growth and development, and high S application (Randle and Galmarini, 1997).

According to Randle and Armstrong (2001), onion flavours are dominated by primary and secondary products from the enzymatic decomposition of unique sulphur compounds collectively known as S-alk(en)yl cysteine sulfoxides (ACSO). The synthesis of the ACSOs occurs when sulphur is absorbed by the plant and metabolized through the ACSO biosynthetic pathway. The metabolism of sulphur through the biosynthetic pathway leading to the synthesis of the ACSOs, and the decomposition of the ACSOs are highly variable among and within onion cultivars. Understanding this variation may be used to develop strategies for flavour manipulation with the intent to provide specific flavour intensities or flavour profiles. Breeding opportunities which may affect final flavour include: differential sulphate absorption influenced by root morphology or S-permease concentration; total bulb sulphur accumulation; the sulphate to organic sulphur ratio influenced by cysteine synthase or other enzymes specific to the ACSO biosynthetic pathway; differential accumulation of individual ACSOs; or alliinase manipulation. Onion flavour is strongly influenced by the growing environment, and by the time of product consumption.

Hamilton *et al.* (1998) reported from USA that pungency and sugar content cannot be manipulated by choosing soil type or by applying extra sulphur (s) to a soil containing sufficient S or irrigated with high S content water.

Lachrymatory factor was detected only when three components, namely purified alliinase, PRENCSO and lachrymatory factor synthase, were present in the reaction mixture. Omission of the synthase from the reaction mixture resulted in an increased yield of thiosulphinate, the condensation product of 1-propenylsulphenic acid. These results indicate that it might be possible to develop a non-lachrymatory onion by suppressing the lachrymatory factor synthase gene while increasing the yield of thiosulphinate. Thiosulphinate is responsible for the flavour of fresh onion, and is converted to compounds reported to exert hypolipodaemic and antiplatelet aggregation effects (Imai *et al.*, 2002).

3.2.0 Uses

Onion is an important and indispensable item in every kitchen as condiment and vegetable, hence commands, an extensive internal market. The green leaves, immature and mature bulbs, and immature inflorescence are used as vegetable. Onion is used to enhance flavour of different recipes.

On account of its special characteristics of pungency, it is valued much. Onions are used in soups, sauces and for seasoning foods. Attention is given to crispness, juiciness, pungency and keeping quality of the bulb. The small bulbs and shallots are pickled in vinegar or brine. Dehydrated bulb or onion powder is in great demand which reduces transport cost and storage losses. Dried onion flakes can be reconstituted by cooking in water.

With regards to its global consumption, Libyans are the one who consumes the highest amount of onion, which accounts for an average of 30 kg annually per capita followed by the Americans (16 kg) (FAO, 2015).

Nadkarni (1927) reported many medicinal properties of onion. Onions are diuretic, applied on bruises, boils and wounds. It relieves heat sensation. Bulb juice is used as smelling on hysterial fits in faintness. It is used to relieve insect bites and soar throat. Results of investigation suggested that onions in the diet may play a part in preventing heart diseases and other ailments and the medicinal uses of onions have been reviewed by Hanley and Fenwick (1985) and Augusti (1990).

Onion is a spice with characteristic flavor and aroma as well as having pronounced medicinal importance. Antibacterial, anti-inflammatory and antioxidant properties are the various diverse functions of onions, which have sound effects on human health. Antioxidant compounds (flavonoids and phenolics), have protective effects against different degenerative pathologies are present in high amount in onions. Onion consider as a good antioxidant additive for food due to Cysteine derivatives. Onion antioxidant activity is reduced after cooking, so onion is efficient in its raw form. Fructo-oligosaccharides are present in abundant quantity in it due to which the growth of potentially harmful bacteria in the colon is retarded by the oligomers thus reducing the risk of emerging tumors in the colon and also initiates the growth of healthy bifidobacterium. Onions are being used since primeval times as a traditional medicine. By having rich photochemistry, these reduce the peril of cardiovascular diseases, diabetes, cancer and atherosclerosis (Arshad *et al.,* 2013).

Onion has antispasmodic, hypoglycaemic, antimicrobial, hypotensive, antiasthmatic, anticholesterolaemic, anticancer, and antioxidant properties (Ashwini *et al.,* 2013). Allium vegetable consumption may have a strong effect for the prevention of stomach cancer. Stomach cancer may be reduce with high consumption of onions >0.5 onion/day. Higher intake (>16 times/week) of onions showed a significant decrease in risk of breast cancer. Variety of sulfides rich in onion extracts, provide protection against tumor growth (Arshad *et al.,* 2013).

Apart from its culinary uses, onion has been reputed in the indigenous knowledge of medicine for ages. The essence of onion proliferated into ancient Greece where it was used as a blood purifier for athletes. Onion is commonly taken raw or as a decoction for treating infectious diseases. It is also used in a wide variety of preparations for internal and external use to relieve several ailments including digestive problems, skin diseases, metabolic disease, insect bites and others. Onions possess a wide range of pharmacological properties including antimicrobial, antioxidant, analgesic, anti-inflammatory, anti-diabetic, hypolipidemic, anti-hypertensive, and immunoprotective effects. Besides, onion also displayed hypoglycemic effect. Sulphur compounds have proven to be the principal active antimicrobial agent present in onion. Onion also possesses other antimicrobial phenolic compounds including protocatechuic, p-coumaric, ferulic acids, and catechol. Quercetin and kaempferol have been found as significant contributors to this activity. Moreover, onion extracts are potent against fungal species, and its essential oil inhibits the dermatophyte fungi. *Aspergillus niger* and *Fusarium oxysporum* were strongly inhibited by the ethyl alcohol extract of dehydrated onion (Teshika *et al.,* 2019). The antibacterial activity of the red variety of onion extract was found to be higher compared to yellow and white varieties (Sharma *et al.,* 2017).

3.3.0 Value Added Products

Onion is an important commodity that is used in everyday food preparation in the Indian kitchens. Onion is semi-perishable in nature; however the postharvest losses are quite high due to improper handling and lack of storage facilities coupled with long period of storage. Onion processing industry is flourishing day by day due to high domestic demand and worldwide markets. Onion is mainly exported in the form of dehydrated onion, canned onion and onion pickle. Product diversification is also commercially visisble. Some of the processed products have been listed below:

Minimally Processed Products

Onion peeling and cutting is time taking, cumbersome and lachrymatory effect. Availablility of minimally processed onion for ready to cook or ready to use reduce overall food preparation time. Therefore demand for fresh-cut, value-added, and ready-to-eat onion in households, as well as large-scale uses in retail, food service, and various food industries, increasing mainly due to the end-use convenience. Minimally processed onion is prepared by peeling and cutting which retain its freshness. Peeling and cutting can be done either manually or mechanically. Intensive discoloration, microbial growth, softening, and off-odor are the typical deteriorations that need to be controlled through the application of suitable preservation methods. Modified atmosphere packaging and freezing are the techniques applied to increase the shelf life of minimally processed onion.

Dehydrated Flakes

Dehydration is one of the most common processes used to improve food stability. Dehydration in to flakes would decrease bulk to store and transport besides increasing shelf life. Flakes are prepared by peeling, cutting, pre-treating and drying using different drying techniques. Pre-traetment before drying improves product quality, prevents browning, accelerates drying rate and also retains volatile compounds.

Powder

Onion powder can be prepared by grinding the dehydrated flakes. Powder dissolves very easily and reconstitute quickly compared to flakes. Powder offers convenience to add to preparation of baked products, soups and also for spicing up different grilled products. It is highly hygroscopic, thus required proper packing to avoid moisture absorption. Powder can further be converted to different products like, soup mix, curry mixes *etc.* by adding different spices in optimized ratio.

Paste

Onion paste contents high moisture. It can be prepared by peeling and grinding. Proper packaging and storage conditions are required to store for longer duration without any microbial spoilage. Addition of preservatives or thermal treatment can be done for longer storage life. Packaging material plays a crucial role in storage of the paste. Temperature and storage duration affect the colour, microbial growth and nutritional changes of onion paste.

Pickle

Pickling can be done in two ways: vinegar based pickling and oil based pickling. Oil based pickle can be be prepared by adding different spices as per desired taste. Addition of pH regulator is important to keep the pH below 4, which extend the storage life and prevent microbial growth.

Oil

Onion oil can be extracted by solvent extraction method or super critical fluid extraction technique. Oil is used as a flavouring substance, preservative and also as medicine. Encapsulation of oil is also done to reduce the pungent flavour while consuming.

Onion can be processed in to onion vinegar, onion wine, and onion beverage and onion sauce. Processed products of onion reduce transport cost and storage losses (Augusti, 1990). Small bulbs of onions are pickled in vinegar and are used for making various seasonings, sauces and soups (Arshad *et al.*, 2013).

Uses of Onion

Onion Oil   **Onion Flakes**

Canned Onion   **Onion Pickles**

Dehydrated Onion Powder   **Onion Vinegar**   **Onion Sauce**

French Fry (Left), Salsa (Middle), and Paste (Right) from Onion

Results of experiments on dehydration of 3-mm thick rings of onion in an air cabinet drier (55 + 2°C) for 9 h with an air flow rate of 1.46 m/second and loading of 4.3 kg/m² to 5-6 per cent (wet basis) moisture content, indicated that cultivars Jalgaon White, Phule White, Decco and Red Creole were suitable for dehydration with good physicochemical and organoleptic qualities. The extent of loss of pungency during dehydration ranged from 67.3 to 74.4 per cent. Decco cultivar retained maximum pungency, followed by Red Creole (Sakhale *et al.*, 2001). Sehrawat *et al.* (2017) determined the effect of combined low pressure superheated steam drying (LPSSD) followed by vacuum drying (VD) on quality of onion slices at different temperature (60-80°C~10 kPa). In the first stage, the moisture content of onion slices was reduced to 50 per cent (w.b.) from 88-90 per cent (w.b.) using LPSSD; and then it was further reduced up to 7-8 per cent (w.b.) by VD. Among the various drying temperatures studied, retention of colour, rehydration ratio, thio-sulphinate content and antioxidant capacity of dehydrated onion was better at 60°C, followed by 70°C, whereas total phenol contents were higher at 80°C. Attkan *et al.* (2017) studied on drying temperature, slice thickness and KMS (potassium meta-bisulphate) concentration using Box-Behnken design of experiments. The optimum operating conditions using selective quality parameters for KMS concentration, slice thickness and drying temperature was found 0.5 per cent, 3 mm and 57.31°C respectively. Drying temperature followed by thickness of the slice had the most significant effect on all the product responses. Optimum values of the response parameters, in particular the drying time, color change, pyruvate content and rehydration ratio acquired were 671.6 mins, 3.53, 12.93 µmol/g and 4.98, respectively. At drying air temperature of 60°C and 3 mm slice thickness, dried onion slices retain maximum color, flavor, and taste and have maximum overall acceptability.

4.0 History and Origin

4.1.0 History

Alliums are among the oldest cultivated plant species. References to edible onion can be found in the Bible, Koran and in the inscriptions of ancient civilizations of Egypt, Rome, Greece and China. They are mentioned as a source of food for the builders of the great pyramid of King Cheops and the Israelites wandering in the desert after the exodus from Egypt bemoaned the lack of appetizing onions.

Onion has got a Sanskrit equivalent 'Plandu', mentioned in *Apastamba Dharma Sutra*-I (dated 800 B.C. to 300 B.C.), which signifies its very early introduction in India.

4.2.0 Origin

According to Vavilov (1950), the primary center of origin for the garden onion lies in northwestern India and the Russian Tadjikistan, Uzbekistan and western Tian-Shan. There are two secondary centers of origin to the west. The first comprises Iran, the Transcaucasus, and eastern and central Turkey. The second borders the Mediterranean.

Jones and Mann (1963) gave an account of origin, domestication and distribution of onion. Onion had been domesticated independently in several places. It occupies a large area in Western Asia, extending perhaps from Palestine to India. Although there is still doubt as to whether *A. cepa* exists today as wild plant, most botanists believe that it had its origin in areas including Iran and Pakistan.

While the place of origin of *A. cepa* is still a mystery, there are many documents from very early times, which described its importance as a food and its use in art, medicine and mummification in ancient Egypt. One of the most important testimonials to the use of onion as a food in ancient Egypt is from the Bible.

The onion must have been grown in India from very ancient times, as it is mentioned in *Charaka-Samhita*, famous early medical treatise of India. Many virtues are attributed to the onion and garlic by Charaka. Reference to onion as food, medicine or religious objects was date back to 3200 B.C. in Egypt. Other early references (600 B.C. to A.D. 79) are made by Indian, Greek and Roman authors. McCollum (1976) is of the view that the primary centre of origin of onion is central Asia and Mediterranean, a secondary centre for larger types of onion.

Meer *et al.* (1997) reported that nearly all *Allium* crops originate from the main centre of *Allium* species diversity which stretches from the Mediterranean basin to central Asia and beyond. Onion (*A. cepa*), garlic (*A. sativum*), shallots (*A. ascalonicum*), leeks (*A. porrum*) and Japanese bunching onions (*A. fistulosum*) are very old crops with a worldwide distribution. Onions and shallots originate from a group including the *A. oschaninii* alliance. The ancestors of garlic, leeks and Japanese bunching onions must have been closely related to *A. longicupus*, *A. ampeloprasum* and *A. altaicum*, respectively. The following Allium crops : rakkyo (*A. chinense*), ciboule vivace, great headed garlic (*A. ampeloprasum*), ever ready onion (*A. perutile*), *A. grayi* and Grise de la Drome, which are only grown in limited areas, have unknown ancestors. Some other small *Allium* crops (*A. proliferum* and *A. wakegi*) are the result of ancient (spontaneous) crosses, notably *A. cepa* × *A. fistulosum* and *A. ascalonicum* × *A. fistulosum*, respectively. Finally, other minor edible *Alliums* are essentially wild species, notably: Chinese chives (*A. tuberosum*), common chives (*A. schoenoprasum*), *A. triquetrum*, *A. nutans*, *A. oleraceum*, *A. pskemense*, *A. aflatunense*, *A. hookeri*, *A. macrostemon*, *etc.* Some of them are not cultivated but mainly collected from the wild, *e.g.*, *A. nutans* and *A. macrostemon*. The lesser known species in question have, if possible, been characterized following standards like morphological description, geographical distribution, variability, *etc.* Generally, the genus *Allium* seems to be a very promising field of exploration because of the following reasons: the very high number (more than 600) of available species; the highly appreciated specific culinary properties of many *Allium* crops; therapeutic and medicinal properties.

According to Friesen *et al.* (2006) the primary evolution center of the genus extends across the Irano-Turanian biogeographical region, and the Mediterranean basin and western North America are secondary centres of diversity.

A low-density genetic map of onion reveals a role for tandem duplication in the evolution of an extremely large diploid genome (King *et al.*, 1998).

5.0 Botany

Alliums are bulbous biennial or perennial herbs which give off a distinctive and pungent odour when the tissues are crushed. The onion smell and taste are important diagnostic features of the genus. Other characteristics include the presence of bulbs formed by the attachment of swollen leaf bases to the underground part of the stem, inflorescences in the the form of umbles with numerous small flowers, a single spathe which splits into segments as the young flowers emerge, and a superior ovary above the six stamens and the petal like perianth. Not all the species of the genus possess all these features, a few lacking in the volatile compounds that produce onion or garlic smell, some have very long sheathing leaf bases and some have their storage organs in the form of rhizomes or storage roots.

The botanical description of onion as prepared by Nanda and Agrawal (2011) has been given below:

Habit: Biennial glabrous herb, usually grown as an annual from seed or bulb, up to 100 cm tall.

Stem: Real stem very short, formed at the base of the plant in the form of a disk, with adventitious roots at base; bulbs form by the thicking of leaf-bases a short distance above the true stem, solitary or in clusters, depressed globose to ovoid or oblate, up to 20 cm in diameter, variously coloured.

Leaf: Leaves 3 to 8, distichously alternate, glaucous, with tubular sheath; blade D-shaped in cross section, up to 50 cm long, acute at apex.

Inflorescence: Inflorescence a spherical umbel up to 8 cm in diameter, on a long, erect terete, hollow scape up to 100 cm long, usually inflated below the middle; umbel initially surrounded by a membranous spathe splitting into 2-4 papery bracts.

Flower: Flowers bisexual, stellate; pedicel slender, up to 4 cm long; tepals 6, in 2 whorls, free, ovate to oblong, 3-5 mm long, greenish white to purple; stamens 6; ovary superior, 3-celled, style shorter than stamens at anthesis, later elongating.

Fruit: Fruit is a globular capsule 4-6 mm in diameter, splitting loculicidally, up to 6 seeded. Seeds 6 mm × 4 mm, black.

Pollination: Onion is a faccultative cross pollinator, the percentage of selfing accounting to 10-20 per cent. The flowers are protandrous. Pollination is by bees, bumble-bees or flies. When mature the fruit dehise, allowing shedding of the seeds.

The underground bulb, the edible portion, varies in shape, colour, size, firmness, keeping quality, period of maturity and flavour. It is potentially a biennial crop producing large bulbs and hollow leaves during first year, and produces flower and seeds in the subsequent year (Rana, 2015). Onion Flowers are borne in simple umbels at the apex of floral stem which is commonly hollow when mature, round in cross section and somewhat swollen at the middle or near the base. The number of seed stems vary from 1-20 or more depending upon the cultivar and the seed stalks may vary from 0.9-1.2 m in height. The umbel before it is fully expanded is enclosed in a papery spathe, consisting of 2-3 bracts, which are split, open by the pressure of developing flowering buds. The number of flowers may vary from 50-2000 depending upon the species and cultivars, time of planting, size and storage condition of mother bulbs.

The umbels are aggregates of cymes of 5-10 flowers each and the flowers open in definite sequence. The flowers are white to bluish in colour. The perianth segments are 6 in 2 whorls spreading, reflexed, free and ovate. The stamens are also 6 in 2 whorls, anthers bilocular; ovary superior, 3 locules with nectary at base. Each locule has 2 ovules. The anthers of inner whorl dehisce first and usually it occurs between 9 a.m. and 5 p.m. The style which is approximately 1 mm long, when the flower first opens is not receptive, until it elongates to a length of 5 mm. This requires a day or two after anthers have all dehisced. An inflorescence may continue opening for two weeks or more and a plant may be in bloom for more than 30 days. Each individual flower contains 6 stamens, 3 carpels united into one pistil and 6 perianth segments. The pistil contains 3 locules each of which has 2 ovules.

In onion, 7 stages of umbel development were distinguished by Klein and Korzonek (1999) with regard to the size of umbels, flower buds and developmental stages of male and female gametophytes. At the beginning of flowering, the stage at which buds are most often collected for induction of gynogenesis, flower bud length ranged from 2.0 to 5.0 mm and the stages from archesporial cell to mature embryo sac were observed in the ovules microspores and bicellular pollen grains were observed in the anthers. Flower bud size was not an accurate indicator of pollen development. Buds of the same size but originating from umbels at different developmental stages were characterized by different stages of pollen development.

Karak and Hazra (2012) studied the floral biology of several short day Indian genotypes. They reported that dichogamous self-pollination control mechanism in the bisexual flowers of onion was manifested by differential maturity of the style and filaments at anthesis, the average length of the style (1.43 mm) being half of that of the filament (3.10 mm) at anthesis. It took on an average 4.75 days for the style to mature and become of the same size as that of filament to reach the vicinity of its own pollens in the anthers. Pollen grains which were small and more or less oblong to oval in shape showed both high viability and germinability with significant variation among the genotypes.

Popandron (1998) noted distinct differences between male sterile and normal plants with respect for length of flower stem, number of flower per inflorescence and stamen weight.

It is pollinated chiefly by honeybees, which visit the nectaries. Though cross pollinated, selfing also takes place as the insects visit many flowers in a single umbel and also umbels of the same plant before leaving for other plant. In artificial pollination for breeding purposes, selfing is accomplished by bagging all the umbels of the same plant together and by shaking or by introducing house flies. In Brazil, *Trigona spinipes* is used as a pollinator in onion breeding programmes. It is easily found, the nests are simple to collect, and colonies can be well managed so that the bees are highly effective pollinators of onion when placed in isolation cages (Nascimento *et al.*, 1998).

Pollen of cv. Aurora can be stored for one year with good viability, either in liquid nitrogen (-196°C) or inside a desiccator with sulfuric acid in the freezer (-18°C). The best condition for one-year maintenance of cv. Petrolini pollen was recorded in liquid nitrogen (Gomes *et al.*, 2000).

5.1.0 Cytogenetics

Jones and Mann (1963) have indicated that the basic chromosome number in the genus *Allium* is 7 or 8 with a few speices having 9 chromosomes. The majority of the species that have been cytologically studied are diploids with x = 8, and in this basic chromosome number group are included all the edible species. The next commonest group is x = 7, and there are only a few specieswith x = 9 (Jones, 1991).

A. cepa is known only as diploid (2n = 16). Vakhtina *et al.* (1977) studied the haploid nuclei of 40 species of the genus *Allium* and found one species with n = 7, 32 species with n = 8, one species with n = 9, two species with n = 10 and four species with n = 8, 16. The basic number of chromosomes, x = 7, 9, 10, 11 has also reported by Fritsch and Astanova (1998) and Fritsch and Abbasi (2013).

Sen (1976) has suggested that *Allium* (x = 7) might have originated from ancestor in which x = 6. Singh *et al.* (1967) reported a natural triploid cv. Pran from India.

The commercial onion has eight pairs of chromosomes. The eight chromosomes can be placed in four groups. The satellite chromosome has an index of 0.366. Three have indexes ranging from 0.546 to 0.559. Three more range from 0.752 to 0.786 and one chromosome with an index of 0.851. Chromosome and chromatid breaks are reported in onion root tips two to four mm long, from seeds. The aberrations occur either in the somatic cells of the embryo of dormant seeds or in the prophase stage in seeds just starting to germinate. Most of these aberrations are soon lost, presumably through failure of the cells containing them to continue to divide. The proportion of aberrations, especially of the chromosomal type, increases with increased age of the seed (Yarnell, 1954). Tetraploid cells were found to occur regularly in the mesophyll of the cotyledon and in the cortex of the transition region between root and shoot in seedlings ranging 10 to 40 mm in length (Berger and Witkus, 1946). Mensinkai (1939) regards onion as a secondarily balanced diploid. The second nucleolar pair of chromosomes has a secondary submedian constriction. *Alliums* are well known for their symmetrical and uniform karyotypes. Across a wide range of species the chromosomes are regularly metacentric or at most submetacentric (Jones, 1991, Salmasi *et al.*, 2019). The chromosomal length of long arms showed a variation from 5 to 14 µm while short arms have 2 to 10 µm at the early and late stages of mitotic metaphase (Okumuo and Hassan, 2000). Sirajo and Namo (2019) reported that the centromere position and variation in complement length of chromosomes accounted for 92.71 per cent of the total variations among the studied onion genotypes.

Two viviparous strains of common onion (*A. cepa*) denoted as *A. cepa* var. *viviparium* (syn. *A. proliferum*), traditionally cultivated in the coastal regions of Croatia, were found to be diploid (2n = 2x = 16) and triploid (2n = 2x = 24), respectively. The triploid, known in Croatian as ljutika, has not been previously reported in Europe (Puizina and Papes, 1996). In Iran, Dolatyari *et al.* (2018) karyologically investigated eighteen species and subspecies of *Allium* sect. Acanthoprason and 11 species belonging to other subgenera and sections of *Allium*. They observed 47 diploid accessions, three tetraploid accessions of two species and a mix of di- and triploid individuals in *A. bisotunense* accession. A basic chromosome number of x = 8 was confirmed for all investigated members of subg. Melanocrommyum and subg. *Allium*, and x = 9 for *Allium tripedale* of subg. Nectaroscordum. Chromosomal aberrations were rarely observed. The karyotypes showed low variation in inter- and intra-chromosomal asymmetry especially inside of the taxonomic groups. Six diferent types of satellites were recognized, two of them were newly described: Type P was prevalent in subg. Melanocrommyum, and type O in sect. Codonoprasum. They noted distinct karyological characters for the subgenera. Salmasi *et al.* (2019) investigated the karyotypes of 10 species of Iranian *Allium*. Except *A. giganteum* (x = 7), other accession showed basic chromosome number x = 8. Karyotypes of 14 taxa of *Allium* were diploid with 2n = 16; only *A. macrochaetum* was tetraploid with 2n = 32. Satellite chromosomes were seen in *A. asarense*. All karyotypes were symmetrical, consisting of metacentric and submetacentric chromosome pairs. Only *A. caspium* and *A. stipitatum* had subtelocentric chromosomes. The longest chromosome length was detected on *A. asarense*, *A. elburzense*, *A. giganteum*, *A. rotundum* and *A. stipitatum* (17.9-19.7 µm), while *A. ampeloperasum* demonstrated the shortest value (8.2 µm).

Kaul (1977) observed complete seed sterility in some translocated onion heterozygotes, though pollen fertility was ranged from 9-22 per cent, which indicated the existence of male and female sterility in those onion clones.

Deogade *et al.* (2017) studied the karyotypic of onion cv. Brown onion, Phule safed and Baswant 780. The chromosome number of *Allium cepa* L. and its two varieties were diploid with (2n = 16) and Phule safed was diploid with (2n = 20). The chromosome types were detected as mostly metacentrics (m) and submetacentrics (sm). Karyotype formula of *Allium cepa* L. and its varieties were found 14 Am + 2 Asm, 12 Am + 2 Bm + 2 Asm and 10 Am + 6 Asm, 10 Am + 8 Am + 2 Bsm.

The cytoplasm of *A. vavilovii* is closely related to N cytoplasm and may be the progenitor of cultivated onion (Havey, 1997).

Onion is among the most used test organisms in plant bioassays. The genotoxic level of the test agent is reflected by structural changes of the chromosomes and their changed numbers.The changes in morphology varied from a single distortion of a single chromosome up to several morphological changes observed on many chromosomes. Peter and Amon (2014) identified 15 categories of morphological aberrations which are classified into three groups: chromatid damage (CtD), centromere damage (CmD) and chromosome damage (CsD). CtD includes: single break chromatid, double break chromatid, isochromatid break, multiple break chromatid, gap chromatid, centric ring chromatid, acentric ring chromatid and triradial chromosomes. CmD includes: break centromere, gap centromere, single break centromere, double break centromere and multiple break centromere. CsD includes: ring chromosomes and dicentric chromosomes. Sometimes also the chromosome number changed which occurred as aneuploidy with monosomy 2n = 15 (2n = 16–1) and euploidy, increased number of the basic chromosome number (2n = 6x to 8x). Based on mitotic index, frequency of chromosomal abnormalities, DNA content and other observations Santos *et al.* (2017) reported that 'Baia Periforme', 'Crioula' and 'Vale Ouro' varieties of onion were most suitable for phytotoxicity and cytotoxotoxicity studies. Lead is polluting agent and very toxic for plants (Heggestad, 1968). Nitrate of lead has a strong inhibitory effect on mitotic division of onion. It causes diverse chromosomial aberrations. It has total inhibitory action at mitotic shaft, perturbeing the behavior of the chromosomes in the same manner as the effect of colchicines (Padureanu, 2005). Onion is sensive to salinity. Salinity showed a significant inhibitory effect on the seed germination and seedling growth of *Allium cepa* (Cavusoglu, 2020). At high salt stress up to 150 mM, the onion roots are capable to rapidly activate antioxidant defence system to resist the salt-induced oxidative stress, but could not control the cytogenetical activities. The recovery is possible at physiological and cytogenetical level by retaining chromosomal and DNA integrity (Singh and Roy, 2016). Salinity reduced significantly the mitotic index in onion root tip cells and increased the chromosomal abnormalities and micronucleus which is the simplest indicator, the most effective of cytological damage. Althought the detrimental effects of salinity on the seed germination, mitotic activity, seedling growth and chromosomal aberrations were alleviated flashilly in varying degrees by propolis application, it was ineffective in reducing of salt damage on the micronucleus frequency (Cavusoglu, 2020). Ahirwar and Verma (2016) isolated two mutants (one translocation heterozygote and one inversion heterozygote) in a population raised from 0.2 per cent EMS. At anaphase-I, the translocation heterozygote displayed various abnormalities such as laggards, unequal distribution, micronuclei, *etc*. On the other hand, the inversion heterozygote plant displayed various types of chromosomal configurations at anaphase/telophase-I and II in meiosis. However, pollen fertility was found to be very low in both the mutants. Grey mould (*Botrytis allii*) fungus has the ability to cause a large number of mitotic abnormalities to onion affecting the growth and development of plants. Elena *et al*. (2018) reported that the mitotic index decreased in onion cell affected by the fungus. The main types of chromosome aberrations and nuclear abnormalities were sticky and laggard type chromosomes, fragments of chromosomes, as well as cells with nuclear erosion. The cytogenetic effects of infection with *Botrytis allii* can probably be similar to those produced by the action of a mutagenic agent.

Uneven distribution of recombination frequencies along the chromosome has been reported in the *Alliums*. Integrating the genetic and chromosomal maps Romanov *et al*. (2016) analyze the distribution of recombination events along onion chromosomes. They found highest recombination frequency found in the interstitial region. The recombination frequency between the markers located in the proximal region was over 20 times less than between markers located in the interstitial region.

6.0 Genetics

6.1.0 Male Sterility

Production of hybrid onions became possible with the discovery of male sterile cytoplasm in the onion cultivar 'Italian red' (Jones and Emsweller, 1936). Jones and Clarke (1943) reported male sterility from an interaction between recessive nuclear gene (chromogene) and cytoplasmic factor (cytogene). It is presumed that there are two types of cytoplasm: normal (N) and sterile (S). All plants with N cytoplasm produced functional pollen and all male sterile plants have S cytoplasm. A recessive gene for male sterility (*ms*) influences pollen development when carried in plants with S cytoplasm, but has no effect when carried by plants with

N cytoplasm. Consequently, the male sterile plants have the genotype *Smsms*. The plants with N type cytoplasm are always male fertile, and may be either *NMsMs*, *NMsms* or *Nmsms*. The plants with genetic constitution *SMsMs* and *SMsms* will always be fertile, because they carry the dominanat gene *Ms*. The sterility factor S is inherited only through the egg and not through the male parent. Neither the genetic factor *ms*, nor the genetic cytoplasmic factor S alone is enough to cause male sterility; plants must have both these factors to be male sterile. Besides modifier genes, environmental factors also cause variation in expression of male sterility (Nieuwhof, 1970; Khaisin, 1975). Yakovlev (1985) reported that male sterility in onion is controlled by 2 complementary genes ms_1ms_1 and ms_2ms_2 in conjunction with male sterile (S) cytoplasm. Another form of cytoplasm conferring male sterility *i.e.* CMS-(T) was discovered in the onion cultivar 'Jaune paille des Vertus' (Berninger, 1965). It was reported that one independent and two complementary genes were involved in fertility restoration of the male sterility (Schweisguth, 1973).

6.2.0 Time of Maturity

This character is conditioned by multiple genes (Jones and Mann, 1963). Hosfield *et al.* (1977) reported that in all F_1 studied, the additive gene action was more important than non-additive gene action. Todorov and Sharma (1950) reported that all F_1s studied indicated shorter growth period than mid parental value. Some of the hybrids showed heterosis for earliness.

6.3.0 Bulbing

Taylor *et al.* (2010) reported that *AcG1* and *AcFKF1* are the key genes involved in the photoperiodic control of bulb initiation. Lee *et al.* (2013) identified six *FT* like genes and suggested that different *flowering locus T* genes regulate flowering and onion bulb formation. They opined that *AcFT1* promotes bulb formation and *AcFT4* prevents upregulation of *AcFT1* and thereby inhibits bulbing. Manoharan *et al.* (2016) identified eight *FT* like genes (*AcFT*) encoding PEBP (phosphatidylethanolamine binding protein) domains in onion. They also reported another gene (*AcFT7*) which showed highly conserved region with *AcFT6* and yet another (comp106231) with low similarity to MFT protein, but containing a PEBP domain. The expression levels of *AcFT4*, *AcFT5* and *AcFT6* increased as it got closer to a condition in long days in association with the onion bulbing (Tagashira and Kaneta, 2015). According to Rashid *et al.* (2016), the *FT-like protein 1* and *FT-like protein 2* genes, similar to *AcFT6* and *AcFT5*, might have influence on bulb formation. In short day onion, Dalvi *et al.* (2016) reported that *AcFT6* expression level was high during bulb initiation stage and suggested that it might be involved in bulb initiation.

The expression of five of the six genes studied (*AcFT1, AcFT3, AcFT4, AcFT5* and *AcFT6*) was noted highest at the bulbing stage (60 DAT) in SD onion variety Pusa Riddhi, suggesting their role in bulbing in SD onions (Lyngkhoi *et al.*, 2019). However, the expression of *AcFT2* was found lowest at bulbing stage, which indicating that the downregulation of this gene promotes bulbing in short day variety. Contrary to observations made in short day variety, expression of all the six genes was comparatively very low at bulbing stage (60 DAT) in Brown Spanish in Delhi condition. It can be assumed that the specific photoperiod conditions required for induction of these genes and in turn bulb formation in long day onion. Twenty-two partial cDNAs representing genes potentially involved in onion bulbing have been identified and isolated by Rashid *et al.* (2016). Eight of these were shown to be differentially expressed in bulb and leaf tissue and with respect to photoperiod. A total of 13665, 12604, 484 and 964 significantly differential expressed transcripts were detected in short day (SD) leaf vs. bulb, long day (LD) leaf vs. bulb, SD leaf vs. LD leaf and SD bulb vs. LD bulb, respectively.

Transition to bulbing in long-day onions is associated with reduction in expression of *AcFT4* which represses, and increase in *AcFT1* which promotes bulbing. Another family member, *AcFT2* is expressed in basal shoot meristems of bulb following cold treatment (McCallum *et al.*, 2016).

6.4.0 Bolting

Baldwin *et al.* (2014) reported that a QTL conditioning an adaptive trait in bulb onion which involved in photoperiod and vernalization physiology. Genome regions on chromosomes 1, 3 and 6 associated with bolting. QTL analysis located a major gene conditioning bolting on chromosome 1 (*AcBlt1*) (McCallum *et al.*, 2016).

6.5.0 Bulb Colour

Among commercial onions there are 4 colour classes: white, yellow, red and brown. The bulbs accumulate a range of flavonoid compounds, including anthocyanins (red), flavonols (pale yellow), and chalcones (bright yellow). Onion bulbs with red, white, yellow, golden, pink, chartreuse, *etc.* colour are available and this variation is due to the mutations in structural and regulatory genes of the flavonoid biosynthesis pathway (Khandagale and Gawande, 2019).

Clarke *et al.* (1944) showed that a dominant basic colour factor *C* is necessary for either yellow or red colour. All plants with *cc* have white bulbs regardless of the presence of other colour factors. Dominant *R* in the presence of *C* gives a red bulb colour. The recessive with *C* produces yellow bulb, the colour inhibiting factor *I* is incompletely dominant over *i*. The cultivars with genotypic constitution *II* produce white coloured bulb, regardless of the presence or absence of *C* and *R* factors. The bulbs with genotype *IiCCrr* are cream or buff in colour. Thus cultivar that is homozygous for red has *iiCCRR*, genotype for yellow is *iiCCrr* and recessive white can be *iiccRR*, *iiccRr* or *iiccrr*. Jones and Peterson (1952) reported complementary factors for light red bulb colour. Complementary factors for yellow and light red bulb colours are present in many recessive white cultivars: coloured bulbs appear in the F_1 when recessive white cultivars are crossed.

El-Shaifi and Davis (1967) after study of the F_1, F_2, F_3 and back cross progenies of several cultivars and lines, concluded that five major genes *I*, *C*, *G*, *L* and *R* (each with two alleles) interact and segregate independently for the 4 different colours, white, yellow, red and brown. The five genes act in a specific order on a biochemical pathway that leads to pigment formation. They proposed a pathway showing the action of each gene. Four of these genes, *I*, *C*, *L* and *R* were reported previously by other workers and the fifth gene *G* has been proposed by them. According to them the results of several crosses cannot be explained without its presence.

However, density of golden yellow and red colours is apparently controlled by several microgenes and it is quantitatively inherited. Patel *et al.* (1971) showed that four factors control skin colour, *viz.*, two basic factors *C1* and *C2* and two duplicate complementary factors R_1 and R_2. C_2 requires presence of both R_1 and R_2 for expression of red colour.

The variation found in shades of reds and browns in onion skin is due to genetic colour-intensifying factors which are probably quantitatively inherited (Currah and Proctor, 1990). Khar *et al.* (2008) reported that onion bulb color is controlled by at least five major loci (I, C, G, L, and R) and seedcoat color by one locus (B). The B and C loci were linked to SSRs on chromosomes 1 and 6, respectively. For all of three families, SNPs in DFR (dihydroflavonol 4-reductase) cosegregated with the R locus conditioning red bulb color. In the family from B2246 × B11159, red bulbs versus yellow bulbs were controlled by DFR and a locus (L2) linked at 6.3 cM to ANS (anthocyanidin synthase). The authors propose that yellow bulb onions have been independently selected numerous times and that yellow populations carry independent mutations in structural or regulatory genes controlling the production of red bulb color in onion.

R2R3-MYB transcription factors associated with the regulation of distinct branches of the flavonoid pathway were isolated from onion. These belonged to sub-groups (SGs) that commonly activate anthocyanin (SG6, MYB1) or flavonol (SG7, MYB29) production, or repress phenylpropanoid/flavonoid synthesis (SG4, MYB4, MYB5). MYB1 was demonstrated to be a positive regulator of anthocyanin biosynthesis by the induction of anthocyanin production in onion tissue when transiently overexpressed and by reduction of pigmentation when transiently repressed *via* RNAi (Schwinn *et al.*, 2016). The bulb color of the F_1 hybrids of Santero (yellow-colored downy mildew resistant cultivar) and OT803 (Yellow breeding line) became light pink, suggesting involvement of complementation between the *DFR-A* and *ANS* genes in the onion anthocyanin biosynthesis pathway. Santero contained active *DFR-A* and inactive *ANS* alleles, OT803 was assumed to harbor active *ANS* and inactive *DFR-A* alleles. However, some yellow-colored individuals of OT803 were shown to contain the homozygous genotype of the active DFR^{AR4} like allele. The nucleotide sequences of the DFR^{AR4} and DFR^{AR4} like alleles were identical except for a single nucleotide deletion in the last exon. This new *DFR-A* mutant allele was designated DFR^{APS2} (Kim *et al.*, 2017).

6.6.0 Bulb Weight and Bulb Diameter

In onion among yield components, bulb diameter and bulb weight had the maximum contribution towards onion bulb yield (Singh, 2001). Mohanty (2001) found that bulb weight contributed predominantly to total divergence in onion. Onion bulb weight had high genetic advance over mean and is be governed by additive gene effects (Aditika *et al.*, 2017). However, Pavlovic *et al.* (2015) reported super-dominance and dominance are the mode of inheritance for bulb fresh weight in onion.

These traits have low or zero value for heritability (McCollum, 1971). Patil *et al.* (1986) reported high genetic advance for bulb weight. Rajalingam and Haripriya

(1998) recorded very high heritability estimates coupled with high genetic advance for the characters like weight of plant, bulb length, bulb diameter and volume of bulb. Ibrahim *et al.* (2000) noticed that bulb production was significantly and highly correlated with equatorial bulb diameter, quantity of marketable bulbs and polar bulb diameter. It also had a positive relationship with the number of rings per bulb and neck thickness. Bolting, sprouting and split percentage in bulb production were negative correlated with bulb yield, quantity of marketable bulbs, equatorial bulb diameter and polar bulb diameter.

6.7.0 Bulb Shape

Salem (1966) reported that onion bulb size, shape, and dimensions appeared to indicate that such characters were quantitatively inherited. The ratios determining onion bulb shape showed that the flat shape was partially dominant to the high globe shape which, in turn, was partially dominant to the torpedo shape. Bulb shape is governed by multiple genes (Jones and Mann, 1963) and moderate to high value for heritability (McCollum, 1971).

6.8.0 Yield

It is a polygenic character and both additive and non-additive type of gene actions is important (Joshi and Tandon, 1976; Hosfield *et al.*, 1977). High heritability with high genetic advance for bulb yield/plant has been reported by Rajalingam and Haripriya (1998). Kumar *et al.* (2000) reported that the yield of onion was positively correlated with plant height, number of leaves per plant, fresh weight of leaves per plant, number of roots per plant, total soluble solid content and total sugar content of onion bulbs.

From path analysis study, Gurjar and Singhania (2006) reported that plant height, number of leaves per plant, bulb neck thickness, bulb weight, equatorial and polar bulb diameter of onion had high positive direct effect through each other on yield. Marketable yield of onion was positively and significantly correlated with plant height, bulb diameter, bulb size index, weight of 20 bulbs and gross yield and negatively correlated with bolters (Singh and Dubey, 2011). Total yield was significantly and positively correlated with bulb fresh weight, vertical bulb diameter, reducing sugars and bulb weight (Ashok *et al.*, 2013). Sharma *et al.* (2015) reported that marketable bulb yield, average weight of bulb, number of leaves/ plant, collar height and leaf length showed positive direct effect on total yield. Bharti *et al.* (2011), Dewangan and Sahu (2014) and Solanki *et al.* (2015) observed high heritability coupled with high genetic advance for bulb yield/plant and total yield (q/ha). Aditika *et al.* (2017) found high genetic advance over mean for yield per plot and number of marketable bulbs per plot. These characters assumed to be governed by additive gene effects.

Santra *et al.* (2017) opined that superior genotypes like Agrifound Dark Red, Gota and Baswant 780 having potentiality to perform in both the locations Kalyani and Bankura of West Bengal, India for Kharif season and can bring a new era of Kharif onion cultivation in Eastern part of India specially in West Bengal. From a study on evaluation of onion genotypes for growth, yield and quality traits under

Variability in Shape and Color of Onion Bulb

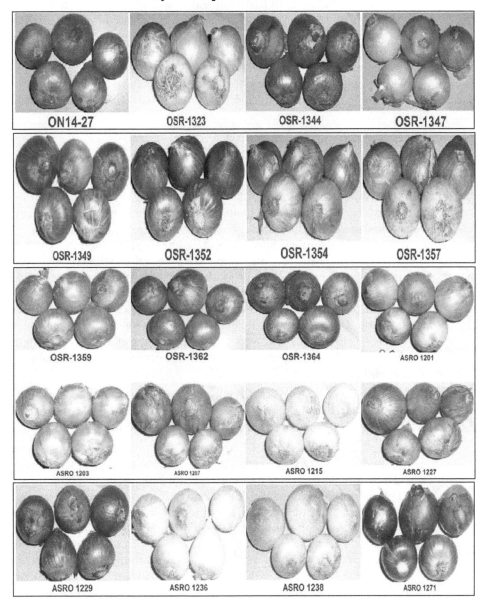

gangetic-alluvial plains of West Bengal, Bal *et al.* (2020) concluded that genotype Sukhsagar was best concerning plant height at 30 DAT and 90 DAT; the number of leaves at 30 DAT, 60 DAT and 90 DAT; equatorial diameter, total soluble solids, dry matter (per cent), total sugar (per cent), vitamin C (mg/g), pyruvic acid (μmole/g). The genotype Arka Kirtiman was found superior in terms of total yield and marketable yield. Punjab Naroya had the highest phenol content. The genotypes Bhima Raj, HO-3, Arka Bheem, Arka Niketan exhibited superiority in terms of number of bulbs per plot, number of days to maturity, polar diameter, number of scales per bulb, respectively. For future onion improvement, these well performing genotypes can be considered for selection or can also be used as parents in hybridization programme to obtain varieties/hybrids with desirable quality and higher yield.

6.9.0 Bio-chemical Characters

Pungency and dry matter are polygenic characters (Jones and Mann, 1963). Total solids had high heritability together with a high genetic advance indicating the additive gene effects (Padda *et al.*, 1973). Warid (1952) reported that Soluble solids (SS) was governed by four to ten gene pairs and partial dominance in nature. Owen (1961) postulated that cumulative gene action and a relatively small number of genes were involved in SS inheritance. Pal and Singh (1988) reported that pyruvic acid (PA) was controlled by additive and dominant gene action, with additive gene action more important. (Lin *et al.*, 1995) found that SS and PA are expressed and inherited in a quantitative manner. Generation means, frequency distributions, deviation from midparent value, and estimates of gene effects all indicated that inheritance of SS and PA was additive. Estimates of broad-sense heritability ranged from 48 per cent to 53 per cent for PA and 8 per cent to 56 per cent for SS. Phenotypic correlations between PA and SS estimated from the F_2 generations of two crosses, were moderate and positive ($r = 0.50$ and 0.42).

Allium fistulosum, shallot (*A. cepa* Aggregatum group) monosomic addition lines (FF+1A to FF+8A) accumulated non-reducing sugars in winter leaf blades, except for FF+2A. Hang *et al.* (2004) concluded that genes related to non-reducing sugar metabolisms are located on the 2A and 8A chromosomes.

Ananthan and Balakrishnamoorthy (2007) observed higher estimates of genotypic and phenotypic co-efficient of variation for bulb weight, economic dry matter yield/ha, chlorophyll a, chlorophyll b and sulphur contents in related to economic dry matter yield. Bulb weight, economic dry matter yield and sulphur content were highly heritable. Bulb weight, total dry matter production, chlorophyll a, net assimilation rate and total soluble solids had high positive and significant associations with economic dry mater yield/ha. Further, in a storage study, Ananthan (2010) observed high GCV and PCV for bulb weight, reducing sugars, non reducing sugars, total sugars, total loss and sulphur content. High heritability coupled with high genetic advance was registered for bulb weight, total loss, reducing sugars and sulphur content of the stored bulbs. In association analysis, reducing sugars, protein and moisture content of the bulbs recorded high positive and significant associations with storage loss; while path co-efficient analysis indicated that the reducing sugars, protein and total loss had the strongest positive direct effect on storage loss. Chattopadhyay *et al.* (2013) indicated that quality traits (TSS, vitamin C, phenol and pyruvic acid contents) represent either weak or negative association with bulb yield of onion.

Studying Total flavonol (TF) content in onion cultivar 'Lady Raider', Smith *et al.* (2003) concluded that TFs are governed by more than one gene.

Galmarani *et al.* (2001) observed significant correlation among soluble solids content (SSC), total dry matter, pungency and onion-induced in vitro antiplatelet activity. A chromosome region on linkage group E accounted for a significant amount of the phenotypic variation for all of these traits. The correlation among these traits may be due to pleiotropy of genes controlling solid content.

McCallum *et al.* (2007) identified genomic regions affecting pungency in the cross 'W202A' × 'Texas Grano 438'. Linkage mapping revealed that genes encoding plastidic ferredoxin-sulfite reductase (SiR) and plastidic ATP sulfurylase (ATPS) are closely linked (1-2 cM) on chromosome 3. Broadsense heritability of pungency was estimated 0.78-0.80. QTL analysis revealed significant associations of both pungency and bulb soluble solids content with marker intervals on chromosomes 3 and 5 which have previously been reported to condition pleiotropic effects on bulb carbohydrate composition. Single-locus models suggested that the partially dominant locus associated with these candidate genes controls 30-50 per cent of genetic variation in pungency in these pedigrees.

6.10.0 Seed Coat Colour

Jirik (1972) identified plants with a brown seed coat marker. Simple monogenic recessive control was observed. Brown seeds are smaller and lighter in weight and have a smoother surface than black ones. The brown seed trait is potentially useful as a marker character for hybrid seed production (Fustos, 1985). Khar *et al.* (2008) reported that onion seed coat color is controlled one locus (B).

6.11.0 Seed Stalk

Horobin (1986) suggested that a major recessive gene is responsible for seed stalk dwarfness; however, substantial minor gene and environmental variation is also involved.

6.12.0 Disease and Insect Resistance

Jones and Mann (1963) reported sources which have been found to be resistant to important diseases like pink root, *Fusarium*, downy mildew, purple blotch, onion smut, black mould, neck rot and similarly, Jones and Mann (1963) and Pawar *et al.* (1975) recorded sources resistant to thrips. These workers have proposed the morphological and chemical basis of resistance in some instances. Waxy foliage and seed stems are more resistant to the purple blotch fungus than the non-waxy or less waxy ones. Coloured bulbs are more resistant to mould than white ones. Glossy foliage plants were more resistant to thrips than non-glossy plants (Molenaar, 1984). Yellow stripe virus in *A. ampeloprasum* was indicated to be controlled by a single recessive gene (Cheremushkina *et al.*, 1989). Kofoet *et al.* (1990) reported that the resistance against downy mildew, caused by *Peronospora destructor*, is most probably based on a single dominantly inherited locus originating from the nuclear genome of *Allium roylei*. They proposed to designate this locus Pd_1, the recessive counterpart being pd_1. However, they can not exclude the presence of a second resistance locus segregating independently from Pd_1.

The inheritance of resistance to neck rot, incited by *Botrytis allii* Munn, was quantitative and mostly additive with a small amount of dominance for susceptibility. Estimates of gene effects also indicated that additive effects predominated, with some dominance and no epistasis (Lin *et al.*, 1995).

Chand *et al.* (2018) suggested that the Purple blotch, caused by *Alternaria porri* (Ellis) Cifferi, resistance is controlled by a single dominant gene (ApR1). Molecular mapping revealed that the SSR marker AcSSR7 and STS marker ApR-450 were closely linked to the ApR1 locus in coupling at distances of 1.3 and 1.1 cM, respectively.

Research suggests that a single gene, two genes, or multiple genes govern resistance to FBR (Cramer, 2000). Bacher (1989) and Bacher *et al.* (1989) reported that resistance to Fusarium basal plate rot (FBR), caused by *Fusarium oxysporum* f. sp. *cepae*, was controlled by two partially dominant genes, designated as *Foc1* and *Foc2*. They hypothesized that the interaction between loci appeared to be additive. Dominant alleles must be present at both loci in order for plants to be resistant to FBR. Tsutsui (1991) reported that resistance to FBR was controlled by a single dominant gene. However, he proposed some variable expression in this gene and the possibility of additional genes for FBR resistance. In addition to single major gene inheritance, polygenic inheritance of FBR resistance has been suggested (Kehr *et al.*, 1962; Lorbeer and Stone, 1965; Holz and Knox-Davies, 1974). Villanueva-Mosqueda (1996) reported high narrow-sense heritability (0.80) of FBR resistance in a short-day, open-pollinated onion population, and noted it was highly correlated (r = 0.9) with resistance to pink root disease.

7.0 Growth and Development

The commercial product of the onion plant is the bulb, formation of which governed by many factors like temperature, photoperiod, light quality and intensity, plant size, nutrition, growth substances, *etc.*

According to Kato (1963), the onset of bulbing is characterized by rapid elongation of leaves due to extension of neck region of the sheath, which is followed by a lateral expansion rather than the division. As bulbing progresses, leaf blades cease to form, and scale leaf initials, in which the blade is much reduced in comparison to the sheath, are differentiated from the apex of disc (Heath and Hollies, 1965). These scale leaves swell to form inner storage of the bulb. The innermost leaves are thickened to a lesser extent than the outer scales. Finally, as the bulb matures, two or three foliage leaf initials are laid down at the apex. These initials elongate to produce leaf blades in the following season, when the bulb sprouts (Abdalla and Mann, 1963).

On average, a dry matter content of 13 per cent has been found with a variation from 11 to 14 per cent. Higher dry matter contents are seen in old cultivars of Stuttgarter or new breedings of shallot cultivars (*A. cepa* var. *ascalonicum*). Dry matter content increases during the period of bulb development. Harvesting later than 80-90 per cent top fall reduced the dry matter content and storage ability. Cutting off the onion leaves before harvest and lifting the crop for artificial drying instead of wind-rowing decreases dry matter content and total dry matter production. Increasing the nitrogen supply above 200 kg N per ha decreases dry matter content in mature bulbs. During the storage period from November to July the dry matter content decreases slowly as a result of respiration loss (Henriksen *et al.*, 2001).

7.1.0 Photoperiod

Onion is normally regarded as a long day plant and the bulb formation is promoted by long day conditions. The cultivars differ greatly in day length requirement and there are some cultivars like Conjumatlan (Mexican cultivar) where bulb formation may start under short day condition. Correctly speaking, it is not also a short day cultivar and apparently it may appear so, since the bulb formation becomes effective under relatively shorter photoperiodic condition. In fact, phtoperiodic requirement in onion is a quantitative character and each cultivar needs a minimum day length for bulb formation, which is known as critical value. This critical value in combination with temperature influences bulb initiation. The so-called short day cultivars, when pass the critical value limit, will initiate bulb formation and development is continued under long day conditions. Conversely, a long day cultivar will not be able to initiate bulb formation unless its longer critical value of photoperiod is reached. Aoba (1964) reported that under long day and high temperature, leaf growth was inhibited and ceased completely within 25-30 days of the start of long day treatment. When short day (8 hours light) treatment was given to the plants for 30 days, before the long day treatment, growth was accelerated and large bulbs were formed. Kato (1964) found that any effect produced by long photoperiod during bulb formation was neutralized if long day treatment was interrupted by a period of short days and the bulb forming phase was reverted to vegetative phase. It was experimentally possible to reverse the bulb formation phase, even after the tops have fallen over.

Onion adapted to latitudes far away from the equator start to initiate bulbing when day length of 14 to 16 hours are reached and onion of these regions are commonly referred to as long day onions. Short day cultivars can initiate and form bulbs under photoperiods of 13 hours or less, equivalent to those found in latitudes which are < 24° from the equator (Currah and Proctor, 1990).

Interesting findings of Heath and Hollies (1965) showed that if onion plants were kept under short day condition they continued growing for over 3 years without bulbing. Kononkov *et al.* (1969) tried some cultivars from Europe, Israel and USA in Cuba and found that the only cultivars from the countries approximately of same day length of Cuba, produced bulbs. Those of far northern region did not produce any bulb. Similarly, Sinnadurai (1970) while testing suitability of long day cultivars in Ghana, found that only a few cultivars like Excel, Eclipse, Crystal Wax and Texas Grano, performed fairly well in the upper region where day length was slightly longer than the south. Austin (1972) showed that the response of different cultivars to day length remained in the same order, relative to each other over a wide range of photoperiodic treatments.

The bulbing behaviour of 21 onion cultivars was studied by Mettananda and Fordham (1997) under short (12 h) and long (16 h) photoperiods in controlled environmentals in the UK and under field conditions in Sri Lanka with the objective of developing simple and rapid screening techniques for identifying cultivars for tropical regions. There was a high correlation between the time taken for bulbing of individual cultivars under 16 h and 12 h photoperiods in both environments as indicated by a bulbing ratio of 2.0, bulb scale initiation, and bulb maturity. The sequence in which the cultivars bulbed remained constant under the 2 photoperiods but the time taken was reduced by 50-70 per cent under the long 16 h condition. There was also a high correlation between the time taken to achieve a bulbing ratio of 2.0 (which can be determined non-destructively) under an extended 17 h photoperiod, and the time to bulb maturity under tropical field conditions where the prevailing day length was approximately 12 h. These findings are discussed in terms of their application to screening of onion germplasm and cultivars for use in different regions. The cultivars Early Lockyer Brown, Superex and Agrifound Rose were very early maturing followed by an intermediate group consisting of cultivars Galil, H489, Agrifound Light Red, Red Creole-C5 and H-226, all of which are suitable for production in the tropics. Wickramasinghe *et al.* (2000) also noted that the tropical cultivars like Red Creole and Agrifound Dark Red needed at least 12 h for bulb formation. Further, they reported that plants grown under 0, 25 per cent, 50 per cent and 75 per cent shading treatments (12 h photoperiod), in only plants receiving 0 and 25 per cent shading bulbed. The results suggest that Red Creole was more sensitive to shorter photoperiods, bulbing earlier than Agrifound Dark Red. Molecular and biochemical changes associated with the initiation and development of bulbing were studied in onion cv. Robusta by Thomas *et al.* (1997). Bulb initiation was induced in two-week-old onion seedlings by transfer from 8 h short days (SD) under fluorescent lighting at 70 W m^{-2} and 25°C to 16 h long days (LD) at the same temperature and irradiance but with 4 × 100 W lamps supplementary tungsten lighting. An increase in the bulbing ratio was detected within 7 days of starting the LD treatment. The LD treatments resulted in higher levels of glucose in pseudostem tissues and sucrose in leaf tissues. Otherwise sucrose, glucose and fructose levels in leaves and pseudostem were similar in LD and SD seedlings. Sucrose synthase levels were increased in LD pseudostems and leaves compared to SD plants but sucrose phosphate synthase was higher in SD than LD pseudostems. A cDNA library from bulbing pseudostem tissues was differentially screened with cDNA from bulb and leaf tissues. More than 20 clones were isolated and placed into 8 complementation groups. ABE 1-7 were expressed strongly in pseudostems but not leaves and ALE 1 was strongly leaf enriched. ALE 1-7 was also expressed to variable degree in roots. Yamasaki *et al.* (2000) concluded that low temperature and a short day photoperiod complementarily induce flower initiation in Japanese bunching onion. Varietal differences exist in the requirement of low temperature and a short day photoperiod: the primary requirement in Kincho is low temperature and that in Asagikujo is a short day. After flower initiation, the early stage of flower development is day-neutral, and after the floret formation stage, a long day photoperiod promotes flower development and elongation of the seed stalk.

In a study on cultivar Pusa Red having long juvenility phase and absence of vernalization requirement for transition to flowering, Tarakanov (1994) observed that regulation of flowering was truly inductive and required a photoperiodic stimulas for its initiation only, while bulbing was photoperiod dependent until maturity.

7.2.0 Leaves and Carbohydrate

Leaf is the vital point of the plant which receives the external stimulus of photoperiod from the environment and thus enhances the physiological activity. Kato (1965) elucidated that role of leaves in bulb formation and development by using method of defoliation. He found that severe defoliation inhibited bulb formation. Similar results had also been reported by Baker and Wilcox (1961) and Vilet and Scheele (1966). Terabun (1971) found that on plants with 4 leaves, exposure of just some leaf to continuous light was sufficient to promote bulbing and the effect was not nullified, even if the remaining leaves were cut off or were exposed to short days. Comrie (1986) reported that defoliation of leaves either one month before transplanting (75 per cent defoliation) or at transplanting (25 per cent defoliation) decreased marketable yields, mainly by decreasing mean bulb weight.

Along with day length, it may be necessary to know certain other important factors associated with initiation and bulb development. Heath and Hollies (1965) showed that removal of leaf blade delayed bulbing until after the emergence of new leaf. It was further shown that swelling of the seedling section could be induced in the presence of 1 per cent sucrose, within a few days even in darkness. On the basis of these observations they concluded that the response of long days in relation to bulb formation would be available, only in presence of sufficient accumulation of sugars in plants both for maintenance of root system and bulb formation. Kato (1965) found that carbohydrate accumulation in leaf sheath and their concentration reached a maximum in the early stage of bulb development and then decreased gradually towards the end of bulb thickening, though total amount of carbohydrate increased. Bertaud (1986) reported that in short days all photosynthate is partitioned to storage carbohydrate resulting in low leaf area and small bulbs that mature early as growth of foliage ceases. Hansen (1999) observed that the stage of maturation at the time of harvest significantly influenced the weight of the onions, the dry matter content in individual onions and the fructose content.

7.3.0 Light Quality

Several reports showed that an extension of day length with incandescent light was more effective in promoting bulbing than extension with fluorescent light (Butt, 1968; Woodbury and Ridley, 1969; Austin, 1972). Terabun (1965) found that photoperiodic extension using farred light or blue light, led to accelerated bulbing, while red light suppressed bulbing. Terabun (1970) further reported that bulbing is a phytochrome induced response. Shading of leafy onion plants by green plants reduces the ratio of red to farred wavelength of light and this gives a signal, detected by the onion plant via phytochrome system, which hastens the onset of bulbing (Mondal *et al.*, 1986). Bosch Serra and Casanova (2000) observed that the normalized difference vegetation index (NDVI) relationship with the fraction of intercepted photosynthetically active radiation (f par) gave a better prediction of crop growth estimation than the ratio of near-infrared to red reflectance (RVI).

7.4.0 Light Intensity

Brewster (1975) observed that in identical light quality, photoperiod temperature, higher the light intensity more rapid was bulbing and earlier ripening of bulb resulted. Lercari (1983) showed that the photoperiodic aspect of the control of bulbing is a high irradiance response of the phytochrome system. This may explain why bulbing in onion occurs more rapidly in season or areas of high light intensity.

Wickramasinghe *et al.* (2000) also observed that low light intensity decreased the number of true scales and increased the number of sheath scales. Again, dormant leaf initials decreased with decreasing temperature while the number of secondary meristems significantly increased.

7.5.0 Temperature

Although photoperiod seems to play the major role in the formation of bulb in onion, temperature also plays a very important part. Other factors being equal, onion will bulb more quickly at higher temperature than at lower temperature, Thompson and Smith (1983) found that critical day length of bulbing decreased with increase in temperature indicating an interaction between temperature and day length. Thus it would appear that it might be difficult to specify photoperiodic requirement of a cultivar, without specifying the temperature. For a North European cultivar, growing in controlled environment at constant temperature, Butt (1968) found the most rapid bulb growth and earliest onset of decline of leaf area indicating maturity at temperatures between 25 and 30°C. Lower temperatures of 20, 15 and 10°C gave successively less rapid bulbing and maturity. However, Abdalla (1968) observed that the bulb development was significantly retarded when the maximum day temperature reached 40-45°C. Thus their studies showed that for bulbing temperature above 15°C would be helpful and that a late cultivar gave better performance at somewhat cooler climate, while early cultivars at warmer climate. Terabun (1981) reported that cv. Kaizuka-Wase could form bulbs under short day conditions irrespective of low (10°C) or high (20°C) temperature treatments once the bulb initiation started under long day condition of 24 h for 14 days. In another cultivar Sapporo-ki, the bulb formation was observed at higher temperature (20°C) only under similar treatments. Thus the cultivars differ in their temperature requirement. Wiles (1989) compared a series of controlled temperature treatments from 14/6°C up to 34/26°C (day/night) at a 14 h day length, and found that cv. Granex bulbed most rapidly at 30/22°C and more slowly at lower or higher temperature suggesting that growth was reduced by stress condition. In a four temperature regimes experiment conducted in a growth room indicated that plants under the 29-34°C treatments bulbed within two weeks and matured within six weeks (Wickramasinghe *et al.*, 2000). Zambrano *et al.* (1997) observed an increase in TSS and a decrease in acidity in bulbs when exposed to low temperature for 4 weeks. In onion bulb, Lancaster *et al.* (1996) noticed that diameter at bulbing was related to thermal time accumulated prior to bulbing. The dual thresholds of a minimal thermal time of 600 degree-days and a photoperiod of 13.75 h were necessary for bulbing to occur. Mathematical relationships were developed between leaf number, sowing date, bulbing date and bulb growth and maturity. Bulb size at bulbing and number of leaves produced after bulbing could be used to predict final bulb size. Bulb maturity date could be predicted from the number of leaves produced after bulbing.

7.6.0 Plant Size

According to results of Butt (1968), young plants do not respond to bulbing stimulus as rapidly as the older ones. Robinson (1971) reported that physiological age was important in triggering initiation of bulbing in mild winter for so-called short day cultivar. Thus it appears that age of plant has considerable effect on bulb formation. In an experiment with dry onion sets, transplants and seed of the same cultivar when planted/sown at the same time, they started to bulb and mature in the order named. It appears that age, size or stored food, possibly all the three play some unknown roles in triggering the mechanism that initiates bulbing in proper day length and temperature conditions. Brewster *et al.* (1986) found experimentally that yield of onion as highly dependent on the Leaf Area Index (LAI) of the crop during the period between onset of bulbing and harvest, a period known as duration of bulbing. Sahu (2016) observed that the age of seedlings significantly influenced the growth and yield of onion. He recorded maximum plant height, number of leaves per plant, bulb weight, bulb diameter and yield with minimum days to maturity under 20 and 30 days old seedlings.

7.7.0 Growth Substances

Levy and Kedar (1970) and Levy *et al.* (1973) reported that Ethephon promoted early and rapid bulb growth under non-inductive condition. However, high dosages led to retarded leaf growth and low dry matter in the bulbs. They also found that 500-1000 ppm of Ethephon spray was effective. Terabun (1965) recorded that foliar application of 500 ppm of MH induced bulbing in normally non-inductive condition but a spray of 2, 4-D (20 ppm) retarted the bulb formation. The growth regulators like MH, abscisic acid induced leaf senescence and reduced bulb size. However, growth promoters like IAA, GA_3 or kinetin alone or in combination prevented leaf blade senescence. Mathur (1971) reported that growth substances like NAA, IAA, IBA at 100, 200 or 3000 ppm increased leaf number and weight of bulbs. Omran *et al.* (1984) obtained the best combination with regard to bulb weight generally with IAA at 200 ppm + $ZnSO_4$ at 0.05 per cent applied three times at 15 days intervals starting from 40 days after transplanting.

Shukla and Namdeo (2000) reported from India that application of 2000 ppm MH on cv. Pusa Red gave the maximum leaf, scale leaf area, dry weight, LAI, SLA, SLW, leaf chlorophyll content, net photosynthesis, stomatal resistance at all the growth stages with highest bulb yield. In contrast, the use of MH (60 per cent w/w) had no effect on yield and bulb size of 6 onion cultivars in field experiments at Akhelia, Cyprus (Gregoniov, 1998). Bhardwaj (1991) obtained male sterility in onion by spraying of MH and GA_3 at the initiation of first floral bud and after a 15 days interval. Ranpise *et al.* (2000) concluded that to get maximum onion yield with minimum bolting, spray of Ethephon 4000 ppm or cycocel 2000 ppm should be given 85 days after transplanting. Exogenous application of salicylic acid (SA) significantly increased the better vegetative growth in terms of plant height, collar thickness with higher level of chlorophyll content of leaves. Application of SA at 30 DAS, 30 DAT and/or 45 or 60 DAT increased the vegetative growth and also bulb yield in onion variety ALR (Pradhan *et al.*, 2016). Singh *et al.* (2017) recorded maximum bulb dry weight (8.96 g), protein (18.32 mg/g), sugar (8.61 per cent) nitrogen (0.64 per cent) and acid content (0.29 per cent) under treatment 125 g a.i./ha of mepiquat chloride at 35 DAT. However, treatment 62.5 g a.i./ha of mepiquat chloride at 35 DAT had significant effect on TSS content (13.670 Brix).

7.8.0 Nutrition

It has been found that at photoperiod above the critical value nitrogen nutrition did not influence bulbing response. However, when the plants were grown near the critical photoperiod, a deficiency of nitrogen supply would have the same effect as lengthening the photoperiod, while a high N supply at the same critical photoperiod would have the effect of shortening of length of day. Therefore, at critical photo-period, deficiency of N will hasten bulbing and excess of N will slow down the bulbing process. Kato (1964) observed that heavy application of N under long day treatment (20 hours) resulted in smaller bulbs though bulb formation was not delayed. Geetha *et al.* (1999) noticed that combined application of farmyard manure and muriate of potash at 25 t/ha and 200 kg K_2O/ha, respectively, resulted in higher dry matter yields and K uptake at various stages of growth.

7.9.0 Bulb Development in Relation to Climatological Conditions

In the light of the enumeration of several factors favouring bulb development, growing of onion in different traditional regions of India is examined, in respect of bulb development in relation to climatological conditions. In the North Indian States, Bihar, Uttar Pradesh, Madhya Pradesh, Rajasthan, Haryana and Punjab, which grow onion mostly in *rabi* season, (December-January to May-June), the temperature prior to bulb initiation is low, and with the onset of bulb formation, days become longer and temperature rises, reaching maximum around 40°C in May-June. Then senescence sets in, characterized by sheath fall, which indicates the complete maturity of bulb. The main cultivars grown in this season are Patna Red, Pusa Red, *etc.* which are light pink in colour with flat bulb shape. Of late, there has been development of growing onions in *kharif* season in North India by transplanting the seedlings (or sets produced in spring-summer) in early to middle August. The bulb formation is initiated in late September and bulbs become ready for lifting in November-December. The cultivars suitable for this season have been found to be N-53 and Nasik Red from Maharashtra. The bulb formation during September in N-53 cultivar takes place under comparatively short day condition and Pusa Red cultivar does not form bulbs when grown in *kharif* season. This brings out the differential response of the cultivars and also the differences in respect of photoperiod requirement of the cultivars.

In western India, onion is grown under comparative warmer temperature, in three well-defined season, *kharif* August-December, early *rabi* (*rangda*) November-March and *rabi* during January-May. Maharashtra commands a large area in *kharif* and *rabi* seasons and Gujarat grows in larger area in *rabi* season. The *kharif* cultivars are N-53 and Nasik Red and for *rabi* and early *rabi* seasons the cultivars are N-2-4-1, and Pusa Red. N-53 can be grown in all the three seasons. The *kharif* production is characterized by somewhat shorter day length with moderate temperature. In the *rabi* season, the days are comparatively longer and bulb development takes place at a higher temperature. In Karnataka, Bellary region grows onion both in *kharif* and *rabi* seasons and the same cultivar Bellary Red grows well in both the seasons. Possibly, the difference in day length in both the seasons, is not marked and hence the same cultivar is adapted to both the seasons.

In Tamil Nadu, mutiplier onion, belonging to *aggregatum* is grown in the foot hills of Dindigul-Palni regions of Madurai district, almost round the year without any appreciable difference in day length. The *aggregatum* has been preferred, because it grows well under high temperature and nearly equal day length conditions and also due to easy vegetative propagation.

The bulb development and maturity in *kharif* and *rabi* seasons provide an interesting contrast, especially in North and Western India. In the *rabi* season, full maturity of bulb is reached and due to senescence and sheath fall (in most of the cultivars), the storage quality of bulbs is improved. In *kharif* season, on the other hand, the full maturity of bulb is probably reached, but due to lack of senescence the vegetative growth continues, affecting the storage quality. This is particularly observed in N-53 and Nasik Red cultivars, in which the bulbs reach maturity in November, when the temperature is low and days are shorter. Hence there is no cessation of growth and senescence does not occur.

8.0 Cultivars

In India, red or pink skinned and pungent type onions are preffered essentially for cooking, due to its strong flavour. In contrast, yellow skinned, mild flavoured and sweet onion types are preffered in western countries for salad purpose. These types are not liked in India and have no demand in Asian market in general.

Onion varieties are notified either by the Central Varietal Release Committee (CVRC) or State Variety Release Committee (SVRC) for commercial cultivation. Onion breeding materials are extensively evaluated for their performance in multi-location trials under the AICVIP/AINRPOG, which play a key role in testing, identification and release of new varieties. Onion is a biennial and cross pollinated crop. It takes almost 12-14 years to purify or develop a new variety. Onion varieties were developed by different SAUs and ICAR institutes and tested under coordinated or network project for release at national level for different agro-climatic conditions. About 65 onion varieties, including 2 F_1 hybrids and 6 multiplier types have been developed and released from public sectors in India. However there is enough scope to develop onion cultivars for processing (white varieties with high TSS suitable for dehydration), salad purpose (more sugar and less pungency), export (short day yellow colour varieties) and for resistance to various biotic and abiotic stresses. The onion cultivars developed can be grouped on the basis of colour of skin (Light red, Dark red, White and Yellow), types (common, rose and multiplier), location (short or long day), growing season (*kharif*, late-*kharif* and *rabi*), maturation (early and late), pungency (pungent and mild) *etc*. Development of varieties suitable to ecological zones of India and for non-traditional areas also required to increase the area coverage under onion.

Table 9: Common Onion Cultivars of India

Sources	Cultivars
ICAR-IARI, New Delhi	Pusa Red, Pusa White Round, Pusa White Flat, Pusa Madhavi, Pusa Ridhi, Early Grano and Brown Spanish
ICAR-NBPGR, New Delhi	Pusa Ratnar
ICAR-IIHR, Bangalore	Arka Niketan, Arka Pitambar, Arka Pragati, Arka Kalyan, Arka Lalima (F_1), Arka Kirthiman (F_1), Arka Bheem, Arka Yojith, Arka Sona and Arka Swadista
ICAR-DOGR, Rajgurunagar	Bhima Super, Bhima Red, Bhima Raj, Bhima Dark Red, Bhima Kiran, Bhima Shakti, Bhima Light Red, Bhima Shweta, Bhima Subhra and Bhima Safed
ICAR-VPKAS, Almora	VL Piaz 3, VL Piaz 67 (F_1)
NHRDF, Nasik	Agrifound Dark Red, Agrifound Light Red, Agrifound White, Agrifound Red, NHRDF Red, NHRDF Red 2, NHRDF Red 3 and NHRDF Red 4
MPKV, Rahuri	Phule Samarth, Phule Safed and Phule Suvarna
PDKV, Akola	Akola Safed
Agril. Dept., Maharashtra	Baswant 780, N-2-4-1, N-257-9-1 and N-53
PAU, Ludhiana	Punjab Naroya, Punjab Round Red, Punjab Selection, Punjab 48 (S 48), Punjab White, PRO 6, PRO 7, PWO 2, PYO 1
HAU, Hisar	Hisar 2, Hisar Onion 3 and Hisar Onion 4
MPUAT, Udaipur	Udaipur 101, Udaipur 102 and Udaipur 103
GAU and JAU, Junagadh	GWO 1, GJRO-11, GJWO-3
Anand Agricultural University, Anand	GAWO 2
CSAUA&T, Kanpur	Kalyanpur Red Round
CSKHPAU, Palampur	Palam Lohit

Table 10: Rose Onion Cultivars

Source	Variety
ICAR-IIHR, Bangalore	Arka Bindu, Arka Vishwas
NHRDF, Nasik	Agrifound Rose
Local, Bangalore	Bangalore Rose

Table 11: Multiplier and Bunching Onion Cultivars

Source	Cultivar
Multiplier onion	
ICAR-IIHR, Bangalore	Arka Ujjawal
NHRDF, Nasik	Agrifound Red
TNAU, Coimbatore	CO 1, CO 2, CO 3, CO 4, CO (On) 5 and MDU 1
Bunching onion	
ICAR-IARI, New Delhi	Pusa Soumya

Table 12: Classification of Common Onion Cultivars According to Skin Colour

Bulb Colour	Cultivars
Red	Pusa Red, Pusa Ratnar, Pusa Madhavi, Pusa Ridhi, Arka Pragati, Arka Niketan, Arka Kalyan, Arka Bindu, Arka Lalima (F_1), Arka Kirthiman (F_1), Arka Bheem, Arka Akshay, Arka Vishwas, Bangalore Rose, Bhima Super, Bhima Raj, Bhima Red, Bhima Dark Red, Bhima Kiran, Bhima Shakti, Agrifound Dark Red, Agrifound Light Red, Agrifound Rose, Hisar 2, Hisar 3, Hisar 4, Punjab Naroya, Punjab Red Round, Punjab Selection, N 53, N-2-4-1, NHRDF Red, NHRDF Red 2, Baswant 780, Udaipur 101, Udaipur 103, Phule Samarth, Kalyanpur Red Round, VL 67, Palam Lohit
White	Pusa White Round, Pusa White Flat, Agrifound White, N-257-9-1, Punjab 48 (S48), Punjab White, PWO 2, Udaipur 102, Bhima Sweta, Bhima Shubra, Phule Safed, Akola Safed, Arka Swadista
Yellow	Early Grano, Brown Spanish, Bermuda Yellow, Arka Pitambar, Arka Sona, Phule Swarna, PYO 1

Table 13: Salient Features of Onion Cultivars Developed in India

Pusa Red

Recommended for *rabi* season, can be grown both late kharif and rabi season in Maharashtra; Developed by mass selection from indigenous material; Released in 1978 (CVRC) for Punjab, U.P., Bihar, Rajasthan, Gujarat, Haryana, Delhi, Karnataka, Tamil Nadu and Kerala; Short to intermediate day length variety; Wider adaptability; Bulbs are flattish round, medium-sized (70-80 g), red colour, less pungent with TSS 12-13 per cent ; Matures in 135-140 days after transplanting; Average yield is 300 q/ha; Very good for storage.

Pusa Ratnar

Recommended for *rabi* season; Developed by selection from segregating population of an F_1 hybrid 'Granex' introduced from USA; Released in 1978, throughout India; Bulbs red in colour, obvate to flat, globular, less pungent and neck drooping, TSS 11-12 per cent ; Mature in 125 DAT; At maturity, bulbs are more exposed above ground; Average yield 300-400 q/ha; Fair in storage; Less pungent, suitable for salad.

Pusa White Round

Recommended for *rabi* season; Developed by the selection from local material; Released in 1975, Throughout India; White colour bulb, roundish flat with TSS 12-13 per cent, drying ratio 8 : 1; Maturity 125-130 days; Average yield 300 q/ha; Very good in storage; Suitable for dehydration purposes and green onion.

Pusa White Flat

Recommended for *rabi* season; Selection from local material; Released in 1977 (CVRC) for Northern plains; Medium to large bulbs; flattish round with attractive white colour, T.S.S. 12-14 per cent, drying ratio 9 : 1; Maturity 120-130 days; Average yield 300 q/ha; Very good for green onions and storage; Suitable for dehydration purposes.

Pusa Madhvi

Recommended for *rabi* season; Selection from indigenous material of Muzaffarnagar; Released in 1989 (CVRC) for Madhya Pradesh and Maharashtra; Medium to large sized bulbs, roundish flat, light red, T.S.S. 11-13 per cent ; Maturity 130-135 DAT; Average yield 300-350 q/ha; Good for storage.

Pusa Riddhi

Suitable for kharif and rabi season; Released in 2013 by SVRC, Delhi for cultivation in Delhi and NRC; Bulbs are compact, flat globe, dark red in colour; Average bulb weight 70-100 g, average equatorial bulb diameter 4.8-6.3 cm; Pungent and rich in antioxidant (quercetin 107.42 mg/100 g); Average yield 316.6 q/ha; Suitable for storage and export.

Pusa Soumya

This is the first bunching onion variety proposed for commercial cultivation in India; Released in 2013 by SVRC, Delhi for cultivation in Delhi and NCR; Multi-cut variety and suitable for round the year green onion production; Produces bluish green leaves and clumps can be separated for further multiplication; Least affected by pets and diseases; Average yield potential is 260 q/ha from single harvest.

Early Grano

Suitable for *kharif* and *rabi* season; Developed through an introduction from USA; Bulbs yellow colour, globular, average weight 150 g, 7-8 cm in size, TSS 6-7 per cent, very mild pungency; Maturity 95-110 days; Average yield 500-600 q/ha; Suitable for green onions and salad; Poor in storage.

Brown Spanish

Suitable for growing in 1000 m or higher regions above sea level; Medium long day type; Bulbs are round, reddish brown colour, size 2-3.5 cm, thick skin, less pungent, TSS 13-14 per cent ; Maturity 160-180 days; Average yield 250-300 q/ha; Good for storage up to 6 months.

Arka Kalyan

Recommended for *kharif* season; Very popular variety and well adopted variety for many agro ecological zones of the country; Developed through vigorous mass selection from IIHR-145; Bulbs are attractive red, flat globe, weighing 100-140 g, 4-6 cm across, pungent with TSS 11-13 per cent ; Deep red coloured outer scales and fleshy succulent internal scales; Maturity 110-110 DAT; Average yield 336 q/ha; Moderately resistant to purple blotch disease.

Arka Niketan

Recommended for *rabi* season, can be grown in late-*kharif* in maharashtra; Developed through mass selection from a local collection IIHR-153; Bulbs globular with thin neck, light red colour, medium size, average wight 100-180 g, bulb diameter 4-6 cm, TSS 12-14 per cent, high in pungency and dry matter; Maturity 145 DAT; Average yield 340 q/ha. Good storability for 5 months under room temperature.

Arka Pragati

Recommended for *rabi* and *kharif* seasons in the South; Developed through mass selection of a local collection IIHR-149 from Nasik; Medium size, globe shaped bulbs with thin neck, average bulbs weight 100-160 g, deep pink coloured outer scales and high pungency; Maturity 140-145 DAT; Average yield 200 q/ha.

Arka Pitambar

Suitable for *kharif* and *rabi* seasons; Short day type; Developed through pedigree selection from the cross U.D. 102 × IIHR-396; Bulbs uniform yellow, average weight 80 g, medium size (5.2-6.0cm), globe shape and thin neck; Less pungent with TSS 11 per cent and total sugar 9.81 per cent ; Duration 140 days; Yield 350 q/ha; Tolerance to purple blotch and basal rot diseases and thrips. Long storage life (3 months); Suitable for export market (Europe, Australia and America).

Arka Bindu

Recommended for *rabi* season; Developed by mass selection from the local material collected from Chickballapur (Karnataka); Bulbs are attractive shining dark red, flat globe, small bulb size (2.5-3.5 cm), highly pungent, high TSS of 14-16 per cent ; Maturity 100 DAT; Free from early bolting and splitting; Average yield 250 q/ha; Export oriented rose onion variety. Exported to Malaysia, Singapore, Indonesia and Bangladesh.

Arka Lalima

Suitable for *kharif* and *rabi* seasons; Recommended for eastern dry zone of Karnataka; F_1 hybrid (MS-48 × Sel. 14-1-1), developed through heterosis breeding (MS based); Medium to big sized bulbs with globe shape and firm texture, red in colour with bulb weight 120-130 g; Duration 130-140 days; Bulb yield 470 q/ha; Tolerance to purple blotch, basal rot and thrips; Long storage (4-5 months).

Arka Kirthiman

Suitable for *kharif* and *rabi* seasons; Recommended for eastern dry zone of Karnataka; F_1 Hybrid (MS-65 × Sel. 13-1-1), developed through heterosis breeding (MS based); Medium to big sized bulbs with globe shape and firm texture; Bulbs red in colour with average weight 120 -130 g; Duration 125-130 days; Yield 470 q/ha; Bulbs Tolerance to purple blotch, basal rot and thrips; Long storage (4-5 months).

Arka Bheem

A tri-parental synthetic variety with red to pinkish red elongated globe shaped bulbs; Average bulb weight is 120 g; Yield 470 q/ha in 130 days.

Arka Akshay

A tri-parental synthetic variety with dark red globe shaped bulbs; Average bulb weight is 115 g; Yield 450 q/ha in 130 days.

Arka Yojith

Bulbs white in colour, flat globe shape, weight 60-80 g, bulb diameter 4.5-5 cm, TSS 18-20 per cent, dry matter content 18-20 per cent ; Bulb yield 250-300 q/ha with duration 110-120 days; Suitable for dehydration.

Arka Sona

Suitable for growing in *rabi* season; Developed by pedigree method; Yellow bulb colour, globe shape, big size, bulb weight 140.0 g and bulb diameter 6.5-7 cm, low TSS (10 °Brix); Duration 120 days; Bulb yield 450 q/ha; Developed for export market for Asian countries,

Arka Vishwas

Rose onion suitable for growing in *rabi* season; Developed through selection; Bulb colour dark red, flat globe shape, small size (3-4 cm), bulb weight 40 g, TSS 16 °Brix; Duration 115 days; Bulb yield 300 q/ha; Developed for export market (Asia).

Arka Ujjwal

True seed multiplier onion developed through pedigree breeding; Uniform bright dark red bulb color, compound bulb with flat shape, bulb size 4-5 cm, number of bulblets/bulb 3-5, bulb weight 40-45 g, TSS 16-18 per cent, dry matter content 14-16 per cent ; Bulb yield 20-25t/ha in 85 days; Developed for export market.

Arka Swadista

Developed through pedigree breeding; Uniform white color, bulbs are oval globe in shape, TSS 18-20 per cent, dry matter content 15-18 per cent, edible bulb 98 per cent, bulb weight 35-40 g, small in size (3-3.5 cm); Bulb yield 160-180 q/ha in 105 days; White onion variety for fermented preservation has suitable for bottle preservation.

Bhima Super

Recommended for *kharif* and late-*kharif* season in Chhattisgarh, Delhi, Gujarat, Haryana, Karnataka, Madhya Pradesh, Maharashtra, Odisha, Punjab, Rajasthan and Tamil Nadu; Bulbs are attractive red with globe shape; Average bulb weight 70-80 g, TSS 11-12 °Brix and less than 5 per cent double bulbs and bolters; Average yield 200-220 q/ha in *kharif* and 400-500 q/ha in late-*kharif*; Bulbs attain maturity within 100-105 DAT in *kharif* and 110-120 DAT in late-*kharif*; Bulb storability 30-45 days in *kharif* and 3 months in late-*kharif*; produces mostly single centered bulbs; field tolerant to thrips and foliar diseases.

Bhima Red

Recommended for *rabi* season for Maharashtra and Madhya Pradesh, and *kharif* and late-*kharif* season in Delhi, Gujarat, Haryana, Karnataka, Maharashtra, Punjab, Rajasthan and Tamil Nadu *i.e.* can be grown in all three seasons; Bulbs are medium red and flat globe shape; Average bulb weight 70-80 g, TSS 10.5-11.5 °Brix and less than 5 per cent double bulbs and bolters; Average yield 190-210 q/ha in *kharif*, 450-500 q/ha in late-*kharif* and 300-320 q/ha in *rabi*; Bulbs attain maturity within 105-110 DAT in *kharif* and 110-120 DAT in late-*kharif* and *rabi*; Bulb storability 30-45 days in *kharif* and 3 months in late-*kharif* and *rabi*; Field tolerant to thrips and foliar diseases.

Bhima Raj

Recommonded for *kharif* and late-*kharif* season in Gujarat, Karnataka and Maharashtra and can be cultivated in *rabi* for immediate market in Delhi, Gujarat, Haryana and Rajasthan *i.e.* can be grown in all three seasons; Bulbs are dark red with globe shape; Average bulb weight 60-70 g, TSS 10-11°Brix and less than 5 per cent double bulbs and bolters; Uniform neck fall in maturity; Average yield 240-260 q/ha in *kharif*, 400-450 q/ha in late-*kharif* and 250-300 q/ha in *rabi*; Bulbs attain maturity within 100-105 DAT in *kharif* and 115-120 DAT in late-*kharif* and *rabi*; Bulb storability 30-45 days in *kharif* and 3 months in late-*kharif* and *rabi*; Field tolerant to thrips and foliar diseases.

Bhima Dark Red

Recommended for *kharif* season in Chhattisgarh, Delhi, Gujarat, Haryana, Karnataka, Madhya Pradesh, Maharashtra, Odisha, Punjab, Rajasthan and Tamil Nadu; Bulbs are dark red with flat globe shape; Average bulb weight 60-70 g, TSS 11-12 °Brix and less than 5 per cent double bulbs and bolters; Average yield 220-240 q/ha; Bulbs attain maturity within 100-110 DAT; Bulb storability 2 months; Field tolerant to thrips and foliar diseases; It is first red onion variety recommended at national level for *kharif* season.

Bhima Kiran

Recommended for *rabi* season in Maharashtra, Karnataka, Andhra Pradesh, Delhi, UP, Haryana, Bihar and Punjab; Average bulb weight 60-70 g, TSS 11.5-12.5 °Brix and less than 5 per cent double bulbs and bolters; Average yield 280-320 q/ha; Bulbs attain maturity within 125-135 DAT; Bulb storability 5-6 months; Uniform neck fall, good in storage; Field tolerant to thrips and foliar diseases.

Bhima Shakti

Recommended for *rabi* season, can be grown in late-kharif in Gujarat, Karnataka and Maharashtra; Bulbs are attractive red with globe shape; Average bulb weight 60-80 g, TSS 12-13 °Brix and less than 5 per cent double bulbs and bolters; Average yield 280-300 q/ha in *rabi* and 350-400 q/ha in late-*kharif*; Bulbs attain maturity within 125-135 DAT in *rabi* and late-*kharif*; Bulb storability 5-6 months; Uniform neck fall during *rabi* and >70 per cent neck fall in late-*kharif*; Moderately tolerant to thrips and field tolerant foliar diseases.

Bhima Light Red

Recommended for *rabi* season in Karnataka and Tamil Nadu; Bulbs are light red with globe shape; Average bulb weight 70 g with thin neck, TSS 12-14 °Brix and almost free from double bulbs and bolters; Average yield 360-400 q/ha; Bulbs attain maturity within 110-120 DAT; Bulb storability 5-6 months; Uniform neck fall and good in storage; Total weight losses after four months of storage was less than 25 per cent.

Bhima Shweta

Recommended for *rabi* and *kharif* season; Bulbs are attractive white with flat globe shape; Average bulb weight 60-70 g, TSS 11-12 °Brix and less than 5 per cent double bulbs and bolters; Average yield 180-200 q/ha in *kharif* and 260-300 q/ha in *rabi*; Bulbs attain maturity within 100-105 DAT in *kharif* and 110-120 DAT in *rabi*; Uniform neck fall; Bulb storability 30-45 days in *kharif* and 3 months in *rabi*; Field tolerant to thrips.

Bhima Shubra

Recommended for *kharif* season in Chhattisgarh, Gujarat, Karnataka, Madhya Pradesh, Maharashtra, Odisha, Rajasthan and Tamil Nadu and late-*kharif* season in Maharashtra; Bulbs are attractive white with globe shape; Average bulb weight 70-80 g, TSS 10-12 °Brix and less than 5 per cent double bulbs and bolters; Average yield 180-200 q/ha in *kharif* and 360-420 q/ha in late-*kharif*; Bulbs attain maturity within 110-115 DAT in *kharif* and 120-130 DAT inlate-*kharif*; Bulb storability 30-45 days in *kharif* and 3 months inlate-*kharif*; Moderately tolerant to thrips and having capacity to tolerate environmental fluctuation; First white variety for kharif season recommended at national level which can fill the gap for processing from October-February.

Bhima Safed

Line 'NRCWO-3' has been christened as 'Bhima Safed'; Recommended for *kharif* season in Chhatisgarh, Gujarat, Karnataka, Madhya Pradesh, Maharashtra, Odisha, Rajasthan, and Tamil Nadu; Bulbs are white with globe shape; Average bulb weight 70-80 g, TSS 11-12 °Brix and less than 5 per cent double bulbs and bolters; Average yield 180-200 q/ha; Bulbs attain maturity within 110-120 DAT; Bulb storability 30-45 days; Moderately tolerant to thrips and foliar diseases.

VL *Piaz* 3

Recommended for *rabi* season. Released in 1991 (SVRC) for Uttarakhand Hills; This is the first open pollinated synthetic onion variety for long day conditions of Uttarakhand Hills; Plant height 50-70 cm, bulbs globular, red colour, medium size, pungent; Less bolting character; Maturity 145 days; Field tolerance of purple blotch and thrips, and a reduced tendency to bolting; Average yield 332.9 q/ha.

VL *Piaz* 67

First hybrid onion developed in India; Released in 1973 (SVRC) for Uttarakhand and plains; Large size bulbs, red colour; Maturity 130-140 days; Yield 350 q/ha.

Agrifound Light Red

Recommended for *rabi* season throughout the country; Developed through mass selection from a local stock of *rabi* onion grown in Dindori area of Nasik; Bulbs are attractive light red, globular round in shape with reddish thick inner scales, 4-6 cm size, 13 °Brix TSS, 14-15 per cent dry matter and12.20 micro mole/g pyruvic acid; Maturity 110-120 DAT; Yield of 300-350 q/ha; Less bolting habit and good keeping quality.

Agrifound Dark Red

Recommended for *kharif* season in the plains of Sutlej-Ganga; Developed by selection from a local stock of *kharif* onion grown in Nasik; Bulbs are dark red, globular round in shape, 4-6 cm in size with tight skin and moderately pungent, TSS 11-12 °Brix, 13-14 per cent dry matter and 10.07 micro mole/g pyruvic acid; Maturity 95-110 DAT; Average yield of 300 q/ha. Moderate keeping quality.

Agrifound White

Recommended for growing in late *kharif* and *rabi* seasons in Maharashtra, Madhya Pradesh, Gujarat and Rajasthan; Develop through selection from a local stock of white onion grown during *rabi* season in Nimad area of Madhya Pradesh; Bulbs are attractive white, globular in shape with tight silvery skin, 14-15 °Brix TSS, 15-16 per cent dry matter and 12.0 micro mole/g pyruvic acid; Maturity 110-120 DAT; Yield 250-300 q/ha; Medium to good keeping quality; Suitable for dehydration.

Agrifound Red

Multiplier onion; Recommended for growing in *kharif* and *rabi* seasons in Karnataka, Tamil Nadu and Kerala; Aveage size of clusters is 7.15 cm, 5-6 bulblets/cluster, light red colour, average weight 65-68 g/clump and 8.85 g/bulblet, TSS 17-19 °Brix, 18-20 per cent dry matter and 10.13 micro mole/g pyruvic acid; Maturity 65-67 days; Average yield 180-200 q/ha; Resistant to lodging and good keeping quality.

Agrifound Rose

A pickling type cultivar; Recommended for growing in *kharif* season in Cuddapah (AP) and in all the three seasons in Karnataka; Bulbs are deep scarlet red,flattish round in shape with 2.5-3.5 cm in diameter, 16-18 °Brix TSS, 17-19 per cent dry matter and10.27 micro mole/g pyruvic acid; Maturity 95-110 DAT; Average yield 250-300 q/ha; Suitable for export and dehydration.

NHRDF Red

Recommended for growing in *rabi* season in Northern, Central and Western India; Bulb are attractive dark red, globular round in shape with reddish thick inner scales, 13-14 °Brix TSS, 14-15 per cent dry matter and 12.0 micro mole/g pyruvic acid; Maturity 110-120 DAT; Yield 250-300 q/ha; Medium keeping quality

NHRDF Red 2

Recommended for growing in *rabi* season Delhi,Uttar Pradesh, Haryana, Gujarat, Maharashtra, Karnataka and Andhra Pradesh; Bulbs are attractive lightred in colour, globular round in shape with thin neck and 5.0-6.0 cm in diameter, 13-14 °Brix TSS, 14-15 per cent dry matter and12.0 micro mole/g pyruvic acid; Maturity 100-120 DAT; Yield of 350-375 q/ha; Good keeping quality.

NHRDF Red 3

Recommended for growing in *rabi* season in Maharashtra, Madhya Pradesh, Gujarat, Delhi, Uttar Pradesh, Bihar, Punjab, Haryana and Rajasthan; Bulbs are light bronze in colour,globular round in shape with thin neck and 5.5-6.0 cm in diameter, 12-13 °Brix TSS, 13-14 per cent dry matter and 12.50 micro mole/g pyruvic acid; Maturity 120-130 DAT; Yield 350-400 q/ha; Good keeping quality.

NHRDF Red 4

Recommended for growing in *rabi* season in Uttar Pradesh, Bihar, West Bengal and Imphal; Bulbs are dark red in colour, globular round in shape with thin neck and 5.5-6.25 cm in diameter, 12-14 °Brix TSS, 13-14 per cent dry matter and 13.0 micro mole/g pyruvic acid; Maturity 110-120 DAT; Yield 350-400 q/ha; Good keeping quality.

Baswant 780

Recommended for *Kharif* planting in western Maharashtra; Developed by mass selection of an indigenous material; Released in 1986; Bulbs globose in shape with crimson red colour, average bulb weight 90-100 g, 13 per cent TSS; Less percentage of bolting; Duration 100-110 days; Yield 250 q/ha; Good for storage.

N-2-4-1 (Nasik Red)

Recommended for *rabi* season for zone IV and VII; Released in 1987; Bulbs are globose in shape with brick red colour which improves in storage, bulb diameter 4-6 cm, TSS 12-13 per cent, pungent; Resistant to premature bolting; Duration 120-130 days; Yield 300-350 q/ha; Tolerant to *Alternaria* blight and thrips; Excellent storage ability; Suitable for export.

N-257-9-1

Suitable for *rabi* season; Developed through selection from a local material; Bulbs are globular, white colour; Maturity 140 days; Yield 300 q/ha; Good keeping quality and suitable for dehydration.

N 53 (Niphad 53)

Recommended for *kharif* season in Western as well as Northern India, now popularly grown all over the country; Recommended by Department of Agriculture during 1960 and released at National level during 1987; Bulbs medium size, flatish round and red in colour, average bulb weight 70-80 g, TSS 11-12 per cent, mildly pungent; Duration 90-100 days; Yield 250-300 q/ha. Poor keeping quality.

Phule Safed

Recommended for *kharif* and *rabi* season; Developed through selection from a local material collected from Kagal; Release in 1994; Bulbs are globular in shape with attractive white colour, TSS is 13 per cent ; Duration 120-125 days; Moderately resistant to purple blotch and thrips under field conditions; Yield 250-300 q/ha; Suitable for dehydration.

Phule Suvarna

Recommended for *kharif* and *rabi* season, but useful for round the year plantation; Released in 1996; Bulbs are yellow coloured, pungency mild, TSS 11.5 per cent ; Duration 120-125 days; Moderately resistant to purple blotch and thrips, under field conditions; Yield 240 q/ha; Excellent keeping quality (4-6 months); Suitable for export to Europe, Australia and America.

Phule Samarth

Recommended for *kharif* and late-*kharif* season; Developed through selection from a germplasm from Sangamner; Released in 2004; Controlled vegetative growth coupled with rapid bulb devilment; dark red lustrous bulb colour, Globular bulb shape, thin bulb neck; early maturity, (75-90 days), natural foliage top fall; Duration 90-100 days; Moderately resistant to purple blotch and thrips under field conditions; Yield 300 q/ha; Excellent keeping quality (2-3 months).

Akola Safed

Suitable for *rabi* season of Vidarbha region (hot and dry climate); Developed through mass selection from local white onion; Released in 2005; White colour and medium size round shaped bulb, necck thickness 0.97 cm, bulb diameter 4.99 cm, average bulb weight 72.12-128.0 g, bulb size index 20.10, TSS 11.68-12.30 per cent, bolting percent-1-1.66 per cent ; Duration 183-185 days; Yield 240-260 q/ha; Tolerant to early blight and thrips.

Punjab Naroya

Developed by mass selection from the indigenous material collected from Maharashtra; Bulbs are red, medium to large, round with tight neck; Maturity 125 DAT; Yield 350-370 q/ha; Tolerant to purple blotch disease both in bulb and seed crop, and tolerant to thrips and *Heliothis*.

Punjab Red Round

Recommended for *rabi* season; Developed by mass selection from the local material; Medium to large sized bulbs (50-70 g), globular with tight neck, shining red colour, TSS 12.7 per cent ; Free from premature bolting; Maturity 140-150 DAT; Average yield 300 q/ha; Good keeping quality.

Punjab Selection

Recommended for *rabi* season; Plant height 55-65 cm with 6-9 leaves/plant; Bulbs red in colour, globular and firm, 5-6 cm in diameter, weighing 50-70 g, TSS 14 per cent, moderately pungent; Less bolting habit; Maturity 160-165 DAT; Average yield 200 q/ha. Field tolerance to purple blotch disease and thrips; Good keeping quality; Suitable for dehydration.

Punjab 48 (S 48)

Recommended for *rabi* season, suitable for sub-humid plains of Sutlej-Ganga; White colour bulb, flattish round shape, very good texture and flavour; Maturity 140 days; Average yield 250-300 q/ha; Good keeping quality; Suitable for dehydration.

Punjab White

Bulbs are large, round, white colour, TSS 15 per cent ; Yield 250-300 q/ha; Suitable for dehydration.

PRO 6

Developed by mass selection from an exotic material EC 383494. Bulbs are medium to large, round with thin and tight neck, deep red colour, TSS 12 per cent ; Maturity 125 DAT; Yield 400 q/ha; Good keeping quality.

PRO 7

Plants are medium tall, leaves green and upright, bulbs are round, medium large and red; Duration 120 days; Yield 397 q/ha; Tolerance to bolting and good keeping quality.

PWO-2

Recommended for *rabi* season; Released in 2019; plants are medium tall, leaves green and upright. Bulbs are round, medium large and white, TSS 14-16 per cent ; Maturity 140 days; Yield 388 q/ha; Tolerance to bolting and good keeping quality; Used to make dehydrated products such flakes, powder, rings and granules.

PYO 1

Plants are medium tall, leaves green and upright, bulbs are globular, large and yellow; Maturity 141 days; Yield 410 q/ha; Tolerance to bolting and good keeping quality.

Hisar 2

Recommended for *rabi* season in Haryana and Punjab; Bulbs bronze-red in colour, flat-globular, sweet but pungent, TSS 11.5-13.9 per cent ; Maturity 165 DAT; Average yield 200-300 q/ha.

Hisar Onion 3

Recommended for *rabi* season in Haryana, Delhi, Rajasthan and Gujarat; Bulbs are globular, thin necked and single hearting, light bronze color and off white flesh.

Hisar Onion 4

Bulbs are globular shape, rose red colour with high TSS 14.2 per cent ; Less bolting (5.4 per cent) and dark green leaves.

Udaipur 101

Recommended for winter *rabi* cultivation in Rajasthan and other adjoining States; Bulbs are flattish-globular, red colour, sweet with less pungency, TSS 12-14 per cent ; Maturity 150-160 days; Average yield 200-300 q/ha; Good for salad.

Udaipur 102

Recommended for *rabi* season; Bulbs round to flat, 4.5-6.5 cm in size, TSS 12 per cent ; white colour; Maturity 120 DAT; Average yield 300-350 q/ha; Suitable for dehydration.

Udaipur 103

Recommended for *rabi* season in Rajasthan and other adjoining States; Bulbs are oblate globular, red colour, sweet but slightly pungent, TSS 10.5-13 per cent ; Maturity 150-160 days; Average yield 250-300 q/ha; Good for salad.

Gujarat White Onion 1 (GWO 1)

Released in 2000; Developed through mass selection; Bulbs are big in size, bulb weight 57-60 g, TSS 15.42 per cent ; Maturity 133-135 days; Yield 431.45 q/ha.

Gujarat Junagadh Red Onion 11 (GJRO-11)

Recommended for growing in Gujarat; Released in 2015; Bulbs are medium in size, flat globe shape and red in colour, TSS 12.94 per cent, less pungent (pyruvic acid 1.22 μM/g); Less affected by purple blotch disease (12.67 per cent) and less population of thrips (5.7/leaf); Bulb yield 336.29 q/ha.

Gujarat Junagadh White Onion 3 (GJWO 3)

Recommended for *rabi* season in Gujarat; Released in 2019; Bulbs are medium size, flat globe, white colour, TSS 13.15 per cent, total carbohydrates 7.97 per cent, total protein 1.19 per cent, ascorbic acid 3.19 mg/g, total phenol 18.23 mg/100 g, pyruvic acid 11.35 and total soluble sugar 1.04 per cent ; Less jointed bulbs (1.70 per cent) and bolting; Tolerance against purple blotch disease (13.88 per cent) and less number of thrips per leaf (6.08); Yield 398.03 q/ha; Preferred by industry.

Gujarat Anand White Onion 2 (GAWO 2)
Recommended for *rabi* season under irrigated condition in middle Gujarat; Released in 2014; Bulbs are globe shape and big in size, white skin colour, high TSS, pyruvic acid, reducing sugar, non-reducing sugar and total carbohydrates; Low percentage of bolting and jointed bulb; Less affected by purple blotch disease and thrips; Yield 595 q/ha,and yield potential 703.7 q/ha.
Kalyanpur Red Round
Recommended for *rabi* season. Popular in U.P.; Bulbs are bronze red in colour, globular in shape, moderately sweet and moderately pungent, TSS 13-14 per cent ; Maturity 140-145 DAT; Average yield 250-300 q/ha; Good keeing quality.
Palam Lohit
Bulbs are round with attractivedeep purple colour, narrow neck; Bulb yield 450q/ha; Moderately resistant to downey mildew.
Bangalore Rose
Bulbs are small (2-2.5 cm), flattish-round shape, deep red colour, TSS 19-20 per cent ; Maturity 110-120 DAT; Yield 180-200 q/ha. Grown near Bangalore exclusively for export.
CO 1
Multiplier onion; Developed through clonal selection from CS 450, an onion type introduced from Manachanallur, is adapted throughout Tamil Nadu; Medium-tall plants, 7-8 bulblets/plant, red colour, average weight 55-60 g/clump, fairly pungent with medium TSS; Maturity 90 days; Average yield 100 q/ha; Good keeping quality.
CO 2
Multiplier onion; Developed through selection from CS 911, is adapted throughout Tamil Nadu; Medium-tall plants with light green cylindrical leaves, moderately bigger sized 7-9 bulblets/plant, crimson colour, average weight 60-65 g/clump, pungent with high TSS; Maturity 65 days; Average yield 120 q/ha; Moderately resistant to purple blotch diseases and thrips; Good storability.
CO 3
Multiplier onion; Developed through clonal selection from open pollinated progenies of CS 450; Taller than CO 1 and CO 2, 8-10 bulblets/plant, pink colour, 75 g/clump, bigger sized bulbs, TSS 13 per cent ; Maturity 65 days; Average yield 160 q/ha; More consumer preference; Moderately resistant to thrips; Good keeping quality up to 120 days without sprouting.
CO 4
Multiplier onion; Developed through clonal selection from hybrid of AC 863 × CO 3; Taller plants, 8-10 bulblets/plant, light brown colour, 90 g/clump, bulbs are bold; Maturity 60-65 days; Average yield 180 q/ha. Moderately resistant to thrips; Can be stored more than 150 days without sprouting in well-ventillated storage.
CO (On) 5
Multiplier onion; A free flowering type with seed producing ability, can be propaged through seeds as well as bulblets; High yielding, attractive pink and bold bulblets; Bulb yield 189 q/ha, seed yield 2.5-3.0 q/ha; Seed crop takes 95-100 days.
MDU-1
Multiplier onion; Well adapted to southern districts of Tamil Nadu; Uniform, big sized 10-11 bulblets/plant, bright red colour, 75 g/clump; Maturity 60-75 days; Average yield 150 q/ha; Tolerant to lodging due to thick erect leaves; Good keeping quality.

National Horticultural Research and Development Foundation (NHRDF), Nashik, collected good number of germplasm and evaluated their performance regarding different attributes.

Table 14: Advance Lines of Onion Developed by NHRDF, Nashik

Genotype	Characters
Advance Line 863	The line is recommended for cultivation in *kharif* and Late *kharif* season; Bulbs are dark red, round in shape, 4.0-5.5 cm diameter, 12-13 per cent TSS, 11-12 per cent dry matter and 12.5-13 micro mole/g pyruvic acid; Crop maturity 80-85 days; Average yield 280-300 q/ha; Medium keeping quality.
Advance Line 883	The line is recommended for cultivation in *kharif* and early *kharif* seasons all over India; Bulbs are dark red in colour with shiny skin, round in shape, 4.5-5.5 cm diameter, 12-13 per cent TSS, 13-14 per cent dry matter and 12.0 micro mole/g pyruvic acid; Crop maturity 85-90 days after transplanting and 80-85 days in direct seeding; Average yield 300-325 q/ha.

At NHRDF (Nashik while at Karnal centre), Singh *et al.* (2020) observed highest gross yield (453.79 q/ha) and marketable yield (297.95 q/ha) in advance lines-824 and 807 in *rabi* season, respectively. At Nashik they noted high TSS (14.11 per cent) in lines-703 and dry matter content (15.43 per cent) in line 825. At Karnal, they recorded the highest gross yield (365.51 q/ha) and marketable yield (333.47 q/ha) in advance line-844. They suggested utilizing these advance lines for onion breeding programme for development useful cultivars for different agroclimatic condition.

During the last few decades, a number of onion varieties have been released. Even though, large area under onion cultivation is covered by released varieties, substantial area is covered by local varieties. In onion, local genotypes play important role in development of new cultivars. Some of the region specific local varieties have been presented in Table 15.

Table 15: Local Cultivars of Onion Grown in India*

State	Local Cultivars
Andhra Pradesh	Rampuram Red, K. P. Onion
Bihar	Patna Red, Patna White
Gujarat	Saurashtra Local, Pillipatti Junagarh, Talaja-Red
Haryana	Panipat Local, Rajpur Red, Hisar Local, Bahadurgarh Local
Karnataka	Bellary Red, Telgi Red, Bangalore Rose, Chikballapur Local, Krishnapuram Local, Kempgende
Madhya Pradesh	Nimar Local
Maharashtra	Niphad, Nasik Red, Nasik White Globe, Poona Red, Bombay Red, Bombay White, Mathewad, Shirwal, Kagar, Peth, Dhulia, Fangira, Lohru, Fursungi Local
Punjab	Ludhiana Local
Tamil Nadu	Cuddalorae, Podusu, Dindigul Red, Mutlore, Natu, Pallu, Vengayam
Uttar Pradesh	Kalyanpur Red, Kalyanpur Lalgol
West Bengal	Sukhsagar

* Swarup (2006).

The two onion cultivars 'Sona 40' and 'Sandeep Pyaz' have been shortlisted, recognized, and incubated for their unique qualities by the National Innovation

Foundation (NIF), Department of Science and Technology. 'Sandip Pyaz' (Average bulb weight 100-150g, firm bulbs, 8-10 number of rings per bulb, retention of outer skin, minimum degree of splitting, bulb yield 38 t/ha and keeping quality 40-45 days) and 'Sona-40' (Average bulb weight 80-90g, 7-9 number of rings per bulb, lower incidence of purple blotch disease, bulb yield 34 t/ha and good keeping quality) were developed through selection from local cultivars. The varieties were validated and evaluated at Dr. Balasaheb Sawant Konkan Krishi Vidyapeeth, Maharashtra and were found significantly superior to checked varieties Agrifound light red and Phule Samarth for yield and their keeping quality or storage life.

Farmers' varieties are mostly developed from local old landraces or cultivars and are therefore genetically diverse, climate-resilient, region specific, high yielding with high storage life/keeping quality, and good nutritional value. Encouraging such novel varieties can help preserve genetic diversity, provide climate-resilient options for farmers, and boost nutritional security.

Some of the cultivars developed in different countries are described below:

Yubilei 50

This Bulgarian cultivar produces large, almost spherical bulbs weighing 90-130 g with a compact internal structure. The yellow-brown outer scales are firm and the white inner scales are fleshy and slightly pungent. Yield is 30-48 t/ha (Todorov *et al.*, 1984).

Sweet Sandwich

It is a mild-flavoured cultivar suitable for fresh use, particularly after storage for 3 months or more. It is adapted to areas of New York and the upper Midwest (Peterson *et al.*, 1986a).

Numex Sunlite

It possesses high level of bolting and pink root resistance and gives high yield (Corgan, 1988).

In recent years, the open-pollinated Texas Early Grano Stock has been subjected to intensive hybridization programme with a Sweet Spanish type cv. Ben Shemen by Pike and his associates in USA and developed a series of cultivars mentioned below.

Texas Grano 1015 Y

It is a yellow onion with slightly flattened globe bulb shape, mild pungency, short-day onion with greater resistance to pink root (Pike *et al.*, 1988a).

Texas Grano 1025 Y

A yellow onion, globe-shaped, resistance to pink root, and shipping and good storage quality (Pike *et al.*, 1988b).

Texas Grano 1030 Y

It is a late maturing, globe-shaped yellow onion with mild pungency and good resistance of pink root (Pike *et al.*, 1988c).

Texas Grano 1105 Y

It is a late maturing, yellow globe onion with medium length storage and good resistance to pink root (Pike *et al.*, 1988d).

Galmarini (2001) reviewed the onion breeding programme of Argentina. For decades, onion-breeding programs have been developing short, intermediate, and long-day cultivars in Argentina. The cultivars released by La Consulta Experiment Station onion-breeding program account for more than 80 per cent of the onions cultivated in Argentina. Most Argentine cultivars trace back to local landraces, known as "Criollas," that are becoming rare. Most of the cultivars used in Argentina originate from domestic breeding programs. Using local 'Valenciana' population improved bolting resistance and keeping quality synthetic cultivars 'Valenciana Synthetics 1 and 2' and later 'Valenciana Synthetic 14' and 'Valenciana Synthetic 15' were developed. Several male sterile and maintainer lines, and the F_1 hybrid 'Híbrido Industria INTA' were developed from 1969 until 1976. The fresh-market cultivars Cobriza INTA, Navideña INTA, and Antártica INTA were developed based on the programme to develop open-pollinated populations and hybrids with high solids content. Valcatorce INTA, Valuno INTA, Cobriza INTA, and Antártica INTA were developed through recurrent selection from 'Valenciana' populations. 'Torrentina' populations, which have been known for more than 100 years in Argentina and may have originated in southern Spain, were used to develop 'Navideña INTA'. 'Refinta 20' originated from a selection from 'Southport White Globe' populations S_1 family selection was the common strategy used for the development of 'Navideña INTA', 'Cobriza INTA' and 'Refinta 20', whereas mass selection was used in the case of 'Valcatorce INTA', 'Valuno INTA' and 'Antártica INTA'. Some of the cultivars developed are described below:

Valcatorce INTA

This variety was developed by selection from 'Valenciana Synthetic 14', requires 14 h photoperiods to produce bulbs, is a moderate to late-maturing cultivar, and has globe-shaped bulbs of medium size, two to three dark-brown scales, pungent and firm flesh, bolting resistance, and excellent keeping quality. It is sown from August to September and harvested from January to February. It is the most commonly grown cultivar in Argentina, and is also grown in Chile and Uruguay.

Valenciana Synthetic 1

It requires 14 h photoperiods to produce bulbs; is a moderate to late-maturing cultivar; has a lighter yellow color; larger bulbs, and shorter keeping time than 'Valcatorce INTA'. It is sown from August to September and harvested from January to March.

Cobriza INTA

Developed by selection from a 'Valenciana' population using S_1 family selection. It has photoperiodic requirements for bulb formation similar to those of 'Valcatorce INTA', medium-sized, globe-shaped bulbs, with four to five dark-brown scales, pungent and firm flesh, bolting resistance, and excellent keeping quality. The yields and production cycles are similar to those of 'Valcatorce INTA'. 'Cobriza INTA' may have a great impact on the domestic onion market, because it yields as well as 'Valcatorce INTA', but is of better quality.

Navideña INTA

It was selected from local 'Torrentina' populations from the north of Mendoza and San Juan provinces. It requires 13 h photoperiods for bulb formation and has brown globe bulbs with two to three colored scales. The bulbs are less pungent than those of the 'Valenciana' types. At Mendoza, it reaches maturity at the end of December.

Antártica INTA

It was selected from white bulbs segregating in 'Valenciana' populations. These bulbs were combined to develop a white population from which the new cultivar was selected. It has photoperiodic requirements for bulb formation similar to those of 'Valcatorce INTA' and good keeping quality. The production cycle is also similar to 'Valcatorce INTA'.

Refinta 20

It was originated from 'Southport White Globe' after mass selection for high solids. It has white, pungent bulbs and an average total solids content of 20 per cent, more than 3 per cent higher than the totals solids of the populations currently used by the Argentine dehydration industry. It is sown from August to September and harvested from January to February. An Argentine dehydration company has exclusive use of this cultivar for 20 years.

Onion cv. Bianca di Guigno of Italy has been reported to be early, gives high yield. Other good cultivars are Bianca Precocissima di Barletta, Bianca Precocissima di Pompei, Bianca di Aprile and Borettana (Miccolis and Montemurro, 1988). In Egypt, the best known cultivars are Giza-6 and Beheri (Meer, 1986). In the dry climate of Egypt these onions can be stored for 8-10 months. In Israel, cultivars have been developed from Egyptian types, for example, cv. Haemek (Currah and Proctor, 1990). A new onion cv. Sekihoku with good bulb quality has been reported from Japan (Miyaura *et al.*, 1985). It stores well, gives high yield and resistant to *Fusarium oxysporum* f. sp. *cepae* and *Erwinia carotovora*.

Based on skin colour, productivity for export and consumer demand, the cultivar Cream Gold (Australia) is highly rated and can be commercially grown at latitude of 30-35°N. Other promising cultivars are Bola Precoce, Creol Red and Violet De Galmi. Among the hybrids, HA 950 was found to have the highest yields, and other promising hybrids are H-226, Elad, HA-870, Rojo, Rio Raji Red, Cardix and Suprex (Singh *et al.*, 2001).

Arzamas is one of the oldest onion cultivars of amateur breeding which has been grown in the Nizhnii Novgorod region, Russia, for 300 years (Matveev, 2001).

In Japan cvs. grown in fall or spring season (August-September) are Marushinoki, Unzen-Gokuwase, Hikari, Shugyo Ku, Inazuma, OA-Ki, Kin-Kyu, OX-Ki, Senshu-Nakadaka, Satuki, Mamiji, Hormer, *etc.* while cvs. Kitamomiji,

Tsukihikari, Ohoutuku-1, Reo, Sapporoki, Getsurin, *etc.* are grown in spring season (February-March) (Yakuwa, 1994).

Yield attributes of 12 short-day onion cultivars assessed at Hermosillo, Mexico by Warid and Loaiza (1996) revealed that high yielding cultivars were Red Creole PRR and Rio Bravo. Low yield characterized cultivars were RAM 710, Red Creole PRR and Red Creole, the bulbs of which were light brown or red. All the yellow cultivars had high yields and so did the white cultivar NM 899. Asencio (1997) evaluated 7 cultivars and 4 lines of medium-day onion and one short-day cultivar as control in the South of Puerto Rico for marketable yield and bulb size. Line PSX13489 had the shortest growth cycle (112 days). Early Harvest, Spano, Candy and H222 had the highest marketable yields (6.055-6.316 kg/ha). The majority of genotypes had medium-sized bulbs, except Texspan P.R.R. and lines PSX13489 and H688, which had small bulbs. In Egypt, El-Sayed and Atia (1999) obtained highest marketable yields with exotic cultivars Ori (Israel), G.6 M (Egypt) and Perla (USA) in both seasons.

In Belgium, highest yields as ripe onions were obtained from cv. Ocean, Buffalo, Swift and Radar (Vanparys, 2000b). When harvested immature, Buffalo, Imai Early Yellow, Meteor, Ocean and Siberia gave the best results (Vanparys, 1996). In Dominican Republic the cultivars with the best attributes of yield and bulb quality were Texas Grano 438. Among white onion crops, Diamante and ontessa were significantly better than Sebaquena and White Hawk, in terms of early development and yield. For the red crops, Red Creole was significantly inferior to Hibrido Rojo and Sivan in bulb grade and yield (Morales *et al.*, 2000).

In Nepal, Early Red (yield of 46.11 t/ha) produced 8.9 and 11 per cent more bulbs than Red Creole and Mallajh Local, respectively. Farmers preferred Early Red since it exhibited less splitting and bolting and it produced good quality bulbs (Jaiswal *et al.*, 1997b).

Onion cv. Melkam, originally from India, has performed well in trials in Ethiopia and was released in 1998 for lowland irrigated production (Aklilu and Dessalegne, 1999).

In Brazil, cultivar EPAGRI 362-Crioulo Alto Vale was derived by selection from 5 local populations. It has a growth period of 180-200 days from sowing to harvest (110-120 days from transplanting to harvest) and is resistant to bolting. EPAGRI 363-Superprecoce, obtained from 4 early maturing populations derived from the cultivar Baia Periforme, has a growth period of 170-190 days from sowing to harvest (or 110-120 days from transplanting to harvest) and is resistant to bolting. Tanaka *et al.* (2000) reported new onion male-sterile line, 'S7946A' and its maintainer line, 'S7946B' from Japan.

New onion seed parent Kitamikou 25 was described by Tanaka *et al.* (1999). It was produced by hybridization of WTN875-07B, AOPFA and PRCX03. The yield of Kitamikou 25 was 57.5 t/ha compared with 51.0 t for Tsukisatsupu, 61.1 t for Kamui and 54.9 t for Rantaro. Disease resistance was good and bulb quality high. Seed yield was 1.5 g/plant and the 1000 seed weight 5.62 g. In the off-season trial in Nepal, cultivars Agrifound Dark Red, Nasik Red and Agrifound Light

Red were tested. Bulb production was poor for all 3 cultivars owing to premature bolting, non-bulbing and splitting. Agrifound Dark Red and Agrifound Light Red produced similar marketable bulb yields, outyielding Nasik Red by 55 and 62 per cent, respectively. However, Nasik Red was the best cultivar for green onion production, producing 12 and 23 per cent more biomass yield than Agrifound Dark Red and Agrifound Light Red, respectively. Farmers preferred to sell the onions green rather than keeping the plants for bulb production (Jaiswal and Subedi, 1996). In normal-season trial, cultivars Early Red, Red Synthetic, Red Creole and Mallajh Local were evaluated in Nepal. Early Red consistently produced the highest yields at all sites, from low to high areas. It out-yielded Red Synthetic, Mallajh Local and Red Creole by 20, 25 and 31 per cent, respectively across 5 sites (700-1600 m asl). Farmers preferred Early Red for its high marketable bulb yield and good quality bulbs (Jaiswal and Subedi, 1996). Derived from NuMex Starlite, NuMex Dulce is a low-pungency, short-day, yellow, grano-type onion that matured from late May to early June when planted in autumn in southern New Mexico. It yielded 64.3-120.2 t/ha depending on sowing date in field trials. It showed high resistance to pink root (*Phoma terrestris*), and was resistant to bolting (Wall and Corgan, 1988). NuMex Snowball is a new onion cultivar derived from a cross between cv. Ringmaster and a pink-root-resistant selection from cv. New Mexico White Grano. NuMex Snowball is resistant to pink root (caused by *Phoma terrestris*) and is for spring sowing in New Mexico, USA (Cramer and Corgan, 2001).

Originating from an intercross of Texas Grano 1015Y, Excel 986B and a population of Texas Grano 502PRR (bolting resistant), 'New Mex Sweetpak' onion cultivar is a low pungency, early maturing, open-pollinated, short-day, yellow grano type. It is ideal for fresh consumption and matures in late May when planted in autumn in southern New Mexico (Wall and Corgan, 1999). The cultivars released by La Consulta Experiment Station (INTA, Argentina) have good keeping ability and early maturity. Newly released cultivars Valcatorce INTA, Valuno INTA, Cobriza INTA and Antartica INTA originated from Valenciana populations; Navidena originated from Torrentina populations; and Refinta 20 from cultivar Southport White Globe (Galmarini, 2000). Some of the new cultivars developed in Poland are Kristine, Efekt, Fiesta, Grabowska, Supra, Pino, Dawidowska and Niagara F_1.

In New South Wales, Australia Red Shine and Sweet Sensation produced total, marketable and exportable yields which were significantly higher than those of other cultivars. Red Shine had the highest (8.57 per cent) Brix value (sugar content). River Brown matured one week earlier than Rio Xena and Sun Shine, which matured few days earlier than Sweet Sensation and Red Shine (Quadir and Boulton, 2001). Arctic, Swift and Glacier are recommended for use as bunching onions in Belgium (Vanparys, 1998). Trials conducted at Gumah Agric. Research Station, Sultanate of Oman showed that onions cv. Giza 20, Nucleus 961 and Giza 6 Mohassan out-yielded a local Omani cultivar (yield 15.12 t/feddan) by 56.5, 46.9 and 24.1 per cent, respectively (El-Aweel *et al.*, 2000).

In Brazil cv. EPAGRI 362-Crioulo Alto Vale was derived by selection from 5 local populations. It has a growth period of 180-200 days from sowing to harvest (110-120 days from transplanting to harvest) and is resistant to bolting. EPAGRI

363-Super-precoce, obtained from 4 early maturing populations derived from the cultivar Baia Periforme, has a growth period of 170-190 days from sowing to harvest (or 110-120 days from transplanting to harvest) and is resistant to bolting (Gandin *et al.*, 1998).

Hybrids

Heterosis breeding provides an opportunity for improvement in productivity, earliness, uniformity and yield attributing characters. Most of the area under onion cultivation in India is covered by open pollinated varieties; in contrast, the areas under hybrid have increased in other countries. At present, development of F_1 hybrids has been increased in both public and private sector in India. An emphasis should also be given for developing F_1 hybrids for yield and quality.

Table 16: List of F_1 Hybrids Developed through Male Sterile Lines*

Institute	F_1 Hybrid	Remarks
ICAR-IIHR	Arka Kirthiman	High bulb yield and quality
	Arka Lalima	High bulb yield and quality
ICAR-IARI	Hybrid 63	High bulb yield and quality
	Hybrid 35	High bulb yield and quality
ICAR-DOGR	DOGR Hy-1	Light red, flat globe, 41.30 t/ha marketable yield which is 42.84 per cent higher than check Bhima Kiran
	DOGR Hy-2	Dark red, globe, 34.96 t/ha marketable yield which is 20.91 per cent higher than check Bhima Kiran.
	DOGR Hy-3	Red, globe, 34.32 t/ha marketable yield during *kharif* which is 16.22 per cent higher than best check Bhima Super.
	DOGR Hy-4	Attractive dark red, flat globe, 34.10 t/ha marketable yield during *kharif* which is 15.81 per cent higher than best check Bhima Super.
	DOGR Hy-5	Dark red, globe, 38.73 t/hamarketable yield during *kharif* which is 20.11 per cent higher than best check Bhima Dark Red.
	DOGR Hy-7	Dark red, flat globe, 40.71 t/ha marketable yield during *rabi* which is 34.88 per cent higher than best check Bhima Red.
	DOGR Hy-8	Red, flat globe, 54.44 t/ha marketable yield during late *kharif* which is 23.57 per cent higher than best check Bhima Shakti.
	DOGR Hy-50	Attactive dark red, globe, 37.46 t/ha marketable yield during *rabi* which is 24.11 per cent higher than best check Bhima Red.

* Gupta *et al.* (2017).

Many F_1 hybrids have been developed in advanced countries by using male sterile lines. Singh *et al.* (1992) reported that red F_1 hybrid VL Paize-67 was developed in India using male sterile line BYG 2207-A and as pollen parent VL line 31. CMS line VL 31-1 A (Female), VL 31-1 B (Maintainer) was developed and being used in heterosis program of onion.

Tsukihikari is a hybrid cultivar of Japan with the parentage W202A × OPP1. The cultivar is high yielding, produces large firm bulbs of good quality, mid-late in maturity and resistant to *Fusarium oxysporum* f. sp. *cepae* (Takayanagi, 1987). Toyohira

Cultivars of Red Onion

| Arka Sona | Arka Bheem | Arka Akshay |

| Arka Vishwas | Arka Kalyan | Arka Bindu |

| Bhima Dark Red | Bhima Raj | Bhima Shakti |

| Bhima Red | Bhima Super | Bhima Kiran |

Cultivars of White Onion

Bhima Shweta

Bhima Shubra

Bhima Safed

Agrifound White

Pusa White Flat

Arka Swadista

Cultivars of Yellow Onion

Arka Pitamber

Early Grano

Hybrids of Onion

Arka Lalima (F₁)

Arka Kirthiman (F₁)

is a F$_1$ hybrid resulting from a cross of the seed parent, 2935A, with CHS3-12. 2935A is a cytoplasmically male sterile line and was introduced with the maintainer, 2935B, from the USDA to the Hokkaido National Agricultural Experiment Station in 1984. Bulb qualities such as pungency, bulb firmness and scale thickness are much improved in Toyohira. Enzymatically formed pyruvic acid (EFPA) content, which is a reliable index of pungency, is 12.7 per cent lower in Toyohira than that in the control cultivar Tsukihikari. The bulbs of Toyohira are 22.2 per cent softer, and the scales are 25 per cent thicker than those of Tsukihikari. These traits are good in bulbs used for salads and for part cooking. The storage abilities of Toyohira are as high as those of Tsukihikari. Toyohira produces larger bulbs and 6-19 per cent higher marketable yields than Tsukihikari (Sato *et al.*, 1999).

9.0 Soil and Climate

9.1.0 Soil

Onions are grown on all types of soils such as sandy loam, silt loam and heavy clay soils. In heavy soils bulb development is restricted and the crop matures late compared to light soils. Sandy soils need frequent irrigation and favours early bulb maturity. Soil should be rich in humus and withhold sufficient moisture for proper growth. For successful onion production, deep friable soils are desirable. Onion is very sensitive to water logging conditions. *Kharif* season crop thus require better dranage condition.

Onion is sensitive to high acidity and alkalinity. The favourable soil pH range is 5.8 to 8.0. Under given level of salinity onion yield was reduced by 80 per cent in hot dry conditions but only 10 per cent in the cooler environment. Yields were reduced by 50 per cent when EC was 4.1 mmhos/cm or approximately 3000 ppm (Russel, 1974).

Sharma *et al.* (2000) reported that under alkaline stress condition onion bulb yield declined by 41 and 49 per cent at p^H 9.2 and 9.45, respectively, compared with the non-stress control, and dropped to negligible level at pH 9.7. On average, over different alkaline levels, Amrawati gave maximum bulb yield followed by Agrifound Light Red. Agrifound Light Red recorded minimum reduction both in seedling fresh weight and final bulb yield up to pH 9.45. Further, under salinity stress, seedling fresh weight declined by 13 and 26 per cent while mean bulb yield (at harvest) declined by 54 and 65 per cent at ECe 3.5 and 5.2 dS/m, respectively. Highest mean bulb yield obtained under salinity was by Pusa White Flat followed by Amrawati amongst the different genotypes tested. In greenhouse, Yadav *et al.* (1998) found that highest salinity (16 dS/m) reduced plant height, number of leaves/plant, leaves fresh and dry weight and bulb yield by about 40-50 per cent as compared to the control (0.3 dS/m). Reduction was of higher magnitude at early growth stages. Bulb maturity was hastened by a week due to increasing levels of salinity. Pusa Red performed best under salinity. Malik *et al.* (1978) observed that salt concentration above 4 mmhos/cm inhibited vegetative growth of most of the onion cultivars except Hissar-2 and Punjab Selection which had better salt tolerance even up to 7.5 mmhos/cm salt concentrations. It yielded better at a pH range of 5.8-6.5.

9.2.0 Climate

The onion grows in mild climate without extremes of high or low temperatures even though it can be grown under a wide range of climatic conditions. The plants at early stage can withstand the freezing temperature. Lovato and Amaducci (1965) reported that 20-25°C temperature was optimal for onion seed germination. For vegetative growth lower temperatures and short photoperiod are required, while relatively higher temperatures and long photoperiod are needed for bulb development. It requires 13-21°C for vegetative growth before bulbing and 15.6-25.1°C for bulb development. However, it is well adapted to higher temperatures. Temperature is more important than the day length for seed stalk development.

Very low temperatures in the initial stages of growth support bolting, while sudden rise in temperature causes early maturity in winter crop (*rabi*) and ultimately reduces the size of bulbs. Moderate temperature favours pollination and seed setting. Karak and Hazra (2012) worked on short-day tropical type of onion in West Bengal condition. They observed that flowering stalks were induced even at 25.35°/11.74°C day/night temperature during last week of December to last week of January.

Onion cannot thrive well in places when average rainfall exceeds 75-100 cm particularly in the monsoon period. Relative humidity should be around 70 per cent for good growth of the crop (Rao and Purewal, 1957). For better germination, twenty per cent available moisture or more is required. Rainy and cloudy weather coupled with high humidity favours disease development. High rainfall and wet soil affect plant stand.

Onion cultivars differ in day length requirement. Almost all the varieties grown in Indian plains are short day type. Intermediate or long day type onions are restricted in hilly areas only. Long day varieties if grown in short day conditions do not forms bulbs. However, short day varieties can produce bulb in long day condition. It is therefore advocated to grow locally acclimatized varieties only for better production. According to growing season Indian onion varieties are grouped as, *kharif*, late *kharif* and *rabi* season varieties. Varieties grown in *kharif* season require short photoperiod (10-11 hr); while *Rabi* season varieties have 12-13 hr photoperiodic requirement.

Onion is a biennial crop. In first phase it requires short day couple with low temperature (13-20°C) to complete its vegetative stage. In second phase it need long day couple with high temperature for bulb development.

Red onions, especially those selected under tropical conditions, sometimes perform better than brown, yellow or white onions in the more equatorial environments, suggesting that some of these red cultivars are better adapted to grow and produce bulb under high temperature and humidity (Currah *et al.*, 1997).

10.0 Cultivation

Onion plants are usually grown by various means like transplanting of seedlings, planting bulbs, direct sowing of seed in the main field by broadcasting, crop by sets, *etc*. Among all those, the cultivation of onion through transplanting of seedlings is more commonly practised for an irrigated crop by the Indian farmers.

10.1.0 Seed Rate

Onion requires average 8 to 10 kg/ha and 12-15 kg/ha seed for transplanting in *rabi* and *kharif* season. In *kharif* season, due to high seedling mortality and diseases more seeds are required for planting a hectare of land. Generally onion is not sown directly in the field. However, for direct sowing it required 20-25 kg seed per hectare. Onion seed lose viability very fast. Therefore sowing of fresh seed is recommended. Onion can also be grown by planting 10-12 q/ha bulbs. However, planting bulbs are costly and shelf lives of the bulbs harvested from bulb plantings are short.

10.2.0 Raising of Seedlings

For raising a crop for bulb production, onion seeds are sown on nursery beds to raise seedlings for transplanting in the field. Raised beds of about 3-6 metre long, 1 metre width and 10-15 cm above the ground level is prepared. About 70 cm distances are kept between two beds to carry out operations of watering, weeding, *etc*. The surface of beds should be smooth and well levelled. To avoid mortality of seedlings due to damping off, drenching of the beds with copper fungicides 25-30 g or Bavistin 15-20 g per 10 litres of water is effective. The seeds are treated with thiram 2-3 g per kg of seed. Thiram is also applied in soil and used for drenching beds to protect seedlings against damping off. The effects of benomyl + captan, soil solarization for 3 and 4 weeks, and compost applied at 25 t/ha were tested for control of root pathogens on onion seedlings in Bagaces, Guanacaste province, Costa Rica. Despite the wet weather conditions, the compost treatment gave the highest survival rates of seedlings and the best plant health. The use of compost is recommended as an adequate alternative for production of onion seedlings (Navarro and Umana, 1997). Gabriel *et al.* (1997) observed that seed germination of the onion cv. Valcatorce increased linearly as seed diameter increased from 2.00 to 2.75 mm. Basu *et al.* (1975) found that soaking of stored onion seeds in water or sodium phosphate solution for 2-6 hours followed by drying was beneficial for vigour and viability. Verma *et al.* (2000) observed that hydration-dehydration treatment enhance the germination of onion and vigour in onion seedlings (in cultivars like Agrifound Light Red and Agri-found Dark Red). Irradiation of seeds of cv. Behary with gamma rays at 6.5 Krad significantly stimulated germination, increased yields and improved pungency (Abd-El-Gawad *et al.*, 1986). Gonzalex *et al.* (2001) indicated that X-rays irradiation of seeds improved quality, total yield and percentage of bulbs of top quality (diameter of 7.6-11 cm). Increasing level of stress caused a gradual and significant decrease in germination and seedling growth in onion. Singh and Jai Gopal (2019) reported that water stress reduced germination by 22 to 83 per cent and it was delayed by 2.19 to 4.69 days and salt stress also showed similar trend and prolonged the germination time by 0.46 to 4.68 days and reduced the germination by 3 to 57 per cent. Palaniappan *et al.* (1999) noted that the cv. Arka Niketan and Arka Kalyan were the most tolerant of salinity (6-7.5 dS/m) and exhibited good germination (92.3-97.9 per cent) at the highest salinities investigated. Of all the cultivars, Arka Niketan exhibited the best seedling growth at high salinities and recorded highest salinity threshold levels. Arka Pragati was the most susceptible to salinity. Seeds are sown on well prepared beds in lines at a spacing of 5-7 cm and are covered with soil. Beds are irrigated regularly with the help of watering can. Seed sowing is done in October-November for winter (*rabi*) crop and May-June for rainy season (*kharif*) crop. The beds are kept weed free and nursery is protected from pests and diseases. Seedlings will be ready in 6-7 weeks for rainy season planting and in 8 weeks for winter (*rabi*) planting. Seedlings of 0.8 to 0.9 cm in diameter and about 20-35 cm in height are ready for transplanting. Over-aged seedlings result in bolting and take longer time to start new growth, whereas underaged seedlings do not establish well. From the results of the investigations it was observed that 8-10 weeks old seedlings established well after transplanting and with stood adverse weather (Patel *et al.*, 1958; Maiti *et al.*, 1968). Rathore and Yadav (1971) obtained the highest yield from planting of 8-week-old seedling.

Cultivation of onions in Korean peninsula is restricted to the southern region, mainly due to inadequate air temperatures for onion growth. Soo *et al.* (2000) examined the possibility of extending the onion production region by adopting the plug seedling system. The effects of plug seedling age, transplanting date, plug cell volume, and cultivar on yield were investigated. Transplanting date (15 March or April) did not affect yield. However, 60-day-old seedling produced significantly greater bulb yield (107.2 t/ha) than 30 or 90-day-old seedlings. Bulb yields of seedlings raised in 128 and 72 cell trays were 125 and 120.8 t/ha, which were significantly greater than those in other cell trays. Yield of seedlings raised in 406 cell trays was the lowest. Seedling of mid-late Chenjuwhang yielded 130.5 t/ha, which was significantly higher than the other cultivars. It was suggested that seedlings should be raised for 60 days and transplanted on 15 March for the production of marketable bulbs; the earliest cultivar was Yongbongwhang.

10.3.0 Raising of Set

Onion sets are small bulbs, usually between 10 mm (5/16 inches) and 25 mm (1 inch) in diameter that after a period of storage of 6 to 8 months, are mostly used by home gardeners as the propagation material to produce either green onions or mature bulbs. Use of onion sets is considered advantageous for a number of reasons such as the bulbs grown from sets mature 3 to 4 weeks earlier than those grown from seeds, there may be a price advantage if the bulbs are sold early in the market; it is also possible to have more uniform spacing using sets than seeds which results in more uniform bulbs and it saves time from nursery raising of seed (Walker *et al.*, 1946; Jones and Mann, 1963; Voss, 1979).

Walker *et al.* (1946) and Jones and Mann (1963) in their experimental studies in Illinois and Colorado it was established that the size of bulbs to be produced, when photoperiod and temperature are favourable, depends on the amount of seed sown per unit area, and later, on the resultant stand of plants. The optimum seeding rates to obtain the maximum amount of sets in the range 10 mm and 25 mm in diameter are between 56 and 112 kg/ha. Voss (1979) found out that in Californian conditions, seeding rates from 67 to 84 kg/ha are recommended to obtain the size of sets between 10mm and 25mm in diameter. Gupta *et al.* (1999) reported that highest yield of large-sized setts (> 1.75 cm diameter) was recorded for a seed rate of 15 g/m square and the lowest for a seed rate of 45 g/m square. Rajesh Kumar *et al.* (2003) determine the effect of seed rates (5.0, 7.5, 10.0, 12.5, 15.0, 17.5, 20.0, 22.5 and 25.0

g/m^2) on the production of onion (cv. Agri-fond Dark Red) sets in Haryana, India. They reported that total number of sets and sets of up to 2 g (small) size increased with increase in seed rate. However, the number of >5-10 g (large) and >10 g (very large) sets decreased with increase in the seed rate. They noted that total yield of sets was highest with the seed rate of 5.0 g/m^2.

Jones and Mann (1963) reported that in the most traditional areas of onion set production the planting time usually extends from mid-April to early May. Voss (1979) reported that in California, planting for set production can start as early as in February. Gupta *et al.* (1999) reported that total yield of onion (cv. Agrifound Dark Red) sets was significantly influenced by sowing dates and it was 2268.5, 1736.5, 522 and 305 g/sqm for sowing dates 25 December, 15 January, 5 and 25 February, respectively. Bhutani *et al.* (2000) concluded that for the higher production of better quality sets, seed may be sown @ 15 g/m^2 in the last week of December under Hissar (Haryana), India conditions. Cheema *et al.* (2003) observed that environment affected set size not the number of sets. However, date of sowing significantly affected set size and number of sets. Sowing dates of 2nd and 4th week of January gave desired set size with 262.92-461.60 sets/kg, they reported. Rajesh Kumar *et al.* (2003) studied the effects of sowing date (1, 11, 21 or 31 December; 10, 20 or 30 January; and 10 February) on set production by onion cv. Agri-fond Dark Red. They reported that highest yield of large sets (1496.0 g/m^2) was obtained with sowing on 1st December.

Mandal *et al.* (2018a and b) reported the varietal differences in onion set production. They reported that the cultivars, Arka Kalyan, Bhima Shweta, Bhima Raj, Baswant-780, Sukhsagar and Bhima Shrubha recorded maximum set diameter, average set weight and number and weight of total sets. The sets/bulblets can be produced for green onion or mature bulb production during *kharif* season in West Bengal condition.

10.4.0 Transplanting and Spacing

Seedlings are usually transplanted in flat beds. Transplanting on ridge is better for rainy (*kharif*) crop. Flat beds of 2-3 m wide and 3-5 m in length depending upon level of land, soil type and irrigation method, are prepared. Kaur (2019) reported that all the growth and yield contributing parameters *viz.*, plant height, number of leaves per plant, neck thickness, bulb weight and bulb yield were significantly better in bed planting as compared to flat planting system.

In certain areas, pruning of top to the extent of 25 per cent is followed to obtain higher yields. Maiti and Sen (1968) found that partial pruning of seedlings at the time of transplanting augmented the stand of crop and increased the size of bulbs. Rathore and Kumar (1974) also reported similar result. Nevertheless, pruning of top at best may add to the convenience in transplanting and has no effect on yield of bulbs.

Regional variation exists in planting time of onion as it is very much sensitive to photoperiod and temperature. It starts from April - June in Karnataka, Tamil Nadu, Andhra Pradesh as early crop, July - August in Maharashtra, Gujarat, Rajasthan and August-September in Odisha and West Bengal (Samra *et al.*, 2006). Singh and

Pal (2001) summarized seasons of onion growing practised in different parts of India (Table 17).

Table 17: Season of Onion Growing Practised in different Parts of India*

Season	Time of Sowing	Time of Transplanting	Time of Harvesting
Maharashtra and part of Gujarat			
Rainy (*kharif*)	May-June	July-August	October-December
Late rainy (*kharif*) or early winter (*rabi*)	August-September	September-October	January-March
Winter (*rabi*)	November-December	December-January	April-June
Tamil Nadu, Karnataka and Andhra Pradesh			
Early rainy (*kharif*)	April-May	May-June	August
Rainy (*kharif*)	May-June	July-August	October-November
Winter (*rabi*)	September-October	November-December	March-April
Rajasthan, Uttar Pradesh, Haryana, Bihar, Punjab, West Bengal and Odisha			
Rainy (*kharif*)	May-June	July-August	Nov.-December
Winter (*rabi*)	October-November	December-January	May-June
Hills			
Winter (*rabi*)	September-October	October-November	June-July
Summer (Long-day-type)	October end to early November	March	August-September

* Singh and Pal (2001).

In Maharashtra, India, early rainy season (*kharif*) crop is transplanted by July end or beginning of August, whereas the usual rainy season onion is transplanted by mid-October. Winter (*Rabi*) onion is transplanted in December or 1st week of January. Singh and Singh (1974) observed that planting on 16th October gave highest bulb yield (341.89 q/ha). In North India, July-August is better time for transplanting onion in rainy season. In South India, planting of cv. Rampur Local in the 2nd and 3rd week of November gave the highest bulb and seed yield (Krishnaveni *et al.*, 1990). Trials conducted at Regional Research Station, Nasik, India during winter (*rabi*), 1999-2000 and 2000-2001 revealed that the cultivar Agrifound Dark Red transplanted on 1st October gave better bulb development and yield (Anon., 2002). Sharma *et al.* (2009) revealed highest plant height, maximum number of leaves, maximum polar diameter of bulbs was observed in 1st August planting whereas average bulb weight and total marketable bulb yield was found maximum in 16th August planting. However, further delay in planting towards 1st September resulted into a significant decline in bulb production. Khodadadi (2012) observed significant effect of the planting date on plant emergence, final height of plant and seed yield per hectare in onion. They reported that 6th November planting with mother bulb size of 65 to 80 mm recorded maximum seed yield/ha. Mohanta *et al.* (2017) suggested planting of kharif onion varieties during September among the four planting dates (15th August, 30th August, 15th September and 30th September) to get vigorous plants so that the growers may sell the plants as green onion in the market under Red and

Laterite Zone of West Bengal. Mohanta and Mandal (2014) recorded increased value in growth and yield parameters of onion as planting dates progressed from August to September. The highest yield was produced from 30th September planting. Rugi (2017) reported that, between two transplanting dates (10th and 25th August), 25th August proved best for growth, yield and economic returns in onion. Sharma and Dogra (2017) observed that second fortnight of August produced the highest bulb yield. Sharma and Jarial (2017) studied effect of dates of planting with four onion cultivars under low hill conditions of Himachal Pradesh at six dates of planting at ten days interval starting from 5th July to 25th August. They recorded continuous decrease in plant height with delay in planting. Planting date also produced significant effect on bulb yield. The highest average yield was obtained from 25th July planting. Dhar *et al.* (2019) studied for off-season onion cultivation and observed delay in planting from August to September significantly increases the growth and yield parameters of onion varieties and September planting given higher marketable yield.

Babrikov *et al.* (1997) assessed annual onion cultivars Top Keeper, Keep Well, Omega, Alix, Aldobo, Mayon, Radar and Siberia in the autumn in Bulgaria when seeds were directly sown on 9, 16, 25 or 30 September and overwintered also. Seeds sown on 9, 16 or 25 September germinated 10 to 14 days after sowing and formed 2-3 leaves by the beginning of winter. However, most plants sown on 30 September died. Top Keeper, Alix and Omega showed the best adaptability to the conditions with the highest yields and were ready for harvesting as early as May-June of the following year. They thus concluded that cultivar selection is essential for successful annual production of autumn-sown onion bulbs. Massiha *et al.* (2001) studied the effect of direct sowing (DS), row direct sowing (RDS), and the transplanting method (TM) for two onion cultivars, Azar-Shahr (red skin onion) and Horand (yellow-brown skin onion) in Iran. TM had higher values of total yield, marketable yield, percentage of class I and III, bulb weight, homogeneity of bulb weight, bulb diameter, homogeneity of bulb diameter, bulb length, homoge-neity of bulb length, bulb diameter at the neck and base, number of centres and percentage of bulbs containing multiple centres than the other methods. In Bologno, Italy the cv. Borettana gave the best bulb yield when sown direct; transplanting of cv. Dorata di Bologna polystyrene cellular trays (PT) plants resulted in a yield increase of 76 per cent compared with the yield obtained from direct sowing and it was achieved 20 days earlier. In Borettana, the percentage of bolting plants was very high, especially from PT (68 per cent) while in Dorata di Bologna it was very low (2 per cent for PT plants). The 80 plants/m^2 gave the highest mean yield (31 t/ha), but bulb weight was lowest for this treatment (Dellacecca *et al.*, 2000). In light sandy loam soil in Belgium, cultivars sown on 2 September 1997 were harvested on 13 May 1998 for marketing as young onions with a diameter of > 0.5 cm. Highest yields were obtained from cvs. Swift and Radar (Vanparys, 1999a). Further twelve onion cultivars were compared by Vanparys (1999b) on a light sandy loam soil. Cultivars were sown on 1 September 1997, and plants were harvested on 7 August 1998 (after ripening of foliage). Bolting at harvesting occurred most in Wapiti (8.47 per cent) and Continental (5.86 per cent), while no bolting occurred in Glacier and Yellow Stone. Highest crop yields were obtained by Swift (240.3 kg/100 sq.m) and Radar (224.4 kg/100 sq.m), and lowest by Glacier (19.7 kg/100 sq.m). Onions with a diameter of > 4 cm were mainly found in Buffalo (58.5 per cent), Yellow Stone (49.8 per cent) and Flix (49.8 per cent). Cultivars Radar and Swift were recommended. In NWFP, Pakistan, 15 February was the best sowing time at lower altitude, while 15 March was the suitable time for transplanting onion seedlings at elevation above 1524.30 m. The best transplanting time at elevations between 1067.10 and 1524 m was 1 March (Tariq and Ali, 2001). In Beligum, cultivars were sown in August-September, and harvested as ripe onions in July-August (Vanparys, 2000a). Further, for harvesting as young onion, cultivars were sown in August-September, and harvested as young (unripe) onions in April-May. Highest yields were obtained for cvs. Juno, Imai Early Yellow, Ocean, Buffalo, Arctic, Swift and Glacier and Radar (Vanparys, 1999c). In Korea Republic the main period of seed sowing was between 1 and 10 September. Seedlings were grown for 46-55 days and irrigated 3 or 4 times. Most farms planted onions in the field between 20 October and 10 November (Tai *et al.*, 1996). In south-eastern Queensland, Australia for early season plantings (February-March) locally developed open-pollinated lines, including Early Lockyer White and Brown and Golden Brown, were the highest yielders. Wallon Brown, Golden Brown and Cavalier hybrid were the most successful cultivars in April plantings. In May, the hybrids Diamond White and Cavalier and the open pollinated Wallon Brown was dominant. Hybrid cultivars, including Gladiator, Omega and SPS 846, were outstanding in June plantings. Average yields for the best three cultivars in the early and late season plantings were 54-56 t/ha, while average yields of 70-80 t/ha were observed for the best three cultivars in the April planting and 60-63 t/ha for May planting (Jackson *et al.*, 2001). In Egypt, the highest total and marketable yields of all 3 cultivars (Shandawel 1, Giza 6 Mohassan and single-centred Shandawel) were obtained from bulbs planted on 20 September. Mean total yield of the 3 cultivars was in the order Giza 6 Mohassan > Shandawel 1 > Shandawel, though the differences were only significant in the first season. However, Shandawel gave the highest marketable yield when the cultivars were shown on 20 September (13.90 t/0.42 ha) (Gamie *et al.*, 1996). In southwestern Korea Republic, the main cultivars were Kinkyu (early maturing cultivar) and Bonganhwang (late maturing cultivar). The main period of seed sowing was between 1 and 10 September. Seedlings were grown for 46-55 days and irrigated 3 or 4 times. Most farms planted onions in the field between 20 October and 10 November. Transparent vinyl was used as a mulch and chemical control of diseases and pest insects were conducted 3-5 times. Early maturity cultivars were harvested between 10 and 20 May and marketed immediately, while late maturing cultivars were harvested between 10 and 20 of June and were stored for 2-3 months before marketing (Ibrahim *et al.*, 2000). According to Madisa (1994), mid-March planting gave the highest yield of 43.0 t/ha compared with 23.6 t/ha for February planting and 31.2 t/ha for April planting. The February planting produced 24 per cent bolting plants, whereas, March planting gave 7 per cent and April planting gave no bolters. In a spacing trial with cultivars Granex 33 and Texas Grano, the treatments had no effect on yield, but influenced bulb size. In Chile, sowing on 21 July produced the highest yields for cultivars Valenciana, Valenciana-INIA and Valcatorce (Gonzalez *et al.*, 1997). In Poland, cvs. Keepwell, Takii's Senshyu Yellow Globe, Top Keeper and Presto showed promise for autumn cultivation in terms of yield, resistance to bolting and earliness of maturity (July).

The optimum sowing date was 15th August (Doruchowski and Krawczyk, 1987). In Denmark, the best results were achieved by sowing different cultivars of Welsh onion on 1st June. About 80 per cent of the harvested onions were marketable within 105 days with a yield of about 35 t/ha (Grevsen, 1989).

The onion seedlings are transplanted at a distance of 15 × 10 cm. The field is to be irrigated immediately after transplanting. At certain places transplanting is followed after irrigation. Das and Dhyani (1956) in an experiment at Benaras concluded that 24 cm spacing were the best for market value and 10 cm spacing for the highest production. Gruda (1987) from Albania reported that white onion cv. Sukthi transplanted at a spacing of 20 cm × 10 cm (50 plants/m²) produced the highest yield. Lokesh *et al.* (2000) in a study at Bangalore, India, inferred that higher mother bulb yield is produced in wider spacing of 15 × 10 cm in Arka Niketan followed by Arka Pithambar and Arka Bindu. The yield increase was mainly attributed to the difference in bulb weight, bulb diameter, neck thickness and number of centres/bulb. Muthuramalingam *et al.* (2002) studied the three spacing, *i.e.* 45 cm × 5 cm, 45 cm × 10 cm, and 45 cm × 15 cm. They observed that increase in length of bulb and shape index were highest at closer spacing of 45 cm × 5 cm, while the widest spacing of 45 cm × 15 cm recorded the greatest bulb diameter. Kumari and Kumar (2017) reported that onion cultivar Agrifound Light Red spaced at 15 × 10 cm² recorded the maximum plant height, number of leaves, girth thickness, yield per hactare and minimum days for maturity. Rumpel *et al.* (2000) noticed that yield of large bulbs decreased with density and was highest at 20-40 plants/m² in contrast to the yield of small bulbs, which was highest at the highest density of 140 plants/m². The yield of medium bulbs increased with density in the same way as the total marketable yield. Plant density hastened maturity of onions and at a density of 140 plants/m², the leaf fall-over occurred 8-10 days earlier compared with at a density of 20 plants/m². The highest yielding cultivars were Mercato F_1, Spirit F_1 and Armstrong F_1 in Italian condition. The highest yield of marketable bulbs was achieved in Brazil by sowing at the lowest density of 25 seeds/m. Sowing directly rather than the traditional method of transplanting seedlings reduced the growth cycle by 30 days with a reduction in production costs (Guimaraes *et al.*, 1997).

Singh and Singh (2002) studied the effects of the size of mother sets (1.0-1.5, 1.6-2.0 and 2.1-2.5 cm) and planting date (21 August, 1 September and 11 September) on the growth and yield of onion cv. N-53. They reported that largest sets planted on 21 August resulted in the greatest growth and yield. In Botswana, Madisa (1994) compared three sizes of onion sets for four onion cultivars Granex 33, Pyramid, De Wildt and Texas and was of the view that 1.2-1.7 cm size gave 30.6 t/ha while 2.5-2.8 cm recorded 37.6 t/ha. However, he opined that optimum size was 1.7-2.5 cm with a yield of 45.5 t/ha. From Pakistan, Khokhar *et al.* (2001) reported that bulbs developing from large sets (21-25 mm) showed significantly higher percentage of bolting (40 per cent) than medium (29.4 per cent) or small sets (20.8 per cent). Medium sets (16-20 mm diameter) and large sets (21-25 mm diameter) produced significantly higher bulb yield than small sets. The average bulb yield in Phulkara was significantly higher (40.05 t/ha) than rest of cultivars. The yields obtained from medium (43.8 t/ha) and large size sets (42.13 t/ha) of Phulkara were statistically *at par* with each other. However, medium size sets of Phulkara were recommended for planting in autumn season as relatively smaller amount of sets would be required to plant per unit area compared with large size sets. In Egypt, Koriem and Farag (1996) concluded that sets of 14-26 mm of cv. Giza 6 Mohassan and young or medium aged sets are best for high yield production. Patil *et al.* (2009) demonstrated a successful kharif onion cultivation via set plantation with reasonably good bulb yield potential (~180 q/ha) from the shortest span of 62-73 days which otherwise normally required for seedling nursery. They studied the effect of various set sizes (equatorial bulb diameter from 0.5 to 1.0 cm to 2.1 to 2.5 cm) and cultivars (Baswant-780 and Phule Samarth) on kharif onion bulb production. They noticed almost progressive increasing bulb yield trend with the bigger set size. In Zimbabwe, Matimati *et al.* (2006) studied the effect of set size and cultivar on marketable yield of onion (*Allium cepa*). Their study evaluated productivity of common onion cultivars (Arad, Dina, Grano, Pyramid, Shahar and Texas Grano) grown from different set sizes, namely large (3.6-5 mm), medium (2.6-3.5 mm) and small (2-2.5 mm). Their study recommends the use of medium to small sets for cultivars Arad, Dina and Grano. Small set sizes are recommended for cultivars Pyramid, Shahar and Texas Grano. Use of large sets tripled unmarketable bulb weight for Shahar and Texas Grano when compared with small sets, they reported. In Islamabad, Pakistan, Khokhar *et al.* (2001) examine the effects of three set sizes: 10-15, 16-20 and 21-25 mm diameter on bulb size, bulb yield, maturity and bolting response in five cultivars of onion namely Burgundy, Texas Early Grano, White Creole, Swat No.1 and Phulkara. They reported that bulbs developing from large sets (21-25 mm) showed significantly higher percentage of bolting (40 per cent) than medium (29.4 per cent) or small sets (20.8 per cent). However, medium sets (16-20 mm diameter) and large sets (21-25 mm diameter) produced significantly higher bulb yield than small sets.

Sharma and Dogra (2017) reported that the yield of onion was significantly affected both by variety and transplanting time. Agrifound Dark Red transplanted around second fortnight of August produced the highest bulb yield in Chamba, Himachal Pradesh. On the other hand, Singh *et al.* (2020) concluded that Agrifound Dark Red cultivar with transplanting on 23 July was the best combination for the farmers of Shahdol district, Madhya Pradesh, India. Mohanta and Mandal (2014) observed significant differences in growth and yield parameters of onion with different cultivars (Agrifound Dark Red, Arka Kalyan, Arka Niketan, Indam Marshal and Red Stone) in *kharif* season. They recorded highest yield in Agrifound Dark Red and suggested the cultivar to grow in *kharif* season in lateritic belt of West Bengal. In Punjab condition, Kaur (2019) observed that planting of bulb size (2.7-3.1 cm) on 16 July by bed planting gave higher bulb yield (223.800 q/ha) in kharif season. Rathore and Chaturvedi (2018) developed of an onion bulblet planter for vertisol. The performance indicated that the field efficiency was maximum 83.33 per cent with minimum seed damage 10.2 per cent. Multiple index was 5.1 per cent, missing index was 2.2 per cent and bulb to bulb spacing was 10.66 cm in chisel type furrow opener at 1.8 km/h speed with moisture content 17.2 per cent.

10.3.0 Manuring and Fertilization

The soil for onion growing should be liberally manured and fertilized. About 20-25 tonnes FYM/ha is considered adequate. In substitution of FYM, Yawalkar *et al.* (1962) recommended 25 t/ha pig manure or 35 t/ha night soil compost or green leaf compost or tank silt. In Brazil, Boff *et al.* (2000) observed that thermophilic compost (produced from ground *Pennisetum purpureum,* cattle manure and rotten onion bulbs) increased emergence and survival of onion cv. Crioula plantlets compared with mineral and organo-mineral fertilizers. The requirement of nutrients will depend on soil type, region of growing and removal of major nutrients. Maynard and Hill (2000) noticed that leaf compost-amended plots produced a greater percentage of very large sized onions in all cultivars. Repeated compost additions also reduced the incidence of soft rot disease, especially in susceptible cultivars in years with higher than average precipitation.

10.3.1 Primary Nutrients

Yawalkar *et al.* (1962) reported that onion crop yielding 300 quintals per hectare removed 73 kg nitrogen, 36 kg phosphorus and 68 kg potash. Duque *et al.* (1989) recorded that for a yield of 2.5 t/ha, uptake levels were 38.8 kg N, 38.6 kg P and 71.3 kg K. In Japan, 150-250 kg nitrogen, 200-250 kg phosphorus and 150-250 kg potassium per hectare are normally applied to onion fields. About 30 per cent nitrogen, 100 per cent phosphorus and 50-100 per cent potassium are usually incorporated into the soil prior to transplanting, with additional nitrogen and potassium applied during crop growth (Yakuwa, 1994).

Deficiency symptoms of common nutrients in onion (Pandey, 1993) are presented below:

Nitrogen

Leaves become yellowish green, curled, wilted and dwarf. At the beginning of the maturity, the tissue above the bulbs becomes soft. Higher doses of nitrogen during the growing period results excessive top-growth and thick neck.

Phosphorus

Growth of the plant is slow and maturity is delayed. The colour of leaves becomes light green and bulbs have a few dried outer scales.

Potassium

The deficiency lead to develop leaves tip burn, dark green and erect. The plant also becomes susceptible to diseases due to the decline in vigour. Timely application promotes bulb formation and improves the quality by raising the total sugar content. Deficiency also increases the bolting tendency.

Copper

Scales become thin and of pale yellow colour. Judicious application of copper sulphate increases thickness of the scales and changes the colour from pale yellow to a brilliant brown. Deficiency also leads to lack of firmness and solidity of bulb.

Boron

Plants become characteristically stunted and distorted appearance. Leaf colour varies from dark grey to deep blue green, the youngest leaves develop conspicuous yellow and green mottlings. Shrunken areas appear followed by ladder like transverse crack on the upper sides of the basal leaves. Leaves become stiff and brittle.

Magnesium

Appearance of irregular elliptical shaped areas almost white in colour near the ends of the leaves are visualised. Soils that are highly calcareous, overlimed or acidic, leachable and sandy will probably require more frequent applications of this nutrient for successful production.

Jyotishi and Pandey (1969) suggested application of 30-35 tonnes of FYM or compost along with 71-112 kg/ha each of nitrogen and phosphorus for optimum yield, depending on agro-climatic conditions. Nitrogen had pronounced effect on bulb yield but it was decreased at excessive rates. The effect of P and K was noticeable only in the soil deficient in these nutrients. Rizk (1997) found that increasing the NPK rate increased all vegetative parameters with economic yield. The best application method for NPK was 2 equal doses applied 30 and 60 days after transplanting. In South Aukland, Sher (1996) recommended 3 post-emergence NPK applications (approx. 10-20, 50-60 and 80-95 days after germination) with cv. Pukekohe Early Longkeeper. According to him, onion crop at harvest absorbs 218 kg N, 32 kg P, 204 kg K, 54 kg S, 108 kg Ca and 13 kg Mg/ha. Singh *et al.* (1997) reported that application of FYM (25 t/ha) with 100 kg N + 25 kg P + 25 kg K/ha increased significantly gross and marketable yield and highest net return was obtained.

Purewal and Dargan (1962), Srivastava *et al.* (1965), Kashyap *et al.* (1967) and Singh and Kumar (1969) reported that application of 100-115 kg nitrogen per hectare increased the yield of onion, while no significant response was observed due to the application of phosphorus and potash. Treatment with high dose of N caused earlier deterioration of bulbs in storage (Singh and Batra, 1972). Sharma (1998) found that plant height, number of leaves per plant and mean bulb yield increased significantly as seedling age at transplanting increased up to 6 weeks old in cv. Pusa Red. Plant height and number of leaves per plant increased as N application rate increased, but the differences were not significant beyond 100 kg N/ha. Bulb yield increased significantly up to 150 kg N/ha. Kashi and Frodi (1998) found highest yield and bulb mean weight at 120 kg N/ha.

The increase in yield and mean bulb weight was correlated with the increase in length and weight of the aerial leaves. The length and width of the bulbs, as well as the thickness of the bulb neck were affected by the N rate. N had positive and significant effects on the total soluble solids (TSS) and dry weight (DW) of the bulbs as quality factors and increasing the N rate up to 120 kg/ha increased the TSS and DW of the bulbs and aerial leaves. N at 160 and 200 kg/ha decreased TSS and DW. N had negative effects on the firmness of the bulbs and weight loss percentage during storage and applying more than 40 kg N/ha resulted in a decrease in firmness and increase in sprouted bulbs and weight loss. In Nigeria, Ekpo (1991) noticed high bulb yield (460.2 q/ha), average bulb weight (197.8 g), per cent contribution of grade A bulb to total bulb and bulb development rate/day in cvs. Bama and Pnanday with applied N at 120 kg/ha 61-90 days after transplanting. Hussaini and Amans (2000) observed that application of N fertilizer at 110 kg/ha substantially increased bulb yield, average bulb weight and number of large bulbs per plot during dry seasons at Kadawa, Nigeria. Further N had no effect on per cent storage loss up to 19 weeks after harvest. Stone (2000) from U.K. opined that starter fertilizer offer clear opportunities for reducing N inputs, while maintaining yield and quality of onion crop. Meena *et al.* (2007) studied the effect of nitrogen doses on growth and yield attributes of onion cv. Nasik Red. They found higher number of leaves per plant when onion was supplemented with 150 kg nitrogen per ha with comparison of lower doses of nitrogen *i.e.* 50 and 100 kg per ha. Mozumder *et al.* (2007) observed that increase in yield was reduced beyond 150 kg/ha nitrogen application and also high risk of bulb storage losses by promoting sprouting and decay of onion was observed. The study of Kumari and Kumar (2017) revealed that the maximum plant height, maximum number of leaves and maximum girth thickness at 45 and 90 days after transplanting (DAT) as well as quality traits like bolting percentage, maximum bulb diameter, minimum days taken for maturity, maximum average weight of bulb and maximum total soluble solids of onion was recorded with treatment 150 kg N/ha. Singh *et al.* (2018) observed that N significantly affected plant height, number of leaves, diameter of stem, fresh weight of plant, postharvest observations, fresh weight of bulb, length of individual bulb, diameter of individual bulb, volume of bulbs, specific gravity of bulb, total soluble solids (°Brix), number of scales and yield per hectare. Nitrogen at the rate of 100 kg/ha gave the best results.

Singh *et al.* (2000a) observed in Meerut, India that application of 75 kg P gave the best seedling survival, plant spread, bulb diameter, vertical thickness of bulb, weight per bulb and per plot, and length of the longest root, and the lowest bolting percentage, while yield and other yield parameters were similar to values for 100 kg. In Egypt, El-Rahim (2000) noticed that total and marketable yields, and average bulb weight increased, while percentage of missing plants, doubles and bolters decreased as a result of increasing rate of P fertilization from 0 to 60 kg P/feddan (0.42 ha). Jha *et al.* (2000) from IARI, New Delhi, reported that all the growth parameters and bulb yield showed significant linear increase over control with increasing doses of P. They opined that P stimulates the root systems and thus better vegetative growth resulting in higher accumulation of assimilates which gave to large-sized bulbs and consequently higher yield. Morphological characters have been found positively correlated with yield as reported earlier by Singh and Joshi (1978) and Pal *et al.* (1988). Further, data on radio-chemical analysis showed a steady and significant increase in per cent Pdff as a result of increased P application. They opined that the extra amount of added fertilizer phosphorus lead to increase in pool of available P, which is proportionately absorbed by plants. Per cent utilization of added P declined steadily and significantly with the increasing P levels. Pot experiment conducted to study the effect of rock phosphate-elemental sulphate granule (RP-SO G) on the yield attributes and yield of onion (cv. Co. 4) revealed that application at the rate of 60 kg P/ha produced the highest yield (Ammal *et al.*, 2000).

With increase in application from potassium fertilizer rate from 0 to 525 kg potassium sulphate/ha, bulb weight increased from 231 g to 324 g and yield from 69.4 t/ha to 97.1 t/ha (Shun *et al.*, 1998). Chatterjee and Dev (1998) studied the effect of K on onion (cv. N-53) grown in refined sand at 3 levels, 0.05 (acutely K deficient), 1.0 (moderately K deficient) and 4 (adequate K) meq/l. At days 25, owing to low potassium (0.05 meq K/l supply), plants showed marked depression in growth as compared to the plants grown at adequate K supply. At 1 meq K/l, the symptoms of potassium deficiency appeared later and were mild as compared to those observed at 0.05 K/l supply. As a result of low K, the bulb formation was affected more at 0.05 meq K/l than at 1 meq K/l supply and the bulb was very small and the scales were loosely arranged due to which the bulb was very soft to touch. The formation of bulb was also affected at 1 meq K/l supply and they were reduced in size compared to those at adequate K supply. Owing to low 0.05 meq K/l supply, the depression in dry weight of plants was very remarkable. As a result of low K, the bulb formation was markedly affected and the weight of bulbs decreased significantly, the decrease being more pronounced in acute K deficiency (84 per cent). At days 109 and 141, compared to 7.9 to 8.5 per cent K in old leaves at adequate K, its concentration ranged from 0.7 to 1.3 per cent in K dificiency. Desuki *et al.* (2006) studied the response of onion plants cv. Giza-20 to the different dose of potassium in the form of potassium sulphate @ 0, 50, 75, 100 kg per ha. Results indicated that the highest nutrient uptake was recorded due to higher dose of sulphate of potassium. Aisha and Taalab (2008) carried out the experiment to study the response of onion to the application of potassium at forms and different rates. Their study indicated that the highest plant height was due to application of potassium as sulphate of potassium. Singh *et al.* (2018) observed that potassium caused formidable impact on various growth and yield characters but not significantly affected.

The results of the experiment on the effect of N, P and K on bulb yield of onion cv. Patna Red indicated that NPK at 40-40-0 was best for seedling growth up to 30 days after transplanting. This showed that P was required for root development at early stage and additional K is not needed at this stage. At 90 and 120 days, 80-0-10 and 80-0-20 levels of NPK maximized growth, whereas the best yield was obtained with 80-0-0 fertilizer combination. It was also found that phosphorus was not utilized at later stage of the crop (Singh and Jain, 1959). Kashyap *et al.* (1967) also suggested 115 kg each of N, P, K/ha for higher yield. Chowdappan and Morchan (1971) recommended 150 kg each of N and P and 80 kg K per hectare in addition to 5 tonnes FYM. Chauhan and Sekhawat (1971) observed that N, P, K at the rate of 135, 45 and 22 kg/ha, respectively was beneficial. Overall the best yield and bulb quality were obtained with N : P : K at 120 : 60 : 50 kg/feddan in trials with cv. Giza 6 Improved in Egypt (Haggag *et al.*, 1986). N : P : K at 90 : 135 : 90 kg/ha applied at ploughing before sowing at 9 kg seed/ha, gave the highest yield of marketable bulbs in Ukraine (Britvich and Goncharenko, 1986). Singh and Mohanty (1998) recommended that rates for commercial onion production in and around Bhubaneswar (Odisha), India, are 160 kg N, 60 kg P and 80 kg K. In Ranchi, India, Singh (2000) recorded that application of 100 kg N + 60 kg K/ha significantly improved growth and economic yield. Singh *et al.* (2000) concluded that *kharif* (rainy season) onion productivity could be enhanced considerably by application of 100 kg N, 30.8 kg P and 83 kg K/ha. The highest yield (18.37 t/ha, compared with 16.59 t/ha in control) was obtained with 45 kg N + 45 kg P. Al-Moshileh (2001) conducted experiments in two different locations in a sandy soil at Al-Qassim area in Saudi Arabia with cv. Giza-6 Improved. A significant increment was obtained in the total onion yield, marketable yield and the mean bulb weight, as a result of N-application at the rate of 200 kg/ha during 1997-98 and 100 kg/ha in 1998-99. No significant effects were observed with the application of P and K on the total yield, marketable bulb yield and the mean bulb weight. Muthuramalingam *et al.* (2002) studied the 3 levels of N (20, 40 and 60 kg/ha), 3 levels of P (20, 40 and 60 kg/ha) and constant level of K (30 kg/ha), along with farmyard manure (FYM) at 25 tonnes/ha, *Azospirillum* at 2 kg/ha and phosphobacterium at 2 kg/ha. Application of 60:60:30 kg NPK/ha, FYM, *Azospirillum* and phosphobacterium was recorded the longest bulb diameter (5.70 cm) and highest yield. Singh and Pandey *et al.* (2006) studied integrated nutrient management (INM) for sustainable production of onion. They evaluated 9 different INM modules involving NPK (50, 75 and 100 per cent recommended dose), 10 tonnes farmyard manure (FYM) and Azotobacter (200 g culture/litre solution dipping before transplanting). They concluded that the combined use of FYM, fertilizers and biofertilizers (75 per cent NPK + 10 tonnes FYM/ha + Azotobacter) would be the optimum integrated nutrient management practices for higher yield, nutrient uptake and fertility status of soil. Nasreen *et al.* (2007) conducted an experiment on the effect of nitrogen (0, 80, 120, and 160 kg/ha from urea) and sulphur (0, 20, 40, and 60 kg/ha from gypsum) fertilization on N and S uptake and yield performance of onion. They observed significantly responded to the application of nitrogen and sulphur. The highest yield of onion and the maximum uptake of N and S were recorded by the combined application of 120 kg N and 40 kg S/ha with a blanket dose of 90 kg P_2O_5, 90 kg K_2O, and 5 kg Zn/ha plus 5 tons of cowdung/ha. Dilruba *et al.* (2010) studied the effect of different levels of nitrogen and potassium on yield contributing bulb traits of onion cv. BARI Peaj-1. The results showed that the treatment 100 kg N with 120 kg K_2O ha^{-1} was found to be the best from overall considerations. Jahan *et al.* (2010) studied the effect of nitrogen (N) and potassium (K) fertilizers levels on the growth, yield and nutrient concentration of onion. Four levels of nitrogen *viz*: 0, 60, 120 and 180 kg N ha^{-1} and three levels of potassium *viz*: 0, 60 and 120 kg K_2O ha^{-1}. The highest bulb yield (8.53 ton ha^{-1}) was obtained when plants were grown with nitrogen at 120 kg/ha. The lowest yield (6.70 ton ha^{-1}) was recorded in the control treatments. Application of potassium at 60 kg K_2O ha^{-1} produced the highest bulb yield (7.77 ton ha^{-1}). The effect of interaction between nitrogen and potassium was statistically significant. The combination of 120 kg N and 60 kg K_2O ha^{-1} gave the highest bulb yield (9.40 ton ha^{-1}). Mahala *et al.* (2019) reported that application of 75 recommended doses of NPKS along with poultry manure @ 5 t/ha is better for realizing better bulb diameter and yield of onion. Jeevitha *et al.* (2017) obtained higher yield and superior bulb quality in onion by the application of 80 per cent of RDF through drip fertigation spaced at 20×15 cm^2 or 10×5 cm^2.

Effect of application of NPK on bolting was studied by Paterson *et al.* (1960), Singh and Kumar (1972) and Singh and Batra (1972) who reported that application of relatively high amount of nitrogen reduced the development of premature seedstalk formation. Further, Verma *et al.* (1972) and Hassan and Ayub (1978) found that application of high dose of nitrogen reduced the percentage of bolting; potash, however, did not influence bolting.

Onion bulbs are often stored for different periods, depending on the necessity. Hence, the doses of NPK should also be optimum so that quality of bulb does not deteriorate and spoilage is not enhanced. In this respect, application of sulphur is found to improve quality and pungency of onion bulbs. It was also observed that application of higher doses of sulphur caused reduction of bolting in onion.

10.3.2 Secondary Nutrients

Application of sulphur (S) invariably increases the yield and quality of onion. Mishra and Prasad (1966) reported beneficial effect of sulphur on the yield of onion. Singh and Batra (1972) quantified the dose and reported that the yield of onion bulbs was increased by the application of sulphur up to 25 kg/ha. Jana and Kabir (1990), however, recorded highest plant height (48.62 cm) of onion (cv. Nasik Red) with the application of 30 kg sulphur per hectare. They also reported that highest root length (13.78 cm) and yield of onion bulb 30.69 t/ha in the treatment where sulphur was applied @ 30 kg/ha. Randle *et al.* (1994) raised 5 short-day cultivars (Z 238, Granex 33, Granex 429, Rio Bravo and Sweet Tex) of onion in a greenhouse at low or high S fertility (supplied with nutrient solution containing 0.1 or 4.0 me/l S, respectively) where leaf S concentration varied between cultivars at each of 8 sampling dates during growth and development, but the pattern of S accumulation in the leaves of cultivars was similar. Leaf S concentration increased during early plant development while still in a non-bulbing photoperiod, but that decreased as bulbing progressed to maturity at both high and low S fertility. The decline in leaf S concentration during bulbing was more severe with low than with high S fertilization. Leaves that were left to dry on the mature bulb, most of their S, especially at 0.1 me/l S resulted to have implications in the final flavour intensity of the bulb. Correlations between leaf S concentration and final bulb S concentration or pungency (in terms of pyruvic acid formation) were generally poor. Randle *et al.* (2002) reported that sequentially reducing sulphate fertility from 1 mM to 0.05 mM during onion growth and development affected bulb flavour at harvest. He also found that leaf and bulb S responded significantly to the sequential reduction of SO_4 fertility during plant growth and development. Leaf and bulb S increased linearly as SO_4 reduction was delayed during growth and development. Singh and Singh (2000a) reported that for maximizing the bulb yield, application of 200 kg/ha nitrogen combined with 50 kg sulphur/ha is best.

Kumar and Sahay (1954) studied the effect of S fertilizer on the pungency of onion and found that volatile sulphur content of bulb contributing pungency increased with the increase in the elemental sulphur fertilizer. Balasubramonium *et al.* (1978) reported that the amino acid, cystine and methionine contents increased with increase dosage of sulphur. The results also indicated that there was an increase in pyruvic acid content due to sulphur application which might be due to increase synthesis of volatile compounds.

The pungency and sugar content of onions cannot be manipulated by choosing soil type or by applying extra S to a soil containing sufficient S or irrigated with water with high S content (Hamilton *et al.*, 1998). Significant variation in the pyruvic acid content among clones was recorded by Hamilton *et al.* (1997) in the high S treatment, but not in the low S treatment. Total sugar content was higher in the low S treatment than the high S treatment (45.2 vs. 43.1 mg/l) with a large variation among the clones under both treatments. Bulb weight was significantly less at low S rates (311 vs 504 g) with a great variation among the clones.

Nagaich *et al.* (1998) reported from Gwalior, India that bulb yields of cv. Nasik Red increased with S rate and were highest at an intermediate K rate (80 kg/ha) under a field experiment. Chatterjee *et al.* (1999) observed that sulphur deficiency resulted in a decreased biomass, loosely arranged bulb formation, lowered bulb yield, reduced concentration of sulphur and protein, lowered activity of starch phosphorylase and ribonuclease. Jaggi and Dixit (1999) also noted positive responses to S application in terms of growth and bulb yield of onion under Palampur, India, conditions.

Sharma *et al.* (2002) reported that onion bulb diameter increased with the increase in S rate up to 30 kg/ha in heavy-textured soils and up to 45 kg/ha in light-textured soils. Rizk *et al.* (2012) reported that bulb and neck dimensions (length and diameter) was significant when onion plant supplied by sulphur at rate of 400 kg S/fed. The interaction between nitrogen plus phosphorus fertilizer and sulphur had a significant difference effect on bulb diameter. Tripathy *et al.* (2013) study the effect of sources and levels of sulphur on growth, yield and bulb quality in onion. They reported that sulphur application influenced the bulb equatorial and polar diameter in onion. Among the levels of sulphur, irrespective of sources, sulphur @ 30 kg/ha recorded significantly higher polar diameter (5.17 cm), equatorial diameter (5.17 cm). Meher *et al.* (2016) reported that yield, yield attributes and other traits of onion cv. Agrifound Dark Red was response favourably to the sulphur application in a range of 40 to 60 kg/ha. Graded level of sulphur application linearly increased the yield up to 50 kg/ha with bulb yield of 35.5 tonnes. Maximum pyruvic acid content in onion bulbs was noticed with sulphur application at 40 and 50 kg/ha. However, application of sulphur did not affect number of scales and TSS content of onion bulb. Mondal *et al.* (2020) found that yield attributes responded favourably to the sulphur application in the range of 40-50 kg/ha. Maximum bulb yield (28.52 t/ha) and nutrient uptake by bulb was recorded with 40 kg S/ha. Maximum pyruvic acid, reducing sugar and total sugar was registered with 40, 40-50 and 60 kg/ha S application, respectively.

The effect of sulphur nutrition on bulb quality was also investigated by Lancaster *et al.* (2001) using onion bulbs grown in hydroponic culture. At lower sulphur supply the per cent S in cell walls was reduced. Bulbs grown at low sulphur supply had reduced firmness and pungency. Therefore, storage may be adversely affected at low sulphur supply.

10.3.3 Micronutrients

Application of micronutrients improved the quality of bulbs. Rao and Deshpande (1971) reported appreciable increase in yield when copper and boron were applied at the rate of 13.4 and 1.8 kg per hectare, respectively. Increased yield due to the application of copper was reported by Arnon and Stout (1949) and Knott (1956). Onion plants grown under B-deficient nutrient solution exhibited thick, brittle, mottled, blue-green colour and eventually necrotic colour of leaves. The bulb scales become hard and rough and the internal scales became necrotic. After 54 days of storage, 70 per cent of B-deficient bulbs had rotted (Calbo *et al.*, 1986).

Boron and molybdenum also had beneficial effect on yield of onion. However, the response of boron is more pronounced than that to molybdenum (Mukhopadhyay and Chattopadhyay, 1999).

Schmidt (1963-64) reported better performance of several vegetable crops including onion with zinc application. Even seed treatment with zinc improved the yield of onion sets and hastens their maturity (Paridan, 1966). Tibabishew (1972) reported that pre-sowing seed treatment of onion with 0.02 per cent zinc produced early and uniform seedling emergence, vigorous leaf development and increase bulb yield. Gallagher (1969) reported that the application of zinc sulphate was effective in restoring the normal growth and appearance of onion plants. Gupta *et al.* (1983) studied the effect of zinc application on the yield and the concentration of the element in the bulbs and found that the yield of onion bulbs increased significantly with zinc over the control irrespective of the levels of application. Midan *et al.* (1986) reported that the number of flower heads of onion increased due to spraying with 0.05 per cent zinc sulphate. The result further revealed that spraying with IAA and either Zn-sulphate or Mn-sulphate gave the highest yield of onion seed with a higher percentage of germination. The increase in seed yield of onion with foliar zinc spraying might be due to an improvement in photosynthesis. Maurya and Lal (1975) studied the chemical compositions of the bulb as affected by Zn application and found that the values of total soluble solids (TSS) increased with Zn concentration up to 3 mg/l. The same trend was also noted with regard to the reducing, non-reducing and total sugars. The improvement in various constituents of bulb was associated with physiological activities in plant for which Zn was the responsible factor. Application of Zn alone (10 kg/ha as Zn-EDTA) or in combination with S (30 kg/ha) reduced rotting, sprouting and physiological weight loss during storage (Kumar *et al.*, 2000). Singh and Dhankar (1988) reported that the application of zinc not only increased the growth and yield of onion significantly but also reduced bolting, increased neck thickness and growth when the element was provided @ 25 kg/ha in combination with 100 kg/ha potash. The observations were in agreement with those of Pandey and Mundra (1971), Islam and Hoque (1977). Singh and Dhankar (1989) also reported that different levels of N, P and Zn significantly affected the growth, yield and quality of onion. Jawaharlal *et al.* (1988) reported that the soil application of zinc, antagonised uptake of phosphorus and potassium by the onion plant. Soil application of 5 kg Zn/ha through $ZnSO_4$ as basal placement with other fertilizer produced highest bulb yield with highest benefit : cost (B:C) over control. Foliar application of 1.5 kg at 40 and 60 days after transplanting had the similar beneficial effects (Sharma *et al.*, 2000). Among the different micronutrients tested (Fe, Mn, Zn or Cu), Sliman *et al.* (1999) reported from Egypt, significant increase in dry yield when $ZnSO_4$ was applied as foliar spray.

The soil application of Cu and Mn @ 5 kg/ha, individually increased the bulb yield of onion over control. There was no significant effect of iron on bulb yield of onion (Kothari *et al.*, 2000). Pre-harvest application of copper oxychloride (250 g/100 l) at bulbing stage induced an increase in skin thickness, decreased weight loss and enhanced colour in cultivars. Dark skin cultivars had a better response to pre-harvest treatments in relation to light skin cultivars. Firmness was not affected by pre-treatments (Ferreira and Minami, 2000).

Goyel *et al.* (2017) registered maximum bulb yield (394.86 q/ha), highest net return (Rs 3, 45, 355/ha) and high benefit cost ratio (7.98) by the foliar application of B (0.25 per cent) in cv. Agrifound Dark Red during kharif season. Foliar application of Zn + Mn + B + Cu (0.5 + 1.0 + 0.25 + 1.0 per cent) also gave good bulb yield (333.32 q/ha). More *et al.* (2017) found that the application of micronutrient enhanced yield in onion. The application of zinc 25 kg/ha resulted in significantly higher bulb yield up to (564.70 q/ha), net income (Rs. 494621.68/ha) and B: C ratio 7.06. Iron (20 kg/ha) and boron (5 kg/ha) application registered 547.58 and 533.93 q/ha bulb yield. Mandal *et al.* (2020) reported that the application of Sulfur, Boron and Zinc alone or in combinations with FYM and recommended dose of fertilizer increased the plant height, neck diameter and bulb diameters in onion.

10.3.4 Biofertilizers

The beneficial role of biofertilizers on onion production has been reported by several scientists. In India, Bhonde *et al.* (1997) observed that Azotobacter-1, 50 per cent N gave the highest market yield (230.62 q/ha) of onion (cv. Agrifound Light Red). Mohd-Mostakim *et al.* (2000) obtained highest fresh (132.24 g) and dry (13.52 g) weight of the bulbs and yield (575.8 q/ha) with the 5 per cent azotobacter treatment. The application of Azospirillum or the phosphobacteria to the seed bulbs (cv. CO-4) and the soil gave in increased yield of 18.3 per cent and saved 25 per cent inorganic fertilizer input. Shape index, bulb colour and pyruvic acid content increased through the application of 45 : 45 : 30 NPK/ha along with *Azospirillum* or the phosphobacteria (Thilakavathy and Ramaswamy, 1999). The results obtained by Singh *et al.* (2000b) indicated that *Azotobacter* inoculation with 75 per cent of recommended dose of nitrogen gave bulb yield *at par* with full dose of recommended nitrogen alone, thus showing a net saving of fertilizer nitrogen. *Azotobactor* inoculation helped the crop growth in several ways. Sai Reddy *et al.* (2000) also suggested that application of 75 per cent of recommended dose of nitrogen along with biofertilizer is recommended for obtaining higher bulb yield in onion. Mahanthesh *et al.* (2008) investigated the influence of integrated use of bio-fertilizers with NPK on growth and yield character of onion (cv. Bellary Red) under rainfed conditions. Plants provided with Azospirillum + 100 per cent N + PK (125 : 50 : 125 kg/ha) produced maximum plant height, more number of leaves, neck thickness, bulb diameter, bulb weight and also highest bulb yield. Prajapati *et al.* (2017) reported that application of Azospirillum (5 kg/ha) at 60 days after sowing and Salicylic acid (250 mg/L) at 30 days after sowing and 30, 45, 60 days after transplanting was performed significantly superior over control in different growth and yield characters. Talwar *et al.* (2017) found that biofertilizers improved the microbial content and nutrient uptake of onion stover. They registered increase in bulb and stover dry mass by the application of Azospirillum or Azotobacter along with recommended fertilizer dose. They noted that application of Azospirillum along with recommended fertilizer dose resulted in significantly higher nitrogen uptake (210.3 kg/ha^{-1}) and application of Azotobacter along with Vesicular-Arbuscular Mycorrhizae and recommended fertilizer dose resulted in highest phosphorus uptake (21.5 kg ha^{-1}). They observed the highest bacteria (27.2 × 10^6) and actinomyctes (34.0 × 10^4) count in FYM @ 20 t/ha treated plots; while highest fungal count (28.2 × 10^3) was noted by the application of Azotobacter along with recommended fertilizer dose at the time of harvesting. They also found that organic manures improved the organic carbon status of soil.

Two year pooled data revealed that application of 75 per cent chemical fertilizer with 10 tonnes vermicompost/ha gave significantly higher bulb yield over recommended dose of NPK (100 : 50 : 100) fertilizers and FYM (30 t/ha). A further addition of 2.5 tonnes vermicompost/ha can save 25 per cent chemical fertilizers without significant reduction in the yield (Lal *et al.*, 2000). By the application of different chemicals–bio-agents and vermicompost, Mathur and Sharma (2000) observed that average plant population was significantly higher in thiophanate methyl and *T. harzianum*. The weight of sets was also significantly superior in thiophanate methyl and *T. harzianum* over check.

Tawaraya *et al.* (1999) found that in Welsh onion (*A. fistulosum*) cultivars with short roots were more responsive to mycorrhizal colonization than those with long roots. This suggests that root length should be considered in mycorrhizal-plant symbiosis. Ramananda *et al.* (2000) reported that onion cultivars N-53 and Bellary Red responded well to inoculation with different AM fungi. Percentage root colonization, spore count and populations of free living N$_2$-fixers and P-solubilizers were significantly higher in N-53 inoculated with *Gigaspora margarita* compared with the other AM fungi screened. Bellary Red responded to inoculation with *Acaulospora laevis*. Shoot P concentration, plant growth and yield were significantly higher in N-53 and Bellary Red onion seedlings inoculated with *G. margarita* and *A. laevis*, respectively. A field experiment was conducted to evaluate the benefit to growth of *Allium cepa* of inoculation and treatment with 2 P rates (25 and 50 kg P/ha). Inoculation significantly increased mycorrhiza formation over that caused by native AM fungi present at the site. At harvest, all inoculated onion cultivars showed higher values for bulb diameter, fresh weight, shoot dry matter, shoot P content and bulb yield than in inoculated plants. However, the magnitude of AM response for yield in a given onion cultivar was found to be different at different rates of P (Sharma *et al.*, 2000). Tawaraya *et al.* (1996) also suggested that selection of suitable fungal species and optimal phosphate application were important for improvement of growth using arbuscular mycorrhiza, because species differences in the rate of infection, increased P uptake, and growth have been observed at each P level.

10.4.0 Irrigation

Irrigation is depending on soil and climate. The cylindrical leaf shape of onion is effective for dissipating heat in air since it has the highest heat transfer per unit surface area (Slatyer, 1967). Stomata closes at low value of potential evapotranspiration compared with other species. Thus the narrow, vertically oriented, cylindrical leaves are well adapted to dissipate energy as heat without any excessive rise in leaf temperature. The root system in onion is normally restricted to top 8 cm and roots penetrate seldom deeper than 15 cm. The water requirement of the crop at the initial growth period is less. It depends on crop growth, soil type and planting season. Irrigation should be stopped 15-20 days before attaining maturity for improving the keeping quality of bulb. Frequent irrigation delays maturity. In onion, Water input per kg of product is 147 and productivity kg/m^2 is 6.826.

In India, results of investigations on irrigation of onion at different locations revealed marked increase in yield when the crop was irrigated at 13 days interval during November-December, at 10 days interval during January, 7 days interval during February-March with 2 acre inches of water (Mande and Arakeri, 1956). Patel (1958) reported that about 13 irrigations were needed during the entire growth period. Murthy and Rao (1964) obtained the highest yield with irrigation at 5 days interval. In Delhi, the optimum soil moisture requirement for onion was found to be 0.65 bar tension at 8 cm soil depth (Joshi, 1963; Narang and Dastane, 1969). Further, it was observed that bulb formation and enlargement stages (60-110 days after planting) were critical in their demand for water (Dastane *et al.*, 1969). Narang and Dastane (1971) recommended application of 30 mm water corresponding to cumulative pan-evaporation of 43 mm. In Maharashtra, Bhonde *et al.* (1996) recorded the highest yield of quality seed of onion cv. Agrifound Light Red, at irrigation 10 days intervals with 80 kg N/ha in split application. Drip fertigation with 50 per cent of the recommended solid fertilizer dose was the best treatment for promoting growth (Chopade *et al.*, 1997). Drip irrigation with 100 per cent RDF and fertigation of 9 lit/ha root booster at weekly interval for eight weeks is the best treatment for improved growth and yield of *rabi* onion cultivated in semi-arid region of Maharashtra state (Ali *et al.*, 2019).

Rana and Singh (1997) recorded highest onion bulb yields with 10 irrigations (each of about 3.6 cm) at intervals of 20 days (up to 100 days after transplanting) during the winter months and then every 10 days during summer. In Coimbatore, Ramamoorthy *et al.* (2000) observed that water use efficiency was greatest when onions were irrigated at IW/CPE of 1.2 and were given 90 kg N/ha.

Hegde (1988) observed that irrigation at 0.45 to 0.65 bar soil water potential resulted in maximum bulb yield and water use efficiency, but the quality of bulbs was not much affected by irrigation treatments. Singh and Alderfer (1966) also reported that the most critical stages for water need were periods of bulb formation and enlargement. Hassan (1984) reported that irrigation of 10 days interval with 90 kg N/ha gave the best yield (12.52-12.94 t/ha) of high quality bulbs.

Chung (1989) noted that application of irrigation throughout the growing period increased the total bulb yield from 52 to 84 t/ha while withholding the last 2 irrigations prior to maturity reduced the 51-70 mm bulb yield from 57 to 44 t/ha.

Zimmerman *et al.* (1970) reported that onion yield increased with sprinkler irrigation, particularly in dry weather and it depended on the time of application of nitrogen, whereas Abrol and Dixit (1972) recorded significant increase in yield and water use efficiency in drip irrigation than conventional methods of check basin which reduced surface evaporation also. In the Mesilla Valley of southern New Mexico, the maximum irrigation water use efficiency (IWUE) values for onion using the subsurface-drip and furrow irrigation were 0.059 and 0.046 t/ha/mm of water applied, respectively (Al-Jamal *et al.*, 1999). Erdem and Kayhan (2019) obtained highest onion yield in the treatment on which lateral was buried in 20 cm depth and the amount of irrigation water was applied based on 125 per cent of Class A pan evaporation. They concluded that the total marketable onion yield increases as the amount of lateral depth and applied irrigation water increases. High onion yield could be achievable using a drip system compared to sprinkler system but drainage water was more in drip than sprinker system. Nouri (1990) from Sudan, noted that water stress resulted in a consistently highly significant reduction in yields and high yield could be realized by irrigation at every eight days intervals.

Koriem *et al.* (1999) calculated average crop coefficient (Kc) value as 0.66 under the conditions of Mallawi, Egypt. IWUE was highest when irrigation was withheld, followed by irrigation after depletion of 30 per cent of available soil moisture. Muldoon *et al.* (1999) in Australia reported that onions absorbed soil moisture to a depth of 50 cm. Al-Jamal *et al.* (1999) from USA developed a Kc (crop coefficient) curve for a particular yield which is simple to schedule irrigation events for onion. Shock *et al.* (2000) recorded that in onion evapotranspiration from emergence to the last irrigation totalled 681 mm in 1997 and 716 mm in 1998. Hussaini and Amans (2000) reported from Nigeria that irrigation at 7 days intervals produced higher yield with larger bulbs. In Argentina, highest total yield 24.5 t/ha with 467 mm of water applied in 1994-95 and 38.9 t/ha with 612 mm of water in 1995-96 were obtained when irrigation was continued until 8 and 7 days before harvest in the first and second years, respectively (Gaviola *et al.*, 1998). Mohamed and Gamie (2000) investigated the effects of irrigation treatments after a depletion of 35-40, 55-60 and 75-80 per cent in the available soil water (designated as wet, semi-wet and dry, respectively) on onion cultivars (Giza-6M, Giza 20, Shandwel-1, Assiut Glob, El-Bostan and Beheary No Pink) in Egypt. The total, marketable and exportable yields, and average bulb weight, significantly increased, while total soluble solids (TSS) declined with increasing available soil moisture. Giza 20 produced the highest total and marketable yields, while El-Bostan had the highest average bulb weight. Water consumptive use and water use efficiency were highest in Beheary No Pink and Giza 20, respectively.

In Finland, Suojala *et al.* (1998) observed that irrigation during warm and dry periods was essential to achieve the maximum yield potential and did not impair the storage quality of onions. Onion is extensively grown under furrow irrigation in the western United States. The application of wheat straw (630 to 900 kg/ha) mechanically to the bottom of irrigation furrows increased yield of sweet Spanish onions in commercial onion fields in Oregon, USA. Yield improvements are attributed to decreased water run-off and increased lateral movement and soil moisture (Shock *et al.*, 1999). Kgopa *et al.* (2018) assessed the distribution of heavy metals in edible and non-edible parts of plants irrigated with treated wastewater under field conditions. They noted that relative to root tissues, the accumulation of heavy metals increased (32-35 per cent) and decreased (20-87 per cent) in leaf and bulb tissues, respectively; and thus, onion bulbs were relatively safe for consumption in plants irrigated with treated wastewater.

In Raipur, Banjare *et al.* (2019) studied on enhancement in water use efficiency and economic feasibility of mulching roll prepared from reusable plastic materials for growing *rabi* onion. The maximum water use efficiency (1.13 q/ha/mm) and the benefit : cost ratio (2.28) was observed under BPM (black plastic mulch). They concluded that low cost mulching roll can be prepared from reusable plastic material and can be utilized as mulches for enhancing water use efficiency particularly for small and marginal farmers under Badi cultivation.

10.5.0 Weed Control

Onion is closely planted and a shallow rooted crop. Thus hand-weeding is difficult which may also damage the crop. Manual weeding is also becoming expensive. Therefore, it is suggested to use chemical weedicides along with one hand-weeding at critical stage. Bhagchandani *et al.* (1973) reported that 2.5 kg of teneran/ha 3-5 weeks after transplanting gave very promising result. Basalin 1 l/ha as pre-emergence and Lorox 0.5 kg/ha, after three weeks from transplanting also proved effective. TokE-25 @ 4-6 litre and basalin 0.5-1 litre per hectare with one hand weeding controlled weeds (Bhagchandani and Pandey, 1983). In trials with cv. Nasik Red, Singh *et al.* (1986) reported that nitrofen at 1 kg/ha applied pre-planting gave good weed control and the highest yields of 282.41 q/ha compared with 193.21 q/ha in the control.

Treatment of onion cv. Red Creole immediately after transplanting with various herbicides showed that Goal (oxyfluorfen) 2 l/ha was most effective in controlling broad leaved weeds and grass spp. (Aristy, 1981). Kerefov and Shkhagumov (1985) reported that application of 4 litres treflan (trifluralin) presowing + 10 kg ramrod (propachlor) pre-emergence + 2 kg afalon (linuron) post emergence/ha was most effective against grassy weeds, decreased weed population in onions by 96 per cent. Patel *et al.* (1986) recommended that for overall weed control efficiency 0.9 kg fluchlo-ralin/ha applied as pre-planting + weeding at 40 days after sowing (92.1 per cent). Application of 100-400 g fluroxypyr (starane)/ha twice controlled weeds effectively without hampering bulb size or yield in U.K. (Pollak *et al.*, 1989).

Perez-Moreno *et al.* (1996) reported that hand weeding was most effective followed by oxadiazon (3.0 l/ha) and oxyfluorfen (1.5 l/ha) treatments. According to Oliveira *et al.* (1997) application of oxyfluorfen should be made at a time when weeds are still susceptible but the onion crop has become sufficiently tolerant to the herbicide. In Denmark, Melander and Hartvig (1997) reported that hoeing close to the row, leaving only a 5 cm unfilled strip, has the potential of saving labour costs for hand weeding in non-herbicidal growing systems for onions. In Polands, application of propachlor in onion significantly decreased the number of weeds (Konopinski, 1997).

Verma and Singh (1997) found that weed population and weed dry weight/m² were lowest in plots treated with 1.5 kg pendimethalin a.i./ha. Nagagouda *et al.* (1998) also noted that pendimethalin at 0.75 and 1.00 kg/ha higher benefit : cost ratios than the weed free control. Singh and Singh (2000) obtained the maximum net profit with treatments pendimethalin 1.5 kg/ha and pendimethalin 1.5 kg/ha + one hand weeding 45 DAT. Pandey (2000) also recorded the maximum yield of bulb (227.76 q/ha) with the application of pendimethalin @ 1.0 kg a.i./ha which was significantly higher as compared to the yield obtained under weed check (163.22 q/ha). Singh *et al.* (1997) observed that the greatest uptake of N, P and K resulted from treatment by 0.25 oxyfluorten + hand-weeding with greatest crop yield along with benefit: cost ratio when compared with the weed control.

According to Satao and Dandge (1999) Trifluralin applied at 1.08 kg/ha was the most effective herbicide, followed by 0.96 kg/ha of triflurin during *rabi* season under Maharashtra, Indian conditions. In California, USA, Bell and Boutwell (2001) observed that onion yields in fields treated with bensulide and pendimethalin were comparable to that fields treated with DCPA (chlorthal-dimethyl). Minz *et al.* (2018) recorded highest plant height (66.67 cm), number of leaves (5.98) and neck thickness (4.76 cm) in hand weeding at all the growth stages, which was at par with pendimethalin (1.00 kg/ha) immediately after transplanting and pendimethalin (1.00 kg/ha) one week before transplanting. Hand weeding recorded 122 per cent more yield (23.71 t/ha) than weedy check (10.68 t/ha). The study of Chaurasiya *et al.* (2018) revealed that the growth and yield characters and bulb yield were higher and equally effective under weed free check, PRE-Oxyfluorfen @ 250 g a.i./ha + one hand weeding at 35 DAT followed by PRE-Oxyfluorfen @ 250 g a.i./ha + Oxyfluorfen @ 250 g a.i./ha at 35 DAT and PRE-Pendimethalin @ 1000 g a.i./ha + Oxyfluorfen @ 250 g a.i./ha at 35 DAT. However, the quality parameters *viz.*; total soluble solids, total sugar (g/100 g) and dry matter content of bulb were not influenced by the various treatments of weed management.

Sahoo and Tripathy (2019) obtained significantly highest marketable bulb yield (19.19 t/ha) and total bulb yield (22.80 t/ha) with a higher percentage 'A' grade bulb (34.04 per cent) and an average bulb weight (69.83 g) by the application of oxyfluorfen before planting + one hand weeding at 45 days after transplanting. The same practice recorded highest 66.09 per cent of weed control efficiency along with the lowest number of monocot weeds (26.33/m²), minimum fresh weight of weeds (40.08 g/m²) and minimum dry weight of weeds (14 g/m²).

Tripathy *et al.* (2019) registered the lowest values of total weed density (18.6/m²), dry weight (6.8 g/m²), weed index (6.2 per cent) and the highest values of weed control efficiency (89.2 per cent), bulb yield (21.89 t/ha) and benefit: cost ratio (2.84) with sequential application of pendimethalin @ 750 g/ha at 3 DAT + oxyfluorfen @ 100 g/ha at 20 DAT. They reported that weed competition resulted in about 54.8 per cent yield loss of onion in weedy check plot and weeds removed 23.9 kg N, 3.8 kg P and 27.6 kg K/ha.

In seed crop also weedicides are effective. Sandhu and Randhawa (1980) reported that oxadiazon (2 and 3 kg/ha, post emergence), basalin + chloroxuron 1.2 + 0.68 kg/ha at pre-plant and post emergence, respectively, proved to be the most promising treatments for controlling the weed population and resulting in significantly higher seed yield.

11.0 Harvesting, Curing, Yield and Storage

11.1.0 Harvesting

Harvesting of onion bulbs at appropriate stage of maturity is a very important factor in deciding storage life of onion as the bulbs may be stored for about six months. The bulbs reach maturity when the plants cease to produce new leaves and roots. In onion, neckfall is the indication of maturity. Davis (1943) suggested that the best time to harvest is when 15-25 per cent of tops have broken over. However, bulbs which are to be stored should be harvested after the tops have begun to break over but before the foliage has completely dried. If left in the grounds until the tops are dead, the bulbs are likely to develop roots (Thompson and Kelly, 1957). Onions for storage should be fully developed. Thickneck bulbs which results due to premature harvesting do not store well (Shoemaker and Teskey, 1939). Late harvesting leads to increase respiration, subsequent susceptibility to diseases and excessive sprouting during prolonged storage. Bhonde *et al.* (1983) reported that onions could safely be harvested one week after 50 per cent crop showed neckfall and 3 days field curing was desirable to improve the storage life of onions. They further reported that wind-row method of field curing and cutting the bulbs leaving 2 cm neck which helped close neck, reduced losses as much as 12 per cent during 5-month storage. The weather at the time of harvesting should be warm and dry. In traditional method of harvesting onions in New Zealand, onions are lifted at 60-80 per cent top down, the bulbs are field-cured, and the foliage is removed after curing, which is the simplest method and best compromise to ensure postharvest onion quality and successful storage (Wright *et al.*, 2001). Timing of harvest is an essential factor affecting the quality and storability of onion yield. An experiment was conducted in Finland by Suojala (2001) to determine the relationship between maturity stage and yield development in onion cv. Sturon. Generally little, if any, yield increase was recorded after plants reached 100 per cent maturity, but in some cases, bulb growth continued after complete fall-down of leaves. On the other hand, harvesting before 100 per cent maturity resulted in a yield loss of 0-45 per cent of final yield. Weight loss, and thus energy consumption during drying, was still reduced after complete leaf fall-down. Therefore, it may be concluded that delaying harvest up to 100 per cent maturity, or even longer, ensures that the highest yield and lowest drying costs. However, Henriksen *et al.* (2001) reported that harvesting later than 80-90 per cent top fall over reduces dry matter content and storage ability. Cutting off the onion leaves before harvest and lifting the crop for artificial drying instead of wind-rowing decreases dry matter content and total dry matter production. Increasing the nitrogen supply above 200 kg N per hectare decreased dry matter content in mature bulbs. During the storage period from November to July the dry matter content decreases slowly as a result of respiration loss. Wall and Corgan (1999) suggested that undercutting of onions should not be performed until just prior to harvest and that harvest should not be delayed for 15 days after maturity.

In India, the bulbs are usually harvested by hand pulling in small area, if soil is light. These are also harvested by hand implements or modified potato digger. Rainy season (*kharif*) onion is harvested when leaves are still green. The tops and roots are removed, and bulbs sent for market. In such crop, since the growth continues forced toppling should be taken up to stop the growth 15 days before harvesting. Onions along with tops are windrowed and kept in field for 2-3 days when tops are cut either in field or in shade if strong sun is there. In the Mediterranean and temperate climates, large-scale mechanical lifting and handling methods used for onion harvesting are suitable only for hard onions and short-day varieties of the Grano/Granex type are usually pulled and bagged by hand in the USA (Currah and Proctor, 1990). In Belgium, plants are usually harvested after ripening of foliage to obtain mature onion (Vanparys, 1999). According to Bottchear (1999) best time to harvest onions was 90 per cent bending when 40-50 per cent leaves were dried.

Tai *et al.* (1996) reported that in Republic of Korea, early maturing cultivars were harvested between 10 and 20 May and were marketed immediately while late maturing cultivars were harvested between 10 and 20 June and stored for 2-3 months before marketing. Onions were harvested at various dates by a traditional harvesting and field curing methods and by a cutting and direct harvesting methods, and dried at various temperatures and humidities in experiments in southeastern Norway by Solberg and Dragland (1999). The occurrence of translucent and leathery scales, leaf sprouting and other postharvest problems was examined after eight months storage. Late harvesting, high drying temperature, and a long drying duration produced the highest incidence of translucent scales. Direct harvesting methods with cuffing the leaves gave higher number of leaf sprouting than the methods with field curing and no leaf cuffing. Late harvesting and high drying temperature raised the number of leaf sprouting.

11.2.0 Curing

Purpose of curing (drying) is the removal of excess moisture from the outer skin and neck of onion which helps in reducing the infection of diseases. This also helps in minimizing shrinkage due to removal of moisture from the interior. Further, curing is an additional measure for the development of skin colour. It is also practised to remove field heat before onion bulbs are stored. Therefore, onion should be adequately cured. Bulbs are cured either in field or in open shades or by artificial means before or in storage. Onions are considered cured when neck is tight and the outer scales are dried until they rustle. The time length required for curing operation largely depends on the weather condition. In winter season, when temperature is low during harvesting, especially in North India, thorough curing is essential for 2-3 weeks alongwith tops. In winter crop, bulbs are cured in field for 3-5 days in windrow method. In rainy season, curing is done in sun. Onion bulbs are artificially cured by passing hot air at 46°C for 16 h when temperature is low. In Maharashtra, in rainy season crop, field curing is also done by windrow method. Bulbs are then placed in shade and cured for 7-10 days to remove field heat. This shade curing improves bulb colour and reduces losses during storage. This system of curing may be practised for multiplier onion also. Artificial curing by hot air blowers is useful in the case of continuous rains or low temperature.

Gorreapti *et al.* (2017) stated that curing is one of the most important postharvest management practices that are required for long term storage of onion with minimum losses. Curing can be done naturally on the field or artificially by using hot air. During curing physiological weight loss occurs to certain extent, however proper curing helps in improving storage life of onion by reducing postharvest losses that occur due to physiological weight loss, rotting and sprouting. Bhattarai and Subedi (1998) of Nepal recorded that average weight loss during storage was lower with curing (31.9 per cent) than without (43.9 per cent). They also noted that there was little difference in weight loss in cured onions irrespective of storage methods.

In New Zealand, Wright and Grant (1997) reported that additional water applied during field curing increased the proportion of bulbs with stained skins and rots. They also noted that heated air curing of bulbs reduced the incidence of rots regardless of harvest methods. Forced air drying also reduced skin staining in most harvest treatments. Solberg and Boe (1999) reported that in Norway, the loss due to watery scales were 2-10 per cent. Years with high precipitation in August and September had the highest frequency of watery scales. A positive correlation between rainfall during the 3 last days of field curing and watery scales was found. High air humidity during the drying process could also be related to high occurrence of watery scales. Solberg and Dragland (1999) noted that high humidity (R.H. 80-100 per cent) during artificial drying gave the highest frequency of translucent scales. No effect of humidity was observed on leathery scales, sunken spots, bacterial rots, leaf sprouting, and other parameters. High drying temperature (14 to 30°C) increased the number of translucent scales, sunken spots, bacterial rots and leaf sprouting. A long field curing period raised the average frequency of onion with leathery scales but not the frequency of translucent scales. The occurrence of onions with bacteria rot and leaf sprouting was not influenced by the duration of the field curing period.

Quality of onion bulbs was significantly affected by curing methods, storage conditions and duration. Curing with foliage resulted in significantly lower weight loss, sprouting and rotting (Nabi *et al.*, 2013). Eda and Kukanoor (2016) studied various methods for curing of onion bulbs (curing under forced hot air dryer, polytunnel, 35 per cent shade and 100 per cent shade with foliage and without foliage). Curing under 35 per cent shade with foliage had maximum marketable bulb, total soluble solids, ascorbic acid retension, reducing sugars, non reducing sugars and total sugar.

11.3.0 Yield

In general, in India the yield of winter (*rabi*) onion is around 25-30 tonnes per hectare and in rainy season (*kharif*) it is comparatively low. However, India's average productivity is low in comparison with China, USA, Netherlands, Korea, Spain and Japan. Average yield of some leading onion producing countries is presented in Table 18.

11.4.0 Storage

Onion is produced mainly in one season in temperate region and one or two or three seasons in tropical region as per the climatic conditions. Thus, a sizable quantity of onion is stored all over the world to fulfill the daily requirements of onion (Tripathi and Lawande, 2019). However, onion is a delicate and perishable commodity difficult to store for a long duration at room temperature, especially in tropical and subtropical countries due to its high water content and many other factors associated with it.

Table 18: Average Yield (t/ha) of Onion Producing Countries in 2018

Country	Yield (t/ha)
Guyana	73.11
USA	62.33
Republic of Korea	57.56
Taiwan	54.16
Australia	54.14
Spain	53.54
Chile	48.80
Canada	44.26
Japan	44.08
Sweden	43.43
Switzerland	40.51
Austria	40.03

FAOSTAT (2020).

In temperate climate, the environmental factors are more favourable for the storage of onion bulbs, since these are harvested in autumn and stored during low temperature period. In these countries, onions are also stored in cold storage especially for the purpose of seed production. But in tropical and subtropical countries like India, the crop is harvested in summer and stored for 4-6 months in ambient condition of high temperature and high humidity due to monsoon rains. During this period, the bulbs sprout very easily besides rotting due to fungal and bacterial diseases. In these countries, proper storage of bulbs is necessary both for consumption and also for seed production. In India, bulbs of rainy season crops are harvested in early winter and consumed during winter and summer months. Hence the problem of storage is not that acute as in the case of winter crop. Besides, the bulbs of rainy season crop after harvesting can be planted immediately for seed production.

The postharvest period of onion bulb has been separated into two well defined stages like rest period, when bulbs cannot sprout even under conditions favourable for growth and a dormant period when sprouting will occur under such conditions like high humidity and suitable temperature. Thomas and Isenberg (1972) have found from their experiments that sprout activation is controlled by a complex natural inhibitor-promoter interaction but sprout extension is controlled by natural gibberellins. Thus, there is evidence that a growth inhibitor may be a prime factor involved in dormancy process. Kato (1966) suggested that it is produced in the leaves during bulbing and translocated into the bulbs during senescence since defoliated onions sprouted much earlier in storage than those in which senescence occurs naturally in the field. It has been observed that this growth inhibitor was present in the bulbs after harvesting but it decreased gradually until the onset of sprouting, while both auxins and gibberellins increased around the beginning of sprouting period.

The storage losses in tropical region are very high (30-40 per cent) due to improper pre-and postharvest management and poor storage environment coupled with climatic conditions (Tripathi and Lawande, 2019). Ranpise *et al.* (2004) reported 5.2 to 32.4 per cent and from 6.8 to 54.8 per cent storage loss in onion grown during *rangda* season after 30 and 60 days of storage, respectively. Ilic *et al.* (2009) noted that prolonged storage in ambient conditions caused a significant decrease in marketable bulbs, up to 40-60 per cent, and an increase of the amount of sprouted bulbs, up to 30-50 per cent. Nabi *et al.* (2013) observed that the weight loss, sprouting and rotting percentage increased with increasing storage duration.

The varieties, nutrient management, time, quality and quality of irrigation, time and methods of harvesting, field and shade curing are some important pre-harvest factors effecting storability. Postharvest treatments such as irradiation, fumigation and storage environment such as type of structure, ventilation pattern, type of construction material, design, stake dimensions, temperature and humidity regime, packing material, season of storage are postharvest factors effecting storability (Tripathi and Lawande, 2019). In addition to these, there are many other factors which affect storage directly or indirectly. Dry matter content, neck thickness, quality of the outer scales, hardness of the bulb, its shape, *etc.*, are important. Among the diseases in the store, black mould (caused by *Aspergillus* sp.) and *Penicillium* rot are the most important. Other important storage diseases of onion are basal rot (*Fusarium oxysporum* sp. *cepae*), smudge (*Colletotrichum aeruginosa*) and bacterial soft rot (*Erwinia orotovora*).

Satish *et al.* (2002) stated that water loss, sprout loss and disease loss were the major storage loss variables responsible for onion losses during storage. They reported that during the 4 months of storage, water loss recorded a significant negative association with storage period and temperature, and significant positive correlation with relative humidity. They also noted that sprouting was negatively affected by storage period and temperature, and positively affected by relative humidity. However, they did not notice any clear association between disease loss and storage conditions.

Onion is generally stored at two temperature and humidity regimes, *i.e.* 0-2°C and 65-70 per cent RH and 25-30°C and 65-70 per cent RH. In tropical region, onion is stored at ambient conditions in different types of structures (Tripathi and Lawande, 2019). Yoo *et al.* (1996) concluded that onions could be stored for 4 months at 4°C in air or with 1 per cent O_2 or at 11°C with 1 per cent O_2 without excessive sprouting and flowering for planting in September or early October. Bulbs could also be stored up to 7 months for crossing or general seed production when kept at 4 or 11°C with 1 per cent O_2. Onions (cv. Walla Walla) stored in 0.5 per cent O_2 and 0.7 per cent O_2 were of better quality and had caused less neck rot immediately following CA storage (Silton *et al.*, 1997). Low storage temperature (0°C) greatly suppressed sprouting, but had no effect on root development. An additional one month of storage at a higher temperature (10-14°C) increased sprouting and decreased the quality of onions. All late maturing cultivars stored better than early maturing cultivars (Adamicki, 1998). Four onion varieties were tested for storability under 2 storage regimes: cold (–1 to –3°C, 70-80 per cent RH) and warm (18 to 20°C, 50-60 per cent RH). Storability of onion was better under the cold regime (–1 to –3°C, 70-80 per cent RH) and storability for 7 months was best when bulbs were stored in trays rather than in wooden boxes. Lowest storage losses under cold storage in trays was observed in cv. Ispanskii 313 (marketable yield after storage was 79.6 per cent) (Pirov, 2000). Okesh *et al.* (2015) observed that in ambient temperature (35°C ±2) and <75 per cent relative humidity the sprouting and rotting could not develop in onion bulbs. They suggested onion should be stored in well ventilated room. Depending on the cultivar, Ilic *et al.* (2009) found 1 to 6 per cent of sprouted bulbs on refrigerated cool storage at 0-2°C during 6 months of storage from November to April.

Kale *et al.* (1991) considered cultural factors such as nutrition and irrigation are critical for successful onion storage. Kato *et al.* (1987) reported that rotting of bulbs stored at room temperature from harvest until August increased with increasing application of N fertilizers. Rotting of bulbs at low temperature (2°C) from August to March, showed a similar trend but the rate of N application was not the main factor related to rotting under this condition. Shinde *et al.* (2016) reported that the total storage losses increases as nutrients levels increases and it was lowest in absolute control. Sankar *et al.* (2009) observed that the total loss of stored bulbs increased steadily as the period of storage was extended. They noted that the organic treatment combination of 3 per cent panchagavya + 50 per cent FYM + 50 per cent poultry manure registered the lowest total loss *viz.*, 30.57 and 32.71 per cent in the variety N 2-4-1 during crop I and crop II, respectively at 120 days after storage. They further reported that the inorganic treatment consisting of 100 per cent recommended dose of NPK fertilizers significantly varied from organic treatment and exhibited the highest total losses in both crops. They observed similar kind of response in sprouting and rotting per cent.

Marketable bulbs (per cent) of cv. Granex 33 after 1 month of storage at 25 to 28°C were greater with maturity with hand harvest than with machine harvest but was significant only at the 52 per cent maturity level. Storability tended to decrease with bulb maturity (Doyle and Bryan, 1988).

Onion contains various organosulfur compounds which have antibiotic and anticarcinogenic properties and flavonoid like quercetin which is a valuable natural source of antioxidants. The sugar concentration is associated with dormancy and storage life of onion, occurring as decrease in glucose, fructose and fructan, particularly towards the end of storage (Cho *et al.*, 2010). Cultivars with a high dry matter percentage and those with a larger number of dry scales showed potential for a longer storage period (Agic *et al.*, 1997). Shekib *et al.* (1986) worked on Egyptian onion cvs. Giza No. 6 and Beheri and observed that pungency increased with storage until sprouting occurred. Chang *et al.* (1987) reported that bulbs of cultivars with high dry matter and pyruvic acid contents and a low N content have the best bulb storage quality. Horbowicz and Grzegorzewska (2000) observed a negative correlation between sugar content and storability of onion. Price *et al.* (1997) observed that apart from a 50 per cent loss of quercetin 4'-o-monoglucoside during the initial drying process, little change in content and composition over 6 months of storage. Sharma *et al.* (2014) investigated the post-storage deterioration of onion under ambient conditions (temperature 20-25°C and relative humidity 60-80 per cent), subsequent to a cold storage period of 8 months, with an emphasis on changes in chemical composition of flavonols, sugars and phenylalanine. They also studied the total phenolics, total flavonoids and antioxidant activity of onion bulbs. They reported that at the 4th week of post-storage, the total quercetin content was three times [2.725 + or -0.097 micro mol/g fresh weight (FW)] than its initial post-storage value (0.877 + or -0.085 micro mol/g FW) and they noted visible signs of sprouting and decay. However, they found the highest content of total quercetin in the onion bulb (3.209 + or -0.350 micro mol/g FW) at the 8th week, and by the 10th week, the onion bulbs were severely decayed and were reduced to roughly half of the initial size. They observed that the fructose and glucose content decreased continuously during post-storage from 0.174 + or -0.007 to 0.024 + or -0.004 mmol/g FW and 0.368 + or -0.04 to 0.014 + or -0.006 mmol/g FW, respectively, whereas the sucrose content was almost constant. They reported that the total phenolics, total flavonoids and antioxidant activity were increased and correlated well with total quercetin content during the post-storage period. Gorrepati *et al.* (2018) reported that during storage, rate of respiration and total phenol content increased up to 60 days of storage and then decreased up to 90 days. Ilic *et al.* (2009) reported that the sugars content (4.5-10.5 per cent) and vitamin C contents (12.4 to 14.9 mg/100 mg) slightly decreased after long term storage (depends of storage temperatures and cultivars). However, they observed a little change between the initial levels and the levels after 6 months in dry matter content. Nabi *et al.* (2013) found maximum dry matter (17.5 per cent) and TSS (11.5 per cent) in bulbs cured with foliage as compared to 15.7 per cent DM and 9.36 per cent TSS with curing without foliage accordingly. They noted maximum DM (21.2 per cent) and TSS (14.9 per cent) in cold stored bulbs was followed by 15.65 and 13 per cent DM with 9.44 and 6.65 per cent TSS in mud and cemented room storage accordingly.

In Argentina, Piccini *et al.* (1987) irradiated onion cv. Valenciana Sintetica 14 with X-ray (0.03 KGy) at 30 days after harvest and stored under cover in an open shed for 268 days. They observed 5.8 per cent sprouting in irradiated bulbs compared to 80 per cent in untreated control. The Malate content increased and reached at peak in irradiated onions (2.57 g/100 g) on 179th day and then declined, while in control onions it continued to increase to 2.72 g/100 g at 268 days. Postharvest application of B as borax is recommended to minimize losses and sustain quality of onions in storage (Alphonse, 1997). Forced air pre-drying for 15-20 days at room temperature is an essential procedure to reduce freezing injury and sprouting, then onion bulbs can be stored at 0°C for 6 months to control sprouting and decay. Bacterial soft rot caused by *Erwinia* and *Pseudomonas* is the main postharvest disease when the bulbs are infected with the bacteria and stored at room temperature (Cho *et al.*, 2010).

In Yugoslavia, cultivar Moldavski showed the greatest potential for storage (Agic *et al.*, 1997). In storage tests in Germany, cultivars Copra, Staccato, Vitesso and Athos showed the best storage ability (Arold, 1998). Ahmad *et al.* (2005) recorded maximum sprouting (99.0 per cent) and total losses (58.48 per cent) in Swat-1 cultivar at the 3rd month of storage. They noted increase in rotting of bulbs in Afghan white from 0 to 17.38 per cent during three months storage period. They recorded maximum weight loss (14.93 per cent) at the 3rd month of storage in Swat-1; while in Afghan white maximum they noted weight loss of 9.42 per cent at the 2nd month of storage.

In India, storage study under room temperature from June to October indicated that bulb rotting, sprouting and weight loss increased with increasing storage duration. Rotting was highest in Arka Niketan, followed by PW-1, while it was lowest in Hisar-2. No sprouting was observed in any of the cultivars up to 60 days of storage, and sprouting was recorded only after 90 days of storage. The lowest sprouting percentage was in IIHR-Yellow, followed by RHR-White (Batra *et al.*, 2000). Seven cultivars of onion were stored in nylon-netted bags at room temperature for 2 months and storage losses were studied by Sakhale *et al.* (2001). The losses due to rotting were found to be highest in 'Decco' cultivar, and lowest in Parbhani Local followed by Tadole Local, Pusa Selection and Red Creole. Mohanty *et al.* (2002) reported that Pusa Madhavi, Arka Niketan, Punjab Red Round, Agrifound Dark Red, Arka Pitamber and Agrifound Light Red produced small to medium sized bulb with thinner neck depicting better storage quality. They suggested Agrifound Dark Red and Arka Niketan for commercial cultivation in rainy season owing to their better storage quality, medium bulb and moderately high yield. In Maharashtra (India), Ranpise *et al.* (2004) evaluated 198 local strains and varieties along with Baswant-780 (local control) and identify the best genotype with the best storability and bulb quality during rangda season. They found that Selection Nos. 175, 151, 176, 186 and 168 recorded significantly minimum storage losses of 28.0, 30.0, 32.0, 32.0 and 33.0 per cent, respectively, after 120 days of storage compared with other selections and Baswant-780 (35.0 per cent). Gorrepati *et al.* (2018) evaluated onion varieties (Bhima Kiran, Bhima Raj, Bhima Red, Bhima Shakti, Bhima Shubra, Bhima Shweta and Bhima Super) for storage losses during *rabi* season. All the varieties were grown under similar conditions and stored at ambient conditions in a modified bottom and top ventilated storage structure. Total losses were found significantly less in Bhima Kiran (26.66 per cent) and Bhima Shakti (35.87 per cent) after four months of storage. Narayan *et al.* (2019) recorded minimum weight loss after 2 months of storage in genotypes oN-1388 (3.33 per cent).

Satodiya and Singh (1997) recorded that bulb neck thickness was significantly correlated with dry matter content, physiological weight loss and percentage of marketable bulbs, but significantly negatively correlated with percentage of rotten bulbs. Sprouting and rooting were the main causes of loss during the long-term storage of onions.

In India, different storage methods are practised by the farmers. These are bulk stored in special houses with thatched roof and sides covered by bamboo sticks with provision for good air circulation. In North India, the side are also covered with gunny cloth. Onions are stored in these sheds by spreading them on dry and damp proof floor and/or racks. Periodical turning of bulbs or removal of rotted, damaged and sprouted bulbs should be done. Well-ventilated room with racks or tiers having two or three layers of bulbs would be desirable for proper storage. Maximum retention of quality up to 100 days at ambient temperature in onion was observed in the two-tier wire mesh structure, thin conventional bamboo structure or single tier by Maini *et al.* (1997). The study of Kumar *et al.* (2005) revealed that, farmers store onions on kutcha floors, pucca floors, and bamboo mats; the cost of storage is highest on marginal farms and on bamboo mats; the highest percentage losses are highest on bamboo mats; and the producer's margin and the marketing efficiency are highest in the direct marketing channel. Tripathi and Lawande (2013) observed highest storage loss in the method of traditional storage where as minimum physiological and rotting loss in top and bottam ventitated storage. In Peshawar, Nabi *et al.* (2013) noted lowest rate per month weight loss (1.95 per cent), sprouting (2.4 per cent) and rotting (0.4 per cent) in cold stored bulbs, while observed maximum weight loss (59.3), sprouting (59.5) and rotting (31.3 per cent) in bulbs stored cemented room. They obtained lowest weight loss, sprouting and rotting (0 per cent each) in cold store during the 1st month storage duration, while it was found highest percentage (13, 2.5 and 1.67, respectively) in cemented room. Similarly, after four months storage, they noted the minimum percentage of weight loss (6 per cent), sprouting (9.6 per cent) and rotting (1.7 per cent) in cold stored bulbs and maximum weight loss (98 per cent), sprouting (100 per cent) and rotting (70 per cent) in bulbs stored in cemented room.

Gupta *et al.* (1991) advocated the following steps for reducing the onion storage losses:

1. Varieties with good storage qualities should be grown.
2. An improved package of practices should be adopted to produce the best possible quality of onion.
3. Harvesting of onion should be done at correct maturity stage of the crop.
4. Proper curing of the onions should be done before storage.
5. Tops should be cut 2 to 2.5 cm above the bulbs so that the neck is properly closed.
6. Sorting and grading should be done before the bulbs are stored.
7. Bulbs should be stored in well-ventilated structures.

Similarly, Tripathi and Lawande (2019) opined that use of recommended varieties, production technologies, curing, ventilated well-designed permanent structure and irradiation may reduce these losses up to 20-25 per cent with economic feasibility. Amit Kumar *et al.* (2017) examine the nature and extent of postharvest losses in onion supply chain in the Ambala district which is major onion district of Haryana. Various losses at field levels were doubles, bolters, rotted bulbs, drying, bulbs injuries, de-topping, improper packing, *etc.* Similarly, the postharvest losses were due to faulty storage, lack of adequate transportation, drying, improper handling of the produce at the time of marketing, rotted bulbs and poor packing facilities. Total losses in the supply chain were estimated to be 10.86 percent. Sharma *et al.* (2017) examine the economics of storage in onion supply chain in the Jhunjunu

Seedling Raising and Field Growing of Onion

Seed Sowing and Seed Germination

Set Sowing and Set Germination

Transplanting of Onion Seedlings

Field View of Onion

Irrigation, Mulching and Harvesting of Onion

Different Irrigation Methods Practices in Onion Field

Mulching in Onion

Harvesting of Onion

Curing, Grading, Packaging and Storage of Onion

district, which is a major onion growing district of Rajasthan. They reported that the onion growers stored 54.66 per cent of the total quantity of marketed surplus at farm level, of which 14.68 per cent of onion is lost during storage period owing to postharvest losses. Onion growers received maximum return from onion marketing during October month (34.24 per cent) due to shortage of produce in market. They obtained an overall average profit of 9.98 per cent during six months storage period. Majority of the respondant said that the major reasons for storing onion by farmers were for home consumption (97.6 per cent) and to reap benefits of higher prices (92.7 per cent). Most of the farmers (58.5per cent) adopted improved methods of storage. However, about 90.2 per cent farmers reported that lack of knowledge about proper scientific methods for storage of onion was the major problem faced by farmers.

12.0 Diseases and Pests

Onion is affected by many diseases and pests which cause considerable damage to the crop unless they are controlled immediately after detection.

12.1.0 Diseases

Onion is attacked by many diseases which cause yield losses and result in lowering the quality and export potential of the produce.

Fungal Diseases

12.1.1 Purple Blotch

This is caused by *Alternaria porri*. Ajrekar (1920-21) reported this disease from Bombay caused by *Macrosporium* species, but in the following year, he again reported that same diseases, now attributing to *Alternaria* sp. Pandotra (1964, 1965) and Gupta *et al.* (1983) confirmed the pathogen as *Alternaria porri*.

The common symptoms occur on leaves, scapes and pseudostem as small, sunken, whitish flecks with purple coloured centres. Further large purple area develops forming dead patches. The intensity was observed from 5-25 per cent on bulb crop and 10-45 per cent on seed crop. Affected plants showed delayed bulb formation and maturation (Salunkhe *et al.*, 2017). Lakra (1999) reported from Hisar, India, that purple blotch of onion (cv. Hisar 2) is a serious constraint in onion seed production. The intensity is epidemic every year during March. The severe infection reduced the seed yield/plant, compared to healthy plant. The disease occurs under favourable environmental conditions (temperature 28-30°C and relative humidity 70-90 per cent). The disease appears from middle of February to April. The leaves and stem fall down from point of attack. Quadri *et al.* (1982) reported highest incidence during *kharif* (62 per cent) as against 38 per cent in *rabi* season. The pathogen survives in the crop debris as a dormant mycelium and in the form of chlamydospores and can remain viable for 12 months. Spore formation increases during humid nights and leaf wetness periods greater than 12 hours. Spores dispersed by wind, rain, sprinkler irrigation and spraying (Salunkhe *et al.*, 2017).

Three summer ploughing reduced the disease severity by 72 per cent with 24.4-29.8 per cent higher bulb yield, compared with unploughed plots. The early sown crop had lower disease index and higher bulb yield, compared with the crop sown later than 15th September in India (Gupta and Pathak, 1987).

Kushal *et al.* (2015) evaluated tolerant/resistant onion varieties purple blotch disease caused by *Alternaria porri*. They reported that Bellary Red, L-819 and Arka Kalyan as resistant whereas Anand-2 and Agrifound Dark Red were susceptible variety to purple blotch.

Gupta *et al.* (1981) reported that dithane M-45 (1000 ppm) significantly inhibited the growth of pathogen. Quadri *et al.* (1982) and Gupta *et al.* (1987) also reported that dithane M-45 at 0.25 per cent concentration as effective in controlling the disease. According to Georgy *et al.* (1986) from Egypt, the Ridomil group especially ridomil MZ (metalaxyl + mancozeb) proved most effective in reducing disease severity of downy mildews and purple leaf blotch. Srivastava *et al.* (1999) noted that mancozel @ 0.25 per cent was the most effective in reducing purple blotch disease incidence and intensity as well as increasing yield and gave a higher cost benefit ratio. Ghure *et al.* (2000) reported that along with mancozel @ 0.25 per cent at 15 days interval cuprous oxide 75 WP @ 0.25 per cent was also effective in controlling the disease and increasing bulb yield. Aujla *et al.* (2013) found Score 25 EC (difenoconazole) and Nativo-75WG (trifloxystrobin 25 per cent + tebuconazole 50 per cent) were most effective as these completely inhibited the mycelial growth at 0.1 per cent. Rao *et al.* (2015) reported that cymoxanil 8 per cent + mancozeb 64 per cent WP @ 2500 ppm and mancozeb 70 per cent WP @ 2500 ppm were effective in reducing the disease severity by 54.86 and 52.88 per cent with high yield. Gangwar *et al.* (2016) observed that chlorothalonil 75 WP (1.5 g/l) and mancozeb 75 (2.0 g/l) when sprayed along with surfactant (0.5 ml/l) showed minimum per cent disease index (5.38 and 7.96 per cent, respectively) and disease incidence (11.80 and 15.47 per cent, respectively) for purple blotch. Mandi *et al.* (2020) reported that seed treatment with Vitavax power @ 0.2 per cent along with foliar application of Tebuconazol 25 EC @ 1 ml/l was most effective in reducing purple blotch disease in onion. Maximum disease control was recorded in Tebuconazol 25 EC @ 1 ml/l with PDC of 56.12 per cent.

According to Mohan *et al.* (2002) leaf blight disease of onion caused by *Alternaria porri* can be managed by ecofriendly method without using any chemical. Two sprays of palmarosa oil (0.1 per cent) first at the time of appearance of the leaf blight disease and the second 15 days later was found to check the disease besides increasing the yield.

12.1.2 Stemphylium Blight

This disease is observed on onion leaves as well as on flower stalk. It is a devastating disease limiting the quality and quantity of both bulb and seed. Stemphylium blight caused by *Stemphylium vesicarium* (Wallr.) Simmons has become an economic threat, especially in Northern and Eastern India (Mishra and Singh, 2017). It is frequestly occurs with purple blotch disease. Gupta *et al.* (1983) reported that the range of intensity of the disease was from 20 to 90 per cent in seed crop and 5 to 40 per cent in bulb crop. Further, they observed that infections occur on radical leaves of transplanted seedlings at 3-4 leaf stage during late March and early April. Infection appears as small yellow to pale orange spots or streaks in the middle of leaves/flowerstalk on one side. The symptoms of this disease includes light yellow to tan, water soaked lesions formed which turn into elongated spots that often extended towards leaf tips. The spots frequently coalesce into pathches resulted in the blighted appearance of leaves. As conidiophores and conidia of the pathogen develop on lesions it turn into light brown to tan purple at the centre and later to dark olive brown to black. The disease can prematurely defoliate the crop which can deteriorate the bulb quality and make the crop more susceptible to secondary diseases like storage rots. The pathogen overwinters in pseudothecia produced on both diseased and symptomless leaves left in and on the soil. Secondary spread of the disease takes place by conidia released from leaf lesions. The disease development is favourable by warm temperature (18-25°C) and high humidity (78-91 per cent) conditions and long period of leaf wetness (16 hours or more) (Salunkhe *et al.*, 2017).

According to Efath *et al.* (2016), 1 November sowing of crop resulted in significantly lower disease intensity of major disease including *Stemphylium vesicarium*. The field studies indicated that Dithane M-45 at 0.25 per cent along with sticker triton can control the disease (Gupta *et al.*, 1983). In greenhouse experiment, application of Ridomil gold plus on onion plants resulted in significant reduction of disease severity percentage (Hussein *et al.*, 2007). Mishra and Singh (2007) reported that combination products azoxystrobin 25 per cent + flutriafol 25 per cent SC and fluopyram 20 per cent + tebuconazole 20 per cent SC can be recommended for the management of *Stemphylium* blight of onion under field conditions.

Hussein *et al.* (2007) reported that the highest inhibition of *Stemphylium vesicarium* mycelial growth was achieved by *P. fluorescens, B. subtilis* and *T. harzianum*. The in vitro study revealed that *Trichoderma harzianum* strain PBAT-21 gave highest mycelial inhibition (60 per cent), followed by *Pseudomonas fluorescens* strain PBAP-27 (57.45 per cent). The field experiments, showed that *T. harzianum* was most effective, giving almost 50 per cent disease control over check after spray at 45 days after transplanting, followed by *P. fluorescens*. The yield and proportion of A-grade bulbs were also higher in the *T. harzianum* treatment (Mishra and Singh, 2017).

12.1.3 Basal Rot

Rotting of bulbs is caused by *Fusarium* fungus and bacteria. Basal rot caused by *Fusarium* is a widespread disease. It is an economically significant disease has been causing major yield loss in the most onion cultivating regions of the world. Gupta *et al.* (1983) reported that it is common in Bihar State near Nalanda district.

The causal organism was identified as *F. oxysporum* f. sp. *cepae* Merlatt. Mathur and Shukla (1963) reported a similar disease from Udaipur. Common symptoms of the disease are wilting and rapid dying-back of leaves from tip as the plants approach maturity and roots turn pinkish. Later on most of the roots rot off. The bulbs become soft and when cut, a semi-watery decay is found advancing from base of the scales upward.

Cramer (2000) reported that fusarium basal plate rot (FBR) is an important soil-borne disease of onion worldwide. The causal organism infects the basal stem plate of the bulb and eventually kills the entire plant through degradation of the basal plate. *F. oxysporum* f. sp. *cepae* infections in dormant bulbs during storage allow secondary infections to occur. The primary method of infection by *F. oxysporum* f. sp. *cepae* is through direct penetration of the basal stem plate. Infection can also occur through wounded tissue particularly roots and basal portions of bulb scales.

The most cost-effective methods of control are crop rotation and host plant resistance. Five-year rotation with unrelated crops is recommended as the pathogen survives in the soil (Gupta *et al.*, 1983). Tanaka *et al.* (1996) reported that in Japan the pathogenicity and survival of *F. oxysporum* propagules in flooded soil decreased more rapidly than non-flooded soil after 2 weeks. It is concluded that satisfactory control can be achieved by flooding in the non-growing season.

Sokhi *et al.* (1974), after screening 47 cultivars, found FBR-resistant in some cultivars, *viz.*, Telgi Red, White Large, Poona Large, Patna Red, N-257-7-1, Udaipur 103, *etc.* Taylor *et al.* (2013) observed that cvs Ailsa Craig Prizewinner and White Lisbon had the highest levels of resistance to *Fusarium oxysporum*. Gupta *et al.* (1983) from laboratory tests indicated that bavistin and benlate were at par and significantly checked the fungal growth. Carbendazim, mancozeb (63 per cent) + carbendazim (12 per cent) and mancozeb were found to be the most effective with complete inhibition of mycelial growth of the fungus at 500, 1000, 1500 and 2000 ppm followed by trifloxystrobin (25 per cent) + tebuconazole (50 per cent). Among the oil cake extracts neem was found the most effective in inhibiting the mycelial growth of *F. oxysporum* followed by mustard (Saravanakumari *et al.*, 2019).

Malathi (2015) reported that biological control by *Trichoderma* sp. and *T. harzianum* (TH 3) gave the greatest (83 per cent) inhibition and *Pseudomonas* sp. (Pf 12) exerted significantly the greatest (75 per cent) reduction of mycelial growth of *F. oxysporum* f. sp. *cepae*. Priti *et al.* (2018) tested the biopesticide formulations of *Bacillus pumilus* (BPS4) and *Pseudomonas monteilii* (PMS2) in combination with *Trichoderma harzianum* (IIHR-TH2) against *Meloidogyne incognita* (Kofoid and White) Chitwood and *Fusarium oxysporum* f. sp. *cepae* Schlecht (FOC) infecting onion under nursery and field conditions. Combination treatments of biopesticides were more effective and reduced the disease incidence of FOC (63.5 per cent to 72.0 per cent), in the field experiment. A pathogenic fungus, *Fusarium proliferatum*, known to infect a wide range of crop and ornamental plants, is causing bulb rots disease in onion. This pathogenic fungus has lethal effect on onion plantlets. The study revealed that *T. harzianum/H. lixii, T. viride* and *T. asperellum* used in combination restrained the infection and enhanced the onion yield Geetika and Ahmed (2019).

12.1.4 Downy Mildew

This disease is caused by *Peronospora destructor*, especially in high humid locations. In India, it is limited to temperate regions mainly in Jammu and Kashmir and Uttarakhand. This disease can cause bulb yield losses up to 12-75 per cent. The primary infection of this disease is through infected planting materials. The secondary spread is due to wind-borne conidia. The fungus may also overwinter in volunteer plants as oospores. Downy mildew can develop an epidemic form very quickly if humidity and temperature conditions (1.5 to 7 hours of leaf wetness and 6 to 27°C) are favourable. Spores can travel long distances in moist air but are quickly killed by dry conditions (Salunkhe *et al.*, 2017).

On the surface of leaves or flowerstalk violet growth of fungus is noticed which later becomes pale-green yellow and finally the leaves or seedstalk collapse. In storage, bulbs of infected plants become soft, shrivelled and outer fleshy scale becomes amber in colour. Gonzalez *et al.* (2011) identified crop rotation, crop vigour and plant density as major factors affecting downy mildew variation in the field. This disease significantly reduced the seed yield of onion. Late planting with low density had significantly lowered the incidence and severity.

The control measures outlined by Rudolph and Wolf (1986) are (i) a crop rotation with a 4-year break in onion cultivation, (ii) an appropriate choice of site, (iii) good weed control and field hygiene, (iv) removal of primary infected onion plants, (v) planting the mother bulbs in spring rather than in autumn, (vi) postharvest thermal treatment of mother bulbs, and (vii) chemical measures, which are considered particularly important in onion set and mother bulb crops.

Yarwood (1943) recommended heating for four hours at 41°C to destroy the fungus. Jones and Mann (1963) could not control the disease by spray or dust. Doorn *et al.* (1954) reported that the disease can be controlled by spraying with 0.2 per cent zineb applied at the rate of 550 l/ha. Varietal differences exist in the degree of susceptibility (Choudhury, 1967). A preliminary model, able to forecast primary infection of onion downy mildew (ONIMIL) caused by *Pernospora destructor* was developed by Battilani *et al.* (1995) in Italy. System analysis was applied to elucidate the relationships between the factors influencing the establishment of primary infection. This model was able to determine for each day the probability of *P. destructor* establishing an infection on onion and its infectivity level, compared with the maximum. Forecasting model DOWNCAST (for downy mildew due to *P. destructor*) is also used in Netherlands (Meier, 2000).

12.1.5 Onion Smut

Onion smut is caused by fungus *Urocystis cepulae*, which lives in soil. It is observed in the area where temperature remains below 30°C. Since the fungus remains in soil, disease appears on the cotyledon of the young plant soon after it emerges. Smut appears as elongated dark, slightly thickened areas at the base of seedling. The black lesions appear near the base of the scales on planting. The affected leaves bend downward abnormally. On older plants numerous raised blisters occur near the base of the leaves. The lesions on plant at all stages often expose a black powdery mass of spore (Walker, 1969). The incidence of visible black mould caused by *A. niger* on the neck, middle and base of onion bulbs was evaluated by Sinclair and Latham (1996) in the brown-skinned fresh market cultivar Creamgold, and the white-skinned dehydrating cultivar Southport White Globe. Black mould occurred all over the bulbs, the incidence was lower in the neck region. Infection levels were consistently higher in Creamgold (77.5 per cent) than in Southport White Globe (46.6 per cent). Infection increased with increasing bulb size. N topdressing was evaluated in a situation where N levels were already high. Rates up to 150 kg N/ha had no clear effects on black mould incidence and no significant influence on bulb yield or soluble solids concentration. The fungus becomes inactive at soil temperatures of 26.7°C and above (Nath, 1976).

The disease can be controlled by treating the seeds with 55-85 g of arsan per 4.5 kg of seed before sowing (Thompson and Kelly, 1957). Application of granule formulation containing fungicides in the furrow with the seed have been successful (Newhall and Brann, 1960). Treatment of the sets with thiram along with methocel sticker proved useful (Larson and Walker, 1953).

12.1.6 Black Mold

The disease is caused by the fungus *Aspergillus niger*. This fungus cause high storage loss. High humid and warm weather favours the disease development. The fungus requires an optimum temperature range of 28-34°C for its growth. The fungus survives in air, soil, plant debris and in infected seed. In onion, the initial infection always starts through neck tissues. Infected bulbs show black discolouration around the neck and affected scales get shrivelled. Masses of powdery black spores develop as streaks along veins on and between outer dry scales. Infection may advance from neck into central fleshy scales (Salunkhe *et al.*, 2017).

The occurrence of black mould due to variables (i) plot characteristics, (ii) crop characteristics and management, (iii) postharvest handling, and (iv) disease indices was reported by Konijnenburg *et al.* (1997) from Argentina. They suggested that some crop management patterns could modify the incidence of the disease.

12.1.7 Blue Mold

The disease is caused by *Penicillium* spp. The fungus can be found in soil, plant and animal debris or in senescing tissues. The pathogen grows well at 21-25°C and under moist conditions. Ifection of bulbs is usually through damaged tissues caused by bruishing, freezing injury or sunscaid. Initially pale yellowish lesions and watery soft spots develop on bulbs. The affected areas are soon covered with characteristics blue-green spores. Fleshy scales may show water-soaking and a light tan or grey colour when affected bulbs are cut open. As the decay continues, bulb may become soft and tough or may develop a watery rot. Severely affected bulb produces a musty odour (Salunkhe *et al.*, 2017).

12.1.8 Botrytis Neck and Bulb Rot

This disease is caused by fungus *Botrytis allii* and *B. squamosa*. The fungus persists in the form of sclerotia on dead onion as well as in the soil. The sclerotia germinate in moist weather and produce airborne conidia, which land on tissue, germinate, and infect when conditions are favourable. The greatest incidence of infection occurs when cool (10 to 24°C) and moist weather prevails. Botrytis bulb rot generally appears during storage, though infection originates in the field. Symptoms generally begin at the neck. The affected tissue softens, becomes water soaked, and turns brown. In humid weather, a grey fungal growth appears on rotted scales and mycelia may develop between scales. Eventually sclerotia developed in neck region and sometimes between scales (Salunkhe *et al.*, 2017).

12.1.9 Onion Smudge

This disease is caused by the fungus *Colletotrichum circinans*. The fungus survives in the soil on the remains of onion leaves. In favourable season it produces spores to infect the next crop. Damp soil and and soil temperature above 20°C favoures the disease development. Humid weather play key role in fungus spore germination. Onion smudge can be developed in field, storage and transit. Dark green to black spots of variable size and shape are developed on the outer scale. The spots may be homogeneous in appearance or may consist of numerous individual stromata scattered miscellaneously or arranged in concentratic rings. The disease is most common in white varieties of onion (Salunkhe *et al.*, 2017).

Bacterial Disease

12.1.10 Bacterial Soft Rot

This disease is common in stored bulbs. Bacteria responsible for the soft rot disease of onion includes: *Erwinia carotovora, E. chrysanthemi, Pseudomonas gladioli* and *Enterobacter cloaceae*. The pathogens are soil borne and may be spread through irrigation water. Bacteria enter through wounds or senescent leaves in the bulb. Free water is essential for entry and spread of the bacteria. High temperature (>25°C) and humid condition favours the growth of the bacteria. Soft and water soaked area form in one or more of the inner fleshy scales of the bulb. Affected tissues are yellow initially, turing brown as the disease progesses and the whole interior of the bulb may break down giving out foul smell (Salunkhe *et al.*, 2017). The rot begins at the neck of the bulb, later it gives offensive smell through the neck when squeezed. The onion bulbs with green neck and not properly cured are often affected, particularly in humid weather. Therefore, proper curing and rapid drying at harvest time are essential or all the bulbs showing damage should be discarded before storing.

O'-Garro and Paulraj (1997) screened 19 commercial onion genotypes for leaf blight resistance and recorded 2 cultivars H-942 and H-508, free of symptoms and had restricted bacterial growth in plant. Sendhilvel *et al.* (2001) studied the varietal reaction of onion bulbs on the incidence of soft rot and reported that there was a direct correlation between quantity of pyruvic acid content and per cent disease incidence due to *Erwinia carotovora* var. *carotovora*. The pyruvic acid content (7.53 µmole/g) was more in cv. Patna White, which showed severe disease incidence (35.98 per cent). In case of country onion the pyruvic acid content was very less (2.19 µmole/g) and the disease incidence was also very less (5.03 per cent).

12.1.11 Brown Rot

This disease is caused by *Pseudomonas aeruginosa*. It is a storage disease of onion. This disease has a characteristics dark brown discolouration in bulb scale. The rot developed below the epidermis and adjacent cells. Browing of the inner scale along with rotting is the main symptom of the disease. Rotting starts from inner scale and spread to outer scale. Apparently the infected bulb seems to be healthy. However on pressing white ooze comes out from the neck.

12.1.12 Slippery Skin

It is a storage disease of onion caused by *Pseudomonas allicola*. The internal scale show brown water soaked cooked appearance on cut opening. When such bulbs are pressed from the bottom, the central core may eject. Infection progess downward and bulb become soft and rotten. At room temperate, the mature bulb may rot completely within 10 days of infection.

Viral Disease

12.1.13 Onion Yellow Dwarf Virus

The Onion yellow dwarf virus (OYDV) is a member of family Potyviridae, genus *Potyvirus*. It is a monopartite, single-strand positive sense RNA genome encapsulated in flexuous filamentous particle. It is transmitted by aphid (*Myzus persicae* and other) in a non-persistant manner or mechanically (Gawande and Singh, 2017). The diagnostic symptoms are severe stunting of the plants, dwarfing and twisting of the flowerstalk. The affected leaves and stems change their normal green colour to various degrees of yellow and leaves tend to flatten and crinkle, and as a result bend over (Gupta *et al.*, 1983). The highest percentage incidence of OYDV positives were observed from Maharashtra (96 per cent) followed by Gujarat and Madhya Pradesh. To control vector of the disease, spraying of rogor or malathion is useful. Healthy bulb should be taken for seed production and healthy seed should be sown.

12.1.14 Iris Yellow Spot Virus

Iris yellow spot virus (IYSV) causes significant losses in bulb and seed crops of onion. The yield reduction may be up to 100 per cent. The diagnostic symptoms are appearance of chlorotic spindle or diamond shaped lesions on the leaves and scapes. The size of the lesions increased and formed necrotic patches on leaves and scapes, as the disease progresses. The disease is more harmful to the seed crop. It is transmitted by thrips and *Thrips tabaci* is the major vector. The Asian and European IYSV formed separate clusters (Gawande and Singh, 2017). The crop needs to monitor strickly for incidence of vectors. Fipronil, Acepahate, Dicofol, Carbosulphan and Profenophos are commonly used to manage the vectors.

12.1.15 Other Diseases

Both commercial cultivars and lines form the USDA breeding programme were evaluated for pink root (*Phoma terrestris*) incidence early in the season and for pink root severity later in the season by Coleman *et al.* (1997) in organic soils. The cultigens with fewest symtoms were Sweet Sandwick, Keepswett II, Spartan Banner 80 and inbred line MSU6785B.

Patale (2019) isolated and identified fungi infected samples of onion bulbs. Those included *Aspergillus niger, A. flavs, Alternaria porri, Fusarium oxysporium, Botrytis allii, Penicillium* sp., *Macrophomina phaseolina, Cladosporium alli, Rhizopus stolonifer* and *Chaetomium* sp. All of them were pathogenic. *Trichoderma* sp. was observed to as microbial agent against the pathogens of onion and garlic.

Due to infection by stemphylium blight, purple blotch and colletotrichum blight pathogens, Srivastava and Verma (2000) observed that chlorophyll contents and minerals were utilized by the pathogens during pathogenesis.

Hayden and Maude (1997) observed that pre- and postharvest treatments (singly or in combination) were useful in order to control the most important fungal pathogens (neck rot black mould) of stored temperate onion bulbs. Seed treatments

Diseases of Onion

Purple Blotch Infection   **Onion Smut infection**

Black Mold Infection   **Blue Mold infection**

Botrytis Neck (Left), and Bulb Rot (Right) Infection

Bacterial Soft Rot Infection

Diseases of Onion

Slippery Skin Infection **Stemphylium Leaf Blight Infection**

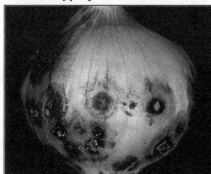

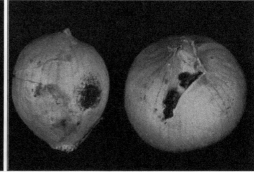

Onion Smudge Infection

Onion Yellow Dwarf Virus Infection

Iris Yellow Spot Virus (IYSV) infection

with fungicides control seed-borne *Botrytis* neck rot and damping-off diseases and seed furrow treatments with fungicides control *Fusarium* and *Pythium* damping off onion smut as reported by Lorbear (1999) from USA. Abdel-Momen *et al.* (2000) observed in field experiments at Malawi Research Station, Egypt that the synthetic lucerne saponin was the most effective saponin source in reducing white rot (59.16 per cent efficacy). The highest yield was recorded with saponin treated plants while the lowest from non-treated control. Chethana *et al.* (2000) reported that dithane M-45 (0.25 per cent) proved to be the most effective in controlling the leaf blight disease of onion followed by difencozole (0.20 per cent). While, on the basis of studies conducted for five years by Rahaman *et al.* (2000) it is evident that the leaf blight disease of onion can be actively controlled by giving 4 sprays of mancozel (@ 0.03 per cent) coupled with monocrotophos (@ 0.05 per cent) at 15 days interval commencing from 45 days after planting with a cost benefit ratio of 1 : 8 : 9.

Patil *et al.* (2016a) surveyed the major onion growing areas of Karnataka to know the incidence and severity of twister disease of onion. Their study revealed that severity of this disease varied from locality to locality. The severity of disease was also dependent on cropping pattern, environmental conditions prevailing in different localities and genotype grown. They observed *Colletotrichum* spp. in 78 fields followed by *Fusarium* sp. in 50 fields and in combination of both *Fusarium* spp and *Colletotrichum* spp. with 34.5 fields; whereas, they also observed *Sclerotium* sp. (16 per cent) and *Meloidogyne* spp. (6 per cent). Patil *et al.* (2016c) characterized the causal agents of twister disease of onion. *C. gloeosporioides* varies with color cream, pink and grey; where as *C. acutatum* color pink, brown, yellow and cream and regular to irregular margin and fluffy course to smooth growth with or without sectoring producing good sporulation; where *F. oxysporum* had white, cream and pink color exhibiting their variability. Patil and Nargund (2016) studied the metabolomic changes in onion incited with twister disease. They reported that the reducing sugar, nonreducing sugar and total sugar were higher in 5th grade (8.74, 6.68 and 15.54 mg/g of leaf tissue) whereas, total proteins were higher in 1st grade (9.92 mg/g of leaf tissue). IAA content was higher in 3rd grade diseased leaves (5.14 mg/g of leaf tissue) whereas, lowest in healthy seedling stage (2.38 mg/g of leaf tissue). Similarly, the GA content was higher in 3rd grade diseased leaves (4.17 mg/g of leaf tissue) whereas lowest in healthy leaves (3.16 mg/g of leaf tissue). Suresh *et al.* (2016a) observed that the genotype Arka Kalayan has shown infection in the range of 25 to 65 per cent followed by Local varieties (0 to 55 per cent); whereas fewer incidences was observed in private hybrids (5 to 10 per cent). Hegde *et al.* (2015) reported that integrated treatment involving seed treatment with carbendazim @ 2 g/kg + soil application of neem cake @ 5q/ha + *Trichoderma harzianum* @ 2 kg/ac + dipping the seedlings in *Pseudomonas fluorescens* @ 10g/l + spray with boron @ 2g/l + multi K @ 5 g/l + spray with hexaconazole @ 0.1 per cent was superior treatments in reducing the twister disease incidence (5.49 per cent) with maximum yields of 279.8 q/ha. Patil *et al.* (2016b) reported that the seed treatment with vitavax power (2 g/kg) + seedling dip with Pseudomonas sp. (10 g/l) + spray of Hexaconazole + Profenophos @ 0.1 per cent, Bacillus subtillus 10 g/l at 15 days resulted low PDI. Patil *et al.* (2016d) evaluated various plant extracts and reported that the extracts of *Azadirachta indica*, *Prosopis julifera*, *Clerodendron inerme* and *Parthenium hysterophorus* in 1 : 1 : 1 : 1 inhibited maximum mycelial growth of *C. gleosporoides* (46.66 per cent). They also observed that four species of *Trichoderma* inhibited mycelial growth (78.00 per cent) of *C. gleosporoides* and *F. oxysporum*.

Ganeshan and Sinha (2000) concluded that damping-off of onion could be effectively controlled by bioagent. Srivastava and Tiwari (2000) reported that highest germination of seeds (83.78 per cent) and lowest damping off disease incidence (24.67 per cent) were recorded in soil solarization with white polythene for 30 days before sowing followed by seed treatment and soil application of *Trichoderma viride* (80.78 per cent and 28.56 per cent) compared to control (73.33 per cent and 45.33 per cent, respectively). Kanwal *et al.* (2016) observed that postharvest rot fungi were significantly inhibit growth by using of *Azadirachta indica, Psidium guajava, Aloe vera, Ocimum basilicum, Eucalyptus camaldulensis, Carica papaya* and *Taraxacum officinale*. The highest percentage growth inhibition was achieved with Aloe-vera (65.5 per cent) against *A. niger* followed by guava leaf (43.8 per cent).

12.2.0 Pests

Onions are attacked by numerous pests all over the world both in field and storage. In India, the following insect pests cause severe damage.

12.2.1 Thrips

The following species of thrips are reported on onion from India; 1. *Thrips tabaci*, 2. *Heliothrips indicus*, 3. *Caliothrips indicus* (Saxena, 1971), 4. *Scirtothrips dorsalis* and 5. *Aeolothrips collaris*.

Thrips tabaci is worldwide distributed polyphagous pest infesting approximately 140 plant species (Pal *et al.*, 2019). It is the major injurious pest of onion and garlic. Besides, it is cosmopolitan in nature, polyphagous, transmit plant pathogen, have high reproductive rate, asexual mode of reproduction (parthenogenesis), high survival *via* criptic (non-feeding prepupa) instars and develop insecticide resistance. With this thrips become a global pest with increasing concern in commercial onion cultivation (Soumia *et al.*, 2017). In India, thrips are widely distributed in all onion growing regions. This pest is active throughout the year and breeds on onion and garlic from November to May.

Both nymphs and adults attack all stages of onion plant. They puncture the leaves and stem and suck the exuding sap. Onions infested with this thrips develop spotted appearance on the leaves which turn into pale white blotches due to drainage of sap. As infestation progress, damage tissues coalesces exhibiting blast like appearance. The tip of the leaf gets dried and eventually falls. Severe infestation may result in distorted and undersized bulbs. Besides, it also acts as a vector for viral disease like Iris Yellow Spot Virus (IYSV).

Mote (1978) reported about 50 per cent losses of onion crops due to thrips attack. Seed production and viability are also hampered due to attack of thrips (Lall and Singh, 1968). The adults hibernate in soil, on grass and other plants in the onion fields (Shirck, 1972). The thrips also overwinter in bulbs and act as source of infestation in the following year (Bagnar and Shanab, 1969). Srinivas and Lawande (2000) noted that different abiotic factors played an important role in population dynamics of thrips. Maximum day temperature was positively corrected with thrips population build up. Other weather parameters *viz.*, minimum temperature, relative humidity and rainfall showed a negative effect on thrips population. El-Gendi (1998) however, reported that maximum temperature and R.H. had little effect on population density of thrips. Thrips multiply in large number during March and April both on seed and bulb onions in the northern parts of India, whereas at Nasik areas (India) thrips incidence was observed even in January and February (Gupta *et al.*, 1983). Kumar *et al.* (2015a) recorded 4th standard week of January and continued up to 18th standard week of May, 2012 with maximum thrips population (63.30 thrips/plant). In Brazil, the population peak of *T. tabaci* occurred between the end of October and mid-November. Cultivars transplanted early tended to escape from high population densities. Populations of *T. tabaci* peaked during bulb formation and decreased thereafter (Goncalves, 1997). Numbers of generation onion thrips is 2-9 per year depending on environment. Spreading of population was faster when temperature rises between 20.8 to 27.7°C and decreased with rainfall, relative humidity and wind speed of 13.7 kph (Pal *et al.*, 2019). The mean maximum temperature ($r = 0.721**$) and minimum temperature ($r = 0.768**$) had positive and highly significant effect on thrips population; while the mean relative humidity (morning) ($r = -0.627*$) and rainfall ($r = -0.393*$) had negative significant effect (Ganai *et al.*, 2018).

Pawar *et al.* (1975) reported a plant with glossy stiff leaves with wider angle between leaves covering neck region are resistant to thrips. Host selection and preference of thrips is more on cultivars with narrow angles of leaves, dark and waxy coating leaves with high sugar content. The attack is more on older plant where outer and upper section of leaves receives 95 per cent of eggs approximately. Population dispersion depends on abiotic condition, planting date, variety and cropping pattern (Pal *et al.*, 2019). Flight activity of the onion thrips in the Central Jordan Valley by using water and sticky traps showed that there were two main activity periods. The first period was in early December to February and the second one was in March to May where average temperatures were 16 and 22.3°C, respectively (Mustafa and Turaikhim, 2001).

Destruction of weeds harbouring adults, larvae and pupae, and burning of debris was suggested by Bailov (1930). Staggered planting of onion and balanced nutrients results in less colonization of onion thrips (Pal *et al.*, 2019). Adoptation of crop rotation (Franseen and Ranteuran, 1933) and use of resistant cultivars such as Nasik Red are effective measures for control of this pest. Lal and Verma (1959) reported Spanish White cultivar as resistant to thrips on account of loose arrangement and light green colour of leaves. El-Gendi (1998) reported that cv. Giza 6 and cv. Behiri was the least susceptible to thrips. Tripathy *et al.* (2013) observed that NRCRO-3, NRCWO-3, NRCWO-4 and VG-19 showed tolerance to thrips and cultivars, Bhima Super, NRCWO-3, NRCRO-4 and the control, Arka Niketan produced significantly high total bulb yield having better tolerance to thrips. Patil *et al.* (2015) noted Bellary Red, L-819 and Arka Kalyan in *kharif* season and NHRDF Red3, Bellary Red, Bheema Red and Bheema Shubra during *rabi* were moderately resistant to thrips.

Introduction in the Dutch leek culture of a diapause strain of the phytoseiid *Amblyseius cucumeris* may give satisfactory control of thrips population (Vierbergen and Ester, 2000).

Application of malathion or nuvacron has proved effective. Bhatnagar *et al.* (2000) found that the use of malathion as insectcide in onion did not pose any residue hazard if green onion were consumped by human after three days of insecticide spray even at the double recommended rates of field application. Chiag (1973) suggested application of phorate @ 1 kg/ha. Rao and Kumaraswami (1986) reported that carbosulfan (0.1 per cent) and endosulfan (0.07 per cent) were very effective in reducing the incidence of *Thrips tabaci*. Naik *et al.* (1986) found that monocrotophos (0.05 per cent) was the most effective for controlling *Thrips tabaci*. Among synthetic pyrethroids, cypermethrin at 50 g a.i. per ha has also been found an economic insecticide for thrips control (Gupta *et al.*, 1991). Bhargava and Singh (2000) reported lowest thrips incidence by treatment with deltamethrin 2.8 EC (0.73 thrips/plant) which was statistically at par with nagada (0.93 thrips/plant) as compared with untreated check (10.80 thrips/plant). Kumar *et al.* (2015b) reported three round spraying of profenofos 50 EC @ 0.01 percent applied at fortnightly intervals starting from 30 days after planting in minimizing thrips population. Wagh *et al.* (2016) reported that fipronil @ 0.0075 per cent was most effective as it recorded least (4.00-6.00 thrips/plant) and maximum (25.67 t/ha) yield of onion bulbs and also with carbosulfan @ 0.025 per cent, thiamethoxam @ 0.0075 per cent and lambda cyhalothrin @ 0.005 per cent were equally effective and recorded less number of thrips and good yields. Ganai *et al.* (2018) observed thiamethoxam 25 EC (0.100 per cent) was the most effective against thrips followed by imidacloprid 200SL (0.008 per cent), methyl-o-demeton 25 EC (0.030 per cent), carbosulfan 250 EC (0.003 per cent), neem oil 5 per cent (0.050 per cent), novaluron 10 EC (0.100 per cent) and bifenthrin 10 EC (0.050 per cent). The study of Wale *et al.* (2018) revealed that the treatment with betacyfluthrin 90 + imidacloprid 210 OD @ (45 + 105) g a.i./ ha was most effective for the control of thrips and recorded highest yield of onion bulbs *i.e.* 498.74 q/ha as against 375.93 q/ha in untreated control. Betacyfluthrin 90 + imidacloprid 210 OD @ (36 + 84), (27 + 63) g a.i./ha and lambda-cyhalothrin 5 per cent EC @ 15 g a.i./ha were found next best treatments for the control of thrips and recorded 488.00; 477.69 and 458.17 q/ha yield of onion bulbs, respectively. Foliar spray applications of these treatments did not showed any adverse effect on natural enemies. Thangavel *et al.* (2018) studied the Bio-Efficacy of Alika 247 ZC on onion and reported that three rounds foliar application of Alika 247 ZC @ 100 to 150 ml/ ha starting from 30 days after planting was effective in reducing the nymphs and adults of Thrips tabaci (89.4 to 86.2 per cent). Thiamethoxam 25 WG @ 100 g/ha and dimethoate 30 EC @ 660 ml/ha were next best which were at par with each other. Alika 247 ZC at all the doses did not cause adverse effect on population of spiders in onion ecosystem and recorded the highest bulb yield of 18.72 t/ha, while in untreated check it was 11.34 t/ha. Yadav *et al.* (2020) observed fipronil (0.01 per cent) to be most effective against thrips, followed by clothianidin (0.01 per cent). The next most effective insecticides were acetamiprid (0.004 per cent) and thiamethoxam (0.025 per cent). On the other hand, Karuppaiah *et al.* (2020) found profenofos @ 160 g a.i./ha was the most effective insecticide with reduction of 86.8-90 per cent of thrips population, low leaf curing and a maximum bulb yield (20.8-24.2 t/ha). Cyantraniliprole @ 120 g a.i./ha was the next best with 73.879 per cent decrease of thrips population.

Altaf Hussain *et al.* (1999) observed that botanical insecticides pandaphose 60 SL and tobacco leaf infusion were most effective in reducing the population of thrips followed by neem leaf infusion during the 1995-96 seasons while neembokil 60 EC reduced thrips populations significantly during the 1996-97 season. Pawar *et al.* (2000) reported that two spray of the crude plant extracts of *Mentha arvensis* 5 per cent (4.0) and *Nerium oleander* 5 per cent (7.5) were effective in reducing the thrips population than untreated (42.4). In a study on management of onion thrips with insecticides and vermicompost, Pawar *et al.* (2000) observed significantly lowered number (5.36) of thrips in the plot treated with cypermethrin (0.01 per cent) than vermicompost (27.5) and untreated control (49.3). While Kulkarni *et al.* (2000) reported that MICP 0.1 per cent (6.49 thrips/plant) was the most effective treatment however it was at par with neembecidine 2.5 l/ha (7.02 thrips/plant) spray.

12.2.2 Head Borer (*Heliothis armigera*)

Gram pod borer is a polyphagous pest of tropics, subtropics and warm temperate regions of the world. Samarjit *et al.* (1979) reported high incidence of the pest at Burari farm (Delhi) on seed crop. It has also been observed on bulb crop feeding on leaves (Bhardwaj *et al.*, 1983). Invade the umbel and feeds on seeds. As a result complete drying of flowers and complete loss of seed occurs. The larva cuts the pedicel of the flower and feeds on stalk and which on maturation pupates inside scape (Soumia *et al.*, 2017). It also pupates in the soil. Spray application of endosulfan or zolone at 2-3 ml/l has been recommended to control the pest.

12.2.3 Onion Maggot (*Delia antiqua*)

Singh *et al.* (1976) reported that the adult appears like housefly. Maggots are small white and devoid of legs. Eggs are laid on the base of plants in cracks in the soil. Maggots enter the bulbs through roots and attack the tender portions. Infested plants turn yellowish brown and finally dry up. The affected bulbs rot in storage as infestation leads to secondary infection by pathogenic organisms. Good sanitation is extremely important in controlling the onion maggot. Removal of the culls from the field is advised. Crop rotation should be followed and application of thimet is beneficial. Application of 50 EC actelic [pirimiphos-methyl] at 1 litre/ha (biological efficiency of 77.4 per cent) and 2.5 EC decis [deltamethrin] at 0.3 l/ha (biological efficiency of 56.4 per cent) were the most effective treatments against onion maggot and onion bulb fly pests in onion (Paulauskyte, 1999). Direct application of insecticides to the root zone is considered the most effective means for controlling maggot damage. Spray should be directed to the base of the plant.

12.2.4 Mites

Red spider mite (*Tetranychus cinnabarinus*), Eriophyid mite (*Aceria tulipae*) and Bulb mite (*Rhizoglyphus robini*) affect onion cultivation. The upper surface of the leaves becomes stippled with little dots that are the feeding punctures of red spider mites. Silk webbings are also visible. Eventually leaves turn bleached and wither.

Eriophyid mite causes stunting, twisting, curling and discolouration of foliage. Yellow mottling is seen mostly on the edge of the leaves. In storage bulbs dry and desiccate. Bulb mites suck sap turning plants pale with sickly appearance. Causes reduced plant stand, stunting and premature bulb rot in storage. In nursery, radical developed from seeds are cut off (Soumia *et al.*, 2017). Sandhu (1979) reported that infested bulbs should be exposed to sun for two days. In field, dusting of sulphur @ 22 kg/ha is recommended.

12.2.5 Nematodes

Lorbeer (1997) reported that a 3-4-year rotation to lettuce was eliminated the stem and bulb nematode from soils infested by pathogen. Soil fumigants control nematode diseases and pink rot.

An experiment was conducted to evaluate the potential of use of bioagents of root-knot nematode *viz.*, *Paecilomyces lilacinus* and *Pseudomonas fleuroscens* along with neem cake suspension of 10 per cent and 20 per cent for the sustainable management of *Meloidogyne incognita* on rose onions. The results clearly indicated that *P. lilacinus* and *P. fleuroscens* can be used effectively along with neem cake suspension for the management of *Meloidogyne incognita* on onions (Rao and Reddy, 2000).

Priti *et al.* (2018) suggested combination treatments of *B. pumilus*, *P. monteilii* and *T. harzianum* in the production of healthy onion seedlings under nursery conditions whereas increased yield and the management of *Meloidogyne incognita* (Kofoid and White) Chitwood and *Fusarium oxysporum* f. sp. *cepae* Schlecht (FOC) complex under field conditions. It reduced the nematode infestation (47.6 per cent to 71.4 per cent) in the field experiment.

12.3.0 Physiological Disorders

Some physiological disorders hamper onion production. These are as follow:

12.3.1 Sprouting

This is an important storage disorder of onion that causes a great loss to growers and traders. It is more common in white cultivars than pink or purple cultivars. Sprouting is associated with excessive soil moisture at maturity and supply of nitrogen. The control measures are, (1) harvesting should be done at dry period, (2) before maturity irrigation should be stopped, (3) balance fertilizer application, particularly nitrogen and (4) spray Chlorpropham or Isopropyl N-(Chlorophenyl) Carbamate (CIPC) @ 2 per cent at 75 days after transplanting (Choudhary, 2014).

12.3.2 Bolting

Bolting refers to premature immergence of seed stalk. It hampers bulb development and affect yield. Early (August) and late transplanting (end of December-January) induces bolting in *kharif* and *rabi* onion, respectively. Rabi onion starts to bolt at 20-25°C. In early planting, onion plants completed their basic vegetative phase and get cold stimulation before favourable condition of bulb formation. This encourages the plants for bolting. Early sowing of seeds before mid November in North Indian plains favours the development of seed stalk in onion

Pests of Onion

Onion Thrips Infestation

Onion Maggot Infestation

Pests of Onion

Head Borer Infestation

Mite Infestation

Nematodes Infestation

bulb crop. Bolting also increases by planting big size sets. Poor vegetative growth sometime leads to bolting. Bolting is considered undesirable in warehouse onion crop as it has shortened shelf life. Bolting also reduces the quality as the bulbs become fibrous and lightweight. Genetic factors, poor seed quality, spacing, seedling size, poor soil condition, temperature variation, photoperiod and cultural practices are influence this process. The control measures are, (1) planting should be done at proper time, (2) non-bolter/less-bolter types should be selected for cultivation, (3) aged seedlings should be avoided for planting, (4) apply proper nutrition doses and (5) remove/cut the seed stalk at early stages in bulb production field (Choudhary, 2014; Rana and Hore, 2015).

12.3.3 Tip Burning

This is necrosis at the margin of young developing leaves. This trait is both genetically controlled and environment influenced. The main cause of tip burning is the deficiency of potassium. High temperature, high light intensity, high soil pH and repeated use of brakish water may cause tip burning. Kapoor *et al.* (2000) reported that tip blight does not appear to be associated with any micro-organisms and was the result of host-weather interaction. Seedlings subjected to hardening a week prior to transplantation are less prone to the disorder. The spraying of micronutrient in the crop raised from hardened seedlings, employing need-based irrigation and fertilizer resisted the disorder up to 89 per cent resulting two fold increase in yield.

12.3.4 Thick Neck

Excessive vegetative growth without forming bulbs leads to thick necked bulbs. The interaction between day length and temperature regulate the bulbing of onion. Prolonged low temperature at bulbing phase may increase the neck thickness. Excessive nitrogenous fertilizer may produce bulb with thick neck. *Kharif* onion often produced thick neck. Long day varieties may fail to start bulbing and keep their growth continues until particular photoperiod came and may produce thick neck.

12.3.5 Premature Bulbing

It is the formation of bulbs very rapidly, soon after transplanting of seedlings in the field. Late in the season transplanting often leads to premature bulbing without completing the basic vegetative growth. When onion seedlings are transplanted in high temperature and relatively long photoperiod the vegetative growth suppresses and bulbing starts. Dense plant population or very close spacing also cause premure bulbing in onion. Premature bulbing give much lesser yield in comparison to timely planted crop.

12.3.6 Sun Scald

Sun scald generally occurs when the onion bulbs left for field curing after harvest. However, sometimes it may occur even befor harvesting in standing crop particularly when the bulbs are exposed to sun. High temperature coupled with low humidity and low soil moisture favous sun scald development. The tissues expose to such weather and soil condition become soft and slippery. The field should be given frequent light irrigation to keep the soil moist and cool. Avoid very shallow planting of bulbs. Under high temperature and dry condition bulbs should not be left in the field for field curing after harvest, rather it should be shed cured to avoid sun scalding. The affected bulbs are poor in storage (Rana and Hore, 2015).

12.3.7 Bulb Splitting and Doubling

Bulb splitting occurs due to the presence of multiple growing points in a single bulb. It is a genetically governed trait. Adverse growing condition and imbalance nutrition may also cause this disorder. Growing of some cultivar under high temperature and short day condition produced lateral shoots. Sometime injury to the plant during cultural operations may lead to bulb splitting. Water stress condition followed by irrigation or rainfall at initial growth phase results in doubling of bulbs, as long dry spell breaks the dormancy of lateral buds. Use of undecomposed manures also leads to bulb splitting.

12.3.8 Watery Scales

This disorder can be seed in store onions. The symptoms are the development of thick and leathery outer most skin, which when removed reveals coatery glossy storage scales. This is highly susceptible to fungus and bacterial infection. Onion is highly susceptible to carbon dioxide. In controlled atmosphere storage, high carbon dioxide may present in internal scales or external atmosphere. Onion bulbs exposed to high temperature after harvest for long time produce more carbon dioxide. Carbon dioxide concentration >13 per cent causes watery scales. In controlled atmosphere storage, carbon dioxide concentration of 10 per cent cause internal breakdown. In storage, higher concentration of carbon dioxide is more harmful than low oxygen content (Rana and Hore, 2015).

12.3.9 Freezing Injury

Onion bulbs stored at very low temperature may cause freezing injury. Cultivar variation exists for this trait in onion. The cultivars having high TSS content are less sensitive to freezing injury, as the bulbs of such cultivars have very low freezing point. The sensitivity of onion bulbs to freezing injury depends also on water content. The bulbs with less water content have more resistance against freezing injury. Some onion cultivars can be stored even at -2°C without any freezing injury. However, below this temperature freezing injury is developed and bulbs start rotting. Injured bulbs can not be stored for longer time.

12.3.10 Chemical Injury

It is sometime known as alkali scorch. It is occur in stored onion. Chemical injury may developed if onion is stored in alkali impregnated/printed jute bags. Higher concentration of ammonia gas in storage may cause injury to onion bulbs. The symptoms of ammonia injury are the development of dark brown to black spots mostly in yellow and red skin varieties.

13.0 Seed Production

Onion seed is usually produced in the temperate and subtropical countries. In regions where high temperature prevails almost throughout the year, only the

Physiological Disorders of Onion

Tip Burning

Sun Scauld

Freezing Injury

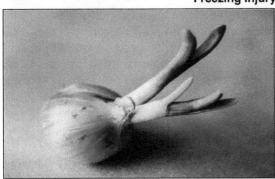

Sprouting of Onion

Bulb Splitting

early bolting type of onion, requiring relatively little low temperature exposure, can produce seed. Onion is a biennial crop for the purpose of seed production. In one season bulbs are produced from seed and in the second season bulbs are replanted to produce seed. Onion seeds are poor in keeping quality and lose viability within a year. Therefore, it is essential to produce seeds freshly and use the same for bulb production. It is a highly cross-pollinated crop which is facilitated by protandrous nature of flowers. Cross pollination is effected by honeybees.

In India, the main reasons for low onion yield are short day length, suboptimal standards of cultivation, weather vagaries, non-availability of quality seed and use of low yielding varieties, *etc*. It is evident that productivity could be increased by 20-30 per cent if quality seed of improved varieties along with improved management practices are followed. In tomato, cabbage, cauliflower and some other vegetable crops, use of quality seeds and F_1 hybrids have brought considerable improvement. Many varieties though they are higher yielding and have better storage qualities are not popular in view of non-availability of adequate quality seeds (Pandey, 2000). He also pointed out that constraints in quality seed productions are due to (i) non-availability of statistics on variety wise area and demand of seed, (ii) marketing of seed collected from premature bolters, (iii) without maintaining isolation, (iv) unawareness about the suitable seed production pockets, (v) slow spread of improved varieties, (vi) inadequate facility for storage of bulbs, (vii) suboptimal standards of seed production and storage, (viii) inadequate facility for seed processing and storage, and (ix) lack of education of farmers in follow up of improved production techniques as also on production planning for onion seed.

Availability of the quality seed material plays major role in increasing onion productivity. In India about 1000 tonnes of onion seed require annually. The seed requirement for *rabi* and *kharif* season is about 60 per cent and 40 per cent, respectively. The organised sector (Government agencies and private firms) contribute about 40 per cent of total requirement. Rest 60 per cent seed is produced by the farmers without following any standard for isolation and varietal purity (Mahajan *et al.*, 2017).

13.1.0 Factors Influencing Seed Yield

Seed yield is influenced by many factors, among which cultivars, bulb weight, soil and climate, spacing, fertilizer application and dates of planting are important. Mittal and Srivastava (1965) reported that seed yield is related to the number of seedstalks per bulb. It appears that a cultivar may have an optimum seedstalk or seedstalks per bulb. The optimum seedstalk number in cv. Pusa Red appeared between 6-8 and 12-15 stalk group, while in cv. Early Grano it was 3-4. Therefore, Pusa Red yields more seed. Increase in seedstalks beyond optimum number did not show proportional increase in yield of both seed and bulb.

Sandhu and Korla (1976) in an investigation on 40 onion cultivars tried to correlate the seed yielding capacity with important plant characters. Seed yield per plant was positively and significantly correlated with number of seedstalks per plant and weight per umbel. Even then, they further observed that seed yield is a quantitative character and thus greatly influenced by fluctuation in environmental conditions.

Varieties like Agrifound Dark Red, Baswant 780 and Arka Pragati emerged as most promising for producing both early *rabi* onion bulb as well as their seeds in the Gangetic alluvial plain of West Bengal (Karak and Hazra, 2012).

Barman *et al.* (2013) evaluated the influence of planting date and spacing on growth and earliness parameters in onion seed crop. Planting of bulbs at a closer spacing of 45 × 15 cm resulted with maximum plant height at 90 and 120 DAP over other spacing levels and maximum number of leaves on 90 DAP but maximum number of leaves resulted in wider spacing of 60 × 30 cm at 120 DAP.

Haile *et al.* (2017) revealed that significantly highest value of days to 50 per cent bolting, days to 50 per cent flowering, days to 50 per cent maturity, number of leaves per plant, flower stalk height, flower stalk diameter, umbel diameter, number of umbels per plant, number of umbels per plot, 1000-seed weight and seed germination percentage were obtained from 25 cm intra-row spacing.

Ashagrie *et al.* (2014) studied on the effects of planting time and mother bulb size on onion (*Allium cepa* L.) seed yield and quality. They observed that significant interactions between mother bulb size and planting time on days to 50 per cent flowering, scape diameter (cm), seed yield per plant (g), seed yield per hectare (kg) and germination index and also reported that early planting (October 25) of large bulbs (4.1-5 cm) can be used for high yield and better quality of onion seeds. Bhakare and Fatkal (2008) repoted that 100 per cent water soluble fertilizer through fertigation was the best system for growth, yield, quality and economically viable for onion seed production with saving of water and fertilizer.

Information about yield components which are less affected by environment is of much help in enhancing the efficiency of selection. Selection based on seedstalks per umbel and weight of seed per umbel accounted for 67.2 per cent variation in seed yield. Flower clusters pollinated by bees produced a mean of 132.85 and 224.52 more seeds than artificially pollinated and control groups, respectively, which represented respective yield increases of 25 per cent and 45 per cent. Absolute seed weight and seed germination percentage were on average 10.57 and 3.67 per cent higher when pollinated by bees than when pollinated with a brush (Martinovski *et al.*, 1997).

13.1.1 Climate and Soil

In general, onion requires cool weather during the early development of bulb crop and again prior to and during early growth of the seedstalk. Later, a moderately high temperature and a dry atmosphere are favourable for the maturing of bulb crop and also for the seed crop during the second year. Temperature throughout the year and day length during the growing season set broad limits to the areas that are suitable for seed production. Temperatures between 4.5 and 14°C are favourable for stored onion before using for seed production. The optimum temperature for flowering is between 15 to 25°C. High temperature early in the season or at the time of year when the young scapes are forming causes many plants to produce bulbs instead of flower stems and may greatly reduce the number of flowers per plant.

Higher temperature (≥35°C) affects the growth of pollen tube and stigma receptivity, which results in poor seed set. The higher temperature is required at the time of seed maturity and ripening. High relative humidity, fog and frost are not favourable for seed production. Low temperature at bulb planting (November-December), moderate temperature during seed stalk development, flowering and seed setting (January-February) and warm temperature at seed maturity and drying (March-April) with bright sunshine and free of cloudy weather is considered optimum for onion seed production in Indian plains. Penninsular and central India is more suitable area for onion seed production.

Well-drained, medium to heavy soil are good for onion seed production. The pH of the soil should be 6.5 to 7.0. Light soil can also be used for onion seed production after application of ample well decomposed organic matter.

13.1.2 Bolting

Bolting is a physiological disorder that reduces quality and marketable yield of onion and having poor storability (Gupta *et al.*, 2018). Seedstalk initiation and development come under the general term 'bolting', which also includes premature bolting found in certain specific conditions. Onions is essentially a biennial crop, requiring one season for bulb development and another for seed production *i.e.*, they normally flower in the second season of their development. This biennial nature is very specific and pronounced in case of long-day cultivars grown especially in temperate regions and also in *rabi* season in subtropical regions of North India. However, under favourable conditions, some sensitive varieties bolt in the first season itself. The so-called short-day cultivars are grown during late summer to early winter in subtropical regions of India (Maharashtra) for bulb production and followed by a seed crop from winter to early summer, both the crops being completed within a year. This can be construed as an annual crop, even though the two specific seasons covering a year follow successively without bulb storage.

Bolting is a result of complicated interaction between genotype, plant age and environmental factors. Bolting is essential for seed production, but undesirable during bulb production. Bolting varies with genotype susceptibility and resistance, and is clearly influenced by temperature or cultivar or both. This problem occurs mostly in late kharif due to variation in temperature during vegetative growth under Indian plains (Gupta *et al.*, 2018).

The growth of onion is characterized by bulb development in the first year, followed by storage and seed production in second year. In case of N-53 or Nasik Red, it appears that both these stages have been sacrificed to produce the seeds, by growing rainy (*kharif*) crop for bulbs and successively followed by winter (*rabi*) seed crop.

The initiation of floral formation is dependent on somewhat lower temperature. In temperate countries, the onions are normally stored at a temperature of 0-2.2°C but it was found that bulbs kept at this temperature did not produce sufficient seed-stalks compared to bulbs which were stored at 7.2-12.8°C. On the basis of these results, it may be suggested that the bulbs when stored at 0-4.5°C, should be lifted and put at a temperature of 10°C before planting for seed production. It is apparent that low temperature around 10-12°C is needed for initiation of the floral primordia. Under North Indian conditions this temperature treatment is received by the bulbs, during their growing period in December-January, when the seedstalks are initiated. However, it appears that this temperature requirement varies amongst the cultivars, since cultivars like Bellary Red or N-53 and N-2-4-1 in Maharashtra produce seed-stalks comparatively at higher temperature.

Holdsworth and Heath (1950) concluded that day length did not directly influence floral initiation and that under low temperature, long days promoted the subsequent emergence of flowers and elongation of scape. The multiplicity of effects of temperature at different seasons makes the flowering behaviour in onion variable rather unpredictable in the field. Paterson *et al.* (1960) and Cleaver and Turner (1975) reported that high levels of N fertilizer reduced flowering in the field in over-wintered onion crops. Stuart and Griffin (1946) reported that a phase of low N nutrition during January and February, promoted subsequent flowering in plant, grown in culture solutions in greenhouse. Cuocolo and Barbieri (1988) from Italy reported that N fertilization increased the number of umbels and seed weight per umbel, resulting in greater seed production. Seed yield increased 0.22 q/ha for every 10 kg/ha N applied.

Inflorescence initiation is favoured by large set size (Heath, 1943) or plant size (Holdsworth and Heath, 1950). Heath and Mathur (1944) suggested that an onion plant should form at least 12-14 leaves before the appearance of flower but the evidence was not conclusive. However, in case of premature bolting in the bulb crop in India, it is generally observed that some percentage of plants show seedstalk formation before the bulb development. As has been pointed out earlier, the seedstalk development is the result of interaction of several factors, like temperature, cultivar, age of seedlings, *etc.* Hence, it is necessary that the time of planting of seedlings in a particular cultivar in a specific location should be worked out to avoid premature seedstalk development in the bulb crop. However, direct bolting, without bulb formation can be utilized for testing and isolation of a line for maintaining male sterility in one year itself. Treatment of bulbs with GA_3 at 300 ppm induced better flower initiation in onion, while the auxins like IAA, IBA, NAA and ethephon also affected flowering. Karak and Hazra (2012) studied on 14 short-day tropical onion genotypes in which flowering stalks were induced even at 25.35/11.74°C day/night temperature during last week of December to last week of January.

13.2.0 Methods of Seed Production

There are two methods of seed production. Most commonly used method of seed production is bulb-to-seed method. Another method is seed-to-seed.

13.2.1 Bulb-to-Seed Method

For seed production, bulbs harvested during warm weather should be selected and stored carefully till middle of October. The storage temperature influences seed yield. Temperature ranging from 4.5 to 14°C with an optimum of about 12° is the best for storage of mother bulbs which are to be planted for seed production.

Plants from such bulbs produce early and heavy yield than those grown from bulbs which have been stored at higher or low temperature. The roots of the bulbs should be left intact after harvest. Bulbs selected for replanting should be free from disease infection. Doubles and long thick-necked bulbs are discarded and only true-to-type bulbs are selected. Such selected bulbs are planted in well prepared field. The growing portion of bulb is cut to the extent of 1/4 to 1/3 for easy and quick sprouting of more growing buds. The lower portion with disc-like stem and roots is used for planting. To avoid rotting due to fungal infection of bulbs in the field, bavistin @ 20 g in 10 litres of water is used for dipping the bulbs before planting. The treated bulbs are planted along one side of the ridge. Kalavathi *et al.* (1990) noted that bulbs with 1/4 top cut produced better quality seed compared with whole bulbs. Cutting the bulbs, however, had a negative effect on yield attributes and seed yield (Nehra *et al.*, 1989). Thiruvelavan *et al.* (1999) observed that soaking of bulbs in 100 ppm GA_3 24 hours before sowing showed early sprouting, umbel emergence and flowering including better umbel character and higher seed yield compared to control.

Kanwar *et al.* (2000) reported that bulb to seed method produced considerably higher number of leaves and umbels per plant and higher per cent bolter as compared to seed to seed method. Bulb to seed method produced seed yield 7.99 q/ha as compared to 2.98 q/ha under seed to seed method. Among the genotypes, Punjab Red Round with seed yield of 4.13 q/ha and 10.86 q/ha in seed to seed method and bulb to seed method, respectively. Close spacings of 25 × 10 cm under seed to seed and 45 × 20 cm under bulb to seed method gave highest seed yield of 3.41 q/ha and 9.39 q/ha, respectively. It is concluded that maximum seed yield of onion can be obtained through bulb to seed method with genotypes Punjab Red Round at the spacing of 45 × 20 cm.

Planting of bulbs of Baswant-780 in Ruhuri, India on 15th October gave significantly higher seed yield over all other dates of planting except 1st and 15th November. The seed yield of all these three dates of planting was at par with each other. The seed quality parameters *viz.*, 1000-seed weight and germination percentage were superior in 15th October planting (Muspade *et al.*, 2000).

13.2.1.1 Size and Weight of Bulb

Size and weight of bulbs used for planting have markedly influenced the seed production in onion. Jones and Emsweller (1939) reported that increase in size and weight of bulb increased seed yield per plant and the optimum bulb weight was 51-60 g. Soloman and Patel (1959) also observed that bulbs weighing 81-100 g each produced the maximum seed yield of 760 kg/ha. Ahlawat and Rathore (1957), however, observed that medium-sized bulbs of 1 to 1 1/2 inch (2.5-3.7 cm) in diameter were better for seed production. Vander Merr and Van Bennekom (1968) recorded less number of flowering stalks from small-sized bulbs. Kulkarni (1963) concluded that for economy and convenience, the bulbs weighing more than 40 g with a diameter up to 3 cm were the best suited for raising the seed. Rajput (1973) in an experiment reported that bulbs weighing 61-80 g produced more seeds per plant as compared to those weighing 30-40 g. Lal *et al.* (1987) observed that maximum number of scapes/plant and highest seed yield/ha were produced by 90 g bulbs planted at 30 × 30 cm. Gill and Singh (1989) obtained highest number of bolting stems/plant and seed yield with bulb diameter less than 5 cm. Toman *et al.* (1989) found that graded mother bulbs of cv. Pogarskit with diameters of 50.1, 60.1-70.0 and 70.1-80.9 mm produced 27, 47 and 62 per cent higher yields of seeds respectively than ungraded controls. Pall and Padda (1972) suggested to plant bulbs weighing 50 g for higher seed production. Although an increase in weight and size of bulbs results in higher seed yield, very large-sized bulbs, if used will need a very high seed rate. A bulb size of 2.5-3 cm diameter needs 1500 kg to plant a hectare and may yield about 815 q of seed. Large-sized bulbs of 3-4 cm diameter will need more seed bulb and may yield 10 q/ha (Choudhury, 1967).

Singh *et al.* (1997) observed that planting of bigger bulbs (80-100 g) in the last week of October at a spacing of 45 × 30 cm gave high seed yields. Singh and Sachan (1999) reported from India that the largest bulb size 4-5 cm with the widest spacing 30 × 45 cm gave the highest seed yield/plants, although the smallest bulbs 2.5-3 cm produced the biggest umbels on average. The closest spacing 15 × 30 cm with the largest bulb size gave much the greatest seed yield per ha. Kalyanpur Round Red was superior to Nasik Red. Therefore, for commercial seed production medium-sized bulbs (2.5-3 cm in diameter) may be used economically.

Singh and Singh (2000) advocated that onion seed production potential (11 q/ha) could be better exploited by using the 3.0 cm diameter bulbs and planting them at 30 × 20 cm inter and intra row spacings respectively in Ladakh, the cold desert region of Jammu and Kashmir.

Thapa *et al.* (2005) reported that the treatment with larger size bulb (50-65 g) had marked effect on seed yield and yield attributing characters compared to the treatment with smaller size bulb (30-45 g).

Geetharani and Ponnuswamy (2007) reported that use of larger sized bulbs and allowing three umbels/plant will be more suitable for production of highly valuable seeds of breeder and nucleus seeds. Khokhar (2008) reported that flowering and seed production were influenced by set-size. The seed yield per umbel increased with increasing set-size from 12.5 mm to 22.5 mm diameter and inflorescence development (*i.e.*, emergence and floret opening) occurred earlier in plants grown from larger onion sets than from smaller sets.

Shukla *et al.* (2008) studied seed production of onion (*Allium cepa* L.) cv. Agrifound Dark Red as influenced by bulb size and nitrogen levels. The large size bulb gave better results in respect of plant height, number of leaves, number of umbels (8.45), seed weight (93.43 g) per plant and seed yield (789.41 kg/ha).

Ud-Deen (2008) studied the effect of mother bulb size and planting time on growth, bulb and seed yield of onion. The mother bulb size showed significant influence on growth, bulb and seed yield of onion. The large mother bulb were favourable for getting higher bulb and seed yields

Ashrafozzaman *et al.* (2009a) studied the effect of paclobutrazol and bulb size on onion seed production and reported that the plant height, number of leaves per plant, length of scape, effective fruits per umbel, percentage of fruit set and seed yield were positively influenced by the bulb size of onion. Ali *et al*. (2015) reported that the plant height, number of leaves per plant, length of scape, number of umbels per plant, umbel diameter, seed yield per plant, germination percentage and electrical conductivity were positively influenced by bulb size of onion. However, the bulb size did not affect seed yield per one seed stalk, but it did affect the seed yield obtained from the entire plant in bigger bulb size.

Mollah *et al*. (2015) reported that the optimum bulb size is 15 g for production true seeds of onion at Bogra region in Bangladesh. Manna *et al*. (2016) observed that planting of large size bulb produced significantly better results than other medium and small size bulb in plant height, number of leaves, number of flower stalks per plant and number of seeds per umbel, seed yield per plant and seed yield per hectare, quality characteristics of freshly harvested seeds *viz*., germination. They concluded that large bulb size was suitable for quality seed production under New Alluvial Zone of West Bengal.

Haile *et al*. (2017) obtained significantly highest value of days to 50 per cent bolting, days to 50 per cent flowering, days to 50 per cent maturity, number of leaves per plant, flower stalk height, flower stalk diameter, umbel diameter, number of umbels per plant, number of umbels per plot, 1000-seed weight and seed germination percentage from bulbs of 6.1 to 7 cm size.

13.2.1.2 Cultivation

Planting of bulbs for seed production is done in October-November. Higher seed yield is obtained from planting in the first fortnight of October. Arakeri and Patel (1956) obtained the highest seed yield by planting bulbs in October at a distance of 30 cm in a row. Singh *et al*. (1974) and Nehra *et al*. (1989) also reported the highest seed yield from planting on 15th October. Seed yield declined as planting was delayed from 16th October until 15th November (Nehra *et al*., 1989). Yadav *et al*. (2002) observed that the larger bulb (5.5-6.5 cm) when planted from October 24 to November 13 at close spacing of 45 × 30 cm was given highest seed yield. Rohini and Paramgur (2016) obtained highest seed and residual bulb yield (807 kg/ha and 8000 kg/ha, respectively) at September planting.

In Korea Republic, plant growth and seed set were highest for plants planted on 25 September with transparent PE film mulch. Seed yields in Paechong-Joseng and Changnyeong-Daego were 40 and 44 per cent higher, respectively, than in non-mulched controls (Hwang *et al.*, 1996). In Argentina, Gaviola (1997) found that bulb to seed method of onion seed production of cv. Valcaforce INTA, planted in March produced the highest yield during the 3 years (average 121.5 g/m^2) resulting in good seed quality.

Planting distance depends upon the size of bulbs. Rajput (1973) also obtained high seed yield from planting at 30 × 30 cm. Singh *et al*. (1974) and Singh and Rathore (1973) reported the highest seed yield from planting at the closest spacing of 30 × 10 cm. An increase of plant density from 4 to 12 bulbs/m^2 gave a linear increase of overall seed yield from 7 to 12 q/ha (Cuocolo and Barbieri, 1988).

Asaduzzaman *et al.* (2012) reported that the highest seed yield (776.67 kg) per hectare was obtained from the large bulb (15 ± 2g) with the closest spacing of 25 × 15cm followed by small bulb size of same spacing. The maximum number of flowers per umbel (371.39), seed weight per umbel (0.80g) and 1000-seed weight (3.92g) were obtained from the largest bulb size (15 ± 2g) with widest (30 × 20 cm) planting spacing. They suggested that for onion seed production large bulb size with closest plant spacing were taken.

Application of fertilizers was found to increase the seed production in onion. Naga-raju *et al.* (1986) obtained highest seed yield (411 kg/ha) with spacing of 60 × 15 cm and NPK rate of 60 : 30 : 60 kg/ha. Singh *et al.* (1988) reported that highest seed yield of cv. Pusa Red was obtained from plants spaced at 45 × 30 cm and receiving 40 kg N/ha. Planting bulbs with a fertilizer rate of 150 kg N + 60 kg P + 20 kg K/ha and spacing at 45 × 15 cm gave the highest seed yields (Bhatia and Pandey, 1991). In Punjab, Nandpuri *et al.* (1968) recorded that 50 kg N and 25 kg P per hectare gave the highest seed yield. Pall and Padda (1972) also obtained higher seed yield with application of 40 kg N/ha. In another experiment, the highest seed yield in onion was recorded by treatment with 120 kg N/ha (Chakrabarty *et al.*, 1980). Yadav *et al.* (2000) reported that use of bio-fertilizer had caused better crop stand, crop health on seed production variety Hisar-2 due to various useful traits of Azotobacter like nitrogen fixation, production of plant growth promoting substances and antibiotic like substances.

Manna *et al.* (2017) suggested that application of nitrogen @ 175 kg/ha with potassium @ 120 kg/ha is optimum for quality seed production with higher economic return of onion cv. Sukhsagar under new alluvial zone of West Bengal.

Onion seed crop sprayed with zinc @ 5 per cent and boron @ 1 per cent at vegetative stage, prior to flowering and at seed set stage significantly enhanced plant growth, yield contributing characters and seed yield (Dogra *et al.*, 2019). Study revealed that FYM played an important role in releasing the micronutrients from applied nutrients as well as from soil and enhancing the seed yield of onion. The direct contribution of FYM was highest (40.9 per cent) followed by applied N, P_2O_5 and K_2O. Contribution of soil nutrients were negligible in FYM treated plot, while it was the highest (FN: 21.3 per cent, SK: 13.9 per cent and SP: 7.2 per cent) in no-FYM plot. Therefore, application of fertilizers with FYM as per yield target enhanced the onion seed production, increase use efficiency of added nutrients, maintain soil health and avoid the mining of soil nutrients (Shinde *et al.*, 2017).

Pohare *et al.* (2018) studied the effect of INM on growth and yield on soybean-onion (for seed) crop sequence. It was concluded that, combined application of 50 per cent RDN (25 kg/ha) through CF+50 per cent RDN through VC (19.5 t/ha) or FYM (46 t/ha) to kharif soybean followed by 100 per cent RDN (100 kg/ha) through CF to rabi onion (for seed) along with recommended dose of phosphorus and potassium to both the crops is necessary for getting higher yield and soil health in soybean-onion (for seed) crop sequence.

Manna *et al.* (2016) suggested that the soil application of recommended dose of fertilizers (N : P : K : S :: 125 : 80 : 100 : 40 kg/ha) + Boron (B) @ 1 kg/ha may be followed in New Alluvial Zone of West Bengal for getting quality seed yield with higher economic return of onion cv. Sukhsagar.

Under Maharashtra condition, Bhalekar and Chalak (2016) studied the best combination of plant nutrients for onion seed production cv. Phule Samarth and concluded that 150 per cent of recommended dose of fertilizer (20 t FYM/ha + 100 : 50 : 50 kg NPK/ha) accompanied with biofertilisers (Azospirillum 12.5 kg/ha, PSB 12.5 kg/ha and VAM 16 kg) is essential for better seed yield of onion.

Discontinuing irrigation when the seeds reached the milk stage gave high yields good quality seeds with economy of labour and water (Globerson *et al.,* 1987). In Maharashtra, India irrigation of 10-day intervals with 80 kg N/ha in split applications gave the highest yield of quality seed of onion cv. Agrifound Light Red (Bhonde *et al.,* 1996). Singh *et al.* (1997) observed that planting of bigger bulbs (80-100 g) in the last week of October, at a spacing of 45 cm between the rows and 30 cm within the rows gave high seed yields. Herbicide treatment (pendimethalin and fluchloralin at 2.5 l/acre) and fungicide sprays (0.3 per cent Dithane M-45 [mancozeb] + 0.5 per cent monocrotophos) at 10 days intervals to control purple blotch also increased yield and quality.

Agarwal *et al.* (2010) observed significant effect of bulb size and planting geometry on seed yield. 175 g bulbs at 40 cm × 40 cm planting geometry appeared most economic with high returns, they reported.

In order to detect and eliminate different plant types, roguing should be started before the bulbs are harvested. It is easier to remove late maturing bulbs at this stage. After the bulbs are harvested, they may carefully be rogued for colour and off-types as thick-necks, doubles, bottle-necks, under and over-sized bulbs and any other character which do not conform to the varietal type. Bulbs are stored for 4-6 months; hence it is necessary to remove rots or sprouts before planting.

13.2.2 Seed-to-seed Method

A considerable amount of onion seed is produced by seed to seed method, which avoids the use of mother bulbs. Bulb-to-seed not only takes two years for seed production but is more expensive as large quantities of bulbs are to be stored at planting in the second year and losses in storage may also be considerable. This method produces good yield of seed and the time of ensure 100 per cent bolting of the plants is also eliminated. If some of the plants bolt, the seeds that are harvested will include those from the easy bolting individuals. Seed from such plants will not be useful for commercial seed production when bolters are undesirable. Yield of seed by this method is often much higher than that obtained from a bulb-to-seed method because of the greater number of plants and seed head per unit area. Seed from seed-to-seed crop should not be used again for seed production.

In India, seed-to-seed method of production has not been popular. Sometimes bolters do appear in the bulb crop of onion, and the seeds produced from these direct bolters are considered unsuitable for raising the crop. Bhagchandani *et al.* (1970) reported that results of investigations on seed-to-seed method. They observed that more than 80 per cent of the plants produced the seed stalks in Early Grano in

October planting and it was reduced significantly when planted in November. In Pusa Red only 26.5 per cent seed stalks were formed when planted on 5th October. In cultivar Early Grano, a seed yield of 650 kg was obtained when planted on 5th October, while Pusa Red did not produce economic yield. There was no significant difference in the performance of crop from seed produced by the two methods of seed production in both the cultivars. In Hungary, onion seed production is achieved by planting sets in autumn or spring. Seeds are harvested in August with only a few days' difference due to the different dates of planting. Both laboratory and field trials, however, showed better quality seeds from spring planting (El-Emery, 1985).

Mohanta and Mandal (2016) studied the onion seed production under Red and Laterite Zone of West Bengal following seed-to-seed method. They reported that the cultivar Agrifound Dark Red (3.9 g) recorded highest seed weight per umbel and closely followed by Indam Marshal (3.7 g). The highest seed yield per plant was recorded in Indam Marshal (11.3 g), followed by Agrifound Dark Red (10.1 g).

Advantages of this method are, high seed yield, production of seed in poor keeping cultivars without storage, and less cost of production by eliminating expenses on harvesting and replanting bulbs.

13.2.3 Rouging

Onion seed production field should be visited regularly and off type plants like yellow, lanky and abnormal plants with varied umble height, diseased plants should be removed before opening of flowers. Generally four rouging are done: before flowering, during seed stalk development, during flowering and at harvest for getting pure seed crop. If bulb crop is cultivated around the seed crop, care should be taken to remove the bolters before flower opening.

13.2.4 Isolation

The location of onion seed fields should be planned well in advance, so that adequate isolation can be maintained. The greater the distance between onion fields, less will be the extent of crossing. Other factors such as the direction of wind and weather conditions at pollination time would influence the extent of crossing. Since it is a cross pollinated crop, two cultivars may be kept about 1000 m apart to produce pure seeds (Anon., 1971).

13.2.5 Insect Pollinators

Under natural conditions, onion umbels are visited by number of insect species. Onion blossoms are highly attractive to both pollen and nector collecting insects and are good sources of minerals and sugars (Karuppaiah *et al.*, 2017).

Karuppaiah *et al.* (2018) observed eleven insect species *viz.*, *Apis florea*, *A. cerana*, *A. dorsata*, *A. mellifera*, *Tetragonula* sp., *Xylocopa* sp., *Vespa* sp., *Pieris rapae*, *Danais chrysippus*, *Eristalis* sp. and *Musca domestica* pollinating onion. Of these 98 per cent of forage visits were of hymenopterans and *A. dorsata* was the dominant. Most of the foraging happened during 1330-1430 hrs. Maximum foraging was by *A. cerana* (3.17 umbels/min) followed by *A. dorsata* (3.0 umbels/min) and the least with *A. florea* (2.0 umbels/min). Time spent/flower was the maximum with *Tetragonula* sp.

(27.50 sec/umbel) followed by *A. florea* (18.83 sec/umbel) and *A. dorsata* (17.83 sec/umbel) and the least with *A. cerana*. The dwarf honey bee *Apis florea* L. was the most abundant flower visitor and comprised more than 94 per cent of the total flower visitors (Abrol, 2011). Honey bees started visiting onion crop at 8.00 h, population was high during 13.00-16.00 h, declined slowly during 16.00-18.00h. The intensity of *A. dorsata* was more between 12-14 h (2.66 bees/m^2/min) (Mupade *et al.*, 2009).

Gupta and Singh (2010) suggested keeping of bee colonies in onion seed field for proper pollination and seed setting. Chandel *et al.* (2004) reported that induced bee pollination increased onion seed yield by 2.5 times and the seeds from induced pollination field resulted in 90 percent germination compared to 69.5 percent germination from the control. Karuppaiah *et al.* (2017) reported that for open pollinated varieties, systemically placing 4 to 6 bee hives per acre of *Apis mellifera* or *Apis cerana* would enhance bee activity and pollination success rate satisfactorily. Hives are placed when 10 per cent flowers are opened. Hybrid seed production field require 10 to 12 bee hives per acre. Hives are to be placed in and around the field.

Spraying of jaggery solution 10 per cent and sugar solution 10 per cent attracted maximum number of *Apis dorsata* up to 3 days after first spray (at 10 per cent flowering). As regard *Apis florea*, jaggery solution 5 and 10 per cent, sugar solution 5 and 10 per cent were efficient in attracting more bees up to 5th day of first spray. In case of *Apis cerana* spraying of jaggery solution 10 per cent attracted more bees up to 5th day of first spray and 3rd day of second spray (Kulkarni *et al.*, 2017). In Bee-Q sprayed plots in general the intensity of all wild bees was increased as compared to open pollination (Mupade *et al.*, 2009). Abrol (2011) reported that the insect pollinated plots produced significantly more seeds with heavier weights than those isolated from insect visits in onion.

The neighbouring crops around the onion seed crop should be non-attractive to bees. Generally grains and tomatoes are advised (Karuppaiah *et al.*, 2017). However, pollinator attractant crops like Fennel and/or Coriander can be planted in a sequence of two rows for every alternate 10 rows of onion including border for higher seed setting and obtaining higher seed yield of onion (Uddin *et al.*, 2015). These attractant crops should be cut down/harvested and no more stand of trap crop is permitted when the umbel initiation is started in seed onion. Balanced fertilizer application and timely irrigation are required to sustain adequate nector production in onion.

13.2.6 Harvesting, Curing and Seed Yield

The maturity of seed ready for harvest is indicated when fruits open and expose the black seed. Harvesting of seeds at proper maturity is essential. Only fully ripe seed should be harvested. All the umbels do not mature at the same time. A field is considered ready for harvesting when about 10 per cent of the heads gave black seeds exposed. It is desirable to harvest seed at intervals. Three to four harvesting of dried umbels is required. The seed heads are cut with 10-15 cm of stem attached. While cutting, the umbel is supported in the palm of the hand and held between the fingers to avoid seed shattering.

Sarati *et al.* (1987) from Russia reported that CCC and camposan M (ethephon), especially when applied in combination, reduced plant height and increased

Seed Production Stages of Onion

Seed to Seed Method

Bulb to Seed Method

Seed Production Stages of Onion

the resistance of the peduncles to bending, thereby facilitating mechanical seed harvesting.

The harvested umbels are heaped (not more than 30 cm height) for a few days for drying before threshing the seed. Regular turning of heaps is followed to avoid dumping of lower layers and for uniform drying. This helps in proper curing of seeds. After threshing, seeds are cleaned and graded by air screen machine and further graded by gravity separator.

A well managed seed plot can yield about 500 to 800 kg seed per hectare. As much as 1000 to 1200 kg seed per hectare can be obtained under best management and climatic conditions. Farmers obtained average seed yield of 3.0 to 4.5 quintal per hectare (Dhar *et al.*, 2018). Manna *et al.* (2016) reported seed yield per hectare as high as 11.39 quintal.

Table 19: Onion Seed Quality Parameters

Parameter	Seed Standard for	
	Foundation Seed	Certified Seed
Seed purity	98.0 per cent	98.0 per cent
Inert matter (Maximum)	2.0 per cent	2.0 per cent
Other crop seed (Maximum)	5/kg	10/kg
Weed seed (Maximum)	5/kg	10/kg
Germination (Minimum)	70 per cent	70 per cent

For packing in porous container (normal cloth bag or jute bag), seeds are dried to about 8 per cent moisture, while they are dried to 5-6 per cent moisture for moisture proof packing like aluminium foil. Seed viability decreased rapidly in paper packing, while it remains viable for 3-4 years in sealed containers. Under cold storage at 12-15°C and 35-45 per cent humidity seed can be stored for 3-4 years provided initial seed moisture is maintained at 6 per cent level. Small seed lots of onion breeding lines often stored in desiccators. Zeolite beads can be used successfully to reduce seed moisture for seed storage in sealed containers under ambient conditions (Bhonde, 2016).

In Germany, Rudolph (1987) investigated four methods of umbel drying to allow after ripening of seeds. Forced air drying using (i) air at ambient temperature, (ii) hot air at 29 ± 3°C or (iii) air at ambient temperature for 10 days followed by hot air for another 10 days or (iv) drying on slatted shelves without forced ventilation had no differential effect on 1000 seed weight, germination capacity or rate of emergence. Dried umbels (moisture content 9-11 per cent) were threshed manually or with the E 512 machine with a drum rotation of 700 to 1200 revolutions per minute for 15 minutes. Drum revolutions of 1000 per minute impaired seed quality and reduced germination capacity, conductivity and emergence rate compared with hand-harvested seed. An average seed yield of 8-10 q can be expected from a hectare. Higher yields even up to 15 q are also obtained. For nucleus and foundation seed production, the bulb-to-seed method should be followed because it provides opportunity for proper selection and roguing. However, the seed-to-seed method will produce higher seed yield.

Seed production on onion was investigated at Kundule (1100 m) and Keware (1100 m) in Nepal by Jaiswal *et al*. (1997a). Regal PVP (213 kg/ha) produced 263 per cent more seeds than Early Red and required 216 days to complete development (46 days earlier than Early Red). In Nepal, seed yield of a new elite onion cv. Red Synthetic of 619 kg/ha was obtained at an altitude of 1200 m. The crop took 223 days to produce mature seeds (Jaiswal *et al*., 1997).

In Cuba, Perez *et al*. (1996) obtained highest seed yield of cv. Caribe with bulbs 4.0 cm in diameter. Locality of origin of different cultivars did not influence seed yield.

Verma *et al*. (2000) reported that in Himachal Pradesh, India, the average cost of production per kg seed was Rs. 523.59 with cv. Brown Spanish excluding charges for land, irrigation and cost of mother bulb with a net profit of 34 per cent. The highest net return was obtained under the treatment combination medium bulb size (3.6-4.5 cm diameter) with closest spacing (30 × 30 cm). The best cost benefit ratio (1 : 2.57) was recorded under the treatment combination of medium bulb size (2.6-3.5 cm diameter) with closest spacing (30 × 30 cm).

13.2.7 Seed Storage

Rapid loss in viability of onion seeds during seed storage is a major problem. Seed moisture or storage relative humidity has a major role in determining the longevity.

Trigo *et al*. (1999) recorded that germination rate and percentage germination at optimum and suboptimum temperatures (20 and 10°C, respectively) were better in treated seeds than in controls, with treatment with KNO_3 for 24 h giving the best results. Treated seeds maintained their physiological quality for up to 6 months in storage.

Doijode (2000) reported that onion seeds remain viable for shorter under ambient conditions. High seed viability (cv. Arka Kalyan) was maintained for three years up to 30 per cent, thereby only seeds at 10 per cent RH exhibited high viability and the moisture reduced from 9.0 to 4.3 per cent. Seedlings characteristics were preserved with ultra low moisture. Ultra low moisture or preserving seeds at low RH is beneficial in maintaining high seed viabiity and vigour during storage.

Paul *et al*. (2018) harvested the seeds of onion cv. Punjab Naroya in the month of June and studied it storage behaviour. The refrigerated stored seeds showed maximum germination percentage (94 per cent) while seeds accelerated aged for twelve days gave least germination percentage (62 per cent). As the duration of ageing increased, there was a marked reduction in seedling length, fresh and dry weight. All the ageing treatments resulted in membrane damage, reduction in activities of peroxidase, catalase and-amylase. There was a reduction in contents of total starch and total soluble proteins whereas, the total soluble sugars and total free amino acids increased with accelerated ageing. The amount of ascorbic acid, α-tocopherol and DNA also reduced with accelerated ageing.

In Himachal Pradesh, Himangini *et al.* (2018) studied the influence of drip fertigation in combination with N and K fertilizers along with boron foliar spray on quality of onion seed during storage. The seeds of the treated plants were harvested and stored for 8 month in air tight plastic jars at ambient temperature under shade. Drip fertigation along with boron foliar spray gave significant results on quality of seeds in storage. Maximum germination percentage (96.5 per cent), emergence index (79.5), mean germination time (5.34), germination speed (9.73) and germination energy (92.11) was recorded in $N_{100}K_{100}B_1$. Onion seeds packed either in glass container, polythene bag 700 gauges or aluminium foil and stored under cold storage condition maintained better seed quality over a period of six months of storage (Manna *et al.*, 2020).

14.0 Crop Improvement

14.1.0 Breeding Objective

In an improvement programme, the objectives for the breeding should be specific so that correct methodology can be adopted to develop a cultivar with the desired characteristics. Onion breeders focus primarily on bulb characteristics such as color, shape, soluble solids content, pungency and flavour, storage ability, and health enhancing attributes, as well as plant characters such as resistances to diseases, pests, and bolting. Important characteristics for seed production include uniform flowering, straight seed stalks, stable expression of male sterility, and seed yield (Havey, 2018).

Hari Har Ram (1998) outlined the following specific breeding objectives of onion : (i) high yield, (ii) longer bulb storage life, (iii) resistance to diseases (purple blotch, basal rot, stemphylium blight, bacterial storage rot), (iv) resistance to insect pest (thrips), (v) resistance to abiotic stresses (moisture stress, high temperature, salinity, alkalinity), and (vi) bulb quality (size, shape, colour, pungency, firmness, dormancy, amount of soluble solid).

Dormancy is important because onions are normally stored for longer time. High TSS is important for dehydration industry producing onion chips and powder. The amount of 8-9/kyl cysteine sulfoxide precursors and the enzyme allinase contribute to the yield of sulphur compounds that constitute the pungency of the onion bulb.

According to van der Meer (1994) low priority should be given to breeding onion hybrids for the tropics while high priority should be given to the following:

1. Collection, multiplication, and evaluation of a wide spectrum of tropical onion strains;

2. Selection for improved characters in promising strains;

3. Combining complementary promising strains and subsequent selection in resulting composites;

4. Addition of individual superior characters from nonadapted varieties and closely related species; and

5. Improvement of multiplication methods for the central tropics (from 10°N to 10°S).

However, he suggested that well-equipped institutions in or close to the subtropics (*e.g.*, AVRDC and/or Rahuri) could pave the way for local hybrid breeding in the near future by: analyzing tropical strains for frequencies of S, N, ms, and Ms genes; and testing the male sterility in question for its thermostability in tropical conditions.

Lawande (2000) indicated the thrust areas are (i) development of varieties/ hybrids for kharif season tolerant to colletotrichum and purple blotch, (ii) development varieties/hybrids for rangda season with low bolting and twins with better keeping quality and resistant to stemphylium blight and thrips, (iii) development of varieties/hybrids for rabi season having dark red as well as light red colour with very good storabiity and resistance to stemphylium blight and thrips, (iv) development of yellow onion varieties/hybrids for export in European countries, (v) development of white onion varieties/hybrids with high TSS (> 20 per cent) for dehydration, and (vi) standardization of protocol for anther/ovule culture for development of homozygous parental lines used in heterosis breeding programme.

Foliage color in green onion and shape and colour of onion bulb are most important characteristics to help customers in choosing cultivars in the market (Solanki *et al.*, 2015). On the other hand, colour, size, shape, number, thickness and adhesion of skins, storage abilities, solids content, quercetin levels, pungency and sweetness are all traits that breeders work on. Local onion growers and traders can be included in the decision making process during selection of new variety. Consumer preference (size and outer scale colour), disease index, crop duration, storability and seeding ability should be taken care of before adopting a new variety for any region.

14.2.0 Selection and Hybridization

Onion is essentially a cross-pollinated crop, which influences greatly the breeding methodology adopted for improvement. This crop plant shows abrupt inbreeding depression, especially after two generations of selfing, which makes it impossible to go in for pure line selection or pedigree method of breeding. Hence, mass selection is the common method which has been adopted in several countries for its improvement. Even in seed production, mass selection of bulbs is necessary, since in seed production without bulb selection, rapid deterioration of pure cultivars usually occurs.

Aklilu *et al.* (2001) found that seed yield per plant had a high, significant correlation with number of flower stalks per plant, number of seeds and flowers per umbel and umbel size. Bolting and flowering period had a significant negative correlation with seed yield per plant. From the path analysis results, the number of flower stalks per plant, bolting period, thousand seed weight, flower stalk diameter and umbel size had a high direct positive effect on seed yield per plant. Since the direct and indirect effects through these components on seed yield are high and positive, selection should concentrate on these characters for high seed yield in onion cultivars. Since these components were found to affect seed yield they could be used for developing varie-ties for the growing onion industry in the country.

Rajalingam and Haripriya (1998) recorded very high heritability estimates coupled with high genetic advance for the characters weight of plant, bulb length, bulb diameter and volume of bulb. Mohanty (2000) observed moderate to high heritability, genotypic coefficient of variation and genetic gain were for the number of seed stalks per plant, flower per umbel, seed yield, and diameter of umbel and 1000 seed weight, which could be improved by simple selection. The seed yield showed positive and significant genotypic and phenotypic correlation with the number of seed stalks per plant, flower per umbel, diameter of umbel and 100 seed weight. Path analysis showed that 1000 seed weight and the number seed stalks per plant had high positive direct effect, while each of these characters had relatively high and positive indirect effect, on seed yield. This suggests that emphasis should be given on such traits while imposing selection for amenability in seed yield of common onion.

Mohanty (2001) further recorded moderate to high estimates of heritability, genetic coefficients of variation and genetic gain for neck thickness, weight of bulb and number of leaves/plants, which could be improved by simple selection. Phenotypic and genotypic associations of bulb yield were significantly positive for plant height, number of leaves per plant and diameter and weight of bulb, but were significantly negative for neck thickness. Path analysis showed that the number of leaves per plant had the highest positive direct effect on yield. Other characters also exerted high positive indirect effects through this trait on yield, suggesting that emphasis should be given to this trait independently or in combination with neck thickness in selecting for higher bulb yield of onion.

Narayan *et al.* (2019) observed that bulb weight was positively associated with polar diameter and length of bulb. And the average bulb weight and yield/m^2 is positively correlated with bulb yield (q/ha). He advocated the generation of genetically broad base population using diverse genotypes in breeding programme for improving bulb yield component traits in onion.

Heritabilities of the pungency and single centre traits were estimated in onion breeding populations using selection response and half-sib family analyses by Wall *et al.* (1996). Pungency was determined indirectly by measuring enzymatically produced pyruvic acid in individual bulbs. After one generation of selection, pungency was lowered by 8.1 and 8.9 per cent in the population 90-61-1 and 89-69-8, respectively, and realized heritabilities of 0.21 and 0.51 were estimated. Selection had no effect in lowering the pungency of population 90-62. Heritability estimates calculated through half-sib progeny analysis were 0.53, 0.48 and 0.25 for pungency in the populations 90-61-1, 90-62 and 89-69-8, respectively. The number of single centred onions was increased by 19 and 22 per cent in populations 90-62 and 89-69-8, respectively, after one generation of selection, and the realized heritability estimates were 0.37 and 0.34, respectively.

The breeding programme carried out in Brazil by Franca *et al.* (1997) has its main goals the development of short-day, yellow cultivars tolerant of *C. gloeosporioides* and *T. tabaci*. Among the released cultivars, Composto IPA-6 has good adaptation during the whole year, high yield capacity, and is being used in 30 per cent of the cropping area. Recently, Blem IPA-9 was released, having as main characteristics, good yield performance, adaptation throughout the year, and a remarkable tolerance to *C. gloeosporioides*. Soluble solids comprise most of onion bulb dry mass, and dehydrator onion cultivars and developed from breeding populations that have high dry mass content. Realized and narrow-sense heritabilty estimates were obtained for the soluble solids content (SSC) trait in two open-pollinated dehydrator onion breeding populations (BP) using response to selection and half-sib family analysis. Parental populations, designated as BP9335-U and BP9243-U, were derived from two-way crosses of lines (Ben Shemen Southport White Globe and Yellow Grano selection × Southport White Globe, respectively), advanced as open pollinated (OP) populations to the F_7 or F_6 generation, respectively. Significant differences in half-sib family performance in the advanced groups BP9335-S and BP9243-S demonstrate that progency testing was effective for evaluating phenotypic selections.

In order to minimize the adverse effects of inbreeding, Jones and Mann (1963) recommended alternate selfing and sibmating. The selected individual bulbs are planted in isolation and selfing is done in each plant. In the second year, progeny rows are grown and considering the typical characters of the cultivar or characters under selection, the progeny rows are selected strictly to the conformity of the character under selection. Further selection of bulbs is done within the progeny. The selected bulbs are planted in isolation for sibmating among the selected progeny. The next cycles of selection are repeated till the uniformity in the characters is obtained. Even in this method, a certain amount of loss in vigour in noticed.

According to Lawande (2000), systematic breeding programme in India was started as early 1960 at Pimpalgaon Baswant, Nasik and later on at IARI, New Delhi, India. The early varieties developed through selection *viz.*, N-2-4-1 and Pusa Red are still dominating.

Most of the Indian varieties of onion are the result of mass selection from the locally adapted material. Arka Kalyan, Arka Pargati and Arka Niketan are high yielding cultivars developed by mass selection from local collection from Maharashtra state. Another high yielding cultivar Pusa Ratnar has been developed by selection from an imported hybrid cv. Red Granex from USA.

Theresa and Armstrong (2001) reported about the development of Polish cultivars. According to him first original Polish Wolska cultivar was bred in 1912 by crossing Zytawska with landrace Wolska. The resultant Wolska cultivar has large globe-rhombic shaped bulbs, straw yellow dry skin, 13.5 per cent dry matter, is suitable for long term storage and was the main cultivar used in production and export for many years. Using the original Wolska cultivar as parent material, 7 further Wolska-type cultivars have been bred, differing amongst themselves with respect to vegetation period, shape and size of bulbs, colour, skin thickness and adherence of dry skin, yield and storage ability. The Wolska type cultivars on the list of National cultivars are Czerniakowska, Warzawska (medium-late),

Sochaczewska, Wolska and Kutnowska (late). Other cultivars are Rawska (early-medium) bred in 1938 with globular, slightly elongated bulbs, brown-yellow dry skin, dry matter content of 15-16 per cent, suitable for storage; Dako bred in 1948 for fresh consumption, has flattened, straw-yellow, sweet tasting and large bulbs; Zytawska selected in 1950 has flat bulbs, straw-yellow dry skin. Landraces that are grown on a small scale, include Szczebrzeszynska, Lubartowska, Drazgowska, Przybyszewska and others. The above-mentioned cultivars landraces were used as the source material for breeding programmes to develop new cultivars with improved quality and productivity.

Bulbs of 9 varieties and lines were tested in the computerized video system by Martinovich and Felfoldi (1996) to develop a method to store true-colour video pictures of onion samples where individual colours and shapes could be measured objectively, as well as establishing homogeneity and genetic distance in lots (varieties and lines).

O'-Garro and Paulraj (1997) screened 19 commercial onion genotypes for leaf blight resistance and recorded 2 cultivars, H-942 and H-508, were generally free of symptoms and had restricted bacterial growth in plant. The Hungarian onion cv. Makoi CR has been selected from Makoi for resistance to bolting under unfavourable conditions (Szalay, 1986).

Mako is a traditional Hungarian onion variety from which a number of new varieties have been derived. In order to develop a variety which could be harvested in April-May for marketing in June-July, several lines and varieties derived from Mako were evaluated from 1991 for ability to over-winter as sets. Individual selection for frost tolerance, earliness and short growth period resulted in the release in 1998 of Makomi. This new variety is adapted to autumn cultivation, needs no irrigation, does not bolt, is early and can be used for bunching or export. It can be harvested at the end of April. In appearance it is similar to Moko onions grown from sets in spring (Barnoczki, 1998). In USA cv. NuMex Sunlite was developed by selection from Texas Early Grano 502 PRR and is resistant to bolting (Corgan, 1988a). Recurrent selection for non-bolting bulbs from Ben Shemen produced two new cultivars, NuMex Sundial and NuMex Suntop (Corgan, 1988b).

Rexas Grano 1015 Y was derived from an original single bulb selection from Texas Early Grano 951 Y were derived from the cross Texas Early Grano 502 × Ben Shemen (Pike *et al.*, 1988b, 1988d). Texas Grano 1030 Y was developed from an F_2 late maturing selection of Texas Early Grano 502 × Ben Shemen (Pike *et al.*, 1988c). In Bulgaria, cv. Yibilei 50 has been selected from a cross between Mako and Bernsteinfarbige (Todorov *et al.*, 1989).

While breeding for summer stress tolerance in onion for development of high-yielding lines for the tropics, 200 onion lines were evaluated during summer in Taiwan. Out of the 30 summer stress lines selected the lines AC429, AC325-1 and AC47 performed best with yields of 25.4, 20.5 and 17.2 t/ha, respectively, compared with only 5 t/ha for Granex 429. Plant survival was > 90 per cent in most of the SST lines compared with 10 per cent in Granex 429 (Pathak *et al.*, 1996).

Onion cv. Rossa di Tropea is grown on a narrow coastal strip from Capo Vaticano to north of the Galfo di S. Eufemia and is renowned for its sweetness and tenderness. Three ecotypes exist: (i) the earliest and sweetest is sown and transplanted in autumn and harvested in April-May; (ii) a less pungent type sown and transplanted as (i) but harvested in May-June; and (iii) the most pungent is sown and transplanted in winter and harvested in June-July. In a breeding programme, designed to remedy a mingling of the ecotypes, 30 diplo-haploid lines were obtained. Improved lines matured more uniformly and one week earlier than the original population. Shape was also improved and yields were 10 per cent higher, due in part to the production of more marketable bulbs (Schiavi and Annibali, 1998).

Brazil import large amounts of onions mainly from Argentina. Brazilian consumers prefer the variety Valenciana 14, which is a Sweet Spanish type, has an excellent, multiple skin with good retention, and a bronze tanned coloration. However, this is a long-day variety and does not produce bulbs under natural conditions when cultivated in Brazil. Cardoso and da Costa (2003) tried crossing this variety with local cultivars in a genetic breeding program to obtain new adapted populations, by incorporating its retention and skin coloration qualities. They selected seventeen half sib progenies for early maturity and twenty five for late maturity, from the intervarietal triple cross [Crioula × (Pira Ouro × Valenciana Sintetica 14)], along with the triple cross itself and the cultivars Pira Ouro (short-day), Crioula (intermediate-day) and Armada (long-day) to study the effect of selection for bulb maturity. The progenies selected for earliness had cycles from 67 to 83 days, whereas those selected for lateness had cycles of 85 to 103 days. They obtained high heritability estimates for all characters and they varied from 0.65 (thick neck percentage, in the late selection) to 0.80 (average bulb weight, in the early selection). Progenies of higher bulb weight and maturity similar to the standard cultivars were obtained. They concluded that the selection for maturity was highly efficient and the population selected for early maturity has potential to originate adapted cultivars, with bulb yield and quality superior to the available cultivars.

14.3.0 Heterosis and Hybrid Seed Production

Being a cross-pollinated crop, heterosis breeding can be taken up profitably, both to avoid ill effects of inbreeding depression and to bring about uniformity in all the economic characters, especially in maturity. F_1 hybrids also allow the breeder to have exclusive ownership of the parent material, which is a commercially important factor (Eady, 1995).

The exploitation of heterosis in this crop was made possible commercially in other countries due to the isolation, maintenance and easier production of cytoplasmic male sterility. Even though there may be some difference of opinion in respect of superiority of F_1 hybrids over open-pollinated cultivars among onion breeders of Europe and the USA (Dowker and Gordon, 1983), the usefulness of F_1 hybrids in developing countries like India, may be more due to several factors. The open-pollinated improved cultivars, at present in cultivation, are by the large, highly heterozygous with at least 15-20 per cent off-types in respect of maturity, bulb size and shape, *etc.* Further, in the short period available for vegetative growth and bulb development, especially in the *rabi* onion crop of North India, which follows an early winter potato crop, the F_1 hybrids would likely to prove more advantageous for quicker bulb development and maturity.

Pathak (1999) reviewed onion hybrid production with special emphasis to floral characteristics, male sterility, production of hybrid onion and method for onion seed production. Heterosis estimates were most often significant for yield and soluble solid content, less often for pungency, storage ability and bulb size, and not significant for water loss in storage. Overall significant GCA estimates indicated that superior onion inbreds and populations may be developed using recurrent-selection strategies that increase the frequency of desirable alleles with additive effects (Havey and Randle, 1996).

To assess the potential of breeding onion for hybrid seed reproduction, mother bulbs of 7 onion cultivars, namely Desi Red, Red Imposta, P K 10321, Rubina, Phulkara, Faisal Red and Dark Red, were planted in the field, in Faisalabad, Pakistan. Seeds obtained from selfed and open-pollinated umbels were used to estimate the inbreeding depression at the seedling stage. There were significant differences among the cultivars for different characteristics. Further, a considerable and varying degree of inbreeding was found in different parameters among the onion cultivars (Khan *et al.*, 2001).

Genetic analysis of seed yield in onion indicated that relative seed yields of individual bulbs after self-pollination cannot be used to predict seed yields of progeny families. However, the seed yield of inbred lines of onion may reflect the potential seed yield of F_1 male-sterile lines (Mosqueda and Havey, 2001).

The production of hybrid onions became possible with the discovery of male sterile cytoplasm in the onion cultivar 'Italian red' by Jones and Emsweller (1936). They propagated this male sterile bulb clonally and used it in crosses with different pollen parents and demonstrated that these F_1 showed heterosis for yield. In their classical paper on inheritance of male sterility, Jones and Clarke (1943) showed that sterility was determined by the combination of a cytoplasmic factor 'S' together with a recessive nuclear gene in its homozygous form *ms/ms*. This hypothesis enabled them to devise practical methods for the commercial production of F_1 hybrid seeds using cytoplasmic male sterile lines and for the regeneration of the male sterile lines by pollination with a maintainer line of constitution N*msms*. Later on, another form of cytoplasm conferring male sterility *i.e.*, CMS-(T) was discovered in the onion cultivar 'Jaune paille des Vertus' (Berninger, 1965). It was reported that one independent and two complementary genes (three recessive genes *a*, *b* and *c*) were involved in fertility restoration of the male sterility (Schweisguth, 1973). There are also reports that the temperature (very low) effect might account for unexpected segregations for male fertility in male sterile lines. Of the two systems, CMS-S system is most widely used because of its stability in various environments (Havey, 2000).

Peterka *et al.* (1997) reported that modern onion breeding is mainly based on hybrid seed production by means of CMS induced by the interaction between S or T cytoplasm and one or three nuclear genes. This work is aimed at producing hybrid containing S-cytoplasm for use in CMS breeding work. Interspecific hybrids between

Allium cepa and *A. ampeloprasum* were generated as a first step for the introduction of S-cytoplasm from onion into leek. Pre-zygotic barriers of crossability were observed after the arrival of pollen tubes at the end of the style when entering the cavity. Nevertheless, micropyle penetration of pollen tubes and the formation of hybrid embryos were also observed. After accomplishing *in vitro* culture of ovaries and ovules successively, triploid hyrbid plants with 24 chromosomes were obtained. Their hyrbid nature was confirmed by RAPD analysis, genomic *in situ* hybridization, and morphological analysis. Southern hybridization with a cytoplasmic probe indicated the transfer of unaltered S-cytoplasm into the hybrid plants.

Normal (N) fertile and S cytoplasms are distinguishable by restriction-enzyme (RE) analysis of the chloroplast genome. RE analysis of the chloroplast DNA established that S cytoplasm has been introgressed into OP onion cultivars since its discovery in 1925. 'Valencia Grano' (released in 1927), 'New Mexico Early Grano' (1931), 'Texas Early Grano (TEG) 502' (1947), and 'Temprana' "(1979) are in N cytoplasm; S cytoplasm was introduced into 'TEG 502 PRR' (1960s), and subsequent selections ('NuMex BR1' and 'NuMex Sunlite') are now exclusively in S cytoplasm (Havey and Bark, 1994). A new source of sterility has been backcrossed into onions from *A. galanthum* (Havey, 1998a).

Sheemar and Dhatt (2015) used a mitochondrial DNA based marker cytochrome b (cob), which used to identify S and N cytoplasms in onion, to cytoplasmically characterize Indian onion populations. The oligonucleotides amplified 414 bp DNA for S-cytotype and 180 bp for N-cytotype in open-pollinated populations, namely, Punjab Naroya, Punjab Selection, Punjab White, PRO-6, PBR-4 and P-266. Out of the 200 seedlings screened in each population, 19 in Punjab Naroya, 15 in Punjab Selection and 5 in Punjab White showed amplification of 414 bp revealing S-cytoplasmic proportion of 9.5, 7.5 and 2.5 per cent, respectively. While PRO-6, PBR-4 and P-266 showed amplification of 180 bp only, indicating the absence of S-cytoplasm. The results suggest that Punjab Naroya, Punjab Selection and Punjab White can be exploited for development of male sterile (A) and maintainer (B) lines, while PRO-6, PBR-4 and P-266 can be utilized to isolate maintainer and restorer (R) lines. The study establishes a rapid method of cytoplasm identification in onion, which is pre-requisite to developing CMS lines. In contrast to 4-8 years taken by conventional approach used in India, this technique identifies cytoplasm within 15 days of sowing and proves its practical efficacy for future onion F_1 breeding in the country. Malik *et al.* (2017) tried marker assisted selection (MAS) using mitochondrial DNA based marker cytochrome b (cob) integrated with phenotypic evaluation to isolate male sterile and maintainer lines from open-pollinated onion varieties adapted to North Indian agro-climatic region. They determined the cytotype (N/S) by cob marker followed by morphological and microscopic study of pollen discovered male sterile plants (Smsms) at frequencies of 0.015 in Punjab Naroya, 0.020 in Punjab Selection, and 0.006 in Punjab White. The progeny scoring of test-crosses between male sterile and N-cytoplasmic plants isolated the maintainers (Nmsms) at frequencies of 0.133 in Punjab Naroya, 0.231 in Punjab Selection and 0.182 in Punjab White, they reported. As a novel approach, they used Trait Recovery Programme to reduce the population size required to recover a male sterile plant by 91.08 per cent in Punjab Naroya, 92.99 per cent in Punjab Selection and 97.66 per cent in Punjab White. For recovering a maintainer, they calculated 10 per cent reduction in Punjab Naroya and 9.10 per cent in Punjab Selection. It validated that in a randomly mating onion population, frequency of recessive ms allele squared is equal to the frequency of male sterile plants among S-cytotype and frequency of maintainers among N-cytotype (fms2 = fSmsms/fS = fNmsms/fN).

Male sterility located in the tropical onion line AC26 from India appeared to be controlled by strong cytoplasmic genes. Backcrossing is being carried out to transfer this male sterility trait to different genetic backgrounds for further use in heterosis breeding (Pathak *et al.*, 1996). For genetic improvement in onion 5 red-skinned open-pollinated population as males were crossed to 4 exotic yellow-skinned cytoplasmic male sterile inbred lines by Mani *et al.* (1999). The results indicated positive and significant heterosis over the better parent for bulb yield in the cross inbred 13 × L43. In this study, the F_1 between In-13 × L43 was used as a base material, advanced to the F_2 and subjected to 3 cycles of mass selection for bulb yield, skin colour, shape and size. An improved, high-yielding onion strain was thus developed and designated VL Piaz 3.

Pal (1980) found that on the basis of heterosis studies, hybrids Sel. 102 × Sel. 126 performed well. The heterosis ranged from 20 to 29 per cent over the best parent. He further observed that the selection on the basis of three characters, *viz.*, bulb weight, bulb diameter (horizontal) and total soluble solids will be useful to the breeders for onion improvement. Further, Pal *et al.* (1999) reported that the hybrids 75 Smsms × Early Grano and (102 × 106) Smsms × 26-3-8-2 × P 5 gave more uniform bulbs than any other F_1 or the control. For total soluble solids the hybrids (102-1 × 106 Smsms) × S_I 13 showed 6.0 per cent heterosis over the top parents. For storage Pusa Red Smsms × S_I 13 gave heterosis of 31.42 per cent over the top parent.

According to Aguiar *et al.* (2001) combinations Red Creole × Red Creole C-5, Caribe-71 Sebaquena and Sebaquena × Red Creole C-5 had the greatest heterosis. Various workers have attempted to induce male sterility by using chemical gameto-cides, and Vander Meer and Van Bennekom (1976) demonstrated the effectiveness of GA_4 and GA_7 as gametocides. However, higher concentrations used gave virtually complete male sterility but severely reduced seed yield.

Keller *et al.* (1996) reported interspecific crosses of onion with distant *Allium* species and characterization of the presumed hybrids by means of flow cytometry, karyotype analysis and genomic *in situ* hybridization. The viable hybrid of the most distant cross resulted from crossing *A. cepa* with *A. sphaerocephalon*. *Allium fistulosum* possesses a number of traits which would be desirable in *A. cepa*. So far, no commercial *A. cepa* cultivars have been released which harbour *A. fistulosum* traits. Peffley and Hou (2000) first demonstrated bulb-type onion introgressants possessing *Allium fistulosum* L. genes recovered from interspecific hybrid backcrosses *between A. cepa and A. fistulosum*.

Onion hybrids are derived mainly from partial inbred lines. Partial inbred lines require 14-16 years to be developed due to the biennial life cycle of the crop. Onion breeding lines can be selfed only for two or three generations due to high inbreeding depression. Thus, with conventional breeding it is difficult to obtain homozygous inbreds for complete genetic and phenotypic uniformity in the resultant hybrid. Doubled haploid (DH) production is an alternative strategy for complete homozygosity and phenotypic uniformity to obtain inbred lines in onion (Bohanec, 2002). Therefore, induction of di-haploids can greatly reduce the time and resources required for developing inbreds (Bohanec *et al.*, 1995). Haploid development in onion is only feasible through *in-vitro* gynogenesis, which is influenced by genotype, geographic origin, genetic constitution, physiological stage, growth conditions and cultural conditions. Whole basal plant as an explant and amiprofos-methyl as a chromosome doubling agent has been found to be safe and effective for for doubled haploids (DHs) induction (Khar *et al.*, 2019). DH technology deployment leads to the development of quicker and more economical homozygous onion lines than conventional breeding procedures (Bohanec, 2002; Alan *et al.*, 2003, 2004).

14.4.0 Resistance Breeding

14.4.1 Breeding for Disease Resistance

In onion the major disease problems are purple blotch and *Stemphylium* rot in the field and black mould and *Penicillium* rot in the storage. In some of these, disease resistant sources have been reported. The onion improvement programme at AVRDC aims to alleviate the production constraints on onions caused by biotic and abiotic stressed prevailing in the tropics. Among the diseases, *Stemphylium vesicarium* (leaf blight), *Alternaria porri* (purple blotch) and *Colletotrichum gloeosporioides* [*Glomerella cingulata*] (anthracnose) were given priority, with the intention of developing resistant varieties. Sources of resistance to leaf blight were identified in *A. fistulosum*, and research programme is under way to transfer this trait to onion lines.

Pathak *et al.* (2001) conducted onion breedings in Taiwan for stemphylium leaf blight (SLB; caused by *Stemphylium vesicarium*), which is one of the major onion diseases of the tropics. Five Welsh onion (*Allium fistulosum*) and 106 *A. cepa* lines were screened in field and in laboratory. All the *A. fistulosum* lines were resistant or moderately resistant to SLB, whereas all the *A. cepa* lines were susceptible. Crosses were successfully made between five *A. fistulosum* and 29 *A. cepa* lines to introgress SLB resistance into onion lines. A total of 48 crosses thus produced were used for further evaluation. All the F_1 hybrids were resistant or moderately resistant to SLB. Segregation in the F_2 generation was not expected to follow a Mendelian ration for disease reaction because of high sterility of the F_1 hybrids; however, the results indicate possible dominant gene control of the resistance trait. Moderate pollen fertility (> 30 per cent) and low seed set (> 5 per cent) was observed in four crosses: CF16 (AC15 × TA198), CF19 (AC50 × TA198), CF19R (TA × AC50), and CF52 (AC49 × TA204). These crosses were further evaluated for diseases resistance and fertility in the F_2 and F_3 generations. In the F_2 generation, there was a marginal increase in fertility among the four crosses. The crosses CF16 and CF19 had seed set up to 30 per cent in some plants, which were also resistant to SLB. In the F_3 generation, resistant plant with high pollen fertility (40-80 per cent) and seed set (20-60 per cent) were obtained in two crosses (CF 16 and CF19). In both these crosses *A. cepa* was used as a female parent. CF19 progenies also had fairly well developed bulbs. The F_3 progenies thus generated in this programme combine the traits of both *A. cepa* and *A. fistulosum* parents. These progenies are now being used in our breeding programme to develop SLB resistant onion lines.

Fusarium basal plate rot (FBR), caused by *Fusarium oxysporum* f. sp. *cepae*, is an important soil-borne disease of onions worldwide. The causal organism infects the basal stem plate of the bulb and eventually kills the entire plant through degradation of the basal plate. Numerous intermediate- and long-day onion hybrids possess moderate levels of resistance to FBR. Although this resistance is not absolute, losses to FBR can be significantly reduced through the use of FBR-resistant cultivars. In USA, under field conditions in southern New Mexico, USA, two fall-planted cultivars, 'NuMex Dulce' and 'NuMex Vado' exhibited moderate resistance to FBR. Among spring-planted cultivars, 'Dawn', 'Impala', 'La Nina', 'Navigator', 'NuMex Casper', 'NuMex Centric', 'Riviera', and 'Utopia' showed high levels of FBR resistance while 'Aspen' and 'Frosty' showed moderate levels of resistance when grown in fields infested with FBR (Cramer, 2000). Dutch variety Daytonna possessed tolerance of the root rot, but was susceptible to basal rot and storage decay (Villanueva-Mosqueda, 1996). Resistant cultivars are available for intermediate and long-day onions but not available for short-day onions. However, Ganeshan *et al.* (1998) reported that three onion lines, IIHR-141, IIHR-506, and Sel 13-1-1 were resistant to FBR in both laboratory and field screenings replicated over years. In addition, Indian breeding lines, 'Hybrid-1', 'IIHR Yellow', and 'Sel. 29' were resistant to FBR from seed and bulb infection (Somkuwar *et al.*, 1996). In Brazil, cultivars, 'Bola Precoce', 'Roxa do Barreiro', 'Cebola de Verao', 'Crioula', 'Monte Alegre', 'Pera IPA 3', 'Roxa IPA 3', and 'Texas Grano 502' were considered resistant to FBR at harvest after inoculation of transplants. After bulbs were stored for 90 days, only 'Cebola de Verao' was considered resistant to FBR. Latent infections of bulbs in the other cultivars reduced bulb yield during storage (Stadnik and Dhingra, 1996).

A gene encoding a non-specific lipid transferase that has activity against 12 types of pathogenic fungi has been isolated from onion (Phillippe *et al.*, 1995) and PGIP protein with antifungal activity has also been identified (Favaron *et al.*, 1997). Potyviruses (*e.g.* onion yellow-dwarf virus, leek yellow-stripe virus), carlaviruses (*e.g.* garlic latent virus, shallot latent virus) (Walkey, 1990) and garlic and shallot virus X (Song *et al.*, 1998) are the most devastating *Allium* viruses. While *in vitro* elimination is possible (Fletcher *et al.*, 1998), inbuilt resistance would provide a simpler solution. Researchers have isolated and sequenced coat-protein gene sequences from *Allium* carla virus (Tsuneyoshi *et al.*, 1998b) and potyvirus types (Kobayashi *et al.*, 1996; Tsuneyoshi *et al.*, 1998a; van der Vlugt *et al.*, 1999).

Screening for resistance to the virus was undertaken with 30 Welsh onion (*Allium fistulosum*) varieties under artificial inoculation and more than 80 varieties under natural field infection by Mei *et al.* (1999) revealed that varieties with dark coloured leaves or with a thicker leaf wax layer were considered most resistant. Of 7 virus-resistant varieties screened, 4 showed stable resistance from the seedling to the flowering stage.

14.4.2 Breeding for Insect-Pest Tolerance

Hamilton *et al*. (1999) suggested that greater genetic gains for thrips resistance in onion can be achieved by selection on a family basis rather than using single plant selection. The study results of Patil *et al*. (2015) revealed that Bellary Red, L-819 and Arka Kalyan during *kharif* and NHRDF Red3, Bellary Red, Bheema Red and Bheema Shubra during *rabi* were moderately resistant to thrips.

Tripathy *et al*. (2013) reported that onion genotypes, NRCRO-3, NRCWO-3, NRCWO-4 and VG-19 showed tolerance to both thrips (25.91 to 28.28 thrips/plant) as well as purple blotch disease (PDI of 42.83 to 51.66 per cent). They identified Bhima Super, NRCRO-4, NRCWO-3 and Arka Niketan for high yield potential with tolerance to both onion thrips and purple blotch disease in Odisha condition.

14.4.3 Breeding for Stress Tolerance

Onion crop requires about 350-500 mm of water during its growing season. The occurance of water deficit stress in onion bulb crop results into split, double centric bulbs, sometime skin cracking which ultimately resulting into tremendous yield losses. Onion crop is highly sensitive to moisture deficit stress due to its shallow root system, which restricts its water absorption capacity zone to only about 18-25 cm of the top soil. This root architechture often makes the onion crophighly sensitive to water stress.

Bulb yield of onion was not severely affected under 10-30 per cent of drought stress. However drought stress beyond 30 per cent of onion crop during any of its growth phase reduces the bulb yield by 15-30 per cent. A water defit stress during early bulb initiation and development stage did not hamper the onion bulb yield and size. However, water deficit stress during bulb enlargement stage resultrd into significant yield losses with poor size onion bulbs. Postharvest losses were also higher with 50 per cent water deficit stress at this stage (Ghodke *et al*., 2017a)

Similarly, in onion seed crop, the reproductive stage is highly sensitive to drought stress limiting the seed production and quality. Water deficit decrease the amount of sugary nector produced by onion florests that decrease the number of pollinator visit. Drought stress also reduces the amount of viable pollen grains per flower, increase floret abortion and finally contributes to the lower seed yield. Bolting and anthesis stage are the most sensitive for drought stress in seed crop. Therefore proper irrigation management is necessary for getting quality seed and higher yield.

Identification of drought tolerance genotypes is the prerequisite in stress breeding programme. In addition, stay-green characters like high chlorophyll content with better photosynthesis efficiency and less leaf senescence rate are considered valuable traits for water deficit stress toletance. Likewise induced ROS scavenging mechanism with increased antioxidant enzyme activityfor maintaining the osmoregulationwithin the plant that confer the drought tolerance further reduces the risk of extreme drought situation. ICAR-DOGR identified onion genotypes W-397 (white bulb colour) and Accession 1656 (red bulb colour) which can perform better under limited moisture condition. Bhima Shubhra has moderate drought tolerance ability (Ghodke *et al*., 2017a).

Onion is a relatively salt sensitive crop. Salts influence bulbing and the quality of the bulbs harvested. Under saline soils conditions the reduced water supply of crops is the most critical growth factor. The permeability of cell walls of onion plants to solutes and water is deferentially reduced by stresses such as salinity. It is postulated that short root hairs of onion contribute to a lower salt tolerance, whereas long root hairs of other crop enhance water uptake from saline soils and crop salt tolerance (Schleiff, 2008). Salts cause alterations in key physiological processes, due to the salt-induced osmotic stress and the specific ionic effect, which in turn result in water deficit, ionic toxicity and plant nutritional imbalances (García *et al.*, 2020).

Identification of onion cultivars which able to maintain productivity at low or moderate levels of salt stress may provide a cost-effective solution to this situaltion. Cultivar variation in onion for salt concentration was reported by Correa *et al.* (2013) and Hussein and El-Faham (2018). Sudha and Riazunnisa (2015) identified Agrifound white as tolerant one and LINE - 28 as susceptible to salt stress. García *et al.* (2020) suggested that the lower sensitivity to salt in some onion genotypes could be related to a better performance of the antioxidant machinery under salinity conditions. They elaborated that as salt stress being first perceived by the root system, the root apoplastic antioxidant defenses were enhanced in the salt-resistant genotype compared to the salt-sensitive genotype. El-baky *et al.* (2003) compared the degree of lipid peroxidation, CAT, SOD and POX enzymes behaviour and glutathione content in leaves of three onion varieties under saline conditions and they found that the most tolerant cultivar showed an increased antioxidant capacity.

In India, heavy rainfall or soil flooding is a serious environmental consequence that severely limits onion production during *kharif* season. The overall damage in onion bulb yield due to heavy rainfall in standing crop ranges from 50-80 per cent and sometime prolonged waterlogged conditioncauses the total crop failure. Onion being a shallow rooted crop is extremely sensitive to excess moisture stress condition. Leaf yellowing due to premature leaf senescence and wilting are the visual morphological symptoms of excess moisture stress. Flooed soils also suffer from low light stresses that affect photosynthesis. Additionally, it limits the other gas exchange processes like transpiration and respiration in both leaves and roots. Excess soil moisture causes the negative impact on plant growth and development by reducing the plant inorganic nutrient absorption efficiency. Heavy rainfall and high atmospheric moisture during *kharif* season favours various diseases out break both in nursery and standing crop. Identification of excess moisture stress and waterlogging tolerant genotypes is required to address this problem. W-208 (white onion) and KH-M-2 (red onion) were identified as promising lines which can withstand frequent to prolonged excess moisture condition. Additionally, Bhima Shweta, Bhima Shubhra and Bhima Dark Red have moderate waterlogging tolerance ability (Ghodke *et al.*, 2017a).

14.5.0 Breeding for Quality

Pungent and sweet taste, texture, and several other qualities are required for onions to be suitable as processing ingredients (Narayan *et al.*, 2019). In the case of processing cultivars, the most important characters for dehydration have been listed by Sethi *et al.* (1973). The quality of the product is mainly dependent on uniform white colour, globe shape for convenient handling, high solids soluble and insoluble preferably more of non-reducing sugar, high degree of pungency (pyruvic acid 0.50-0.70 per cent) and its better retention after dehydration. In addition, they suggested for better recovery, the root zone should be small and the cultivar should be free from doubles. Mitra *et al.* (2012) listed some essential characteristics that should be present in onion cultivar suitable prior to drying. White coloured flesh with total solid content 15-20 per cent and having high pungency is strongly recommended for drying. Insoluble solid should be high whereas, ratio of reducing to non-reducing sugar should be low to lessen discolouration and browning during drying.

Pal and Singh (1987) reported that pyruvic acid content is positively associated with storage and drying quality was controlled by both additive and dominance gene effects, the former being more important. While breeding, attention should be given on good sized bulbs having high solids, since these characters are usually negatively correlated. Develop of non-lachrymatory onion can be possible by suppressing the lachrymatory factor synthase gene while increasing the yield of thiosulphinate (Imai *et al.*, 2002).

Many countries prefer sweet and less- or non-pungent salad type onions. Current "tearless" onion cultivars (*e.g.* 'Vidalia') are achieved through deficient uptake and partitioning of sulfur and/or growth in sulfur-deficient soils. However, Eady *et al.* (2008) genetically manipulated the sulfur secondary metabolite pathway of onion using RNAi. They successfully decreased endogenous LFS (lachrymatory factor synthase) production through a single simple transformation event. This unambiguously lowered LFS activity in the plant, and the consequence of this was a significant reduction in the production of the irritant LF (lachrymatory factor). This reduced lachrymatory synthase activity by up to 1,544-fold, so that when wounded the onions produced significantly reduced levels of tear-inducing lachrymatory factor. They noted that this silencing had shifted the trans-S-1-propenyl-L-cysteine sulfoxide breakdown pathway so that more 1-propenyl sulfenic acid was converted into di-1-propenyl thiosulfinate. A consequence of this raised thiosulfinate level was a marked increase in the downstream production of a nonenzymatically produced zwiebelane isomer and other volatile sulfur compounds, di-1-propenyl disulfide and 2-mercapto-3,4-dimethyl-2,3-dihydrothiophene, which had previously been reported in trace amounts or had not been detected in onion. LF is the cause of the unpleasant pungent aroma associated with cut onions and the cause of heat and pungency flavor notes. Disulfide and dihydrothiophene compounds are associated with the sweeter aroma of cooked or fried onions. Trans- and cis-zwiebelanes are associated with sweet raw onion sulfur tastes and sweet brown saute flavor notes, respectively.

Yoo *et al.* (2020) studied the recurrent selective breeding program to produce low-pungency lines in short-day type onions. They tested two parent populations, 'T81079' (n = 754) and 'T81082' (n = 347), with mean pungencies of 4.1 and 3.5 mM pyruvic acid, respectively, for pungency. Genetic shifts towards the selection differential were observed. The fourth cycle of selection from a population with a mean pungency of 1.5 mM pyruvic acid was observed. The genes controlling pungency became fairly homogeneous after four generations. The recurrent selection of low pungency shifted the pungency distribution toward lower pungency levels, and therefore developing low-pungency (1.5 mM pyruvic acid) lines will be feasible in about 10 years in short-day onions. A selection pressure of 1-3 per cent is necessary for fast genetic progress.

Onion cultivars not only vary in pungency and sweetness, but also in flavour profile which is reflected by differences in flavour precursor ratios and the type of soluble sugar accumulated (Randle and Galmarini, 1997). Barbieri *et al.* (1998) evaluated volatile components contents in onion cultivars and found that cv. Density had more aroma-giving volatile and sulphurous compounds. Kandoliya *et al.* (2015) studied the nutritional quality along with various parameters contributing antioxidant activity from onion of different red (AGFL Red, Pillipati, JDRO-07-13, Talaja Red) and white (JWO-05-07, GWO-1, PWF-131) type local varieties. They recorded 58.14 to 77.67 per cent DPPH value, comparable amount of flavanoids (0.422 to 1.232 mg/g) and anthocyanine content along with total phenol (8.96-18.23 mg/100 g), Pyruvic acid (1.09 to 1.33 mg/g), ascorbic acid (1.18 to 3.89 mg/100 g), protein (0.79 to 1.27 per cent) and titrable acidity (0.34 0.75 per cent). JDRO-07-13 of Red variety and GWO-1 of white nutritionally found better due to its higher antioxidant property, proteins, carbohydrates and reducing sugar. Aggarwal *et al.* (2016) analyzed the bioactive compounds in seven genotypes of onions (Punjab Naroya, Punjab White, PRO-6, P-305, POH-1, POH-2 and CLX-18021). PRO-6 contained maximum anthocyanin (2.32 mg/100 g), ascorbic acid (11.40 mg/100 g) and total phenols (283.57 mg/100 g). P-305 contained maximum total carotenoids (86.43 µg/100 g) and β-carotene (13.72 µg/100 g). The genotype PRO-6 had maximum antioxidant activity (13.63 per cent). Dangi *et al.* (2018) observed significant variability in the evaluated morphological and biochemical traits of 58 onion genotypes. Pusa White Flat (PWF) was highest yielder and maximum TSS was recorded in 106BS2 (15.30 ± 3.03°B). Arka Kalyan had highest dry matter (15.70 ± 0.57 per cent), Red Creole3 had highest pyruvic acid (6.32 ± 0.17 µmol/ml) and total phenolic content was recorded highest (36.96 ± 2.00 mg/g FW) in Juni. They also identified genotypes with waxy leaves and firm skin. They observed highly significant correlation between TSS and dry matter ($r = 0.45$), pyruvic acid and dry matter ($r = 0.35$) and significant and negative correlation between TSS and TPC (total phenolic content) ($r = -0.32$).

Due to the potential health benefits of quercetin, its importance in onion, and varietal differences, quercetin is a trait of interest in onion breeding programs. Smith *et al.* (2003) studied total flavonol (TF) content in 'Lady Raider' (Bulbs are pearlized, red in colour, grano shaped, and are in the short-day onion class). They registered 79 to 431 mg/kg TF in the parent population and grouped the bulbs by TF concentration into high (>232 mg/kg), medium (203-223 mg/kg), or low (<203 mg/kg). These three populations were sib-pollinated which formed three Sib-one (S1) populations which designated as Sib-one high (S1H), Sib-one medium (S1M), and Sib-one low (S1L) based upon the parent population from which it was generated. They noted different TF ranges in the sibs. TF values of the S1 populations ranged from 228 - 675 mg/kg. S1H has the highest TF flavonol mean (456 mg/kg), S1M mid-range (329 mg/kg), and S1L the lowest (286 mg/kg). Flavonol content in the S1 generation segregated into classes similar to those of parental populations. They concluded that quercetin in onion is likely highly heritable and selection for TFs for increased levels can be achieved.

14.6.0 Breeding for Adaptation

Bulbing in onion is a photoperiodically driven process. Besides day length requirement, temperature is yet another pivotal factor which governs flowering and bulbing in onion. Both of these physiologically antagonistic phenomena are driven by the same group of genes. Economic production of bulb onion requires adaptation to photoperiod and temperature such that a bulb is formed in the first year and a flowering umbel in the second. 'Bolting', or premature flowering before bulb maturation, is an undesirable trait strongly selected against by breeders during adaptation of germplasm.

In short day onion, very low (<10°C) temperature during bulb development leads to bolting of onion. On the other hand, excessive high temperatures (>42°C) at the time of maturity during April-May leads to reduction in bulb sizes. Many varieties have been developed worldwide but are restricted for cultivation to specific season or climate and sensitive to climate change. Looking to over sensitiveness of varieties available, any one of the variety cannot be cultivated through the world. In India short day varieties were developed but that too are restricted for different seasons. *Kharif* or late-*kharif* season varieties cannot be cultivated in *rabi* season and vice-versa. Efforts were initiated to develop varieties which can sustain environmental vagaries and cultivar 'Bhima Super', 'Bhima Raj', and 'Bhima Red' were identified, those can be grown in all the three seasons *viz. kharif* (May to October), late-*kharif* (August to February) and *rabi* (October to April). Similarly in white onion Bhima Shubhra and Shweta is also giving encouraging results. These varieties have less effect of photoperiod and temperature and can sustain up to greater extent in changing climate (Mahajan and Gupta, 2016).

Baldwin *et al.* (2014) conducted linkage mapping and population genetic analyses of candidate genes, and QTL analysis of bolting using a low-density linkage map. They performed tagged amplicon sequencing of ten candidate genes, including the FT-like gene family, in eight diverse populations to identify polymorphisms and seek evidence of differentiation. They observed low nucleotide diversity for most genes, which were consistent with purifying selection. Significant population differentiation was observed only in AcFT2 and AcSOC1. Genotyping in a large 'Nasik Red × CUDH2150' F_2 family revealed genome regions on chromosomes 1, 3 and 6 associated with bolting. Two F_2 families grown in two environments confirmed that a QTL on chromosome 1, designate as AcBlt1, consistently conditions bolting susceptibility in this cross.

14.7.0 Mutation Breeding

In onion, mutation breeding was attempted by workers with a view to improving specific characters. Low doses of some mutagens induced stimulatory effects on seedling height and bulb weight in onion. However, the increasing doses of mutagens decreased germination, seedling height, survival, bulb weight and seed setting. The damages due to mutagens were not similar to all the onion varieties. The variety White Warangal suffered more damages due to reduction in respect of bulb weight and plant survival which were scored in later stages of plants indicating the presence of some recovery mechanism in onion. Of the total number of macromutations detected in M_2, chlorophyll mutations accounted for more than 60 per cent in the three varieties. Sectorial mutations or chimerism in true sense was absent in the types of chlorophyll mutations. Spontaneous chlorophyll mutations frequented in all the varieties and the irradiation of seeds with 12.5 kR gamma-rays increased the induced mutations rate to 14 times. A total of 29 different types of morphological mutations affecting foliage, bulb formation, flowering, seed and growth habit were recorded in three varieties. The order of mutation frequency they observed in three varieties was Pusa Red > Pusa Ratnar > White Warangal (Kataria and Singh, 1989a and b). Kataria *et al.* (1990) tried chemical mutagens *viz.*, ethyl-methane-sulphonate, N-methyl-N-nitrosourea and ethyl-imine along with gamma rays and reported that TSS can be increased from 12.12 to 14.01 per cent in Pusa Red and 10.88 to 13.07 per cent in White Warangal. Changes in morphological characters have been reported by many Japanese and Russian workers.

Kim *et al.* (2004a) identified a new locus (Pink; P) responsible for a pink trait in onions resulting from natural mutations of anthocyanidin synthase. These unusual pink onions were found in haploid populations induced from an F_1 hybrid between yellow and dark red parents and in F_3 populations originating from the same cross. Segregation ratios of red to pink in F_2, backcross, and F_3 populations indicated that this pink trait is determined by a single recessive locus. The transcript level of anthocyanidin synthase (ANS) was significantly reduced in the pink line. Reduced transcription of the ANS gene caused by mutations in a cis -acting element is likely to result in the pink trait in onions. Unlike the five loci determining qualitative color changes, this locus is the first reported gene whose mutant phenotype results in reduced intensity of a color in onions. P locus is likely to be the gene encoding anthocyanidin synthase (ANS), one of the late genes in the anthocyanin pathway. Kim *et al.* (2004b) also observed a natural mutation that yield an unusual gold-colored onion from a F_3 family originating from a cross between US type yellow and Brazilian yellow onions. The gold onions contained a significantly reduced amount of quercetin. They identified a premature stop codon and a subsequent single base-pair addition causing a frameshift in the coding region of the CHI gene in the gold onions.

Inactivation of the gene (DFR-A) coding for dihydroflavonol 4-reductase (DFR) involved in the anthocyanin biosynthesis pathway results in a yellow bulb color in onion. Phylogenetic analysis of DFR-A alleles showed that all inactive alleles were independently derived from four different active alleles. In addition, the close relatedness and diversity of DFR-A mutants implied that all these mutations might have occurred after domestication of onions and had probably been maintained by artificial selection. Song *et al.* (2014) opined that at least nine independent natural mutations of the DFR-A gene are responsible for appearance of yellow onions from red progenitors.

Two genes (DFR-A and ANS) encoding dihydroflavonol 4-reductase (DFR) and anthocyanidin synthase (ANS) enzymes in the anthocyanin biosynthesis pathway, respectively, are complementarily involved in anthocyanin production in onion. Eleven inactive DFR-A alleles have been reported, with only a single inactive ANS allele previously identified. Kim *et al.* (2016) identified two novel mutant ANS alleles responsible for inactivation of anthocyanidin synthase and failure of anthocyanin production in onion. A mutant ANS allele containing a 4-bp insertion at the end of exon1 was identified from yellow bulbs of the F_2 population in which the DFR-A genotype was homozygous for an active allele. The 4-bp insertion caused a frame-shift mutation and resulted in creation of a premature stop codon at the start of exon2. This mutant ANS allele was designated ANS[PS] allele. They identified another inactive ANS allele from the light-red F_1 populations showing complementation between DFR-A and ANS genes. A critical amino acid change of the strictly conserved serine residue into leucine was found in this mutant allele designated ANS[S188L].

Radiation breeding technique usually makes use of γ-ray, laser, ion beam and space to radiate plant. Radiation breeding technique can shorten breeding cycle, break the linkage of disadvantageous characters and improve the rate of mutation. It has become one of important ways for modern crop breeding (Huang and Li, 2008). Pan *et al.* (2012) studied the biological effect of laser-induced mutation on fibrous roots of yellow skin onion. They reported that laser radiation showed stimulating effect on root activity of onion. The variability of net assimilate rate, net photosynthetic rate, respiratory efficiency, protein content, catalase, chlorophyll and sugar content in leaf and stem in onion L_1 generation induced mutation by He-Ne laser was larger than induced mutation by CO_2 laser (Pan *et al.*, 2000; Ren and Li., 2002; Pan, 2014).

Li (2004) developed a red onion variety "Chang Ji 99-3a" by exposing wet seeds of onion in He-Ne laser. It gives high yield, good quality, multiple resistances and more suitable to nature. Another new red peel onion variety "Xicong No.2" was developed by Li *et al.* (2004) using CO_2 and He-Ne laser radiation. This is a high quality early-ripening and low bolting-rate variety developed from variant offsprings of Xichang local onions. A new yellow-peel onion variety "Changji 07-9" with high yield, high quality, multi-resistance and stronger adaptability was bred out from the variation progenies of Japanese yellow-peel onion "96203", irradiated by laser. "Changji 07-9" had the characteristics of strong growth potential, spicy tasteless, good quality, larger bulb, high yield, strong resistance and long storage without bulb separation (Li *et al.*, 2009). Yang *et al.* (2017) isolated the glossy, bright green BianGan Welsh onion (*Allium fistulosum* L.) (GLBG) which was derived from a natural variation in the field, and it was tentatively demonstrated that the GLBG or glossy phenotype is conditioned by a single recessive gene. An analysis of the physiological characteristics related to cuticle wax showed that the density and shape of wax crystals on the leaf surface of GLBG clearly differed from those of the waxy BianGan Welsh onion (BG) and that its epicuticular wax coverage was 50 per cent less than that of BG. Compared to BG, GLBG had poorer growth but higher net photosynthetic rate and carotenoid content.

14.8.0 Breeding for Increased Shelf-life

The bulbing and yield of dry onion cultivars vary, but they depend on the contents of high dry matter and non-structural carbohybrates (fructooligo saccharides) which contribute to keeability.

Dhotre *et al*. (2010) studied the genetic variability, character association and path coefficients in red onion involving 14 genotypes. They recorded high heritability with moderate to high GCV and genetic advance for storage losses due to rotting, sprouting and total loss denoting their possibility of improvement with simple selection. They observed positive and significant association of bulb yield with TSS and number of rings per bulb, and neck thickness was significantly correlated with rotting and total storage loss.

14.9.0 Breeding for Organic Farming

At present organic farmers depend on varieties bred for conventional high-input farming systems. As organic farming refrains from high and chemical inputs, it needs reliable varieties better adapted to organic growing conditions to improve the yield stability and quality of crops.

Gupta *et al*. (2019) screened thirty advancelines/varieties of onion were to assess the suitability of cultivars under organic farming during late kharif under short day conditions. Maximum marketable yield was recorded in DOGR-1168 (47.59 t/ha) followed by EL1044 (45.44 t/ha) and DOGR-595 (45.35 t/ha) and these lines performed better under organic conditions. These lines also recorded more than 85 per cent marketable yield, more than 95 g average bulb weight, less than 6 per cent doubles and less than 10 per cent bolters and were found highly suitable for organic farming.

15.0 Biotechnology

Onion breeding faces some inherent difficulties. The species has a biennial habit and many cultivars are open pollinated; and high levels of heterozygosity must be maintained. Interspecific hybrids are sterile or have poor fertility causing problems in the transfer of characteristics between varieties or species. There is an urgent need for introduced resistance to pests and diseases, because gene pools of most cultivated alliums lack sources of resistance.

However, due to the biennial generation time of the onion, the development of value added populations and hybrids is a time consuming and expensive process. And the work with wild species is rather painstaking and complicated. The use of biotechnological approaches, such as marker aided selection, production of doubled haploids, gene editing, and cytoplasmic conversions, offers great promise for population improvement and hybrid development addressing changes in consumer preference and production environments (Havey, 2018). Genetic modification using biotechnological tools has a significant role to play in overcoming a number of technical problems encountered in *Allium* crop improvement. Traits that might be altered in *Alliums* include their sulphur biochemistry, pigmentation, fructan metabolism, and susceptibility to environmental conditions and to specific pests and diseases.

Onion biotechnology has made rapid strides and regeneration has been successful from various explant sources. Luther and Bohani (1999) have been able to induce direct organogenesis. These protocols can be very successfully used for *Agrobacterium* transformation procedures. Production of gynogenic lines for induction of haploidy from the female reproductive organs has helped in the development of genetically uniform pure onion lines. Besides, refinement of molecular studies will enable the taxonomy of *Allium* more evident (Klass, 1998).

15.1.0 Tissue Culture

All commercially important *Allium* species have been to some extent cultured *in vitro*. Apart from minor modifications to suit particular cultivars or clones, they generally respond to media and growth regulators in a typical manner (forming callus on media containing high auxin levels and shoots on media containing high cytokinins). In vitro techniques have been used with *Allium* for: clonal propagation; (2) production of disease-free clones; (3) germplasm conservation; (4) development of new cultivars through somaclonal variation and molecular genetics; and (5) secondary metabolism studies (Eady, 1995).

15.1.1 Micropropagation

Bulblet formation, dormancy of plantlets, vitrification of tissue and decreasing regeneration ability are the main factors limiting the efficiency of onion micropropagation (Kamstaityte and Stanys, 2004).

Shoot regeneration ability of *Allium cepa* was investigated using meristems and shoot-tips (Barringer *et al.*, 1996), basal plate slices, shoot (Mohamed-Yasseen *et al.*, 1995), inflorescence (Barringer *et al.*, 1996), root-tips, zygotic embryos (Zheng *et al.*, 1998) and inner leaf tissues of dormant bulbs. Disease-free bulbs of onion were produced using meristem culture (Dijke *et al.*, 1995).

In vitro regeneration of callus, which requiring totipotency for regeneration, has been developed for onion. The most efficient of the callus systems are probably basal plate callus and embryo/seedling derived callus. Field-grown onion bulbs are, however, very difficult to surface sterilise which restricts the amount of material readily available for transformation studies. To improve the chances of obtaining transformants there must be a high proportion of cells within a particular callus that can both receive DNA and regenerate. Unfortunately, most onion callus usually has a low frequency of regeneration (Eady, 1995).

Hussey and Falavigna (1980) observed that the adventitious shoots were induced firstly on twin scales cut from small bulbs and subsequently on split *in vitro* shoots used assecondary explants, on media containing 6-benzylaminopurinewith or without 1-naphthaleneacetic acid according to thecultivar. They found that more shoots were induced in 16 h than in 8 h days, but day length had negligible effect on the growth of *in vitro* shoots, dormancy being delayed slightly in short days. Adventitious shoots were consistently initiated from at least two tissue layers, the epidermis and hypodermis, on the abaxial surface of leaves and scales close to the basal plate where the frequency of endopolyploid cells was lowest. They suggested that the multicellular origin of the shoots ensures the genetic stability essential for multiplication and storage of elite material.

Viterbo *et al.* (1992) studied plant regeneration from callus cultures of *Allium trifoliatum* subsp. *hirsutum* fertile accession F 370. They obtained best proliferation on modified BDS medium supplemented with (mg/1): 0.75 picloram, 2.0 benzyladenine, and 900 casein hydrolysate on *in vitro* basal leaf explants. Shoot and root organogenesis were obtained in 3 to 5 month old subcultured calli, on BDS or MS medium supplemented with (mg/1): either 0.03 picloram or no auxin, 2 BA or 2 isopentenyladenine, and 900 casein hydrolysate. They observed direct bulb formation, without shoot elongation, occurred on BDS medium with 10 mg/1 IBA. However, under these conditions, callus formation and organogenesis were not obtained with *A. trifoliatum* subsp. *hirsutum* var. sterile, a male sterile genotype. Later, Viterbo *et al.* (1994) developed a procedures for *in vitro* clonal propagation and for cold-storage of propagules for fertile and male-sterile genotypes of *Allium trifoliatum* subsp. *hirsutum*, var. *hirsutum* and var. *sterile*, respectively. They achieved highest rate of shoot multiplication from basal leaf and umbel explants on a modified BDS medium supplemented with 9 mg/l benzyladenine. However, they observed that naphthalene acetic acid reduced the propagation rate, and was not required for shoot multiplication. The resulting shoots were rooted in an indole butyric acid-supplemented medium, and bulbing occurred upon exposure to a 16 h photoperiod. The small dormant bulbs were transplanted into potting mixture and sprouted after termination of dormancy, resulting in phenotypically-normal plants.

Jeong *et al.* (1997) cultured shoot-tips, including 1-2 leaf primordia of the cv. Samda on liquid BDS medium containing both auxins and cytokinins under continuous lighting (2000-8000 lux) in vertically rotating gyration drum with a speed of 2 revolutions/min. Induction of shoot primordia was dependent on the kind and concentration of plant growth regulators. The best induction was obtained using a combination of picloram (0.5 mg/l) or 2, 4-D (1 mg/l) + BA (1 mg/l). Shoot formation in response to thidiazuron (TDZ), a potent cytokinin was studied (Mohamed-Yasseen and Splittstoesser, 1991). Rodrigues *et al.* (1997) standardized the culture conditions and sterilization of *in vitro* micropropagation using basal disc and stem bud explants of onion cultivars. The highest regeneration frequency (60 per cent) was obtained using 2.0 mg/l BA and 1.0 mg/l IBA. The highest shoot number per explant (5.15) was obtained with 4.0 mg/l BA and 1.0 mg/l NAA with a multiplication rate of 20.60 shoots per bulb. For cv. Pira Ouro stem bud explants, the highest multiplication rate was 6.5 using a combination of 2.0 mg/l BA and 0.5 mg/l IBA, while for cv. Baia Piriforme this was reduced to 2.6, using 1.0 mg/l BA and 1.0 mg/l IBA. Rodriguez *et al.* (1996) cultured sections of bulbils of onion cultivars 90 days after culture in MS medium supplemented with 2.0 or 4.0 mg/l BAP in combination with 0.25 g or 0.5 mg/l NAA. Shoot number was highest in cultivars when cultured with lower concentrations of growth regulators and placing explants vertically increased the percentage of regeneration. Jeong *et al.* (1998) standardized a method for the *in vitro* production of onion plantlets involving shoot induction from flower buds and plant regeneration. The best medium was BDS supplemented with BA (2 mg/l), 2, 4-D (2 mg/l), myoinositol (100 mg/l), sucrose

(100 g/l) and agar (6 g/l). Pike and Yoo (1990) developed a tissue culture technique to obtain large numbers of plants rapidly from immature flower buds taken from unopened spathes and cultured on modified MS medium containing 0.5 mg/l NAA and 5.0 mg/l BA. About 10 per cent of the flower buds gave rise adventitiously to shoots and 70 per cent developed into deformed flowers which produced a greater number of shoots than those which arose adventitiously. Kamstaityte and Stanys (2004) took onion cvs. 'Lietuvos didieji', 'Stutgarten Riesen' and 'Centurion' F_1 for micropropagation experiment. They used MS medium, supplemented with 1 mg l^{-1} naphthaleneacetic acid, 0.9 to 13.1 μM concentrations of 6-benzylaminopurine (BAP), kinetin (1.1 to 5.8 μM) and 30 g l^{-1} sucrose for their study. The highest numbers of microshoots (1.8 to 2.4 microshoots per explant) were formed by 'Centurion' F_1 and 'Lietuvos didieji' explants, containing stem dome plus basal plate. Experiments with growth regulators showed that the number of microshoots increased (from 1.0 to 2.1 microshoots per explants) when the BAP concentration was raised from 0.9 to 4.4 μM, respectively. The highest micropropagation frequency using kinetin (1.9 to 2.1 microshoots per explant) was obtained at a moderate (10.6 μM) concentration. The regeneration intensity (output of microshoots) was 68 per cent higher using kinetin in comparison with BAP.

Jeong *et al.* (1997) reported regeneration of haploid plants when unpollinated flower buds were used as explants rather than ovules or ovaries. The most effective stage of megaspore development was 6-10 days prior to anthesis. Explants from different-sized flower buds of *Allium cepa* cultivars Wolska and Kutnowska (greenhouse grown) and Wolska, Czerniakowska and Rawska (grown in the field) were cultured on various culture media at 3 different temperatures (Michalik *et al.*, 1997). The highest percentage of ovaries producing embryos was obtained from mature flower buds (4.3-4.5 mm diameter) collected 1 or 2 days before opening. Of the regenerants obtained, 58 per cent were diploid, 33 per cent were haploid and 9 per cent were polyploid.

Keller *et al.* (1996) made interspecific crosses by hand-pollination of *Allium cepa* with pollen of 19 species belonging to nine sections of two subgenera of the genus *Allium*. In all cases they obtained the viable plantlets ovary culture. The efficiency depended on the relationship of the pollen donor to *A. cepa*. Hybrids were confirmed for 18 new species combinations. The viable hybrid of the most distant cross resulted from crossing *A. cepa* with *A. sphaerocephalon*.

Onion is multiplied through seed or sets and requires two-years to complete one seed cycle necessitating the dependence on huge resources and the involvement of high risks (Sidhu *et al.*, 1992). Poor seed viability, very high out crossing, bulblet formation, dormancy in plantlets, vitrification of tissues and decrease in regenerability for natural vegetative multiplication are some of the limitations in propagation of onion in the open field. Reports are available on *in vitro* callus induction, shoot regeneration and micropropagation in long day onion, but information on short day onion is lacking.

Passi *et al.* (2018) studied the *in vitro* micropropagation of short day tropical varieties of onion 'Agrifound Dark Red', 'Punjab Naroya' and 'PRO-6', which were exposed to different concentrations and combinations of growth hormones.

They reported that pre-sterilization of basal plate of onion in 0.5 per cent solution of bavistin followed by treatment with 0.1 per cent mercuric chloride for 10 min. produced the highest rate of survival of explants (47.9 per cent). Survival was further enhanced to 53.17 per cent with the addition of 750 ppm cefotaxime in MS medium. They noted varietal differences for *in vitro* establishment, multiplication and root induction. Among various combinations of growth hormones, MS medium supplemented with 4.0 mg/l BAP + 0.5 mg/l NAA, 2 mg/l BAP and 0.5 mg/l NAA and half MS carrying 1.0 mg/l IBA + 0.5 mg/l NAA produced the highest *in vitro* establishment (53.07 per cent), multiplication (64.46 per cent) and rooting (66.37 per cent) respectively, in 'Agrifound Dark Red'.

Patena *et al.* (1998) developed seed production systems for garlic and shallot *via* the *in vitro* shoot multiplication technique in either Gamborg's B5 and Shenck and Hildebrandt media (for garlic) or MS (Murashige and Skoog) medium (for shallot), subsequent *in vitro* bulblet (G0) formation and finally, field production of generation 1 (G1) to generation n (Gn) bulbs.

15.1.2 *In vitro* Bulb Formation

Onion belongs to a family of which several tissues express a high regeneration potential (Hussey, 1976), therefore *in vitro* technique has been introduced to it for vegetative multiplication. Bulbing in onion has been studied for years by agronomist and plant physiologists in onion. Bulb formation with micropropagatd plants also allows conversation possibilities in onion growing area. Even so onion tissues were capable of shoot regeneration for only a limited length of time, bulbiet formation, plantlet dormancy, and a decrease in regenerative ability or in multiplication rate have been previously reported as limiting factors to micropropagation of onion.

A successful multiplication process for several authors from basal parts of one bulb have described a single cycle of regeneration cultivated *in vitro* on MS medium supplemented with NAA and BA or kintein (Khalid *et al.*, 2001). Mohamed-Yasseen *et al.* (1993) cultured inflorescence explants of onion on MS medium containing various concentrations of TDZ (0.0, 0.005, 0.01, 0.05 or 0.1 μM). Regenerated shoots from explants were cultured on medium with BA and trans-ferred to bulb induction medium after four weeks. Multiple shoots (10.6) were obtained from explants cultured on BA-containing medium. Further, they found that for shoot regeneration hormonal supplementation was not essential since plantlets were regenerated in the basal medium (MS). The number of bulbs per explant from indirect and direct bulb formation was 9.7 and 6.5 respectively and they could be transferred to soil without acclimatization. In this system, normal viable plants were obtained from bulbs induced *in vitro*. Hence, this method has potential to propagate disease-free onion bulbs in relatively less time. Mohamed-Yasseen *et al.* (1995) described the procedure for bulb formation from onion explants. They culture the explants from cut stem bases in shoot induction medium composed of MS medium with or without N6-benzyladenine. Shoots produced were then transferred to bulb induction medium composed of MS medium containing 5 g/liter activated charcoal and 120 g/liter sucrose under a long-day photoperiod and 28°C. They reported that bulbs were also produced from onion directly, without passing through shoot formation, when explants were cultured in the bulb induction medium. Bulbs were transferred to soil without acclimatization and produced viable plants. Keller (1993) induced *in vitro* cultivated seedlings of onion and leek to form bulblets by increase in sucrose concentration (30, 50, 150 g/1), and addition of benzyladenine (BA-0, 12.5 mg/1), or ethephon (0, 5, 20 days). He obtained highest bulbing ratios within combinations of sucrose and ethephon treatments. BA caused not only bulb swelling but also an increase of multiple adventitious shoot formation. Khalid *et al.* (2001) studied the effect of growth regulators on plantlet regeneration and bulbing of onion *in vitro*. They found that shoot induced from twin scale in 0.1 mg/l NAA was with a percentage of 93 per cent and per explant shoot rate of 0.84. A combination of 0.1 mg/I NAA + 2 mg/I 4PU30 induced 89 per cent shoot regeneration from twin scale. The highest per explant shoot (1.25), however, was observed in 0.1 mg/I NAA + 1 mg/I 4PU30. The highest root percentage (89 per cent) in twin scale was obtained with a combination of 0.1 NAA + 3 mg/I BA, and highest per explant root (1.06) was observed in 0.1 mg/I NAA + 1 mg/1 BA treatment. The highest bulb percent (84 per cent) was induced in 0.8 mg/I 4PU30. The mean bulb circumference at the same concentration was 2.6 cm after six weeks of culture. In sucrose treatment, the highest bulb percentage (81 per cent) was observed in 100 g/I sucrose. Malla *et al.* (2015) studied rapidly growing and maintainable callus formation using the basal meristem plate and twin scale leaves as explants in Bellary and CO-3 varieties for callus induction and *in vitro* bulb formation. They found maximum frequency of callus induction on MS medium fortified with B5 vitamins supplemented with 0.5 and 1 mg/L picloram. Regeneration of plantlets from the callus was observed on MS supplemented with 0.5 mg/L each of BAP and KIN and 0.1 mg/L NAA. They observed bulbing along with whorl formation of scales leaves in presence of equal concentrations of BAP and TDZ.

In vitro plant production by direct organogenesis from immature flower heads is an ideal approach for clonal propagation of onions. This technique ensures genetic stability, high propagation rate, and maintains donor plant of explants with anadvantage over other means of in vitro regeneration. According to Marinangeli (2012), for *in vitro* culture, the mature onion bulbs weed to be induced to reproductive phase by vernalization and forced to inflorescence initiation. Immature umbels were dissected from bulbs or cut directly when they appear from the pseudo stem among the leaves. Disinfected inflorescences were cultivated in BDS basal medium supplemented with sucrose (30 g/L), naphthalene acetic acid (0.1 mg/L), N –benzyladenine (1 mg/L), and agar (8 g/L) in pH 5.5 and under 16 h photoperiod white fluorescent light (PPD: 50–70 µmol/ms) for 35 days. Then the regenerated shoot clumps were divided and subculture under the same conditions. For bulbification phase, the individual shoots were cultured in BDS basal medium containing 90 g/L sucrose, without plant growthregulators, pH 5.5, under 16 h photoperiod. The observed that the microbulbs can be directly cultivated *ex vitro* without acclimation.

Kahane *et al.* (1992) studied the bulbing in long day onion cultured *in vitro*. According to them the bulb formation was mainly dependent either on sucrose concentration in the culture medium, or on light spectral composition. Raising the sucrose concentration from 40 to 120 g/l increased plant basal swelling and stopped further vegetative development. These plants were not dormant. When fluorescent light was enriched in incandescence during a long day period, bulbs were obtained in two months and underwent a consecutive dormancy. Similarly, Kahane *et al.* (1997) obtained basal swelling of onion without light induction, only with high sucrose concentrations (> 80g/1) or with ethephon (> 2 mg/1). The addition of incandescent lamps to fluorescent light induced bulbing in onion. Combinations of long photoperiod (16 h), low R : FR ratio (<2) and low temperature (3-4°C for 4 months) could lead to bulbing in any of the tested cultivars of onion (Kahane *et al.*, 1997). Kastner *et al.* (2001) studied the influences of light conditions, sucrose and ethylene on *in vitro* formation of onion bulblets in various accessions. The found that light, sucrose and ethylene influenced bulb formation. The bulbing process was characterised by changes in bulbing ratio, leaf length, number of leaves and leaf development time.

Onion bulbing is a storage process controlled by daylength and phytohormones, with Gas playing a key role. Le Guen – Le Saos *et al.* (2002) observed bulbing of shallot plants (*A. sativum* var. *aggregatum* Group) occurred under a 16 hr photoperiod with fluorescent + incandescent light and 30-50 g/1 sucrose in culture medium. They found that exogenous gibberellins (10 μM GA$_3$) inhibited leaf and root growth and bulbing. However, three inhibitors of gibberellins biosynthesis (ancymidol, flurprimidol and paclobutrazol) promoted bulb formation and percentage of bulbing when added to the media. Application of ancymidol causes 66 per cent decrease in sucrose contentin leaf bases but greatly increase the glucose, fructose and fructan contents by 188, 274 and 131 per cent respectively. The addition of ancymidol to the culture medium would reinforce the carbohydrate accumulation induced by the addition of far-red light to a 16 hr photoperiod. This requirement of light enriched in far red radiation indicates that the bulbing response depends on phytochrome action.

15.1.3 Gynogenesis

Gynogenesis refers to the use of unpollinated female gametophyte for haploid development. Therefore, in other words it is the induction of maternal haploids. Bohanec and Jakse (1999) reported that haplody via gynogenesis has great potential for onion breeding since due to its biennial flowering habit, the process of inbreeding is slow. Attempts to produce haploid plants via androgenesis have been a failure. In onion, haploid development through gynogenesis was first reported by Campion and Azzimonti (1988) followed by Muren (1989) and subsequently by Campion and Alloni (1990). A high rate of success in onion through gynogenesis was observed by Luthar and Bohanec (1999) and Bohanec (2009). Factors that govern haploid induction and subsequent regeneration in onion are genotype of donor plant, stage of flower/ovule development, pre-treatment, culture medium and cultural conditions. To date, only gynogenesis has been successfully utilised for development of DHs in onion.

Developmental stage of explant is one of the crucial factors determining the success of gynogenic haploid induction. In onion, haploid plants have been produced

Production of Doubled Haploid Onion Plants by Gynogenesis

(A) Onion umbels at the time of harvest. (B) Optimal stage of flower bud development for culture (3.5–4.5 mm length, flowers tagged with an asterisk). (C) Flower after 7 days of culture. (D) Flowers from cultivar Recas after 45 days of culture. (E) Gynogenetic embryo from Fuentes de Ebro emerging from the ovary at 90 days of culture. (F) Isolated embryo from Rita. (G) Onion embryo treated with APM in liquid medium. (H) OH-1 plant regeneration in Eco2box. (I) Plants during acclimation in greenhouse. (J) Bulb formation in a shade house. (K) Seed production from Rita (Fayos *et al.*, 2015).

from ovules, ovaries or whole flower buds. Gynogenesis can be achieved with *in vitro* culture of various parts from un-pollinated flower such as ovules, placenta with ovules attached, ovaries, or whole flower-buds (Bohanec, 2002; Dhatt and Thakur, 2014). Gynogenesis of four onion cultivars was induced starting from ovules and ovaries, using a 2-step culture procedure by Bohanec *et al.* (1995). They found that the induction frequency was much higher from ovaries and was strongly affected by genotype. Keller (1990) observed that ovule culture was the most laborious and yielded the lowest number of embryo regenerants. Flower bud culture is the simplest way of inducing gynogenic haploids in onion. The most efficient method is to plate whole onion flowers, without sub-culture, to induce haploid plants from cells of the female gametophyte (Bohanec *et al.*, 2003).

Flower buds 3-5 days prior to anthesis were superior to either older or younger ones (Muren, 1989). Musial *et al.* (2001) suggested that embryo sacs at early stages of development which are capable of more parthenogenesis, might be more suitable for haploid induction than mature megagametophytes. According to Alan *et al.* (2004) the most ideal stage to culture the flowers is between 3 to 5 days before anthesis. Musial *et al.* (2005) reported that medium and large flower buds gave better response than the small flower buds. Fayos *et al.* (2015) opined that flowers of 3.5-4.5 mm in length are excellent material for doubled haploid production. Michalik *et al.* (2000) observed that small young buds of 2.8-3 mm length produced significantly fewer embryos than older ones of 3.5-4.5 mm length, while displaying genotype specificity. Alan *et al.* (2004) observed that the small flower buds (<2-2.5 mm) responded poorly, whereas medium-sized buds (2.25-4.5 mm) gave the best results.

Genotype and type of explants are the main determining factors for gynogenesis. According to Bohanec *et al.* (2001) the genetic make-up of donor onion plant and growth conditions plays the most important roles to succeed at gynogenesis. Muren (1989) classified genotypes into high, medium and low response groups based on their haploid induction efficiency and reported similar response of short day materials and long day materials. Genotypic differences for haploid induction in onion depend on day length, geographic origin and genetic constitution. Geoffriau *et al.* (1997) tested variable genetic material from different regions across the world. They found that only two out of 18 onion cultivars showed a high gynogenic potential. Bohanec and Jakse (1999) analyzed the long day onion accessions from Europe, Japan and North America. In terms of geographic origin, genetic material bred in America was on average almost five and nine times more responsive than European and Japanese material, respectively. Very high variability was found among cultivars, and even within inbred lines. Michalik *et al.* (2000) reported a maximum of 10 per cent embryo yield in a breeding line out of 11 Polish cultivars and 19 breeding lines studied. Crossing responsive with non-responsive onion lines resulted in increased gynogenic ability in the hybrid progeny (Bohanec and Jakse, 1999). Geoffriau *et al.* (1997) reported that among genetic structures, inbreds regenerated significantly better than synthetics. Regenerants from inbreds were the most vigorous, whereas, synthetics were confirmed to be good donor material for quality embryos. Jakse *et al.* (1996) compared the influence of different culture media on gynogenic regeneration using four onion cultivars. Marked variation

Production of Onion Haploid Plants with *In vitro* Gynogenesis

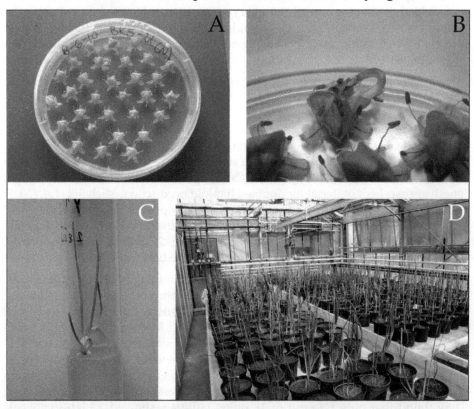

(A) *In vitro* culture of un-pollinated flower buds on BDS medium (Dunstan and Short, 1977) supplemented with 500 mg/l myo-inositol, 200 mg/l proline, 2 mg/l BAP, 2 mg/l 2,4-D, 100 g/l sucrose and 7 g/l agar; (B) germination of haploid embryos after 60 to 180 days in culture; (C) elongation of haploid plantlets and (D) acclimatization of haploid plants in the greenhouse (Murovec and Bohanec, 2012).

in regeneration of somatic embryos was also observed (46-152 days), which was particularly pronounced in genotypes with lower regeneration capacity. Isozyme patterns of regenerants showed that all analyzed regenerants of the cultivar with a high regeneration capacity were homozygous. Ebrahimi and Zamani (2009) investigated the effects of polyamines (putrescine and spermidine) on *in vitro* gynogenesis of two Iranian onion cultivars and found that polyamines could improve onion gynogenesis. Cultivar 'Sefid- e- Kurdistan', showed a higher embryo generation capacity (6.90 per cent) than cultivar 'Sefid- e-Neishabour' (3.33 per cent). They also observed that the medium containing 0.01 mM BA, 0.01 mM 2, 4-D, 2 mM putrescine and 0.1 mM spermidine produced the highest number of gynogenic embryos in both cultivars. About 73.6 per cent of regenerated plants were haploid.

Theoretically, for haploid induction in onion, a maximum of 600 per cent frequency can be expected. However, in practice, yields are low. Jakse *et al.* (2010) used individual plants from the inbred long day onion lines, populations and hybrids and reported gynogenic efficiency in the range of 0.03-0.82 per cent while the responsiveness of the selfed DH lines ranged from 0.00-0.63 per cent. In India, Sivalingam *et al.* (2014) reported gynogenic efficiency in short day onion with a range of 0.9 per cent (Bhima Shweta) to 4.5 per cent (Bhima Shubhra) whereas Khar *et al.* (2018) observed that open pollinated varieties and hybrids were more responsive towards embryo induction followed by exotic and landraces. Fayos *et al.* (2015) reported variation in gynogenic responsiveness in Spanish onion germplasm and found that Valencia type commercial variety 'Recas' had the highest percentage (2.09 per cent) of embryogenesis and the sweet cultivar 'Fuentes de Ebro' had the lowest percentage (0.53 per cent).

Chromosome doubling is the most important step, after obtaining haploids, to get sufficient number of DH lines of adequate fecundity. The spontaneous doubling of gynogenic plants is a rare event in the bulb-onion (Bohanec, 2002). The major problem in genome doubling in onion is inaccessibility of the apical meristem of adult field-grown plants. Therefore, chromosome doubling of haploid onion plantlets should be attempted during *in vitro* propagation. Three different types of explants have been used in chromosome doubling treatments: intact plantlet, split basal, and whole basal explant. Different antimitotic agents like colchicine, dinitroaniline, oryzalin, trifluralin and phosphoric amide herbicide amiprophos-methyl (APM) at various concentrations and for various periods of exposure are being used for chromosome doubling.

May'-yakhina and Polumordvinova (1989) reported polyploidization of interspecific hybrids to be the most practical in breeding work. Tissue with high meristematic potential (the inflorescence or roots at the base of the bulb) was treated with colchicine, which was introduced into the nutrient medium. Subsequent micropropagation combined with selection stabilized ploidy and reduced the percentage of chimaeras (over 75 per cent of regenerates contained 80-100 per cent polyploid cells). Geoffriau *et al.* (1997) assessed the ploidy stability in gynogenic clones. In order to induce chromosome doubling in haploid forms, explants and plantlets of three clones were treated *in vitro* with colchcine (0.625-12.5 mM) and oryzalin (10-200 µM) for 24 h (4°C in the dark or 24°C with a 16/8-h day/night photoperiod). Diploidization performed well (64 per cent) during micropropagation of explants treated with 0.625 mM colchicine under 24°C, 16/8 day/night conditions. Song *et al.* (1993) demonstrated that *in vitro* colchicine treatment of regenerating calli of interspecific F_1 hybrids was effective in recovering tetraploids. Regenerating calli of *A. fistulosum* × *A. cepa* interspecific F_1 hybrids were treated *in vitro* with colchicine. Shoot production of regenerating calli following *in vitro* colchicine treatment decreased with increasing colchicine concentration and treatment time. Bohanec *et al.* (1995) reported that Thidiazuron (2 mg/l) had a positive effect on regeneration and subsequent *in vitro* culture of regenerants. Alan *et al.* (2004) reported the use of colchicine (200-400 mg/l in liquid medium for 48 h) to the whole basal explants from 2- to 4-month-old *in-vitro* haploid plants and the plants became ready for transfer to the greenhouse in 1-3 months and following this method, they recovered fecund DH lines implying their possible use in further research/breeding programme. Bohanec and Jakse (1997) tested the effect of oryzalin and colchicine on halved basal shoots. They opined that diploidization with oryzalin (67 per cent) was better than with colchicine (21 per cent). Grzebelus *et al.* (2004) reported oryzalin, trifluralin and APM as better agents than colchicines for *in vitro* chromosome doubling in onion tissue. However, APM is recommended due to its low toxicity. Alan *et al.* (2007) recommended APM (100 or 150µM) due to its low toxicity and ability to yield results comparable to that with colchicine (750 or 1000µM). Fayos *et al.* (2015) recovered the highest number of doubled haploid plants through application of 25 µM APM in a solid medium for 24 h. Jakse *et al.* (2003) also found that amiprofos-methyl (APM) was superior to oryzalin on the basis of a lower toxicity. A 2 day treatment in liquid media supplemented with 50 mM APM was the most successful with respect to chromosome doubling. They reported that 36.7 per cent of the plants were diploid. Havey and Bohanec (2007) released 'Onion haploid OH-1' to be used as a responsive control for the extraction of gynogenic haploids of onion.

The DH lines are completely homozygous and develop in a single generation whereas the traditionally bred inbreds are nearly homozygous with some residual heterozygosity which results in minor to major heterogeneity in the hybrids developed therein. Besides, the time, space and cost associated with onion inbred development are substantially higher than the doubled haploid production due to biennial generation time and high inbreeding depression owing to accumulation of recessive lethal alleles. Higher uniformity, reduced space requirement production and maintenance cost as well as higher yield makes production of DH lines a viable option for hybrid development (Khar *et al.*, 2019).

15.1.4 Protoplast Culture

The transfer of desirable traits from related species by interspecific crossing has been hampered by sexual incompatibility within the genus *Allium*. Successful crosses between species within the *A. ampeloprasum* complex have been achieved; however successful gene transfer between species of different sections or subgenera is not possible or very di ¦ cult. In this respect, somatic hybridization may be a viable alternative that will enable breeders to combine the genomes of incompatible species and to transfer nuclear or cytoplasmic traits from one species to another. The number of reports on successful realization of somatic hybridization and cybridization of monocotyledons, to which the genus *Allium* belongs, is very limited, in comparison to dicotyledon species. This is due to recalcitrance of the protoplasts to regeneration (Buiteveld *et al.*, 1998).

Plant cell protoplasts are valuable experimental objects for developmental and physiological investigations. Guard cells are of great interest because of their distinctive morphogenetic and physiological properties. Zeiger and Hepler (1976) report on the production and use of protoplasts from the guard cells of the stomatal complex in onion. They reported on the isolation and culture of guard cell protoplasts using microchambers, which permit observation of the entire sequence of wall degradation, protoplast release, and the response of the protoplasts to different osmotic and culture conditions. The digestion course was quantified by following under polarized light the loss of retardation of the birefringent cellulose of the guard cells. Osmotic conditions were crucial for GCP survival. Onion guard cells fragment in the presence of strong plasmolyticum (>0.45 M) indicating cytoplasmic connnections between neighboring guard cells and/or cytoplasmic attachments to the wall. Onion GCP were obtained by digesting paradermal slices with 4 per cent (w/v) cellulysin in 0.23 M mannitol. Protoplasts can be osmotically released by replacing the enzyme solution with 0.23 M mannitol at early stages of digestion. They are also available after prolonged digestion (6-12 hours). Paradermal slices also yield mesophyll and epidermal cell protoplasts but they can be selectively washed away if a pure preparation of GCP is desired. Onion GCP have been kept alive in a simple culture solution for up to 10 days.

In working with the culture of protoplasts attention must be paid to two major aspects. First, effective isolation procedures must be achieved, involving cell plasmolysis followed by enzymatic degradation or partial breakdown of the cell wall leading to protoplast release; second, cultural conditions must be found for maintenance of the protoplasts and provision of the appropriate osmotic conditions, nutrients, and stimuli for cellular proliferation. Bawa and Torrey (1971) isolated the protoplasts were isolated from root callus tissues of onion grown in agar or liquid culture by cellulase treatment. Greatest protoplast release could be affected by the addition of 0.4 M sorbitol to the nutrient medium. They were maintained for over 3 weeks in liquid medium containing either 0.4 1 sorbitol or a mixture of sorbitol and 1 per cent sucrose. A few protoplasts exhibited nuclear divisions on the second day of culture. About 4 per cent -5 per cent became binucleate, and 2 per cent became three- to four- nucleate. Hansen *et al.* (1995) cultured protoplasts in modified K8P liquid medium. Microcalli recovery depended on the number of weeks the cell suspension has been in culture with highest recovery from 4- to 5-month-old cell suspension. Microcalli were moved to semi-soil media when they were approximately 2 mm in diameter. After 4-6 weeks, embryogenic calli thus recovered were moved to variations of standard onion regeneration media containing picloram and BA. Elongating shoots were obtained from up to 88 per cent of the microcalli of one line and 40-50 per cent of the shoots were further multiplied in culture. Karim and Adachi (1997) prepared callus culture from two onion cultivars, one Bangladeshi cultivar 'Manikgonj' and one Japanese cultivar 'Kairyo-Unzenmaru'. For plant regeneration, they took 3-5 mm (diameter) cell clumps from liquid suspension culture, which were placed onto solid BDS regeneration medium supplemented with kinetin alone or in conjunction with ABA at 25°C and 16 h of light per day from cool white fluorescent lamps. Then Protoplastswere isolated from mesophyll cells of aseptic 2-3 weeks old seedlings and 4 to 6 months old suspension cultures. In protoplast culture, they examined the effects of different combinations of 2, 4-D, NAA, BAP and Zeatin on protoplast division and colony formation. Cultivar differences were evident in the ability to produce shoots. They observed very frequent budding of protoplast both in liquid and in agarose bead culture.

Regeneration from protoplasts is very dependent upon genotype (Eady, 1995). Successful plant regeneration has been observed only at low frequency using embryonic suspension derived protoplasts of *A. porrum* (Buitveld and Creemers-Molenaar, 1994).

Buiteveld *et al.* (1998) reported on the production and characterization of somatic hybrids between leek (*Allium ampeloprasum*) and common onion (*A. cepa*). They carried out both symmetric and asymmetric protoplast fusions using a polyethylene-based mass fusion protocol. Asymmetric fusions were performed using gamma ray-treated donor protoplasts of *A. cepa* and iodoacetamide treated *A. ampeloprasum* protoplasts. However, they noted that the use of gamma irradiation to eliminate or inactivate the donor DNA of *A. cepa* proved to be detrimental to the development of fusion calli, and thus it was not possible to obtain hybrids from asymmetric fusions. The symmetric fusions yielded a high number of hybrid calli and regenerated plants. The analysis of the nuclear DNA composition using interspecific variation of rDNA revealed that most of the regenerated plants were hybrids. Flow cytometric analysis of nuclear DNA showed that these hybrid plants contained a lower DNA content than the sum of the DNA amounts of the parental species, suggesting that they were aneuploid. A shortage of chromosomes in the hybrids was confirmed by genomic in situ hybridization. Chromosome counts in metaphase cells of six hybrids revealed that these plants lacked 2-7 leek chromosomes. One hybrid showed also the loss of onion chromosomes. The hybrids had an intermediate phenotype in leaf morphology. Shimonaka *et al.* (2002) produced interspecific somatic hybrid plants between Japanese bunching onion (*A. fistulosum* L.) and bulb onion (*A. cepa* L.) *via* electrofusion of protoplasts. IOA treated protoplasts of Japanese bunching onion and untreated protoplasts of bulb onion were mixed together and fused. Four hundred and seventeen colonies reported to form from fusion products after 45 days culture, of which approximately 80 per cent developed into calli. Plant regeneration was achieved in 33 out of 325 (10.1 per cent) calli. They noticed many abnormalities such as depressed root formation, twisted leaves and slow growth. They successfully transplanted two regenerants in a greenhouse, which were found amphidiploids (2n = 4x = 32). Another three regenerants possessed the nuclear genome of Japanese bunching onion, whereas, their chloroplasts were from bulb onion.

15.1.5 Somatic Embryogenesis

Establishing a high-frequency regeneration system through tissue culture is an important prerequisite for genetic modification. Different pathways for onion in vitro regeneration exist, including organogenesis, somatic embryogenesis and haploid culture. Somatic embryogenesis is a promising approach to plantlet regeneration. Immature or mature seeds can be used as initial inocula to induce somatic embryogenesis in *Allium* species. Regeneration system using onion immature embryos is limited by the time of harvesting immature embryos. Also, the plantlet regeneration rate was relatively low. Compared with immature embryos, mature seeds are easy to obtain and do not have seasonal limitations (Wu *et al.*, 2015).

Organogenic responses have been reported in onion by many workers. Scale bases from bulbs (Kahane *et al.*, 1992) and floral parts like receptacles (Matsubara and Hihara, 1978) and immature flower buds (Luthar and Bohanec, 1999) have been used to induce direct somatic organogenesis. Callus has also been induced from seedling radicle, seedling leaf sheath, immature embryos, unfertilized ovules and basal plates. Phillips and Hubstenberger (1987) standardized procedures for micropropagation and plant regeneration from callus of specific crosses *A. cepa* × *A. fistulosum, A. cepa* × *A. galanthum* and *A. cepa* × *A. oschaninii*. All the genotypes studied gave rise to shoots and most to regenerated plants. Luthar and Bohanec (1999) also developed a novel method for direct organogenesis in onion wherein multiple shoot structures were induced on mature flower buds or ovaries in a two-step culture procedure. Flowers were cultured on an induction medium containing 2 mg/l 2, 4-D and 2 mg/l BAP.

Van der Valk *et al.* (1992) studied the regenerative ability of zygotic embryo-derived callus cultures of 12 *A. cepa* varieties and accessions. Addition of abscisic acid to the regeneration medium stimulated the formation of both somatic embryos and shoots in a number of varieties. These cultures had potential to regenerate plantlets up to one year and proved to be target tissues in genetic transformation of useful traits in *Allium*. Eady *et al.* (1998) obtained somatic embryos from immature embryos of onion and regenerated plantlets. Supplementation of 5 mg/l picloram and culturing 0.5-1.5 mm size immature embryos resulted in increased rate of somatic embryogenesis. The production of somatic embryos was significantly affected by the addition of auxin, embryo size and cultivar. Further, they observed that somatic embryos were formed either directly on the surface of embryos or from compact cultures. Saker (1998) reported a reliable protocol for the regeneration of onion through repetitive somatic embryogenesis. Embryogenic callus was derived from mature seeds on MS medium supplemented with 2, 4-D 2 mg/dm^3. Somatic embryos formed on the surface of calli cultures and produced plantlets after the removal of 2, 4-D or after it was substituted with kinetin 1 mg/dm^3. Hailekidan *et al.* (2013) developed an efficient protocol for *in vitro* regeneration of shallot. They took two local shallot varieties (Huruta and Minjar) where basal discs were used as explants. Maximum callus induction was observed in genotype Huruta (81.11 per cent). In combined effect both genotypes Huruta and Minjar showed highest callus induction (74.44 per cent) from basal discs placed in medium supplemented with 1 mg/l 2,4-Dichlorophenoxyacetic Acid. Maximum callus fresh weight of 1.26 and 1.20 g were achieved with 1 mg/l 2, 4-Dichlorophenoxyacetic Acid and α-naphthaleneacetic acid combined with 1 mg/l 6-Benzylaminopurine, respectively. They obtained the regenerated plants via somatic embryogenesis and organogenesis. Murashige and Skoog medium supplemented with 5.0 mg/l 6-Benzylaminopurine

Somatic Embryogenesis of Onion

a, Shoot apex explants cultured on MS medium b, Initiation of callus from shoot apex explants after 15 d of incubation in the dark on MS medium supplemented as in (a). c, Enlargement of callus after 3 wk of incubation in the dark on MS medium supplemented as in (a) (arrow shows primary embryogenic callus at globular stage). d, Formation and maturation of somatic embryos after 5 wk of incubation in dark on MS medium (arrow indicates embryogenic callus at globular stage). e, Germination and shoot regeneration of somatic embryos after 2 wk of culture under a 16-h photoperiod on MS medium f, Shoot elongation from germinated embryo after 2 wk of culture under a 16-h photoperiod on MS medium g, Root induction from elongated shoots after 2 wk of culture on MS medium. h, Plants maintained in paper cups covered with polythene bags for hardening. i, Plant maturation in soil after hardening. j, Maturation and bulb setting of soil-grown plants (arrow indicates bulb) (Ramakrishnan *et al.*, 2013).

+ 0.1 mg/l α-naphthaleneacetic acid showed higher percentage of shoot regeneration (91.11 per cent). Whereas, 1.5 mg/l indole-3-butyric acid + 2 mg/l 6-Benzylaminopurine was the optimum concentration giving 86.66 per cent of rooted plantlets. They noted satisfactory survival rate of transferred regenerated plantlets, 66.6 per cent and 60.0 per cent for Minjar and Huruta.

Ramakrishnan *et al.* (2013) developed an efficient somatic embryogenesis and regeneration system for the first time in onion using shoot apex explants. They initiate callus in MS medium supplemented with 4.0 mg/l 2, 4-dichlorophenoxyacetic acid using these explants, which has 85.3 per cent induction frequency of primary callus. Transferring the primary calli onto medium supplemented with 2.0 mg/l 2, 4-dichlorophenoxyacetic acid and folled by two biweekly subcultures formed the embryogenic callus. They reported that the inclusion of a low concentration of 6-benzylaminopurine in the subculture medium promoted the formation of embryogenic callus. The addition of 2.0 mg/l glycine, 690 mg/l proline, and 1.0 g/l casein hydrolysate also increased the frequency of callus induction and embryogenic callus formation. However, they obtained the highest frequency of embryogenic callus (86.9 per cent) and greatest number of somatic embryos (26.3 per callus) by the further addition of 8.0 mg/l silver nitrate. They found that somatic embryos formed plantlets on regeneration medium supplemented with 1.5 mg/l 6-benzylaminopurine; addition of 2.0 mg/l glycine to the regeneration medium promoted a high frequency of regeneration (78.1 per cent) and plantlet formation (28.7 plants per callus). The regenerated plantlets were then transferred to half-strength MS medium supplemented with 1.5 mg/l indole-3- butyric acid for root development. They noted the maximum frequency of root formation was 87.7 per cent and the average number of roots was 7.6 per shoot. The regenerated plantlets were successfully grown to maturity after hardening in the soil.

Wu *et al.* (2015) established a high frequency *in vitro* regeneration culture system for onion using mature zygotic embryos. They found the highest induction rate of embryogenic callus (91 per cent) when the NH_4 +/NO_3 ratio in Murashige and Skoog (MS) medium was adjusted to 30/30 mM/mM. They observed well-proliferated embryogenic calli in a 3 L airlift balloon-type bioreactor. During bioreactor culture, the continuous immersion culture was better. Inoculation density of 5 g/L and 0.1 vvm (air volume/culture medium volume per min) air volumes were optimal for embryogenic callus proliferation. In addition, they examined the effects of sucrose and thiamine (VB1) to obtain mature somatic embryos. According to them the selection of an appropriate concentration of sucrose (20 g/L) or VB1 (10 mg/L) is essential for somatic embryo maturation. In such case the rate of somatic embryo germination reached 93.3 per cent. They obtained maximum survival rate (93.8 per cent) in the substrate of 1/2 vermiculite and 1/2 perlite during plantlet acclimatisation.

15.2.0 Biochemical and Molecular Markers

Isozymes were identified as biochemical markers for onion during the 1980s. Different isozymes have been characterized in onion seeds and roots (Table 20). Isozymes have been used primarily to compare other *Allium* species to *A. cepa*, common onion, and to identify the origins of chromosomal regions in interspecific hybrids (Cramer and Havey (1999).

Table 20: Biochemical Markers of Onion*

Isozyme	References
Alcohol dehydrogenase (*Adh*-1)	Hadacova *et al.* (1981); Peffley *et al.* (1985)
Acid phosphatase (*Aps*-1,2)	Peffley *et al.* (1985)
Alkaline phosphatase	Hadacova *et al.* (1981)
Catalase	Hadacova *et al.* (1981)
Cholinesterase	Hadacova *et al.* (1981)
Esterase	Hadacova *et al.* (1981); Peffley *et al.* (1985); Cooke *et al.* (1986)
Glucose-6-phosphate dehydrogenase	Hadacova *et al.* (1981)
Glutamate dehydrogenase	Hadacova *et al.* (1981)
Glycerate dehydrogenase (*Gdh*-1)	Peffley *et al.* (1985)
Glutamate oxaloacetate transaminase (*Got*-1,2,3)	Peffley *et al.* (1985)
Isocitrate dehydrogenase (*Idh*-1)	Peffley *et al.* (1985)
Lactate dehydrogenase	Nakamura and Tahara (1977)
Malate dehydrogenase (*Mdh*-1)	Peffley *et al.* (1985)
NAD+-glyceraldehyde-3-phosphate dehydro-genase	Hadacova *et al.* (1981)
NADP+-glyceraldehyde-3-phosphate dehydro-genase	Hadacova *et al.* (1981)
NAD+-malate dehydrogenase	Hadacova *et al.* (1981)
NADP+-malate dehydrogenase	Hadacova *et al.* (1981)
NADH2-tetrazolium reductase	Hadacova *et al.* (1981)
NADPH2-tetrazolium reductase	Hadacova *et al.* (1981)
Peroxidase (*Prx*-1)	Peffley *et al.* (1985)
6-phosphogluconate dehydrogenase (6*Pgdh*-1,2)	Peffley *et al.* (1985)
Phosphoglucoisomerase (*Pgi*-1)	Peffley *et al.* (1985)
Phosphoglucomutase (*Pgm*-1)	Peffley *et al.* (1985)
Superoxide dismutase (*Sod*)	Hadacova *et al.* (1981)

* Rabinowitch (1988) and Cramer and Havey (1999).

Two alleles for *Adh*-1 were characterized in *A.cepa*, while third and fourth putative alleles were identified in *A. fistulosum* L. (Peffley and Orozco-Castillo, 1987; Peffley *et al.*, 1985). *Mdh*-1 formed fast band in *A. cepa* and a slow band in *A. fistulosum* (Ulloa-Godinez *et al.*, 1995). Cryder *et al.* (1991) suggested that *Idh*-1 and *Pgi*-1 were linked in backcross progeny between *A. cepa* and *A. fistulosum*. Segregation analyses for *A. fistulosum* suggest that 6-PGDH is governed by two loci, *6-Pgdh-1* and *6-Pgdh-2*, that have two alleles (1 and 2) each (Magnum and Peffley, 1994). Peffley and Currah (1988) used alien chromosome addition breeding lines and placed the location of *Adh*-1 to the sub-telocentric chromosome 5 and *Pgm*-1 to the sub-metacentric chromosome 4 in *A. fistulosum*. The isozyme investigation in large *Allium* collection were performed by MaaB and Klaas (1995) in 300 accession of *A. sativum*; Pooler and Simon (1993) in 110 accession of *A. sativum*; Arifin and Okubo (1996) in 189 accession of *A. cepa var. ascalonicum*; Preffley and Orozco-Castillo (1987) in 118 and 29 accession of *A. cepa* and *A. fistulosum* respectively; and Rieseberg *et al.* (1987) in *A. douglassi*. Polymorphisms for six enzyme systems were analysed in the top onion, *Allium proliferum* (Maass, 1997). Five multilocus in isozyme genotypes were found. The banding pattern of top onions was compared with those of *A.* × *wakegi, A. cepa, A. fistulosum, A. altaicum*, and artificial hybrids between these three species. One top onion type and one artificial hybrid had identical banding patterns. Shallots and *A. altaicum*, the wild progenitor of *A. fistulosum* could not be distinguished from the common onion and suggesting that these species are potential contributors to the top onion's gene pool. The major limitation of isozyme analysis is the small number (15 or less) of suitable enzyme systems, of which usually only a subset will exhibit sufficient variability.

The genomic research in onion is still inadequate. This is partly due to its biological nature and enormous genome size (16.3 GB per 1C Nucleus) which is characterised by high frequency of duplication of recessive lethal alleles that remains masked in heterozygous state and maintenance of bulb population by open pollination (Havey 1993). Havey *et al.* (1996) stated that the development of a low-density genetic map for onion has been a challenge due to its relatively few polymorphisms as compared to other outcrossing diploid species, its biennial generation time, and its extremely large genome.

Onion possesses two main challenges to the development of molecular marker. The first is the huge nuclear genome of onion. This enormous amount of DNA causes some PCR based marker system such as, RAPDs and AFLPs, to produce too many fragments that may not resolve well in electrophoretic gels. The second challenge is that onion inbeds and OP populations often retain high level of heterozygosity, which with low levels of allelic diversity among onion populations causes molecular markers, especially dominant ones, to appear mono-morphic among hybrid or OP cultivars (Jakse *et al.*, 2005).

Bradeen and Havey (1995) evaluated 580 decamer primers and were able to establish the genetic bases of only 14 repeatable RAPDs. King *et al.* (1998a) constructed a low-density genetic map using 58 F_3 families generated from a cross between inbreds 'Brigham Yellow Globe (BYG) 15-23' and 'Alisa Craig (AC) 43'. They identified 128 segregating loci (112 RFLPs, 14 RAPDs, 2 morphological) and constructed a map consisting of 114 loci distributed over 11 linkage groups, one linked pair, and 12 unlinked markers. The predominance of dominant and duplicated RFLP loci observed in the low-density genetic map indicated that tandem, intrachromosomal duplications may have contributed to the large size of the onion genome (King *et al.*, 1998b).

Havey (1991) reported that phylogenetic relationships among the cultivated *Allium* are not well understood and taxonomic classification are based on relatively few morphological characters. Chloroplast DNA is highly conserved and useful in determining phylogenetic relationships and its size was estimated at 140 kb and restriction enzyme sites were mapped for KpnI, PstI, PvuII, SaII and XhoI. Out of 189 restriction enzyme sites detected with 12 enzymes, 15 mutations were identified and used to estimate phylogenetic relationships. There are several reports on molecular markers particularly RAPD and RFLP markers in *Allium* assessing variability among species (Bark *et al.*, 1994; Bradeen *et al.*, 1995) and cultivars (Bradeen and Havey, 1995). Wilkie *et al.* (1993) studied RAPD analysis in onion (*Allium cepa*) and other *Allium* species in order to assess the degree of polymorphism within the genus. Seven cultivars of *A. cepa* including shallot, and single cultivars of Japanese bun-ching onion (*A. fistulosum*), chive (*A. schoenoprasum*), leek (*A. ampeloprasum*) and a wild relative of onion (*A. royle*) were evaluated for variability using a set of 20 random 10-mer primers. Seven out of the twenty primers revealed scorable polymorphisms between cultivars of *A. cepa*. Linne von Berg *et al.* (1996) tried to structure the genus Allium by molecular markers taking 48 species. The position of the genus with the Alliaceae was investigated by Fay and Chase (1996), through a phylogenic analysis of plastid DNA sequences coding for the large subunit of ribulose-1, 5-biphosphate carboxylase (rbcL). The data set comprised of 52 species. According to them, *Nectaroscordum siculum* should be included in the genus *Allium*. They also highlighted that *Milula spicata* is the closest relative to the genus *Allium*. Mes *et al.* (1998) conducted a phylogenetic study taking 29 species of Allium and 7 species of related genera using RFLP data from PCR amplified cpDNA.

Commercial onion production is based on open-pollinated (OP) and hybrid cultivars. Grow-out test is required in appropriate environment to identify specific seed lot, which are costly and can cause significantly delay in the movement of seed to market. Theus, the commercial production of onion inbreds, hybrids and OP cultivars would benefit from a robust set of molecular markers that confidently distinguish among elite germplasm. Large scale DNA sequencing has revealed that single nucleotide polymorphisms (SNPs), short insertion-deletion (indel) events, and simple sequence repeats (SSR) are relatively abundant class of codominant DNA markers. Phylogenic analysis of simple matching and Jaccard's coefficients for SSRs produced essentially identical trees and relationship were consistant with known pedigrees and previous marker evaluations (Jakse *et al.*, 2005). Bardeen and Havey (1995) identified RAPD between two inbred onion lines, demonstrated their Mendelian inheritance and tried to distinguish and examine changes in independently maintained, publicly released inbred lines of onion. They observed poor agreement between data sets based on genetically characterized and uncharacterized RAPD markers and analyses used only genetically characterized RAPD markers and revealed that contamination in addition to drift and/or selection likely contributed to differences among independently maintained, publicly released inbreds. King *et al.* (1998) used nuclear RFLPs to estimate relationships among 14 elite commercial inbred of bulb onion (*A. cepa*) from Holland, Japan and the United States. Variability for known alleles at 75 RFLP loci and 194 polymorphic fragments revealed by 69 anonymous cDNA probes and a clone of alliinase were scored to yield genetically characterized and uncharacterized data sets, respectively. RFLPs confidently distinguished among elite inbreds within a specific market classes. Tanikawa *et al.* (2002) examined 22 onion cultivars using RAPD markers. A total 88 fragments were produced by 17 primers, of which 35 were polymorphic among the cultivars. All the 22 cultivars could be distinguished by the combinations of polymorphic bands generated by various primers. The similarity values were range from 0.836 to 0.979, which indicated low genetic diversity among the onion cultivars. Based on the evaluation of polymorphism in 21 SSR loci, Mitrova *et al.* (2015) established a panel of 15 easy to score SSR markers that differentiated 16 commercial onion cultivars in the Czech Republic. The polymorphism ranged from 2-3 alleles per locus, indicating a low level of diversity among the cultivars. Molecular genetic markers, as RFLP, RAPD, AFLP, SNP and SSR are most frequently used in the characterization and evaluation of genetic diversity within and between species and populations. However, the most used types of genetic markers are microsatellites or simple sequence repeats (SSR). Baldwin *et al.* (2012) estimated within- and amongpopulation heterozygosity in a set of onion populations using genomic simple sequence repeat (SSR) markers developed by genomic skim sequencing aiming to develop quantitative estimates of diversity. Primer sets (166) designed to flank SSR motifs identified were evaluated in a diverse set of lines, with 80 (48 per cent) being polymorphic. The 20 most robust single copy markers were scored in 12 individuals from 24 populations representing short-day to long-day adapted material from diverse environments. The assessed onion populations were distinct with moderate to large population differentiation but also had high within-population variation (Fst = 0.26). Assessment of within- and among-population diversities revealed that observed heterozygosities were significantly lower than expected. The clustering based on the principal component and Bayesian model-based analyses using Structure led to similar population groupings based primarily on the geographical location. This marker resource will be applicable for population structure analysis for association mapping, DNA fingerprinting for cultivar identification, and testing identity and hybridity of commercial seedlots. An important potential application for these markers in genetic diversity studies will be to aid in clarifying the relationships between onion and other *Alliums*. Karic *et al.* (2018) evaluated the genetic diversity of five most common Bosnia and Herzegovina (BiH) onion genotypes, including the sample of unknown origin. Seven SSR markers for genetic similarity analysis were used to address the genetic backgrounds. They noted sufficient dissimilarity characteristics which reflect significant genetic diversity among onion cultivars. A total of 30 alleles were detected, out of which 17 bands (56.7 per cent) were polymorphic. The phylogenetic analysis revealed that the Konjic onion shares similarity with the Ptujska rdeæa, indicating regional cohesion, whereas all onions in this study are clearly separated from the Majski Srebrenjak onion, an Italian origin onion. Using competitive allele-specific PCR (KASP) genotyping technology, Villano *et al.* (2019) explored the genetic variation of 73 onion accessions (including wild species, commercial, and local varieties) from different areas of the world. The SNP dataset inspection returned 375 polymorphic loci with a very low percentage of non-calling sites (0.03 per cent). Eight-nine percent of the onions amplified all polymorphic loci and were considered for a population structure analysis. The method suggested four populations and enabled the identification of genepools, reflecting the geographical origin of the samples. The SNP set successfully revealed population-specific alleles and potential candidates for use in future breeding programs. They highlighted that 74 loci were associated with phenotypic traits (bulbing photoperiod, bulb shape, or bulb color), and 3 loci were identified as putative targets of selection associated with onion improvement.

Marker assisted selection (MAS) has a great potential in the genetic enhancement of onion and several breeders expressed the hope to fasten breeding by skipping several breeding cycles during the segregating generations and condense timelines (Mazur, 1995). The greatest benefits of MAS for the possibilities to achieve the same breeding progress in a much shorter time than through conventional breeding, pyramid combinations of genes that could not be readily combined through other means and to assemble target traits more precisely, with less unintentional losses (Xu and Crouch, 2008) and to finally having found a tool to control the allelic variations for all genes of agronomic importance (Peleman and Van Der Voort, 2003).

Onion is a biennial crop. It takes four to eight years to determine if maintainer lines (*Nmsms*) can be extracted from population or segregating family. Identification of sterile plants on the basis of colour of anthers is relatively easy in comparison to the identification of maintainer line (*Nmsms*) where the flowers produce viable pollen and there is no visual marker. Identification of maintainer line takes at least two years to confirm the plants to be the maintainers. Selection of one wrong type of plant can contaminate the whole population which makes identification of the maintainer line very difficult. It is now possible to identify/confirm a maintainer line within one year (one season) using marker assisted selection. Identification of molecular markers tightly linked to the nuclear (*Ms*) locus greatly aided the development of maintainer lines. These markers could allow the breeders to identify the plants carrying recessive *ms* alleles and help in identifying the maintainer lines in the first year only. Molecular markers distinguishing normal (N) versus sterile (S) cytoplasm (Xu *et al.*, 2008; Sato, 1998) and T cytoplasm (Havey 1995; Engelke *et al.*, 2003) have been reported in onion. Identification of *ms* allele in dominant, heterozygous and recessive condition has been estimated by employing various types of markers. Gokce and Havey (2002) developed a RFLP based PCR marker to identify the *Ms* locus which was later on converted into PCR marker (Bang *et al.*, 2011; Huo *et al.*, 2012), dominant SCAR makers (Yang *et al.*, 2013), SNP marker (Havey, 2013) and codominant PCR markers (Kim, 2014) linked to *Ms* locus. In order to identify sterile (S) and normal (N) cytoplasm, PCR was performed using the markers anchoring in the upstream region of the mitochondrial gene *cob* (Sato, 1998), referred as 5' *cob* markers, and also by primers designed for *orf*A501 gene (Engelke *et al.*, 2003). Two markers linked to putative oligopeptide transporter (OPT) and photosystem I subunit (*PsaO*) reported by Bang *et al.* (2011) were used to identify fertility restorer gene (*Ms/ms*). Saini *et al.* (2015) identified male sterile and maintainer lines using molecular markers in three long day onion populations. Molecular markers, 5'*cob* and *orf*A501 were able to distinguish effectively normal (N) and sterile (S) cytoplasm in three populations. The observed frequency of S cytoplasm in VL Piaz 67 (100 per cent), VL Piaz 3 (86.4 per cent) and KR1 (90 per cent) was higher than N cytoplasm.

Out of the two PCR markers *viz.*, OPT and PsaO used to determine the nuclear fertility restorer locus (*Ms* locus), OPT was found better than PsaO. They observed increased frequency of dominant homozygous alleles (87.4 per cent) followed by heterozygous alleles (8.6 per cent) and homozygous recessive alleles (4.0 per cent). They noted only 12 plants (2.15 per cent) as completely male sterile in VL Piaz 3 and KR1 population, whereas there was no male sterile plant observed among the population in VL Piaz 67. Selfing and test crossing of plants having normal (N) cytoplasm led to the development and identification of maintainer lines in VL Piaz 3 and KR1 population. This was the first example of deploying DNA markers for identification and purification of male sterility and hybrid development in long day onion in Indian population.

Very limited polymorphic and cross-transferable markers are available in onion. There is an urgent need to develop polymorphic markers in Allium to expedite and introgress desirable traits from wild relatives (which are rich bioresource of various biotic and abiotic resistance genes) to *A. cepa*. Intron Length Polymorphism (ILP) makers are among the popular molecular markers due to their sequence specificity, co-dominance and random distribution. Moreover, high polymorphism in intron regions together with higher conservation in primer binding exonic sites make them superior markers for diversity as well as cross-species transferability studies (Jayaswall *et al.*, 2019). They successfully utilized 20,204 ESTs (3750 contigs and 8364 singletons), of *A. cepa* for identification of over 2689 intron length polymorphic (ILP) markers. A set of 30 markers was tested for polymorphism in onion and cross-transferability in garlic and related wild species. Among these, eighteen markers amplified at least one of the accessions of *A. cepa*. Transferability of these ILP markers was ranged from 21.7 to 95.7 per cent in *Allium* spp. They found that all the 23 varieties/accessions were clustered under three groups. All the varieties of *A. cepa* were clearly clustered separately under one group. However, there was intermixing of varieties/accessions of *A. sativum* L. and wild relatives, which may possibly be due to less number of markers validated for cross-transferability.

Single nucleotide polymorphisms (SNPs) play important roles as molecular markers in plant genomics and breeding studies. Genotyping-by-sequencing (GBS) offers a greater degree of complexity reduction followed by concurrent SNP discovery and genotyping for species with complex genomes. Jo *et al.* (2017) developed a genome-wide SNP resource in onion using a GBS de novo approach on an F_2 population, derived from a cross between 'NW-001' and 'NW-002,' as well as multiple parental lines. A total of 56.15 Gbp of raw sequence data were generated and 1,851,428 SNPs were identified from the de novo assembled contigs. Stringent filtering resulted in 10,091 high-fidelity SNP markers. Robust SNPs that satisfied the segregation ratio criteria and with even distribution in the mapping population were used to construct an onion genetic map. The final map contained eight linkage groups and spanned a genetic length of 1,383 centiMorgans (cM), with an average marker interval of 8.08 cM. These robust SNPs were further analyzed using the high-throughput Fluidigm platform for marker validation. This is the first study in onion to develop genome-wide SNPs using GBS.

15.3.0 Genetic Transformation

Genomic resources for onion and garlic are limited because of their large, extremely complex, repetitive, and often polyploid genomes and long generation times. *Alliums* have been initially considered recalcitrant to *in vitro* regeneration and not amenable to *Agrobacterium*-mediated genetic transformation. Subsequently, however, several approaches have been developed (Cardi *et al.*, 2017). Callus induction, regeneration and transformation of *A. cepa* by various explants were reported. Some reported direct DNA delivery (Klein *et al.*, 1987; Eady *et al.*, 1996; Barandiaran *et al.*, 1998; Scott *et al.*, 1999), while some others reported *Agrobacterium* transformation using mature (Joubert *et al.*, 1995; Eady *et al.*, 1996; Zheng *et al.*, 2001) and immature embryos (Eady *et al.*, 2000). However the use of mature embryos is tedious, while immature embryo leads to contamination. Reddy *et al.* (2006) used seedling radicle as an explant. Khar *et al.* (2005) reported that in onion, callus proved to be the best explant source for genetic transformation, followed by shoot tip and root tips.

Table 21: Availability of *in vitro* Regeneration/Genetic Transformation Protocols and of Genomic Resources for *Alliums**

Cultivated Species	Estimated Genome Size (Mb)	Regeneration and Transformation	Genomic Resources
Allium cepa	16000	various protocols available	information available
Allium sativum	15901	protocols available	Little information available
Allium porrum	28607	protocols not available	Little information available

* Cardi *et al.* (2017).

Eady *et al.* (1996) studied particle bombardment and *Agrobacterium*-mediated DNA delivery into immature embryos and microbulbs to investigate the expression of the *uidA* gene in *in vitro* onion cultures. Both methods were successful in delivering DNA and subsequent *uidA* expression was observed. Optimal transient beta-glucuronidase activity was observed in immature embryos that had been precultured for three days and bombarded at a distance of 3 cm from the stopping plate, under 25 in Hg vacuum, using 900-1300 psi rupture discs. The CaMV35S-*uidA* gene construct gave five-fold higher transient GUS activity than the *uidA* gene construct regulated by any of the four other promoters initially chosen for high expression in monocotyle-donous tissues.

A repeatable protocol for transformation of onion was reported by Eady *et al.* (2000). They used immature embryos as explant source of *Agrobacterium*-mediated transformation. Transgenic plants were recovered from the open pollinated onion cultivar Canterbury Longkeeper at a maximum transformation frequency from immature embryos of 2.7 per cent. The method took 3-5 months from explant to primary regenerant entering the glasshouse. The binary vector used carried the *nptII* anti-biotic resistance gene and *mgfp 5-ER* reporter gene. The transgenics were

Genetic Transformation in Onion

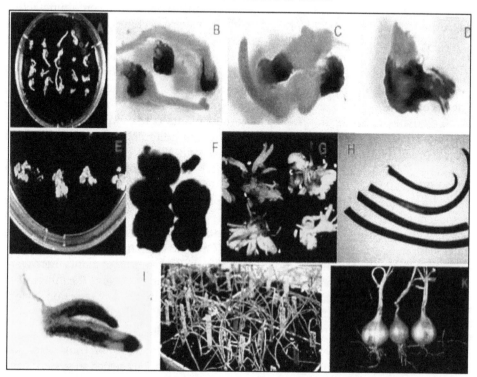

A. Three-week old callus derived from mature embryos. B. Transient expression of GUS in two-week old callus derived from mature embryos of cv. Bawang Bali; infection with LBA4404 (pTOK233) after four days of co-cultivation. C. Transient expression of GUS in two-week old callus derived from immature embryos of cv. Bawang Bali; infection with LBA4404 (pTOK233) after four days of co-cultivation. D. Transient expression of GUS in three-week old callus derived from mature embryo of Kuning; infection with EHA105 (pCAMBIA1301) after four days of co-cultivation. Callus was chopped and the elongated part of the cotyledon was removed. E. Hygromycin-resistant callus of cv. Kuning after two months of growth on selective medium. F. Stable and uniform expression of GUS in hygromycin-resistant callus of cv. Kuning. G. Plant regeneration of cv. Kuning. The photograph was taken four weeks after the hygromycin-resistant callus had been transferred to regeneration medium with hygromycin. H. Expression of GUS in the leaves of a transformant. I. Expression of GUS in the root of a transformant. J. Transgenic onion and shallot plants in the greenhouse. K. Bulbs from transgenic shallot plants (Zheng *et al.*, 2001).

confirmed by Southern blot analysis. Zheng *et al.* (2001) developed a reproducible *Agrobacterium tumefaciens* mediated transformation system both for onion and shallot with young callus derived from mature embryos with two different *Agrobacterium* strains.

Transgenic onion plants (*Allium cepa*) tolerant to herbicides containing active ingredients glyphosate and phosphinothricin were recovered from immature embryos of open pollinated and hybrid parent onion lines at a maximum transformation frequency of 0.9 per cent. Transformants of different onion cultivars, grown on different selective agents and confirmed by Southern analysis, thrived with no apparent ill effects when sprayed with the respective herbicides at double the recommended field dosage for weed eradication (Eady *et al.*, 2003a). Eady *et al.* (2003b) produced the transgenic onion plants containing the Cauliflower mosaic virus 35s promoter (*CaMV35s*) and *gfp* gene construct encoding the visual green fluorescent reporter protein from *pBINmgfp ER* and the *CaMV35s-bar* gene construct encoding resistance to the herbicide phosphinothricin from pCAMBlA3301 by *Agrobacterium*-mediated transformation. These plants were grown to maturity and selfed in order to determine the expression and inheritance of the transgenes. *CaMV35s* regulation in onion, as observed by GFP expression, was essentially constitutive, and profiles of regulation were typical of those observed in dicotyledonous plants. Both the expression of GFP and tolerance to phosphinothricin appeared to be inherited in a Mendelian fashion. In the majority of plants there were no obvious detrimental phenotypic effects caused by the transgene, the integration event, or somaclonal variation. Aswath *et al.* (2006) devised a new selection system for onion transformation that does not require use of antibiotics or herbicides, using *Escherichia coli* gene that encodes phosphomannose isomerase (*pmi*).

Reddy *et al.* (2006) developed a new selection system for onion transformation that does not require the use of antibiotics or herbicides. The selection system used the *Escherichia coli* gene that encodes phosphomannose isomerase (*pmi*). Transgenic plants carrying the *manA* gene that codes for *pmi* can detoxify mannose-6-phosphate by conversion to fructose-6-phosphate, an intermediate of glycolysis, *via* the *pmi* activity. Six-week-old embryogenic callus initiated from seedling radicle was used for transformation. Transgenic plants were produced efficiently with transformation rates of 27 and 23 per cent using *Agrobacterium* and biolistic system, respectively. Untransformed shoots were eliminated by a stepwise increase from 10 g/l sucrose with 10 g/l mannose in the first selection to only 10 g/l mannose in the second selection. Integrative transformation was confirmed by PCR, RT-PCR and Southern hybridization.

Resveratrol (RV) is a natural polyphenolic compound found in certain plant species including grapes. RV is well known for its nutraceutical properties and to assuage several disease conditions. Naini *et al.* (2019) developed nutraceutical onion by engineering RV biosynthetic pathway. A codon-optimized grapevine synthetic stilbene synthase gene (VvSTS1) was synthesized using native grapevine sequence. Six-week-old healthy yellowish compact nodular calli were co-cultivated with *Agrobacterium tumefaciens* harbouring pCAMBIA1300-hpt II-CaMV35S-VvSTS1-nos. LC–ESI-HRMS analysis revealed the accumulation of variable quantities of

RV (24.98-50.18 µg/g FW) and its glycosylated form polydatin (33.6-67.15 µg/g FW) in both leaves and bulbs, respectively, indicating the successful engineering of RV biosynthetic pathway into onion. The transgenic onion accumulating RV and polydatin may serve as a potential nutraceutical resource.

Transgenic onions can be developed and applied to some of the problems presently affecting onion breeding. These are: (1) the relatively simple secondary metabolite pathway leading to the production of onion flavour could be manipulated by antisense or antibody technology. Onions with precise flavour characteristics could be produced; (2) the insertion of the anthocyanin genes into onions could be used to alter bulb colour; (3) Altering maturity rates in onions to meet early markets. This is not feasible at present because this characteristic is regulated by precise daylength and temperature requirements and is likely to be under the control of several gene loci; (4) Hybrid seed selection and post-emergence weed control could be solved simply by introducing a herbicide resistance gene; (5) Transgenic plants can be produced to overcome bacterial disease by the introduction of genes encoding antimicrobial peptides or the transfer of genes encoding lysozyme. Resistance to fungal disease such as downy mildew may prove more difficult to overcome. Introduction of synergistically acting chitinase and P-l, 3-glucanase genes may confer increased resistance as has been demonstrated in tomato against *Fusarium oxysporium*. Genes encoding artificial antibodies raised against specific pathogen antigens can, when introduced into plants, confer a high degree of protection. This approach may also be successful for onions; (6) Thrips (*Thrips tabacii*) are a major pest of onions that may be controlled by the introduction of anti-insect genes (Eady, 1995).

In onion, Using RNA interference silencing, Eady *et al.* (2008) able to develop "Tear-free Onion" by suppressing the lachrymatory factor synthase gene, through a single genetic transformation. Transgenic shallot and garlic plants containing *Bt* resistance genes have been produced which confer resistance to beet armyworm (*Spodoptera exigua*) (Zheng *et al.*, 2004).

15.4.0 *In vitro* Conservation

There are two approaches for conservation of plant genetic resources, namely *in situ* and *ex situ* conservation. In situ conservation involves maintaining genetic resources in the natural habitats where they occur, whether as wild and uncultivated plant communities or crop cultivars in farmers' field as components of the traditional agriculture system. On the other hand, ex situ conservation involves conservation outside native habitat and is generally used to safeguard populations in danger of destruction, replacement or deterioration. Approaches to ex siu conservation include methods like seed storage, DNA storage, pollen storge, in vitro conservation, field gene banks and botanical gardens. Wild species of Allium that are endemic, rare/threatened/vulnerable; or at risk of genetic erosion due to habitat destruction or over exploitation need to be conserved through suggested *ex-situ* approach (Sancir *et al.*, 1989).

In bulbous plants, such as onion, induction and storage of *in vitro* bulblets could enable long term maintenance of special genotypes. Keller (1993) cultured *in vitro* seedlings of onion (cv. Stuttgarter Riesen) which formed bulblets under increased sucrose concentration (30, 50, 150 g/l) in the medium and with the addition of BA (0, 12.5 mg/l) or ethephon (33 mg/l for 0, 5 or 20 days). Kastner *et al.* (2001) reported that the storability of onion bulblets was primarily enhanced by a high sucrose concentration (100 g/l) in the culture medium.The viability of bulbs after 1 year of *in vitro* storage at low temperatures was determined by their growth reaction in subsequent subcultures, growth after transfer into the greenhouse and tetrazolium staining. Sufficient sprouting of bulblets previously stored at −1 °C demonstrated the possibility of storing them in a low-temperature, slow-growth culture. Keller (1993) suggested increasing the sucrose concentration and treatment with ethephon to obtain bulblets for *in vitro* storage under conditions of slow growth. Patena *et al.* (1998) developed *in vitro* conservation systems for garlic and shallot *via* the slow growth technique using G0 bulblets in controlled atmosphere.

Stanwood and Sowa (1995) described methods for long term storage strategies for seed germplasm to assure preservation of diminishing plant genetic resources. Cryopreservation in liquid nitrogen (−196°C) is being studied as a potential method to reduce the rate of seed deterioration and thus increased the storage life of the seed. Onion seed was stored for 10 years at 5, −18 and −196°C (liquid nitrogen). Average germination of seed stored at −18 and −196°C did not decline over the 10-year period, while germination of seed stored at 5°C dropped from 94 to 68 per cent.

In vitro culture is a complex of methods widely adopted for application in plant breeding. *In vitro* culture is important for germplasm preservation because of the wide distribution of seed sterile forms in the genus *Allium*. Besides this, it can help to safeguard the maintenance of other forms possessing uniquecharacters allogamous species. Keller (1991) established a collection of *in vitro* clones derived from gynogenetic haploids, diploidregenerants, and their mother plants. Keller and Hanelt (1992) induced 115 gynogenetic haploids of onion and characterized the maintened haploid material in the genebank. They were maintaining 163 gynogenetic clones, 64.9 percent of which being haploid. They obtained about 96 partially multiple plantlets in *Allium tuberosum* and *A. ramosum* through culture of unfertilized ovaries and flower buds. In cultures of ovaries having been previously fertilized by alien pollen, regenerants were obtained from crosses with *A. altaicum*, *A. ledebourianum*, *A. schoenoprasum*, and *A. senescens*. A method of slow growth storage basing on combinations of bulblet induction *in vitro* by media rich in sucrose and treatment by ethephon, cold storage and micropropagation by cutting and culture on media containing cytokinins were followed. Similarly, the long term maintenance of this material comprises a cycle consisting of a slow growth phase at 4°C and a multiplication phase using cuttings of *in vitro* bulblets and cultivation on media with benzylaminopurine (BAP). And the maintenance of the haploid constitution is repeatedly checked by stomata guard cell measurements (Keller, 1991; Keller and Hanelt, 1992). Viterbo *et al.* (1994) stored the basal leaf explants of *in vitro* grown plantlets at 4-6°C for up to 16 months. They used standard medium, or modified media containing 0.4 to 10.0 per cent sucrose or 1 to 10 mg/l paclobutrazol for storage study. They obtained eighty to 100 per cent and 70 per cent survival rates after 8 and 16 months cold-storage, respectively, in a 10 per cent sucrose medium. They assessed the genetic stability of cold-stored explants using isozyme polymorphism of

13 proteins. Considerable differences in zymograms were found between the fertile and the sterile varieties. However, no differences in isozyme profiles were detected between the control field-grown plants, and those which were established from *in vitro*-stored leaf base explants, except plantlets exposed to a high paclobutrazol concentration in storage.

Vitrification-based cryopreservation techniques are operationally less complex than classical ones since they do not require the use of a programmable freezer. In addition, since ice formation is avoided during freezing, they are more adapted for freezing complex organs such as apices or embryos which contain a variety of ceU types, each with unique requirements under conditions of freeze-induced dehydration. *Allium sativum* and *Allium wakegi* apex were tried to conserve using vitrification technique (Engelmann, 1997). Wang *et al.* (2019) attempted to develop an efficient droplet-vitrification cryopreservation method for shallot '10603' shoot tips. They excised the shoot tips (2.0-3.0 mm in length) from 4-week-old stock shoots and step wise precultured with increased sucrose concentrations from 0.3 to 0.5 M, each concentration for 1 day. The precultured shoot tips were then loaded for 20 min with a solution composed of 2 M glycerol and 0.5 Msucrose, before exposure to PVS3 for 3 h at room temperature. Dehydrated shoot tips were transferred onto aluminum foils (2 × 0.8 cm), prior to direct immersion into liquid nitrogen (LN) for cryostorage. For thawing, they moved the frozen aluminum foils from LN and immediately transferred into unloading solution composed of liquid MS containing 1.2 M sucrose. After incubation at room temperature for 20 min, shoot tips were post-cultured on solidified MS medium containing 0.3 M sucrose for 2 days and then transferred onto a recovery medium for shoot regrowth. With this procedure, they noted, 94 per cent shoot tips survived, and 58 per cent shoot tips regenerated into shoots following cryopreservation.

Kim *et al.* (2012) review the establishment procedure of cryopreserved *Allium* germplasm collection at the genebank of the National Agrobiodiversity Center, Republic of Korea. The systematic approach to *Allium* cryopreservation included by them: (1) revealing the most critical factors that affected regeneration after cryostorage; (2) understanding the mechanisms of cryoprotection by analyzing the thermal behavior of explants and cryoprotectant solutions using DSC and influx/ efflux of cryoprotectants using HPLC; (3) assessing genetic stability of regenerants; and (4) revealing the efficiency of cryotherapy. They found that the bulbil primordia, *i.e.* asexual bulbs formed on unripe inflorescences was the most suitable material for conservation of bolting varieties due to high post-cryopreservation regrowth and lower microbial infection level, followed by apical shoot apices from single bulbs and cloves. During 2005-2010, they cryopreserved a total of 1,158 accessions of garlic as well as some *Allium* species using the droplet-vitrification technique with a mean regeneration percentage of 65.9 per cent after cryostorage.

Wild *Allium* relatives maintained in MTS and FGB in ICAR-NBPGR are: *Allium fistulosum, A. tuberosum, A. schoenoprasum, A. sativa* var. *ophioscordon, A. ampeloprasum, A. angulosum, A. oschaninii, A. ascalonicum, A. ledebourianum, A. proliferum* and *A. altaicum*. In cryopreservation experiments using vitrification, encapsulation-dehydration or droplet vitrification technique led to varying degree of pre- and/

or post-freezing success in *Allium* spp. Subculture frequency and protocol applied depended on species/genotype. In 2014-15 they procured a total of 63 accssions including *Allium ampeloprasum* (1), *A. cepa* var. *aggregatum* (1), *A. chinense* (1) *A. hookeri* (1) and *A. sativum* (59) from various sources. One accession each of *A. chinense* and *A. hookeri* were added in the *in vitro* Genebank. In *A. ampeloprasum* and *A. cepa* var. *aggregatum*, contamination-free cultures were established following sterilant treatment and only single shoot per explant was obtained using bulbous shoot bases as explants. In *A. chinense, A. ramosum* experiments revealed that *in vitro* shoot base explants, excised from mother cultures (maintained at low temperature on high sucrose medium), pregrown at low temperature on high sucrose, exhibited 30 per cent and 20 per cent regrowth following LN freezing, with 30 min PVS2 dehydration, respectively. In *A. hookeri*, explants isolated from mother cultures maintained on shoot multiplication medium (SM) at 25°C and pregrown on high sucrose at 50°C exhibited only 20 per cent post-thaw survival with 20 min PVS2. However, preconditioning of mother cultures on SM medium at low temperature (50°C) led to 20 per cent post-thaw regrowth of explants which were pregrown on high sucrose at 25°C or 50°C (ICAR-NBPGR, 2015).

16.0 References

Abdalla, A.A. (1968) *Proc. 17ᵗʰ Int. Hort. Congress.*

Abdalla, A.A. and Mann, L.K. (1963) *Hilgardia,* **35** : 85-112.

Abd-El-Gawad, A.A., El-Tabbakh, A.M., El-Habbal, M.S. and Thabet, E.M.A. (1986a) *Ann. Agric. Sci.,* **31** : 1021-1031.

Abd-El-Gawad, A.A., El-Tabbakh, A.M., El-Habbal, M.S. and Thabet, E.M.A. (1986b) *Ann. Agric. Sci.,* **31** : 1011-1019.

Abdel-Momen, S.M., Omar, S.A., Hanafi, A.A. and Abdel Rahman, T.M. (2000) *Bull. Facul. Agric.,* Univ. of Cairo, **51** : 365-377.

Abrol, D.P. (2011) *Ind. J. Entomol.,* **73** : 207-212.

Abrol, J.P. and Dixit, S.P. (1972) *Exp. Agric.,* **8** : 171-175.

Adamicki, F. (1998) *Biuletyn Warzywniczy,* **48** : 89-100.

Aditika, P., Dod, V.N. and Sharma, M. (2017) *Int. J. Farm Sci.,* **7** : 123-126.

Agarwal, A., Gupta, S. and Ahmed, Z. (2010) *Indian J. Agric. Sci.,* **80** : 333-334.

Aggarwal, P., Rajput, H., Dhatt, A.S. and Kaur, A. (2016) *Agric. Res. J.,* **53** : 97-100.

Agic, R., Gjorgjievaska, M.C., Martinovski, G. and Lazic, B. (1997) *Acta Hortic.,* **462** : 565-570.

Aguiar, F.N.A., Perez, A., Fundora, Z., Gonzalez, M., Prats, A., del. C.M. Alonso-Rodriguez (2001) *Alimentaria,* **38** : 79-82.

Ahirwar, R. and Verma, R.C. (2016) *Cytologia,* **81** : 149-153.

Ahlawat, M.R. and Rathore, D.S. (1957) *Indian Fmg.,* **6** : 17.

Ajrekar, S.L. (1920-21) *Ann. Rept. Bomay Dept. Agric.,* pp. 102-104.

Aisha, H, Ali and Taalab, A.S. (2008) *Res. J. Agric. Bio. Sci.*, **4** : 228-237.

Aklilu, S. and Dessalegne, L. (1999) *AgriTopia*, **14** : 6.

Aklilu, S., Dessalegne, L. and Currah, L. (2001) *Acta Agronomica Hungarica*, **49** : 175-181.

Al-aghabary, K., Guo, D. and Zhu, Z. (2001) *Pak. J. Biol. Sci.*, **4** : 374-377.

Ali, M.A., Hossain, M.M., Zakaria, M., Naznin, A. and Md. Islam, M. (2015) *Int. J. Agron. Agric. Res.*, **6** : 174-180.

Al-Jamal, M.S., Ball, S. and Sammis, T.W. (1999) *Amer. Sco. Agril. Eng.* (ASAE), p. 15.

Al-Jamal, M.S., Sammis, T.W., Ball, S. and Smeal, D. (1999) *Appl. Eng. Agric.*, **15** : 659-668.

Al-Jamal, M.S., Sammis, T.W., Ball, S. and Smeal, D. (2000) *Agril. Water Manag.*, **46** : 20-41.

Al-Moshileh, A.M. (2001) *Assiut J. Agric. Sci.*, **32** : 291-305.

Alphonse, M. (1997) *Alexandria J. Agril. Res.*, **42** : 171-183.

Altaf Hussain, H., Shakeel, M., Khan, J., Iqbal, M., Khan, S. (1999) *Sarhad J. of Agric.*, **15** : 619-624.

Ammal, U.B., Mathan, K.K. and Mahimairaja, S. (2000) *J. Trop. Agri.*, **38** : 55-58.

Ananthan, M. (2010) *Progressive Agric.*, **10** : 260-264.

Ananthan, M. and Balakrishnamoorthy, G. (2007) *Agric. Sci. Digt.*, **27** : 190-193.

Andargie, B.H.M. and Assefa, K. (2013) *Res. Plant Sci.*, **1** : 45-52.

Anonymous (1971) *Indian Minimum Seed Certification Standards, Central Seed Committee, Min. Food Agric.*, Comm. Dev. Co-op., New Delhi.

Anonymous (1973) *FAO Production Year Book.*

Anonymous (1981) *FAO Production Year Book.*

Anonymous (1998) *FAO Production Yearbook*, **52** : 135-136.

Anonymous (1999) *FAO Yearbook, Trade and Commerce*, **53** : 135-137.

Anonymous (2002) *Annual Report (2000-2001)*, National Horticultural Research and Development Foundation, Nashik.

Aoba, T. (1964) *J. Jap. Soc. Hort. Sci.*, **33** : 46-52.

APG (2009) *Botanical Journal of Linnean Society*, **161** : 105-121.

Arakeri, M.R. and Patel, S.S. (1956) *Indian J. Agron.*, **1** : 75-80.

Arifin, N.S., Ozaki, Y. and Okubo, H. (2000) *Euphytica*, **222** : 23-31.

Aristy, J.D. (1981) *In 17th Annual Meeting, Caribbean Food Crops Society*, pp. 34-38.

Arnon, D.J. and Stout, P.R. (1949) *Pl. Physiol.*, **14** : 599-602.

Arold, G. (1998) *Gemuse-Munchen*, **34** : 456-459.

Arrgusti, K.T. (1990) In : *Onions and Allied Crops, Vol. III*, (eds. J. L. Brewster and H. D. Rabinowitch). Boca Raton, Florida, CRC Press, pp. 93-108.

Asaduzzaman, M., Hasan, M., Moniruzzaman, M. and Howlander, M. H.K. (2012) *Bangladesh J. Agr. Res.*, **37** : 405-414.

Asencio, C.I. (1997) *J. Agril. Univ. Puerto Rico*, **81** : 75-77.

Ashagrie, T., Belew, D., Alamerew, S. and Getachew, Y. (2014) *Int. J. Agric. Res.*, **9** : 231-241.

Ashok, P., Sasikala, K. and Pal, N. (2013) *Int. J. Farm Sci.*, **3** : 22-29.

Ashwini, M., Balaganesh, J., Balamurugan, S., Murugan, S.B., and Sathishkumar, R. (2013). Antioxidant activity. In: *In Vivo* and *in Vitro* cultures of onion varieties (Bellary and CO 3). *Food and Nutrition Sciences, 4,* 6, Article ID: 36052.

Aswath, C.R., Mo, S.Y., Kim, D.H. and Park, S.W. (2006) *Plant Cell Rep.*, **25** : 92-99.

Attkan, A.K., Raleng, A. and Alam, M.S. (2017) *Vegetos*, **30** : 11-17.

Augusti, K.T., Brewster Ed., J.L. and Rabinowith, H.D. (1990) Therapeutic and medicinal values of onion and garlic. In: *Onion and Allied Crops*, Baco Ratani, Florida, **3** : 93-108.

Aujla, I.S., Amrate, P.K., Kumar, P. and Thind, T.S. (2013) *Plant Disease Research*, **28** : 171-173.

Austin, R.B. (1972) *J. Hort. Sci.*, **47** : 492-504.

Aykroyd, W.R. (1963) *ICMR Special Rept. Series* No. 42.

Babrikov, T.D., Gueorguiev, V.T., Meranzova, R.A. and Lazic, B. (1997) *Acta Hortic.*, **463** : 553-555.

Bacher, J.W. (1989). Inheritance of resistance to *Fusarium oxysporum* f. sp. *cepae* in cultivated onions. MS Thesis, Michigan State Univ., East Lansing.

Bacher, J.W., Pan, S. and Ewart, L. (1989). Inheritance of resistance to *Fusarium oxysporum* f.sp. *cepae*in cultivated onions. In: L. Jensen (Ed.), Proc. 1989 Natl. Onion Res. Conf. Boise, pp. 85-91.

Bagnar, S. and Shanab, L.M. (1969) *Acta Phytopath. Aca. Sci. Hung.*, No. 2-3, pp. 153-161.

Bailov, D. (1930) *Res. Inst. Rech. Agron.*, Bulgaria, **4** : 4-5.

Bajaj, K.L., Kaur, G., Singh, J. and Gill, S.P.S. (1980) *Qualit. Plant.*, **30** : 117-122.

Baker, R.S. and Wilcox, F.E. (1961) *Proc. Amer. Soc. Hort. Sci.*, **78** : 400-405.

Bakr, A.A. and Gawish, R.A. (1998) *Nahrung*, **42** : 94-101.

Bal, S., Maity, T. K. and Maji, A. (2020). *Int. J. Chem. Stud.*, **8** : 2157-2162.

Balasubramaniam, A.S., Raman, G.V. and Krishna Moorthy, R.K. (1978) *Agric. Res. J.*, Kerala, **17** : 138.

Baldwin, S., Pither-Joyce, M., Wright, K., Chen, L. and McCallum, J. (2012) *Mol. Breed.*, **30** : 1401-1411.

Baldwin, S., Revanna, R., Pither Joyce, M., Shaw, M., Wright, K., Thomson, S., Moya, L., Lee, R., Macknight, R. and McCallum, J. (2014) *Theor. Appl. Genet.*, **127** : 535-547.

Banjare, C., Sinha, J. and Sahu, K. (2019) *J. Soil Water conserv.*, **18** : 401-405.

Barandiaran, X., Pietro, A.D., Martin, J. and Di Pietro, A. (1998) *Plant Cell Rep.*, **17** : 734-741.

Barbieri, G., Bolzoni, L., Siviero, P. and Macchiavelli, L. (1998) *Informatore Agrario*, **54** : 43-46.

Bark, O.H., Havey, M.J. and Coirgan, J.N. (1994) *J. Amer. Soc. Hort. Sci.*, **119** : 1046-1049.

Barman, D., Mulge, R., Madalageri, M.B. and Das, S.C. (2013) *Int. J. Agric.l Sci.*, **9** : 72-75.

Barnoczki, A. and Barnoczki-Stoilova, E. (1998) *Zoldsegtermesztesi Kutato Intezet Bulletinje*, **28** : 13-17.

Basu, R.N., Bose, T.K., Chattopadhyay, K., Gupta, N.D., Dhar, N., Kundu, C., Mitra, R., Lal, P. and Pathak, G. (1975) *Indian Agric.*, **19** : 91-96.

Batrta, V., Singh, J. and Singh, V. (2000) *Haryana J. Hort. Sci.*, **29** : 124-125.

Battilani, P., Rossi, V., Racca, P. and Giosue, S. (1996) *Bull. OEPP*, **26** : 567-576.

Bawa, S.B. and Torrey, O.G. (1971) *Bot. Gaz.*, **132** : 240-245.

Bell, C.E. and Boutwell, B.E. (2001) *Calif. Agric.*, **55** : 35-38.

Berger, C.A. and Witkus, E.R. (1946) *Am. Jour. Bot.*, **33** : 785-787.

Bhagchandani, B.S., Gupta, R.P. and Pandey, U.B. (1983) *Indian Hort.*, **28** : 23-27.

Bhagchandani, P.M. and Pandey, V.B. (1983) *Rept. National Workshop on Onion*, pp. 10A-30A

Bhagchandani, P.M., Pal, N. and Choudhury, B. (1970) *Progressive Hort.*, **2** : 15-23.

Bhagchandani, P. M., Pal, N., Choudhury, B. and Seshadri, V. S. (1973) *Indian J. Hortic.*, **30** : 121-124.

Bhagchandani, P. M., Pal, N., Singh, N. and Choudhury, B. (1989) *Indian Hort.*, **24** : 7-9.

Bhakare, B.D. and Fatkal, Y.D. (2008) *J. Water Manag.*, **16** : 35-39.

Bhardwaj (1991) *New Agriculturist*, **1** : 139-142.

Bhardwaj, B.S., Gupta, R.P. and Pandey, U.B. (1983) *Indian Hort.*, **28** : 23-27.

Bhargava, K.K. and Singh, S. (2000) *Nat Symp. on Onion and Garlic, Production and Postharvest Management, Challenges and Strategies*, Nashik, Nov. 19-21, 2000, Abst., p. 216.

Bharti, N., Ram, R.B., Meena, M.L. and Yogita. (2011) *Ann. Hortic.*, **4** : 171-175.

Bhatia, A.K. and Pandey (1991) *Res. Develop. Rept.*, **8** : 10-16.

Bhatnagar, A., Gupta, A. and Singh, B. (2000) *Natl. Symp. Onion and Garlic, Production and Postharvest Management, Challenges and Strategies*, Nashik, Nov. 19-21, 2000, Abst., p. 213.

Bhattarai, S.P. and Sabedi, P.P. (1998) *Working paper - Lumle Agril. Res. Centre*, No. 98/20, p. 5.

Bhonde, S. (2016) Onion seed storage for buffer stock. In: 2nd National Symposium on Edible Alliums: Challenges and future strategies for sustainable production, 7-9th November, 2016, Jalna, Maharashtra, India, pp. 171-175.

Bhonde, S.R., Ram, L. and Pandey, V.B. (1983) *Rept. National Workshop on Onion*, pp. 13B-22B.

Bhonde, S.R., Mishra, V.K. and Chougule, A.B. (1996) *Newsl. Nat. Hort. Res. Dev. Found.*, **16** : 4-7.

Bhonde, S.R., Sharma, S.B. and Chougule, A.B. (1977) *News letter, Nat. Hort. Res. and Dev. Foundation*, **17** : 1-3.

Boff, P., Debarba, J.F., Silva, E., Werner, H. and Niggli, U. (2000) *Proc. 13th International IFOAM Scientific Conference, Basel, Switzerland, 28-31 August, 2000*, p. 56.

Bohanec, B. (2002) Doubled haploid onions. In: *Allium* crop science: Recent advances. Rabinowich, H. and Currah, L. (eds.), CABI Publishing House, Wallingford, UK. pp. 145-148.

Bohanec, B. and Jakse, M. (1999) *Plant Cell Rep.*, **18** : 737-742.

Bohanec, B., Jakse, M., Ihan, A. and Javornik, B. (1995) *Pl. Sci.*, **104** : 215-224.

Bosch Serra, A.D. and Casanova, D. (2000) *Acta Hortic.*, **519** : 53-59.

Bottcher, H. (1999) *Gartenbauwissenschaft*, **64** : 220-226.

Bozzini, A. (1964) *Caryologia*, **17** : 459-464.

Bradeen, J.M. and Havey, M.J. (1995) *J. Amer. Soc. Hort. Sci.*, **120** : 752-758.

Bradeen, J.M., Havey, M.J. and Piscataway, N.J. (1995) *Amer. J. Bot.*, **82** : 1455-1462.

Brewster, J.L. (1975) *Ann. Rept. Nat. Veg. Res. Sta.*, p. 54.

Brewster, J.L. (1987) *Ann. Appl. Biol.*, **111** : 463-467.

Brewster, J.L., Mondal, E.M. and Moris, G. E. L. (1986) *Ann. Bot.*, **58** : 221-233.

Britvich, M.D. and Goncharenko, V. Yu. (1986) *Ovochivnitstvo i Bashtannitstvo*, No. 31, p. 12-14.

Brown, B.D., Hornbacher, A.J. and Naylor, D. V. (1988) *J. Amer. Soc. Hort. Sci.*, **113** : 864-869.

Buiteveld, J., Suo, Y., Campagne, M.M.L. and Creemers-Molenaar, J. (1998) *Theor. Appl. Genet.*, **96** : 765-775.

Buitveld, J. and Creemers-Molenaar, J. (1994) *Pl. Sci.*, **100** : 203-210.

Butani, R.D., Gupta, R.S. and Khurana, S.C. (2000) *Nat Symp. on Onion and Garlic, Production and Postharvest Management, Challenges and Strategies*, Nashik, Nov. 19-21, 2000, pp. 102-106.

Butt, A.M. (1968) *Meded. Landbouw.*,Wageningen, **68** : 1-211.

Cabrera-Asencio, I. (1997) *J. Agri. Univ. Puerto Rico*, **81** : 75-77.

Calbo, M.E.R., Monnerat, R.H., Shimoya, C. (1986) *Revista Ceres*, **33** : 174-180.

Cardi, T., D'Agostino, N. and Tripodi, P. (2017) *Front. Plant Sci.*, **8** : 241.

Cardoso, A.I.I and da Costa, C.P. (2003) *Sci. Agric.*, **60** : 59-63.

Çavuþoðlu, D. (2020) *Bulg. J. Agric. Sci.*, **5** : 26-31.

Chakrabarty, A.K., Choudhury, B. and Singh, C. (1980) *Seed Res.*, **8** : 1-4.

Chand, S.K., Nanda, S. and Joshi, R.K. (2018) *Mol. Breed.*, **38** : 109.

Chandel, R.S., Thakur, R.K., Bhardwaj, N.R. and Pathania, N. (2004) *Acta Hortic.*, **631** : 79-86.

Chandrakar, O., Dixit, A., Sharma, P., Rangare, S.B. and Yadav, V. (2015) *J. Progress. Agric.*, **6** : 67-70.

Chang, W.N., Chang, Y.H., Su, A.Y. and Kuo, F.Y. (1987) *FFTC/ASPAC Book Series*, No. 36, p. 166-176.

Chatterjee, C. and Dev, G. (1998) In : *Potassium deficiency symptoms in important vegetable croips in India* (Ed. Dev, G.), Potash Phosphate Institute of Canada, Gurgaon, India, pp. 15-16.

Chatterjee, C., Jyoti, G. and Krirane, N. (1999) *Indian J. Hortic.*, **56** : 155-158.

Chattopadhyay, A., Sharangi, A.B., Dutta, S., Das, S. and Denre, M. (2013) *Vegetos*, **26** : 151-157.

Chauhan, K.S. and Sekhawat, J.S. (1971) *Fertl. News*, **16** : 45-47.

Chaurasiya, J., Verma, R.B., Verma, R.K., Panwar, G.S., Patel, V.B. and Saha, B.C. (2018) *Indian J. Hortic.*, **75** : 717-722.

Cheema, K.L., Saeed, A. and Habib, M. (2003) *Int. J. Agric. Bio.*, **5** : 185-187.

Chenna, R.A., Sung, Y.M., Doo, H.K. and Park, S.W. (2006) *Plant Cell Rep.*, **25** : 92-99.

Cheremushkina, N.P., Shishkina, T.S. and Lukonina, E.I. (1989) *Soviet Agriculture Sciences*, **6** : 41-44.

Chethana, B.S., Kochapur, M.R. and Hosamani, R.M. (2000) *Nat Symp. on Onion and Garlic, Production and Postharvest Management, Challenges and Strategies*, Nashik, Nov. 19-21, 2000, Abst., p. 213.

Chiag, L. C. (1973) *J. Taiwan Agric. Res.*, **22** : 285-89.

Cho, J.G., Bae, R.N. and Lee, S.K. (2010) *Korean J. Hortic. Sci. Technol.*, **28** : 522-527.

Chopade, S.O., Bansode, P.N., Hiwase, S.S. and Bhuyae, R.C. (1997) *Ann. Plant Physiol.*, **11** : 45-48.

Choudhary, B.R. (2014) *Vegetables*. Kalyani Publishers, Ludhiana. pp. 141-142.

Choudhury, B. (1967) *Vegetables*, National Book Trust, India, New Delhi.

Chowdappan, S.R. and Morchan, Y.B. (1971) *Madras Agric. J.*, **58** : 88-91.

Christopher, S.C. and Michael J.H. (1999) *HortScience*, **34**(4) : 589-593.

Chung, B. (1989) *Acta Hortic.*, No. 247, p. 233-237.

Clarke, A.E., Jones, H.A. and Little, T.M. (1944) *Genetics*, **29** : 569-575.

Cleaver, T.J. and Turner, M.K. (1975) *Ann. Rept. Nat. Veg. Res. Sta.*, Wellesbourne for 1974, p. 58.

Clifford, M. (2000) *J. Sci. Food Agric.*, **80** : 1063-1072.

Coleman, P.M., Ellerbrock, L.A. and Lorbeer, J.W. (1997) *Pl. Dis.*, **81** : 138-142.

Comrie, A.G. (1986) *Acta Hortic.*, No. 194, pp. 125-132.

Cooke, R.J., Smith, T.M. and Morgan, A.G. (1986) *J. Matl. Inst. Agr. Bot.*, **17**: 209-218.

Corgan, J.N. (1988a) *HortScience*, **23** : 432-434.

Corgan, J.N. (1988b) *HortScience*, **23** : 421-422.

Correa, N.S., Bandeira, J.M., Marini, P., Borba, I.C.G., Lopes, N.F. and Moraes, D.M. (2013) *Acta Bot. Bras.*, **27** : 394-399.

Cramer, C.S. (2000) *Euphytica*, **115** : 159-166.

Cramer, C.S. and Corgan, J.N. (2001) *HortScience*, **36** : 1337-1338.

Cramer, C.S. and Corgan, J.N. (2001) *HortScience*, **36** : 1339-1340.

Cramer, C.S. and Havey, M.J. (1999) *HortScience*, **34**(4): 589-593.

Cuocolo, L. and Barbieri, G. (1988) *Rivista di Agronomonia*, **22** : 195-202.

Currah, L. and Proctor, F.J. (1990) *Onion in tropical regions. Natural Resources Institute Bulletin*, NO. 35, pp. 232, NRI, U.K.

Currah, L., Green, S.M. and Galmarini, C.R. (1997) *Acta Hortic.*, **433** : 197-205.

Daljeet-Singh, Singh, D. and Armstrong, J. (2001) *Acta Hortic.*, **555** : 141-145.

Daljeet-Singh, Singh, D. and Galmarini, C.R. (1997) *Acta Hortic.*, **433** : 75-79.

Dangi, R., Khar, A., Islam, S. and Kumar, A. (2018) *Indian J. Hortic.*, **75** : 226-236.

Das, B.C. and Dhyani, K.C. (1956) *Phyton.*, **6** : 47-56.

Dastane, N.G., Parasher, K.S. and Yusuf, M.Y. (1969) *Ann. Rept. IARI*, New Delhi.

Davis, G.N. (1943) *Onion Production in Calif. Circ.* No. 357.

Dellacecca, V., Lovato, A.F.S. and Damato, G. (2000) *Acta Hortic.*, **533** : 197-203.

Deogade, D., Sukeshini, K.P. and Nasare, N. (2017) *Bioinfolet*, **14** : 275-281.

Desuki, M.M., Abdel-Mouty and Ali, A.H. (2006) *J. Appl. Sci. Res.*, **2** : 592-597.

Dewangan, S.R. and Sahu, G.D. (2014) *Agric. Sci. Digt.*, **34** : 233-236.

Dhar, M., Mandal, J. and Mohanta, S. (2017) In: *Issues in Sustainable Development in India: Present Problems and Future Perspective* (Eds. P.K.Chattopadhyay, and D.S. Kushwaha), New Delhi Publishers, New Delhi. pp. 19-28.

Dhar, M., Mandal, J., Maity, T.K. and Mohanta, S. (2019) *J. Pharmacogn. Phytochem.*, **8** : 1317-1321.

Dharam Singh, Chakravarty, A.K. and Sinha, S. N. (1976) *Seed Tech. News*, **6** : 1-5.

Dhatt, A.S. and Thakur, P. (2014) *J. Hort. Sci.*, **9** : 107-112.

Dhotre, M., Allolli, T.B., Athani, S.I. and Halemani, L.C. (2010) *TAJH*, **5** : 143-146.

Dilruba, S., Alam, M.M., Rahman, M.A. and Hasan, M. F. (2010) *Int. J. Agric. Res.*, **5** : 430-435.

Dogra, R., Thakur, A.K. and Sharma, U. (2019) *Int. J. Econ. plants*, **6** : 136-139.

Doijode, S.D. (2000) *Nat Symp. on Onion and Garlic, Production and Postharvest Management, Challenges and Strategies*, Nov. 19-21, 2000, Abst., p. 225.

Dolatyari, A., Mehrvarz, S.S., Fazeli, S.A.S., Naghavi, M.R. and Fritsch, R.M. (2018) *Plant Syst. Evol.*, **304** : 583-606.

Doorn, A.M., Vankoert, J. L. and Vilet, M.V. (1954) *Rev. Appl. Mycol.*, **34** : 273.

Doruchowski, R.W. and Krawczyk, J. (1987) *Biuletyn Warywniczy*, **30** : 7-28.

Dowker, B.D. and Gordon, G.H. (1983) *Heterosis, Theor. Apl. Genet. Monogr.*, **6** : 220-233.

Doyle, A.S. and Bryan, W.M. (1988) *HortScience*, **23** : 141-143.

Du, W.F., Wang, X.H. and Galmarini, C.R. (1997) *Acta Hortic.*, **433** : 179-184.

Dunstan, D.I. and Short, K.C. (1977) *Physiol. Plant.*, **41** : 70-72.

Duque, M.C.M., Perdomo, G.C.E. and Jaramillo, V.J. (1989) *Acta Agron.*, **39** : 45-53.

Eady, C.C. (1995) *N. Z. J. Crop Hortic. Sci*, **23** : 239-250.

Eady, C., Davis, S., Farrant, J., Reader, J. and Kenel, F. (2003b) *Ann. Appl. Biol.*, **142** : 213-217.

Eady, C.C. and Lister, C.E. (1998) *Plant Cell Rep.*, **18** : 177-221.

Eady, C.C., Kamoi, T., Kato, M., Porter, N.G., Davis, S., Shaw, M., Kamoi, A. and Imai, S. (2008) *Plant Physiol.*, **147** : 2096-2106.

Eady, C.C., Lister, C.E., Suo, Y., Schaper, D. and Suo, Y.Y. (1996) *Plant Cell Rep.*, **15** : 958-962.

Eady, C.C., Reader, J., Davis, S. and Dale, T. (2003a) *Ann. appl. Biol.*, **142** : 219-224.

Eady, C.C., Weld, J.J. and Lister, C.E. (2000) *Plant Cell Rep.*, **19** : 376-381.

Ebrahimi, R. and Zamani, Z. (2009) *American-Eurasian Journal of Sustainable Agriculture*, **3** : 71-74.

Eda, R. and Kukanoor, L. (2016) *Progress. Hortic.*, **48** : 106-109.

El-Aweel, M.A.T., Ghobashi, A.A. and El-Kafory, A.K. (2000) *Assiut J. Agri. Sci.*, **31** : 89-100.

El-baky, A., Hanaa, H., Hussein, M.M. and Amal, M.A. (2003) *Asian J. Plant Sci.*, **2** : 633-638.

El-Emery, M. (1985) *Kertgazdasag*, **17** : 81-84.

Elena, B., Ioan, S. and Irina, P. (2018) *Annals of the University off Craiova- Agriculture,* XL (VIII) : 38-43.

El-Gendi, S.M. (1998) *Arab Univ. Agri. Sci.,* **6** : 267-276.

El-Rehim, G.H.A. (2000) *Assiut J. Agri. Sci.,* **31** : 115-121.

El-Sayed, A.M. and Atia, A.A.M. (1999) *Egyptian J. Hort.,* **26** : 67-75.

El-Shaifi, M.W. and Davis, G.N. (1967) *Hilgardia,* **38** : 607-622.

Engelmann, F. (1997) *In vitro* conservation methods. In : Callow J.A., Ford-Lloyd B., Newbury H.J. (ed.) Biotechnology and plant genetic resources: conservation anduse. Wallingford: *CAB International,* **19** : 119-161.

Erdem, T. and Kayhan, A. (2019) *Indian J. Hortic.,* **76** : 57-465.

FAO. (2015) Food and Agriculture Organization Statistical Pocketbook on world food and agriculture. ISBN 978-92-5-108802-9

Favaron, F., Castiglioni, C., Dovidio, R. and Alghisi, P. (1997) *Physiol. Mol. Plant Pathol.,* **50** : 403–417.

Fayos, Oreto, Valles, María, Garcés-Claver, Ana, Mallor, Cristina and Castillo, Ana. (2015) *Front. Plant Sci.,* **6**. 10.3389/fpls.2015.00384.

Feng, D.W., Hua, W.X., Du, W.F., Wang, X.H. and Burba, J.L. (1997) *Acta Hortic.,* **433** : 179-184.

Ferreira, M.D. and Minami, K. (2000) *Sci. Agric.,* **57** : 693-701.

Ferreres, F., Gil, M.I., and Tomás-Barberán, F.A. (1999) *Food Res. Int.,* **29** : 389.

Firbas, P. and Amon, T. (2014) *Caryologia,* **67** : 25-35.

Fletcher, P.J., Fletcher, J.D. and Lewthwaite, S.L. (1998) *New Zeal. J. Crop Hortic. Sci.,* **26** : 23-26.

Fossen, T. and Andersen, O.M. (2003) *Phytochem.,* **62** : 1217-1220.

Fraga-Aguiar, N.A., Perez, A., Fundora, Z., Gonzalez-Chavez, M., Prats, A., Alonso-Rodriguez, M. del C. Alonso-Rodriguez, M. (2001) *Alimentaria,* **38** : 79-82.

Franca, J.G.E. de, Candeia, J.A., Menezes, J.T. de, Maranhao, E.A.A. de, Menezes, D., Franseen, C.J.H. and Rantteuran, W.C. (1933) *Korte medel Inst. Pl. Zoxket No. 18,* P. 20.

Friesen, N., Fritsch, R.M., Blattner, F.R. (2006) *Aliso,* **22** : 372-395.

Fritsch, R.M. and Abbasi, M. (2013) A Taxonomic Review of *Allium* subg. *Melanocrommyum* in Iran. IPK Gatersleben, Germany.

Fritsch, R.M. and Astanova, S.B. (1998) *Feddes Repertorium,* **109** : 539-549.

Fritsch, R.M. and Friesen, N. (2002) Evolution, domestication and taxonomy. In: Rabinowitch HD and Currah, L. (eds.). *Allium* Crop Science. CABI Publishing, Wallingford, U.K. pp. 5-30.

Fritsch, R.M., Blattner, F.R. and Gurushidze. M. (2010) *Phyton.*, **49** : 145-220.

Fritsch, R.M., Farideh-Matin, Klaas, M. (2001) *Genet. Resour. Crop Evol.*, **48** : 401-408.

Fustos, Z. (1985) *Zoldsegtermesztesi Kutato Intezet Bulletinji, Hungary,* **18** : 71-74.

Gabriel, E.L., Makuch, M.A. and Piccolo, R.J. (1997) *Acta Hortic.*, **433** : 573-581.

Gallagher, P.A. (1969) *Fm. Res. New.*, **10** : 74.

Galmarani, C.R., Goldman, I.L. and Havey, M.J. (2001) *Mol. Genet. Genomics*, **265** : 543-551.

Galmarini, C.R. (2000) *HortScience*, **35** : 1360-1362.

Galmarini, C.R., Della-Gaspera, P., Fuligna, H. and Armstrong, J. (2001) *Acta Hortic.*, **555** : 259-261.

Galmarini, C.R., Goldman, I.L. and Havey, M.J. (2001) *Mol. Genet. Genom.*, **265** : 543-551.

Gamie, A.A., El-Rhim, G.H.A., Imam, M.K. and Abdoh, E. (1996) *Assiut J. Agri. Sci.*, **27** : 101-110.

Ganai, S.A., Ahmad, H., Kour, D., Sharma, R., Norboo, T. and Khaliq, N. (2018) *Indian J. Entomol.*, **80** : 563-566.

Gandin, C.L., Thomazelli, L.F., Zimmermann-Filho, A.A., Neto, J.S., Oliveira, S.O. de, Rosset, V., Biasi, J., Garcia, A., Neto, J.A.Z. and Debarba, J.F. (1998) *Agropecuaria Catarinense*, **11** : 5-7.

Ganeshan, G. and Sinha, P. (2000) *Nat Symp. on Onion and Garlic, Production and Postharvest Management, Challenges and Strategies*, Nov. 19-21, 2000, Abst., pp. 142-149.

Ganeshan, G., Pathak, C.S. and Gowda, B.V. (1998) *P K V Res. J.*, **22** : 53-54.

Gangwar, R.K., Agarwal, N.K., Singh, T.P. and Sharma, R.K. (2016) *Research on Crops*, **17** : 555-561.

García, G., Clemente-Moreno, M.J., Díaz-Vivancos, P., García, M. and Hernández, J.A. (2020) *Antioxidants (Basel)*, **67** : 1-17.

Gaviola, J.C. (1997) *Horticultura Argentina*, **16** : 40-41.

Gaviola, S., Lipinski, V. and Galmarini, C.L. (1998) *Ciencia del Suelo*, **16** : 115-118.

Gawande, S. J. and Singh, M. (2017) *Indian Hort.*, **52** : 70-72.

Geetha, K., Raju, A.S. and Shanti, M. (1999) *J. Res., ANGRAU*, **27** : 18-23.

Geetharani, P. and Ponnuswamy, A.S. (2007) *Int. J. Plant Sci.*, Muzaffarnagar, **2** : 12-15.

Geetika, A.M. (2019) *Journal of eco-friendly agriculture*, **14** : 93-101.

Geoffriau, E., Kahane, R., Bellamy, C. and Rancillac, M. (1997) *Plant Sci. Limer.*, **122** : 201-208.

Georgy, N.I., Radwan, A., Mohamed, H.A. and Shahabi, A.E. (1986) *Agric. Res. Rev.*, **61** : 25-41.

Ghodke, P.H., Thangasamy, A., Mahajan, V. Gupta, A.J. and Singh, M. (2017a) *Indian Hort.*, **62** : 36-38.

Ghodke, P.H., Thangasamy, A., Mahajan, V. Gupta, A.J. and Singh, M. (2017b) *Indian Hort.*, **62** : 33-35.

Gill, S.S. and Singh, H. (1989) *Seed Res.*, **17** : 11-15.

Global Trade (2019) https: //www.globaltrademag.com/

Globerson, D., Levy, M., Huppert, H. and Eliassy, R. (1987) *Acta Hortic.*, **215** : 17-24.

Gomes, P.R., Garcia, A., Raseira, M. do C.D., Silva, J.B. da and da Silva, J.B. (2000) *Agropecuaria Clima Temperado*, **3** : 19029.

Goncalves, P.A.S. (1997) *Anais da Sociedade Entomologica do Brazil*, **26** : 365-369.

Gonza´lez, P.H., Colnago, P., Peluffo, S., Gonza´lez, I.H., Zipý´trý´a, J. and Galva´n, G.A. (2011) *Eur. J. Plant Pathol.*, **129** : 303-314.

Gonzalex, L.M., Ramirez, R., Licea, L., Garicia, B., Porra, E. and Perez, A. (2001) *Alimentaria*, **38** : 125-128.

Gonzalez, M.I. and Galmarini, C.R. (1997) *Acta Hortic.*, **433** : 549-554.

Gorrepati K., Thangasamy, A., Bhagat, Y. and Murkute, A.A. (2017) *Indian Hort. J.*, **7** : 8-14.

Gorrepati, K., Murkute, A.A., Bhagat, Y. and Gopal, J. (2018) *Indian J. Hortic.*, **75** : 314-318.

Govaerts, R., Kington, S., Friesen, N., Fritsch, R., Snijman, D.A., Marcucci, R., Silverstone-Sopkin, P.A. and Brullo, S. (2013) http: //apps.kew.org/wcsp/

Goyal, R., Uike, V. and Verma, H. (2017) *Agric. Science Digt.*, **37** : 160-162.

Gregoniov, S. (1998) *Reports - Agril. Res. Inst.*, Ministry of Agriculture and Natural Resources (Nicosia), No. 71, p. 51.

Grevsen, K. (1989) *Acta Hortic.*, **242** : 319-324.

Gruda, N. (1987) *Buletini i Skencave Bujquesore*, **26** : 27-33.

Guimaraes, D.R., Torres, L. and Diltrich, R.C. (1997) *Agropecuaria Catarinense*, **10** : 57-61.

Gupta, A. J. and Mahajan, V. (2019) *Veg. Sci.*, **45** : 226-231.

Gupta, A.J., Mahajan, V. and Lawande, K.E. (2019) *Veg. Sci.*, **46** : 107-113.

Gupta, A.J., Mahajan, V. and Lawande, K.E. (2018) *Veg. Sci.*, **45** : 92-96.

Gupta, A.J., Mahajan, V. and Singh, M. (2017) *Indian Hort.*, **62** : 24-27.

Gupta, R.B.L. and Pathak, V.N. (1987) *Zentralblatt fur Mikrobiologie*, **142** : 163-166.

Gupta, R.P. and Singh, R.K. (2010) Technical Bulletin No. 9, NHRDF, Nasik, Maharashtra, pp. 46.

Gupta, R.P., Sharma, V.P., Singh, D.K. and Srivastava, K.J. (1999) *Newsl. Nat. Hort. Res. Dev. Found.*, **19** : 7-11.

Gupta, R.P., Srivastava, K.J. and Pandey, U.B. (1991) *Onion Newsletter for the Tropics*, No. 3, pp. 15-17.

Gupta, R.P., Srivastava, P.K. and Pandey, U.B. (1981) *Pesticides*, **15** : 16.

Gupta, R.P., Srivastava, P.K., Pandey, U.B. and Mehta, U. (1983) *Rept. National Workshop on Onion*, pp. 48-55.

Gupta, R.P., Srivastava, V.K. and Pandey, U.B. (1987) *Pesticides*, **21** : 33-34.

Gupta, R.S., Bhutani, R.D., Khurana, S.C. and Thakral, K.K. (1999) *Veg. Sci.*, **26** : 137-139.

Gurjar, R.S.S. and Singhania, D.L. (2006) *Indian J. Hortic.*, **63** : 53-58.

Gvozdanovic-Varga, J., Vasic, M., Panajotovic, J. and Lazic, B. (1997) *Acta Hortic.*, **462** : 557-563.

Hadacova, V., Klozova, E., Pitterova, K. and Turkova, V. (1981) *Biol. Plant* (Prague), 23: 442-448.

Haggag, M.E.A.; Rizk, M.a., Hagras, A.M. and Aboel-Hamad, A.S.A. (1986) *Annals Agric. Sci.*, Ain Shams University, **31** : 989-1010.

Haile, A., Tesfaye, B. and Worku, W. (2017) *Afr. J. Agric. Res.*, **12** : 987-996.

Hamilton, B.K., Pike, L.M. and Yoo, K.S. (1997) *Sci. Hort.*, **71** : 131-136.

Hamilton, B.K., Pike, L.M., Sparks, A.N., Bender, D.a., Jones, R.W., Candeia, J. and Franca, G. de (1999) *Euphytica*, **109** : 117-122.

Hamilton, B.K., Yoo, K.S., Pike, L.M. and Yoo, K.S. (1998) *Sci. Hort.*, **74** : 249-256.

Hang, T.T.M., Shigyo, M., Yaguchi, S., Yamauchi, N. and Tashiro, Y. (2004) *Genes Genet. Syst.*, **79** : 345-350.

Hanley, A.B. and Fenurick, G.R. (1985) *J Plant Food,* **6** : 211-238.

Hansen, E.E., Hubstenberger, J.F. and Phillips, G.C. (1995) *Plant Cell Rep.*, **15** : 8-11.

Hansen, S.L. (1999) *Soil Plant Sci.*, **49** : 103-109.

Hari Har Ram (1998) *Onion : In Vegetable Breeding - Principles and Practices*, Kalyani Publishers, India, pp. 309-321.

Hassan, M.S. (1984) *Acta Hortic.*, **143** : 341-348.

Hassan, M.S. and Ayub, A.T. (1978) *Sudan Gezira Exp. Agric.*, **14** : 29-32.

Havey, M.J. (1991a) *J. Hered.*, **82** : 501-503.

Havey, M.J. (1991b) *Theor. Appl. Genet.*, **81** : 752-757.

Havey, M.J. (1997) *Genet. Res. Crop Evol.*, **44** : 307-313.

Havey, M.J. (1998a) In: *Proceedings of the 1998 National Onion (and Other Allium) Research Conference, 10-12 December, Sacramento, California, USA*. University of California, Davis, California, pp. 35-38.

Havey, M.J. and Bark, O.H. (1994) *J. Amer. Soc. Hort. Sci.*, **119** : 90-93.

Havey, M.J. and Randle, W.M. (1996) *J. Amer. Soc. Hort. Sci.*, **121** : 604-608.

Havey, M.J., King, J.J., Bradeen, J.M. and Bark, O. (1996) *HortScience*, **31** : 1116-1118.

Hayden, N.J. and Maude, R.B. (1997) *Acta Hortic.*, **433** : 475-479.

Heath, O.V.S. (1943) *Ann. Appl. Biol.*, **30** : 208-220.

Heath, O.V.S. and Hollies, M.A. (1965) *J. Exp. Bot.*, **16** : 128-144.

Heath, O.V.S. and Mathur, P.V. (1944) *Ann. Appl. Biol.*, **31** : 173-186.

Hegde, D.M. (1988) *Singapore Journal of Primary Industries*, **16** : 111-123.

Hegde, G.M., Nargund, V. B, Nayak, G.V. (2015). *Int. J. Plant Prot.*, **43** : 90-93.

Heggestad, H.E. (1968) *Phytopatology*, **58** : 1089-1098.

Henriksen, K., Hansen, S.L. and Armstrong, J. (2001) *Acta Hortic.*, **555** : 147-152.

Himangini, Kanwar, H. and Thakur, A. (2018) *J. Hill Agric.*, **9** : 300-303.

Holdsworth, M. and Heath, O.V.S. (1950) *J. Exp. Bot.*, **1** : 353-355.

Holz, G. and Knox-Davies, P.S. (1974) *Phytophylactica,***6**: 153-156.

Horbowicz, M. and Grzegorzewska, M. (2000) *Folia Hortic.*, **12** : 65-75.

Horobin, J.F. (1986) *Ann. Appl. Biol.*, **108** : 199-204.

Hosfield, G.L., Vest, G. and Peterson, C.E. (1977) *J. Amer. Soc. Hort. Sci.*, **102** : 56-61.

Huang, X.J. and LI, J.Z. (2008) *J. Southern Agric.*, 06.

Hussaini, M.E. and Amans, F.B. (2000) *Trop. Agric. (Trinidad)*, **77** : 145-149.

Hussein, M.M. and El-Faham, S.Y. (2018) *Egypt. J. Agron.*, **40** : 285-296.

Hussein, M.A.M., Hassan, M.H.A., Allam, A.D.A. and Abo-Elyous, K.A.M. (2007) *Plant Pathol. Egypt. J. Phytopathol.*, **35** : 49-60.

Hussey, G. (1976) *J. Exp. Bot.*, **97** : 375-382.

Hussey, G. and Falavigna, A. (1980) *J. Exp. Bot.*, **31** : 1675-1686.

Hutton, R.C. and Wilson, G.J. (1986) *New Zealand J. Exp. Agric.*, **14** : 453-457.

Huygens, D. (1997) *Proeftuinnieuws*, **7** : 14-15.

Hwang, H.J., Suh, J.K., Ha, I.J. and Ryu, Y.W. (1996) *RDA J. Agri. Sci., Hortic.*, **38** : 640-647.

Ibrahim A.S. (1966) The inheritance of onion bulb shape and its component measurements. *Retrospective Theses and Dissertations.* 2914. *Iowa State University.* https: //lib.dr.iastate.edu/rtd/2914

Ibrahim, M., Ganiger, V.M. and Naik, B.H. (2000) *Karnataka J. Agri. Sci.*, **13** : 975-977.

ICAR-NBPGR (2015) Annual Report of the ICAR-National Bureau of Plant Genetic Resources 2014-2015, ICAR-NBPGR, Pusa Campus, New Delhi, India, 210+x p.

Ilic, Z., Milenkovic, L., Djurovka, M. and Trajkovic, R. (2009) *Acta Hortic.*, **830** : 635-642.

Imai, S., Tsuge, N., Tomotake, M., Nagatome, Y., Sawada, H., Nagata, T. and Kumagai, H. (2002) *Nature*, **419** : 685.

Islam, M.T. and Haque, M.A. (1977) *Bangladesh Hort.*, **5** : 5.

Jackson, K.J., Duff, A.A., O'-Donnell, W.E. and Armstrong, J. (2001) *Acta Hortic.*, **555** : 239-242.

Jaggi, R.C. and Dixit, S.P. (1999) *Indian J. Agric. Sci.*, **69** : 289-291.

Jahan, M. A., Hossain, M. M., Yesmin, S. and Islam, M. S. (2010) *Int. J. Sust. Agric. Technol.*, **6** : 1-6.

Jaiswal, J.P. and Subedi, P.P. (1996) *Working Paper*, Lumle Regional Agricultural Research Centre, No. 96-14, p. 8.

Jaiswal, J.P., Bhattarai, S.P. and Subedi, P.P. (1997) *Working Paper*, Lumle Regional Agricultural Research Centre, No. 97-48, p. 8.

Jakse, M., Bohance, B. and Ihan, A. (1996) *Plant Cell Rep.*, **15** : 934-938.

Jakse, M., Havey, M.J. and Bohanec, B. (2003) *Plant Cell Rep.*, **21** : 905-910.

Jana, B.K. and Kabir, J. (1990) *Crop Res.*, **3** : 241.

Jaske, J., Martin, W., McCallum, J. and Havey, M. J. (2005) *J. Amer. Soc. Hort. Sci.*, **130** : 912-917.

Jasol, F.S.K.R. (1989) *Indian Hort. (Veg. Spl.)*, **33 and 34** : 79-85.

Jawaharlal, M., Sundararajan, S. and Veeraragavathatham, D. (1998) *South Indian Hort.*, **36** : 308.

Jayaswall, K., Sharma, H., Bhandawat, A., Sagar, R., Yadav, V.K., Sharma, V., Mahajan, V., Roy, J. and Singh, M. (2019) *Genet. Resour. Crop Evol.*, **66** : 1379-1388.

Jayeeta, M., Shrivastava, S.L. and Rao, P.S. (2012) *J. Food Sci. Technol.*, **49** : 267-277.

Jeevitha, D., Manohar, R.K., Madhuri, R.K. and Vasnathakumari, R. (2017) *J. Hill Agric.*, **8** : 250-252.

Jeong, H.B. and Park, H.G. (1997) *J. Korean Soc. Hort. Sci.*, **38** : 123-128.

Jeong, H.B., Ha, S.H. and Kang, K.Y. (1996) *RDA J. Agri. Sci. Biotech.*, **38** : 295-301.

Jha, A.K., Pal, N. and Singh, N. (2000) *Indian J. Hortic.*, **57** : 347-350.

Jiang, Q.S., Zhang, H.L. and Ai, G.H. (1998) *China Veg.*, **4** : 38.

Jirik, J. (1972) *Bulletin*, Vyzkummy Ustav Zelionarsky Olomovc, **16** : 36-44.

Jo, J., Purushotham, P.M., Han, K., Lee, H.R., Nah, G. and Kang, B.C. (2017) *Front. Plant Sci.*, **8** : 1606.

Joaheer, D. T., Aumeeruddy, M.Z., Toorabally, Z., Gokhan, Z., Kannan, R.R., Rengasamy, S., Karutha, P. and Mahomoodally, M.F. (2019) *Crit. Rev. Food Sci. Nutr.*, **59** : 39-70.

Jones H.A. and Mann L.K. (1963). Onion and their allies Botany, cultivation and utilization. *Leonard Hill (Books) Ltd.* London, New York.

Jones, H.A. and Clark, A.E. (1943) *Proc. Amer. Soc. Hort. Sci.*, **43** : 189-194.

Jones, H.A. and Emsweller, S.L. (1936) *Proc. Amer. Soc. Hort. Sci.*, **34** : 582-585.

Jones, H.A. and Emsweller, S.L. (1939) *Bull Calif. Agric. Exp. Sta.*, **628** : 14.

Jones, H.A. and Mann, L.K. (1963) *Onion and their Allies, Botany, Cultivation and Utilization*, World Crops Books, Leonard Hill (Books) Ltd., London.

Jones, H.A. and Peterson, G.E. (1952) *Proc. Amer. Soc. Hort. Sci.*, **59** : 457.

Jones, R.N. 1991. Cytogenetics of Alliums. In: Chromosome Engineering in Plants: Genetics, Breeding, Evolution, Part B edited by Suchiya, T.T. and Gupta, P.K. Elsevier Science Publishers B.V., The Netherlands. pp. 215-228.

Jones, R.N. (1983) *Cytogenetics of Crop Plants* (Swaminatha, M.S., Gupta, P.K. and Sinha, U.K. eds.), Macmillan India Ltd.

Joshi, H.C. and Tandon, J.P. (1976) *Indian J. Agric. Sci.*, **46** : 88-92.

Joshi, R.S. (1963) *M.Sc. Thesis* submitted to P.G. School, IARI, New Delhi.

Joubert, P., Sangwan, R.S., Aouad, M.E.A., Beaupere, D. and Snagwan, N.B.S. (1995) *Phytochemistry*, **40** : 1623-1628.

Jyotishi, R.P. and Pandey, R.C. (1969) *Punjab Hort. J.*, **9** : 192-197.

Kahane, R., De La Serve, B.T. and Rancillac, M. (1992) *Ann. Bot.*, **69** : 551-555.

Kahane, R., Rancillac, M. and Teyssendier de la Serve, B. (1992) *Plant Cell Tissue Organ Cult.*, **28** : 281-288.

Kahane, R., Schweisguth, B., Rancillac, M., Burba, J.L. and Galmarini, C.R. (1997) *Acta Hortic.*, **433** : 434-443.

Kalavathi, D., Vanangamudi, K. and Ramamoorthy, K. (1990) *Madras Agric. J.*, **77** : 61-63.

Kale, P.N., Warade, S.D. and Jagtap, K.B. (1991) *Onion Newsletter for the Tropics*, No. 3, pp. 25-27.

Kamala, V., Gupta, A.J., Rajput, A.S., Sivaraj, N., Pandravada, S.R., Sunil, N., Varaprasad, K.S. and Lawande, K.E. (2014) *Indian J. Hortic.*, **71** : 499-504.

Kamstaityte, D. and Stanys, V. (2004) *Acta Universitatis Latviensis, Biology*, **676** : 173-176.

Kandoliya, U.K., Bodar, N.P., Bajaniya, V.K., Bhadja, N.V. and Golakiya, B.A. (2015) *Int. J. Curr. Microbiol. App. Sci.*, **4** : 635-641.

Kanwal, G., Lal A. A. and Simon, S. (2016) *Ann. Plant Protec. Sci.*, **24** : 335-338.

Kanwar, J.S., Sidhu, A.S. and Singh, S. (2000) *Natl. Symp. Onion Garlic, Production Postharvest Management, Challenges and Strategies*, Nov. 19-21, 2000, pp. 43-48.

Kapoor, K.S., Kallo, G., Pandey, K.K. and Pandey, P.K. (2000) *Natl. Symp. on Onion and Garlic, Production and Postharvest Management, Challenges and Strategies*, Nov. 19-21, 2000, Abst. pp. 211-212.

Karak, C. and Hazra, P. (2012) *Veg. Sci.*, **39** : 161-164.

Karic, L., Golzardi, M., Glamoclija, P. and Sutkovic, J. (2018) *Genet. Mol. Res.*, **17** : gmr16039870.

Karim, M.A. and Adachi, T. (1997) *Plant Cell, Tissue and Organ Culture,* **51** : 43-47.

Karuppaiah, V., Soumia, P.S. and Singh, M. (2020) *Indian J. Entomol.,* **82** : 195-199.

Karuppaiah, V., Soumia, P.S. and Wagh, P.D. (2018) *Indian J. Entomol.,* **80** : 1366-1369.

Kashi, A. and Frodi, B.R. (1998) *Iranian J. Agri. Sci.,* **29** : 589-597.

Kashyap, P., Jyotishi, R.P. and Srivastava, S.R. (1967) *Punjab Hort. J.,* **7** : 119.

Kastner, U., Klahr, A., Keller, E. and Kahane, R. (2001) *Plant Cell Rep.,* **20** : 137-142.

Kataria, A.S. and Singh, N. (1989a) *Indian J. Hortic.,* **46** : 199-203.

Kataria, A.S. and Singh, N. (1989b) *Indian J. Hortic.,* **46** : 395-400.

Katayoon, O. S., Hamideh, J. and Seied, M. M. (2019) *J. Plant Physiol. Breed.,* **9** : 115-127.

Kato, T. (1963) *J. Japanese Soc. Hort. Sci.,* **32** : 229-237.

Kato, T. (1964) *J. Japanese Soc. Hort. Sci.,* **33** : 53-61.

Kato, T. (1965) *J. Japanese Soc. Hort. Sci.,* **34** : 51-57.

Kato, T. (1966) *J. Japanese Soc. Hort. Sci.,* **35** : 142-152, 295-303.

Kato, T., Yamagata, M. and Tsukahara, S. (1987) *Bulletin of the Shikoku National Agricultural Experiment Station,* No. 48, pp. 26-49.

Kaul, M.L.H. (1977) *Cytologia,* **42** : 681-689.

Kaur, A. (2019) *Int. J. Farm Sci.,* **9** : 112-115.

Kehr, A.E., O'Brien, M.J. and Davis, E.W. (1962) *Euphytica,* **11**: 197-208.

Keller, E.R.J. (1993) *Genet. Resour. Crop Evol.,* **40** : 113-120.

Keller, J. (1991) *Acta Hortic.,* p. 289.

Keller, E.R.J., Schubert, I., Fuchs, J. and Meister, A. (1996) *Theor. Appl. Genet.,* **92** : 417-424.

Keller, J. and Hanelt, P. (1992) In: Hammer, K. and Knupffer, H. (Eds.), Proceedings International Symposium on the Genus Allium - Taxonomic Problems and Genetic Resources, Gatersleben, Jun 11-13, 1991, Germany DE. pp. 137-152.

Kerefov, K.N. and Shkhagumov, K.K. (1985) *Zashchita Rastenii,* No. 5, p. 39.

Keusgen, M., Kusterer, J. and Fritsch, R.M. (2011) *J. Agric. Food Chem.,* **59** : 8289-8297.

Kgopa, P.M., Mashela, P.W. and Manyevere, A. (2018) *Res. Crops,* **19** : 62-67.

Khaisin, M.F. (1975) *Stink,* pp. 70-81.

Khan, S.A., Muhammad-Amjad and Khan, A.A. (2001) *Inter. J. Agri. Biol.,* **3** : 498-500.

Khandagale, K. and Gawande, S. (2019) *J. Hortic. Sci. Biotechnol.,* **94** : 522-532.

Khar, A., Bhutani, R.D, Yadav, N. and Chowdhury, V.K. (2005) *AkdenizUniversitesi Ziraat Fakultesi Dergisi,* **18** : 397-404.

Khar, A., Islam, S., Kalia, P., Bhatia, R.and Kumar, A. (2019) *Indian J. Agric. Sci.,* **89** : 396-405.

Khar, A., Jakse, J. and Havey, M.J. (2008) *J. Amer. Soc. Hort. Sci.*, **133** : 42–47.

Khodadadi, M. (2012) *Int. J. Agric. Res. Rev.*, **2** : 324-327.

Khokhar, K.M. (2008) *J. Hortic. Sci. Biotechnol.*, **83** : 481-487.

Khokhar, K.M., Hussain, S.I., Tariq-Mahmood, Hidayatullah and Bhatti, M.H. (2001) *Sarhad J. Agri.*, **17** : 355-358.

Kim, B., Cho, Y. and Kim, S. (2017) *Plant Breed. Biotech.*, **5** : 45-53.

Kim, H., Popova, E., Shin, D., Yi, J., Kim, C., Lee, J., Yoon, M. and Engelmann, F. (2012) *Cryo Letters*, **33** : 45-57.

Kim, S., Binzel, M.L., Yoo, K.S., Park, S. and Pike, L.M. (2004a) *Mol. Gen. Genomics*, **272** : 18-27.

Kim, S., Jones, R., Yoo, K.S.and Pike, L.M. (2004b) *Mol. Gen. Genom.*, **272** : 411-419.

King, J.J., Bradeen, J.M. and Havey, M.J. (1998a) *J. Amer. Soc. Hort. Sci.*, **123** : 1034-1037.

King, J.J., Bradeen, J.M., Bark, O., McCallum, J.A. and Havey, M.J. (1998b) *Theor. Appl. Genet.*, **96** : 52-62.

Klein, M. and Korzonek, D. (1999) *Acta Biologica Cracoviensia Series Botanica*, **41** : 185-192.

Klein, T.M, Wolf, E.D., Wu, R. and Sanford, J.C. (1987) *Nature*, **327** : 70-73.

Knott, J.E. (1956) *Agric. Exp. Sta. Bull.*, **650** : 20.

Ko, M.J., Cheigh, C.I., Cho, S.W., and Chung, M.S. (2011) *J. Food Eng.*, **102** : 327-333.

Kobayashi, K., Rabinowicz, P., Bravo Almonacid, F., Helguera, M., Conci, V., Lot, H. and Mentaberry, A. (1996) *Arch. Virol.*, **141** : 2277-2287.

Kofoet, A., Kik, C., Wíbïsxih, W.A. and De Vries, J.N. (1990) *Plant Breed.*, **105** : 144-149.

Konijnenburg, A. Van and Ardizzi, M.C.P. (1997) *Acta Hortic.*, **433** : 635-638.

Kononkov, P.F., Ustimenko, G.W. and Perez, A.P. (1969) *Beitr. trop. subtrop. Landw. trop. vet Med.*, **7** : 183-192.

Konopinski, M. (1997) *Ann. Universitatis Mariae Curie Sklodowska Sectio EEE, Horticultura*, **5** : 167-177.

Kopsell, D.A. and Randle, W.M. (1997) *Euphytica*, **96** : 385-390.

Koriem, S.O. and Farag, I.A. (1996) *Assiut J. Agri. Sci.*, **27** : 107-118.

Koriem, S.O., El-Koliey, M.M. and El-Sheekh, H.M. (1999) *Assiut J. Agril. Sci.*, **30** : 75-84.

Krishnaveni, K., Subramanian, K.s., Bhaskaran, M. and Chinnasami, K.N. (1990) *South Indian Hort.*, **38** : 258-261.

Kulkarni, C. (1963) *Indian Fmg.*, **10** : 11.

Kulkarni, S.R., Gurve, S.S. and Chormule A.J. (2017) *Ann. Plant Protec.*, **25**(1): 78-82.

Kumar, A., Kumar, R., Choudhary. R., Kumar, D. and Kumar, A. (2017) *Progress. Agric.*, **17** : 272-275.

Kumar, J., Singh, P.P., Paidi, S., Rani D. and Devika. (2015b) *Biosci. Trends*, **8** : 1535-1539.

Kumar, J., Singh, P.P., Paidi, S., Rani, D. and Devika. (2015a) *Biosci. Trends*, **8** : 1515-1518.

Kumar, K. and Sahay, R.K. (1954) *Curr. Sci.*, **11** : 368.

Kumar, M., Das, D.K., Bera, M.K., Chattopadhyay, T.K. and Kumar, M. (2000) *Environ. Ecol.*, **18** : 302-310.

Kumar, R., Khurana, S.C., Bhutani, R.D. and Bhatia, A.K. (2003) *Haryana J. Hort. Sci.*, **32** : 282-285.

Kumari, S. and Kumar, R. (2017) *Int. J. Bio-resource Stress Manag.*, **8**(3) : 424-428.

Kushal, Patil, M.G., Amaresh, Y. S., Patil, S.S., Kavitha, K. (2015) *Biosci. Trends*, **8** : 4374-4377.

Kwon, B.S. (1996) *RDA J. Agri. Sci. Hort.*, **38** : 454-461.

Lakra, B.S. (1999) *Indian J. Agric. Sci.*, **69** : 144-146.

Lal, B.S. and Singh, L.M. (1968) *Indian J. Econ. Ent.*, **61** : 676-679.

Lal, B.S. and Verma, S.K. (1959) *Sci. Cult.*, **24** : 575-576.

Lal, S., Malik, Y.S. and Pandey, U.C. (1987) *Haryana J. Hort. Sci.*, **16** : 264-268.

Lancaster, J.E., Farrant, J., Shaw, M., Bycroft, B., Brash, D. and Armstrong, J. (2001) *Acta Hortic.*, **555** : 111-115.

Lancaster, J.E., Triggs, C.M., Gandar, P.W. and De Ruiter, J.M. (1996) *Ann. Bot.*, **78** : 423-430.

Larson, R.H. and Walker, J.C. (1953) *Phytopathology*, **43** : 596-597.

Lawande, E.K. (2000) *Natl. Symp. Onion Garlic, Production Postharvest Management, Challenges and Strategies*, Nov. 19-21, 2000, Abst. pp. 2-10.

Lee, E.T., Choi, I.H., Oh, Y.B., Kim, J.K. and Kwon, B.S. (1996) *RDA J. Agri. Sci.*, Horticulture, **38** : 454-461.

Lee, J.S., Seong, K.C., Sin, Y.A., Ro, H.M. and Um, Y.C. (2000) *Korean J. Hort. Sci. Technol.*, **18** : 9-13.

Leighton, T., Ginther, C., Fluss, L., Harter, W. K., Cansado, J. and Notario, V. (1992) Molecular characterization of quercetin and quercetin glycosides in Allium vegetables: Their effects on malignant cell transformation. In *ACS symposium series (USA)*.

Lercari, B. (1983) *Photochem. Photobiol.*, **38** : 219-222.

Levy, D. and Kedar, N. (1970) *HortScience*, **5** : 80-82.

Levy, D., Kedar, N. and Karacinque, R. (1973) *HortScience.*, **8** : 228-229.

Li, C.Z., Xia, M.Z., Cai, G.Z., Ren, Y.H., Tian, C. and Shan, C.H. (2004) *J. Xinchang Agric. Coll.*, pp. 202.

Li, C.Z. (2004) *Acta Laser Biology Sinica*, p. 4.

Li, C.Z. (2009) *J. Anhui Agric. Sci.*, p. 25.

Lin, M., Watson, J.F. and Baggett, J.R. (1995) *J. Amer. Soc. Hort. Sci.*, **120** : 119-122.

Lin, M., Watson, J.F. and Baggett, J.R. (1995) *J. Amer. Soc. Hort. Sci.*, **120** : 297-299.

Liu, H.M., Zhang, Q.P. and Wei, Y.Y. (1999) *J. Shandong Agri. Univ.*, **30** : 31-36.

Lokesh, K.L., Kalapa, V.P. and Veere Gowda, R. (2000) *Natl. Symp. Onion Garlic, Production Postharvest Management, Challenges and Strategies*, Nov. 19-21, 2000, Abst. p. 224.

Lombard, K., Peffley, E., Geoffriau, E., Thompson, L., and Herring, A. (2005) *J. Food Comp. Ana.*, **18** : 571-581.

Lorbeer (1997) *Acta Hortic.*, **433** : 585-591.

Lorbeer, J.W. and Stone, K.W. (1965) *Plant Dis. Rep.*, **49** : 522-526.

Lovato, H.A. and Amaducci, M.T. (1965) *Proc. Int. Seed Test. Ass.*, **30** : 803-820.

Lulkrni, S.R., Darekar, K.S. and Ranapise, S.A. (2000) *Natl. Symp. Onion Garlic, Production Postharvest Management, Challenges and Strategies*, Nov. 19-21, 2000, Abst. p. 214-215.

Luthar, Z. and Bohanec, B. (1999) *Plant Cell Rep.*, **18** : 797-802.

Lyngkhoi, F., Khar, A., Mangal, M., Gaikwad, A.B. and Thirunavukkarasu, N. (2019) *Indian J. Genet.*, **79** : 77-81.

Maass, H.I. (1997) *Plant Syst. Evol.*, **208** : 35-44.

Macha, M. (1998) *South African J. Sci.*, **94** : 454-456.

Madisa, M.E. (1994) *Acta Hortic.*, **358** : 353-357.

Mahajan, V. and Gupta, A.J. (2016) *Acta Hortic.*, **1143** : 61-68.

Mahajan, V., Thangasamy, A., Gupta, A.J., Gawande, S.J. and Singh, M. (2017) *Indian Hort.*, **52** : 39-43.

Mahala, P., Chaudhary, M.R. and Garhwal, O.P. (2019) *Indian J. Hortic.*, **76** : 312-318.

Mahamed-Yasseen, Y., Urbana, I.L., Barringer, S.A. and Splittstoesser, W.E. (1995) *In vitro Cell. Dev. Biol. Plant*, **31** : 51-52.

Mahanthesh, B., Sajjan, M.R.P. and Harshavardhan, M. (2009) *Mysore J. Agric. Sci.*, **43** : 32-37.

Mahanthesh, B., Sajjan, M.R.P., Thippeshappa, G.N., Harshavardhan, M. and Janardhan, G. (2008) *Environ. Ecol.*, **26** : 381-384.

Maini, S.B., Sagar, V.R., Chandan, S.S. and Rajesh Kumar (1997) *Veg. Sci.*, **24** : 73-74.

Maiti, R.G., Singh, S.M. and Singh, R.P. (1968) *B.V.G. Agri. Sci. Res.*, **6** : 8-13.

Maiti, S.C. and Sen, P.R. (1968) *Curr. Sci.*, **37** : 566-567.

Malathi, S. (2015) *Asian J. of Bio-sci.*, **10** : 21-26.

Malik, G., Dhatt, A.S. and Malik, A.A. (2017) *Vegetos,* **30** : 94-99.

Malik, Y.S., Kirti Singh and Pandita, M.L. (1978) *Haryana J. Hort. Sci.*, **7** : 61-67.

Malla, A., Balamurugan, S., Balamurugan, S. and Ramalingam, S. (2015) *J. Crop Sci. Biotech.*, **18** : 37-43.

Mandal, J., Acharyya, P., Bera, R. and Mohanta, S. (2020) *Int. J. Curr. Microbiol. Appl. Sci.*, **9** : 1137-1144.

Mandal, J., Acharyya, P., Singh, W.D. and Mohanta, S. (2018) *J. Allium Res.*, **1** : 28-31.

Mandal, J., Sangma, N.S., Mohanta, S. and Ajgalley, R. (2018) *J. Crop and Weed*, **14** : 97-104.

Mandal, S., Saxena, A., Cramer, C.S. and Steiner, R.L. (2020) *Horticulturae*, **6** : 26.

Mandi, M., Nayak, B.S., Sahoo, B.B., Prasad, G. and Khanda, C. (2020) *Int. J. Curr. Microbiol. App. Sci*, **9** : 1970-1976.

Mandke, D.V. and Arakeri, H.R. (1956) *Indian J. Agron.*, **1** : 115-122.

Manna, D. and Maity, T.K. (2016). *J. Plant Nutr.*, **39** : 438-441.

Manna, D., Santra, P., Maity, T.K. and Basu, A.K. (2016) *Agri Res and Tech: Open Access J.*, **2** : 1-7.

Manna, D., Santra, P., Maity, T.K. and Basu, A.K. (2016) Influence of micronutrients on quality seed production in onion (*Allium cepa* L.). *In 2nd National Symposium On Edible Alliums: Challenges and future strategies for sustainable production, 7-9th November, 2016*, Jalna, Maharashtra, India, pp. 237.

Manna, D., Maity, T.K. and Basu, A.K. (2017) *Res. J. Chem. Environ. Sci.*, **5** : 38-45.

Manna, D., Maity, T.K. and Basu, A.K. (2020) *The Asian J. Hort.*, **15** : 15-25.

Marianna, L., Judita, B., Tomáš, T. and Miroslava, H. (2016) *J. Cent. Eur. Agric.*, **17** : 1119-1133.

Marinangeli, P. (2012) In: Lambardi, M., Ozudogru E., Jain, S. (Eds), Protocols for Micropropagation of Selected Economically-Important Horticultural Plants. Methods in Molecular Biology (Methods and Protocols), vol 994. Humana Press, Totowa, NJ.

Marringer, S.A., Mohammed-Yasseen, Y., Schloupt, R.M. and Splittstoesser, W.E. (1996) *J. Veg. Crop. Prodn.*, **2** : 7-33.

Martinovich, L. and Felfoldi, J. (1996) *Hort. Sci.*, **28** : 69-75.

Martinovich, L. and Felfoldi, J. (1998) *Zoldsegtermesztesi Kutato Intezet Bulletinje*, **28** : 67-75.

Martinovski, G., agic, R., Martinovski, D. and Lazic, B. (1997) *Acta Hortic.*, **462** : 103-109.

Massiha, S., Motallebi, A. and Shekari, F. (2001) *Acta Hortic.*, **49** : 169-174.

Masthanareddy, B.G. and Sulikeri, G.S. (1998) *Karnataka J. Agri. Sci.*, **11** : 533-534.

Mathew, B. (1973) *Dwarf Bulbs, B.T. Brasford*, London, pp. 29-39.

Mathur, B.L. and Shukla, H.C. (1963) *Curr. Sci.*, **9** : 420.

Mathur, K. and Sharma, S.N. (2000) *Natl. Symp. Onion Garlic, Production Postharvest Management, Challenges and Strategies*, Nov. 19-21, 2000, Abst. p. 199-200.

Mathur, M.M. (1971) *Indian J. Hortic.*, **28** : 296-300.

Matimati, I., Murungu, F.S., Handiseni, M., Dube, Z.P. (2006) *J. New Seeds*, **8** : 61-70.

Matsubara, S. and Hihara, H. (1978) *J. Japanese Soc. Hort. Sci.*, **46** : 479-486.

Matveev, A.A. (2001) *Kotofel i Ovoshchi*, **4** : 22.

Maurya, A.N. and Lal, S. (1975) *Punjab Hort. J.*, **15** : 61-67.

Maynard, A.A. and Hill, D.E. (2000) *Compost-Science and Utilization*, **8** : 12-18.

McCallum, J., Pither-Joyce, M., Shaw, M., Kenel, F., Davis, S., Butler, R., ScheVer, J., Jakse, J. and Havey, M.J. (2007) *Theor. Appl. Genet.*, **114** : 815–822.

McCallum, J.A., Grant, D.G., McCartney, E.P., Scheffer, J., Shaw, M.L. and Butler, R.C. (2001) *New Zealand J. Crop Hort. Sci.*, **29** : 149-158.

McCallum, J., Baldwin, S., Thomson, S., Pither-Joyce, M., Kenel, F., Lee, R., Khosa, J.S. and Macknight, R. (2016) *Acta Hortic.*, **1110** : 71-76

McCollum, G.D. (1971) *J. Hered.*, **62** : 101-104.

McCollum, G.D. (1976) *Evolution of Crop Plants* (Simmonds, N.W. rf.), Longman, London and New York, pp. 186-190.

Meena, P.M., Vijai, K., Umrao, A.K. and Kumar, R. (2007) *J. Maharashtra Agric. Univ.* **9** : 53-56.

Meer, Q.P. Van Der (1986) *Onion World*, **2** : 9-12.

Meer, Q.P. Van Der and Galmarini, C.R. (1997) *Acta Hortic.*, **433** : 17-31.

Meher, R., Mandal, J., Saha, D. and Mohanta, S. (2016) *J. Crop Weed*, **12** : 86-90.

Meier, R. (2000) *PAV-Bulletin Akkerbouw*, **4** : 25-30.

Mensinkai, S. W. (1939) *J. Genet.*, **39** : 1-45.

Mettananda, K.A. and Fordham, R. (1997) *J. Hort. Sci.*, **72** : 981-988.

Miccolis, V. and Montemurro, P. (1988) *Acta Hortic.*, **220** : 149-157.

Michael, J. Havey (2018) In: Goldman, Irwin (Ed.), Plant Breeding Reviews, Wiley Online Library. https://doi.org/10.1002/9781119521358.ch2.

Michalik, B., Adamus, A. and Nowak, E. (1997) *Acta Hortic.*, **447** : 377-378.

Midan, A.A., El-Sayed, M.M., Omran, A.E. and Fattahallah, M.A. (1986) *Seed Sci. Technol.*, **14** : 519.

Ministry of Agriculture and Farmers' Welfare, (2018) *Horticultural statistics at a glance 2018. Horticulture Statics Division*, Department of Agriculture, Cooperation and Farmers' Welfare, Ministry of Agriculture and Farmers' Welfare, Government of India. p. 458. (www.agricoop.nic.in)

Minz, A., Horo, P., Barla, S., Upasani, R.R.and Rajak, R. (2018) *Indian J. Weed Sci.,* **50** : 186-188.

Mishra, B. and Singh, R.P. (2017) *Biosci. Biotechnol. Res. Asia,* **14** : 1043-1049.

Mishra, B. and Singh, R.P. (2017) *J. Hill Agric.,* **8** : 325-328.

Mishra, N.M. and Prasad, K. (1966) *Fert. News,* **11** : 18.

Mitrova, K., Svoboda, P. and Ovesna, J. (2015) *Czech J. Genet. Plant Breed.,* **51** : 62-67.

Mittal, S.P. and Srivastava, G. (1965) *Indian J. Hortic.,* **21** : 264-269.

Miyaura, K., Shinada, Y. and Gabelman, W.H. (1985) *HortScience,* **20** : 769-770.

Mohamed, K.M. and Gamie, A.A. (2000) *Assiut J. Agri. Sci.,* **31** : 115-127.

Mohamed-Yasseen, Y. and Splittstoesser, W.E. (1991) *Plant Growth Reg. Soc. Amer.,* **19** : 41-45.

Mohamed-Yasseen, Y., Barringer, S.A. and Splittstoesser, W.E. (1995) *In VitrᵒCell Dev Biol.- Plant,* **31 :** 51-52.

Mohamed-Yasseen, Y., Splittstoesser, W.E. and Litz, R.E. (1993) *HortScience,* **28** : 1052.

Mohan, K., Ebenezer, E.G. and Seetharaman, K. (2002) *NHRDF News Lett.,* **22** : 11-14.

Mohanta S., Mandal, J. and Dhakre D.S. (2017) *HortFlora Research Spectrum,* **6** : 262-267.

Mohanta, S. and Mandal, J. (2014) *HortFlora Research Spectrum,* **3** : 334-338.

Mohanta, S. and Mandal, J. (2016) *Veg. Sci.,* **43** : 274-275.

Mohanty, B.K., Prusti, A. M. and Bastia, D.K. (2002) *JNKVV Res. J.,* **35** : 73-74.

Mohanty, B.K. (2000) *Indian J. Hort.,* **57** : 329-333.

Mohanty, B.K. (2001) *Indian J. Hort.,* **58** : 260-263.

Mohanty, B.K. (2001) *J. Trop. Agri.,* **39** : 17-20.

Mohd-Mostakim, Ahmad, M.F. and Singh, D.B. (2000) *Appl. Biol. Res.,* **2** : 35-37.

Molenaar, N.D. (1984) *Disser. Abs. Internat.,* B (Science and Engineering), **45** : 1075 B.

Mollah, M.R.A, Ali, M.A, Ahmad, M., Hassan, M.K. and Alam, M.J. (2015) *European J. Biotechnol. Biosci.,* **3** : 23-27.

Mondal, M.F., Brewster, J.L., Morris, G.E.L. and Butter, H.A. (1986) *Ann. Bot.,* **58** : 197-206.

Mondal, S., Ghosh, G.K. and Mandal, J. (2020) *Int. J. Curr. Microbiol. Appl. Sci.,* **9**(4) : 2858-2866.

Moore, H.E.Jr. (1954) *Baileya,* **2** : 103-113.

Morales, J.P., Navarro, F., Rondon, F., Baez, C. and Cenao, R. (2000) *Agronomia Mesoamericana,* **11** : 115-121.

More, S.G, Varma, L.R. and Joshi, H.N. (2017) *Curr. Hortic.,* **5** : 53-55.

Mosqueda, V.E. and Havey, M.J. (2001) *J. Amer. Soc. Hort. Sci.,* **126** : 575-578.

Mote, U.N. (1978) *Pesticides,* **12** : 42-43.

Mozumder, S.N., Moniruzzaman, M. and Halim, G.M.A. (2007) *J. Agric. Rural Dev.,* **5** : 58-63.

Muhammad, S.A., Muhammad, S., Muhammad, N., Farhan, S., Ali, I., Ahsan, J., Zaid, A. and Syeda M.B. (2017) *Cogent Food Agric.,* **3** : 1280254.

Muhammad-Tariq and Nawab-Ali (2001) *Sarhad J. Agric.,* **17** : 537-540.

Mukhopadhyay, T.P. and Chattopadhyay, S.B. (1999) *Hort. J.,* **12** : 71-76.

Muldoon, D., Hickey, M. and Hooger, R. (1999) *Farmers' Newsletter, Horticulture,* **183** : 15-16.

Mupade, R.V., Kulkarni, S.N., Kamte, G.S. (2009) *Int. J. Plant Prot.,* **37** : 83-86.

Muren, R. (1989) *HortScience,* **24** : 833-834.

Murthy, K.S.N. and Rao, M.P. (1964) *Andhra Agric. J.,* **11** : 214-223.

Murovec, Jana and Bohanec, Borut (2012) In: *Haploids and Doubled Haploids in Plant Breeding,* pp. 87-106, DOI: 10.5772/29982.

Mustafa, T.M. and Turaikhim, M.S. (2001) *Arab J. Plant Prot.,* **19** : 49-51.

Muthuramalingam, S., Muthuvel, I., Sankar, V. and Thamburaj, S. (2002) *News Letter National Horticultural Research and Development Foundation,* **22** : 1-6.

Nabi, G., Rab, A., Sajid, M., Abbas, F.S.J. and Ali, I. (2013) *Pak. J. Bot.,* **45** : 455-460.

Nadkarni, K.M. (1927) *Indian Material Media,* Nadkarni and Co., Bombay.

Nagagouda, B.T., Honyal, S.C., Malabasari, T.A., Pattar, P.S. and Aski, S.G. (1998) *World Weeds,* **5** : 131-134.

Nagaich, K.N., Trivedi, S.K. and Rajesh Lekhi (1998) *South Indian Hort.,* **46** : 266-271.

Nagaraju, A.P., Kurdikeri, C.D. and Rao, M.R. (1986) *J. Fmg. Sys.,* **2** : 44-46.

Naik, R.L., Pokharkar, D.S., Patil, B.D., ambekar, J.S. and Pokharkar, R.N. (1986) *Curr. Res. Rept.,* Mahatma Phule Agric. Univ., **2** : 309-310.

Naini, R., Pavankumar, P., Prabhakar, S, Kancha, R.K., Rao, K.V. and Reddy, V.D. (2019) *Plant Cell Rep.,* **38** : 1127-1137.

Nakamura, S. and Tahara, M. (1977) *J. Jap. Soc. Hort. Sci.,* **46**: 233-244.

Nanda, J.S. and Agrawal, P.K. (2011) *Botany of vegetable crops.* Kalyani Publishers, Ludhiana. pp. 356-373.

Nandpuri, K.S., Madan, S.P.S., Singh, A. and Singh, S. (1968) *J. Res. Punjab Agric. Univ.,* Ludhiana, **5** : 478-479.

Narang, R.S. and Dastane, N.G. (1969) *Indian J. Hortic.,* **26** : 176-180.

Narayan, R., Singh, D. B., Kishor, A.and Singh, M. M. (2019) *Progressive Hort.,* **51** : 92-98.

Nascimento, W.M., Pessoa, H.B.S.V. and Araujo, M. de T. (1998) *J. Appl. Seed Prod.,* **16** : 47-49.

Nasreen, S., Haq, S.M.I. and Hossain, M.A. (2003) *Asian J. Plant Sci.*, **2** : 897-902.

Nath, P. (1976) *Vegetables for the Tropical Region*, ICAR, New Delhi.

Navarro, J.R. and Umana, G. (1997) *Agronomia Mesoamericana*, **8** : 107-111.

Nehra, B.K., Malik, Y.S. and Yadav, A.C. (1989) *Haryana Agric. Univ. J. Res.*, **19** : 225-229.

Newhalb, A.G. and Brann, J.L. (1960) *Plant Dist. Reptr.*, **44** : 269-272.

Nieuwhof, M. (1970) *In La sterilite male chez les plantes horticole*, Versailles, France, Eucarpia, **8** : 385-403.

Nouri, A.H. (1990) *XXIII Int. Hort. Cong.*, Italy, Abst. No. 3176.

O'-Garro, L.W. and Paulraj, L.P. (1997) *Plant Dis.*, **81** : 978-982.

Okumuo, A. and Hassan, L. (2000) *Pak. J. Biol. Sci.*, **3** : 613-614.

Oliveira, R.S., De, J.R., Silva, J.F.D.A. and Ferreira, L.R. and Reis, F.P. (1997) *Revista Ceres.*, **44** : 1-16.

Omran, A.F., El-Sayed, M.M., Midan, A.A. and Fatthalla, M.A. (1984) *Minufiya J. Agril. Res.*, **8** : 385-403.

Owen, E.W. (1961) The inheritance of dry matter in onion bulbs. M.Sc. thesis. Univ. Idaho, Moscow.

Ownbey, M. and Aase, H.C. (1955) *Res. Stud. State Coll.*, Washington, **24** : 1-106.

Padda, D.S., Singh, G. and Saimbhi, M. S. (1973) *Indian J. Hortic.*, **30** : 391-393.

Padureanu, S. 2005. *Analele stiintifice ale Universitatii, Alexandru Ioan Cuza", Genetica si Biologie Moleculara*, TOM V.

Pal, N. (1980) *Ph.D. Thesis* submitted to P.G. School, IARI, New Delhi.

Pal, N. and Singh, N. (1987) *Curr. Sci.*, **56** : 719-720.

Pal, N. and Singh, N. (1988) *Plant Breeding Abstr.*, 379.

Pal, N., Singh, N. and Choudhury, B. (1988) *Indian J. Hortic.*, **45** : 295-299.

Pal, N., Singh, Narendra and Singh, N. (1999) *Indian J. Agri. Sci.*, **69** : 826-829.

Pal, S., Wahengbam, J., Raut, A.M. and Banu, A. N. (2019) *J. Entomol. Res.*, **43** : 371-382.

Palaniappan, R., Yerriswamy, R.M. and Varalakshmi, L.R. (1999) *Agri. Sci. Digest Karnal*, **19** : 31-34.

Pall, R. and Padda, D.S. (1972) *Indian J. Hortic.*, **29** : 185-189.

Pan, T.C. (2014) *Northern Hortic.*, 6.

Pan, T.C., Li, C., Shan, C. and Xia, M. (2012) *J. Anhui Agric. Sci.*, **40**(13) : 7763–7765.

Pan, T.C., Li, C., Shan, C. and Xia, M. (2000) *Acta Laser Biology Sinica.*

Pandey, A.K. (2000) *Natl. Symp. Onion Garlic, Production Postharvest Management, Challenges and Strategies*, Nov. 19-21, 2000, Abst. p. 196

Pandey, P.C. and Mundra, R.S. (1971) *Indian J. Agric. Sci.*, **41** : 107.

Pandey, U.B. (1993) *Assoc. Agric. Develop. Found.*, Nasik, India, pp. 12-13.

Pandey, U.B. (2000) *Natl. Symp. Onion Garlic, Production Postharvest Management, Challenges and Strategies*, Nov. 19-21, 2000, Abst. pp. 43-48.

Pandey, U.C. and Ekpo, U. (1991) *Res. Dev. Rept.*, 8 : 5-9.

Pandotra, V.R. (1964) *Proc. Indian Acad. Sci.*, 11 : 336-340.

Pandotra, V.R. (1965) *Proc. Indian Acad. Sci.*, 12 : 229-233.

Paridan, O.G. (1966) *Acad. Nauk.*, USSR. 106 : 900.

Passi, R., Dhatt, A.S. and Sidhu, M.K. (2018) *Bangladesh J. Bot.*, 47 : 961-967.

Patale, S.S. (2019) *Bioinfolet*, 16 : 116-118.

Patel, C.L., Patel, Z.G., Patel, R.B. and Naik, A.G. (1986) *Indian J. Agron.*, 31 : 414-415.

Patel, J.A. (1958) *Farmer*, 9 : 29-32.

Patel, J.A., Deokar, A.B. and Maslekar, S.R. (1971) *Res. J.*, Jawaharlal Nehru Krishi Viswavidyalaya, 2 : 92-93.

Patel, J.A., Waghmare, D.R. and Patel, A.K. (1958) *Poona Agric. Coll. Mag.*, 49 : 83-86.

Pateña, L.F., Rasco-Gaunt, S.M., Chavez-Lapitan, V.P., Bariring, A.L. and Barba, R.C. (1998) *Acta Hortic.*, 461 : 503-508.

Paterson, D.R., Blackhurst, H.T. and Siddiqui, S.H. (1960) *Proc. Amer. Soc. Hort. Sci.*, 76 : 460-467.

Pathak, C.S. (1999) *J. New Seed.*, 1 : 89-108.

Pathak, C.S., Black, L.L., Cherng, S.J., Wang, T.C. Ko, S.S. and Armstrong, J. (2001) *Acta Hortic.*, 555 : 77-81.

Pathak, C.S., Black, L.L., Ko, S.S., Cherng, S.J. and Wang, T.C. (1996) *Onion Newsletter for the Tropics*, 7 : 12-16.

Pathak, C.S., Ko, S.S. and Cherng, S.J. (1996) *TVIS Newsletter*, 1 : 19.

Patil M.G., Kushal, Patil S.S., Hosamani, A. and Kavita, K. (2015) *Biosci. Trends*, 8(16) : 4415-4418.

Patil, D.G., Dhake, A.V., Sane, P.V. and Subramaniam, V.R. (2012) *Acta Hortic.*, 969 : 143-148.

Patil, J.D., Desale, G.Y. and Kale, D.N. (1986) *J. Maharashtra Agril. Univ.*, 11 : 281-283.

Patil, M.G., Kushal, Patil, S.S., Hosamani, A. and Kavita, K. (2015) *Trends Biosci.*, 8 : 4415-4418.

Patil, R.S., Yevale, H.V. Kolse, R.H., Bhalekar, M.N. and Asane, G.B. (2009) *Adv. Plant Sci.*, 22 : 129-131.

Patil, S. and Nargund, V.B. (2016) *Adv. Life Sci.*, 5 : 2278-4705.

Patil, S., Nargund, V.B. and Machenahalli, S. (2016c) *Adv. Life Sci.*, 5 : 3983-3987.

Patil, S., Nargund, V.B., Gurudath, H., Dharmatti, P. R. and Ravichandran, S. (2016a) *Adv. Life Sci.*, 5 : 3547-3554.

Patil, S., Nargund, V.B., Hegde, G., Ravichandran, S. (2016b) *Adv. Life Sci.*, **13** : 614-618.

Patil, S., Nargund, V.B., Ravichandran, S. (2016d) *Adv. Life Sci.*, **5** : 3629-3632.

Paul, A., Bedi, S. and Singh, R, (2018) *Indian J. Agric. Biochem.*, **31** : 79-81.

Paulauskyte, K. (1999) *Sodininkyste Darzininkyste*, **18** : 228-234.

Pavloviæ, N., Cvikiæ, D., Zdravkoviæ, J., Ðorðeviæ, R., Zdravkoviæ, M., Varga, J.G. and Moravèeviæ, D. (2015) *Ratar. Povrt.*, **52**(1) : 24-28.

Pawar, B.B., Patil, A.V. and Sonone, H.N. (1975) *Res. J.*, Jawaharlal Nehru Krishi Viswavidyalaya, **6** : 152-153.

Pawar, D.B., Warade, S.D., Patil, S.K. and Brave, H.S. (2000) *Natl. Symp. Onion Garlic, Production Postharvest Management, Challenges and Strategies*, Nov. 19-21, 2000, Abst. p. 214.

Peffley, E.B. and Hou, A. (2000) *Theor. Appl. Genet.*, **100** : 528-534.

Peffley, E.B., Corgan, J.N., Horak, K.E. and Tanksley, S.D. (1985) *Theor. Appl. Genet.*, **71** : 176-184.

Perez Gregorio, R.M., Garc, M.S., Simal-Gandara, J., Rodrigues, A.S. and Almeida, D.P. (2010) *J. Food Compos. Anal.*, **23** : 592- 598.

Perez-Moreno, L., Jimenez-Floresalarorre, H.E. and Grana-Luna, M. (1996) *Onion Newslwtter for the Tropics*, No. 7, pp. 67-77.

Peterka, H., Budahn, H. and Schrader, O. (1997) *Theor. Appl. Genet.*, **94** : 383-389.

Peterson, C.E., Simon, P.W. and Ellerbrock, L.A. (1986) *HortSci.*, **21** : 1466-1468.

Phillippe, B., Cammue, B.P.A., Thevissen, K., Hendriks, M., Eggermont, K., Goderis, I.J., Proost, P., van Damme, J., Osborn, R.W., Guerbette, F., Kader, J.C. and Broekaert, W.F. (1995) *Plant Physiol.*, **109** : 445-455.

Phillips, G.C. and Hubstenberger, J.F. (1987) *HortScience*, **22** : 124-125.

Piccini, J.L., Evans, D.R. and Quaranta, H.W. (1987) *J. Food Sci. Technol.*, **24** : 91-93.

Pike, L.M. and Yoo, K.S. (1990) *Sci. Hortic.*, **45** : 31-36.

Pike, L.M., Horn, R.S., Anderson, C.R., Leeper, P.W. and Miller, M.E. (1988a) *HortScience*, **23** : 634-635.

Pike, L.M., Horn, R.S., Anderson, C.R., Leeper, P.W. and Miller, M.E. (1988b) *HortScience*, **23** : 635-636.

Pike, L.M., Horn, R.S., Anderson, C.R., Leeper, P.W. and Miller, M.E. (1988c) *HortScience*, **23** : 636-637.

Pike, L.M., Horn, R.S., Anderson, C.R., Leeper, P.W. and Miller, M.E. (1988d) *HortScience*, **23** : 638-639.

Pirov, T.T. (2000) *Kartofel i Ovoshch*, No. 3, 27, 32.

Pohare, V.B., Thawal, D.W., Shinde, L.D. and Kamble, A.B. (2018) *Int. J. Econ. Plants.*, **5**(2): 65-70.

Pollak, R.T., Green, M. and Prince, B. (1989) In : *Proc. of the Brighton Crop Protection Conference*, Weeds, **3** : 1027-1032.

Polunin, O. (1969) *Flowers of Europe*, Oxford University Press.

Popandron, N. (1998) *Anale Institutul de Cercetari pentru Legumicultura si Floricultura*, Vidra, **15** : 51-59.

Pradhan, M., Tripathy, P., Mandal, P., Sahoo, B. B., Pradhan, R., Mishra, S. P. and Mishra, H. N. (2016) *Int. J. Bio-resource Stress Manag.*, **7** : 960-963.

Prajapati, S., Jain, P.K. and Singh, O. (2017) *J. Func. Environ. Bot.*, **7** : 50-56.

Prakash, D., Singh, B.N., and Upadhyay, G. (2007) *Food Chem.*, **102** : 1389-1393.

Prats-Perez, A., Munoz de Con, L. and Fundora-Mayor, Z. (1996) *Onion Newsetter for the Tropics*, No. 7, pp. 25-32.

Price, K.R., Bacon, J.R. and Rhodes, M.J.C. (1997) *J. Agril. Food Chem.*, **45** : 938-942.

Puizina, J. and Papes, D. (1996) *Plant Syst. Evol.*, **199** : 203-215.

Puizina, J. and Papes, D. (1999) *Acta Botanica Croatica*, **58** : 65-67.

Purewal, S.S. and Dargan, K.S. (1962) *Indian J. Agron.*, **7** : 46-53.

Quadir, M.A. and Boulton, A. (2001) *Farmers Newsletters, Horticulture*, No. 185, pp. 38-39.

Quadri, S.M.H., Srivastava, K.J., Bhonde, S.R., Pandey, U.B. and Bhagchandani, P.M. (1982) *Pesticides*, **16** : 11-16.

Kahane, R. Schweisguth, B. and Rancillac, M. (1997) *Acta Hortic.*, 433.

Rabinowitch, H.D. (1988) In: Proc. EUCARPIA 4th *Allium* Symp., Inst. Hort. Res., Wellesbourne, Warwick, U.K. pp. 57-70.

Rahman, M.A., Chitranjeevi, C.H. and Reddy, I.P. (2000) *Natl. Symp. Onion Garlic, Production Postharvest Management, Challenges and Strategies*, Nov. 19-21, 2000, Abst. pp. 147-149.

Rajalingam, G.V. and Haripriya, K. (1998) *Madras Agri. J.*, **85** : 248-250.

Rajput, S.G. (1973) *M. Sc. (Agri.) Thesis*, submitted to Marathwada Agric. Univ., Parbhani.

Ramakrishnan, M., Ceasar, S.A., Duraipandiyan, V., Daniel, M.A. and Ignacimuthu, S. (2013) *Cell. Dev. Biol. Plant*, **49** : 285-293.

Ramamoorthy, K., Salvaraj, K.V. and Velayuthana, A. (2000) *Madras Agril. J.*, **86** : 140-141.

Rana, M.K. (2015) Technology for vegetable production. Kalyani Publishers, Ludhiana. pp. 305-334.

Rana, M.K. and Hore, J.K. (2015) In: Technology for vegetable production. Kalyani Publishers, Ludhiana. pp. 326-329.

Rana, R.S. and Singh, A.K. (1997) *Veg. Sci.*, **24** : 20-22.

Randhawa, K.S., Mohan Singh and Kooner, K.S. (1987) *Punjab Vegetable Grower,* **22** : 15-17.

Randle, W.M. and Armstrong, J. (2001) *Acta Hortic.,* **555** : 57-61.

Randle, W.M. and Galmarini, C.R. (1997) *Acta Hortic.,* **433** : 299-311.

Randle, W.M., Bussard, M.L. and Warnock, D.F. (1994) *J. Amer. Soc. Hort. Sci.,* **118** : 762.

Ranpise, S.A., Patil, B.T. and More T.A. (2004) *J. Maharashtra Agric. Univ.,* **29** : 104-106.

Rao, A.S., Girija, G., Ramachandra, Y.L. and Chethana, B.S. (2015) *Int. J. Agric. Environ. Biotechnol.,* **8** : 89-95.

Rao, H. and Purewal, S.S. (1957) *Farm Bull.,* ICAR, New Delhi, India, No. 3, pp. 18-22.

Rao, J.M.V. and Deshpande, R. (1971) *Indian J. Agric. Res.,* **5** : 257.

Rao, M.S. and Reddy, P. (2000) *Natl. Symp. Onion Garlic, Production Postharvest Management, Challenges and Strategies,* Nov. 19-21, 2000, Abst. p. 215.

Rao, P.V.S. and Kumaraswami, T. (1986) *Pesticides,* **20** : 61-62.

Rashid, M.H.A., Massiah, A.J. and Thomas, B. (2016) *Acta Hortic.,* **1143**: 2.

Rasooli, M.A., Kale, K.D., Shinde, M.G. and Shinde, S.D. (2019) *Bioinfolet,* **16** : 25-128.

Rathore, F. and Chaturvedi, S. (2018) *Agric. Sci. Digt.,* **38** : 183-187.

Rathore, S.V.S. and Kumar, R. (1974) *Progress. Hort.,* **6** : 105-111.

Rathore, S.V.S. and Yadav, B.S. (1971) *Progress. Hort.,* **3** : 29-33.

Ray, P. and Gupta, M.N. (1980) *'Caraka Samhita' Sci. Synop.,* Indian Natl. Sci. Acad., New Delhi.

Ray, P., Gupta, M.N. and Roy, M. (1980) *Sunsruta Samhita Sci. Synop.,* Indian Natl. Sci. Acad., New Delhi.

Reddy, P.N. and Madalgeri, B.B. (1978) *Curr. Sci.,* **2** : 147.

Ren, Y.B. and Li, C.Z. (2002) *Acta Laser Biology Sinica,* **4** : 102-121.

Rizk, F.A. (1997) *Egyptian J. Hort.,* **24** : 219-238.

Rizk, F.A., Shaheen, A.M., El-Samad, E.H.A. and Sawan, O.M. (2012) *J. Appl. Sci. Res.,* **8** : 3353-3361.

Ro, H.M. and Um, Y.C. (2000) *Korean J. Hort. Sci. Tech.,* **18** : 9-13.

Robinson, J.C. (1971) *Rhodesian J. agric. Res.,* **9** : 31-38.

Rodrigues, B.M., Pinto, J.E.B.P. and De Souza, C.M. (1997) *Agrotecnologia,* **21** : 343-352.

Rodrigues, B.M., Pinto, J.E.B.P., Maluf, W.R. and De Souza, C.M. (1996) *Bragantia,* **55** : 19-28.

Rodriquez, S.N., Belmar, N.C. and Valenzuelap, A. (1999) *Agricultura Technica (Santiago),* **59** : 122-132.

Rohini, N. and Paramaguru, P. (2016) *Int. J. Farm Sci.,* **6** : 174-183.

Romanov, D.V., Kirov, I.V., Razumova, O.V., Khrustaleva, L.I., 2016, **2** : 75-79.

Rudolph, M. (1987) *Archiv fur Gartenbau*, **35** : 217-223.

Rudolph, M. and Wolf, P. (1986) *Nachrichtenblatt fur den Pflanzenschutz in der DDR*, **40** : 190-193.

Rugi, M. (2017) Effect of varieties and transplanting dates on growth, yield and quality of *kharif* onion (*Allium cepa* L.) M. Sc. Thesis. Department of Vegetable Science. Rajmata Vijayaraje Scindia Krishi Vishwa Vidyalaya, Gwalior, College of Horticulture, Mandsaur (MP).

Rumpel, J., Felczynski, K. and Damato, G. (2000) *Acta Hortic.*, **533** : 179-185.

Russel, E.W. (1974) *Soil Conditions and Plant Growth* (10th edn.), Longman Group Co., London.

Saghir, A.R.B., Ownbey, M., Mann, L.K. and Berg, R.Y. (1966) *Amer. J. Bot.*, **53** : 477-484.

Sahoo, B.B. and Tripathy, P. (2019) *Indian J. Weed Sci.*, **51** : 308-311.

Sahu, S. (2016) *Indian Hortic. J.*, **6** : 349-351.

Sai Reddy, C., Chiranjeevi, C.M. and Reddy, J.P. (2000) *Natl. Symp. Onion Garlic, Production Postharvest Management, Challenges and Strategies*, Nov. 19-21, 2000, Abst. p. 205.

Saini, N., Hedau, N.K., Khar, A., Yadav, S., Bhatt, J.C. and Agrawal, P.K. (2015) *Indian J. Genet.*, **75** : 93-98.

Saker, M.M. (1998) *Biol. Plant.*, **40** : 499-506.

Sakhale, B.K., Kulkarni, D.N., Pawar, V.D. and Agarkar, B.S. (2001) *J. Food Sci. Technol.*, **38** : 412-414.

Salunkhe, V., Gawande, S.J. and Singh, M. (2017) *Indian Hort.*, **62** : 52-54.

Salunkhe, V., Gawande, S.J. and Singh, M. (2017) *Indian Hort.*, **62** : 67-69.

Rai, S., Jotwani, M.G. and Gupta, R.P. (1979) *Bull. Ent.*, **20** : 109-110.

Samra, J.S., Ramakrishna, Y.S., Desai, S., Subba Rao, A.V. M., Rama Rao, C.A., Reddy, Y.V.R., Rao, G.G.S.N., Victor, U.S., Vijaya Kumar, P., Lawande K.E., Srivastava K.L. and Krishna Prasad, V.S.R. (2006) Information Bulletin - Hyderabad : Central Research Institute for Dryland Agriculture (ICAR) pp. 22-24.

Sancir, C., Pistrick, K., Hanelt, P. and Zur Nutzung, V. (1989) *Kulturpflanze*, **37** : 133-143.

Sandhu, J.S. (1979) *Sci. Cult.*, **52** : 221-222.

Sandhu, J.S. and Korla, B.N.B. (1976) *Indian J. Hortic.*, **33** : 170-172.

Sandhu, K.S. and Randhawa, K.S. (1980) *Trop. Pest Manag.*, **26** : 41-44.

Sankar, V., Veeraragavathatham, D. and Kannan, M. (2009) *Asian J. Hortic.*, **4** : 16-20.

Santos, S.C.D., Oliveira, U.A.D., Trindade, L.D.O.R., Assis, M.D.O., Campos, J.M.S., Salgado, E.G. and Barbosa, S. (2017) *Pak. J. Bot.*, **49** : 2201-2212.

Santra, P., Manna D., Sarkar, H.K. and Maity, T.K. (2017) *J. Crop Weed.*, **13** : 103-6.

Sarati, S., Aleksandrova, I.M., Tarakanov, I.G. and Onushko, T.N. (1987) In : *Intensif. Vozdel. Polev. Kul'tur i Morfol. Osnovy Ustoichivosti Rast*, USSR, pp. 50-56.

Saravanakumari, K., Thiruvudainambi, S., Ebenezar, E.G. and Senthil, N. (2019) *Int. J. Farm Sci.*, **9** : 93-96.

Satao, R.N. and Dandge, M.S. (1999) *Crop Res. (Hissar)*, **18** : 480-481.

Sato, Y., Nagai, M., Ito, K., Tanaka, M., Yoshikawa, H. Uragami, A. and Muro, T. (1999) *Res. Bull. Hokkaido Nat. Agri. Exp. Sta.*, No. 168, pp. 47-57.

Satodiya, B.N. and Singh, S.P. (1997) *Agric. Sci. Digt.* (Karnal), **17** : 123-125.

Satoh, Y., Nagai, M., Tanaka, M., Yoshikawa, H. and Uragami, A. (1996) *Res. Bull. Hokkaido Nat. Agri. Exp. Sta.*, No. 164, pp. 49-59.

Saxena, R.S. (1971) *Indian J. Ent.*, **33** : 342-345.

Schiavi, M. and Annibali, S. (1998) *Informatore Agrario*, **54** : 41-42.

Schleiff, U. (2008) *J. Agron. Crop Sci.*, **194** : 1-8.

Schmidt, W.A. (1963-64) *Proc. Carib. Reg. Amer. Soc. Hort. Sci.*, **7** : 17.

Schulz, H., Kruger, H., Liebmann, J. and Peterka, H. (1998) *J. Agri. Food Chem.*, **46** : 5220-5224.

Schweisguth, B. (1973) *Annals AMEL Plant*, **23** : 221-253.

Schwinn, K.E., Ngo, H., Kenel, F., Brummell, D.A., Albert, N.W., McCallum, J.A., Pither-Joyce, M., Crowhurst, R.N., Eady, C. and Davies, K.M. (2016) *Front. Plant Sci.*, 1865.

Scott, A., Wyatt, S., Tsou, P.L., Robertson, D. and Allen, N.S. (1999) *BioTechniques*, **26** : 1125, 1128-1132.

Sehrawat, R., Nema, P. K. and Chandra, P. (2017) *J. Agric. Eng.*, **54** : 32-39.

Sen Nag, O. (2017) https://www.worldatlas.com/

Sen, S. (1976) *Indian J. Hortic.*, **8** : 41-50.

Sendhilvel, V., Palaniswami, A., Thiruvudainambi, S., Rajappan, R. and Raguchander, T. (2001) *NHRDF News Lett.*, **21** : 4-8.

Sethi, V., Anand, J.C., Netra Pal and Bhagchandani, P.M. (1973) *Indian Food Packer*, **27** : 5-8.

Shahnaz, E., Razdan, V.K., Kumar, A. and Bandey, S. (2016) *Int. J. Farm Sci.*, **6** : 102-107.

Sharma, A., Chandrakar, S. and Thakur, D.K. (2015) *Biosci. Trends*, **8** : 1473-1476.

Sharma, A.K., Bhatia, R.S. and Raina, R. (2009) *Veg. Sci.*, **36** : 74-76.

Sharma, D. and Dogra, B.S. (2017) *Indian J. Hortic.*, **74** : 405-409.

Sharma, D.P. (1998) *Adv. Plant Sci.*, **11** : 237-239.

Sharma, K., Mahato, N. and Lee, Y.R. (2017) *J. Food Drug Anal.*, **26** : 518–528.

Sharma, K., Assefa, A.D., Kim, S., Ko, E.Y. and Park, S.W. (2014) *New Zeal. J. Crop Hort.*, **42** : 87-98.

Sharma, M.P., Alok-Adholeya and Adholeya, A. (2000) *Bio. Agri. Hort.*, **18** : 1-14.

Sharma, M.P., Singh, A. and Gupta, J.P. (2002) *Indian J. Agri. Sci.*, **72** : 26-28.

Sharma, P.C., Mishra, B., Singh, R.K., Singh, Y.P. and Tyagi, N.K. (2000) *Indian J. Agri. Sci.*, **70** : 674-678.

Sharma, S., Jain, R., Leua, A. and Shukla, R. (2017) *Indian J. Econ. Dev.*, **13** : 696-700.

Sheemar, G. and Dhatt, A.S. (2015) *Res. Crops*, **16** : 133-138.

Shekib, I.A., Shehata, A.A.Y. and El-Tabey, A. (1986) *Alexandria J. Agril. Res.*, **31** : 167-174.

Sher, D.J. (1996) *Commercial Grower*, **51** : 10, 12-13.

Shimonaka, M., Hosoki, T., Tomita, M. and Yasumuro, Y. (2002) *J. Japan. Soc. Hort. Sci.*, **71** : 623-631.

Shinde, K.G., Warade, S.D. and More, T.A. (2001) *J. Maharashtra Agric. Univ.*, **26** : 12-18.

Shinde, K.G., Bhalekar, M.N. and Patil, B.T. (2016) *Veg. Sci.*, **43** : 91-95.

Shinde, S.D., Nimbalkar, C.A., Wani, V.S., Kadlag, A.D. and Pharande, A.L. (2017) *J. Indian Soc. Soil Sci.*, **65**(3) : 316-320.

Shirck, F.H. (1972) *J. Econ. Ent.*, **44** : 1029-1031.

Shock, C.C., Feibert, E.B.G. and Saunders, L.D. (2000) *HortScience*, **35** : 63-66.

Shock, C.C., Jensen, L.B., Hobson, J.H., Seddigh, M., Shock, B.M., Saunders, L.D. and Stieber, T.D. (1999) *Hort. Tech.*, **9** : 251-253.

Shoemaker, T.S. and Teskey, N.J.E. (1939) *Practical Horticulture*, Wiley Eastern Pvt. Ltd., pp. 212-213.

Shrivastava, R.K., Verma, B.K., Mehta, A.K. and Dwivedi, S.K. (1991) *Veg. Sci.*, **26** : 170-171.

Shukla, A.P. and Namdeo, K.N. (2000) *Crop Res.*, Hisar, **20** : 93-96.

Shukla, N., Mehta, N., Singh, B. and Trivedi, J. (2008) *Orissa J. Hortic.*, **36** : 102-105.

Sidhu, A.S., Kanwar, J.S. and Chadha, M.L. (1992) *Seed Tech. News.*, **22** : 23.

Sinclair, P.J. and Letham, D.B. (1996) *Australasian Plant Path.*, **25** : 8-11.

Singh, A.K., Janakiram, T., Singh, M. and Mahajan, V. (2017) *Indian Hort.*, **62** : 3-8.

Singh, A.K. (2000) *Natl. Symp. Onion Garlic, Production Postharvest Management, Challenges and Strategies*, Nov. 19-21, 2000, Abst. p. 200.

Singh, A.K. and Singh, V. (2002) *Ann. Agric. Res.*, **23** : 654-658.

Singh, D. and Roy, B.K. (2016) *Braz. J. Bot.*, **39** : 67-76.

Singh, D., Dhiman, J.S., Brar, S.S. and Saimbhi, M.S. (1997) *Acta Hortic.*, **433** : 323-328.

Singh, D., Dhiman, J.S., Sidhu, A.S. and Hari Singh (1992) *Onion Newsletter for the Tropics*, No. 4, p. 43.

Singh, D., Singh, D. and Armstrong, J. (2001) *Acta Hortic.*, **555** : 141-145.

Singh, D.N. (2001) *Environ. Ecol.*, **19** : 980-982.

Singh, D.P. and Joshi, M.C. (1978) *Veg. Sci.*, **5** : 1-3.

Singh, D.P. and Singh, R.P. (1974) *Indian J. Hortic.*, **31** : 69-74.

Singh, Foja, Ved Brat, S. and Khoshoo, T.N. (1967) *Cytologia*, **32** : 403-407.

Singh, J. and Dhankar, B.S. (1989) *Indian Agric.*, **32** : 163-170.

Singh, J.R. and Jain, N.K. (1959) *Indian Hortic.*, **16** : 31-38.

Singh, K. and Batra, R.B. (1972) *Proc. 3rd Int. Symp. Subtrop. Trop. Hort.*, Bangalore, p. 144.

Singh, K. and Kumar, S. (1969) *J. Res. Punjab, Agric. Univ.*, Ludhiana, **6** : 764-768.

Singh, K. and Kumar, S. (1972) *Plant Sci.*, **1** : 181-182.

Singh, K.P., Singh, K. and Jaiswal, R.C. (1988) *Veg. Sci.*, **15** : 120-125.

Singh, L., Bhonde, S.R. and Mishra, V.K. (1997) *News Letter, Nat. Hort. Res. and Dev. Foundation*, **17** : 1-3.

Singh, M., Sharma, A., Chouhan, D. and Tripathi, P.N. (2020) *Int. J. Farm Sci.*, **10** : 50-52.

Singh, N. (2001) In : Chadha, K.L. (Ed.), *Handbook of Horticulture*, ICAR, New Delhi, pp. 428-430.

Singh, N. and Pal, N. (2001) In : Thamburaj, S. and Singh, Narendra (Eds.), *Textbook of Vegetables, Tubercrops and Spices*, ICAR, New Delhi,

Singh, N. and Singh, B. (2000a) *Natl. Symp. Onion and Garlic, Production Postharvest Management, Challenges and Strategies*, Nov. 19-21, 2000, Abst. pp. 225-226.

Singh, N., Malik, Y.S., Nehra, B.K., Yadav, K.S. and Lakshminarayana (2000 b) *Natl. Symp. Onion and Garlic, Production Postharvest Management, Challenges and Strategies*, Abst., p. 202.

Singh, P. and Gopal, J. (2019) *Indian J. Hortic.*, **76** : 368-372.

Singh, P.K., Singh, D., Singh,S. K., Singh, M. K., Singh, A. K. and Singh, U. (2018) *Vegetos*, **31** : 28-32.

Singh, R. and Alderfer (1966) *Soil Sci.*, **101** : 69-80.

Singh, R., Kohli, U.K. and Dobhal, V.K. (1997) *Crop J. Hill Res.*, **10** : 69-71.

Singh, R., Sinha, A.P. and Singh, S.P. (1986) *Haryana J. Hort. Sci.*, **15** : 76-82.

Singh, R.K. and Dubey, B.K. (2011) *Progress. Hort.*, **43** : 116-120.

Singh, R.K., Singh, S.K. and Tailor, A.K. (2020) *J. Pharmacogn. Phytochem.*, **9** : 963-967.

Singh, R.P., Jain, N.K. and Poonia, B.L. (2000a) *Indian J. Agric. Sci.*, **70** : 871-872.

Singh, S. and Singh, T. (2000a) *Natl. Symp. Onion Garlic, Production Postharvest Management, Challenges and Strategies*, Abst., p. 206.

Singh, S.P. and Mohanty, C.R. (1998) *Orissa J. Hort.*, **26** : 70-71.

Singh, S.P., Laddha, S. and Pal, S.K. (2017) *Veg. Sci.*, **44** : 66-69.

Singh, S.R. and Sachan, B.P. (1999) *Crop Res. Hisar*, **17** : 351-355.

Singh, S.S. and Rathore, S.V.S. (1973) *Allahabad Fmr.*, **47** : 348.

Singh, V and Pandey, M. (2006). *J. Indian Soc. Soil Sci.*, **54** : 365-367.

Singh, V., Singh, V. and Singh, I.J. (1974) *Balwant Vidyapeeth J. Agric. Sci. Res.*, **16** : 32-37.

Sinnadurai, S. (1970) *Ghana J. Agric. Sci.*, **3** : 13-15.

Sirajo, S.A. and Namo, O.A.T. (2019) *Afr. Crop Sci. J.*, **27** : 253-265.

Slatyer, R.O. (1967) *Plant Water Relationship*, Academic Press, London.

Sliman, Z.T., Abdelha Kim, M.A. and Omran, A.A. (1999) *Egyptian J. Agril. Res.*, **77** : 983-993.

Smith, B.M., Crowther, T.C. and Galmarini, C.R. (1997a) *Acta Hortic.*, **433** : 317-321.

Smith, C., Lombard, K.A., Peffley, E.B. and Liu, W. (2003) *Texas J. Agric. Natl. Resour.*, **16** : 24-28.

Sokhi, S.S., Sohi, M.S., Singh, D.P. and Joshi, M. (1974) *Indian J. Mycol. Pl. Path.*, **4** : 214-215.

Solanki, P., Jain, P.K., Prajapati, S., Raghuwanshi, N., Khandait, R.N. and Patel, S. (2015) *Int. J. Agric. Environ. Biotechnol.*, **8** : 783-793.

Solberg, S.O. and Boe, E. (1999) *J. Veg. Crop Prod.*, **5** : 31-42.

Solberg, S.O. and Dragland, S. (1999) *J. Veg. Crops Prod.*, **4** : 23-34.

Solomon, S. and Patel, J.A. (1959) *Poona Agric. Coll. Mag.*, **50** : 30-33.

Somkuwar, R.G., Gowda, R.V., Singh, T.H. and Pathak, C.S. (1996) *Madras Agr. J.*, **83** : 273-275.

Song, P., Kang, W. and Ellen, P.B. (1993) *HortScience*, **28** : 498.

Song, S., Kim, C.W., Moon, J.S. and Kim, S. (2014) *Mol. Breed.*, **33** : 173-186.

Song, S.I., Song, J.T., Kim, C.H., Lee, J.S. and Choi, Y.D. (1998) *J. Gen. Virol.*, **79** : 155-159.

Song, Z., Zhang, Q.P., Li, J.R., Wei, Y.Y., Zhang, S., Zhang, Q.P., Li, J.R. and Wei, Y.Y. (1997) *J. Shandong Agril. Univ.*, **28** : 134-140.

Soo, L.J., Cheol, S.K., An, S.Y., Myong, R.H., Cheol, U.Y., Lee, Y.S., Seong, K.C., Sin, Y.A., Soumia, P.S., Karuppaiah, V. and Singh, M. (2017) *Indian Hort.*, **62** : 55-56.

Srinivas, P.S. and Lawande, K.E. (2000) *Natl. Symp. Onion Garlic, Production Postharvest Management, Challenges and Strategies*, Nov. 19-21, 2000, Abst. p. 218.

Srivastava, K.J. and Tiwari, B.K. (2000) *Natl. Symp. Onion Garlic, Production Postharvest Management, Challenges and Strategies*, Nov. 19-21, 2000, Abst., p. 211.

Srivastava, K.J. and Verma, L.R. (2000) *Natl. Symp. Onion Garlic, Production Postharvest Management, Challenges and Strategies*, Nov. 19-21, 2000, Abst. p. 211.

Srivastava, P.K., Tiwari, B.K. and Srivastava, K.J. (1999) *News Letter - National Horticultural Res. and Dev. Foundation*, **19** : 10-11.

Srivastava, R.P., Agarwal, N.C. and Srivastava, K.L. (1965) *Fertl. News*, **10** : 27-33.

Stadnik, M.J. and Dhingra, O.D. (1996) *Fitopatol Bras*, **21** : 431–435.

Stanwood, P.C. and Sowa, S. (1995) *Crop Sci.*, **35** : 852-856.

Stearn, W.T. (1944) *Herbertia*, **11** : 11-34.

Stone, D.A. (2000) *Soil Use Manag.*, **16** : 285-292.

Stuart, N.W. and Griffin, D.M. (1946) *Proc. Amer. Soc. Hort. Sci.*, **48** : 398-402.

Sudha, G.S. and K. Riazunnisa, (2015) *Inter. J. Plant Anim. Environ. Sci.*, **5** : 125-128.

Suojala, T. (2001) *J. Hort. Sci. Biotechnol.*, **76** : 664-669.

Suojala, T., Salo, T. and Pessala, R. (1998) *Agri. Food Sci. Finland*, **7** : 477-489.

Susek, A., Javornik, B. and Bohanec, B. (2002) *Plant Cell Tissue Organ Cult.*, **68** : 27-33.

Szalay, F. (1986) *Zoldsegtermesztesi Kutato Intezet Bulletinje*, **19** : 11-18.

Tai, L.E., Huk, C.J., Bee, O.Y., Kwang, K.J., Sun, K.B., Lee, E.T., Choi, I.H., Oh, Y.B., Kim, J.K. and Kwon, B.S. (1996) *RDA J. Agric. Sci. Hort.*, **38** : 454-461.

Takayanagi, K. (1987) *Japanese J. Breed.*, **37** : 109-112.

Talwar, D., Singh, K. and Walia, S.S. (2017) *Int. J. Agric. Environ. Biotechnol.*, **10** : 289-294.

Tanaka, K., Terao, M., Kubota, E., Teramoto, S. and Motomura, T. (1996) *Marine and high land Bio-science Centre report*, **3** : 39-43.

Tanaka, S., Iritani, M., Komai, F., Nakano, M., Mori, N., Miyaura, K., shiga, Y., Tanaka, A., Sato, M., sako, K. and Nishimura, N. (1999) *Res. Bull. Hokkaido Nat. Agri. Exp. Sta.*, No. 77, pp. 17-21.

Tanaka, S., Miyaura, K., Shinada, Y. and Nakano, M. (2000) *Bull. Hokkaido Prefectural Agril. Expt. Sta.*, No. 79, pp. 85-88.

Tanikawa, T., Takagi, M. and Ichii, M. (2002) *J. Japan Soc. Hort. Sci.*, **71** : 249-251.

Tarakanov, L.G., Karlov, G.I. and Khrustalava, L.L. (1994) *XXVI th Int. Hort. Cong.*, Kyoto, Japan, Abst. No. p. 15.

Tariq, M. and Ali, N. (2001) *Sarhad J. Agric.*, **17** : 537-540.

Tawaraya, K., Imai, T. and Wagatsuma, T. (1999) *J. Plant Nutr.*, **22** : 589-596.

Tawaraya, K., Kinebuchi, T., Watanabe, S., Wagatsuma, T. and Suzuki, M. (1996) *Japanese J. Soil Sci. Plant Nut.*, **67** : 294-298.

Taylor, A., Vagany, V., Barbara, D.J., Thomas, B., Pink, D.A.C., Jones, J.E. and Clarkson, J.P. (2013) *Plant Pathol.*, **62** : 103-111.

Terabun, M. (1965) *J. Japanese Soc. Hort. Sci.*, **34** : 186-204.

Terabun, M. (1970) *J. Japanese Soc. Hort. Sci.*, **39** : 325-330.

Terabun, M. (1971) *J. Japanese Soc. Hort. Sci.*, **40** : 50-56.

Terabun, M. (1981) *J. Japanese Soc. Hort. Sci.*, **50** : 53-59.

Thangavel, K., Baskaran, R.K.M. and Suresh, K. (2018) *Annals Plant Protec. Sci.*, **26** : 16-20.

Thapa, U., Patti, M.K., Chattopadhay, S.B. and Mandal, A.R. (2005) *Research on Crops*, **6** : 55-57.

Theresa, K. and Armstrong, J. (2001) *Acta Hortic.*, **555** : 231-237.

Thilakavathy, S. and Ramaswamy, N. (1998) *Newsl. Nat. Hort. Res. Dev. Found.*, **18** : 18-20.

Thilakavathy, S. and Ramaswamy, N. (1999) *Veg. Sci.*, **26** : 97-98.

Thiruvelavan, P., Thamburaj, S., Veeraragavathatham, D. (1999) *South Indian Hort.*, **47** : 223-224.

Thomas, B., Hornby, P. and Partis, M.D. (1997) *Acta Hortic.*, **433** : 375-380.

Thomas, T.H. and Isenberg, F.M. (1992) *Exp. Hort.*, No. 23, pp. 48-51.

Thompson, H.C. and Kelly, W.C. (1957) *Vegetable Crops*, Tata McGraw-Hill Publishing Co. Ltd., New Delhi, pp. 362-363.

Thompson, H.C. and Smith, O. (1938) *Bull. Cornell Agric. Exp. Sta.*, No. 708, p. 21.

Tibabishew, N.K. (1972) *Referactung Zhurnal*, **55** : 553.

Todorov, I., Gencheva, D. and Minkov, I. (1984) *Gradinarska i Lozarska Nauka*, **21** : 63-66.

Todorov, S.H. and Sharma, R.C. (1950) *Science*, **121** : 595-558.

Toman, S.I., Pavlov, L.V., Gorchakova, N.O. and Rudenok, V.I. (1989) *Kartofel' i Ovoshchi*, No. 6, pp. 20-21.

Traub, H.P. (1968) *Plant Life*, **24** : 147-163.

Trigo, M.F.O.O., Nedal, J.L. and Trigo, L.F.N. (1999) *Scientia Agric.*, **56** : 1059-1067.

Tripathi, P.C. and Lawande, K.E. (2013) *Indian J. Hortic.*, **70** : 455-458.

Tripathy, P., Priyadarshini, A., Das, S.K., Sahoo, B.B. and Dash, D.K. (2013) *Int. J. Bio-resource Stress Manag.*, **4** : 561-564.

Tripathy, P., Sahoo, B.B., Patel, D., Das, S.K., Priyadarshini, A. and Dash, D.K. (2013b) *Indian J. Entomol.*, **75** : 298-300.

Tripathy, P., Sahoo, B.B., Priyadarshini, A., Das, S.K. and Dash, D.K. (2013a) *Int. J. Bio-resource Stress Manag.*, **4** : 641-644.

Tripathy, S.K., Mohapatra, S., Mohanty, A.K., Panigrahy, N. and Lenka, S. (2019) *Indian J. Agron.*, **64** : 248-252.

Tsuneyoshi, T., Matsumi, T., Deng, T.C., Sako, I. and Sumi, S. (1998b) *Arch. Virol.*, **143** : 1093-1107.

Tsuneyoshi, T., Matsumi, T., Natsuaki, K.T. and Sumi, S. (1998a) *Arch. Virol.*, **143**: 97-113.

Tsutsui, K. (1991) Inheritance of resistance to *Fusarium oxysporum* in onion. M.Sc. Thesis, Univ. Wisconsin, Madison.

Uddin, M.K., Kamal, M.M., Akand, M.M., Hasan, M.M. and Chowdhury, M.N.A. (2015) *Int. J. Appl. Sci. Biotechnol.*, **3** : 737-743.

V., Biasi, J., Garcia, A., Neto, J.A.Z. and Debarba, J.F. (1998) *Agropecuaria Catarinense*, **11** : 5-7.

Vakhtina, L.I., Zakirova, R.O. and Vakhtin, Y.B. (1977) *Botanicheskii Zhurnal*, **62** : 677-684.

van der Meer, Q. P. (1994) *Acta Hortic.*, 358.

Van der Valk, P., Scholten, O.E., Verstappen, F., Jansen, R.C. and Dons, J.J.M. (1992) *Plant Cell Tissue Organ Cult.*, **30** : 181-191.

van der Vlugt, R.A.A., Steffens, P., Cuperus, C., Barg, E., Lesemann, D.E., Bos, L. and Vetten, H.J. (1999) *Phytopathology*, **89**: 148-155.

Vander Meer, Q.P. and Van Bennekom, J.L. (1968) *Euphytica*, **17** : 216-219.

Vander Meer, Q.P. and Van Bennekom, J.L. (1976) *Euphytica*, **25** : 293-296.

Vanparys, L. (1996) *Provinciaal Onderzoek en Voorlichtingscentrum voor Land en Tuinbouw*, Beitem Roeselare, No. 374, p. 4.

Vanparys, L. (1998) *Proeftuinnieuws*, **8** : 26-27.

Vanparys, L. (1999a) *Proeftuinnieuws*, **9** : 22-23.

Vanparys, L. (1999b) *Proeftuinnieuws*, **9** : 33-34.

Vanparys, L. (1999c) *Provinciaal Onderzoek en Voorlichtingscentrum voor Land en Tuinbouw*, Beitem Roeselare, No. 412, p. 4.

Vanparys, L. (2000a) *Provinciaal Onderzoek en Voorlichtingscentrum voor Land en Tuinbouw*, Beitem Roeselare, No. 416, p. 4.

Vanparys, L. (2000b) *Proeftuinnieuws*, **10** : 26-27.

Vassilev, N., Taro, M., Vassilev, M., Azcon, R. and Barea, J.M. (1997) *Biosource Technol.*, **61** : 29-32.

Vavilov, N.I. (1950) *Chron. Bot.*, **13** : 1-364.

Vavilov, N.I. (1951) *Chronica Botanica*, 13, Wal ham, Mass.

Verma, J.P., Rathore, S.V.S. and Ram, V. (1972) *Progress. Hort.*, **4** : 57-68.

Verma, L.R., Srivastava, K.J. and Pandey, U.B. (2000) *Nat Symp. on Onion and Garlic, Production and Postharvest Management, Challenges and Strategies*, Nov. 19-21, 2000, Abst. p. 226.

Verma, S.K. and Singh, T. (1997) *Indian J. Agron.*, **42** : 540-543.

Vierbergen, G. and Ester, A. (2000) *In : Proc. 52nd International Symposium on Crop Protection*, Gent, Belgium, 9 May, 2000, Part. I, **65** : 335-342.

Vilet, M. Vander and Scheele, J. (1966) *Verolagen Van het Landbouwkundig Onderzoek*, No. 669, p. 27.

Villano, C., Esposito, S., Carucci, F., Iorizzo, M., Frusciante, L., Carputo, D. and Aversano, R. (2019) *Mol. Breeding*, **39**: 5.

Villanueva-Mosqueda, E. and Havey, M.J. (2001) *J. Amer. Soc. Hort. Sci.*, **126** : 575-578.

Villanueva-Mosqueda, E., 1996. Onion heritability for pink root resistance, *Fusarium* basal rot resistance, and bolting traits. M.Sc. Thesis. New Mexico State Univ., Las Cruces.

Villevieille, M. (1996) *Acta Hotanica Gallica*, **143** : 109-115.

Viterbo, A., Altman, A. and Rabinowitch, H.D. (1994) *Genet. Resour. Crop. Evol.*, **41** : 87-98.

Viterbo, A., Rabinowitch, H.D. and Altman, A. (1992) *Plant Breed.*, **108** : 265-273.

Voss, R.K. 1979. Onion production in California, Division of Agricultural Science, University of California. Publication 4097. p. 49.

Vvdensky, A.I. (1935) *Flora of the USSR* (Komarow, V.L. ed.), Leningrad Academy of Sciences, USSR.

Wagh, K.D., Pawar, S.A., Datkhile, R.V., Bhalekar, M.N. (2016) *J. Life Sci.*, **13** : 282-285.

Wale, S.D., Pawar, S.A. and Datkhile, R.V. (2018) *Bioinfolet*, **15** : 256-260.

Walker, J.C. (1969) *Plant Pathology*, Tata McGraw-Hill Publishing Co. Ltd., New Delhi, p. 819.

Walker, J.C., Edmondson, W.C. and Jones, H.A. (1946). Onion Set Production. Farmers Bulletin. U.S. Department of Agriculture. pp. 1-2.

Walkey, D. (1990) In: Rabinowitch, H.D. and Brewster, J.L. (Eds), *Onions and Allied Crops*, Vol. II. *Agronomy, Biotic Interactions, Pathology, and Crop Protection.* CRC Press, Boca Raton, Florida, pp. 191-212.

Wall, A.D. and Corgan, J.N. (1999) *Euphytica*, **106** : 7-13.

Wall, M. and Corgan, J. (1998) *HortScience*, **33** : 762-763.

Wall, M. and Corgan, J. (1999) *HortScience*, **34** : 1303-1304.

Wall, M.M., Mohammad, A. and Corgan, J.N. (1996) *Euphitica*, **87** : 133-139.

Wanderley, L.J. da G. and Galmarini, C.R. (1997) *Acta Hortic.*, **433** : 285-289.

Wang, M.R., Zhang, Z., Haugslien, S., Sivertsen, A., Rasmussen, M., Wang, Q.C. and Blystad, D.R. (2019) *Acta Hortic.*, 1234.

Warade, S.D., Kale, P.N., Jagtap, K.B., Desale, S.B. and Shinde, K.G. (1996) *J. Maharashtra Agri. Univ.*, **21** : 52-54.

Warid, W.A. (1952) Inheritance studies in the onion, *Ph.D. Thesis*, Louisiana State Univ.

Warid, W.A. and Loaiza, J.M. (1996) *Onion Newsletter for the Tropics*, No. 7, pp. 22-24.

Wickramasinghe, U.L., Wright, C.J. and Currah, L. (2000) *J. Hort. Sci. Biotech.*, **75** : 304-311.

Wiedenfeld, R.P. (1986) *HortScience*, **21** : 236-238.

Wiles, G.C. (1989) *Ph.D. Thesis*, Wye College, University of London, UK, p. 268.

Wilkie, S.E., Isaae, P.G. and Slater, R.J. (1993) *Theor. Appl. Genet.*, **86** : 497-504.

Woodbury, C.W. and Ridley, J.R. (1969) *J. Amer. Soc. Hort. Sci.*, **94** : 365-367.

Wright, P.J. and Grant, D.G. (1977) *J. Crop Hort. Sci.*, **25** : 353-358.

Wright, P.J., Grant, D.G. and Triggs, C.M. (2001) *New Zealand J. Crop Hort. Sci.*, **29** : 85-91.

Wu, W., Yang, F., Piao, F.C., Li, K.H., Lian, M.L. and Dai, Y. (2015) *New Zealand J. Crop Hort. Sci.*, **43**(4) : 249-260.

Yadav, D., Khinchi, S.K., Sharma, P and Jangir, H. (2020) *Ann. Plant Protec. Sci.*, **28** : 12-14.

Yadav, S.S., Narendra-Singh, Yadav, B.R. and Singh, N. (1998) *Indian J. Hort.*, **55** : 243-247.

Yadav, V.K., Singh, B., Srivastava, J.P. (2002) *Progress. Agric.*, **2** : 148-150.

Yakovlev (1985) *Genetika*, USSR, **21** : 595-596.

Yakuwa, T. (1994) In : *Horticulture Organizing Committee*, Japan (ed), XXIV Int. Hort. Cong., Pub. Committee, Japan, pp. 87-88.

Yamasaki, A., Miura, H. and Tanaka, K. (2000) *J. Hort. Sci. Biotech.*, **75** : 645-650.

Yang, L., Liu, Q., Wang, Y. and Liu, L. (2017) *Sci. Hortic.*, **225** : 122-127.

Yarnell, S.H. (1954) *The Botanical Review*, **XX** : 323-325.

Yarwood (1943) *Hilgardia*, **14** : 595-596.

Yawalkar, K.S., Jakate, P.N. and Srivastava, M.M.P. (1962) *Commercial Fertilizers in India*, Agri-Horticultural Publishing House, Nagpur.

Yoo, K.S., Pike, L.M. and Too, K.S. (1996) *HortScience*, **31** : 875.

Yoo, K.S., Pike, L.M., Patil, B.S. and Lee, E.J. 2020 *Sci. Hortic.*, 266, 109269.

Zambrano, J., Ramirez, H. and Manzano, J. (1997) *Acta Hortic.*, **433** : 543-547.

Zeiger, E. and Hepler, P. K. (1976) *Plant Physiol.*, **58** : 492-498.

Zeven, A.C. and Zhukovsky, P.M. (1975) *Dictionary of Cultivated Plants and their Centres of Diversity*, Wageningen, Netherlands.

Zhang, S., Zhang, Q.P., Li, J.R. and Wei, Y.Y. (1997) *J. Shandong Agric. Univ.*, **28** : 134-140.

Zheng, S.J., Henken, B., de Maagd, R.A., Purwito, A., Krens, F.A., Kik, C. (2005) *Plant Biotechnol. J.*, **14** : 261-272.

Zheng, S.J., Henken, B., Sofiari, E., Jacobsen, E., Krens, F.A. and Kik, C. (1998) *Plant Cell Tissue Organ Cult.*, **53** : 99-105.

Zheng, S.J., Khrustaleva, L., Henken, B., Sofiari, E., Keizer, E., Jacobsen, E., Kik, C. and Krens, F.A. (2001) *Mol. Breed.*, **7** : 101-115.

Zimmerman, H., Engel, E. and Banholzer, G. (1970) *Btsche Gartenb.*, **17** : 192-193.

GARLIC

N.N. Shinde, D. Sanyal, M.B. Sontakke,
R. Chatterjee, T.K. Maity and P.K. Maurya

1.0 Introduction

Garlic is the second most widely used cultivated *Allium* after onion. It has long been recognized all over the world as a valuable spice for foods and a popular remedy for various ailments and physiological disorders. The word garlic derives from old English word 'garleac', meaning gar (spear) and leek, as a 'spear-shaped leek'.

It is grown throughout the plains of India and consumed by most of the people. It is used practically all over the world for flavouring various vegetarian and non vegetarian dishes. In America, about 50 per cent of the entire output of fresh garlic is dehydrated and sold to food processors. In India and other Asian and Middle-East countries, it is already being used in several food preparations, notably in *chutneys*, pickles, curry powders, curried vegetables, meat preparations, tomato ketchup *etc*. In medieval Europe, it was widely used for disguising the smell and flavour of salted meat and fish. In the Philippines, much of Eastern Asia and in parts of tropics, the green tops as well as bulbs of garlic are used. Dehydrated garlic in powdered or granulated form is replacing the fresh bulbs for industrial and home-use in many countries.

In 2016, the world production of garlic bulbs was reported to be around 27 million tonnes, where China contributed nearly 80 per cent of the global production to the extent of 21 million tonnes (FAOSTAT, 2016). The important garlic producing countries in the world are China, India, Bangladesh, Egypt, South Korea, Russia, *etc*. In India, Rajasthan is the leading state in terms of production share of garlic (52.87 per cent) followed by Uttar Pradesh (17.13 per cent), Gujarat (13.72 per cent), Punjab and Assam (NHB, 2019).

2.0 Composition and Uses

2.1.0 Composition

Garlic has been considered as a rich source of carbohydrates, proteins and phosphorus. Ascorbic acid content was reported to be very high in green garlic Pradan *et al*. (1977). Composition of fresh peeled garlic cloves (bulblets) and dehydrated garlic powder as reported by Pruthi (1979) is given in Table 1. The vitamin C content in garlic varies with cultivars, mean content being 315 mg/kg fresh matter (F.M.), cv. Vladan showed the lowest amount (273 mg/kg F.M.), while the highest one was found in cv. Vekan (403 mg/kg F.M.) (Pokluda and Petrikova, 2001). Shankaracharya (1974) reported that garlic contained about 0.1 per cent volatile oil. The chief constituents of the oil are : diallyl disulfide (60 per cent), diallyl trisulfide (20 per cent), allylpropyl disulfide (16 per cent), a small quantity of diethyl disulfide and probably diallyl polysulfide. Diallyl disulfide is said to possess the true garlic odour.

Table 1: Composition of Garlic*

Particulars	Fresh Peeled Garlic Cloves (Bulblets)	Dehydrated Garlic Powder
Moisture per cent	62.80	5.20
Protein per cent	6.30	17.50
Fat per cent	0.10	0.60
Mineral matter per cent	1.00	3.20
Fibre per cent	0.80	1.90
Carbohydrates per cent	29.00	71.40
Calcium per cent	0.03	0.10
Phosphorus per cent	0.31	0.42
Potassium per cent	-	1.10
Iron per cent	0.001	0.004
Niacin per cent	-	0.70
Sodium per cent	-	0.01
Vitamin A (I.U.)	0.0	175.00
Nicotinic acid (mg/100 g)	0.40	-
Vitamin C (mg/100 g)	13.00	12.00
Vitamin B (mg/100 g)	-	0.68
Vitamin B2 (mg/100 g)	-	0.08

* Pruthi (1979).

Most of the available cultivars of garlic are white coloured, but sometimes pink or red types are also found. Kenmochi and Katayama (1975) reported that the anthocyanin pattern of garlic and onion were very similar and the major pigment in each was cyanidin-3-glucoside. The uninjured bulb contains a colourless, odourless water soluble amino acid 'alliin'. On crushing the garlic bulb, the enzyme alliinase breaks down into alliin to produce allicin of which the principal ingredient is the odoriferous diallyl disulfide. The precursor alliin does not possess bactericidal properties. Allicin (an allyl thiosulfinate), the main bioactive compound produced by garlic, is thought to have therapeutic activity, particularly in the reduction of elevated plasma cholesterol (Sterling *et al.*, 2001). Allicin content of garlic is influenced by a number of factors like genotype, environment, soil type, sulphur fertilization, relative water content, harvesting date of bulb, and light spectrum (Huchette *et al.*, 2007; Yang *et al.*, 2005). The allicin content in garlic leaf was observed to be increased slightly during maturation and increased rapidly in developing bulbs during bulb maturation (Cho and Lee, 1974).

2.2.0 Uses

2.2.1 Medicinal Properties

Since prehistoric time garlic was used as a staple in the Mediterranean region, as well as a frequent seasoning in Asia, Africa, and Europe. It was also known to ancient Egyptians, as ingredient used for both culinary and medicinal purposes (Sripradha *et al.*, 2014). According to the *Unani* and *Ayurvedic* systems as practised in India, garlic is carminative and is a gastric stimulant and thus aids in digestion and absorption of food. It is also used in flatulence. Augusti (1977) reported that allicin, which has a hypocholesterolaemic action, is present in the aqueous extract of garlic and reduces the cholesterol concentration in human blood. This antimicrobial action could be applied in the prevention of dental caries. *In vitro* studies have shown that garlic extract has an inhibitory effect on periodontopathic and cariogenic bacteria (Masaadeh *et al.*, 2006; Chen *et al.*, 2009). The inhalation of garlic oil or garlic juice has generally been recommended by doctors in cases of pulmonary tuberculosis, rheumatism, sterility, impotency, cough and red eyes (Pruthi, 1979).

2.2.2 Insecticidal Action

Garlic extract possesses insecticidal action and repellent properties (Bhuyan *et al.*, 1974). A formulation containing 1 per cent garlic extract gave protection to persons against mosquitoes and blackfly for about 8 hours. According to Sukul *et al.* (1974), extracts of garlic, chilli and ginger showed strong nematicidal action of killing *Meloidogyne incognita* and other species of soil nematodes in less than 40 minutes. Gupta and Sharma (1993) isolated and tested allicin against *M. incognita* infesting tomato. It was found that allicin at 25 ppm or 5 minutes as a root-dip treatment for tomato seedlings is effective against *M. incognita*. Dhillon *et al.* (2019) reported that garlic crop can be opted in vegetable based cropping systems to manage the root knot nematode in infested soils. They found that the inoculum load of root knot nematode and carry over population to the next crop in rotation decreased significantly by opting garlic in the sequence and the suppressive effect was more pronounced when the rotation with garlic was followed for three years. Debkirtaniya *et al.* (1980) recorded that garlic extracts showed larvicidal properties against the larvae of *Culex* sp., *Spodoptera litura* and *Euproctis* sp.

2.2.3 Fungicidal Action

Fungicidal action has also been attributed to garlic by some workers. Appleton and Tansey (1975) and Barone and Tansey (1977) reported that there is inhibition of the growth of many zoopathogenic fungi by aqueous garlic bulb extract. Toxicity of garlic juice to a number of plant fungi has also been reported by Pordesimo and Ilag (1976) and Agarwal (1978). Use of garlic extract as seed treatment has been indicated by Russel and Mussa (1977) for the control of foot rot of French bean. Moore and Atkings (1977) showed that aqueous extract of garlic cloves had inhibitory and fungicidal effects on a number of medically important fungi. Slusarenko *et al.* (2008) reported that the volatile antimicrobial substance allicin is very effective against a range of plant pathogenic fungi. Allicin effectively controlled seed-borne *Alternaria* spp. in carrot, *Phytophthora* leaf blight of tomato and tuber blight of potato as well as *Magnaporthe* on rice and downy mildew of Arabidopsis.

2.2.4 Bactericidal Action

Bactericidal property of garlic against *Staphylococcus aureus* has been demons-treated by Borukh (1975). Sharma *et al.* (1977) reported that a crude extract of garlic

Different Uses of Garlic

| Roasted Garlic Clove | Garlic Soup | Garlic Powder |

| Garlic Paste | Garlic Toast |

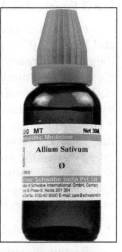

| Garlic Tablets | Garlic Oil | Homeopathy Medicine |

clove had antibacterial activity against gram negative and gram positive bacteria. Mantis *et al.* (1979) observed that colonies of food poisoning bacteria (*Clostridium perfringens*) were unable to grow if concentration of garlic extract in the medium was more than 1-2 per cent. According to Purseglove (1975), allicin of garlic has bactericidal properties. Studies by Hovadik (1981) indicated that garlic constituents like allicin, allistalin, garlicin, diallyl disulfide, diallyl trisulfide and essential oils were active against certain bacteria and other microorganisms.

3.0 Origin

Original abode of garlic is said to be central Asia and southern Europe, especially Mediterranean region (Thompson and Kelly, 1957). Garlic has long been known as a cultivated plant. It is among the most ancient of cultivated vegetables giving pungency of the genus *Allium*.

Some authorities consider that *Allium longicuspis*, which is endemic to central Asia, is the wild ancestor and spread in ancient times to the Mediterranean region. It was known in Egypt in predynastic times, before 3000 B.C. and also to ancient Greeks and Romans. It has long been grown in India and China. Garlic was carried to the western hemisphere by the Spanish, Portuguese and French (Purseglove, 1975). Garlic was known to ancients and is said to have been disliked by Romans on account of strong odour and hence it was fed to their labourers and soldiers. It was used in England as early as the first half of the 16th century.

The early domestication of garlic took quite a different turn from that of the largely seed propagated leek and onion; it became exclusively vegetatively propagated by cloves, and in those cultivars that still bolt, by inflorescence bulbils. Some modern cultivars may produce flowers mixed with the bulbils but the flowers never set seeds. Garlic thus presents an interesting problem as to the origin of the many cultivars, differing in maturity, bulb size, clove size and number, scale colour, bolting, scape height, number and size of inflorescence bulbils and presence or absence of flowers.

It is not known how much variation due to bud mutation has arisen after garlic became vegetatively propagated.

Nearly all *Allium* crops originate from the main centre of *Allium* species diversity which stretches from the Mediterranean basin to central Asia and beyond. The ancestors of garlic, must have been closely related to *A. longicupus* (Meer *et al.*, 1997).

4.0 Botany

Garlic (*Allium sativum* L.) belongs to the family Alliaceae (Liliaceae). It is a frost hardy bulbous perennial, erect herb of 30-100 cm in height with narrow flat leaves and bears small white flowers and bulbils (Janick, 1979). It is a herbaceous annual for bulb and a biennial for seed production. The scape of garlic is smooth, round and solid for its entire length unlike onion which is hollow. Many cloves of garlic do not produce flower stalks. The inflorescence may be partially or not at all exerted, its bulbils forming a swelling somewhere within the false stem a few cm above the bulb. The bulb consists of 6-35 smaller bulblets is called 'cloves' and is surrounded by a thin white or pinkish pappery sheath.

In ancient literature garlic was classified as *Alliums* under the tribe Alliaceae of the family Liliaceae, owing to its superior ovary. Few decades back Takhtajan (1967) included garlic under the order Amaryllidales, family Alliaceae. Dahlgren *et al*. (1985) and Walter *et al*. (1999) described the present day garlic as *Alliums* under the family Alliaceae and order Asparagales.

There are mainly two types or subspecies of garlic. Soft neck garlic (*Allium sativum* var. *sativum*) and hard neck Garlic (*Allium sativum* var. *ophioscorodon*).The hard neck garlic is grown in cool climates. It produces flower stalks called scapes. The scapes often topped with cluster of small round propagules called bulbils or top sets. A single ring of cloves surrounds a stiff central stem that curls as it grows. The soft neck garlic is grown close to the equator. It produces large cloves around the outside, and smaller cloves in the middle. Cloves produce strong flavor and store well. It does not have flower stalk (scape). According to Nair *et al*. (2013) garlic species are classified into four groups: *A. longicuspis*, *A. ophioscorodon*, *A. sativum*, *A. subtropical* and *A. pekinense* sub-group. The *longicuspis* group is considered the oldest and it is postulated to be the original group. The *ophioscorodon* group is distributed in Central Asia, the *sativum* group in the Mediterranean zone and the *subtropical* in the south and southeast of Asia. The *pekinense* group comes from the east of Asia.

Garlic cultivated in rural farms of South Italy is often a heterogeneous clone population, which can comprise different cytotypes. A collection of cultivated garlic from the University of Basilicata gene-bank was evaluated for ploidy level and 16 different morphological traits. Out of 50 accessions, 7 were hexaploids (*A. ampeloprasum*) and 43 were diploids (*A. sativum*). Significant differences in yield were observed within and between ploidy levels. Discriminant analysis did show that four characters (leaf basal width, total number of leaves, clove diameter and neck height) were able to correctly discriminate all germplasm accessions between the two species (Figliulo *et al.,* 2001).

It is suggested that chromosome structure altered during species expansion from the centre of origin, which is assumed to be the area around the Shen mountains from western China to Kazakhstan and Kirgizstan (Ethoh and Pank, 1996). The different configurations of chromosome synapsis, such as tetravalents, hexavalents and octovalents, were observed at meiosis of pollen mother cells in 4 garlic varieties. The formation of the multivalents may result from chromosome reciprocal translocation among the non-homologous chromosomes. The karyotype analysis showed that the varieties with multivalents at meiosis also showed variation in the karyotypes (Zhang *et al.,* 1999).

The elephant garlic or great headed garlic (*Allium ampeloprasum;* hexaploid 2n = 6x = 48) is not a true garlic, but a type of leek that produces larger bulb with fewer larger cloves (4-8). Several small cormels may also develop on the bulb near root zone (5-10). Plants have flat and thin leaves and produced a large scape and umbel (flowering head) but no seeds. It gives milder flavor which is intermediate between garlic and onion. Allicin content of great headed garlic is less than common garlic. The coves are largely used for medicinal purpose.

Diffrent Species of Garlic

A. longicuspis **A. ophioscorodon**

A. sativum

A. ampeloprasum

Purseglove (1975), Cobley and Steele (1976), and Pruthi (1979) described the plant in somewhat greater detail. Accordingly, it has superficial adventitious root system. The bulb is composed of a disc-like stem, thin dry scales which are the bases of foliage leaves and smaller bulbs or cloves which are formed from axillary buds of the younger foliage leaves. The cloves are enclosed by the dry outer scales. Each clove consists of a protective cylindrical sheath, a single thickened storage leaf sheath and a small central bud. If very small cloves are planted or if grown under poor conditions, a single solid clove (round) is produced. Leaf blade is linear, flat, solid, 2.5 cm or less in width, folded length wise. Flowers are variable in number and sometimes absent, seldom open and may wither in the buds. Flowers are borne on slender pedicels, anthers and style exerted, seed seldom, if ever produced.

Histogenesis of seedstalk of inflorescence was studied by Kothari and Shah (1974). It was observed that the seedstalk bears a terminal inflorescence which in turn bears bulbils instead of flowers. The shoot becomes flat and finally aborts after the development of bulbils in the inflorescences. Etoh and Ogura (1977) studied the formation of morphological abnormal flowers and reported that they were sterile to a great extent. Even though floral parts develop normally, pollen sterility occurs after meiosis (Konvicka *et al.*, 1978). In the studies by Etoh (1982), tapetum was observed to have reached maximum development at uninucleate pollen stage but nearly all such pollen grains were disintegrated before pollen mitosis.

5.0 Growth and Development

Chun and Soh (1980) studied the histology of bulb development which indicated that the cloves had differentiated from primordia at the axil of the terminal leaf. Each clove primordium then differentiated into 3-4 cloves as the leaves developed further over the next 25 days. Branching in garlic is a physiological process affecting bulb quality due to product deterioration. Branched plants produce rough bulbs with an extremely high number of small cloves. It is well known that the thermal conditions before and after planting, day length, high availability of water and nitrogen, and sparse plantations are factors involved in branching expression. Because thermal conditions and photoperiod are the main factors that affect process induction, it was postulated that branching could be a good indicator of adaptation of the genotype to the environment in Argentina. Delayed planting resulted in an increase in branching intensity. There should be a physiological stage for the effective reception of the environmental signal in the white garlic type (Portela and Armstrong, 2001). Portela (1998) also reported that thermal conditions before and after planting, day length, high availability of water and nitrogen, and sparse planting are factors involved in branching expression.

Stahlschmidt *et al.* (1997) identified three growth phases; an initial one of slow growth, a second phase where growth was logarithmic and a third phase where growth was linear. During the first phase, relative growth rate (RGR) was similar for all 3 cultivars. During the second phase whole plant RGR was higher in cvs. Blanco and Colorado than in Rosada because of a higher leaf area ratio (LAR) in the former 2 cultivars. In this phase, bulb RGR in Blanco and Colorado was 43 per cent higher than in Rosada. In the third phase, RGR decreased in all cultivars. Bulb NAR was very similar for all cultivars implying that the lower bulb DM in Rosada was due to a lower LAR.

Among the environmental factors, photoperiod and temperature may be the key factors to regulate the garlic bulb development. Exposure to prolonged daylight and temperatures higher than 20°C improves garlic bulb production and quality (Grubben and Denton, 2004). In addition, vernalization accomplishment provided with prolonged daylight and high temperature may enhance garlic growth and development (Wu *et al.*, 2015). Thus growth and bulb development in garlic is affected considerably by variations in photoperiod and temperature thereby influencing its morphology, physiology, and nutritive quality (Atif *et al.*, 2019).

Garlic bulbing is affected by the day length and temperature to which the dormant cloves are subjected before the start of bulbing (Bandara *et al.*, 2000). The competition for resources by the developing bulbs and inflorescence sinks regulates the fate of stalk elongation and bulb formation (Etoh and Simon, 2002). It has been suggested that impact of photoperiod and temperature on bulb growth should be considered in the background of the immediate but viable bulb growth (Mathew *et al.*, 2010). Environmental restrictions may result because of the interaction between photoperiod and temperature, and the warm conditions of late spring accelerate bulbing response in short-day cultivars thereby imposing plants to develop smaller bulbs, however colder winter temperatures can induce bolting. The time spell from planting to inflorescence progressively curtails with increasing photoperiod or temperature (Gaskell *et al.*, 1998). The impact of external factors like photoperiod and temperature on plant evolution and growth has been described in numerous garlic genotypes (Takagi, 1990). Different combinations of long or short photoperiod with low or high growth temperatures produce changes in plant development. Nevertheless, long photoperiod and high temperatures induce bulb growth once bulbing has started. Influence of environmental effects on garlic bulb development has been studied and the internal physiological mechanisms causing alterations in productivity between cultivars under the same ecological conditions have been unidentified (Kamenetsky, 2007).

5.1.0 Temperature

Borgo *et al.* (1994) reported that bulbs derived from the cool area had deeper dormancy than those from a warmer area. Consequently, cloves of bulbs from the warmer area showed a higher percentage of sprouting. Nunez *et al.* (1997) reported that the inductive stage of bulbing lasted 90 days from sowing to bulb initiation. At that point bulbing index declined, isoenzymes activity increased significantly with changes in the isoenzyme patterns. The morphological stage started at the end of the inductive period, 80 days after sowing and continued until 170 days after sowing. Soluble protein content was not an indicator of morphological changes.

Plantlets of 2 local garlic cultivars, grown *in vitro* at 15°C, bulbed when transferred to 20-25°C. Pre-incubation at 4°C enhanced but did not directly induce bulbing. The cultivar Taicang White (late maturing) bulbed only under daylengths longer than 12 h. Photoperiod did not affect bulbing of the early maturing cultivar Xuzhou White. The analysis of endogenous growth regulators during bulbing showed that ABA and IAA levels increased, while IPA (isopentenyladenosine) and GA decreased. At the beginning of bulbing, the IPA/ABA ratio decreased but the IAA/ABA ratio increased. The IPA/ABA ratio was sensitive to photoperiod in Taicang White. It was concluded that high temperature was one of the major factors inducing bulbing (Liu *et al.*, 1997).

Recent study conducted by Atif *et al.* (2019) revealed that longer photoperiod (14 h or 16 h) and higher temperature (25°C or 30°C) treatments significantly improved the garlic bulbing imparting maximum bulb diameter, height, bulbing index, and the shortest growth period, whereas, 12-h photoperiod had maximum bulb weight. In addition, total soluble solids (TSS), contents of soluble protein, soluble sugar, total sugar, glucose, sucrose, fructose, starch, total phenols, and total flavonoids increased significantly because of 14-h photoperiod and 30°C temperature condition, however exhibited decline with 8 h photoperiod and lowest temperature (20°C).

5.1.1 Temperature on Dormancy

Rakhimbaev and Solomina (1980) studied the activity of endogenous cytokinins in bulbs stored for 8 weeks at 4°C. Low temperature stimulated the accumulation of cytokinins in the tissue and the bulbs stored at low temperature sprouted earlier and grew more rapidly than bulbs stored at 22°C. In an experiment by Chang and Park (1982), dormancy was broken earlier in local garlic bulbils treated for 40 days at 0-5°C before planting than the untreated ones, but clove differentiation was unaffected. Cold storage (6 ±1°C) for 30 and 40 days before planting reduced the dormant period and increased field emergence (Silva and Casali, 1987).

5.2.0 Light

Rahim and Fordham (1994) demonstrated that bulb formation was promoted by long day treatment and the longer the day the sooner was the start of bulb formation. Bulb yield was higher under 12 h day length followed by 8 h and 16 h. There was a quantitative response of bulb initiation, growth and maturity to light intensity. Bulb initiation was earlier, development of bulbs was more rapid, and maturity early under more complete light conditions.

5.3.0 Changes on Growth and Chemical Constituent

In India, Sood *et al.* (2000) studied the changes on growth, protein, pungency and flavour characteristics in five varieties (HG-1, HG-6, HG-17, HG-19 and G-1) of garlic bulb beginning 150 days after planting at an interval of 15 days till maturity (195 days). They observed that fresh and dry weights, diameter, height, diameter : height ratio of garlic bulbs, total, non-enzymatic and enzymatic pyruvate [pyruvic acid], and specific activity of allinase increased up to 180 days after planting and declined thereafter. The mean sulphur and protein contents increased throughout bulb development. Cultivars HG-6 and HG-17 had better size after 195 days of planting.

In plants, synthesis of phenolics and flavonoids is influenced by environmental factors (Hemm *et al.*, 2004). Plant growth and development is not only regulated by light but also the biosynthesis of primary and secondary metabolites (Liu *et al.*, 2002). Phenolic biosynthesis is modulated by light, and flavonoid biosynthesis is categorically light dependent (Xie and Wang, 2006). Bulbing at the end of crop cycle is a very vital stage in development where bulb becomes the greatest sink organ of photo assimilates which are altered to fructooligosaccharides and high-molecular weight fructans till plant senescence happens (Rizzalli *et al.*, 2002). The quantity of fructans like other metabolites is sturdily prejudiced by genetic and ecological aspects, and growth factors like photoperiod, temperature, humidity and fertilizers, microbes and insects, UV radiation, heavy metals and pesticides (Orcutt and Nilsen, 2000).

6.0 Cultivars

Most of the available cultivars of garlic are white coloured, but sometimes pink or red types are also found. Two distinct types namely Fawari and Rajalle Gaddi with slightly bigger bulbs are grown in the Bellary District of South India. Breeding for production of new cultivars with large bulbs of white colour, uniform and medium to large shape and size, compact cloves, high pungency and yield, and resistance to pests and diseases, is being done at Punjab Agricultural University, Ludhiana, India.

Some of the local known clones in addition to Fawari and Rajalle Gaddi are Madrasi, Tabiti, Creole, Eknalia, T-56-4, Jamnagar, *etc.* The cultivar Jamnagar was reported to be the biggest and best, giving the highest recovery of dehydrated peeled garlic and garlic powder of good pungency and antibacterial activity (Pruthi, 1979). In an evaluation trial, Pandey and Singh (1989) recorded best cultivar HG-I for yield, number of cloves/bulb and weight of bulbs. It also had lowest incidence of infection by purple blotch (*Alternaria porri*) and *Stemphylium*.

In a trial with 5 garlic genotypes (G-41, G-1, DARL-52, G-283 and Pithoragarh Local) in India, Bhatt *et al.* (1998) observed that the cv. G-41 reached maturity earliest (203 days) followed by G-1 (207 days). Highest vegetative growth (73.3 cm) and number of cloves per bulb (23.16), and lowest number of bulbs/kg (26.6) were recorded from Pithoragarh Local. The highest bulb yield (149.91 q/ha), ascorbic acid content (10.35 mg/100 g) and total minerals (1.52 per cent) were recorded from DARL-52. Highest TSS (38.31 per cent) was recorded from G-431, followed by DARL-52 (37.11 per cent), while the lowest TSS value (33.31 per cent) was recorded from G-283. Singh and Dubey (2015) evaluated 300 indigenous genotypes including 19 advanced lines and four checks Yamuna Safed (G-1), Agrifound White (G-41), Yamuna Safed-2 (G-50) and Yamuna Safed 3 (G-282) for yield, quality and storage life of garlic bulb.The result revealed that the genotypes G-189 and G-324 were promising for higher yield with quality bulbs and good keeping quality. The advanced line, G-200 was suitable for longer storage life.

Seven promising accessions out of the 88 held at Nasik were evaluated for yield components. G-14 produced the highest bulb size index (12.41 cm^2), gross yield

(7.25 t/ha), clove size index (2.42 cm^2), 10-bulb weight (273 g) and marketable yield (6.6 t/ha) (Singh *et al.*, 1993). The bulbs of fifteen white and purple white colour garlic lines were analysed for quality parameters for dehydration purposes. It was revealed that garlic collection G-189 recorded significantly highest TSS (43.42 per cent), dry matter (44.24 per cent), drying ratio (2.26 : 1), pyruvic acid content (28.35 micro mole/g) and rehydration ratio (1:5.06). The performance was at par with collection G-176 (43.07 per cent, 43.90 per cent, 2.28 : 1 and 27.38 micromole/g and 1:4.64, respectively). It was concluded that garlic collection G-189 was best for dehydration purpose on account of better recovery and colour of dehydrated product (Singh and Gupta, 1998).

Some 23 promising advance lines screened from the garlic germplasms were evalua-ted for productivity and storability characters and compared with check variety Yamuna Safed-2 (G-50) at National Horticultural Research and Development Foundation, Karnal, India. The lines G-4 performed better in respect of bulb size (3.45 cm), clove size (0.67 cm), gross yield (131.94 q/ha) and marketable yield (126.57 q/ha) and minimum storage losses (6.42 per cent) (Gupta and Singh, 1998).

Seven genotypes were evaluated for 8 yield components at Lembucherra (Tripura, India). Genotypes from Guwahati, Meghalaya, Sikkim and Tripura performed better than those from Delhi, Varanasi and Jaunpur (Sharma *et al.*, 1997). On the basis of pooled data for 3 seasons, it was observed that G-282 was best for size of bulbs and cloves (diameters of 4.32 and 1.3 cm, respectively), produced the lowest number of cloves/bulb (14.1) and gave good yields (gross and market yields of 15.4 and 14.4 t/ha, respectively) (Pandey *et al.*, 1996a). On the basis of their high yields and good storability, G-1 and G-50 are recommended for cultivation under Karnal conditions, India (Pandey *et al.*, 1996b).

The performance of 16 garlic cultivars was evaluated by Saraf *et al.* (2000) on clay loam soils at Sagar, India. Results revealed that ARU-52 produced the highest bulb yields. Cultivars G-50, G-282, G-1, Sel-2 and LEC-1 performed next best (in descending order). Bulb diameter, weight of bulb and plant height showed positive relationships with bulb yield, while number of cloves/bulb exhibited a negative association. Singh *et al.* (2014) reported that Agrifound Parvati-2 (G-408) is a long day type variety of garlic developed by NHRDF for mid and higher hills of India and highly suitable for cultivation in zone-I (Jammu Kashmir and Himachal Pradesh). The maximum average yield (151.56 q/ha) was 165.34 per cent, 81.34 per cent 82.65 per cent and 56.41 per cent higher than the checks variety G-41, G-282, G-189 and VL Garlic-1, respectively.

Akimov *et al.* (1973) reported two cultivars of garlic namely Odeskii-13 and Imeruli-23 from different regions of the former U.S.S.R. which had very high essential oil content. Of the 14 cultivars studied by Chavchanidze (1976), Maikopskii-2, Snezhok and Ukrainskii Belyi were observed to be relatively resistant to *Acceria tulipae*, which were characterized by small clove and compact bulbs. Marchesi and Fuochi (1979) reported that continued mass selection by farmers had led to the isolation of a white garlic biotype, in which bulb weight was 50.24 ± 4.82 g.

Taicangbai garlic from China has large white bulbs and large, uniform bulbils, 6-9 per bulb. It has a strong taste and a delicate aroma. The cultivar is midlate with a growth period of 260 days. The mature bulbs have a relatively long dormancy period and sprout late. Yields of dry bulbs are 30-35 per cent higher than those of common garlic (Chen, 1986). Prikindel is a midseason Moldavian cultivar of the non-bolting type. It had a growth period of 85-107 days and forms bulbs weighing 28-32 g and have 7-15 cream- coloured thick cloves with pungent flavour and fairly firm flesh (Khaisin, 1988).

In Brazil, highest yields of commercial grade air-dried bulbs were obtained from the cultivars Selecado Jetiba, Gigante Roxao, Caltura and Selecao Regional (Zimmerer, *et al.*, 1988). In trials at the Universidade Federal de Lavras, Minas Gerais, Brazil, cultivars Gigante Curitibanos, Gigante Roxo and Amarante were considered promi-sing for the Lavras region, based on marketable yield, bulb weight, number of cloves/bulb and clove weight. Cultivars Dourados, Dourados de Castro, Gigante Lavinia and Cara had an excessive number of cloves/bulb and low clove weights (Blank *et al.*, 1998). In Havana, Cuba-7 clones were compared; growth period ranged from 110 days in HO-V-I and Martinez to 155 days in Guadalupe 15 and Guadalupe 25. The highest yeilds (7.6-8.0 t/ha) were obtained with Guadalupe 15, Guadalupe 25, Sancti Spiritus 3, HO-V-I and Martinez (Ayala and Savon, 1986).

In Bangladesh, an evaluation of cultivars revealed that GCO23 and GCO17 were most adaptable and recommended for cultivation by Ahmed and Hoque (1986).

Cultivar Timis R is adapted to the conditions of the South-West Romania. Yield is good, bulbs contain 11-15 bulbils, each bulbil weighing over 3.5 g (Suciu *et al.*, 1988).

Three garlic cultivars were evaluated by Tariq-Mahmood *et al.* (2001) for high yield and other characteristics. Chinese cultivar (Exotic) gave significantly higher yield as compared to local cvs. Lehson Gulabi and GA-I. The cultivar GA-I matured earlier followed by Lehson Gulabi and Chinese.

In an effort to solve the seed quality problem and to begin a programme of selection and identification of high yield clones suitable for internal and external market consumption in SE Buenos Aires, 14 red garlic clones were evaluated during 1993 and 1994. Bulb weight and bulblet weight and colour and roughness were the variables which best indicated differences among clones. Colorado Espanol, 203 and Colorado A-1 clones were the best (bulb weight higher than 45 g, regular shape, bulblet weight of 4 g and nil roughness). Positive and significant correlations between yield components, weight, size and number were obtained for all clones, except Colorado Espano. The genetic variation (total genetic variance/phenotypic variance) for yield components of clones were 7 per cent, so clonal differences were mainly yield × environment effects (Saluzzo and Rattin, 1998).

Five local cultivars of northern type (Euiseong, Yeochun, Danyang, Jungsun, and Yongin) garlic and one of the southern type (Namdo) were collected, and their growth, developmental characteristics and productivity were investigated in order to select a cultivar well adapted to the Euiseong region, Korea. Incidence of bolting was low in Jungsun and Yongin, while high in Euiseong and Yeochun. Considering yield factors, Yeochun was the most promising northern type cultivar in the Euiseong region, showing the highest bulb weight and largest cloves with a high growth rate. Danyang, Jungsun, and Yongin had a relatively low productivity in the Euiseong region (Ha *et al.*, 2000).

In Nepal, co-ordinated varietal trial of garlic was conducted by Shrestha *et al.* (2000) at three AER sites (Bhakimle 1650 m asl, Chhahara 1350 m asl and Maduwa 850 m asl) and Lumle Station (1600 m asl) during 1998 for identifying early maturing and high yielding genotypes in the middle hills. Genotype Bhote lasun produced the highest marketable bulb yield (19.5 t/ha) followed by Mallaij Local (13.7 t/ha) and Pakhribas Local (11.4 t/ha). Bhote Lasun was later in maturity (221 days after sowing) than the other genotypes.

Thirty-six different garlic ecotypes, formerly collected in Southern Italy and in other Mediterranean areas, were grown at the Chiancalata Experimental Farm (Matera, Italy). The ecotypes Gravina 3 and Altamura 2 showed the highest bulb yields (9 t/ha). Diallyl trisulfide and diallyl disulfide were the 2 major volatiles found in the oils from all the selected garlic ecotypes (Avato *et al.*, 1998). A 3-year trial at Mussomeli, Italy, showed that yields were highest from cv. Lucchese Nostrale followed by Putignano and Rosid di Napoli. These cultivars also produced the largest bulbs (Caruso *et al.*, 1994). A total of 21 garlic cultivars (19 European, 1 Asiatic and 1 Chinese) was evaluated in France. Good results (emergence, size and weight) were obtained with Messidrome, Thermidrome and Germidour (autumn cultivars), Iberose and Goulurose (rose stick cultivars), and Moulinor and Artop (spring cultivars) (Lavigne and Audubert, 1998).

There is wide diversity in germplasm of non-bolting garlic in Bulgaria, and especially of early genotypes. Several breeding programmes have been undertaken during the last few decades and, as a result, new cultivars have been released including White June N20, Nora, Zahorsci. The most stable results have been achieved by the new clonal cultivar Zdravets, which is one of the central sterile leaf types (Datshvarov *et al.*, 1997).

Seven clones derived from local populations by clonal selection were compared with the variety Piros over 3 years to evaluate bulb morphology and weight, clove number and weight, dry matter percentage, and yield. The highest mean yield over 3 years was given by clone L4/2 (5.46 t/ha). Clones L3/2, L8/1 and L35/4 gave yields greater than 5 t/ha as against 4.14 t/ha in the standard. All the clones had significantly greater bulb weight than Piros and also exceeded it in clove weight and number (Gvozdanovic-Varga *et al.*, 1994).

The cultivars of garlic developed by different research institutes of India are given in Table 2.

Table 2: Commercial Cultivars of Garlic in India*

Cultivar	Features
Yamuna Safed (G-1)	Bulbs are compact, silvery white with creamy flesh. Diameter of bulb is 4.0-4.5 cm. Cloves are sickle shaped and 25-30 in number per bulb. Bulb contains 38-40 °Brix total soluble solids, 40 per cent dry matter and 29 micro mole/g pyruvic acid. Crop matures in 140-150 days. It gives an average yield of 150-175 q/ha. The variety is recommended for all over India.
Yamuna Safed-2 (G-50)	Bulbs are compact, attractive with white creamy flesh. Diameter of bulb is 3.5-4.0 cm. The number of cloves is 35-40 per bulb. Bulb contains 38-40 °Brix total soluble solids, 41 per cent dry matter and 26 micro mole/g pyruvic acid. The crop matures in 140-160 days. It gives an average yield of 150-200 q/ha. The variety is recommended for North India, Punjab, Haryana and Madhya Pradesh.
Yamuna Safed-3 (G-282)	Bulbs are creamy white, bigger in sized and 4.5-5.5 cm in diameter. The number of cloves is 15-16 per bulb. Bulb contains 38-42 °Brix total soluble solids, 42 per cent dry matter and 25 micro mole/g pyruvic acid. The crop matures in 120-140 days. It gives an average yield of 175-200 q/ha. Suitable for export. The variety is recommended for Madhya Pradesh, Maharashtra, Haryana, Gujarat, Punjab, Rajasthan, Uttar Pradesh and Chhattisgarh.
Yamuna Safed-4 (G-323)	Bulbs are silvery white and 3.5-4.0 cm in diameter. The number of cloves is 20-25 per bulb. Bulb contains 40-42° Brix total soluble solids, 44.5 per cent dry matter and 25 micro mole/g pyruvic acid. The crop matures in 140-150 days. The variety gives an average yield of 175-200 q/ha. The variety is recommended for recommended for North India and Central India.
Yamuna Safed-5 (G-189)	Bulbs are creamy white and 4.5-5 cm in diameter. The number of cloves is 22-30 per bulb. Bulb contains 40-42 °Brix total soluble solids, 44 per cent dry matter and 26 micro mole/g pyruvic acid and are suitable for processing. The crop matures in 140-160 days. Suitable for dehydration purpose. The variety gives an average yield of 150-180 q/ha. The variety is recommended for Delhi, Uttar Pradesh, Haryana, Bihar and Punjab, Rajasthan, Gujarat, Maharashtra, Karnataka and Andhra Pradesh.
Yamuna Safed-6 (G-324)	Bulbs are compact, white in colour with creamy white flash and better keeping quality. Diameter of bulb varies from 4.50 to 4.80 cm and 27-30 cloves per bulb. Bulbs contain 39- 40 °Brix total soluble solids, 42 per cent dry matter and 25.5 micro mole/g pyruvic acid. Crop matures in 155-165 days after sowing. The average yield ranges between 145 and 150 q/ha. The variety is recommended for growing in kharif season in Maharashtra, Gujarat and Madhya Pradesh.
Yamuna Safed-7 (G-378)	Plants are straight with dark green leaves. Bulbs are compact, bold white in colour. Diameter of bulb varies from 4.50 to 4.80 cm and number of cloves 20-25 per bulb. Bulbs contain 38-39 °Brix total soluble solids, 41 per cent dry matter and 25.0 micro mole/g pyruvic acid. Crop matures in 145-150 days after sowing. The average yield ranges between 155 and 165 q/ha. The variety is recommended for growing in hilly areas.
Yamuna Safed-8 (G-384)	Bulbs are solid, white, flash attractive and dark green plants. Diameter of bulb varies from 4.5 to 5.0 cm, 22-30 cloves per bulb. Bulbs contain 41° Brix total soluble solids, 42 per cent dry matter and 26 micro mole/g pyruvic acid. Crop matures in 150-160 days after sowing. The average yield of variety ranges from 175 to 200 q/ha. The variety is recommended for growing in Jammu, Punjab, Delhi, Haryana and Rajasthan.

Cultivar	Features
Yamuna Safed-9 (G-386)	Bulbs are solid, light purple skin, attractive creamy white flash and dark green plants. Diameter of bulb varies from 4.80 to 5.50 cm, cloves 22-30 per bulb. Bulbs contain 40 °Brix total soluble solids, 43 per cent dry matter and 25 micro mole/g pyruvic acid. Crop matures in 150-160 days after sowing. The average yield of variety ranges between 180 and 225 q/ha. The variety is recommended for growing in Jammu, Punjab, Delhi, Haryana and Rajasthan.
Yamuna Purple-10 (G-404)	Bulbs are compact, attractive light purple in colour with creamy flesh. Diameter of bulb varies from 4.8 to 5.5 cm and bigger elongated cloves 25-30 in number per bulb. Bulbs contain 40 °Brix total soluble solids, 42.9 per cent dry matter and 26.8 micro mole/g pyruvic acid. Crop matures in 230-250 days after sowing. The average yield of variety ranges between 200 and 225 q/ha. The variety is recommended for Delhi, Rajasthan, Haryana, Jammu and Kashmir, Punjab, Tarai region of Uttar Pradesh, Uttarakhand, Bihar and Jharkhand.
Agrifound White (G-41)	Bulbs are compact, silvery white with creamy flesh. Diameter of bulb is 3.5 to 4.5 cm. Cloves bigger elongated and 20-25 in number/bulb. They contain 41 °Brix total soluble solids, 43 per cent dry matter and 27 micro mole/g pyruvic acid. Crop matures in 140-150 days. It gives an average yield of 130 q/ha. The variety is recommended for growing in Maharashtra and Madhya Pradesh.
Agrifound Parvati (G-313)	Bulbs are bigger, creamy white in colour with pinkish tinge. Diameter of bulb varies from 5.0 to 6.5 cm, 10-16 bigger cloves per bulb. Bulbs contain 36 °Brix total soluble solids, 38.5 per cent dry matter and 23 micro mole/g pyruvic acid. Crop matures in 230-250 days after sowing. The average yield of variety ranges between 200 to 225 q/ha. The variety is recommended for Jammu Kashmir, Himachal Pradesh and Uttarakhand.
Agrifound Parvati-2 (G-408)	Bulbs are bigger, creamy white in colour. Diameter of bulb varies from 5.0 to 6.0 cm, cloves bigger 12-14 in number per bulb. Bulbs contain 37 °Brix total soluble solids, 39 per cent dry matter and 23 micro mole/g pyruvic acid. Crop matures in 240-260 days after sowing. The average yield of variety ranges between 200 to 225 q/ha. The variety is recommended for Jammu and Kashmir, Himachal Pradesh and Uttarakhand.
Bhima Omkar	Compact white coloured medium sized bulb. Crop matures in 120-135 days after sowing. The average yield of variety ranges between 80-140 q/ha. The variety is recommended for Gujarat, Haryana, Rajasthan and Delhi.
Bhima Purple	Bulbs are attractive and purple skinned. Crop mature in120-135 days after sowing. The average yield of variety ranges between 60-70 q/ha. The variety is recommended for Delhi, Karnataka, Maharashtra, Punjab, Haryana and Uttar Pradesh.
VL Garlic 1	Bulbs are white in colour. Bulbs mature in 180-190 days after sowing. The average yield of variety ranges between 140-150 q/ha in Hills and 90-100 q/ha in plains. The variety was developed by ICAR-VPKAS, Almora. The variety is recommended for Bihar, Himachal Pradesh, Jammu and Kashmir, Punjab, Uttarakhand and Uttar Pradesh.
VL Lahsun 2	Bulbs are white in colour. Bulbs mature in 190-200 days after sowing. The average yield of variety ranges between 140-160 q/ha in mid hills and 240-260 q/ha in above mid hills. The variety was developed by ICAR-VPKAS, Almora. The variety is recommended for Himachal Pradesh, Jammu and Kashmir and Uttarakhand.

* Anonymous (2020).

Types and Cultivars of Garlic

Hardneck and Softneck Type

Single Clove Type

Agrifound White

Yamuna Safed (G-1)

Yamina Safed-2 (G-50)

Yamuna Safed-3

Agrifound Parvati (G-313)

Yamuna Safed-4 (G-323)

Bhima Omkar

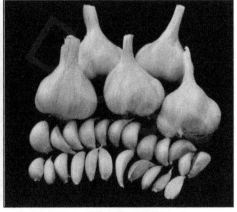

Bhima Purple

7.0 Soil and Climate

7.1.0 Soil

It can be grown on variety of soils but thrives better on fertile, well-drained loamy soils. Heavy clay soils may result in mis-shapen bulbs and make harvesting difficult. Kim *et al.* (1977) reported that a pH range between 5 and 7 had little effect on growth and yield of garlic.

The effects of different concentrations of A1C13 on root growth, cell division and on nucleoli in root tip cells of *A. sativum* were studied. A1C13 at 10-2 and 10-4 M inhibited root growth and caused a decrease in mitotic index. At 10-2 M A1C13, the effect was mainly observed on nucleoli in root tip cells while 10-4 M A1C13 caused chromosome abnormalities during mitotic cell division (Kaymak, 1996).

In the lowlands, garlic forms insufficient leaf area resulting in low yields and small bulbs. It is suggested that successful adaptation of garlic to lowland conditions is more dependent on selection of suitable heat-tolerant cultivars than on adjustment of cultural practices (Grubben and Midmore, 1994).

In a trial on yield and quality of salt-stressed garlic, its relative bulb yield was reduced by 14.3 per cent with each unit increased in soil salinity > 3.0 dS/m. Increasing soil salinity significantly reduced all yield components (*i.e.*, bulb weight and diameter, plants per unit area). Percentage of solids in the bulb was significantly reduced as soil salinity increased. Leaf tissue accumulated significantly higher Cl, Na and Ca concentrations than did bulb tissue (Francois, 1994).

7.2.0 Climate

Garlic is adapted to tropical and sub-tropical conditions. According to Purseglove (1975), bolting does not seem to be influenced by temperature and some clones never produce flowers.

7.2.1 Temperature

Garlic survives well in areas with 600-1200 mm annual rainfall with temperatures ranging between 5-25°C to 25-40°C. It is a frost hardy plant requiring cool and moist period during growth and relatively dry period during maturity of bulbs. Garlic tolerates morning and evening relative humidity in the range of 15-80 per cent and 15-70 per cent, respectively. High relative humidity favour diseases build up. Bulbing takes place during longer days and at high temperatures, exposure to low temperature subsequent to bulb formation favours the process. Adequate vegetative growth promotes bulb formation. Excessively hot and long days are not conducive to proper bulb formation. In the studies by Abdel (1973), garlic bulbs planted on west side of north-south ridges where the temperatue was lower, were larger than those on the east. Yield and survival were also higher at lower temperatures. Further, the survival rate, bulb weight and yield of garlic planted on north side of west-east ridges was higher than that of the plants grown on the south side or on the flat beds. Moravec and Kvasnicka (1975) evaluated the world collection of garlic and noted that during cool growing seasons the yield differences between the cultivars were more conspicuous, whereas warm wet wheather reduced it. These results showed that low temperature is a prerequisite for higher yields in garlic.

Low temperature pretreatment and night interruption generally increased bulb weight and reduced neck: bulb diameter ratio in all 4 cultivars (3 southern ecotypes – Shanghi, Shenyang and Jejuione, 1 northern ecotype – Euisung). Of the cultivars, Shenyang had the heaviest bulbs (56.0 and 26.6 g in field and protected winter cultivation, respectively) (Hahn, 1994).

The cause of poor bulbing in the late planted crops in Bangladesh was probably due to increasing air and soil temperatures at the end of the season. Both air and soil temperatures above 20°C had adverse effects on growth and development. Results also demonstrated the possibility of accelerating crop development by either pre-cooling or with the use of growth regulators. Seed cloves treated either at 5 or 10°C for 15 to 30 days before planting, accelerated initiation, development and maturity of bulbs relative to those of cloves stored at 15 and 20°C (Rahim *et al.*, 2001).

The sprouting and emergence of garlic clones California, Chol-Chol, Rosado-INIA and Rose de Lautrec were investigated at 6 constant temperatures ranging from 5 to 30°C in darkness and also under field conditions. The rates of sprouting and emergence, defined as the inverse of the time taken to reach 25, 50 or 75 per cent of those stages, were related to temperature. Base temperature (Tb) for 50 per cent sprouting and 50 per cent emergence ranged between –0.1 and 2.2°C and between 2.8 and 5.8°C, respectively. Clone California showed the highest values for Tb. In the field, rates of emergence were similar to those found under controlled conditions (Barrera *et al.*, 1998).

7.2.2 Photoperiod

Moon and Lee (1980) noted that short days of 8 hours resulted in suppression of growth, poor bulb formation and induction of secondary growth. They concluded that the activities of ABA like substance and high total sugar levels are major factors related to bulb formation, whereas gibberellin-like substances play a relatively minor part. Contrary to the above findings, Kim *et al.* (1980) showed that long-day conditions necessary for growth and bulb formation could be replaced by night interruption with incandescent, red or far-red light. The interruption was most effective when applied in the middle of dark period for 60 minutes or longer. Excessively hot and long days are not conducive to proper bulb formation.

7.2.3 Photoperiod and Temperature

Mann (1952) also found that long days and high temperature encouraged bulb developments in garlic. As soon as bulbing commences leaf initiation ceases. Because of this reason and high yields, garlic should be planted early to promote vegetative growth under short photoperiod and cool temperature. The yield potential depends on the amount of vegetative growth made before bulbing commences. Late planting adversely affects vegetative growth and thereby results in lower bulb yield. Zhila (1978) noted that exposure to 10 h day stimulated branching in both bolting as well as non-bolting cultivars. This treatment retarded bolting in the bolting cultivars and reduced bulb formation. Park and Lee (1979) reported that increase in day length from 8 to 12 h increased the bulb weight, bulb diameter and number of cloves produced and reduced the number of secondary leaves. The critical daylength for bulbing was 12 h. Short days promoted secondary growth and suppressed bulbing. Treatment of long day (16 h light) markedly accelerated bulbing and senescence.

Flowering percentage was influenced most by clone, although interactions with photoperiod, growth temperature and cold storage occurred. Clone R81 flowered equally well in all conditions, whereas flowering percentage of clones D129, D130 and PI485592 was reduced by cold (4°C) storage of either bulbs or plants, long (16 h) photoperiod and growth at 18°C rather than 10°C. The highest flowering percentage in all clones (13-43 per cent) was achieved by growing plants at 10°C under short (9- to 10-h) photoperiod with no cold storage of bulbs before planting (Pooler and Simon, 1993a).

8.0 Cultivation

In general, the cultural practices followed in onion are similar for garlic, execpt certain requirements which are specific for this crop.

8.1.0 Planting Material

Garlic is propagated vegetatively by single clove, but bulbils are also used occasionally. In addition to the above, some recent methods have also given encouraging results. Practical utility of aerial bulbils as planting material was investigated by some workers. Ku *et al.* (1974) reported that the highest yield of bulbs was obtained when the bulbils were 2-year-old or more. Trippel and Chubrikova (1976) observed that the plants raised from aerial bulbils were more productive than those raised from cloves. A yield to the extent of 96.10 q/ha was obtained by Om and Srivastava (1977) at Chaubattia from aerial bulbils when planted at a distance of 10 × 20 cm. The maximum yield of 129.06 q/ha was, however, obtained with cloves at the same spacing. These results clearly indicate that the aerial bulbils can also be used as a planting material for garlic when closer spacing needs to be followed. Virus-free garlic seed cloves having average weights of 1.51 to 2.32 g planted with space between plants of 12.5 and 15.0 cm generate plants with bulbs of larger diameter and higher titratable acidity, pungency, and industrial index, enabling the production of better-quality bulbs with good prospects for industrialization (Lima *et al.*, 2019).

Apart from bulbs, garlic has been successfully propagated by tissue culture which has been discussed in biotechnology.

8.2.0 Time of Planting

In regions having temperate climate, garlic is often planted in late autumn to facilitate top growth at higher temperature in spring. It is planted from August to November in Maharashtra, Karnataka and Andhra Pradesh and from September to November in the northern plains of India. Proper season of planting in the hills is March-April. Om and Srivastava (1974) reported that the planting from September 6 to October 3 was significantly superior to other sowing dates for getting better growth and yield of bulbs at Chaubattia, in the hilly region of U.P.

According to Mathur *et al.* (1975), the optimum time for planting of garlic was 25th September. Bulb yield was highest from planting on 20th September and declined with delayed planting on 30th October in Garhwal, U.P., India (Singh and Phogat, 1989). In an experiment to standardize sowing date at Parbhani in Maharashtra, India it was observed that the bulb yield was significantly reduced at both the levels of significance as planting was delayed from 10th November (122.09 q/ha) to 25th November (101.45 q/ha) and 10th December (80.77 q/ha) (Anon., 1982-83). On the other hand, in northern districts of West Bengal, 11th November planting gave the highest yield (Das *et al.*, 1985a). In Ludhiana, Punjab, Bhathal and Thakur (1986) observed with three different cultivars (56-4, G-11 and G-15) that mean yield was highest (133.1 q/ha) from 15th October planting and declined to 100.6 q/ha when planted on 16th November. The crop was harvested on 24 April. In Bihar, India bulb yield was highest with direct planting of cv. DG 78-2 in the field on 15 October followed by transplanting on 15 November. Harvesting was delayed with further delay in transplanting (Maurya, 1987). Choudhary and Choudhary (2018) evaluated four dates of sowing of garlic *viz.*, 10 October, 25 October, 10 November and 25 November under semi-arid region of Rajasthan, India. Sowing of garlic crop on 25 October resulted in maximum yield and net returns from garlic cultivation. Chattopadhyay *et al.* (2006) recorded maximum bulb yield of garlic by planting the cloves on November 15 in presence of mustard cake (5 t/ha) and inorganic fertilizers (NPK - 60: 60: 120 kg/ha) in the Gangetic plains of West Bengal, India. The effect of planting material quality (clove weights of > 4 g, 1-4 g, and < 1 g) on the growth and yield of 2 garlic cultivars (Mexican and Local) was investigated in the field and in plastic tunnels. Mexican cultivar exhibited higher fresh and dry weights, length of leaves, bulb diameter and yield than Local cultivar. Regarding planting material quality, better growth and higher yield were associated with heavier cloves at planting. Similar trends were observed in the field and in plastic tunnels, although the fresh weight of leaves and bulbs tended to be higher for plastic tunnel-grown plants compared with field-grown plants (Hafidh, 2000).

In Bangladesh, the highest bulb yield of garlic was recorded when planted on 30 October (Rahim *et al.*, 2003).

Garlic cultivars (Swat Local, Tarnab Peshawar and Italian) were planted on 3 different dates : 22 September, and 7 and 22 October, at Mingora, Swat, Pakistan. For all planting dates, the highest yields were produced by cv. Italian. For all cultivars, the yields were highest for the early planting, followed by the 2nd and 3rd planting dates (Humayun-Khan *et al.*, 1997). At Faisalabad, Pakistan biological yield, leaf area, bulb weight, clove weight, neck diameter, bulb diameter, clove maximum length, clove minimum length, number of clove per bulb, clove volume, fresh weight, dry weight and total phenolic contents were highest when garlic was sown on October 30 beyond which the growth and yield of garlic was reduced (Ahmad *et al.*, 2018).

In Germany, the most suitable cultivars for field cultivation were Stamm, Burgenland, Thuringer, Mako and Ungarischer. Cloves should be planted out from the middle of September to the middle of October at a depth of 7-8 cm and a density of 500 000 cloves/ha (row spacing of 25 cm and a planting distance of 8 cm). Bulbs should be harvested the following year, producing a yield of 160-240 dt/ha fresh bulbs (80-120 dt/ha superficially dried bulbs) with an allicin content of 0.32-1.01 per cent (Bomme *et al.*, 1996).

The effect of planting time on the incidence of *A. porri* was evaluated using 11 cultivars at 3 planting times. The cultivars Dourada and Centenario were the most resistant, and the 3rd planting time (end of May) was the most effective for reduction of yield loss (Ferreira *et al.*, 1995). Planting in May resulted in the highest yields. Yields were highest in cultivars Gigante Lavinia, Gigante Inconfidentes and Sao Lourenco, planted in May (15.45, 13.57 and 13.00 t of cured bulbs/ha, respectively). Gigante Lavinia and Gigante Inconfidentes had short cropping cycles (160 days when planted in May) and Sao Lourenco had a longer cropping cycle (196 days) (Trevisan *et al.*, 1996).

Relative growth and yield of eight garlic lines were evaluated in autumn- and spring-planted trials in Saskatchewan. The best adapted lines (California Late and Mexican) produced commercially acceptable yields of high-quality garlic, while other lines proved poorly adapted to Saskatchewan growing conditions. Autumn planting enhanced yields over spring planting by an average of 59 and 25 per cent in the two test years; however, some lines were only suited to spring planting. California Late was the highest yielding variety in both autumn-planted trials but its yields were relatively poor in the spring-planted trials. In contrast, the Jamaican variety yielded poorly when planted in autumn, but gave good yields in the spring-planted trials. Planting date had little effect on overwinter survival or yields of autumn plantings (Waterer and Schmitz, 1994).

Time required for 50 per cent shoot emergence was shortened as planting dates were delayed. Good shoot emergence was observed at earlier planting dates following treatment with K_3PO_4. Sowing date and chemical treatment did not significantly affect the differentiation of cloves, and in all treatments, differentiation was >85 per cent. Average bulb weight was higher at earlier planting dates (Cho *et al.*, 1998).

8.3.0 Spacing and Seed Size

For planting, cloves or sometimes bulbils are used. They are carefully detached from the composite bulbs without any damage or injury for higher germination. The planting distance as well as the seed size varies in different regions of India and other countries. Similarly, the quantity of planting material required for an area is also variable. Purewal and Dargan (1961) in Punjab obtained the best yield at a spacing of 15 × 18 cm. According to Nath (1976), the cloves should be dibbled at a spacing of 15 cm between rows and 8 cm between plants for getting good yield. About 350-500 kg of cloves are required at this spacing to plant a hectare of land. Highest yield of bulbs was obtained by Menezes *et al.* (1974) when planting was done at a distance of 20-30 cm between rows and 7.5 cm between plants. Purseglove (1975) recommended a spacing of 30 × 15 cm.

The highest yield (102 centners/ha) of bulb was obtained by Bogatirenko (1976) by planting cloves 16-21 mm in diameter at the optimum spacing of 45 × 4 cm. Similarly cloves over 11-13 mm was recommended for bulb production for consumption, whereas smaller cloves could be used for further propagation (Burba *et al.*, 1982). Duimovic and Bravo (1980) recommended clove weight and plant population of 4-5 g and 333000 plants/ha, respectively. Garlic cloves weighing 5.8, 4.6 and 3.6 g were planted by Lucero *et al.* (1982) at densities of 160000 or 200000/ha. Bulb yields were greatest (74 q/ha) from 5.8 g cloves at the highest planting density and produced higher proportion of large bulbs (46-55 mm diameter). Further, it was seen that only 5.8 g cloves produced bulbs with diameter over 55 mm. Rahman and Talukdar (1986) reported that early planting on 15 November with high density of 3 × 15 cm gave the highest yield of 11.28 t/ha but bulb weight was only 6.42 g. It increased with spacing to 9.66 g at widest spacing of 12 × 15 cm but the yield decreased to 8.16 t/ha.

Large bulblets from bulbs of different sizes were tested by Kusumo and Widjajato (1973) at spacing of 15 × 15 cm and 15 × 10 cm. Best yield and quality were obtained with bulblets from medium-sized bulbs at wider spacing. Aliudin (1980) planted the garlic cloves graded as large (weight > 1.5 g), medium (1-1.5 g) and small (>1 g) at 6 spacings from 20 × 20 cm to 10 × 7.5 cm. It was observed that the plant weight, bulb weight and bulb diameter decreased with closer planting. The closest spacing gave the highest yield (17.25 t/ha) of small bulbs suitable for use as planting material and the widest spacing gave the lowest yield (8.8 t/ha) of large bulbs suitable for consumption. Mishra and Pandey (2015) obtained highest bulb yield (248.21 q/ha) of great headed garlic by adopting commercial seed cloves (6.0–6.5 g) and closer planting distance (15 × 7.5 cm).

Nourai (1994) noted that high seed rates influenced bulb size. High seed rates decreased the percentage by weight of large bulbs and reduced the bulb weight, number of cloves per bulbs, clove weight, bulb diameter and bulb length.

In former Czechoslovakia, for vegetable production direct sowing to give a density of 250000 to 300000 plants/ha is recommended (Luzny, 1987). Orlowski and Rekowska (1989) from Poland reported highest yield with 5.0 g cloves planted at 30 × 6 cm, whereas a spacing of 30 × 8 cm gave the best return.

Distance between rows, size of the planting material and site had significant effects on marketable yield and average size of harvested cloves. The highest yield was obtained with 45-50 cm between rows. Planting cloves smaller than 20 mm produced smaller bulbs, but yield was not affected by whether the clove came from the periphery of the bulb or elsewhere (Landry and Khanizadeh, 1994).

Naruka (2000) planted garlic cultivars at row spacing of 10 × 7.5 cm (narrow) and 15 × 7.5 cm (wide) in a field experiment in Rajasthan, India. Higher plant height (50.65 cm), leaf chlorophyll content (1.02 mg/g) and leaf fresh weight (24.92 g/plant) were observed at wide spacing compared to narrow spacing. The number of leaves per plant and maturity period was not affected by row spacing treatments. Ramniwas *et al.* (1998) observed that increasing the planting distance significantly increased bulb yield (83.57 q/ha at 20 cm, compared with 59.27 and 41.78 q/ha at 15 and 10 cm, respectively). Bulb FW and DM, number of cloves/bulb, and length and diameter of bulb and clove were also highest at the lowest planting density (20 cm). Bulb yield also increased with increasing clove size. Bulb weight, DM and diameter were higher with larger cloves. Spraying with GA at 45 days after sowing had no significant effect on growth and yield.

A field experiment was conducted in Punjab, India, to determine the effect of clove size on garlic cv. LLC-1 yield. Cloves of different sizes (0.71, 1.66 and 3.30 g) were sown on 18 October. Bulb diameter and bulb weight per 10 bulbs increased with increasing clove size. The maximum bulb weight per 10 bulbs (287.50 g) was obtained with the sowing of the largest clove size (3.30 g). The highest bulb yield (20.92 t/ha) was obtained with the sowing of the largest clove (Brar and Gill, 2000). The trials on the effect of bulb size and plant density on the yield of garlic indicated that yield was more affected by plant density than by the size of the planting material (Xu *et al.*, 1999).

Bernard *et al.* (1993) opined that emergence was always significantly higher when planted vertically with the apex uppermost and lowest when the apex pointed downwards. Planting horizontally with the convex side uppermost gave proper emergence particularly when planting was shallow. It was also found that mechanical planting with a pneumatic selector could provide a viable and cheaper alternative to manual planting provided that soil and propagules were well-prepared, that working parts were accurately adjusted and that the speed of advance was suitable. Garlic (strain G1) seeds were sown at spacings of 12 × 5 cm, 12 × 10 cm or 12 × 15 cm on 28 November on a loamy soil. The closest spacing (12 × 5 cm) resulted in the tallest plants, longest leaves, highest number of leaves per plant and highest yield (73.95 q/ha), compared with 56.82 and 41.73 q/ha at the middle and widest spacings, respectively (Singh *et al.*, 1995). The highest bulb yield (19,014 kg ha^{-1}) was recorded in Hamedan with the spacing 12.5 × 12.5 cm, while the lowest bulb yield (7572 kg ha^{-1}) was detected in Tarom with the spacing 27.5 × 27.5 cm.

Significant differences between both planting spaces and cultivars in all variables were assessed by Fakhar *et al.* (2019). They recorded cultivar Hamedan had a higher yield and yield components than Tarom and the spacing 12.5 × 12.5 cm had a higher yield than the other spacing.

Castillo *et al.* (1996) studied densities of 104,000 to 400,000 plants/ha under Mediterranean conditions and reported that the favorable effect of high plant stand on garlic yield occurs as a result of an increase in leaf area index (LAI) and its duration. They also recommended handling plant densities from 140,000 to 180,000 plants/ha to ensure a good bulb diameter, given that bulb quality diminishes at high populations. From an economic point of view, the best planting densities for fresh garlic production were between 300,000 to 420,000 plants/ha, given that large sized bulbs with a greater market value are achieved (Castellanos *et al.*, 2004). The best seed size ranged from 3.6 to 6.5 g/clove, corresponding to classes 5 and 7. Only under production conditions that lead to a very high yield potential (above 30 t/ha) can very large seed size (up to 10 g/clove) be expected to have economic advantages. Larger sized seed or higher planting densities may produce significantly greater yields, but with lower market classes, and the lower priced bulbs that decrease overall profit. The best planting method, according to yield, quality and profit, was spaced equidistant by hand with apex up (Castellanos *et al.*, 2004).

8.4.0 Preplanting Storage of Bulbs

Preplanting storage of cloves markedly influence the plant growth, yield and quality of bulbs. Starikova (1977) observed that spring planted garlic stored at –2 to 2°C gave an early crop but the yield was low and of inferior quality. Similarly, bulbs stored at 18-20°C produced plants which grew and developed more slowly but the yield and bulb quality were appreciably higher than from cold stored bulbs. Ledesma *et al.* (1980) reported that the number of cloves from which shoots had emerged 15 days after planting was highest in cloves stored for 2 months at 10°C or 1 month at 5°C and lowest in controls (20°C). Initial growth was more rapid from cloves stored at low temperature than in controls, but it decreased later after bulb initiation giving shorter plants. At harvest, bulb weight and yield were lower in plants from cold stored cloves than from controls and were lowest from those stored at 5°C for 2 months. Contrary to the above findings, low temperature (–2 to 0°C) storage of garlic bulbs resulted in good growth and development of plants which gave high yields, whereas high tempe-rature (18°C) storage resulted in a prolonged vegetative period, reduced maturation and decreased yields (Polishchuk *et al.*, 1982).

Cloves of tropical cv. Bangladesh Local were stored at 5°C (cold treatment) or 20°C (control) for 30 days before planting during summer in the south-east of England. The cloves primordia were initiated 21 and 35 days after planting and differentiation and development were completed in 87 and 117 days in cold treated and control plants, respectively (Rahim and Fordham, 1988). Park and Lee (1989) observed that emergence following storage of seed bulbs of northern and southern ecotypes of garlic in Cheju, Korea was more rapid at low temperature (0-5°C), specially in the southern ecotype. Bulb weight was highest with storage for 30 days at both temperatures of 30-35°C and 0-5°C but number of cloves was highest with storage at high temperature (30-35°C) for 60 days.

Garlic bulbs from 37 clones were stored at 15°C and at room temperature (< 32°C). The visual index of dormancy (VID) was measured, and at 70 per cent, the cloves were planted. Date of 50 per cent sprouting was correlated with harvest date. Yields were lower for clones with shorter growth periods. The specific relative weight of bulbs changed with length of storage (according to genotype), and this can be used as selection index for germplasm. Correlations between dormancy versus bulb diameter and weight and specific relative weight indicated that with long dormancy periods, planting dates were delayed (which in turn reduced the net assimilation period), and as a consequence bulb yields were reduced significantly (Burba *et al.*, 1997). Storage of the bulbs at 4°C for 4 weeks reduced the time from planting to sprout emergence in most clones. If the bulbs were stored at 12-15°C, only 3 clones formed bulbs. In the field, the periods of planting to sprouting and sprouting to bulb initiation were shorter with the later sowing dates in all clones. The time from bulb initiation to maturity showed little variation with planting date (Pozo *et al.*, 1997).

Rendon *et al.* (1997) obtained the highest yield (bulbs with an average dry weight of 28.91 g) with white garlic stored at 15°C and planted early. The effect of storing cultivars Omani, Iranian, Indian and Chinese for 0, 7 or 25°C for 30 days before planting in the field on leaf initiation, and clove differentiation, development and yield was investigated by Satti and Lopez (1994). All cultivars, except Omani, showed the lowest leaf initiation at 25°C than at lower storage temperatures. Bulb diameter was greater at 0 and 7°C for all cultivars. Clove differentiation was highest in Omani and lowest in Indian. Single clove bulbing increased with temperature, reaching 80 per cent at 25°C. Mean bulb weight was highest after storage at 0°C, whilst mean number of days to reach maturity was lower at 9 and 7°C. Bulb yield decreased with the increasing storage temperature. All cultivars, except Indian, showed a 3-fold increase in bulb yield at 0°C compared to 25°C.

In a greenhouse study, California Early and California Late cloves were planted after low-temperature treatment at 4°C for 0 (control), 30, 45, 60 or 75 days. In field studies, cloves from greenhouse-grown bulbs of 3 cultivars (an unnamed local selection, California Early and California Late) were planted, and chilling treatments were similar to those for the greenhouse study. Pre-plant chilling treatment of cloves produced significant increases in cloving and bulb yield for all cultivars. In general, chilling treatment periods exceeding 30 days (for field) and 45 days (for greenhouse) resulted in improved cloving, bulb diameter and bulb yield per plant, particularly in greenhouse-grown garlic. Pre-plant chilling was not a pre-requisite for bulb formation, but it was essential for cloving (Bandara *et al.*, 2000).

The effect of freezing time on the growth and yield of 4 garlic cultivars that originated from tissue culture (Jureia, Gigante Roxo, Gigante de Lavinia and Gravata) and one cultivar that originated from conventional propagation (Gigante Curitibanos), was evaluated by Silva *et al.* (2000). Bulb growth increased with increasing time in the freezer, resulting in early plant production. There was a linear production decrease and an increase in the percentage of non-commercial and oversprouted bulbs as a result of increasing time in the freezer.

The study of Bandara *et al.* (2000) showed that better cloving and higher bulb yields were obtained if garlic cloves were treated at 4°C for 45 or 60 days prior to field or greenhouse planting, respectively. Pyo *et al.* (1979) indicated that the most striking effect of vernalization in garlic was the increase in earliness, especially in cultivars with a greater low-temperature requirement for development (Motaz *et al.*, 1971). Siddique and Rabbani (1985) reported that the treatment of garlic cloves at 6 °C for 50 days before planting increased the bulb size and yield. Nevertheless, much lower temperatures tend to initiate the secondary growth of garlic, which is defined as the lateral bud differentiation into secondary plants on the axils of the outer layer leaves of the primary plant or the continuous growth of the primary plant's clove or bulbil buds into secondary plants instead of turning dormant at the end of the season (Cheng *et al.*, 1991). The secondary growth of the garlic plant significantly affects the commercial performance of fresh scapes and bulbs.

8.5.0 Manuring and Fertilization

Garlic responds very well to organic manure. Nath (1976) and Hazra *et al.* (2001) recommended application of 15-20 tonnes of FYM/ha. It should be applied at the time of field preparation and mixed thoroughly in the soil. In India, a fertilizers dose comprising of nitrogen-phosphorus-potash (100:50:50 kg/ha) along with sulphur at 30-50 kg/ha and zinc sulphate at 20 kg/ha has been recommended for enhancement of garlic yield and quality (Anonymous, 2020). Bogatirenko (1976) recorded a yield of 91 centners/ha by using 40 tonnes of FYM as compared to 79 centners from the untreated field. In Brazil, Pereira *et al.* (1987), however, obtained highest yield (7067 kg/ha) with 20 t compost/ha and it decreased to 6800 kg/ha at 50 t compost/ha.

Rao and Purewal (1957), Purewal and Dargan (1961) and Mathur *et al.* (1975) suggested application of 100 to 115 kg N/ha. Singh and Tewari (1968) obtained the best result in respect of clove number and yield by treatment with 50-100 kg N/ha. Ferrari and Churata (1975) found that the rates of N from 25-75 kg/ha were optimum for bulb size and yield, while application of 67-90 kg N/ha gave better yield (Purse-glove, 1975). Sotomayor (1975a) obtained higher yield by increasing the N rates and plant population but bulb size was best at 15 cm^2 with 256 kg N/ha. Further, no adverse effect of high N rates on storage of garlic was noted. In an investigation on nitrogen levels and plant population, Aljaro and Gacitua (1976) obtained high yields and optimum bulb size from a population of 1000000 plants/ha and by treatment with 150 kg N. Arboleya *et al.* (1994) obtained the marketable bulb yield on irrigated plots ranged from 3.76 t/ha at 112000 plants/ha to 8.1 t/ha at 500000 plants/ha, and from 4.66 t/ha with zero N to 8.04 t/ha at the highest N rate. A similar trend was observed on non-irrigated plots also. Usman *et al.* (2016) recorded significantly higher plant height (cm), number of leaves per plant, leaf area, number of bulbs, and the total yield of the garlic by application of 50 kg N/ha.

A study was conducted in Mexico to determine the appropriate phosphorus (P) amount for garlic cv. Tacascuaro under fertigation. Furrow irrigation was used during 1997-98 and fertigation during 1998-99. P_2O_5 at 80 kg/ha was given for all experimental years, while 240, 285 and 405 kg N/ha was given in 1997, 1998 and 1999-2000 experiments, respectively. Crop densities were 300000 for the furrow-irrigation experiment and 380000 plants/ha for the fertigation experiment in 1998 and 1999. Garlic took up very little P during the first 50 days after planting, but uptake greatly increased after that date. Garlic grown under furrow irrigation took up 64 kg P_2O_5/ha, while under fertigation, the crop took up 89 kg P_2O_5/ha. Their respective crop yields were 19.1 and 29 t/ha. Thus, the higher yield potential of the crop under fertigation increased P demand by the plant by almost 50 per cent (Castellanos *et al.*, 2001).

Potassium (K) is a vital nutrient for increasing garlic yields. Proper application rates and timing are critical for generating a yield or quality response. As crop yield increase, the amount of K required also increases, along with all other nutrients (Mendez *et al.*, 2001). Garlic demand for K ranges from 125 to 180 kg K_2O ha^{-1} (Bertoni, 1988). Total bulb yield was increased with increasing level of K application (Linx and Niwuzhang, 1997; Jiang *et al.*, 1998 and Melzer *et al.*, 1999). Moreover, Al-Moshile (2001) reported that potassium application at 150 kg fed^{-1} reduced the unmarketable bulb yield as compared to control. On the other hand, Sharma (1992) reported that the application of a moderate level of K (40 kg fed^{-1}) significantly increased bulb yield but adding the high level of K (80 kg fed^{-1}) had no further effect. Potassium helps in the root development and increasing the efficiency of leaf in the manufacture of sugar and starch. It is essential for the translocation of sugars. It exerts a balancing role on the effect on both nitrogen and phosphorus.

Recent study revealed that the yield of garlic increased with increasing potassium fertilizer, where K application at 200 kg ha^{-1} is suitable for obtaining maximum yield (Jiku *et al.*, 2020).

Besides the sole effect of NPK, higher yield has also been reported by the combined application of N, P and K. Joshi (1961) observed that a supply of 75 kg N, 85 kg P and 55 kg K along with 50 tonnes of FYM/ha produced high yield. Singh *et al.* (1961) recommended 90 kg each of N and P/ha for increasing the yield of garlic, whereas Mathur *et al.* (1975) did not find any response to phosphorus application. Bogatirenko (1976) reported that 120, 60 and 120 kg of N, P and K, respectively, with 40 tonnes of FYM/ha gave the highest yield of bulbs and aerial bulbils (91 centners/ha). In West Bengal, India the highest yield of multiple clove garlic was obtained by the application of 60 kg each of N and P, and 120 kg K per hectare (Das *et al.*, 1985b). Setty *et al.* (1989) obtained highest yield (7.91 t/ha) and largest bulb diameter (3.67 cm) with application of 100 kg N, 50 kg P_2O_5 and 50 kg K_2O/ha. In Ukraine, Borabash and Kochina (1987) observed highest returns of underground bulbs with 90 kg N, 90 kg P_2O_5 and 90 kg K_2O/ha. Singh *et al.* (2014) obtained maximum number of cloves and highest bulb yield of garlic by using organic manure in the form of vermicompost (15 tonnes/ha) along with inorganic fertilizers (NPK:: 60:36:60 kg/ha). The use of specialty fertilizers (water soluble fertilizers) as NPK :: 19:19:19 at 1 per cent at 30, 45 and 60 DAP followed by NPK::13:0:45 at 1 per cent at 75, 90 and 105 DAP have the potentiality to increase the yield and enhance the storability of garlic bulb (Anonymous, 2020). Priyanshu *et al.* (2019) assessed the effect of integrated nutrient management on growth parameters of garlic. The results showed that application of organic manures in the form of farmyard manure(3 t/ha) and vermicompost (1 t/ha) along with 75 per cent RDF (N:P:K :: 100: 50: 50 kg/ha) and sulphur (40 kg/ha) in presence of the biofertilizer *Azotobacter* and Phosphate solubalizing bacteria emerged best for plant height (80.07 cm), number of leaves per plant (8.37), fresh weight of plants (48.43 g), oven dry weight of plants (7.59 g) and earliness in maturity (127.07 days).

The effect of compound fertilizer has also been studied on garlic. The normal fertilizer rates of 240 kg N, 60 kg P and 200 kg K/ha (control) were compared with 5 other fertilizer treatments by Kusumo and Widjajanto (1973). The best of these five treatments was a compound fertilizer, Rustica Blue supplemented with urea to give 232 kg N, 120 kg P and 170 kg K/ha. Contrary to these results, Aliudin and Suminto (1978) could not find significant differences on growth and yield when compound fertilizers used were Rustica Yellow (15-15-15) at 16 q/ha, Rustica Blue (12-12-17 + 2) at 16 q/ha and NPK (10-15-15) at 20 q/ha, where control plants were given the recommended rates of 240 kg N, 60 kg P and 200 kg K/ha. It is suggested that complete dose of FYM should be broadcasted in the field before planting. At the time of planting half of nitrogen along with full dose of P and K should be used, the remaining half of nitrogen can be applied 30-45 days after planting.

In an experiment on the uptake of nutrients, Ramirezh *et al.* (1973) observed the highest amount of P, K, Ca and Mg in the plants at 50-60 days after planting. Therefore, they suggested fertilizer application before this stage. It was further observed that the best yield corresponded to 0.40 per cent P level in the leaves.

Depriving plants of N at the beginning of bulb growth accelerated and increased the redistribution of organic N from leaves and roots to the bulbs (Bertoni *et al.*, 1992). Cho and Lee (1974) reported that large amount of Ca was found in the leaf during maturation. Potassium was present in similar amount in the leaf and bulb, and decreased during bulb development and increased during maturation. The Mg content was lower than that of Ca and K and it was highest in the leaf during maturation. In the plants grown without Mg there was greater uptake of N, P and K. Calcium had no effect on uptake of N, P and K when Mg was supplied but decreased their uptake in the absence of Mg (Kim *et al.*, 1988). The dry bulb yield of 5.7 t/ha corresponded to the removal per hectare of 134 kg N, 28 kg P, 117 kg K, 46 kg S, 85 kg Ca and 7 kg Mg in the bulbs and leaves (Bertoni *et al.*, 1988). Lazzari (1982) concluded that garlic removed more N from urea than from ammonium sulphate, which indicated preference of garlic to source of nitrogen through urea.

Bulb yield, and uptake rates and bulb concentrations of N, P and S increased significantly with increasing rate of applied S. Amongst the S sources, gypsum and potassium sulphate gave the highest yields and nutrient uptake rates. Sodium sulphate gave the poorest results (Singh *et al.*, 1995). The contents of K, Ca and Mg in the leaves were influenced by the potassium applied at sowing and at 81 days after emergence (DAE); an antagonistic effect being observed in relation to potassium and contents of Ca and Mg in the leaves. The increase of soil potassium level at the sowing stage led to increase of bulb yield, and this was not influenced by potassium rates applied at 81 DAE (Bull *et al.*, 2001). Thangasamy and Chavan (2017) assessed the dry matter accumulation and nutrient uptake pattern of garlic and reported that garlic plants accumulated 84.7 and 84.6 per cent of total nitrogen (N) and potassium (K) from planting to 75 days after planting, whereas, the total phosphorus (P), sulphur (S), zinc (Zn), iron (Fe), manganese (Mn) and copper (Cu) uptake accounted for 59.5–66.7 per cent of total uptake from planting to 75 days after planting which coincided the bulb initiation and development stages.

The effect of S application (0, 40 or 80 kg/ha) on the yield and quality of garlic (cultivars Useong and Namdo) was investigated in the field on soils low in available S in Korea Republic. Application of S at 40 kg/ha significantly increased the yield of garlic. In general, the content of allyl disulfide and 5-methyl-1,2,3-thiodiazole tended to be higher in Namdo than in Useong. At the rate of 40 kg S/ha, the content of allyl disulfide and 5-methyl-1,2,3-thiodiazole was increased 6.5-times in Namdo and 2.1-times in Useong, compared with the controls. Application of K_2SO_4 increased the content of S-containing aromatic compounds by 2- to 3-times when compared with the application of KCl (Park *et al.*, 1997). Patidar *et al.* (2017) obtained maximum bulb yield, TSS and sulphur content of bulb when garlic was raised using vermicompost (4.0 t/ha) and sulphur (50 kg/ha).

Garlic (cv. Peshawar Local) cloves were separated into 2 sizes, <1.0 and >2.3 g, and planted at 8 × 15 cm distance. Controls received no fertilizer; the others were supplied with N (120 kg/ha) and P (90 kg/ha) alone or in combination. The P, as single superphosphate, was applied in full immediately before planting; half the N (as urea) was applied before sowing and half one month later. Yield/ha was not significantly different between plants from large and small cloves but it was higher with than without fertilizer, particularly with N + P (14.23 t/ha) and P alone (12.38 t/ha) (Grad *et al.*, 1993). Das *et al.* (2014) recorded maximum leaf number, bulb weight and yield of garlic by using nitrogen fixing biofertilizers (*Azospirillum lipoferum*), phosphate solubilizing biofertlizers (*Bacillus polymixa*) and potassic solubilizer (*Fraturia aurantea*) inoculated farm yard manure in presence of 100 per cent recommended inorganic fertilizers.

Micronutrients also play an important role in increasing the yield of garlic. Melnik (1973) observed that spraying with $MnSO_4$ at 0.1 per cent, boric acid at 0.02 per cent, $CuSO_4$ at 0.02 per cent or $ZnSO_4$ at 0.02 per cent stimulated dry matter accumulation in the cloves. Soaking cloves in $ZnSO_4$ or $MnSO_4$ solutions induced more vigorous growth of the aerial parts. $MnSO_4$ at 0.01 per cent and $ZnSO_4$ at 0.1 per cent produced the highest yield (Likhatskil, 1987). Singh and Singh (1974) reported that insufficient supply of boron inhibited the growth of plant and reduced the number of cloves per bulb, the optimum concentrations varied from 0.5-1.0 ppm. Ferrari and Churata (1975) found that the rates of borax up to 10 kg/ha increased bulb size and yield, and higher rates proved toxic. In sand culture experiment, Francois (1991) observed that relative yields of garlic were reduced by 2.7 per cent with each unit (mg/l) increase in soil solution of B above 4.3 mg/l. Lad *et al.* (2013) found that application of gypsum (2 t/ha) along with recommended dose of fertilizers produced maximum bulb yield of garlic. Gypsum application increased 26.7 per cent bulb yield over control under Navsari, Gujarat condition. Singh (2008) compared sulfer-95 and gypsum in garlic under sulphur deficient soils of Ranchi, India. Sulfer-95 at 40 kg/ha along with recommended dose of NPK fertilizers emerged best for bulb yield and sulphur uptake by the plants. Chanchan and Hore (2014) obtained maximum bulb yield of garlic by foliar spray of 0.2 per cent Borax with sticker at 45, 60 and 75 days after planting. Rani *et al.* (2018) recorded maximum number of cloves per bulb (37.17) and total bulb yield (155.51 q/ha) of garlic (cv. HG-17) when 0.5 per cent zinc sulphate in combination with 1.5 per cent urea were sprayed along with sticker at 30, 45 and 60 days after planting. Micro minerals (Zn and Fe @ 5 kg ha[-1]) supplementation played a vital role in enhancing the growth, yield and quality of garlic, especially in case of mixture (1:1 ratio) of both Zn and Fe proved to be superior and was very effective to enhance growth and yield of garlic (Alam *et al.*, 2019).

Chanchan *et al.* (2018) reported that under alluvial plains of West Bengal inorganic fertilizers (NPK:: 150: 125: 150 kg/ha) along with bioferlizers *Azospirillum brasilense* and vesicular arbuscular mycorrhizae (*Glomus fasciculatum*) recorded the maximum plant height (74.32 cm), leaf number (11.86), bulb weight (26.34 g) and plot yield (2.76 kg/3 m²) of garlic.

8.5.1 Nutritional Deficiencies

Symptoms due to deficiency of different nutrients as reported by Satyagopal *et al.* (2014) are presented below:

Nitrogen

Leaves become yellowish green erect and upright curled, wilted and dwarf. At maturity tissue above bulbs become soft.

Correction measure: Foliar spray of urea 1 per cent or DAP 2 per cent twice at weekly intervals.

Phosphorus

Slow growth, maturity blazed. Leaf colour becomes light green and bulbs have few dried outer peals. Tip burn in older leaves.

Correction measure: Soil application of recommended dose of phosphorous should be applied at the time of sowing or planting. Foliar spray of DAP 2 per cent twice at fortnightly intervals.

Potassium

Since potassium is very mobile within the plant, symptoms only develop on young leaves in the case of extreme deficiency. Tip burn symptoms, leaves become dark green and erect. Bolting promoted. Older leaves become yellow and necrotic.

Correction measure: Foliar application of K_2SO_4 @ 1 per cent twice at weekly interval.

Sulfur

The leaves show a general overall chlorosis. The yellowing is much more uniform over the entire plant including young leaves.

Correction measure: Foliar spray of K_2SO_4 or $CaSO_4$ @ 1 per cent twice at fortnightly interval.

Manganese

Leaves show tip burn, light coloured and curling. Growth restricted. Bulbing delayed with thick necks.

Correction measure: Foliar spray of $MnSO_4$ @ 0.3 per cent twice at fortnightly interval.

Zinc

Growth restricted. The leaves show interveinal necrosis. In the early stages of zinc deficiency the younger leaves become yellow and pitting develops in the interveinal upper surfaces of the mature leaves. As the deficiency progress these symptoms develop into an intense interveinal necrosis but the main veins remain green.

Correction measure: Foliar spray of $ZnSO_4$ @ 0.5 per cent twice at fortnightly interval.

Iron

Complete yellowing of young leaves. The most common symptom for iron deficiency starts out as an interveinal chlorosis of the youngest leaves, evolves into an overall chlorosis, and ends as a totally bleached leaf. Because iron has a low mobility, iron deficiency symptoms appear first on the youngest leaves. Iron deficiency is strongly associated with calcareous soils, anaerobic conditions, and it is often induced by an excess of heavy metals.

Correction measure: Foliar spray of $FeSO_4$ @ 0.5 per cent.

8.6.0 Irrigation

In general, garlic needs irrigation at an interval of 8 days during vegetative growth and 10-15 days during maturation. Irrigation requirement for green garlic was studied by Dimitrov (1974). The optimum soil moisture for emergence was 80-100 per cent of field capacity. Plants grew fastest, were ready for consumption earlier and produced the highest yield when moisture was maintained at 80-90 per cent. However, keeping quality was poorer than that of plants grown at a lesser soil moisture, because of the large cells and thinner cuticle which led to higher transpiration. For a high yield of good quality garlic, 2 irrigations (20 mm) in August and September, and 3 irrigations (30 mm) in October and November were found optimum (Donnari *et al.*, 1978). Choi *et al.* (1980) reported that garlic yields were increased by 64 and 84 per cent when 30 mm of water were applied at 10-day interval in 1977 and at 5-day interval in 1978, respectively. In the garlic production zone of India, irrigation at 1.5 CPE (Cumulative Pan Evaporation) was recommended for higher yield and quality of garlic bulbs (Anon., 2020). Shiwani *et al.* (2017) compared the effect of withholding irrigation before harvesting on yield contributing characters of garlic cv. HG-17 and found that withholding irrigating 7 days before harvesting produced maximum yield of garlic. For Malwa plateau of Madhya Pradesh, India 100 per cent CPE (cumulative pan evaporation) with three days' irrigation interval was found best in order to get higher marketable and gross bulb yield of garlic (Gupta *et al.*, 2017). On sandy soil, weekly irrigations of an amount equal to the crop evapotranspiration plus losses due to irrigation inefficiencies should be applied (Hanson *et al.*, 2003). On fine-texture soil with the soil-moisture profile at field capacity in February, deficit irrigations can occur without reducing yield.

The water requirement for garlic during the spring irrigation period was estimated to be around 425 mm, of which approximately 25 per cent was contributed by rainfall and the remainder through irrigation (Ayars, 2008). An additional 30 to 40 mm was used from stored soil water, 150 mm of rainfall occurred during the winter, and 115 mm was applied for germination. Statistically, at the 100 per cent water replacement levels, there was no difference in yield between the three types of irrigation systems used in this experiment. Subsurface drip was effective in maintaining soluble solids regardless of the imposed irrigation treatment, which was not the case for the other irrigation systems. The peak crop coefficient was estimated to be approximately 1.3 to 1.4, which is larger than that given in FAO 56 (Allen *et al.*, 1998).

Carrijo *et al.* (1982) studied the effect of different evaporation pan factors (0.4, 0.7, 1.0 and 1.3) under drip irrigation on the productivity in garlic. The highest yield of top grade bulbs in cv. Jureia (58.69 q/ha) and Gigante de Lavinia (78.37 q/ha) were obtained at factors of 1.0 and 1.3, respectively.

In Paraiba State, Brazil, garlic is usually irrigated by sprinkler system using water with varying salt concentrations that may cause damage to plants. It was found that garlic plants were relatively tolerant to salinity at the bulb formation stage and initial growth up to 30 days. During the final stage (90-120 DAP), wetting of the leaves affected the growth of aerial parts and the number of garlic cloves. The salinity levels started affecting aerial parts during the period 30-60 DAP while the bulb was affected only between 60 and 90 DAP. The most sensitive phase of bulb growth to salinity was the last 30 days of the crop cycle (Amorim *et al.*, 2002). In India, Sankar *et al.* (2001) observed that among the different irrigation methods and levels tested, drip irrigation at 100 per cent PE recorded the highest yield of garlic (147.8 q/ha) followed by 75 per cent PE at the same system. Drip irrigation at 100 per cent PE recorded the tallest plant (79.3 cm). Up to 44 and 41 per cent water were saved in the drip and sprinkler systems, respectively. Plants treated with either drip or sprinkler irrigation had higher yield and yield contributing characters than those treated with surface irrigation.

The effects of 4 irrigation frequencies (drip irrigation at 3-, 4-, 5- or 6-day intervals) and 4 N rates (0, 20, 40 or 80 kg N/ha) on garlic cv. Roxo Perola de Cacador, growing in a dark-red podzolic, eutrophic soil in Ilha Solteira, Sao Paulo, Brazil, were evaluated. The bulbs were pretreated for 60 days at 4°C, and N and irrigation were applied from 54 days after planting. Neither irrigation nor N application had a significant effect on commercial yields (up to 1115.68 g/1.6 m^2). The combination of irrigation applied every 3 days plus 20 kg N/ha resulted in the greatest bulb weight (18.40 g) and decreased the percentage of small bulbs (Seno, 1997). In a field experiment in 1991-94 at Junagadh (Gujarat, India), garlic was irrigated at irrigation water : cumulative pan evaporation (IW : CPE) ratios of 1.0, 1.2 or 1.4, and given 25, 50 or 75 kg each of N and P_2O_5/ha. Irrigation at IW: CPE ratio of 1.4 produced the highest mean bulb yield of 4.93 t/ha and the highest net returns. Bulb yield increased up to 50 kg N (4.44 t) application, whereas P application did not significantly affect yield (Sadarai *et al.*, 1997a). They further suggested that higher water availability at 1.2 and 1.4 than at 1.0 IW : CPE increased nutrient availability, and therefore increased growth and productivity. The different N treatments tested had no significant effects on bulb yield, and the effects of P treatments were not clear (Sadaria *et al.*, 1997b). Tripathi *et al.* (2017) stated that adoption of drip irrigation can increased the bulb yield (15–40 per cent), bulb size and storability of bulbs. Besides saving water it reduced weed population, disease infection and labour requirement. In field trials on a dystrophic dark red latosol in Lavras, Minas Gerais, Brazil in 1993, garlic cv. Gigante de Lavinia planted in late April was given 3 rates of applied N (0-120 kg/ha) and 4 rates of K (0-160 kg K_2O/ha) with irrigation corresponding to 60 per cent, 100 per cent or 140 per cent of maximum evapotranspiration (401.5-716.5 mm). Rainfall during the growing season was relatively high and no significant effects of irrigation were observed. Although various effects of N and K on emergence and morphology were noted, total and marketable bulb yields were affected only by N. The highest yield was obtained with 70-76 kg N/ha (2400-4440 kg/ha) (Carvalho *et al.*, 1996). In a trial on a deep sandy loam soil, yields of coloured garlic increased by up to 35 per cent when irrigation frequency was increased from 10 to 25 times during the growing period. Total and commercial yields increased with increase in planting clove size from < 5 g to > 5 g (Lipinski *et al.*, 1995).

In Brazil, Melo *et al.* (1999) observed that plant height increased with water level, reaching 44.0 and 46.0 cm 60 days after planting (DAP), and 45.0 and 47.5 cm 90 days DAP, for 85 per cent and 100 per cent soil water availability, respectively. The greatest bulb weight (14.85 g) was obtained with 95.15 per cent available soil water, while total production was greatest (4604 kg/ha) with 97.93 per cent available soil water. Each percentage increase in available water in the soil corresponded to an increase of 80.35 kg/ha of commercial bulbs and a reduction of 10.71 kg/ha of non-marketable bulbs. Neither the use of bovine manure, nor its interaction with levels of water in soil significantly influenced the crop growth and bulb weight. Silabut *et al.* (2014) studied the response of garlic cultivars to irrigation levels in Mandsaur Madhya Pradesh, India. Result showed garlic cultivar G-323 along with 15 irrigation levels was found superior in terms of plant height, number of leaves, number of cloves per bulb, weight of cloves, fresh bulb weight, dry bulb weight, bulb yield and sulphur content of bulb.

Lipinski *et al.* (1994) observed that 25 irrigations increased yield by 35 per cent over the treatment with 10 irrigations, and the use of large cloves gave a 15 per cent yield advantage over small cloves. N applications gave only a 10-12 per cent yield increment in a soil with an initial total N content of 800-900 mg/kg. The highest commercial yield of 990, 10-kg boxes was obtained with a combination of the large cloves, 25 irrigations and 80 kg N/ha. In trials with garlic (cv. G1) on a clay loam soil at the Regional Research Station, Karnal, India, 5 levels of irrigation (0.5, 0.75, 1.00, 1.25 and 1.50 ID/CPE) and 3 levels of N (50, 100 and 150 kg N/ha) were compared. Half the N was applied as a basal dose and the rest was applied 30 days after planting while P_2O_5 at 50 kg/ha and K_2O at 50 kg/ha were applied as a basal dose. Yields were highest with irrigation at 1.5 ID/CPE (156.0 q/ha) and with N at 150 kg/ha (157.33 q/ha) (Pandey and Singh, 1993).

In semi-arid regions of the world, where limited water reservoirs are getting more valuable due to climate change and increasing demand, implementation of irrigation management techniques such as the optimized regulated deficit irrigation (ORDI) are to be explored (Sanchez-Virosta *et al.*, 2019). ORDI distributes the total available water, based on the needs at each growing stage [in the case of garlic: establishment = Ky(i'), crop development = Ky (i''), bulbification = Ky (ii) and ripening = Ky(iii)].

Their study supports the idea that characterization and monitoring of key physiological traits with fast and low-cost effective methods along the crop cycle can assist regulated deficit irrigation methods and improve irrigation management of garlic in areas with restricted water availability. The specific methodology ORDI was successfully calibrated earlier for purple garlic under the semi-arid conditions of Castilla-La Mancha region (Spain) (Domínguez *et al.*, 2013). Available evidence has shown that impact of water deficit on garlic, depends on its intensity and the growth stage in which it occurs. SangSik *et al.* (2007) stated that both the leaf development and the size of the bulb are affected by water deficit. If the crop suffers water deficit during the formation of the bulb, its size tends to decrease (Fabeiro

Cortés *et al.*, 2003). At very early stages, water deficit causes dehydration of the bulbils. Yet overwatering either at early or late stages can lead to bulb/clove rot (De La Cruz, 2007).

8.7.0 Interculture

Since garlic is a closely planted crop, manual weeding is tedious, expensive and often damages the plants. In East Java, two weedings at 30 and 60 days after planting led to the highest yield of fresh garlic (Aliudin, 1979). Mollejas and Mata (1973) reported that treatment with Simazine-Diuron (0.75-1 kg/ha) as post-emergence and Simazine-Nitrofen (0.75-3 kg/ha) as pre-emergence was 20 per cent more profitable than cultivating the land twice.

Formigoni (1972) observed that the application of Methazole at 3 kg/ha as pre-emergence controlled weeds up to harvest to the extent of 79 per cent. According to Delahousse *et al.* (1973), application of 1 kg oxadiazon/ha gave effective weed control. Treatment with 1.5 kg a.i. oxadiazon/ha resulted in the maximum average plant height, number of leaves per plant at harvest, cloves per bulb, bulb weight and bulb yield in both seasons (Nandal *et al.*, 2001).

Trials conducted by Armellina and Dall (1976) showed that weeds were controlled most effectively by post-emergence application of linuron, prometryne or ioxynil at normal rates. Janyska (1986) reported best weed control and highest yields by post-emergence application of Prope 75 W.P. (methazole) at 3 kg/ha followed by Fusilade W (fluazifopbutyl) at 4 kg/ha about 10 days later. For pre-emergence treatments higher rates were needed (Linuron 2.25 kg/ha, Prometryne 2.8 kg/ha) to control weeds over a long time. However, the high rate of Prometryne caused yield reductions. Linuron 50 EC at 1.50 and 1.00 kg/ha tested during 1995-96 gave comparable results (Sandhu *et al.*, 1997).

Deuber and Fornasier (1980) found that pre-emergence treatments on the day of planting with Pendimethalin at 1.5 kg/ha, Linuron at 1 kg/ha and Oxadiazon at 1 kg/ha gave excellent selective control of both grasses as well as broad leaved weeds, 15 days after treatment. Kumar *et al.* (2013) stated that untreated check or weedy plots reduced 72.5 per cent bulb yield of garlic against the application of pendimethalin (0.75 kg/ha) along with three hand weeding 30, 60, 90 days after planting.

In a study at Jammu, India Sampat *et al.* (2014) recorded highest economic return in weed free plot followed by the combination of oxadiargyl @ 90 g/ha as pre-emergence and *fb* quizalofop-ethyl as post-emergence @ 50 g/ha applied at 2-3 leaf stage of weeds in small and large segmented garlic. Patil *et al.* (2017) reported that application of oxyfluorfen @ 0.150 kg a.i./ha and quizalofop ethyl 0.05 kg a. i./ha as post emergence herbicide recorded lowest weed index (5.12 per cent) and maximum weed control efficiency (84.17 per cent) along with highest bulb yield of garlic. In the irrigated areas of Punjab, Pakistan it is recommended to use Stomp (pendimethalin) 455 G/L CS @ 2000 ml/ha or Hadaf (oxyflourfen) 24EC @ 750 ml/ha as pre-emergence herbicide followed by a hand weeding after 70 days of planting for getting better weed control and higher bulb yield (Ali *et al.*, 2017). In garlic growing zones in India, use of pendimethalin at 3.5 l/ha along with one hand weeding or oxyfluorfen at 0.25 Kg a.i/ha along with one hand weeding are recommended for effective control of weeds in garlic (Anonymous, 2020).

Qasem (1996) reported that the shorter the initial weed-free period, or the longer weeds remained in plots before removal, the greater the reduction in garlic yield and quality. Average reduction in bulb yield over the 3 years was up to 85 per cent. None of the weed-free periods gave a total garlic FW comparable with the weed-free control. Sharma *et al.* (2011) studied the persistence and bioaccumulation of oxyfluorfen residues in soil and garlic crop. Soil samples and garlic bulbs were analyzed by Gas Liquid Chromatography method for oxyfluorfen residues. Results revealed that pre-emergence application of oxyfluorfen dissipated completely in soil at the time of crop harvest and bulb residues were below the prescribed MRL (0.05 mg/kg).

8.8.0 Mulching

Garlic is sensitive to moisture stress and high temperature and found about 60 per cent reduction in yield when it was associated with water stress (Miko *et al.*, 2000). Walters (2008) compared garlic produced on bare soil during the winter and wheat (*Triticum aestivum*) straw mulch in the spring to black plastic. Black plastic provided greater winter protection for garlic (95 per cent survival rate) compared with bare soil (85 per cent survival rate). Greater marketable weights and bulb diameters resulted when garlic was grown in black plastic compared with the bare soil/wheat straw mulch treatment. Islam *et al.* (2007) reported that the effect of black polyethylene and water hyacinth mulch were almost similar on the growth and yield of garlic. They proposed that the water hyacinth and black polyethylene mulch were suitable for increasing garlic production. Karaye and Yakubu (2006) indicated that the number of leaves/plant, weed growth and cured bulb yield responded significantly to mulching and they based on their results for optimum bulb yield in garlic proposed the using of 9 t/ha mulch. The use of water-hyacinth root, rice straw and dried grass as mulches was evaluated for their effects on the growth and yield of late planted garlic (Baten *et al.*, 1995). Plants treated with any kind of mulches under study significantly increased plant height, number of leaves per plant, length of leaf, length of pseudo stem, number of roots per plant, bulb and neck diameter over the control (Baten *et al.*, 1995). These mulches significantly influenced both on chlorophyll-a and chlorophyll-b contents. Bulb length, bulb diameter, clove length, clove diameter, clove number per bulb, 100 clove weight and yield were also significantly higher in plants treated with mulches. All mulches provided good weed control as well.

In Brazil, mulching of cv. Chines with *Paspalum nbotatum* gave a maximum yield of 10.30 t/ha with an average bulb weight of 31.22 g compared with 6.06 t/ha and 19.01 g, respectively, for the unmulched control (Sumi *et al.*, 1986). In Korea Republic, Shin *et al.* (1988) noticed that cv. Namdo gave the best yield when planted with polyethylene film mulch during the last ten days of September. In Bangladesh, Haque *et al.* (2003) compared the efficiency of natural and synthetic mulches on garlic and found that dry water hyacinth mulch enhanced the fresh and dry weight of bulb as well as recorded highest yield of garlic bulb. Najafabadi *et al.* (2012) conducted experiment at the National Rice Research Institute, Rasht, Iran using three kinds of mulches (Transparent and black PE and rice straw). Results showed that garlic total yield, bulb ash percent, TSS, vitamin C and flavonoids content were affected by mulching. Although mulching could improved some quality indices in garlic but no effect on forcing was observed. They recommended the usage of rice straw in rainy and cool season and plastic mulch in low rain fall and warm season for increasing garlic quality as second crop in rice field.

8.9.0 Use of Growth Substances

Bio-regulators play essential roles in plant development by influencing various biochemical and physiological responses. They have been implicated in the enhancement of garlic yield (Memane *et al.*, 2008, Gautam *et al.*, 2014; Singh *et al.*, 2014; Chattopadhyay *et al.*, 2015). Takagi and Aoba (1976) reported that foliar application of GA_3 at 200-400 ppm or treatment of seed bulbs in GA_3 solution at 50-800 ppm stimulated formation of lateral buds. Application of GA_3 at the inflorescence stage enhanced bulblet formation and increased the number of cloves per bulb but delayed leaf formation in storage and inhibited its development. Application of GA_3 at 25 ppm at 45 and 60 days after planting have been recommended to enhance the yield of garlic and improve the quality under Nashik condition of Maharashtra and Karnal conditions of Haryana, India (Anonymous, 2020). Further, it was observed that benzyladenine (BA) at 50 ppm also induced lateral bud formation but its effect was less pronounced than that of GA_3. BA at 50-100 ppm failed to increase bulb and clove weight compared to control. Ethrel at 960-1920 ppm also inhibited plant height and storage leaf formation but increased leaf width. NAA at 50-800 ppm also inhibited the formation and development of storage leaves. Foda *et al.* (1979) obtained the highest yield of garlic bulbs (44 kg/plot of four 5 m long rows) by soaking the cloves for 24 hours before planting in Cycocel at 1000 ppm. Further, they noted the least storage weight loss for 6 months in bulbs from plants treated 3 times (3, 12 and 18 weeks after planting) with cycocel. Patel *et al.* (2019) recorded highest TSS (44.77 per cent) and maximum ascorbic acid (17.02 mg/100 g) content of bulb of garlic variety GC-4 when plants were sprayed with 100 ppm citric acid along with 300 ppm thiourea. It also recorded the maximum storage life (179.33 days) of bulbs. Ram *et al.* (2019) recorded significant increase in growth and yield parameters of garlic as well as reduced PLW of bulbs in storage when cycocel (1000 ppm) was sprayed at 90 days (bulb development stage) and 150 days (50 per cent neck fall stage) of the plants. Meena *et al.* (2016) obtained maximum plant height, number of leaves per plant, bulb yield, dry matter and pungency level of garlic (cv G-282) by soaking of cloves followed by foliar spray of salicylic acid at 200 ppm, where as number of cloves per bulb, total soluble solids (TSS), ascorbic acid, nitrogen and protein content were found maxium by soaking of cloves followed by foliar spray of ethrel at 200 ppm. Abd Elwahed *et al.* (2019) concluded that pre-sowing treatments of garlic cloves by IBA at 100 ppm enhanced the quantity and quality of garlic bulb yield.

Gautam *et al.* (2014) recorded highest bulb yield of garlic (cv. Agrifound Parvati) by planting 3.1–3.5 g clove and foliar spray of 1000 ppm Cycocel. Kumara *et al.* (2014) obtained highest bulb yield of garlic (cv. Vannur Local) by spraying

1000 ppm Cycocel. Singh *et al.* (2008) recorded maximum bulb weight and yield by spraying 600 ppm Cycocel at 50 and 80 days after sowing. Aqueous solutions of both paclobutrazol at 5 ppm and cycocel (CCC) at 1000 ppm had been reported to be effective in accelerating crop development while at the same time increasing bulbs yields (Rahim and Fordham, 1994). It was further concluded that growth retardants could shorten the growth cycle of garlic.

Injecting garlic plant with GA_3 solution induced axillary meristem formation and increased the number of cloves per bulb (Liu *et al.*, 2019). Moreover, soaking seed cloves in 1 mmol L^{-1} GA_3 solution for 24 h not only promoted axillary bud formation and secondary plant growth (equal to tillering or branching), but also slightly increased the number of cloves per bulb and changed bulb structure with a low yield and marketable quality (Hong-jiu *et al.*, 2020). Their findings also suggest that the level of endogenous plant hormone (GA_3, IAA and ZR) cooperates with the content of sugar (sucrose and fructose) in leaf and stem to regulate axillary bud outgrowth in garlic.

9.0 Harvesting and Yield

9.1.0 Harvesting

The crop is ready for harvesting when the tops turn yellowish or brownish and show signs of drying up and bend over. The bulbs begin to mature in about 4-6 months after planting depending on the cultivar, soil, season, *etc.* The bulbs are lifted, cleaned and the leaves are tied at the top. The bulbs are dried for a week or so under shade. Green garlic is harvested like green onion. According to Hinykh (1975), optimum after-ripening period for garlic was 1-3 days. Pruthi (1979) reported that the bulbs should be cured for 3-4 days in the shade before storing them in an ordinary room.

9.2.0 Yield

In India the average yield of garlic varieties ranged between 150 and 225 q/ha. Recovery of clove in the bulbs ranges from 86-96 per cent. Fifty-three genotypes of garlic were evaluated for 7 yield and morphological traits at Ludhiana, India (Thakur *et al.*, 1997). Yields ranged from 7.69 to 16.71 t/ha and bulb weights ranged from 9.9 to 20.8 g. Yield was significantly and positively correlated with weight per bulb, plant height, leaves per plant, and leaf length and width

Bulbs of 10 garlic cultivars were planted at the Federal University of Santa Maria, Brazil, on 27 April, 18 May or 14 July. In all cultivars, marketable yields and the percentage of high quality bulbs were greatest with bulbs planted on 18 May. Marketable yields were highest in cultivars Sao Lourenco, Gigante Inconfidentes, Gigante Lavinia, and Quiteria (12 756, 11 888, 11 854 and 11 212 kg/ha, respectively) (Trevisan *et al.*, 1997). Baiday and Tiwari (1995) reported that the genotypes G61 had the highest bulb yield (6.84 t/ha) and IC25599 the lowest (1.87 t/ha). Yield was highly correlated with bulb weight, bulb diameter, neck diameter and plant height.

Genotype G-282 performed best for bulb yield (258.8 q/ha) and sulphur content (0.896 per cent). Pantnagar Selection-1 performed second highest (215.5

Field Growing of Garlic

Planting of Clove

Irrigation of Garlic

Interculture and Mulching of Garlic

q/ha) for bulb yield. The highest protein content (9.0 per cent) was observed in Pantnagar Selection-6 while the total soluble solids remained highest (40.1 per cent) in Pantnagar Selection-10. The volatile oil content was recorded highest (0.12 per cent) in Rampur Selection followed by G-282 and HG-1 (0.10 per cent in each) (Singh and Tiwari, 1995).

Shinde *et al.* (1999) reported that date of planting was significantly correlated with bulb yield. Bulb weight was the most important morphological factor affecting garlic production and it was significantly positively correlated with garlic yield. Maximum humidity, rainfall, number of leaves per plant and plant height were also significantly correlated with garlic yield. It is concluded that a cool, humid climate in winter is favourable for garlic production.

10.0 Postharvest Management

Many operations are performed for getting mature and quality bulbs from the field to the consumer. About 15-50 per cent losses occur if proper postharvest management practices are not followed. These practices differ from place-to-place. Proper curing, sorting and grading, transportation and storage are essential to minimize these looses. Drying and curing are very essential. Drying is done to remove excess moisture from outer skin and neck to reduce storage rot, while curing is an additional process of drying to remove the excess moisture and to allow the colour development and help the bulbs to become compact and go into dormant stage. It is done for about a week in the field for drying. The method and period of curing vary depending on weather at the time of harvesting. Bulbs are covered along with their tops to avoid damage to bulbs from sun. These are also cured for 7-10 days in shade either with tops or after curing the tops by leaving 2.5 cm above the bulbs and removing the roots. Harvesting at 100 per cent neck fall and curing by windrow method have been recommended. The curing in field till foliage turns yellow should be done. Artificial curing can be done by passing hot air at 27.35°C through the curing room. It takes about 48 hours for complete curing process if humidity is between 60 and 75 per cent.

Garlic bulbs after curing are run over a grader or graded manually before their storage or marketing. The thick- necked, splitted, injured, and diseased or bulbs with hollow cloves are sorted out. Size grading is done after sorting. It is very much necessary for getting better price and to minimize losses on account of drying and decay. Government of India has prescribed certain grade designations for different qualities of garlic for export. The grade designations and definition of different qualities of garlic have been prescribed.

The effect of processing methods (crushing, boiling, microwave treatment) on the major functional components of garlic was investigated by Hong *et al.* (1999). Total organic acid content was higher in the Danyang cultivar compared with the Namdo cultivar. Pyruvic acid content was lowest in samples subjected to boiling and microwave treatment. Alliin content was high in samples subjected to boiling and microwave treatment. Fructan was little affected by processing methods.

The occurrence of spotty symptoms, its cause and prevention were determined for garlic cultivars Euisung and Namdo, collected from 3 major growing areas in

Different Stages of Harvesting and Curing of Garlic

Korea Republic. Spotty symptoms on cloves were classified into 3 types in terms of size, shape and colour. Discolouration appeared as red, brown and purple spots, possibly caused by fungal pathogens. Red-coloured spots were small (hardly detectable by the naked eye); brown and purple spots were larger. Red spots were usually observed at harvest, whereas brown spots appeared during curing period and storage. Purple spots were mostly observed after storage. The incidence of the symptoms differed with cultivar and growing area, and were in the range 9-27 per cent at harvest and 38-41 per cent after storage for 8 months. The incidence of spotty symptoms was reduced by shortened curing period and rapid refrigeration storage (Park and Park, 1997).

10.1.0 Storage

In India, almost 80 per cent of garlic produce is stored by farmers and businessmen for domestic supply throughout the year (Tripathi *et al.*, 2009). Traditional methods of garlic storage lead to physiological loss and deterioration due to pests and insects.

The total losses during storage are of 25–40 per cent under ambient conditions (Tripathi and Lawande, 2006). The postharvest handling, curing, storage conditions are important factors that affect the storage of garlic. These storage problems can be overcome to a great extent by the adoption of improved postharvest technologies and storage structures. Prolonging the storage life of garlic will result in better remunerative prices to both producers and traders. Hence, scientific information on the storage of garlic bulbs is required to reduce postharvest losses.

Proper storage is highly essential for augmenting steady and continuous supply to domestic as well as overseas marketing because of increasing demand for quality garlic products. The main pre- and postharvest factors that affect storage of garlic are very well documented by Petropoulos *et al.* (2017). They reported that genotype, irrigation, fertilizer, chemical agents, and harvest stage are main preharvest factors and bulb handling, curing, storage conditions, and treatments during processing are main postharvest factors. The influence of different storage conditions and parameters on quality characteristics of garlic have been analyzed and reported by Vazquez-Barrios *et al.* (2006) and Pellegrini *et al.* (2000).

For extension of storage life of bulbs, packaging singly in polyethy-lene bags, selection of small-sized bulbs, storage at 0-1.6°C and irradiation with 6 krad of cobalt-60 gamma rays have been recommended by Habibunnisa *et al.* (1971). Use of gamma irradiation with controlled atmosphere condition *i.e.* only less than 0.5 per cent of oxygen or combined with 5–10 per cent carbon dioxide and the spray of maleic hydrazide before harvesting was reported as a viable measure to control sprouting and increase in shelf life (Madhu *et al.*, 2019). Application of irradiation practices has been also extensively studied mostly by considering their capacity to expand the shelf life of garlic (Pellegrini *et al.*, 2000; Perez *et al.*, 2007; Dhall and Ahuja, 2013). Gamma radiation doses of 10 Gy considerably decreased the sprouting in garlic and stopped the process of mitosis (Pellegrini *et al.*, 2000). Perez *et al.* (2007) studied the influence of gamma rays in garlic bulbs and revealed that a 60 Gy dose aggravated a significant decrease in contents of fatty acid and lipids with a simultaneous decrease in the occurrence of sprouting. Irradiation process also used to reduce or avoid the microbial infection during the period of storage, along with to substitute use of chemical fungicides for the duration of the postharvesting period (Thomas, 1999).

Directorate of Onion and Garlic Research (DOGR), India recommended that the garlic bulbs treated with irradiation ranging from 2 to 6 Krad of cobalt 60 gamma rays minimize the sprouting during storage (http://www.dogr.res.in). Properly cured garlic bulbs treated with 60–90 Gy gamma rays can overcome 100 per cent sprouting even after cold storage. Irradiating bulbs before starting of sprouting or in the early 8 weeks of storage can greatly reduce the sprouting, decreases the physiological weight loss and increase the shelf life of the bulbs up to 1 year (http://www.dogr.res.in; Dhall and Ahuja, 2013). Irradiation doses higher than 10 Kr decreases the content of diallyl disulphide which provides garlic flavour (Dhall and Ahuja, 2013). Applying 0.1 per cent carbendazim before harvesting and using a clean and dry place for storing and handling the bulbs also greatly decrease the postharvest losses, especially decay loss (http://nhrdf.org/en-us/pPostHarvestTech_G).

Thoroughly cured garlic bulbs keep fairly well in ordinary well-ventilated rooms. The temperature of 13–18°C and humidity between 40 and 60 per cent has been suggested as ideal for storage of garlic bulbs as pests and diseases are less active at this temperature (https://www. gourmetgarlicgardens.com). Naresh *et al*. (2013) reviewed the influence of storage duration and conditions on bulb losses in onion and garlic. Many of the garlic and onion growers in Khurda and Ganjam districts of Odissa store by binding and hanging them in bunches in the house (Naresh *et al*., 2013). Storage of garlic bulbs after removing tops with nylon-netted bags increases the storage period up to 6–8 months at Nasik and Karnal, India hence the same can be recommended for storage to reduce the postharvest losses of garlic (http://nhrdf. org/en-us/pPostHarvestTech_G). Tripathi *et al*. (2009) studied storage environment and packing method for the storage of garlic bulb and recorded lowest losses of bulb under top and bottom ventilated storage structure.Again heap storage of bulb was found better than the storage in hessian cloth bags.

Garlic can be stored for 1–2 months at 20–30°C (*i.e.*, at room temperature) (Tripathi *et al*., 2009). The garlic bulbs in the ambient storage lose their stiffness or firmness, turn into elastic (spongy) and change in color because of the water loss. Dormancy of bulbs stored at temperature ranging from 5 to 18 °C ends rapidly (Cantwell, 2004). Miedema (1994) revealed that the storage temperatures from 10 to 20 °C initiated sprouting and Takagi (1990) included that the rate of respiration was more noteworthy at temperatures 5, 10, and 15 °C than storage at either 0 or 20 °C temperatures. Well cured and healthy garlic bulbs can be stored at temperature 0 °C and relative humidity ranging from 65 to 70 per cent for 6–7 months with low storage loss. The best temperature and relative humidity for garlic with good quality maintenance are –1 to 0°C and 60–70 per cent relative humidity and can be stored for more than 9 months under these conditions with the provision of good airflow to prevent any moisture accumulation (Cantwell and Suslow, 2002; http://www.cargohandbook. com/index.php/Garlic). Bulbs stored in a cold store at temperatures ranging from –1 to 0°C with 60–70 per cent relative humidity can have long term storage, (http://www.dogr.res.in). Cantwell (2004) also reported that the optimum temperature for garlic storage is from –1 to 0°C.

Volk *et al.* (2004) reported that the properly cured garlic bulbs, when stored at –3°C can be spring planted and utilized around the year. Cold storage of garlic bulbs can be possible at a temperature ranging from 0 to 4°C with 60–70 per cent relative humidity. The losses of garlic bulbs during storage are reportedly 12.5 per cent at temperature ranging from 1 to 5°C with 75 per cent relative humidity compared to 42.4 per cent losses in garlic stored at ambient temperature (http://nhrdf.org/en-us/AreaandProductionReport). Garlic stored under refrigeration condition should not be used for seed purposes. Storage of planting stock at temperatures below 5°C results in rough bulbs, side-shoot sprouting and early maturity, while storage above 18°C results in delayed sprouting and late maturity. Garlic bulbs stored at temperatures ranging from 5 to 18°C can cause most rapid internal sprout development. Ideal storage of garlic bulb for seed is at 10°C temperature with 65–70 per cent relative humidity (Cantwell and Suslow, 2002).

The proper storage system plays an important role in the postharvest handling of garlic bulbs. Storage structures are crucial in maximizing of the shelf life of garlic with desirable quality. There are many local and traditional storage structures or methods at present, but most of these structures have disadvantages like lack of good ventilation facilities. An economically feasible storage structure for small scale farmers to enhance the shelf life of garlic bulbs was developed by the Kota Agricultural University, Rajasthan, India. This storage structure costs around Rs 1 lakh and has the capacity of 90 tons of garlic bulbs in a ventilated chamber separated into eight tiers, with the dimensions of 15 × 30 feet and 600 sq ft area. A similar structure built with cement concrete can cost up to Rs 10 lakhs. This structure can store the bulbs up to 8 months and is made with bamboos in the form of hut (http://www.ukrup.com.ua/en/indian-scientists-have-developed-a-cheap-garlic-storagesystem/).

A low-cost storage structure with bottom and side ventilated single rows has been developed by ICAR-DOGR for small scale garlic growers. It was built with the bamboo or wooden frame with an aeration facility at the bottom side. The side and bottom walls were made with bamboo and the roof was made with thatch from dried leaves sugarcane or grasses. It was estimated that the storage loss of garlic in this structure was up to 35 per cent for 4 months of storage (Lawande, 2018; Tripathi *et al.*, 2009; Tripathi and Lawande, 2006). A double row storage structure with bottom and side ventilation was developed by ICAR-DOGR, MPKV, and NHRDF, India. This storage structure is a permanent or semi-permanent type built with capacity ranging from 25 to 50 tons. These structures are 30–50 ft long, 12 ft wide and consist of two rows. It is a facility with 4 ft free space for walking along the aisles between the rows. The length of the structure should be limited to 50 ft as rotting percentage increases with the increases in length. For providing bottom ventilation these structures are built at 2 ft above the ground level supported by the reinforced cement concrete pillars (Dhall and Ahuja, 2013). Green shed net is used to cover the door and opposite sides, thus protecting them from rain. Asbestos sheets are used to construct the roof of the structure; unfortunately, asbestos is an anachronistic material causing the disease called asbestosis which has led plenty of people to death. The storage losses of garlic bulbs can range from 30 to 40 per cent in 4 months of storage period (Tripathi and Lawande, 2006; Tripathi *et al.*, 2009; Dhall and Ahuja, 2013; Lawande, 2018).

Investigation was carried out at Parbhani, India to find out the effect of planting date and storage conditions on shelf-life of garlic (Sonkamble *et al.*, 2000). Bulbs obtained from corms planted on 20 October exhibited the lowest weight loss during storage. Bulbs could be stored for longer in perforated polythene bags compared with non-perforated bags. Storage decay was reduced if bulbs were stored intact (with leaves)

In a storage study with 13 cultivars for 16 weeks, Horcin and Simekova (1986) observed that respiration rate was negatively correlated with storability and positively correlated with weight loss in storage. Storability was also affected by enzyme activity and the cultivars most suitable for storage had generally low ascorbate and polyphenol oxidase activity. Ascorbic acid content was always higher in irradiated bulbs and it showed a constant level of enzymatic pyruvate whereas the level in non-irradiated garlic dropped by 10 per cent during storage. Iglesias *et al.* (2000) studied the effects of storage conditions [room temperature (21-35°C) with forced ventilation or 4-11°C] and time (up to 11 months) on the texture of irradiated (gamma-radiation, 60-80 Gy) and non-irradiated garlic. Garlic cv. Creole could be preserved at room temperature with forced ventilation for 11 months. In general, the irradiated treatments showed better texture than controls in all storage conditions. Texture decreased with storage period.

Mihailescu *et al.* (1979a, 1979b) noted that storage of garlic at 1.5°C at RH < 75 per cent reduced the storage losses to 12.2 per cent over 150 days from 42.4 per cent at ambient temperature. Treatment with UV light for 30 minutes and then storage at 1.5°C further reduced the loss to 8.6 per cent. Lin *et al.* (1987) reported that after 200 days of storage at –2 to 0°C in an atmosphere of approximately 6 per cent CO_2 and 2-5 per cent O_2, 90 per cent of bulbs was still salable and retained their fresh colour and flavour.

Om and Awasthi (1977) concluded that the storage life of garlic cloves could be prolonged and loss in weight reduced by spraying 3000 ppm MH three weeks before harvest. Omar and Arafa (1979) also applied 2500 or 5000 ppm MH as foliar spray 2 weeks before harvest and observed inhibition of sprouting in storage up to 300 days without any appreciable adverse effects on yield. The treatment also reduced the loss in weight of the bulb during storage. Kumari *et al.* (2013) studied storage life of garlic (cv. Agrifound Parvati) and reported that twice spraying of borax (1000 ppm) at one month and two months before harvesting of the garlic minimized the physiological loss of bulb weight, sprouting incidence, rotting incidence and hollow bulb occurrence during storage of garlic.

Jourdain and Lavigne (1987) reported that bulbs may be stored for 2-3 months at –0.5°C and 80 per cent RH or at 18°C and 70 per cent RH. They, however, found that the use of irradiation or the application of maleic hydrazide before harvest did not prolong the storage period. The limiting factor was development of the tunic, which lowered commercial quality. Madhav *et al.* (2016) studied the quality of modified atmosphere packed minimally processed garlic cloves of two Indian varieties, *viz.*, Yamuna Safed (G-1) and Yamuna Safed-4 (G-323) stored at 10°C and 75–85 per cent RH for 28 days. The result revealed that garlic cloves stored with 1–2 per cent oxygen level and 5 per cent carbon dioxide level emerged best for retaining firmness, colour, total antioxidant capacity, total phenolic content and pyruvic acid of the samples and minimizing the physiological loss in weight and respiration rate throughout the storage period of 28 days irrespective of variety used.

11.0 Diseases and Pests

Most of the common diseases and pests of onion also attack garlic. The control measures that are followed for onion are applicable to garlic.

11.1.0 Diseases

11.1.1 Purple Blotch

Purple blotch (*Alternaria porri*) first appears as small, whitish sunken lesions. Almost immediately, the spots turn brown, enlarge, and become zoned, somewhat sunken, and more or less purplish. The lesions occur on the leaves, flower stalks, and floral parts. The lesion borders are reddish and surrounded by a yellow "halo." If conditions are favorable for disease development, the lesions quickly girdle the leaves and seed stems. Affected leaves and stems may turn yellow, die back, collapse, and die within several weeks after the first lesions appear. In moist weather, diseased tissues are covered with a dense, dark purplish black mold composed of large numbers of microscopic, dark multi-celled spores (conidia). Free moisture, in the form of rain, persistent fog, or dew, is required for infection and spore production. Mycelial growth of the fungus occurs over a temperature range of 6° to 34°C (optimum 25° to 27°C) at a relative humidity of 90 per cent.

The effect of the planting time on the incidence of *Alternaria porri* was evaluated using 11 cultivars at 3 planting times (Ferreira *et al.*, 1995). The cultivars Dourada and Centenario were the most resistant, and the 3rd planting time (end of May) was the most effective for reduction of yield loss. Iprodione and metalaxyl + mancozeb were superior to chlorothalonil, copper oxychloride, mancozeb and zineb in providing control of garlic purple blotch in Himachal Pradesh (Sugha, 1995).

El-Ganaieny (1998) recorded purple blotch and downy mildew (*Peronospora destructor*) was more pronounced on Chinese garlic than on cultivar Balady. Susceptibility of garlic cultivars was decreased when plants were irrigated at long intervals compared with irrigation at short intervals. The highest disease severity was recorded when plants were irrigated at 5-10 days intervals, while the least infection and best growth were recorded when plants were watered at 20-30 days intervals. Ammonium nitrate at 400 kg/feddan resulted in more infection by either purple blotch (42.88 per cent) or downy mildew (45.25 per cent). By contrast, the same rate (400 kg/feddan) of calcium superphosphate reduced infection of purple blotch (13.75 per cent) and downy mildew (15.63 per cent). Least infection by either purple blotch (7 per cent) or downy mildew (4 per cent) and the best yield were expressed when plants were treated with potassium sulphate at 100 kg/feddan and irrigated every 30 days (El-Ganaieny, 1998).

Foliar spray of mancozeb at 0.25 per cent or Ziram at 0.3 per cent with sticker Triton at 0.06 per cent at fortnightly intervals after appearance of disease have been recommended to control the purple blotch (*Alternaria porri*) and *Stemphylium* blight (*Stemphylium vesicarium*). Again 15th October planting of garlic cloves in Nashik, India was found suitable for reducing the prevalence of diseases in garlic varieties Yamuna safed, Yamuna safed-2 and Yamuna safed-3 (Anon., 2020). Kolte and Patale (2019) evaluated different plant extracts for management of purple blotch and *Stemphylium* blight of garlic. Extracts of *Azadirechta indica, Datura stramonium, Moringa oleifera, Annona squamosa* and *Ocimum sanctum* leaves were tested. The leaves extracts of *Azadirechtia indica* showed maximum inhibition against Purple blotch (71.21 per cent) and *Stemphylium* blight (70.06 per cent) followed by the leaves extracts of *Ocimum sanctum*.

11.1.2 *Stemphylium* Blight

The disease appears as small, yellow to orange flecks or streaks on leaf which soon develop into elongated, spindle-shaped to ovate elongate, diffused spots, often reaching the leaf tips. They usually turn gray at the centre, brown to dark olive brown with the development of conidiophores and conidia of the pathogen. The spots frequently coalesce into extended patches blightening the leaves and gradually the entire foliage.

Stemphylium vesicarium was isolated from lesions on garlic leaves showing purple leaf blotch symptoms in commercial garlic farms at Tenterfield, New South Wales and at Waikerie, South Australia. This is the first record of *S. vesicarium* on garlic in Australia. Disease incidence at Waikerie, on three commercially grown garlic cultivars, was 100 per cent with severity of 19-30 per cent leaf area diseased 1 week before harvest. The disease epidemic was influenced by prolonged periods (8 h) of leaf wetness (Suheri and Price, 2000). Pooja *et al.* (2020) studied the effect of predisposing factors like leaf wetness duration, plant age, date of sowing, temperature, relative humidity, total rainfall and number of rainy days on the severity of the *Stemphylium* blight of garlic. They observed that with an increase in the leaf wetness duration and age of plant, corresponding decrease in the incubation period and increase in the rate of disease progress. They recorded the highest disease severity at 25°C (63.11 per cent) followed by 30°C (56.00 per cent) and 20°C (49.78 per cent) and the disease progress ceased altogether at 15°C. The rate of disease severity was found minimum in the late sown crop (16th Nov.) in comparison to early sown crop (10th Oct. or 26th Oct.). Katoch and Kumar (2017) reported that a minimum of 24 hrs of leaf wetness decreased the incubation period while increased the disease severity of *Stemphylium* blight of garlic. Katoch and Kumar (2017) tested the efficacy of dfferent fungicides on *Stemphylium* blight of garlic. Result showed that seed treatment with tebuconazole (Raxil) followed by two foliar sprays of difenoconazole (Score) was found most effctive in *Stemphylium* blight inhibituion under field condition of Himachal Pradesh, India. Sonawane *et al.* (2017) explored different fungicides against stemphylium blight (*Stemphylium vesicarium*) of garlic. Four sprays of Tebuconazole (0.1 per cent) at ten days interval with the onset of disease minimized the leaf blight of garlic (PDI 1.90 per cent) with maximum percent disease control (PDC 89.29 per cent). Khosla *et al.* (2007) found that schedule spray of Companion (carbendazim 12 per cent + mancozeb 63 per cent) - Score (difenconazole 25 EC-) - Indofil M-45 (mancozeb 75 WP) was most effective with highest per cent disease control (91.6 per cent) and maximum economic return (cost benefit ratio-1:3.12).

11.1.3 Cercospora Leaf Blight

Caused by *Cercospora duddiae*, this disease appears on leaves as small, ash-coloured and irregular shaped spots scattered on leaf lamina which later converted to brownish or dark brown and finally black. The spots coalesce gradually and results in blightening of foliage. Spraying of ziram or captan @ 2.0 g/litre of water or copper oxychloride @ 3.0 g/litre of water at fortnightly intervals gives good control.

11.1.4 Powdery Mildew

Powdery mildew is caused by *Leveillula taurica*. Distinct pale-yellow patches of variable size on abaxial surface of leaf associated with powdery mass are its main characteristics. Sulphur fungicides @ 2.0 g/litre of water if sprayed at regular intervals of 15 days after disease appearance gives good control.

11.1.5 Downy Mildew

The symptoms of downy mildew are quite distinct: a whitish, furry growth will appear on the leaves, along with yellow discoloration. It can kill younger plants and stunt the growth of older ones. Diseased leaf tips and other tissues will eventually collapse. Bulbs in storage will have a blackened neck, be shriveled, and outer scales will become water-soaked. Some bulbs may sprout prematurely. Foliar spraying of Zineb (0.2 per cent), Karathane (0.1 per cent) or Tridemorph (0.1 per cent) also gives good control of the disease.

11.1.6 Mosaic

Garlic plants infected with mosaic virus show typical symptoms of chlorotic mottling and strips on first emerging leaf followed by pale-yellow broken stripes, resulting in typical mosaic pattern on matured leaves. Yellowish dots on leaves, whitish leaf margin or twisting of leaves are also recorded on a few cultivars. Generally symptoms are mild on younger leaves than on matured leaves. Bulbs harvested from mosaic affected plants after maturity remains smaller in size and cloves are fewer in number. Since virus is transmitted through aphids, spraying of imidacloprid @ 3.0 ml/10 litres of water or acetamiprid @ 1.5 g/litre of water is useful.

In Darjeeling hills, India a new garlic virus which infects onion was reported by Ahlawat (1974), the incidence being about 30 per cent. Experiments conducted in Siberia demonstrated that virus infection and unfavourable weather conditions resulted in single clove garlic bulbs and a reduction in bolting. Of the cultivars tested, K 2848, 2818 and 2702 were most resistant (Orzhekhovskaya and Starikova, 1990).

In India, most garlic cultivars are infected by filamentous viruses causing mosaic disease. A total of 115 germplasm lines were evaluated during the winter for resistance to mosaic disease. One month after sowing of cloves, disease evaluations were performed at 20-day intervals. A disease incidence of 0-25 per cent was recorded in 8 lines, while line 15 showed a disease incidence of 25-50 per cent (Ghosh and Ahlawat, 1998). The extent of infection by garlic virus X (GVX) of garlic plants was analysed by Northern or immunoblot analyses of individual garlic plants cultivated in different regions. These results showed that almost all of the garlic plants tested from 40 different regions including the USA, China, Japan and Korea Republic were infected with GVX (Song *et al.*, 1997).

Using electron microscopy and ELISA, mixed infections with the following viruses were found in garlic cv. Ptujski Spomladanski : onion yellow dwarf potyvirus (OYDV, garlic strain), leek yellow stripe potyvirus (LYSP, garlic strain), garlic common latent carlavirus, carnation latent carlavirus and mite-borne filamentous viruses (Ravnikar *et al.*, 1996).

Results of a survey on garlic fields at Tegal and Ciwidey, Java, demonstrated symptoms of striping due to OYDV-garlic strain and leek yellow stripe potyvirus-garlic strain (LYSV-G). Mean virus incidences on garlic were 86.5 and 18.5 per cent for OYDV-G and LYSV-G, respectively (Sutarya *et al.*, 1994). Viruses infecting garlic in Venezuela were isolated and characterized. Onion yellow dwarf potyvirus (OYDV) was present in the majority of field-grown garlic clones Criollo venezolano showed to infection of leek yellow stripe potyvirus (LYSV) and OYDV (Marys *et al.*, 1994).

Gawande *et al.* (2013) reported the presence of Onion Yellow Dwarf virus (OYDV) on garlic and related *Allium* spp. in parts of India. Maximum OYDV positives were recorded from Maharashtra (96 per cent) followed by Gujarat (75 per cent) and Madhya Pradesh (75 per cent).

11.1.7 Other Diseases

Nakov *et al.* (1974) detected a new disease of garlic for the first time in Bulgaria caused by *Helminthosporium alliicampamile*. A new leaf blight of garlic was reported to be caused by *Cladosporium delicatulum* (Pal and Basuchaudhury, 1976) and *Stemphylium botryosum* (Singh and Sharma, 1977). A new white rot of garlic disease caused by the fungus *Sclerotium cepivorum* was reported from Himachal Pradesh, India (Gupta and Bharat, 2017). Patale (2019) isolated a number of pathogenic fungi from the infected samples of garlic bulbs. Those included *Aspergillus niger, A. flavs, Alternaria porri, Fusarium oxysporium, Botrytis alli, Penicillium sp., Macrophomina phaseolina, Cladosporium alli, Rhizopus stolonifer* and *Chaetomium sp.*

In Venezuela, a disease characterized by plant wilting and basal rot of the stem has been reported. The causal agent was identified as *Fusarium oxysporum* f. sp. *cepae* and patho-genicity was confirmed by inoculation (Martinez *et al.*, 1996). The symptoms of basal rot are slow to develop. Often, they are seen as a yellowing and eventual dieback of the leaves. Sometimes one can also see white fungal growth at the bulb base, which will lead to both pre and postharvest rotting. Postharvest rotting can include single, several or all of the cloves in the garlic bulb. A moderate temperature of 22 to 28°C favours disease development. Seed treatment with thiram (3 g/kg of seed) and soil application of carbendazim, thiophanate methyl (Topsin-M) or benomyl at 0.1 per cent is effective in the controlling the disease (Mishra *et al.*, 2014).

White rot caused by *Sclerotium cepivorum* is an important of garlic disease first reported in Italy (Walker, 1924). The symptoms of white rot may look almost identical to basal rot, with the exception that the process of disease initiation to plant death is more rapid. Early symptoms include white, fluffy fungal growth on the stem that extends around the bulb base. Small, dark, over-wintering structures called sclerotia form in the decayed tissue. The tolerance and adaptation response to this disease differs among cultivars. Chileno San Javier 1000-7, Hermo-sillo 500-1, Pocitas 500-13 and Blanco de Egipto 750-7 showed the highest tolerance. Pocitas 500-13, Hermosillo 500-7 and Pocitas 750-4 showed the highest yields and Hermosillo 500-6 and Chileno San Javier 750-1 produced the lowest number of cloves per bulb (Perez-Moreno *et al.*, 1995). The incidence of white rot disease in fields at 3 locations in Korea during 1988-91 was up to 85 per cent. Infection was severe on cv. Daeseo and cv. Namdo but not on native cultivars. The optimum temperature for growth was 20°C. Pathogenicity to garlic bulbs was confirmed by inoculation (Cho *et al.*, 1994).

The genotypes IC-49373 and IC-49382 showed high yield potential and moderate disease development and might be suitable for use in crop improvement programmes (Bisht *et al.*, 1993). Lesions appeared early on the highly susceptible cultivars. Rapid progress of disease development was noticed during the last 3 weeks before bulb maturity. Peak severity at crop maturity was significantly higher on highly susceptible genotypes. Individual garlic leaves became more susceptible to purple blotch as they aged and emerging leaves were more susceptible the closer they emerged to bulb maturity. Percentage loss in bulb weight and bulb volume was significantly higher on highly susceptible genotypes. It is proposed that 4 weeks before bulb maturity should be the action threshold for initiation of fungicidal application to prevent damaging levels of disease (Bisht and Agrawal, 1993). White rot (*S. cepivorum*) is a major disease of garlic and onions in the Lockyer Valley in southeastern Queensland, Australia where garlic is a minor industry. The percentage of surviving plants at harvest for the 0.75 and 1.0 treatments of tebuconazole were 91.1 and 92.5 per cent, respectively, in cv. Glenlarge, and 84.2 and 85.8 per cent, respectively, in cv. Southern Glen. Highest marketable yields were obtained from the highest rate of tebuconazole in Glenlarge (7.7 t/ha) and Southern Glen (8.6 t/ha), but in both cases this was not significantly higher than the 0.75 rate (Duff *et al.*, 2001).

Patil and Rane (1973) obtained good bulb weight when cloves infected with white rot of garlic (*Sclerotium cepivorum*) were treated with PCNB or agrosan GN at 6 g/kg, whereas Duchon (1979) recommended dipping of garlic cloves in Ronilan at 0.3 kg/q for the same. Seed treatment with thiram (3 g/kg of seed) and soil application of carbendazim, thiophanate methyl (Topsin-M) or benomyl at 0.1 per cent is effective in the controlling the disease (Mishra *et al.*, 2014).

Geetika and Ahmed (2019) reported that consortium of *Trichoderma* spp was found very effective in controlling the bulb rot (*Fusarium proliferatum*) disease besides increasing the garlic yield. Aguilar (1986) found that mancozeb 80 w.p. at 1.92 kg/ha gave the best control of garlic rust (*Puccinia allii*) and the use of spreader-stickers was suggested for increasing its efficacy (Aguilar *et al.*, 1986). Occurrence of three new rot diseases of stored garlic has been reported by Roy *et al.* (1977). Rath and Mohanty (1987a) recommended the treatment of garlic bulbs with 2 per cent boric acid immediately after harvest to avoid or minimize the storage loss due to *Macrophomina* infection. The best results for the control of *Rhizopus* rot of stored garlic were obtained in another study by Rath and Mohanty (1978b) by treating the cloves with mercuric chloride at 0.1 per cent (if stored for seed) or formalin at 0.3 per cent (if stored for table use). Prasad *et al.* (1986) detected that due to *Aspergillus niger* infection, the bulb fails to emerge or emerges with scanty leaf with decaying bulbs. Treating the bulbs with various fungicides including Dithane Z-78, Dithane M-45, Difolatan, Brassicol and Captan can control the disease. Fusarium rot of stored garlic caused by *F. oxysporum* and *F. solani* can be best controlled by fumigation with formalin (Rath and Mohanty, 1986).

In high humidity region most of the stored garlic is contaminated by the *Penicillium* fungus known as blue mold or green mold resulting in water-soaked spots, followed by white patches of mycelium which turn blue or green in colour. *Penicillium hirsutum* and *Penicillium allii* are two predominant species identified for blue mold.

Several abnormalities of physiological disorders have also been observed in garlic. The sprouting of garlic bulbs in the field before harvest as found by Garcia (1980) was more common in white than in purple cultivars. Further, the sprouting was associated with excessive water and nitrogen supply and was reduced by dense planting.

11.2.0 Pests

11.2.1 Mite

The garlic mite, *Aceria tulipae* (Keifer), also known as the dry bulb mite, tulip mite or onion leaf mite, was described from specimens collected from tulip bulbs in Sacramento by W. B. Carter in 1937, which was originated in Holland (Keifer, 1938). Later, it was found in bulbs of garlic and onion (Keifer, 1952).

The damage signs due to mite attack appeared as twisting and curling of leaves, which did not open properly, creating a micro-environment on the upper leaf surface where all the biological stages of the mite, namely eggs, nymphs and adults, colonized along the mid-rib (Debnath and Karmakar, 2013). The infested leaves typically arched downwards with the tip tucked into the next young leaves (Channabasavanna, 1966). In heavy infestations, leaves showed yellow streaks, most often along the mid-rib and leaf margin. Yamashita *et al.* (1996) and Koo *et al.* (1998) demonstrated that mosaic and streak symptoms were induced by the garlic mite-borne mosaic virus of the genus *Rymovirus* (Potyviridae) vectored by *A. tulipae*. Mite is also known to cause severe crop losses to garlic in all production areas

Diseases of Garlic

 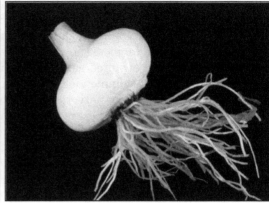

White Rot (Left), and Pink Root Rot (Right) infection

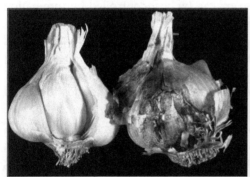 

Black Mold (Left), and Blue Mold (Right) Infection

Damping Off Infection in Nursery Bed

Diseases of Garlic

Stemphylium Leaf Blight Infection

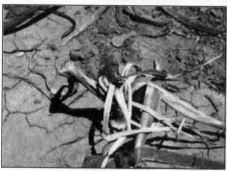

Diamond shaped viral lesions

Bacterial Soft Rot   **Irish Yellow Spot Virus**

Botrytis Neck Rot

Diseases of Garlic

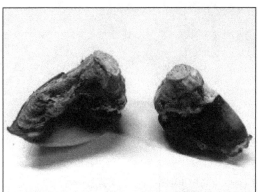 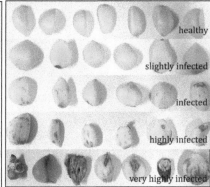

healthy

slightly infected

infected

highly infected

very highly infected

Fusarium Bulb and Basal Rot

Rust Infection in Garlic

Downy Mildew Infection   **Garlic Yellow Dwarf Virus**   **Embellisia Skin Blotch**

around the world, reducing yields up to 23 to 32 per cent (Larrain, 1986; Debnath and Karmakar, 2013). The results of the 2-year study on the varietal responses to garlic mite revealed that the mite initiated its population during the second fortnight of December and attained the maximum level during February before dwindling (Debnath and Karmakar, 2013). Very few mite populations were encountered during the second week of March because of the prevalence of adverse weather conditions and also due to the crop maturity. All the varieties were found infested by the mite; Goldana, Gangajali and G-323 were most susceptible, and Katki and G-50 were found to be the most tolerant. Pawar *et al.* (1990) also conducted on varietal susceptibility and reported that out of the seven varieties tested, G-41, IC-49383 and G-1 were the most tolerant to *A. tulipae* infestations.

Treatment of garlic cloves a day before planting with Ekatin was observed to be useful against mites (Doreste, 1965). Cindea (1982) treated the garlic bulbs before planting for 15 minutes with Hostathion at 0.05 per cent or Nemafos at 0.15 per cent which gave very good control of garlic bulb mites (*Rhizoglyphus echinopus*) to the extent of 94.4-97.2 per cent. The garlic yield from treated bulbs was 7.6-7.7 t/ha compared to 4.6 t/ha in the control.

Of 6 pesticidal treatments tested against *Aceria tulipae* on garlic in Maharashtra, India, dipping of seed in wettable sulphur (0.3 per cent) + dimethoate (0.03 per cent) resulted in the minimum number of mites ($5.0/cm^2$ leaf) in a pre-sowing experiment. Post-sowing, the same pesticides applied as foliar sprays resulted in the lowest number of mites of $4.7/cm^2$ leaf. These treatments also gave the highest yields (138.7 q/ha pre-sowing and 159.2 q/ha post-sowing) (Katkar *et al.*, 1998). Ten days after spraying dicofol (2.5 ml/l) was found to be the most effective acaricide against garlic mite and very low population (1.40 mites per cm^2) was found in this treatment closely followed by ethion @ 1.0 ml/l (3.0 mites per cm^2) (Bala *et al.*, 2015).

11.2.2 Aphid

In an experiment, Nadejde (1977) found that malathion at 500 g a.i./ha, dimefox at 350 g a.i./ha and pirimicarb at 250 g a.i./ha were effective to control peach aphid on garlic without any phytotoxicity.

11.2.3 Storage Pests

Stored garlic was observed to be infested and the bulblets eaten from inside by larvae of *Ephestia elutella* (Bhardwaj and Thakur, 1974). Cosenza *et al.* (1981) treated garlic bulbs by fumigation with posphine at 1-4 tablets/m3 for 12 hours. The treatment every 2 months had no adverse effect on clove sprouting and gave good control of five storage pests.

10.2.4 Thrips

Thrips (*Thrips tabaci* Lindman) is a serious and major biological constraint in garlic production causing heavy economical loss, if infestation starts at bulb initiation stage (Patel and Patel, 2012). Thrips prefers to feed on newly emerged leaves in the center of neck, therefore, majority of thrips are found at the base of the youngest leaves in the lower center of the neck. In case of severe infestation, the bulbs remain undersized and distorted (Butani and Verma, 1976). The pest is responsible for curling of leaves, low yield, and poor quality of bulbs. Leaf curling reduces the activity of photosynthesis and thus reduces the crop yield. According to Changela (1993), losses of 15.35 to 46.82 per cent in garlic bulb yield was recorded due to infestation of this pest.

The susceptibility of 10 garlic genotypes to attack by *T. tabaci* was studied in the field in Sao Paulo, Brazil. The results showed high variability among the genotypes, with Piaui being the least infested and Dourados, Franca and Jureia, the most infested. Correlation analysis showed that the genotypes with the highest medium number of leaves and biggest leaf area had the highest thrips population (Bortoli *et al.*, 1995). Hossain *et al.* (2014a) recorded the lowest thrips population and higher bulb yield in genotype GC0034 in Bangladesh.

In Pakistan, it has been reported that the infestation of *T. tabaci* began on 2 December and continued to the end of April varying from 0.02 to 2.50 thrips/plant. The peak infestation was recorded from 28 March to 28 April. The variety Kund and Akora attracted the least number of thrips, while CZ 49315 and Yogo 383821 attracted the greatest number. Methamidophos was the most effective insecticide for control followed by dicrotophos and endosulfan while cypermethrin and mono-crotophos were the least effective (Hussain *et al.*, 1997).

The incidence of *T. tabaci* was studied in 3 varieties grown in Jaboticabal, Sao Paulo, under 4 nitrogen regimes (0, 50, 100 and 150 kg/ha) with/without mulching and with/without application of deltamethrin (30 ml/1 water). Dourados, which had the highest thrips population (5.32 thrips/plant), even with the pesticide treatment, differed significantly in this respect from the other varieties, Cabaceiras and Centenario, with respectively 1.35 and 0.85 thrips/plant. Mulching did not affect the thrips population, which was greatest at around 100 kg N/ha (Oliveira *et al.*, 1995). They further reported that the thrips population was related to the leaf insertion angle; the greater the insertion angle the lower the thrips population on the plant. Cabaceiras and Centenario showed greater insertion angles (23.08o and 25.7o, respectively) and lower thrips population densities, compared with 17.86o for Dourados. It is suggested that the smaller angles may offer greater protection to the thrips against bad weather conditions (*e.g.*, rain, wind and strong sunlight) (Oliveira *et al.*, 1995b). Gajera *et al.* (2009) evaluated the myco-insecticides on against *Thrips tabaci* infesting garlic. Mortality of *Thrips tabaci* was found maximum with *Verticillium lecanii* (7 g/l) followed by *Beauveria bassiana* (7 g/l).

Bajpai *et al.* (2014) recorded lowest infested of thrips (7.9 per cent) and mites (3.75 per cent) in garlic plants by split application of potash (100 kg/ha), 50 per cent as a basal dose and remaining 50 per cent at 60 days age of the crop.

Installation of sticky white trap along with spraying of spinosad (Tracer 45 SC) @ 0.4 ml/l may be recommended for effective management of thrips in garlic (Hossain *et al.*, 2014b).

11.2.5 Nematodes

The stem and bulb nematode (*Ditylenchus dipsaci*) is a serious pest of commercial garlic. The infections can arise from planting in nematode-infested soil or more

Pests of Garlic

Leek Moth Infestation

Blister Beetle

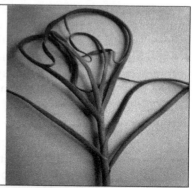

Damage Caused by Garlic Mite, *Aceria tulipae*

commonly from planting infected garlic seed cloves. Planting clean nematode-free seed into non-infested soil is the best option to avoid the nematode. A four years crop rotation with non-susceptible crops can be practised to suppress the nematode population in the soil. Again dipping of infested cloves in a hot water at 49°C for 20 minutes also helps to kill the nematodes present in cloves.

Barber (1977) described the symptoms of damage by nematode, *Ditylenchus dipsaci*. The first report of stem and bulb nematode in garlic was documented in New Mexico by French *et al.* (2017). According to Cucchi *et al.* (1986), nematodes in garlic can be controlled by immersing the bulbs in solution of bidrin, phosdrin or mixture of both or metasystox. More than 98 per cent control of this nematode was obtained by Cindea (1979, 1980) when cloves were treated with nemafos 46 at 0.15 per cent before planting or vydate at 0.2 per cent for 15 minutes, but heat treatment had an adverse effect on growth. In a four-year trial by Romascu (1977), excellent nematode control was obtained with several nematicides like temik and nemafos each at 30 kg/ha applied to the soil before or at planting.

The effect of pigeon manure, poultry manure and sawdust (all at 900 kg/ha 2 weeks before sowing of garlic bulbs) on the population density of *Helicotylenchus indicus*, *Hoplolaimus seinhorsti* and *Merlinius brevidens* was determined in field plots in Karachi, Pakistan. Pigeon and poultry manures were markedly more effective in controlling the nematodes than sowdust (Aly-Khan *et al.*, 1997).

12.0 Seed Production

Since this crop is mainly propagated by cloves, well grown, uniform composite bulbs of a particular cultivar are selected and the cloves are sorted out. Healthy and uniform cloves are selected and used for planting. The planting details including cultural practices followed for the multiplication of bulbs are the same as that for the production of garlic bulbs for consumption. However, Nath (1976) has recommended that the cloves may be planted 7.5-12.5 cm in rows which are spaced 40 cm apart for seed production. In former Czekoslovakia, for seed production, over-wintering in the field and a density of 100000 to 134000 plants/ha were recommended (Luzny, 1987).

Garlic is known as a sterile plant. However, following an expedition to former Soviet Central Asia in 1983, several of the clones collected were found to be fertile. One clone collected in Kirgiz showed male sterility, and was found to produce the highest number of seeds among the collected clones. The seeds produced from this male sterile clone in the following year also showed high germination rates in different crosses. More than half of the F_1 seedlings between this male sterile clone and other fertile clones also showed pollen sterility (Etoh and Galmarini, 1997).

Pooler and Simon (1994) reported that despite a long history of obligate vegetative propagation, selected garlic clones can produce sexual seeds. By removing vegetative topsets from the inflorescence and cutting inflorescences from the underground bulb, 63 germinable seeds were produced from 11 garlic clones. Isoenzyme analysis of the seedlings confirmed their syngamic origin. The generation of new recombinants through sexual reproduction could have a major impact on garlic production worldwide.

Nematode Infestation in Garlic

From left to right, two garlic plants exhibiting severe symptoms of *Ditylenchus dipsaci* infection compared with healthy plant on the right. The plants exhibit leaf chlorosis and necrosis with a progressive loss of functioning roots. Close-up photos of bulb symptoms shown at right (French *et al.*, 2017).

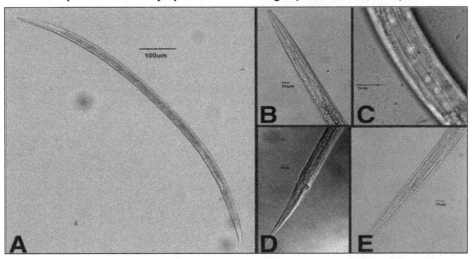

Microscopic photos of representative nematodes recovered from symptomatic garlic plants: (A) female; (B) esophagus with metacorpus and stylet; (C) 4 incisures in lateral field; (D and E) male and female tail shapes, respectively.

13.0 Crop Improvement

Garlic is a diploid (2n=2x=16) predominantly cross-fertilizing species. Garlic however has lost its blooming potential and fertility already millenia ago and thus vegetative propagation is the sole production system used. Consequently garlic breeding has been limited to the selection of the pre-existing genetic variability and increase in garlic variability was attempted via mutation breeding and *in vitro* techniques, however with very limited successes. In recent years increase in garlic variability is sought via sexual hybridization and genetic transformation. In all cases, an efficient mass propagation system is needed for multiplication of selected elite material for commercial purposes.

Since garlic is propagated exclusively by vegetative methods clonal selection of local existing types in various regions is the important method of improvement. In the hill districts of Uttar Pradesh, cv. Chauballci was found to· be a very promising clonal selection in respect of bulb size and yield. In different regions of India good number of local strains is available. They vary in number of cloves, ranging from 16-50/bulb and the size of bulb (Nath, 1976). There is very little variation in shape, whereas most of the strains have white colour, few have reddish tinge. Special efforts are needed for the improvement of these strains.

Due to apomictic nature of garlic, exploration and clonal selection have been the most widely implemented breeding strategies in India. The Directorate of Onion and Garlic Research (DOGR), Rajgurunagar; Vivekananda Parvatiya Krishi Anusandhan (VPKAS), Almora; Central Institute of Temperate Horticulture (CITH), Srinagar; Mahatama Phule Krishi Vidyapeeth (MPKV), Rahuri and National Horticulture Research and Development Foundation (NHRDF), New Delhi have been instrumental in developing and identify clones suitable for different day length conditions of the country after rigorous evaluation through All India Network Research Project on Onion and Garlic (AINRPOG).

The National Bureau of Plant Genetic Resources (NBPGR), India has collected over 220 accessions of onion and garlic, as well as the wild relatives *A. ampeloprasum*, *A. auriculatum, A. ascalonicum, A. carolinianum, A. chinense, A. wallachii, A. tuberosum* and *A. rubellum* (Singh *et al.*, 1994). The Bureau has introduced over 1100 *Allium* accessions, including improved cultivars, germplasm collections and related species from over 40 countries. These valuable collections, both indigenous and exotic, have been established, multiplied and are in the process of evaluation and characterization at different locations. Some of them have been identified as tolerant/resistant to diseases such as purple blotch (*Alternaria* sp.), *Stemphylium* blight and garlic mosaic carlavirus (Singh *et al.*, 1994).

Despite the importance of crop, limited breeding work has been done so far. As a first step of systemic breeding programme, collection and evaluation of germplasm is required. The adequacy of germplasm collection is determined by the amount of genetic variability present in the germplasm. Assessment of variability present in these genotypes is helpful in selection of suitable genotype. Correlation estimates between bulb yield and its components are useful in developing suitable selection criteria for selecting desired plant type or developing high yielding varieties. Path analysis is helpful in choosing the character(s) that have direct effect on yield. Improvement techniques like polyploidy, mutation breeding and tissue culture have proved beneficial in increasing yield of garlic.

13.1.0 Horticultural Diversity and Characterization of Genetic Resources

During a long cultivation history, garlic plants were grown in diverse climatic and biogeographic regions. They exhibit wide variations in bulb size, shape and color, number and size of cloves, peeling ability, maturity date, flavor and pungency, bolting capacity, and numbers and sizes of topsets and flowers in the inflorescence (McCollum, 1976; Astley *et al.*, 1982; Astley, 1990; Hong and Etoh 1996; Lallemand *et al.*, 1997; Kamenetsky *et al.*, 2005; Meredith, 2008). A strong interaction between genotype and environment has led to a variety of phenotypic expressions (Lallemand *et al.*, 1997; Portela, 2001; Kamenetsky *et al.*, 2004b; Meredith, 2008).

Based on temperature and day-length response, garlic has been classified as having long-day and short day varieties. Depending on the ability to develop a flower stem, garlic producers distinguish between softneck and hardneck varieties (Engeland, 1991; Meredith, 2008). Hard-neck varieties bolt and flower but these flowers are usually sterile, while soft-neck varieties do not flower at all. Hard neck varieties cannot be braided for storage whereas softneck varieties can be braided and stored. Hard neck (long-day varieties) is characterized by big bulbs, less number of cloves (10-15), ease of peeling and, generally, has low storage life. Typical examples are Agrifound Parvati and Chinese garlic. Because of big size, their productivity is higher and these fetch a good price in local and international markets. Soft-neck (short-day) varieties are characterized by small bulbs, more number of cloves (20-45), more aroma and are, generally, good storers *e.g.*, Indian garlic varieties G41, G1, G50, G282, *etc.* (Lawande *et al.*, 2009).

However, from a physiological point of view, the terminology *bolters* and *nonbolters* is more accurate. Depending on the traits of scape elongation and inflorescence development, garlic varieties were classified by Takagi (1990) as: (i) *nonbolters*, which normally do not form a flower stalk or produce cloves inside an incomplete scape; (ii) *incomplete bolters*, which produce a thin, short flower stalk, bear only a few large topsets, and usually form no flowers; and (iii) *complete bolters*, which produce a long, thick flower stalk, with many topsets and flowers.

Based on morphological and physiological phenotype, worldwide garlic cultivars were classified into several horticultural groups, reflecting the broad diversity of the crop. The group named Purple Stripe, which includes bolting hardneck cultivars, is considered to be genetically closest to the origin of garlic. The other groups include the Artichoke, Asiatic, Creole, Glazed Purple Stripe, Marbled Purple Stripe, Middle Eastern, Porcelain, Rocambole, Silverskin, and Turban types (Meredith, 2008).

Knowledge about the morphological characteristics of garlic germplasm is most important for selection of the appropriate varieties for appropriate place and achievement of high yield. Besides, characterization is an important aspect for documentation of the performance of the studied cultivars, which subsequently will help to introduce, select and improve the existing varieties (Brar and Gill, 2000; Dhakulkar *et al.*, 2009; Dyduch and Najda, 2001; Chand *et al.*, 2010; Kumar, 2015; Salahuddin *et al.*, 2019). Prajapati *et al.* (2018) evaluated thirty diverse garlic genotypes for morphological characters. The genotype JAS 28 recorded the highest value for most of the quantitative traits such as, leaf length (36.33 cm), leaf width (1.74 cm), pseudostem diameter (0.84 cm), average weight of bulb (27.0 g), marketable yield (117.0 q/ha) and total yield (118.0 t/ha). However the maximum plant height (62.20 cm), number of cloves/bulb (23.66) and average weight of bulb (27.0 g) were recorded by the genotype JAS 16. Seven local bolting ecotypes of winter garlic were assessed by Nurzynska-Wierdak (1997a) for morphological and quality properties. Significant differences were observed for plant height, number of leaves, plant weight, weight of bulbs and weight of inflorescences. Ecotypes R and X had the largest bulbs, cloves, topsets and total yield of bulbs. They further reported that Ecotypes P, Y and A were notable for their high contents of sugars, L-ascorbic acid and mineral components. Ecotypes X, Y and Z showed a low tendency to accumulate heavy metals compared to other ecotypes, and were therefore recommended for cultivation in polluted regions of Poland (Nurzynska-Wierdak, 1997b). The number of cloves per bulb (CPB) in garlic shows a normal distribution in each cultivar and gives a rather constant average and modal value per cultivar. On the basis of this knowledge, CPB was used as the main selection criteria with Chileno-type. It was found that use of CPB as a changing value according to the average of every new generation was a very effective method to reduce the number of CPB and hence improve the quality (Heredia-Zepeda *et al.*, 1997a). After a few cycles of selection using number of cloves per bulb (CPB) as the main selection criterion in Taiwan-type garlic, reduction was more than desired (2.55 CPB), so that procedure previously used for the Chileno-type was modified. From the original Taiwan population, only bulbs with less CPB than the average of the original population were selected during the three first cycles of selection. There were sufficient differences in emergence, plant height, and bulb shelf-life in eight of the selections to justify their acceptance as new varieties. Tacatzcuaro, Tinguindin, Tocumbo, Texcoco, Huerteno, Chapingo-94, INIFAP-94 and Celayense were released to growers (Heredia-Zepeda *et al.*, 1997b). At INIFAT a breeding programme was developed with the aim of obtaining germ-plasm adapted to Cuban conditions : latitude 19-23oN; mean monthly temperature during the growing period of 21-23°C, and day length of 11 h at the beginning of the bulb development period. A clone collected in Sancti Spiritus province with some degree of adaptation to the country, one from Celaya, Mexico, and another introduced from Guadalupe were employed as germplasm. Best results were obtained with the selection of bulbs with the greatest diameter. With this method it was possible to increase mean yields from 3 to 7 t/ha (Munoz de Con *et al.*, 1997). Kumar *et al.* (1994) estimated the phenotypic stability for 20 genotypes by growing them under 3 artificially created environments (by changing the dose of N) in Dholi, Bihar, India and evaluating 5 economic characters. Significant genotype × environment interac-tion was observed for all the characters. The genotypes 78-13, 78-15, 78-4, 80-5 and 78-5 had regression value of more than 1.5 in respect of average weight of cloves. However, for yield (weight of bulb), the genotypes 78-5, 78-15, 43, 78-13, PG20 and 78-20 had higher potential irrespective of environment.

Characterization of germplasm collections based on their bulb concentration of phytochemicals and traits associated with garlic nutraceutical properties is important for the identification and selection of garlic clones with high functional value (Barboza *et al.*, 2020). This is particularly relevant in this species, for which conventional breeding strategies by means of sexual reproduction are not readily feasible and, therefore, its improvement largely relies on the identification and selection of new variants followed by their asexual propagation. They (Barboza *et al.*, 2020) found significant and broad variation among 73 accessions for all three phytochemicals (pyruvate, total phenolocs and total solids), and garlic genotypes with high levels of these compounds were identified. Their results also suggest that it will be possible to select for mild and pungent garlics, as well as for garlics with high functional value and longer shelf-life both for fresh consumption and for industrial dehydration. They also observed significant variation for phytochemicals content among different classes of botanical variety, flowering behavior and ecophysiological group.

Gupta and Mahajan (2019) stated that DUS test (Distinctness, Uniformity and Stability) are essential for registering new varieties of garlic under Protection of Plant Varieties and Farmers Rights Act 2001 (PPV and FRA) in India. The variety must be clearly distinguishable by one or more essential characters from any other existing varieties. DUS test guide lines for garlic was notified by PPV and FRA in 2009 in India with 32 DUS test characters for garlic varieties. The candidate variety was tested for DUS characters in two seasons at two locations; whereas farmers' variety was tested only in one season at two locations.

13.2.0 Selection Indices

Studies on genetic variability in garlic indicated high phenotypic coefficient of variation (PCV) and genotypic coefficient of variation (GCV) for average weight of cloves, plant height, bulb weight, number of cloves/bulb (Shaha *et al.*, 1990; Agrawal, 1999). The variation for different morphological characters, yield and yield components among different genotype of garlic have been reported by Vijay (1990) and Kohli and Fageria (1992). High heritability along with high genetic advance was observed for width of leaf, neck thickness of bulb, plant height, average weight of clove, number of cloves per bulb, bulb diameter and bulb weight (Korla *et al.*, 1981; Shaha *et al.*, 1990; Agrawal, 1999; Kohli and Prabal, 2000; Singh and Chand, 2004; Khar *et al.*, 2005; Singh *et al.*, 2018).

The genetic potentialities of yield contributing characters and their interrelationship should properly be assessed for improving the crop. In garlic, Moravec *et al.* (1974) found positive correlation between bulb and clove weights. The number of cloves/bulb showed significant positive correlation with bulb weights. Trippel and Chubrikova (1976) observed a direct correlation between bulb yield and bulb size, in which they studied more than 700 clones. More than double yield was obtained when cloves utilized for planting were from large and medium-sized bulbs than from small bulbs. Plant height, bulb weight, average weight of clove and bulb diameter showed significant positive correlation with bulb yield (Korla and Rastogi, 1979; Kalloo *et al.*, 1982; Shaha *et al.*, 1990; Agrawal, 1999). Number of cloves per bulb showed significant positive correlation with yield (Shaha *et al.*, 1990; Kohli and Mahajan, 1993; Agrawal, 1999). Kohli and Mahajan (1993) reported that dry leaf weight was positively and significantly correlated with yield. In an experiment, Selvaraj *et al.* (1997) observed that plant height was significantly positively correlated with bulb yield and was the greatest direct contributor. Its indirect effect on bulb yield was mainly via bulb diameter, followed by number of cloves and leaf length. The existence of genetic variability in "white" garlic was confirmed along with the possibility of further selection for clove number and bulb weight, and also further increases in bulb equatorial diameter through indirect selection for bulb weight (Lopez-Frasca *et al.*, 1997).The correlation between yield, leaf size and bulb diameter was positive and highly significant (Kohli and Prabal, 2000). Singh *et al.* (2017) observed that bulb yield is positively associated with plant height, number of leaves per plant and number of cloves per bulb, whereas bulb diameter, bulb weight and number of cloves were positively correlated with each other, therefore selection of plants having bigger cloves, thick, broad and flat leaves will helps to achieve maximum yield efficiency of garlic.

Shaha and Kale (1999) identified polar bulb diameter and weight of 50 cloves were the major yield-contributing characters in garlic. Kalloo *et al.* (1982) observed direct contribution of bulb weight, clove number and plant height during 1978 and those of plant height, bulb diameter, weight and number of cloves per bulb during 1979 towards yield. Agrawal (1999) reported that bulb weight had the highest positive direct effect on the bulb yield followed by bulb diameter, plant height and weight of 10 cloves. Singh (1987) reported that clove weight and leaf length had the maximum positive direct effect on bulb weight. It is suggested that genotypic selection for improving bulb yield in garlic should be based on clove weight and leaf length. These informations may be of immense utility for further crop improvement programme. Agarwal and Tiwari (2009) recorded that bulb weight, clove length and leaf area index have positive direct effect on bulb yield, hence selection of genotypes bearing bigger and heavier bulbs and cloves along with broad leaves will be beneficial.

13.3.0 Genetic Divergence

Controversy on the origin of garlic cultivars and clones obtained through selection from germplasm collected in Southern Brazil led to an analysis of electrophoretic variants for proteins and 5 enzyme systems in 35 native and introduced clones from Uruguay and Argentina. Patterns of leaf protein, root and leaf esterase, and root peroxidase allowed the classification of the cultivars and clones into 8 groups. These groups were divided further into subgroups based on morphological and agronomic characteristics. Clones introduced from Uruguay and Argentina showed a high degree of genetic similarity with the most important garlic cultivars grown in Southern Brazil (Augustin and Garcia, 1993).

Ten quantitative characters were studied by multivariate analysis in 64 local [Korean] and 32 foreign cultivars of garlic (Hwang, 1993). The cultivars were classified into 3 groups by bolting habit, namely completely bolting, incompletely bolting and non-bolting types. By principal component analysis, 3 main components accounted for 83.5 per cent of total variation. By single linkage cluster analysis, local cultivars could be classified into 4 or 8 groups and the foreign cultivars into 8 groups. These cultivar groups corresponded with geographical distribution (Hwang, 1993).

Menezes-Sobrinho *et al.* (1999) reported that principal components analysis identified the following traits as best differentiating the clusters : plant height with erect leaves at 60 days; plant height with normal leaves at 60 days; leaf insertion angle at 90 days; bulb colour; number of bulblets from sieves 1, 2 and 4; weight of bulblets from sieve 2; average weight of bulbs at harvest; and average bulb weight at threshing. The most important features for distinguishing group genotypes by canonic variables analysis were : plant height with erect leaves at 120 days; plant height with normal leaves at 60 days; leaf insertion angle at 90 days; number of leaves at 60 and 90 days; number of bulbs at harvest; total number of bulbs; average weight of plants on day 1 and day 60; and average weight of bulbs. A multivariate study based on 17 phenological characteristics of 66 garlic clones was carried out by Matus *et al.* (1999) to evaluate individual and group variations and to identify the most relevant characters for distinguishing them. Four principal components (PCs) were identified which together explained 63.8 per cent of the variation. The first PC (29.9 per cent of the variation) was associated principally with duration of the growth period, quality characteristics of the bulb (diameter and weight of bulb) and plant characteristics (height and number of green leaves 150 days after planting). The second PC (13.5 per cent of the variation) was associated only with percentage of plants with secondary growth of cloves.

Cluster analysis and principal component analysis were used to investigate the genetic diversity of 40 garlic germplasms analyzed with 23 sequence-related amplified polymorphism (SRAP) primer combinations (Chen *et al.*, 2013). Cluster analysis revealed that the 40 germplasms could be divided into 3 groups. The results of principal component analysis were consistent with those of unweighted pair-group method with arithmetic averages (UPGMA) clustering analysis. The Shannon-Weaver information index ranged from 0.2419 to 0.4202, indicating that the garlic germplasms had high genetic diversity. In most of the previous studies the grouping pattern of genoptypes suggested no parallelism between genetic divergence and geographical distribution of genotypes (Sheikh and Khandy, 2008; Swaroop, 2010; Singh and Dubey, 2011; Singh *et al.*, 2012; Yadav *et al.*, 2018).

13.4.0 Fertility Restoration and Production of Seeds

The sexual sterility of garlic inhibits or markedly reduces possibilities of improvement of economically important traits, including pest and disease resistance, yield, and quailty, through breeding. Restoration of fertility and, therefore, of sexual reproduction would permit genetic studies and classical breeding in garlic. In addition, fast propagation of desired genotypes *via* true seeds would be expected to result in reduction of storage costs and fewer injuries caused in the production field by viruses, diseases and pests transmitted by infected propagules. Many researchers attempted to restore fertility in garlic (Novak and Havranek, 1975; Katarzhin and Katarzhin, 1978; Konvicka, 1984; Etoh, 1985; Etoh *et al.*, 1988; Etoh, 1997; Etoh and Simon, 2002; Kamenetsky *et al.*, 2004a, 2004b; Simon and Jenderek, 2004).

Under the same environmental conditions, garlic accessions differ significantly in morphological and physiological traits, including leaf number prior to bolting, flowering date (date of spathe opening), final stem length, as well as in flower/ topset ratio and pollen viability (Etoh, 1985, 1986; Takagi, 1990; Pooler and Simon, 1993a, 1993b; Engeland, 1995; Etoh and Simon, 2002; Simon, 2003; Kamenetsky *et al.*, 2004b; Simon and Jenderek, 2004; Kamenetsky *et al.*, 2005). Marked variation among garlic clones with regard to flowering ability and flower to topset ratio led Etoh (1985) to propose that garlic undergoes a process of transition from sexual to asexual reproduction. Accordingly, ancestral garlic had normal meiosis, was fertile, and developed numerous flowers in the long-scaped umbel. Compared with modern clones, ancestral plants probably had greater adaptation to a variety of climatic conditions, a larger numbers of foliage leaves, and diverse maturation dates. The status of garlic fertility prior to domestication is at present unclear. However, fertility restoration by decapitation, removal of topsets and/or environmental manipulation clearly indicates that the genetic cascade coding for flowering remained intact and is not impaired. In some garlic genotypes gradual accumulation of chromosomal mutations during millennia of vegetative propagation resulted in complete sterility, shorter scapes and fewer flower buds and topsets. Domestication and subsequent cultivation and sterility of garlic have probably been accelerated by selection for larger bulbs.

In the early 1980s, expeditions to Central Asia, collected a number of garlic accessions in Uzbekistan, Tadjikistan, Kirgizistan, and Kazakhstan (Etoh *et al.*, 1988). The collected plants were grown at Kagoshima, Japan, and following topset excision, 17 clones developed fertile flowers with over 3,000 viable seeds. One of the clones was male sterile. Later, more fertile garlic plants were found in Armenia, Georgia, and Sin-Kiang (Etoh *et al.*, 1991). For induction of flowering, Tizio (1979) suggested that gibberelic acid with adenine or biotin could stimulate normal development of some flowers on pieces of garlic flower-stalk growth *in vitro*, while inhibiting formation of aerial bulbils on the inflorescence. However, no seeds were produced. Pooler and Simon (1994) improved floral production and seed set by scape decapitation and removal of topsets, but seed germination was low and ranged between 10 and 12 per cent. Later, Inaba *et al.* (1995) and Jenderek (1998) obtained 50,000 and 1.2 million garlic seeds, respectively. They identified 27 clones as highly fertile, producing over 400 seeds per umbel, with seed germination of 67 to 93 per cent. Removal of topsets was necessary only in the early generations, as the strong selection pressure for blooming and seed production resulted in improved seed set. Fertile accessions were identified in the USDA garlic collections (Jenderek and Hannan, 2000, 2004), producing 0-85 seeds per umbel in the first propagation cycle. Major blooming traits, including flower stem appearance, spathe opening, umbel shape, and the number of flowers per umbel were stable and similar across populations evaluated.

The first step towards marker-assisted selection of fertile garlic clones was made by Etoh and Hong (2001) using RAPDs. They screened twelve pollen fertile and sterile clones using 60 10-mer primers and found two RAPD markers, which amplified only in the pollen fertile clones. These two markers were further tested on 60 clones, and found to be correlated to pollen fertility.

A search for genes involved in the control of flowering in garlic, resulted in identification of garlic *LEAFY/FLO* homologue, *gaLFY* which is expressed in both flowering and non-flowering genotypes, however further comparative analyses of gene expression revealed two *gaLFY* transcripts, differing in 64 nucleotides, with clear splicing borders (Rotem *et al.*, 2007). The short one was common in both genotypes throughout their development, whereas the long variant appears in the flowering genotype only during the reproductive phase. Thus, the phenotypic differences in garlic, with regard to flowering, may be associated with the efficacy of the splicing process.

The discovery of fertile plants on one hand and the development of a sophisticated environmental system (Kamenetsky *et al.*, 2004b) for fertility restoration, thus enable in depth studies on the genetics of garlic, on flowering and fertility in plants, as well as classic and molecular breeding work.

Production of true seed in garlic (*Allium sativum* and *A. longicuspis*) has been known with the discovery of fertile clones by Etoh (1986) that efforts were started to induce flowering and seeds in garlic. With the advent of flowering garlic, Jenderek and Hannan (2004) were able to evaluate reproductive characteristics and true seed production in garlic germplasm and were successful at producing S1 bulbs in a few fertile clones. This represented valuable material for studies on garlic genetics (Jenderek, 2004). Jenderek and Zewdie (2005) studied within and between family variability for important bulb and plant traits and observed that bulb weight, number of cloves, and clove weight were the main factors contributing to yield, and concluded that vegetative propagation of garlic over the centuries had produced highly heterozygous plants. Koul *et al.* (1979) studied prospects for garlic improvement in the light of its genetic and breeding systems and Simon and Jenderek (2004) made a comprehensive review about flowering, seed production and genesis of garlic breeding.

Garlic accessions in the Gaterleben, Germany collection were surveyed for their poolen fertility. Of the 35 accessions in which pollen was counted, 15 accessions showed some degree of pollen fertility, 14 of which originated from Central Asia, Caucasus or Russia. These accessions are useful for garlic breeding. Meiosis was observed in only 5 accessions. All of them showed regular chromosome pairing at meiosis and contained 8 bivalent chromosomes. Of these, 2 accessions came from Europe, Italy and France. It was confirmed that there exist some garlic clones with regular meiosis in Europe (Gtoh *et al.*, 2001). The different configurations of chromosome synapsis, such as tetravalents, hexava-lents and octovalents, were observed at meiosis of pollen mother cells in 4 garlic varieties. The formation of the multivalents may result from chromosome reciprocal translocation among the non-homologous chromosomes. The karyotype analysis showed that the varieties with multivalents at meiosis also showed variation in the Karyotypes (Zhang *et al.*, 1999).

Kamenetsky *et al.* (2004) evaluated 115 accessions of garlic collected from Central Asia for bolting, umbel formation, pollen viability, stigma receptivity, embryo viability and seedling development. Five accessions were found to fulfill all above criteria of being classified as fertile accessions for use in creating genetic variability and act as virus free garlic propagation material. In addition, sexually reproducing accessions that set true seed have also been developed but are mainly being used for research purpose as of yet (Etoh and Ogura, 1997).

Most garlic accessions and commercial cultivars either do not flower or have sterile pistils and/or stamens or exhibit floral abortion. This restriction also necessitates the use of cloves asb propagating material, which is cumbersome and cost intensive approach to garlic propagation and production. The various possible morphological, physiological, genetic, anatomical and molecular reasons of apomixes have been attributed to floral abnormalities (Khar, 2012), sterile hybridity from cross between two ancestral parents (Koul and Gohil, 1970), competition of floral buds with vegetative topsets (Novak, 1972), tapetal degeneration (Konvicka and Die, 1973), interference of degenerative like diseases with sexual reproduction (Konvicka *et al.*, 1978; Pooler and Simon, 1994) *etc.*

Bozzini *et al.* (1991) identified an Italian fertilie tetraploid (2n = 32) line of garlic which produced a floral scape and fertile seeds.

Cultivated garlic, being non-flowering, has limited variability. Breeders depend upon natural clonal mutations and selection of superior clones from the germplasm. There is considerable interest in improving garlic cultivars but, until the causes of sterility are determined and overcome (Koul and Gohil, 1970), the only feasible breeding method is the slow one of selecting spontaneous or induced mutations and somaclonal variation to broaden germplasm.

13.5.0 Mutagenesis and Somaclonal Variation

Garlic being a sterile and vegetatively propagated plant, genetic variation can be induced only by somaclonal variation, induced mutations or genetic transformation (Novak, 1990; Kondo *et al.*, 2000). Natural occurring variation originates from two sources namely genetic recombination and mutation. Domestication of garlic involved the selection of genotypes with strong vegetative development and bulb production. This selection process decreased generative reproduction and resulted in the development of almost only asexually propagating garlic cultivars. How long this situation is already occuring is not known, but it can most probably safely be stated that genetic recombination did not contribute much to the increase of variation in the garlic genepool over the last thousands of years. No information is available on the effects of natural occuring mutations in garlic. Reports dealing with mutation breeding research in garlic are limited.

Choudhary and Dnyanansagar (1982) exposed to various doses of X-ray on cloves of garlic and a range of concentrations of ethyl methones sulphonate (EMS), diethyl sulphate (DES) and ethyleneimine (EI). In the second and third vegetative generations 16 types of morphological mutant were recorded at various frequences. The most effective agent was X-ray followed by EI, EMS and DES. Gohil and Koul (1984) found that irradiation of garlic with different doses of X-rays, reduced mitotic activity and induced chromosome aberrations. Dedul *et al.* (1985) observed that as the X-rays dose increased, the rate of shoot production declined. At low doses (1-3 Gy) there were little differences between irridiation plants and control in growth rate and seed production. A-12 Gy dose proved lethal for garlic before their observations were recorded in M_1 and M_2 generations. Garlic clone INIFAT RM-2 was obtained by gamma-radiation-induced mutation from commercial clone Guadalupe 15. It has better yield characteristics than its parent variety, tolerance to *A. porri* and good postharvest storage (Talavera *et al.*, 1999). Perez-Talavera (2001) determined natural variability of Guadalupe-15 garlic clone and radio-induced variability in MV3 plants irradiated with 6 Gy gamma radiation. Low natural variability was observed, while irradiation resulted in an increase in variability. Small (<8 mm) and large (>8 mm) cloves of 5 garlic varieties were irradiated with 4, 8, 12, 16 and 20 Gy and plant height was evaluated 25 days after planting. Radiosensitivity varied significantly between cloves of different sizes (Talavera and Cervantes, 1999). Al-Safadi *et al.* (2000) showed that via the use of gamma irradiation (4-7 Gray) the resistance to white rot (caused by *Sclerotium cepivorum*) could be improved. In cultivar 'Kisswany' the infection by white rot could be reduced to 3 per cent as compared to 29 per cent in the control, in cultivar Yabroudy these figures amounted less than 5 per cent white rot infection for mutant lines versus 20 per cent infection for the control. Recently, Taner *et al.* (2005) determined the effective dosis for Cesium-137 around 4.5 Gray for mutation breeding in garlic.

Induction of somaclonal variation in garlic and its use in plant breeding was scarcely carried out. *In vitro* culture was primarily aimed at the rapid propagation of elite material (Novak, 1990). Novak (1990) concluded that direct organogenesis in the absence of a differentiating callus stage in *in vitro* culture of various garlic tissues, like bulb plate, inflorescences and apical meristems, resulted in genetically uniform plants. A callus phase on the other hand often leads to somaclonal variation. In this context, Al-Zahim *et al.* (1999) showed that basal plate callus cultures generated significant somaclonal variation of five garlic cultivars as determined by RAPD and karyotype analyses. Via RAPD analyses it was found that the frequency of variant RAPD fragments was around 0.63 per cent. Karyotype analyses indicated that 16 per cent of the regenerants were abnormal (non-diploids). Comparing these data with data from other plant species indicated that stress due to *in vitro* conditions has a considerable impact on the genome constitution of garlic (Al-Zahim *et al.*, 1999). An *in vitro*-selected somaclonal variant of garlic was characterized by increased vigour and high productivity, both unchanged after five cycles of field cultivation. Somaclonal variant metaphase cells had the same chromosome number (2n=16) as those of a control; however, their Feulgen-DNA values in metaphases and interphase nuclei were different from those of the control. A 4.5 per cent increase in DNA content, possibly due to gene amplification, was observed in the somaclonal variant metaphases (Vidal *et al.*, 1993).

13.6.0 Disease Resistance

Garlic is highly susceptible to a wide range of viruses that include Onion Yellow Dwarf Virus (OYDV), Leek Yellow Stripe, Iris Yellow Spot Virus (IYSV), shallot latent viruses and Garlic mosaic viruses. Common fungal and bacterial diseases that are Stemphylium leaf blight (*Stemphylium vesicarium*), downy mildew (*Peronospora destructor*), purple blotch (*Alternaria porri*), Fusarium basal plate rot (*Fusarium oxysporum* f. sp. *cepae*), white rot (*Sclerotium cepivorum*), Colletotrichum blight/anthracnose (*Colletotrichum gleosporoides*), black mould (*Aspergillus* sp.), bacterial soft rot (*Erwinia carotovora* and others), bacterial brown rot/slippery skin

(*Pseudomonas aeruginosa*) and the major insect pests are onion thrips (*Thrips tabaci*) and onion maggots (*Delia antiqua*). However, the prevalence of viruses in long day conditions is lesser but it is nevertheless affected by Stemphylium leaf blight, downy mildew and occasionally purple botch with onion thrips being its major insect pest. Significant laboratory success has been achieved in elimination of viruses and pests by tissue culture techniques like meristem culture, somaclonal variations, and thermotherapy. Proper utilization of resistant or clean end products from these experiments will be helpful in developing virus free/resistant cultivars in future (Bruna, 1997; Ebi *et al.*, 2000; Walkey *et al.*, 1987; Rout *et al.*, 2014).

Indian workers have succeeded in characterizing and analyzing candidate gene linked to *Fusarium* basal plate rot disease which helps in developing molecular markers linked to this disease and screening resistant plants/cultivars at seedling stage (Gupta *et al.*, 2013). ICAR-DOGR, Rajgurunagar has developed commercial rapid detection RT-PCR kits for *Iris Yellow Spot Virus* and *Onion Yellow Dwarf Virus* useful for rapid and high through put detection of diseases in garlic crop samples. Few reports on discovery of new viruses (*Leek Yellow Stripe Virus*) infecting garlic in India have also been reported (Gawande *et al.*, 2014). Reaction of garlic genotypes to purple blothch in South India revealed that Bhima Purple, Yamuna Safed-2, Yamuna Safed-3, Yamuna Safed-4, Yamuna Safed-5, Yamuna Safed-8, Yamuna Safed-9, Baram Local-06, Jamnagar Local, Mandsaur Local, Ranebennur Local, Ooty Local, GRS-1330, GN14-25, GN14-15, DWG-2 and DWG-1 were found resistant against purple blotch (Asiya Kowser *et al.*, 2019). In Pakistan, cultivar Alladher local was found resistant to rust while it displayed moderate susceptibility to purple blotch. None of the cultivars except Hazro showed acceptable resistance levels to both diseases and prior recommending it for wider cultivation; this cultivar should be evaluated at multi-location hot spots in order to determine its true genetic potential (Alam *et al.*, 2007).

13.7.0 Polyploidy *In vitro*

Attempts were made to induce polyploids by colchicine treatment in clones of Namdo-garlic. Calli were induced by culturing shoot apexes in MS medium supplemented with 2.0 mg l-1 2,4-D and 30 mg l-1 sucrose. They were suspension-cultured with liquid media containing different concentrations of colchicine 0.1, 1.0, 10, 20 or 50 mM for different period of 2, 4, 6, 8 and 10 days. The optimum concentration and period of colchicine treatment for inducing polyploidy was 1.0 mM for 6-10 days. After acclimatization, initial growth of polyploid was slower than that of diploid but its growth speed recovered in the late stage (Jang *et al.*, 2000). A simple protocol for garlic *in vitro* management was described by Barandiaran *et al.* (1999). It comprised of a single medium for all developmental stages and 23 genotypes tested. The immature bulbs were used as source of axillary buds. Although genetic variability existed among the different accessions tested, for both the multiplication rate and bulblet size, acceptable values of multiplication were reached for all the accessions, and values of bulb formation approached 100 per cent in the shoots produced.

14.0 Biotechnology

Because of the difficulties in conventional breeding methods, garlic has attracted the attention of biotechnology. Virus elimination has been main consideration in micropropagation. Recently, developments have been made in DNA marker technology and genetic transformation.

14.1.0 Tissue Culture

14.1.1 Micropropagation and *In vitro* Bulb Formation

Kehr and Schaeffer (1976) could successfully produce differentiated plantlets by tissue culture from growing points of bulbs. Bhojwani (1979) obtained healthy garlic plants to the extent of 70 per cent by shoot proliferation with the help of NAA. In another study, Bhojwani *et al.* (1982) produced virus-free garlic plants from micropropagation of shoot tips. Regeneration of virus free plantlet by shoot tip culture was observed to be best in a medium containing BA, IBA and GA_3. Rooting was better with Gamborg and Skylark nutrients (Bertaccini *et al.*, 1986).

Makowska and Kotlinska (2001) found that the best kind of explant giving the highest rate of regeneration was shoot tips isolated from garlic bulbils. The best medium in the first step of micropropagation was MS, supplemented with 0.1 mg IAA/dm3 and 0.1 mg kinetin/dm3 and in regeneration step, MS supplemented with 0.1 mg NAA/dm3 and 0.5 mg 2iP/dm3.

Xu *et al.* (2000) investigated the various factors affecting the efficiency of rapid multiplication of virus-free garlic by inflorescence meristem culture and multi-bulbil induction. A total of 5000 multi-shoots or multi-bulbils were obtained from one pedicel per year through 4 cycles of inflorescence meristem culture. Garlic genotype and culture medium significantly affected the number of multi-shoots, while the daylength affected multi-bulbil induction. Xiong *et al.* (2000) presented the *in vitro* propagation of virus-free *A. sativum* plantlets. The effects of explant (rachis) developmental stage, cultivar, pH, BA and NAA on plant growth and proliferation rate were investigated. The best medium for rachis culture was B5 supplemented with BA at 2 mg/l and NAA at 0.1 mg/l at pH 6.5. Plant growth and proliferation varied with cultivar. Suitable subculture conditions included MB medium supplemented with BA at 2 mg/l, NAA at 0.1 mg/l, gibberellic acid at 0.05 mg/l, and sucrose at 20 g/l at pH 6.2.

Garlic bulbils obtained from Poland were grouped according to size : large (0.8 cm in diameter) – accession numbers (AN) 242, 357 and 333; medium-size (0.6 cm) – AN 230, 183 and 189; and small (0.4 cm) – AN 180, 330 and 171. Shoot apices with three primordial leaves from garlic bulbils were isolated and placed in modified MS 1 medium supplemented with 0.1 mg/dm3 IAA and kinetin with 3 per cent sucrose at 20°C for 3 months. The plants were transferred into a modified MS medium (MS2) containing 0.1 mg/dm3 NAA, and 0.5 mg/dm3 isopentenyladenine. The regeneration factor for each clone was determined. Root formation was evident after 2 weeks in explants placed in MS1. AN 180 gave the highest survival of explants (100 per cent), followed by AN 230 (90 per cent). AN 333 developed the highest number of shoots (9.3) (Makowska and Kotlinska, 2000).

Root tips culture from aseptically sprouted cloves of garlic produced 40 explants per clove. Shoots were induced from 95 per cent of explants. An average 7.5 bulblets were formed per explant. The bulblets sprouted and formed plants when transferred to soil. The present protocol seems to be an effective alternative of seed clove system for rapid multiplication of virus-free garlic clones through continuous bulblet formation (Haque *et al.*, 2000). Virus-free materials of *A. sativum* were obtained from *in vitro* shoot apex culture. The results indicated that the best medium for inducing callus was MS + BA 0.2 mg/l + NAA 0.5 mg/l. The best medium for multiplying clumpy buds was B5 + BA 0.5 mg/l + IAA 0.2 mg/l. Virus-free materials were observed repeatedly under the electron microscope (Gao *et al.*, 2000). Successful plant regeneration of garlic through root tip culture was reported by Robledo *et al.* (2000). In the protocol, root apices from *in vitro* cultured garlic cloves were used as axenic explants for organogenic callus production and plant regeneration.

Resende *et al.* (1998) reported that five garlic clones, numbered 8, 9, 14, 15 and 28, of cv. Gigante Roxo derived from tissue culture exhibited degeneration after 4 generations of standard propagation in the field. In the first field reproduction, the tissue culture clones were superior to the standard for all the traits analysed. After the fourth generation, however, only clones 8, 14 and 15 were superior to the standard for height and average number of leaves/plant, whereas clones 8 and 28 were equal to the standard for those traits. In the first generation, clones 28, 14, 15 and 8 gave yields which were 64 per cent, 49 per cent, 52 per cent and 44 per cent higher than that of the standard, but this decreased to 27 per cent, 44 per cent, 38 per cent and 19 per cent lower than the standard by the fourth generation. Only clones 9 showed no change in yield performance throughout the experiment. Identical behaviour was observed for average bulb weight. They further evaluated growth and production of micropropagated plants (apical meristem culture) compared with plants conventionally multiplied. As for tissue culture multiplication, cultivars Gravata and Gigante Roxao were the most productive, while for conventional multiplication the cultivars Gravata and Lavinia presented the highest production (Resende *et al.*, 2000).

Single shoot development from isolated garlic cv. Ptujski Jesenski meristems occurred on a medium without plant growth regulators. After thermotherapy, the percentage of successful development of meristems into plantlets was c. 55 per cent and the yield of onion yellow dwarf potyvirus (OYDV) and carnation latent carlavirus (CLV)-free plants was 88-100 per cent. The largest increase in bulb weight was observed in virus-free garlic, which reached commercial size in the same vegetative season (Ravnikar *et al.*, 1994). Screening for virus indicated that scape tip culture provides an effective method for elimination of garlic mosaic carlavirus and could be used for eliminating viruses from vegetatively propagated species (Ma *et al.*, 1994).

Mohamed-Yasseen *et al.* (1994) described a procedure to regenerate shoots and bulbs *in vitro* with high frequency from shoot-tips of garlic using BA or TDA. Regenerated shoots were induced to form bulbs on MS medium containing 5 g activated charcoal and 120 g sucrose/l under a long-day photoperiod. Bulbs formed *in vitro* were transferred to soil without acclimatization and produced viable plants.

This method could be useful to produce low cost bulbs which are easy to handle and store until needed. Conditions for bulblet formation vary between genotypes (Naga-kubo *et al.*, 1993). Matsubara *et al.* (1990) reported that virus-free garlic bulbs yielded 93.5 g bulblets/bulb compared with 64.4 g from virus-infected bulbs and also produced more bulblets with a high average bulblet weight. Zel *et al.* (1997) studied the effect of sucrose, jasmonic acid (JA) and darkness on bulb formation of garlic cv. Ptujski Jesenski. For bulb induction, explants with developed shoots were transferred onto medium with 3 per cent or 8 per cent sucrose in the presence or absence of 5 µM JA. Sucrose (8 per cent) significantly increased the proportion of shoots which formed bulbs by 86-90 per cent, bulb diameter and the number of bulbs per basal plate.

A novel procedure for rapid and cyclic clonal propagation of garlic through *in vitro* bulblet formation was developed by Haque *et al.* (1998b). In contrast to previously employed micropropagation methods that make use of a single shoot tip, basal plate or bulb scale explant per clove, the procedure described, used more than 40 root tips per clove of the garlic cv. White Roppen as starting material. Use of roots as explants allowed continuous bulblet formation. This method has the potential to produce 5838 bulblets from a single clove after 2 cycles completed within 10 months. Yun *et al.* (1998) reported that the most effective medium for induction and multiplication of shoots was MS medium supplemented with 0.3 mg IAA, 2 mg 2iP and 30 g sucrose/l. Bulblets were obtained on MS medium supplemented with 0.5 mg IAA, 0.1 mg 2iP and 70 g sucrose/l. Low concentrations of GMV were found in shoot-tip-derived plants.

Roksana *et al.* (2002) depicted a procedure to regenerate bulblets *in vitro* from garlic cloves using MS medium plus 0.5 mg/l of 2-iP along with 0.25 mg/l of NAA. Multiple bulblets formation was induced by repeated (fourth) subculture of regenerated plantlets on both liquid and semi-solid MS medium. The regenerated bulblets were planted successfully in soil.

Barandiaran *et al.* (1999) have developed an efficient *in vitro* shoot regeneration protocol of garlic. In the system, axenic root tips were used in presence of light. The process significantly improved the explants regeneration ability and 250 shoots per gram of callus were produced in a two months period.

14.1.2 Somatic Embryogenesis

Interspecific hybrids between *A. cepa* and *A. sativum* were obtained using the fertile clone *A. sativum* as the male parent. The nascent embryos which formed shortly in interspecific hybridization between *A. cepa* and *A. sativum* were rescued by ovule culture at an early stage. The zygotes or proembryos developed in MS medium containing 5.7×10^8 M IBA. Once developed, the embryos were taken out of the ovule and cultured on embryo culture medium where they regenerated into whole plants. The hybrids produced not only S-propenyl-L-cysteine sulfoxide, which is the major flavour precursor in *A. cepa*, but also S-allyl-L-cysteine sulfoxide (alliin), which is characteristic of *A. sativum* (Ohsumi *et al.*, 1993).

Root tips measuring 2-3 mm were excised and cultured on agar-solidified MS medium containing different concentrations of 2,4-D for callus and embryo

In vitro Shoot Development of Garlic through Indirect Organogenesis

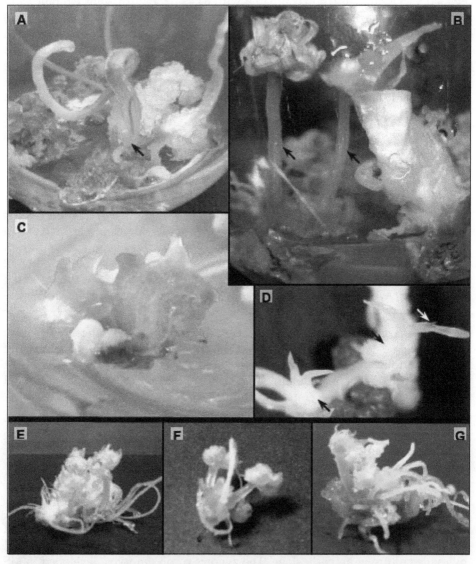

Regeneration on medium A (Kondo *et al.*, 2000) (A-D) and medium B (Zheng *et al.*, 2003) (E-G). (A) Developing shoots from cv. 'Amarante-Embrapa' (arrow); (B) cv. 'Jonas' with two shoots (arrows); (C) cv. 'Cajuru 2315' with initial shoot development; (D) Leaf primordia developing from calli of cv. 'Jonas' callus; (E-G) Developing roots and aerial organs from calli of: (E) cv. 'IAC 75-Gigante de Curitibanos'; (F) cv. 'IAC 63-Mexicano Br' and (G) cv. 'Amarante-Embrapa'. Piracicaba, CENA/USP, 2011.

formation. The optimum concentration of 2,4-D was 0.5 µM and higher concentration (0.1 µM) inhibited callus and embryo formation. Embryos germinated and formed rooted plantlets on MS solid medium containing 5 µM kinetin. The number of plantlets regenerated/root tip explant depended on the concentration of 2,4-D in the callus initiation medium. Plantlets were established in the soil after acclimatization in a growth cabinet (Haque *et al.*, 1998a). The formation of globular bodies by leaf callus of *A. sativum* cv. Jiading White was influenced by gamma-radiation (1-10 Gy); the number of globular bodies decreased at higher radiation doses. Globular bodies were capable of differentiating into plantlets (Zhen and Zhen, 1998).

Plants were generated by somatic embryogenesis from long-term callus cultures derived from 5 garlic cultivars. Thirty-five of these plants were subjected to RAPD analysis. The frequency of variation was found to be cultivar dependent : approximately 1 per cent in Solent White and California Late and around 0.35 per cent in Chinese, Long Keeper and Madena. No association could be shown between the rate of variation for molecular and cytological characters either by comparing cultivars or examining individual regenerants (Al-Zahim *et al.*, 1999).

Fereol *et al.* (2002) developed a scheme for somatic embryogenesis and plant regeneration in garlic using young leaf or root explants. Result showed that the embryogenic potential was higher in callus proliferated from young leaves in B5 medium supplemented with 0.1 mg/l of 2, 4- D and 0.5 mg/l of Kn.

14.2.0 Biochemical and Molecular Markers

In the contemporary research on the genetic diversity of garlic, molecular markers such as RAPD (Restriction fragment length polymorphism), AFLP (Amplified fragment length polymorphism), SSR (Simple sequence repeat), SRAP (Sequence related amplified polymorphism) and ISSR (Inter-simple sequence repeat) are largely uses to assess the diversity and relationship among the different garlic genotypes as the molecular markers are extremely sensitive and neutral to environment effects and crop management practices.

RAPD variation of 30 garlic clones from the primary centre of origin (Central Asia) was compared by Etoh *et al.* (2001) with that of 30 garlic clones from the western most area of distribution (Iberian Peninsula). Central Asian garlic clones were complete-bolting type, and some of them were fertile clones. On the other hand, Iberian garlic clones showed incomplete-bolting type, and all of them were sterile clones. The genetic similarity among the Iberian clones was high, and poor genetic diversity was estimated among the clones from Spain and Portugal. On the other hand, genetic similarity among the central Asian clones was comparatively low, rather there was greater genetic diversity. It is confirmed that the RAPD technique provides a useful and efficient tool in classification and identification of garlic genetical resources (Xu *et al.*, 2001).

Studies were conducted to evaluate genetic similarities in 20 Australian garlic cultivars. Using DNA fingerprinting, the cultivars were differentiated into bolting, non-bolting or intermediate bolting types and into early-, mid- and late-season types. The range of genetic similarity between the 20 garlic cultivars varied from 58 to 97 per cent, confirming a high degree of genetic variation (Bradley *et al.*, 2001).

Al-Zahim *et al.* (1997) performed RAPD analysis to assess genetic variation in 27 cultivars of garlic and to examine relationships between cultivated garlic and the wild progenitor *A. longicuspis*. Using 26 ligonucleotide primers, a total of 292 bands were detected, of which 63 were polymorphic. Cluster analysis revealed groupings that in part reflected patterns of morphological variation. All bolting forms (including wild and cultivated) grouped separately from non-bolting cultivars. *A. longicuspis* and *A. sativum* var. *ophioscorodon* were grouped together, indicating close taxonomic affinity.

Twenty-three accessions of garlic from different geographical parts of India and two accessions from Argentina were used for RAPD profiling and bioactivity evaluation. In RAPD analysis, a total of 2998 bands were observed; out of which, 2459 (82.02 per cent) bands were polymorphic and 52 (1.73 per cent) were unique bands. Similarity and variation among the garlic accessions was observed by cluster analysis and a dendrogram was constructed and later compared with the dendrogram constructed from the morphological characters (Shasany *et al.*, 2000).

Garlic reproduces only by vegetative propagation yet displays considerable morphological variation within and between cultivars. Twenty common Australian garlic cultivars were analysed by the RAPD technique using 20 random decamer primers. The amplification products of 5 of these primers resulted in 65 clear polymorphic bands. The genetic similarities were used to perform a cluster analysis and produce a dendrogram grouping the cultivars. Bolting and intermediate/ non-bolting types could be differentiated from each other. These could be further subdivided into 4 groups based on length of growing season (Bradley *et al.*, 1996).

In garlic, molecular markers associated with the trait of pollen fertility were determined using RAPD. Twelve pollen fertile and sterile garlic clones were screened using 60 RAPD primers to find the DNA fragments related to pollen fertility. With the 60 arbitrary 10-mer primers, 625 discrete fragments of DNA were amplified, and 397 (64 per cent) of them showed polymorphisms among the 12 clones. Two RAPD markers, OPJ121300 and OPJ121700, were amplified only in the pollen fertile clones with the primer OPJ12 (5'-GTCCCGTGGT-3') (Etoh *et al.*, 2001).

Isoenzyme and morphological characters were analysed by Pooler and Simon (1993b) for 110 diverse clones of garlic and the proposed progenitor species, *A. longicuspis*. The clones displayed 17 different electrophoretic phenotypes, which were associated with morphological traits. The lack of unique isoenzyme and morphological characters of *A. longicuspis* suggests an artificial species separation.

Some 65 garlic cultivars from 25 countries were evaluated for several morphological and physiological characters and by isoenzyme electrophoresis using 20 enzyme systems. These characters allowed the delimitation of 6 varietal groups in the Western world (Europe, Africa, America). These groups were homogeneous in isoenzyme banding patterns and there were only 4 polymorphic isoenzymes : esterase, phosphoglucomutase, malate dehydrogenase and diaphorase. Groups I and II share the same enzymatic banding pattern while groups III to VI are characterized by unique banding patterns. The Asiatic cultivars investigated were not fully representative of the whole continent. These studies provided evidence of the narrow genetic base in cultivated garlic (Lallemand *et al.*, 1997).

Molecular markers associated with bulb characters of garlic were searched using RAPD technique (Song *et al.*, 2001). Eighty-four accessions were subjected to RAPD analysis using 18 oligonucleotide primers. The clove adherent types were divided into four types. The majority was of type A collected from East Asia, types B and C were collected from Europe and type D was collected from Nepal. Four RAPD markers, WF701400, WG85980, WF641050, and WF701700 were amplified accessions of clove adherent types A, B, C and D, respectively. High correlation was shown between bolting and clove adherent types. Types C and D were exhibited non-bolting while types A and B showed bolting. An additional RAPD marker, WF821220 was amplified from accessions with white bulb colour. RAPD analysis was used to evaluate genetic diversity among eight garlic mutants resistant to white rot disease (*S. cepivorum*). Twelve of the 13 synthetic random primers were found to identify polymorphism in amplification products. Mutants charactertized with moderate resistance to white rot were closely related to the control using cluster and correlation analyses. On the other hand, highly resistant mutants were quite distant from the control with low correlation coefficients. The banding patterns produced by primer OPB-15 (GGAGGGTGTT) with highly resistant mutants may be used as genetic markers for early selection of resistant plants (Nabulsi *et al.*, 2001). Gogia *et al.* (2014) studied the cloning and molecular characterization of LECASAI lectin gene from garlic via PCR, cloning, characterization and bioinformatics analysis. Results from In-vitro analysis indicated a higher homology of LECASAI lectin gene of garlic with those of insecticidal lectins carrying insecticidal activity to several insects pests, apart from being non-toxic for man, mammals and birds.

14.3.0 Genetic Transformation

Despite recent progress in restoration of garlic fertility, the availability of a reliable genetic transformation system remains highly valuable due to lack of variability in desired traits, or because of insurmountable species barriers. Untill 1998, no report on garlic transformation was published. Barandiaran *et al.* (1998) reported transfer of *uidA* gene into different garlic tissues, including regenerable calli, through biolistic particle delivery. Garlic tissues showed a high endogenous nuclease activity, preventing exogenous DNA expression. Since then, genetic transformation in garlic has been reported through indirect (Kondo *et al.*, 2000) as well as direct (Sawahel, 2002) transformation system. Park *et al.* (2002) were the first to generate transgenic plants resistant to chlorsulfuron, a sulfonylurea herbicide.

As monocotyledons, the *Allium* species were predisposed to be recalcitrant to transformation. It has, therefore, been relatively under-studied with respect to application of biotechnology. Using *Agrobacterium tumefaciens*, Eady *et al.* (2000, 2003, 2005) developed a stable transformation protocol using immature embryos of *A. cepa*, *A. porrum* and *A. sativum*. Successful transformation of one onion cultivar, mediated by *Agrobacterium tumefaciens* was reported by Eady *et al.* (2000) using immature embryos as inoculated explants. Kondo *et al.* (2000) used highly regenerative calli from shoot primordial-like tissues to produce transgenic garlic plants by *Agrobacterium*-mediated gene transfer. Zheng *et al.* (2001) developed a reproducible *Agrobacterium tumefaciens* mediated transformation system both for onion and shallot with young callus derived from mature embryos with two

Genetic Transformation in Garlic

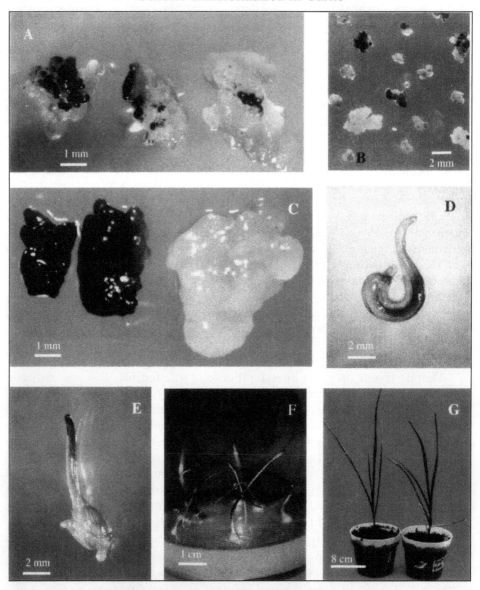

Genetic Transformation of Garlic Mediated by Particle Bombardment with the Plasmid pWRG1515.

(A) Transient expression of β-glucuronidase activity in calluses 24 h after bombardment. (B) Transgenic clones after 4 months on selective medium. (C) Stable expression of β-glucuronidase in transgenic clones. (D and E) Regenerated plantlets showing gusA gene expression. (F) Microbulb development from transgenic plantlets of garlic. (G) Transgenic plants of garlic established under soil conditions (Robledo *et al.*, 2004).

different *Agrobacterium* strains. In India, Khar *et al.* (2005b) reported that in onion, callus proved to be the best explant source for genetic transformation, followed by shoot tip and root tips. Aswath *et al.* (2006) devised a new selection system for onion transformation that does not require use of antibiotics or herbicides, using *Escherichia coli* gene that encodes phosphomannose isomerase (*pmi*). Through a single genetic transformation in onion, Eady *et al.* (2008) were able to develop "Tear-free Onion" by suppressing the lachrymatory factor synthase gene, using RNA interference silencing. For the first time, transgenic garlic resistant to beet armyworm (*Spodoptera exigua*) using *cry1Ca* and *H04* resistance genes from *Bacillus thuringiensis* were developed (Zheng *et al.*, 2004). The *cry1Ca* transgenic garlic plants developed normally in the greenhouse to maturity producing normal bulbs, but none of the transgenic *in vitro H04* garlic plants survived the transfer to the greenhouse. Transgenic *cry1Ca* garlic plants proved completely resistant to beet armyworm in a number of *in vitro* bio-assays, thus providing good perspectives for the development of new garlic cultivars resistant to beet armyworm.

A biolistic particle delivery system was used to introduce DNA containing a beta-glucuronidase (*gus*) reporter gene under the control of the CaMV35S promoter in three different garlic tissues : embryogenic calluses, leaves and basal plate discs. Following bombardment, high levels of beta-glucuronidase (*gus*) were found without the need for treatment to block previously reported putative endogenous nuclease activity (Ferrer *et al.*, 2000). The similar system was used to transfer the *uidA* gene into different garlic tissues, including regenerable calluses. These tissues showed a high endogenous nuclease activity preventing exogenous DNA expression. After submerging the tissues for 24 h in a 2 mM solution of the nuclease inhibitor aurintricarboxylic acid, transient expression of the *uidA* gene was detected (Baran-diaran *et al.*, 1998). Transgenic garlic plants were developed directly from the juvenile leaf tissue through selective culture subsequently *Agrobacterium tumefaciens* mediated gene transformation and regeneration (Kenel *et al.*, 2010).

14.4.0 Cryopreservation

Keller (2002) strongly advocated the cryopreservation of garlic as vegetative propagation is predominantly followed in garlic and vegetative maintenance of genotypes is necessary. Meristem culture allows production of virus-free lines. It circumvents the problem of explants as long as the materials are kept permanently *in vitro* storage through cryopreservation. In vitro storage is, therefore, preferable for this vegetatively propagated material. An experiment performed with a total of 11 garlic accessions showed that the success of cryopreservation depended on the size of the bulbils from which apices had been dissected. The following trend was observed as regards regrowth of cryopreserved apices : the highest regrowth frequency was achieved with apices taken from large bulbils, whereas no regrowth was noted with apices from small bulbils, and regrowth of apices from medium size bulbils was intermediate. Apices from cloves had higher survival and regrowth frequencies than those from bulbils (Makowska *et al.*, 1999).

15.0 References

Abd Elwahed, M.S.A., Mahdy, H.A., El-Saeid, H.M. and Abouziena, H.F. (2019) *Middle East J. Appl. Sci.*, **9**(1): 17-24.

Abdel, A. Z. E. (1973) *Acta Hortic.*, **33** : 43-49.

Abou-Hussein, M. R., Fadl, M. S. and Wally, Y. A. (1974) *Egyptian J. Hort.*, **1** : 113-125.

Agarwal, M.K. (1999) Variability, correlation and path coefficient analysis in garlic. *M.Sc. Thesis. Submitted to Rajasthan Agricultural University*, Bikaner, India.

Agrawal, P. (1978) *Trans. British Mycol. Soc.*, 70 : 439-441.

Aguilar, J. A. E. (1986) *Horticultura Brasileira,* **4** : 26-28.

Agullar, J.A.E., Reifschnelder, F.J.B. and Cordeiro, C.M.T. (1986) *Fitopatologia Brasileira,* **11** : 237-240.

Ahlawat, Y. S. (1974) *Sci. Cult.,* **40** : 466-467.

Ahmad, M., Din, S., Sabar, K., Siddique, M., Salahuddin, M., Hafeez, M. and Imran, F. (2018) *Nature Sci.,* **16** : 17-26.

Ahmed, N. U. and Hoque, M. M. (1986) *Bangladesh Hort.,* **14** : 19-24.

Akimov, Y. A., Ponurova, N. P. and Chipiga, A. P. (1973) *Trudy VNII Efirnomaslichnykh Kultur,* **6** : 197-199.

Alam, S., Mahmood, I., Nawab, N., Haider, S. and Aamir, S. (2019) *Food Biol.,* **8** : 13-16.

Alam, S.S., Ahmad, M., Alam, S., Usman, A., Ahmad, M.I. and Naveedullah (2007) *Sarhad J. Agric.,* **23** : 149-152.

Ali, N., Ashiq, M. and Ameer, H. (2017) *J. Environ. Agric. Sci.,* **12** : 19-24.

Aliaro, U. A. and Gacitua, M. E. (1976) *Agric. Tecnica (Chile),* **36** : 63.

Aliudin and Suminto, T. J. (1978) *Bull. Penelitian Hortikultura,* **6** : 3-10.

Aliudin, (1979) *Bull. Penelitian Hortikultura,* **7** : 23-48.

Aliudin, (1980) *Bull. Penelitian Hortikultura,* **8** : 3-11.

Allen, R.G., Pereira, L.S., Raes, D. and Smith, M. (1998) Crop evapo-transpiration: guidelines for computing crop water requirements, Irrigation and Drainage Paper 56. United Nations FAO, Rome, p. 300.

Al-Moshile, A.M. (2001) *Assiut. J. Agric. Sci.,* **32** : 291–305.

Al-Safadi, B., Mir-Ali, N. and Arabi, M.I.E. (2000) *J. Genet. Breed.,* **54** : 175-181.

Aly-Khan, Shaukat, S.S. and Khan, A. (1997) *Appl. Entomol. Phytopathol.,* **66** : 13-18.

Al-zahim, M., Newbury, H.J. and Ford-Lloyd, B.V. (1997) *HortScience,* **32** : 1102-1104.

Al-Zahim, M.A., Ford Lloyd, B.V. and Newbury, H.J. (1999) *Plant Cell Rept.,* **18** : 473-477.

Amorim, J.R. de A., Fernandes, P.D., Gheyi, H.r. Azevedo, N.C. de, de A Amorim, J.r. and de Azevedo, N.C. (2002) *Pesquisa Agropecuaria Brasileira,* **37** : 167-176.

Anonymous (1988) *FAO Production Year Book,* **53** : 137.

Anonymous, (1976) *FAO Production Year Book.*

Anonymous, (1979-80) *Agricultural Situation in India,* **34.**

Anonymous (2020) Report of the ICAR-Directorate of Onion and Garlic Research, Pune, Maharashtra, India

Appleton, J. An. and Tansey, M. R. (1975) *Mycologia,* **67** : 882-885.

Arafa, A. E., Abdel, G. M. G. and Eman, Y. T. (1977) *Agric. Res. Rev. Hort.,* **55** : 159-164.

Arboleya, J., Garcia, C. and Burba, J.L. (1993) *III Curso taller sobre produccion, comercializacion, e industrializacion de ajo,* Argentina, 22-25 June, 1993, pp. 261-266.

Armellina and Dall, A. A. (1976) *ASAM Mardel plants,* **3** : 73-78.

Asiya Kowser, R., Amarananjundeswara, H., Doddabasappa, B., Aravinda Kumar, J.S. and Pannure Aarti (2019) *J. Entomol. Zool. Stud.,* **7**(1): 63-66.

Astley, D. (1990) In: Rabinowitch, H.D. and Brewster, J.L. (Eds.), Onions and Allied Crops. I. Botany, Physiology, and Genetics. CRC Press, Boca Raton, FL, pp. 177–198.

Astley, D., Innes, N.L. and Van der Meer, Q.P. (1982) Genetic Resources of *Allium* Species: A Global Report, IBPGR, Rome.

Aswath, C.R., Mo, S.Y., Kim, D.H. and Park, S.W. (2006) *Plant Cell Rept.,* **25** : 92-99.

Atif, M.J., Amin, B., Ghani, M.I., Hayat, S., Ali, M., Zhang, Y. and Cheng, Z. (2019) *Agronomy,* **9**(12) : 879.

Augusti, K. T. (1977) *Indian J. Exp. Biol.,* **15** : 489-490.

Augustin, E. and Garcia, A. (1993) *Horticultura Brassileira,* **11** : 10-13.

Avato, P., Miccolis, V. and Tursi, F. (1998) *Adv. Hort. Sci.,* **12** : 201-204.

Ayala, A. and Savon, J. R. (1986) *Ciencia y Tecnica en la Agricultura, Hortalizas,* Papa, Granos y Fibras, **5** : 73-80.

Ayars, J.E. (2008) *Am. Soc. Agric. Biol. Eng.,* **51**(5) : 1683-1688.

Bacdrati, G. and Cangna, D. (1970) *Ind. Conserva,* **45** : 125-130.

Badshah, N. and Umar, K. (1999) *Sarhad J. Agric.,* **15** : 431-436.

Baiday, A.C. and Tiwari, R.S. (1995) *Rec.Hortic.,* **2** : 117-123.

Bajpai, N.K., Jeengar, K.L., Singh, D.K., Gupta, I.N. and Meena, S.N. (2014) *Curr. Adv. Agric. Sci.,* **6** : 201-202.

Bala, S., Karmakar, K. and Ghosh, S. (2015) *Int. J. Sci. Technol.,* **4** : 1365-1372.

Bandara, M.S., Krieger, K., Slinkard, A.E. and Tanino, K.K. (2000) *Can. J. Plant Sci.,* **80** : 379-384.

Barandiaran, X., Di Pietro, A. and Martin, J. (1998) *Plant Cell Rept.,* **17** : 737-741.

Barandiaran, X., Martín, N., Rodriguez, C.M., Pietro, D.A. and Martín, J. (1999) *HortScience,* **34** : 348-349.

Barber, C. J. (1977) *New Zealand Commercial Grower,* **32** : 11.

Barboza, K., Salinas, C., Acuña, C., Bannoud, F., Beretta, V., Garcia Lampasona, S., Burba, J., Galmarini, C. and Cavagnaro, P. (2020) *Sci. Hortic.*, **261**, htpps://10.1016/j.scienta.2019.108900.

Barone, F. E. and Tansey, M. R. (1977) *Mycologia,* **69** : 793-825.

Barrera, C., Gonzalez, M.I. and del Pozo, A. (1998) *Agro-Ciencia,* **14** : 207-215.

Baten, M.A., Nahar, B.S., Sarker, S.C. and Khan, M.A.H. (1995) *Pak. J. Sci. Indus. Res.*, **38** : 138-141.

Bernard, J.K., Gracia, C., Torregosa, A., Val, L. and Burba, J.L. (1993) *3rd Curso taller sobre produccion, commercializacion, e industrialization de AJO Proyecto A JO INTA, Mendoza, Argentina,* pp. 381-389.

Bertaccini, A., Marani, F. and Borgia, M. (1986) Rivista della Ortoflorofrutticoltura Italiana, **70** : 97-105.

Bertoni, G. (1988) *Agrochimica,* **32**(5-6) : 519–530.

Bertoni, G., Morard, P. and Espagnacq, L. (1988) *Agrochimica,* **32** : 518-530.

Bertoni, G., Morard, P., Soubielle, C. and Llorens, J. M. (1992) *Sci. Hort.,* **50** : 187-195.

Bharadwaj, A. K. and Thakur, J. R. (1974) *Curr. Sci.,* **43** : 419-420.

Bhathal, G. S. and Thakur, J. C. (1986) *Punjab Vegetable Grower,* **21** : 29-31.

Bhatt, R.P., Biswas, V.R., Narendra-Kumar and Kumar, N. (1998) *Veg. Sci.,* **25** : 95-96.

Bhojwani, S. S. (1979) *Sci. Hort.,* **12** : 47-52.

Bhojwani, S. S., Cohen, D. and Fry, P. R. (1982) *Sci. Hort.,* **18** : 39-43.

Bhuyan, M., Saxena, B. N. and Rao, K. M. (1974) *Indian J. Exp. Bot.,* **12** : 575-576.

Bisht, I.S. and Agrawal, R.C. (1993) *Ann. Appl. Biol.,* **122** : 31-38.

Bisht, I.S., Agrawal, R.C. and Venkateshwaran, K. (1993) *Indian Phytopath.,* **46** : 89-91.

Blank, A.F., Pereira, A.J., Arrigoni-Blank, M. de F. and de Souza, R.J. (1998) *Ciencia e Agrotecnologia,* **22** : 5-12.

Bockish, T., Saranga, Y., Altman, A. and Ziv, M. (1997) *Acta Hortic.,* **447** : 241-242.

Bogatirenko, A. K. (1976) *Referativny i Zhurnal,* **9** : 55-62.

Bomme, U., Regenhardt, I. and Schilling, W. (1996) *Zeitschrift fur Phytotherapie,* **17** : 22-29.

Borabash, O. Yu and Kochina, T. N. (1987) In : Puti Intensifikatsii Ovoshchevodstva. Kiev, 12-16.

Borgo, R., stahlschmidt, O., Leuzzi, L., Gauna, C., Canatalla, C. and Burba, J.L. (1994) *III Curso taller sobre produccion, commercializacion e industrialization de ajo,* Mendoza, Argentina, 22-25 June, pp. 135-140.

Bortoli, S.A. de, Castellane, P.D., Terasaka, F. and De Bortoli, S.A. (1995) *Cientifica Jaboticabal,* **23** : 353-366.

Borukh, I. F. (1975) *Pishcheraya Teknologiya,* **1** : 21-23.

Bovo, O. A. and Mroginski, L. A. (1985) *Phyton, Argentina,* **45** : 159-163.

Bozzini, A., Casoria, P. and Luca, P. De. (1991) *Econ. Bot.,* **45** : 436-438.

Bradley, K., Rieger, M., Collins, G. and Armstrong, J. (2001) *Acta Hortic.,* **555** : 159-160.

Bradley, K.F., Rieger, M.A. and Collins, G.G. (1996) *Australian J. Exp. Agri.,* **36** : 613-618.

Brar, P.S. and Gill, S.P.S. (2000) *Haryana J. Hortic. Sci.,* **29** : 1-2.

Brodnitz, M. H. and Pascale, J. V. (1971) *J. Agric. Food Chem.,* **19** : 273-275.

Bruna, A. (1997) *Acta Hortic.,* **433** : 631-634.

Bull, L.T., Boas, R.L.V., Fernandes, D.M. and Bertani, R.M.A. (2001) *Scientia Agricola,* **58** : 157-163.

Burba, J. L., Fontan, H. M., Lansranconi, L. and Beretta, R. (1982) *Revista de Ciencias Agropecuarias,* No. 3, pp. 37-48.

Burba, J.L., Riera, P.G. and Galmarini, C.R. (1997) *Acta Hortic.,* **433** : 151-164.

Butani, D.K. and Verma, S. (1976) *Pesticides,* **10**(11) : 33-35.

Cantwell, M. and Suslow, T. V. (2002) In : Kader, A. A. (Ed.), Postharvest technology of horticultural crops (Publication 3311). Postharvest Technology Center, University of California. Davis, CA.

Cantwell, M. I. (2004) Retrieved from http: /postharvest.ucdavis/Produce/Produce facts.

Carrijo, O. A., Olitta, A. f. L. and Minami, K. (1982) *Pesquisa Agropeararia Brasileira,* **17** : 783-790.

Caruso, P., D'-Anna, F., Palazzolo, E. and Panno, M. (1994) *Colture Protette,* **23** : 79-85.

Carvalho, L.G. de, Silva, A.M. da, Souza, R.J. de, Carvalho, J.G. de, Abreu and A.R. de (1996) *Ciencia e Agrotecnologia,* **20** : 249-251.

Castellanos, J.Z., Ojodeagua, J.L., Mendez, F., Villalobos-Reyes, S., Badillo, V., Vargas, P. and Lazcano, F. (2001) *Better Crops International,* **15** : 21-23.

Castellanos, J.Z., Vargas-Tapia, P., Ojodeagua, J.L., Hoyos, G., Alcantar-Gonzalez, G., Mendez, F.S., Álvarez-Sánchez, M. and Gardea, A.A. (2004) *HortScience.,* **39** : 1272-1277.

Castillo, J.E., López-Bellido, L., Fernández, E.J. and López, F.J. (1996) *J. Hort. Sci.,* **71**(6) : 867–879.

Chanchan, M. and Hore, J. K. (2014) *Res. Crops,* **15** (3) : 701-704.

Chanchan, M., Thapa, P. and Hore, J. (2018) *Res. Crops,* **19** : 127-131.

Chand, P.K., Chattopadhyay, S. and Hasan, M.A. (2010) *Indian J. Hort.,* **67**(3) : 201.

Chang, J. I. and Park, Y. B. (1982) *J. Korean Soc. Hort. Sci.,* **23** : 179-187.

Changela, N.B. (1993) Bionomics, population dynamics and chemical control of thrips (*Thrips tabaci* L.) on garlic. M. Sc. (Agri.) thesis submitted to Gujarat Agricultural University, Sardarkrushinagar, India, pp. 82-83.

Channabasavanna, G.P. (1966) A contribution to the knowledge of Indian eriophyid mites (Eriophyoidea: Trombidiformes, Acarina). University of Agricultural Sciences, Bangalore, p. 60.

Charchar, J.M., Aragao, F.A.S. and de Menezes Sobrinho, J.A. (1999) *Horticultura Brassileira*, **17** : 96-101.

Chattopadhyay, N., Lalrinpuii, F. and Thapa, U. (2015) *J. Crop Weed*, **11**(2): 67- 71.

Chavchanidze, T. M. (1976) *Referativny i Zhurnal*, **6**.55.656.

Chen, J. B., Kwong, K. H. and Chiu, Y. M. (1976) *Rept. Taiwan Sugar Res. Inst.*, pp. 131-141.

Chen, S., Zhou, J., Chen, Q., Chang, Y., Du, J. and Meng, H. (2013) *Biochem. Syst. Ecol.*, **50** : 139–146.

Chen, Y. (1986) *Seed World*, China, No. 10, p. 19.

Chen, Y.Y., Chiu, H.C. and Wang, Y.B. (2009) *J. Food Drug Anal.*, **17**(1): 59-63.

Cheng, Z.H., Lu, G.Y. and Du, S.L. (1991) *Acta Hortic*. Sinica, **18** (4) : 345–349.

Cho, S. Y. and Lee, S. W. (1974) *J. Korean Soc. Hort. Sci.*, **15** : 1-6.

Cho, W.D., Kim, W.G., Lee, Y.H. and Lee, E.J. (1994) *RDA J. Agri. Sci., Crop Prod.*, **36** : 327-330.

Cho, Y.C. and Jeong, B.R. (1998) *Korean J. Hort. Sci. Technol.*, **16** : 219-221.

Choi, J. K., Ban, C. D. and Kwon, Y. S. (1980) *Res. Rept. Hort. and Seri.* (Korea), **22** : 20-23.

Choudhary, A.D. and Dnyanansagar, V.R. (1982) *J. Indian Bot. Soc.*, **61** : 85: 90.

Choudhary, Kavita and Choudhary, M.R. (2018) *Int. J. Agric. Sci.*, **10** : 5791-5793.

Choudhury, S. K. (1960) *Curr. Sci.*, **29** : 251-252.

Chun, K. B. and Soh, W. Y. (1980) *J. Korean Soc. Hort. Sci.*, **21** : 119-125.

Cindea, E. (1979) *Analele Institutului de Cercetari pentru Legumicultura si Floricultura,***5** : 267-273.

Cindea, E. (1980) *Bull. de I Academic des Sci. Agricoleset Forestieres*, No. 9, pp. 95-100.

Cindea, E. (1982) *Productia Vegetla, Horticultura*, **31** : 17-18.

Cobley, L. S. and Steele, W. M. (1976) *Introduction to the Botany of Tropical Crops*, Longman, London, pp. 151-152.

Cosenza, G. W., Menezes, S. J. A. D. E., Regina, S. M. and Gontizo, V. P. M. (1981) *Pesquisa Agropeararia Brasileira*, **16** : 199-203.

Cucchi, N. J., Puiatti, A. E. and Salvarredi, A. U. (1966) *Idia*, No. 230, pp. 54-58.

Curzio, O. A., Croci, C. A. and Ceci, L. N. (1986) *Food Chemistry*, **21** : 153-159.

Dahlgren, R.M.T., Clifford, H.T. and Yeo, P.F. (1985) The Families of Monocotyledons – Structure Evolution and Taxonomy. Springer – Verlag, Berlin Heidelberg, New York, pp. 191-198.

Das, A. K., Sadhu, M. K., Som, M. G. and Bose, T. K. (1985a) *Indian Agric.*, **29** : 177-181.

Das, A. K., Som, M. G., Sadhu, M. K. and Bose, T. K. (1985b) *Indian Agric.*, **29** : 183-189.

Das, S., Chanchan, M. and Hore, J.K. (2014) *Res. Crops*, **15** : 912-915.

Datshvarov, S. and Galmarini, C.R. (1997) *Acta Hortic.*, **433** : 137-140.

De La Cruz, J. (2007) *Garlic: Post- Harvest Operatio. Agric. Food Eng. Technol. Serv.*, pp. 2–40.

Debkirtaniya, S., Ghosh, M. R., Chaudhury, A. N. and Chatterjee, A. (1980) *Indian J. Agric. Sci.*, **50** : 507-510.

Debnath, P. and Karmakar, K. (2013) *Int. J. Acarol.*, **39** : 89–96.

Dedul, F.A., Novak, F.J. and Havel, L. (1985) *Agropromizdat*, pp. 151-154.

Delahousse, B., Deloraine, J. and Guillot, M. (1973) In compte Rendu de la 7 conference du Columa (France), pp. 715-726.

Deuber, R. and Fornasier, J. B. (1980) Resumos 13th Congresso Brasileira de Herbicidas Ervas Daninhas, Bahia, Brazil.

Dhall, R.K. and Ahuja, S. (2013) *Natl. Sem. High-Tech Cult. Veg. Postharvest Manag.*, NHRDF, Karnal, India, pp. 117–121.

Dhillon, N.K., Kaur, S., Sidhu, H. and Chaudhary, A. (2019) *Indian J. Hortic.*, **76** : 472.

Dimitrov, Z. (1974) B'lgarski Plodove Zelenchutsi i Konservi, No. 3, pp. 9-11.

Domínguez, A., Martínez-Romero, A., Leite, K.N., Tarjuelo, J.M., de Juan, J.A., López and Urrea, R. (2013) *Agric. Water Manag.*, **130** : 154–167.

Donnari, M. A., Rosell, R. A. and Torre, L. (1978) *Turrialba*, **28** : 331-337.

Doreste, S. E. (1965) *Proc. 12th Ann. Mtg. Carib. Reg. Amer. Soc. Hort. Sci.*, **8** : 39-43.

Duchon, D. J. (1979) *Pepinieristes Hort. Maraichers*, No. 200, pp. 33-43.

Duff, A.A., Jackson, K.J., O'-Donnell, W.E. and Armstrong, J. (2001) *Acta Hortic.*, **555** : 247-250.

Duimovic, M. A. and Bravo, M. A. (1980) *Ciencia e Investigacion*, **6** : 99-103.

Dyduch, J. and Najda, A. (2001) *Horticultura*, **9** : 295-299.

Eady, C., Davis, S., Farrant, J., Reader, J. and Kenel, F. (2005) *Ann. Appl. Biol.*, **142** : 213–217.

Eady, C.C., Kamoi, T., Kato, M., Porter, N.G., Davis, S., Shaw, M., Kamoi, A. and Imai, S. (2008) *Plant Physiol.*, **147** : 2096-2106.

Eady, C.C., Reader, J., Davis, S. and Dale, T. (2003) *Ann. Appl. Biol.*, **142** : 219-224.

Eady, C.C., Weld, R.J., Lister, C.E. (2000) *Plant Cell Rept.*, **19** : 376-381.

Ebi, M., Kasai, N. and Masuda, K. (2000) *HortScience*, **35** : 735-737.

El-Ganaieny, R.M.A., Osman, A.Z. and Mohamed, H.Y. (1998) *Assiut J. Agric. Sci.*, **29** : 45-58.

Engeland, R.I. (1991) Growing Great Garlic. Filaree Productions, Okanogan, WA.

Etoh, T. (1982) *Pleme. Faculty Agric. Kagoshima Univ.*, **18** : 75-84.

Etoh, T. (1985) *Mem. Fac. Agr. Kagoshima Univ.*, **21** : 77-132.

Etoh, T. (1986) *J. Japanese Soc. Hortic. Sci.*, **55** : 312-319.

Etoh, T. (1997) *Acta Hortic.*, **433** : 247-255.

Etoh, T. and Galmarini, C.R. (1997) *Acta Hortic.*, **433** : 247-255.

Etoh, T. and Hong, C.J. (2001) *Acta Hortic.*, **555** : 209-212.

Etoh, T. and Ogura, H. A. (1997) *Mem. Fac. Agr. Kagoshima Univ.*, **13** : 77-88.

Etoh, T. and Pank, F. (1996) *Beitrage zur Zuchtungs forschung Bundesanstalt fur Zuchtungs forschung an Kulturpflanzen*, **2** : 108-115.

Etoh, T. and Simon, P.W. (2002) In : Rabinowitch, H.D. and Currah, L. (Eds.), *Allium* Crop Science: Recent Advances, CAB International: Wallingford, UK, pp. 101–117.

Etoh, T., Hong, C.J. and Armstrong, J. (2001) *Acta Hortic.*, **555** : 209-212.

Etoh, T., Kojima, T. and Matsuzoe, N. (1991) In: Hanelt, P., Hammer, K. and Knupffer, H. (Eds.), Proc. Int. Symp. The Genus *Allium* - Taxonomic Problems and Genetic Resources, Gatersleben, Germany, pp. 49-54.

Etoh, T., Noma, Y., Nishitarumizu, Y. and Wakamoto, T. (1988) *Mem. Fac. Agr. Kagoshima Univ.*, **24** : 129-139.

Fabeiro Cortés, C., Martín de Santa Olalla, F. and López Urrea, R. (2003) *Agric. Water Manag.*, **59** : 155–167.

Fakhar, F., Biabani, A., Zarei, M. and Ali, M. (2019) *Italian J. Agron.*, **14** : 108-113.

FAOSTAT. (2016) Available at: http: //faostat.fao.org/faostat/collections? subset=agriculture.

Fereol, L., Chovelon, V., Causse, S., Michaux, F., Ferriere, N. and Kahane, R. (2002) *Plant Cell Rept.*, **21** : 197-203.

Ferrari, V. A. and Churata, M. G. C. (1975) *Cientifica, (Brazil)*, **3** : 254-262.

Ferreira, P.V., Silva, W.C.M. da and Da Silva, W.C.M. (1995) *Summa Phytopathologica*, **21** : 181-183.

Ferrer, E. Linares, C. and Gonzalez, J.M. (2000) *Agronomie*, **20** : 869-874.

Figliuolo, G., Candido, V., Logozzo, G., Miccolis, V. and Zeuli, P.L.S. (2001) *Euphytica*, **121** : 325-334.

Foda, S. A., Saleh, H. H. and Shahein, A. H. (1979) *Agric. Res. Rev. Hort.*, **57** : 171-177.

Formigoni, A. (1972) *Proc. 6th Int. Velsicol Symp.* p. 7.

Francois, L. E. (1991) *HortScience*, **26** : 547-549.

Francois, L.E. (1994) *HortScience*, **29** : 1314-1317.

Freeman, G. G. (1975) *J. Sci. Food Agric.*, **26** : 471-481.

French, J.M., Beacham, J., Garcia, A., Goldberg, N. P., Thomas, S. H. and Hanson, S. F. (2017) *Plant Health Progress*, **18** : 91-92.

Fujime, Y., Ono, M.M., Kudou, R. and Midmore, D.J. (1994) *Acta Hortic.*, **358** : 199-203.

Gajera, R.C., Kapadia, M.N. and Jethva, D.M. (2009) *Agric. Sci. Digest.*, **29**(4) : 294-296.

Garcia, A. (1980) *Pelotas Res.*, **9** : 3.

Gaskell, M.; Cantwell, M.; Nie, X.; Smith, S.; Faber, B.; Voss, R. (1998) In : Proc. Natl. Onion (and other Allium) Res. Conf., Sacramento, CA, USA, 10–12 December 1998; pp. 337–341.

Gautam, N., Kumar, D., Kumar, R., Kumar, S., Sharma, S. and Dogra, B. (2014) *Int. J. Farm Sci.*, **4**(3); 49-57.

Gawande, S.,Chimote, K., Gurav, V. and Gopal, J. (2013) *Indian J. Hortic.*, **70** : 544-548.

Gawande, S.J., Gurav, V.S., Ingle, A.A. and Gopal, J. (2014) *Plant Dis.*, **98**(7) : 1015.

Ghosh, D.K. and Ahlawat, Y.S. (1998) *Indian J. Agric. Sci.*, **68** : 380-381.

Goa, S.L., Jin, Y.A., Cai, Z.H. and Liu, L.J. (2000) *J. Plant Res. Environ.*, **9** : 15-18.

Gogia, N., Kumar, P., Singh, J., Rani, A., Sirohi, A. and Kumar, P. (2014) *Int. J. Agric. Environ. Biotechnol.*, **7** : 1-10.

Gohil, R.N. and Koul, A.K. (1984) *Allium News No. 1*, pp. 70-74.

Grad, Z.Y., Nawab-Ali and Hussain, S.A. (1993) *Sarhar J. Agric.*, **9** : 313-316.

Grubben, G.J.H. and Denton, C.M. (2004) In : *Plant Resources of Tropical Asia Foundation*; Bakhuys Publishers: Leiden, The Netherlands, p. 59.

Grubben, G.J.H. and Midmore, D.J. (1994) *Acta Hortic.*, **358** : 333-339.

Gupta, M. and Bharat, N.K. (2017) *Int. J. Farm Sci.*, **7** : 47-50.

Gupta, N., Prabha, K., Islam, S. and Baranwal, V.K. (2013) *J. Plant Pathol.*, **54** : 69-77.

Gupta, R. and Sharma, N.K. (1993) *Int. J. Pest Manage.*, **39** : 390-392.

Gupta, R., Mishra, K. P. and Hardaha, M. K. (2017) *Int. J. Pure App. Biosci.*, **5**(6) : 1334-1340.

Gupta, R.P. and Singh, D.K. (1998) *Newsl. - Nat. Hort. Res. and Dev. Foundation*, **18** : 13-18.

Gvozdanvic-Varga, J., Takac, A. and Vasic, M. (1994) *Selekcija i Semenarstvo*, **1** : 65-68.

Ha, H.T., Hwang, J.M. and Park, Y.M. (2000) *Korean J. Hort. Sci. Technol.*, **18** : 499-502.

Habibunnisa Mathur, P. B. and Zakia Bano (1971) *Indian Fd. Pack.*, **25** : 10-13.

Hafidh, F.T. (2000) *Dirasat. Agric. Sci.*, **27** : 458-464.

Hahn, S.J. (1994) *J. Korean Soc. Hort. Sci.*, **35** : 559-573.

Hanson, B., May, D., Voss, R., Cantwell, M. and Rice, R. (2013) *Agric. Water Manage.*, **58**(1): 29-43

Haque, M.S., Wada, T. and Hattori, K. (1998a) *Plant Prod. Sci.*, **1** : 216-222.

Haque, M.S., Wada, T. and Hattori, K. (1998b) *Breed. Sci.*, **48** : 293-299.

Haque, M.S., Wada, T., Hattori, K. and Klerk, G.J. de (2000) *Acta Hortic.*, **520** : 45-52.

Hasan, M. and Suhardi, A. (1979) *Bull. Penelitian Hortikultura,* **7** : 21-30.

Hemm, M.R., Rider, S.D., Ogas, J., Murry, D.J. and Chapple, C. (2004) *Plant J.,* **38** : 765–778.

Heredia-Zepeda, A. Heredia-Garcia, E., Laborde, J.A. and Galmarini, C.R. (1997b) *Acta Hortic.,* **433** : 271-277.

Hong, C.J. and Etoh, T. (1996) *Breed. Sci.,* **46** : 349–353.

Hong, G.H., Jang, H.S. and Kim, Y.B. (1999) *J. Korean Soc. Hort. Sci.,* **40** : 23-25.

Hong-jiu, L., Cai-ping, H., Pei-jiang, T., Xue, Y., Ming-ming, C. and Zhi-hui, C. (2020) *J. Integ. Agric.,* **19**(4): 1044–1054.

Horcin, V. and Simekova, E. (1986) *Sbornik UVTIZ, Zahradnictyi,* **13** : 154-160.

Hossain, A.M., Mostofa, M., Alam, M.N., Sultana, M.R. and Rahman, D.M. (2014a) *Bangladesh J. Vet. Med.,* **12** : https: //10.3329/bjvm.v12i1.20453.

Hossain, M.M., Khalequzzaman, K.M., Fahim, A.H.F., Ahmed, R.N. and Islam, M.S. (2014b) *Int. J Expt. Agric.,* **4**(3) : 30-33.

Hovadik, A. (1911) *Bull. Vyzkumny Sleihitielsky Ustav,* No. 25/26, pp. 107-118.

Hredia-Zepeda, A. and Galmarini, C.R. (1994) *Acta Hortic.,* **433** : 265-270.

Huchette, O., Kahane, R., Auger, J., Arnault, I. and Bellamy, C. (2005) *Acta Hort.,* **688** : 93-99.

Humayun-Khan, Azim-Khan, Derawadan, Mazullah-Khan and Abdul-Majeed (1997) *Sarhad J. Agri.,* **13** : 357-361.

Hussain, T., Iqbal, M., Farmanullah and Anwar, M. (1997) *Sarhad J. Agric.,* **13** : 175-180.

Hwang, J.M. (1993) *J. Korean Soc. Hort. Sci.,* **34** : 257-264.

Iglesias-Enriquez, I., Hombre, R. de and de Hombre, R. (2000) *Alimentaria,* **37** : 57-59.

Ilinykh, Z. G. (1975) *Trudy uralskogo NII,* **15** : 161-171.

Inaba, A., Ujiie, T. and Etoh, T. (1995) *Breed. Sci.,* **45** : 310.

Islam M.J., Hossain, A.K.M.M, Khanam F., Majumder, U., Rahman, M.M. and Rahman, M. (2007) *Asian J. Plant Sci.,* **6** : 98-101.

Jang, Y.S., Oh, Y.B., Choi, I.H., Song, Y.S. and Park, J.H. (2000) *J. Korean Soc. Hort. Sci.,* **41** : 157-160.

Janick, J. (1979) *Horticultural Science, Freeman and Co. San Francisco,* p. 544.

Janyska, A. (1986) *Bulletin, Vyzkumny A Slechtitelsky Ustav Zelinarsky, Olomoue,* No. 30, p. 3-16.

Jenderek, M.M. (1998) *Sesja Naukowa* (in Polish), **57** : 141-145.

Jenderek, M.M. and Hannan, R.M. (2000) Proc. Third Intl. Symp. Edible Alliaceae, Athens, Georgia USA, pp. 73-75.

Jenderek, M.M. and Hannan, R.M. (2004) *HortScience,* **39** : 485-488.

Jenderek, M.M. and Zewdie, Y. (2005) *HortScience*, **40** : 1234-1236.

Jiang, Q., Zhang, H., Guanghui, A. I., Jiang, Q. S., Zhang, H. L. and Gh, A.I. (1998) *China Vegetables*.

Jiku, M. A., Alimuzzaman, M., Singha, A., Rahaman, M., Ganapati, R. K., Alam, M. and Sinha, S. (2020) *Bull. Natl. Res. Cen.*, **44** : 9.

Joshi, S. N. (1961) *Punjab Hort. J.*, **9** : 192-197.

Jourdain, J. M. and Lavigne, D. (1987) *Infos, Centre Technique Interprofesscionel des Fruits et Legumes, France*, No. 34, pp. 23-27.

Jousseayme, C. (1982) *Herbicidas en Hortofruticultura*, p. 9.

Kalloo, Pandey, U. C., Lal, S. and Pandita, M. L. (1982) *Haryana J. Hort. Sci.*, **11** : 97-101.

Kamenetsky, R. (2007) *Hortic. Rev.*, **33** : 123.

Kamenetsky, R., London Shafir, I., Baizerman, M., Khassanov, F., Kik, C. and Rabinowitch, H.D. (2004b) *Acta Hortic.*, **637** : 83–91.

Kamenetsky, R., London Shafir, I., Khassanov, F., Kik, C., van Heusden, A.W., Vrielink van Ginkel, M., Burger Meijer, K., Auger, J., Arnault, I. and Rabinowitch, H.D. (2005) *Biodivers. Conserv.*, **14** : 281–295.

Kamenetsky, R., London Shafir, I., Zemah, H., Barzilay, A. and Rabinowitch, H.D. (2004a) *J. Am. Soc. Hortic. Sci.*, **129** : 144-151.

Karaye, A.K. and Yakubu, A.I. (2006) *Afr. J. Biotechnol.*, **5** : 260-264.

Katarzhin, M.S. and Katarzhin, I.M. (1978) *Bulleten' Vsesoyuznogo Ordena Lenina i Ordena i Druzby Narodov Instituta Rastenievodstva Imeni N. I. Vavilova* (in Russian), **80** : 74-76.

Katkar, S.M., Pawar, D.B., Warade, S.D., Kulkarni, S.S. (1998) *Insect Environ.*, **4** : 103.

Katoch, S. and Kumar, S. (2017) *Indian Phytopathol.*, **70** : 294-296.

Kaymak, F. (1996) *Turkish J. Biol.*, **20** : 139-145.

Kehr, A. E. and Schaeffer, G. W. (1976) *HortScience*, **11** : 422-423.

Keifer, H.H. (1938) Eriophyid studies. Bulletin of the California, Department of Agriculture, **27** : 181–206.

Keifer, H.H. (1952) The eriophyid mites of California (Acarina: Eriophyidae). Bulletin of the California Insect Survey, **2** : 1–123.

Keller, E.R.J. (2002) In: Towill, L.E. and Bajaj, Y.P.S. (Eds.), Biotechnology in agriculture and forestry, Vol. 50, Cryopreservation of plant germplasm II. Springer, Heidelberg.

Kenmochi, K. and Katayama, D. (1975) *Nippon Shokuhin Kogyo Gakkaishi*, **22** : 598-605.

Khaisin, M. F. (1988) *Kartofel'i Ovoshchi*, No. 1, p. 48.

Khan, A. M., Siddiqui, Z. A., Alam, M. M. and Szena, S. K. (1976) *Indian J. Agric. Sci.*, **46** : 439-441.

Khar, A. (2012) *Acta Hortic.*, **969** : 289-295.

Khar, A., Asha Devi, A. and Lawande, K.E. (2005b) *J. Spices Arom. Crops*, **14** : 51-55.

Khar, A., Mahajan, V., Devi, A.A. and Lawande, K.E. (2005) *J. Maharastra Agric. Univ.*, **30** (3): 277-280.

Kim, B. W., Lee, B. jY., Moon, W. and Pyo, H. K. (1980) *J. Korean Soc. Hort. Sci.*, **21** : 5-18.

Kim, K. S., Pyo, H. K. and Lee, B. Y. (1977) *J. Korean Soc. Hort. Sci.*, **18** : 162-172.

Kohli, U.K. and Fageria, M.S. (1992) *J. Res., APAU.*, **20** : 164-168.

Kohli, U.K. and Mahajan, N. (1993) *Haryana J. Hortic. Sci.*, **22** : 163-165.

Kohli, U.K. and Prabal (2000) *Haryana J. Hort. Sci.*, **29** : 209-211.

Kolte, A.R. and Patale, S.S. (2019) *BIOINFOLET*, **16** : 201-203.

Kondo, T., Hasegawa, H. and Suzuki, M. (2000) *Plant Cell Rept.*, **19** : 989-993.

Konvicka, O. (1973) *Biol. Plant.*, **15** : 144-149.

Konvicka, O. (1984) *Allium Newsl.*, **1** : 28-37.

Konvicka, O., Niehaus, F. and Fischbeck, G. (1978) *Zeitschrift fur Pflanzenzuchtung*, **80** : 265-276.

Konvicka, O., Nienhaus, F. and Fischbeck, G. (1978) *Z. Pflanzenzucht.*, **80** : 265-276.

Koo, B., Chang, M. and Choi, D. (1998) *Korean J. Plant Pathol.*, **14**(2) : 136–144.

Korla, B. N. and Rastogi, K. B. (1979) *Punjab Hort. J.*, **19** : 89-92.

Korla, B.N., Singh, A.K. and Kalia, P. (1981) *Haryana J. Hortic. Sci.*, **10** : 77-80.

Kothari, I. L. and Shah, J. J. (1974) *Phytomorphology*, **24** : 42-48.

Koul, A. K., Gohil, R. N. and Langer, A. (1979) *Euphytica*, **28** : 457-464.

Koul, A.K. and Gohil, R.N. (1970) *Cytologia*, **35** : 197-202.

Ku, Y. S., Nho, S. P., Lee, G. J. and Chung, D. S. (1974) *Res. Rept. Rural Dev. Hort. Agric.*, **16** : 99-106.

Kumar, A., Prasad, B., Saha, B.C., Sinha, R.P. and Maurya, K.R. (1994) *J. Appl. Biol.*, **4** : 23-26.

Kumar, M. (2015) *J. Plant Dev. Sci.*, **7** (5) : 473-474.

Kumar, S., Rana, S.S., Chander, N. and Sharma, N. (2013) *Indian J. Weed Sci.*, **45**(2) : 126–130.

Kumara, B.R., Shankargoudapatil, Hegde, N.K. and Gangadharappa, P.M. (2014) *Trends Biosci.*, **7**(12): 1323-1326.

Kumari, S., Sharma, M.K., Kumar, D. and Singh, S.P. (2013) *Ann. Hortic.*, **6** (2) : 326-330.

Kusumo, S. and Widjajanto, D. D. (1973) *Bull. Horticultura*, 'Tjahort' No. 10, pp. 16-29.

Lallan-Singh, Dubey, B.K., Pandey, U.B. and Singh, L. (1993) *Newsletter Associated Agricultural Development Foundation*, **13** : 8-10.

Lallemand, J., Messian, C.M., Briand, F., Etoh, T. and Galmarini, C.R. (1997) *Acta Hortic.*, **433** : 123-132.

Landry, B.S. and Khanizadeh, S. (1994) *Canadian J. Plant Sci.*, **74** : 353-356.

Larrain, S.P. (1986) *Agricultura Tecnica* (Chile), **46**(2) : 147–150.

Lavigne, D. and Audubert, A. (1998) *Infos Paris*, **140** : 48-52.

Lawande, K. E. (2018) *J. Allium Res.*, **1**(1) : 1–6.

Lawande, K.E., Khar, A., Mahajan, V., Srinivas, P.S., Sankar, V. and Singh, R.P. (2009) *J. Hortic. Sci.*, **4** (2) : 91-119.

Lazzari, M. A. (1982) *Plant and Soil*, **67** : 187-191.

Ledesma, A., Reale, M. I. and Racca, R. (1980) *Phyton, Agrentina*, **39** : 37-48.

Lee, W. S. (1974) *J. Korean Soc. Hort. Sci.*, **15** : 119-141.

Lee, Y.H., Lee, W.H. and Lee, D.K. (1988) *Korean J. Plant Pathol.*, **14** : 594-597.

Li, Z.Y., Luo, H.L. and Guo, Y. (1999) *J. South China Agri. Univ.*, **20** : 54-58.

Likhatskil, V. I. (1987) In : *Puti Intensifikatsii Ovoshchevodstva. Kiev*, 52-57.

Lima, Mayky Francley Pereira De, Lopes, Welder De Araújo Rangel, Negreiros, Maria Zuleide De, Grangeiro, Leilson Costa, Sousa, Hiago Costa De and Silva, Otaciana Maria Dos Prazeres Da. (2019) *Rev. Caatinga*, **32**(4) : 966-975.

Lin, G. *et al.* (1987) *Journal of the Chinese Association of Refrigeration*, No. 2, pp. 34-44.

Linx, A. and Niwuzhang, Z.Y. (1997) *Acta Agricultural Zhe Jiangensis*, **9** : 143–148.

Lipinski, V., gaviola de Heras, S. and Filippini, M.F. (1995) *Ciencia del Suelo*, **13** : 80-84.

Lipinski, V., Gaviola de Heras, S., filippini, M.F. and Burba, J.L. (1994) *III Curso taller sobre produccion, comercializacion, e industrializacion de ajo*, Argentina, 22-25 June, 1993, pp. 235-245.

Liu, C.Z., Guo, C., Wang, Y.C. and Ouyang, F. (2002) *Process Biochem.*, **38** : 581–585.

Liu, G.Q., Li, S.J. and Zhang, X.P. (1997) *Acta Hortic.*, **24** : 165-169.

Liu, H., Deng, R., Huang, C., Cheng, Z. and Meng, H. (2019) *Sci. Hortic.*, **246** : 298–306.

Lopez-Frasca, A. Rigoni, C., Silvestri, V., Burba, J.L. and Galmarini, C.R. (1997a) *Acta Hortic.*, **433** : 279-284.

Lucero, J. c., Andreoli, C., Reyzabal, M. and Larregui, Y. (1982) *Anales de Edafologia Agrobiologia*, **40** : 1807-1814.

Luzny, J. (1987) *Metodiky ProZavadeni Vysledku Vyzkumu do Zemedelske Praxe, S-IV*, pp. 1-24.

Ma, Y., Wang, H.L., Zhang, C.J. and Kang, Y.Q. (1994) *Plant Cell Rep.*, **14** : 65-68.

Madhav, J.V., Sethi, S., Kaur, C. and Pal, R.K. (2016) *Indian J. Hortic.*, **73** : 274-278.

Madhu, B., Mudgal, V.D. and Champawa, P.S. (2019) *J. Food Process. Eng.*, https://doi.org/10.1111/jfpe.13177.

Makowska, Z. and Kotlinska, T. (2000) *Veg. Crops Res. Bull.*, **53** : 17-21.

Makowska, Z. and Kotlinska, T. (2001) *Veg. Crops Res. Bull.*, **54** : 19-23.

Makowska, Z., Keller, J. and Engelmann, F. (1999) *Cryo-Letters*, **20** : 175-182.

Mann, L. K. (1952) *Hilgardia*, **21** : 195-231.

Mantis, A. J., Koidis, P. A., Karaionnoglou, P. G. and Panetsos, A. G. (1979) *Labensmittel-Wissenschaft Technologie*, **12** : 330-332.

Marchesi, G. and Fouchi, A. (1979) *Sementi Elette*, **26** : 23-25.

Martinez, Beringola, M. L. and Garcia, A. A. (1979) *Protection Vegetas*, No. 9, pp. 33-43.

Martinez, G.E., Arcia, A., Subero, L., Albarracin, M. and De Albarracin, N. (1996) *Agronomia Tropical Maracay*, **46** : 265-274.

Marys, E., Carballo, O. and Izaguirre-Mayoral, M.L. (1994) *J. Phytopath.*, **142** : 227-234.

Masaadeh, H.A., Hayajneh, W.A. and Momani, N.M. (2006) *J. Med. Sci.*, **6**(4): 650-653.

Mathew, D., Forer, Y., Rabinowitch, H.D. and Kamenetsky, R. (2010) *Environ. Exp. Bot.*, **71** : 166–173.

Mathur, R. B. L., Arora, P. N. and Parshad, M. (1975) *Indian J. Agron.*, **20** : 46-47.

Matus, I., Gonzalez, M.I. and Pozo, A. del (1999) *Plant Genety. Reso. Newslet.*, **117** : 31-36.

Maurya, K. R. (1987) *Progress. Hort.*, **19** : 132-134.

McCollum, G.D. (1976) In: Simmonds, N.W. (Ed.), Evolution of Crop Plants. 9th Edition, Longman, London, pp. 186–190.

Meena, B., Arvindakshan, K., Singh, P. and Patidar, D. K. (2016) *The Ecoscan*, **9** : 615-619.

Meer, Q.P. van der and Galmarini, C.R. (1997) *Acta Hortic.*, **433** : 17-31.

Melnik, T. K. (1973) *Referativny i Zhurnal*, **9**.11.55.

Melo, J.P.L., Oliveira, A.P. de and de Oliveira, A.P. (1999) *Horticultura Brasileira*, **17** : 11-15.

Melzer, O., Alt, D., Ladebusch, H., Burns, I.G., Bending, G.D. and Mulholland, B. (1999) *Acta Hortic.*, **506** : 29–36.

Memane, P.G, Tomar, R.S., Kakade, D.K., Kulkarni, G.U. and Chovatia, R.S. (2008) *Asian J. Hort.*, **3**(1) : 82-86.

Méndez, V., Lok, R. and Somarriba, E. (2001) *Agroforest. Syst.*, **51** : 85-96.

Menezes, J. A. D. E. S., Novais, R. F. D. E., Dos, Santos, H. L. and Sans, L. M. A. (1994) *Revista Ceres*, **21** : 203-212.

Meredith, T. (2008) The Complete Book of Garlic: A Guide for Gardeners, Growers, and Serious Cooks. Timber Press, Portland, OR.

Miedema, P. (1994) *J. Hortic. Sci.*, **69**(1) : 29–39.

Mihailescu, N., Iordachescu, C. and Ivan, M. (1979a) *Productia Vegetala Horticultura, Romania*, **23** : 19-23.

Mihailescu, N., Ivan, M., Ioradachescu, C. and Ungureanu, A. (1979b) *Fructelor Bucharest, Romania,* **10** : 47-53.

Miko, S., Ahmed, M.K., Amans, E.B., Falaki, A.M. and Ilyas, N. (2000) *J. Agric. Environ.,* **1** : 260-264.

Mishra, A.C. and Pandey, V. (2015) *Indian J. Hortic.,* **72** : 514.

Mishra, R.K., Jaiswal, P.K., Kumar, D., Saabale, P.R. and Singh, A. (2014) *J. Plant Breed. Crop Sci.,* **6** : 160–170.

Mollejas, J. F. and Mata, R. H. (1973) *Bol. Tech. Estac. Exp. Agric.,* **6** : 1-13.

Moon, W. and Lee, B. Y. (1980) *J. Korean Soc. Hort. Sci.,* **21** : 109-118.

Moore, G. S. and Atkins, R. D. (1977) *Mycologia,* **69** : 341-348.

Moravec, J. and Kvasnicka, S. (1975) *Sbornik UVITY Zahradnictyi,* **2** : 117-123.

Moravec, J., Kvasnicka, S. and Velicka, O. 91974) *Bull. Vyzkumny Ustav Zelinarsky Olomouse,* No. 18, pp. 15-23.

Motaz, B.M., Omar, F., Abd, E.E., Shiaty, M., Imam, A.A., Galm, G., Shahin, H. and Zein, A. (1971) *Agric. Rev.,* **49** : 157-172.

Munoz de Con, L. and Galmarini, C.R. (1997) *Acta Hortic.,* **433** : 257-263.

Nabulsi, I., Al-Safadi, B., Mit-Ali, N. and Arabi, M.I.E. (2001) *Ann. Appl. Biol.,* **138** : 197-202.

Nadejda, M. (1977) *Analele Inst. de Cercetari pentru Protectia Plantelor,* **13** : 247-267.

Nair, A., Khar, A., Hora, A. and Malik, C. (2013) *Int. J. Life Sci.,* **2** : 72-89.

Najafabadi, M., Peyvast, G., Asil, M., Olfati, Jamal-Ali and Rabiee, M. (2012) *Int. J. Plant Prod.,* **6**(3) : 279-290.

Nakov, B., Kotetsov, P. and Angelov, D. (1974) *Restitelna Zashchita,* **22** : 33.

Nandal, T.R., Ankur-Vermani, Ravinder-Singh, Vermani, A. and Singh, R. (2001) *Res. Crops,* **2** : 159-161.

Naresh, B., Srivastava, S. K. and Agarwal, S. (2013) *Indian J. Trad. Knowl.,* **12**(3) : 518–523.

Naruka, I.S. (2000) *Curr. Agric.,* **24** : 137-138.

Nath, Prem, (1976) *Vegetables for the Tropical region, ICAR,* New Delhi.

NHB. (2019) National Horticulture Board, Ministry of Agriculture and Farmers Welfare Government of India, Gurugram - 122015 (Haryana), India.

Nogueira, F. D., Ferreira, F. A., Andrade, H. and Gualberto, V. (1982) *Horticultura Brasileira,* **1** : 28-32.

Novak, F. J. (1990) In: Rabinowitch, H. D. and Brewster, J. L. (Eds.), Onions and Allied Crops., Vol. II, CRC Press, Boca Raton, Fl., U.S.A., pp. 233-250.

Novak, F.J. (1972) *Experientia,* **28** : 363-364.

Novak, F.J. and Havranek, P. (1975) *Biologia Plantarum* (Praha), **17** : 376-379.

Nunez, S.B., Ledesma, A., Milanesi, E., Dubois, M.E., Cerana, M.M., Arguello, J.A. and Galmarini, C.R. (1997) *Acta Hortic.*, **433** : 395-403.

Nurzynska-Wierdak, R. (1997a) *Folia Hortic.*, **9** : 67-75.

Nurzynska-Wierdak, R. (1997b) *Folia Hortic.*, **9** : 77-83.

Ohsumi, C., Kojima, A., Hinata, K., Etoh, T. and Hayashi, T. (1993) *Theor. Appl. Genet.*, **85** : 969-975.

Oliveira, A.P. de, Castellane, P.D. and Bortoli, S.A. de (1995b) *Agropecuaria Catarinense*, **8** : 5-6.

Oliveira, A.P. de, Castellane, P.D. and Bortoli, S.A. de (1995a) *Horticultura Brasileira*, **13** : 202-205.

Oliveira, A.P. de, Castellane, P.D., Bortoli, S.A. de and Banzatto, D.A. (1995c) *Horticultura Brassileira*, **13** : 202-205.

Oliveira, A.P. de, Castellane, P.D., Bortyoli, S.A. de, De Oliveira, A.P. and De Bortoli, S.A. (1995d) *Agropecuaria Catarinense*, **8** : 506.

Om, H. and Awasthi, D. N. (1977) *Progress. Hort.*, **9** : 63-67.

Om, H. and Srivastava, R. P. (1974) *Progress. Hort.*, **6** : 71-76.

Om, H. and Srivastava, R. P. (1977) *Indian J. Hort.*, **34** : 152-156.

Omar, F. A. and Arafa, A. E. (1979) *Agric. Res. Rev. Hort.*, **57** : 215-221.

Orcutt, D.M. and Nilsen, E.T. (2000) The Physiology of Plants under Stress: Soil and Biotic Factors, John Wiley and Sons: New York, NY, USA.

Orlowski, M. and Rekowska, S. (1989) *Biuletyn Warzywniczey, Suppl. III*, 85-89.

Orzhekhovskya, T. E. and Starikova, D. A. (1990) *sibirskit Vestnik Sel'skokhozyaistvennoi Nauki*, No. 3, pp. 30-34.

Pal, A. K. and Basuchaudhury, K. C. (1976) *Curr. Sci.*, **45** : 739.

Pandey, U. C. and Singh, J. (1989) *Agric. Univ. Res. J.*, **19** : 69-71.

Pandey, U.B. and Singh, D.K. (1993) *Newsl. - Nat. Hort. Res. Dev. Found.*, **13** : 10-12.

Pandey, U.B., Chauhan, K.P.S. and Singh, D.P. (1996) *Newsl. - Nat. Hort. Res. Dev. Found.*, **16** : 10-11.

Pandey, U.B., Chauhan, K.P.S. and Singh, D.P. (1996b) *Newsl. - Nat. Hort. Res. Dev. Found.*, **16** : 10-11.

Pandey, U.B., Gupta, R.P. and Chougule, A.B. (1996a) *Newsl. - Nat. Hort. Res. Dev. Found.* **16** : 10-11.

Parashar, K. S., Arora, P. N. and Sharma, R. P. (1979) *Indian Sugar*, **29** : 217-223.

Park, H.M., Kim, J.O., Kang, U.G., Kang, H.W. and Park, K.B. (1997) *RDA J. Agro Environ, Sci.*, **39** : 35-39.

Park, M.Y., Yi, N.R., Lee, H.Y., Kim, S.T., Kim, M., Park, J., Kim, J., Lee, J.S., Cheong, J.J. and Choi, Y.D. (2002) *Mol. Breed.*, **9** : 171–181.

Park, Y. B. and Lee, B. Y. (1989) *Abstract of communicated papaers, Korean Soc. Hort. Sci.*, **7** : 76-77.

Park, Y.M. (1997) *J. Korean Soc. Hort. Sci.*, **38** : 704-708.

Patale, S. S. (2019) *BIOINFOLET*, **16** : 116-118.

Patel, P.B. and Patel, J. J. (2012) *AGRES*, **1** : 256-262.

Patel, Z., Bhalerao, P.P., Gaikwad, S.S. and Patel, N. (2019) *Curr. Hortic.*, **7** : 43-45.

Patidar, M., Shaktawat, R.P.S. and Naruka, I.S. (2017) *J. Krishi Vigyan*, **5** : 54.

Patil, B. G. and Rane, M. S. (1973) *Mag. Coll. Agric.*, Nagpur, **45** : 86-89.

Patil, B.T., Siddhu, G.M., Shinde, K.G. and Handal, B.B. (2017) *Veg. Sci.*, **44** : 71-74.

Pawar, D.B., Lawande, K.E. and Kale, P.N. (1990) *J. Maharashtra Agric. Univ.*, **15**(3) : 352–353.

Pellegrini, C.N., Croci, C.A. and Orioli, G.A. (2000) *Rad. Phys. Chem.*, **57** : 8–11.

Pereira, E. B., Fornazier, M. J., Souza, J. L. De., Ventura, J. A. and Nogueira, F. D. (1987) *Horticultura Brasileira*, **5** : 36-38.

Perez, M. B., Aveldaño, M. I. and Croci, C. A. (2007) *Postharvest Biol. Technol.*, **44**(2), 122–130.

Perez-Moreno, L., Salinas-Gonzalez, J.G. and Sanchez-Pale, J.R. (1995) *Revista Mexicana de Fitopatologia*, **13** : 18-25.

Perez-Talavera, S. (2001) *Alimentaria*, **38** : 75-78.

Petropoulos, S., Ntatsi, G. and Ferreira, I. (2017) *Food Rev. Int.*, **33** : https://10.108 0/87559129.2015.1137312.

Pokluda, R. and Petrikova, K. (2001) *Zahradnictvi Horticultural Science*, **28** : 112-115.

Polishchuk, S. F., Gorkutsenko, A. V. and Zhila, E. D. (1982) *Selektsiya i Semenovodstvo*, No. 50, pp. 66-76.

Pooja, B., Pramanik, G. and Pramanik, T. (2020) *Res. J. Pharm. Tech.*, **13**(1) : 152-156.

Pooler, M.R. and Simon, P.W. (1993a) *HortScience*, **28** : 1085-1086.

Pooler, M.R. and Simon, P.W. (1993b) *Euphytica*, **68** : 121-130.

Pooler, M.R. and Simon, P.W. (1994) *Sexual Plant Reproduction*, **7** : 282-286.

Pordesimo, A. N. and Ilag, L. L. (1976) *Kalikasan Philippine J. Biol.*, **5** : 251-258.

Portela, A. (1998) *Advances in Horticulture*, **3** : 30-42.

Portela, J.A. (2001) *Acta Hortic.*, **555** : 175–178.

Portela, J.A. and Armstrong, J. (2001) *Acta Hortic.*, **555** : 175-179.

Pozo, A. del, Gonzalez, M.I., Barraza, C., and Galmarini, C.R. (1997) *Acta Hortic.*, **433** : 389-393.

Prasad, B. K., Sinha, T. S. P., Shanker, U. and Kumar, (1986) *Indian Phytopath.*, **39** : 622-624.

Priyanshu, A., Singh, E.M., Malik, S., Kumar, M., Kumar, V., Tripathi, S. and Shahi, U. (2019) *Progress. Agric.*, **19** : 242-246.

Prodan, G. E., Florescu, Mihalache, M., Visarion, M., Baciu, E., Dorobantu, N. and Tudor, T. (1977) *Nicolae Balcescu Hort.*, **17** : 7-15.

Pruthi, J. S. (1979) *Spices and Condiments*, National Book Trust, India, New Delhi, pp. 125-132.

Purewal, S. S. and Dargan, N. G. (1961) *Indian J. Agron.*, **5** : 262-268.

Purseglove, J. W. (1975) *Tropical Crops : Monocotyledons*, ELBS Longman, London, pp. 52-56.

Pyo, Hk. Lee, By; Moon, W. and Woo, Jk. (1979) *J. Korean Soc. Hortic. Sci.*, **20** : 19-27.

Qasem, J.R. (1996) *J. Hort. Sci.*, **71** : 41-48.

Rahim, M.A., Chowdhury, M.N.A, Anwar, H.R.M.M. and Alam, M.S. (2003) *Asian J. Plant Sci.*, **2** : 171-174.

Rahim, M.A., Fordham, R. and Armstrong, J. (2001) *Acta Hortic.*, **555** : 181-188.

Rahm, M. A. and Fordham, R. (1988) *Sci. Hort.*, **37** : 25-38.

Rahman, A. K. M. M. and Talukdar, M. R. (1986) *Bangladesh J. Agric.*, **11** : 19-26.

Rakhimbaev, I. R. and Olshanskaya, R. V. (1976) *Fiziologiya Rastenii*, **23** : 76-79.

Rakhinbaev, I. R. and Solomina, V. F. (1980) *Referativny i Zhurnal*, 7.55.307.

Ramirez, H. V. E., Lopez, G. C. A. and Loria, M. W. (1973) *Boletin Teck, Facult. Agromomia, Univ. de Costa Rica*, **6** : 14.

Ramniwas-Goutam, Singh, D.K. and Goutam, R. (1998) *Orissa J. Hort.*, **26** : 72-74.

Rani, P., Batra, V.K., Bhatia, A.K. and Shiwani. (2018) *Int. J. Curr. Microbiol. App. Sci.*, **7**(12): 2946-2951.

Rath, G. C. and Mohanty, G. N. (1978a) *Indian Phytopath.*, **31** : 256-257.

Rath, G. C. and Mohanty, G. N. (1978b) *Sci. Cult.*, **44** : 467-468.

Rath, G. C. and Mohanty, G. N. (1986) *Indian Phytopath.*, **39** : 614-515.

Rathore, D. N., Kadian, V. S. and Panwar, B. S. (1981) *Haryana Agric. Univ. J. Res.*, **11** : 355-360.

Ravnikar, M., Mavric, I., Ucman, R., Ivanovic, S., Kus, M. and Zel, J. (1996) *Novi izzivi v poljedelstvu '96, Zhornik simpozija*, Ljubljana, Slovenia 9-10 December, 1996, pp. 189-193.

Ravnikar, M., Plaper, I., Ucman, R., Zel, J. and Kreft, I. (1994) *Proc. International Colloquium on Impact of Plant Biotechnology on Agriculture*, Rogla, Slovenia, 5-7th December, pp. 97-102.

Rendon, M.C., Balzarini, M., Burba, J.L. and Galmarini, C.R. (1997) *Acta Hortic.*, **433** : 507-518.

Resende, F.V., Gualberto, R., Souza, R.J. de and de Souza, R.J. (2000) *Scientia Agricola*, **57** : 61-66.

Resende, F.V. and Souza, R.J. de (1998) *UNIMAR-Ciencias,* **7** : 81-88.

Rizzalli, R.H., Villalobos, F.J. and Orgaz, F. (2002) *Eur. J. Agron.,* **18** : 33–43.

Robledo, P.A., Villalobos arambula, V. and Jofre-Garfias, A. (2000) *In Vitro Cell. Dev. Biol. Plant.,* **36** : 416-419.

Robledo, P.A., Cabrera-Ponce, J., Villalobos arambula, V., Herrera-Estrella, L. and Jofre-Garfias, A. (2004) *HortScience,* **39** : 1208-1211.

Roksana, R., Alam, M.F., Islam, R. and Hossain, M.M. (2002) *Plant Tissue Cult.,* **12**(1) : 11-17.

Romascu, E. (1977) *Analele Inst. De Cercetari pentru Protectia Plantelor,* **13** : 159-162.

Rotem, N., Shemesh, E., Peretz, Y., Akad, F., Edelbaum, O., Rabinowitch, H.D., Sela, I. and Kamenetsky, R. (2007) *J. Exp. Bot.,* **58** : 1133-1141.

Rout, E., Nanda, S., Nayak, S. and Joshi, R.K. (2014) *Physiol. Mol. Plant Pathol.,* **85** : 15-24.

Roy, A. N., Sharma, R. B. and Gupta, K. C. (1977) *Curr. Sci.,* **46** : 716-717.

Russell, P. E. and Musa, A. E. A. (1977) *Ann. Appl. Biol.,* **86** : 369-372.

Sadaria, S.G., Malavia, D.D., Khanpara, V.D., Dudhatra, M.G., Vyas, M.N. and Mathukia, R.K. (1997a) *Indian J. Agron.,* **42** : 653-656.

Sadaria, S.G., Malavia, D.D., Khanpara, V.D., Dudhatra, M.G., Vyas, M.N. and Mathukia, R.K. (1997b) *Indian J. Agri. Sci.,* **67** : 402-403.

Salahuddin, M., Rahim, M.A., Alam, S.M.J.B., Rahman, M.M. and Rahman, J. (2019) *Malaysian J. Halal Res.,* **2** : 46-52.

Saluzzo, J.A. and Rattin, J.E. (2000) *Horticultura Argentina,* **17** : 42-43.

Sampat, Chopra, S., Kumar, A. and Samnotra, R.K. (2014) *Indian J. Weed Sci.,* **46**(2): 146–150.

Sánchez-Virosta, Á., Léllis, B.C., Pardo, J., Martínez, A., Sánchez-Gómez, D. and Domínguez, A. (2019) *Agric. Water Manag.,* **228**. 105886. 10.1016/j. agwat.2019.105886.

Sandhu, K.S., Singh, Daljit, Singh, Jaswinder, Singh, D. and Singh, J. (1997) *Veg. Sci.,* **24** : 7-9.

SangSik, N., InHu, C., SangKyung, B. and JinKi, B. (2007) *Korean J. Hortic. Sci. Technol.,* **25** : 169–173.

Sankar, V., Qureshi, M.A., Tripathi, P.C. and Lawande, K.E. (2001) *South Indian Hort.,* **49** : 379-381.

Saraf, R.K., Kurmvanshi, S.M., Sharma, R.S. and Parihar, M.S. (2000) *Crop Res.,* Hisar, **19** : 149-151.

Satti, S.M.E. and Lopez, M. (1994) *Pak. J. Bot.,* **26** : 161-165.

Satyagopal, K., Sushil, S.N., Jeyakumar, P., Shankar, G., Sharma, O.P., Boina, D., Sain, S.K., Ram Asre, Kapoor, K.S., Arya, S., Kumar S., Patni, C.S., Chattopadhyay, C.,

Pawar, S.A., Shukla, A., Bhale, U., Basanagoud, K., Mishra, H.P., Ekabote, Suresh D., Thakare, A.Y., Halepyati, A.S., Patil, M.B., Sreenivas, A.G., Sathyanarayana, N. and Latha, S. (2014) AESA based IPM package for garlic, p. 46.

Sawahel, W. (2002) *Cell. Mol. Biol. Lett.*, **7** : 49-59.

Selvaraj, N., Irulappan, I. and Vedamuthu, P.G.B. (1997) *South Indian Hort.*, **45** : 75-77.

Seno, S. (1997) *Cultura Agronomica*, **6** : 29-40.

Setty, B. S., Sulikeri, G. S. and Hulamani, N. C. (1989) *Karnataka J. Agric. Sci.*, **2** : 160-164.

Shaha, S.R. and Kale, P.N. (1999) *J. Maharashtra Agri. Univ.*, **24** : 92.

Shaha, S.R., Kale, P.N., Dhankhar, B.S. and Shirsath, N.S. (1990) *Haryana J. Hortic. Sci.*, **19** : 313-317.

Shankaracharya, N. B. (1974) *Symposium on Spice Industry in India, AFST, CFTRI,* Mysore, pp. 24-36.

Sharma, C.K., Barman, D., Singh, I.P. and De, L.C. (1997) *Indian J. Hill Fmg.*, **10** : 113-114.

Sharma, N., Sehgal, S. and Kumar, S. (2011) *Pesticide Res. J.*, **23** : 14-17.

Sharma, R.P. (1992) *Indian J. Agron.*, **37** : 268–284.

Sharma, V. D., Sethi, M. S., Kumar, A. and Rarotra, J. R. (1977) *Indian J. Exp. Biol.*, **15** : 466-468.

Shasany, A.K., Ahirwar, O.P., Sushil-Kumar, Khanuja, S.P.S., Kumar, S. and Singh, A.K. (2000) *J. Medicnal Aromatic Plant Sci.*, **22** : 586-592.

Sheikh, M.Q. and Khandy, B.A. (2008) *J. Ornament. Hortic.*,**11**(3): 216-219.

Shin, K. H., Park, J. C., Lee, K. S., Han, K. Y. and Lee, Y. S. (1988) *Research Reports of the Rural Development Administration, Horticulture,* Korea Republic, **30** : 41-52.

Shinde, B.N., Pawar, B.R., sonkamble, A.M., Wankhede, S.D. and Khandare, V.S. (1999) *J. Soils Crops*, **9** : 111-113.

Shiwani, Lal, M., Raj, D. and Kumar, V. (2017) *Int. J. Curr. Microbiol. App. Sci.*, **6**(12): 560-567.

Shrestha, R.L., Karki, T. and Pun, A. (2000) *Working Paper Lumle Agricltural Research Centre,* No. 2000-I, pp. 3-5.

Siddique, M. A. and Rabbani, M. G. (1985) *Bangl. J. Bot.*, **14** : 41–46.

Silabut, N., Naruka, I.S., Shaktawat, R.P.S., Verma, K. and Seyle, A. (2014) *Indian J. Hortic.*, **71** : 354-359.

Silva, E.C. da, Souza, R.J. de, Santos, V.s., da Silva, E.C. and de Souza, R.J. (2000) *Ciencia e Agrotechnologia*, **24** : 939-946.

Silva, N. F. and CAsali, V. W. D. (1987) *Horticultura Brasileira*, **5** : 29-30.

Simon, P.W. and Jenderek, M.M. (2004) *Plant Breed.*, **23** : 211-244.

Singh, B. M. and Sharma, Y. R. (1977) *Indian Phytopath.*, **30** : 272-273.

Singh, B.P., Rana, R.S. and Midmore, D.J. (1994) *Acta Hortic.*, **358** : 181-188.

Singh, D.K. and Gupta, R.P. (1998) *Newsl. - Nat. Hort. Res. Dev. Foundation*, **18** : 19-22.

Singh, G., Ram, C.N., Singh, A., Shrivastav, S.P., Maurya, P.K., Kumar, P. and Sriom. (2018) *Int. J. Curr. Microbiol. App. Sci.*, **7**(02) : 1362-1372.

Singh, H.D., Maji, S. and Kumar, S. (2008) *Int. J. Agric. Sci.*, **10** : 510-524.

Singh, H.D., Maji, S. and Kumar, S. (2014) *Int. J. Agric. Sci.*, **10**(2): 546-549.

Singh, J. R. and Tewari, J. (1968) *Indian J. Hort.*, **25** : 191-195.

Singh, J.V., Kumar, A., Sirohi, H.S. and Kumar, A. (1995) *Indian J. Agric. Res.*, **29** : 153-156.

Singh, M.C. and Tiwari, R.S. (1995) *Haryana J. Hort. Sci.*, **24** : 46-49.

Singh, R.K. and Dubey, B. K. (2015) *Curr. Hortic.*, **3**(1) : 41-48.

Singh, R. K. and Dubey, B. K. (2011) *Indian J. Hortic.*, **68** : 123–127.

Singh, R. N. and Singh, J. R. (1974) *Indian J. Hort.*, **31** : 255-258.

Singh, R. P. (1987) *J. Res. Assam Agril. Univ.*, **5** : 181-183.

Singh, R. V. and Phogat, K. P. S. (1989) *Progress. Hort.*, **21** : 145-147.

Singh, R.K., Dubey, B.K. and Gupta, R.P. (2014) *J. Spices Aromatic Crop*, **21** (2): 136-144.

Singh, S. (2008) *Agric. Sci. Digt.*, **28** : 189–191.

Singh, S.P. and Dwivedi, V. K. (2002) *New Agriculturist*, **13** : 5–7.

Singh, V., Singh, A., Mehta, V.S., Singh, V. and Singh, A. (1995) *Fert. News*, **40** : 47-49.

Singh, Y. and Chand, R. (2004) *Hortic. Soc. Haryana*, **33** : 146-147.

Slusarenko, A., Patel, A. and Portz, D. (2008) *Sust. Dis. Manag. European Cont.*, **121** : 313-322.

Sonawane, J.M., Yadav, A., Ghosh, P.C. and Adeloju, S.B. (2017) *Biosens. Bioelectron.*, **90** : 558-576.

Song, J.T., Chang, M.U., Lee, J.S. and Choi, Y.D. (1997) *Molecules and Cells*, **7** : 705-709.

Song, Y.S., Choi, I.H., Jang, Y.S., Choi, W.Y. and Park, J.H. (2001) *J. Korean Soc. Hort. Sci.*, **42** : 305-309.

Sonkamble, A.M., Shinde, B.N., Khandare, V.S., Wakle, P.K., Sonakamble, M.M. and Pawar, B.R. (2000) *J. Soils Crops*, **10** : 86-89.

Sood, D., Chhokar, V., Singh, J. and Singh, J. (2000) *Veg. Sci.*, **27** : 180-184.

Sotomayor, R. I. (1975a) *Agric. Tecnica (Chile)*, **35** : 175-178.

Sotomayor, R. I. (1975b) *Investigacion Progreso, Chile*, **7** : 34-44.

Sripradha S., Murthykumar, K., Soundarajan, S. and Naveed, N. (2014) *Res. J. Pharm. Tech.*, **7**(6): 727-729.

Stahlschmidt, O., Cavagnaro, J.B., Borgo, R. and Galmarini, C.R. (1997) *Acta Hortic.*, **433** : 427-434.

Starikova, D. A. (1977) *Referativny i Zhurnal,* **8**.55.599.

Stavìlíková, H. (2008) *Hort. Sci.* (Prague), **35**(3): 130–135.

Sterling, S.J., Eagling, D.R. and Armstrong, J. (2001) *Acta Hortic.,* **555** : 63-73.

Suciu, Z., Berar, V., Negrau, G., Ivascu, R., Ivascu and Radulescu, A. (1988) *Productia Vegetala, Horticultura,* **37** : 6-7.

Sugha, S.K. (1995) *Indian J. Agri. Sci.,* **65** : 455-458.

Suheri, H. and Price, T.V. (2000) *Aust. Plant Pathol.,* **29** : 192-199.

Sukul, N. C., Das, P. K. and De, G. C. (1974) *Nematologica, Bengal (India),* **20** : 187-191.

Sumi, S., Castellane, P., Bellingieri, P. and Churatamasca, M. G. C. (1986) *Horticultura Brasileira,* **4** : 32-34.

Sutarya, R., Van-Dijk and Sudarsono (1994) *Acta Hortic.,* **369** : 134-143.

Swaroop, K. (2010) *Indian J. Agric. Sci.,* **80**(8): 742-745.

Takagi, H. (1990) In : Rabinowitch, H.D. and Brewster, J.L. (Eds.), Onions and Allied Crops, Volume 3, Biochemistry, Food Science, and Minor Crops, CRC Press: Boca Raton, FL, USA, pp. 109–157.

Takagi, H. and Aboba, T. (1976) *Yamagata Agric. Forestry Soc.,* No. **33** : 39-50.

Takhtajan, A. (1967) Diversity and Classification of Flowering Plants. Columbia University Press, New York, pp. 501–504.

Talavera, S.P. and Cervantez, M.L. (1999) *Alimentaria,* **300** : 101-102.

Talavera, S.P., Lezcano, A.P., Cervantes, M.L., Enriquez, I.I., Nunez, L.M.G., Tabares, F.P. and Fraga, R. (1999) *Alimentaria,* **300** : 103-105.

Taner, Y., Kunter, B., Besirli, G. and Yanmaz, R. (2005) *Bahce,* **33** : 95-99.

Tariq-Mahmood, Khokhar, K.M., Hussain, S.I. and Bhatti, M.H. (2001) *Sarhad J. Agri.,* **17** : 209-212.

Thakur, J.C., Bhathal, G.S. and Gill, S.P.S. (1997) *J. Res. Punjab Agri. Univ.,* **34** : 40-44.

Thangasamy, A. and Chavan, K. (2017) *Indian J. Hortic.,* **74** : 80-84.

Thomas, P. (1999) *Int. Conf. Ensuring Safety Quality Food Rad. Process.,* pp. 24–26.

Thompson, H. C. and Kelly, W. C. (1957) *Vegetable Crops, McGraw-Hill Book Co., Inc.* New York, 368-370.

Tizio, R. (1979) Florasion in vitro de l'ail (*Allium sativum* L.), C. R. Academy of Science. Paris, **289** : 401-404.

Trevisan, J.N., Martins, G.A.K. and Santos, N.R.Z. dos (1997) *Ciencia Rural,* **27** : 7-11.

Trevisan, J.N., Martins, G.A.K., Zamberlan dos Santos, N.R. and Dos-Santos, N.R.Z. (1996) *Ciencia Rural,* **26** : 29-32.

Tripathi, P. and Lawande, K.E. (2006) *Tech. Bull. No. 15,* pp. 1-8.

Tripathi, P., Sankar, V. and Lawande, K.E. (2017) *Curr. Hortic.,* **5** : 3-14.

Tripathi, P., Sankar, V. and Lawande, K.E. (2009) *Indian J. Hortic.*, **66** : 511-515.

Trippel, V. V. and Chubrikova, L. P. (1976) *Referativny i Zhurnal*, 8.55.771.

Trunkenboltz, M. (1975) *Pepinieristes Hort. Maraichers (France), No.* **155** : 49-51.

Usman, M.G., Fagam, A. S., Dayi, R.U. and Isah, Z. (2016) *Int. J. Agron.*, **16** : 1-9.

Vázquez-Barrios, M., López-Echevarría, G., Mercado-Silva, E., Castaño-Tostado, E. and Gonzalez, F. (2006) *Sci. Hortic.*, **108** : 127-132.

Vidal, B.C., Mello, M.L.S., Illg, R.D. and Campos-Vidal, B. de (1993) *Revista Brasileira de Genetica*, **16** : 347-356.

Vijay, O.P. (1990) *Indian J. Hortic.*, **47** : 431-433.

Volk, G. M., Rotindo, K. E. and Lyons, W. (2004) *HortScience*, **39**(3) : 571–573.

Walkey, D.G.A., Webb, M.J.W., Bolland, C.J. and Miller, A. (1987) *J. Hortic. Sci.*, **62**(2) : 211-220.

Walter, S.J., Camphell, C.S., Kellog, E.A. and Stevens, P.F. (1999) Plant Systematics–A Phylogentic Approach, Siananer Associates Inc Publishers Sunderland, Massachussetts, USA, pp. 189-190.

Walters, A. (2008) *HortTechnol.*, **18** : 286-289.

Wang, H.L., Kang, Y.Q., Zhang, C.J., Ma, Y. and Wang, T.K. (1998) *Biologia Plantarum*, **41** : 49-55.

Waterer, D. and Schmitz, D. (1994) *Can. J. Plant Sci.*, **74** : 611-614.

Whitaker, J. R. (1976) *Adv. Food Res., California Univ., Davis.* (USA), **22** : 73-133.

Wu, C., Wang, M., Dong, Y., Cheng, Z. and Meng, H. (2015) *Sci. Hortic.*, **194** : 43–52.

Xie, B.D. and Wang, H.T. (2006) *Nanjing For. Univ.*, **30** : 51–54.

Xiong, Z.Q., Li, S.J., Liu, G.Q. and Huang, B.J. (2000) *J. Nanjing Agri. Univ.*, **23** : 25-28.

Xu, K., Yang, J.H. and Liu, H.Q. (1999) *J. Shandong Agric. Univ.*, **30** : 409-412.

Xu, P.W., Srinives, P. and Yang, C.L. (2000) *Thai J. Agri. Sci.*, **33** : 11-20.

Yadav, S., Pandey, V.P., Maurya, R., Sriom and Kumar, S. (2018) *J. Pharmacogn. Phytochem.*, **7**(6): 1625-1630.

Yamashita, K., Sakai, J. and Hanada, K. (1996) *Ann. Phytopathol. Soc. Japan*, **62**(5): 483–489.

Yang, F., Liu, S. and Wang, X. (2005) *Sci. Agricultura* Sinica, **38** : 1011-1016.

Yun, J.S., Hwang, S.G., Song, I.G., Lee, C.H., Yun, T., Jeong, I.M. and Peak, K.Y. (1998) *RDA J. Hort. Sci.*, **40** : 14-19.

Zhang, C.H., Shen, S.X. and Wang, M. (1999) *Acta Hortic.*, **26** : 268-270.

Zhen, H.R. (1998) *Acta Agriculturae Shanghai*, **14** : 21-23.

Zheng, S.J., Khrustaleva, L., Henken, B., Sofiari, E., Jacobsen, E., Kik, C. and Krens, F.A. (2001) *Mol. Breed.*, **7** : 101-115.

Zheng, S., Henken, B., Krens, F. and Kik, C. (2003) *In Vitro Cell. Dev. Biol. Plant*, **39** : 288-292.

Zheng, S., Henken, B., Ahn, Y., Krens, F. and Kik, C. (2004) *Mol. Breed.*, **14** : 293-307.

Zhila, E. D. (1978) *Fiziologiya i Biokhimiya Kulturnykh Rastenii*, **10** : 190-193.

Zimerer, A. T., Ribeiro, L. G. and Coelho, R. I. (1988) *Horticultura Brasileira*, **6** : 25-26.

Zu, P.W., Liu, H.Y., Gao, Z.T., Peerasak-Srinives and Armstrong, J. (2001) *Acta Hortic.*, **555** : 213-220.

LEEK

T.K. Maity, D. Sanyal and P. Hazra

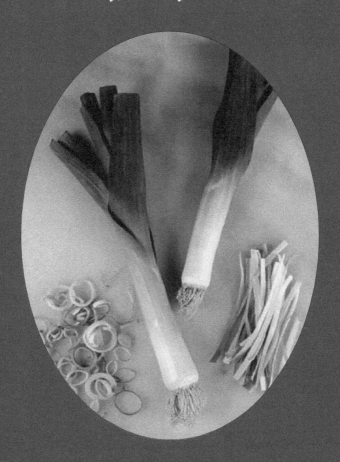

1.0 Introduction

Leek is a herbaceous, biennial and non-bulb forming member of the onion family and is grown for its blanched stem and leaves. It is a self-compatible outbreeding tetraploid (2n = 4x = 32) belonging to the diverse species complex *Allium ampeloprasum L.* which includes wild and cultivated forms with ploidy levels of 4x, 5x and 6x, tetraploidy being the most common (Koul and Gohil, 1970). Leeks (*Allium porrum* or *A. ampeloprasum* var. *porrum*), sometimes called "the gourmet's onion" are related to onions (*A. cepa*) and garlic (*A. sativum*) but have flat leaves instead of tubular and relatively little bulb development. Leek has been cultivated in Western Europe since the middle ages. The greatest production of the leek is in Indonesia and Turkey, with France and Belgium being major producers in Europe. According to the most recent data of the United Nations Food and Agriculture Organization, worldwide leek production was approximately 2,096.067 tonnes in 2016 harvested from131.766 hectares (Hanci *et al.*, 2018). Commercial cultivation of leek is not followed in India and wherever it grows, it is on a home scale, mainly in the kitchen garden as a favourite vegetable.

2.0 Composition and Uses

2.1.0 Composition

Leek has a more delicate flavour and is sweeter than the onion. It is packed with essential vitamins, minerals, antioxidants and dietary fibre. It contains an impressive amount of flavonoids, particularly kaempferol, and considerable amounts of sulphur. The characteristic flavour of the leek is associated with the presence of S-alkyl-cysteine sulphoxides, propyl- and methyl-cysteine sulphoxides being predominant in modern leeks.

Table 1: Composition of Leek (per 100 g of edible portion)*

Moisture	78.9 g	Phosphorus	70 mg
Protein	1.8 g	Iron	2.3 mg
Fat	0.1 g	Vitamin A	30 I.U.
Carbohydrates	5.0 g	Thiamine	0.23 mg
Minerals	0.7 g	Vitamin C	11mg
Fibre	1.3 g		

* Aykroyd (1963).

Bulb of cultivated leek contains very low (below 20 mg/kg fresh weight) or traces (below 2 mg/kg FW) of quercetin and kaempferol (Horbowicz *et al.*, 2000).

2.2.0 Uses

It is grown for its etiolated pseudostem formed by the leaf sheaths. Its tolerance of low temperatures, together with the absence of photoperiod requirements or a

Uses of Leek

Different Cooked and Raw Dishes of Leek

Leek Soup   **Leek Pickles**

Leek Flakes   **Leek powder**

bulb resting stage, make leek extremely adaptable and it is grown as a vegetable crop worldwide. Leek has a milder and more delicate flavour than onion, though of coarser texture. For centuries, the leek has been widely used in preparing many delicious recipes all around the globe. The edible part of the leek plant is a bundle of leaf sheaths which is sometimes called a stem or stalk. The stem is pseudostem. When tender, it is eaten raw. It is also cooked with other vegetables or used for flavouring in soups and stews.

Leek has been known and used for centuries for its medicinal properties. The popularity of leek is connected not only with its high nutritional value but also with its wide spectrum of biological activities, primarily due to high antioxidants content. In the traditional medicinal system, leeks were used for curing many diseases.

3.0 Origin and Taxonomy

Leek probably originated in the eastern Mediterranean area (Jones and Mann, 1963) where it has been in cultivation since prehistoric times. This vegetable has been known as a food for over 4000 years in the Middle East. Researchers have discovered traces of leeks near Egyptian pyramids. It is thought that the vegetable spread across Europe and to the British Isles by the Romans. European settlers brought the vegetable to North America. It had spread to Europe by the Middle Ages (Helm, 1956) and from there it was introduced to North America by the early European settlers. Nowadays leeks are of minor importance in the USA and Canada but are widely grown in Europe, particularly in France which is the largest producer of leeks in the world. It is not grown in India on a commercial scale but is a favourite vegetable in a kitchen garden.

Leek, *Allium porrum*, belonging to the family Alliaceae (Liliaceae), is a tall hardy biennial with white, narrowly ovoid bulbs and broad leaves. It resembles the green onion but it is larger. Leek, a cultivated form of *Allium ampeloprasum*, is a tetraploid with 2n=32 (McCollum, 1976). According to Traub (1968), leek and garlic are put together in section Allium. Leek closely resembled wild *A. ampeloprasum*, differing mainly in a lesser tendency to form bulbs. Leek was known in Europe in the Middle Ages and is still popular there, many cultivars having been selected for long, white edible leaf-based and green tops, winter hardiness, and resistance to bolting. Leek and kurrat (*Allium kurrat*) are interfertile and it may be supposed that they would cross readily with wild *A. ampeloprasum* of the same (4x) ploidy level (McCollum, 1976). It is a self-compatible out-breeding tetraploid (2n = 4x = 32) species of the onion family, Alliaceae (Koul and Gohil, 1970).

4.0 Cultivars

Leek cultivars differ significantly in growth habit which affects the final product. They vary from long, green narrow-leaf types with long slender white stems to long wide-leaf types with thicker shorter white stems and blue-green leaves. Apart from the vegetatively propagated tropical types, old introductions from temperate countries, there are no tropical leek cultivars.

In temperate parts of the world, where it is grown commercially the varieties are grouped in two *viz.* Early Season and Late Season. Early season leeks are ready in about 50 to 100 days after planting, depending on the variety. In comparison to late-season leeks, early season leeks tend to be smaller, more sensitive to frost and have a milder flavour. They are easier to grow than late-season leeks as they do not have any blanching requirement.

Depending on the variety, late-season leeks mature about 120 to 180 days after germination, The varieties in this group have a stronger flavour, develop wider stalks and require blanching for optimum quality.

Very little work has been done on the improvement of cultivar of leek in India. What so ever the cultivars grown in India are introduced ones. The cultivars of interests in India are London Flag and American Flag (Choudhury, 1976). In recent times, two cultivars developed through selection are Palam Paushtik and PPL-1.

Palam Paushtik

It is an alternative of green onion, suitable for salad, soup and cooking which matures in about 140-150 days with an average yield of 300-350q/ha.

PPL-1

It is a selection from exotic germplasms having light green leaves, swollen stem without forming any bulb. It is a good replacement for green onion and matured in 150-160 days.

Leek varieties of commercial importance are London Flag, American Flag, Elephant, Gennivillies, St. George, Copenhagen Market, Bluevetia, Borde Pearl, Belgium Winter Giant, Large Musselburgh, _etc._ (Rana and Jatav, 2018). The characteristics of some of these varieties as suggested by them are as follows:

Renova

A variety developed through a selection from the variety Regions produces marker leaves, longer stalk and fewer tendencies to form bulbs and is more uniform in size and shape.

Miner

An autumn or early winter variety derived from St. George produces long stem and high greens yield.

Splendid

This variety is suitable for cultivation throughout the season.

London Flag

It is most generally cultivated variety with tall and broad leaves.

Musselburgh

It is very long and hardy variety. Its leaves are broad and tall and stem long and thick.

Large Roven

Its leaves are very dark green, broad and thick and stem rather short but remarkably thick. It grows as thick as a man's arm. It is found to be the best kind of forcing, as it requires a sufficient thickness of stem sooner than any other varieties.

Lincoln

It is early Bulgarian giant type leek, which gives uniform long and thick stems. It is suitable for Rabi season planting and matures in 80 days after transplanting.

Carlton

This hybrid variety is very fast in growth and leaves are erect and green in colour. Its long shaft nodes have excellent uniformity.

Roxton

This variety possesses long shaft length with excellent uniformity. The growth of this hybrid is very fast with green to dark green erect leaf colour.

Parton

Medium growth of this hybrid produces medium shaft length with excellent uniformity. Leaves are dark green.

The characteristics of some the cultivars/hybrids grown across the world are given below.

American Flag

It matures within 90-120 days, classic, home garden cultivar.

Giant Musselburg

It matures within 105 days, Scottish heirloom, best winter hardy.

King Richard

It matures within 75 days, grows to full size in summer; not winter hardy.

Lancelot

It matures within 70 days, bolt resistant, virus tolerant.

Lincoln

It matures within 50-100 days, early variety with long, white shanks.

Almera

An autumn type cropping from mid July to September. Long slender stems/ mid green semi-upright leaves.

Atlantic

Very good frost tolerance for winter cropping well into the new year.

Axima

Long strong shaft without bulbing. Dark green erect foliage.

Below Zero (F₁)

British breeding has combined the vigour of an F_1 Hybrid with extreme cold tolerance to produce quality leeks which will withstand the harshest of weathers. Dark leaves, pure white stems with no bulbing, long standing ability and bolting and rust tolerant.

Blue Solaise

A traditional French variety with deep blue-purple leaves. Superb flavour and very hardy.

Carentan

Large thick stems with blue-green foliage. Crops late October to Early January.

Giant Winter

Excellent late variety with heavy thick stems. It will stand in the ground for a long time.

Hannibal

Fast growing variety for summer and autumn cropping.

Jolant

Very early variety with a mild flavour. Use for mini-veg or grow on to harvest form August onwards. Grows vigorously and gives a high-density stem.

Mammoth Blanch

A superior exhibition variety with extra long white blanch and thick, broad flag. Sow mid-January to early March at 15°C. Do not overwater seedlings as this may cause damping off. Harden off and plant out from early May.

Oarsman (F₁)

Medium to dark flag leaf, the plants remain virtually free from bolting even when direct drilled. Second early to mid season maturity slot. Shows good resistant to bolting.

Pot

A true exhibition variety producing very large, heavy leeks.

Prizetaker/Lyon

Uniform habit produces long thick white stems. As the name implies is ideal for the show bench. Matures from early autumn onwards.

Tadorna

Medium length, very upright habit and extremely winter hardy. Crops from December to March.

Cultivars/Hybrids of Leek

Palam Paushtik

Almera

Atlantic

Blue Solaise

Below Zero (F₁)

Carentan

Hannibal

Cultivars/Hybrids of Leek

Jolant

Giant Winter

Mammoth Blanch

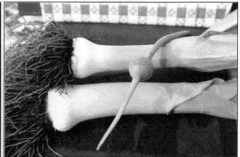

Musselburgh

Oarsman (F₁)

Prizetaker

Pot

Tadorna

Cultivars/Hybrids of Leek

American Flag

 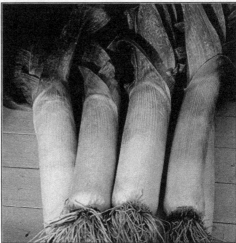

Early Giant   Autumn Giant

Lancelot   Tornado

The cultivars listed by American seedsmen include London Flag, American Flag, Elephant, or Monstrous Carentan, Giant Musselburgh, and the Lyon. But these cultivars are not very distinct (Thompson and Kelly, 1978). Parijs (1983) reported that cultivars like Acadia, Derrick and Electra are suitable for autumn cultivation in Belgium while Alberta, Blizzard and Carina are for winter. In Canada, promising cultivars mentioned by Maurer (1982) are Longa, Odin, Kilima, Goliath, Siberia and Artico. Carina and Alberta (Rijbroek and Riepma, 1982) and Arkansas (Aalbersberg, 1988) were considered the best cultivars in the Netherlands. Kanters (1991) recorded the highest yield from cultivar Rami followed by Albana and Alma in summer cultivation in the Netherlands. The other recommended cultivars are Alma-Norda, Enak, E 8903 and Jolant.

In west-central Florida, Majnard (1989) obtained high yield from cultivars Kind Richard, Verina, Tivi, Kazan, Kilima and Albana. Perko (1989) reported that in high altitude, cv. Alaska out yielded other cultivars and along with cultivars Arca and Labrador, it was able to withstand the weight of a snow cover most effectively, owing to its leaves being borne lower down on the stem. In Norway, high yielding cultivars are Rami and Prelina (Flones, 1990). Kirnosova (1990) from Russia described K 2321, K 2368, K 2042, Suttons Prizetaker, K 2245 and Topaz as promising cultivars.

In a trial in Belgium for early autumn cultivation, Vanparys (1996) reported that the cultivars Alora, Carolina, Casarca, Ginka, Lancelot and Porvana gave the best results. Based on the results of a field study in respect of susceptibility to disease (*Fusarium culmorum, Phytophthora porri*), pests (*Thrips*), frost and yield. Vanparys (1998a) recommended cultivars like Arkansas, Atlanta, Farinto, Latina, Suprina and Vrizo for loamy sand soil in Belgium. A total of 32 leek cultivars were evaluated in Belgium. Cultivars Carolina, Natan and Porvite showed the best crop yield, crop quality and colour, while Alora, Landina, Rami and Ramona were promising (Vanparys, 1998b). Callens *et al.* (2000) evaluated 18 leek cultivars (including 12 hybrids) at 3 sites in Belgium and 2 sites in the Netherlands. Parton F_1, Rami, RX 10511 F_1, Tadorna, AP 9705, Logan F_1, Upton F_1, Breugel F_1, Nun 7492 F_1, Sultan F_1 and Cadet F_1 were susceptible to *Alternaria*, while Davinci F_1 was farily resistant to this disease. Highest yields were obtained for Logan F_1, followed by Rami, Breugel F_1 and Upton F_1, and Nun 7492 F_1. The best quality was obtained for Apollo F_1 and Parton F_1.

The findings of Rooster *et al.* (2000) were in agreement with the previous study. F_1F_1, Breugel F_1, Cadet F_1, and Gavia were susceptible to Alternaria, while Angelos, Apollo F_1, Aristos, Arkansas, Bluetan F_1, Flextan F_1, Farinto, Tara and Prospecta were fairly resistant to Alternaria. Highest yields were obtained for RS 94007 F_1, followed by Parker F_1 and Breugel F_1. The good quality was obtained for Apollo F_1, Avidia F_1, Breugel F_1, Parker F_1 and Parton F_1. Fourteen leek cultivars were compared for summer cultivation at 3 sites, Sint-Katelijne-Waver, Kruishoutem and Aartrijke, in Belgium. Upton F_1 gave high yields of uniform quality with a dark green colour, while Roxton F_1 also produced high yields of uniform quality, but the colour was paler with brown strips. Bolting was a problem for Atal and Apollo F_1 (Reycke *et al.*, 2000). Based on the yields, average leek size and other characteristics, Vanaerde *et al.* (1997) considered the cvs. Ginka, Davina, Tara, Rami and Baikal as the best in Belgium.

Fourteen leek cultivars and hybrids were compared by Reycke *et al.* (1999a) at 3 sites, Kruishoutem, Sint-Katelijne-Waver and Leuven, in Belgium. Highest yields were obtained for Logan F_1 and Upton F_1, but they had a pale colour and were difficult to store. Parton F_1 also gave high yields and showed better crop and storage quality. Apollo F_1 was also high-yielding with a dark colour and good resistance to disease (*Alternaria*). Cultivars Remi and Angelos were also promising. Twenty-one leek cultivars (Rami, Alesia, Angelos, Davina, Firena, Porbella, Prenora, Alcazar, Alexis, Alesus, Alvito, Casarca, Arena, Carolina, Landina, Natan, Porvite, Ramona, Tara, Parton F_1 and Upton F_1) were compared for the crop, harvest and disease resistance characteristics at 4 sites in Belgium. They were sown during 5-7 March, planted during 27-29 May and harvested between 15 September and 14 October. The best results were found for Rami, Davina, Angelos, Alesia, Casarca, Landina, Ramona. Both hybrids (Parton and Upton) had good uniformity and yields (Reycke *et al.*, 1998).

Vanparys (1998c) further noted that bolting occurred mostly in cv. Enak (33.6 per cent), and was least in Pandora (5.6 per cent), Almiros (6.7 per cent) and Casarca (6.9 per cent) in 1995. The highest yields were seen in Ginka (368 kg/acre), Pandora (359 kg/acre) and Casarca (349 kg/acre). In 1996, Enka and Rami showed the highest percentage of bolting (5.7 and 4.0 per cent, respectively). The highest yields were seen in Carlton (516 kg/acre) and RS 92628 (510 kg/acre). In 1997, a slight infection with *A. porri* occurred in July, while bolting was almost absent (2.2 per cent). The highest yields were seen in RS 92628 (557 kg/acre), Marina (464 kg/acre), Alora (458 kg/acre) and Carlton (442 kg/acre). They recommended cultivars Alora, Carlton, Marina and RS 92628 for initial raising under heated glass.

Kolota *et al.* (2001) conducted a varietal trial in Poland, to evaluate the suitability of new leek hybrids (Upton F_1, Carlton F_1, Stanton F_1, Norton F_1 and Roxton F_1) and open- pollinated cultivars (Kilima, Varna, Alita, Rami, Albana and Kamush) for early cropping. Results of the study showed that delay of harvest from 30 June to 1 August increased the marketable yield of leeks by 107.4 per cent. Plants harvested in later stages of maturity contained higher amounts of dry matter, while a lower level of vitamin C and nitrates. The maximum yield of early harvested leeks was produced by Norton F_1 and Carlton F_1. Irrespective of the date of harvest, Carlton F_1, Stanton F_1 and Albana accumulated the lowest, while Kamush and Kilima, the highest quantity of nitrates. Small and rather variable differences were found in dry matter and vitamin C contents among the tested cultivars.

Reycke *et al.* (1996b) evaluated sixteen leek cultivars at 2 sites (Kruishoutem and Sint-Katelijne-Waver) in Belgium for late autumn cultivation. Highest yields and good quality at both sites were obtained for Parton F_1. Tadorna, Idaho, AP 9805, Tara and Arena also showed good quality, but lower yields. However, Arkansas, Farinto and Vrizo were slightly disappointing due to the weather. In a similar study, Vanparys (1998d) obtained the highest percentages of marketable leeks in Columbus and RS 89029. Columbus, Ramona, Alesia, Carolina and E70010 produced the highest percentage of leeks with a diameter > 3 cm.

Twenty-two leek cultivars (Arkansas, Arena, Astor, Carolina, Davina, Farinto, Ginka, Libra, PX 89029, Tadorna, Tara, Alcoy, Angelos, AP 9701, AP 9703, E 70010, Gavia, Monzie, Parton, Porbella, Vrizo and Winora) were compared for crop yield and disease resistance (*Alternaria*) characteristics in Belgium. Cultivars were sown on 10 March, planted on 24 June and harvested at the end of November. The best results were obtained with Arkansas, Farinto, Gavia, Tadorna, Tara, Vrizo, Arena, E 70010, PX 89029, Angelos and AP 9701. The hybrid Parton was better for early cultivation (Rooster and De Rooster, 1998a). Heine (1998) reported that marketable yields were in the range 61.4 t/ha for the F_1 hybrid Upton to 42.5 t/ha for Conora; F_1 hybrid Parton gave the 2nd highest yield (55.6 t/ha). Upton had good stem length (24.6 cm) and thickness (3.7 cm), comparable to those of normal cultivars.

Six leek cultivars (Alamos, Angelos, Apollo F_1, Logan F_1, Parton F_1 and Upton F_1) were direct sown on 23 April 1998 and harvested on 6 January 1999. Hybrid Upton F_1 produced the highest yield but was very light-coloured. The best result was obtained for hybrid Parton F_1 (Rooster *et al.*, 1999). Reycke *et al.* (1999) reported that cultivars Hiberna, Latina, Aristos, AP 9501 gave high yields followed by Prospecta in Belgium. The best quality was obtained with Prospecta and Idaho, but Idaho gave a low yield. Vanparys (1999) recommended the cultivars Arkansas, Atlanta, Farinto, Latina, Suprina and Vrizo for growing in Belgium.

Seventeen leek cultivars were compared at 3 sites (Kruishoutem, Beitem and Sint-Katelijne-Waver) in Belgium. At Sint-Katelijne-Waver, hybrid Carlton F_1 showed an early harvest with high yields and was of good quality, while hybrid Upton and cultivars Rami, Firena and Alesia also showed good yields and quality. Hybrid Upton showed the best yield and quality at both other sites, while Landina had high yields, but lower quality. AP 9803 scored well at Beitem, but did not perform well at Sint-Katelijne-Waver (Callens *et al.*, 1998). Cultivars Gavia and RX 93017 were the least susceptible to *Phytophthora*, thus resulting in the greatest crop yields (353 and 381 kg/acre, respectively), while Janssens was more susceptible to Phytophthora, but showed the shortest flower stem (5.1 cm) and the most compact shafts (Rooster and De Rooster, 1998c).

Rooster *et al.* (2000) reported that cvs. Parton F_1, Atlanta, Avidia F_1, RS 94007 F_1, AP 9706, Sultan F_1, Cadet F_1, and Gavia were susceptible to *Alternaria*, while Angelos, Apollo F_1, Aristos, Arkansas, Bluetan F_1, Flextan F_1, Farinto, Tara and Prospecta were fairly resistant to *Alternaria*. Highest yields were obtained for RS 94007 F_1, followed by Parker F_1 and Breugel F_1. The good quality was obtained for Apollo F_1, Avidia F_1, Breugel F_1, Parker F_1 and Parton F_1.

Golubkina *et al.* (2018) assessed yield, quality indicators, antioxidants and elemental composition of nine leek cultivars grown in greenhouses under organic or conventional systems in the Moscow region. The management system did not affect yield, which attained the highest value with the cultivar Giraffe and the lowest with Premier and Cazimir. Pseudo-stem dry matter and sugars were higher with organic management, whereas nitrate concentration was higher with conventional management. The cultivars Vesta and Summer Breeze showed the highest dry matter and total sugar content, whereas Goliath had the highest antioxidant, selenium and potassium concentrations. Among the antioxidants, ascorbic acid attained higher values with organic management.

5.0 Soil and Climate

5.1.0 Soil

Leek may be grown successfully on a wide range of soil types but in heavy, clay soils harvesting may be delayed. It grows luxuriantly on medium, well-drained soils which are rich in plant nutrient and organic matter with a pH between 6.0 and 7.0. Deep ploughing is recommended so that a longer shaft can be developed. Stone and Rowse (1982) recommended a thorough loosening of the sandy clay loam subsoil which has helped in easy penetration of roots up to 70 cm and thereby increased the yield. The most favourable range of soil pH for leek is 6.0-8.0 (Chauhan, 1972).

5.2.0 Climate

Leeks grow best in a cool to moderate climate. Although leek is a cool-season crop, it can withstand heat and cold better than the onion. They need a minimum of eight hours of bright sunlight daily. In India and Sri Lanka, it thrives well at higher altitudes. Seeds are produced in India at higher altitudes in the hills.

A temperature range of 11–23°C is optimum for better germination of seeds and it is reduced above 27°C. The ideal temperature for rapid vegetative growth is 21–24°C. Leek does not form bulbs under the coolest conditions, however, bulb formation increases under long-day conditions with a temperature range of 15°C to 18°C as the accumulation of dry matter is noticed more at this temperature. Fresh as well as the dry matter of plant is increased with the rising in temperature up to 21°C and day length. The number of adventitious roots is found more in increased temperature and day length (Rana and Jatav, 2018).

6.0 Propagation

6.1.0 Seed Propagation

Leeks are mainly propagated by seeds. Seeds are sown from August to October in the nursery bed and the seedlings are ready to plant when they attain a height of 15 cm. Seeds take 5-12 days for germination. About 5-7 kg of leek seeds is required to be sown per hectare, but leeks are not commonly direct-seeded. Hardening off seedlings is often practised for better stand in the field. The seed rate is somewhat less than that of onion as greater spacing is maintained in leek. Leek seed, like other Alliums, has very limited useful viability (less than 2 years) unless stored under ideal conditions.

Seeds of leek (cvs. Arial and 791), were subjected to freeze drying and vacuum drying. For freeze drying seeds were frozen at –80°C for 3 h, transferred to a freeze drier at –25°C for 1 h rising to room temperature during the drying period. For vacuum drying seeds were dried in a vacuum desiccator with silica gel at 22°C for 2-3 days. After drying seeds were vacuum sealed in aluminium foil bags and stored at 5, 20 and 30°C. There was a slight reduction in germination rate of some dried seeds but not enough to affect seed quality (Zheng *et al.*, 2001).

Propagation through tissue culture has been summarised in biotechnology.

7.0 Cultivation

7.1.0 Planting

Leeks are transplanted about 5-15 cm deeper in trenches or furrow to assist subsequent blanching of a base of the developing pseudostem since the length of a blanched portion of pseudostem influences its market value. Seedlings are generally spaced 10 cm apart in rows and 40 cm between rows. By increasing the plant density from 196000 to 256000 plants per hectare, the yield was increased by 10-13 per cent.

Paschold and Gottlick (1982) grew 25000 plants per hectare as minimum plant population. In an evaluation trial in Netherlands, leek seeds were sown in early April, seedlings were planted in early July at 50 × 12 cm and the crop was harvested between January and early April depending on weather (Aalbersberg, 1988).

Planting in furrows and earthing up gave lower yields, fewer high-quality leeks and lower average plant weight for crops harvested in autumn but gave better results than flat cultivation for crops harvested in spring. The cultivars Platina, Winter Riesen, Alaska and Wila were the most resistant cultivars to frost damage. In general, during overwintering in the field, dry matter and total sugar contents decreased and reducing sugar contents increased (Kolota *et al.*, 1996).

Parys *et al.* (1997) conducted field studies with 16 (at Kruishoutem), 15 (at Sint-Katelijne-Waver) and 14 (at Rumbeke-Beitem) leek cultivars in Belgium. Leeks were sown between 27 December and 10 January, planted between 12 March and 10 April, and harvested between 12 June and 28 July. The best results at Kruishoutem were obtained with cv. Upton, followed by Carlton, Rami, Alora and Firenza and RS 92628, Porvite and Columbus (all pale colour and requiring prompt harvesting). The early cultivation at Sint-Katelijne-Waver resulted in very good quality and yields of Upton and Rami, and lower yields with Casarca and Firenza. The best results at Rumbeke-Beitem were with Rami and Carlton, followed by Alora (disease problem), RS 92628 (pale colour) and Marina (pale colour and disease problem).

Seven leek hybrid cultivars (Upton, Parton, Newton, Angelos, Davina, Firena and Prenora) were direct sown on 2 April on loamy sand in Belgium and were harvested during November-December. Crop yields were lower than predicted by the initial plant density due to attacks by leafminer flies (Agromyziae). Newton had the greatest crop yield (379 kg/ha), while Parton had the best quality (Rooster and de Rooster, 1998 b and d).

It appeared that removing light plants decreased the number of dead plants, improved uniformity at harvest and slightly increased yield. Heavier plants resulted in higher yields, better uniformity and fewer dead plants. Grading and separate planting out of the different grades improved uniformity (Embrechts, 1996).

Earthing up, planting into furrows or into 20 cm deep holes increased the length and weight of the blanched part of the shaft, compared with traditional planting (Sasa *et al.*, 1987). Kaniszewski *et al.* (1989) studied 4 growing methods of leek, *viz.* (i) traditional planting at a depth of 5 cm, (ii) planting as above followed

Different Stages of Field Growing of Leek

by earthing-up, (iii) planting into 15 cm deep furrows, levelled during the growing season, and (iv) planting into 20 cm deep holes using cultivars Alaska, Darkal, Jolant and Nebrasak. Planting into 20 cm deep holes reduced the yield, compared with the other 3 treatments which gave similar yield.

7.2.0 Manuring and Fertilization

The plants of leek being larger than those of onion, the requirement of manures and fertilizers are high. A yield of 30 tonnes of leek per hectare removes 100 kg nitrogen, 65 kg phosphorus and 130 kg potash from the soil (Choudhury, 1976).

The soil should be prepared with green manure plough down or farmyard manure to enhance organic content and provide nutrients and the extra moisture-holding ability for the crop. Compost is widely used to increase soil fertility, usually practised by incorporating the compost into the upper soil layer. However, the addition of compost as mulch resulted in a significantly higher quality leek, including more first-class leeks, longer and thicker shafts, and a generally better appearance over incorporation into the upper soil layer due to higher availability of plant nutrients (Reeh and Jensen, 2002). Mulching helps in the retention of moisture and suppression of weeds. Decomposed mulch releases nutrients into the soil which are absorbed by the crop.

The slurry obtained from the anaerobic digestion of filter cake had a positive effect on leek plants, increasing production by 64.7 per cent compared to the control (Valdes-Mendez *et al.*, 1999). However, Bath and Ramert (2000) reported that the amounts of plant-available N from slurry or compost were too small to support a leek crop with a high N demand towards the end of the growing season.

Leeks require more nitrogen than phosphorus and potassium. Application of nitrogen should preferably be in three splits - one-third pre-plant incorporated, one-third as a side dressing, and one-third as a top dressing. Phosphate requirements of leeks are not very substantial. Potash requirements are also low and the application of sulfate of potash is suggested.

The diameter and length of the bulb are increased by nitrogen fertilization. The highest yield of the crop was obtained by the application of 200 kg nitrogen per hectare (Venter, 1982). Results of the study conducted by Karic *et al.* (2005) demonstrated that increasing level of nitrogen to 200 kg ha^{-1} resulted in greater number of leaves per plant (14.4), maximum leaves weight (194.6 g plant^{-1}), higher pseudo-stem diameter (36.3 mm), maximum pseudo-stem weight (146.5 g) and highest total yield (91.98 t ha^{-1}). There was also slight evidence that higher nitrogen amount decrease dry matter. Furthermore, increase in nitrogen levels had no appreciable effect on chemical composition (total and reducing sugars, vitamin C) of leek. In southern Ethiopia, application of 138 kg N ha^{-1} and intra row spacing of 5 cm was found to be promising for maximum yield of leek (Kiffo, 2016). Different nitrogenous fertilizers had little effect on yield. In dry years in Poland, the highest yields were observed by Kaniszewski (1986) with the pre-planting application of 200 kg N/ha under both irrigated and non-irrigated conditions, and in wet years with the split application of 600 kg N/ha. Under the plastic tunnel, Bohec *et al.* (1996) obtained a gross yield of 87.6 t/ha with a total N uptake of 240 kg/ha. Most of the

N uptake (200 kg/ha) occurred over a period of a month and a half between the removal of the plastic at the 4-leaf stage and harvest. It is recommended that the soil N content should be measured at 3 points: when the plastic is removed, and then 30 and 45 days later. N fertilizer can then be administered in accordance with the phenological stage of the plants and the soil N content. The nitrate content of leek increased with increasing regional N use, especially when used in the form of nitrate (Karaman *et al.*, 2000).

Biczak *et al.* (1999a) obtained the highest marketable yield of leeks (1.73 kg/m²) with 75 mg N + 75 mg P_2O_5 + 100 mg K_2O per kg soil (compared with 1.16 mg/m² in the unfertilized control). Fertilizer application (particularly N) had a great effect on acid phosphatase activity and it increased enzyme activity by up to 36 per cent (Biczak *et al.*, 1996b). They further noted that increasing the N, P and/or K content of the soil affected catalase activity: the changes were related to the rates and kind of fertilizer, to the plant species and its age. Nitrogen had the greatest effect on increasing catalase activity by up to 21 per cent (Biczak *et al.*, 1999c). Increasing the soil content of N, P and K stimulated peroxidase activity in leaves. The greatest effect (an average of 26 per cent increase) was achieved by the combined application of N, P and K (the highest rate of each) (Biczak *et al.*, 1999d).

In pot experiments with greatly differing rates of N, P, S, K and Ca, dry matter (DM) yields of leek cv. Imperial stems varied from 25 to 164 g/pot. Total N and NO_3-N concentrations ranged from 1.18 to 3.56 per cent and from 10 to 15 ppm in DM, respectively. Both N applications and P and K deficiency greatly increased total N and NO_3-N concentrations. S applications increased total S concentration from 0.047 to 0.359 per cent in DM, of which between 25 and 100 per cent was found in methionine + cystine. Total N: total S ratio decreased from 57 to 6 with the highest S rate. P and K application increased their respective concentrations in DM 2- and 3-fold. Severe Ca deficiency reduced Ca concentration from 0.495 to 0.045 per cent. Fe, Zn, Mn and Ca deficiency were 33-69, 14-26, 11-34 and 3.1-5.7 ppm in DM, respectively (Eppendorfer and Eggum, 1996).

The uptake of N in direct-sown leeks was not increased with the placement of N and drought stress during growth significantly depressed plant growth, N uptake and N uptake efficiency (Sorensen *et al.*, 1996). Field studies in France showed that N needs of leek plants depended on the growth stage and N levels in soil (Berry and Thicoipe, 1998). Reycke and de-Reycke (1998) did not find any significant differences between N treatments in leek cv. Rami. However, crop quality varied, with the palest colour and lowest nitrate residue in untreated plants. No differences were found between K treatments. Differences were found for P treatments, with the greatest yields in ammonium polyphosphate-treated plots, followed by triple superphosphate.

Fertigation with N or with Polyfeed compound fertilizer significantly affected growth, yield and N content of leeks as compared to broadcast N application. Total and marketable yield was highest with 1/3 of the N applied pre-planting and 2/3 with trickle irrigation (a total rate of 200 kg N/ha). Reducing the N rate to 125 kg N/ha, applied in several drip fertigations, caused an insignificant reduction in yield. Fertigated plants had greater leaf area and produced more fresh and dry matter, as compared to plants given only the broadcast N. Leaf NO_3-N content in leek plants decreased during the growing season and was highest under surface fertigation, intermediate under subsurface fertigation and lowest with only broadcast N application (Kaniszewski *et al.*, 1999).

After its late harvest, a leek crop will normally leave large quantities of N in the soil which may be subjected to leaching. A possible method to reduce N leaching losses is to establish a catch crop before the harvest of the leek. It is possible to establish a catch crop (chicory) in leek thereby depleting soil N and reduce the risk of N leaching without jeopardizing crop yield (Nielsen *et al.*, 2001).

Rooster and De Rooster (1998e) conducted 5 field studies at 3 sites (Putte, Sint-Katelijne-Waver and Branst) in Belgium with phosphorus fertilizer added to the irrigation water for cultivars Rami, Upton, Angelos and Casarca. Treatment included ammonium polyphosphate (APP) at 0.5, 1.0 or 2.0 litres/a, or 1.0 litre APP/a + Topsin M (thiophanate-methyl) at 80 ml/a, all applied in 150 litres; the 1-litre APP treatment being equivalent to 48 kg P_2O_5/ha. Control treatments were the basal application of triple superphosphate (1.2 kg/a), and no supplementary phosphate application. All treatments (except at Putte) resulted in better growth and greater crop yields.

In Belgium, in a trial with leek cvs. Alora and Daving, yield/acre and average leek weight were the highest with treatments of ammonium polysulfate at 1.05 litres/are and potassium dihydrogen phosphate at 0.96 kg/are, each applied with irrigation at 200 litres/are, with no significant difference between them.

Pre-trans plant inoculation of leeks with VA mycorrhizal fungi attained marketable weights 25 days earlier than un-inoculated leeks from untreated soil, and their final dry matter yields were 5.7 and 1.5 times as high as those of uninoculated leeks from Dazomet-disinfected and untreated soil, respectively (Sasa *et al.*, 1987). Ducsay *et al.* (2000) conducted a pot experiment using orthic luvisol soil. NPK fertilizer was applied alone or with concentrated or dilute preparation of *Desulfovibrio desulfuricans* bacteria, vermicompost extract, vermicompost or NaOH (to adjust soil pH) to leeks. Control plants were not given any fertilizer or amendment. Biomass production and macroelement uptake increased with NPK fertilizers. The yield of edible leek increased by 88.4 per cent compared with the control. Adding the concentrated *D. desulfuricans* preparation further increased yield (by 143.8 per cent compared with the control).

Rolin *et al.* (2001) studied leek plants colonized by the arbuscular mycorrhizal fungus *Glomus etunicatum* under conditions where the symbiosis is actively affecting host plants' growth and carbon metabolism compared to uncolonized plants grown either with (NM+P) or without (NM–P) supplemental phosphate. Total amino acid levels in roots were lower in colonized plants (M) than in uncolonized plants without NM–P plants and this effect was also induced partially in NM+P plants by supplemental phosphate. The relative proportions of different amino acids were neither affected by the arbuscular mycorrhizal (AM) symbiosis nor were amino acid levels in leaves affected by either colonization or increased phosphate supply. There were no indications of the effects of mycorrhizas upon relative enrichments in different amino acids. More rapid turnover of amino acids was noted in the M plants roots, however, this was probably due to the lower concentration of amino acids found in the plants. These results suggest that there are indeed effects of AM symbioses on overall nitrogen handling by the host but that these are secondary effects of the improved phosphate supply to the host and do not cause significant shifts in levels or fluxes through different amino acids. The effects of inoculation by arbuscular mycorrhizal fungi (AMF) and indigenous AMF populations on biomass production and P uptake by leeks and on its utilization of P from Kola apatite and bone meal were studied in the field in Finland. *Arbuscular mycorrhiza* did not enhance P uptake from Kola apatite, which had no clear effect on P uptake within 1 growing season, either in the field or in pots. AM increased P uptake from bone meal by 62 per cent in a pot experiment. Inoculation did not increase the effect in unsterile soil compared with the indigenous AMF populations. The results demonstrated a key role for AM in the utilization of organic P. It was suggested that regular inoculation is not an appropriate AM utilization strategy in a sustainable farming system (Kahiluoto and Vestberg, 1998).

The effectivity of arbuscular mycorrhizal spores in promoting the growth of leek cv. Musselburgh was tested for inocula from three soil series, under long term organic or intensive, conventional grass and grass arable rotations. Mycorrhizal treatments generally increased growth of leek. In the least fertile soil, inocula from organic farms were more effective than those from conventional farms. In the least fertile soil, there was evidence of more efficient uptake of phosphorus in plants inoculated with spores from organic farms. Inoculating leeks with AMF root fragments produced a plant response similar to that obtained when spores were used, confirming that spore viability was not the sole factor influencing AMF effectivity in earlier experiments (Scullion *et al.*, 1998).

A correct balance between macronutrients and micronutrients is essential to obtain the best results from a crop. A deficiency of any single nutrient is enough to limit crop yield and the availability of each nutrient needs to be related to the crop requirements.

Nitrogen Deficiency

The young leaves maintain a green but paler color and tend to become smaller in size. As the deficiency progresses, the older leaves gradually change from their normal characteristic green appearance to a much paler green colour and become uniformly chlorotic.

Phosphorous Deficiency

Stunted growth is the major visual symptom. Root growth is also poor.

Potassium Deficiency

Some leaves show marginal necrosis and/or tip burn, while others at a more advanced deficiency status show necrosis in the interveinal spaces between the main veins along with interveinal chlorosis.

Copper Deficiency

Deficiency may also be expressed as a light overall chlorosis along with the permanent loss of turgor in the young leaves. Leaves are curled and their petioles bend downward. Matured leaves show netted, green veining with areas bleaching to a whitish gray. Some leaves develop sunken necrotic spots and have a tendency to bend downward.

Zinc Deficiency

The younger leaves become yellow in the early stages of zinc deficiency and pitting develops in the interveinal upper surfaces of the mature leaves. These symptoms develop into an intense interveinal necrosis but the main veins remain green as the deficiency progress.

7.3.0 Irrigation

Leeks being a monocotyledonous plant have a fibrous root system and these roots develop from the base of the plant. Regular watering should be done to encourage the formation of a large number of roots and for best production. Depending on weather conditions, a post-planting irrigation is desirable to ensure rapid establishment. Soils need to be maintained near field capacity. Moisten the soil thoroughly to a depth of 45 cm every 7 days. Water needs are critical since rooting depth in leeks is shallow. Drought stress during growth will decrease yield. When plants are four months old the roots will have penetrated to a depth of 45–60 cm.

In Poland, sprinkler irrigation from mid-June to mid-September significantly improved both yield and quality. Planting in groups (2-3 plants/planting hole) increased yields compared with single planting, but the crop was less uniform and mean plant weight was lower (Kolota, 1986).

Effective utilization of saline water could have significant potential for agricultural development in many areas, particularly in water-scarce regions. Kiremit and Arslan (2016) studied the effect of salinity on growth and yield of leek (*Allium porrum* L.). Increases in irrigation-water salinity caused decreases in plant height; stem diameter; leaf, stem and root fresh weights; and leaf, stem and root dry weights. Water-use efficiency, leaf area and chlorophyll content also decreased with increases in irrigation-water salinity; however, leaf number was not significantly affected. Soil salinity and drainage-water salinity both increased with increases in irrigation-water salinity. They suggested that if appropriate leaching and drainage systems are applied, slightly saline water can be used for irrigation with little or no soil damage and minimal decreases in plant yield, thus saving large amounts of water for the cultivation of more salt-sensitive crops as well as for industrial and domestic usage.

From a pot experiment on response of leek to different irrigation water levels under rain shelter, Kiremit and Arslan (2018) observed that decreases in irrigation water resulted in decreases in plant height, stem diameter, leaf and stem fresh weights, leaf and stem dry weights and leaf area, but did not significantly affect leaf number or chlorophyll content. A yield-response factor of 1.26 was obtained, implying that the leek crop was sensitive to water stress caused by deficit irrigation. Comprehensive analysis of yield, water use efficiency, and evapotranspiration, the 75 per cent times of consumed water (I75) treatment can be suggested for leek production in water-scarce regions.

7.4.0 Blanching

Blanching is important in leek cultivation. It is done by covering the plants to a certain height with soil so as to bleach them, which improves the quality of the crop. For this purpose, plants are put in up to their centre leaves in trenches or pits which are heavily manured and to earth up soil as they grow (Katyal, 1977). Care should be taken not to earth up soil too early when the plants are young.

The white part of the leek is the most desirable for cooking and eating, being tender and less fibrous than the green. Therefore, gardeners often hill the dirt around the shaft to prevent that part of the plant from making chlorophyll and turning green.

7.5.0 Weed Control

Leeks are poor competitors with weeds and therefore the field should be kept weed-free to avoid the losses attributed to weed infestation. Control weeds through regular cultivation but root damage are to be avoided that slows plant growth by damaging shallow roots. Weed control is particularly important during the first 2 months of growth when plants are growing slowly and compete poorly. Mulching with compost, grass clippings or leaves will conserve water and smoother weeds.

The manual weeding of intra-row weeds in direct-sown leek crop grown organically can be very labour-intensive. Generally, pre-emergence flaming plus brush weeding gave the highest intra-row weed control (92 per cent and 87 per cent, respectively) in the two leek experiments (Melander and Rasmussen, 2001). Time consumption for hand weeding after the different treatments was linearly related to the remaining number of intra-row weeds, with no significant influences of the experimental factors on the general relationship. Generally, the cultural methods had no significant influence on the effects of physical weeding in terms of their effect against intra-row weeds. They did not affect the tolerance or robustness of the crop plants against negative impact from the physical control methods. However, generally, seed priming and cultivar choice improved the yield. Application of 1 kg bladex (cyanazine) +1.5 kg tribunilk (methabenzthia-zuraon)/ha showed potential in weed control (Alofs, 1988). From a study on critical period for weed control in leek, Tursun (2007) suggested that the crop should be kept weed free between 7 days after transplanting and 85 days after transplanting to avoid yield losses in excess of 5 per cent. From a study on the effects of herbicides and mulching (barley straw) on weed flora, growth and yield of a leek crop, Karkanis *et al.* (2012) reported that mulching and oxyfluorfen application provides satisfactory control of weeds. The use of mulching is an option for the weed management in organic leek crop. The herbicides providing the most consistent season-long weed control in leek were metolachlor, oxyfluorfen, prodiamine, and pendimethalin (Gilreath *et al.*, 2008). Plots treated with metolachlor, prodiamine, and pendimethalin had the highest yields and plant vigor and were comparable to those for the weed-free control plots. Pendimethalin @ 3 l/ha proved to be the best herbicide in controlling weeds in leek at Tehran, Iran (Sadeghi *et al.*, 2010).

7.6.0 Intercropping and Cover Cropping

In an intercropping system with leek and celery, weed suppression was improved through increased canopy light interception. Intra- and interspecific competition in the system, however, affected the performance of the crops with respect to yield and quality. Relative yield totals around one showed that with respect to biomass production, no yield advantage was found in the crop mixture. Effects of intra- and interspecific competition resulted for both crops in a reduction of the quality. Nitrogen utilization efficiency was generally poor in all crop stands, particularly at a high N application rate (Baumann *et al.*, 2001).

The potential effect of two sowing dates for ryegrass, *Lolium perenne* var. Elke, interplanted in winter leek, on weed control, soil nitrogen allocation and crop yield was studied at two sites in Switzerland. In order to reduce potential competition with the ryegrass, row application of the herbicide methazole (75 per cent a.i., as additional treatment) and increased fertilizer doses in spring (as split-plot treatment) were used, and compared with a weed-free control. Under the experimental conditions, inter-sowing ryegrass 6 weeks after planting resulted in crop yield similar to that in the control plots. Crop quality variates were increased and the overall production system was environmentally more advantageous. Approximately 20 kg N/ha could be retained for being washed out over the winter, and 50 kg N/ha was stored in the inter-plants up to harvest. The inter-plants were incorporated into the soil after the harvest to serve as a N-source for the subsequent crop. Surplus fertilizer doses in spring increased N-allocation to leek in the control plot (Muller, 1996).

Bath (2000) examined the time of incorporation of red clover as a means to improve the availability of mineralized N to leek. Strips of red clover, left between the rows of leek, were incorporated at planting, or 2 or 4 weeks after planting. The difference in N-uptake was small between treatments with incorporation of red clover strips between rows at planting and 2 weeks after planting. With the latest time of incorporation the N-uptake was lower. The inter-row clover strips assimilated and transferred inherent soil-N as well as N mineralized from red clover incorporated in the leek rows. This kept mineral-N values at a low level until the time of incorporation. Low growth rate made it difficult for the leek crop to exploit N mineralized between rows and thereby decreased N-uptake and increased the leaching risk after harvest. Results from this experiment showed that intercropping, beyond the crop establishment phase, could not be recommended. However, if other benefits such as decreased pressure from weeds and pests, and improved soil structure were taken into account, intercropping might prove to be economically justifiable.

Field/pot experiments in the Netherlands showed that in the range of fertilizer dosages tested (0, 50 and 150 kg/ha), suppression of *Thrips tabaci* in leek intercropped with clover was not affected by nitrogen concentration in the plant or size of the plant (expressed as leaf area index). In contrast to the response of other pest species to plant quality as mediated by nitrogen application, *T. tabaci* population dynamics appeared not to be influenced by this factor. It was concluded that attractiveness or nutritional quality of the leek plant for *T. tabaci* was reduced as a direct result of the interaction of leek with clover (Belder *et al.*, 1998).

Due to earlier and faster canopy light interception, the critical period for weed competition was shorter in the intercrop compared with a pure stand of leeks. Relative soil cover of weeds was reduced by 41 per cent, the biomass of *Senecio vulgaris* by 58 per cent and the number of seedlings which emerged as offspring by 98 per cent in the intercrop. Relative yield total of the intercrop exceeded that of pure stands by 10 per cent due to better resource utilization (Baumann *et al.*, 2000).

7.7.0 Mulching

Mulching the soil surface can reduce weed problems by preventing weed seed germination or by suppressing the growth of emerging seedlings. Mulching decrease the numbers of hand hoeing and mechanical cultivations for remove of weeds. The key factors that make straw mulch attractive are low cost and easy in availability and application.

Weber *et al.* (1997) studied the effect of under sowing of *Trifolium subterraneum* (cv. Gerladton) or *T. fragiferum* and straw mulches on the population dynamics of *T. tabaci* in leeks. Both clover under sowing significantly reduced the mean number of *T. tabaci*, whereas straw mulch did not decrease population densities. The basal diameter of plants was regressed against mean numbers of thrips.

A field study was carried out in Belgium in leek (cv. Arkansas) to investigate the effect of drip irrigation and the use of white plastic film mulch on plant development. Crop growth of mulched plants was better and resulted in greater plant weights in August (300 g compared with 125 g for unmulched plants). This was probably caused by improved light reflection and soil moisture. Supplementary irrigation resulted in a 58 per cent higher plant weight in August, compared with plants irrigated according to the Makkink formula; the latter were 31 per cent heavier than non-irrigated plants. Correspon-ding values for 3 September were 23 and 9 per cent respectively. It was recommended that the water supply should be reduced from September onwards to compensate for evapotranspiration losses (Benoit and Ceustermans, 1998).

Mulching increased crop weights by 16 per cent compared with non-mulched plots. The different colours did not affect the trapping of thrips, probably due to the strips being covered by leek leaves (Benoit and Ceustermans, 1998). On the other hand, growing of perennial ryegrass and white clover as living mulches considerably reduced the marketable yield of leeks by up to 45.1 per cent (Kolota and Adamczewska- Swinska, 2002).

While studying the impact of two mulching materials (barley straw mulch and spent mushroom compost) on weed infestation and yield of leek cv. 'Bulgarian Giant', Gerasimova and Yordanova (2015) observed that mulching with barley straw mulch and spent mushroom compost have a significant depressing effect on weeds, especially on *Echinochloa crus-galli* L., *Setaria glauca* (L.) Beauv., *Galinsoga parviflora* Cav., *Polygonum lapathifolium* L. and *Portulaca oleracea* L. over two control variants : non-mulching, but weeding control and non-mulching and non-weeding control. The yields were increased from 3.7 to 4 times when the leek was grown with mulches.

8.0 Harvesting, Yield and Storage

Leek is harvested just like green onion and marketed in bunches. Most leek varieties are fully developed when the stems grow to be a minimum of one inch in width. Some smaller varieties, however, mature at one-half to three-fourths of an inch in diameter. Four leek cultivars (Arial, Atal, HFI Charlton and San Giovanni) were collected every 3 weeks for 5 months from 4 November with the aim of evaluating the correct harvesting date. Comparison of the mean values of methyl propyl disulfide, yield and dry matter indicated that the optimal harvesting date could occur between the 6th and 15th week after the first harvest (Cesare *et al.*, 1999). In Italy, harvesting in December-January is recommended for all cultivars for production with the best organoleptic properties, despite lower yields (Ferrari *et al.*, 1999).

Success has been achieved in lifting the crop by a complete harvester and packed into Lattice crates, before being washed and cooled to 2°C in a cool-room (Schow, 1998). Mechanical cleaning and packing have been attempted in Germany. The new machine was installed in 1996 and is operated by 25 workers. The leeks are mechanically transferred from the harvesting machine to the machine's conveyor belt, manually orientated on a second belt and passed in front of up to 16 work places for trimming of the leaves and roots. They then pass through a mechanical washing tunnel before manual packing into bundles of 1 kg or 5 kg (Geyer, 1997).

Yield depends upon several factors. However, an average yield of 30 tonnes per hectare can be obtained from a well managed crop.

After harvesting, they should be kept under shade and cool conditions. The green crop can be stored for 1-3 months at a temperature of 0°C with a relative humidity of 85-90 per cent (Thompson and Kelly, 1978).

In France, storage of leek at temperatures below freezing point (-2.2°C) for 2 to 5 months depending upon cultivar has been reported by Bohec and Chombart (1989). At 0.5°C, storage duration was 6-7 weeks in CA condition in 10 per cent CO_2 and 11 per cent O_2 compared with 4 weeks in ordinary cold storage. The storage period could be extended to 9 weeks at -1°C in CA storage (Embrechts, 1988). The storage of a total of 22 leek cultivars in different types of crate was studied in Poland. The leeks were stored in universal (U-type), half-size universal or specially designed leek crates at 0 or -1.5°C. The specially designed crates were 600 × 400 × 435 mm in size and held 5 kg of leeks stored vertically. The quality of leeks was similar following storage in the universal and specially designed crates. Storage was better at -1.5°C than at 0°C (Grazegorzewska and Bakowski, 1996).

9.0 Bolting

Results of several French studies revealed that cultivars Albana, Carlton, Nemina and Rami were the most susceptible to bolting (assessed in terms of length of the flower stalk). Bolting was greater for the early planting date, plant receiving 1310 h <12°C, compared with 770 h and for bare-rooted transplants. Finally, the effects of the growth regulators fazor (maleic hydrazide, MH), cycocel C5 (chlormequat + choline chloride), cyter (chlormequat + mepiquat), terpal (mepiquat + ethephon) and tomathrel (ethephon) were investigated for cv. Carlton crop direct sown on 16 October, 1996 under a small tunnel (removed on 8 April) and harvested on 27 May, or a cv. Roxton crop planted on 4 March, covered by a small tunnel until 2 May and harvested on 8 July. The mean length of the flower stalk and the percentage difference were compared with the untreated control plants. In the direct-sown crop, fazor and cyter (chlormequat + mepiquat) reduced flower stalk length, early application being the most effective. Tomathrel, particularly later application at a higher rate, was also effective. Terpal increased bolting. With the transplanted crop, all treatments reduced the flower stalk length (by up to 52 per cent), except of one of the cycocel treatments tested (Bohec *et al.*, 1997). For early leeks in UK, it is suggested that rather than planting early, with subsequent air temperatures near to the optimum for vernalization, which will result in high levels of bolters, it would be better to delay transplanting. Plants can then be maintained at a rising temperature that is higher than the optimum for vernalization and transplanted when there is a strong probability of air temperatures being higher (Wurr *et al.*, 1999).

10.0 Diseases and Pests

Leeks do not have many pests but most insects and diseases that attack onions can also affect leeks.

10.1.0 Diseases

They are mostly the same as those of onion. However, diseases like downy mildew, purple blotch, pink root, smut, smudge, *Heterosporium* blight and white tip are the common diseases of leek.

Damping-off (*Fusarium* spp.)

Seedlings that grow slowly and then wilt and die may be a sign of damping off. Root tips of affected plants may be tan, yellow, black, or pink.

Planting of disease-free seed, seed treatment with fungicide, crop rotation with cereals or grasses to reduce levels of pathogen in soil, steam treatment or fumigation of soil can help reduce levels of *Fusarium* in the soil

Downy Mildew (*Peronospora parasitica*)

Pale spots or elongated patches on leaves; gray-purple fuzzy growth on leaf surface; leaves turning pale then yellow; leaf tips collapsing may be a sign of downy mildew.

Avoidances of infected sets planting, crop rotation with non-allium species for 3-4 years, planting in well-drained areas and maintenance of proper spacing; destruction of all infected crop debris; application of appropriate foliar fungicides taking care to apply thoroughly to waxy leaves will be the better management options for the disease.

Purple Blotch (*Alternaria porri*)

Purple blotch presents on the flag or stalk with small, water-soaked lesions with white centers. These blotches enlarge and become brown or purple, with yellow edges. Foliage may die.

Cultural controls include long rotations with non-hosts and the reduction of leaf wetness by planting in well-draining soil and timing irrigation to allow plants to dry adequately during the day; some fungicides are effective at controlling the disease but should be rotated for optimal control.

Pink Root (*Phoma terrestris*)

Stunted plants with undersized shafts may be a sign of pink root, which also causes roots to turn light pink and then darken to purple. The roots may be transparent and water-soaked.

Disease is most severe when alliums are planted continuously. Growing of resistant varieties; solarization and/or fumigation can help reduce the levels of pathogen in the soil.

Botrytis Leaf Blight (*Botrytis squamosa*)

This fungus is characterized by small white lesions ringed in light green which may expand slightly as they grow; in prolonged periods of moisture fungus may develop rapidly and cause leaf blighting.

Planting of leeks in single rows allowing at least 30 cm between plants to promote good air circulation and quick drying of foliage after rain; timely irrigation to allow plants time to dry out sufficiently; spraying of suitable fungicide when plants have at least five true leaves and early symptoms of disease are better management practices of the disease.

White Rot (*Sclerotinia cepivorum*)

Yellowing of older leaves; stunted growth; death of all leaves; fluffy white growth on base of bulb which spreads up bulb to storage leaves are common symptoms of the disease.

Seed treatment with hot water prior to planting; a long term rotation with non-allium crops; application of appropriate fungicides may be effective at controlling this disease.

Rust (*Puccinia porri*)

It is a fungal disease that presents as orange pustules on leaves. It can reduce market value and yield of the crop severely.

White tip of leek is known for a long time in the British Isles. Onion is not affected by this disease. The disease is caused by the fungus *Phytophthora porri*. The tips of leaves become yellow die and turn white for a distance up to 15 cm. Most affected leaves turn backward. Water-soaked areas in the vicinity of the midrib may appear half way down the leaf or near the base. Young severely infected plants

Diseases of Leek

Rust · White Tip (*Phytopthora porri*)

Purple Blotch

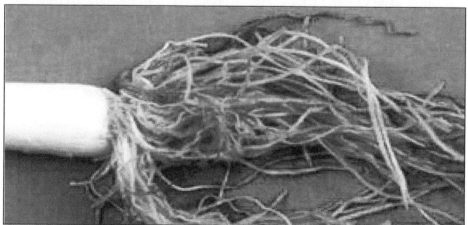

Pink Root

remain stunted. When large plants are attacked severely, they break over the soil line. The disease can be effectively controlled by spraying with Bordeaux mixture.

Reaction of leek cultivars to infection by *Puccinia allii* in the U.K. revealed that 'slow rusting' cultivars Agria and Winterreuzen showed sufficiently high agronomic quality to be considered for commercial production (Uma and Taylor, 1991). The effect of reduced doses of propiconazole and different initial inoculation levels on leek rust caused by *P. allii* was studied by Clarkson *et al.* (1997). All propiconazole doses tested reduced rate of plant to plant spread compared with an unsprayed treatment but lower doses had little effect on new pustule appearance. Rust spread occurred at a maximum rate unless incidence was extremely low. Reduced doses of propiconazole also suppressed spore germination of *P. allii* and the appearance of rust pustules in inoculated plants in the glasshouse.

10.2.0 Pests

Leaf Miners (*Lyriomyza* spp.)

These tiny, 1/8-inch flies are generally yellow, dark gray, or black, or a combination thereof. They leave thin, white, winding trails on leaves. The fly lays its eggs in the leaf, and the larvae feed on the leaf's interior. White blotches on leaves may also be seen.

Use of Neem oil is suggested to get rid of the pest.

Thrips (*Thrips tabaci*)

Thrips (*Thrips tabaci*) cause considerable damage to crop. Thrips are small winged insects that damage plants by sucking leaf sap. Discoloured, distorted tissue; scarring of leaves; severely infected plants may have a silvery appearance.

Belder *et al.* (1999) observed that thrips densities were lower in the intercropped (with strawberry) leek plots than in the monocropped plots. The number of thrips recorded were not correlated with the amounts of nitrogen present in the foliage or shaft of the leek plants. Sowing clover (*Trifolium fragiferum*) between rows in leek resulted in significant and sufficient thrips population suppression and good yields. Similar results were achieved by full field sowing of clover simultaneous with leek crop planting (Theunissen and Schelling, 1998). In an experiment in the Netherlands, leek seeds were sown at distances of 1 cm (seed bed) and 4 cm (direct drilling). On plants sown at the 4 cm seed distance, significantly more *T. tabaci* (immatures and adults) were present in comparison to the plants from the seeds sown at 1 cm. Also, the damage caused by thrips at the 4 cm spacing was more severe than at 1 cm, in both treated (with fipronil and thiram) and untreated seeds (Ester *et al.*, 1998).

In a study to investigate the potential reduction of chemical sprays on vegetables, a variety of insect nets (woven polytehylene with different mesh sizes and spun-bonded polypropylene) were evaluated. The dense mesh or spun-bonded polypropylene reduced growth and harvest weight but the former provided the best protection against pests and diseases. The quality of the leek plants was mainly affected by thrips, so the dense mesh produced the best quality (Beyenburg *et al.*, 1998).

Pests of Leek

Leek Moth Infestation

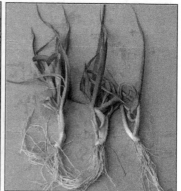

Leaf Miner Infestation

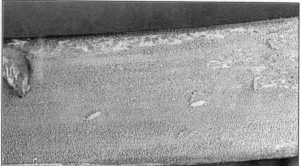

Aphid (Left), and Thrips (Right) Infestations

Onion Maggots (*Delia antiqua*)

Stunted or wilting seedlings; plant will commonly break at soil line if an attempt is made to pull it up; if infestation occurs when plants are bulbing, bulbs will be deformed and susceptible to storage rots after harvest; adult insect is a greyish fly which lays white, elongate eggs around the base of the plant; the larvae that emerge from the eggs are tiny and white and bore into the onion plant; mature larvae are about 1 cm (0.4 in) long with feeding hooks. The damaged plants are also more susceptible to other pathogens, such as bacteria.

Management of onion maggots is heavily reliant on good sanitation; all bulbs should be removed at the end of the season as maggots will die without a food source. Spraying of suitable insecticides is suggested. Crop rotation is the best way to prevent the infestation.

Leek Moth (*Acrolepiopsis assectella*)

The leek moth is a lepidopteran pest that attacks Allium crops, including onion, leek, and garlic. Eggs are laid on leaves and emerging larvae may cause extensive damage by mining leaves, feeding on leaf surfaces and feeding directly on bulbs. It is a serious pest in Italy as reported by Scaltriti and Rezzadore (1982). The moth had 5 overlapping generations in a year. The third generation which occurred in July was the most injurious. Two applications of chemical insecticides against each of the last 3 generations were recommended.

11.0 Seed Production

Leek does not usually seed in the plains as the mature plants should be exposed to low temperatures for a period before they form flowering stalks. The seed is produced at higher altitudes in the hills. Like onion, there are two methods of seed production; one is seed to seed method and the other one is Bulb to seed method. Both the methods have advantages and disadvantages. However, mostly bulb to seed method is followed because it permits selections of true to type and healthy bulbs for seed production with comparatively higher seed yield.

Leeks are cross and insect pollinated and will cross with elephant garlic, kurrat and other leek varieties. They may also self fertilize but they won't cross with onion.

They are biennial flowering in their second year. Ideally a distance of up to 1600 meters should be observed between two different flowering leek crops. Other agronomical practices are similar to onion.

Any plants which fail to grow well, which look weak or bolt too early should be pulled out and discarded before flowering. The leek flowers later than onions producing usually just one seed head and will take longer to mature delaying harvest.

The effects of development of leek seeds at 20/10°, 25/15° and 30/20°C (day/night) and drying of seed harvested at different developmental stages on subsequent performance were examined in each of 3 years by Gray *et al.* (1992). An increase in temperature from 20/10° to 30/20°C reduced mean seed weight from 2.90 to 2.55 mg as a result of a reduction in the duration of seed growth from 80 to 55 days;

seed growth rate was unaffected. Seed moisture content reached a minimum, up to 35 days after the attainment of maximum seed dry weight and 115, 90 and 70 days after anthesis at 20/10°, 25/15° and 30/20°C, respectively. Germination was little affected by temperature of seed development. Drying immature seeds increased percentage germination. Growing seeds at 30/20°C and drying at 35°C and 30 per cent RH raised the upper temperature limit of germination compared with growing at 20/10°C and drying at 15°C and 30 per cent RH.

Leek probably has an obligatory vernalisation requirement whereas the effects of day length are quantitative only. There is a juvenile phase that ends at a minimum size of about 2 g or five visible leaves. Optimum vernalisation temperature is at 5°C, the inductive temperature ranges between 0 and 18°C. Temperatures above 18°C cause devernalisation. Bolting is promoted by short days during and long days after vernalization (Wiebe, 1994).

Once around 60 per cent of seed heads have matured manual harvesting can take place over several weeks cutting the entire plant with 30 – 45 cm of stem to allow the sap to continue to ripen the seeds. Seed heads are to be kept in a well aired space to dry and turn the stems every few days. An indication of seed maturity can be seen by seeds turning black. Threshing can be done by hand by rubbing the seed heads gently and sieving out the trash materials. Leek seeds do not have a long storage life and therefore may not keep for longer than 2 years.

12.0 Crop Improvement

Major breeding objectives in temperate countries are yield, uniformity and disease resistance. Leek can be considered an allogamous autotetraploid with 20–30 per cent self-pollination in seed-production fields. Nowadays seed producing sectors either go on maintaining traditional open-pollinated cultivars under permanent mass selection pressure, or offer F_1 hybrids. Plant vigour loss by selfing is about 30 per cent for the first generation and 40 per cent between S_1 and S_2. F_1 hybrids are based on male sterility. F_1 hybrid leek has 20 per cent higher yields and is more uniform. The rationale behind this is the exclusion of progeny resulting from self-pollination. F_1 cultivars can have a different genetic background but they always have the male sterile parent in common. This line is vegetatively propagated via in-vitro multiplication and/or formation of bulbils, obtained in the umbels by early flower ablation and exposition to photoperiods of 20 hours. Cytoplasmic male sterility has recently been transferred from onion. No special breeding work for tropical countries has been reported.

Kampe (1980) and Pink and Innes (1984) listed the following objectives for leek breeding in the UK

☆ Winter hardiness

☆ Resistance to bolting

☆ Dark green leaf blade

☆ Freedom from bulbing

☆ Long shaft (the blanched pseudo-stem formed by the young folder leaf blades) length, particularly for processing purpose

☆ High yield

☆ Resistance to leek rust and yellow stripe virus

☆ Uniformity

In Canada, Belgium and the Netherlands some promising cultivars have been identified. In India, practically no improvement work on this crop has been done.

Leek shows considerable inbreeding depression during inbreeding. According to review of Currah (1986), inbreeding depression was about 35 per cent in I_1, and nearly 40 per cent in I_2 generation. This depression was equally strong on seed production capability. Therefore, mass selection has been the most effective breeding method of cultivar improvement. Male sterility is also encountered in I_1 generation. Crosses between two varieties and I_1 lines often show heterosis. F_2 generations of I_1 x I_1 crosses could also be used as segregating source populations for selection of desirable genotypes which can again be crossed to reverse the drastic effect of inbreeding. Use of male sterility seems to be a possibility for developing hybrid cultivars of leek, but this shall not be easy to employ as done in onion because of tetraploid nature of leek.

Allium porrum (cultivated leek) (2n = 4x = 32) is a fertile tetraploid that forms bivalents with pericentric chiasmata at metaphase I. The karyotype appears to be autotetraploid. Thus, the patterns of crossing over and partner trades promote balanced disjunction and high fertility in autotetraploid. Rare quadrivalents observed at metaphase I must be due to infrequent partner trades between recombination nodules (RNs). Polycomplexes, unusual in their number and sizes, were observed during zygonema (Stack and Roelofs, 1996). A detailed analysis of metaphase I of meiosis in pollen mother cells has confirmed that despite frequent quadrivalent pairing (71 per cent) at prophase I and a high mean chiasma frequency (3.5 per tetrasome), the majority of quadrivalents are resolved into bivalent associations at metaphase I. However, neither chiasma localization nor quadrivalent resolution are as prevalent as most previous reports have indicated. It appears that meiosis in the leek is less regular than previous reports indicated. The implication for fertility, stability and uniformity are not trivial and ought to be considered in the context of leek production and breeding (Jones *et al.*, 1998).

Crosses between 14 cultivars of leek, kurrat (*A. kurrat*) and preianak were carried out with the wild relatives *A. ampeloprasum, A. commutatum, A. bourgeaui* and *A. atroviolaceum*. Mitochondrial and chloroplast DNA was isolated and subjected to restriction analyses. The results showed and mtDNA variation in the cultigens leek, kurrat and prei-anak is limited compared to that of their wild relatives in the *A. ampeloprasum* complex. The phylogenetic relationships among these cultigens and their wild relatives were quite close, with the majority of the species clustering within one mitochondrial clade. The presence in leek of an extra-mitochondrial genetic element was noted. Analysis of crossability showed that all species were interfertile with leek. It is suggested that the genetic variation present within the *A. ampeloprasum* complex could be exploited in order to broaden the genetic basis of leek (Kik *et al.*, 1997).

Breeding methods, for improving this crop are like that of onion.

12.1.0 Hybridization Techniques

Pink (1993) described the hybridization technique in leek. The inflorescences (umbles) of both the female and male parents are enclosed in a muslin cloth bag. The individual flowers (protandrous in nature) are emasculated as they open. Pollination is generally carried out a day or two after emasculation when the stigma is feathery. As flowers open sequentially from the bottom to the top of the umbel, emasculation is done daily to prevent self-pollination of emasculated flowers by pollen from later-opening flowers. Once sufficient pollinations are made, remaining unopened flowers are trimmed off. Leek can also be propagated -vegetatively by the induction of top-sets and plantlets in the umbel or from the basal cloves formed after flower stalk formation.

Kirnosova (1990) observed that the yield of marketable stem was positively and highly correlated with mean stem weight and stem height. For commercial production of hybrid seeds in leek, Schweisguth (1970) has proposed the use of genetic male sterile line.

A. tricoccum var. *burdickii*, commonly called wild leek, is a rare species in Nova Scotia, Canada, growing in isolated populations. Genetic variation was investigated in three populations using cellulose acetate gel electrophoresis. The results showed that all 13 enzyme loci were polymorphic. An excess of heterozygotes was found in all populations. A total of 29 genotypes were detected in the populations. Very few genotypes were common to two populations and none to the three populations. Occasional sexual reproduction and gene flow might occur to maintain high level of variation among populations. Environmental conditions could also influence population genetic structure as they occur in highly different habitats (Vasseur, 2001).

Leek is mainly grown for the fresh market and varieties of different earliness are demanded. Leijon and Olsson (1999) identified some aspects of breeding of leek. Medium-stalked types with a large leaf mass give high total yield and are thus desired by the food industry. The fresh market prefers long-stalked types with a small leaf mass. The thickness of the stalk is also important for the economic outcome and thick stalks are often more crispy. Plants with blue-green leaves which are much keeled have better winter hardiness than plants with light-green, flat leaves. Plants should have an upright growth habit and no bulb formation. A higher dry matter content favours the cooking characteristics but can reduce crispness. Resistance against rust (*P. allii*) is an important breeding objective. Western Europe is the world's greatest leek producing and consuming region. Clercq *et al.* (1999) evaluated the breeding potential of the Belgian landraces. Clustering analysis revealed that 4 of the Belgian landraces were grouped with 6 commercial varieties to form the most winter hardy group of leek. For the important characteristics leaf colour and senescence resistance, the Belgian landraces performed better than most of the commercial varieties; however, for the economically more important characteristics shaft length and some disease resistance traits they scored low. The Belgian landraces were high yielding, especially in the late season.

A. ampeloprasum ssp. *porrum*, the leek, shows extensive and frequent quadrivalent formation during prophase I of meiosis, and that despite the resolution of most of these into bivalents some quadrivalents persist to metaphase I; in addition, a substantial number of univalents are found at this stage. Khazanehdari and Jones (1997) proposed that the majority of aneuploid seedlings originate from meiotic irregularity, especially univalent missegregation. It is very likely that the presence of this level of aneuploidy in the seedling populations contributes to the problems of phenotypic non-uniformity experienced by the leek crop, although its relative importance is difficult to assess. Major volatile components detected in leek oils were dipropyl trisulfide, dipropyl disulfide and (E)-propenyl propyl disulfide.In accordance with the higher amount of leek chromosomes in the cell nucleus, the percentages of the measured sulphur volatiles in the hybrid material corresponded to the leek flavour profile (Schulz *et al.*, 1998).

Within winter leek, five resistance classes to *P. porri* were defined on the basis of average field scores of 21 plants. Clones from these plants were tested with the immersion technique (inoculation by 24 h-immersion of leek plantlets in the 3-6 leaf stage in a suspension of ca. 100 zoospores/ml). A selection experiment yielded a strong response to selection for resistance (53-97 per cent) but no response to selection for susceptibility. This may indicate that resistance is due to few recessive genes in the studied winter leek. Crosses between landraces and winter leek were analyzed by means of F_2 (selfed F_1) and BC1 progenies. This analysis indicated the presence of few loci with dominant genes for resistance in accession CGN873243, and additive polygenes in accession PI368351 (Smilde *et al.*, 1997).

Disease resistance of several leek genotypes against leek rust was evaluated by Smith *et al.* (2000) in separate trials in the UK. Sixteen cultivars representing a wide range of leek germplasm were evaluated during the 1991 trial, while the cultivar Agria (previously identified as the best slow-rusting genotype), some breeding lines and 7 commercial cultivars were evaluated during 1994. Breeding lines [including half-sibs (HS), S1, and single seed descent (SSD) lines] were developed from the 1991 and 1994 materials and evaluated for leek rust resistance. Wintra was ranked with the more susceptible genotypes while Agria was not significantly different from the majority of the commercial cultivars. The breeding line C407 was significantly better than the other cultivars and test lines at all recordings. Comparison between Wintra and the partially resistant genotype AP96146 showed that AP96146 scored better than Wintra in terms of slowing rust spread between plants by 21 per cent, increasing the time to 50 per cent infection (from 46.98 to 87.94 days), and slowing the rate of new pustule appearance by 44 per cent.

12.2.0 Inter-specific Hybridization

In leek (*Allium ampeloprasum*) several important breeding traits are missing that are present in other cultivated *Allium* species. The high crossing barriers between leek and species of subgenus *Rhizirideum* have been overcome to introgress new traits from other cultivated Alliums. Interspecific hybrids between *A. cepa* and *A. ampeloprasum* were generated as a first step for the introduction of S-cytoplasm from onion into leek. Pre-zygotic barriers of crossability were observed after the arrival of pollen tubes at the end of the style when entering the cavity. Nevertheless, micropyle penetration of pollen tubes and the formation of hybrid embryos were also observed. After accomplishing *in vitro* culture of ovaries and ovules successively, triploid hybrid plants with 24 chromosomes were obtained. Their hybrid nature was confirmed by RAPD analysis, genomic *in situ* hybridization (GISH), and morphological analysis. Southern hybridization with a cytoplasmic probe indicated the transfer of unaltered S-cytoplasm into the hybrid plants (Peterka *et al.*, 1997; 2005). GISH and FISH techniques as well as molecular markers have been applied to characterize the mitotic chromosomal constitution and the meiotic behaviour of hybrids and their derivatives. To open the chance for a CMS system in leek, a sexual hybrid between a cytoplasmic male-sterile onion and leek was used to transfer the alien cytoplasm into leek. The phenotype of alloplasmic leek is modified in comparison to euplasmic leek suggesting some incompatibility between T-cytoplasm of onion and leek genomes. Addition of onion chromosome 1C conferred normal growth of the alloplasmic leek. The hybridization between Bunching onion and leek could be useful both for transferring a new cytoplasm for CMS induction and to introgress genomic components transmitting resistance. An *A. fistulosum* × *A. ampeloprasum* hybrid (FAA, 2n=24) was produced by means of embryo rescue and the hybrid chromosome number was doubled with colchicines and hybrid plants from crosses *A. schoenoprasum* × *A. ampeloprasum* were obtained (Peterka *et al.*, 2005).

13.0 Biotechnology

The major achievements in biotechnology of leek are regeneration of plants *in vitro*, mainly from somatic embryogenesis and success in protoplast culture and development of somatic hybrids.

13.1.0 Tissue Culture

13.1.1 Micropropagation

Tissue culture of leek has been perfected for induction of somatic embryogenesis, organogenesis and protoplast culture. Gorecki and Gorecka (1989) suggested micro-propagation, anther culture; ovule culture and cybridization as various means of *in vitro* manipulations for leek. However, Dore (1988) presented one of the first suitable methods of an *in vitro* propagation method for continuous, large-scale production of plants. It involved stimulating the latent meristems located at the basal part of the plantlet, multiplication in successive cycles and storage either in a medium containing no growth substances or bare-rooted in a sealed jar. Various explants such as meristems, ovaries, ovules, flower buds *etc.* have been used for tissue culture studies. Ionescu and Popandron (1995) cultured ovules, ovaries and intact flower buds of leek on 12 different media. The ovary and flower-bud cultures developed callus and regenerated plants on media A6 and A11, which contained 1-5 mg BA, 0.5 mg NAA and 1 mg IBA/l.

In vitro cultivated seedlings of leek (cv. Poros) were induced to form bulblets by an increase in sucrose concentration (30, 50, 150 g/l) of the medium, and the addition of BA (0, 12.5 mg/l) or ethephon (33 mg/l for 0, 5 or 20 days). The highest bulbing ratios were obtained with combination of sucrose and ethephon treatments.

BA caused not only bulb swelling but also an increase in multiple adventitious shoot formation (Keller, 1993).

13.1.2 Somatic Embryogenesis

In an investigation on anther and ovule culture of some species and hyrbid of *Allium,* Keller (1990) reported strong callus development on media with equimolar concen-trations of 2,4-D and BA (10^{-6} mol/l). Species, lines and individual plants differed significantly from one another in rate of callus formation, the highest rates being recorded for *A. porrum.* Some haploid regenerates were obtained by culturing unfertilized ovules. High frequency plant regeneration via somatic embryogenesis was induced by Hong *et al.* (1995) from *in vitro* shoot-base cultures of seedlings of garden leek. The presence of 2,4-D in the medium and light conditions were shown to be essential for nodular callus induction and somatic embryogenesis. Abscisic acid was not a prerequisite for somatic embryogenesis, but it significantly increased the frequency. Plant regeneration system via organogenesis and somatic embryogenesis was established with callus cultures derived from mature zygotic embryos of different leek genotypes (*A. ampeloprasum*). Relatively high concentrations of the auxin, 2,4-D reduced callus weight and subsequent shoot regeneration and primordia formation of the callus. Shoot regeneration and primordia formation of the callus decreased after prolonged subculture on media containing 2,4-D. Differences between leek genotypes in callus type, callus weight, shoot regeneration and primordia formation were observed (Silvertand *et al.,* 1996).

Compact callus was induced on embryo and leaf explants of 8 (*A. porrum*) cultivars and 4 accessions on three media (Buiteveld *et al.,* 1993). The highest frequency of compact callus formation (up to 90 per cent) was obtained when mature, zygotic embryos were cultured on MS medium containing 30 g sucrose and 1 mg 2,4-D/l. Regeneration occurred through somatic embryogenesis on MS medium supplemented with 1 mg kinetin/l. Schavemaker and Jacobsen (1995) reported formation of somatic embryos and callus from embryos excised from seeds resulting from reciprocal crosses involving 3 cultivars and 2 from selfing of 2 cultivars in MS medium supplemented with 2 mg BA/l. The average number of somatic embryos produced per callus form zygotic embryo varied between 8 and 39.4. The average of 10 genotypes for the percentage of somatic embryos forming rooted plantlets from the second to the fifth cycles of somatic embryogenesis declined from 68 to 52 per cent.

In vitro propagation method offers possibilities for large scale micro-propagation of selected leek plants. A detailed study was carried out by Kaska *et al.* (2016) to determine the effects of media on somatic shoot production from six open-pollinated standard leek genotypes. They cultured a total of 28515 unopened flower buds collected from six donor leek lines in various tissue culture media. Bud cultures established in BDS based media provided somatic shoot regeneration at higher frequencies than those cultured in MS-based media. The highest frequency of somatic shoot regeneration (~6 per cent) was obtained from the buds cultured in BDSC medium (BDS medium containing 2 mg L^{-1} BAP and 100 g L^{-1} sucrose). A selection line (PAUPR2) showed the highest somatic regeneration (~18 per cent) in BDSC medium. Flow cytometry analysis showed that all somatic regenerants were tetraploids. Somatic regenerants transferred to greenhouse showed normal development similar to their donor lines. Somatic plants showed higher vigor and uniformity compared to those grown from seeds.

13.1.3 Protoplast Culture

Buiteveld and Creemers-Molenaar (1994) studied regeneration-competent protoplasts obtained from an embryogenic suspension culture that was initiated from friable, embryogenic callus derived from immature embryos. The generally low plating efficiency was increased by embedding the protoplasts in Ca-alginate, compared to culturing the protoplasts in liquid or agarose-solidified medium. Upon transfer of the protoplast-derived callus on agarose-solidified BDS medium, morphologically different callus types proliferated. After transfer to regeneration medium, compact or friable embryogenic callus produced somatic embryos and plantlets at a frequency of up to 80 per cent.

Plant regeneration from protoplasts of leek was reported by Schum *et al.* (1994) from leaf tissue of *in vitro*-cultured seedlings of leek cultivars Alberta, Artico, Carina, Farinto and Ekkehart, from the hypocotyls of cultivars Abila, Arkansas, Carina, Farinto and Gavia, from callus of cv. Alaska with shoot initials, and from cell suspension cultures of cv. Alaska. Sustained cell divisions were obtained only from protoplasts obtained from cell suspension. Elongated leaves arose on solid MS medium, and shoots developed in the leaf axils after further subculture. Rooted plantlets were obtained 30 weeks after isolation of protoplasts.

13.1.4 Somatic Hybrids

Buiteveld *et al.* (1998a) analysed organelle composition of somatic hybrid plants, produced by symmetric protoplast fusion between leek (*A. ampeloprasum*) and cytoplasmic male sterile onion (*A. cepa*). Mitochondrial DNA (mtDNA) analyses were performed using PCR amplification. Of the 55 hybrids analysed with PCR, only 3 had mtDNA restriction fragment patterns of both parents, while the remaining hybrids displayed a pattern identical to *A. ampeloprasum*. Detailed analysis of 18 of these hybrids with Southern hybridization using 5 mitochondrial probes revealed that 15 hybrids possessed a rearranged mitochondrial genome of both parents, but with a predominance of mtDNA fragments of *A. ampeloprasum*.

Buiteveld *et al.* (1998b) also reported the production and characterization of somatic hybrids between *A. ampeloprasum* and *A. cepa*. Both symmetic and asymmetric protoplast fusions were carried out using a polyethylene-based mass fusion protocol. Asymmetric fusions were performed using gamma ray-treated donor protoplasts of *A. cepa* and iodoacetamide-treated *A. ampeloprasum* protoplasts. However, the use of gamma irradiation to eliminate or inactivate the donor DNA of *A. cepa* proved to be detrimental to the development of fusion calluses, and thus it was not possible to obtain hybrids from asymmetric fusions. The symmetric fusions yielded a high number of hybrid calluses and regenerated plants. The analysis of the nuclear DNA composition using interspecific variation of rDNA revealed that most of the regenerated plants were hybrids. Flow cytometric analysis of nuclear

DNA showed that these hybrid plants contained a lower DNA content than the sum of the DNA amounts of the parental species, suggesting that they were aneuploid.

13.2.0 Genetic Transformation

There is very little information on the genetic transformation in leek. Schavemaker (2000) achieved a cyclic somatic embryogenesis regeneration system with flower stalk for transformation purposes. Particle gun bombardments were carried out resulting in chimeric leek plants.

Transgenic leek plants have been recovered by the selective culturing of immature leek embryo via *Agrobacterium*-mediated transformation by Eady *et al.* (2005). This method involved the use of a binary vector containing the m-gfp-ER reporter gene and nptII selectable marker, and followed the protocol developed previously for the transformation of onions with only minor modifications pertaining to the post-transformation selection procedure which was simplified to have just a single selection regime. Transgenic cultures were selected for their ability to express the m-gfp-ER reporter gene and grown in the presence of geneticin (20 mg/l). The presence of transgenes in the genome of the plants was confirmed using TAIL-PCR and Southern analysis. This is the first report of leek transformation. It now makes possible the integration of useful agronomic and quality traits into leek.

14.0 References

Aalbersberg, W. (1988) *Groenten en Fruit*, **44** : 62-63.

Alofs, W. J. (1988) *Groenten en Fruit*, **43** : 54-55.

Aykroyd, W. R. (1963) *ICMR Special Rept. Series*, No. 42.

Bath, B. (2000) *Biol. Agri. Hort.*, **18** : 243-258.

Bath, B. and Ramert, B. (2000) *Acta Agriculturae Scandinavica*, **49** : 201-208.

Baumann, D.T., Bastiaans, L. and Kropff, M.J. (2001) *Crop Sci.*, **41** : 764-774.

Baumann, D.T., Bastiaans, L., Kropff, M.J. and Niggli, U. (2000) *Proceedings 13th International IFOAM Scientific Conference*, Basel, Switzerland, 28 to 31 August, 2000, p. 190.

Belder, E. den, Elderson, J., Den Belder, E. and Brunel, E. (1999) *Bulletin OILB-SROP*, **22** : 151-156.

Belder, E. den, Elderson, J., den Belder, E. and Francke, P.J. (1998) *Proc. Sec. Exp. Appl. Entomol. Netherlands Entomol. Soc.*, **9** : 123-127.

Benoit, F. and Ceustermans, N. (1998) *Proeftuinnieuws*, **8** : 36-38.

Berry, D. and Thicoipe, J.P. (1998) *Infos Paris*, **140** : 39-42.

Beyenburg-Weidenfeld, H., Lenz, F. and Whitehead, P. (1998) *International Seminar*, Warszawa, Poland, 10-15 June, 1997, pp. 9-10.

Biczak, R., Gurgul, E., Karczmarczyk, S. and Herman, B. (1999a) *Folia Univ. Agriculturae Stetinensis, Agricultura*, **73**: 45-52.

Biczak, R., Gurgul, E., Karczmarczyk, S. and Herman, B. (1999b) *Folia Univ. Agriculturae Stetinensis, Agricultura*, **73** : 37-44.

Biczak, R., Gurgul, E., Karczmarczyk, S. and Herman, B. (1999c) *Folia Univ. Agriculturae Stetinensis, Agricultura*, **73** : 29-35.

Biczak, R., Gurgul, E., Karczmarczyk, S. and Herman, B. (1999d) *Folia Univ. Agriculturae Stetinensis, Agricultura*, **73** : 19-27.

Bohec, J. Le and Chombart, B. (1989) *Infos, Centre Technique Interprofessionnel des Fruits et Legumes*, France, No. 48, pp. 5-8.

Bohec, J. le and Fouyer, L. (1997) *Infos Paris*, **137** : 34-38.

Bohec, J. le, Fouyer, L., Routhiau, O. and Le Bohec, J. (1996) *Infos Paris*, **122** : 37-39.

Buiteveld, J. and Creemers-Molenaar, J. (1994) *Plant Sci., Limerick*, **100** : 203-210.

Buiteveld, J., Suo, Y., Lookeren-Campagne, M.M. van and Vreemers-Molenaar, J. (1998b) *Theor. Appl. Genet.*, **96** : 765-775.

Buiteveld, J., Valk, P. van der, Jansen, J., Creemers-Molenaar, J., Colijn-Hooymans, C.M. and Van der Valk, P. (1993) *Plant Cell Rep.*, **12** : 431-434.

Buiteveld, J., Kassies, W., Geels, R., Lookeren-Campagne, M.M. van, Jacobsen, E., Creemers-Molenaar, J. and Van Lookeren-Campagne, M.M. (1998a) *Plant Science, Limerick*, **131** : 219-228.

Callens, D., Rooster, L. de and Reycke, L. de (1998) *Proeftuinnieuws*, **8** : 35-37.

Callens, D., Rooster, L. de, Reycke, L. de and Visser, J. (2000) *Proeftuinnieuws*, **10** : 40-41.

Cesare, L.F. de, Giombelli, R., Senesi, E. and Ferrari, V. (1999) *Industrie Alimentari*, **38** : 667-673.

Chauhan, D. V. S. (1972) *Vegetable Production in India* (3rd edn.), Ram Prasad and Sons. Agra.

Choudhury, B. (1976) *Vegetables*, National Book Trust, India, New Delhi, pp. 105-106.

Clarkson, J.P., Kennedy, R., Phelps, K., Davies, J. and Bowtell, J. (1997) *Plant Pathol.*, **46** : 952-963.

Clercq H. de, Baert, J. and Bockstaele, E. van (1999) *Euphytica*, **106** : 101-109.

Dore, C. (1988) *Agronomie*, **8** : 509-511.

Ducsay, L., Lozek, O., Varga, L. and Kona, J. (2000) *Acta Horticulturae et Regiotecturae*, **3** : 29-31.

Eady, C., Davis, S., Catanach, A., Kenel, F. and Hunger, S. (2005) *Plant Cell Rep.*, **24**(4) : 209-215.

Embrechts, A. J. M. (1988) *Groenten en Fruit*, **43** : 52-53.

Embrechts, A.J.M. (1996) *Publicatie - Proefstation voor de Akkerbouw en de Groenteteelt in de Vollegrond, Lelystad*, No. 81B, pp. 59-69.

Eppendorfer, W.H. and Eggum, B.O. (1996) *Plant Foods Hum. Nutr.*, **49** : 163-174.

Ester, A., Evenhuis, A. and Francke, P.J. (1998) *Proc. Sec. Exp. Appl. Entomol. Netherlands Entomol. Soc.*, **9** : 117-122.

Ferrari, V., Cesare, L. di and Schiavi, M. (1999) *Informatore Agrario*, **55** : 55-58.

Flones, M. (1990) *Gartneryrket*, **80** : 18-19.

Gerasimova, N. and Yordanova, M. (2015) *Acta Agric. Serbica*, **20**: 41-49.

Geyer, M. (1997) *Gemuse Munchen*, **33** : 257-258.

Golubkina, N.A., Seredin, T.M., Antoshkina, M.S., Kosheleva, O.V., Teliban, G.C. and Caruso, G. (2018) *Horticulturae*, 4, https: //doi.org/10.3390/horticulturae4040039.

Gilreath, James P., Santos, Bielinski M., Gilreath, Phyllis R. and Maynard, Donald N. (2008) *Crop Protec.*, **27**(3): 847-850.

Golubkina, Nadezhda A., Seredin, Timofey M., Antoshkina, Marina S., Kosheleva, Olga V., Teliban, Gabriel C. and Caruso, Gianluca (2018) *Horticulturae*, **4**: 39, doi: 10.3390/horticulturae4040039.

Gorecki, R. and Gorecka, K. (1989) *Biuletyn Warzywniczy, 1989,* Suplement, I, pp. 39-42.

Grazegorzewska, M. and Bakowski, J. (1996) *Biuletyn Warzywniczy*, **45** : 77-89.

Hanci, Öðr. Üyesi Fatih, Pinar, Öðr. Üyesi Hasan and Uzun, Aydýn (2018) In: *Ejons v – International Conference on Mathematics – Engineering – Natural and Medical Sciences*, November 22-25, Gaziantep/TURKEY, pp. 808-818.

Heine, J. (1998) *Gemuse Munchen*, **34** : 23-24.

Helm, J. (1956) *Allium Kulturpflanze* 4: 130-137.

Hong-Wang, Debergh, P.C. and Hong, W. (1995) *Plant Cell Tissue Organ Cult.*, **43** : 21-28.

Horbowicz, M., Kotlinska, T. and Jager, A. de (2000) *Acta Hortic.*, **517** : 375-380.

Ionescu, A. and Popandron, N. (1995) *Anale Institutul de Cercetari pentru Legumicultura si Floricultura*, Vidra, **13** : 17-22.

Jones, G.H., Khazanehdari, K.A. and Ford-Lloyd, B.V. (1996) *Heredity*, **76** : 186-191.

Jones, H. A. and Mann, L. K. (1963) In: *Onions and their allies*. Leonard Hill, London.

Kahiluoto, H. and Vestberg, M. (1998) *Biol. Agri. Hort.*, **16** : 65-85.

Kampe, R. (1980) *Archiv furZuchtungsforschung,* **10**: 123- 128.

Kaniszewski, S. (1986) *Biuletyn Warzywniczy*, **26** : 95-106.

Kaniszewski, S. Rumpel, J. and Elkner, K. (1989) *Acta Hortic.*, **244** : 229-234.

Kaniszewski, S., Rumpel, J. and Dysko, J. (1999) *Veg. Crops Res. Bull.*, **51** : 39-47.

Kanters, F. (1991) *Groenten en Fruit, Vollegrondsgroenten*, **1** : 29.

Karaman, M.R., Brohi, A.R., Gunes, A., Inal, A. and Alpaslan, M. (2000) *Turkish J. Agri. Fores.*, **24** : 1-9.

Kariæ, Lutvija, Vukašinoviæ, Smiljka and Žnidarèiè, Dragan (2005) *Acta Agric. Slovenica*, **85** : 2.

Karkanis, A., Bilalis, D., Efthimiadou, A. and Katsenios, N. (2012) *Hort. Sci.* (Prague), **39**(2) : 81–88.

Kaska, A., Yildirim, S., Top, B., Celebi-Toprak, F. and Alan, A.R. (2016) *Acta Hortic.*, **1143** : 55-60.

Katyal, S. L. (1977) *Vegetable Growing in India*, Oxford and IBH Publishing Co., New Delhi, pp. 12-22.

Keller, E.R.J. (1993) *Genet. Resour. Crop Evol.*, **40** : 113-120.

Keller, J. (1990) *Archiv fur zuchtungsforschung*, **20** : 189-197.

Khazanehdari, K.A. and Jones, G.H. (1997) *Euphytica*, **93** : 313-319.

Kiffo, Birhanu Lencha (2016) *J. Biol. Agric. Healthcare*, **6** : 21-33.

Kik, C., Samoylov, A.M., Verbeek, W.H.J., Raamsdonk, L.W.D. van and Van Raamsdonk, L.W.D. (1997) *Theor. Appl. Genet.*, **94** : 465-471.

Kirnosova, T. I. (1990) *Sbornik Nauchnykh Trudov po Priklad noi Botanike*, Genetike i Selektsii, **134** : 44-47.

Kolota, E. (1986) *Biuletyn Warzywniczy*, **26** : 65-80.

Kolota, E. and Adamczewska-Sowinska, K. (2001) *Veg. Crop Res. Bull.*, **54** : 29-34.

Kolota, E. and Adamczewska-Sowinska, K. (2002) *Acta Hortic.*, **571** : 103-108.

Kolota, E., Adamczewska-Sowinska, K. and Michalak, K. (1996) *Biuletyn Warzywsnczy*, **45** : 39-49.

Leijon, S. and Olsson, K. (1999) *Sveriges Utsadesforenings Tidskrift*, **109** : 46-52.

Majnard, D. W. (1989) *Proc. Ann. Meet. Florida State Hortic. Soc.*, **101** : 385-389.

Maurer, A. R. (1982) *Res. Rev.*, Res. Sta. Agassiz B.C., pp. 6-7.

McCollum, G. D. (1976) In : Simmonds, N.W. (Ed.), *Evolution of Crop Plants*, Longman, London and New York.

Melander, B. and Rasmussen, G. (2001) *Weed Research Oxford*, **41** : 491-508.

Muller-Scharer, H. (1996) *Crop Protection*, **15** : 641-648.

Nielsen, K.L., Thorup-Kristensen, K and Romheld, V. (2001) *Fourteenth Int. Plant Nutr. Colloq.*, Hannover, Germany, pp. 1010-1011.

Parijs, L. Van. (1983) *Boer en de Tuinder*, **89** : 16.

Parys, L. var, Rooster, L. de and Reycke, L. de (1997) *Proeftuinnieuws*, **7** : 20, 35-37.

Paschold, P. J. and Gottlick, G. (1982) *Gartenbau*, **29** : 321-323.

Perko, J. (1989) *Revue Suisse de Viticulture, d' Arboriculture et d'Horticulture*, **21** : 159.

Peterka, H., Budahn, H. and Schrader, O. (1997) *Theor. Appl. Genet.*, **94** : 383-389.

Peterka, H., Budahn, H. and Schrader, O. (2005) *Acta Hort.*, **688**: 101-107

Pink, D. A. C. and Innes, N. L. (1984) *Plant Br. Abstr.*, **54**: 197.

Rana, M.K. and Jatav, P.K. (2018) In: *Vegetable Crops Science,* CRC Press, p. 8.

Reycke, L. de and de Reycke, L. (1998) *Proeftuinnieuws,* **8** : 30-31.

Reycke, L. de, Callens, D., Rooster, L. de and Janssens, M. (1999) *Proeftuinnieuws,* **9** : 36-37.

Reycke, L. de, Rooster, L. de and Callens, D. (2000) *Proeftuinnieuws,* **10** : 36-37.

Reycke, L. de, Rooster, L. de and Janssens, M. (1999a) *Proeftuinnieuws,* **9** : 14-15.

Reycke, L. de, Vanparys, L., Rooster, L. de and Janssens, M. (1998) *Proeftuinnieuws,* **8** : 32-34.

Rijbroek, V. van, and Riepma, F. (1982) *Groenten en Fruit,* **37** : 56-57.

Rolin, D., Pfeffer, P.E., douds, D.D. Jr., Farrell, H.M. Jr. and Shachar-Hill, Y. (2001) *Symbiosis Rehovot,* **30** : 1-14.

Rooster, L. de and De Rooster, L. (1998a) *Proeftuinnieuws,* **8** : 35-36.

Rooster, L. de and De Rooster, L. (1998b) *Proeftuinnieuws,* **8** : 18-19.

Rooster, L. de and De Rooster, L. (1998c) *Proeftuinnieuws,* **8** : 28-29.

Rooster, L. de and De Rooster, L. (1998d) *Proeftuinnieuws,* **8** : 18-19.

Rooster, L. de and De Rooster, L. (1998e) *Proeftuinnieuws,* **8** : 30-31.

Rooster, L. de, Reycke, L. de and Visser, J. (2000) *Proeftuinnieuws,* **10** : 42-43.

Rooster, L. de, Spiessens, K. and de Rooster, L. (1999) *Proeftuinnieuws,* **9** : 27-28.

Sadeghi, S. , Rhnavard, A. and Ashrafi, Z. Y. (2010) *Int. J. Agric. Technol.,* **6** : 607-614.

Sasa, M., Zahka, G. and Jakobasen, I. (1987) *Plant Soil,* **97** : 279-83.

Scaltriti, G. Pellizzari and Rezzadore, P. (1982) *Informatore Agrario,* **38** : 21515-21522.

Schavemaker, C.M. (2000) *Ph.D. Thesis,* Wageningen Agricultural University, Netherlands, p. 99.

Schavemaker, C.M. and Jacobsen, E. (1995) *Plant Cell Rep.,* **14** : 227-231.

Schow, E. (1998) *Gemuse Munchen,* **34** : 20-22.

Schulz, H., Kruger, H., Liebmann, J. and Peterka, H. (1998) *J. Agri. Food Chem.,* **46** : 5220-5224.

Schum, A., Junge, H. and Mattiesch, L. (1994) *Gartenbauwissenschaft,* **59** : 383-389.

Schweisguth, B. (1970) *Ann. Amellior. Plantes,* **20** : 215-231.

Scullion, J., Eason,W.R. and Scott, E.P. (1998) *Plant and Soil,* **204** : 243-254.

Silvertand, B., Rooyen, A. van, Lavrijsen, P., Harten, A.M. van, Jacobsen, E., Van Rooyen, A. and Van Harten, A.M. (1996) *Euphytica,* **91** : 261-270.

Smilde, W.D., Heusden, A.W. van, Kik, C. and van Heusden, A.W. (1999) *Euphytica,* **110** : 127-132.

Smilde, W.D., Nes, M. van, Reinink, K. and Kik, C. (1997) *Euphytica,* **93** : 345-352.

Smith, B.M., Crowther, T.C., Clarkson, J.P. and Trueman, L. (2000) *Ann. Appl. Biol.*, **137** : 43-51.

Sorensen, J.N. and Wichmann, W. (1996) *Acta Hortic.*, **428** : 131-140.

Stack, S.M. and Roelofs, D. (1996) *Genome*, **39** : 770-783.

Stone, D. A. and Rawse, H. R. (1982) *Proc. 9th Soil Tillage Research Organization*, U. K.

Theunissen, J. and Schelling, G. (1998) *Bi°Control*, **43** : 107-119.

Thompson, H. C. and Kelly, W. C. (1978) *Vegetable Crops*, Tata McGraw-Hill Publishing Co. Ltd., New Delhi, p. 368.

Traub, H. P. (1968) *Plant life*, **24** : 147-163.

Tursun, N., Bükün, B., Karacan, C.S., Ngouajio ve, M. and Mennan, H. (2007) *HortScience*, **42** : 106-109.

Uma, N. U. and Taylor, G. S. (1991) *Plant Pathol.*, **40** : 221-225.

Valdes-Mendez, W., Rodriguez-Perez, S. and Cardenas, J.R. (1999) *Interciencia*, **24** : 264-267.

Vanaerde, H., Reycke, L. de, Vanparys, L., Rooster, L. de and Jnssens, M. (1997) *Proeftuinnieuws*, 7 : 48-50.

Vanparys, L. (1996) *Mededeling Provinciaal Onderzoek en Voorlichtingscentrum voor Land en Tuinbouw*, Beitem Roeselar, No. 363, p. 4.

Vanparys, L. (1998a) *Mededeling Provinciaal Onderzoek en Voorlichtingscentrum voor Land en Tuinbouw, Beitem Roeselar*, No. 399, p. 4.

Vanparys, L. (1998b) *Proeftuinnieuws*, **15** : 34-36.

Vanparys, L. (1998c) *Proeftuinnieuws*, **14** : 34-36.

Vanparys, L. (1998d) *Proeftuinnieuws*, **13** : 33-34.

Vanparys, L. (1999) *Proeftuinnieuws*, **10** : 40-41.

Vasseur, L. (2001) *Plant Syst. Evol.*, **228** : 71-79.

Venter, F. (1982) *Gemuse*, **18** : 402-405.

Weber, A., Hommes, M. and Vidal, S. (1997) *Mitteilungen der Deutschen Gesellschaft fur Allgemeine und Angewandte Entomologie*, **11** : 271-275.

Wurr, D.C.E., Fellows, J.R., Hambidge, A.J. and Fuller, M.P. (1999) *J. Hort. Sci. Biotechnol.*, **74** : 140-146.

Zheng, X.Y., Li, X.Q. and Ma, L.P. (2001) *IPGRI-Newsl. for Asia, the Pacific and Oceania*, **34** : 22-23.

ASPARAGUS

J. Kabir, J.C.Jana, T.K.Maity and S. Chatterjee

1.0 Introduction

Asparagus is a herbaceous perennial cultivated for its tender shoots, commonly known as spears, which has an agreeable flavour after it has been boiled. It is delicious, nutritious, and versatile vegetable. It can be used in a myriad of culinary and medicinal ways. Once the plant starts producing, it continues to give economic yield for 10-15 years. Globally, major asparagus growing countries are China, Peru, Mexico, Germany Spain, Italy, USA, Japan, Thailand and the Netherlands. In that year, China was the biggest producer of asparagus worldwide, producing approximately 7.8 million metric tons of asparagus. The U.S. production of asparagus amounted to about 35,460 metric tons in 2018(FAO, 2020). In India, its cultivation is very negligible and Himachal Pradesh is the largest growing state of asparagus.

A key production factor is the cultivation date. Greece, Spain and Italy being in a favourable position can harvest early varieties, which command the highest prices. The southern part of Italy would be the best site, where high yields could be obtained in February and March when market prices are high (Falavigna, 2000).

2.0 Composition and Uses

2.1.0 Composition

Asparagus is a good source of carbohydrates, proteins, calcium, iron, chromium, dietary fibre, folate, vitamins A, C, E, and K. Besides, asparagus is a great source of antioxidants. It contains high levels of the amino acid asparagine obtained from the juice of young shoots. Asparagus ranks among the top 20 foods in regards to its ANDI score (Aggregate Nutrient Density Index); this score measures vitamin, mineral, and phytonutrient content concerning the caloric content. The edible portion contains:

Table 1: Composition of Asparagus (per 100 g of edible portion)*

Water	91.7 g	Iron	1.0 g
Carbohydrate	5.0 g	β-carotene	540 µg
Protein	2.5 g	Ascorbic acid	33 mg
Fat	0.2 g	Niacin	1.5 mg
Ash	0.69 g	Riboflavin	0.2 mg
Potassium	278 mg	Thiamine	0.18 mg
Phosphorus	62 mg	Calories	26
Calcium	22 mg	Dietary fibre	0.7 g
Sodium	2 mg		

* Pandita and Bhan (1992).

2.2.0 Uses

The plant is famous for its culinary and medicinal uses. The tender shoots or

Uses of Asparagus

Different Recipes of Asparagus

Canned (Left), and Frozen (Right) Asparagus

spears are cooked for various vegetable dishes and preparation of soup. It is also eaten raw. As a processed product it is canned and frozen in large quantities.

As asparagus contains high levels of asparagine, it serves as a natural diuretic, and increased urination not only releases fluid but helps rid the body of excess salts and is also used especially in cardiac dropsy and chronic gout.

3.0 Origin, Domestication and Botany

The postulated centre of origin comprises Eastern Europe, Caucasus, and Siberia, where it supposedly was domesticated (Sturtevant, 1890). Greeks and Romans followed the culture of growing asparagus from eastern nations, from which they took the old-Iranian word 'sparega', which means shoot, rod, spray; becoming 'asparagos' and 'asparagus' in Greek and Latin, respectively. One of the first detailed guides on how to raise asparagus is traced back to about 65 A.D. by the Roman Columella (Lužný, 1979). Romans spread the culture of growing asparagus along with their empire throughout Europe. There is also evidence that crusading troops brought asparagus seeds from Arabian countries to the Rhine valley around 1212 (Reuther, 1984). In entire Europe, except Spain (Knaflewski, 1996), the decline of the Roman Empire brought a decline in its cultivation, which was confined only to some feudal lords and monastery gardens as a medicinal plant, until the Renaissance, when it was rediscovered as an appreciated vegetable (Lužný, 1979).

Asparagus is comprised of about 150 species of herbaceous perennials, tender woody shrubs and vines. Some of them are grown for ornamental purpose, and one of them (*A. officinalis* L.) for food (Bailey, 1942). In India, nearly 17 species are found. Three subgenera, *Asparagus*, *Protoasparagus* and *Myrsiphyllum* have so far been recognized (Clifford and Conran, 1987). The species of the subgenus *Asparagus* are dioecious, while those of the other subgenera are hermaphrodite. Asparagus species are naturally distributed along Asia, Africa and Europe. Many of them have economic value as ornamentals (*Asparagus plumosus, A. densiflorus, A. virgatus*), or for their medicinal properties (*A. racemosus, A. verticillatus, A. adscendens*) (Štajner *et al.*, 2002). Even when the search for young tender shoots as a tasty vegetable of the wild species *A. acutifolius* has cultural roots in Spain and Greece (Ellison, 1986), the only worldwide cultivated species for tender shoots either blanched (white) or light exposed (green) is *A. officinalis*.

Asparagus has evolved from a primitive hermaphrodite form, via an intermediate gynodioecy state, to the actual dioecious or subdioecious populations, depending if andromonoecious plants (those bearing male and bisexual flowers), are observed (Galli *et al.*, 1993). The suggested pathway (Charlesworth and Charlesworth, 1978) involves first a mutation for male sterility, in the case of asparagus this gene being recessive (*X*). This mutation will spread either if the female and hermaphrodite have the same fertility, but some rate of selfing and inbreeding depression occur in the cosexual form, or if no selfing and inbreeding depression are present but some gain in fertility (ovule production) is gained in the female by allocation of resources in comparison to the hermaphrodite state (Carlesworth and Charlesworth, 1978; Charlesworth, 1999). In asparagus, bisexual flowers are mostly self-pollinated (Galli *et al.*, 1993); and some inbreeding depression was shown after continued selfing of andromonoecious plants in comparison to unrelated outbred-cultivars (Ito and Currence, 1965) and in the comparison of a Hybrid F_1 cultivar UC 157 (cross of two heterozygous selected plants) and the so-called F_2 progeny (first generation of full-sibs, F = 0.25) (Farías _et al._, 2004).

Once the gynodioecious state (females and hermaphrodites) has been reached, a second mutation for female-sterility is necessary to reach full dioecy. This mutation, often called modifier, will spread as long as it confers increased pollen production in comparison to the cosexual form. Again there is no clear evidence in asparagus referring the gained pollen output of males _vs._ cosexuals, however it has been noticed that male flowers are longer than cosexual flowers (Lazarte and Palser, 1979), and this feature has been associated in general to an increased male fitness (Lloyd and Webb, 1977). The modifier can arise tightly linked to the previous male sterility loci or to some extent of independence when it behaves hypostatically, that is not affecting females.

Franken (1970) studied several selfed progenies of andromonoecious plants and concluded that a partial dominant gene (modifier, _A_ in his nomenclature) was responsible for the suppression of pistil development (Table 2); and this gene was inherited independently of the _X_ male sterility locus, previously defined by Rick and Hanna (1943) as a simple Mendelian factor mode of gender inheritance.

Table 2: Phenotypes (Genders) and Genotypes as a Model of Inheritance of Sex

Pistillate	Staminate	Andromonoecious
XX AA	XY AA	XY Aa weakly
XX Aa	YY AA	XY aa strongly
XX aa	YY Aa	YY aa

Franken (1970) proposed model is also in concordance with the results of Galli _et al._ (1993) who, after analyzing the length of pistils in some backcrosses, concluded that the factors affecting style length and stigma development (modifiers) are not localized on the chromosome possessing the _X_ locus; moreover, the backcross distribution of style length fitted a model of at least two loci.

Though asparagus used to be a part of the lily family (Liliaceae)-a distinction shared with aloe, agave, and onion-it has since been kicked off into its group, Asparagaceae. Some botanists still place the vegetable with these beautiful flowers, but others contend that the DNA structure differs too greatly.

The perennial part of the plant is the rhizome (crown), which is composed of clusters of buds with primary fleshy (storage) and secondary fibrous (absorbent) attached roots. The buds sprout rendering the edible organ, a tender growing shoot (spear) between 18-25 cm long, either blanched or not. Once harvests are discontinued, shoots continue to grow becoming the aerial part of the plant (fern), which is responsible for the replenishment of carbohydrates and accumulation until the next harvest season is conducted. The full expanded stems, between a few to 50 or more per plant depending on age, sex and cultivar, have long internodes and can vary in height between 30 and 200 cm. Each stem contains primary and secondary branches where flexuous cladodes (10-25 mm) are disposed in whorls. In normal temperate field conditions flowering starts at the second year from seed germination, but some plants can flower at the end of the first year. Flowering is in flushes, with up to three flushes per year. Anthesis in each stem begins once it is almost expanded, following an apical direction in the main stem and in primary and secondary branches. Depending on air temperatures, blossom in each stem can last up to two weeks. Normal sex ratio in out-bred populations and cultivars is 1:1 staminate (male) to pistillate (female) plants; however some hybrid cultivars are composed of entirely male plants. Yellow-reddish male flowers (5-6 mm) and yellow-greenish female flowers are disposed in bundles of two or three flower per node, rarely mixed with cladodes (Valdes, 1980). Natural pollination is conducted by bees and bumble bees, and normally male plants flower earlier and produce many more flowers than females. Fruits, reddish berries when mature, bear up to ten (normally 6-8) round black seeds. A 3-year-old or older female plant can produce more than 2000 flowers and has the potential to produce more than 10,000 seeds (Machon *et al.*, 1995).

4.0 Cultivars

Green asparagus or green coloured spears are more popular and produced mainly for the fresh market while the light green or white asparagus is mainly used for processing. Selection of cultivar is important for its successful crop production. There are different types of asparagus cultivars available globally. Cultivars are broadly divided into two groups; i. green coloured spears: more popular and mainly used in the fresh market, ii. white or light green coloured spears: mainly used for processing. Older cultivars like Martha Washington, Mary Washington, *etc.* are a mixture of both female and male plants (dioecious). Female plants are identified by their larger spears and produce berries. All-male hybrids are resistant to rust and *Fusarium* crown rot diseases and give uniform-sized spears with improved productivity.

General features of some common cultivars of asparagus are:

Perfection

It is early, spears large, green, succulent, heavy cropper and uniform in maturity with an average yield of 80-100 q/ha, recommended by IARI, New Delhi.

Conovers Colossal and Mammoth White

They produce light green or whitish spears. They are high yielder and commonly grown for marketing purpose.

Palmetta, Argentenil and Mary Washington

These are the most important cultivars and produce a very high yield. The tips of the spear before exposer to the light are purplish and spears become dark green in the sunlight.

Diversity of Asparagus

Types of Asparagus

Violet Touch

It is the oldest cultivar which may be regarded as the form from which all the modern types have been derived.

White German

It is selected in Germany and widely grown by Germans. It has rounded tips and good quality spears.

Palmetto

It has a little difference in appearance than other varieties and considerably resistant to rust and extensively grown throughout America.

Schewetzinger

It is a commercial cultivar and has a greater vigour when planted by clones.

Paradise

It is a selection of Mary Washington developed in California are said to be superior to the standard strain in some regions.

Jersey Giant

An early producer and one of the oldest developed cultivars, does better in cool weather and yields large, thick spears.

Jersey Knight

A popular variety with a wide variability of soils (including clay) and tolerates warmer climates.

Jersey King

Much like Jersey Knight, wide variability and performs well in warm climates.

Jersey Supreme

Great producer and earlier than Giant, but pickier about being grown in sandy soil. Prefers cool weather.

Jersey Queen

A clone cultivar selected from Mary Washington. It produces high-quality spear and maximum yield.

Lucullas 1883 and Limbrass 126 are the best white type cultivars and **Limbrass 22** is the best green type cultivar grown in Poland.

Argenteuil Early: A selection from the Dutch purple type.

Argenteuil Late: Dutch type.

Connivers Colossal: An American light green type.

Marther Washington: An American variety with roundish tipped bud.

Palmetto: Argentevil type pea-green in colour.

Sutton's Giant Fresh: Argenteuil type with stout buds.

Selection-841: Bush type, medium, uniform plants with an average yield of 90-110q green spears/ha. The spears are 15-20 cm long, succulent, tender, green with better flavour and suitable for soup preparation.

Despite these cultivars, **UC-72, UC-66** and **Sel-831** are also grown in Kashmir, India

Hybrids

Gijnilm: This asparagus hybrid is very early in production with medium, thick, straight spears of excellent quality. A very good total yield by means of 100 per cent male plants gives 20 per cent higher production compared to female plants and gives the advantage of no production of seeds and seedlings under normal conditions.

Lara and Mira: These hybrids are early and high yielding evolved through breeding

Faribo: It is a high yielder and produces good quality spears.

Aneto and Desta: These are inter-clone hybrids, high yielding and less heterogeneous. Evolved through *'in-vitro'* tissue culture.

UC 711: It is a better hybrid than others and gives maximum yield and high return.

Thirty-nine introductions of asparagus cultivars/strains from the collections of genetic stock were evaluated at Regional Research Laboratory farm Sanat Nagar, Srinagar, India.

Four distinct complexes were recognized on the basis of a number of spears/ plant and weight of spear/plant. Majority of the introductions were in the complex IV. Marked differences were observed between complex I and complex II. The introductions SL 17, SL 21, SL 29 were among top ranking for the weight of spear/ plant and spear yield/ha (Pandita and Bhan, 1996).

Newly bred F_1 hybrids are **Diego, Ringo, Argo, Golia, Eros, Sirio, Marte** and **Gladio** (Falavigna, 1995). Knaflewski *et al.* (2014) recorded the highest mean total and marketable yield from Dutch cultivars **Gijnlim** and **Cumulus** and German cultivar **Mondeo**. The lowest yields were obtained in New Zealand cultivars **Pacific Challenger** and **Pacific 2000**.

All these described cultivated forms so far are diploids (2n = 20). There are some tetraploid local landraces as Morado de Huetor in Spain (Moreno *et al.*, 2006), Violetto d'Albenga in Italy (Falavigna and Fantino, 1985), Cereseto and Poire in Argentina (López Anido *et al.*, 2000), and derived cultivars as Purple Passion (Benson *et al.*, 1996) and Purple Pacific (Fallon and Andersen, 1999) which could be considered as a distinct (tetraploid) varietal group. Their spears are recognized by showing intense anthocyanin pigmentation when grown exposed to sunlight, and, as suggested by Kay *et al.* (2001), they are not simple autotetraploid derivatives from diploids *A. officinalis*, but probably derived wholly or partially from tetrapoids *A. maritimus* or *A. prostratus*.

5.0 Soil and Climate

Asparagus prefers deep, well-drained, sandy, silt loam or alluvial soils but a high level of organic material in the soil is essential. Commercial plantings of asparagus should not be made in soil that is heavier than a sandy loam. A pH of 6.0-6.7 is suitable.

Generally, cool regions are better for asparagus cultivation. Asparagus can be grown successfully in temperate to sub-tropical climates. It requires a cooler and humid climate for growth and yield of spear production. The upper portion dries up during winter and again sprouts with the onset of the spring season. The temperature should not be lower than 8°C for better germination of seeds and the temperature of 15-25°C is required for good spear development. A lower temperature is necessary for 60-90 days of the year when the plants are in a dormant state. Asparagus tolerates frost to some extent but continuous freezing temperatures, however, are harmful. It should be grown over 1000 m elevation for economic yields (Rice *et al.*, 1990).

Chang *et al.* (1987) suggested that temperature was more important than photoperiod for influencing precocious flowering in asparagus.

6.0 Cultivation

6.1.0 Sowing of Seeds and Transplanting

Asparagus is planted either from seed or from one to two years old crowns. Starting from seed requires an extra year before harvest. Seeds are usually soaked in water for 24 hours before sowing in a well-prepared nursery bed in rows 30 cm apart with 4-6 cm between seeds at a depth of 3-4 cm.

Seedling emergence was more rapid from primed seeds in –0.6 MPa polyethylene glycol 8000 at 20°C than from water-soaked ones, but increased drying before planting led to slower emergence (Evans and Pill, 1989). In the cold region, seeds should be soaked before sowing and should be covered with plastic film after sowing (Liu *et al.*, 1991). Gupta and Singh (1999) suggested that seed scarification at –20°C for 48 hours was most effective for achieving the highest germination.

To break seed dormancy and improve their germination in *Asparagus racemosus*, a study was conducted by Prabha *et al.* (2018). Mature seeds of Asparagus were exposed to three different treatments: warm (35°C) and cold (50°C) stratification and no stratification for one month followed by four different soaking treatments; water at room temperature (25°C), KNO_3 1 per cent, GA_3 50 ppm and NAA 100 ppm. Maximum germination was observed when the seeds were warm stratified followed by soaking in 1 per cent KNO_3 (96.66 per cent) and 50 ppm GA_3 (90 per cent), which was rapid and uniform. After cold stratification germination was reduced (8.3 per cent). Seeds without stratification but with different soaking treatments also showed improvement in germination. Cold stratification leads to secondary dormancy in the seeds which was released after soaking treatments but still the germination was less. Results indicate that the warm stratification followed by soaking were the most effective treatments, which significantly improved germination and removed seed dormancy impediment.

Optimal carbamate treatment conditions were studied for flower induction in seedlings of asparagus cultivar Mary Washington 500 W by Ozaki *et al.* (1999). Flower induction was most accelerated by soaking the seeds in 50 mg/l carbamate solution for 12 days at 25°C under fluorescent light. Longer exposure to carbamate over 12 days induced a higher percentage of seedlings to flower and a higher percentage of seedlings were male.

Aneja *et al.* (1999) advocated a rapid and accurate method of identifying the percentage of females in "male" cultivars. Imbibing seeds in water for 5 days and then treating germinated seeds with 0.4 mM n-propyl N-3, 4-dichlorophenyl carbamate (NPC) for 5 days after radicle emergence, with seedling aeration in the light, resulting in the production of the flowering seedling from > 90 per cent of the treated seeds. For seven male asparagus cultivars, chemical induction of flowering in seedlings with NPC produced sex similar to that of field-grown plants, demonstrating that NPC induces flowering without altering floral differentiation of sex expression. Flower formation in 1-month-old seedlings of asparagus cv. Mary Washington 500 W was induced by soaking the seeds in solutions of triazines, carbamates or thiocarbamates at 200 mM for 8 days. The most effective chemicals were carbamates which induced flowering in 81 per cent of seedlings (Abe *et al.*, 1996).

Seeds required for sowing one hectare of land are 2-3 kg. In Kashmir, the best time of sowing is May and June. Germination is best at a temperature of 25-30°C. Ren *et al.* (1997) reported that the optimum temperature for germination was 20-25°C. Alternating temperature gave better germination than constant temperature. Day and night temperature regime of 25 and 20°C respectively gave the best results. The seed is sown in early spring and takes 3 to 4 weeks to germinate. The seedlings are allowed to grow in a nursery bed for about 7-12 months and then transplanted to ridges about 75-90 cm apart and 50-60 cm between plants in the row. Chen *et al.* (1987) suggested a quick method of raising asparagus seedlings in peat pots. Sowing the seeds in peat pots promoted germination and seedling growth, shortened. The seedling production period by 3 months and subsequently increased the spear yield.

Crowns of one-year-old plants are divided to provide a source of planting materials and are planted at a depth of 15 cm with the same spacing as seedlings. Best time of transplanting in Kashmir is mid-October to Mid-November. Knaflewski *et al.* (1994) obtained the highest cumulative total and marketable yields from beds started with 1-year-old crowns.

In Poland, green asparagus was grown in the greenhouse at 19000 plants/ ha and white asparagus at 1800 plants/ha at a depth of 10-15 cm or 20-25 cm (Knaflewski, 1988).

Feher (1994) reported that satisfactory seedlings can be raised, under heated glass, in Jiffy cylinders (compressed peat) of 30 or 47 mm diameter even in winter. At an average of 20-22°C and light intensity of 18-24 Klx transplantable seedlings can be obtained in 8-9 weeks. In peat cylinders or plastic cups of 100-200 cu cm capacity, seedlings can be raised in 45-50 days when environmental conditions are more favourable.

Madalageri and Ramanjinigowda (1991) recorded that a few plants produced spears of harvestable size within 6 months from transplanting under Bangalore condition from open-pollinated seeds collected at the University of California.

6.2.0 Manuring and Fertilization

Asparagus is a medium-heavy feeder. According to Choudhury (1967) yield of 17.5 quintals of asparagus removes 44 kg of nitrogen, 25 kg of phosphorus and 40 kg of potash. He recommended 50-60 kg of nitrogen, 25 kg of phosphorus and 50 kg of potassium per hectare every year in two split doses. Pandita and Bhan (1992) advocated that 20 tonnes of farmyard manure or compost, 60 kg nitrogen, 100 kg phosphorus and 80 kg potash per hectare should be incorporated in the field a couple of days before transplanting. Mehwald (1988) opined that in soil with pH 5.7-6.0, fertilizer application to asparagus should not exceed 100-120 kg N/ha, 50 kg P_2O_5/ha and 250-350 kg K_2O/ha. Krarup (1989) indicated that P was the most important element as regard yield of asparagus spears. Espejo *et al.* (1996) found that yield and spear diameter were positively influenced by P fertilizer application, but other spear quality parameters (fibre, protein content, *etc.*) were unaffected. Summer foliage of asparagus after the 6th year of the harvest was found to contain N 3.74 per cent, P 0.28 per cent, K 1.95 per cent, Ca 0.63 per cent, Mg 0.28 per cent, S 0.29 per cent, Fe 190 ppm, Zn 41 ppm, Cu 17 ppm, Na 408 ppm, Mn 66 ppm, Co 25 ppm and B 41 ppm. Analysis of fresh spring spears gave N 4.29 per cent, P 0.78 per cent and K 3.98 per cent (Krarup, 1990).

Flori *et al.* (1997) from Brazil, suggested that for aiming a yield of 5.0 t/ha and maintaining soil fertility, annual fertilization of 300 g/ha of N, 100 kg/ha of P_2O_5 and 300 kg/ha K_2O, complemented with organic fertilization for the replacement of micronutrients. According to Hikasa (2000) for good production plants required 200 kg N, 60 kg P_2O_5 and 120 kg K_2O/ha annually. Further to achieve high productivity, methods to promote good growth and accumulation of sugars in roots should be conducted in tandem with adequate soil management and fertilization. The spear should not be harvested too long a period, otherwise, plant vigour and production in subsequent can be reduced.

Hossain *et al.* (2009) investigated the performance of *Asparagus racemosus* grown by the application of different forms and doses of nitrogen fertilizer at Bangladesh. The number, length, diameter and both fresh and dry weight of tuberous roots were found higher with super granule urea than that of prilled urea. Root protein content was found to be 25.20 per cent higher in super granule urea treated plants compared to prilled urea. The rates of nitrogen also had a significant effect on plant height, leaves number and number, length, diameter and both fresh and dry weight of tuberous roots when compared to 0 kg N kg ha^{-1} to the rest of the rates. Root protein content was 21.87, 12.5 and 14.06 per cent higher than the control at the 100, 200 and 300 kg N ha^{-1} concentrations, respectively. Therefore, application of 100 kg N ha^{-1} as super granule urea was found to be sufficient for the sustainable production of tuberous roots of Asparagus.

Vijay *et al.* (2009) studied the influence of N, P and K on chlorophyll, carbohydrate, proteins and sapogenin contents of *Asparagus racemosus* (wild). The treatment consisted different concentrations of nitrogen (N 20, N 40, N 80 and N 160 mg/kg), phosphorus (P 20, P 40, P 80 and P 160 mg/kg) and potassium (K 20, K 40, K 80 and K 160 mg/kg) in the form of Urea, superphosphate and muriate of potash. A significant increase in the chlorophyll content was recorded with all applications of N, P and K. Root protein and carbohydrate content were found linearly increase with K treatment while a slight decline was found with higher dose of N. Root sapogenin content was 1.66, 1.87 and 1.75 folds higher than the control with N, P and K, respectively. Application of phosphorus was found to be best for growth and biochemical contents of root tuber.

Considering the medicinal and economic importance of *A. racemosus*, study was conducted on the effect of different organic manures, biofertilizers and harvesting schedule on its growth and yield by Thakur *et al.* (2016). The maximum plant height (53.73 cm), number of branches/plant (4.5), number of roots/plant (31.15), fresh root weight (63.85 g), dry root weight (13.42 g), fresh root yield (20.58 q/ha) and dry root yield (4.97 q/ha) were recorded when the field was applied with a combination of FYM 5 tons/ha + vermicompost 2 tons/ha + PSB compared to other treatments. Harvesting 24 months after transplanting in the main field resulted in maximum yield attributes compared to the 12 and 18 months. Economic analysis revealed that highest cost of cultivation (Rs 37120/ha), total cost of production (Rs 53092/ha) and maximum gross return was calculated for FYM 5 tons/ha + vermicompost 2 tons/ha + PSB and minimum for control (Rs 15120 and 31092 per hectare, respectively) but maximum benefit/cost ratio (4.12) was found for FYM 5 tons/ha + PSB.

From a study on effect of organic manures and biofertilizers on growth and yield of asparagus, Palande *et al.* (2017) reported maximum plant height (177.20 cm), number of branches/plant (19.5), root length (24.8 cm), fresh root yield (206.10 q/ha) and dry root yield (41.22 q/ha) were recorded when FYM (5 tons/ha) + Vermicompost (2 tons/ha) + PSB (5 kg/ha) was applied. Harvesting at 18 months after transplantation resulted in maximum yield.

Brandenberger *et al.* (2014) reported that 70 pounds of nitrogen per acre about two weeks before the end of the harvest season (about May 15) should be broadcasted after first year. At five-year intervals, a soil test should be conducted and if recommended, any needed lime, phosphate and/or potash fertilizers in addition to the annual nitrogen application can be applied.

6.3.0 Irrigation

Adequate moisture should be maintained for good germination and early seedling growth. Asparagus roots can penetrate up to 1.0 meter to obtain soil water if not restricted but their greatest water uptake occurs from the top 15 to 60 cm of rooting zone. The plant should not be under water stress situation particularly during establishing a root system after two months of planting. Water stress during this early stage can reduce yields. After the root system is established, irrigation is needed only during extreme drought. To maintain healthy fern development, soil moisture during this period should not be allowed to deplete more than 50 to 60 per cent of the soil's water holding capacity in the active rooting zone.

Two irrigations during the cutting season and two afterwards are sufficient. Light, frequent irrigation applications should be avoided during fern growth to minimize foliage disease development. On the other hand, over irrigation should also be avoided as it may cause some of the applied nitrogen to be leached below the plant's root zone and possibly into the ground water. Takatori *et al.* (1970) obtained the highest yield in terms of weight and number of spears with medium irrigation of 52.5 cm water annually.

Sterrett *et al.* (1990) grown New Jersey 'Syn 4' asparagus (*Asparagus officinalis*, L.) on a sandy loam soil to compare plant survival and yield of asparagus grown from crowns and transplants under four irrigation treatments *i.e.* sprinkler (SPR), surface trickle (ST), subsurface trickle (SST) and no irrigation (NI). While plant survival of crowns was not appreciably influenced by any irrigation treatment, survival of transplants was significantly increased by SST. Total and marketable yields from crowns and transplants were similar in the first harvest season (year 3). However, in years 4 and 5, the yield of crowns was higher than that of transplants. Subsurface trickle increased yield from transplants in years 4 and 5 and increased yield from crowns in year 5. All irrigation methods significantly increased both spear production (spear/ha) and average spear weight. Subsurface trickle irrigation resulted in the largest increase over NI in total yield and spear production.

Asparagus is cultivated on desert sand in Peru and can be harvested all year round due to the favourable temperatures, but sophisticated irrigation systems are essential. The high temperatures also inhibit asparagus maturation which has to be induced by reduced irrigation 3 weeks before the expected harvest (Ziegler and Riedel, 2000).

Rolbiecki and Rolbiecki (2008) studied the effect of surface drip irrigation on the possibilities of cultivation of some European and non European asparagus cultivars for white spear production. The field experiment was carried out in the first three years of cultivation (2003-2005) at Kruszyn Krajenski near Bydgoszcz on a sandy soil (soil type Hapludolls). The water reserve to 1 m soil depth at field capacity was 87 mm and the available water quantity 67 mm. The first factor was irrigation used in two variants [O – non-irrigated plots (control), D – drip-irrigated plots]; the second factor was 10 European or 14 non European cultivars. The drip irrigation system influenced the investigated features of the asparagus cultivars positively. The most significant effects were obtained for German and Dutch cultivars and 'Purple Passion'. Yields obtained from irrigated plots were higher than those from nonirrigated plots. Byl (2013) reported that yield of Asparagus increased from 6 to 21 per cent with trickle and overhead irrigation treatments in cultivars 'Guleph Millenium' (GM) and 'Jersey Supreme' (JS). With supplemental irrigation, a significant increase in stem number, light interception, fern height, root carbohydrates, cladophyll weight, and dry fern weight occurred. Cultivar responses to irrigation treatments differed depending on drought stress severity and plant growth stage. Increased yields for GM were attributable largely to increased weight per spear, rather than increased spear number as seen in JS.

6.4.0 Interculture

Cultivation methods of asparagus vary depending on regions and their climatic conditions. It is grown mainly for green spears in North and South Americas (Benson, 2008), while in Europe for white asparagus (Knaflewski *et al.*, 2014). White asparagus is produced when spears are grown in the absence of light and brings a higher market price for its milder flavour than green asparagus. The traditional practice for blanching asparagus is to mound up soil or straw over the asparagus row. Simple row tunnels covered with black, opaque plastic are also used to produce white asparagus. Asparagus should be kept weed-free for obtaining maximum yield. Since it is a perennial crop and bed-width increases every year, cultivation in the row during spear harvest and the following harvest during fern production, is difficult. Weeds are controlled through a combination of cultural, mechanical and biological control techniques. Weeds between the rows can be kept under control by proper shallow cultivation. Hand hoeing may be desirable to keep down the weeds during the cutting season. To suppress weeds use of mulches, cover crops, weeder geese and flaming with propane flamers are also practised. Pre-emergence application of Metribuzin (1 kg/ha) in direct sown asparagus and early post-emergence application gave good control of broad-leaved weeds (Freeman and Maurer, 1989).

Mounding the soil over the rows is practised to blanch the young spears. After harvesting the green asparagus for fresh market, it is a common practice to mound the row with soil to blanch the asparagus for canning.

Spraying on 9-month-old seedlings with dikegulac-sodium (Atrinal) solutions with 300-500 ppm increased the number of new shoots. The emergence of new shoots, after the initial shoot growth of the plants had been cut off, was equally affected by dikegulac especially at 300-500 ppm (Mahotiere *et al.*, 1989).

Araki and Tamura (2008) reported that it is not possible to intercultivate asparagus (*Asparagus officinalis*) fields with large agricultural machines because the crop grows in the field for a long time in the same field. Therefore, soil improvement and weed control are difficult. So, they conducted an experiment for weed control and field management with barley living mulch in an asparagus field in Hokkaido, the cool and snow cover region in Japan. Barley cultivar 'Temairazu' with a strong winter habit for heading was used. When it is sown in spring, heading does not occur for lack of low temperature and only leaves emerge, so living mulch is formed. After sowing on April, barley living mulch covered the ground-surface after mid-May and reduced emerged weed quantity to 18 per cent of the control, bare soil without barley, on July 15th. Such living mulch continued to grow during the summer. In their observation, seeding density was more than 60-80 kg/ha for complete weed control. When sown after mid-May, the barley growth decreased the effect of weed control compared to the sowing date of April 15th, 2 or 3 weeks before the beginning of the asparagus harvest. Weed control effect with living mulch of barley sown on May 15th, 2004, showed a little reduction compared to the treatments of conventional methods, herbicide and tillage. Serious problems were not observed for spear harvest and it was possible to add organic matter into the soil in barley living mulch system.

Martelloni *et al.* (2013) reported that weed competition during spears harvesting reduces asparagus yields. The application of herbicides during this period is illegal and alternative non-chemical practices are needed. They tested the effectiveness and efficiency of a custom-built combined flamer-cultivator to control weeds (both in the inter and intra spears production bands) during the spears harvest season. It also analyzed the effects of various liquefied petroleum gas (LPG) doses on total asparagus yield, mean spear weight and total number of marketable spears. The asparagus spears were generally not damaged by flame weeding using LPG doses of between 43 and 87 kg/ha. The same LPG doses were effective in controlling weeds, showing the same total marketable yields as the weed-free control. At high LPG doses (*e.g.* 130 and 260 kg/ha), yields decreased as a consequence of the damage caused to the spears, resulting in a lower number of marketable spears. Flaming did not affect the mean spear weight and can be applied repeatedly during harvesting to maintain the weeds at a level that does not lead to a yield reduction. The repeated use of the combined flamer-cultivator (every seven days) led to higher yields than plots where weed control was not conducted. The new machine can be used in a period when herbicides are not possible. Flaming could be introduced by asparagus producers as an alternative or in addition to herbicides applied in the pre-emergence and postharvest of spears.

7.0 Harvesting, Yield and Storage

Harvesting period of asparagus can be extended by planting at different depths, periodic removal of mulch materials, part wise harvesting and allowing the foliage to grow for the rest of the season. The harvesting period lasts about eight to nine weeks. Spears are hand harvested when they are six to eight inches long. In commercial scale, self-propelled harvest aids are also used to cut the spears 1 to 2 inches below the soil surface. For high perishability, it requires pre-cooling to remove field heat after harvest before shipment. Different hydro-cooling like flooding, spraying or immersing in chilled water are used for the purpose.

In Kashmir, the spears are ready for harvest in the last week of March to the first week of April. Harvesting begins more or less 13 to 20 months from the establishment of the seedlings. Spears of 10 to 15 cm are generally harvested. Male plants give a higher total yield while female plants produce larger individual spears. Yields of 2.5-4.0 t/ha or 2.5-4 kg/10 m^2 have been reported (Rice *et al.*, 1990).

Spears of asparagus can be stored up to 3 weeks at 0°C. Asparagus will turn flaccid and dull gray-green if kept for more than 10 days at 0°C. Exposure to ethylene gas should be avoided as ethylene can cause toughening of spears. High relative humidity (95-100 per cent) should be maintained with good ventilation to reduce carbon dioxide and ethylene builds up. Asparagus can be stored for 2-3 weeks at 95 per cent relative humidity and 0-2°C (Rice *et al.*, 1990). Spears stored in wet tissue paper looked fresh and firm after 13 or 16 days of storage (Poll, 1991). Hurst *et al.* (1993) reported that the shelf life of asparagus spears declines during the season. Late-season spears have only half the shelf life of early-season spears and this is strongly associated with a decline in spear weight. It is possible to predict the shelf life of spears harvested at any time during the season and this should be valuable to the asparagus industry in its marketing strategy.

Seedling Raising and Field Growing of Asparagus

Harvesting and Packaging of Asparagus

Bhowmick *et al.* (2002) harvested asparagus spears (*Asparagus officinalis* L. cv. 'Welcome') from a greenhouse using the mother stalk cultivation method, from March to October and held at 1.5°C to evaluate the effect of an extended harvest season on storage quality and shelf life. Both storage quality and shelf life declined over the harvest season. The deterioration of textural quality during storage took place for all harvest months, in general being higher for cooler months and less at the top portion of the spear than at the bottom. During the harvest season, the soluble sugars and organic acid content of freshly harvested spears significantly declined up to 53 per cent after a 7-day storage period and the losses were greater in the top portion than in the bottom portion. Shelf life declined by 50 per cent from a maximum of 2.7 weeks.

Renquist *et al.* (2005) investigated the likelihood of interactive effects of storage of asparagus spears in controlled atmosphere (CA) along with spear feeding solutions, including 2 per cent sucrose. Standing the spears in a feeding solution extended shelf-life in air but conferred little additional benefit in CA. Feeding the spears with 2 per cent aqueous sucrose reduced asparagine accumulation and protein loss, but gave no visual benefit over water alone. Spears in the feeding solutions gained weight, particularly during the first 2 days after harvest, but weight gain was slower in CA than in air. Sensory assessment indicated that spears held in CA for 6 days had similar flavour and acceptability to spears held in air for 1 day. Spear quality was more strongly influenced by CA than by feeding solutions. Both approaches could assist in asparagus quality retention where a good refrigerated cool chain is not available, but these technologies are technically challenging to apply to air-freighted asparagus.

Asparagus plants need stored nutrients and time to recover from harvests. They Weed-free environments, moderate soil fertility, and adequate moisture to build up food reserves in their crowns are also needed. Neglecting asparagus fields after harvest is a more significant contributor to poor yields in subsequent years than insect or disease damage.

8.0 Diseases and Pests

8.1.0 Diseases

Several important diseases can cause significant losses to asparagus.

Rust (*Puccinia asparagi*)

It is the most serious disease of asparagus and appears on all aerial plant parts. Orange circular lesions are formed on stems and leaves. The first symptoms appear in spring and then turn to black pustules in summer. Rust becomes more severe in heavy rain, high humidity or abundant dew. Increasing plant spacing and row orientation towards the prevailing summer wind also help in preventing disease build-up. The use of rust-resistant cultivars has been suggested. Jersey Giant and UC 157 are slow rusting cultivars (Johnson and Lunden, 1992). By elimination of the infected plants, cultivating resistant varieties, ensuring good drainage of the field and treating plants with Thiovit Jet, Polrram DF, Score 250 EC the disease can be managed. It can be controlled by dusting fine sulphur @ 24 kg/ha even before infection takes place. Spraying of benlate or bavistin @ 0.1 per cent at early stage of the plant growth has also been suggested.

Fusarium Crown, Root and Lower Stem Rot (*Fusarium oxysporum* f. sp. *asparagi/Fusarium proliferatum*)

Infection commonly occurs when *Fusarium* spp. enters the roots and spreads throughout the plant. Symptoms of asparagus crown rot include wilting of mature plants during hot summer weather, stunting, yellowing, seedling blight, and death. Infected areas of the crown turn brownish as cells that transport water and nutrients become clogged due to the infection. Affected plants reveals dark, reddish-brown colored decay of lower stems, crowns, and roots. Later, portions of the crown begin to dry up until the entire plant dies. *Fusarium spp.* survives in crown and stem lesions of diseased, old asparagus plantings. Fungal spores are spread by air currents and on the surface of contaminated asparagus seeds.

Fusarium diseases are extremely difficult to manage once the fungus is established in an asparagus field. Primary controls are choosing healthy, Fusarium-tolerant varieties of plants obtained from a reputable source, and planting in fields not previously used for asparagus. Fusarium-resistant varieties for Minnesota growers include Jersey Giant, Jersey, Knight, Jersey Prince, and Viking KB3. The seed treatment and drenching of the plants with brassicol (0.2 per cent) or bavistin (0.1 per cent) help in controlling this disease.

Phytophthora Crown, Root and Spear Rot (*Phytophthora asparagi*)

Infection is more likely when soils are wet. Spear rot begins as soft, water soaked lesions and/or shriveling occurring slightly above or below the soil line. Infected crowns and roots show water-soaked lesions and/or shriveling, but the tissue remains firm at the lesion site. Fern of infected plants show yellowing leading to crown death in nursery and commercial fields. It can shorten the lifespan of production fields by 50 per cent despite good cultural practices. Healthy planting materials, good soil drainage and treatment with Merpan 50WP or Folpan 80 WDG can prevent and combat the disease.

Purple Spot (*Stemphylium vesicarium/Pleospora herbarum*)

Sunken, purple, oval-shaped lesions that develop on asparagus spears. Epidemics may affect 60-90 per cent of the spears. Tan to brown lesions on the fern, including the needlelike leaves (cladophylls). May expand, coalesce and cause defoliation.

Controlling purple spot enhances fern vigor and may aid in managing soil borne pathogens. Burning of crop debris in late fall or winter is also suggested although not feasible in large acreage due to human and environmental safety concerns

Cercospora Blight (*Cercospora asparagi*)

Symptoms first appeared on lower portions of the ferns are small, oval, and gray to tan lesions (spots) with reddish brown borders on the needles and small branches. Spores of the fungus are produced on the lesions and are dispersed by wind and rain. Development of the disease depends on rainfall and humidity levels.

Diseases of Asparagus

Rust Infection

***Fusarium* Rot Infection**   **Purple Spot Infection**

***Phytophthora* Rot and Shepherd's Crook with Shrivelling**

Pests of Asparagus

Spotted Asparagus Beetle

Pests of Asparagus

Common Asparagus Beetle Infestation

Asparagus Miner infestation

The disease results in reduced photosynthesis of affected ferns leading to yield loss due to reduced crown vigor caused by the early defoliation.

Elimination of affected plants, collection and destruction crop residues after harvest and foliar sprays of Ortiva 250 Sc, Bravo 500 Sc or Score 250 EC are practised to overcome its infestation.

Insect Pests

Common Asparagus Beetle (*Crioceris* sp. L.)

Extensive damage is caused by feeding the crop by two species of beetles, the common asparagus beetle (*Crioceris asparagi* L.) and the 12-spotted asparagus beetle (*Crioceris duodecimpunctata* L.). Injury by both adults and larva may be reduced by cutting all shoots every 3 or 5 days. This will remove all eggs deposited on them. The occurrence of asparagus a leaf beetle (*Crioceris quatuordecimpunctata*) has been reported from Japan (Nishijima and Saito, 1988). Adults and larvae damaged the shoots, leaves, flowers and seeds of asparagus in May-October. Sanitation helps in suppressing these pests. Natural predators like chalcid wasp and lady beetle larvae may be used.

Spraying with 1.80 kg of 4 per cent rotenone in 100 gallons of water is considered more effective than dusting rotenone (Thompson and Kelly, 1972).

Garden Centipede (*Scutigerella immaculata*)

The insect makes a large number of small, round holes in the spear below the surface of the ground. The fibres around the holes harden and make the spears unfit for canning. Flooding the field is the only practical means of control where infestations are scattered throughout the field.

Asparagus Miner (*Ophiomyia simplex* Loew)

It damages the tender portion of the plants and creates problems for the crop. Spraying of suitable insecticide to control the fly is suggested. In case of severe infestation, spraying should be repeated at 7 days interval.

9.0 Seed Production

The production of hybrid seed with high germination, vigour, and genetic purity is important for a high value perennial crop such as asparagus. Walker *et al.* (1999) and López Anido and Cointry (2008) described a complete guideline for seed production of asparagus.

9.1.0 Site Selection

Ideal location should have frost-free growing season of more than 180 days, 250-500 mm annual rainfall, low frequency of severe storms, hails and high winds. It also must be free from difficulties in the control of perennial weeds and not previously planted with asparagus. The isolation from commercial asparagus fields should be at least 2 km and from any volunteer plant at least 300 m (Ellison, 1986).

9.2.0 Establishment of Parental Material

The arrangement recommended is a single male row intercalated every four rows of female plants in a plantation grid of 1.5 to 1.8 m between rows and 0.6 to 0.9 m within plants in the row.

9.3.0 Pollination Management

Spears of the earliest parent can be removed in order to maximize blossom coincidence and seed production. The placement of honeybees at a rate of 2-4 hives per acre since first female flowers open is recommended. Hives should be removed in the late summer to avoid pollination in the last stalk flushes, which may not complete berry and seed development prior to the first frost.

9.4.0 Seed Genetic Purity

In all-male seed production blocks contamination could be from pollen of volunteer male plants or eventually from seeds of volunteer female established by seeds brought in by birds. In dioecious cultivar seed production blocks volunteers could be of either sex. In all instances efforts should be taken in eliminating volunteers as early as they are emerging. Selective herbicides as linuron have excellent control in seedling stage only. If volunteers escape this stage the may be difficult to rogue out unless mechanically. During fruit harvests caution should be placed in avoiding berry dropping. In all-male hybrids seed lot samples, multiplication of the percentage of female plants by two provides an indication of the percentage of genetic impurity. For certain dioecious and all-male materials specific markers as isoenzymes (Lallemand *et al.*, 1994), RFLPs and RAPDs (Roose and Stone, 1996) and RAPDs (Khandka *et al.*, 1996; Jaag *et al.*, 1998) were found and could be used to check genetic purity.

9.5.0 Pest Management

Effective weed and insect control is essential in seed blocks along years. Caution should be taken in the insecticide used to avoid bee mortality. If frequent rains or heavy dew occur preventive fungicides application should be conducted.

9.6.0 Seed Harvest

Large-scale seed harvesting can be conducted by a modified grain combine harvest machine adjusted to remove berries from the foliage in the field or by stationary threshing equipment, which requires hand cutting and transporting. The former has the disadvantage of depositing some seed back onto the field. Berries have high moisture at harvest and caution should be taken because respiration heat can

be elevated when held in piles. Harvested berries are machine crushed, and the pulp/seed mixture is water washed, and floated low-density pulp separated from seeds. Seed moisture should be lowered to 12-13 per cent prior to cleaning, storing and packing.

For seed production of asparagus one row of male plants is needed for every set of four female rows. Planting of different rows is arranged in such a way in the

Seed Production of Asparagus

field that it provides one male plant for every four female plants to ensure good pollination.

A spacing of 50 × 60 cm is recommended for seed production (Ellison, 1986).

Ellison (1986) suggested the following important points for seed production as asparagus:

i Spears should not be harvested because it reduced stored reserves for maximum stalk growth and seed production.

ii) A large population of bee is necessary for good pollination.

iii) Asparagus beetles should be controlled.

iv) Red ripe fruit should be harvested because seeds from ripe berries are superior in germination to that from less mature bronze coloured fruit.

Seeds from 4 to 8-year-old plantation measuring 3 mm in diameter are regarded best. Machine grading is quite effective in removing lighter and immature seeds of low germination. After harvest seed should be separated from skins and pulp and dried.

10.0 Crop Improvement

Being dioecious, asparagus produces male and female florets on separate staminate and pistillate plants. The staminate (male) plants give rise to more spears and consequently higher yield than the pistillate (female). The pistillate spears generally are heavier individually but fewer in number than in the staminate plants. For this reason, breeding efforts are concentrated on developing super males that will produce higher, more stable yields of spears. The normal ratio of staminate to pistillate plants is 1:1. Besides, there are occasional hermaphroditic plants, which may someday become the type of plant most desired. The release of all-male asparagus hybrids has been a major objective of the DSIR Crop Research (CR) asparagus breeding programme, New Zealand which commenced in 1968 (Falloon 1982). This work started following the observations of Sneep (1953), Franken (1970), and Moon (1976), which predicted that in populations differing only in sex, the population of males out-yields the dioecious population. Other reported advantages of all-male hybrids are: the elimination of asparagus seedlings in production beds; the greater longevity of male plants, which may be related to better disease resistance of males compared to females and the seed supplier has complete control of the seed supply because there is no seed production in all-male hybrids. Hence F_2 seed cannot be saved by growers. The male parent used in all-male hybrids may be produced either via doubled haploids (Falavigna *et al.*, 1989) or by using hermaphroditic seed bearing male plants (andromonoecious plants) and a test cross procedure to identify the plants that are homozygous (YY) for sex, so called supermales. Although both methods have been tried at DSIR Crop Research (CR) asparagus breeding programme the andromonoecious (hermaphroditic male) plant method has proved most successful. If more supermales are produced and selection pressure imposed for the traits associated with saleable yield, then better supermale parents should result. To ensure a wide genetic base, supermales should be derived using the test-cross procedure, as well as from andromonoecious supermales (Nikoloff and Falloon,1990).

Asparagus rust was a serious problem for asparagus cultivation in the USA and J.B. Norton was the pioneer worker in breeding for rust resistance. Norton (1913) adopted different methods for breeding asparagus for rust resistance. This has led to the development of improved cultivars Washington, Martha, Martha Washington (the first generation cross of Martha × Washington) and Mary Washington (the first generation cross of Mary × Washington). Unfortunately, due to lack of effort to maintain the pure stocks of Martha or Mary Washington asparagus the genetic qualities of original cultivar have lost over the years. In USA and Canada, since 1930 every strain of asparagus has been a selection out of Martha and Mary Washington.

The cultivars Apollo, Atlas and Grande were bred in the early eighties from previously adapted cultivars in California to improve yield and quality. Apollo, Atlas and Grande produced deep green spears with very tight heads and little anthocyanin pigmentation. These cultivars showed a 25-50 per cent yield increase over UC 157. Purple Passion was bred from 2 generations of single plant selections from Violet d'Albinga from Italy. These tetraploid cultivars produced large diameter, tight headed purple spears. Spear yields from Purple Passion were similar to UC 157 when planted at the same plant density (Benson *et al.*, 1996).

The main objectives of asparagus breeding are the development of male hybrids, Fusarium and rust resistance, improved spear quality, polyploidy breeding and high yielding ability and adaptability.

Male asparagus plants yield more and live longer than female plants and comparatively more disease resistant. Certain male hybrid produces larger spears than dioecious lines and produces no seedling weeds. Sneep (1953) outlined a scheme for inbreeding andromonoecious asparagus plants for the production of uniform male hybrids.

Wolyn (1996) reported that male asparagus plants yield 20 to 25 per cent more than male-female plants. Supermale mutants are obtained by tissue culture or from rare berries on male plants where female reproductive organ develop to allow self-pollination. All-male hybrids were produced by crossing supermale mutants with female plants. In two seasons, the all-male hybrid G 24 × G 305 yielded 40 per cent more than the control Jersey Giant, due to increased numbers of spears per plant.

In France, the first commercial hybrids released by Corriols-Thevenin (1979) were double-crossed hybrids. It permitted large scale seed production and hybrids Diane, Junon, Minerve and Larac produced 30 per cent increased after the commercial tissue culture was perfect, further increased the yield over the standard by 75 per cent. In New Zealand, dioecious hybrids, Astora and Tarmea have been released by DSIR Crop Research Division after extensive trials (Anon., 1991). Both these hybrids have been developed from crosses between selected female and male plants from cv. Mary Washington 500 W. Falavigna *et al.* (1998) reported that of the 7 male hybrids identified as suitable for northern Italy. Eros, Marte, Gladio and Golia showed particular adaptability to this region combined with resistance to rust (*Puccinia asparagi*). The cultivars 86-22, Ida-lea, 86-21 and 96-13 are reasonable resistance to asparagus aphid (*Aphis gossypii*) (Li, 1995).

Polyploidization appeared promising for use in breeding and it is recommended that hybrids be produced from lines and clones of a male plant at tetraploid level (Skieber *et al.*, 1989). To develop clone-based hybrids, in particular male 4x cultivars, generally divergent diploid ideotypes were subjected to meiotic polyploidization using colchicine (Skiebe *et al.*, 1991).

Falloon *et al.* (1999) made 39 pair crosses between 9 female and 6 male plants with purple spears selected from a small population of VA. The hybrids were compared to two commercial diploid cultivars (UC 157 and JWC 1) and a tetraploid cultivar selected VA that segregates for colour (Purple Passion). One experimental hybrid produced significantly higher yields than Purple Passion and no hybrids produced significantly higher yields than UC 157 and JWC 1.

Difficulties were experienced in tissue culture of the tetraploid species. Seeds of a Pure Purple cultivar are now being produced from a poly cross-block established with plants identified in a progeny test as producing seedlings with only purple spears. This variety has been named "Pacific Purple" (Falloon *et al.*, 1999).

Intra and interspecific crosses of *Asparagus officinalis* and *A. densiflorus* cv. Sprengeri revealed that two types of internal barriers are acting. Cross-incompatibility at the pollen-stigma and pollen-style levels and stronger post-stylar barriers (Marcellan and Camadro, 1996). Chiusa *et al.* (1993) suggested interspecific crosses of ornamental species *A. sprengeri* and *A. plumosus* with *A. officinalis* for breeding for resistance to *Stemphylium vesicarium*.

In India, the improvement work of this crop began in 1982. A cultivar SL 831, derived from hybridization of Mary Washington and UC 66 gave higher yield than open-pollinated seed in cultivation (Pandita and Bhan, 1992). The spears produced were of superior size with tight tips and less fibre content. In Kashmir, India, genotypes SL-9, SL-25, SL-27, SL-29 and SL-32 appeared stable and may be used in breeding cultivars for spear yield (Pandita *et al.*, 1998). Variability patterns and correlation studies with 39 diverse strains/varieties of asparagus revealed that the number of spears/plant was closely associated with the weight of spears/plant. Fairly high heritability coupled with high genetic advance indicated that yield may be controlled by additive gene action (Pandita and Bhan, 1999).

11.0 Biotechnology

Tissue culture has opened avenues of modern asparagus breeding that otherwise would be closed. Most of the new improved asparagus cultivars from the United States and Europe are clonal hybrids and the efficient way of vegetative propagation is by tissue culture.

11.1.0 Tissue Culture

11.1.1 Micropropagation

The plants showed very poor response for rapid multiplication through conventional methods by dividing the crowns into smaller parts with subsequent regeneration to the whole plant. *In vitro* technique was found suitable for rapid multiplication. Complete plantlets could be regenerated using meristem and shoot-tips. Several authors reported successful multiplication using varied concentration of kinetin (0.3 to 1.4 mM) and NAA (0.5 to 1.6 mM) (Matsubara, 1973; Greiner, 1974; Yang, 1977) in MS medium.

Chin (1982) reported propagation of asparagus by culturing single-node spear segments on MS medium. Incorporation of ancymidol accelerates the production of plantlets. Gunawan *et al.* (1944) cultured shoot-tips on MS medium supplemented with 0.5 mg/l 2iP and auxin (IAA, IBA or NAA at 0.25 or 0.5 mg/l). The maximum number of shoots was obtained with 2iP + 0.5 mg/l IAA.

Kohmura *et al.* (1994) obtained an effective micropropagation system using embryogenic calli induced from bud clusters in *Asparagus officinalis*. They reported induction of compact bud clusters from an excised shoot apex cultured in MS liquid medium supplemented with 10 mg/l ancymidol.

Araki *et al.* (1996) reported acceleration of aerial crown-like body formation through nodal segment culture in medium supplemented with 2 mg/l ancymidol and 0.1 M sucrose.

Micropropagation of Asparagus by In vitro Shoot Culture (Stajner, 2012).

Regalado *et al.* (2018) extracted rhizome buds of *A. macrorrhizus*, disinfected, and then cultured on Asparagus Rhizome Bud Medium (ARBM) consisting of MS medium supplemented with 0.3 mg l⁻¹ NAA, 0.1 mg l⁻¹ KIN, 2 mg l⁻¹ ancymidol and 6 per cent sucrose. A percentage of 69.7 ± 8.0 per cent of the rhizome buds developed shoots, but only 17.4 ± 7.9 per cent of them rooted. To increase this low rooting rate, the shoots were cultured on Macrorrhizus Rooting Media (MRM) supplemented with three different concentrations of IBA. The highest rooting rate (55.0 ± 7.9 per cent) was reached when shoots were incubated in MRM-2 consisting of MS medium supplemented with 2 mg l⁻¹ IBA and 4 per cent sucrose. The acclimatization rate of micro propagated plantlets was 90 per cent. The method developed in this study allows the micropropagation of *A. macrorrhizus*, offering a new option to preserve this almost extinct species.

11.1.2 Anther Culture

Pelletier *et al.* (1972) reported a culture of anther on a basal medium of the following composition: MS macro elements, microelements according to Heller without FeCl$_3$, 0.1 mM FeSO$_4$, 7H$_2$O, 0.4 mM Na$_2$ EDTA, a mixture of vitamins according to Morel, 0.058 M sucrose and 5 g`agar. Qiau and Falavigna (1990) developed an improved *in vitro* anther culture method to obtain double haploid clones. Commercial gelling agents were used to solidify the A1 medium on which anthers were cultured. The proliferation of morphogenetic calli and development of somatic embryos were obtained in T1 medium. Production of double haploid male and female clones was also reported by Falavigna *et al.* (1990). Ancymidol was found to enhance the development of haploid asparagus embryo from a liquid culture of anther derived calli (Feng and Walyn, 1993).

A high percentage of micro calli and micro-derived embryos were also reported (Zhang *et al.*, 1994) from isolated asparagus microspores on culturing them on MS medium supplemented with 1 mg/l 2,4-D and 0.5 mg/l BA.

Anthers of cv. Mary Washington 500 W were cultured on MS medium supplemented with BA and NAA at 0.1 and 0.3 mg/l respectively (low concentration medium LCM) or 1 and 3 mg/l respectively (high concentration medium HCM) for callus induction. Calli about 5 mm in length was divided into 2 species and transferred to LCM supplemented with 0 or 0.1 mg/l ABA. Shoot differentiation under these conditions was better for calli produced on LCM than those produced on HCM.

Asparagus anthers containing microspores at the early uninucleate to early binucleate stages were successfully cultured by Shen *et al.* (1995) on half-strength MS medium (without FeNa-EDTA) supplemented with 1.0 mg 2,4-D, 0.5-2.0 mg NAA and 0.5-2.0 mg BA/l for callus formation. The highest frequency of stem and pseudo leaf induction (76.96 per cent) was obtained on MS medium with 0.1 mg/l NAA and 0.1 mg/l BA. A few calli developed shoots and roots. Peng and Wolyn (1999a) established an efficient asparagus microspore culture system by pre-treating genotype (G 459) flower buds at 4°C for 7-9 days, coculturing anthers with shed microspores for 14 days and including 6 per cent sucrose, 2 mg NAA and 1 mg BA/l in the culture medium.

Anther culture is used to develop 'super-male' (di-haploids) in asparagus, which can be used to develop 'all-male' varieties, by crossing them with suitable females; their progenies will be formed only by males which is advantageous for producers. This report described a new anther culture protocol adapted to 'Morado de Huétor', a Spanish tetraploid landrace. Regalado *et al.* (2016) studied the different factors involved in callus proliferation success from anther explants such as the microspore development stage, or the type of stress used to induce the symmetric division of the microspores, to obtain a high success rate (90 per cent). For plantlets regenerates from anther culture (PRACs) regeneration they developed a proliferation media supplemented with a combination of pCPA and BA able to induce callus proliferation and plantlet regeneration in the same step in a 50 per cent of calli, simplifying the procedure. The high percentage of heterozygous male recovery originated from somatic cells, is an important problem in the anther culture, and to elucidate the origin of PRACs they combined different tools: ploidy analysis, characterization with the linked sex-marker Asp1-T7 and with EST-SRRs. they could establish that 50 per cent of PRACs obtained in the work were regenerated from diploid microspores of 'Morado de Huétor', regenerating diploid, di-diploid and tetra-diploid plantlets. The di-diploids males *(MMmm)* would generate a ratio male: female of 5:1 (83.3 per cent) and the tetra-diploid males *(MMMMmmmm)* a ratio male: female of 69:1 (98.6 per cent), so the tetra-diploid males could be considered "super-males" and be used to develop 'all-male' varieties of 'Morado de Huétor'.

11.1.3 Protoplast Culture

Kong and Chin (1988) reported a culture of asparagus protoplast on porous polypropylene membrane. Elmer *et al.* (1989) used the donor callus cells for protoplast from mature plants of crowns. Optimum protoplast was obtained after 10-20 days of culture from these callus cells and 65 to 75 per cent of the isolated protoplast was viable. Callus was induced on a solidified MS medium containing NAA (1 mg/l), 2, 4-D (1.2 mg/l), BA (0.9 mg/l) and 3 per cent sucrose. Shoot generated at 28 per cent efficiency after transfer and culture for an additional 4 to 5 weeks on solidified MS medium with NAA (0.2 mg/l) and kinetin (1 mg/l). Regeneration of plants from callus-derived protoplast was also reported by Sink *et al.* (1990), Hsu *et al.* (1990) and Dan and Stephen (1991).

May and Sink (1996) regenerated plants via protoplast-derived embryos for all 4 genotypes studied. The best treatment was protoplasts cultured without plant growth regulators derived from 2-month-old NAA suspensions of Rutgers 22 (> 40 per cent of embryos regenerated into plants).

Guangyu *et al.* (1997) conducted experiments to maximize the isolation and purification of viable protoplasts from shoot cultures of asparagus (*Asparagus officinalis* L.). Important factors for high yield of viable protoplasts included: the use of in vitro etiolated shoots as source material; 0.6 M glucose as an osmoticum in a modified KM medium; a combination of pectinase, cellulase, and hemicellulase, each at 1 per cent (w/v) for enzymatic digestion of cell walls; and physical factors such as the volume of enzyme solution and speed of gyratory shaking. Protoplasts were purified by suspending digested etiolated shoot tissue in 0.6 M sucrose, overlaid with KMG medium and centrifugation at 650 g. The asparagus genotype had a marked influence on protoplast yield, with some genotypes yielding up to 18.4×10^6 protoplasts/g fresh etiolated shoot tissue with 90 per cent viability.

11.1.4 Somatic Embryogenesis

Various aspects of somatic embryogenesis with different explants of asparagus have been reported earlier (Gorter, 1965; Wilmar and Hellendoorn, 1968; Harada, 1973).

Ghosh and Sen (1994) reported somatic embryogenesis and plantlet formation from callus derived from the sub-apical region of spears of asparagus. Callus was obtained in MS medium supplemented with NAA and kinetin. Embryo formation could be obtained during subsequent subculture with a higher concentration of

KNO$_3$. The embryos underwent rapid multiplication on transfer to a medium containing a different source of nitrogen and low level (0.01 mg/l) of GA$_3$. Media containing zeatin or GA$_3$ led to the formation of complete plantlet from an embryo.

Li and Wolyn (1995) concluded that ancymidol and ABA were more effective in conversion to plantlet from somatic embryo than from uniconazol and paclobutrazol.

Highly efficient somatic embryogenesis in asparagus was reported by Saito (1999) from suspension cultures by elevating the concentration of gelrite in the regeneration medium in combination with an aseptic ventilation filter as a capping material of the culture vessel. Kohmura*et al.* (1996) concluded that micropropagation of superior asparagus clones by somatic embryogenesis is an economical means of producing uniform spears of high quality and yield.

Mamiya and Sakamoto (2001) proposed a method to produce encapsulatable units for synthetic seeds in *Asparagus officinalis* L. Encapsulatable units with high conversion ability in non-sterile soil were produced from somatic embryos by a pre-encapsulation culture. The synthetic seeds containing somatic embryos without the pre-encapsulation culture did not germinate in soil. When the pre-encapsulation culture medium did not contain growth regulators, the roots elongated too much to accomplish encapsulation. Several growth regulators were studied and indole-3-acetic acid was considered to be optimum at 28.5 µM. The pre-encapsulation culture medium with indole-3-acetic acid inhibited the growth of roots during the pre-encapsulation culture and produced compact encapsulatable units. The growth of roots was promoted when plants were produced from the encapsulatable units. The percent conversion of the synthetic seeds with these encapsulatable units was 72 per cent in non-sterile soil. This is the first report on synthetic seeds in *Asparagus officinalis* L.

Mousavizadeh *et al.* (2017) used spear buds of *Asparagus breslerianus* to obtain *in vitro* somatic embryos and, subsequently turn them into plantlets. Calli were developed on callus induction media (MS + 0.88 µM BA + 1.07 µM NAA). Then, they were transferred for four weeks to somatic embryogenesis induction phase media with different concentrations of 2,4-D (0, 4.53, 9.05 and 13.57 µM) and BA (0, 4.44, 8.88 and 13.32 µM). 2,4-D and BA were eliminated from induced calli media in the realization phase. As a result globular, bipolar and mature embryos were observed six weeks later. Calli with compact or friable structures either green or yellowish produced embryogenic calli. Once the mature embryos developed shoots, they were transferred to root initiated media consisting of MS +3.9 µM Ancymidol and 1.07 µM NAA. Two months later, storage roots were produced on *in vitro* well-regenerated plantlets. The survival of asparagus plantlets in the acclimation phase was improved by the development of storage roots.

11.2.0 Biochemical and Molecular Markers

11.2.1 Isozymes

Roux and Roux (1983) characterized the hybrids of heterozygous clones with the help of peroxidase and acid phosphatase. Eleven enzymes were examined by Qiu and Zhou (1995) in callus, roots and stems of 1, 2 and 7-year-old asparagus plants derived from anther culture to identify polymorphic loci which could be useful as genetic markers.

Peroxidase isoenzymes from different organs and tissues of male and female plants of asparagus were analyzed using PAGE (Fan and Song, 1995). Differences in the banding pattern were observed between male and female plants. Male plants had one less band than the female plants. The differences between isozymes could be thus used to identify the sex of asparagus plants.

Gonzalez-Castanon (1999) used isoenzyme markers to classify hybrids and cultivars.

Altintas *et al.* (2019) analyzed three species of genus *Asparagus* native to Lake Van Basin of Turkey using an internal transcribed spacer (ITS) and cpDNA trnL intron sequence and screened for their antioxidant activity and total phenolic and flavonoid contents.

11.2.2 Molecular Markers

RAPD and RFLP techniques were used to distinguish F_1 seed from F_2 or open-pollinated seed. RAPD variation among the parental clones was screened using 60 random decamer primers (Roose *et al.*, 1996). Caporali *et al.* (1996) constructed a linkage map of asparagus through RFLP and RAPD analysis. To identify molecular markers associated with sex-determining genes of asparagus, segregation of RFLP and RAPD polymorphism was investigated in BC progeny of 6 families derived from double haploid parents. Nine linkage groups were identified integrating 23 RFLP and 2 RAPD markers.

The molecular linkage map was also constructed using combined RFLP and RAPD markers by Jiang *et al.* (1997). Spada *et al.* (1998) reported the construction of an integrated genetic map of the asparagus based on RFLP, RAPD, AFLP and isozyme markers. The segregation analysis of the polymorphic markers was carried out on the progeny of five different crosses between male and female double haploid clones generated by anther culture. A total of 274 markers have been organized to ten linkage groups spanning 721.4 cM. A total of 33 molecular markers (13 RFLPs, 18 AFLPs, 2RAPDs and 1 isoenzyme) have been located on chromosome 5.

Many workers (Irshad *et al.*, 2014; Idrees *et al.*, 2018; Chen *et al.*, 2020) evaluated genetic diversity among asparagus species and its cultivars studied using SSRs markers. The findings elucidated asparagus germplasm genetic background and determined hybrid parents, which will facilitate the optimal application of asparagus germplasm resources and provide additional data for genetic improvement. Kapoor *et al.* (2020) characterized 48 accessions belonging to 10 different species of *Asparagus* using 24 SSR markers. Of these, eight SSR markers were newly developed in this study. All SSR markers were polymorphic and revealed high genetic diversity in studied accessions of different *Asparagus* species. Their results showed that there are 2 genetic stocks contributing to the genetic makeup of all the 10 species. *A. adscendens* and *A. pyramidalis* were found closely related. The findings of this research work can be useful in identification of promising genotypes for large-scale production and for initiating new improvement programs in *Asparagus* species to meet the industries demands.

11.3.0 Genetic Transformation

The possibility to transform and obtain transgenic plants will enable us to explore the potential of genome editing for basic and applied research in asparagus. In asparagus there is not yet a reliable approach, although asparagus was the first transformed monocotyledon (Bytebier *et al.*, 1987). Asparagus transgenic plants were regenerated after co-cultivation with disarmed *Agrobacterium tumefaciens* strains (Bytebier *et al.*, 1987; Delbreil *et al.*, 1993) or after particle bombardment (Cabrera-Ponce *et al.*, 1997; Li and Wolyn, 1997). However, the transmission of transgenes to the progeny was established in asparagus only by Limanton-Grevet and Jullien (2001).

Delbreil *et al.* (1993) discussed the *Agrobacterium*-mediated transformation of asparagus. A total of 23 independent karamycin resistant lines were obtained after cocultivation of long-term embryogenic cultures of three genotypes with an *Agrobacterium* strain harbouring *gus* and *npt II* genes. All the lines showed GUS activity after histological staining, DNA analysis by southern blots confirmed the integration of the T-DNA.

Mukhopadhyay and Desjardins (1994) demonstrated a suitable technique using callus derived protoplast as a source for direct gene transfer to obtain transgenic plants. Chen *et al.* (1996) described a protocol for the selection of transformed asparagus protoplasts and the optimization of electroporation parameters for direct gene transfer.

The development of asparagus transformation systems, based on microprojectile bombardment and direct DNA uptake into protoplast was discussed by Conner *et al.* (1996) and Cabrera-Ponce *et al.* (1997). Li and Wolyn (1997) reported the recovery of transgenic asparagus plants by particle bombardment of somatic cells.

An effective method for consistent regeneration of transgenic asparagus plants from electroporated protoplasts is described by Mukhopadhyay *et al.* (2002). Transgenic plants containing β-glucuronidase (*GUS*) and neomycin-phosphotransferase (*NPT II*) genes were obtained by electroporating callus-derived protoplasts of *Asparagus officinalis* L. Embryogenic callus tissue and plants from four kanamycin resistant lines expressed P-glucuronidase activity, as revealed by histological staining. The amplification of genomic DNA by polymerase chain reaction revealed the presence of both *GUS* and *NPT II* genes in transformed callus tissue and plants. Southern hybridization confirmed the integration of these genes into the asparagus genome.

Two different explants were used by Sala *et al.* (2018) to transform asparagus, cut stems and androgenic calli. They were tested two different *Agrobacterium tumefaciens* strains (AGL1 and C58 GV2260); each one was co-cultivated singly with the two types of explants in different times of incubation (15, 25 and 60 min) for AGL1 and only 60 min for C58 GV2260. Their preliminary data showed greater regeneration capacity of androgenic calli in comparison to the cut stems and this enabled to obtain four putative transgenic shoots.

12.0 References

Abe, T., Yeo, D. and Yoshida, S. (1995) *Asparagus Res. Newslett.*, **12** : 36-37.

Abe, T., Yoshida, S., Kameya, T. and Swain, D. (1996) *Acta Hortic.*, **415** : 405-410.

Altýntaþ, S., Pakyürek, M., Þensoy, S., Emre Erez, M. and Behcet Ýnal (2019) *Pol. J. Environ. Stud.*, **28**(4): 2049-2055.

Aneja, M., Gianfagna, T.J., Garrison, S.A. and Durner, E.F. (1999) *HortScience*, **34** : 1090-1094.

Anonymous (1991) *New Zealand Growers*, **46** : 17-18.

Araki, H. and Tamura, H. (2008) *Acta Hortic.*, **776** : 51-54.

Araki, H., Watanabe, S., Harada, T., Yukuwa, T. and Nichols, M. (1996) *Acta Hortic.*, **415** : 209-214.

Bailey, L. H. (1942) In: *The Standard Cyclopedia of Horticulture*, Macmillan Publishing Co., New York, pp. 406-407.

Benson B.L., 2008. *Acta Hort.*, **776**: 496–507.

Benson, B.L., Mullen, R.J., Dean, B.B. and Swain, D. (1996) *Acta Hortic.*, **415** : 59-65.

Bhowmick, P.K., Matsui, T., Ikeuchi, T. and Suzuki, H. (2002) *Postharvest Biol. Tech.*, **26** : 323-328.

Brandbenberger, l., Shrefler, J. and Damicone, J. (2014) *Oklahoma Cooperative Extension Service.*, HLA-6018-1-HLA-6018-8.

Byl, B. (2013) *M.Sc. (Horticulture) Thesis*, Submitted to Michigan State University.

Bytebier, B., Deboeck, F., De Greve, H., Montagu, M.V. and Hernalsteens, J.P. (1987) *Proc. Natl. Acad. Sci. U.S.A.*, **84** (15): 5345–5349.

Cabrera-Ponce, J.L., Lopez, L., Assad-Garcia, N., Medina-Arevalo, C., Bailey, A.M. and Herrera-Estrella, L. (1997) *Plant Cell Rep.*, **16** : 255-260.

Caporali, E., Carboni, A., Spada, A., Maziani, G.P., Biffi, R., Restivo, F., Tassi, F. and Nichols, M. (1996) *Acta Hortic.*, **415** : 435-440.

Chang, T.L., Shii, C.T. and Hung, L. (1987) *Memoirs College Agric.*, National Taiwan University, **27** : 15-21.

Chang, D.C.N., Peng, K.H. and Nichols, M. (1996) *Acta Hortic.*, **415** : 411-416.

Chen, H., Guo, A., Wang, J., Gao, J., Zhang, S., Zheng, J., Huang, X., Xi, J. and Yi, K. (2020) *Physiol. Mol. Biol. Plants*, **26:** 305–315.

Chen, Y.W., Yen, Y.F. and Sun, W.C. (1987) *Asparagus Res. Newslett.*, **5** : 65.

Chin, C.K. (1982) *HortScience*, **17** : 590-591.

Chiusa, G., Stancanelli, G., Rossi, V. and Falavigna, A. (1993) *Agricoltura Ricerca*, **15** : 49-54.

Choudhury, B. (1967) *Vegetables*, National Book Trust, New Delhi.

Clifford, H. T. and Conran, J. G. (1987) In: *Flora of Australia* (Ed. A. S. George), Australian Government Publishing Service, Canberra, pp. 159-164.

Conner, A.J., Abernethy, D.J. and Nichols, M. (1996) *Acta Hortic.*, **415** : 51-58.

Corriols-Thevenin, L. (1979) *In Eucarpia, Section Vegetable, Proceedings of the Fifth International Asparagus Symposium* (Ed. Reuther), pp. 8-20.

Dan, Y. and Stephens, C.T. (1991) *Plant Cell Tissue Organ Culture,* **27** : 321-331.

Delbreil, B., Guerche, P. and Jullien, M. (1993) *Plant Cell Rep.*, **12** : 129-132.

Ellison, J.H. (1986) In : *Breeding Vegetable Crops,* (Ed. M.J. Bassett), pp. 521-569.

Elmer, W.H., Ball, T., Volokita, M., Stephens, C.T. and Sink, K.C. (1989) *J. Amer. Soc. Hort. Sci.*, **114** : 1019-1024.

Espejo, J.A., Tejada, M., Benitez, C. and Gonzales, J. (1996) *Asparagus Newsletter,* 13 (1), y 2.

Evans, T.A. and Pill, W.G. (1989) *J. Hort. Sci.*, **64** : 275-283.

Fadanelli, L. and Meroni, E. (1999) *InformatoreAgrario,* **55** : 33-37.

Falavigna, A. (1995) *Informatore-Agrario,* **51** : 29-33.

Falavigna, A. (2000) *InformatoreAgrario,* **56** : 45-48.

Falavigna, A., Casali, P.E. and Tacconi, M.G. (1990) *Acta Hortic.*, **271** : 39-46.

Falavigna, A., Casali, P.E., Palumbo, D., Materazzo, G., Cerbino, D., Rosa, L. de and de Rosa, L. (1998) *InformatoreAgrario,* **54** : 51-55.

Falavigna, A. and Fantino, M. G. (1985) In: *Proceedings Sixth International Asparagus Symposium* (Eds. E. C. Loughedd and H. Tiessen), University of Guelph, Guelph, p. 398.

Falavigna, A., Palumbo, D., Arcuti, P. and Cerbino, D. (1997) *Informatore-Agrario,* **53** : 39-42.

Fallon, P.G., Andersen, A.M. and Benson, B. (1999) *Acta Hortic.*, **479** : 109-113.

Fan, S. and Song, X. (1995) *Acta AgriculturaeBorealiSinica,* **10** : 67-71.

FAO, 2020 http: //www.fao.org/faostat/en/#data/QC

Feher, E. (1994) *Asparagus Res. Newslett.*, **11** : 1-4.

Feng, X.R. and Walyn, D.J. (1993) *Plant Cell Rep.*, **12** : 281-285

Flori, J.E., Resende, G.M. de, Faria, C.M.B. and De Resende, S.M. (1997) *Hortic. Brasil.*, **15** : 116-118.

Freeman, J.A. and Maurer, A.R. (1989) *Canadian J. Pl. Sci.*, **69** : 265.

Gasperetti, L. (1996) *Informatore-Agrario,* **52** : 59-62.

Ghosh, B. and Sen, S. (1994) *Plant Cell Rep.*, **13** : 381-385.

Gonzalez-Castanon, M.L. (1999) *Acta Hortic.*, **479** : 77-84.

Gorter, C.J. (1965) *J. Hort. Sci.*, **40** : 177-179.

Greiner, H.D. (1974) *Gartenbauwissenschaft.*, **39** : 549-554.

Guangyu, C., Conner, A. J., Christey, M. C., Fautrier, A. G. and Field, R. J. (1997) *Int. J. Plant Sci.*, **158**(5) : 543-551.

Gunawan, L.W., Sidharta, A.A. and Harjadi, S.S. (1994) *Acta Hortic.*, **369** : 226-235.

Gupta V. and Singh B.B. (1999) *Indian J. Plant Genet. Resources,*12: 117-118.

Han, P.L., Chen, L.M. Shen, G.Z., Kiang, L., Zhou, Z.L. and Lu, S.H. (1994) *Acta Agriculturae, Shanghai*, **10** : 85-88.

Harada, H. (1973) *4eme Reunion sur la Selection de l'Asperge* (Ed. L.Thevenin), pp. 163-172.

Hikasa, Y. (2000) *Report of Hokkaido Prefectural Agricultural Experiment Stations,* **94** : 72.

Hossain, K.L., Rahman, M.M., Banu, M.A. and Ali, M.S. (2009) *Asian J. Plant Sci.*, **5** : 1012-1016.

Hsu, J.Y., Yeh, C.C., Yang, T.P., Lin, W.C. and Tsay, H.S. (1990) *Acta Hortic.*, **271** : 135-143.

Hung, L. Chen, Y.W. and Swain, D. (1996) *Acta Hortic.*, **415** : 115-118.

Hurst, P.L., Borst, W.M. and Hannan, P.J. (1993) *NZ. J. Crop Hort. Sci.*, **21** : 229-233.

Idrees, M., Irshad, M., Pathak, M. L., Tariq, A. and Naeem, R. (2018) J. *Biodivers. Conserv. Bioresour. Manag.*, **4**(2): 21-32.

Irshad, M; Idrees, M.; Saeed, A: Muhammad and Naeem, R (2014). *Int. J. Biodiver. Conserv.*, **6**(5): 392-399.

Jiang, C., Lewis, M.E. and Sink, K.C. (1997) *Genome*, **40** : 69-76.

Johnson, D.A. and Lunden, J.D. (1992) *Plant Dis.*, **76** : 84-86.

Kapoor Manish, Mawal Pooja, Sharma Vikas and Gupta Raghbir Chand (2020) *J. Genet. Engin. Biotech.*, **18**: 50, doi.org/10.1186/s43141-020-00065-3.

Knaflewski, M. (1988) *Gemuse*, **24** : 238-239.

Kay, Q.O.N., Davies, E.W. and Rich, T.C.G. (2001) *Bot. J. Linnean Soc.*, **137**: 127-137.

Knaflewski, M. (1996) *Acta Hortic.*, **415**: 87-91.

Knaflewski, M., Ka³u¿ewicz, A., Chen, W., Zaworska, A., Krzesiñski, W. (2014) *J. Hortic. Res.*, **22**(2): 151-157.

Knaflewski, M. and Sadowski, C. (1990) *Acta Hortic.*, **271** : 383-387.

Knaflewski, M. and Nichols, M. (1996) *Acta Hortic.*, **415** : 393-398.

Knaflewski, M., Konys, E. and Babik, I. (1994) *Acta Hortic.*, **371** : 175-181.

Kohmura, H., Chokyu, S. and Harada, T. (1994) *J. Japanese Soc. Hort. Sci.*, **63** : 51-59.

Kohmura, H., Ito, T., Shigemoto, N., Imoto, N. and Yoshikawa, H. (1996) *J. Japanese Soc. Hort. Sci.*, **65** : 311-319.

Kong, Y. and Chin, C.K. (1988) *Plant Cell Rep.*, **7** : 67-69.

Krarup, A. (1989) *Asparagus Res. Newslett.*, **6** : 1-2.

Krarup, A. (1990) *Asparagus Res. Newslett.*, **7** : 14-16.

Krarup, H.A. (1991) *Agro Sur.*, **19** : 88-93.

Lai, P.C., Hsu, Y.J., Yeh, C.C. and Tsay, H.S. (1991) *J. Agric. Res. China*, **40** : 94-101.

Li, B.C. and Wolyn, D.J. (1995) *Plant Cell Rep.*, **14** : 529-533.

Li, B.C. and Wolyn, D.J. (1997) *Plant Science, Limeick*, **126** : 59-68.

Limanton-Grevet, A. and Jullien, M. (2001) *Mol. Breed.*, **7** (2): 141–150.

Li, X., Zhu, X.S. and Yuang, F. (1995) *J. of Hebei Agri. Univ.*, **18** : 53-58.

Liu, G.Z., Li, M.R. and Jin, Z.F. (1991) *Modernizing-Agriculture*, **11** : 20-21.

López Anido, F. (1996) *Master Thesis*, Rosario's National University, Zavalla, p. 104.

Lužný, J. (1979) In: *Proceeding of the 5th International Asparagus Symposium* (Ed. G. Reuther), Eucarpia, Geisenheim, pp. 82-86.

Machon, N., Deletre-Le Boulc'h, V. and Rameau, C. (1995) *Can. J. Bot.*, **73**: 1780-1786.

Madalageri, B.B. and Ramanjinigowda, P.H. (1991) *Curr. Res.*, Univ. Agri. Sci., Bangalore, **20** : 106.

Mahotiere, S., Johnson, C. and Howard, P. (1989) *HortScience*, **24**: 468-469.

Mamiya, K. and Sakamato, Y. (2001) *Plant Cell, Tissue Organ Cult.*, **64** : 27-32.

Marcellan, O.N. and Camadro, E.L. (1996) *Canadian J. Bot.*, **74** : 1621-1625.

Martelloni, L., Fontanelli, M., Frasconi, C., Raffaelli, M., Pirchio, M. and Peruzzi, A. (2017) *Span. J. Agric. Res.*, **15** : 1-10.

Matsubara, S. (1973) *J. Japanese Soc. Hort. Sci.*, **42** : 142-146.

May, R.A. and Sink, K.C. (1996) *Acta Hortic.*, **415** : 237-248.

McGrady, J. and Tilt, P. (1990) *Asparagus Res. Newslett.*, **7**: 12-13.

Mehwald, J. (1988) *Gemuse*, **24** : 232, 234.

Moreno, R., Espejo, J. A., Cabrera, A., Millán, T. and Gil, J. (2006) *Genet. Resour. Crop Evol.*, **53** : 729-736.

Mousavizadeh, S. J., Mashayekhi K, Hassandokht M R, (2017) *Sci. Hort.*, **226** : 184-190.

Mukhopadhyay, S. and Desjardins, Y. (1994) *Plant Cell Rep.*, **13** : 421-424.

Mukhopadhyay, S., Overney, S., Yelle, S. *et al.* (2002) *J. Plant Biochem. Biotechnol.*, **11**: 57–60.

Mullen, R.J., Viss, T.C., Chavarria, R., Reeder, R.K., Whitely, R.W. and Swain, D. (1996) *Acta Hortic.*, 415 : 93-96.

Nishijima, Y. and Saito, O. (1988) *Annual Rep. Soc. Plant Protec.*, North Japan, **39** : 237-238.

Norton, J.B. (1913) *U.S. Dep. Agric. Bur. Plant. Ind. Bull.*, p. 263.

Ozaki, Y., Kurahashi, T., Tashiro, T. and Okubo, H. (1999) *Euphytica*, **110**: 77-110.

Palande A., Deokar C. D., Gaykawd R. T. and Mali M. D. (2017) *Bioinfolet - A Quarterly J. Life Sci.*, **14**: 172-174

Pandita, P.N. and Bhan, M.K. (1992) *Indian Hort.*, **36** : 23-26.

Pandita P.N. and Bhan M.K. (1996) *Indian J. Plant Genet. Resour.*, **9**: 97-103.

Pandita, P.N. and Bhan, M.K. (1999) *J. Medicinal and Aromatic Plant Sci.*, **21** : 1051-1053.

Pandita, P.N., Bhan, M.K. and Kaul, B.L. (1998) *J. Medicinal Aromatic Plant Sci.*, **20** : 1026-1027.

Paschold, P.J., Hermann, G. and Arelt, B. (1999) *GemuseMunchen*, **35** : 588-592.

Paschold, P.J., Hermann, G. and Artelt, B. (1999) *GemuseMunchen*, **35** : 261-266.

Pelletier, G., Raquin, Ch. and Simou, G. (1972) *C. R. Acad. Sci. Paris*, **274** : 848-851.

Peng, M. and Wolyn, D.J. (1999a) *Acta Hortic.*, **479** : 357-363.

Peng, M. and Wolyn, D.J. (1999b) *Plant Cell Rep.*, **18**: 954-958.

Perko, J., Coppey, G. and Berthouzoz, F. (1996) *Revue Suisse de Viticulture,-d'Arboriculture et d'Horticulture*, **28** : 385-387.

Poll, J.T.K. (1991) *Asparagus Res. Newslett.*, **9** : 6-9.

Prabha D., Bhutia N. T., Chauhan J.S.and Negi Y. K. (2018) *Vegetos- An Int. J. Plant Res.*, **31**: 162-166

Qiau, Y.M. and Falavigna, A. (1990) *Acta Hortic.*, **271** : 145-150.

Qiu, D.Y. and Zhou, W.Y. (1995) *Acta Hortic.*, **402** : 398-402.

Reamon-Buttner, S.M. and Jung, C. (2000) *Theor. Appl. Genet.*, **100** : 432-438.

Regalado, J.J., Carmona-Martín, E., López-Granero, M., Jiménez, Ana, Castro, Patricia, Encina, C. (2018) *Plant Cell Tiss Organ Cult*, **132:** 573–578.

Regalado, J.J., Martín, E.C., Madrid, E., Moreno, R., Gil, J. and Encina, C. L. (2016) *Plant Cell Tiss. Organ Cult.*, **124:** 119–135.

Ren, A.X., Zhou, Q., Wang, Y.M., Ren, A.X., Zhu, Q. and Wang, Y.M. (1997) *China Veg.*, **1** : 11-14.

Renquist, A.R., Lill, R.E., Borst, W.M., Bycroft, B.L., Corrigan, V.K. and Odonoghue, E.M. (2005) *New Zeal. J. Crop Hort.* **33** : 269-276.

Reuther, G. (1984) In: *Handbook of Plant Cell Culture*, (Eds. W.R. Sharp, D.A. Evans, P.V. Amminato, and Y. Yamada), Vol. 2, Macmillan Publishing Co., New York, pp. 211- 242.

Rice, R.P., Rice, L.W. and Tindall, H.D. (1990) *Fruits and Vegetable Production*, Macmillan Publishers Ltd., London and Basingstoke, pp. 234-235.

Rolbiecki, R. and Rolbiecki, S. (2008) *Acta Hortic.*, **776** : 45-50.

Roose, M.L., Stone, N.K. and Nichols, M. (1996) *Acta Hortic.*, **415** : 129-135.

Roose, M.L., Stone, N.K. and Benson, B. (1999) *Acta Hortic.*, **479** : 101-107.

Roux, L. and Roux, Y. (1983) *Agronomie*, **3** : 67-74.

Saito, T. (1999) *Bull. Nat. Res. Inst. Vegetables, Ornamental Plants and Tea*, **14** : 105-164.

Sala, T., Losa, A., Ferrari, L., Casali, P.E., Campion, B., Schiavi, M. and Rotino, G.L. (2018) *Acta Hortic.*, **1223**, doi 10.17660/ActaHortic.2018.1223.10.

Sanders, D.C. and Benson, B. (1999) *Acta Hortic.*, **479** : 421-425.

Shen, S.X., Zou, D.Q., Zhang, C.H. and Liu, S.X. (1995) *Acta Hortic.*, **402** : 299-305.

Sink, K.C., Boll, T., Volokita, M., Stephens, C.T. and Elmer, W.H. (1990) *Acta Hortic.*, **271** : 117-127.

Skiebe, K., Gottwald, J. and Stein (1989) *ArchivfiirZuchtungsforschung*, **19** : 367-376.

Skiebe, K., Stein, M., Gottwald, J. and Walterstorff, B. (1991) *Plant Breeding*, pp. 99-106.

Sneep, J. (1953) *Euphytica*, **2** : 89-95.

Spada, A., Caporali, E., Marziani, G., Portaluppi, P., Restivo, F.M., Tassi, F. and Falavigna, A. (1998) *Theor. Appl. Genet.*, **97** : 1083-1089.

Štajner N. (2012) In: *Protocols for Micropropagation of Selected Economically-Important Horticultural Plants. Methods in Molecular Biology (Methods and Protocols)*, (Eds. M. Lambardi, E. Ozudogru, S. Jain), Vol. 994, Humana Press, Totowa, NJ. https://doi.org/10.1007/978-1-62703-074-8_27.

Štajner, N., Bohanec, B. and Javornik, B. (2002) *Plant Sci.*, **162** : 931-937.

Sterrett, S.B., Ross, B.B and Savage, C.P Jr. (1990) *J. Amer. Soc. Hort. Sci.*, **115** : 29-33.

Takatori, F.W., Cannell, G.W. and Asbell, C.W. (1970) *Calif. Agric.*, **24** : 10-120.

Thakur, U., Sood, M. and Bhardwaj, R. (2016) *Intl. J. Farm Sci.*, **6**: 281-291

Thompson, H.C. and Kelly, W.C. (1972) In: *Vegetable Crops*, Tata McGrow-Hill Publishing Co. Ltd., New Delhi.

Vijay, N., Kumar, A. and Bhoite, A. (2009) *Res. J. Environ. Sci.*, **3** : 485-491.

Wilmar, C. and Hellendoorn, M. (1968) *Nature*, **217** : 369-370.

Wolyn, D.J. (1996) *Agri-Food Research in Ontario*, **19** : 12-15.

Wu, L.R.and Feng, X.T. (1998) *Adv. Hort.*, **2** : 613-617.

Yang, H.J. (1977) *HortScience*, **12** : 140-141.

Zhang, C.J., Wang, H.L., Ma, Y. and Kang, Y.Q. (1994) *Plant Cell Rep.*, **13** : 637-640.

Ziegler, J. and Riedel, W. (2000) *GemuseMunchen*, **36** : 31-33.dan

⑥

GLOBE ARTICHOKE

J. Kabir, J.C. Jana, T.K. Maity,
S. Banerjee and A.A. Malik

1.0 Introduction

The globe artichoke, commonly known as artichoke is not a well-known vegetable in India. But in European countries it is an important perennial vegetable crop. Globe artichoke is widely distributed all over the world (126,429 ha) with a production of 1,386,848 tonnes and productivity of 10.9 tonnes/ha, although it is concentrated in the Mediterranean regions (FAO, 2008). Italy is the leading producer (40 per cent) followed by Spain, France and Egypt.

2.0 Composition and Uses

2.1.0 Composition

The composition of edible portion is given in Table 1.

Table 1: Composition of Globe Artichoke (per 100 g of edible portion)*

Moisture	77.3 g	Protein	3.6 g
Fat	0.1 g	Minerals	1.8 mg
Fibre	1.2 g	Other carbohydrates	16 mg
Calories	79 g	Calcium	120 mg
Phosphorus	100 mg	Iron	2.3 mg
Vitamin A	63 I.U.	Thiamine	0.23 mg
Riboflavin	0.01 mg	Vitamin C	Nil

* Choudhury (1967).

Artichokes, whatever one might think of their looks and taste, are actually quite good in nutritive values and have among the highest levels of antioxidants found in vegetables. An excellent source of dietary fiber, they are also packed with magnesium, phosphorous, and sodium, and are a good source of many of the B vitamins as well as vitamins C and K.

According to Khan *et al.* (1999), globe artichoke is rich in nutritive value containing 1.45 per cent minerals, 13.09 per cent carbohydrates, 0.82 per cent fat, and 1.46 per cent fibre.

Miceli *et al.* (1996) reported that its seeds contained 20.5 per cent oil. Physico-chemical characterization showed the oil is high content of polyunsaturated fatty acid requires purification before consumption as edible oil.

The globe artichoke possesses both food value and medicinal properties. The main active components of this plant are mono- and di-caffeoyquinic acids, flavonoids and sesquiterpenes (Dranik *et al.*, 1996).

Dabbou *et al.* (2017) analysed the contents of fatty acids (FAs), volatiles, and phenolic derivatives, the receptacle (the edible part) of two globe artichoke cultivars ('Violet d'Hyéres' and 'Blanc d'Oran'), including their antioxidant activities. The FA profiles of the receptacles exhibited that the most abundant acids were linoleic and palmitic ones. Forty-one volatile compounds, accounting for 97.6-96.3 per cent of the receptacle aroma emission, were identified. Sesquiterpenes hydrocarbons represented the main chemical class; in particular β-selinene, followed by β-caryophyllene, reached the highest levels in 'Violet d'Hyéres' (48.7 per cent and 14.5 per cent, respectively). Total phenols and tannins ($p \leq 0.05$), o-diphenols and flavonoids ($p \leq 0.01$) were significantly higher in the receptacles of 'Violet d'Hyéres' cultivar (73.0, 17.2, 15.4, and 12.2 mg 100 g^{-1} of fresh weight, respectively). The antioxidant activity was notably phenolic derivatives content-dependent, where higher values were observed in the 'Violet d'Hyéres' cultivar.

Biel *et al.* (2020) analysed leaf extracts from the Green Globe cultivar of artichoke to determine the levels of basic nutrients, selected macroelements (K, P, Ca, Mg, and Na) and microelements (Zn, Fe, Mn, Cr, Pb, Cd, and Ni), and their ratios. They also analyzed antioxidant activity (aa) of the extract by using ABTSÿ⁺ and DPPHÿ⁺ radicals and the ferric reducing antioxidant power assay (III) (FRAP). The macroelement concentrations in the artichoke leaf extract were present in descending order as follows: K > P > Ca > Mg > Na. Microelement content in the extract was as follows: Zn > Fe > Cr > Mn. Mean total phenolic content in artichoke leaf extracts was high 2795 mg CAE/100 g dry matter (DM). The ABTSÿ⁺ assay showed a very high ability of artichoke extract to scavenge free radicals (79.74 per cent), and the antioxidant capacity measured at 1060.8 Trolox/1 g DM.

2.2.0 Uses

Artichokes are grown primarily for consumption and can be eaten raw or cooked. They may also be canned or pickled or processed to make tea, liqueur or for extraction of secondary metabolties such as cynarin and chlorogenic acid which can be used in the preparation of alcoholic beverages. The globe artichoke has one of the highest antioxidant contents among vegetables and is therefore valued as a health food with roles in lowering cholesterol and regulating blood sugar levels.

The tender inner bracts and receptacle, commonly known as 'heart' of globular immature flower head or the fleshy bases of the bracts of flower bud are used as vegetable. The small heads are eaten raw or cooked while large heads are eaten only after cooking. The thick receptacle known as heart is used for canning.

Several biochemical studies reported high content of polyphenolic compounds in particular hydroxycinnamates (chlorogenic acid, cynarine and caffeoylquinic acid) and flavonoids (apigenin and luteolin and their glycosides) in its flower heads and leaves (Maurizio *et al.*, 2010). Its root and flower heads contain an important oligosaccharide 'ínulin' that is used as a prebiotic ingredient in functional foods (Raccuia and Melilli, 2004). The leaves are bitter and considered to be diuretic and useful in dropsy and rheumatism. Dranik *et al.* (1996) reported that medicinal preparations from the leaves clearly possess choleretic and diuretic activity and lower the level of cholesterol in the blood. The most suitable raw material for the manufacture of medicinal preparations is fresh leaves gathered in the first-year sowing. It could be used for making soap, hair shampoo, alkyd resin and shoe polish (Miceli *et al.*, 1996).

Uses of Globe Artichoke

Different Recipes of Globe Artichoke in Indian Cuisine

The antibacterial properties of the dichloromethane and ethanol extracts of leaves of *C. scolymus* (collected from Brazil) were investigated by Mossi *et al.* (1999). The dichloromethane extract (5 mg/ml) completely inhibited growth of *Staphylococcus aureus, Bacillus cereus* and *B. subtilis* with bactericidal effect. Fractions exhibiting antibacterial properties were isolated, but the active chemical compounds were not identified. Artichoke is recognized as a medicinal plant where leaves and heads are rich source of polyphenolic compounds. It is recommended for the treatment of gallstones, liver disease or damage and poor liver function. The pharmaceutical properties of artichoke are linked to their special chemical composition, which includes high levels of polyphenols such as cynarin, along with its biosynthetic precursor chlorogenic acid (Bekheet and Sota, 2019). These health-promoting properties are mainly linked to the high content of polyphenolic compounds, which include mono- and dicaffeoylquinic acids and flavonoids (Lattanzio *et al.,* 2009; Pandino *et al.,* 2010).

The fleshy and developed artichoke root system could be used as a source of inulin. Inulin is an oligosaccharide known to have pre-biotic effects through enhancing the activity of bifidobacteria and lactic acid bacteria in the large intestine, with positive effects on bowel habits. Inulin also increases the absorption of Ca, has positive effects on lipid metabolism, and has anti-cancer properties (Lopez-Molina *et al.,* 2005; Sonnante *et al.,* 2007; Lattanzio *et al.,* 2009). Inulin has also been recognized as a beneficial food ingredient since it is used in yoghurt and ice cream preparations (Raccuia and Melilli, 2004).

3.0 Origin and Taxonomy

The globe artichoke is said to have originated from the Mediterranean region (Foury, 1989). Wild form and wide range of variation exists in Southern Europe and Northern Africa.

Rottenberg and Zohary (1996) reported that the genetic affinities between the cultivated globe artichoke *Cynara cardunculus* var. *scolymus* (*C. scolymus*) and its wild relatives were tested by means of a crossing programme. The following wild taxa were involved: (1) wild cardoon *C. carcundulus* var. *sylvestris;* (2) *C. syriaca;* (3) *C. cornigera;* (4) *C. algarbiensis;* (5) *C. baetica* (= *C. alba*); and (6) *C. humilis.* Only the wild cardoon was found to be fully cross-compatible but all the other five wild *Cynara* species turned to be almost cross-incompatible with the crop, and the few interspecies F$_1$ hybrids recovered were partly or almost fully sterile. These findings establish wild cardoon as the wild ancestor of cultivated globe artichoke.

The botanical name of the plant originates from the Greek "skolymos" meaning pointed stake, because of its spines, whereas "Kynara" possibly comes from the name of an Aegean Island where it was grown or from the recommendation of Columella to fertilize this crop with ashes (Cineres) (Chevallier, 1996). The common name of artichoke comes from the Arabic "al Quarshuff".

It belongs to family Compositae and genus *Cynara.* There are about 12 species in this genus, of which *Cynara scolymus* L. is valued as a vegetable crop.

According to De Vos (1992), the adjective "globe" refers to the immature flower head of globular shape, which is the harvestable part of globe artichoke. Globe artichoke is a thistle-like perennial herb, 1-1.5 m high. The shoots produce a rosette of deeply cut pinnatifid, wooly leaves, 5 to 8 cm long. The flowers are borne terminally on the main stem and the laterals. The edible portion consists of immature flower buds and bracts with fleshy lower end. The flower head is large and composed of many florets which are all fertile. The immature florets at the bud stage are hair-like, and composed of a fleshy receptacle with multiple rows of bracts in the external part and florets in the internal part (Ciancolini, 2012). When the bud matures and opens, 800-1,200 purple or blue florets expand. Flowering is centripetal. Cross-pollination is largely assured by protandry, and the principal pollination vector is the honeybee (Basnizki, 2007). The fruits are achenes of elliptical shape, dark coloured, with a prominent pappus that contributes to wind dispersal (Ciancolini, 2012).The achenes of globe artichoke show no dormancy. In a Mediterranean environment, globe artichoke plants sprout in September or October. Winter leaf rosettes develop in November, main stem elongation takes place in April-May, full bloom occurs in June, ripe fruit are available in July, and drying of the aerial biomass occurs in August (Ciancolini, 2012). In a cold climatic zone, globe artichoke is cultivated as an annual, and seedlings are transplanted in the middle of May, and flowering lasts from July to September (Sa³ata, 2006).

The above ground portion dies every year during winter and the crown of individual plant decays after a year's growth. New growth emerges in spring renewed by offshoots from the rootstocks.

4.0 Cultivars

Porceddu *et al*. (1976) have suggested four distinct germplasm groups, namely: 1) the 'Spinosi' (for example 'Spinoso di Palermo' and 'Spinoso Sardo'), containing cultivars with long sharp spines on bracts and leaves; 2) the 'Violetti' (for example 'Violetto di Toscana' and 'Nostrano'), with medium-sized violet-coloured heads; 3) the 'Romaneschi' (for example 'Castellamare' and 'Tondo di Paestum'), to which belongs cultivars with spherical or sub-spherical shape; 4) the 'Catanesi' (for example 'Violetto di Sicilia' and 'Violet de Provence'), with relatively small and elongated heads. Another classification is adopted in France, and includes two groups: Breton (with large green heads) and Midi (with smaller pigmented heads) (De Falco *et al.*, 2015). However, the number of commercially important varieties is only twelve (*i.e.* 'Green globe', 'Blanca de Espãna', 'Opera', 'Violetto di Sicilia', 'Violet de Provence', 'Camus de Bretagne', 'Tema', 'Spinoso sardo', 'Opal', Madrigal', 'Romanesco', 'Madrigal'). Apart from these major groupings, many traditional landraces were cultivated in small holdings. They typically yield less than the commercial varieties, but are well suited for specific final uses, tolerant of environmental stress and adapted to a low input farming system (Mauro *et al.*, 2009).

Cultivars Green Globe and Purple Globe are very popular; Violet de Provence, Catanese and Spinosa Sarda are suitable for processing. Frost resistant cultivars are Violetto de Provenza and Bull. Some other cultivars are Tudella, E 15, Brindisino, Romanesco, *etc*.

Green Globe (Open pollinated, 90-100 days)

It is the original globe type artichoke that produces 3-4 large, 3-5" (8-13 cm), heavy heads over the course of the summer. Globe artichokes have thick, delicious, creamy hearts.

Imperial Star (Open pollinated, 85 days)

It is bred for annual production the first year from seed. Produces 6-8 mature buds, each 3-4" (8-10 cm) in diameter. Plants are 1-1.2 meters tall and as wide.

Tempo (F$_1$ hybrid, 100 days)

It produces the first season from seed. Plants produce 3-4 main buds, each 4-5" (10-13 cm) in diameter, followed by 10-15 secondary 2-3" (5-8 cm) buds. Bracts have a purple tinge.

Violetto (Open pollinated, 85-100 days)

It is an Italian heirloom artichoke that produces elongated, 3" (8 cm) wide by 5" (13 cm) long artichokes, tinged with violet on the bracts. Produces 6-8 main buds, then dozens of 'baby' chokes later in the season. Very little choke if harvested young and tight.

Omaha

The dense and rotund omaha artichoke (up to six inches wide) owes its striking appearance to its sharply tapered red and green leaves. The omaha is less bitter than many artichoke varieties.

Mercury

The petite mercury, with its red-violet hue and distinctive rounded top, is sweeter than many other artichokes and is usually three and a half inches in diameter. Like the baby anzio, the mercury is derived from the Italian romanesco.

Siena

The oblong, about four inches in diameter and born of a breeding program in central Italy, has a small choke and a wine red color. Slow to mature and still grown in relatively small quantities, this small artichoke usually weighs less than a pound and has a heart tender enough to be eaten raw.

Baby Anzio

Light red and only roughly one inch in diameter when fully grown, the purple baby anzio is a relative of the romanesco artichoke of the Lazio region of Italy. Like many baby artichokes, baby anzios can be cooked and eaten whole.

Chianti

It is a classically shaped, four-inch-wide green artichoke with a touch of maroon on the leaves, also (like the mercury) traces its lineage to the iconic Italian romanesco.

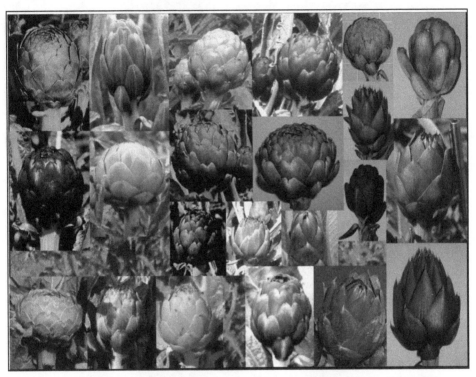

Genetic Variation for Head Shape and Colour in Globe Artichoke Germplasm Collected at University of Catania (Lombardo *et al.*, 2018)

Cultivars/Hybrids of Globe Artichoke

Green Globe

Imperial Star

Sangria Artichoke

Tempo

Siena Artichoke

Violetto Artichoke

Big Heart

It was developed in the mid-1980s by a California grower named Rusty Jordan, the big heart is aptly named. It is endowed with a large, fleshy base and weighs in at over a pound. This green, 3 ½-5 ½" giant—the first patented annual artichoke grown from seed—is excellent for stuffing.

King

The blocky and vividly coloured king has distinctive green spots at the tips of its leaves. Usually four inches in diameter and bred from romanesco varieties mixed with other Italian artichoke strains, the king typically weighs more than a pound in peak season.

Fiesole

The two-inch-wide fiesole artichoke has a fruity flavor and a deep wine color that does not fade with cooking. Bred from the violetta de provence, a purple variety native to southern France, the fiesole has a comparatively tender stalk that can be quickly steamed and eaten.

Traditional Cultivars (Vegetatively propagated)

- ☆ **Green, big:** Camus de Bretagne and Castel (France), Green Globe (USA).
- ☆ **Green, medium-sized:** Blanca de Tudela (Spain), Argentina and Española (Chile), Blanc d'Oran (Algeria), Sakiz and Bayrampasha (Turkey).
- ☆ **Purple, big:** Romanesco and C3 (Italy).
- ☆ **Purple, medium-sized:** Violetto de Provenza (France), Brindisino, Catanese and Niscemese (Italy), Violet d'Algerie (Algeria), Baladi (Egypt).
- ☆ **Spined:** Spinosa Sarda (Italy), Criolla (Peru).

Seed Propagated Cultivars

1. **For industry:** Madrigal, Lorca, A-106 and Imperial Star
2. **Green:** Symphony Harmony
3. **Purple:** Concerto, Opal and Tempo

In an evaluation trial of exotic cultivars, Pandita *et al.* (1988) obtained highest yield with Green Globe and F_1 Salanquet. Green Globe also had the highest reducing sugar content (3.9 per cent) and a high percentage of water-soluble polysaccharides (34.3 per cent). Cultivars di Teramo and Spinaso Sardo had the highest content of water-soluble polysaccharides (38.2 per cent) and total N/crude protein (15.8 per cent) respectively.

Terma 2000 is a perpetually flowering Italian cultivar selected from cv. Terom. The plant is moderately vigorous, reaches a height of 80-90 cm. It is an early cultivar, has good frost resistance and yields 11 heads/plant (Tesi, 1994).

In a trial on an alluvial soil near Zaragoza, Spain, the performance of 3 clones of cv. Tudela (ITG, INIA-D and CNA 303) propagated by offshoots was compared with that of INIA-D plants derived from 2 different stages of *in vitro* propagation (first-year plants and second-year cuttings) and with cv. Imperial Star plants propagated by seed. Imperial Star performed best in terms of survival rate, yield, earliness and total yield (Gil-Ortega *et al.*, 1995).

In Uttarakhand, India cv. Green Globe plants could be harvested 225 days after planting. Mean yield/plant and mean yield/m^2 were 475 g and 1.55 kg, respectively (Khan *et al.*, 1999).

Globe artichokes are also best suited for annual production. 'Imperial Star' is a green artichoke variety specifically bred for annual production. It produces well-developed artichokes the first year from seed. 'Colorado Star' variety is also for annual production, but it produces purple buds.

5.0 Soil and Climate

Artichokes are deep-rooted plants adapted to a wide range of soil types, but will perform best in well-drained, fertile, and deep soil with a pH between 6.0 and 8.0. The extremes of heavy clay and light sandy soils should be avoided. Growing in raised-bed is recommended where drainage is problem, as it results in warmer soil temperatures in the spring and faster establishment.

It is cool-season crop, raised in places where winter is mild and moist. The ideal day and night temperatures are 24°C and 13°C, respectively. To initiate bud development, plants require sufficient chilling exposure or "vernalization," which is generally 250 to 500 hours of temperatures below 10°C. Therefore, bud formation must be artificially induced to produce artichokes in Florida. One vernalization technique is the use of gibberellic acid (GA), a plant hormone that can induce the expression of the same genes activated by cold weather (Catalá *et al.*, 2012). Plants can tolerate light freezes, but buds may suffer freeze damage and show a blistered, whitish appearance. After a few days the blistered skin turns dark; this does not impair the eating quality of the artichoke but does make it more difficult to market. When temperatures fall below -2°C, plants can suffer significantly, and yield may be reduced. High temperatures above 30°C reduce the tenderness and compactness of the "heart" and cause buds to open quickly. The roots can tolerate and survive freezing temperatures, white it is injurious to aerial portions. A hot dry climate resulted artichoke buds to open quickly and destroys the tenderness of the edible parts.

6.0 Propagation

The plant is propagated by seeds and by suckers or offshoots from the old rootstock or by dividing the old crown into pieces with a stem and a portion of the crown. The suckers or off-shoots are removed when they are 30-45 cm high. Plants raised from seeds are highly variable and thus less popular. More uniform crops can be raised from suckers or off-shoots obtained from selected plants. The seeded plants, however, produce better quality flower heads than those from suckers. Now, micropropagation of different plant parts has been successful and widely used as source of planting materials. Germination of artichoke seeds is variable depending on the cultivar and generally ranges from 50 per cent to 90 per cent (Basnizki and Mayer, 1985). After sowing, it normally takes 6 to 7 weeks for seedlings to be ready for transplanting. The optimal size of seedlings is 5 to 6 inches in height with 4 to 5 true leaves.

In Zaragoza, Spain the globe artichoke cultivar Imperial Star, grown from seed, was compared with 3 cvs. Tudela clones (ITG, INIA-D and CAN-303) vegetatively propagated from stumps. Micro- propagated plants of Tudela clone INIA-D were also included. In the first season, spring and total yields were clearly superior for Imperial Star (17.1 and 33.8 t/ha, respectively) than for Tudela clones (for which the highest spring and the total yields were 13.6 and 25.3 t/ha, respectively). In the second season, autumn yields were greater for the vegetatively propagated Tudela clones than for Imperial Star or the micro- propagated INIA-D plants, although total yields were still greatest for Imperial Star. Seed grown Imperial Star heads were less uniform than those of the Tudela clones, however almost all were considered marketable and they were slightly larger than those of Tudela (Gil-Ortega *et al.*, 1998).

Vetrano *et al.* (2000) reported about the propagation of globe artichoke from underground stem sections. Mature stumps of cv. Romanesco were horizontally cut with a saw at the basal, middle and apical positions, to give sections each with one bud which was rooted in plastic pots. Two rooting dates (27 July and 11 August, 1994) were tested. The plants were transplanted in the field in Sicily, Italy, on 18 October, 1994. Almost 90 per cent of the pieces developed into plants with a well-developed fibrous root system. Plant emergence, scored 1, 3, 5 and 7 weeks from planting, was affected by the position of the bud on the stump. After one week, 8, 26 and 27 per cent plant emergence were observed with sections obtained from the apical, middle and basal parts of the stump, respectively. After 3 weeks, plant emergence from buds situated on the apical part of the stump did not exceed 37 per cent, whereas section from the middle and basal part of the stump, respectively, reached 56 and 62 per cent. Regardless of the position of the stump, the total numbers of heads per plant were 2.0 and 1.9 (end of April) and 9.0 and 8.0 (end of May) for the first and second rooting dates, respectively. Irrespective of the rooting dates, section position on the stump had a low influence on total yield as a mean number of 9.0, 8.9 and 9.0 heads per plant were harvested from plants originating from the apical, middle and basal parts of the stump, respectively.

Dawa *et al.* (2012) used meristem tips of good young offshoots from selected globe artichoke plants as explants. Dipping shoot tips (1-2 cm long) in 70 per cent ethanol for 5-10 seconds followed by 0.1 per cent $HgCl_2$ (W/v) for 2 minutes and then in 3 per cent sodium hypochlorite for 20 minutes was found to be the most effective sterilizing and disinfectant treatment for surviving the majority of meristem tip cultures after 5 weeks of culturing. In the establishment stage, adding 0.5 mg TDZ/l to MS basal medium gave the best values for both shoots and leaves number but registered middle values for average shoots length compared to the other treatments. Among the different cytokinin types and concentrations used such as 2iP, BA and Kin (at 0.5, 2.5 and 5.0 mg/l for each) and TDZ (at 0.05, 0.5 and 1.0 mg/l) in addition to control treatment (hormone-free MS medium), 2iP with all concentrations recorded the highest values in this respect in terms of shoots and leaves number and average shoots length. The multiplication rate of proliferated globe artichoke shoots and average shoots length were markedly increased with increasing subcultures number on MS basal medium augmented with 5.0 mg 2iP/l + 1.0 mg IAA/l till the fourth subculture then declined thereafter during the fifth one.

Three plant cuttings management methods were tested by Riahi *et al.* (2017): summer ovoli (T0); spring offshoots nursery's cuttings forced to pass a vegetative rest period by stopping irrigation (T1); and offshoots nursery's cuttings not forced (T2). The result exhibited that the T1 nursery plants produced the heaviest primary heads, 7 per cent and 23 per cent higher than T2 and T0, respectively. T1 plants exhibited the highest yield during the harvest season, with 17.7 per cent and 12.2 per cent higher compared to T0 and T2, respectively. T0 and T1 showed the highest total antioxidant capacity and inulin content. They concluded that the T1 is a viable and sustainable alternative to the traditional one that does not heavily impact on growing costs and improves yield and quality of artichoke.

Soliman *et al.* (2019) conducted experiments to study the performance of globe artichoke plants as affected by propagation methods and spraying with gibberellic acid. Three propagation methods (explants) were used, included stumps cuttings (crown pieces), offshoots, and a mixture between the both explants (stumps cuttings + offshoots) of the same aforementioned globe artichoke plants 'Balady' cv. were joined tightly on planting. Each explant was of the same constant weight, more or less, ca. 160 g/explant. It was planted at 0.5 m apart and 1.0 m width of ridge. So, the total number of plants/feddan was approximately 8400 plants (2 plants/ m^2). Also, four concentrations of dissolved GA_3 were used in the present study (0, 25, 50 and 75 mg/l) as a foliar spraying. The spraying was applied twice; the first one was carried out after 65 days from planting (at the stage when the plants had approximately 12-15 leaves), the second application was 20 days after the first one (or when plants had approximately 20 leaves). The obtained results indicated, generally, that the interaction between any propagation method (explants) and GA_3 at its higher level (75 mg/l) recorded the highest average values; nevertheless, the offshoots + stumps cuttings upon interacted with GA_3 at 75 mg/l ; resulted in the highest average values and might be considered as an optimal treatment for the production of high yield and good quality of globe artichoke plants under the environmental conditions of Behiera Governorate of Egypt and other similar regions.

7.0 Cultivation

7.1.0 Seed Sowing and Planting

Seeds are sown by broadcasting on well-prepared seed beds in plains during August-October and in the hills during March-May. Seedlings are transplanted when 10-12.5 cm in height in rows at a distance of 120 × 90 cm.

Suckers or off-shoots are removed when they are 30 to 45 cm high and planted at 240 × 180 cm spacing during fall and winter in California (Thompson and Kelly, 1972). Plant spacing of 100 × 60 cm generally increased the total weight and the number of buds per unit area but reduced the number and weight of buds per plant compared with spacing of 100 × 100 cm and 100 × 200 cm (Shanmugavelu, 1989). Recommended depth of planting is 10 cm.

Globe artichoke requires vernalization of the plants, either through cold treatment of transplants or from natural temperature conditions in the spring. Rangarajan *et al.* (2000) reported under New York conditions, vernalization treatment increased the number of plants producing buds and the marketable yields, when transplants were set after 15 May. Natural vernalization was achieved and cold treatment before transplanting did not improve yields of plants established in early May. At later planting dates, vernalizing transplants increased the number of plants producing apical buds (largest) by about 20 per cent, yet over 57 per cent of non-vernalized plants of each cultivar produced buds within the season. Welbaum (1994) investigated regarding annual culture from seed in Virginia. Globe artichoke are usually propagated vegetatively because plants grown from seed lack uniformity. Furthermore, in much of the USA, only a small percentage of plants grown from seed flower during the first season due to insufficient chilling for vernalization. Artichokes cannot be grown reliably as perennials without winter protection where temperatures are consistently below −10°C. The new cultivars Imperial Star (IS) and Talpiot (TP) reportedly produce uniform plants from seed and a high percentage of flower heads in the first year with minimal chilling. IS and TP were compared with the standard seed propagation cultivars Green Globe Improved (GG) and Grand Beurr (GB). Plants of each cultivar were tested over a 3-year period in Blackburg, Virginia, or for one year in 3 other locations. Essentially all IS and GG plants flowered after receiving 1356 h of chilling at < 10°C. With 205 h of chilling, 83 per cent of IS plants flowered compared with 25 per cent for GG. No TP or GB plants flowered after 528 h of chilling. In the mountains of western Virginia, only IS plants transplanted into the field in early May received sufficient chilling to produce flower heads during the late summer and early autumn; June transplants did not flower. In warmer areas of central and eastern Virginia, autumn transplanting for spring harvest may yield a higher percentage of flowering plants than spring planting and summer harvest.

Temperature treatments of globe artichoke off-shoots were studied by Iapichino and Vetrano (1994). Dormant off-shoots of globe artichoke cv. Violetto spinoso di Menfi harvested in July were kept at 30°C for 0, 1, 2, 3, 4 or 5 weeks followed by 5°C for 5, 4, 3, 2, 1 or 0 weeks, respectively. Controls were kept at 27-28°C for 5 weeks. Keeping at 5°C for 5 weeks and at 5°C for 3 or 2 weeks reduced flower bud abortion. All treatments reduced yields compared with controls. Calabrese *et al.* (1994) studied yield and quality of new artichoke cultivars propagated by seed. Seeds of the globe artichoke cultivars 044, 137, 271, 307, 374, 386-lb, 386-2b and 386-a were sown in June and transplanted 37 days later at a density of 12000 plants/ha. Heads were harvested 19 times between Nov. and May, by which time cultivars 044, 271, 307, 386-lb, 386-2b and 386-a had produced the greatest number of heads (209000 heads/ha, compared with only 143000 heads/ha for cultivars 137 and 374). Mean yield time (the time from 1 Nov. to when the last head was harvested) and time to 20 per cent yield were shortest for cv. 374 (147 and 42 days, respectively) and longest for cv. 044 (189 and 179 days, respectively). The inflorescence pappus was 2 mm long in cultivars 044, 307, 374 and 386-2b and 4 mm long in cultivars 137 and 271.

In southern Italy, best sowing time was summer, when the yield was about twice that in autumn. Spring sowing, on the other hand, produced no yield in the same year. Increasing the plant density reduced the number of heads/plant, while the yield per unit of area increased. Plant densities above 2.5 plants/m² reduced the number of productive plants, indicating that this density appeared to be the upper limit (Elia *et al.*, 1991). Elia *et al.* (1994b) further conducted trials on plant spatial arrangement and production using the new seed-propagated cultivars 044, 137 and 271. At transplanting, seedlings were arranged in single rows at 120 × 70 cm or in double rows at 170 × 70 × 70, 210 × 60 × 60 or 280 × 50 × 50 cm, giving a planting density of 12000 plants/ha. In the first year off-shoot removal (0, 1 or 2) was carried out with very little effect. The double row spacing of 170 × 70 × 70 cm and the single row spacing resulted in the highest, yields, but the wider-spaced double-row beds may be preferable for access by cultivation machinery. Cultivar 271 produced the highest and earliest yield of small heads and cv. 044 had the shortest harvesting season.

In the Metaponto area of southern Italy, Elia *et al.* (1998) recommended that seed propagated artichoke cultivars should be sown in late July with a density of 12000 plants/ha and N rate of 150 kg/ha.

7.2.0 Manuring and Fertilization

The soil should be well-manured for artichoke cultivation and immediately after stalks are cut, 10-12 tonnes of organic matter per acre may be applied (Thompson and Kelly, 1972).

The nutrients removal in southern Italy by plant populations of 6,900 plants per ha in a single cropping cycle was recorded to be 286 kg N, 19 kg P, 305 kg K, 157 kg Na, 179 kg Ca, 5.2 kg Fe, 0.29 kg Zn, 0.17 kg Cu and 0.64 kg Mn per ha (Magnifico and Lattanzio, 1981).

Moulinear (1980) recommended application of 120-140 kg N, 100 kg P_2O_5 and 400-500 kg K_2O per ha in autumn, with one or two further split applications of N in spring. Fertilizer application is made in California after the plants have been cut back at the end of the previous cropping period or when a new planting is being established. A schedule of 150-280 kg N, 43 kg P and 199-415 kg K has been recommended for south-eastern France.

At Policoro, in south Italy with a soil rich in N, the application of 150 kg N/ha increased the yield by 3 t/ha, no further increase was observed with 300 kg N/ha (Elia *et al.*, 1991).

Elia and Santamaria (1994) concluded that nutrient solutions of 130 ppm N + 100 ppm P + 250 ppm K produced high quality globe artichoke seedlings in a vermi-culite : perlite : peat medium low in N, P and K.

Globe artichoke cv. Balady plants were grown by Abd-El-Fattah (1998) on newly reclaimed calcareous soil at Nabaria Research Station, Egypt to study the influence of soil dressing with mineral P at 0, 30, 60, 90 or 120 kg P_2O_5/fed (1 fed = 0.42 ha), alone or in combination with phosphorien (a biofertilizer produced by the Egyptian Ministry of Agriculture and containing active bacteria capable of converting tri-calcium phosphate to mono-calcium phosphate) at 5 kg/fed, on plant growth, head yields, head trait and chemical constituents in head receptacles (edible parts). In general, all the studied parameters were better in plants receiving mineral P than in unfertilized plants. Increasing the applied mineral P rate from 30 to 120 kg P_2O_5/fed significantly increased most parameters. Soil inoculation with biofertilizer resulted in significant increases in most parameters compared with plants grown on non-inoculated soil, in both seasons. Positive interactions between mineral P and phosphorien inoculation were often observed. The best results were obtained using 90 kg P_2O_5/fed in the presence of phosphorien. These values were similar to those obtained with 120 kg P_2O_5/fed without phosphorien. Therefore, it was concluded that soil inoculation with biofertilizer (phosphorien) reduced the need for mineral P manuring by up to 25 per cent, thereby reducing costs and environmental pollution problems.

Elia *et al.* (1996) suggested from results of his study that NO_3 is the form of N preferred by artichokes. Pomares *et al.* (1993) observed no yield response to N above 200 kg N/ha. P or K had no effect on yield indicating that available P levels of 26-33 ppm and available K levels of 250-282 ppm in the soil were adequate for normal growth. N, P and K rates had no effect on the proportion of heads harvested early in the season (Oct.-Dec.).

Morzadec *et al.* (1998) reported that black spot of globe artichoke is a calcium-deficiency disorder.

Ierna *et al.* (2006) studied the effects of nitrogen and phosphorus fertilization on yield and harvest time of globe artichoke in order to evaluate the possibility of its rationalizing and improvement. The following experimental treatments were compared: a) two P_2O_5 fertilization rates, 50 and 200 kg P_2O_5 ha^{-1}; b) four N fertilization rates, 0, 150, 300 and 450 kg N ha^{-1}; two cultivars 'Violetto di Sicilia', vegetatively propagated, and 'Orlando F_1', seed propagated. The results of this study revealed that yield significantly increased with increasing nitrogen rate from 0 to 300 kg ha^{-1} and did not vary with 450 kg N ha^{-1}. The results suggest that optimum nitrogen application rate was 300 kg ha^{-1} and that 200 kg P_2O_5 ha^{-1} improved the yield response of globe artichoke to nitrogen fertilization. The yield response to phosphorus varied with cultivars. In fact, irrespective of nitrogen rate, head yield of 'Violetto di Sicilia' was improved by 200 kg P_2O_5 ha^{-1}, while yield of 'Orlando F_1' was improved by 50 kg P_2O_5 ha^{-1} (Ierna *et al.*, 2006).

Two field experiments were carried out by Shaheen *et al.* (2007) during two successive seasons of 2004/2005 and 2005/2006 in the newly reclaimed soil to investigate the response of artichoke plant for different fertilization rates of ammonium sulphate 20.6 per cent N (80, 100 and 120 N-units/fed.) and agricultural sulphur (0, 150 and 300 kg/fed.), and they found that the addition of ammonium sulphate as a nitrogen source at rate within 100-120 N-units/fed., gained the best plant growth parameters. However, addition of sulphur at rate of 150 up to 300 kg/fed. resulted the vigor growth of artichoke plant.

Application of beneficial symbionts such as arbuscular mycorrhizal fungi (AMF) is an environment friendly and efficient strategy to enhance plant biosynthesis of secondary metabolites with health promoting activities (Maurizio *et al.*, 2010).

Five treatment combinations between two doses of nitrogen (100 and 150 kg) N/feddan (4200 m²) and two doses of potassium (50 and 100 kg) K/feddan were applied in five equal constant doses (control) or dynamic doses (variable doses with plant growth cycle) (Saleh *et al.*, 2016). Results exhibited that the dynamic application doses of nitrogen (start by high nitrogen dose at the beginning then decreasing application doses during plant growth cycle) and potassium (start by less potassium dose at the beginning then increasing application doses during plant growth cycle) increased vegetative growth characters of artichoke plants and improved quantity and quality of artichoke heads compared to constant applications. The dynamic mode combination between doses of 150 kg N/feddan and 100 kg K/feddan led to increase all vegetative growth characters of artichokes (the height of plant, leaf numbers, leaf fresh and leaf dry weights and leaf area as well as chlorophyll content) and increased head yield, and even more enhanced earliness and head quality (Saleh *et al.*, 2016).

The results of Deligios *et al.* (2017) showed for the first time that long-term biannual rotation with cauliflower coupled with cover crop use could optimize nutrient fluxes of conventionally grown globe artichoke.

7.3.0 Irrigation and Weed Control

Artichokes are large plants with deep roots and should evenly be watered throughout the growing season. Low soil moisture and plant stress during bud formation will result in loosely formed, tough, poor-quality buds that do not size well. Over-irrigation and long-term saturation should be avoided, especially on heavier soils. The amount of normal rainfall must be taken into consideration, especially in bare-ground plantings. After each irrigation the soil between the rows is disked or harrowed to kill the weeds and level the surface.

Artichokes are sensitive to too little or too much moisture in the soil. Prolonged periods of drought result in small numbers of tiny, stunted buds with weak stems. Bracts often develop a brown to black discoloration without enough water. When exposed to soggy soils or standing water, artichoke roots suffocate and begin to rot. Excessive moisture also opens the gate for fungal diseases. Monitor the soil moisture constantly to ensure a proper balance. Artichokes should be planted in well-draining soil and water thoroughly when the top 2 to 3 inches of soil become dry.

Cosentino and Mauromicale (1990) observed that values of leaf transpiration and stomatal conductance of seed propagated cv. Talpiot plants were 46.1 and 49.2 per cent higher than those of vegetatively propagated cv. Violetto di Sicilia plants, irrespective of water regime. Leaf water potential measured just before irrigation on a representative day of the irrigation season averaged 6.9 and −12.1 bar in Talpiot and Violetto di Sicilia, respectively; there was no significant difference between water regimes. A double-emitter source (DES) irrigation system using trickle irrigation can be used for precise application of varying salts or nutrients. Artichoke yield declined by 60 per cent due to salinity when EC increased from 1.5 to 6.2 ds/m (Malach *et al.*, 1996).

Shinohara *et al.* (2011) investigated three irrigation [50 per cent, 75 per cent, and 100 per cent crop evapotranspiration (ETc)] regimes rates under subsurface drip

Seedling Raising and Field Growing of Globe Artichoke

irrigation on growth and yield of globe artichoke. Marketable yields significantly increased at 100 per cent *etc.* compared with 75 per cent and 50 per cent *etc.*, whereas a 20 per cent to 35 per cent yield reduction occurred at 50 per cent *etc.* across seasons. This yield reduction was associated with a decrease in both number of marketable heads and head weight and with reductions in plant physiological responses.

Weeds can be managed through a combination of both pre-emergence and post-emergence materials. Mulching with polyethylene film is the better option for weed management when the crop is grown in the beds, while bare-ground production requires weed management over the entire field. Weeds can be mechanically managed by tillage as well. Once plants reach a certain size, the large leaves will shade the soil, discouraging weed growth.

Trifluralin at 1.2 kg/ha plus diuron at 1 kg/ha, linuron at 0.35 kg/ha and metobro-muron at 1 kg/ha before planting gave good control of broad-leaved and grass weeds without affecting the crop (Shanmugavelu, 1989). Elia *et al.* (1998) recommended pendimethalin, propyzamide and metolachlor for weed control in direct sown artichokes. Pyraflufen ethyl (Evolution) showed excellent control of *Brassica nigra, Convolvulus arvensis, Senecio vulgaris, Sonchus arvensis, Diplotaxis erucoides, Fumaria officinalis, Galium aparine* and *Stellaria media* in globe artichoke field (Montemurro *et al.*, 2016). In particular, this herbicide was more effective on *D. erucoides, F. officinalis* and *G. aparine* than oxyfluorfen and linuron and not different from diquat. They concluded that Evolution provides good control of weeds between rows of artichoke. It also showed high selectivity, both when applied alone and in mixture.

7.4.0 Effect of Growth Substances

Gibberellic acid has been shown to promote earliness in artichokes and is thus useful in timing the harvesting thus is interesting for timing harvest. However, its effectiveness is influenced by cultivar, duration and concentration of treatment, climatic conditions and other factors (Halter *et al.*, 2005). Gibberellic acid (GA_3 or GA_{4+7}) foliar application enhances earliness by several weeks and improves uniformity of flowering. Foliar Application of 30 ppm GA_3 is applied at 5-7 weeks after transplanting when plants are 45-60 cm in diameter.

Elia *et al.* (1994a) reported that duration of harvest was reduced with delay in sowing date and by gibberellic (GA_3) treatment, more so in cv. 271 than the other 2 F_1 hybrid cultivars 137 and 223. Of the 4 cultivars, cv. 271 had significantly smaller heads than the others. Elia *et al.* (1994b) also noticed that single spray of GA_3 at 90 or 120 DAS or double sprays at 90 + 120 DAS resulted in earlier harvest dates and significantly increased yield compared with controls in seed sown globe artichoke cv. 044 directly sown in the field in August. In the following year, seeds of cultivars 044, 137, 223 and 271 were sown in June, July or August in a cold greenhouse and the seedlings planted out in the field 40 DAS and given 3 spray applications of 60 ppm GA_3 at 3-week intervals beginning at the 10- to 12-leaf stage (100 DAS). The number of heads per hectare decreased as sowing date was delayed (except for cv. 137). GA_3 treatment tended to increase yield, although less so for later sowing dates.

Mauromicale and Ierna (2000a) reported that combination of sowing date and GA$_3$ application enables an uninterrupted harvest of globe artichoke heads from October/November to April. A two-year study was carried out in Sicily (Southern Italy) to check the effects of harvest period (November-April), sowing date (from 1 July to 20 August) and GA$_3$ treatment (0, 1, 2 or 3 applications per plant) on characteristics of flower heads of Orlando, a new F$_1$ seed-grown hybrid of globe artichoke. Anticipating the harvest time from April to November, early sowing or 2 or 3 consecutive GA$_3$ applications, generally decreased the head weight, increased the head length/width ratio, and the stalk length.

In Sicily, the effect of GA$_3$ (0, 1, 2 or 3 plant applications at 60 ppm, at the 8, 8+15 or 8+15+25-leaf stages, respectively) and sowing date (from 1 July to 10 August) on the timing of production and head yield of F$_1$ hybrids cv. Orlando was investigated by Mauromicale and Lerna (1995). Regardless of sowing date, untreated plants of Orlando produced heads the next spring, confirming that in the Mediterranean environment seed-grown cultivars have to be exposed to the winter season or part of it to meet their cold requirement for flower initiation. GA$_3$ application replaced this cold requirement, allowing autumn production. GA$_3$'s effectiveness, however, was more evident in early sowing than in later ones. A combination of early sowings (1 and 10 July) and GA$_3$ application (2 or 3 times) resulted in a pattern of head production of Orlando similar to that of Violetto di Sicilia (VS), a typical, early, vegetatively propagated cultivar, with harvesting starting at the end of October and continuing until mid-May. In addition, the total cumulative yield at the end of cycle was significantly higher in Orlando than in VS. When sowing was delayed until 10 August, harvesting of GA$_3$ treated plants was delayed until the early spring (February/March). In another experiment cultivars Orlando and Violetto di Sicilia were planted out on 10 or 22 August or 11 September 1991, sprayed with 60 ppm GA$_3$ twice or 3 times at the 10- to 11-leaf stage and irrigated with salt water of electrical conductivity 3840 æS/cm. Orlando matured later than Violetto di Sicilia but produced higher yields (8 heads/plant compared with only 4.4). Planting Orlando early (10 August) and GA$_3$ treatment resulted in earlier maturation and higher yields than Violetto di Sicilia from mid-March onwards. Heads of Orlando tended to be heavier than those of Violetto di Sicilia (Mauromicale *et al.*, 1996).

Ways of increasing the profitability of globe artichoke crops in Argentina have been discussed by Garcia *et al.* (1998). The most rapid ways are by reducing production costs, mainly by omitting sucker removal. Another alternative is to aim for earlier harvests, achieving higher market prices. This could be achieved either by protected culture, or by exogenous applications of gibberellic acid. Moreover, additional revenue can be obtained by selling leaves to industry for the extraction of pcynarine, a pharmacological compound.

Kocer and Eser (1999) reported that when offshoots at the vegetative growth stage were used for planting, higher early and total yields were obtained than from using offshoots at the generative growth stage. Applying 30 ppm GA$_3$ improved early yields, with a double application being more effective than a single application.

The production of globe artichokes in the Horticultural Rosary Belt (33°1'S and 60°59' W) in Argentina is concentrated from August to October. Garcia *et al.* (1999)

investigated the use of GA_3 sprays to bring crop maturity forward to months of higher prices. Twenty-one cultivars were used, grouped according to their precocity. In 1994 and 1995, GA_3 was applied at 50 ppm in April and at 25 ppm in May. The application of GA_3 in 1994 brought production forward 52 days for the group I cultivars (the earliest), 6 days for group II and only 3 days for group III (the latest), extending the harvesting period by 60, 8 and 3 days, respectively. A similar trend was observed in 1995. There were no significant differences in other variables (yield parameters, head characteristics, *etc.*). Application of GA_3 generated an increase in gross revenue, which varied depending on year and cultivar group.

Abou-Hadid *et al.* (1996) studied the effect of shading and GA_3 on yield and storage of the artichoke. Globe artichoke plants (of a local cultivar) were subjected to 60 per cent shading (with nets) and/or 100 ppm GA_3 treatment in the field. The shade + GA_3 treatment resulted in the greatest yields (weight and number of heads/plant) at harvest and best storage quality (lowest weight loss and decay and highest TSS and DM contents).

Garcia *et al.* (1994) reported that in cultivar Nato, early (April/May) foliar applications of GA_3 (one spray of 50 ppm or one spray of 50 ppm followed by one of 25 ppm a month later) significantly increased early yield compared with control and advanced harvesting date by an average of 20 days. A 50 ppm spray in April followed by a 25 ppm spray in each of the next 4 months had similar effects. However, sprays applied later (June/July) had no such effects. Lin *et al.* (1991) noticed that application of GA_3 at 45 ppm in February and March was the most effective treatment in terms of bolting earliness, bolting percentage and flower bud yield and they concluded that GA_3 could be used to compensate for the insufficient chilling conditions in February/March in Taiwan.

Hot fall weather during September and October suppresses plant growth and causes premature flowering, which lowers yield and average bud size. Schrader (2005) evaluated the effect of plant growth regulator (PGR) treatments in annual artichokes to determine if they could reduce the adverse effects of hot weather during September and October. Treatments included multiple applications of Apogee (gibberellin inhibitor), Retain (ethylene inhibitor), Apogee + Retain, Cytokinin, and untreated control plots. Numbers of harvestable buds were counted as a measure of earlier flowering induced by hot weather. Apogee and cytokinin reduced heat stress during hot fall artichoke production. Other plant growth regulator treatments increased the number of harvestable buds compared to control plots.

Two field experiments were carried out by Saif Eldeen (2015) during two seasons to investigate the effect of foliar spray of natural plant extracts with garlic extract rates (1, 2 and 3 ml/L) and moringa leaf extract rates (1, 2 and 3 ml/L) on earliness, productivity and quality as well as chemical constituents of globe artichoke (*Cynara scolymus* L.) plants cv. Balady. The results revealed that increasing moringa leaf extract and garlic extract rates up to 3 ml/L of each were accompanied with significant increases in vegetative growth characters (plant height, Leaf area of leaves, dry weight of leaves and number of shoots/plant) and yield distribution (early, middle, late yield and total yield/fed.) compared with control. Application of moringa leaf extract and garlic extract rates improved most head quality characteristics (head weight, length, diameter, receptacle fresh and dry matter). Dry matter, N, P, K, protein, total sugars and total carbohydrates contents in receptacle were positively and significantly responded to moringa leaf extract and garlic extract rates in the two seasons.

Hatem *et al.* (2016) conducted two field experiments to investigate the effect of foliar spraying of some growth regulators on two globe artichoke cultivars, *viz.*, Violetto Di Chioggia and French Heryous on early and total yield and yield components. The results indicated that no significant differences between the two cultivars were recorded in most characters except, plant height, inulin content and dry matter in early and total edible part. However, cv. Violetto Di Chioggia recorded higher values than cv. French Heryous. Foliar application with TDZ at 2 or 5 ppm plus GA_3 at 50 ppm exhibited the highest values of most head and receptacle qualities. Moreover, spraying with PP_{333} at 25 or 50 ppm had positive effect on number of leaves and inulin content. The highest early and total yield were obtained by spraying plants with TDZ at 2 ppm plus GA_3 at 50 ppm without significant differences with TDZ at 5 ppm plus GA_3 at 50 ppm or PP_{333} at 25 and GA_3 at 50 ppm.

8.0 Harvesting, Yield and Storage

8.1.0 Harvesting and Yield

Plants grown from seeds come into bearing in about 8 months while plants raised from suckers flower earlier. The flower heads usually appear from February to May in North India. Globe artichoke flower buds should be harvested before they begin to open. Small, immature heads are tender than older heads close to opening. The buds are harvested before they become loose and fibrous and the involural bracts start opening. The buds are cut from the plant leaving a 2.5 to 7.5 cm section of stem with each bud. Once all buds have been harvested from the plant, the plant should be cut down to the ground otherwise once established; the plant continues to produce for 3-4 years. Artichoke buds left too long on the stem have a bitter, pungent taste and contain more woody tissue than soft, edible tissue. Mature buds will change from bright green to dull green in colour and develop a stiffer stem.

During the harvest period (from November to March), the head characteristics of three early Mediterranean cultivars (Violetto di Sicilia, Violet de Provence and Violet Margot) of globe artichoke grown in the field at Siracusa, Italy was evaluated by Mouromicale *et al.* (2000). The delay in harvesting caused a linear increase in head weight and width and a slight decrease in length for both Violet de Provence and Violet Margot, and a linear decrease in head weight, width and length for Violetto di Sicilia. The head length : width ratio, receptacle weight : head weight ratio and scape length, for the three cultivars, decreased with delay in harvesting from November to March.

The average yield is 10-12 t/ha (Veeraragavathatham, 2000).

8.2.0 Storage

Good quality globe artichokes can be stored in the refrigerator for about two weeks without loss of quality. Highest quality is maintained by storing near 0°C

Harvesting, Packaging and Marketing of Globe Artichoke

with over 90 percent relative humidity. After washing, buds are placed in a plastic bag and refrigerated immediately to preserve quality.

In artichoke pre-cooling of buds at 5° C is practised. Pre-cooled buds showed less weight loss, discolouration and incidence of decay than buds not pre-cooled (Shanmugavelu, 1989). Hydro-cooling is commonly employed in California.

Kaynas and Simsek (1992) determined the effect of storage on chemical changes and respiratory patterns of 9 cultivars (Bayrampasa-1, Bayrampasa-21, Sakiz-4, Sakiz-9, 1.3, 1.6, Camus de Bretange, Violet de Province (Violet de Provence) and Bologne). During storage at 0°C for 30 and 60 days, invert sugars, sucrose, total sugars, ascorbic acid, total N and crude protein contents decreased. Soluble solids, titratable acidity and fibre content gradually increased during storage. The artichokes showed no climacteric pattern of respiration. Based on the results, the cultivars tested can be stored for approximately 30 days at 0°C and 90-95 per cent RH. Optimum storage properties were observed in cultivars Bayrampasa-1, Sakiz-4 and Violet de Provence.

Lattanzio *et al.* (1994) reported that cv. Catanese could be kept in good condition for 1-2 days when stored in sealed polyethylene bag at 4° C. He proposed mechanism of browning phenomena in cold-stored, non-mechanically damaged artichoke heads. According to this mechanism, low-temperature storage induces an increase in phenolics, especially chlorogenic acid, as a result of increase phenylalanine ammo-nialyase activity. Chlorogenic acid (synthesized in chloroplasts) releases Fe^{2+} from ferritin (stored in chloroplasts) and under oxidizing conditions (O_2 and/or the quinones formed by cytoplasmic catechol oxidase when cold-induced membrane modification causes partial cell decompartmentalization) this leads to the formation of a grey-blue chlorogenic acid/Fe^{2+}complex and then to browning.

Abou-Hadid *et al.* (1996) harvested heads (of a local cultivar) at maturity, graded, packed in boxes (8/box) and stored at 22° C and 65-70 per cent RH for 6 days.

Sabi *et al.* (2013) conducted a trial of two postharvest treatments consisting of a) stored in refrigerated chamber at 1°C and 90 per cent RH, and b) stored in non refrigerated chamber at room temperature. Both treatments were stored for 14 days. Weight loss, colour and organoleptic characteristics (visual quality, external browning, pilosity, bracts opening) were measured during storage period. Genotypes showed no significant differences in weight loss percentage at seven days. Significant differences were found in weight loss due to storage temperature. All artichoke cultivars stored at room temperature reached no more than seven days of shelf life. In contrast, hybrids stored at 1°C maintained the visual quality for two weeks.

Helaly *et al.* (2016) carried out an experiment in which fresh–cut heads of globe artichoke cultivar "French Hyrious" were dipped in solutions of Gibberellic acid (GA_3) at 100, 200 and 300 ppm, Potassium sorbate (PS) at 1, 2 and 3 per cent and edible coating Chitosan at 1, 2, and 3 per cent for 5 minutes then placed in polyethylene trays that covered with polypropylene film which closed thermally and stored at 0°C and 95 per cent R.H. The result showed that GA_3 exhibited comparatively the most obvious results during storage in lessening the deterioration of loss in weight, unmarketable percentage, T.S.S, inulin besides lowering the activity of polyphenol oxidase enzyme. The maximum T.S.S per cent was obtained from gibberellic acid at 200 ppm, potassium sorbate 1 per cent and chitosan 3 per cent. The application of gibberellic acid led to keep the highest total sugars content at 200 ppm followed by potassium sorbate at 1 per cent and finally chitosan at 2 per cent. The maximum inulin content happened due to the application of gibberellic acid at 200 ppm followed by potassium sorbate 3 per cent and at least was chitosan at 3 per cent. The highest ascorbic acid was obtained from the application of GA_3 at 200 ppm, 3 per cent potassium sorbate and 2 per cent chitosan respectively. Gibberellic acid at 200 ppm followed by potassium sorbate 1 per cent exhibited the least enzyme activity while chitosan at 1 per cent reflected the highest activity during the various storage periods.

Haggag *et al*. (2017) found that storing heads under cold storage conditions at 0°C and 95 per cent R.H. were much better than those stored at 5°C and 95 per cent R.H as it possessed the lower loss in weight and unmarketable percentage besides maintaining the higher visual quality and colour in addition to contain more concentrations of T.S.S., total sugars, inulin, ascorbic acid and lower respiration rate. The heads stored in plastic boxes either lined with perforated polypropylene or unlined at 0°C or 5°C and 95 per cent R.H exhibited that lined plastic boxes existed lower loss in weight, unmarketable percentage, higher visual quality, greater colour retention and kept more concentrations of T.S.S., total sugars, inulin and ascorbic acid.

9.0 Diseases and Pests

9.1.0 Diseases

The artichoke plant is susceptible to a number of diseases.

Damping Off (*Pythium* spp.)

Young seedlings are prone to damping off, which is a soil borne disease that causes seedlings to wilt and die. Root and crown tissue is discolored and decayed. Severely infected seedlings rarely recover and will die. Commencing from 10-15 days after seed sowing, the seedlings may be protected by spraying with chlorothalonil (0.2 per cent) + carbendazim (0.1 per cent). Rhizosphere bacteria belonging to the fluorescent *Pseudomonas* spp. and *Bacillus subtillis* are receiving increasing attention for the protection of plants against soil borne fungal pathogens.

Verticillium Wilt (*Verticillium dahliae*)

This soil-borne fungus is one of the main constraints in artichoke production worldwide. It causes wilting, chlorosis and stunting of plants. Diseased plants produce smaller buds and the plants may collapse in severe infections. All the available varieties are susceptible to this disease.

In California, Bhat *et al*. (1999) first reported *Verticillium* wilt in artichoke. The plants were stunted with chlorotic, drooping and dried leaves near the bottom and middle of the plants. Diseased plants produced smaller edible buds and, in severe cases, buds were discoloured with dried outer bracts. Roots exhibited the characteristic vascular discolouration of Verticillium infection. In one part of the infested field, globe artichoke was near to harvestable stage with 85 per cent of plants showing wilt symptoms with vascular discolouration whereas the other part had a 60-day-old crop with 98 per cent of plants infected. Yield in the field was reduced by as much as 50 per cent. *V. dahliae* was isolated from infected plants on NP-10 medium and identity was confirmed on the basis of colony morphology and formation of micro-sclerotia. Annual artichoke can be rotated with broccoli to help reduce the inoculums levels and manage the disease.

The important disease control measures for the management of *Verticillium* wilt in artichokes are: (i) use of pathogen-free planting material; (ii) site selection to avoid planting into high-risk soils; (iii) reduction or elimination of *V. dahliae* inoculum in soil; (iv) protection of healthy planting material from infection by residual inoculum in soil; and (v) use of resistant cultivars (Cirulli *et al.*, 1994; Amenduni *et al.*, 2005).

Treatments with 1,3-dichloropropene (Telone II emulsifiable) at 18 g m^{-2} followed by metham sodium at 72 g m^{-2}, the mixture of 55.4 per cent 1,3-dichloropropene and 32.7 per cent chloropicrin (Agrocelhone emulsifiable) at 40 g m^{-2}, was suggested by Cebolla *et al.* (2004). All applied with irrigation water, and methyl bromide at 30 g m^{-2} under Virtually Impermeable Film, effectively controlled *Verticillium* wilt and increased bud yield in the first growing season compared with the control.

Botrytis Rot or Gray Mold (*Botrytis cinerea*)

Infected areas typically turn brown and soft on the outside, with gray mould spores visible on the inner surfaces of the bracts.

Crop should be raised in light, well-drained, fertile soils. Sufficient spacing is to be maintained to avoid overcrowding of plants and seeds are also not to be planted too deeply. It is suggested not to wet foliage when watering; plants are to be watered at base. Crop debris should be removed from soil after harvest.

Powdery Mildew (*Leveillula taurica, Erysiphe cichoracearum*)

Two types of powdery mildew infect artichokes. *Leveillula taurica* is more commonly found and primarily colonizes the undersides of older leaves. Careful examination of leaf undersides reveals spores produced singly or in very short chains; however, the profuse white hairs of the leaf may obscure this sign. Severely infected leaves will turn yellow, then brown. With time the brown leaves may collapse and dry up. *Leveillula* infects only the older leaves; the younger leaves escape infection until they mature.

The protective effect of 4 fungicides against powdery mildew of artichoke was studied at the Agricultural Experiment Station of Cairo University by Agwah *et al.* (1990) using the local cv. Balady. The fungicides used were metalaxyl (ridomil MZ 58 w.p. at 0.025 per cent), triadimefon (bayleton 25 per cent at 0.5 g/litre), chlorothalonil (daconil 75 per cent at 0.25 per cent) and fenarimol (rubigan at 0.01 per cent). Early and total yields were increased significantly by metalaxy and triadimefon application every 2 weeks, starting 2 months after planting and these 2 fungicides and also rubigan increased head and edible portion. The inulin content of the edible portion was relatively higher in the early yield and was increased by metalaxyl and triadimefon.

Elsisi and Shams (2019) found significantly decreased percentage of disease incidence and disease severity of powdery mildew by spraying some essential oils *i.e.* garlic, camphor, black seed and clove each at 0.5 per cent three times with 15 days interval in combination with DL- β-aminobutyric acid (BABA). A reduction of the disease incidence by DL- βaminobutyric acid (BABA) was occurred in addition to an increase of vegetative characters when used with essential oils. They also noticed spraying any of the tested oils alone does not have a significant effect on growth, yield and quality unless it is associated with the addition of BABA compared with the fungicide only one 40 per cent EC. The spraying of garlic oil + BABA has achieved the highest rates of increase in the early and total yield.

Artichoke Leaf Spot (*Ramularia cynarae*)

It is characterized by circular, brown lesions on both upper and lower surfaces of leaves. If disease is severe, lesions will coalesce and the entire leaf will turn brown and dry up. White spores of the fungus will usually develop in leaf lesions. On flower bracts, brown patchy lesions will also form, causing the bracts to curl and dry out. Removal and destruction of the infected leaves will limit the spread of the disease. Wetting of leaf surface should be avoided.

Bacterial Crown Rot (*Erwinia chrysanthemi*)

Stunted plant growth; wilted leaves in high temperatures; plant collapse; new leaves do not expand and turn brown and dry; crown tissue becomes soft and rots; black discoloration when cross-section of stem taken.

The pathogen is spread to other plants through cutting machines, or by cutting tools. The bacterium survives on both plant tissue and on dead organic matter. Infected crowns should not be used as planting material; better to start plants from seed or disease free transplants. Annually grown seed planted artichokes or transplants may not develop this disease.

Artichoke Curly Dwarf Virus (ACDV)

Plant growth reduced; plant lacking vigor; leaves may be distorted with dark necrotic spots and/or patches; deformed buds.

Only certified planting materials are to be used; removal and destruction of infected plants will limit the spread of the disease.

Paradies *et al.* (2000) observed that globe artichoke cv. Brindisino plants grown in the province of Foggia, Italy, were found to harbour mixed infections of artichoke latent virus (ALV), cucumber mosaic virus (CMV), and tomato spotted with virus (TSWV). Plants were severely stunted and showed strong deformation of the leaves. Flower heads were few and heavily damaged by marginal necrosis of the scales. In general, symptoms closely resembled those elicited by TSWV. This is the first report of CMV on artichoke in Italy. Application of 2 per cent mineral oil has been reported to reduced ALV transmission by aphids.

Diseases of Globe Artichoke

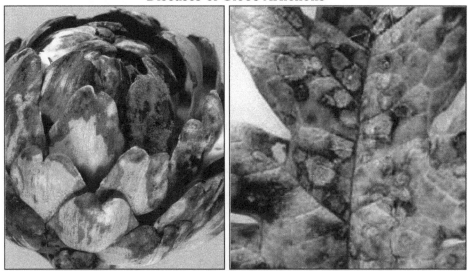

Botrytis Rot or Gray Mold (Left), and Leaf Spot (Right) Infection

Foliar Symptoms of *Verticillium* Wilt in Stump-propagated Artichokes
A: Chlorosis and necrosis in basal leaves, B and C: Intense yellowing on half of
blade or leaflets of young leaves, and D: Severe leaf wilting and necrosis.

Many globe artichoke ecotypes have remained neglected and unnoticed for a long time and have been progressively eroded by several causes, which include a poor phytosanitary status. Sanitation of such ecotypes from infections of vascular fungi and viruses may be a solution for their ex situ conservation and multiplication (Spano *et al.*, 2018). Sanitized stocks should be maintained ex situ in nursery structures adequately protected to prevent reinfection and must be subjected periodically to phytosanitary controls.

Table 2: Viruses Infecting Globe Artichoke in Nature: Taxonomic Allocation, Epidemiology, and Disease Symptoms*

Viruses	Genus Family	Vector/Mode of Transmission	Disease Symptoms
Viruses with Isometric particles			
Artichoke Aegean Ringspot virus (AARSV)	Nepovirus (Subgroup A) Comoviridae	Unknown	Yellow blotches and mild mottling. Often symptomless
Artichoke Italian latent virus (AILV)	Nepovirus (Subgroup B) Comoviridae	*Longidorus apulus L. fasciatus* Seed coat and expanded cotyledons	Mostly symptomless; patchy chlorotic stunting observed in some cvs
Artichoke mottled crinkle virus (AMCV)	*Tombusvirus Tombusviridae*	Contact Soil-borne	Severe deformations, mottling and crinkling of the leaves. Severe distortion of flower capitula. Plant death. New foliage emerging from underground buds develops poorly and often shows bright chrome-yellow discoloration
Artichoke yellow ringspot virus (AYRSV)	Nepovirus (Subgroup C) Comoviridae	Pollen and seed transmission in tobacco and onion	Bright yellow blotches, ring spots and line patterns sometime followed by necrosis
Artichoke vein banding virus (AVBV)	Cheravirus Secoviridae	unknown	Chlorotic discolorations along the leaf veins
Broad bean wilt virus (BBWV)	Fabavirus Comoviridae	*Capitophorus horni* and other aphid species	Yellow mottle, mosaic, or line patterns
Cucumber mosaic virus (CMV)	Cucumovirus Bromoviridae	several aphid species	Probably symptomless. Usually found in mixed infection with ArLV and/or TSWV in plants showing severe stunting
Pelargonium zonate spot virus (PZSV)	Anulavirus Bromoviridae	several thrips species	Moderate stunting and leaf chlorotic mottling. Chlorotic spots and line patterns after inoculation under experimental conditions
Tobacco streak virus (TSV)	Ilarvirus (Subgroup 1) Bromoviridae	several thrips species	Stunting and leaf deformation
Tomato black ring virus (TBRV)	Nepovirus (Subgroup B) Comoviridae	*Longidorus attenuatus*	Mild leaf mottling

Viruses	Genus Family	Vector/Mode of Transmission	Disease Symptoms
Viruses with Rod-Shaped particles			
Tobacco mosaic virus (TMV)	*Tobamovirus*	Contact	Symptomless
Tobacco rattle virus (TRV)	*Tobravirus*	*Thrichodorus Christiei*	Leaf bright yellow discoloration
Viruses with Filamentous particles			
Artichoke curly dwarf virus (ACDV)	*Potexvirus* Flexiviridae	Unknown	Stunting, leaf distortion and extended vein necrosis, delay in the development of flower capitula
Artichoke degeneration virus (ADV). Putative potyvirus	Non-classified	Unknown	Leaf mottling curling and crinkling
Artichoke latent virus (ArLV)	*Macluravirus* Potyviridae	*Myzus persicae, Brachycaudus cardui, Aphis fabae*	Symptomless
Artichoke latent M virus (ArLMV)	*Carlavirus* Flexiviridae	Unknown	Symptomless. Often in mixed infection with AILV, ArLV, AMCV
Artichoke latent S virus (ArLSV)	*Carlavirus* Flexiviridae	Aphids	Symptomless
Bean yellow mosaic virus (BYMV)	*Potyvirus* Potyviridae	Several aphid species	Leaf yellowing, yellow flacking, line patterns, often in mixed infection with ArLV
Filamentous viruses and globe artichoke disease	unknown	Unknown	Foliar mottling, leaf deformation and crinkling, stunting, and decreased yield. Progressive degeneration
Potato virus X (PVX)	Potexvirus Flexiviridae	Unknown	Mosaic, leaf deformation and narrowing, stunting
Ranunculus latent virus (RaLV)			Mostly symptomless. Often in mixed infection with cynara 42 virus (Cy42)
Turnip mosaic virus (TuMV)	*Potyviridae* Potyvirus	Several aphid species	Symptomless. Often in mixed infection with ArLV
Tomato infectious chlorosis virus (TICV)	Crinivirus Closteroviridae	*Trialeurodes vaporariorum*	Symptomless or mild interveinal yellowing
Unnamed putative carlavirus	Non-classified	Unknown	Symptomless
Viruses with Enveloped particles			
Cynara virus (CraV). Putative cytorhabdovirus	unassigned *Rhabdoviridae*	Unknown	Progressive degeneration. Often in mixed infection with AILV, ArLV, BBW, BYMV
Tomato spotted wilt virus (TSWV)	*Tospovirus* Bunyaviridae	Several thrips species	Severe stunting and deformation, generalized chlorosis and bronzing of the apical leaves, distortion of the head stalk, and necrosis of portions of the inner and outer scales

* Spano *et al.* (2018)

9.2.0 Pests

Artichokes are susceptible to a number of insect pests. Some of the organisms that can harm an artichoke plant include sucking insects, such as aphids, mites, and thrips. These pests not only aid in transmitting diseases that can harm an artichoke plant, they also slow down the rate at which the plant grows.

Artichoke Plume Moth (*Platyptilia carduidactyla*)

Larvae, which cause damage, are small and worm like. Larvae may chew holes in choke bracts, new stems, and leaves and bore into stem and buds. Larvae may also bore into the crown below the soil surface. All infested buds should be picked at harvest and destroyed; plant stems above ground are to be cut, shredded and incorporated into soil; Application of *Bacillus thuringiensis* or suitable insecticide is suggested.

Bacillus thuringiensis was effective in limiting the infestation of young sprouts and developing artichokes. The effectiveness of this bacillus is dependent on its high virulence for the first-instar larva, which dies within 1-2 days after infection, usually before it can mine into the plant tissue. Using pathogens against insect pests with a low economic threshold, such as in the case of *P. carduidactyla*, in which a single larva can cause the loss of an artichoke suggested by Tanada and Eeiner (1960).

Artichoke Aphid (*Capitophorus elaeagni*)

Leaves curling and turning yellow; reduced plant growth; small, deformed buds; stalks cannot support weight of buds and droop; sooty mold growing on plants due to honeydew deposits secreted by insect; insect is small, soft-bodied and pale green to yellowish green in color.

Plants immediately after harvest should be destroyed to prevent population spread; washing aphids from plants with a strong stream of water; insecticidal soaps or oils such as neem or canola oil are effective organically acceptable methods of control. Parasites of aphids like wasps of the genera *Aphidius* and *Lysiphlebus*, predators such as ladybird beetles, hoverfly larvae and lacewings can be used to control aphids.

Armyworms (Beet Armyworm, Yellow Striped Armyworm) *Spodoptera exigua, Spodoptera ornithogalli*

Singular, or closely grouped circular to irregularly shaped holes in foliage; heavy feeding by young larvae leads to skeletonized leaves; shallow, dry wounds on fruit; egg clusters of 50-150 eggs may be present on the leaves; egg clusters are covered in a whitish scale which gives the cluster a cottony or fuzzy appearance; young larvae are pale green to yellow in color while older larvae are generally darker green with a dark and light line running along the side of their body and a pink or yellow underside.

Organic methods of controlling armyworms include biological control by natural enemies which parasitize the larvae and the application of *Bacillus thuringiensis*; there are chemicals available for commercial control but many that are available for the home garden do not provide adequate control of the larvae.

Flea Beetle (Palestriped Flea Beetle) *Systena blanda*

Small holes or pits in leaves that give the foliage a characteristic "shothole" appearance; young plants and seedlings are particularly susceptible; plant growth may be reduced; if damage is severe the plant may be killed; the pest responsible for the damage is a small (1.5-3.0 mm) dark coloured beetle which jumps when disturbed; the beetles are often shiny in appearance.

In areas where flea beetles are a problem, floating row covers may have to be used prior to the emergence of the beetles to provide a physical barrier to protect young plants; plant seeds early to allow establishment before the beetles become a problem - mature plants are less susceptible to damage; trap crops may provide a measure of control - cruciferous plants are best; application of a thick layer of mulch may help prevent beetles reaching surface; application on diamotecoeus earth or oils such as neem oil are effective control methods for organic growers; application of insecticides containing carbaryl, spinosad, bifenthrin and permethrin can provide adequate control of beetles for up to a week but will need reapplied.

Loopers (Cabbage Looper, Alfalfa Looper) *Trichoplusia ni, Autographa californica*

Large or small holes in leaves; damage often extensive; caterpillars are pale green with a white lines running down either side of their body; caterpillars are easily distinguished by the way they arch their body when moving; eggs are laid singly, usually on the lower leaf surface close to the leaf margin, and are white or pale green in color.

Looper populations are usually held in check by natural enemies; if they do become problematic larvae can be hand-picked from the plants; an organically acceptable control method is the application of *Bacillus thuringiensis* which effectively kills younger larvae; chemical sprays may damage populations of natural enemies and should and should be selected carefully.

Spider Mites (Two-spotted spider mite) *Tetranychus urticae*

Leaves stippled with yellow; leaves may appear bronzed; webbing covering leaves; mites may be visible as tiny moving dots on the webs or underside of leaves, best viewed using a hand lens; usually not spotted until there are visible symptoms on the plant; leaves turn yellow and may drop from plant.

In the home garden, spraying plants with a strong jet of water can help reduce build up of spider mite populations; if mites become problematic apply insecticidal soap to plants; certain chemical insecticides may actually increase mite populations by killing off natural enemies and promoting mite reproduction.

Slugs and Snails

They often rasp off the outer surface of the artichoke buds and eat jagged holes in the leaves and stems. You may wish to use chemicals to help control pests. Chemicals safe for home use are available at your local nursery or garden supply centre. Before purchasing make sure that the pesticide that you choose is recommended for the pest that you want to destroy, and that it is safe to use in the

Pests of Globe Artichoke

Aphid Infestation

Plume Moth

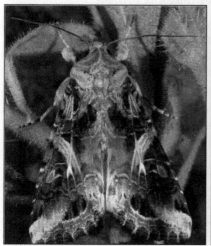

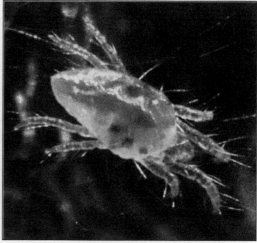

Adults of Yellow Striped Armyworm (Left), and Two Spotted Spider Mite (Right)

vegetable garden. Find out how long after applying the pesticide you should wait before harvesting.

Disorder

Black tip of buds was noted on some plants. This may be due to a physiological response related with reduced calcium uptake. The outside bracts of smaller buds were affected, turning them brown to black. It was primarily noted under conditions of moisture stress and high temperatures affecting plant water uptake and transpiration balance.

10.0 Crop Improvement

Commercial production is mainly based on perennial cultivation of vegetatively propagated clones that are highly heterozygous and segregate widely when progeny-tested. Genetic variation among artichoke clones belonging to the same varietal type was in some cases higher than that found among accessions differently named and coming from different areas. The lowest Jaccard's Similarity Index found within a varietal type can be considered as a threshold for the identification of accessions which share an analogous genetic background. This will enable the selection of representatives in order to develop and manage a germplasm 'core collection' as well as the identification of suitable material for future artichoke breeding efforts (Lanteri *et al.*, 2004).

Mauromicale and Ierna (2000a) overviewed recent research in Italy on globe artichoke breeding covering, genetic resources; classification and characterization of Italian germplasm; and methods, advantages and limits of breeding; clonal selection in local populations and in progeneis from selfing and hybridization; constitution of varieties propagated by seed; and varietal choice.

Several types of globe artichoke are currently cultivated in Europe using vegetative propagation. Breeding programmes focus on the development of new cultivars from seed (Collet, 1999).

Lopez-Anido *et al.* (1998) evaluated 23 globe artichoke clones and recorded that heritability was intermediate to high : 0.83, 0.82, 0.94 and 0.48 for yield (YD), number of artichokes harvested (AN), height/diameter ratio (MR) and harvest period (HP), respectively. The expected selection advance was related to the magnitude of the genetic coefficient of variation, and was over 30 per cent of the mean for weight of secondary artichokes, bottom weight, harvest period and yield. Characters with highest association with yield were the number of artichokes and weight of secondary artichokes (W2) (rg -0.98 and 0.99, respectively). Harvest period and days to harvest were negatively correlated (rg = -0.92). DH was positively correlated with YD, W2, W1 and BW, while the opposite was true for HP, indicating that high yielding clones with good head size should be searched for in late maturing materials.

The background for breeding of seed planted globe artichoke is reviewed by Basnizki and Zohary (1994). Information is presented in sections covering : a rationale for developing seed-planted globe artichoke; genetic resources, including the cultivated and wild gene pools; reproductive biology, including floral biology, seed yields, bolting and germination; inheritance of economically important traits, including polygenic inheritance, traits controlled by major genes, male sterility, consequences of inbreeding and heterosis; and seed-planted hybrid cultivars.

The trait earliness in globe artichoke is probably polygenic and response to selection depends on the materials that breeders are handling as suggested by Gil and Villa (2004).

Phenotypic variability experiment in globe artichoke population resulted the distance between individuals varied from 3.0 to 50.9, revealing high intra-population variability. The greater relative contribution characters for genetic divergence were the primary head fresh mass (79.88 per cent) and bottom fresh mass (8.43 per cent). This indicates the possibility of plant selection for head quality in this population. The clustering analysis through UPGMA method, based on quantitative characters, allowed the formation of five groups. For multi-categoric traits, the similarity among individuals varied from 1.22 per cent to 100 per cent (Costa *et al.*, 2014).

Significant differences between the mean values of several traits among both diallel sets were found by Cravero *et al.* (2004). Most of the variables evaluated were primarily controlled by additive genetic effects. They suggested simple recurrent selection would be effective for increasing the mean value of these variables.

Crippa *et al.* (2011) found a positive genotypic correlation of number of heads and the length, diameter and weight of heads with total yield. They also noticed number of head and their weight had direct positive contribution towards total yield per plant. Indirect effect of plant diameter and length and weight of head, on number of heads plant height was revealed.

López Anido *et al.* (2010) carried out an experiment to determine the gene actions, heterosis, and genetic variance present in reciprocal F_2 populations obtained by combining divergent artichoke cultivars. The means value of F_1 generations were significantly different. The intermediate values between parents for precocity and length-diameter ratio indicate the presence of additive gene action, while for weight of the main head, total and marketable yield dominance effects were detected. For these traits the absolute mean parent heterosis is important but for marketable yield only the best parent heterosis is expressed (13.1 per cent). F_1 reciprocal crosses were not significantly different in any case however, for the reciprocals F_2 all the comparisons indicate unexpected differences between them. The higher genetic variance developed in reciprocal F_2 facilitated the selection.

11.0 Biotechnology

Application of biotechnology on globe artichoke has resulted in marked improvement in propagation through tissue culture and identification of genotypes using molecular markers.

11.1.0 Tissue Culture

11.1.1 Micropropagation

A micropropagation method was developed starting from seeds of the cultivar Green Globe by Lauzer and Vieth (1990). Cultures were initiated with *in vitro* seedlings and with their shoot tips. Shoot multiplication occurred through a proliferation of axillary leaf buds and the best multiplication rates were obtained when different BA concentrations and 0.5 mg NAA/l were combined. During the multiplication stage, vitrification was reduced by avoiding contact of explants' axillary leaf buds with the culture medium. Almost 65 per cent of the *in vitro* shoots produced roots after 2 months in the presence of 1.0 mg NAA/l. The rooted shoots were easily hardened-off and transferred to soil.

Tissue cultures were established on shoot apices of globe artichoke. Sterilization of plant material was one of the most problematic stages because accessions of the same plant species specifically responded to the concentration of sterilization solution as well as to the time of sterilization. Optimal proliferation medium for cv. Istra was MS medium supplemented with 10 mg of kinetin and 50 mg L-tyrosine per litre. Treatmemt with B9 (0.010 g/l) ameliorated rooting capacity rate (Erzen-Vodenik and Baricevic, 1996). El-Bahr *et al.* (1993) reported that regeneration from shoot tips of cv. Herious was best on MS medium supplemented with 0.5 mg IAA + 10 mg kinetin/l. Percentage shoot proliferation was highest (85 per cent) when 4 mm lengths of shoot tips were cultured. Explant browning was best prevented by immersing sterilized shoot tips in a solution of 100 mg ascorbic acid/l. Micropropagated plantlets regenerated from 4 mm long shoot tips, were successfully transferred to the greenhouse in a mixture of peatmoss and sand (El-Gizawy *et al.*, 1993).

Ordas *et al.* (1990) induced shoot regeneration in bracts collected from very young (1.5-2 cm in length) capitula of cv. Romanesco. Bract cultures were maintained on MS medium containing growth regulators in different combinations. The most rapid and the most prolific callus response was obtained using 5 mg NAA and 2 mg BA/l. Shoot differentiation occurred after about 5 weeks, starting from the lower region of the bract explant. Shoots elongated after transfer to MS medium containing 0.5 mg IBA and 0.5 BA/l.

Microrosettes of globe artichoke were subcultured in a medium with NAA and kinetin. For rooting, shoots were placed on an auxinic rooting medium (2.68 æM NAA) supplemented with or without GA_3. Treatment with 2.9 and 14.4 æM GA_3 did not significantly change the rooting rate produced by auxin alone (50 per cent). Moreover, the number of roots per shoot significantly increased for plants grown on rooting media containing 2.9 and 14.4 æM GA_3. Stimulation of rhizogenesis by GA_3 was accompanied by a large increase in leaf growth (Morzadec and Hourmant, 1997).

Shoot apex explants treated with thidiazuron, and intact mature zygotic embryos, were used for adventitious shoot regeneration of globe artichoke by Kanakis and Demetriou (1993). Adventitious shoot regeneration from shoot-apex explants of cv. Argos occurred on a basic cultural medium (MS supplemented with 10 mg ascorbic acid, 0.1 mg IAA, 40 g sucrose and 7 g agar/l) to which thidiazuron was added at concentrations between 1 and 5 mg/l. Thidiazuron was as effective as the commonly used cytokinins kinetin and BA. The explants initially treated with thidia-zuron continued to produce adventitious shoots when subcultured on to cytokinin-free or thidiazuron-free medium, but this did not occur when kinetin or BA was initially applied to the explants. Intact mature zygotic embryo explants,

Micropropagation of Globe Artichoke

A) Shootlet cultures of globe artichoke proliferated on MS medium supplemented with 2 mg/l BA, (B) Callus induced from leaf explant on medium contained 5mg/l kin + 0.5 mg/l IAA, (C) callus grown on medium contained 5 mg/l kin + 0.5 mg/l IAA + 3 mg/l picloram, (D) Direct shootlet proliferation from leaf explants, (E) Indirect, organogenesis, (F) Shoot bud multiplication (Bekheet *et al.*, 2014).

cultured like the shoot apices, produced a high number of adventitious shoots when the medium was supplemented with 10 mg kinetin/l or 5, 10, 12 mg BA/l.

In Tunisia, Ghorbel *et al.* (1993) reported that globe artichoke plants multiplied *in vitro* were free of ALV (artichoke latent potyvirus). These ALV-free plants displayed higher vigour and yield when grown in the field than plants produced by traditional methods. However, some recontamination of these plants was observed after 1 year in an open field.

A study was carried out by Bedini *et al.* (2012) to examine the possibility of *in vitro* propagation of four most common cultivars in Tuscany (central Italy): *Terom, Violetto di Toscana, Chiusure* and *Empolese.* Explants were cultured on an induction medium (IM), which is a modified MS medium consisting of nitrate concentrations reduced by one quarter, 0.8 mg L^{-1} 6-benzylaminopurine (BA) and 0.2 mg L^{-1} 3-indole butyric acid (IBA). Explants were then transferred to a proliferation medium (PM) consisting of the same basal medium together with 0.03 mg L^{-1} BA and 0.05 mg L^{-1} gibberellic acid (GA_3). A rooting double-phase was then established. The pre-rooting medium (PRM), consisting of a basal MS medium with half strength nitrate concentrations, 0.5 mg L^{-1} indole-3-acetic acid (IAA) and 1 mg L^{-1} paclobutrazol (PBZ) was used for two weeks. Over the next four weeks, a rooting medium (MR) was used, consisting of a basal MS medium with 2 mg L^{-1} β-cyclodextrin and 2 mg L^{-1} α-naphthaleneacetic acid sodium salt (NAA). The cv *Empolese* provided the highest number of proliferated explants and rooted plantlets using the method described.

Influences of different doses of growth regulators have been investigated by Ercan (2016) on the rooting of globe artichoke cultivars Sakiz and Bayrampaoa. In this experiment indole-3-butyric acid (IBA), naphthalene acetic acid (NAA), gibberellin (GA_3) in combination with/without activated charcoal (AC), putrescine acid and jasmonic acid were added to Murashige and Skoog (MS) basal medium (*Cynara scolymus* L.) to study the rooting percentage of globe artichoke. Results showed that the highest rooting percentage of 63.35 per cent and 36.6 per cent was obtained in Sakiz and Bayrampaoa, respectively, from the 5 mg/L GA_3 with 1 g/L AC added to MS medium.

11.1.2 Anther and Ovule Culture

Several artichoke cultivars were screened for response to *in vitro* culture of anthers and unfertilized ovules by Motzo and Deidda (1993). Anther culture showed marked genotypic differences and only 3 out of 5 genotypes produced calli. The highest percentage (2.6 per cent) of callusing anthers was recorded in cv. Castellammare on a medium supplemented with 2 mg 2,4-D/litre. Anther calli grew rapidly and turned green when transferred to a proliferation medium, but never produced roots or shoots when transferred on to different regeneration media. A total of 5080 ovules of different ages from the cultivars Masedu and Spinoso Sardo were cultured on 6 induction media, where some of them turned green and produced calli either on the funicle break or inside the ovule integuments. These calli became dark and necrotic after 30-40 days of culture despite being transferred to different proliferation media. Medium composition greatly affected the rate of growth and the percentage of green ovules. Concentrations of BA greater than 3 mg/litre showed a detrimental effect on both ovule growth and percentage of green ovules.

11.1.3 Protoplast Culture

Ordas *et al.* (1991) isolated viable protoplasts from suspension culture of callus initiated from bract explants. Optimum protoplast yield (3.0×10^6 ml PCV) was obtained from a cell suspension at a density of 12.1 cells $\times 10^5$/ml and 10-15 days old, with an enzyme solution that included pectinase. An initial protoplast density of 2 $\times 10^5$/ml was required to stimulate subsequent cell divisions which were observed after 4-6 days of culture. Upon transfer to solid medium the cell colonies gave rise to proliferating green calluses although only those sub-cultured on to MS medium containing 0.5 mg BA and 0.5 mg IBA/l developed globular structures and these failed to regenerate shoots.

11.2.0 Molecular Markers

The magnitude of genetic differences among and heterogeneity within globe artichoke cultivars is unknown. Variation among individual heads (capitula) from 3 artichoke cultivars and 2 breeding populations were evaluated using RAPD markers by Tivang *et al.* (1996). One vegetatively propagated cultivar (Green Globe), 2 seed-propagated cultivars (Imperial Star and Big Heart) and 2 breeding populations were examined. Two to 13 polymorphic bands were observed for 27 RAPD primers, which resulted in 178 scored bands. Variation was found within and among all cultivars and breeding populations, indicating that all 5 groups represent heterogeneous populations with respect to RAPD markers. The genetic relationship among individual genotypes was estimated using the ratio of discordant bands to total bands scored. Multidimensional scaling of the relationship matrix showed 5 independent clusters corresponding to the 3 cultivars and 2 breeding populations. The integrity of the 5 clusters was confirmed using pooled chisquares for fragment homogeneity. Average gene diversity (Hs) was calculated for each population sample, and a one-way analysis of variance showed significant differences among populations. Big Heart had an Hs value equivalent to the 2 breeding populations, while clonally propagated Green Globe and seed propagated Imperial Star had the lowest Hs values. The RAPD heterogeneity observed within clonally propagated Green Globe was consistent with phenotypic variability observed for this cultivar. Overall, the results demonstrated the utility of the RAPD technique for evaluating genetic relationships and contrasting levels of genetic diversity among populations of artichoke genotypes.

Portis *et al.* (2005) assessed the level of genetic variation, *via* AFLP fingerprinting, present in autochthonous globe artichoke germplasm in Sicily, thought to be one of the origins of its domestication.

Martin *et al.* (2008) studied with 118 F_2 plants to detect molecular markers linked to two agronomic traits, colour head and precocity of production. They used Bulked Segregant Analysis (BSA) and Sequence-related amplified polymorphism (SRAP) to identify molecular markers linked to the traits. They detected one band/marker that may be linked to the green colour head and one band/marker that may be associated with precocity in head production. The result shows that the BSA and SRAP analyses are useful to identify molecular markers associated with agronomic traits and could be used in molecular marker aided-selection (MAS) for breeding programs in globe artichoke.

Boury *et al.* (2012) extracted genomic DNA from 556 genotypes belonging to the artichoke germplasm collection and analysed using different types of molecular markers (*i.e.* ISSR, SSR, AFLP). The different markers were able to detect different levels of polymorphism, with AFLP markers detecting higher levels of variation. The cluster analysis grouped the accessions into 6 clusters.

ISSR techniques were efficient markers in showing the occurrence of the genetic changes that occur during the micropropagation process of globe artichoke (Rey *et al.*, 2013).

De Felice *et al.* (2016) used 10 SSR microsatellite markers to assess genetic variation between cultivated and wild species of globe artichoke. Specific molecular markers showed efficient introgression of such traits in cultivated as well as wild artichokes species. Cluster analysis discriminated all 30 accessions and classified cultivated and wild species in distinct groups. Results from PCoA analysis showed that artichoke genotype contains a higher number of unique alleles.

11.3.0 Genetic Transformation

Kchouk *et al.* (1997) studied genetic transformation of globe artichoke cells. However, the percentage of transformed cells is low, with the regeneration process being the problematic step towards obtaining transgenic plants. *In vitro* organogenesis was investigated in artichoke cotyledons and leaves. Cotyledons gave better rates of neoformation than leaves, and in a shorter time. A model system was developed for transferred genetic resistance to artichokes using *Nicotiana benthamiana* as a systemic host for artichoke viral diseases. Mutagenized squences of the replicase gene of artichoke mottled crinkle tombusvirus (AMCV) were used as transferable genetic material for resistance induction. Transgenic lines of *N. benthamiana* were produced, which showed reduced symptoms when inoculated with AMCV.

Menin *et al.* (2012) developed an efficient protocol for callus induction from leaf explants of three globe artichoke genotypes of the varietal type 'Romanesco'; after just one week culture callus formed with efficiency close to 100 per cent. Leaf explants were transformed with *Agrobacterium tumefaciens* Agl0 01-124 strain, harbouring the binary vector pCAMBIA 2301 with the gene marker GUS, under the control of CaMV 35S promoter. After two weeks, about 30 per cent of the calli obtained from infected leaf explants were positive to GUS assay.

Scaglione *et al.* (2016) described the first genome sequence of globe artichoke. The assembly, comprising of 13,588 scaffolds covering 725 of the 1,084 Mb genome, was generated using ~133-fold Illumina sequencing data and encodes 26,889 predicted genes. Re-sequencing (30×) of globe artichoke and cultivated cardoon (*C. cardunculus* var. *altilis*) parental genotypes and low-coverage (0.5 to 1×) genotyping-by-sequencing of 163 F_1 individuals resulted in 73 per cent of the assembled genome being anchored in 2,178 genetic bins ordered along 17 chromosomal pseudomolecules.

12.0 References

Abd-El-Fattah, A.E. (1998) *Assiut. J. Agri. Sci.*, **29** : 227-240.

Abou-Hadid, A.F., Beltagy, A.S., Abdel-Rahman, S.Z. and Gaafer, S.A. (1996) *Egyptian J. Hort.*, **22** : 41-48.

Agwah, E.M.R., El-Fadaly, K.A. and El-Hassan, E.A.A. (1990) *Bull. Fac. Agric.*, **41** : 767-775.

Amenduni, M., Cirulli, M., D'Amico, M., and Colella, C. (2005) *Acta Hortic.*, **681** : 603-606.

Ashour, H.M., Ismail, S.A., Ibrahem N.M. and Abd El-Latif A.A. (2016) *American-Eurasian J. Agric. Environ. Sci.*, **16**(8) : 1484-1497.

Basnizki, J. and Zohary, D. (1994) *Plant Breeding Reviews*, **12** : 253-269.

Basnizki, Y. (2007) *Ital. J. Agron.*, **4** : 373-376.

Basnizki, Y. and Mayer, A.M. (1985) *Agronomie*, **5** : 529-532.

Bedini, L., Lucchesini, M., Bertozzi, F. and Graifenberg A. (2012) *Eur. J. Biol.*, **7**(4) : 680-689.

Bekheet, S.A., Gabr, A.M.M. and El-Bahr, M.K. (2014) *Int. J. Acad. Res.*, **6**: 297-303.

Bhat, R.G., Subbarao, K.V. and Bari, M.A. (1999) *Plant Dis.*, **83**: 782.

Biel, W., Witkowicz, R., Pi¹tkowska, E. and Podsiad³o, C. (2020) *Biol. Trace Elem. Res.*, **194** : 589-595.

Boury, S., Jacob, A.M., Egea-Gilabert, C., Fernandez, J.A., Sonnante, G., Pignone, D., Rey, N.A. and Pagnotta, M.A. (2012) *Acta Hortic.*, **942** : 81-88.

Calabrese, N., Elia, A., Sarli, G. and Rumpel, J. (1994) *Acta Hortic.*, **371** : 189-193.

Catalá, R., Díaz, A. and Salinas. J. (2012) *Molecular responses to extreme temperatures*, In: P.M. Hasegawa (ed.). Plant Biotechnology and Agriculture. Academic Press, San Diego, pp. 287-307.

Cebolla, V., Navarro, C., Miguel, A., Llorach, S., and Monfort, P. (2004) *Acta Hortic.*, **660** : 473-478.

Chevallier, A. (1996) *The encyclopedia of medicinal plants. DK Publishing*, New York, pp. 96-97.

Choudhury, B. (1967) *Vegetables, National Book Trust*, New Delhi.

Ciancolini, A. (2012). Characterization and Selection of Globe Artichoke and Cardoon Germplasm for Biomass, Food and Biocompound Production. *Ph.D.Thesis*. University of Toulouse, Toulouse, France. P. 224.

Cirulli, M., Ciccarese, F. and Amenduni, M. (1994) *Plant Dis.*, **78** : 680-682.

Collet, J.M. (1999) *Infos. Paris*, **156** : 29-31.

Cosentino, S. and Mauromicale, G. (1990) *Acta Hortic.*, **278** : 261-270.

Cravero, V.P., Lopez Anido, F.S., Asprelli, P.D. and Cointry, E.L. (2004) *NZ. J. Crop Hortic. Sci.*, **32**(2) : 159-165.

Crippa, I., Martin, E.A., Espösito, M.A., Cravero, V.P., Lopez, A.F.S. and Cointry, E.L. (2011) *Electron. J. Plant Breed.*, **2**(1) : 151-156.

da Costa, A.R. Basso, S.M.S., Grando, M.F. and Cravero, V.P. (2014) *Cienc. Rural.*, **44**(11) : 2003-2009.

Dabbou, S., Dabbou, S., Flamini, G., Peiretti, P.G., Pandino, G. and Helal, A.N. (2017) *Int. J. Food Prop.*, **20**(1) : 810-819.

Dawa, K.K., El-Saady, W.M., El-Denary, M.E. and Abo-Elglagel, I.M. (2018) *J. Plant Prod.*, **3**(7) : 2237-2248.

De Falco, B., Incerti, G., Amato, M. and Lanzotti, V. (2015) *Phytochem. Rev.*, **14** : 993–1018.

De Felice, B., Borra, M., Manfellotto, F., Anna, S., Biffali, E. and Guida, M. (2016) *Genet. Resour. Crop Evol.*, **63** : 1363-1369.

De Vos, N.E. (1992). *HortTechnol.*, **2** : 438-444.

Deligios, P.A., Tiloca, M.T. Sulas, L., Buffa, M., Caraffini, S., Doro, L., Sanna, G., Spanu, E., Spissu, E., Urracci, G.R. and Ledda, L. (2017) *Agron. Sustain. Dev.*, **37**(6) : 54

Dranik, L.I., Dolganenko, L.G., Slapke, J., Thoma, H., Toma, kh., and Slapke, Yu (1996) *Rastitel'nye Resursy*, **32** : 98-104.

El-Bahr, M.K., El-Gizawy, A.M., Moursy, H.A., Bekheet, S.A. and Francis, R.R. (1993) *Ann. Agric. Sci.*, **2** : 751-763.

El-Gizawy, A.M., El-Bahr, M.K., El-Oksh, I.I., Bekheet, S.A. and Francis, R.R. (1993) *Ann. Agric. Sci.*, **2** : 765-776.

Elia, A. and Santamaria, P. (1994) *Agricoltura Mediterranea*, **124** : 106-111.

Elia, A., Calabrese, N. and Bianco, V.V. (1994a) *Acta Hortic.*, **372** : 347-354.

Elia, A., Calabrese, N. and Bianco, V.V. (1998) *Annali della Facolta di Agraria*, Universita di Bari, **35** : 137-146.

Elia, A., Calabrese, N., Sarli, G., Losavio, F. and Rumpel, J. (1994b) *Acta Hortic.*, **371** : 195-199.

Elia, A., Paolicelli, F. and Bianco, V.V. (1991) *Adv. Hort. Sci.*, **5** : 119-122.

Elia, A., Santamaria, P. and Serio, F. (1996) *J. Plant Nutrition*, **19** : 1029-1044.

Elsisi, A.A. and Shams, A.S. (2019) *Middle East J. Appl. Sci.*, **9**(2) : 443-455.

Ercan, N. (2016) *J. Agric. Sci. Technol.*, **6**(5) : 335-340.

Erzen-Vodenik, M. and Baricevic, D. (1996) *Novi izzivi v polijedelstvu '96. Zbornik simpozija, Ljubljana, Slovenia 9-10 decembra 1996*, pp. 201-206.

FAO Statistical database (2008) http: //www.fao.org/faostat/en/#data/QC

Foury, C. (1989) *Acta Hortic.*, **242** : 155-166.

Garcia, S.M., Firpo, I.T., Anido, F.S.L. and Cointry, E.L. (1999) *Pesquisa-Agropecuaria Brasileira*, **34** : 789-793.

Garcia, S.M., Firpo, I.T., Lopez-Anido, F.S. and Cointry, E.L. (1998) *Advances en Horticultura*, **3** : 43-48.

Garcia, S.M., Panelo, M.S. and Nakayama, F. (1994) *Horticultura Argentina*, **13** : 77-82.

Ghorbel, A., Chabbouh, N., Draoui, N., Cherif, C. and Kchouk, M.E. (1993) *Agricoltura Mediterranea*, **123** : 133-138.

Gil, R. and Villa, F. (2004) *Acta Hortic.*, **660** : 35-37.

Gil-Ortega, R., Villa, F., Arce, P. and Macua, J.I. (1995) *HF-HortoinFormacion*, **6** : 58-60.

Gil-Ortega, R., Villa, F., Arce, P. and Macua, J.I. (1998) *ITEA-Produccion Vegetal*, **94** : 13-18.

Haggag, I.A.A., Shanan, S.A, Abo El-Hamd, A.S.A., Helaly, A.A. and El-Bassiouny, R.E.I. (2017) *Adv. Plants Agric. Res.*, **6**(5) : 138-145.

Halter, L., Habegger, R. and Schnitzler, W.H. (2005). *Acta Hortic.*, **681** : 75-82.

Hatem M.A., Shadia A.I., Nadia M.I. and Amany A.A. (2016) *American-Eurasian J. Agric. and Environ. Sci.*, **16**(8) : 1484-1497.

Helaly, A.A, Haggag I.A.A., Shanan, S.A., Abbo El–Hamd, A.S.A. and El-Bassiouny R.E.I. (2016) *Adv. Plants Agric. Res.*, **5**(2) : 474-481.

Iapchino, G. and Vetrano, F. (1994) *Colture Protette*, **23** : 85-88.

Ierna, A. Mauromicale, G. and Licandro, P. (2006) *Acta Hortic.*, **700** : 115-120.

Kanakis, A.G. and Demetrious, K. (1993) *J. Hort. Sci.*, **68** : 439-445.

Kaynas, K. and Simsek, G. (1992) *Doga, Turk Tarim ve Ormancilik Dergisi*, **16** : 571-580.

Kchouk, M.L., Mliki, A., Chatibi, A. and Ghorbel, A. (1997) *Mol. Cell Biol.*, **43**: 399-408.

Khan, I.A., Verma, G.S. and Pandey, H.K. (1999) *Agric. Sci. Dig.*, **19** : 55-58.

Kocer, G. and Eser, B. (1999) *Turk J Agric For.*, **23** : 325-332.

Lanteri, S., Saba, E., Cadinu, M. Mallica, G.M., Baghino, L. and Portis, E. (2004) *Theor. Appl. Genet.*, **108** : 1534-1544.

Lattanzio, V., Cardinali, A., Venere, D. di, Linsalata, V., Palmieri, S. and Di-Venere, D. (1994) *Food Chem.*, **50** : 1-7.

Lattanzio, V., Kroonb, P., Linsalatac, V. and Cardinalic, A. (2009) *J. Functional Foods.*, **1** : 131-144.

Lauzer, D. and Vieth, J. (1990) *Plant Cell Tissue Organ Cult.*, **21** : 237-244.

Lepez-Anido, F.S., Firpo, I.T., Garcia, S.M. and Cointry, E.L. (1998) *Euphytica*, **103** : 61-66.

Lin, T.C., Chuan, S.H. and Hong, S.T. (1991) *Bull. Taichung Dist. Agric. Improv. Sta.*, **32**: 11-15.

Lombardo, Sara, Pandino, Gaetano and Mauromicale, Giovanni (2018) *Sci. Hort.*, **233** : 479–490.

López Anido, F.S., Cravero, V.P., Martín, E.A., Crippa, I. and Cointry, E.L. (2010) *Span. J. Agric. Res.*, 8(3) : 679-685.

Lopez-Molina, D., Navarro-Martinez, M.D., Melgarejo, F.R., Hiner, N.P., Chazarra, S. and RodriguezLopez, J.N. (2005) *Phytochem.*, **66** : 1476-1484.

Magnifico, V. and Lattanzio, V. (1981) *Proc. 3rd Intern. Congr.*, Artichoke Studies, pp. 283-294.

Malach, Y. de, Ben-Asher, J., Sagih, M., Alert, A. and De-Malach, Y. (1996) *Agron. J.*, **88** : 987-990.

Martin, E., Cravero, V., Espósito, A., Anido, F.L., Milanesi, L. and Cointry, E. (2008) *Aust. J. Crop Sci.*, **1**(2) : 43-46.

Maurizio, C., Piero, P., Luca, M., Cristiana, S. and Manuela, G. (2010) *Mediterr. J. Nutr. Metab.*, **3** : 197-201.

Mauro, R.P., Portis, E., Acquadro, A., Lombardo, S., Mauromicale, G. and Lanteri, S. (2009) *Conserv. Genet.*, **10** : 431–440.

Mauromicale, G. and Ierna, A. (2000a) *Agronomie*, **20** : 197-204.

Mauromicale, G. and Ierna, A. (2000b) *Informatore Agrario*, **56** : 39-45.

Mauromicale, G. and Lerna, A. (1995) *Agronomie*, **15** : 527-538.

Mauromicale, G., Ierna, A., Donzella, G. and Assenza, M. (1996) *Sementi Elette*, **42** : 43-50.

Mauromicale, G., Raccuia, S.A. and Damato, G. (2000) *Acta Hortic.*, **533** : 483-488.

Menin, B., Moglia, A., Comino, C., Lanteri, S., Van Herpen, T.W.J.M. and Beekwilder, J. (2012) *Acta Hortic.*, **961** : 34.

Menin, B., Moglia, A., Comino, C., Lanteri, S., Van Herpen, T.W.J.M. and Beekwilder, J. (2012) *Acta Hortic.*, **961** : 267-271.

Miceli, A., Leo, P. de and De-Leo, P. (1996) *Bioresour. Technol.*, **57** : 301-302.

Montemurro, P., Capella, A., Cazzato, E., Fusiello, R. and Guastamacchia, F. (2016) In: *Atti, Giornate Fitopatologiche, Chianciano Terme* (Siena), 8-11 March 2016, pp.511-515.

Morzadec, J.M. and Hourmant, A. (1997) *Sci. Hort.*, **72** : 59-62.

Morzadec, J.M. and Hourmant, A. (1997) *Sci. Hort.*, **72** : 59-62.

Morzadec, J.M., Hourmant, A., Corre, J. Romancer M. le, Cottignies, A., Migliori, A. and Le-Romancer, M. (1998) *J. Phytopath.*, **146** : 79-82.

Mossi, A.J., Echeverrigaray, S. and Acevedo, C. (1999) *Acta Hortic.*, **501** : 111-114.

Motzo, R. and Deidda, M. (1993) *J. Genet. Breed.*, **47** : 263-266.

Moulinear, H. (1980) *Compt. Rend. Sean Acad. Agri.*, France, **66** : 527-531.

Nurgül, E. (2016) *J. Agricul. Sci. Technol.*, **6** : 335-340.

Ordas, R.J., Tavazza, R. and Ancora, G. (1990) *Plant Sci.*, **71** : 233-237.

Ordas, R.J., Tavazza, R. and Ancora, G. (1991) *Plant Sci.*, **77** : 253-259.

Pandino, G., Courts, F.L., Lombardo, S., Mauromicale G. and Williamson, G. (2010) *J. Agri. Food Chem.*, **58** : 1026-1031.

Pandita, P.N., Ogra, R.K. and Archana-Kaul (1998) *Indian J. Agric. Sci.*, **58** : 724-726.

Paradies, F., Sialer, M.M.F., Franco, A. di, Gallitelli, D. and di Franco, A. (2000) *J. Plant Path.*, **82** : 244.

Pomares, F., Tarazona, F., Estela, M., Bartual, R. and Arciniaga, L. (1993) *Agrochimica*, **37** : 111-121.

Porceddu, E., Dellacecca, V. and Bianco, V.V. (1976) In: *Proceedings of the II International Congress on Artichoke* (Ed. Medica, Minerva), Nov. 22–24, Bari, pp. 1105–1119.

Portis, E., Mauromicale, G., Barchi, L., Mauro, R. and Lanteri, S. (2005) *Plant Sci.*, **168**(6) : 1591-1598.

Raccuia, S.A. and Melilli, M.G. (2004) *Aust. J. Agric. Res.*, **55** : 693-698.

Rangarajan, A., Ingall, B.A., Zeppelin, V.C. and Rangaajan, A. (2000) *HortScience*, **10** : 585-588.

Rey, N.A., Tavazza, R., Papacchioli, V. and Pagnotta, M.A. (2013) *Acta Hortic.*, **983** : 225-229.

Riahi, J., Nicoletto, C., Bouzaein, G., Sambo, P. and Khalfallah, K.K. (2017) *Agron.*, **7**(4) : 65.

Rottenberg, A. and Zohary, D. (1996) *Genet. Resour. Crop Evol.*, **43** : 53-58.

Rubatzky, V.E., Snyder, M.J. and Sciearoni, R.H. (1976) *Proc. end Intern. Congr.*, Artichoke Studies, pp. 1005-1025.

Sabi, M.N., Logegaray, V. and Chiesa, A. (2013) *Acta Hortic.*, **989** : 363-368

Saif Eldeen U.M. (2015) *J. Product. Dev.*, **20**(3) : 307-324

Saata, A. (2006) *Acta Agrobot.*, **59** : 463-470.

Saleh S.A., Zaki, M.F., Tantawy, A.S. and. Salama, Y.A.M. (2016) *Int.J. ChemTech. Res.*, **9**(3) : 25-33.

Scaglione, D., Reyes-Chin-Wo, S., Acquadro, A., Froenicke, L., Portis, E., Beitel, C., Tirone, M., Mauro, R., Lo Monaco, A., Mauromicale, G., Faccioli, P., Cattivelli, L., Rieseberg, L., Michelmore, R. and Lanteri, S. (2016). *Sci. Rep.*, **6** : 19427.

Schrader, W.L. (2005) *Acta Hortic.*, **681** : 207-208.

Shaheen, A.M., Fatma, A., Rizk, A.M.E. and El-Shal Z.S.A. (2007) *Res. J. Agric. Bio. Sci.*, **3**(2) : 82-90.

Shanmugavelu, K. G. (1989) In: *Production Technology of Vegetable Crops*, Oxford and IBH Publishing Co. Pvt. Ltd., New Delhi.

Shinohara, T., Agehara, S., Yoo, K.S. and Leskovar, D.I. (2011) *HortScience*, **46**(3) : 377-386.

Soliman, A.G.M., Alkharpotly, A.A., Gabal, A.A.A. and Abido, A.I.A. (2019) *J. Adv. Agric. Res.*, **24**(1) : 1-33.

Sonnante, G., Pignone, D. and Hammer, K. (2007) *Annal. Bot.*, **100** : 1095-1100.

Spano, R., Bottalico, G., Corrado, A., Campanale, A., Franco, A. D. and Mascia, T. (2018) *Agric.*, **8** : 36.

Tanada, Y. and Eeiner, C. (1960) *J. Insect Pathol.*, **2**(3) : 230-246.

Tesi, R. (1994) *Informatore Agrario*, **50** : 49-51.

Thompson, H. C. and Kelly, W. C. (1972) *Vegetable Crops,* Tata McGraw-Hill Publishing Co. Ltd., New Delhi.

Tivang, J., Skroch, P.W., Nienhuis, J., Vos, N. de and De Vos, N. (1996) *J. Amer. Soc. Hort. Sci.*, **121** : 783-788.

Veeraragavathatham, D., Natarajan, S. and Azhakiamanavalan, R.S. (2000) In: *Textbook of Vegetables, Tubercrops and Spices* (Eds. Thamburaj, S. and Singh, N.), ICAR, New Delhi, pp. 429-430.

Vetrano, F., Iapichino, G., Guella, V., Stoffella, P.J. and Catliffe, D.J. (2000) *Acta Hortic.*, **533** : 593-596.

Welbaum, G.E. (1994) *HortTechnol.*, **4** : 147-150.

MORINGA

J. Kabir, T.K. Maity, J.C. Jana,
S. Banerjee and D. Mukherjee

1.0 Introduction

Moringa (*Moringa oleifera* Lam.) or drumstick is a fast growing multipurpose tree extensively grown in tropical and subtropical India (Kumar *et al.*, 2000) and Africa (Kokou *et al.*, 2001). It is also commonly known as horse radish or ben oil tree, and well distributed in Indian sub-continent, south-east Asia, West Indies, Cuba, Egypt and Nigeria. In eastern and southern India *Moringa* is a widely used vegetable and grown commercially. *Moringa* is widely cultivated for its edible tender fruits and also leaves. The plants also serve as animal forage, biogas, fuel, domestic cleaning agent, blue dye, fencing, fertilizer, foliar nutrient, green manure, gum, honey and sugar cane juice-clarifier, medicine, bio-pesticide, pulp, rope, tannin, water purifier, *etc*. Various parts of this plant such as the leaves, roots, seed, bark, fruit, flowers and immature pods are used as medicine against multiple human diseases. There are some studies about leaves, root, seed and pod uses for health treatment, including the use as anticonvulsant, antioxidant (for cardiovascular diseases), antidiabetic, anti-nephrotixicity, anti-gram-positive and gram-negative bacterial, antiulcer, antimutagenic, antiurilhiatic, antitumor, anti-inflamation and anti-hypertensive (Choudhari *et al.*, 2013; Coppin *et al.*, 2013; Silva *et al.*, 2014).

2.0 Composition and Uses

2.1.0 Composition

Moringa is nutritious. It is very rich in vitamins and minerals.

2.1.1 Pods

The nutritive value of pod is given in Table 1.

Table 1: Composition of Pod (per 100 g edible portion)*

Moisture	86.9 g	Iron	5.3 mg
Protein	2.5 g	Copper	3.1 µg/g
Fat	0.1 g	Iodine	18 µg/kg
Carbohydrates	3.7 g	Oxalic acid	0.01 g
Fibre	4.8 g	Carotene (as vitamin A)	184 I.U.
Mineral matter	2.0 g	Nicotinic acid	0.2 mg
Calcium	30 mg	Ascorbic acid	120 mg
Phosphorus	110 mg		

* Anon. (1962).

2.1.2 Leaves

The nutritive analysis of leaves is given in Table 2.

Table 2: Composition of Leaves (per 100 g)*

Moisture	75 g	Iron	7 mg
Protein	6.7 g	Copper	1.1 µg/g
Fat (Ether extract)	1.7 g	Iodine	51 µg/kg
Carbohydrates	13.4 g	Carotene (as vitamin A)	11,300 I.U.
Fibre	0.9 g	Vitamin B1	210 µg
Mineral matter	2.3 g	Nicotinic acid	0.8 mg
Calcium	440 mg	Ascorbic acid	220 mg
Phosphorus	70 mg	Tocopherol	7.4 mg

* Anon. (1962).

Six new and three synthetically known glycosides were isolated from the ethylacohol extract of the leaves of *M. oleifera* (collected from Karachi, Pakistan), following bioassay-directed isolation. Most of these compounds, bearing thiocarbamate, carbamate or nitrile groups, were fully acetylated glycosides, which are very rare in nature. Their structures were elucidated from spectral and chemical analyses. The 4 thiocarbamates showed hypotensive activity in anaesthetized normotensive rate (Faizi *et al.*, 1995).

2.1.3 Flower and Seeds

A significant amount of thiamin, riboflavin, nicotinic acid, folic acid, pyridoxine, ascorbic acid, beta-carotene and alpha-tocopherol were detected in seeds and flowers. Flowers were particularly rich in riboflavin, ascorbic acid, nicotinic acid and folic acid. Seeds contained significantly more thiamin, beta-carotene and alpha-tocopherol (Dahot, 1988). The aqueous extract of the mature flowers of *M. oleifera* collected in West Bengal contained the free neutral sugars D-mannose and D-glucose in the ratio 1:5 and 2 unidentified carbohydrate bearing materials along with proteins and ascorbic acid, but no polysaccharide. In contrast, the aqueous extract of pre-mature flowers was composed of the above materials (with varying proportions) and a polysaccharide which on hydrolysis gave D-glucose, D-galactose and D-glucuronic acid in a molar ratio of 1.0 : 1.9 : 0.9 (Pramanik and Islam, 1998).

Seeds of *M. oleifera* contain 18.9 per cent oil (Ahmad *et al.*, 1989). The oil from seeds of a cv. Mbololo (from Kenya) was extracted by Tsaknis *et al.* (1999) using a cold press (CP) method, extraction with n-hexane (H), and extraction with a mixture of chloroform/methanol (50 : 50) (CM). The oil concentration was in the range 25.8 per cent (CP) to 31.2 per cent (CM). The oil contained high levels of unsaturated fatty acids, especially oleic acid (up to 75.39 per cent). The dominant saturated acids were behenic (up to 6.73 per cent) and palmitic acids (up to 6.04 per cent). The oil contained high levels of beta-sitosterol (up to 50.07 per cent), stigmasterol (up to 17.27 per cent) and campesterol (up to 15.13 per cent). Alpha-, gamma- and delta-tocopherols were detected up to levels of 105, 39.54 and 77.60 mg/kg of oil, respectively. The induction period (at 120°C) of *M. oleifera* seeds oil was reduced from 44.6 to 64.3 per cent after degumming. The *M. oleifera* seed oil showed high stability to oxidative rancidity. Oliveria *et al.* (1999) assessed the nutritional potential of raw seeds by analysing the protein, oil and amino acid contents and the presence of lectin trypsin inhibitors and urease. Mature seeds contained crude proteins 332.5 g/ka DM, crude fat 412.0 g/kg DM, carbohydrate 211.2 g/kg DM and ash 44.3 g/kg DM. The essential amino acid profile compared with the FAO/WHO/UNU scoring pattern requirements for different age groups showed deficiency in lysine, threonine and valine. The content of methionine + cysteine (43.6 g/kg protein), however, was exceptionally higher and close to that of human milk, chicken egg and cow's milk. Okuda *et al.* (2001) isolated a coagulant from the NaCl crude extract of *Moringa oleifera* seeds (collected from Philippines).

2.2.0 Uses

Almost every part of a *Moringa* tree is useful to human being. *Moringa* is chiefly valued for its tender pods, which are cut into pieces and use in culinary preparations. The tender leaves and flowers are eaten as pot herb. The leaves are also used for seasoning pickles and for flavouring ghee. Seeds are consumed after frying. *Moringa oleifera* contains essential amino acids, carotenoids in leaves, and components with nutraceutical properties, supporting the idea of using this plant as a nutritional supplement or constituent in food preparation (Abdull Razis *et al.*, 2014). All the parts of the tree are considered to have medicinal properties and used in the treatment of ascites, rheumatism and as cardiac circulatory stimulants. The root of the young tree and also root bark are rubefacient and vesicant. The leaves are rich in vitamin A and vitamin C and are considered useful in scurvy, catarrhal infections and also used as emetic. Pressed juice of the leaves of the plant shows strong antibacterial activity. The oils from the seeds are known as 'Ben' or 'Behen' oil. It is used for edible purposes, illumination, in cosmetics and also as lubricant for the fine machinery. *Moringa* contains a group of unique biochemical compounds containing sugar and rhamnose that are uncommon sugar-modified glucosinolates (Fahey *et al.*, 2001; Fahey, 2005; Amaglo *et al.*, 2010). These compounds were reported to demonstrate some chemopreventive activities by inducing apoptosis (Brunelli *et al.*, 2010). Rao *et al.* (2001) reported *in vivo* radioprotective effect of *Moringa* leaves. They demonstrated that pretreatment with the methanolic leaf extract of *Moringa* confers significant radiation protection to the bone marrow chromosomes in mice and this may lead to the higher 30 days survival after lethal whole body irradiation. Abilgos and Barba (1999) reported of the possibility of incorporating the leaves as source of micronutrients for supplementing rice flat noodle. Utilization of *Moringa* as fresh forage for cattle has been reported in Nicaragua (Mendez, 1999). It is planted at high densities as hedges or in plots planted from stakes. The plant materials are harvested at 35- to 45-day intervals for use as forage. Extracts from twigs and leaves acts as *in vitro* antimalarial activity (Gbeassor *et al.*, 1990). Ghasi *et al.* (2000) demonstrated that *M. oleifera* leaves have definite hypocholesterolaemic acitivity.

Kokou *et al.* (2001) reported that the residual presscakes from oil extraction of seed can be used in natural water purification processes. The uses of *M. oleifera* seed for the sedimentation and decontamination of household water has been demonstrated by Nyein *et al.* (1997). They also highlighted the acceptance of the use of seeds for the sedimentation of turbid water in Myanmar. Seeds of *Moringa*

Uses of Moringa

Moringa Leaf, Plower, Pods and Powder Used in Culinary Preparation

Different Recipes of Moringa in Indian Cuisine

species are used as naturally occurring flocculants for water treatment (Sutherland *et al.*, 1998) and for softening hard water (Muyibi and Evison, 1995).

Moringa is a multipurpose plant and an alternative for medicinal purposes throughout the world. It is used as potential antioxidant, anti-inflamatory, anticancer, antiulcer, antihyperglycemic, antidiabetic and antimicrobial agent (Bhishagratna, 1991; Verma *et al.*, 2009; Hamza, 2010; Arora *et al.*, 2013; Abdul Razis *et al.*, 2014).

In Guatemala, *M. oleifera* is used to treat many disorders, in particular infectious diseases of the skin, digestive system and respiratory tracts. The antimicrobial activities of the leaves, roots, bark and seeds were investigated *in vivo* against bacteria yeast, dermatophytes and helminths pathogenic to man. The fresh leaf juice and aqueous extracts from the seeds inhibited the growth of *Pseudomonas aeruginosa* and *Staphylococcus aureus* and extraction temperatures above 56°C inhibited this activity (Caceres *et al.*, 1991a). Pan *et al.* (1997) reported that the leaf, seed and root powder had some antifungal properties.

Hematological and hepatorenal functions of methanolic extract of *M. oleifera* root has been reported (Mazumder *et al.*, 1999). Anti-inflammatory and antihepatotoxic activities of the roots have been reported by Rao and Mishra (1998).

Caceres *et al.* (1991b) suggested that *M. oleifera* could be used in any reforestation programme in Guatemala because of its use for water purfication and medicinal purposes, and its relatively fast growth, although its fuel wood potential is limited. Dasmana *et al.* (1990) reported pulp yields and strength properties for paper making were satisfactory and comparable to those of conventional raw materials.

Moringa oleifera seeds are a promising resource for food and non-food applications, due to their content of monounsaturated fatty acids with a high monounsaturated/saturated fatty acids (MUFA/SFA) ratio, sterols and tocopherols, as well as proteins rich in sulfated amino acids (Leone *et al.*, 2016).

Moringa is used as potential antioxidant, anticancer, anti-inflammatory, antidiabetic and antimicrobial agent. *M. oleifera* seed, a natural coagulant is extensively used in water treatment. It has also use as a cure for diabetes and cancer (Gopalakrishnan *et al.*, 2016).

3.0 Origin, Taxonomy and Floral Biology

Moringa is found growing wild in the sub-Himalayan tract from the Chenab eastwards to Sarda and in the tarai tract of Uttar Pradesh in India. It is an indigenous tree, native to north-western India and African tropics. Muluvi *et al.* (1999) however, opined that the trees were perhaps introduced to Africa from India at the turn of this century.

Moringa belongs to familly Moringaceae and genus *Moringa*. About 33 species have been reported in the family Moringaceae (Arora *et al.*, 2013). There are about 13 known species in this genus (Khawaja *et al.*, 2010), of which *Moringa oleifera* Lam. (syn. *M. pterygosperma* Gaertn.) is the most valued vegetable crop. Olson and Carlquist (2001) suggested that the most likely sister taxon to *Moringa* is *Cylicomorpha*.

It is a deciduous tree of 8-10 m in height. It is a small or medium-sized tree with corky bark and woody and brittle stem. The roots are long and pungent. The leaves are small, tripinnate and crowded at the end of slender branches. Yellowish white fragrant flowers are produced in the large panicles, pods pendulous, greenish, 30 to 60 cm length with several three winged seeds.

3.1.0 Floral Biology

Moringa oleifera showed diurnal anthesis. Pollen-ovule ratio per flower was found to be variable. Netting and bagging of flowers indicated that the external agents might be required for successful fruit setting which varied depending upon the availability of effective pollinators (Haldar, 2000). At Visakhapatnam, India, *M. oleifera* flowered twice a year, in February to May and in September to November. Xenogamous pollinations gave 100 per cent fruit set, 81 per cent seed set and 9 per cent fecundity, compared with 62, 64 and 6 per cent, respectively for geitonogamous pollinations. Flowers were zygomorphic and of the gullet type; they opened during 03.00 to 19.00 h and were visited by diurnally active insects during 06.00 to 15.00 h. Carpenter bees (*Xylocarpa latipes* and *X. pubescens*) (Jyothi *et al.*, 1990) and honey bees (*Aphis mellifera*) (Arumugam *et al.*, 2017) were the most reliable pollinators.

Floral biology and pollen characteristics of *M. oleifera* genotype AD-4 were studied by Babu and Rajan (1996). Plants were found to flower in August-September and again in December-January, the latter resulting in fruit set. Flowers were protandrous, with flower opening, anther dehiscence and stigma receptivity occurred in a phased manner and flower opened between 14.30 and 09.00 h. In August-September, pollen grains did not disperse due to their very sticky nature which shrivelled and become unviable by the time the stigma became receptive. In early December, pollen stickiness was reduced and they became more powdery. At the same time, activity of black ants and flea beetle peaked, thereby increasing the likelihood of cross pollination and fruit set was achieved.

4.0 Cultivars

Moringa cultivars are broadly classified into two groups: perennial and annual. Perennial types are typically propagated from cuttings. These types have several characteristics that have constrained their use for in commercial production, and favoured development and cultivation of annual varieties: long growing time before reaching maturity for production of pods, limited availability of suitable planting materials (stem cuttings), less resistance to pests and diseases and greater water requirements. The Horticultural College and Research Institute, Periyakulam of Tamil Nadu Agricultural University released Annual *Moringa* Periyakulam-1 (**PKM-1**) variety in 1989 from Pure line selection and Annual *Moringa* **PKM-2** in 2000 a hybrid derivative. PKM-1 and PKM-2 have now replaced most the perennial varieties that previously dominated commercial production in India. Improved **MPKM1** is the best variety for *Moringa* cultivation for leaves and **MOMAX3** the highest seed yielding variety for *Moringa* plantation for seeds/oil. These are seed propagated, offer rapid pod maturation, higher yields and greater adaptability to varied agro-climatic conditions.

Cultivars differ in branching habit and pigmentation of plant, floral parts and fruits. There are a few named cultivars of *Moringa*. A type named Jaffna is grown in parts of South India. Commercial cultivars are **KM1, PKM1, PKM2, GKVK 1, GKVK 2, GKVK 3, Dhanaraj, Bhagya (KDM 01), Konkan Ruchira, Anupama** and **Rohit**. Other cultivars which have become popular are known as **Chavakacheri Murungai, Chem Murungai (red-tipped fruits), Kattu Murungai, Pal Murungai, Moolanoor Murangai, Palamendu Murangai and Kodikal Murungai**.

Recently, a short duration *Moringa* is cultivated in some parts of Tamil Nadu. The tree becomes exhausted and loses its vigour after one year and such cultivars are often treated as annual *Moringa*.

Thamburaj (2001) described some of the local genotypes of South India which are given below :

Jaffna

It is a popular perennial type in South India and said to be introduced from Jaffna. It is suited for coastal tracts of South India. It bears long pods (60-90 cm) of soft flesh of good taste. This type can yield 400 pods per tree per year from second year which increases to 600 pods from third year.

Moolanoor Murangai

It is commercially cultivated in some parts of South India. It produces medium size pod and yields 500-600 fruits per tree per year.

Chavakacheri Murungai, Chem Murungai and Pal Murungai

All these types are perennial and ecotype of Jaffna *Moringa* and produce about 400 to 600 pods per tree per year. Chem Murangai produces flowers and red tipped fruits throughout the year.

Kodikal Murungai

This is cultivated in some districts of Tamil Nadu in India. It is predominantly cultivated in betel vine garden and is useful for training betel vine and provides shade. The trees are shorter in height and propagated by seeds.

KM 1

It is a selection from an annual type propagated by seeds. Plants are of short stature which facilitates easy harvesting of fruits. It comes to bearing in 6 months and the plants can be ratooned for 2-3 years by cutting the trunk at a height of one metre.

PKM 1

It is an annual type evolved at Tamil Nadu Agricultural University. Each tree bears on an average of 200-250 pods/year. Suitable for leaf production in high density cultivation.

PKM 2

It is a recently developed hybrid derived from the cross between line MP 31 and MP 28 at Tamil Nadu Agricultural University, India. The plants are quick growing

Popular Cultivars of *Moringa*

PKM-1   PKM-2

Konkan Ruchira

Bhagya (KDM-1)

reaching a height of 4.8 m in six months. Plants propagated through seeds start flowering in 100-110 days and pods can be harvested in 170-180 days. Each tree yields 220 bigger sized pods weighing 62 kg. As on average each pod weight 280 grams with 125 cm length and 8 cm girth having more flesh (70 per cent) content. Yield per hectare is 98 tonnes. Ratoon crop can also be taken up for three years. It is suitable for growing in most soil types varying from sandy loam to clay loam with good drainage.

Anupama

A high yielding cultivar released from Kerala Agricultural University.

Bhagya (KDM-1)

It is developed by the University of Horticultural Sciences, Bagalkot (UHS Bagalkot 2017). It is resistant to pest and well adapted to growing in the hot, dry region of Chikkodi Taluk. The variety can be grown up to 15-20 years.

Konkan Ruchira

This cultivar was released in 1992 from Dr. Balasaheb Sawant Konkan Krishi Vidyapeeth, Dapoli. Medium size (5.6 m height) tree with dark green foliage, dark green colour, medium long sticks with best quality yield up to 30-35 t/ha.

At Periyakulam, India about 20 seed types were evaluated for 7 yield-related characters by Suthathirapadian *et al.* (1992). Significant difference in yield among the genotypes was observed in the first, second and ratoon crops. Pod yield/tree ranged from 18.2 to 76.3 in the first crop, from 36.4 to 96.8 kg in the second crop and from 92.6 to 284.3 kg in the ratoon crop. The coefficients of variation indicated relatively uniform variation in the productivity of plants in each crop. Pod length ranged from 42.8 to 63.9 cm, pod weight from 15.4 to 72.8 g and number of seeds per pod from 12.8 to 21.3. There were no significant differences for pod girth. MT16, MT11 and MT19 were ranked first, second and third, respectively, based on their yield performance.

In West Bengal, India, two distinct genotypes known as Sajna and Najna are cultivated. Sajna flowers only once in a year (January-February), while Najna flowers 2 or 3 times in a year.

5.0 Soil and Climate

5.1.0 Soil

It grows well in all types of soils with wide range of pH (5.0 to 9.0) (Palada and Changl, 2003) except stiff clays, but performs best in sandy loam soil. A deep loamy, sandy or sandy-loam soil is best for its cultivation.

Salinity affects the growth of *Moringa*. Significant reductions in chlorophyll a and b and the number of stomata were obtained at soil electrical conductivity above 4 dS/m. The rate of transpiration increased significantly with salinity, even at 4 dS/m (Valia *et al.*, 1993a). They (1993b) also reported that significant decrease in plant height, number of leaves, stem diameter, plant spread, total leaf area and root length occurred at exchangeable sodium percentage (ESP) > 41.

5.2.0 Climate

It thrives best under the moist tropical climate. It also grows well under subtropical climate. It can tolerate a wide range of annual rainfall (250 mm to 3000 mm). It is hardy crop and can withstand drought to a great extent. The optimum temperature for growth is 25-35°C and is highly susceptible to frost and high temperature exceeding 40°C causes flower shedding (Thamburaj, 2001). Though, it can withstand summer temperatures of up to 48°C for a limited period (Price, 2000). Seed germination of *M. oleifera* was studied under greenhouse conditions where the higher temperature 20/30°C was found better than 10/20°C (Muhl *et al.*, 2011; Batool *et al.*, 2016). The high temperature increased germination speed but decreased the germination percentage. The low temperature delayed the germination time (Hassanein and Al-Soqeer, 2017).

Tesfay *et al.* (2016) studied the effect of temperature in *Moringa* seed phytochemical compounds and carbohydrate mobilization. They found, 30/20 °C accelerated seed radical emergence with germination occurring within 48 h. Subsequently, germination was observed between 48 h and 72 h at 25/15 °C and after 72 h at 20/10 °C. Similarly, temperature especially 30/20°C also significantly influenced the biosynthesis and accumulation of biochemical compounds in the seeds.

The tree sheds its leaves in December-January and new leaves appear in February-March in North Indian condition, followed by flowering and fruiting in early summer. Sometimes, particularly in South India, flowers and fruits appear twice a year (September and February).

6.0 Cultivation

6.1.0 Propagation

Moringa tree is propagated by seeds and by limb-cuttings.

6.1.1 Seed Propagation

Annual types are propagated by seeds. For planting one hectare of land 625 g of seeds is required (Thamburaj, 2001).

In Togo, the best germination rate was obtained for seeds buried at a 2 cm depth (Kokou *et al.*, 2001). Seeds from Madagascar had the highest germination rate (100 per cent), while those from Togo had lowest (80 per cent), however, seedlings from Togo had far better growth (Kokou *et al.*, 2001).

Influence of ageing on the seed quality of annual *Moringa* was studied by Vijayakumar *et al.* (1999) and one-year-old seeds of *Moringa* stored under ambient conditions were sown at 15-day intervals (365, 380, 395, 410, 425 and 440 days). Percentage germination, germination rate, plant height and number of branches 60 days after sowing decreased as seed age increased. The decrease in percentage of germination was initially slow (96.25 per cent germination in 365 days and 78.15 per cent in 390 days) but accelerated thereafter (51.00 per cent in 440 days). Kumar *et al.* (2000) reported that treatment of seeds with KNO_3 promoted seedlings growth. He also found that plants from black seeds and seeds with coats grew taller than those from white seeds and seeds without coats, while branch production was better in white seeded plants, the presence/absence of the seed coat did not affect branch production. Caceres *et al.* (1991b) reported from Guatemala that germination after storage decreased from 94 per cent during the first month to 78 per cent after 12 months. The best germination was obtained by immersion of seeds in water for 24 h at room temperature. At higher altitude seedling growth and survival were less than lower altitude. In an investigation on seed germination, Sivasubramanium and Thiagarajan (1997) recorded that black seeds of *Moringa* showed 6.7, 51.5 and 24.9 higher average germination percentage than brown, white and ungraded seeds, respectively. Seed vigour decreased gradually during storage, but vigour of black seeds was higher than that of brown, white and ungraded seeds at each stage of storage. Germination percentage was higher when stored in polyethylene bags than when stored in cloth bags, regardless of seed colour.

Hydropriming (8 h) was more effective in improving emergence, shoot vigor, and chlorophyll-b contents, while priming with *Moringa* leaf extract priming (8 h) produced vigorous roots and increased the chlorophyll a and mineral contents of *Moringa* leaves (Nouman *et al.*, 2012).

6.1.2 Limb-cutting

Limb-cutting is generally followed for propagation of perennial types. In this method, cuttings of 1 to 1.5 m long, measuring 14-16 cm across is planted *in situ*, preferably in the months of June to August. The cuttings are planted in pits at a spacing of 3 to 5 m each way. The cuttings produce root in moist soil in short period and grow to sizeable trees within a few months; dehydration of cutting in arid condition adversely affect rooting and survival. Kokou *et al.* (2001) reported that cuttings longer than 1.5 m had the best survival rate and those lignified were more resistant to dehydration than chlorophyllous shoots. It can also be propagated by shield budding and air-layering.

6.1.3 Planting

The limb cuttings are planted in $60 \times 60 \times 60$ cm sized well prepared pits at spacing of 8.5-3 metres for perennial types and annual types are planted $45 \times 45 \times 45$ cm pits are with 2.0×2.5 m or 3.25 m spacing. The pits are filled with a mixture of top soil and 15 kg FYM/compost. Seeds can be either sown in situ in the prepared pits or can be transplanted after raising the seedlings in PE bags of 15×7 cm size. One month old seedlings are ready for planting.

6.2.0 Manuring and Fertilization

Manuring is not always practised. This tree is capable of growing with very little attention of manuring. About 45 : 15 : 30 g NPK/plant along with 25 kg FYM or compost are applied in soil around the base of the plant within a week after cutting back every year to get quality pods. Fertilization with about 125 : 75 : 75 kg of NPK ha^{-1} is practised for leaf production. In Kerala, India ring trenches are dug around the trees and filled with green leaves, farmyard manure and ash during the rainy season. It has been found that the application of 75 kg of farmyard manure and 0.037

Seedling Raising and Field Growing of *Moringa*

Flowering and Pod Bearing of *Moringa*

kg of ammonium sulphate per tree gives three times more yield (Sundararaj *et al.*, 1968). Thamburaj (2001) advocated 100 g each of urea and superphosphate and 50 g of muriate of potash three months after planting. Another 100 g of urea alone has to be applied three months after the first application. Organic fertilization enhanced plant resistance to water stress compared to chemical fertilization (Mahmuod *et al.*, 2018).

Dania *et al.* (2014) conducted a trial to study the effects of NPK, poultry manure, and organo-mineral fertilizer on the growth and nutrient concentration of *Moringa oleifera* leaves. They found that application of poultry manure, NPK, and organo-mineral fertilizer resulted 66 per cent, 62 per cent, and 39 per cent higher in leaf number than the control at eight weeks after planting. The application of poultry manure significantly increased the nutrient content of *Moringa* leaves compared to other sources of fertilizer applied. They concluded that the application of poultry manure significantly improved the growth and nutrient content of *Moringa*. Adebayo *et al.* (2017) found growth values for NPK and compost treated plots were higher than the control in *Moringa*. Poultry manure applied at 30 tons ha^{-1} resulted in highest growth values for plant height, stem girth, stem weight and number of leaves, and leaf biomass.

6.3.0 Irrigation, Intercultural Operation and Pruning

Watering should be done until the plants are well established. Growing plants do not require watering except during the hot weather. Water stagnation should be avoided. The drip irrigation method recorded the highest yield and its components compared to that of flood irrigation method (Mahmuod *et al.*, 2018).

The basin of the tree should be kept weed-free and light hoeing of soil around the tree is useful for proper growth and development. Pinching off the seedling at about 75 cm height or 60 days after sowing is practised to facilitate more branching. *Moringa* responds favourably to annual pruning of branches. It not only reduces the plant height which facilitates harvesting but also promotes axillary branches and increases yield. The management of tree also becomes easier. The trees are pruned after harvesting of fruit to a height of 1-1.5 m from the ground level to regulate flowering and obtain higher yield in commercial ratoon crops (Arumugam *et al.*, 2017).

S.du Toit *et al.* (2020) carried out an experiment to establish an ideal pruning intensity level. They found that light pruning (3 m from the ground) resulted in the highest quantity of flower buds and of flowers. In addition, moderate pruning (2 m from the ground) increased flower number on trees. Highest quantity of fruits was recorded from the moderate pruning treatment. Moderate pruning significantly increased fresh and dry biomass, leaf area index over time and stem circumference of trees.

6.4.0 Intercropping

In high density multi-species cropping system in old coconut garden in Tamil Nadu annual *Moringa* has been profitably cultivated along with other crops (Marimuthu *et al.*, 2001). In the temperate hills of Sikkim, apple orchards are generally poor, so the major returns to farmers are from the intercrops like *Moringa* scattered on farmlands in no definite crop combinations (Singh *et al.*, 1991). Bhindi or cowpea can be cultivated as intercrops in young *Moringa* plantation. Moringa intercropping with seasonal crops is recommended to be done only during the first cropping season, due to severe competition thereafter (Edward *et al.*, 2012). Essien *et al.* (2016) found the treatments moringa + maize and moringa + sweet potato crop combinations produced the highest crop growth and yield of maize, sweet potato and *Moringa oleifera* while sole crops produced the lowest.

7.0 Harvesting, Yield and Packaging

7.1.0 Harvesting and Yield

The plant from cutting comes to bearing in 5-6 months after planting. However, after two years of growth, it gives a yield of 500 to 600 fruits per tree. A good tree is capable of yielding 1000 fruits.

Annual type bears fruits 6 months after sowing, while perennial type bears fruits after 8-9 months of sowing. The fruit are ready for harvest after 60 days of flowering. Initially annual type bears 200-250 fruits/tree, while perennial type bears 80-90 fruits/tree. The yield gradually increases up to 500-600 fruits/tree/ year during fourth and fifth year. In the month of March-June fruits are generally harvested for a period of 2-3 months (Thamburaj, 2001).

Different harvesting time for tender leaves significantly affected the number of branches, shoot dry weight, chlorophyll a and b contents, carotenoid contents, total phenolic contents, K contents and total soluble protein except shoot fresh weight. Maximum growth, nutrient and biochemical attributes of leaves were observed in the month of August.

7.2.0 Packaging

Packaging methods to prolong the shelf life of *Moringa* cv. PKM-1 during transit were studied by Damodaran *et al.* (1999). The highest physiological loss in weight (PLW) of 77.94 per cent was recorded in fruits packed in wooden boxes with dried grass as filling material after 12 days. The lowest PLW (24.24 per cent) was recorded in fruit packed in polyethylene bags, followed by corrugated fibre board boxes (CFB) with coir waste as filling material (27.20 per cent). The highest carotene (182.02 mg/100 g) and ascorbic acid (128.17 mg/100 g) contents were found in fruits packed in polyethylene bags.

Controlled atmosphere storage at refrigerated conditions revealed that the shelf life of *Moringa* pods could be increased to approximately three to four times compare to ambient condition. The best treatment for increasing the shelf life of *Moringa* pods up to 40 days at 14°C was 4 per cent O_2 and 5 per cent CO_2 (Selvi and Varadharaju, 2016).

Suganthi *et al.* (2019) found that quality deterioration was found to be lower in aluminum foil followed by High Density Polyethylene (HDPE) of 400 gauge thickness without vent kept at refrigerated conditions respectively. Aluminum foil wrapped treatment was found to be best in retaining the colour, less reduction

Harvesting of Pods, Seeds and Marketing of *Moringa*

in vitamin C content, protein content and higher moisture content with the higher shelf life of twelve days.

8.0 Diseases and Pests

8.1.0 Diseases

Moringa is less infected by diseases. Root rot caused by *Diplodia* sp., twig canker caused by *Fusarium pallidoroseum*, and fruit rot caused by *Cochliobolus hawaiiensis* are some of the reported diseases (Carbungco *et al.*, 2017; Patricio and Palada, 2017).

Moringa tree is not affected by serious diseases. A rot of the edible pods of *M. oleifera* in Maharashtra was caused by *Drechslera* (*Cochliobolus*) *hawaiiensis*, a previously unreported host. Pathogenicity was confirmed experimentally (Kshirsagar and D'souza, 1989). *Fusarium pallidoroseum* is reported for the first time as the causal agent of twig canker of *M. oleifera*, following its isolation and identification at Wakawali, Maharashtra, India (Mandokhot *et al.*, 1994). Leaf extracts of *M. oleifera* could inhibit germination of fungal spores of *Alternaria alternata* to 35.7 per cent at 4000 ppm (Rashmi *et al.*, 1998). *Fusarium oxysporum* f. sp. *moringae* sp. nov. is described as the causal agent for a new wilt disease of wild *Moringa* (*M. concanensis*) seedlings in nursery (Pande *et al.*, 1998).

8.2.0 Pests

Different caterpillars such as bark eating caterpillar, hairy caterpillar, green leaf caterpillar, budworms, leaf miners, aphids, stem borer, mites and termites damage the crop from time to time. To manage these pests the affected plants may be treated with contact and systematic insecticides. But, the use of highly toxic agrochemicals is currently not acceptable and recommended. Hence the quality of *Moringa* products should be thoroughly monitored before reaching the consumer to prevent health hazards. Two caterpillar pests, hairy caterpillar and leaf (*Eupterote mollifera*) caterpillar (*Noorda blitealis*) cause serious damage. They infest the unopen flower buds which ultimately drop. Hairy caterpillars are found in a swarm on the tree trunk and branches. They also feed on the tender foliage and shoots. Hairy caterpillars can be controlled with a burning torch. A spray of fish oil, resin soap, methyl parathion, chlorpyrifos or quinalphos can also control the pest (David and Ramamurthy, 2016). A type of stem borer is also known to affect the tree. To manage these pests different measures like pesticide sprays biological control, botanical extracts and use of resistant varieties are prescribed (Anjulo, 2009; Patel *et al.*, 2010; Satti *et al.*, 2013; Litsinger, 2014; Kumari *et al.*, 2015; David and Ramamurthy, 2016).

Budworm (*Noorda moringae* Tams.) larvae bore into the flower buds and can damage by shedding of flower buds up to 78 per cent. Spraying of insecticides is suggested to manage its infestation, but extreme caution is needed especially for highly toxic products (David and Ramamurthy, 2016). Drumstick pod-fly (*Gitona distigma* Meigen) is another serious pest and can be managed with poison baits and sprays of pesticides (Math *et al.*, 2014). Bark borer (*Indarbela tetraonis* Moore) larva feeds on the bark, usually at night under a shelter of webs. David and Ramamurthy (2016) suggested injecting chlorpyrifos or profenofos emulsion into the bored holes and then sealing the holes with wet mud.

Disease and Pests of *Moringa*

Stem (Left) and Root Rot Infection

Hairy Caterpillar Infestation

Leaf Caterpillar (*Noorda bitealis*)

Pod Fly (*Gitona distigma*)   **Bud Worm (*Noorda moringae*)**

The population dynamics of *Gitona* sp. on annual *Moringa* were investigated by Murthy and Regupathy (1992) in Coimbatore, India. The annual form of this popular vegetable had recently been introduced to cultivation in southern India and *Gitona* sp., previously regarded as a minor pest, had become more important. *Gitona* sp. damaged fruits were recognised in the initial stages of infestation by the presence of gummy exudates, in association with eggs laid in the grooves between the ridges of the fruits, and by drying of the fruits in later stages. Gummy exudates were also observed following feeding by *Oxycetonia versicolor* and *Anatona stillata*. *Gitona* sp. were most numerous in August-September, when 48-49 per cent of fruits were damaged. Incidence decreased to 13-20 per cent in November-December. A slight increase in January-February, when 23.4 per cent of fruits were damaged, was followed by another decrease in March-June. Incidence of *Gitona* sp. was negatively correlated with maximum temperature and hours of sunshine, and positively correlated with relative humidity and sunshine of the previous months. In field trials with 6 insecticides in Tamil Nadu, India, dichlorvos at 0.04 per cent and fenthion at 0.05 per cent provided the best control of the drosophilid, *Gitona* sp. the pyralid, *Noorda blitealis* and *Aphis craccivora* in *M. oleifera* (Anjaneyamurthy and Regupathy, 1989). Ragumoorthi and Arumugam (1992) obtained best results against the 2 pests *Gitona* sp. and *Noorda blitealis* with 0.04 per cent dichlorvos and fenthion and 1 per cent neem cake extract and neem oil. Kulkarni *et al.* (1996) gave a report of *Ascotis selenaria imparata* [*A. imparata*] as a pest of *Moringa pterygosperma* [*M. oleifera*] from the nursery of the Tropical Forest Research Institute at Jabalpur, Madhya Pradesh.

Logiswaran (1993) found that fenthion (0.4 per cent) in particular significantly reduced incidence of *Gitona* sp. in *Moringa* fruits and resulted in greater yields compared with untreated controls. A field trial in Tamil Nadu, Ragumoorthy *et al.* (1998) indicated that *Gitona distigma* can be managed by adopting IPM measures. This includes application of fenthion 80 EC (0.04 per cent) during the vegetative and flowering stages, application of 150 ppm (0.03 per cent) nimbecidine during 50 per cent fruit set and 35 days later, soil application of neem (*Azadirachta indica*) seed kernel extract 2 l/tree at 50 per cent fruit set and weekly removal of affected fruits.

Nematicidal properties of water extracts obtained from plant leaves against 2nd stage juveniles of *Meloidogyne incognita* were tested by Sosamma *et al.* (1998) and 100 per cent mortality was observed at 1 : 5 dilution after 24 hours.

9.0 Crop Improvement

Important breeding objectives of *Moringa* are to develop dwarf statured plants, varieties suitable for leaf and tender pod production, low seed, low fibre and high flesh content, uniform pod length and green colour, soft even late in harvest and long shelf life in pod and resistant to pest and diseases.

As *Moringa* is a cross-pollinated tree, high heterogeneity in form and yield is expected. Several works reported variability in flowering time (Raja *et al.*, 2013) (from annual type to perennial type), tree nature (from deciduous to evergreen), tree shape (from semi spread to upright), resistance to hairy caterpillar (Mgendi *et al.*, 2011; Raja *et al.*, 2013), flowering time (*i.e.*, some tree flowering throughout the year while others flower in two distinct season) (Ramachandran *et al.*, 1980).

Of 9 traits studied by Suthanthirapandian *et al.* (1989) in a population of 184 seedlings derived by open pollination in *Moringa*, number of flowers/inflorescence, fruit weight and number of fruits/plant showed wide variation, providing ample scope for improvement by selection in the population. Twenty *Moringa* genotypes were grown and data recorded for pod length, pod weight, pods/tree and yield/tree in the first, second and ratoon crops (Pandian *et al.*, 1992). Variability was high for pod weight and yield in all 3 crops and genotypic and phenotypic variances and coefficients of variation were highest for yield in the ratoon crop. Highest heritability was recorded for yield in the ratoon crop (76.5 per cent). The results indicated that selection for pod weight and yield will lead to improved yielding genotypes (Pandian *et al.*, 1992).

Seventy-seven accessions of drumstick germplasm were collected from Tamil Nadu (64), Karnataka (10) and Pondicherry (3), and established in the experimental farm of IIHR, India during 2006 by Varalakshmi and Devaraju (2007). Wide variability has been observed for leaf characters. Petiole color ranged from greenish pink (52) to dark pink (21) and green (4). Leaflet size varied from big (6) to medium (46) and small (25). Out of 77 accessions, 41 flowered. Thirty accessions had white flowers, 8 had pink flowers, and 3 had dark pink flowers. With respect to flower size, 28 had big flowers and 13 medium flowers. Twelve accessions set fruits; fruit color ranged from pink (2) to green (3) and pinkish green (7). All these 12 accessions showed wide range of variability in fruit weight, fruit length and fruit girth. IIHR-26 and IIHR-35 had high fruit weight, fruit length and girth. Likewise, IIHR-39 had high fruit weight and length and IIHR-15 and IIHR-4 had high fruit weight and fruit girth. Promising drumstick germplasm identified were IIHR-35, IIHR-39 and IIHR-26 with regard to their fruit characters. These accessions could either be directly used for commercial cultivation or utilized in the breeding programmes.

Raja *et al.* (2013) identified PKM-2 and MO-1 as stable cultivars among 14 diverse annual genotypes for number of fruits per plant and yield, and PKM-1 for unfavourable environment for commercial cultivation for semi arid region of India. The present work also confirmed that regression coefficient, being a stability parameter, is more important than deviation from regression value in heterozygous genotype like drumstick.

Many ecotypes are present in India: Jaffna (soft and taste fruits), Chavakacheri murungai (similar to Jaffna), Chemmurungai (red tipped fruits), Kadumurungai (small and inferior fruits) Palmurungai (bitter taste), Punamurungai (similar to Palmurungai), Kodikalmurungai (short fruit), Palmurungai, Puna Murungai and Ko≈l Murungai and wild Kadumurunga (Ramachandran *et al.*, 1980; Kumar *et al.*, 2014). Two varieties (PKM-1 and PKM-2) have been developed at Horti Nursery Networks, Tamil Nadu, India, to improve pod production: usually these two varieties are grown as annual; after two harvests the tree is dragged out and new seedlings are planted. At Kerala Agricultural University (India) several varieties have been developed. Outside India there is research centers focused on *Moringa oleifera* improvement across the world: AVDRC (Taiwan), Rural development initiative (Zambia), *Moringa* Philippines foundation (Philippines) *Moringa* community (Zambia).

In spite of the great variability of *Moringa oleifera* no institutions have a germplasm bank or data base with either cultivated or spontaneous accessions. The divergence between genetic variability inherent to the species and poor variability reflected in germplasm banks should be fixed since it represents an obstacle for the progress of breeding programs. *Moringa* cytological studies revealed that *Moringa oleifera* has 2c genome size of 1.2 pg (Ohri and Kumar, 1986) and it is a true diploid with $2n = 28$ (Ramachandran *et al.*, 1980). Since 1999 molecular markers have become standards for the genetic characterization of *Moringa oleifera*. Studies started with the use of dominant markers until the development of co-dominant markers (in 2010) that allow distinction between homozygotes and heterozygotes, which provides optimal genetic information profiles.

Although *Moringa oleifera* shows diversification in many characters and high morphological variability, which may become a resource for its improvement, the major factors that limit productivity are the absence of elite varieties adapted to local conditions and the use of seeds obtained through open pollination from plants in the planted area. A germplasm bank encompassing the genetic variability present in *Moringa* is needed to perform breeding programmes and develop elite varieties adapted to local conditions (Leone *et al.*, 2016).

The correlation studies among 34 *Moringa* genotypes conducted by Joshi (2016) indicated strong association of number of pods/tree, number of seeds/pod, pod length, seed yield/tree, 100-seed weight and seed oil content with oil yield/tree. The highest positive direct effect was recorded by seed yield/tree, followed by seed oil content, number of pods/tree, 100-seed weight, tree height and pod length indicating their relationship and selection based on these traits would be highly desirable.

Karunakar *et al.* (2018) studied genetic variability, correlation and path analysis in *Moringa*. The maximum GCV, heritability and genetic advance as a percentage of mean was recorded for pod weight, number of leaves per rachis, yield per plant, length of pod, number of seed per pod. Yield per plant had significant and positive association with stem girth, fruit setting percentage, length of pod, girth pod, weight and number of pods per plant indicating their usefulness in selection for yield. Among the twelve characters studied, six characters plant height, leaf length, number of leaves per rachis, number of flowers per inflorescence, pod weight and number of seeds per pod showed positive direct effect on yield per plant.

10.0 Biotechnology

10.1.0 Micropropagation

Limited information on biotechnological research works are available in this crop. Recently, due importance has been stressed on micropropagation and characterization of germplasm through molecular markers for its improvement.

A technique was described by Kantharajah and Dodd (1991) for *in vitro* propagation of *M. oleifera* using nodal segments collected from a mature field-grown tree. An average of 22.1 ± 6.3 shoots/plant were developed in the woody plant medium containing 2 per cent sucrose, solidified with 0.8 per cent agar and supplemented with 1 mg BA/l. Root formation was readily achieved using MS basal medium with 0.5 mg NAA/l. The resulting plantlets were transferred to soil and successfully grown in the greenhouse. Tissues from seedlings were found to be less useful as sources of explants for micropropagation than those from mature nodal segments from older trees.

Stem segments from 10-day-old seedlings were transferred on to MS medium supplemented with various combinations and concentrations of auxin (NAA) and cytokinin (kinetin and BAP). Regeneration of plantlets of 100 per cent (with profuse rooting) was obtained after 3 weeks on MS medium supplemented with NAA (0.2 mg/l) + kinetin (0.2 mg/l) + glutamine (100 mg/l) (Mughal *et al.*, 1999). Direct somatic embryogenesis was obtained in immature zygotic embryos of *Moringa* cultured in continuous light in media with GA_3, BA and activated charcoal (Iyer and Gopinath, 1999). Long term, fast-growing callus cultures were established from rapidly elongating epicotyls of *in vitro* plantlets of *Moringa* in media with 2, 4-D, NAA and coconut milk (Iyer and Gopinath, 1999).

Saini *et al.* (2012) attempted to develop the rapid *in vitro* micropropagation protocol of *cv.* PKM-1 from nodal sections of young, aseptically grown seedlings. Benzyladenine (BA) at 4.44 µM was found to be optimal in producing on maximum an average of 9.0 ± 1.0 axillary shoots per explant after 15 days of inoculation. A high multiplication rate was established through routine sub culturing of nodal sections explanted from *in vitro* shoot cultures. *In vitro* rooting of individual shoot culture was maximum (100 per cent) on medium containing indole-3-acetic acid (IAA) at 2.85 µM along with indole-3-butyric acid (IBA) at 4.92 µM. Eighty per cent of the rooted plants survived after being transplanted in the soil, provided that the potted plantlets were covered with clear polythene bags and kept in a shaded greenhouse for 15 days before exposure to ambient conditions. Tissue culture-derived plants were found nutritionally superior over control plants to contain 13.2 and 14.7 per cent higher amount α-tocopherol and total carotenoids, respectively.

Efficient shoot formation and multiplication were obtained when nodal or shoot tip explants were cultured on MS medium supplemented with 2.5 µM 6-benzylaminopurine (BAP) at mild ventilation conditions (Salem, 2016). Adventitious root formation was obtained when shoot cuttings were cultured on MS medium supplemented with 4.92 µM indole3-butyric acid (IBA). Under salinity, *M. oleifera* shoots showed symptoms of vitrification such as retardation in shoot multiplication and growth, stimulated callus formation on the basal part of the cultured explants and increased shoot thickness. *M. oleifera* shoot multiplication and growth were negatively affected by ventilation deficiency (Salem, 2016).

Ravi *et al.* (2019) developed an improved micropropagation protocol facilitating continuous multiplication of elite germplasm of *Moringa oleifera*. Initial culture of nodal explant in MS medium supplemented with 2.5 µM BA resulted in the formation of 12.5 shoots per explant with high frequency of leaf fall (84.3 per cent). In order to reduce leaf fall and improve multiplication, varying concentration of anti-ethylene agent, $AgNO_3$ was incorporated in the medium. Addition of 2.5 µM

Micropropagation of *Moringa*

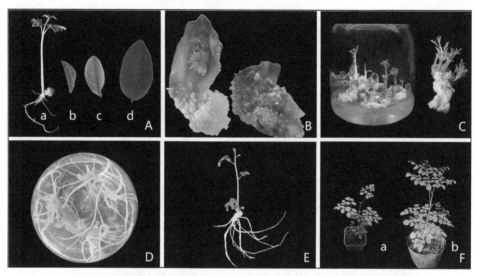

In vitro **Plant Regeneration of Drumstick from Leaf Segments and Acclimatization of the Regenerated Plantlets.**

(A) From a to d, a 2-week-old seedling, stage 1 leaf: justproduced and not fully expanded, stage 2 leaf: fully expanded, light green and immature, stage 3 leaf: completely mature; (B) Numerous shoot bud primordia developedon a leaf segment explant in MS medium with 0.8 mg/L BA, 0.2 mg/L KT and 0.05 mg/L NAA for 20 days; (C) Shoots recovered from leaf segment explants in MS mediumwith 0.8 mg/L BA, 0.2 mg/L KT and 0.05 mg/L NAA for 40 days; (D) Root systems of regenerated shoot-buds in MS medium supplemented with 0.1 mg/L NAA; (E) An intactregenerated plantlet; (F) Acclimatized drumstick plants after transplantation. a, 20 days and b, 40 days (Jun-jie *et al.*, 2017).

AgNO$_3$ in combination with 2.5 μM BA produced maximum number of shoots (17.6) including shoots originated from the base of the explant and shoots from the axillary buds of the primary shoots, where significant reduction in leaf fall (20.6 per cent) was noticed. Microshoots obtained from fourth subculture onwards were used for *ex vitro* rooting and found that by treating 50 μM NAA for 30 s, maximum numbers of microshoots (83.3 per cent) were rooted.

10.2.0 Molecular Markers

In order to facilitate reasoned scientific decisions on its management and conservation and prepare for a selection breeding programme, genetic analysis of seven populations was performed using amplified fragment length polymorphism (AFLP) markers by Muluvi *et al.* (1999). The four pairs of AFLP primers (*pstI/MseI*) generated a total of 236 amplification products of which 157 (66.5 per cent) were polymorphic between or within populations. Analysis of molecular variance revealed significant difference between regions and populations, even though outcrossing perennial plants are expected to maintain most variation within populations.

Interestingly, out of 2857 scientific publications on *Moringa oleifera* in the primary database (Web of Science), only 12 include genetic characterization based on molecular markers. Furthermore, only 77 fragments of DNA and RNA sequences are available (data from NCBI nucleotide database). This means that the genetic approach and its potential application in breeding programs are just at the beginning step. The common aim of almost all the studies is the genetic diversification among different populations and/or accessions: dominant markers are the most used (66 per cent of all papers). In spite of the limited range of dominant markers (heterozygote cannot be distinguished from homozygote specimens), the studies among commercial, cultivated or natural accessions have contributed to the understanding of genetic variability of *Moringa oleifera*. In this context, Amplified Fragment Length Polymorphism (AFLP) and Random Amplified Polymorphic DNA (RAPD) analyses along with Inter-Simple Sequence Repeat (ISSR) and cytochrome P450 were used. Muluvi *et al.* (1999) used AFLP to investigate seven natural populations from India and introduced populations in Malawi and Kenya. Authors found a significant level of population differentiation and separation of genotypes based on geographical origin. Moreover, high portion of genetic variability was within Indian accessions. In line with these findings, the authors argued that Kenya populations presumably came from India, as suggested by the small number of genetically related accessions. Thank to the same molecular markers (Muluvi *et al.*, 2004) the outcrossing rate in *Moringa oleifera* was detected: 26 per cent of selfing in *Moringa* trees. This evidence had a strong impact on the breeding program, as inbreed lines and hybridization allowed improvement of the species. RAPD were used by different authors to investigate cultivated and non-cultivated population of Tanzania (Mgendi *et al.*, 2010), different accessions in Nigeria (Abubakar *et al.*, 2011), accessions present in Embra Cosatal Teblelands Sergipe germplasm bank in Brazil (Da Silva *et al.*, 2012), commercially grown varieties in India (Saini *et al.*, 2013), new genotype developed in different countries (Thailand, USA, India and Malaysia, Tanzania, Taiwan) (Rufai *et al.,* 2013), and further accessions in Nigeria (Popoola *et al.,* 2014). All these studies showed the higher level of genetic diversity in natural population with respect to the cultivated ones. Cultivated accessions present in the considered germplasm banks are genetically close and need to be widen to promote increased diversity and used in breeding programs. Many studies disagree with Muluvi's conclusions that a significant level of population differentiation and separation of genotypes can be based on the geographical origin. Indeed no clusters were found according to geographical origins. This could be due to the planting spread that produced a high rate of gene flow through cross-pollination. Interestingly, Popoola *et al.* (2014) investigated morpho-metric characters along with molecular markers and showed a good correlation between 100 seed weight with pod length, pod weight, number of seeds per lobule and number of seeds per pod. Studies on *Moringa oleifera* with co-dominat markers started in 2010 when Wu *et al.* (2010) developed microsatellite markers. The first successful estimates of genetic diversity and population structure were obtained with Simple Sequence Repeat (SSR) in 2013 by Shahazad *et al.* (2013). These authors' evaluated accessions collected in different locations of Pakistan and accessions from different countries (India, Tanzania, Senegal, Mozambique, Zimbabwe, Florida, Mexico, Haiti, Belize) obtained from Educational Concerns foe Hunger Organization (ECHO). They found high genetic diversity in wild Pakistan accessions, whereas low genetic diversity in ECHO accessions. Moreover, ECHO accessions are more similar to those of a single province of Pakistan. Most likely, British colonialists introduced *Moringa oleifera* in early of twentieth century in Africa from India, while in the 1784 an Englishman took *Moringa oleifera* over to Jamaica (Shahazad *et al.,* 2013). The export pathway was restricted to Indian coastal region (where most movement of goods and people took place) and involved a relatively small number of accessions that belonged to a common or few populations. This explains the low genetic variability within ECHO accessions with respect to the Pakistan ones. Later on, a further investigation on twelve Indian populations, from northern and southern regions of India, was performed through SSR together with morphological markers (Ganesan *et al.,* 2014). In this study too, individuals from various geographical areas were not significantly different genetically, while a large variability exist in Indian populations. Morphological analysis on fourteen quantitative and eleven qualitative characters showed correlation among some quantitative characters, *e.g.,* between tree tallness with fruit girth, trunk girth with tree branching. More SSR were identified in 2014 thanks to EST examination involving several plant species (Haq *et al.,* 2014) and not utilized so far in *Moringa oleifera* genetic investigation. Even if all the reported studies are valuables and have a tremendous importance for conservation, selection and collection of *Moringa oleifera* seeds, same questions are still to be addresses in order to develop an improved *Moringa oleifera* cultivation. Considering the cultivation challenges, some research activities should be prioritized: (i) collection and characterization of world accessions both cultivated and natural in order to obtain a true understanding of the genetic diversity and structure of *Moringa oleifera*; (ii) set a collaborative network among National and International Research Centre, O.N.G, farmers that already work on *Moringa oleifera*.

This will help scientists and producers to:

☆ Have a reliable access to information about genetics and materials to develop better *Moringa* varieties and technologies for farming practices: phenotypic characterization is a priority to evaluated the accessions;

☆ Ensure that *Moringa* production is improved along with best cultivation practice;

☆ Focus research on the association between phenotypic and molecular data within the contest of breeding;

☆ Define maps (both association map and physical map) to identify genes that may confer resistance to biotic and abiotic stress and quantitative traits loci (QTL) for a possible introgression of genes into commercial and cultivated accessions. Today next generation sequencing (NGS) (Davey *et al.*, 2011) is an approachable tool to discovery genome-wide genetic markers. This technique could be applied to species with no existing genome data like *Moringa*. Thank to NGS a saturated genetic map could be obtained within reasonable cost and time, in turn interesting characters could be identified and exploited in breeding programmes.

Hassanein and Al-Soqeer (2018) carried out an experiment with seven genotypes per species and were characterized morphologically using 14 morphological characteristics and genetically using 10 ISSR primers. The studied genotypes were classified according to each characterization, and the correlation between morphological and genetic diversity was investigated. The ISSR molecular markers were effective in the characterization of genetic diversity of *Moringa* where the average of polymorphism across the 14 genotypes was sufficient (90.8 per cent). Dinucleotide repeat $(AC)_n$ primers (UBC825, UBC826 and UBC827) and a trinucleotide primer (UBC864) were the best primers, regenerating the maximum number of polymorphic bands per primer (8-10) and the highest polymorphism level among genotypes (91-100 per cent). Principal coordinate analysis showed similar classification for morphological and molecular data where the two species were separated in two main clusters with three sub-clusters per species. The association analysis showed good correlation, up to a 0.84 determination coefficient, between genetic diversity and morphological variability. The primers UBC826 and UBC827 were the most informative markers, revealing correlations with 12 morphological characteristics.

11.0 References

Abdull Razis, A.F., Ibrahim, M.D. and Kntayya, S.B. (2014) *Asian Pac. J. Cancer Prev.*, **15** : 8571-8576.

Abilgos, R.G. and Barba, C.V.C. (1999) *Philippine J. Sci.*, **128** : 79-84.

Abubakar, B.Y., Wusirika, R., MuA'zu, S., Khan, A.U. and Adamu, A.K. (2011) *Int. J. Bot.*, **7** : 237-242.

Adebayo, A.G., Akintoye, H.A. Shokalu A.O. and Olatunji, M.T. (2017) *Int. J. Recycl. Org. Waste Agricult.*, **6** : 281-287.

Ahmad, M.B., Rauf, A. and Osman, S. M. (1989) *J. Oil Techno. Asso. India*, **21** : 46-47.

Amaglo, N.K., Bennett, R. N., Curto, R.B.L., Rosa, E. A. S., Turco, V.L., Giuffrida, A., Curto, A.L., Crea, F. and Timpo, G.M. (2010) *Food Chem.*, **122** : 1047-1054.

Anjaneyamurthy, J.N. and Regupathy, A. (1989) *South Indian Hort.*, **37** : 84-93.

Anjulo, A. (2009) *Indian J. For.*, **32**(2) : 243-250.

Anonymous (1962) *The Wealth of India*, (Raw Materials), Council of Scientific and Industrial Research New Delhi, **VI** : 426-429.

Arora, D.S., Onsare, J.G. and Kaur, H. (2013) *J. Pharmacog. Phytochem.*, **1** : 193-215.

Arumugam, T., Tamil Selvi, N. A. and Premalakshmi, V. (2017) *Indian Hortic.*, **62** (4) : 49.

Babu, K.V.S. and Rajan, S. (1996) *J. Trop. Agric.*, **34** : 133-135.

Batool, A., Wahid, A. and Farooq, M. (2016) *Int. J. Agric. Biol.*, **18** : 757-764.

Bhishagratna, K.K. (1991) An English translation of Shushrutam Samhita based on the original Sanskrit text, Chowkhamba Sanskrit Series Office, Varanasi, India, 3 : 213-9.

Brunelli, D., Tavecchio, M. and Falcioni, C. (2010) *Biochem. Pharmacol.*, **79** : 1141-8.

Caceres, A., Cabrera, O., Morales, O., Mollinedo, P. and Mendia, P. (1991a) *J. Ethnopharmacology*, **33** : 213-216.

Caceres, A., Freire, V., Giron, L. M., Aviles, O. and Pacheco, G. (1991b) *Econ. Bot.*, **45** : 522-523.

Carbungco, E.S., Pedroche, N.B., Panes, V.A., and De la Cruz, T.E. (2017) *Acta Hortic.*, **1158** : 373-380.

Choudhary, M.K., Bodakhe, S.H. and Gupta, S.K. (2013) *J. Acupunct. Meridian. Stud.*, **6** : 214-220.

Coppin, J.P., Yanping, Xu, Hong, Chen, Min-Hsiung, Pan, Chi-Tang, Ho, Rodolfo, Juliani, Simon, J. E. and Qingli, Wu (2013) *J. Functional Foods*, **5** : 1892-1899.

Da Silva, A.V.C., dos Santos, A.R.F., Lédo, A.D.S., Feitosa, R.B., Almeida, C.S., da Silva, G.M. and Rangel, M.S.A. (2012) *Trop. Subtrop. Agroec.*, **15** : 31-39.

Dahot, M.U. (1988) *Pakistan J. Biotech.*, **21** : 21-24.

Damodaran, T., Anbu, S., Azahakiamanavalan, R. S. and Vennila, P. (1999) *South Indian Hort.*, **47** : 292-293.

Dania, S.O., Akpansubi, P. and Eghagara, O.O. (2014) *Adv. Agric.*, 1-6.

Davey, J.W., Hohenlohe, P.A., Etter, P.D., Boone, J.Q., Catchen, J.M. and Blaxter, M.L. (2011) *Nat. Rev. Genet.*, **12** : 499-510.

David, B. and Ramamurthy, V.V. (2016) *Elt. Eco. Entomol.*, **398** : 147-148.

Dhasmana, B., Madan, R. N. and Dhasmana, B. (1990) *Van-Vigyan*, **28** : 138-140.

Drisya Ravi, R.S., Siril, E.A. and Nair, B.R. (2019) *Physiol. Mol. Biol. Plants*, **25** : 1311-1322.

Edward, E., Shabani, A.O. Chamshama, Yonika, M. Ngaga and Mndolwa, M. (2014) *Afr. J. Plant Sci.*, **8**(1) : 54-64.

Essien, B.A., Essien, J.B. and Eluagu, C.J. (2016) *Niger. Agric. J.*, **46**(2) : 101-108.

Fahey, J.W. (2005) *Trees Life J.*, **1** : 1-15

Fahey, J.W., Zalcmann, A.T., Talalay, P. (2001) *Phytochemistry*, **56** : 5-51.

Faizi, S., Siddiquid, B.S., Saleem, R., Siddiqui, S., Aftab, K. and Gilani, A.U.H. (1995) *Phytochem.*, **38** : 957-963.

Ganesan, S.K., Singh, R., Roy Choudhury, D., Bharadwaj, J., Gupta, V. and Singode, A. (2014) *Ind. Crop. Prod.*, **60** : 316-325.

Gbeassor, M., Kedjagni, A.Y., Koumaglo, K., Souza, C. de., Agbo, K., Aklikokou, K., Amegbo, K.A. and De-Souza, C. (1990) *Phytotherapy Res.*, **4** : 115-117.

Ghasi, S., Nwobodo, E. and Ofili, J. O. (2000) *J. Ethnopharmacology*, **69** : 21-25.

Haldar, P. (2000) *In : Proc. Nat. Seminar on Environmental Biology*, (Eds. Battacharya, A., Mandal, S., Bhattacharya, A., Mandal, S. and Aditya, A. K.) Visva-Bharati Univ., Santiniketan, India, 3-5 April, 1998, pp. 197-204.

Hamza, A.A. (2010) *Food Chem. Toxicol.*, **48** : 345-55.

Haq, S.U., Jain, R., Sharma, M., Kachhwaha, S. and Kothari, S.L. (2014) *Int. J. Genomics*, **2014** : 863948.

Hassanein, A.M.A. and Al-Soqeer, A.A. (2017) *Int. J. Agric. Biol.*, **19** : 873-879.

Hassanein, A.M.A. and Al-Soqeer, A.A. (2018) *Hortic. Environ. Biotechno.*, **59** : 251-261.

Iyer, R. I. and Gopinath, P. M. (1999) *J. Phytol. Res.*, **12** : 17-20.

Joshi J.A. (2016) *Indian J.Hortic.*, **73** (4) : 601- 603.

Jun-jie, Zhang, Yue-sheng, Yang, Meng-fei, Lin, Shu-qi, Li, Yi, Tang, Han-bin, Chen, Xiao-yang, Chen (2017) *Ind. Crops Prod.*, **103**: 59–63.

Jyothi, P. V., Atlura, J. B. and Reddi, C. S. (1990) *Proc. Indian Acad. Sci.*, **100** : 33-42.

Kantharajah, A. S. and Dodd, W. A. (1991) *South Indian Hort.*, **39** : 224-228.

Karunakar, J., Preethi, T.L., Manikanta Boopathi, N., Pugalendhi, L. and Juliet Hepziba S. (2018) *J. Pharmacog. Phytochem.*, **7**(5) : 3379-3382.

Khawaja, T.M., Tahira, M., Ikram, U.K. (2010) *J. Pharm. Sci. Res.*, **2** : 775-781.

Kokou, K., Joct, T., Broin, M. and Aidam, A. (2001) *Cahiers Agricultures*, **10** : 131-133.

Kshirsagar, C.R. and D'Souza, T.F. (1989) *J. Maharashtra Agril. Univ.*, **14** : 241-242.

Kulkarni, N., Shamila Kalia., Sambath, S. and Joshi, K.C. (1996) *Indian Forester*, **122**(11) : 1075-1076.

Kumar, A.R., Prabhu, M., Ponnuswami, V., Lakshmanan, V. and Nithyadevi, A. (2014) *Agric. Rev.*, **35** : 69-73.

Kumar, S., Singh, R. C. and Kumar, S. (2000) *Ann. Agril. Res.*, **21** : 148-151.

Kumari, M.S.B., Kotikal, Y.K., Narabenchi, G., Nadaf, A.M. (2015) *Karnataka Journal of Agricultural Science*, **28**(2) : 193-196.

Lakshmipriya, G., Kruthi, D. and Devarai S.K. (2016) *Food science and human wellness*, **5**(2) : 49-56.

Leone, A., Spada, A., Battezzati, A., Schiraldi, A., Aristil, J. and Bertoli, S. (2016) *Int. J. Mol. Sci.*, **17** : 2141.

Litsinger, J.A. (2014) Pesticide evaluation and safe use practices for USAID and Catholic Relief Services (CRS) Project in Niger on Development Food Aid Program (DFAP). 14 January-28 February, 2014, Niger, P. 148.

Logiswaran, G. (1993) *Madras Agril. J.*, **80** : 698-699.

Mahmuod, M., EL-Sayed, and Mahmoud, A.W.M. (2018) Middle East Journal of Applied Sciences, **08** : 01.

Mandokhot, A.M., Fugro, P.A. and Gonkhalekar, S.B. (1994) *Indian Phytopath.*, **47** : 443.

Marimuthu, R., Athmanthan, U. and Mohan, S. (2001) *South Indian Hort.*, **49** : 34-36.

Math, M., Kotikal, Y.K., Narabenchi, G. (2014) *International Journal of Advances in Pharmacy, Biology and Chemistry*, **3**(1) : 54-59.

Mazumder, U.K., Gupta, M., Chakrabarti, S. and Pal, D. (1999) *Indian J. Exp. Biol.*, **37** : 612-614.

Mendez, M.R. (1999) *FAO Animal Production and Health Paper*, No. 143, pp. 341-346.

Mgendi, M., Manoko, M. and Nyomora, A. (2010) *J. Cell Mol. Biol.*, **8** : 95-102.

Mughal, M. H., Ali, G., Srivastava, P. S. and Iqbal, M. (1999) *Hamdard-Medicus*, **42** : 37-42.

Muhl, Q.E., du Toit, E.S. and Robbertse, P.J. (2011) *Seed Sci. Technol.*, **39** : 208-213.

Muluvi, G.M., Sprent, J.I., Odee, D. and Powell, W. (2004) *Afr. J. Biotechnol.*, **3** : 145-151.

Muluvi, G.M., Sprent, J.I., Soranzo, N., Provan, J., Odee, D., Folkard, G., McNicol, J.W. and Powell, W. (1999) *Mol. Ecol.*, **8** : 463-470.

Murthy, J.N.A. and Regupathy, A. (1992) *South Indian Hort.*, **40** : 43-48.

Muyibi, S.A. and Evison, L.M. (1995) *Water Res. Oxford*, **29** : 1099-1104.

Nouman, W., Siddiqui, M.T., Basra, S.M.A., Afzal, I., Rehman, H.U. (2012) *Turk. J. Agric. For.*, **36** : 227-235.

Nyein, M.M., Aye, T., Khine, W.W., Wai, K.T., Tun, S., Htwe, S.S., Myint, T. and Swe, T. (1997) *Myanmar Health Sci. Res. J.*, **9** : 163-166.

Ohri, D. and Kumar, A. (1986) *Caryologia*, **39** : 303-307.

Okuda, T., Baes, A.U., Nishijima, W. and Okada, M. (2001) *Water Res. Oxford*, **35** : 405-410.

Oliveira, J.T.A., Silveira, S.B., Vasconcelos, I. M., Cavada, B.S. and Moreira, R.A. (1999) *J. Sci. Food Agri.*, **79** : 815-820.

Olson, M.E. and Carlquist, S. (2001) *Bot. J. Linnean Soc.*, **135** : 315-348.

Palada, M.C. and Changl, L.C. (2003) *International Cooperators Guide*, AVARDC, Shanhua, Taiwan, ROC, pp. 1-4.

Pan, S., Deb, G. and Pan, S. (1997) *Indian Agric.*, **41** : 277-285.

Pande, A., Ghate, V. (1998) *J. Econ. Taxon. Bot.*, **22** : 423-425.

Pandian, I.R.S., Sambandamoorthy, S. and Irulappan, I. (1992) *Madras Agril. J.*, **79** : 58-59.

Patel, B.P., Radadia, G.G., Pandya, H.V. (2010) *Insect Environment*, **16**(3) : 135-138.

Popoola, J., Oluyisola, B. and Obembe, O. (2014) *Covenant J. Phys. Life Sci.*, **1** : 43-60.

Pramanik, A. and Islam, S. S. (1998) *Indian J. chem.*, **37** : 676-682.

Price, M.L. (2000) *http: /www.echotech.org/technical/technotes/moringa biomasa.*

Ragumoorthi, K.N. and Arumugam, R. (1992) *Indian J. Plant Protect*, **20** : 61-65.

Ragumoorthy, K.N., Rao, P.V.S., Reddy, P.P., Kumar, N.K.K. and Verghese, A. (1998) *Proc. of the First Nat. Symp. on Pest Management in Hort. Crops : Environmental implication and thrusts*, Bangalore, India, 15-17 October, pp. 140-144.

Raja, S., Bagle, B.G. and More, T.A. (2013) *J. Plant Breed. Crop Sci.*, **5**,: 164-170.

Ramachandran, C., Peter, K.V. and Gopalakrishnan, P.K. (1980) *Econ. Bot.*, **34** : 276-283.

Rao, A.V., Devi, P.U. and Kamath, R. (2001) *Indian J. Exp. Biol.*, **39** : 858-863.

Rao, K.S. and Mishra, S.H. (1998) *Indian J. Pharmaceutical Sci.*, **60** : 12-16.

Rashmi, Yadav, B.P. and Ojha, K.L. (1998) *J. Appl. Biol.*, **8** : 61-64.

Rufai, S., Hanafi, M.M., Rafii, M.Y., Ahmad, S., Arolu, I.W. and Ferdous, J. (2013) *Biomed. Res. Int.*, **2013** : 1-6.

S.du Toit, E., Sithole, J. and Vorster, J. (2020) *South Afr. J. Bot.*, **129** : 448-456.

Saini, R.K., Saad, K.R., Ravishankar, G.A., Giridhar, P. and Shetty, N.P. (2013) *Plant Syst. Evol.*, **299** : 1205-1213.

Saini, R.K., Shetty, N.P., Giridhar, P. and Ravishankar, G.A. (2012) *Biotech.*, **2**(3) : 187-192.

Salem J.M., 2016. *Genetics and Plant Physiology*, **6**(1-2) : 54-64.

Satti, A.A., Nassr, O., Fadelmula, A, and Ali, F.E. (2013). *Int. J. Sci. Nat.*, **4**(1) : 57-62.

Selvi, A., Varadharaju, N. (2016) In: Navarro, S., Jayas, D.S. and Alagusundaram, K. (Eds.), *Proceedings of the 10ᵗʰ International Conference on Controlled Atmosphere and Fumigation in Stored Products (CAF2016)*. CAF Permanent Committee Secretariat, Winnipeg, Canada, pp. 62-66.

Shahzad, U., Khan, M.A., Jaskani, M.J., Khan, I.A. and Korban, S.S. (2013) *Conserv. Genet.*, **14** : 1161-1172.

Silva, M.F., Nishi, L., Farooqi, A. and Bergamasco, R. (2014) *Journal of Medical and Pharmaceutical Innovation*, **1**(3) : 9-12.

Singh, K.A., Rai, R.N. and Pradham, I.P. (1991) *Indian Farming*, **41** : 7-10.

Sivasubramanian, K. and Thiagarajan, C.P. (1997) *Madras Agril. J.*, **84** : 618-620.

Sosamma, V.K., Jayasree, D., Koshy, P.K. and Mehta, U.K. (1998) In : *Proc. of the Third Internatl. Symp. of Afro-Asian Society of Nematologists*, Coimbatore, India, April 16-19, pp. 222-225.

Suganthi, M., Balamohan, T.N., Beaulah, A. and Vellaikumar, S. (2019) *Int. J. Chem. Stud.*, **7**(3) : 483-486.

Sundararaj, J.S., Shanmugavelu, K.G. and Raman, K.R. (1967) *South Indian Hort.*, **15** : 382.

Suthanthirapandian, I.R., Sambandamurthy, S. and Irulappan, I. (1989) *South Indian Hort.*, **37** : 301-302.

Sutherland, J.P., Folkard, G.K. and Grant, W.D. (1989) *Sci. Technol. Dev.*, **7** : 191-197.

Tesfay, S.Z., Modi, A.T. and Mohammed, F. (2016) *S. Afr. J. Bot.*, **102** : 190-196.

Thamburaj, S. (2001) *Drumstick*. In : Text book of Vegetables, Tubercrops and Spices (Ed. S. Thamburaj and N. Singh) Directorate of Information and Publications of Agriculture, ICAR, New Delhi, pp. 400-405.

Tsaknis, J., Lalas, S., Gergis, V., Dourtoglou, V. and Spiliotis, V. (1999) *J. Agril. Food Chem.*, **47** : 4495-4499.

Valia, R. Z., Patil, V. K. and Kapadia, P. K. (1993a) *J. Maharashtra Agril. Univ.*, **18** : 455.

Valia, R. Z., Patil, V. K. and Patel, Z. N. (1993b) *South Indian Hort.*, **41** : 84-90.

Varalakshmi B. and Devaraju (2007) *Acta Hortic.*, **752** : 411-412.

Verma, A.R., Vijayakumar, M. and Mathela, C.S. (2009) *Food and Chem. Toxicol.*, **47** : 2196-201.

Vijayakumar, R.M., Srimathi, P., Vijayakumar, M. and Chezian, N. (1999) *South Indian Hort.*, **47** : 275-277.

Wu, J.C., Yang, J., Gu, Z.J. and Zhang, Y.P. (2010) *HortScience*, **45** : 690-692.

CURRY LEAF

*Priyadarshani P. Mohapatra, T.K.Maity and
Sk Masudul Islam*

1.0 Introduction

Curry leaf (*Murraya koenigii*) is an important perennial tree vegetable. It is a plant found throughout tropical and subtropical East Asia from India and China to New Caledonia and North Eastern Australia (Xie *et al.*, 2006). Its leaves are used mainly to improve the taste and flavour of foods. Leaves are slightly pungent and retain their flavour even after drying. Ground curry leaf with mature coconut kernel and spices forms an excellent preserve. It grows wild in the foothills and plains of the Himalayas from Kumaon to Sikkim. In south India, especially in Tamil Nadu, Kerala and Karnataka, at least one curry leaf plant is found in each homestead. Recently it has gained importance as a commercial crop. It is cultivated commercially in Tamil Nadu and Karnataka. It is also cultivated in West Bengal, Assam and Deccan Plateau. Curry leaf is christened as 'Magical plant of Indian Spice' is extremely popular condiment plants all over the world (Nishan and Subramanian, 2015). In South India it is considered as backyard plants used in various culinary recipes in Indian subcontinents due to its distinctive aroma and taste (Perera and Dahanayake, 2015). In Indian Ayurvedic medicine, it is known as "Krishnanimba" (Ahluwalia *et al.*, 2014). It is a handsome, aromatic more or less deciduous shrub or a small tree up to 6m in height and 15 to 40 cm in diameter. It is an important perennial tree vegetable, found growing in wild form throughout India from time immemorial. Curry leaf plays an important role as a spice and condiment in the culinary preparation of Indian dishes and other tropical countries (Sandip, 2006).

2.0 Composition and Uses

2.1.0 Composition

The composition of edible portion is given in Table 1.

Table 1: Nutritive Value of Curry Leaf per 100 g of Edible Portion*

Constituent	Content	Constituent	Content
Moisture (g)	66.3	Manganese (mg/100 g)	0.15
Protein (g)	6.1	Magnesium (mg/100 g)	44.0
Fat (g)	1.0	Copper (mg/100 g)	0.10
Minerals (g)	4.0	Cr (mg/100 g)	0.005
Fibre (g)	6.4	Chlorine (mg/100 g)	198.0
Carbohydrates (g)	18.7	Carotene (mg)	7,560.0
Energy (k cal)	108.0	Thiamine (mg)	0.08
Calcium (mg/100 g)	810.0	Riboflavin (mg)	0.21
Phosphorus (mg/100 g)	57.0	Niacin (mg)	2.3
Iron (mg/100 g)	0.93	Folic acid (mg) (free)	23.5
Zinc (mg/100 g)	0.20	Folic acid (mg) (Total)	93.9
Sulphur (mg/100 g)	81.0	Vitamin C (mg/100 g)	4.0
Vitamin A (I.U./100 g)	600.0		

* Singh *et al.* (2014).

2.1.1 Chemical Properties

The chemical composition of curry leaf or plant defines the aroma and the taste. Following are the chemical composition or the constituents of the curry leaf:

2.1.2 Aroma Constituents

The aroma components consist of β-caryophyllene, β-gurjunene, β-trans-ocimene, β-thujene a-selinene, β-bisabolene, furthermore limonene, β-elemene, β-phellandrene and β-cadinene.

2.1.3 Essential Oil Components

The essential constituents of curry leaf are: monoterpenes including β-phellandrene, a-pinene, β-pinene. In some species sesquiterpenes is then main constituents.

The fresh leaf has about 2.6 per cent essential oil, but there is a gradual decrease in volatile content with advancing maturity. The essential oils mainly consist of following:

Sabinene: 34 per cent

Alfa-pinene : 27 per cent

Dipentene: 16 per cent with beta-caryophyllene, beta-gurjunene, beta-elemene and beta-phellandrene

2.2.0 Uses

The use of curry leaves as a flavouring for vegetables is described in early Tamil literature dating back to the 1st to 4th centuries AD. Its use is also mentioned a few centuries later in Kannada literature. Curry leaves are still closely associated with South India where the word 'curry' originates from the Tamil 'kari' for spiced sauces. An alternative name for curry leaf throughout India is kari-pattha. Today curry leaves are cultivated in India, Sri Lanka, Southeast Asia, Australia, the Pacific Islands and in Africa as a food flavouring (Singh et al., 2014). The leaf is used in South India as a natural flavouring agent in various curries and chutneys (Sindhu and Arora, 2012). Volatile oil is used as a fixative for soap perfume. The leaves, bark and root of the plant are used in the indigenous medicine as a tonic, stimulant, carminative and stomachic. It prevents the premature greying of the hair (Ramaswamy and Kanmani, 2012 and Parul et al., 2012). The seeds of curry leaf are having some insecticidal or repellent properties (Balaji, 1988 and Nur et al., 2009). Curry powder is a British invention to imitate the flavour of Indian cooking with minimal effort. In Indian cuisine curry leaves are used fresh for some recipes or fried in butter or oil for a short while. Since South Indian cuisine is dominantly vegetarian, curry leaves seldom appear in non-vegetarian food. The leaves have soft texture but are usually removed before serving but if eaten they are harmless. Gautam and Purobit (1974) reported that this essential oil exhibited a strong antibacterial and antifungal activity. An alkaloid, murrayacinine, is also found in this plant (Chakrabarty et al., 1974).

2.2.1 Therapeutic Uses

Its leaves, root and bark are used as medicinal aids in India. The leaves are used to help blood circulation and menstrual problems. The fresh leaves are taken to cure dysentery and diarrhoea (Saini and Reddy, 2013) and an infusion made of roasted leaves stops vomiting (Nayak *et al.*, 2010). It is also recommended for relieving kidney pains. Curry leaves is the richest source of carbazole alkaloids such as koenigine, mahanimbine extracted from the leaves which have anti-cancer and anti-oxidant properties (Kirupa and Kariitha, 2015).

2.2.2 Industrial Uses

The fresh leaves steam distilled under pressure yield a volatile oil (curry leaf oil) which may find use as a fixative for a heavy type of soap perfume. It has a specific gravity (Anon., 1962) of 0.9748 at 25°C, a saponification value of 5.2 and an acid value of 3.8. The edible fruit yields a yellow, volatile oil with neroli-like odour and a pepper-like taste, accompanied by an agreeable sensation of coolness on the tongue. It has a specific gravity of 0.872 at 13°C and a boiling point of 173.74°C. A yellow, clear, and transparent oil known as limbolee oil is extracted from the seeds of curry leaf (Drury, 1978).

Lipid oxidation in foods causes its quality deterioration (rancidity, off flavour, degradation of texture, and colour patterns) in meat and meat products during storage, making them unfit for consumption by humans. *M. koenigii* berry extract has been reported to be an excellent source of antioxidative compounds (tocopherol, β carotene, lutein, flavonoids, and phenolics) to prevent oxidative damage of meat and meat products. The curry leaves incorporated in functional poultry meat finger sticks improved lipid stability and antimicrobial quality of the products, indicating the effective use of *M. koenigii* as an alternative to synthetic food preservatives in functional meat food snacks (Aswathi *et al.*, 2014).

2.2.3 Different Uses of Curry Leaves

While there are many different kinds of curry powders and curry dishes throughout the world, curry leaves come from only one type of tree, the curry leaf tree. However, curry leaves can come in four different forms: fresh, dried, powdered and cooked (Singh, *et al.*, 2014).

Fresh

Fresh curry leaves are the preferred form for cooking. Fresh leaves may be used directly after harvesting from a curry leaf tree. They also may be placed or vacuum-packed in plastic bags and refrigerated or frozen after harvesting, which keeps them fresh from one week to two months. Fresh curry leaves are generally found in the freezer section of stores.

Dried

Curry leaves may be air dried or oven dried, producing leaves that have a longer shelf life. According to Gernot Katzer's Spice Pages, some recipes require the baking or toasting of fresh curry leaves before the leaves are added as a flavoring. Dried leaves are also available commercially.

Powdered

Powdered curry leaves are also called for in some recipes and powdered curry is also available commercially. After being dried, curry leaves can be pulverized, producing a concentrated powder. Powdered curry leaves, though, should not be confused with curry powder. Commercial curry powder is usually a mixture of many spices, while powdered curry leaf is a powdered version of the actual dried curry leaf. It is important to read spice labels for accuracy prior to purchase.

Cooked

Sautéed or fried curry leaves are prepared by the cook or chef prior to or during the cooking process. Some recipes require that fresh curry leaves be cooked before being added as flavouring. Such sautéed or fried curry leaves would not generally be purchased in advance. Instead, curry leaves would be purchased fresh, or perhaps dried, and then cooked in the kitchen.

Fresh (Left), Powdered (Middle), and Dried (Right) Curry Leaves.

3.0 Origin and Distribution

Curry leaf trees are naturalised in forests and waste land throughout the Indian subcontinent except in the higher parts of the Himalayas. From the Ravi River in Pakistan its distribution extends eastwards towards Assam in India and Chittagong in Bangladesh, and southwards to Tamil Nadu in India. The curry leaf is native to India and is found nearly everywhere in the Indian subcontinent (Harish *et al.*, 2012). In India, it is distributed throughout its mainland including Andaman and Nicobar Islands, and found naturally in semi-deciduous to evergreen forests with medium to high rainfall. Foot hills of Himalayas in North, continued through Terai region, Sikkim hills, Darjeeling hills and end up in Khasi-Garo hills in far east, reaching far up to Nilgiri and Annamalai hills in South, covering deciduous and semi-deciduous forests in Middle India and Eastern Ghats, including evergreen forests of Western Ghats are the major diversity regions for curry leaf in India. The plants were spread to Sri Lanka, Malaysia, Indonesia, Burma, Thailand, and South Africa with South Asian immigrants.

Based on several ethno-botanical reports and other floral distribution studies, the germplasm rich regions of curry leaf could be identified into six zones for future exploration and genetic improvement in India (Raghu B.R. 2020)

Region-1

This region is confined to sub-tropical forests running all along the sub-Himalayan foothills from Jammu and Kashmir, Himachal Pradesh, Uttarakhand to Terai regions of Uttar Pradesh and Bihar. In Himachal Pradesh, curry leaf is found in the forest ranges found between Tanda and Shahpur in Kangri district, Nalagarh and Nahan-Paonta ranges in Sirmaur district and some warmer areas of Solan, Shimla, Bilaspr, Hamirpur and Mandi districts. In Uttarakhand, sub-tropical forests in foot hills of Shivalik of Dehradun and Gharwal and Udham Singh Nagar including Rajaji national park and sub-tropical forests of Kumoan regions of Almora, Nainital and Chamavat districts are rich in curry leaf. The Terai regions of Uttar Pradesh covering 15 districts (Saharanpur, Muzaffar nagar, Bijnor, Moradabad, Rampur, Bareilly, Pilibhit, Kheri, Bahraich, Shravasti, Balrampur, Siddharth, Mahrajganj, Kushinagar, Gorakhpur) including Dudhwa national park, and West-Champaran, East-Champaran and Gopalganj districts of Bihar are rich sources of curry leaf.

Region-2

The region is confined to tropical evergreen and semi-evergreen, and tropical moist and dry deciduous forests of North-eastern states. This region covers from foot hills of South Sikkim, Darjeeling, Jalpaiguri and CoochBehar districts of west Bengal. The Assam valley includes Brahmaputra valley, Barak valley, Karbi plateau and Barail hills and parts of South Kamrup, Sibsagar, North Lakhimpur, Cachar, Goalpara, Nowgoan and Darrang districts of Assam. East and West Kameng, lower and Upper Subansiri, Lohit and Tirap districts in Arunachal Pradesh. Southern and Northern slopes of Meghalaya including Northern and North-western slopes of Garo hills. Western and North-western parts of Nagaland. Jiri, Moreh, Vangoi, Tamenglong forest areas and forest areas adjacent to Myanmar in Manipur. Dharam Nagar, Kailashahr, Belonia, Amarpur, Sonamura, Udiapur and Sadar sub-division in Tripura. Northern side forest areas of Kawnpuri, Hortaki, Bhairabi, Kolarib, Vairentee and western parts of Mizoram. Besides, the curry leaf is an integral part of many tribal medicines and hence it could be seen commonly in home-stead gardens in north eastern states.

Region-3

This region is confined to central India, from Sundarban delta to Satpur range covering Chotanagpur plateau, Hazaribagh plateau, Ramgarh hills, Malayagiri, Dandaka ranya and Vindhyan Ranges. Unlike regions 1 and 2, this region is loosely spread over South-western parts of West Bengal, Jharkhand, Northern parts of Odisha, Chhattisgarh, and Northern parts of Maharashtra to Madhya Pradesh including Southern Uttar Pradesh. A pursuance of review of literature of ethno-botanical use and floral and medicinal plants diversity studies of curry leaf in central India indicated that, it has been grown and used in Sonebhadra district of Uttar Pradesh, Jabalpur, Neemuch, Raisen, Rewa, Umaria, Anuppur, Nimar eco region, Satpur plateau of Madhya Pradesh, Bhadrak, Koraput, Jharsuguda, Keonjhar districts, Rourkela, Nandan Kanan wild life sanctuary of Odisha, Mahasamund, Dantewada, Koria, Jashpur, Raipur, Surguja, Ratanpur region of Bilaspur, Raigarh area, Bhoramdeo wild life sanctuary, Kabirdham wild life sanctuary of Chhattisgarh,

Dumka of Jharkhand, Bhagalpur, Banka, Buxar of Bihar and Burdwan, Hoogly, South 24 Parganas, Birbhum, West Rarrh region of West Bengal.

Region-4

This region is confined to Eastern parts of India mainly covering Eastern Ghats. This region starts from Southern parts of Odisha, covering Andhra Pradesh and up to Northern Tamil Nadu.

Region-5

This region is confined to Western parts of India mainly covering Western Ghats. This region starts from Southern parts of Gujarat, covering, Maharashtra, Goa, Karnataka and up to Kerala.

Region-6

This region is confined to Andaman and Nicobar Islands mainly covering the Andaman semi-evergreen forests, Andaman moist deciduous forests and Andaman secondary moist deciduous forests.

4.0 Botany and Taxonomy

It is an unarmed semi-deciduous aromatic pubescent shrub or small tree 3-6m high with slender but strong woody stem and branches covered with dark grey bark and closely crowded crown (Mhaskar *et al.*, 2000). The leaves are imparipinnate, leaf lets 9-25 ovate, lanceolate, almost glaborous. The white small terminal, hermaphrodite flowers are about 1 cm long, fragrant and arising in terminal, corymbose panicles and blooming in the month of April-May. Calyx, 5-lobed, persistent, inferior, green; corolla, white, polypetalous, inferior, with 5 petals, lanceolate; length, 5 mm; androecium, polyandrous, inferior, with 10 stamens, dorsifixed, arranged into circles of five each; smaller stamens, 4 mm. long whereas the longer ones, 5 to 6 mm; gynoecium, 5 to 6 mm long; stigma, bright, sticky; style, short; ovary, superior. Purplish black fruits when ripe occur in close clusters containing 1 or 2 seeds. Fruit indehiscent, a berry. The roots are woody, widely spread and produce many suckers. The seed will be spinach green with 11mm long and weigh about 445mg (Prajapati, 2003, Jain *et al.*, 2012 and Singh *et al.*, 2014).

The taxonomic status and taxonomy of the plant along with M. peniculata was studied by various workers. The presence of plant *M. koenigii* is mainly seen in India and Sri Lanka were as *M. paniculata* is native to deep forests in India, Sri Lanka and related countries.

Kingdom - Plantae

Sub-kingdom - Tracheobionta

Superdivision - Spermatophyta

Division - Magnoliophyta

Class - Magnoliospida

Subclass - Rosidae

Order - Sapindales

Family - Rutaceae

Genus - Murraya J. Koenig ex L

Species - *koenigii, paniculata, exotica*

Chromosome number - 2 n = 18

Mode of pollination - Self Pollination.

5.0 Cultivars

Curry leaves plant is of three different morphological types.

Regular: These grow tall and fast and look most like the curry leaf buy at the grocery store;

Dwarf: These do not grow as tall and the leaves are lighter in colour and longer than the ones you generally buy at the store;

Gamthi: The most fragrant type, grows very slowly and has dark brown thick but smaller leaves.

Curry leaf are of two types *i.e.*, broad leaved and the small leaved, while the small leaved types have fragrance, which is used for the extraction of essential oils.

Farmer prefers local varieties, which have pink midrib. At the Department of Horticulture, University of Agricultural Sciences, Dharwad, two genetically distinct cultivars, *viz.*, DWR-1 and DWR-2, have been identified and are being multiplied.

DWR-1 (Suvasini)

This is a clone obtained from root suckers that have dark green, highly aromatic, shining leaves. It is sensitive to low temperature in the winter season and hence the bud burst is poor during winter. The leaves are dark green (0.1629 mg of chlorophyll content per gram of fresh leaf), shiny and highly quality, and made into dry powder. The plant could be trained to grow as a low bush (up to 90- 120 cm height). This variety is resistant to Cylindrosporium leaf-spot.

DWR-2

This is an open-pollinated seed progeny with pale green leaves with less aroma (4.09 per cent). It is not very sensitive to low winter temperature and is much superior in the number of buds burst and internodal length. It is also 8 times higher in the growth of shoot length and weight of new shoots than DWR-1. Due to its winter tolerance, it can give extra income to planters.

'Senkaampu' is a local cultivar grown in different parts of Tamil Nadu, especially in the Karamadai tract of Coimbatore district. The petiole in this variety is purplish-red in colour. The leaves have a good aroma and flavour due to its high oil content.

6.0 Soil and Climate

It does not require a specific climate and can come up in dry climate too. It can tolerate a maximum temperature ranging from 26° to 37°C. In places where minimum

Morphototype and Cultivar of Curry Leaf

Morphotypes in Curry Leaf

Senkaampu, a Popular Farmer's Variety of Curry Leaf

temperature goes below 16°C, the vegetative buds become dormant arresting the new growth of the plant. It can be grown in places upto an elevation of 1500 m msl.

Curry leaf can be cultivated in a wide range of soils. Although any soil is suitable for the cultivation of the curry leaves, but it is believed that for commercial cultivation red loamy type of soil would be extremely beneficial. The soil which is more suitable for growing curry leaf plants is light textured red loamy soil with good drainage for its normal and fleshy growth which results in a better leaf yield. Heavy clay soil with poor drainage is not suitable for its cultivation.

7.0 Cultivation

7.1.0 Propagation

Curry leaf is mainly propagated through seeds. Well-ripen fruits are collected from high-yielding plants and the seeds after extraction are immediately sown uniformly in well-prepared nursery beds, in lines 10 cm apart at a depth of 1 cm whereas two seeds can also be sown in polythene bags filled with a mixture of 1:1:1 sand, soil and farmyard manure. The seed germinates in about three weeks. The nursery beds should always be kept at optimum moisture level and free from weeds. The healthy seedlings are ready for transplanting in one year. The raised nursery beds are prepared well in advance by mixing equal parts of sand, red soil and farmyard manure. The seeds are available for sowing during July-August. A spacing of 3-4cm has been suggested for planting seedlings (Anon., 1962). Planting in Southernmost parts of India is mostly done just before the arrival of Southern most monsoons in May.

It can also be propagated by root suckers. There are a number of root suckers near its plants. They are separated from the main plant during rainy season and planted immediately in the main field.

7.2.0 Preparation of Field, Planting and Field Growing

Main field should be ploughed repeatedly. Normally, a spacing of 90 to 120 cm is followed on either side. One month before planting, pits of 30 × 30 ×30 cm dimensions are dug out and are filled with top soil mixed with well decomposed farm yard manure at the time of planting. Healthy seedlings are planted in the centre of the pits. Then long furrows are formed connecting all the pits to facilitate easy irrigation. From a field experiment on effect of high density planting on quality of curry leaf was conducted at Department of Spices and Plantation Crops, Horticultural College and Research Institute, Tamil Nadu Agricultural University, Periyakulam during 2011-13, Jagadeeshkanth *et al.* (2017) concluded that high density planting (0.6 × 0.9 m) would results in better quality *viz.*, essential oil (0.156 per cent), crude protein (5.3 per cent), iron (0.156 per cent), calcium content (834.55 mg per 100 g) and phosphorous (0.71 per cent) in curry leaf. Further, the density of planting which would facilitate the cultivation of curry leaf under high density of planting.

Seedling Raising, Field Growing and Harvesting of Curry Leaf

The seedlings are irrigated once in 5 to 7 days up to three years and once in 15 days afterwards. The field should be kept free of weeds. Normally curry leaf plants are not fertilized with inorganic source by the farmers. However, for a better growth and yield, each plant may be fertilized with 20 kg of Farmyard manure besides 150 g nitrogen, 25 g phosphorus, and 50 g of potash per year.

Leguminous crops like cowpea, black gram, green gram or soybean could be raised as inter-crops during the first year.

In the curry leaf tree, the leaves are the economically important part. For the maximum encouragement of leaf production in the plants they have to be pruned first around 100 cm height. The plant terminal buds are removed to encourage lateral branching. A minimum of 5 to 6 branches are kept per plant. It is very important to prune curry leaf plant early in its growth stage. Plants that are older and have an open shape can be pruned early in the period just as growth starts and they will still flower later in the year. It is recommended that pruning should be done annually to keep plant tight, compact and producing the greatest amount of tasty foliage. The additional dosage of N: P: K in required proportions is to be supplied to the plants. Within a few weeks, the plant will generate side shoots and fill out into a mini-grove of numerous bushy stalks with pointed leaves. That is a sign that the curry leaf plants are healthy and thriving.

8.0 Harvesting, Yield and Storage

Curry leaves are picked or harvested 15 months after planting. Commercial harvest can be started from three-year-old plants. The leaves should be plucked at least 20-25 days after spraying. As a routine the leaves are clipped from the young shoots at the end of the first year and a total of four harvests can be made. During the winter season, curry plant's leaves might turn yellow. This generally means that the plant is about to go dormant and might lose its' leaves.

Fully developed curry tree can yield nearly 100 kg of leaves each year. Curry leaves are picked 15 months after planting. Commercial harvest process can be started from 3-year-old plants. With good management, normal yield can be obtained up to the age of 20 to 25 years. The leaves can be harvested at two-and-a-half to three months intervals. The leaf yield ranges from 5 tonnes in the second year to ten tonnes per hectare in the fourth year, and from the fourth year onwards the yield can be around 20 tonnes per hectare.

Curry leaves can be kept in an airtight container in the refrigerator for up to two weeks or freeze them. While dried curry leaves are also obtainable, they have less aroma and flavour.

9.0 Post-harvest Management

The young shoots with tender leaves are harvested, packed in gunny bags or tied in bundles and transported. Water is sprinkled on the bags/bundles. No research has been done on the postharvest management of curry leaf. The leaves are dried and ground into powder and used as curry-leaf powder.

The value-added products of curry leaf are volatile oil and dehydrated curry leaves.

10.0 Diseases and Pests

The curry plant is normally free of pests and diseases. The curry leaves have a strong smell that deters many insect pests and keeps the disease away. However, depending on the weather, there are some pests that create problems to the plant.

10.1.0 Diseases

Leaf-spot (*Phyllosticta murrayae* and *Cylindrosporium* sp.) is a serious disease of curry leaf which causes severe defoliation. Carbendazim (1 per cent) or Mancozeb (2 per cent) could be sprayed 15 days prior to harvest to control this disease.

Sap-rot (*Fomes pectinatus*) and collar-rot of seedlings (*Rhizoctonia solani*) are the other diseases, but the economic losses due to these diseases are minimal. The spraying of Sulphur compounds should be avoided since it results in leaf shedding.

10.2.0 Pests

Among the pests attacking the curry leaf, the caterpillar of citrus butterfly (*Papilio demoleus*) is very important. Early instar larvae are dark with white patches resembling the droppings of birds. When grown up, they turn deep green in colour, stout and cylindrical in shape. Hand picking and destruction of the larvae can be done wherever possible. If the population is heavy, suitable insecticide which has less mammalian toxicity can be sprayed.

The branches affected by citrus psylla (*Diaphorina citri*), aphids and scale insect (*Anoidiella orientalis*) have to be removed immediately. If necessary, a systemic insecticide can be sprayed. But the leaves should not be harvested for at least 20-25 days. The stem (bark) borer (*Indorbela tetraonis*) is reported to cause drying of the branches occasionally and mites are also found to infest the leaves, for which no control measures have been suggested.

The nymphs and adults of *Dasynus antennatus* are reported to infest the growing tips of the branches and cause the drying of young leaves, adversely affecting the growth of the plants. The other pests of curry leaf include black fly (*Aleurocanthus woglumi*), leaf roller (*Tonica zizyphi*), tartoise beetle (*Silana farinosa*), citrus leaf miner (*Phyllocnistis citrella*), leaf weevils (*Myllocerus discolor, M.viridanus*), hairy caterpillars, coreid bugs and termites.

11.0 Crop Improvement

Curry leaf is still regarded as an under-utilized and unexplored vegetable in terms of efforts towards its genetic improvement despite its multiple use and numerous benefits. The genetic improvement of curry leaf is still in infancy stage and no systematic breeding work has so far been done in India. Earlier, few efforts were made to characterize curry leaf germplasm for different morphological and biochemical traits. More often, they are isolated programmes, characterized by limited use of plant genetic resources (PGR), and has narrow genetic base.

Moreover, in majority of the studies, the sampling was done on natural populations, wherein, such samples lack replications and uniform environments. The inferences made from such studies, especially on traits like bio-chemical composition and essential oils content which vary greatly with locations and seasons, often failed to provide comprehensive information regarding extant of genetic base of plant diversity. Further, some other studies have done on nutrient analysis and profiling in curry leaf. However, in those studies, genotype × environment interaction components have not been dissected for nutritional related traits. Thus, it implied that, no comprehensive efforts have been made to collect and conserve curry leaf germplasm covering different geographical regions, and characterize under uniform environmental conditions. Further, an inclusive field gene bank contains broad based germplasm in terms of number, regions and traits, is lacking in curry leaf (Raghu, 2020).

In spite of lack of comprehensive PGR activities in curry leaf, a few and scanty breeding initiatives were made in the past. Towards this end, two improved varieties of curry leaf namely, DWD-1 and DWD-2 (Suhasini) were developed and released by University of Agricultural Sciences (UAS), Dharwad, Karnataka, India. Basically, these two varieties were developed by clonal selections from the germplasm collected from Western Ghats. Both have good aroma and suitable for fresh leaves. Besides, several local landraces and farmers' varieties are being popularly grown by the farmers; these were selected based on appearance, fragrance and response to local environments. Senkaampu is such local landraces or farmers' variety are being popularly grown in many parts of Tamil Nadu. It has pigmented petiole, shiny and leathery leaves with good fragrance. It has huge local demand for fresh leaves. Lately, ICAR-Indian Institute of Horticultural Research, Bengaluru, has initiated comprehensive breeding programme in curry leaf. Till now, more than 150 germplasm were collected from different parts of India covering Himachal Pradesh, Odhisa, Karnataka, Tamil Nadu and Kerala, and successfully established a field gene bank to conserve them. Germplasm characterization with respect to different morphological traits, bio-chemical and nutritional traits including chemical profiling for essential oils are being taken up in different seasons (Raghu *et al.*, 2020). Further, germplasm augmentation is being done through regular exploration programmes concentrating in different states and regions.

Studies on variability and association of characters in curry leaf genotypes were carried out by Shoba *et al.* (2020). Highest phenotypic coefficient of variation and genotypic coefficient of variation was recorded for weight of the matured shoot, the number of matured shoots, the number of compound leaf per matured shoot and fresh leaf yield/plant. Number of matured shoots, length of matured shoots, weight of matured shoots, the number of compound leaves per matured shoot and fresh leaf yield per plant showed high heritability with high genetic advance, suggesting these characters could be improved through selection. The magnitude of genotypic correlation was higher than the phenotypic correlation for all the traits that indicated inherent association between various characters.

Raghu *et al.* (2020) suggested a breeding programme inclusive of development and utilization of different genomic resources in PGR management, trait discovery and allele mining, identification of trait specific genotypes and its successful introgression into elite lines is need of the hour in curry leaf. Being a perennial tree species, different breeding methods like clonal selection, mutation breeding and polyploidy breeding should be integrated in curry leaf. Protection and conservation of such unique germplasm requires novelty, distinctness, uniformity and stability tests conducted through nationally or internationally accepted guidelines. However, there is no internationally or nationally accepted DUS test guidelines available in curry leaf, which is a mandatory requirement to protect unique germplasm under PPV and FRA, New Delhi.

12.0 Biotechnology

The conventional method of propagation of this crop is limited to seeds only, which retain their viability for only a short period of time. Hence, a biotechnological approach might have an advantage edging over traditional breeding as well as the genetic improvement of this crop within a short period. There are only a few reports of *in-vitro* studies of curry leaf which are restricted to in-vitro shoot multiplication from intact seedlings, internode segments and nodal cuttings (Bhuyan *et al.*, 1997; Joshi *et al.*, 2011; Babu *et al.*, 2000). Khatik and Joshi (2018) reported for the first time an efficient protocol for high frequency regeneration of *in-vitro* plants of curry leaf through axillary shoot formation from repeated re-culturing of nodal explants. Rooting (95 per cent) in regenerated shoot occurred in 3-4 weeks of transfer of MS basal medium containing 12.30 µM IBA and 5.13 µM IAA. The hardened *in-vitro* plantlets were successfully transferred to field condition with 82-85 per cent survival.

Green synthesis of silver nanoparticles (SNPs) using plant extracts is an emerging field of research since such photosynthetic method is simple, cost effective, and eco friendly, and like various medicinal and spice plants *M. koenigii* has the capacity to act as reducing agents of Ag^+ to Ag°, thereby forming SNPs that are fairly stable in solution and play a great role in nanomedicine. Since ancient silver has been in use in its various forms such as silver nitrate, silver chloride, *etc.*, in medicine to cure various diseases due to their bioactive properties. The SNPs are excellent antimicrobial agents and are useful against life threatening microbial infection caused due to pathogenic fungus and bacteria; such SNPs (size: 10-100 nm) are stronger and promising when compared with the regular metal, and had synergistic activity against potential human pathogenic bacteria including *E. coli*, *S. aureus* and *P. aeruginosa*, as well as fungus including *C. albicans*. Bonde *et al.* (2012) synthesized SNPs using *M. koenigii* leaf extract that had bactericidal activity against *E.coli*, *S.aureus* and *P.aeruginosa* alone, and showed synergistic interaction when used in combination with conventional antibiotics. It has been reported that the SNPs synthesized with *M. koenigii* extract had antimicrobial efficacy against *A. niger*, *A. flavus*, *Trichoderma*, and *Rhizopus*, and had more efficacy against *Rhizopus* than the test agents: Amphotericin and plant extract. The SNPs, thus, might be the antimicrobial agents of newer generation possessing broad antimicrobial spectrum and can potentially be utilized in the preparation of several antimicrobials.

13.0 References

AESA based IPM-Curry Leaf (2014) Department of Agriculture and Cooperation, Ministry of Agriculture, Government of India

Ahluwalia, V., Sisodia, R., Walia, S., Sati, O. P., Kumar, J. and Kundu, A. (2014) *J. Pest Sci.*, **87**: 341-349.

Amna, U., Halimatussakdiah, P. W., Saidi, N. and Nasution, R. (2019) *J. Adv. Pharm. Technol. Res.*, **10**: 51-55.

Anonymous (1962) Wealth of India, Council of Scientific and Industrial Research, India, pp. 446-448.

Aswathi, P. B., Biswas, A. K., Beura, C. K., Yadav, A. S. and Khatke, P. A. (2014) *Int. J. Meat Sci.*, **4**(1): 15.

Babu, K. N., Anu, A., Remashree, A. B. and Praveen, K. (2000) *Plant cell, Tissue and Organ Cult.*, **61**(3): 199-203.

Balaji, B. R. (1988) *M.Sc. (Ag.) Thesis*, University of Agricultural Sciences, Bangalore, India.

Bhuyan, A. K., Pattnaik, S. and Chand, P. K. (1997) *Plant Cell Rep.*, **16**(11): 779-782.

Bonde, S. R., Rathod, D. P., Ingle, A. P., Ade, R. B., Gade, A. K. and Rai, M. K. (2012) *Methods Nanotechnol.*, **1**(1): 25-36.

Chakraborty, D. P., Bhattacharya, P., Islam, A., Roy, S. (1974) Chemistry and Industry London, pp.165-166.

Dikui, Z. I. B. M. (2009) *Ph.D. Thesis* submitted to Faculty of Chemical Engineering and Natural Resources, University Malaysia Pahang.

Drury, H. C. (1978) In: *The Useful Plants of India*, 2nd ed, p. 78.

Goutam, M. P., and Purohit, R. M. (1974) *Indian J. Pharm.*, **36**(1): 11-12.

Gowdra, K. H. B. (1990) *M.Sc. (Ag.) Thesis*, University of Agricultural Sciences, Dharwad.

Harish, K. H., Pandith, A. and Shruthi, S. D. (2012) *Asian J. Pharm. Clin. Res.*, **5**(4): 5-14.

Husna, F., Suyatna, F. D., Arozal, W., and Poerwaningsih, E.H. (2018) *Drug Res. (Stuttg)*, **68**: 631-636.

Jagadeeshkanth, R. P., Sankaranarayanan, R. and Suresh, V. (2017) *Int. J. Curr. Microbiol. App. Sci.* **6**(12): 2861-2864.

Jain, V., Momin, M. and Laddha, K. (2012) *J. Ayu. Her. Integr. Med.*, **2**(4): 607-627.

Joshi, R., Jat, B. L., Sharma, A., Joshi, V., Bohra, N., and Nandwani, D. (2011) *Funct. Plant Sci. Biotechnol.*, **5**: 75-77.

Khatik, N., and Joshi, R. (2018) *Indian J. Biotech.*, **17**: 379-382.

Kirupa, S. L. S and Kariitha, R. (2015) *Int. J. Pharm. Biol. Sci.*, **6**(1): 507-514.

Kumar, S. R., Loveleena, D. and Godwin, S. (2013) *Int. Res. J. Bio. Sci.*, **2**(9): 80-83.

Mandal, S. (2016) *Int. J. Clin. Exp. Physiol.*, **3**(2): 59-65.

Mhaskar, K. S., Blatter, E., Caius, J. F., Kirtikar, B. (2000) *Indian Medi. Sci. Seri.*, Delhi, India

Mishra, R. K. (2018) *Van Sangyan*, **7**(5): 23-28

Nayak, A., Mandal, S., Banerji, A. and Banerji, J. (2010) *J. Chem. Pharma. Res.*, **2**(2): 286-299.

Nishan, M., and Subramanian, P. (2014) *Int. J. Pharm. Tech. Res.*, **7**(4): 566-572.

Nur, AI., Jamil, R., Hafizah, R. and Sharifah, F. S. M. (2009) Faculty of Chemical and Natural Resources Engineering, University Malaysia Pahang, Lebuhraya TunRazak, Malaysia.

Parul, S., Javed, A., Neha, B., Honey, J., and Anuj, B. (2012) *Asian J. Pharm. Clin. Res.*, **2**(2): 51-53.

Perera, P. C. D and Dahanayake. N. (2015) *J. AgriSearch,* **2**(3): 212-217.

Prajapati, N. D., Purohit, S. S., Sharma, A. K. and Kumar, T. A. (2003a) Agrobios India, P. 256.

Prajapati, N. D., Purohit, S. S., Sharma, A. K. and Kumar, T. A. (2003b) Agrobios, Jodhpur, pp. 352-353.

Raghu, B. R. (2020) *J. Hortic. Sci.*, **15**(1): 1-8.

Raghu, B. R., Aghora, T. S. and Dhananjaya, M. V. (2020) *Indian Hort.*, **65**(3): 127-130.

Raina, V. K., Lal, R. K., Tripathi, S., Khan, M., Syamasundar, K. V., and Srivastava, S. K. (2002) *Flavour frag. J.*, **17**(2): 144-146.

Ramaswamy, L. and Kanmani, M. G. (2012) *Int. J. Ayur. Her. Medi.*, **2**(3): 469-476.

Saini, S. C. and Reddy, G. B. S. (2013a) *Int. J. Pharm. Sci. Lett.*, **2**(10): 13-17.

Saini, S. C. and Reddy, G. B. S. (2013b) *J. Pharm. Biol. Sci.*, **7**(6): 15-18.

Sandip. (2006) NIPER Publication, Punjab, India, pp.101-102.

Shoba, N., Balakrishnan, S., Paramaguru, P., and Vithya, K. (2020) *Electron. J. Plant Breed.*, **11**(02): 694-697.

Sindhu, R. K., and Arora, S. (2012) *J. Appl. Pharm. Sci.*, **2**(11): 12.

Singh, S., More, P. K., Mohan, S. M. (2014) *Indian J. Sci. Res.*, **4**(1): 46-52.

Singh, S., Omreb, P.K. and Madan, M.S. (2014) *Indian J. Sci. Res.*, **4**(1): 46-52

Xie, J., Chang, W., Wang, C., Mehendale, S. R., Li, J., Ambihaipahar R., Ambihaipahar, U., Harry, H. F. and Yuan, C. (2006) *Amer. J. Clin. Med.*, 34: 279-284.

Yankuzo, H., Ahmed, Q. U., Santosa, R. I., Akter, S. F. U and Talib, N. A. (2011) *J. Ethnopharma.*, **135**: 88-94.

Yeap, S. K., Abu, N., Mohamad, N. E., Beh, B. K., Ho, W. Y., Ebrahimi, S., Yusof, H. M., Ky H., Tan, S. W. and Alitheen, N. B. (2015) *Altern. Med.*, **4**: 306.

Index

9 789354 616853